T0377672

Control Systems

Control Systems
Classical, Modern, and AI-Based Approaches

Jitendra R. Raol

Ramakalyan Ayyagari

CRC Press
Taylor & Francis Group
Boca Raton London New York

CRC Press is an imprint of the
Taylor & Francis Group, an **informa** business

MATLAB® is a trademark of The MathWorks, Inc. and is used with permission. The MathWorks does not warrant the accuracy of the text or exercises in this book. This book's use or discussion of MATLAB® software or related products does not constitute endorsement or sponsorship by The MathWorks of a particular pedagogical approach or particular use of the MATLAB® software.

CRC Press
Taylor & Francis Group
6000 Broken Sound Parkway NW, Suite 300
Boca Raton, FL 33487-2742

© 2020 by Taylor & Francis Group, LLC
CRC Press is an imprint of Taylor & Francis Group, an Informa business

No claim to original U.S. Government works

Printed on acid-free paper

International Standard Book Number-13: 978-0-8153-4630-2 (Hardback)

Visit the Taylor & Francis Web site at
http://www.taylorandfrancis.com

and the CRC Press Web site at
http://www.crcpress.com

eResource material is available for this title at
https://www.crcpress.com/9780815346302

Contents

Section I   Linear and Nonlinear Control

Section II　Optimal and H-Infinity Control

Section III   Digital and Adaptive Control

Section IV   AI-Based Control

Section V   System Theory and Control Related Topics

Preface

In recent years, the availability of powerful (with huge memory and very high speed), low-cost computers has given a great impetus to the application of digital and adaptive control to robotics and aerospace vehicles, including micro air vehicles (MAVs) and unmanned aerial vehicles (UAVs). At the same and earlier times (in last four decades), many practical nonlinear control systems have been developed, ranging from digital FBW (fly-by-wire) flight control systems for aircraft to DBW (drive-by-wire) automobiles, to advanced robotic, missiles, and space systems. Thus, the subject of nonlinear systems and control is playing an increasingly important role in automation control engineering and autonomous mobile robotics. Many physical, chemical, and economical systems can be modeled by mathematical relations/equations, such as deterministic and/or stochastic differential and/or difference equations. It is possible to alter these systems' states from one value (of a state vector) at a point in time to another value at any other point by suitable application of some type of external inputs and/or controls (control signals). If this can be done (which is in most cases probable and feasible), then there would be many different ways of doing this task. However, there should be, or would be, one distinct way of doing it, such that it can be called the best way, in some sense (either the optimum or robust): minimum time to go from one state to another state, maximum thrust/power developed by an engine, minimum loss/cost of doing some operation/function/task, or maximum profit in a material-production unit. This is an optimization/optimal control problem. The input given to the system corresponding to this best situation is called the optimal control input, and the state trajectory optimal one. The measure of the best way or performance is called a performance index, cost function or even a loss function. Thus, we have an optimal control system, when a system is controlled in an optimum way satisfying a given performance index, which is either to be minimized or maximized, depending on the type of performance aimed for a given system under given circumstances, called system-environmental combination scenario.

Here, it is important to clarify some terms of the title of the present volume: Classical control is meant in the sense of well proven and/or conventional control, (it is not considered old/outdated, although it is quite there from very previous times, several decades); modern control is meant in the sense of the appearance of the optimal/digital/adaptive control approaches, quite later than the classical approaches; and intelligent control (IC)

is meant to be based on the computational intelligence. IC is mainly treated in this volume from the point of view of applying soft computing paradigms to design control systems. It can be called AI-based intelligent control (AIIC), since the intelligent control is not really 'intelligent' (not even very near to it) as it would/might take several decades for it to develop/evolve to 'near intelligent' control.

The theory of optimal control systems is considered a part of the so-called modern control theory, which has been in vogue for more than the last five/six decades. The interest in theoretical and practical aspects of the subject has been sustained due to its application is diverse fields, such as electrical power (systems/grids), aerospace (though in some limited sense), chemical/industrial plants, and much later to economics, medicine, biology, and ecology. Another fascinating aspect of the modern control is the design of robust control system that might not be optimal in the sense of minimization/maximization of some cost function, but would obtain a robust performance in the presence of some deterministic but unknown/bounded uncertainties. The idea of robustness is that a certain level of (stability and error) performance is obtained, and the (quantifiable) effects of uncertainty and model errors, on the designed (closed loop) control system are kept within a specified bounds, thereby obtaining the robustness. In this spirit, H_∞ (H-infinity/HI) control is being increasingly accepted for design of robust control.

Many dynamic systems, after a long and repeated usage (due to wear and tear), change with time or any other independent variable; in fact, some coefficients or parameters would change and would affect the performance (and even stability) of the overall system. This calls for the concepts and techniques of adaptive control to determine these changes, then adjust the control system parameters (like gains, and time constants, and/or the same of the control filters) to regain the stability and retain the performance of the original control system. In this sense, the adaptive control is a fascinating field for study and research, mainly because of our intention to capture the essence of learning from the presented data of the dynamic systems. The adaptive techniques are being used more and more in industrial control systems and in some aerospace programs, and robotics.

Further, with rapid advances in computing, the computational intelligence, supported by soft computing, is becoming more important in several engineering/non-engineering disciplines including control engineering.

At a higher level of abstraction, the performance index can be multi-objective–simultaneously minimizing/maximizing several metrics (error-criteria). Also, often prior knowledge about the plant may not be completely available and controller design would become complicated. In practice, by experience, (human-) operators know certain rules about the system and its control. This knowledge can be incorporated into a fuzzy logic inference mechanism using control actions that can be designed. Likewise, artificial neural networks, modeled based on the biological neural networks, can be utilized for learning the behavior of the system and, subsequently, for controlling the plant. Other biologically inspired techniques such as genetic algorithms (modeled based on the nature's evolutionary mechanism) can be used to obtain global optimal solutions to optimization problems, which in fact, can give robust designs.

In this composite volume, we present certain important concepts and theoretical results on linear, nonlinear, optimal, robust, digital, adaptive, and intelligent control (essentially based on soft computing). Where appropriate, we use illustrative examples: numerical, analytical, and/or simulation-based; the latter are coded in MATLAB® (MATLAB is the trade mark of MathWorks Ltd. USA); where feasible, these scripts are *reassembled (and then rerun to produce the plots and figures afresh)* from the sources that are mentioned at the appropriate places in the text (these scripts would be available from the web site of the book). The user is expected to have access to PC-based MATLAB and certain toolboxes: signal processing, control system, system identification, optimization, and ANN-FL-GA related.

Interestingly, there are already several good books on control; however, most are quite dated, some are not as comprehensive as expected, and some are special volumes. The treatment in the present volume is comprehensive and composite and covers, in four/five sections, several areas of control: (i) linear, (ii) nonlinear, (iii) calculus of variation, (iv) optimal, (v) model predictive, (vi) robust/H-infinity, (vii) digital, (viii) adaptive, and (ix) intelligent-soft computing based. The approach is not specifically that of explicit Theorem-Proof type; however, where appropriate, several important theoretical and/or analytical results, are presented in various chapters and the appendixes, and the treatment in the entire book can still be safely considered as formal (and proper). The end users of the technology of control systems presented in this book will be aero/mechanical/civil/electrical/electronics and communications-educational institutions, several R&D laboratories, aerospace and other industries, robotics, transportation and automation industry, environmental sciences and engineering, and economics system studies.

MATLAB® is a registered trademark of The MathWorks, Inc. For product information, please contact:

The MathWorks, Inc.
3 Apple Hill Drive
Natick, MA 01760-2098 USA
Tel: 508 647 7000
Fax: 508-647-7001
E-mail: info@mathworks.com
Web: www.mathworks.com

Acknowledgments

This book is dedicated to all those who worked in the very challenging and yet exciting area of control theory and practice and primarily to the pioneers of the discipline. The author (JRR) is very grateful to Prof. Dr. Ranjit C. Desai (emeritus professor, M.S. University of Baroda, Vadodara), and the late Prof. Dr. Naresh Kumar Sinha (emeritus professor, Department of Electrical and Computer Engineering, McMaster University, Canada, and my doctoral thesis supervisor) who taught me major aspects of control theory. I am very grateful to Drs. Mrs. Girija, Gopalratnam, Jatinder Singh, and Abhay A. Pashilakar (senior scientists, FMCD, CSIR-NAL, Bangalore) for their constant support for several years. I am also very grateful to Prof. Radhakant Padhi (Department of Aerospace Science and Engineering, Indian Institute of Science, Bangalore), for giving me an opportunity to work on the modeling and parameter estimation of type I diabetes-patients' data. I (JRR) am very grateful to Jonathan Plant and his entire unique team at CRC Press for their tremendous support, quick responses, alertness, and very helpful/assuring/encouraging nature, during this book project and during all the earlier ones with CRC. Mr. Plant has been a unique editor with tremendous zeal, patience, and a lot of respect to his authors. If not for him, I would not have been able to write *six* books for the CRC press. I am also very grateful to Dr. Mrs. Lakshmi, S. for her contribution on the application of the interval type 2 fuzzy logic to pilot's situation assessment.

I am as ever, very grateful to my own family (my 93-year-old mother, wife, daughter, and son) for their love, affection, care, support, and endurance for more than four decades; and at times, I was/am compelled to learn a few alternative aspects of doing logistics/ management (in personal and technical spheres) during the course of my life's journey, due to criticism from some of them.

This book is very special to me (AR) for several things. First, its title makes me reflect my own five decades of existence–classical, modern, and hopefully, intelligent too. I am deeply indebted to my family–wife Priyadarsini, daughter Sumadyuti, son Virinchi, mother Radha, brother Rajkiran Ayyagari and his family for their confidence in all of my fancy-flight endeavors; in fact, they are my floodlights never allowing me to grope in darkness. I always place my teachers from childhood and the schools, Andhra University and IIT Delhi, aloft on a high pedestal for at least two reasons—the high moral and academic standards I have been brought up in, and their blessings in the form of feedback control to regulate myself, at least, asymptotically. There are innumerous friends who keep me young and jubilant all along, eagerly waiting for this book. Thank you very much everybody. Quite unexpectedly, this has come to me as a wonderful gift from my senior colleague and co-author Dr. J. R. Raol. We have known each other for over two decades; we interacted in various professional occasions; and, I believe we got closer enough since both of us are admirers of (Control and Aerospace) science and technology, in general, and Nature in particular and enjoy prose and poetry; but never had the idea that one day I would co-author a volume like this with him. It was indeed a pleasure working with him. I take this opportunity to sincerely thank him from his kind gesture, and also wish him a long, healthy, and happy life ahead. Serving an academic institution of the order of NIT (National Institute of Technology, Tiruchirappalli, Tamilnadu) is a challenge, an honor, and the best I have. I thank each one of my colleagues for the excellent climate I am assured of here. My students are the backbone of my learning, and without them probably I would have never ventured into writing books. It is a wonderful moment remembering several of my students I taught over the past 22 years. In particular, I place on record the enthusiastic and excellent support, yes in the last minute literally, given by my current doctoral students Mrs. Sharmila Devi JaiGanesh and Mrs. Thilagavathy Giridhar in bringing order to the figures in my chapters. Traditionally, many authors would close their acknowledgements with a line on their favorite God. To me, it is my father Dr. Venkata Rajeswara Prasada Rao Ayyagari, and with all my love I dedicate my part of writing in this book to his evergreen memory.

We are grateful to Mayur J. Raol and his team for the design of the cover for the book. We are also grateful to the teams at CRC Press and Lumina Datamatics for their efficient handling of the book manuscript in its various stages of editing and publication.

Authors

Jitendra R. Raol earned BE and ME degrees in electrical engineering at M. S. University of Baroda, Vadodara, in 1971 and 1973, respectively, and a PhD (in electrical and computer engineering) at McMaster University, Hamilton, Canada, in 1986, where at both the places he was also a postgraduate research and teaching assistant. He joined the National Aeronautical Laboratory (NAL) in 1975. At CSIR-NAL, he was involved in the activities on human pilot modeling in fix- and motion-based research flight simulators. He rejoined NAL in 1986 and retired on 31 July 2007 as Scientist-G (and head, flight mechanics and control division at CSIR-NAL). He has visited Syria, Germany, the United Kingdom, Canada, China, the United States, and South Africa on deputation/fellowships to work on research problems in system identification, neural networks, parameter estimation, multisensor data fusion and robotics, to present technical papers at international conferences, and deliver guest lectures. He has given several guest lectures at many Indian colleges and universities and Honeywell (HTSL, Bangalore). He is a fellow of the IEE/IET (UK) and a senior member of the IEEE (US). He is a life-fellow of the Aeronautical Society of India and a life member of the Systems Society of India. In 1976, Dr. Raol won the K. F. Antia Memorial Prize of the Institution of Engineers (India) for his research paper on nonlinear filtering. He was awarded a certificate of merit by the Institution of Engineers (India) for his paper on parameter estimation of unstable systems. He has received one best poster paper award from the National Conference on Sensor Technology (New Delhi) for a paper on sensor data fusion. He has also received a gold medal and a certificate for a paper related to target tracking (from the Institute of Electronics and Telecommunications Engineers, India). Also, Dr. Raol was one of five recipients of the CSIR (Council of Scientific and Industrial Research, India) prestigious technology shield for the year 2003 for the leadership and contributions to the development of integrated flight mechanics and control technology for aerospace vehicles in the country. The shield was associated with a plaque, a certificate, and the prize of INRs 30,00,000 for the project work. He has published nearly 150 research papers and numerous technical reports. He guest edited two special issues of *Sadhana* (an engineering journal published by the Indian Academy of Sciences, Bangalore) on (1) advances in modeling, system identification, and parameter estimation (jointly with the late Prof. Dr. Naresh Kumar Sinha) and (2) multisource, multisensor information fusion. He also guest edited two special issues of *Defense Science Journal* (New Delhi, India) on (1) mobile intelligent autonomous systems (jointly with Dr. Ajith K. Gopal, CSIR-SA) and (2) aerospace avionics and allied technologies (jointly with Prof. A. Ramachandran, MSRIT). Under his guidance, eleven doctoral and eight masters research scholars have successfully received their degrees. He co-authored an IEE/IET (London, UK) Control Series book *Modeling and Parameter Estimation of Dynamic Systems* (2004), *Flight Mechanics Modeling and Analysis* (2009), and *Nonlinear Filtering: Concepts and Engineering Applications* (2017). He has also written *Multi-Sensor Data Fusion with MATLAB* (2010) and *Data Fusion Mathematics: Theory and Practice* (2015). He edited (with Ajith Gopal) *Mobile Intelligent Autonomous Systems* (2012). He has served as a member/chairman of numerous advisory-, technical project review-, and doctoral examination committees. He has also conducted sponsored research and worked on several projects spanning industry and R&D organizations to CSIR-NAL, all of which had substantial budget. He is a reviewer of a dozen national/international journals and MTech/doctoral theses (from India and overseas). He had been with MSRIT (M. S. Ramaiah Institute of Technology, Bengaluru) as emeritus professor for five years; with the Government College of Engineering, Kunnur (Kerala) as a senior research advisor; and with the Department of Aerospace Engineering (IISc, Bangalore) as a consultant on modeling and parameter estimation for type I diabetes patients' data for a period of three months. Dr. Raol was one of the five recipients of the Chellaram Foundation Diabetes Research Award (2018) for their best paper (Second International Diabetes Summit, March 2018, Pune, India). His main research interests include data fusion, system identification, state/parameter estimation, flight mechanics, flight data analysis, H-infinity filtering, nonlinear filtering, artificial neural networks, fuzzy logic systems, genetic algorithms, and soft technologies for robotics. He has also written a few books as the collection of his three hundred (free-) verses on various facets closely related to science, philosophy, evolution, and life. His new area of study and research is data systems analytics (DaSyA).

 **Ramakalyan Ayyagari** is a professor of instrumentation and control in the Department of Engineering, NIT Tiruchirappalli. He earned a PhD at IIT Delhi for his work on dynamic noncooperative games and robust control for a class of nonlinear systems. He is deeply interested in looking into computational problems that arise out of algebra and graphs in control theory and applications, particularly the NP-hard problems and randomized algorithms. Dr. Ayyagari has several significant papers in international conferences and journals. He was a visiting associate professor at the Institute of Mathematical Sciences, Chennai, 2001–2004. He was a recipient of the Government of India's Young Scientist Award in 2005 for his funded project 'Robust and Efficient Algorithms for Modern Control Systems'. In the same year, he also worked at National Chemical Laboratories at Pune (a constituent of Government of India's Central Scientific and Industrial Research [CSIR]) on 'Density Functional Theory and Quantum Control of Systems', under the aegis of Indian Academy of Sciences. He was among the first UKIERI recipients in 2007 and has successfully completed a collaborative project on unmanned air vehicles (UAVs) together with University of Leicester (UK), Indian Institute of Science (IISc) Bangalore, Indian Institute of Technology (IIT) Bombay, and National Aerospace Laboratories (NAL) Bangalore. Currently, his research and consultancy projects are a fine balance of theory and practice in the areas of model-driven engineering (funded by ABB), traffic scheduling and decongestion (funded by the Government of India), nonlinear control (funded by Bosch), and fault-tolerant control (funded by DRDO, Government of India). Dr. Ayyagari has recently developed a course on circuit theory under the pedagogy project at IIT Kharagpur, based on his textbook *Linear Circuits: Analysis and Synthesis* (Oxford University Press). He visited Texas A&M University during the summer of 2008, University of Leicester in 2008 and again in 2011, and Institut Henri Poincaré, Paris, in 2014. He is a senior member of IEEE and a member of SIAM. He was the founding secretary and present president of Automatic Control and Dynamic Optimization Society (ACDOS), the Indian National Member Organization (NMO) of the International Federation of Automatic Control (IFAC), through which he contributes to controls education and research in the country. He loves Carnatic music, literature, and economics.

Introduction

We start with a brief note on biofeedback/biocontrol/biological control (BFC/BC). It is a treatment in/for which the concerned people are trained to improve their health by using signals/data (as some feedback information) from their own bodies: (i) therapists use BF/C to help stroke-affected patients regain movement in their paralyzed muscles, (ii) psychologists use it to help tensed/anxious patients learn how to relax, (iii) our own body temperature or weight measuring feedback information tells us our body's certain conditions, (iv) when a person raises a hand, or flexes the knee, she controls these voluntary functions, and (v) person's heart rate, skin temperature, and blood pressure are automatically controlled by the human body's nervous system; however, the intentional biocontrol/biofeedback term came in late 1960s [1]. The aim of using biofeedback on a patient is to improve muscle strength and/or function: when a patient sees a flash of a light or hears an audible cue from the monitoring device, she makes an internal adjustment to improve muscle's functionality. The actual biological systems are also required to operate effectively in the presence of internal and/or external uncertainty; say, genetic mutations and temperature changes, respectively [2]. The study of (natural/nature's) evolution has resulted in the widespread research and the use of feedback (concept/mechanism/connections) in systems' biology; however, due to the latter's complexity, the resulting control systems are usually large interconnected regulatory networks; then, in many cases, it is not easy to distinguish the "process" (plant/system-dynamics) from its regulatory component/controller. There are certain merits in separating conceptually the different components of a biological control system into individual functional modules so that their roles are clearly defined and understood; this facilitates subsequent analysis of the system's dynamics. In that case, the role of the network structure, in delivering the required system level performance, would be more specifically and clearly identified. Many biological networks are governed by some exquisitely complex feedback control system mechanisms/connections and certain functions are achieved by up to three distinct negative feedback controllers: (i) genetic regulation, (ii) mRNA attenuation, and (iii) enzyme inhibition [2]. Applications of a kind of parallel or distributed control architectures are widespread in engineering systems as well as in biological networks. The much of the introduction is influenced by Refs. [1–6].

Simple and automatic feedback control systems are believed to have been known and used for nearly more than 2000 years; an earliest example being a water clock, Figure I.1, and described by Vitruvius and attributed to Ktesibios, 270 B.C., (An early nineteenth century illustration of Ctesibius's (285–222 BC) clepsydra from the third century B.C., the hour indicator ascends as water flows in, and a series of gears rotate a cylinder to correspond to the temporal hours); and nearly three hundred years later, Heron of Alexandria had described a range of automata that were based on different feedback mechanisms [3]; and the word feedback is a jargon introduced in the 1920s in order to describe parasitic (dependent) aspect, a feeding back of a signal from the

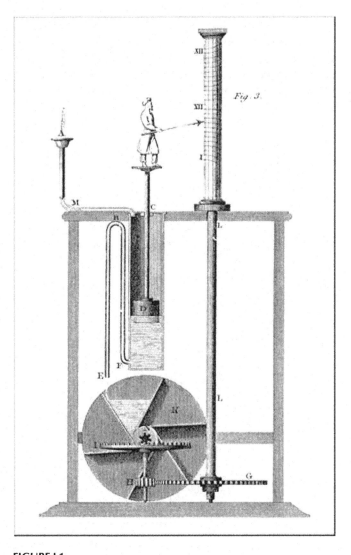

FIGURE I.1
Water-clock. (Source-public domain: https://en.wikipedia.org/wiki/Water_lock)

output to the input of an amplifier circuit. During the period (not exactly known) of early control to 1900, the knowledge of control systems of the Hellenic period is believed to have been preserved within the Islamic culture, and this knowledge was rediscovered in the West toward the end of the Renaissance [3]. The novel inventions and applications of older concepts began to appear during the eighteenth century: (i) Rene-Antoine Ferchault de Reamur had proposed several automatic devices for controlling the temperature of incubators (based on an invention of Cornelius Drebbel), see Figure I.2, (ii) improved temperature control systems were devised by Bonnemain; the sensor and actuator were based on the differential expansion of different metals (today's temperature control in a refrigerator). A wide range of thermostatic devices was invented and sold during the nineteenth century. The most significant control development during the eighteenth century was the steam engine governor, see Figure I.3 with its origin in the lift-tenter mechanism, which was used to control the gap between the grinding-stones in wind/water mills. Later, James Watt adapted it to govern the speed of a rotary steam engine (1788). However, the Watt's governor had some demerits, it: (i) had only proportional control and, hence, exact control of speed

was at only one operating condition; (ii) it operated at a small speed; and (iii) required a good maintenance. Ironically, in the first 70 years of the nineteenth century, extensive efforts were made to improve on the Watt's governor: (a) William Siemens replaced proportional action by integral action and produced floating controllers with no fixed point; (b) the loaded governor of Charles T. Porter could be run at much higher speeds and with greater forces that were developed to operate an actuator; and (c) Thomas Pickering and William Hartnell advanced a spring-loaded governor that operated at higher speeds (than the Watt's governor) with an additional merit of smaller physical size than other governors. In the early years of the nineteenth century, the problems of governors' hunting were observed. Although J. V. Poncelet and G. B. Airy had showed how dynamic motion of the governor could be described with differential equations, still it was difficult to determine the conditions for stable (non-hunting) operation. James Clerk Maxwell showed/derived linear differential equations for governor mechanisms. Around this time, it was known that the stability of a dynamic system could be determined by the location of the roots of the characteristic equation, and when the real part of a complex root became positive, the system would become

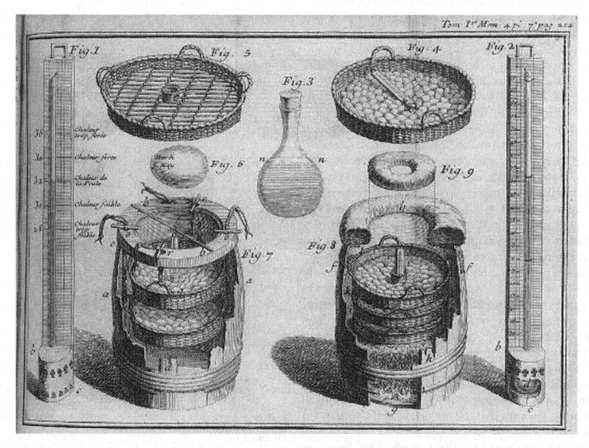

FIGURE I.2
Egg-Incubator (Reaumur's). (Source-public domain: http://www.wikiwand.com/en/Incubator_(egg))

steering engines to assist the helmsman. This was required because, on large ships, the hydrodynamic forces on the rudder were very large. Hence, large gear ratios between the helm and the rudder were required and moving the rudder took a long time. The first powered steering engine, designed by Frederick Sickels in the US, was an open loop system; and the first closed loop steering engine was designed by J. McFarlane Gray for Brunel's steamship, the Great Eastern, see Figure I.4. Around the same time in France, Jean Joseph Farcot designed a range of steering engines and other closed-loop position control systems. He suggested his devices be named as *servo-motcur* or *motcur asservi*; hence, today's prevailing terms are *servomechanisms* and *servomotors*. With the growth in knowledge of electricity and its applications, more uses for control became apparent:

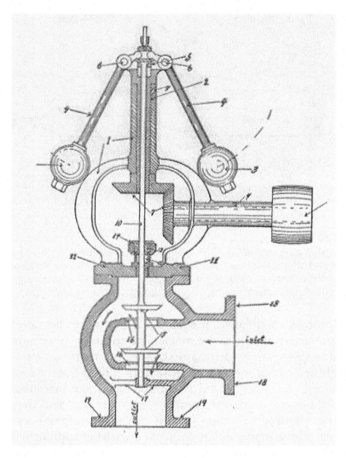

FIGURE I.3
Steam engine speed governor. (Source-public domain: https://en.wikipedia.org/wiki/Centrifugal_governor#/media/File:Centrifugal_governor_and_balanced_steam_valve_(New_Catechism_of_the_Steam_Engine,_1904).jpg)

unstable; the difficulty was how to determine the location of this real part without actually finding the roots of the equation. Maxwell showed, for second- (third- and fourth-) order systems, that the stability of the system could be ascertained by examining the coefficients of the differential equations; and he gave the necessary and sufficient conditions only for equations up to fourth order (fifth-order equations needed two necessary conditions). This problem was further studied by Edward J. Routh, and by now the Routh-Hurwitz stability criterion is well known; the Swiss mathematician Adolf Hurwitz derived the criterion independently (based on some results of C. Hermite). Most of the study of this period was concerned with the basic activities of controlling temperatures, pressures, liquid levels, and the speed of rotating machinery; the main aim was regulation and stability. However, the application of steam, hydraulic, and pneumatic power systems to operate position control mechanisms was prompted due to the growth in the size of ships and naval guns, and use of new weapons (e.g. torpedoes). The engineers in the United States, Britain, and France began to work on devising powered

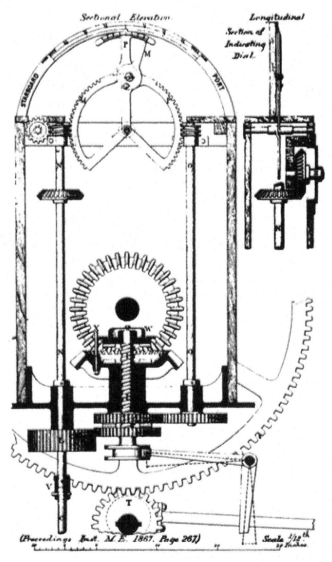

FIGURE I.4
Steering engine. (Source-public domain: https://en.wikipedia.org/wiki/Steering_engine#/media/File: Greysteeringengine.png)

(a) arc lamps required the gap between the electrodes to be kept constant, and it was if either the voltage or the current was kept constant; (b) additional tools came up for measurement, transmission and manipulation of signals, and their actuation, (c) the electric relay that provided high gain power amplification, and (d) the spring biased solenoid that provided (crude) proportional control action, Figure I.5.

The pre-classical period (1900–1935) saw the rapid and widespread application of feedback controllers for: (a) voltage, current, and frequency regulation; (b) boiler control for steam generation; (c) electrical motor speed control; (d) ship and aircraft steering and auto stabilization; and (e) temperature, pressure, and flow control in the process industries. The range of devices manufactured was large; however, but the most were designed without an understanding of the dynamics of these devices; of the system, sensors, and actuators. For simple regulation applications, the lack of understanding was not a serious problem. However, some complex mechanisms had complicated control laws: (a) Elmer Sperry's automatic ship-steering mechanism incorporated PID control and automatic gain adjustment to compensate for the disturbances due to the varying sea conditions, Figure I.6, and (b) in boiler control, both water level and steam pressure have to be controlled, and for efficient combustion, the draught to the boiler has to be controlled.

With the control devices and systems being used in different areas of engineering, some problems became apparent: (a) a lack of theoretical understanding with no common language to discuss technical problems and (b) no simple, easily applicable analysis and design methods. With increased number of applications, engineers faced difficulties, because: (a) controllers that worked satisfactorily for one application/set of conditions, were unsatisfactory to different systems/conditions; and (b) when a change in one sub-system

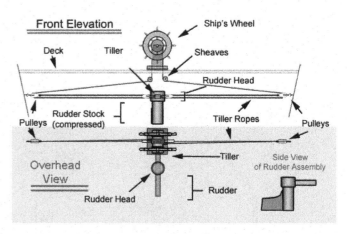

FIGURE I.6
A ship steering mechanism—the operation of a tiller using a ship's wheel tiller poles (a traditional ship's wheel and tiller shows a series of pulleys/sheaves). (Source-public domain: https://en.wikipedia.org/wiki/Ship's_wheel#/media/File:Wheel_and_rudder_assembl-y.gif.)

(process, controller, measuring system, or actuator) affected a change in the major time constant containing this sub-system; causing instability in otherwise a stable system. Some careful observers (Elmer Sperry and Morris E. Leeds) noted that the best human operators did not use an on-off approach to control, instead they used both anticipation (a little advanced action; backing off the power as the controlled variable approached the set-point), and small, slow adjustments (when the error persisted). In 1922, Nicholas Minorsky formulated a PID control law. However, the engineers lacked a suitable linear and stable amplification device to convert the low-power signals obtained from sensors to operate a control actuator. The hydro-mechanical systems had some solutions in slide and spool valves (the early part of the twentieth century), Figure I.7; however, they had dead space and stiction problems. Harold Stephen Black worked on the problem of amplification of signals (early 1920s); and realized that the distortion due to noise/component drift could be reduced, if the amplification of a high-gain amplifier were sacrificed by feeding back a part of the output signal, and eventually sketched a circuit for a negative feedback amplifier. Further he was assisted by Harry Nyquist, whose paper "Regeneration Theory" laid down the foundations of Nyquist analysis (1932). This led to a method of analyzing and designing control systems without requiring the derivation/manipulation of differential equations. This gave a scope of using the experimental data, i.e. the measured frequency responses. This paved the way to assess the degree of stability and the performance of the system. Almost concurrently, Clesson E. Mason developed a pneumatic negative feedback amplifier; Figure I.8. Edgar H. Bristol had invented the flapper-nozzle amplifier (1914). The early versions of the flapper-nozzle amplifier were highly nonlinear, and in 1930, he produced a feedback

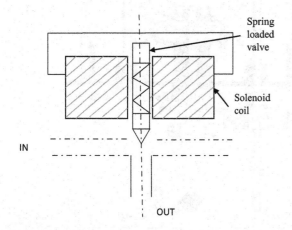

FIGURE I.5
Solenoid. (From https://www.techbriefs.com/media/thumbnails/xt/400/images/stories/NTB/sup/MCAT/ 2015/applications/55593-320_fig1.png.)

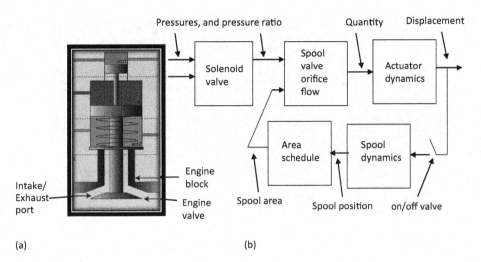

(a)

(b)

FIGURE I.7

Spool valve with circuit. (a) Functional circuit of a spool valve. (b) Block diagram of the spool valve. (From https://www.google.com/search?q=hydro-mechanical+slide+and+spool+valve&client=firefox-b-ab&tbm=isch&tbo=u&source=univ&sa=X&ved=2ahUKEwjkhZGD36HfAhVISxUIHZOVAdMQsAR6BAgEEAE&biw=946&bih=845#imgrc=KUUZrdxmBgnrhM.)

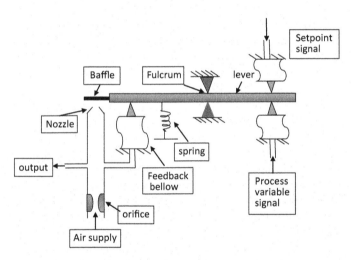

FIGURE I.8

Pneumatic negative feedback amplifier: the spring pulls down on the lever; the signals are in the range of 3–15 psi; the bellow is symbolic and does not represent an actual shape. (Adopted and modified from https://nptel.ac.in/courses/112103174/module6/lec5/3.html.)

circuit that linearized the valve operation. The work on analog calculating machines under the direction of Vanevar Bush resulted in the differential analyzer that provided a means of simulation of the dynamic systems and solving the differential equations; Figure I.9, this led to the study/design of servomechanism by Harold Locke Hazen (1901–1980).

During the period 1935–1940, a part of the classical control period (1935–1950), the advances in understanding of control system analysis and design were pursued independently by several groups in several countries (US, Russia, and Germany); these prompted the ways of extending the bandwidth of communication systems and

obtaining good frequency response characteristics [3]. Hendrik Bode (in 1940) adopted the point (−1,0) as the critical point, rather than the point (+1,0) used by Nyquist, and introduced the concept of gain/phase margins. By 1940, field-adjustable instruments with PID control were available. J.G. Ziegler and N.B. Nichols (1942) described how to find the optimum settings for PI/PID control. The advent of World War II concentrated control system work on the aiming of anti-aircraft guns; the work on this problem brought together mechanical and electrical/electronic engineers, and it was recognized that neither the frequency response approach nor the time domain approach were, separately, effective design approaches for servomechanisms. Arnold Tustin (1899–1994) in England; and R.S. Phillips, W. Hurewicz, L. McColl, N. Minorsky, and George Stibbitz in the US; by the end of the war were studying nonlinear and sampled data systems. Norbert Wiener (1894–1964) studied the problem of predicting the future position of an aircraft based on his work on generalized harmonic analysis (1931). The work lead to Wiener's report "The Extrapolation, Interpolation and Smoothing of Stationary Time Series with Engineering Applications" (1942) (this was known as "the yellow peril" because of its yellow covers and the formidable difficulty of its mathematics). By the end of the war, the classical control techniques (except the root locus method of Walter Evans [1948, 1950]) had been established. The design methodologies were for linear single-input single-output (SISO), and the frequency response techniques (due to Nyquist, Bode, Nichols, and Inverse Nyquist charts) assessed performance in terms of bandwidth, resonance, and gain/phase margins. The alternative approach was based on the solution of the differential equations using Laplace transform, and performance can

FIGURE I.9
Early computer-and-plotter dating to 1944 for solving complex equations. (From https://en.wikipedia.org/wiki/Differential_analyser; *online public image*, from http://ajw.asahi.com/article/sci_tech/technology/AJ201412020060 with original photo taken by Kabata, H. et al. Early computer dating to 1944 solving complex equations again after long 'reboot', The Asahi Shimbun/Technology, 2014.)

be expressed in terms of rise time, percentage overshoot, steady-state error, and damping. The conferences on Automatic Control (in 1951 at Cranfield, England, and 1953 in New York) signified the transition period leading to modern control theory.

Real systems are nonlinear, and measurements are contaminated by noise; and both the process and the environment are uncertain. Ziegler and Nichols had shown how to choose the parameters of a controller to obtain an "optimum" performance of a given PI, or PID controller. In addition to performance criteria based on minimizing some error function, there was interest in minimizing the time to reach a set-point (military target seeking servomechanisms and certain classes of machine tools). The problem was studied by Donald McDonald (1950), Bushaw (1952) and Bellman (1956), and T. P. LaSalle (1960) generalized all the previous results and showed that if optimal control exists it is unique and bang-bang. The progress made in this area was summarized by Oldenburger (Optimal Control, 1966). Design methods for systems with nonlinearities and the theoretical foundations of sampled-data systems were advanced.

In the era of modern control, Richard Bellman (1948–1952) studied the problem of determining the allocation of missiles to targets to inflict the maximum damage, which led to the "principle of optimality" and to dynamic programming. Bellman began working on optimal control theory, at first using the calculus of variations, and later (because of the boundary value problem inherent in the calculus of variations) formulated deterministic optimization problem in dynamic programming setting; the problem can be treated as a multistage decision making process. The generalization of Hamilton's approach to geometric optics by Pontryagin (1956), in the form of maximum principle, laid the foundation

(1956) of optimal control theory. With the advent of the digital computer during (late 1950s), a recursive algorithmic solution was made possible. The presentation of a paper "On the General Theory of Control Systems" by R. E. Kalman, at the Moscow conference showed a duality between multivariable feedback control and filtering. Kalman's work on the concepts of observability and controllability, and Rosenbrock's idea of modal control, led to extensive work on "pole shifting." The Kalman-Bucy filter demonstrated the basic role of feedback (through innovations sequence which is $y(t)-y_m(t)$), in filtering theory. Ironically, the powerful optimal control methods could not be used on general industrial problems because accurate plant models were not available. Astrom, K. J. and Eykoff, P. had remarked that a strength of the classical frequency response method is its "very powerful technique for systems identification (i.e. frequency analysis through which TFs can be determined accurately for use in the synthesis technique)." In modern control, the models used are parametric models in terms of state equations. There was a revival of interest in the frequency-response approach for multivariable systems by Howard Rosenbrock (1966). The main impact during the 1950s and 1960s was to support theoretical investigations and particularly synthesis, and by the early 1960s the digital computer had been used on-line to collect data, for optimization and supervisory control and in a limited number of applications for direct digital control. A leading advocate for the use of the digital computer in the process industries was Donald P. Eckman, who argued in support of "Systems Engineering": i.e. the engineers should have "a broad background across conventional boundaries of the physical engineering/mathematical sciences" and with "an ability to approach problems analytically, and to reduce physical systems to an appropriate math model to which the methods of mathematical manipulation, extrapolation, and interpretation can be applied."

In early 1950s, the activities of the design of autopilots for high-performance aircraft motivated intense research in adaptive control; and in 1958 and 1961, model reference adaptive control (MRAC) was indicated by Whitaker (and coworkers) in order to solve the problem of autopilot control. In 1958, Kalman suggested an adaptive pole placement scheme based on the optimal linear quadratic problem. However, ironically and unfortunately, the lack of stability proofs and the lack of good understanding of the properties of the proposed adaptive control solutions coupled with a disaster in a flight test diminished the interest in adaptive control. The period of 1960s was the most important one for the development of control theory and adaptive control in particular: (i) state space methods and stability theory based on Lyapunov approach were introduced, and (ii) developments in dynamic programming, dual/stochastic control, and system identification/parameter estimation played an important role in the reformulation and redesign of adaptive control. By 1966, Parks and others re-designed the MIT rule-based adaptive laws used in the MRAC schemes using the Lyapunov design approach. Certain advances in stability theory and the progress in control theory (in the 1960s) improved the understanding of adaptive control and a strong renewed interest was kindled (in the 1970s). Also, the simultaneous development and progress in computers and electronics made it feasible to implement the complex controllers, including the adaptive ones, which saw in the 1970s breakthrough results in the design of adaptive control: (i) the concept of positivity was used to develop a wide class of MRAC schemes with well-established stability properties and (ii) several classes of adaptive control schemes were produced for discrete time systems. All these were accompanied by several successful applications; however, these successes were followed by controversies over the practicality of adaptive control; one of the latter being (in as early as 1979), that the adaptive schemes of the 1970s could easily go unstable in the presence of small disturbances. In the 1980s, the non-robust behavior of adaptive control became controversial when more examples of instabilities were published (by Ioannou/Rohrs); and these examples stimulated/prompted many researchers to direct their research towards understanding the mechanism of instabilities and finding ways to counteract them. Thus (by the mid-1980s), several new re-designs and modifications were proposed and analyzed, which lead to robust adaptive control; this trend continued throughout the 1980s; which in turn improved the understanding of the various robustness modifications and led to their unification to a more general framework. For discrete time problem, Praly was the first to establish global stability in the presence of un-modeled dynamics; and by the end of the 1980s, several results were established in the area of adaptive control for linear time-varying plants. The focus of research on adaptive control (in the late 1980s to the early 1990s) was on performance properties and extending the results (of the 1980s) to certain classes of nonlinear systems with unknown parameters; these led to new classes of adaptive schemes (motivated from nonlinear system theory) as well as to adaptive control schemes with improved transient and steady-state performance. These led to the use of new concepts such as adaptive back-stepping, nonlinear damping, and tuning functions to address the more complex problem of dealing with parametric uncertainty in classes of nonlinear systems. In the late 1980s to early 1990s, further developments of neural networks led to the use of online parameter estimators to "train" or update the weights of the neural networks. Adaptive control has a rich literature full of different techniques for design, analysis, performance, and applications; despite this, there is a general feeling that adaptive control is a collection of unrelated technical tools and tricks: (a) the choice of the parameter estimator, (b) the choice of the control law, and (c) the way these are combined leads to different types of adaptive control schemes. The need of the design of autopilots for high-performance aircraft was one primary motivation for active research in adaptive control in the early 1950s.

A brief and concise history of control is given in Tables I.1 through I.4, where we gather the following succinct points [7]: (i) feedback control is an engineering discipline and (ii) its progress was/is closely tied to the practical problems that needed to be solved during various phases of the human history: (a) the Greeks and Arabs were preoccupied with keeping accurate track of time (about 300 B.C. to about 1200 AD); (b) the Industrial Revolution (IR) happened in Europe, (c) the beginning of mass communication and the First and Second World Wars (from about 1910 to 1945), and (d) the beginning of the space/computer age in 1957. In between the IR and the World Wars, control theory began to acquire its written language—the language of mathematics; Maxwell provided the first rigorous mathematical analysis of a feedback control system in 1868. Thus, relative to this written language, the period before about 1868 can be called the prehistory of automatic control. Then, following Friedland (1986), the primitive period of automatic control can be taken from 1868 to the early 1900s; then, the period from then until 1960 the classical period; and the period from 1960 to the present times (may be up to the beginning of the end of the twentieth century) the modern period; and from then onward to the present times the AI-based intelligent control (AIIC), and possibly in future (after several) decades, it might emerge as, or lead to, "nearly" intelligent control (NIC).

TABLE I.1

Brief History of Control-Part 1: Prehistory of Control Theory

300 B.C. to 1200 AD: Greeks/Arabs; and 1600: Industrial evolution (IR)	To keep accurate track of time, and IR in Europe
Water clocks	270 B.C: Greek Ktesibios—a float regulator for a water clock to control the inflow of water through a valve (compare modern flush toilet's ball and clock); 250 B.C.: Philon of Byzantium used a float regulator to keep a constant level of oil in a lamp; 800–1200: Arab engineers used float regulator for water clocks, concept of "on/off" came up, lead to minimum time problem in 1950s; 12th century: a pseudo-feedback control for navigational purpose.
The Industrial Revolution	In Europe→ prime movers, grain mills, furnaces, boiler, steam engines, temperature/pressure regulators, speed control devices; 1712-Newcomen, T. built the first steam engine;1769-J. Watt's steam engine.
The Millwrights	1745: British blacksmith, Lee, E.—fantail to point the windmill continuously into the wind; The millwrights-devices for speed of rotation led to self-regulating windmills sails.
Temperature regulators	1624: Cornelis Drebbel—an automatic temperature control (ATC) for a furnace, in incubator; 1680: Becher, J. J. and in 1754 by Reaumur used in incubator; 1775: Bonnemain—suitable for industrial use for the furnace of a hot-water heating plant.
Float regulators	1746/1775: used in flush toilet; 1758: Bindley, J.—in a steam boiler; 1784: Wood, S. T.—in a steam engine; 1765: Polzunov, I. I.—for a steam engine that drove fans for blast furnaces; 1791: adopted by the firm of Boulton and Watt.
Pressure regulators	1681/1707: Papin, D.—a safety valve for a pressure cooker, on a steam engine; 1799: refined by Delap, R., and Murray; 1803: was combined with a float regulator by Boulton & Watt for steam engines.
Centrifugal governors	1783/1786: the rotary output engine; 1788: Watt—the centrifugal flyball governor for regulating the speed of rotary steam engine.
The *Pendule Sympathique*	1793: French-Swiss A.-L. Breguet—invented a CLFS (closed loop feedback system) to synchronize pocket watches, the *pendule sympathique* used a special case of speed regulation. At a certain time, a pin emerges from the chronometer, inserts into the watch, and begins a process of automatically adjusting the regulating arm of the watchs balance spring.

Adapted and modified from: Lewis, F.L., *Applied Optimal Control and Estimation*, Prentice Hall, Upper Saddle River, NJ, 1992; http://www.uta.edu/utari/acs/history.htm.

TABLE I.2

Brief History of Control-Part 2: Mathematical Control Theory

	Formal Theories
The birth of mathematical control theory	1800s: mathematics was first used to analyze the stability, mathematics is a formal language of automatic control theory.
Differential equations	1840: British astronomer Greenwich, G. B. Airy—a feedback device for pointing a telescope, had led to oscillations, discuss the instability of CLCSs, used differential equations; Newton, I. (1642–1727), Leibniz (1646–1716), Bernoulli (late 1600s and early 1700s), Riccati, G. W. (1676–1754): infinitesimal calculus and related works; Lagrange (1736–1813), Hamilton, W. R. (1805–1865): use of differential equations in analysing the motion of dynamic systems.
Stability Theory	1868: Maxwell, J. C.—analysis of control systems in terms of differential equations, analyzed the stability of Watts flyball governor, a system is stable if the roots of the characteristic equation have negative real parts, the theory of control systems was firmly established; 1877: Routh, E. J.—provided a numerical technique for determining when a characteristic equation has stable roots; 1877: Vishnegradsky, I. I.—analyzed the stability of regulators using differential equations; 1893: Stodola, A. B.—studied the regulation of a water turbine using the techniques of Vishnegradsky, and was the first to mention the notion of the system time constant; 1895: Hurwitz, A.—solved the problem independently; 1892: Lyapunov, A. M.—studied the stability of nonlinear differential equations using a generalized notion of energy; 1892–1898: British engineer Heaviside, O.—invented operational calculus, studied the transient behavior of systems, introducing a notion equivalent to that of the transfer function.
System Theory	Eighteenth and nineteenth centuries: Smith, A.—in economics (*The Wealth of Nations*, 1776), Darwin, C. R.—(*On the Origin of Species By Means of Natural Selection*, 1859), and other developments in politics, sociology, and elsewhere made a great impact on the human consciousness; Early 1900s: Whitehead, A. N.—(1925), with his philosophy of "organic mechanism," and L. von Bertalanffy (1938), with his hierarchical principles of organization, and others had begun to speak of a "general system theory."

(Continued)

TABLE I.2 (*Continued*)

Brief History of Control-Part 2: Mathematical Control Theory

	Formal Theories
Frequency-Domain Analysis	1920s, 1930s: the frequency domain approaches developed by P.-S. de Laplace, J. Fourier, and A.L. Cauchy, were explored and used in communication systems; 1927: Black, H. S.—usefulness of negative feedback; 1932: Nyquist, H.—regeneration theory for the design of stable amplifiers, derived his Nyquist stability criterion based on the polar plot of a complex function; 1938: H.W. Bode—used the magnitude and phase frequency response plots of a complex function, and investigated closed loop stability using the notions of gain and phase margin; 1964~: Groszkowski, P. J.—used the describing function method in radio transmitter design, and formalized by Kudrewicz, J.
The world wars and classical control: ship control, weapons development and gun pointing, MIT radiation laboratory,	1910: Sperry, E. A.—designs of sensors, invented gyroscope; 1922: Minorsky, N.—introduced three-term controller and used PID strategy, also considered nonlinear effects; 1934: Hazen, H. L.-coined the word *servomechanisms* used mathematical theory; 1946: Hall, A. C.—used frequency domain techniques to face the noise effects, and used the approach to design a control system for an airborne radar; 1947: Nichols, N. B.—developed Nichols charts; 1948: Evans, W. R.—root locus technique that provided a direct way of determining the closed loop poles in the s-plane.
Stochastic analysis	1942: Wiener, N.—developed a statistically optimal filter; 1941: Kolmogorov, A. N.—a theory of discrete time stationary stochastic processes.

Source: Lewis, F.L., *Applied Optimal Control and Estimation*, Prentice Hall, Upper Saddle River, NJ, 1992; http://www.uta.edu/utari/acs/history.htm.

TABLE I.3

Brief History of Control-Part 3: The Space/Computer Age and Modern Control

	Optimal/Modern Control
Time-Domain design for nonlinear systems	1948: Ivachenko—investigated a principle of relay control, where the original control signal is switched discontinuously between discrete values; 1955: Tsypkin—used the phase plane for nonlinear controls design; 1961: Popov, V. M.—provided circle criterion for nonlinear stability analysis.
Sputnik-1957	1957: Soviet Union—launched the first satellite, the history of control theory there paved the way.
Navigation	1960: Draper, C. S.—invented inertial navigation system, used gyroscopes to provide accurate information on the position of a moving body in space, e.g. ship, aircraft, spacecraft.
Optimality in Natural Systems	1696: Bernoulli, J.—first mentioned the principle of optimality (all are minimum principles) in connection with the Brachistochrone; 1600s: P. de Fermat—minimum time principle in optics; 1744: Euler, L. and Hamilton's work—a system moves in such a way as to minimize the time integral of the difference between the kinetic and potential energies; 1900s: Einstein, A.—showed that relative to the 4-D space-time coordinate system, the motion of systems occurs such as to maximize the time. Naturally occurring systems exhibit optimality in their motion.
Optimal Control and Estimation Theory	1957: Bellman, R.—applied dynamic programming to the optimal control of discrete time systems, showed that the natural direction for solving optimal control problem is backwards in time; 1958: Pontryagin, L. S.—developed maximum principle, that relies on the calculus of variations of (Euler, L., 1707–1783); 1960–1961: Kalman, R.—LQG, optimal filtering, a recursive solution to the least-squares method of Gauss, C. F. (1777–1855), and a new era of modern control emerged.
Nonlinear Control Theory	1966: Zames, G., 1964: Sandberg, I. W., 1964: Narendra, K. S., and 1965: Desoer, C. A.—extended the work of Popov and Lyapunov in nonlinear stability.

Source: Lewis, F.L., *Applied Optimal Control and Estimation*, Prentice Hall, Upper Saddle River, NJ, 1992; http://www.uta.edu/utari/acs/history.htm.

TABLE I.4

Brief History of Control-Part 4: Computers in Control-Design/Implementation

	Growth of Computers
The development of digital computer	1830: Babbage, C.—introduced modern computer principles (memory, program control, branching); 1948: J. von Neumann—directed the construction of the IAS stored-program at Princeton; IBM built its SSEC stored-program machine; 1950: Sperry—built the first commercial data processing machine, the UNIVAC I; 1960: computers with solid-state technology; 1965: DEC—built the PDP-8; 1969: Hoff, W.-invented the microprocessor.
Digital Control and Filtering Theory	1969: the growth of digital control theory; 1950s: Ragazzini, J. R., Franklin, G., Zadeh, L. A., Jury, E. I., Kuo, B. C.—theory of sampled data systems; Astrom and Wittnmark (1984): the idea of using digital computers for industrial process control; 1940s/1950s: the control of chemical plants, the development of nuclear reactors motivated the exploration of the industrial process control and instrumentation; 1950s: Shannon, C. E.—importance of the sampled data techniques; 1974: Gelb, A.—application of the digital filtering theory.
Personal computer	1983: PC; SW packages—ORACLES, Program CC, Control-C, PC-MATLAB, MATRIX$_x$, Easy5, SIMNON for design of control systems emerged.
Union of modern and classical control	1970s: Rosenbrock, H. H., MacFarlane, A. G. J., Postlethwaite, I.—extended the classical frequency-domain techniques and the root locus to multivariable systems; Horowitz, I.—developed quantitative feedback theory; 1981: Doyle, J., and Stein, G., Safanov, M. G., Laub, A. J. and Hartmann, G. L.—showed the importance of the singular value plots vs frequency in robust multivariable design; 1986: Athans, M.—pursued in aircraft and process control; this led to a new control theory that blends the best features of classical and modern control methods.

Source: Lewis, F.L., *Applied Optimal Control and Estimation*, Prentice Hall, Upper Saddle River, NJ, 1992; http://www.uta.edu/utari/acs/history.htm.

I.1 The Philosophies of Classical and Modern Control

The approach of classical control was and is based on frequency domain and s-plane analyses. It relies on transform methods: Laplace transforms (for LTI systems) and Describing Function method (for nonlinear systems); $G(s) = \frac{k(s+a)}{s(s+b)}$, an open loop TF of a system for which a control law is required. The internal description of the plant is not needed for the classical designs; an I/O description is enough. The designs were made by using hand, with the use of graphical techniques, which led to several possibilities and the resulting control systems were not unique; it was an engineering-art-type approach. The classical theory is natural for the design of the systems that are (inherently) robust to the disturbances to the systems, and noise in observations, such designs were carried out using the ideas of gain and phase margins; simple compensators: PI, PID, lead-lag, washout circuits are used, and these can be tuned easily for meeting certain performance specifications. A basic concept in classical control is the ability to describe closed loop properties in terms of open loop properties; the latter are known or easy to measure: the Nyquist, Bode, and root locus plots are in terms of the open loop transfer functions (OLFTs). Also, the closed loop disturbance rejection properties and steady-state error can be described in terms of the return difference and sensitivity. However, it is difficult to apply to multiple-input

multiple-output (MIMO) control design, due to interactions involved from the other control loops; however many of these limitations were overcome with the development of multivariable frequency-domain and quantitative feedback theory approaches (1970s).

The modern control approach is based mainly on time-domain methods. An exact state-space mathematical model of the system for which a controller is to be designed, is required:

$$\frac{d(x)}{dt} = \dot{x}(t) = Ax(t) + Bu(t); \; y(t) = Cx(t) \qquad (I.1)$$

In (I.1), $x(.)$ is the vector of internal states/variables, $u(.)$ the control input, and $y(.)$ the observed signals (called measurements/observations); the latter contain directly/explicitly or indirectly/implicitly the information on the states, $x(t)$ of the system. The power of the modern control is that the state space models could be SISO or MIMO. The matrices A, B, and C describe the dynamics of the system. The modern control started first for linear systems, then was extended to nonlinear systems. Open loop optimal control designs for the nonlinear system can be obtained by solving nonlinear two-point boundary value problem (TPBVP). Certain open loop properties, like observability/controllability, can give insight on what can be achieved using feedback. To obtain good closed properties, one can use state feedback control

$$u(t) = -Kx(t); \; K - \text{feedback gain matrix/vector} \qquad (I.2)$$

Then, in a linear quadratic regulator (LQR) problem, a performance index (PI) is used and minimized with respect to the gain and optimal value is obtained:

$$J(K) = \sum [x^T(t)R(t)x(t) + u^T(t)Qu(t)]dt \qquad (I.3)$$

In (I.3), $u(.)$ contains gain $K(.)$ as seen from (I.2). This PI is an extension of the SISO PIs to the MIMO systems. The elements of matrices R and Q are the design parameters and are to be tuned to obtain an adequate performance of the designed system. It can be shown that the following control law can be obtained by the minimization of (I.3):

$$K(.) = Q^{-1}B^T P \qquad (I.4)$$

The matrix P can be solved from the associated matrix Riccati equation

$$0 = \dot{P}(t) = A^T P(t) + P(t)A - P(t)BQ^{-1}B^T P(t) + R \qquad (I.5)$$

The features/merits of this approach are: (i) if the system is controllable, and R, Q are properly determined, then the gain K guarantees the stability of the closed loop control system that is given as

$$\frac{d(x)}{dt} = \dot{x}(t) = (A - BK)x(t) + Bu(t); \qquad (I.6)$$

(ii) easy to apply to multiple-input systems; (iii) for obtaining K, computer-aided design approach can be used; (iv) the approach is a formal one and provides a unique answer to the controller design, and it is again an engineering art in modern control (in choosing appropriate Q, R some trial and error approach, or parametric study is needed using the engineering judgement); (v) the gain K is computed in terms of the open loop quantities A, and B (and Q, R), thus, interestingly, both the classical and modern control design approaches determine the closed loop properties in terms of open loop parameters; and (vi) since the complete K is determined, closing of all the feedback control loops happens concurrently. The design as such works, but does not provide much intuition on the properties of the CLCS. Since often all the state variables are not available for the feedback, one can use the output feedback: $u(.) = -Ky(t)$. The state feedback does not offer any structure in the control system; however, the output feedback control law can be used to design a compensator with a desired dynamical structure, thus providing some intuition of the classical approach. The LQR with full state feedback has some robustness properties: (i) infinite gain margin, and (ii) 60° phase margin. However, the LQR design using static or dynamic output feedback has no guaranteed robustness features. The robust modern control techniques are now available (early 1980s), LQR-LTR (LQR/H-Infinity) design incorporates a proper treatment to account for and reduce the effects of modeling uncertainties and disturbances, on the performance of the CLCS, thus designed. With these modern/robust approaches, much of the classical-intuition can now be incorporated in the newer approaches. The major aspect is that the modern/robust control strategies are becoming more suitable and useful for the design of multivariable and complex control systems that can be routinely implemented on microprocessors, in on-board microcomputers for aerospace vehicles, and robots; and these strategies might be much more effective than the classical PID/lead-lag compensators-based approaches, and in some cases where the latter are not even feasible. Of course, there are other developments like matrix-fraction descriptions and polynomial equation design methods that are suitable to represent the MIMO plant in I/O form, and this is an extension of the classical TF approach.

I.2 The Philosophy of AI-Based Intelligent Control (AIIC)

We consider here, the AIIC, since to actually realize an intelligent control in a true sense is nearly impossible, if not so then, at least extremely difficult, since we (and our brains) have evolved to the present state after millions of years of (natural) evolution. Then, the idea of building, at least an AIIC is to help us achieve many complex goals that require very sophisticated and robust approaches to design. The advent of very fast and powerful digital (and parallel) computers and microprocessors/microcomputers/chips has made it possible to implement the techniques of soft computing (artificial neural networks [ANNs]; fuzzy logic [FL]; and Genetic Algorithms [GAs]) for the development of the sophisticated controllers for complex systems. In fact, these approaches are such that the systems are implicitly (but of course actually) controlled and there might not always be a specific structure of a (digital) controller. In most cases, ANN/FL are used for modeling certain uncertainties in the plant modeling and/or the unknown disturbances. In most cases, actually, behavior is modeled rather than the actual plant, so rendering it as a model-free approach. In classical and modern approaches, some explicit model of the plant is required (either TF, or state space). In the ANN-based approach, some learning is involved, and in FL-based control design, If…Then rules can be incorporated, thereby incorporating the (learned) knowledge of the human expert (from the past experience), so it is a knowledge-based approach. Also, in FL, the uncertainty is incorporated via (fuzzy) membership functions. The GAs then can be used for the optimization of the chosen criterion, i.e. performance index, optimization of the learning procedure of ANNs, and learning rules for FL from the available data, if these rules are not available readily.

Thus, the ANN-based models, or the learned ANNs will act as (an implicit) controller in a control loop with the plant, providing a CLCS. In the case of a FL-based controller, the entire fuzzy inference system (fuzzification, application of rules, aggregation, and defuzzification) would act as a controller in a control loop with the plant, providing a CLCS. Interestingly, various suitable combinations of the classical, modern, and AIIC-based control design approaches can be exploited to derive benefits from the good features of each individual approach.

Since some historical notes on these soft-computing paradigms are provided in some chapters of Section IV, it is not dealt with here in the introduction.

Interestingly, there are many good books on control and closely related topics [8–30]; however, a few are dated, some are not as comprehensive, and a few are highly specialized. The treatment in the present volume is relatively comprehensive and composite (not necessarily complete, since control theory has grown too far, and has several facets), and covers several areas of control (linear, nonlinear, calculus of variation, optimal, model predictive, robust, digital, adaptive, and ANN-FL-GA), though the attempt is not to replace any of the equivalent popular volumes. The book will be useful to upper undergraduate (7th/8th Semester classes) and some Masters programs (1st Semester), besides being useful to doctoral research scholars and practicing scientists/engineers.

I.3 Outline of the Sections and Chapters of the Book

In Chapter 1, we primarily present the charm of the transfer function perspective of systems and the intuitive PID controls; a limited dose of the state space approach, which has also become classical by now in some sense, and it appears towards the end of the chapter. Care is taken to strike an adequate balance between the theoretical inputs and numerical examples to make the presentation comprehensive, yet wholesome. In Chapter 2, we briefly introduce the nonlinear systems and present some sort of counterintuitive nature of the systems; for instance, if there is a slight perturbation in the initial condition, the response might undergo an unexpectedly large change. While a general theory, such as the transfer functions, is a bit of a distant dream, we present via

appropriate examples, certain classification of nonlinear systems in Chapter 3. We throw a comprehensive light on certain fundamental concepts, such as existence and uniqueness of solutions, asymptotic stability of systems, and so on. With the ground prepared in Chapters 2 and 3, we move on to the ideas in controller design for nonlinear systems in Chapter 4. Primarily, we focus on a specific class of nonlinear systems, known as the affine systems, which encompass several practical applications, such as robotics, and develop an interesting modern design perspective called feedback linearization. Towards the end of the chapter, we explore a couple of special cases—Backstepping control and Sliding Mode control—of this design methodology.

Section I concludes with a few exercises for the reader to review and learn further through simulations, the bag of ideas contained in all the four chapters. Also, in the Appendixes of this section, we give a brief account of performance error criteria, table of Laplace transforms, describing function analysis of a saturation nonlinearity, and a control design process.

In Section II with four chapters, we cover the main topics of optimal control, including calculus of variation, model predictive control, and the theory of robust control. In Chapter 5 through 8, we mainly cover variational aspects, maximum principle, dynamic programming, differential games, linear quadratic optimal regulator, pole-placement, eigenstructure assignment, minimum-time optimization, and H_2 optimal and H_∞ control. Certain supporting material is presented in the Appendixes of Chapters 5 through 8. Also, some illustrative examples are presented in Appendixes 7B, 8A, and IIC.

In Chapters 9 and 10 of Section III, we present fundamental aspects of discrete time control systems, design of such systems based on root locus method, frequency domain analysis, compensator design, state feedback control, state observers, and optimal control for discrete time systems. In Chapter 11, we present gain-scheduling approach, self-tuning control, and model reference adaptive control, and in Chapter 12, a brief explosion of computer controlled system is given: computers in measurement and control, smart sensor systems, and aspects of implementation of a digital controller. Certain supporting material is presented in the Appendixes of the Chapters 9 and 11. Also, some illustrative examples are presented in Appendix III. Most of the exercises (and their solutions; the manual of which would be available from the CRC press to the instructors who may follow the book) are from several references cited for Sections II and III.

In Section IV of four chapters, we first provide a comprehensive introduction to the three techniques–the artificial neural networks (ANNs), fuzzy logic (FL), and the Genetic Algorithms (GAs), in Chapter 13. The topics covered under ANNs are fairly exhaustive, ranging from its inception as perceptrons, the sigmoidal neurons, through supervised and unsupervised learning methodologies, radial basis and recurrent networks, to the current day deep neural networks. There is a reason for this apparent bias—the mathematically deep rooted theory has certain strongly logical pointers towards the understanding of problems. We present the topics in an intuitive manner, but also provide various directions for further reading as well as research. Generally speaking, fuzzy logic has gained more popularity over ANNs for a variety of reasons including, perhaps, the advantages in the fuzziness of human decision making capabilities. There have also been interesting strides in this school of intelligence, and recently several applications have been reported in the next generation-the type II Fuzzy Logic. The chapter concludes with a fair amount of discussion on the Genetic Algorithms. Today, there are more than a hundred algorithms, largely claimed to have drawn inspiration from Nature's various facets, available in literature. While everything cannot be put tightly in a few pages, we have presented the basic principle behind this school of intelligence and briefly discussed about one of the hundred plus algorithms that has attracted relatively more attention, the particle swarm optimization (PSO). Chapters 14 through 16 are devoted to the showcasing of various success stories employing intelligent control strategies. The case studies have a wider range, including process control systems, aerospace applications, as well as a social-comfort system. We have also presented some of the trends in hybridization such as using Genetic Algorithms to tune the PID controller gains in Chapter 16. Also, in the Appendixes of this section and Chapters, we have given further treatment of some important aspects of use and applications of ANNs, FL, and GAs for the control approaches: (i) adaptive-neuro control, and Lyapunov method based stability analysis of the ANN-based control system; (ii) FL-based gain scheduling, and development of IT2FL-based decision system; and (iii) use of GAs in the design of PID and helicopter controllers. In Appendix IVA, we present the main aspects/features of artificial intelligence, intelligent systems, and intelligent control.

In Section V, some important system theoretic concepts (controllability/observability, stochastic processes, Lyapunov stability results) needed for understanding the material of some chapters of the first four sections, procedure for decentralized control design, and some more illustrative examples are provided.

Disclaimer: Although enough care has been taken to present the theory/concepts and examples of various control approaches in this book, the readers/users should use their own discretion in utilizing these (theories, and illustrative example

algorithms, programs, and software [SW]) for solving their own practical problems. The authors and the publisher of the book do not take any responsibility in the unsuccessful application of the said algorithms/programs/SW to the problem of the readers/users. The same applies to the dates/periods and notes in the brief history of control, concisely given in Tables I.1 through I.4. The Figures I.1 through I.9 are included here for the sake of illustrating certain developmental aspects of early history of control, and the pictures and images are not used/not to be used for any commercial purposes.

Credits: Due to proliferation of MATLAB and its toolboxes for solving the problems of control system analysis and design, various URLs now have MATLAB-based examples. We have followed these in some sections, *re-assembled* certain scripts, and *re-run* these codes to generate *fresh results and plots* presented in certain Appendixes of the present volume, with appropriate citation of the (open-) source references therein. The idea has been that the reader gets a quick overview of these results (after or along with the study of control methods from the chapters of the book) and then can visit, if needed, the original sources for further details. The Figures I.1 through I.9, and the material for the Tables I.1 through I.4 are also from (open-) sources that are appropriately cited along with the titles of these figures and under the table/s.

References

1. Heim, G. Biofeedback. PTA 236 presentation, ppts. http://www.mccc.edu/~behrensb/documents/BiofeedbackGHeim.pdf, accessed May 2017.
2. Cosentino, C., and Bates, D. *An Introduction to Feedback Control in Systems Biology.* CRC Press, Boca Raton, FL; *Feedback control in Systems Biology.* CRC Press, Boca Raton, FL, 2011. https://www2.warwick.ac.uk/fac/sci/eng/staff/dgb/bookmainfinal.pdf, accessed May 2017.
3. Bennett S. A brief history of automatic control. *IEEE Control Systems,* pp. 17–25, June 1996. http://ieeecss.org/CSM/library/1996/june1996/02-HistoryofAuto Ctrl.pdf., accessed May 2017.
4. Anderson, B. D. O. Present developments of control theory. Presentation at the IFAC 50th Anniversary Celebrations, IFAC, September 2006, Heidelberg, Germany. https://www.ifac-control.org/about/ifac-50-lectures/2-Anderson.pdf., accessed May 2017.

5. Fernandez-Cara, E., and Zuazua, E. Control theory: History, mathematical achievements and perspectives. *Bol. Soc. Esp. Mat. Apl.*, pp. 1–62.

6. Ioannou, P. A., and Sun, J. *Robust Adaptive Control.* Prentice-Hall, Upper Saddle River, NJ, 1996.

7. Lewis, F. L. *Applied Optimal Control and Estimation.* Prentice Hall, Upper Saddle River, NJ, 1992.

8. Ogata, K. *Modern Control Engineering.* Prentice Hall of India, New Delhi, India, 1985.

9. Vukic, Z. *Nonlinear Control Systems.* CRC Press, Boca Raton, FL, 2003.

10. Kirk, D. E. *Optimal Control Theory: An Introduction.* Dover Publications, Mineola, New York, 1970, 1998.

11. Kuo, B. C. *Automatic Control Systems* (3rd ed.). Prentice-Hall, Englewood Cliffs, NJ, 1975.

12. Shinners, S. M. *Modern Control System: Theory and Design* (2nd ed.). John Wiley & Sons, New York, 1998.

13. Elgerd, O. I. *Control System Theory.* McGraw-Hill Education-Europe; McGraw-Hill, New York, 1967.

14. Houpis, C. H., Sheldon, C. H., and D'Azzo, J. J. *Linear Control System Analysis and Design* (5th ed.). CRC Press, Boca Raton, FL, 2001.

15. Siddique, N. *Intelligent Control: A Hybrid Approach Based on Fuzzy Logic, Neural Networks and Genetic Algorithms.* Springer International Publishing, Berlin, Germany, 2014.

16. Baocang, D. *Modern Predictive Control.* CRC Press, Boca Raton, FL, 2009.

17. Garcia-Sanz, M. *Robust Control Engineering: Practical QFT Solutions.* CRC Press, Boca Raton, FL, 2017.

18. Saridis, G. N. *Self-organizing Control of Stochastic Systems.* Marcel Dekker Publications, New York, 1977.

19. Lewis, F. L. *Optimal Control.* John Wiley & Sons, New York, 1995.

20. Bryson, A. E., and Ho, Y.-C. *Applied Optimal Control: Optimization, Estimation, and Control.* John Wiley & Sons, New York, 1975.

21. Stengel, R. F. *Optimal Control and Estimation.* Dover (Courier) Publications, New York, 1994.

22. Graham, D., and McRuer, D. *Analysis of Nonlinear Control Systems.* John Wiley & Sons, New York, 1961.

23. Landau, I. D. *Adaptive Control: The Model Reference Approach.* Marcel Dekker Publications, New York, 1979.

24. Narendra, K. S., and Annaswamy, A. M. *Stable Adaptive Systems.* Prentice Hall, Englewood Cliffs, NJ, 1989; Dover Publications, 2004.

25. Sastry, S., and Bodson, M. *Adaptive Control: Stability, Convergence and Robustness.* Prentice Hall, NJ, 1989.

26. Astrom, K. J., and Wittenmark, B. *Adaptive Control.* Addison-Wesley, Boston, MA, 1995.

27. Landau, I. D., Lozano, R., and M'Saad, M. *Adaptive Control.* Springer-Verlag, New York, 1998.

28. Tao, G. *Adaptive Control Design and Analysis.* Wiley-Interscience, Hoboken, NJ, 2003.

29. Goodwin, G. C., and Sin, K. S. *Adaptive Filtering Prediction and Control.* Prentice-Hall, Englewood Cliffs, NJ, 1984.

30. Vidyasagar, M. *Control System Synthesis: A Factorization Approach*, MIT Press, Cambridge, MA, 1985.

Section I

Linear and Nonlinear Control

Feedback is ubiquitous in nature—no wonder it pervades the day-to-day conversation of virtually everyone, and control theory is enriched by the language, questions, and perspectives of fields as diverse as linguistics, animal behavior, neurosciences, and even fine arts. Thus, it is easy to see that control theory helps us understand both natural and man-designed technological systems. All this just happens because generalized models abstracted from Nature give us a mathematical means to connect control theoretic explanations of nature with opportunities in control design.

Practical control mechanisms—as simple as responding to any call of action, either form the Nature or any human—at any point of time since antiquity, have been placing some interesting demands on control theory. Since then, the theory is continually evolving and migrating to practice and, at the same time, the practicing control engineers are both well-trained and imaginative, and new theories are developing, literally in a feedback loop.

Most natural and real-life systems are nonlinear in their behaviour and difficult to analyse; hence, we often deal with linear systems to develop the theories and methods of analysis. The nonlinear systems are often linearized so that the linear control theory can be utilized.

This section focuses primarily on classical control and comprises of four chapters on linear and nonlinear control theory and applications.

1

Linear Systems and Control

This text book is written for students who have loads of energy and enthusiasm to jump into the vortex of **think-work out-execute** exciting projects. It is not necessarily only meant for students of electrical sciences, but also for students from different disciplines, such as economics, who may join hands to build, say, mobile robots. Let us first look into some examples to convince ourselves that this wonderful subject of control systems is already known to us in an informal way, but it is time to learn things in an organized way.

1.1 Dynamic Systems and Feedback Control

1.1.1 Balancing a Stick

This is one of the projects every one of us would have certainly had hands-on experience in our childhood, and even later. Recollect how you were excited to balance a long stick in your palm. Now throw in some mechanics and see how much this fun, simple play can teach us.

Mathematically, this is a pendulum to which we can apply a torque $u(t)$ as an external force, as shown in Figure 1.1a. Let us make a couple of harmless assumptions : friction is negligible, all of the mass m is concentrated at the end, and the rod has unit length.

We might recall Newton's laws for rotating objects. When applying these laws for our case, we get the following equation

$$m\frac{d^2}{dt^2}\theta(t) + mg\sin\theta(t) = u(t) \qquad (1.1)$$

in terms of the variable θ (measured in radians) that describes the counterclockwise angle with respect to the vertical. To reduce the clutter without loss of generality, let us further assume that $m = g = 1$. If we put $u(t) = 0$, it is easy to see that the pendulum simply hangs down motionless. Interestingly, the equation tells us that there is one more such position for the pendulum, $\theta = \pi$ with zero angular velocity. This position corresponds to our stick in the palm, as shown in Figure 1.1b. However, this position results in an "unstable" motion if there is a *small* deviation. Thus, it is this instability that excites

us when playing with the stick. So, let us now ask ourselves what torques to apply to nullify such deviations. If these deviations are indeed small, we may make a further assumption that

$$\sin(\psi) \approx \psi$$

and remove sin(.) from the equation. Thus, with a new variable ψ defined as

$$\psi \triangleq \theta - \pi$$

our equation becomes

$$\frac{d^2}{dt^2}\psi(t) - \psi(t) = u(t) \qquad (1.2)$$

With this equation, our question is clear: What $u(t)$ brings the relative position ψ and the relative angular velocity to zero for any small nonzero initial values in eqn. 1.2? There could be a few more questions, such as figuring out the rate at which we force the convergence, that we might add. However, we shall not do that now. Instead, we shall focus on writing an equation to the strategy we employed while playing with the stick. First, if the stick were to lean towards our right, we would instantaneously (quickly) force the stick to move towards our left; i.e. *opposite* direction. Next, if the deviation is small, then the amount of torque we apply would also be small; otherwise, we tend to force a larger deviation in the opposite direction. In other words, the magnitude of torque is *proportional* to the magnitude of deviation. Putting these two observations together, we get

$$u(t) = -\alpha \cdot \psi(t) \qquad (1.3)$$

where α is some positive real number called the *gain*. Substituting this into our differential eqn. 1.2, we get

$$\frac{d^2}{dt^2}\psi(t) + (\alpha - 1)\psi(t) = 0 \qquad (1.4)$$

Now, we may apply our knowledge of ordinary linear differential equations and examine the auxiliary equation

$$x^2 + \alpha - 1 = 0$$

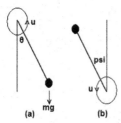

FIGURE 1.1
Balancing a stick.

which suggests that the solution consists of two exponential terms, for example:

$$Ae^{x_1 t} + Be^{x_2 t}$$

But, to our dismay, the stick never rests motionless in the desired vertical position. That this is our experience has been *formally* verified. The problem is not with approximating the sine function; this negative proportional torque does not work even if we retain the sine function. Later in this chapter, we shall see that, in addition to the proportional term, we will need to add a term that acts like a brake. The modified torque function looks like

$$u(t) = -\alpha \cdot \psi(t) - \beta \cdot \frac{d}{dt} \psi(t) \tag{1.5}$$

Perhaps, we did not bother ourselves too much by moving our palm to produce this kind of a torque, and quite possibly it is because of this reason this simple stick balancing game has been entertaining us for generations. We shall leave this example here. We expect the reader to brainstorm and learn more on his own.

1.1.2 Simple Day-to-Day Observations

After a slightly heavier dose of mathematical approach, we shall look into more intuitive examples. The following are some common phrases we come across every moment.

- He has no *control* over his expenditures
- I could not *control* my laughter.
- The law and order situation in the state remains *controlled*.
- Pest *control* for better yield of coconuts.
- The doctor suggested weight *control* with exercises.

Don't we agree?

1.1.3 Position Control System

In this section, we shall once again resume putting our ideas in mathematical language. During this process, we shall also learn, albeit informally, some of the technical terms we choose to live with hereafter.

Let us imagine we are driving an automobile. We start from point **A**, and our destination, point **B**, is 60 km away. Obviously, depending upon the time constraint, we choose to drive at a *pre-calculated* speed. Let us say we need to reach our destination exactly in 60 minutes. Hence, the pre-calculated speed is 60 kmph. We may put this in the following sentence: If we drive the automobile at 60 kmph, then we will reach our destination in exactly an hour.

The antecedent clause in this sentence refers to the **cause**, and the consequent refers to the **effect**. This may be depicted as shown in Figure 1.2.

The cause may be physical, for example, the pressure we apply on the accelerator pedal to maintain the speed of the vehicle at constant 60 kmph. We prefer the word **input** for convenience. Similarly, the effect may be the rate of displacement or the speed of the vehicle; or equivalently, the distance covered by the vehicle. Here, we prefer the word **output** to effect. Thus, we assume that if an input is applied, there will be a unique output. This relationship between the input and the output is essential in defining a system.

Evidently, so far, we have considered an ideal case where it has been assumed that there is absolutely no traffic, no speed-breakers, and so on. Let us now include one such practical factor. Suppose, there is some obstacle, such as a railway crossing, after traveling 10 km, when we need to apply the brakes and wait for 10 minutes. Soon after the railway gate is open, we need to start and drive the automobile with the speed increased to 75 kmph so that the time lost may be *compensated*, and we reach the destination on scheduled time.

What exactly is the process that we accomplished just now? We **measured** the variables—the distance and the time—and accordingly changed or modified the speed. In other words, we **controlled** the automobile according to the difference between the **desired** distance to be covered and the **actual** distance covered so far. This may be depicted as shown in Figure 1.3.

For obvious reasons, we call the scheme in Figure 1.2 an **open loop** system and the one in Figure 1.3 a **closed loop** system. In this illustration, the system we attempted to control is an automobile. Alternatively, the system could be energy, or information, or money.

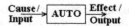

FIGURE 1.2
Schematic of a system.

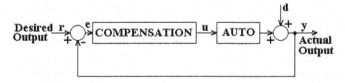

FIGURE 1.3
Simple closed loop system.

In general, any system that is required to be controlled is called a **plant, P**. The input applied to the plant is called the **control signal, u**. This is also called **actuating signal**, or **control law**, which is provided by a **compensator, C**. In the closed loop control system, the plant output **y** is compared with a **reference, r** input and an **error, e** is computed. Accordingly, we also call this system a **negative unity feedback** control configuration. Other configurations also exist (as might be expected), and we shall see them as we proceed through the subsequent chapters. This negative unity feedback configuration, however, is the most used one as it is conceptually quite simple. In our earlier stick-balancing example as well, it is worth verifying that after the torque is expressed in terms of the position and velocity, we have closed the loop with a negative feedback.

The signals $u(t)$ and $r(t)$ are depicted in Figure 1.4 for the above automobile example. Figure 1.4b shows the control signal $u(t)$ in the case of open loop system and Figure 1.4c shows $u(t)$ for the closed loop system.

Although it was mentioned above that the closed-loop system is more *practical*, the control signals shown in Figure 1.4 do not appear to conform. It is *physically* not possible to pick-up (or pick-down) velocities of the order of tens instantaneously. The corresponding delay in time has to be further compensated. However, for all good reasons to follow, we ignore the inertias at the moment.

1.1.4 Temperature Control System

Now we shall turn our attention to a very interesting physical example [1] we are quite familiar with in various guises. Figure 1.5 shows a simple temperature control system—the thermostat. We have a temperature sensing device LM335 to monitor the current temperature. This is compared using the Op-Amp comparator LM339 with the preset temperature, established by the zener LM329 via the resistance R_2. The LM335 is an active reference diode designed to produce a temperature dependent voltage according to $V(T) = T/100$, where T is absolute temperature in Kelvins. For the circuit to work over a range of temperatures, i.e. voltages, the transducer bridge is stabilized by the zener LM329 at about 7 volts, and the rotary button we find on our heaters, for instance, is actually the resistor-potentiometer combination $R_1 - R_2 - R_3$.

The operation of the circuit (although it looks complicated) is simple. If the temperature measured by LM335 is above the preset value, then the high-beta power transistor LM395 is driven into cut-off, which in turn puts off the heater. If the temperature falls below the preset, the transistor saturates turning the heater fully on.

1.1.5 Mathematical Modeling of Systems

The term modeling has its roots in the Latin word *modulus*, meaning "small measure." Thus, a mathematical model is, in some sense, a measure of the behavior of what is under study. It is also implied in this word that the model is smaller than the original, and that there is a similitude in its actions.

It is very difficult to teach someone how to mathematically model physical systems. The most common practice is for college professors to teach modeling indirectly through several examples, derivations, and other classroom material that portray models without

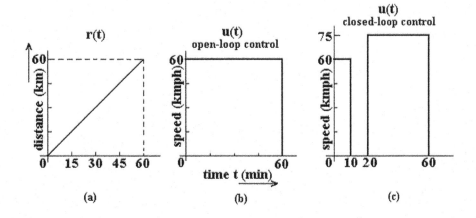

FIGURE 1.4
Elementary Signals.

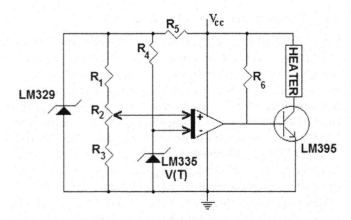

FIGURE 1.5
Temperature Control System. (Adapted from Franco, S., *Design with Operational Amplifiers and Analog Integrated Circuits*, 3rd ed., Tata McGraw-Hill, New Delhi, India, 2002.)

any emphasis on the modeling methods themselves. The actual practice of modeling is seldom taught. This is especially puzzling in the light of the depth to which modeling is embedded into engineering and science. Further, psychologists confirm that people are continually forming their own personal mental models of nature from cradle to grave. The key step in creating a mathematical model is expressing these mental models formed by ideas and concepts into useful *expressions*. From this vantage, mathematics is the language that allows expression of abstract ideas in a concise and complete manner. The prerequisites for applying this *transformation* are a broad knowledge of elementary physics and engineering and a good grasp of basic mathematics. With these tools and experience with practice, anyone can build appropriate mathematical models.

Throughout this textbook, we shall be concerned exclusively with models that are characterized by the following features:

1. Based on one or more physical principles and a set of analysis requirements, a mathematical formulation of the system model can be derived.

2. At an initial time t_o, a set of system variables can be specified.

3. For all time $t \geq t_o$, all inputs to the system can be specified.

This is exactly what we did for the stick-balancing problem, for instance.

The significance of these general features is that they are the necessary and sufficient conditions to establish the system's behavior for all time. The first feature is primarily concerned with the formulation of the equations of motion; the second and third features relate primarily to the solution of those equations for a specified

set of initial conditions and subsequent inputs. An important consequence of these features is that the past history of the model, i.e. for $t < t_o$, has no significance. Differential equations, either ordinary differential equations (odes) or partial differential equations (pdes), will be the primary mathematical descriptions of systems. We shall come to this point again and again in this book.

Mathematical modeling of engineering systems combines analytical reasoning and technical insights that are often disguised in the name of intuition. These are based on both scientific principles and aesthetic notions to explain features that are observed and to reveal features that are to be observed. Mathematical modeling is a complete course by itself. Accordingly, this textbook assumes that the reader has had an introduction to mathematical modeling in one or more earlier courses such as those described next. For electrical engineering students, a course on circuit/network theory provides the tools for honing one's intuition. Similarly, for mechanical engineering students, a course on engineering mechanics provides the tools. It would be a good exercise for the reader to collect a list of books in this area and keep, at least, one of them for ready reference.

Broadly speaking, system models can be developed by two distinct methods: (i) by *analysis* and (ii) by *experiment*. The discussion in this section so far illustrates modeling by analysis. Experimental modeling is the selection of mathematical relationships that seem to fit an observed input-output data. In some situations, the system structure may be based solidly on the laws of physics, and perhaps only a few key parameter values are uncertain. Even these unknown parameters may be known to some degree—upper and/or lower bounds or mean and variance or other probabilistic descriptions may be available at the outset. In other situations, notably in socio-economic and biological systems, the only thing available is an assumed model form, which is convenient to work with and does a reasonable job of fitting observations—for instance, a second order ordinary differential equation. All the coefficients are usually unknown and must be determined in these cases. We also call this by the names *parameter estimation*, *system identification*, and so on.

Good mathematical modeling satisfies the specifications without undue complexity. It is a very valuable engineering undertaking; it is also frequently subtle and difficult. In analyzing mathematical models of systems, the most important *fact* to be kept in mind is: *The model being analyzed is not the real system.*

The purpose of a model is to enhance the understanding of a real physical phenomenon. It is to the extent that a model succeeds in this purpose that the analysis of the model may succeed.

1.1.6 Linear, Time-Invariant, and Lumped Systems [2]

A brief taxonomy of various classes of systems, *vis-a-vis* this particular beginner's class, is presented next. System models are usually classified according to the types of equations used to describe them. The tree is shown in Figure 1.6. In this tree, dashed lines indicate the existence of subdivisions similar to the others shown on the same level.

Distributed parameter systems require *partial differential equations* for their description, for example, a transmission line. Lumped parameter systems are those for which all energy storage or dissipation can be lumped into a finite number of locations. These are described by purely algebraic equations as in the case of simple resistances ($v = R \cdot i$), or by ordinary differential equations as in the case of simple inductances ($v = L \cdot \frac{di}{dt}$). Analogously, one might think of the mechanical quantities—position x, its derivative velocity, and its derivative acceleration.

If the parameters or signals (including the inputs) present in a system are described in a probabilistic fashion (due to ignorance or actual random behaviour), the systems are called stochastic, or random, systems. Otherwise they are deterministic; i.e., we know all the details completely.

If all the signals in a system (and hence all the elemental equations) are defined at every instant of time, then the system is a continuous-time system. On the other hand, if some elemental equations are defined or used only at discrete points in time, we call the system a discrete time system. A typical example of a discrete time system is a flip-flop, which operates at every tick of a clock. As might be expected, continuous-time systems are described by differential equations (ordinary or partial) and discrete time systems by difference equations.

If all the elemental equations are linear, so is the system. If one or more equations are nonlinear, a diode for instance, then the entire system is nonlinear. A nonlinear system may be dubbed as a *not*-linear system.

We call a system time-invariant, or stationary or constant coefficient, when all the parameters of the system are constant with respect to time, for instance a 100Ω resistance is deemed to offer 100Ω forever. If one or more parameters, or the very form of an elemental equation, vary in a known fashion with respect to time, the system is said to be time-varying. A rocket in flight is a time-varying system since its mass includes fuel, which decreases (at a known rate) with time.

Basically, in nature, all the systems are nonlinear, time-varying, distributed parameter, and stochastic systems. However, similar to our modeling process, we start with the simplest systems. Accordingly, in this part of the book, we are primarily concerned with deterministic dynamical systems that are linear, time-invariant, and lumped. We shall hereafter use the symbol linear time-invariant and lumped parameter (LTIL) for these systems. We believe that once we have thoroughly understood these systems, each of the assumptions, such as linearity or time-invariance, can be relaxed one after another and asymptotically understand the behavior of real systems in the rest of the book.

1.2 Transfer Functions and State Space Representations [3,4]

Let us now turn our attention to a few important definitions regarding the description of systems.

1.2.1 Definition: Dynamical Systems

A system is called **static**, or **memoryless**, if its output $y(t_1)$ at some time $t = t_1$ depends only on the input applied at that particular instant; $y(t_1)$ is independent of the input applied prior to and later than $t = t_1$. For instance, an electrical circuit that has only resistors (as in the Ohm's law or Hooke's law).

If the output $y(t_1)$ depends on the input applied prior to $t = t_1$, at the instant $t = t_1$, and (possibly) later than $t = t_1$, then the system is called a **dynamical** system. The system is also said to have **memory**; $y(t_1)$ is dependent on the input applied in the past and in the present as well as on the input to be applied in the future. For instance, an electrical circuit that has capacitors and/or inductors in addition to the resistors.

Several examples may be generated based on the above definition of dynamical systems: the free-wheel of a bicycle, a sequential boolean circuit such as a shift

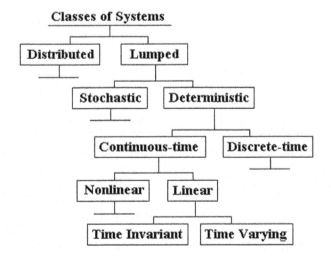

FIGURE 1.6
Classification of systems.

register, shock absorbers in vehicles, automotive alternators, propulsion systems, and so on.

It is easy now to see that simple *algebraic* equations such as

$$v = R \cdot i$$

may be used to describe static systems while *differential* equations such as

$$v = L \cdot \frac{di}{dt}$$

may be used to describe dynamical systems. Notice that we omitted "time" in the algebraic equation.

1.2.2 Definition: Causal Systems

A system is said to be **causal** if its output $y(t_1)$ at $t = t_1$ depends only on the input applied until that instant, but not later. In other words, every "effect" has always a "cause." On the contrary, if the output $y(t_1)$ depends on some input $u(t_2)$ to be applied at a later time $t_2 > t_1$, then the system is said to be **non-causal** or **anticipatory**.

In practice, we would like to have anticipatory systems such as a telephone that does not ring if the caller is an annoying salesman or an audio system that anticipates our mood and gives appropriate music. Clearly, no physical system can be anticipatory in *real time*.

1.2.3 Definition: Linear Systems

A system is said to be **linear** if and only if it satisfies the following two properties:

1. If an input u_j causes an output y_j, then an input that is α *times* u_j causes an output that is also α times y_j. Symbolically,

$$u_j \mapsto y_j \;\Rightarrow\; \alpha \cdot u_j \mapsto \alpha \cdot y_j$$

This property is called "homogeneity," or "scaling."

2. If each of n inputs u_j, $j = 1, 2, ..., n$ causes n different outputs y_j respectively, then an input that is the sum of these n inputs causes an output that is also the sum of the n individual outputs. Again, in symbols,

$$u_j \mapsto y_j \;\; \forall j = 1, 2, ..., n \;\Rightarrow\; u = \sum_{j=1}^{n} u_j \mapsto y = \sum_{j=1}^{n} y_j$$

This property is called "additivity."

These two properties may be combined as follows:

$$u = \sum_{j=1}^{n} \alpha_j \cdot u_j \mapsto y = \sum_{j=1}^{n} \alpha_j \cdot y_j \qquad (1.6)$$

This is known as the principle of "superposition."

Thus, a system is said to be linear **if and only** if it satisfies the principle of superposition. Otherwise, we call the system **nonlinear**. A nonlinear system is *not* a linear system. More about nonlinear systems and control in the next chapter.

Thus far, we have agreed that dynamical systems could be described by differential equations. To this extent we can easily refer to some typical Resistive-Capacitive (RC) and Resistive-Inductive-Capacitive (RLC) circuits. Hereafter, we shall use the words differential equation and system interchangeably.

Once there is a differential equation, its solution, i.e. the system output to a particular input demands a natural question immediately: What are the initial (or boundary, in general) conditions? The answer has two parts—the *number* of initial conditions and the *numerical values* associated. For us, to begin with, the number matters more than the numerical values. We might quickly recall that the differential equation of a simple RC circuit has one initial condition, the initial charge on the capacitor, and that the RLC circuit has two initial conditions, the initial current through the inductor and the initial charge on the capacitor. In general, a *nth* order system has n initial conditions. A system is said to be order n if the order of the highest derivative present in the differential equation is n. We use the adjective *ordinary* for the differential equation if n is finite; i.e., ordinary differential equations (ode) have finite number of initial conditions.

We also have another variety of differential equations, viz., *partial* differential equations (pde). Consider, for instance, thermal systems; while describing heat transfer by radiation or convection, the heat transfer coefficient is a function of fluid flow rate, temperature, and several other system variables. As a result, the dynamics of thermal systems have to be described by partial differential equations, in order to ensure high accuracy of calculations. Another example that involves the application of pde is a communication system in which a pair of (infinitely long) transmission lines, with distributed series inductance $L(x)$ per unit length and distributed shunt capacitance $C(x)$ per unit length, are employed for long distance communications. Yet another example is the longitudinal motion of a system containing slender structural members called rods; these rods undergo uniaxial deformation due to longitudinal forces. Thus, if the dependent variables of the system are functions of spatial coordinates such as length as well as of time, then we have a pde. Analytical solution to the pde is always considered to represent an extremely challenging mathematical problem. Clearly, a system described by a pde has infinite initial/boundary spatial conditions.

Knowledge of the initial conditions at some initial time t_o, plus knowledge of the system inputs after t_o, allows us to determine the *status* of the output at a later time t_1. As far as the output at t_1 is concerned, it does not matter how the initial conditions are attained. Thus, the initial conditions at t_o constitute a complete description of the system prior to t_o, insofar as that history affects future behavior. Henceforth, we prefer to use the term **state** for an initial condition. It is noteworthy that it appears in many other technical as well as nontechnical contexts: "state" of affairs in an organization, financial "state"ment at the beginning of a fiscal year, marital "status," and so on. In these and several other examples, the concept of state is essentially the same.

Ordinary differential equations are, thus, of great importance in systems engineering, because many physical laws and relations appear mathematically in the form of differential equations. We are more interested in the particular solution that satisfies a given initial condition $x(0)$; the initial state prescribes the initial situation at an instant (quite often at $t = 0$) and the solution of the problem shows what is happening thereafter.

While analysis of systems is solving the ode using an appropriate analytical tool, the most systematic way of doing this is using a linear transformation called the **Laplace Transformation.**[1] In fact, the reader should have already been familiar with this in an early course. Accordingly, the treatment is geared to motivate the reader rather than to burden him with mathematics.

The Laplace transformation is formally defined as follows:

$$F(s) \triangleq L\{f(t)\}$$

$$= \int_0^\infty f(t)e^{-st}dt \tag{1.7}$$

The transformation of a function *in the time-domain* into a function *in the s-domain* is expected to simplify our computations significantly.

A first order ode has the following general form:

$$y^{(1)}(t) + a_o y(t) = b_1 u^{(1)}(t) + b_o u(t) \tag{1.8}$$

where the superscript denotes the order of differentiation. Alternatively, we place dots, e.g. $\dot{y}$, above the variable to denote the order of differentiation.

---

[1] Pierre Simon Marquis De Laplace (1749–1827) was a great French mathematician who made important contributions to celestial mechanics, probability theory etc. It is interesting to note that Napolean Bonaparte was his student for a year!

Applying Laplace transformation on both sides of the equation, and remembering that the transformation is linear, we get:

$$sY(s) - y(0) + a_o Y(s) = b_1\left(sU(s) - u(0)\right) + b_o U(s)$$

$$\text{or,} \quad Y(s) = \frac{y(0) - b_1 u(0) + \left(b_1 s + b_o\right)U(s)}{s + a_o} \tag{1.9}$$

Observe that we need the Laplace transformation of the input signal $u(t)$, the initial conditions $y(0)$ and $u(0)$, in addition to the coefficients a_0, b_0, b_1 of the ode to solve for $Y(s)$. Later we may obtain the inverse Laplace transformation to determine the output signal $y(t)$ in time-domain.

Earlier, we agreed that we would apply signals to the systems only for the time $t \geq 0$, and henceforth, we assume that $u(0) = 0$. We shall rewrite the above equation in the following in a more readable way:

$$Y(s) = \underbrace{\frac{y(0)}{s + a_o}}_{\text{natural response}} + \underbrace{\frac{b_1 s + b_o}{s + a_o} \cdot U(s)}_{\text{forced response}} \tag{1.10}$$

If we examine each of the terms—the natural response and the forced response—we understand that the natural response, due to the initial condition $y(0)$, is a decaying exponent in time. Further, if we define

$$H(s) \triangleq \frac{b_1 s + b_o}{s + a_o} \tag{1.11}$$

then we see that the forced response is a *product* of two functions $H(s)$ and $U(s)$, which means that in the time-domain there is a convolution integral to be evaluated in order to solve the ode completely. Indeed, using the method of integrating factors, we can show that the complete solution in the time-domain has a convolution integral. This is the first simplification we seek.

If a transformation could make things simpler for us for a first order ode, then why not for the higher order ones? Let us look at the general form of a second order ode as follows:

$$y^{(2)} + a_1 y^{(1)} + a_o y = b_2 u^{(2)} + b_1 u^{(1)} + b_o u \tag{1.12}$$

Applying the Laplace transformation on both sides and rearranging the terms, we get

$$Y(s) = \frac{\overbrace{(s + a_1)y(0) + y^{(1)}(0)}^{\text{natural response}}}{s^2 + a_1 s + a_o}$$

$$+ \frac{b_2 s^2 + b_1 s + b_o}{\underbrace{s^2 + a_1 s + a_o}_{forced\ response}} \cdot U(s) \qquad (1.13)$$

Once again, we might extend the definition of $H(s)$ as follows:

$$H(s) \triangleq \frac{b_2 s^2 + b_1 s + b_o}{s^2 + a_1 s + a_o} \qquad (1.14)$$

and end up evaluating a convolution integral in the s-domain, this time with some more complicated functions, but still as an algebraic product.

We shall now present the general form of a *nth* order ode as follows:

$$y^{(n)} + a_{n-1} y^{(n-1)} + a_{n-2} y^{(n-2)} + \cdots a_o = b_m u^{(m)}$$

$$+ b_{m-1} u^{(m-1)} + b_{m-2} u^{(m-2)} + \cdots b_o \qquad (1.15)$$

We leave it to the reader to expand this. We simply show that the function $H(s)$ looks like:

$$H(s) = \frac{b_m s^m + b_{m-1} s^{m-1} + \cdots + b_1 s + b_o}{s^n + a_{n-1} s^{n-1} + \cdots + a_1 s + a_o} \qquad (1.16)$$

Example 1.2.1:

Let us solve the following ode

$$\ddot{y}(t) + 3\,\dot{y}(t) + 2y(t) = 2\dot{u}(t) + u(t)$$

subject to the given initial conditions: $y(0) = 1$ and $y^{(1)}(0) = 2$.

We get the following natural response:

$$Y_{natural}(s) = \frac{(s + 3)y(0) + y^{(1)}(0)}{s^2 + 3s + 2} = \frac{s + 5}{(s + 1)(s + 2)}$$

We leave it to the reader to obtain $y_{natural}(t)$ by obtaining the inverse Laplace transformation. The forced response would look like:

$$Y_{forced} = \frac{2s + 1}{(s + 1)(s + 2)} \cdot U(s)$$

Given the input (forcing function) $u(t)$, we may obtain its Laplace transform $U(s)$ and evaluate the forced response by simple multiplication in the s-domain. For instance, if $u(t)$ is a unit ramp function, applying the idea of partial fractions,

$$Y_{forced}(s) = \frac{2s + 1}{s^2(s + 1)(s + 2)}$$

$$= \frac{k_1}{s} + \frac{k_2}{s^2} + \frac{k_3}{s + 1} + \frac{k_4}{s + 2}$$

and hence

$$y(t) = k_1 + k_2 t + k_3 e^{-t} + k_4 e^{-2t} \qquad (1.17)$$

Next, we shall look into the time and frequency response $y(t)$ of the systems.

1.2.4 Time and Frequency Domains

Once we have a mathematical model, i.e. the ode together with the initial state $x(0)$, we wish to obtain the output of the system $y(t)$, i.e. the particular solution to the differential equation, to a given input $u(t)$. This is what we mean by "time response." Broadly speaking, we come across several classes of ordinary differential equations: separable equations, exact equations, Cauchy equations, and so on. In our text, first we are interested in giving a common format to the ordinary differential equations, as given in eqn.1.15, where all the coefficients a_j and b_j are real numbers, and $m \ngtr n$. That $m \ngtr n$ is rather a strict requirement since if m exceeds n, the system is no longer causal.

Thus, the mathematical description of a particular system requires an appropriate choice of the coefficients a_j and b_j and the order n. Next, we are interested in solving this *general* ode. We do this by employing the Laplace transformation. The advantages of this transformation were already elaborated. Applying this transformation to a ode results in an algebraic expression in the complex variable s. The complete response is a *superposition* of two parts—the natural response and the forced response due to an externally applied signal $u(t)$. The natural response is also called the **zero-input response** since it is independent of any external input $u(t)$. Similarly, the forced response is called the **zero-state response** since the initial state does not appear in this term. Notice that the denominator polynomia is common to both the natural and forced responses. In general, the denominator is a *nth* order polynomial that can be written directly from eqn. 1.15 as follows:

$$D(s) = s^n + a_{n-1} s^{n-1} + \cdots + a_1 s + a_o \qquad (1.18)$$

Now we can a give a format to the response of the system as follows:

$$y(t) = \underbrace{y_x(t)}_{zero-input\ response} + \underbrace{y_u(t)}_{zero-state\ response} \qquad (1.19)$$

where the suffixes x and u denote the appropriate portion of the response.

We shall take a quick, but important, detour and shift our interest to the ratio of the transforms of the output and the input, i.e.,

$$H(s) = \frac{Y(s)}{U(s)} \qquad (1.20)$$

and define this as **transfer ratio** or **transfer function**. We have been looking at this ever since we introduced Laplace transformations. Since the Laplace transform of a convolution of two functions in time is the product of the individual Laplace transforms, i.e.,

$$L\{h(t) * u(t)\} = H(s)U(s)$$

one can also immediately identify that the transfer function is no other than the Laplace transform of the impulse response, since $U(s) = 1$ for a unit impulse. All this is to say that, whatever is the description of the system—the convolution integral, or the ode, or the transfer function—essentially the same amount of information regarding the system is available for all $t \geq 0$.

It is easy, but important, to observe that the transfer function is always a **rational** function of two *polynomials* in the complex variable s. Henceforth, we denote the transfer function as

$$H(s) = \frac{N(s)}{D(s)} \qquad (1.21)$$

Notice that eqns. 1.20 and 1.21 are not, in general, identical; there could be some common factors in the numerator and the denominator that get canceled. The reader would experience this as we proceed ahead.

The roots of the numerator polynomial $N(s)$ are called *zeros* of the system since $H(s) = 0$ if s takes any of these values. The roots of the denominator polynomial $D(s)$ are called *poles* of the system since $H(s) = \infty$, and the function "blows up" in the three dimensional representation of the magnitude of $H(s)$ irrespective of the input applied. Poles and zeros are called critical frequencies. At other complex frequencies s, the transfer function has a finite and non-zero value.

Example 1.2.2:

For example, the system

$$H(s) = \frac{s + 3}{s^2 + 3s + 2} = \frac{(s + 3)}{(s + 1)(s + 2)}$$

has a zero at $s = -3$ and two poles at $s = -1$ and $s = -2$. Obviously, the number of poles is equal to the order of the polynomial, which in turn is equal to the order of the system.

Substituting the initial conditions $y(0)$ and $\dot{y}(0)$ and assuming that a unit step input is applied to the system, the complete response can be found to be of the form

$$y(t) = \underbrace{Ae^{-t} + Be^{-2t}}_{zero-input\ response} + \underbrace{Ce^{-t} + Ee^{-2t} + F}_{zero-state\ response}$$

where the coefficients A, B, C, E, and F can be computed using Heaviside's expansion. The reader may refer to the appendix for a quick review of this idea.

Since the response is determined solely by the roots of $D(s)$, i.e. the poles of the system, $D(s)$ is called as the **characteristic polynomial** of the system. This is equivalent to saying that the response of a system is *characterized* by the roots of its characteristic polynomial.

The response $y(t)$ given above may be rewritten in the following way

$$y(t) = \underbrace{Pe^{-t} + Qe^{-2t}}_{transient} + \underbrace{R}_{constant}$$

We may now notice that the exponential terms are prominent in the beginning, i.e. as $t \to 0$, but decay to zero as $t \to \infty$.

Thus, in a slightly different way, the complete response is, once again, a superposition with two components—the **transient response** that dies out soon, and the **steady-state response**, in contrast, that persists forever.

Let us now define another very important parameter—the time-constant τ of the system.

1.2.4.1 Definition: Time-Constant

The time-constant is defined as the interval of time during which an exponential decays by approximately 37%.

In other words, τ is the time at which the exponential drops to about 63% of its initial value.

Example 1.2.3:

In Example 1.2.2,

$$Pe^{-t} = 0.368 \times P \text{ at } t = 1 \quad \text{and}$$

$$Qe^{-2t} = 0.368 \times Q \text{ at } t = 0.5$$

Clearly, the first exponent takes longer to decay than the second one. Since it takes about 1 sec. for the transient to decay by about 37%, the time-constant of this system is 1 sec. Let us then ignore the second exponent and further examine the decay of the first exponent.

$$Pe^{-t} = 0.135 \times P \text{ at } t = 2$$
$$= 0.049 \times P \text{ at } t = 3 \quad \text{and}$$
$$= 0.002 \times P \text{ at } t = 6$$

The reader is encouraged to verify these numbers on his calculator. Thus, the exponent, and hence the transient response, becomes negligible (the magnitude falls below 0.5% of its initial value) within 6 sec.

In general, the transient response appears in the following format

$$y_{tr}(t) = P_1 e^{-\frac{t}{\tau_1}} + P_2 e^{-\frac{t}{\tau_2}} + \cdots + P_n e^{-\frac{t}{\tau_n}}$$

In this case, the time-constant τ may be found to be

$$\tau = \max\{\tau_1, \tau_2, \ldots, \tau_n\}$$

and the transient response becomes negligible within about 6 times τ.

This interval of 6τ may be considered to be **infinite time** in engineering since, theoretically, e^{-at} approaches zero as $t \to \infty$. Typically, electrical systems, i.e. circuits, have a time-constant of the order of $100\,\mu s$; production systems, such as paper production, have a time-constant of the order of several hours. This is the fundamental reason why factories cannot be switched on and off every day; instead, they are run on shifts and a break-down is considered very serious. In the light of this fact, it is important to study the transient response as well as the steady-state response.

The zeros and poles of the transfer function $H(s)$ play a crucial role in the steady-state response. Since $H(s)$ is multiplied by $U(s)$, we get a new set of zeros and poles for the product $H(s)U(s)$. For instance, earlier when we assumed a unit step input in Example 1.2.2, i.e. $U(s) = \frac{1}{s}$, the new set of poles is $\{0, -1, -2\}$. We call this new set the set of **modes**. One may then observe the possibility that one or more poles of $H(s)$ might be canceled by the zeros of $U(s)$. In such a situation, not all the roots of the characteristic polynomial determine the steady-state response. Accordingly, we say that the system is *not* completely characterized. In other words, if the set of modes contains the set of poles, then the system is said to be **completely characterized**.

Example 1.2.4:

Suppose in the previous example, we have applied an input

$$u(t) = -2e^{-3t} + 3e^{-4t}$$

instead of a unit step, then

$$U(s) = \frac{(s+1)}{(s+3)(s+4)}$$

and, the system pole at $s = -1$ and the system zero at $s = -3$ get canceled, rendering the system incompletely characterized.

Next, we shall focus on first-order systems. It turns out that most chemical processes, such as the continuous stirred tank reactors, may be, reasonably well, modeled as first-order systems.

1.2.4.2 First-Order Systems

A typical first order system is given by

$$\tau \dot{y} + y = u \quad where \quad \tau \in \Re \qquad (1.22)$$

Notice that we have not strictly adhered to the earlier format according to the literature, but this is the standard format as far as first-order systems are concerned.

The characteristic polynomial is

$$D(s) = \tau s + 1;$$

hence, the system has only one pole at $s = -\frac{1}{\tau}$. Since $s = \sigma + j\omega$ is a complex variable, let us now look at the *complex plane*, called the **s-plane** whose axis are the real part σ and the imaginary part $j\omega$ of the complex variable s. Figure 1.7 shows this plane. If we assume that the time-constant is always positive (as it happens to be, in general), then the pole at $s = -\frac{1}{\tau}$ can be depicted on the negative real axis in Figure 1.7.

1.2.4.3 The Role of Time-Constant

The zero-input response exhibits an exponential decay to zero with the time-constant τ. Figure 1.8 shows this decay for different values of τ.

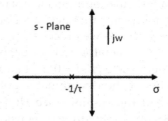

FIGURE 1.7
A single pole on the *s*-plane.

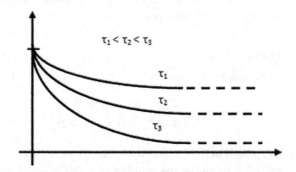

FIGURE 1.8
Role of time-constant.

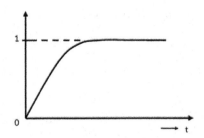

FIGURE 1.9
Typical Step-response.

Clearly, the smaller the time-constant is, i.e. farther the pole away (along the negative real axis) from the origin of the s-plane, the faster the response. If a unit step input is applied, the response will be as shown in Figure 1.9.

The response catches up with the step input within about 6 time-constants, as one might expect. Again, quite obviously, the smaller the time-constant is, the faster the response. We observe that a tangent drawn at $t = 0$ in Figure 1.9 has a slope τ.

Thus, first-order systems are governed by *only one* parameter, the time-constant; this is a positive real number, and the smaller it is, the faster the response.

1.2.5 Response of Second-Order Systems

Let us now look at a second order system, which is typically given by

$$\ddot{y} + 2\zeta\omega_n\dot{y} + \omega_n^2 y = \omega_n^2 u \qquad \zeta, \omega_n \in \Re \qquad (1.23)$$

As we might generalize from the previous section, a second-order system must be governed by *two* time-constants τ_1 and τ_2; after all, the system has two poles now. However, it has been common practice to redesignate these and call them as **figures of merit** $-\zeta$ and ω_n. The reason behind this becomes clear as we go ahead.

The characteristic polynomial of this system is

$$D(s) = s^2 + 2\zeta\omega_n s + \omega_n^2$$

This is a *quadratic expression* in s, and thus, the roots depend upon the discriminant. Let us examine this in greater detail. The two roots of $D(s)$ are

$$s_{1,2} = -\zeta\omega_n \pm \omega_n\sqrt{\zeta^2 - 1}$$

Given this general formula, we see that there are four possibilities enumerated as follows.

- **Case (i):** If $0 < \zeta < 1$, then the roots are complex conjugates with real part equal to $-\zeta\omega_n$. Let us call them

$$-\zeta\omega_n \pm j\omega_n\sqrt{\zeta^2 - 1}$$

The poles in the complex s-plane are shown in Figure 1.10a.

- **Case (ii):** If $\zeta = 1$, then the roots are real and identical. Let us call them

$$s_1 = -p = s_2$$

The poles for this case are shown in Figure 1.10b. Here, there are, apparently, two time-constants, as in the examples illustrated earlier.

- **Case (iii):** If $\zeta > 1$, then the discriminant is positive, and the roots are real and distinct. Let us call them

$$s_{1,2} = -\zeta\omega_n \pm \omega_n\sqrt{\zeta^2 - 1}$$

These poles are shown in Figure 1.10c.

- **Case (iv):** If $\zeta = 0$, then the roots are purely imaginary:

$$s_{1,2} = \pm j\omega_n$$

and conjugate to each other.

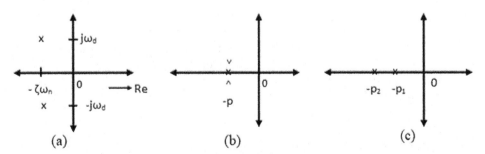

FIGURE 1.10
Possible pole locations.

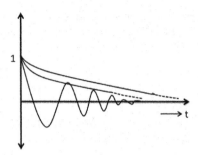

FIGURE 1.11
Natural responses.

The zero-input response of this system is shown in Figure 1.11.

Let us now look at each of these cases in some more details.

1.2.5.1 Underdamped Systems

In the case where $0 < \zeta < 1$, the response will be the sum of two exponents,

$$Ae^{-(\zeta\omega_n + j\omega_d)t} + Be^{-(\zeta\omega_n - j\omega_d)t}$$

This can be reduced, using some trigonometry, to the form

$$Ce^{-\zeta\omega_n t} \times \cos(\omega_d t - \phi) \qquad (1.24)$$

where $\phi = \arctan\frac{\zeta\omega_n}{\omega_d}$. The response is thus a *damped* sinusoid as shown in Figure 1.11a; interestingly, we have not energized the system with any sinusoidal forces. Since the system output has oscillations that are not completely suppressed, we call this system an **underdamped system**. If ζ is very small, the damping is relatively less. The time-constant in this case may be taken as $\frac{1}{\zeta\omega_n}$, and infinite time would be approximately $\frac{6}{\zeta\omega_n}$.

1.2.5.2 Critically Damped Systems

Here $\zeta = 1$, and the response is

$$Dte^{-\zeta\omega_n t}$$

This is a product of t and $e^{-\zeta\omega_n t}$. In the beginning, t dominates, and the response appears to grow. But soon e^{-pt} starts dominating, and the net response is a decaying exponential. This is shown in Figure 1.11b. We call this system a **critically damped system**. Here too, the time-constant is $\frac{1}{\zeta\omega_n}$. The reason for the adjective becomes clear after we learn about the third case.

1.2.5.3 Overdamped Systems

In the case where $\zeta > 1$, the response is the sum of two exponents

$$Ee^{-p_1 t} + Fe^{-p_2 t}$$

One of the exponents is relatively slower since the corresponding pole is closer to the origin, and the total response is sluggish as shown in Figure 1.11c. We call this system an **overdamped response**. This is the slowest response of all the three. The time-constant in this case is

$$\max\left\{\frac{1}{p_1}, \frac{1}{p_2}\right\} = \min\{p_1, p_2\}$$

i.e. the reciprocal of the pole that is closer to the origin. This pole, nearer to the origin, is often called the **dominant pole**, since it corresponds to the time-constant that causes the exponent to take a long time to decay.

The constants $A, B, C, D, E,$ and F, in the above three cases, may be determined from the initial conditions. (How many initial conditions are there?)

Finally, it is quite easy to verify that if $\zeta = 0$, the poles are purely imaginary at $s = \pm j\omega_n$ and the zero-input response is an *undamped sinusoid* of frequency ω_n. This is the fourth possibility, and such systems are called **undamped systems**.

Let us now give appropriate names to the two parameters of second-order systems.

- $\omega_n, \triangleq$ **natural frequency**,

- $\omega_d = \omega_n\sqrt{1 - \zeta^2} \triangleq$ **damped natural frequency**, and

- $\zeta \triangleq$ **damping factor** or **damping ratio**.

The suffixes n and d denote appropriate frequencies.

Thus, the underdamped response is the fastest in the sense that it hits the steady-state value much earlier than the other two responses. For most engineering systems we prefer this kind of a response despite the presence of oscillations. Let us look at the response of an underdamped second order system to a unit step input. We assume zero initial conditions for the sake of simplicity.

Example 1.2.5:

Assume that the parameters are given by

$$\zeta = \frac{1}{\sqrt{2}}(<1), \quad \omega_n = \sqrt{2} \quad \text{so that} \quad \omega_d = 1$$

If the input is a unit step, then applying Laplace transformation and solving the partial fractions, we get

$$y(t) = 1 - \sqrt{2}e^{-t}\cos\left(t - \frac{\pi}{4}\right)$$

This response is shown in Figure 1.12

We notice the significance of this response as follows:

- The oscillations are (more or less) completely damped, and the response *tracks* the input within about 5 to 6 sec. In other words, the output *settles* to its final value (within 1%) in a time called **settling time** t_s.
- At $t_r = \frac{3\pi}{4}$ sec the response just reaches the input, and at $t_p = \pi$, it *shoots up* much above the input. We call t_r the **rise time** and t_p the **peak time**. The response attains its maximum $1 + M_p$ at the peak time. We call this excess the **maximum overshoot**, M_p.
- From eqn. 1.24, we can determine the parameters t_r, t_p, M_p, and t_s as follows. Notice that these quantities are defined only when the system is given a unit step input.
- Since the rise time is defined as the time at which the output touches the input (i.e. $y(t_r) = 1$), this occurs when the sinusoid in eqn. 1.24 equals zero. Hence,

$$t_r = \frac{\frac{\pi}{2} + \phi}{\omega_d}; \quad \phi = \cos^{-1}\frac{\omega_d}{\omega_n} \qquad (1.25)$$

- The output overshoots the input at the peak time. Since this maximum is greater than 1, eqn. 1.24 says that this can happen only when

$$\frac{\omega_n}{\omega_d}e^{-\zeta\omega_n t}\cos(\omega_d t - \phi)$$

- is minimum, i.e.

$$\cos(\omega_d t - \phi) = -\frac{\omega_d}{\omega_n}$$

- We may quickly notice that $\cos\phi = \frac{\omega_d}{\omega_n}$, and hence,

$$t_p = \frac{\pi}{\omega_d} \qquad (1.26)$$

- Using eqn.1.25, we get

$$M_p = e^{-\pi\tan\phi}$$

- It is standard practice to express this is as a percentage, i.e.

$$M_p = 100 \cdot e^{-\pi\tan\phi}\% = 100 \cdot e^{-\pi\frac{\zeta}{\sqrt{1-\zeta^2}}}\% \qquad (1.27)$$

- It is worth noting that the peak overshoot is a function of the damping ratio alone.
- The sinusoid is damped by the exponential $e^{-\zeta\omega_n t}$. Hence, the settling time or the *infinite* time, defined as the time it takes for the output to reach within 98% of the steady-state, can be chosen as

$$t_s \approx \frac{5}{\zeta\omega_n} \qquad (1.28)$$

owing to the idea of exponential decays and infinite time.

The relationship among ζ, ω_n, and ω_d is shown in Figure 1.13.

The (negative) real part is the product $\zeta\omega_n$. Hence, if we would like to design a system that settles down faster, we should see that this product has a large magnitude. The natural frequency ω_n is the radial distance of the complex poles from the origin. If this radial line joining the origin and a complex pole subtends an angle θ with the negative real axis as shown in the figure, then

$$\cos\theta = \zeta \text{ and } \phi = \frac{\pi}{2} - \theta$$

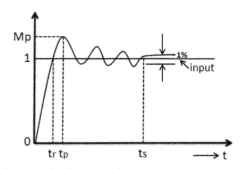

FIGURE 1.12
Step-response of underdamped system.

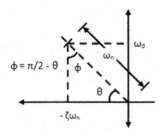

FIGURE 1.13
Second order system parameters in the *s*-plane.

Since eqn. 1.27 tells us that the overshoot is directly related to ζ, it is easy to see that if radial lines are drawn at an angle $\pm\theta$ with respect to the negative real axis, then all the underdamped second-order systems that have poles on these lines exhibit the same peak overshoot. In other words, these radial lines become the loci of poles of all second order systems that exhibit the same peak overshoot. Further, since ζ and θ are related by a cosine function, smaller the θ, smaller the peak overshoot (why?).

Similarly, from eqns. 1.25 and 1.26, the rise time and the peak time shall become progressively smaller, i.e. the response becomes faster, if the poles are farther away from the origin along these radial lines.

For a typical control system, the design requirements on the transient response appear as follows:

- $M_p \leq 10\%$,
- $t_s \leq 5$ units (in terms of the time-constant of the system)
- t_r as small as possible.

For the design purposes, these specifications can be mapped onto the s-plane as shown in Figure 1.14.

The hashed bounded region is usually referred to as the desired pole region D.

The faster response of the underdamped second order system has also a moral for us: *if you demand a smaller rise time, you have to live with a larger overshoot.* If the output of our system is fed as an input to a second system, we need to be careful in the sense that no damage to the second system is done because of the overshoot from the first system. Accordingly, we may have to make a trade-off between the speed of the response and overshoot.

All the while, an inquisitive reader would have been questioning: What is so important about a unit step signal as in input? We end this part of the chapter with a plausible answer to this question.

First, *unit* stands for a normalized amplitude of 1 unit. It could be *one* volt, or *one* foot, or any other physical quantity, such as pound per square inch (psi). A unit step signal has a sudden change in its magnitude from 0 to 1 at time

$t = 0$. If we assume that this signal is the output of an underdamped second order system, then it has

$$M_p = 0\%, \quad t_s = 0 \text{ units}, \quad \text{and} \quad t_r = 0 \text{ units}$$

This sounds too ideal! For instance, imagine that a pilot would like to fly his plane at an altitude of 10,000 feet. Certainly, he cannot accomplish this, i.e. reaching an altitude of 10,000ft and staying there, in *zero time*. Therefore, all that he does is to reach the desired altitude at the *earliest possible* time and stay there. In other words, a step signal serves as a model or "reference" to the fastest response *desirable*. In another example, consider semiconductor transistors. An ordinary Bipolar Junction Transistor (BJT) takes some time called the storage time (akin to our notion of time-constant) to be driven from cut-off to saturation and vice versa. Hence, this is not suitable for high-speed digital switching purposes. On the other hand, the Metal Oxide Semiconductor (MOS) gates serve as excellent digital logic gates. We may look at this example from our systems point of view—some design has been carried out to reduce the (dominant) time constant.

There is another reason for considering the step signal. In laboratories, we can generate periodic square or rectangular waveforms. If the period is large enough, i.e. larger than the settling time of the system, we may consider one portion of the waveform to represent a step signal. We may apply this to our system and record the response. We are interested in linear systems and differentiating this recorded response gives the impulse response since (i) differentiation is a linear operation and (ii) derivative of a unit step is unit impulse. Thus, *it is possible to obtain the mathematical model of a linear system in the laboratory.*

1.2.5.4 Higher Order Systems

Before we establish results on third and higher order systems, let us look back on the idea of time-constant. A time-constant is an index of the exponential decay—larger the time-constant, slower the response. In other words, the closer the pole is on the negative real axis to the origin, the slower the response. We have called such a pole as a *dominant pole*. Also for an underdamped system, if the real part $-\zeta\omega_n$ happens to be closer to the origin, i.e. the complex conjugate poles are closer to the imaginary axis, it takes longer for the sinusoidal oscillations to disappear. In this sense, we call these complex conjugate poles a **dominant pair**. This is why we prefer to place the poles, in the desired pole region, radially far away from the origin.

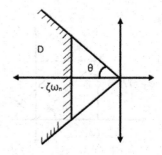

FIGURE 1.14
Desired pole region.

We shall now utilize the idea of dominant pole pair to examine third and higher order systems. Assume that we can factorize the characteristic polynomial as follows:

$$D^n(s) = (s^2 + 2\zeta\omega_n s + \omega_n^2) \cdot D^{n-2}(s) \quad 0 < \zeta < 1$$

where a superscript on D is used to denote the order of the polynomial. Notice, however, that this is not an easy assumption since factoring a *nth* order polynomial is not as trivial as factoring a quadratic polynomial. Further, assume that the $n - 2$ poles are farther away relative to the dominant pair of poles from the origin. Then, we may neglect them and view the system as an approximation of a second order system. This is a general practice, and any arbitrary higher order system can be deemed as a second order system with a pair of dominant complex-conjugate poles. Now the following question arises: Is there any *index* of saying "farther" or "closer?" It is difficult to answer this question precisely. But, we may *heuristically* say that if the magnitude of the real part of any pole is *at least* five times the magnitude of the real part of the dominant pole pair, its contribution is insignificant. Looking back, we borrow the factor five from our notion of (relative) infinite time.

Example 1.2.6:

Consider the characteristic polynomial

$$D(s) = s^4 + 10s^3 + 50s^2 + 80s + 64$$

We may factorize this to see that the poles are

$$s = -1 \pm j1 \quad \text{and} \quad -4 \pm j4$$

The zero-input response shall have two sinusoids with frequencies 1 rad/sec and 4 rad/sec. But the second sinusoid is damped relatively heavily, and clearly, the poles at $s = -1 \pm j1$ dominate the response.

We close this discussion with the following remarks.

1. It is worth noting that the relative location of the poles—closer or away to the origin, and/or real or complex conjugates—dictates the time response of a system. For this reason, we make direct comments on the time response—fast, or slow, or underdamped, etc.—directly from the transfer function/characteristic polynomial without actually obtaining the inverse Laplace transformation; in fact, the Laplace transformation is a 1-1 map. We encourage the reader to cultivate this habbit with more practice. Generally, there is hue and cry on the apparent abstract nature of this subject;

however, once the reader gets into the groove, it would take no time to discover that he is indeed talking about the most practical things.

2. Thus far, we have not considered the zeros of the system. We just remarked that certain inputs may cancel the poles and zeros in the transfer function. Strictly speaking, zeros alter the nature of the response drastically—a zero of the system may cancel one of the poles. We no longer get *nice* damped sinusoids for our underdamped second order systems. In fact, the very notions of rise time, peak time etc. become invalid. It is very difficult to examine the general nature of the response in the presence of zeros. In some sense, this is the price we pay for easier analysis using approximations; in a way this poses severe limitations to the classical control theory. We postpone the discussion on this aspect to later chapters.

1.2.5.5 A Time Response Analysis Example

In this section, we shall illustrate the time response of an important system. We encourage the reader to work out the example and fill-up all the gaps provided to gain hands-on experience.

Example 1.2.7:

Quite often we travel in automobiles. For a comfortable ride, the role of shock absorbers is quite evident. A physical model of the suspension system of an automobile is depicted in Figure 1.15.

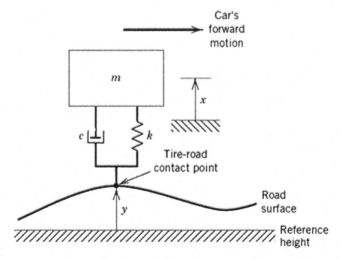

FIGURE 1.15
Typical Shock absorption system. (Adapted from: Palm III, W.J., *Mechanical Vibrations*, John Wiley & Sons, 2006, https://nasportscar.com/springs-and-dampers-and-rods-and-more/.)

Applying elementary mechanics, we obtain the following equation of motion.

$$m \cdot \frac{d^2}{dt^2} y + k_2 \cdot \frac{d}{dt} y + k_1 \cdot y = u$$

Here, we have assumed that the spring and the dashpot to be linear devices. Upon hitting a pot hole, the sudden jolt experienced by the automobile shock absorbing system may be modeled by a impulse signal $\delta(t)$. (What are the poles of the system ?) Accordingly, we get the following transfer function

$$H(s) = \frac{1}{ms^2 + k_2 s + k_1}$$

Case (i): All is well

If we choose the parameters m, k_1, and k_2 so that the system behaves like an underdamped one, the impulse response is a damped sinusoid approaching zero. We do get displaced upon hitting a pothole, but we come back to our original position, rather comfortably. Physically, this phenomenon is called cushioning. In fact, this is the purpose of a shock absorber. From a design point-of-view, we would build a shock absorber so that the parameters obey such a relationship.

Case (ii): Suppose the spring fails

In this case, the transfer function is

$$H(s) = \frac{1}{ms^2 + k_2 s}$$

We direct the reader to notice that the constant term in the characteristic polynomial is missing. The response is of the form

$$y(t) = A + Be^{-t/\tau}$$

This responses tells us that there is an exponential rise in the displacement. Physically, we feel being pushed downwards, and we never return to our original position. Do we need to mention that this is a very unpleasant experience?!

Case (iii): Suppose the dashpot fails

The transfer function in this case is

$$H(s) = \frac{1}{s^2 + k_1}$$

and the poles of the system are purely imaginary. We leave the consequences to the imagination of the reader. The purpose of this example is to show that the mathematical model need not be too complicated as long as the analysis of performance carried out is appropriate for the physical situation. It also gives a way of designing systems; for instance any random combination of the mass-spring-dashpot system is not suitable for the shock absorption—the parameters have to bear a specific relationship.

In the analysis example we saw just now, what exactly could be the difference between Case (i) and the other two cases? In other words, on what basis did we call the first case "All is well?" Apparently, in the latter two cases, one or the other term in the characteristic polynomial is missing. Does this really matter? Before we elaborate, let us look at a simpler example. With a gentle push behind, we enjoy a ride on a bicycle down a hill slope. How about choosing a bicycle without brakes?

The answer to the above question and the like lies in the concept of stability, and we shall learn about it in great detail in the next section.

1.2.5.6 Frequency Response

The response of dynamical systems exhibits transients whenever there is a sudden change, for instance putting a cube of ice from the refrigerator on a hot plate, in the input excitation. Accordingly, we are interested in the step response characteristics such as overshoot, settling time, and the rise time. Moreover, the reference input r may also, in general, be modeled as a step signal (or a linear combination of several step signals). Thus, the specifications in time domain are, generally, the transient response specifications, such as the time it takes for the ice cube to melt. In this part of the chapter, we shall look at the system response at steady-state, i.e. after all the transients (response in terms of e^{-at}) have disappeared. We call this response as **sinusoidal** *steady-state* **response**. The name suggests us the following: the input is a sinusoid, such as $A\cos(\omega t)$, and we study the response after all the transients have died out.

There are several reasons for choosing a sinusoidal excitation. First, nature itself seems to have a sinusoid character: the vibration of a string, the motion of a simple pendulum, and countless other phenomena. As the reader would have learned in an earlier couse, many mathematical functions governing the physical world may be expressed as suitable sums of pure sinusoids, i.e. Fourier series. Since our systems are linear, we may use the principle of superposition to determine the response to any arbitrary signal once we know the response to a sinusoid. In other words, sinusoidal analysis constitutes a far more general analytical tool than it might at first appear.

Secondly, generators supply electrical power in sinusoidal form, which is a very efficient form for

Linear Systems and Control   19

transmission and distribution also. Further, for reasons of efficiency, sinusoidal waves are used in several commercial applications such as electronic communications. Thirdly, since a linear system is governed by differential equations, a sinusoidal input induces a sinusoidal response, with the *same frequency* as the excitation, but differing in *magnitude* and *phase*. Also, the magnitude and the phase are frequency dependent.

Let us then concentrate on the following issue: as we vary the frequency of the input excitation, how does the (steady-state) response vary? There is a related and important issue regarding the stability of the closed-loop system at all frequencies. However, we defer this issue until the next chapter. We shall begin with the following trivial, but important, observation.

$$\cos(\omega t) = Re\ e^{j\omega t} = Re\ e^{st}\,|_{s\,=\,j\omega}$$

so that we need not explicitly show the sinusoidal input. For the next few pages, we will look into the product $L(s) = P(s) \times C(s)$, called the **loop transfer function**. As it turns out, we will be looking into the product of two transfer functions—the plant $P(s)$ and the controller $C(s)$—called the **Loop Transfer Function**.

The frequency response of a system, whose transfer function is $L(s)$, may be obtained directly by letting $s = j\omega$. The response can be *graphically* visualized by plotting the magnitude $|L(j\omega)|$ and the phase $\angle L(j\omega)$ versus ω. For instance, for the magnitude, the curve representing $|L(j\omega)|$ can be obtained as the *intersection* of the surface representing $|L(s)|$ with the vertical plane passing through the imaginary axis.[2] Here, let us use a simple example to illustrate the idea.

Example 1.2.8:

Let

$$L(s) = \frac{1}{s+1}$$

so that

$$L(j\omega) = \frac{1}{j\omega + 1}$$

Although ω is real, $L(j\omega)$ is, in general, complex. If $\omega = 0$ (i.e. dc), $L(j\omega) = 1$. If $\omega = 1$, then

$$L(j1) = \frac{1}{1+j1} = \frac{1}{\sqrt{2}} e^{-j45°}$$

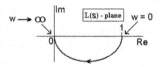

FIGURE 1.16
Polar plot of Example 1.2.8.

or, $|L(j1)| = 0.707$ and $\angle L(j1) = -45°$. Similarly, we can compute $L(j1.2)$, $L(j5)$, $L(j\infty)$, and so on.

We may use this data to plot $L(j\omega)$. If we choose *polar coordinates*, we obtain the plot as shown in Figure 1.16. This is called the **polar plot**, for obvious reasons.

The horizontal and vertical axes are, respectively, the real part Re $L(j\omega)$ and the imaginary part Im $L(j\omega)$. Thus, a point in the $L(j\omega)$-plane is a phasor with magnitude and phase. The polar plot of the transfer function in the above example is a semicircle for all the positive real frequencies $0 \le \omega < \infty$. A polar plot is fairly easier to obtain.

1.2.6 Bode Plots

There is another way of plotting $L(j\omega)$. As we have seen earlier, $L(j\omega)$ contains two components, the magnitude and the phase, which are functions of the frequency. It seems logical to obtain two different graphs—magnitude versus ω, and phase versus ω for all $0 \le \omega < \infty$.

The frequency range we encounter is infinitely wide, and hence, it would save some of our space if we consider a *logarithmic* scale for the frequency axis.[3] We need to make *equally spaced marks* on the axis to represent **decade** frequency intervals, i.e. 1, 10, 100, and so on. This is shown in Figure 1.17.

Given a specific frequency within the decade $10^k \le \omega \le 10^{k+1}$ rad/sec, its location l within the decade is given by

$$l = \log_{10} \frac{\omega}{10^k}$$

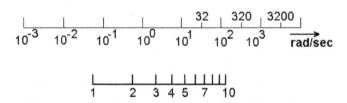

FIGURE 1.17
Semi-log frequency scale.

[2] The reader is expected to have acquired familiarity with these ideas in one or two earlier courses on electrical circuits and/or signals and systems.

[3] To sketch your plots you may purchase *semi-log* graph paper. However, preparing your own logarithmic scale on plain graph paper will make you better appreciate the convenience of this scale. The reader is urged to prepare at least one immediately.

For instance, $\omega = 320$ rad/sec lies approximately half-way between 100 rad/sec and 1000 rad/sec. It is easy to see that $\omega = 32$ rad/sec lies approximately half-way between 10 rad/sec and 100 rad/sec; $\omega = 3200$ rad/sec lies approximately half-way between 1000 rad/sec and 10^4 rad/sec. Figure 1.17b shows the frequencies within the decade $1 \le \omega \le 10$ rad/sec. The advantage of a logarithmic scale is very apparent as it compresses higher frequencies and expands the lower frequencies, thereby allowing us to visualize the response at both extremes with a comparable level of detail.

We further define one more quantity, the **decibel**.

1.2.6.1 Definition: Decibel

The decibel (dB in short) value of a transfer function $L(s)$ is defined as

$$|L|\text{dB} \triangleq 20 \cdot \log_{10} |L| \qquad (1.29)$$

Notice that *unity magnitude* corresponds to 0 dB, *gain* corresponds to positive dB, and *attenuation* corresponds to negative dB. Accordingly, a magnitude of $1/\sqrt{2}$ very closely corresponds to -3 dB, and a magnitude of 2 corresponds to $+6$ dB. Converting the dB gain to ordinary values is trivial.

We may readily verify that given two systems $L_1(s)$ and $L_2(s)$ in cascade,

$$|L_1 \cdot L_2|\text{dB} = |L_1|\text{dB} + |L_2|\text{dB}$$

$$\angle L_1 \cdot L_2 = \angle L_1 + \angle L_2$$

$$|L_1 / L_2|\text{dB} = |L_1|\text{dB} - |L_2|\text{dB}$$

$$\angle (L_1 / L_2) = \angle L_1 - \angle L_2$$

$$|1/L_1|\text{dB} = -|L_1|\text{dB}$$

$$\angle 1/L_1 = -\angle L_1 \qquad (1.30)$$

We shall learn these ideas quickly using the following examples.

Example 1.2.9:

Let

$$L(s) = \frac{10^{-6}s + 1}{10^{-3}s + 1}$$

i.e. the system has a pole at $s = -10^3$ Nepers/sec and a zero at $s = -10^6$ Np/sec. To find the frequency response, we rewrite $L(s)$ as

$$L(j\omega) = \frac{1 + j\dfrac{\omega}{10^6}}{1 + j\dfrac{\omega}{10^3}}$$

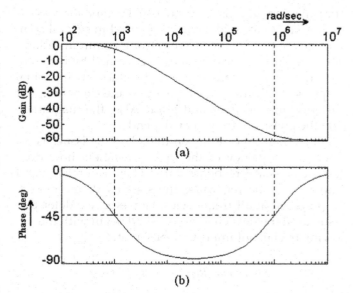

FIGURE 1.18
Bode plot for Example 1.2.9.

Accordingly, we get

$$|L|\text{dB} = 20\log_{10}\left[\sqrt{1 + \left(\frac{\omega}{10^6}\right)^2}\right] - 20\log_{10}\left[\sqrt{1 + \left(\frac{\omega}{10^3}\right)^2}\right]$$

and

$$\angle L = \arctan\frac{\omega}{10^6} - \arctan\frac{\omega}{10^3}$$

Using a small calculator we may evaluate these expressions for several values of ω and plot them point by point on the semi-log paper. The plots are shown in Figure 1.18a and b.

Let us make a few observations on the example. As we vary ω from 0 to ∞, the slopes of the magnitude and phase curves change when ω approaches the frequencies $\omega_{pole} = 10^3$ rad/sec and $\omega_{zero} = 10^6$ rad/sec. In particular, ω_{pole} brings about an overall change of -20 dB/decade in the magnitude curve and $-90°$ in the phase curve. Similarly, ω_{zero} brings about an overall change of $+20$ dB/decade and $+90°$, respectively. For obvious reasons we call these frequencies **corner frequencies**. In Figure 1.18, we observe that the slope of the magnitude curve changes from 0 dB/dec to -20 dB/dec at ω_{pole} and from -20 dB/dec back to 0 dB/dec at ω_{zero}. The phase difference due to the pole starts at $\omega \approx 0.1\omega_{pole}$, reaches $-45°$ at ω_{pole}, and approaches $-90°$ at $\omega \approx 10\omega_{pole}$. Similarly, the phase difference due to the zero

---

[4] It is exciting to note that Hendrik W. Bode and Harry Nyquist were at Bell Laboratories, and their contributions directly followed Harold Black's negative feedback amplifier.

starts at $\omega \approx 0.1\omega_{zero}$, reaches $-45°$ at ω_{zero}, and approaches $0°$ at $\omega \approx 10\omega_{zero}$.

Thus, we see a connection between the location of the zeros and poles in the s-plane, where $L(s)$ is either zero or infinity, and the corresponding corner frequencies on the $\log_{10}\omega$ axis, where both the slope and the phase angle undergo a positive or negative change. It is also interesting to observe that when dealing with the s-plane as a whole, we attribute the units Nepers/sec (Np/sec) to zeros and poles, while we regard the corresponding corner frequencies in radians/sec (rad/sec).

The two curves—magnitude in dB versus $\log_{10}\omega$ and phase versus $\log_{10}\omega$—jointly provide us the information about the transfer function as a function of frequency. Accordingly, it is customary to call both the curves by a singular noun, **Bode plot.**[4] It is recommended that both the curves be plotted on the same graph paper—the magnitude curve above the phase curve—with the same scale along the $\log_{10}\omega$ axis.

1.2.6.2 Construction of Bode Plots

Let us now consider the elementary transfer functions and their Bode plots. The reader is strongly advised to *work out* this section rather than read it.

- **Case (i):** Let

$$L(s) = K \qquad (1.31)$$

This transfer function is independent of frequency. Therefore, the magnitude curve is a horizontal straight line for all frequencies. If K is positive the angle is $0°$, and if K is negative the angle is $180°$ for all frequencies. The Bode plot is shown in Figure 1.19.

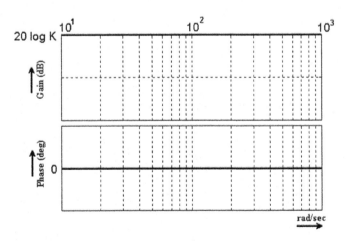

FIGURE 1.19
Bode plot for Case (i).

- **Case (ii):** Let

$$L(s) = s \qquad (1.32)$$

i.e.

$$|L(j\omega)|\,\mathrm{dB} = 20\log\omega$$

$$\angle L(j\omega) = \arctan\frac{\omega}{0} = 90° \qquad (1.33)$$

The Bode plot of this simple zero at origin is shown in Figure 1.20a.

The expression for the magnitude is of the form $y = mx$, where both y and x are in the logarithmic scale. Accordingly, the magnitude curve is a straight line with slope $m = +20$ dB/decade. We say that the magnitude curve **rolls up** at 20 dB/dec. Notice that for every tenfold change in ω, the corresponding change in the magnitude is 20 dB. It is worth noting here that a frequency $\omega = 0$ rad/sec cannot be shown on the semi-logarithmic scale. At most, we can start at a very low frequency, such as $\omega = 0.001$, at the far left end of the frequency axis. Owing to this, our description of the magnitude curve is incomplete without an intercept. We may come out of this problem

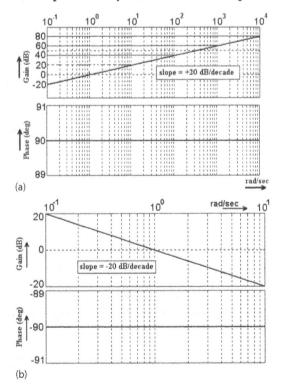

FIGURE 1.20
(a) Bode plot for a simple zero at origin. (b) Bode plot for a simple pole at origin.

by observing that the magnitude is 0 dB at $\omega = 1$ rad/sec. Thus, the magnitude curve is a straight line with slope 20 dB/dec with an intercept of 0 dB at $\omega = 1$ rad/sec.

The phase curve, however, is constant at 90° for all frequencies.

If

$$L(s) = \frac{1}{s} \qquad (1.34)$$

then the magnitude curve **rolls off** at -20 dB/dec with an intercept of 0 dB at $\omega = 1$ rad/sec. The phase curve is constant throughout at $-90°$. This is shown in Figure 1.20b.

- **Case (iii):** Let

$$L(s) = s + \alpha \quad \alpha > 0 \qquad (1.35)$$

We rewrite $L(j\omega)$ as follows.

$$L(j\omega) = \alpha \left(1 + \frac{j\omega}{\alpha} \right) \qquad (1.36)$$

so that

$$|L(j\omega)|\,\mathrm{dB} = 20\log \alpha + 20\log \left[\sqrt{1 + \left(\frac{\omega}{\alpha}\right)^2} \right]$$

$$\angle L(j\omega) = \arctan \frac{\omega}{\alpha} \qquad (1.37)$$

Let us first examine the term

$$20\log \left[\sqrt{1 + \left(\frac{\omega}{\alpha}\right)^2} \right]$$

in the expression for the magnitude. If $\omega \ll \alpha$, then

$$20\log \left[\sqrt{1 + \left(\frac{\omega}{\alpha}\right)^2} \right] \approx 20\log 1 = 0 \qquad (1.38)$$

i.e. the magnitude is almost 0 dB, or, the magnitude curve is a straight line with 0 slope and grazing the frequency axis, for frequencies far less than the *corner frequency* α. If $\omega = \alpha$, then

$$20\log \left[\sqrt{1 + \left(\frac{\omega}{\alpha}\right)^2} \right] = 20\log \sqrt{2} \approx 3\ \mathrm{dB} \qquad (1.39)$$

Finally, if $\omega \ll \alpha$, then

$$20\log \left[\sqrt{1 + \left(\frac{\omega}{\alpha}\right)^2} \right] \approx 20\log \frac{\omega}{\alpha} \qquad (1.40)$$

and the magnitude curve rolls up at $+20$ dB/dec with an intercept of 0 dB at $\omega = \alpha$. The straight lines we obtain for the two cases $\omega \ll \alpha$ and $\omega \gg \alpha$ are called the **asymptotes**. The actual magnitude curve will be very close to these asymptotes passing through 3 dB point at the corner frequency. This is shown in Figure 1.21a below. The maximum deviation of the actual magnitude curve from the asymptote is 3 dB occurring at the corner frequency.

The complete magnitude curve is obtained by adding $20\log \alpha$. This amounts to simply shifting the curve in Figure 1.21a upwards if $\alpha > 1$ or downwards if $0 < \alpha < 1$. This is because you add a constant $20\log \alpha$ at *every* frequency. This is shown in Figure 1.21b.

The phase curve is an inverse tangent function of ω. For $\omega \le 0.1\alpha$, the phase is almost 0° and for $\omega \ge 10\alpha$, the phase is almost 90°. In the range $0.1\alpha \le \omega \le 10\alpha$, the phase curve is almost a straight line passing through 45° at $\omega = \alpha$. This is shown in Figure 1.21c. Again the maximum deviation of the actual phase curve from the straight line approximation is about 6°.

$$L(s) = (s + \alpha)^{\pm m} \quad \alpha > 0,\ m \ge 1$$

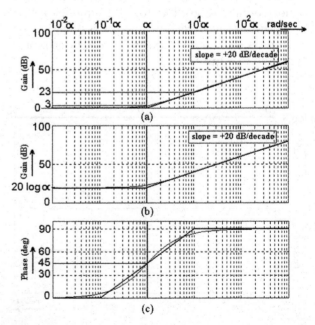

FIGURE 1.21
Bode plot of a first order factor with a gain.

At this point, let us consider an example and learn how to construct the Bode plots in a general case.

Example 1.2.10:

Let

$$L(s) = \frac{6s}{s+2}$$

Notice that this transfer function is a combination of the three cases discussed above, i.e. there is a factor independent of frequency, there is a zero at the origin and there is a factor similar to $s + \alpha$ in the denominator. Let us then rewrite $L(s)$ as

$$L(s) = s \cdot \frac{1}{1+\frac{s}{2}} \cdot \frac{6}{2}$$

Observe the way we have split the factors—the constant K and the factor of the form $20\log\alpha$ have been put at the end. The purpose of splitting $L(s)$ in this fashion becomes obvious if we look at the expressions for the magnitude and the phase.

$$|L(j\omega)|\,\mathrm{dB} = |L_1(j\omega)| + |L_2(j\omega)| + |L_3(j\omega)|$$
$$= 20\log\omega + 20\log\frac{1}{\sqrt{1+\left(\frac{\omega}{2}\right)^2}} + 20\log\frac{6}{2}$$

$$\angle L(j\omega) = \angle L_1(j\omega) + \angle L_2(j\omega) + \angle L_3(j\omega)$$
$$= 90° - \arctan\frac{\omega}{2} + 0°$$

By way of logarithms we have nicely transformed the multiplication × in $L(s)$ to addition + in $|L(j\omega)|$ dB; $\angle L(j\omega)$ is already in the addition form. Now the idea is clear—

- There may be any number of factors in the transfer function $L(s)$. All we need to do is to simply ADD algebraically, at every frequency, the Bode plots of the constituting factors.

Figure 1.22a shows the individual magnitude asymptotes drawn to the same frequency scale.

Figure 1.22b shows the *addition* of the individual asymptotes in Figure 1.22a. For instance, at $\omega = 1$ rad/sec

$$|L(j1)| = 0 + 0 + 20\log 3 = 9.54\,\mathrm{dB}$$

Similarly, at $\omega = 1.5$ rad/sec

$$|L(j1.5)| = 3.52 + 0 + 20\log 3 = 13.06\,\mathrm{dB}$$

In fact, until $\omega = 2$ rad/sec, $|L_2|$ is 0 dB.

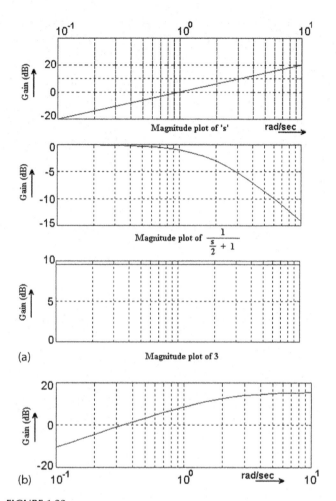

FIGURE 1.22
(a) Magnitude plot of individual factors for Example 1.2.10. (b) Magnitude plot for Example 1.2.10.

At $\omega = 2$ rad/sec, the magnitude is $6 + 0 + 9.54 = 15.54$ dB. For $\omega > 2$ rad/sec, the first term $|L_1|$ rolls up at $+20$ dB/dec and the second term $|L_2|$ rolls off at -20 dB/dec. Therefore, the sum of these two factors is a constant 15.54 dB $\forall \omega > 2$ rad/sec. This is where the slopes of the curves are needed. The actual magnitude curve may be obtained from the asymptotic curve by making a simple error correction at the corner frequency. At the corner frequency 2 rad/sec, the magnitude of the $|L_2|$ is -3 dB. Therefore, it would be sufficient to mark a point 3 dB below 15.54 dB on the magnitude asymptote and draw a smooth curve along the asymptotes and passing through this point. As was mentioned earlier, the maximum deviation is only 3 dB and that too occurs only at the corner frequency. In general, the gain of a control system would be large enough compared to 3 dB. Accordingly, we may consider the errors as negligible at all other frequencies. In other words, asymptotes themselves are reasonably good Bode plots.

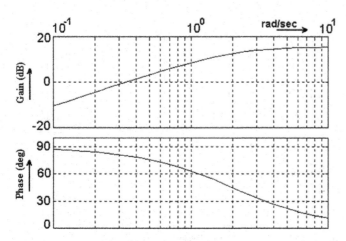

FIGURE 1.23
Bode plot of Example 1.2.10.

With little more practice, the reader would find it easier to add the asymptotes using the slopes for all simpler functions of s and then simply shift the result by the constant arising out of the *last* term.

Figure 1.23 shows the complete Bode plot of the transfer function. We leave it as an exercise to the reader to study the phase curve. Here too, compared to large phase angles, the deviation of $\pm 6°$ may be considered negligible.

- **Case (iv):** Let

$$L(s) = s^2 + 2\zeta\omega_n s + \omega_n^2 \qquad (1.41)$$

Here ζ and ω_n carry the usual meanings—damping ratio and natural frequency. We shall rewrite this function as

$$L(j\omega) = \left[1 - \frac{\omega^2}{\omega_n^2} + j2\zeta\frac{\omega}{\omega_n}\right] \cdot \omega_n^2 \qquad (1.42)$$

so that we clearly identify the corner frequency as ω_n. The magnitude and phase functions can be written routinely as

$$|L(j\omega)| = 20\log\left[\sqrt{\left(1-\frac{\omega^2}{\omega_n^2}\right)^2 + \left(2\zeta\frac{\omega}{\omega_n}\right)^2}\right] + 20\log\omega_n^2$$

$$\angle L(j\omega) = \arctan\frac{2\zeta\frac{\omega}{\omega_n}}{1-\frac{\omega^2}{\omega_n^2}} \qquad (1.43)$$

For the magnitude curve, we shall once again consider only the first term (since the second term contributes just a shift), and look at the asymptotic cases. First, if $\omega \ll \omega_n$, then

$$|L| \approx 0\,\text{dB} \qquad (1.44)$$

Secondly, at the corner frequency $\omega = \omega_n$,

$$|L| = 20\log 2\zeta \text{ dB} \qquad (1.45)$$

i.e. the magnitude is a function of the damping ratio alone. If ζ is very small, the magnitude is very large with a negative sign, if $\zeta = 0.5$, the magnitude is 0 dB, and if ζ is close to 1, the magnitude is close to 6 dB. Thus, the magnitude at the corner frequency can be anywhere in the range $-\infty < |L| < 6$ dB. Lastly, if $\omega \gg \omega_n$, then

$$|L| \approx 40\log\frac{\omega}{\omega_n} \qquad (1.46)$$

This is because the squared term $\frac{\omega^4}{\omega_n^4}$ dominates 1 as well as $4\zeta^2\frac{\omega^2}{\omega_n^2}$. Thus, for frequencies greater than ω_n, the magnitude curve rolls up at +40 dB/decade. This can be expected because a single zero contributes +20 dB/decade; hence, a pair of zeros contribute +40 dB/dec. The asymptotes are shown in Figure 1.24a. Observe that the second asymptote is a straight line with a slope +40 dB/dec with an intercept of 0 dB at the corner frequency. We have added the second term, i.e. $20\log\omega_n^2$ so that the intercept actually appears to be $20\log\omega_n^2$ dB. The correction factor can be determined from the value of ζ given. In Figure 1.24b, we have given the actual magnitude curve for several values of ζ. For clarity, we have taken $\omega_n = 4$.

The phase curve is, once again, an inverse tangent function of ω. For small values of ω, i.e. $\omega \le 0.1\omega_n$, the phase is almost 0°. In the range $0.1\omega_n \le \omega \le 10\omega_n$, the phase depends on the exact value of the damping ratio ζ. The only exception is at $\omega = \omega_n$ where the phase is exactly 90°. And, for large values of ω, i.e. $\omega \ge 10\omega_n$, the phase is almost 180Υ. This is because the argument

$$\frac{2\zeta\frac{\omega}{\omega_n}}{1-\left(\frac{\omega}{\omega_n}\right)^2} \to 0 \text{ as } \frac{\omega}{\omega_n} \to \infty$$

Further, the denominator is always negative suggesting us that the phase approaches the

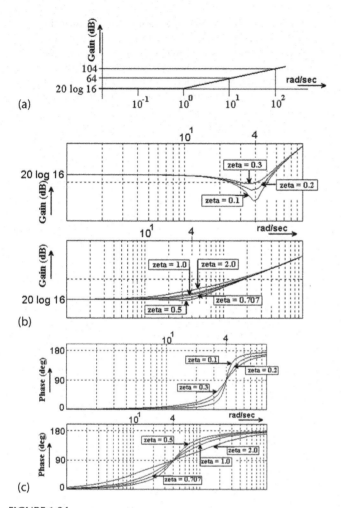

FIGURE 1.24

(a, b) Magnitude plots for complex-conjugate zeros. (c) Phase plots for complex-conjugate zeros.

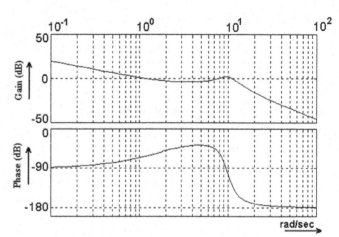

FIGURE 1.25

Bode plot for Example 1.2.11.

This may be factored as

$$L(s) = \frac{1}{s} \cdot (s+2) \cdot \frac{1}{\left(\frac{s}{10}\right)^2 + 2 \cdot 0.2 \cdot \left(\frac{s}{10}\right) + 1} \cdot \frac{50 \cdot 2}{100}$$

where we have identified the quadratic factor in the denominator to have $\zeta = 0.2$ and $\omega_n = 10$. The complete Bode plot is shown in Figure 1.25.

Example 1.2.12:

Let

$$L(s) = \frac{s - z}{s + p}$$

We notice that the zero in this transfer function lies in the right half of the s-plane. Let us look at the magnitude and phase of this factor.

$$|L_z(j\omega)| = 20 \log \sqrt{1 + \left(\frac{\omega}{z}\right)^2}$$

$$\angle L_z(j\omega) = \arctan \frac{\frac{\omega}{z}}{-1}$$

The expression for magnitude does not change whether the zero is in the left half or right half of the s-plane. However, the expression for the phase angle indicates that

$$90° \le \angle L_z(j\omega) \le 180°$$

while the phase angle of a LHS-plane zero would be between 0° and 90°. Therefore, a zero in the RHS-plane contributes to a larger phase angle, and hence, we call it as **non-minimum phase** zero. The physical origin of this non-minimum phase zeros is the time-delays in the systems.

negative real axis from above. The phase curve for several values of ζ is given in Figure 1.24c. Here too, the two poles contribute $2 \times 90 = 180°$ asymptotically.

The deviation from a straight line depends on the value of ζ at hand. Hence, we cannot make any claim here that the asymptotic plots are reasonably good. However, we always obtain the asymptotic plot first and then make necessary corrections for the given damping ratio.

Example 1.2.11:

Let

$$L(s) = \frac{50(s+2)}{s(s^2 + 4s + 100)}$$

In addition to the larger phase, these right half of the s-plane, i.e., the rhp zeros create a different problem in terms of the limitations on *achievable performance*. This idea is beyond the scope of this book.

1.2.7 State Space Representation of Systems

In the previous parts of this chapter we have primarily focused on (i) LTIL single-input single-output (SISO) systems, and (ii) real-rational functions, called the transfer functions, in the complex variable s. Historically, it was more than adequate until World War II. Thereafter, two factors influenced the course of control systems—the availability of detailed physical models (in terms of linear and nonlinear differential equations) together with more accurate measuring instruments, and Poincare's formulation that led to **state-space** framework. In fact, this heralded the era of, what is now popularly called, modern control.

When the paradigm shifts, we need to evaluate its merits. This new state space paradigm has several salient features we briefly describe. First, it captures the models, both SISO and multiple-input multiple-output (MIMO), in a unified framework. As we see in this chapter, it is only the dimension of a certain matrix that serves as an index to SISO or MIMO systems. Secondly, it accommodates *time-varying* systems, i.e. differential equations with time-varying coefficients. Thirdly, it allows us to design an optimal control (out of a large number of possible controls). Next, nonlinear systems can also be considered under this framework since, unlike the transfer functions, we cannot neglect the initial conditions. Last, but not the least, the mathematical elegance that has inspired several best minds to work in this area. The fundamental idea of control theory that the inputs should be computed from the *state* was enunciated and emphasized by Richard Bellman in the mid 1950s. Subsequently, Rudolf Kalman made giant strides and put control theory in a firm *algebraic* setting where we go from a completely abstract consideration to completely practical results, which are, in the end, quite simple.

1.2.7.1 Two Examples

In this section, we shall consider two examples that would motivate our study. The first is the automobile suspension system we considered earlier.

Example 1.2.13:

The second order ode governing the automobile suspension system is given by

$$m \cdot \frac{d^2}{dt^2} y + k_2 \cdot \frac{d}{dt} y + k_1 \cdot y = u$$

where y is the displacement, m is the mass of the auto, k_2 is the dashpot constant, k_1 is the spring constant, and u is the force applied on the suspension system. Since velocity is the *rate of change of displacement*, and acceleration is the rate of change of velocity, let us now define two variables as follows:

$$x_1 \triangleq \text{displacement} = y$$

$$x_2 \triangleq \text{velocity} = \frac{d}{dt} y = \frac{d}{dt} x_1$$

We rewrite the ode in terms of these new variables as

$$m \cdot \frac{d}{dt} x_2 + k_2 \cdot x_2 + k_1 \cdot x_1 = u$$

Apparently, we have a *first order* ode in two variables. A closer inspection reveals that, in fact, there are *two* first order odes:

$$\dot{x}_1 = x_2$$

$$\dot{x}_2 = -\frac{k_1}{m} x_1 - \frac{k_2}{m} x_2 + \frac{1}{m} u$$

We shall pack[5] these pair of equations into vectors and matrices as follows:

$$\frac{d}{dt} \begin{bmatrix} x_1 \\ x_2 \end{bmatrix} = \begin{bmatrix} 0 & 1 \\ -\dfrac{k_1}{m} & -\dfrac{k_2}{m} \end{bmatrix} \begin{bmatrix} x_1 \\ x_2 \end{bmatrix} + \begin{bmatrix} 0 \\ \dfrac{1}{m} \end{bmatrix} u$$

Since the output is the displacement y, i.e. x_1, we shall also write

$$y = \begin{bmatrix} 1 & 0 \end{bmatrix} \begin{bmatrix} x_1 \\ x_2 \end{bmatrix}$$

We are now ready to generalize this new formulation:

$$\Sigma(A, B, C):$$

$$\frac{d}{dt} \vec{x} = A \cdot \vec{x} + B \cdot u \tag{1.47}$$

$$y = C \cdot \vec{x} \tag{1.48}$$

where A is called the **system coefficient matrix**. The variables x_i[6] are called the state variables, and the vector $\vec{x}$ is called the state vector, or

---

[5] This is the *gift* of R.E. Kalman!

[6] Throughout the book the symbol $s = -P$ is reserved for state variables.

simply (hereafter) the **state**. Recall that earlier we have used the word *state* to denote the initial condtion; now, we let it evolve over time as $x(t)$. Equation 1.46 is called **state equation**, since it tells us how the state evolves, and eqn. 1.47 is called the **output equation**, since it tells us what the output is in terms of the state. Notice that the state equation is a differential equation and the output equation is an algebraic one.

In this example, the choice of the state variables was obvious. In a later example we shall show the importance of these variables in terms of compensator design.

Example 1.2.14(a):

Consider the simple series RLC circuit shown in the Figure 1.26.

Applying Kirchhoff's Voltage Law together with the elemental laws, we get,

$$R \cdot i + L \cdot \frac{d}{dt} i + y = u$$

$$C \cdot \frac{d}{dt} y = i$$

In circuit analysis, the general practice is to plug the second equation into the first one and obtain a second order ode. Some manipulation is required to obtain the initial conditions in the desired form. However, in the current paradigm, we shall simply rewrite the above equations as:

$$\frac{d}{dt}\begin{bmatrix} x_1 \\ x_2 \end{bmatrix} = \begin{bmatrix} -\dfrac{R}{L} & -\dfrac{1}{L} \\ \dfrac{1}{C} & 0 \end{bmatrix} \begin{bmatrix} x_1 \\ x_2 \end{bmatrix} + \begin{bmatrix} 1 \\ 0 \end{bmatrix} u$$

$$y = \begin{bmatrix} 0 & 1 \end{bmatrix} \begin{bmatrix} x_1 \\ x_2 \end{bmatrix}$$

Here, again, the choice of the state variables, i.e. $x_1 = i$ and $x_2 = y$, is obvious since we know the *initial conditions* directly in terms of these variables. It is clear that the variables x_1 and x_2 are arbitrary. In other words, the reader is free to choose $x_1 = y$ and $x_2 = i$. What is important to us is the **state model** given in eqns. 1.46 and 1.47.

Before we provide a formal definition of *state*, we shall look at simple extensions of the above example.

Example 1.2.14(b):

Suppose in the RLC circuit, the input u comprises two sources – u_1 and u_2 – in series. Then, we may obtain the state model as:

$$\frac{d}{dt}\begin{bmatrix} x_1 \\ x_2 \end{bmatrix} = \begin{bmatrix} -\dfrac{R}{L} & -\dfrac{1}{L} \\ \dfrac{1}{C} & 0 \end{bmatrix} \begin{bmatrix} x_1 \\ x_2 \end{bmatrix} + \begin{bmatrix} 1 & 1 \\ 0 & 0 \end{bmatrix} \vec{u} \qquad (1.49)$$

$$y = \begin{bmatrix} 0 & 1 \end{bmatrix} \begin{bmatrix} x_1 \\ x_2 \end{bmatrix} \qquad (1.50)$$

where we have defined an **input vector**

$$\vec{u} \triangleq \begin{bmatrix} u_1 \\ u_2 \end{bmatrix}$$

This may be extended to any p number of inputs.

Example 1.2.14(c):

Similarly, if the output of the RLC circuit is the current i through the inductor *as well as* the voltage y across the capacitor, then we may define an **output vector** $\vec{y}$ as

$$\vec{y} \triangleq \begin{bmatrix} y \\ i \end{bmatrix}$$

and the state model becomes

$$\frac{d}{dt}\begin{bmatrix} x_1 \\ x_2 \end{bmatrix} = \begin{bmatrix} -\dfrac{R}{L} & -\dfrac{1}{L} \\ \dfrac{1}{C} & 0 \end{bmatrix} \begin{bmatrix} x_1 \\ x_2 \end{bmatrix} + \begin{bmatrix} 1 & 1 \\ 0 & 0 \end{bmatrix} \vec{u} \qquad (1.51)$$

$$\vec{y} = \begin{bmatrix} 0 & 1 \\ 1 & 0 \end{bmatrix} \begin{bmatrix} x_1 \\ x_2 \end{bmatrix} \qquad (1.52)$$

and this may be extended to any m number of outputs.

Clearly, the format of state model does not change; it is only the dimensions of the matrices and/or vectors that change. Now we are ready for the formal definition of *state* and a few general observations.

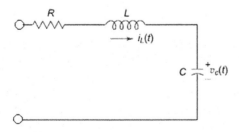

FIGURE 1.26
A standard RLC circuit.

1.2.7.2 Definition: State

The state of a system at a time t_o is the *minimum* set of internal variables which is sufficient to uniquely specify the system outputs given the input signal over $[t_0,\infty]$.

Very often the choice of states is natural and obvious. However, the choice can be arbitrary. For instance, in the example above, we might as well select the voltage across R as a state variable instead of the current through the inductor, i.e. the *linear combination*

$$x_1 \leftarrow R \cdot x_1 + 0 \cdot x_2$$

This flexibility provides freedom to the designer; for instance, the current is now converted to voltage that can be more easily tapped.

We make the following observations.

1. The number n of state variables is equal to the order of the system; the **system matrix** A is a square matrix of size $n \times n$.

2. If there are r number of inputs, $\vec{u}$ is a p-dimensional vector; the size of the **input matrix** B is $n \times p$.

3. If there are m number of outputs, $\vec{y}$ is a m-dimensional vector; the size of the **output matrix** C is $m \times n$.

4. The term state vector *parameterizes* the model and the components of the vector are, in general, a mixture of several physical quantities such as voltages, currents, temperatures, displacements etc. The set of all such possible vectors can be thought of as a *linear* vector space[7] called the **state space** X. In other words, this space satisfies the following property

$$\forall \alpha_i \in F \text{ and } \forall \vec{x}_i \in X, \quad \sum_i \alpha_i \cdot \vec{x}_i \in X$$

where F is any scalar *field*, typically the real field.

5. Certain transformations, for instance a linear combination of the state variables, allow greater insight into the system behaviour and simplify the analysis and design.

In general, the elements of the matrices A, B, and C and the vectors $\vec{x}$, $\vec{u}$, and $\vec{y}$ are real numbers. Having understood that a scalar is a one dimensional vector, we shall drop the accent $\bar{}$ hereafter. In summary, the following is the general state model we consider throughout.

---

[7] The reader is strongly advised to refer to a good book on linear algebra.

$\Sigma(A, B, C, D)$:

$$\dot{x} = Ax + Bu \quad (1.53)$$

$$y = Cx + Du \quad (1.54)$$

with

$$x \in \Re^{n \times 1}, u \in \Re^{p \times 1}, y \in \Re^{m \times 1} \text{ and}$$

$$A \in \Re^{n \times n}, B \in \Re^{n \times p}, C \in \Re^{m \times n}, D \in \Re^{m \times r} \quad (1.55)$$

1.2.7.3 Solution of the State Equation

Now we shall proceed to the solution of the state equation.

Let us first look at the homogeneous system, i.e. the system with no external input.

$\Sigma^d(A, B, C, D)$:

$$\begin{bmatrix} \dot{x}_1 \\ \dot{x}_2 \\ \dot{x}_3 \end{bmatrix} = \begin{bmatrix} -p_1 & 0 & 0 \\ 0 & -p_2 & 0 \\ 0 & 0 & -p_3 \end{bmatrix} \begin{bmatrix} x_1 \\ x_2 \\ x_3 \end{bmatrix} \quad (1.56)$$

The signficance of this canonical form is the mutual independence of the odes governing each state variable. In other words, we have n independent differential equations governing the system. Each of these equations

$$\dot{x}_i = -p_i x_i$$

has the solution

$$x_i(t) = e^{-p_i t} \cdot x_i(0) \quad (1.57)$$

and if we pack the solutions we get

$$x(t) = \begin{bmatrix} e^{-p_1 t} & 0 & \cdots & 0 \\ 0 & e^{-p_2 t} & \cdots & 0 \\ & & \ddots & \\ 0 & 0 & \cdots & e^{-p_n t} \end{bmatrix} x(0)$$

$$= e^{At} x(0) \quad (1.58)$$

where the matrix e^{At} is called the **state transition matrix** and has certain distinction of its own. For instance, it is just not the entire matrix A that is raised to the exponent, rather it is the diagonal elements that are raised. A closer observation reveals that it is the eigenvalues that are raised to the exponent. The name comes from the linear transformation involved, i.e. in the absence of any external input signal u, the initial state $x(0)$ moves to $x(t)$, at a later time t, via the transformation e^{At}.

Let us now obtain the solution of the nonhomogeneous system. The ode governing the ith state,

$$\dot{x}_i = -p_i x_i + b_i u$$

has the solution

$$x_i(t) = e^{-p_i t} \cdot x_i(0) + \int_0^t e^{-p_i(t-\lambda)} \cdot b_i \cdot u(\lambda)d\lambda \qquad (1.59)$$

and if we pack the solution we get

$$x(t) = e^{At}x(0) + e^{At} * Bu \qquad (1.60)$$

where $*$ denotes the convolution operation applied elementwise.

Let us look at an example to understand the above.

Example 1.2.15:

Let

$$A = \begin{bmatrix} -1 & 0 \\ 0 & -2 \end{bmatrix}, \quad \text{and} \quad B = \begin{bmatrix} 1 \\ -1 \end{bmatrix}$$

$$\text{with} \quad x(0) = \begin{bmatrix} -1 \\ 1 \end{bmatrix}$$

Then, for a unit step input,

$$x(t) = \begin{bmatrix} e^{-t} & 0 \\ 0 & e^{-2t} \end{bmatrix} \begin{bmatrix} -1 \\ 1 \end{bmatrix} + \begin{bmatrix} \int_0^t e^{-(t-\lambda)} \cdot u(\lambda)d\lambda \\ -\int_0^t e^{-2(t-\lambda)} \cdot u(\lambda)d\lambda \end{bmatrix}$$

$$= \begin{bmatrix} -2e^{-t}+1 \\ \dfrac{3}{2}e^{-2t}-\dfrac{1}{2} \end{bmatrix}$$

Apparently, the solution of state equation requires the state transition matrix e^{At} to be computed from the system matrix A. In order to explore certain inherent properties of e^{At}, let us now look at the general state equation without reference to any particular canonical form. We shall resort to Laplace transformation to solve the ode

$$L\{\dot{x} = Ax + Bu\}$$

$$\Rightarrow s \cdot x(s) - x(0) = Ax(s) + Bu(s)$$

$$i.e.\,(s \cdot I - A)x(s) = x(0) + Bu(s)$$

$$i.e.\,x(s) = (sI - A)^{-1}\big[x(0) + Bu(s)\big]$$

$$\text{and} \quad x(t) = L^{-1}\big\{(sI-A)^{-1}\big[x(0)+Bu(s)\big]\big\} \qquad (1.61)$$

Notice that we have made use of the fact that the vector $x(s) = I \cdot x(s)$ where I is the identity matrix of suitable dimension. This facilitates us to group similar terms together.

Let us look at the dimensions of the vectors and matrices in the equation above. On the left-hand side, we have a $n \times 1$ vector. Since A is $n \times n$, so is $sI - A$ and its inverse. $x(0)$ is a $n \times 1$ vector and so is $Bu(s)$. Hence the equation is dimensionally satisfactory. We have certain important questions regarding this new matrix $sI - A$ and its inverse. For instance, if it were a scalar first order system, then $(sI-A)^{-1}$ would look like $\frac{1}{s-a}$ and the inverse transform would be e^{at}. Following the same convention, we may call

$$L-1\{(sI - A)-1\} = \Delta\, eAt$$

where A is a matrix.

Example 1.2.16:

Let us consider A:

$$A = \begin{pmatrix} 0 & 1 \\ -2 & -3 \end{pmatrix}$$

It follows that

$$(sI-A)^{-1} = \begin{bmatrix} s & -1 \\ 2 & s+3 \end{bmatrix}^{-1}$$

$$= \frac{1}{(s+1)(s+2)}\begin{bmatrix} (s+3) & 1 \\ -2 & s \end{bmatrix}$$

$$= \begin{bmatrix} \left(\dfrac{2}{s+1}+\dfrac{-1}{s+2}\right) & \left(\dfrac{1}{s+1}+\dfrac{-1}{s+2}\right) \\ \left(\dfrac{-2}{s+1}+\dfrac{2}{s+2}\right) & \left(\dfrac{-1}{s+1}+\dfrac{2}{s+2}\right) \end{bmatrix}$$

and

$$e^{At} = \begin{bmatrix} 2e^{-t}-e^{-2t} & e^{-t}-e^{-2t} \\ -2e^{-t}+2e^{-2t} & -e^{-t}+2e^{-2t} \end{bmatrix}$$

Here comes an interesting issue. If A were a diagonal matrix, e^{At} was simple and straightforward. However, if A is in a different form, then computing e^{At} is a bit involved in terms of finding the eigenvalues, and obtaining partial fractions and the result looks much messier. In fact, in both the examples, the eigenvalues were the same -1 and -2. While we do not bother about the final shape of e^{At}, we are indeed bothered about the computation involved. This is because, in general, an nth order characteristic polynomial has to be solved for the eigenvalues, which is not always an easy task.

1.3 Stability of Linear Control Systems

We learned, until the previous sections of this chapter, about physical systems, their mathematical representation, and their analysis using Laplace transformation; this tool is quite obviously indispensible. Certainly, it was an interesting exercise all through. In principle, we are ready to venture into the design of systems. However, we are at a crossroads and need to investigate yet another important aspect of systems called **stability**. In fact, all the examples we have seen earlier (except perhaps the very first one of an inverted pendulum) have cast a shadow on this aspect; we chose those examples purposefully to drive a point without making the whole thing complex. It is now time for us to briefly unravel this agenda.

Let us begin with some intuitive arguments. Consider the way boats and ships have been designed to stay afloat. The geometrical properties and the specific gravity of the floating bodies determine the so-called *stable configurations*. It was Archimedes (287–212 BC, much before the invention of calculus) who formulated the law of flotation—the buoyant upthrust balances the weight of the floating body. It is not difficult for us to understand that "stability" denotes the ability of a body to retain its position and orientation in space. Even if the body is disturbed *slightly*, it returns to the original configuration. Next, let us look at a contrasting phenomenon. Whenever we attend a public meeting or concert, we do not appreciate the shriek from the public address system. It is not difficult to understand this as a chain of events leading to an *ever-increasing* volume of sound. A little sound is initially picked up by the microphone, gets amplified, put out of the speakers, picked up again by the microphone, gets further amplified, and thus goes the *regenerative* cycle. In this particular case, we refer to the phenomenon as *instability*. Clearly, what we appreciate is a moderate volume; by "moderate" we mean a volume to our taste given the acoustics of the auditorium. Perhaps, had we been referring to our financial investments, we might have called it as a stable phenomenon. As a final example, we may consider the depletion of the ozone layer owing to the ever-increasing emissions of automobiles; we have a feeling that the "limits" have been reached, and that we cannot tolerate any further depletion. We encourage the reader to brainstorm and generate many more examples of this nature.

Let us now study the arguments of this sort in greater detail, albeit formally.

1.3.1 Bounded Signals

Suppose that we have a signal $f(t)$ that satisfies the following condition:

$$|f(t)| \le K < \infty \quad \forall t \in (-\infty + \infty) \quad (1.62)$$

Put in words, if a signal $f(t)$ is finite in magnitude, we call it a *bounded signal*. We may naturally extend this definition to input signals and output signals and discuss about **Bounded Inputs** and **Bounded Outputs (BIBO)**.

For instance, the signal e^t is unbounded on $0 \le t < +\infty$ (why?), while the signal e^{-t} is bounded on the same interval. The given signal $f(t)$ need not hit the bound; rather, we are interested whether the signal is bounded or not. For instance, the signal $f(t) = A + sBe^{-t/\tau}$ is bounded by A, but only gets arbitrarily close to this bound for large t. The reader may quickly recollect the basic definitions of open- and closed-intervals he would have studied in an elementary calculus course.

Next, we move on to study the relationship between bounded-inputs and bounded-outputs.

1.3.1.1 Definition (a): BIBO Stability

A system Σ is said to be **BIBO stable** if, and only if, *every* bounded input produces a bounded output.

Put in symbols,

$$(\forall K_I)(\exists K_O): u(t) \in U \le K_I \mapsto y(t) \in Y \le K_O \quad (1.63)$$

We emphasize that the bound K_O on $y(t)$ depends only on the bound K_I on $u(t)$, and not on any specific input signal $u(t)$.

With reference to our earlier intuitive arguments, a faulty public address system produces a loud and unpleasant shriek, i.e. an unbounded output even when there is a small (bounded) perturbing input; similarly, the total amount of gaseous emission is an unbounded output. Of course, this "unboundedness" is merely a mathematical exaggeration; in practical systems (e.g. amplifiers in the public address system) the output is always limited by a certain amount of saturation. *Stability*, according to the above definition, is a very fundamental idea, and we need to carefully investigate possible instability in the systems we design. While many physical systems, such as amplifiers, are inherently unstable, certain systems (e.g. modern fighter aircrafts) are intentionally designed to be unstable. We may think of designing feedback controllers, like Black's negative feedback networks around the transistors, that stabilize these systems. In this sense, stability overrides all other performance specifications in the overall system design.

Let us now look at an equivalent definition of stability that allows us to talk about systems' stability in terms of their mathematical models.

1.3.1.2 Definition (b): BIBO Stability

A linear system is BIBO stable *if, and only if*, there is a constant $K < \infty$ such that

$$\int_{-\infty}^{+\infty} |h(t-\lambda)| \, d\lambda \; < \; K \;\; \forall t \in (-\infty, +\infty) \qquad (1.64)$$

where $h(.)$ is the impulse response of the system.

In fact, the reader might prove the above statement in a mathematical sense, and we leave it to him as an exercise. Let us now look at investigating stability as was suggested by the definitions above. An immediate question would be: "Do we have to apply every possible bounded input and see if the corresponding output is also bounded?" Certainly, this is a formidable task. Instead, we shall turn around this question and ask: "Assuming that the input is bounded, do we have any characterization of the impulse response for the output to be bounded?" It is easier to address this latter question. In plain english, "In the product $k_i k_h = k_o$, how big is k_h allowed, yet ensuring restricted k_i and k_o?" We might quickly recollect that the characteristic polynomial and its roots (i.e. the poles of the system) affect the system response. For example, if the poles look like

$$s_i = \alpha_i \pm , j\beta_i$$

the response is

$$y(t) = \sum_i K_i e^{\alpha_i t} \cos(\beta_i t + \theta_i)$$

Probing into this little deeper, we observe that

1. If the real part $\alpha_i < 0 \; \forall \, i$, then the system is **stable** since the response decays to zero in a finite amount of time. This is like the stability of boats and ships we were discussing before. We may quickly verify that this is true whether the pole is simple or not.

2. If the system has simple poles only on the imaginary axis, i.e. if $\alpha_i = 0 \; \forall i$, then the response is a summation of sinusoids; hence, generally, oscillatory. If the pole is at origin, then the response is constant with respect to time. If this were the case, we call the system **critically stable**.

3. If the real part $\alpha_i > 0 \; \forall i$, then the response, in general, is a growing sinusoid, i.e. unbounded, and the system is clearly **unstable**.

4. If the pole s_i has multiplicity r, the system is **unstable**.

5. If there are poles of multiplicity r on the imaginary axis, then the response is, once again, a **polynomially growing** sinusoid; hence, the system is **unstable**.

6. All the above statements are true irrespective of the imaginary part β.

The inference of these observations is that the poles must be restricted to lie only in the *open* left half of the complex s-plane (i.e. excluding the imaginary axis) for the system to be stable. Mathematically, we say that the characteristic polynomial must be **Hurwitz**. Although we have mapped our original problem into an equivalent simpler one, still the task appears to be formidable since it is hard to compute the roots of a general *nth* order polynomial. This goads us to investigate how polynomials with all of their roots in the left half of the complex plane look like. Equivalently, we wish to investigate a polynomial and decide that, owing to such and such properties, it does/ does not have roots with positive real parts. In what follows, we shall occasionally use the words "characteristic polynomial" and "system" interchangeably.

1.3.2 Routh-Hurwitz Criterion

This stability problem was posed by Maxwell way back in the year 1867 (after calculus was invented by Leibniz and Newton). He demonstrated that, by examining the coefficients of the differential equations, the stability of the system could be determined. He was also able to give necessary and sufficient conditions for equations up to the fourth order; for fifth-order equations he gave two necessary conditions. However, Edward John Routh in 1877 and Adolf Hurwitz in 1895 solved the problem completely in an interesting competion where Maxwell himself was on the jury. The result involved computing a sequence of determinants from the coefficients of the characteristic polynomial. The "Routh-Hurwitz" criterion can be formally developed as follows.

First, let us consider the characteristic polynomial given in the following form:

$$D(s) = a_o s^n + a_1 s^{n-1} + a_2 s^{n-2} + \cdots + a_{n-1} s + a_n \quad (1.65)$$

Next, we define the following **Routh-Hurwitz (R-H) array** (Table 1.1):

Observe that the elements in s^n and s^{n-1} row of the array are appropriately ordered coefficients of $D(s)$. The elements of the third s^{n-2} row are formed using the preceding two rows as follows

TABLE 1.1

Routh-Hurwitz Array

s^n row:	a_0	a_2	a_4	a_6	$\cdots$
s^{n-1} row:	a_1	a_3	a_5	a_7	$\cdots$
s^{n-2} row:	b_2	b_4	b_6	b_8	$\cdots$
s^{n-3} row:	c_1	c_5	c_7	c_9	$\cdots$
$\vdots$	$\vdots$				
s^1 row:	d_n				
s^0 row:	$e_n + 1$				

$$b_2 = -\frac{1}{a_1}\begin{vmatrix} a_o & a_2 \\ a_1 & a_3 \end{vmatrix} = a_2 - \frac{a_o}{a_1} \cdot a_3$$

and the general formula for $b_k, k = 2, 4, \cdots$ is

$$b_k = -\frac{1}{a_1}\begin{vmatrix} a_o & a_k \\ a_1 & a_{k+1} \end{vmatrix} = a_k - \frac{a_o}{a_1} \cdot a_{k+1}$$

Notice that the first column remains the same in all the determinants.

The fourth s^{n-3} row, similar to the third row, is computed as follows.

$$c_k = -\frac{1}{b_2}\begin{vmatrix} a_1 & a_k \\ b_2 & b_{k+1} \end{vmatrix} = a_k - \frac{a_1}{b_2} \cdot b_{k+1} \quad k = 3, 5, 7, \cdots$$

The reader would have, by now, observed a nice pattern in the computations. Each subsequent row of the $(n+1)$-row Routh-Hurwitz array is formed from its preceding two rows in an analogous fashion. Zeros can be inserted for undefined elements, as required, and any row can be scaled by a positive scalar to simplify the computation of a subsequent row. The following example serves as an illustration.

Example 1.3.1:

Let

$$D(s) = s^6 + 5s^5 + 15s^4 + 55s^3 + 154s^2 + 210s + 100$$

We form the first two rows immediately:

s^6 row:	1	15	154	100
s^5 row:	5	55	210	0

We may divide the s^5 row with 5 and replace it with the new expression. Accordingly, we get the s^4 row as follows:

s^4 row:	4	112	100	0

We may divide this s^4 row by 4. We proceed analogously to compute the subsequent rows, and the completed array is computed as follows (Table 1.2):

We encourage the reader to write a computer program that generates the Routh-Hurwitz array given the coefficients $a_o, \cdots, a_n$ of a polynomial.

The Routh-Hurwitz Criterion:

The number of roots of $D(s)$ with positive real parts is equal to the number of sign changes in the first column of the Routh-Hurwitz array. We may readily infer that the system is stable if, and

TABLE 1.2

Routh Array for Example 1.3.1

s^6 row:	1	15	154	100
s^5 row:	5	55	210	0
s^4 row:	1	28	25	0
s^3 row:	−1	1	0	
s^2 row:	29	25		
s^1 row:	1.862			
s^0 row:	25			

only if, *all* the elements in the first column of the Routh-Hurwitz array have same sign. We call such a polynomial a **Hurwitz polynomial**.

Example 1.3.2:

In the previous example, there are exactly two sign changes in the first column—the first due to the sign change from +1 to −1 and the second due to the sign change from −1 to +29. It follows that the system has two poles with positive real parts; thus, it is unstable.

1.3.2.1 Special Cases

It is quite evident now that the R-H criterion helps us decide the stability of systems, without having to compute the roolts of the characteristic polynomial. As one might (or, might not) suspect, this criterion has certain interesting special cases.

Case (i): If a row has a zero in the first column and has at least one nonzero element, we replace the zero in the first column with ε, an infinitesimally small positive number and continue building the array.

We shall illustrate this in the following example.

Example 1.3.3:

Let

$$D(s) = 2s^4 + 4s^3 + 4s^2 + 8s + 1$$

Note that the necessary condition is satisfied since all of the coefficients are present and have the same sign. The first two rows of the R-H array are

s^4 row:	2	4	1
s^3 row:	4	8	0

TABLE 1.3

Router Array for Example 1.3.3

s^4 row:	2	4	1
s^3 row:	4	8	0
s^2 row:	ε	1	
s^1 row:	$\approx -\dfrac{4}{\varepsilon}$		
s^0 row:	1		

We may compute the s^2 row as

s^2 row:	0	1

We now replace this row with an ε in the place of 0.

s^2 row:	ε	1

The complete Routh-Hurwitz array is as follows (Table 1.3)

Since there are exactly two sign changes in the first column, there are exactly two roots with positive real parts, and hence, the system is unstable. In fact, the roots of this polynomial are −1.9561, −0.1327, and the complex conjugate pair 0.0444 ± j1.3871 with positive real parts.

Case (ii): An entire row of zeros may be encountered. It is interesting to note that this happens only in the row s^{2k+1} corresponding to the odd powers of s. This implies the presence of mirror-image roots relative to the imaginary axis or one or more pairs of imaginary-conjugate roots of the form $\pm j\omega$.

In such a situation, we use the preceding row s^h to form an *auxiliary polynomial* in even powers as follows:

$$A(s) = A_{2k}s^{2k} + A_{2k-2}s^{2k-2} + \cdots + A_0$$

where A_i are the elements of the s^h row. We differentiate $A(s)$ with respect to s, and take the resulting coefficients as the entries of the s^{h+1} row. There is a bonus in this case: the roots of $A(s)$ are also the roots of the original polynomial $D(s)$.

We shall illustrate this in the following example.

Example 1.3.4:

Let

$$D(s) = s^6 + 2s^5 - 9s^4 - 12s^3 + 43s^2 + 50s - 75$$

(notice that in this case even the necessary condition fails).

The s^4 row is as follows:

s^4 row:	−1	6	−25

The s^3 row is then

s^3 row:	0	0	0

At this point, we define the auxiliary polynomial

$$A(s) = -s^4 + 6s^2 - 25$$

We can now replace the entire zero s^3 row with the coefficients of

$$\frac{d}{ds}A(s) = -4s^3 + 12s$$

Thus, the s^3 row is

s^3 row:	−4	12	0

We may continue to build the R-H array.

As has been mentioned earlier, the auxiliary polynomial is a factor of the given polynomial. In fact,

$$D(s) = A(s)(-s^2 - 2s + 3)$$

and the four roots of $A(s)$ are

$$s = \pm 2 \pm j1,$$

which are located symmetrically about the origin.

In control systems design, the Routh-Hurwitz criterion is most often used to ascertain the ranges of certain design parameters that lead to stability. We use the following example to illustrates this idea.

Example 1.3.5:

Consider the following polynomial with an unknown parameter K.

$$D(s) = s^3 + 24s^2 + 53.255s + 88.76K$$

We will be interested in asking: For what values of K is the following system stable? The R-H array is as follows (Table 1.4):

TABLE 1.4

Routh Array for Example 1.3.5

s^3 row:	1	53.255	0
s^2 row:	24	88.76 K	
s^1 row:	53.255−3.69 K		
s^0 row:	88.76 K		

Applying the criterion, we immediately see that

$$53.255 > 3.69K \quad \text{and} \quad 88.76K > 0,$$

which means that

$$0 < K < 14.3997$$

for the system to be stable. The reader would appreciate this example better after learning controller design techniques in the next chapter.

While this Routh-Hurwitz criterion appears quite attractive in deciding stability problems, the proof of this criterion tends to be quite complex. However, it is stimulating to note that there is considerable interest in the research community discussing simpler proofs that could be accessible to undergraduate students.

We end this section with a note on the Routh-Hurwitz criterion. If the system has multiple poles on the imaginary axis, the response would be unbounded and the Routh-Hurwitz criterion will not always reveal this.[8]

1.3.3 Nyquist Criterion

Recall from Example 1.2.8 that the frequency response of a system, whose transfer function is $L(s)$, may be obtained directly by letting $s = j\omega$. The response can be *graphically* visualized by plotting the magnitude $|L(j\omega)|$ and the phase $\angle L(j\omega)$ versus ω. If

$$L(s) = \frac{1}{s + 1}$$

so that

$$L(j\omega) = \frac{1}{j\omega + 1}$$

Although ω is real, $L(j\omega)$ is, in general, complex. If $\omega = 0$ (i.e. dc), $L(j\omega) = 1$. If $\omega = 1$, then

$$L(j1) = \frac{1}{1 + j1} = \frac{1}{\sqrt{2}} e^{-j45°}$$

or, $|L(j1)| = 0.707$ and $\angle L(j1) = -45°$. Similarly, we can compute $L(j1.2)$, $L(j5)$, $L(j\infty)$, and so on. We encourage the reader to do these computations immediately.

1.3.3.1 Polar and Nyquist Plots

We may use these pairs of numbers computed at several frequencies to *plot* the function $L(j\omega)$. If we choose *polar*

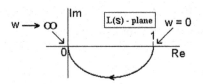

FIGURE 1.27
Polar plot of Example 1.2.8.

coordinates, we obtain the plot as shown in Figure 1.27 (Figure 1.16 replicated). This is called the **polar plot**, for obvious reasons.

The horizontal and vertical axes are, respectively, the real part $\text{Re}L(j\omega)$ and the imaginary part $\text{Im}L(j\omega)$. Thus, a point in the $L(j\omega)$-plane is a phasor with magnitude and phase. The polar plot of the transfer function in Example 1.2.8 is a semicircle for all the positive real frequencies $0 \le \omega < \infty$. As we see it, a polar plot is easier to obtain. And, it possesses information about a transfer function for the range of frequencies -0 to ∞.

Before we discuss another plot of $L(j\omega)$, let us study this polar plot in more detail. Each point on the $L(j\omega)$-plane is a *map of some point* ω on the imaginary axis of the s-plane. More precisely, the polar plot is a map of the *positive* imaginary axis in the $L(j\omega)$-plane. Let us now generalize this idea and look at the maps of *contours* in the s-plane. A **contour** is a closed curve, i.e. it starts and ends at the same point. A **simple contour** is one that does not intersect with itself; otherwise, it is called a complex contour.

Let us look at the following example.

Example 1.3.6:

Let

$$L(s) = \frac{1}{s + 1}$$

as before. Let the contour Γ_s in the s-plane be the simple closed curve ABCD as shown in Figure 1.28a. If we evaluate $L(s = \sigma + j\omega)$ at every point along the contour in the clockwise direction $A - B - C - D - A$, we obtain the contour Γ_L in the $L(s)$-plane as shown in Figure 1.28b.

For the contour Γ_s shown in Figure 1.29a, the corresponding contour Γ_L is shown in Figure 1.29b. We observe that Γ_s in Figure 1.29a *encloses* the pole of $L(s)$, and Γ_L now *encircles* the origin in the *counterclockwise direction* just once.

Example 1.3.7:

Let

$$L(s) = \frac{s + 1}{s + 4}$$

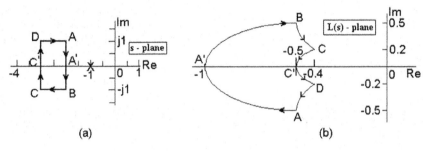

FIGURE 1.28
Polar plot for Example 1.3.6.

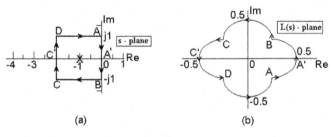

FIGURE 1.29
Polar plot for Example 1.3.6, change of contour.

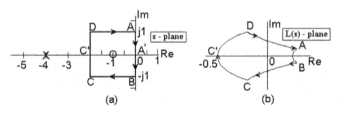

FIGURE 1.30
Polar plot for Example 1.3.7.

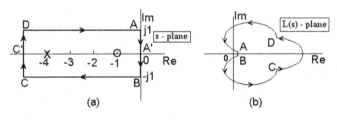

FIGURE 1.31
Polar plot for Example 1.3.7, change of contour.

The contours Γ_s and Γ_L are shown in Figure 1.30a and b, respectively.

We prompt the reader to observe that the contour Γ_L encircles the origin once, but in the same direction as that of Γ_s. Finally, let us see the map in Figures 1.31a and b.

Let us now consolidate all these observations. We get Cauchy's theorem, also called the Principle of Argument, in complex variables.

Theorem: Cauchy's Theorem:

Let Γ_s be a simple closed curve in the s-plane. Let $L(s)$ be a rational function that has neither poles nor zeros on Γ_s. Let Z and P be the number of zeros and poles of $L(s)$ (counting multiplicity), encircled by Γ_s. Let Γ_L be the map of Γ_s in the $L(s)$-plane. Then Γ_L will encircle the origin of the $L(s)$-plane $N = Z-P$ number of times in the same direction as Γ_s.

Let us now dissect this theorem statement and make some concrete observations:

1. Γ_s does not pass through the zeros or poles of $L(s)$. Hence Γ_L does not pass through the origin and is always well defined.
2. $L(s)$ is a continuous function in s and Γ_L is also a closed curve; not necessarily simple.
3. Finally, we shall explain the meaning of $N = Z-I$. We shall make a careful notice that Z and P are the number of zeros and poles, respectively, *encircled by* Γ_s. For clarity, let us assume that Γ_s is in the clockwise direction.
   a. Suppose $P = 0$. Then the number of encirclements of the origin by Γ_L is Z and the direction is same as Γ_s, i.e. clockwise.
   b. Similarly, if $Z = 0$, the number of encirclements is P, but the direction is counterclockwise, as is implied by the negative sign.

Thus, it follows that there shall be Z encirclements in the clockwise direction and P encirclements in the opposite direction. Accordingly,

- If $N>0$, the total number of encirclements is N in the clockwise direction and,
- If $N<0$, the net number of encirclements is N in the counterclockwise direction.

We urge the reader to verify these observations in the earlier examples.

Example 1.3.8:

Let

$$L(s) = \frac{s+1}{s(s+2)}$$

and the contour Γ_s be as shown in Figure 1.32a. The contour Γ_L is shown in Figure 1.32b. There would be two counterclockwise encirclements and one clockwise encirclement of the origin; hence, the contour Γ_L effectively encircles the origin once in the counterclockwise direction.

Example 1.3.9:

Let

$$L(s) = \frac{2}{s-1}$$

and the contour Γ_s be as shown in Figure 1.33a. Since the pole of $L(s)$ is encircled by Γ_s, we see that the contour Γ_L in Figure 1.33b encircles the

origin once in the counterclockwise direction. Notice that the semicircle of infinite radius in Γ_s is automatically considered by letting $\omega \to \infty$. Further, Γ_L is symmetric with respect the real axis.

A polar plot, as in Figure 1.33b, that maps $-\infty < \omega < \infty$ along the imaginary axis and the infinite semicircle enclosing the entire right half of the plane is called the **Nyquist Plot**.

Example 1.3.10:

Let

$$F(s) = 1 + L(s)$$

where $L(s)$ is the transfer function in Example 1.3.9, and the contour Γ_s is shown in Figure 1.33a. Since a real number 1 is added to $L(s)$ to obtain $F(s)$, the contour Γ_F shown in Figure 1.34 is similar to that in Figure 1.33b except for the shift on the real axis to the right by one unit. In other words, the origin of $F(s)$-plane is $(-1, j0)$ in the $L(s)$-plane. Accordingly, we say that the contour Γ_F in Figure 1.34 encircles the origin $(0, j0)$ in the $F(s)$-plane once in the counterclockwise direction, or the point $(-1, j0)$, hereafter called the **critical point**, in the $L(s)$-plane once in the counterclockwise direction.

Example 1.3.10 gives us lots of information. As might be recollected from the earlier chapters, the characteristic polynomial is always of the form $1 + L(s)$, where $L(s)$ is the loop transfer function. Therefore, once the polar plot of $L(s)$ is obtained, the polar plot of the characteristic polynomial is already there. Let us now explore further in this direction.

Let the loop transfer function $L(s)$ be a rational function $\frac{N(s)}{D(s)}$ so that

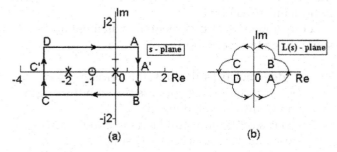

(a)   **(b)**

FIGURE 1.32
Polar plot for Example 1.3.8.

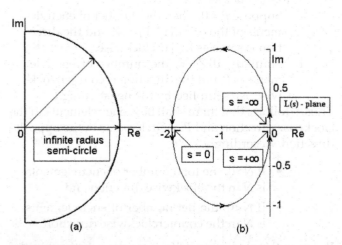

(a)   **(b)**

FIGURE 1.33
Nyquist plot for Example 1.3.9.

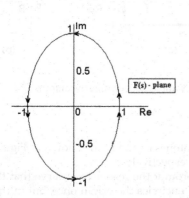

FIGURE 1.34
Nyquist plot for Example 1.3.10.

$$F(s) = 1 + L(s) = \frac{D(s) + N(s)}{D(s)} \qquad (1.66)$$

is the characteristic polynomial. The open loop system's poles are the roots of $D(s)$, while those of the closed loop system are the zeros, i.e. the roots, of $D(s) + N(s)$. Clearly, we do not want the *zeros* of $F(s)$ to lie in the right half of the s-plane, even if there might be a few open loop poles there.

Let us then choose our contour Γ_s to be the infinite semicircle enclosing the entire right half of the s-plane including the $j\omega$ axis. Let us also fix the direction of traversal of Γ_s to be clockwise. Such a contour is considered above in Examples 1.3.9 and 1.3.10 and is shown in Figure 1.33a. The corresponding map Γ_L is called the Nyquist plot. We do not want any of the zeros of $F(s)$ to lie in the rhp, i.e. we need $Z = 0$. There might be a few poles of the open loop system, i.e. P need not be zero. Therefore, we expect that given a loop transfer function, Γ_L encircles the critical point $(-1, j0)$ in the $L(s)$-plane P number of times in the *counterclockwise* direction. Conversely, if the contour Γ_L of the loop transfer function encircles the critical point $(-1, j0)$ in the $L(s)$-plane P number of times in the counterclockwise direction, then we may infer that the *closed loop system is stable*. This observation leads to the following important theorem.

Theorem: Nyquist's Theorem:

Consider a unity negative feedback system with the closed loop transfer function:

$$T(s) = \frac{L(s)}{1 + L(s)} = \frac{L(s)}{F(s)} \qquad (1.67)$$

Let the contour Γ_s in the s-plane be the infinite semicircle enclosing the entire right half of the s-plane including the imaginary axis. Let the direction of traversal be clockwise. The closed loop system is stable **if and only if** the Nyquist plot of $L(s)$ does not pass through the critical point and the number of counterclockwise encirclements of $(-1, j0)$ equals the number of poles of $L(s)$ in the right half of the s-plane.

In most applications, we may come across a situation where

$$T(s) = \frac{k \cdot L(s)}{1 + k \cdot L(s)} \qquad (1.68)$$

in which case, we need to scale the critical point by k and take it as $\left(\frac{-1}{k}, j0\right)$.

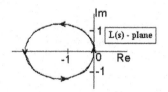

FIGURE 1.35
Nyquist plot for Example 1.3.12.

Example 1.3.11:

Consider Example 1.3.9 again. The closed loop system has the transfer function

$$T(s) = \frac{2}{s + 1}$$

and the closed loop system is stable. This conforms to our observation that the contour in Figure 1.33(b) encircles the point $(-1, j0)$ once in the counterclockwise direction.

Example 1.3.12:

Let

$$L(s) = \frac{8s}{(s-1)(s-2)} \qquad \text{Notice here that } P = 2$$

The polar plot happens to be a circle and, hence, its reflection about the real axis (for $\omega \le 0$) is also a circle. Thus, the complete Nyquist plot consists of the solid and dashed circles as shown in Figure 1.35. It encircles $(-1, j0)$ twice in the counterclockwise direction. Therefore, the closed loop system is stable.

On the other hand,

$$L(s) = \frac{2s}{(s-1)(s-2)}$$

may be shown to be unstable.

Example 1.3.13:

Let

$$L(s) = \frac{s+1}{s^2(s-2)}$$

Observe that the transfer function has poles at the origin. In this case, we need to modify our contour Γ_s so that it does not pass through the origin. This new contour takes a small semicircular detour $\varepsilon e^{j\phi}$ at the origin, as shown in Figure 1.36a.

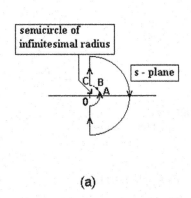

(a)   (b)

FIGURE 1.36
Nyquist plot for Example 1.3.13.

The radius of the semicircle ε is very small and ϕ varies from $-90°$ to $+90°$. The mapping of this small semicircle may be computed using the following approximation

$$L(s) = \frac{s+1}{s^3 - 2s^2} \approx \frac{-1}{2\varepsilon^2} \quad \text{as} \quad s \to \varepsilon$$

If we consider the points A, B, and C on the semicircle, then

$$L(A) = \frac{1}{2\varepsilon^2} \angle 180° \quad L(B) = \frac{1}{2\varepsilon^2} \angle 90°$$

$$\text{and} \quad L(C) = \frac{1}{2\varepsilon^2} \angle 0°$$

where we have used the fact that $-1 = 1\angle 180°$. Notice that the magnitude of $L(s)$ at these points is infinite. The complete Nyquist plot is shown in Figure 1.36b. The number of encirclements is one (equal to the number of poles of $L(s)$ in the rhp), but the direction is clockwise. Thus, the closed loop system is unstable with a closed loop pole in the rhp.

Example 1.3.14:

Let

$$L(s) = \frac{8k}{(s+1)(s^2 + 2s + 2)}$$

The Nyquist plot is shown in Figure 1.37 for the case $k = 1$. Notice that the portion between -0.8 and $+4$ on the real axis is encircled by the plot. Accordingly, if $1/k$ is in this range, the number of encirclements is one and the system becomes unstable. In other words, if

$$-0.8 \leq \frac{1}{k} \leq +4.0$$

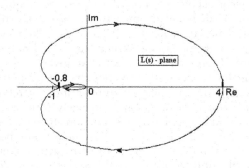

FIGURE 1.37
Nyquist plot for Example 1.3.14.

the system is unstable. Therefore, the stability range of the system is

$$\frac{-1}{4} < k < \frac{5}{4}$$

1.3.3.2 Gain and Phase Margins

While the Routh-Hurwitz criterion gives us an analytical tool, the Nyquist criterion gives us an excellent graphical tool to *decide* whether a given system is stable or not. An added advantage of the latter criterion is that it offers us information on the relative stability: how safe we are from the edge of instability. In this section we shall learn this idea.

Suppose that at some frequency ω_o, the Nyquist plot of a given loop transfer function passes through the point $(-1, j0)$. This means that the characteristic polynomial has a zero, i.e. $1 + L(j\omega_o) = 0$ at this frequency; thus, the closed-loop system is unstable at this frequency. Naturally, the distance between the Nyquist plot and the point $(-1, j0)$ can be used as a *measure* of the closed loop system stability—the larger the distance, the greater the stability. This distance equals the radius of the circle drawn with center at $(-1, j0)$ and touching the Nyquist plot. This is shown in Figure 1.38. Such a distance, however,

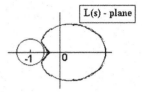

FIGURE 1.38
Illustrating stability margins.

is not convenient to measure. We replace this with "gain margin and phase margin." The word margin should be intuitively clear. For instance, imagine that you are burning a fire cracker with a stick in your hand—the longer the stick, the safer you are.

We now define two important frequencies.

1.3.3.3 Definition: Gain Crossover Frequency

The **gain cross-over frequency** ω_p is that frequency at which

$$|L(j\omega_p)| = 1 \qquad (1.69)$$

At this frequency, since the gain is already unity, we do not want the phase to become $\pm 180°$; otherwise the Nyquist plot passes through the point $(-1, j0)$. Accordingly, we shall compute the **phase margin** as

$$\text{phase margin} \triangleq \pm 180° - \angle L(j\omega_p) \qquad (1.70)$$

1.3.3.4 Definition: Phase Crossover Frequency

The **phase cross-over fresquency** ω_g is the frequency at which

$$\angle L(j\omega_g) = 180° \qquad (1.71)$$

Again, at this frequency, since the phase is already $180°$, we do not want the gain to become unity. Accordingly, we shall compute the **gain margin** as

$$\text{gain margin} \triangleq |-1 - |L(j\omega_g)|| \qquad (1.72)$$

This gain margin, however, as measured on the Nyquist plot is not so convenient. We would rather explore these margins in greater detail in terms of the Bode plot.

1.3.3.5 The Margins on a Bode Plot

We saw in Section 1.2 how a Bode plot can be obtained given the transfer function. Once again, we restrict our discussion to transfer functions that could be factored into terms whose powers are either 0 or 1 or 2.

By definition, the *gain cross-over frequency*, ω_p is the one at which the gain is unity, i.e.

$$20\log\big(|L(j\omega_p)|\big) = 0 \text{ dB}$$

In other words, the frequency at which the magnitude curve crosses 0 dB is the gain cross-over frequency. To obtain the phase margin, we simply read the phase at this frequency and measure how *far away* is $180°$. Let us quickly look at the following example:

Example 1.3.15:

Let

$$L(s) = \frac{50(s+2)}{s(s^2 + 4s + 100)}$$

This may be factored as

$$L(s) = \frac{1}{s} \cdot (s+2) \cdot \frac{1}{\left(\dfrac{s}{10}\right)^2 + 2 \cdot 0.2 \cdot \left(\dfrac{s}{10}\right) + 1} \cdot \frac{50 \cdot 2}{100}$$

where we have identified the quadratic factor in the denominator to have $\zeta = 0.2$ and $\omega_n = 10$. The complete Bode plot is shown in Figure 1.39.

In this Example 1.3.15, the gain becomes unity at $\omega_p \approx 1$ rad/sec. At this frequency, the phase is $-65°$ approximately, and hence, the phase margin is $115°$.

At this point, the reader may raise the following question: Since $+180° = -180°$, i.e. we get the negative real axis either clockwise or counterclockwise, how is the phase margin exactly measured? Recall that we assumed the loop transfer function to be strictly proper with n poles and $m < n$ zeros. Accordingly, the phase angle approaches

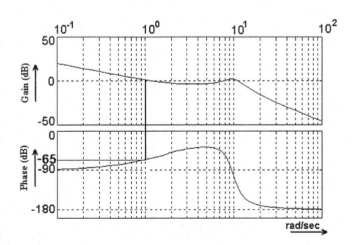

FIGURE 1.39
Bode plot and margins for Example 1.3.15.

$(n - m) \times -90°$ asymptotically. It stands to reason, then, to measure the phase margin as the distance with reference to $-180°$.

Similarly, the frequency at which the phase curve crosses $-180°$ is the *phase cross-over frequency*, ω_g. To obtain the gain margin, we simply read the gain at this frequency and measure how far away is 0 dB. In Example 1.3.15, the gain margin is ∞ dB since the phase curve grazes $-180°$ only at infinity and the gain at this frequency is $-\infty$.

Thus, we see that it is quite simple to *measure* the gain and phase margins on a Bode plot. It is not difficult to observe that these two plots differ only in the coordinates; in fact, both the plots convey the same information, viz., the frequency response. The basic idea behind the margins has to be emphasized.

For the closed loop system to be stable, at the frequency when the magnitude becomes unity, the phase should not be 180°, and at the frequency when the phase becomes 180°, the magnitude should not be unity. Naturally, the larger the margins the better.

If the loop transfer function $L(s)$ has open right half plane poles, still the closed loop system is stable if the Nyquist plot encircles the point $-1, j0$ sufficient number of times. In such a situation, we may find that there are more cross-over frequencies, and hence, the margins may not be unique. The logical way to overcome situations like this is to simply follow the basic idea and see that at no frequency ω_x (or over a range of frequencies)

$$L(j\omega_x) \neq -1 \qquad (1.73)$$

We shall revisit these margins while learning to design controllers.

1.3.4 The Root Locus

In the earlier chapters, we have established that it is the poles of the system that determine the nature of response. Henceforth, it may be expected that any compensation involves altering the roots of the characteristic polynomial. This may be accomplished by altering one or more coefficients of the characteristic polynomial; of course, we are now talking about the closed loop systems. In this chapter, we shall explore a graphical technique, called **Root Locus**, introduced by W.R. Evans in 1948. It allows us to design controllers so that the feedback system meets the desired performance specifications. Needless to say, the design must not perturb the stability of the control system.

1.3.4.1 Definition: Root Locus

It is the locus of each of the roots of a polynomial equation as one of its parameters is varied.

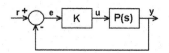

FIGURE 1.40
Feedback control with gain K.

Before elaborating this definition, let us look at the following examples. Most of these examples require solving quadratic equations for a large number of different values of the parameter K. The reader is urged to equip himself well, preferably with a good computer program developed by himself, and work out each of these completely before moving to the next one for a better understanding. Throughout these examples we consider the control system shown in Figure 1.40.

Example 1.3.16:

Let the plant transfer function be

$$P(s) = \frac{1}{s}$$

We may readily see that the overall transfer function is

$$T(s) = \frac{1}{s + K}$$

The characteristic polynomial is now a function of the parameter K, and we understand that the pole is at $s = -K$ for this first-order system. As K is varied from 0 to $+\infty$, the location of the pole is as shown in Figure 1.41 with a thick line.

Thus, the locus of the root of the characteristic equation

$$F(s) = s + K = 0$$

is the entire negative real axis of the s-plane as the parameter K is varied from 0 to ∞.

Example 1.3.17:

In this example, let

$$P(s) = \frac{1}{s(s + 2)}$$

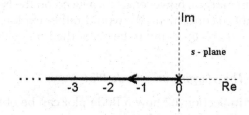

FIGURE 1.41
Root Locus for Example 1.3.16.

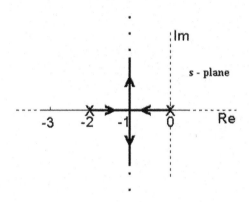

FIGURE 1.42
Root Locus for Example 1.3.17.

The characteristic equation is

$$F(s) = s^2 + 2s + K = 0$$

Figure 1.42 shows the root loci.

In this example, we have a second order system and notice that there are two loci, called the *branches* of the graph, corresponding to the two roots of the characteristic equation. These branches start at the open loop pole locations ($s = 0$ and $s = -2$) for $K = 0$, meet at $s = -1$ for $K = 1$, and thereafter, for $K > 1$, become complex conjugates with real part equal to -1.

Example 1.3.18:

Suppose we have the following plant

$$P(s) = \frac{s + 3}{s^2 + 2s + 2}$$

For this second order system the root loci are shown in Figure 1.43.

In this example, the loci of the two roots, i.e. the two branches, start at the complex conjugate open loop poles and remain complex conjugates

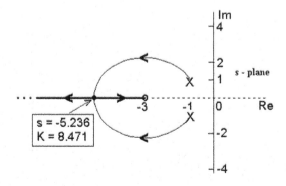

FIGURE 1.43
Root Locus for Example 1.3.18.

until $K = 8.471$, when they meet at $s = -5.236$. Thereafter, the roots are real and distinct. We may notice that one of the roots approaches the open-loop zero at $s = -3$, and the other approaches $-\infty$ (which is also a zero of the system since it is strictly proper).

Let us now generalize these observations and put forth some important properties of the root loci. In what follows, we shall consider the control system depicted in Figure 1.40. We assume that K is a non-negative real number, i.e. $K \geq 0$.

1. Owing to our fundamental requirement on the well-posedness of systems, the plant may be assumed to be strictly proper; hence, the degree of the characteristic equation is equal to the order of the plant. Thus, for each $K \in \Re_+$, there are n roots. Since the roots of any polynomial are continuous functions of its coefficients, the n roots form n continuous *branches* as K varies. Moreover, since we also assume that all the coefficients in $P(s)$ are real, complex roots, if any, should appear as conjugates. Thus,

Property #1

If the order of the plant is n, then the root loci consists of n continuous branches. These branches are symmetric with respect to the real axis.

2. For the rational plant transfer function

$$P(s) = \frac{N(s)}{D(s)}$$

with numerator and denominator polynomials let us examine the characteristic equation

$$F(s) = D(s) + K \times N(s) = 0$$

at the two extremes, i.e. as $K = 0$ and $K \to \infty$. In the former case,

$$F(s) = D(s) = 0$$

and the closed loop poles coincide with the open loop poles. In the latter case,

$$F(s) \to N(s) = 0$$

and the "closed loop" poles approach the open loop zeros. It must be noted that the case $K = 0$ is a limiting case, which means that the system consists of only the plant $P(s)$, and it is meaningless to attempt to compute the overall transfer function with $K = 0$.

Property #2

The root loci start at the open loop poles for $K = 0$ and terminate at the (open-loop) zeros for $K \to \infty$.

Notice that the zeros could be either the finite ones, as in Example 1.3.18, or could be those at infinity, as in Examples 1.3.16 and 1.3.17. In other words, all the n continuous branches that start at n open-loop poles terminate at n zeros, which could be finite or at infinity or both.

3. The general form of the characteristic equation is

$$F(s) = 1 + K \times P(s) = 0$$

where $P(s)$ is a rational function of a complex variable s, i.e. $F(s)$ evaluates to a complex number. If $C^{\#}$ is the set of all the roots of the characteristic polynomial, then $\forall s \in C^{\#}$, the characteristic polynomial should satisfy the following conditions on magnitude and phase angle respectively.

$$|KP(s)| = 1 \quad \text{and} \qquad (1.74)$$

$$\angle P(s) = \pm(2q + 1)180^{o} \qquad q = 0, 1, 2, \dots.$$

Notice that the gain $K \in \Re_{+}$ need not appear in the angle condition. We encourage the reader to ask himself why it is so.

It is often tedious to evaluate the magnitude and the phase of any complex function $P(s)$ at a given point $s^{\#} \in C$ in the complex plane. However, it is far easier (and hence strongly recommended) to evaluate them on the graph itself using a ruler and a protractor. The reader should have had familiarity with this technique in an earlier course on complex variables. For instance, in Example 1.3.17, from Figure 1.42, we see that at

$$s^{\#} = -1 + j1$$

the lines drawn from the poles of $P(s)$ would represent the phasors

$$(s^{\#} + 0), \quad \text{and} \quad (s^{\#} + 2)$$

as shown in Figure 1.44.

This equivalence of a rectangular and polar form for a factor like $(s^{\#} + p_1)$ may be expressed in the equation

$$s^{\#} + p_1 = M_{p1}\, e^{j\theta_{p1}}$$

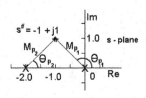

FIGURE 1.44
Graphical computation of gain K.

where M_{p1} represents the magnitude of the phasor and θ_{p1} represents the phase angle. Since $K = 2$ at $s^{\#} = -1 + j1$, $KP(s^{\#})$ may be evaluated as

$$|KP(s)| = \frac{K}{M_{p1} \times M_{p2}}$$

$$= \frac{2}{\sqrt{2} \times \sqrt{2}}$$

$$= 1$$

$$\angle P(s) = -135^{\circ} - 45^{\circ}$$

$$= -180^{\circ}$$

where

$$M_{p1} = M_{p2} = \sqrt{2}, \ \theta_{p1} = 135^{\circ}, \ and \ \theta_{p2} = 45^{\circ}$$

using elementary geometry. Since the poles appear in the denominator of the rational function, we placed the magnitudes in the denominator and we used negative sign for the angles.

In general, a function like

$$\frac{N(s)}{D(s)} = K \frac{(s - z_1)(s - z_2) \cdots (s - z_m)}{(s - p_1)(s - p_2) \cdots (s - p_n)}$$

may be expressed in terms of the magnitude and phase angle quantities as follows:

$$\frac{N(s)}{D(s)} = K \frac{M_{z1} M_{z2} \cdots M_{zm}}{M_{p1} M_{p2} \cdots M_{pn}} \cdot$$

$$e^{j(\theta_{z1} + \theta_{z2} + \cdots + \theta_{zm} - \theta_{p1} - \theta_{p2} - \cdots - \theta_{pn})}$$

From this relationship, the magnitude and phase angle of a function for a given value of s may be found.

It is now easy to establish the following property using the angle condition.

Property #3

On the real axis, the root loci lie to the left of an odd number of poles and zeros counted together.

It is not very difficult to verify this property. From the above expression, for any point

s on the real axis, $\theta = 0$ or $180°$. Therefore, if the point lies to an odd number of poles and zeros counted together, the net angle would $(2q \pm 1)\theta = \pm 180° = 180°$.

4. From the examples cited above and properties 2 and 3, it is clear that there is always a possibility for two of the branches to meet at a point (not necessarily on the real axis always). Let us now determine analytically where such a point, such as s_m, exists. Since s_m is a repeated root of the polynomial $F(s)$, then

$$F(s_m) = 0 \text{ and } \frac{d}{ds}F(s_m) = 0$$

Substituting $F(s) = D(s) + K \times N(s)$ in the above equation and solving using the chain rule of differentiation, we may state the following property.

Property #4

Any two branches of the root loci meet at a point determined as the solution of the equation:

$$\frac{d}{ds}P(s) = 0 \qquad (1.75)$$

Such a point is called, either a **break-away** point (as in Example 1.3.17), or a **break-in** point (as in Example 1.3.18).

It is important to notice that the eqn. 1.75 represents a necessary (but not sufficient) condition; i.e. the solution of equation need not represent a break point, but all break points must satisfy eqn. 1.75.

5. As the gain parameter K is made arbitrarily large, it may be expected that the magnitude of each of the n roots also tends to become large. Thus, as $K \to \infty$ the loci stretch *asymptotically* towards ∞ in the complex s-plane. (What is ∞ in a complex plane?) These asymptotes appear to have a common meeting point on the real axis making certain angles with it. We may look at a map for an analogy. Though a city may be quite large, when it is represented on the map of a country, the scale chosen would not permit more than a point the size of a pin-head to represent the city. In other words, here we are interested in larger scales of s, typically, $s \approx 10000$.

Let us illustrate this. As shown in Figure 1.45, if $P(s)$ has m zeros and n poles, in the limit $s \to \infty$,

$$P(s) = \frac{(s-z_1)(s-z_2)\cdots(s-z_m)}{(s-p_1)(s-p_2)\cdots(s-p_n)}$$

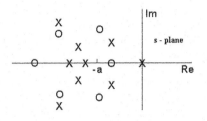

FIGURE 1.45
A typical zero-pole distribution in the s-plane.

$$\approx \frac{1}{(s+a)^{n-m}} \qquad (1.76)$$

i.e. there appears a cluster of $n-m$ poles on the negative real axis at $s = -a$. (This cluster could be on the positive real axis as well). Applying polynomial division (on the lhs of eqn. 1.76) and the binomial theorem (on the rhs of eqn. 1.76), and neglecting only the higher order terms,

$$a = \frac{\Sigma_n(poles) - \Sigma_m(zeros)}{n-m} \qquad (1.77)$$

where $\Sigma(poles) = a_{n-1}$ and $\Sigma(zeros) = b_{m-1}$. Thus, we assume that the asymptotes are straight lines. Now that we have found the point of their intersection, it remains to find the slope. Since there are $n-m$ such asymptotes, each of these contribute equally to the phase angle; hence the slope of each asymptote may be written as

$$\phi = \frac{(2q+1)180°}{n-m} \qquad (1.78)$$

Property #5

For large values of K, the loci follow $n-m$ asymptotic straight lines given in the point—slope form. The point, called the "centroid," is given by eqn. 1.77, and the slope is given by eqn. 1.78.

Observe that there was one asymptote along the negative real axis with centroid at the origin in Example 1.3.16; there were two at right angles to the negative real axis with centroid at $s = -1$ in Example 1.3.17. It is left to the reader to find the asymptotes in Example 1.3.18. The reader should be cautious at this stage to notice that although for some problems, the break-away/break-in point and the centroid coincide (as in Example 1.3.17), and these two indicate two altogether different properties of the loci.

6. So far, we have seen examples wherein the poles and the zeros lie on the real axis. Apparently, the loci leave the real poles at angles 0° or 180°. Similarly, the loci arrive at the real zeros at 0° or 180°. If there are complex poles and/or zeros, how do the loci appear? In particular, how do the branches start and how do they terminate? Again, the answer is provided by the angle condition. Consider the loci in Example 1.3.18 as shown in Figure 1.44. Assume that one of the loci has started from, for example $s = -1 + j1$, and moves by a very small amount ε to s_ε making an angle θ_p with reference to $s = -1 + j1$. The angle condition says that

$$\angle P(s)|_{s_\varepsilon} = \pm(2q+1)180°$$

i.e. the algebraic sum of all the angles measured from the poles and zeros to the pole at $s = s_\varepsilon$ is equal to ±180°. Graphically, assuming $\varepsilon \to 0$, the phasors drawn from the zero and the rest of the poles to s_ε are the same as those drawn to the pole $s = -1 + j1$ itself. Employing this approximation, we can easily compute the "angle of departure" of the locus from $-1 + j1$ as follows:

$$18.43° - \theta_p - 90° = \pm(2q+1)180°,$$

which gives $\theta_p = 108.43°$. Since it has been already established (property #1) that the loci are symmetric with respect to the real axis, the angle of departure θ_p^* from the complex pole at $s = -1 - j1$ may be found easily as $\theta_p^* = 360° - \theta_p$.

If there is a complex zero the "angle of arrival" may be found by applying the same reasoning as above.

In general, let there be n poles with $2R$ of them as complex conjugate pairs, and m zeros with $2Q$ of them as complex conjugate pairs. Further, let $\theta_{z_i}^{p_r}$ denote the m angles measured from the zeros at $s = z_i$, $i = 1, \ldots, m$ to the complex pole at $s = p_r$, and let $\theta_{p_j}^{p_r}$, $j \neq r$, denote the $n-1$ angles measured from the poles at $s = p_j$, $j = 1, \ldots, n-1$ to $s = p_r$. The angle of departure $\theta_{p_r}^{p_r}$ may be computed using

$$\theta_{p_r}^{p_r} = \Sigma_{i=1}^m \theta_{z_i}^{p_r} - \Sigma_{j=1 \neq r}^n \theta_{p_j}^{p_r} \mp (2q+1)180° \quad (1.79)$$

$$\forall r = 1, \ldots, 2R.$$

Similarly, the angle of arrival $\theta_{z_t}^{z_t}$ may be computed using

$$\theta_{z_t}^{z_t} = \Sigma_{i=1 \neq t}^m \theta_{z_i}^{z_t} - \Sigma_{j=1}^n \theta_{p_j}^{z_t} \mp (2q+1)180° \quad (1.80)$$

$$\forall t = 1, \ldots, 2Q.$$

Property #6

The angle of departure from a complex pole p_r may be found using eqn. 1.79 and the angle of arrival at a complex zero z_t may be found using eqn. 1.80.

In fact, it is sufficient to compute only half the number of angles. The remaining half may be obtained easily by noticing that the loci are symmetric with respect to the real axis.

All these six properties may be used to construct reasonably good root loci. This is illustrated in the following examples.

Example 1.3.19:

Let

$$P(s) = \frac{1}{s(s+2)(s^2+2s+2)}$$

1. Property 1 says that there would be 4 branches.
2. Property 2 says that each of the branches starts at the open loop poles at 0, −2, and −1 ± j1, and terminate at infinity.
3. Property 3 says that the negative real axis between the two real open loop poles would form a portion of the loci.
4. Property 4 says that there would be break-away points at $s = -1$. (The solution to $\frac{d}{ds}P(s) = 0$ is $s = -1, -1, -1$.)
5. Property 5 says that the centroid is also at $s = -1$ and there would be 4 asymptotes at angles 45°, 135°, 225°, and 315°, respectively.
6. Property 6 says that the angles of departure from the two complex conjugate open loop poles would be −90° and +90°, respectively.

The exact loci are shown in Figure 1.46. The loci obtained from MATLAB© are shown in Figure 1.47. Owing to certain limitations in numerical computations, the software fails to plot the exact loci. All four loci actually meet at the break-away point, then depart along the asymptotes.

Example 1.3.20:

Let us now consider

$$P(s) = \frac{(s+4)}{(s-1)(s+2)}$$

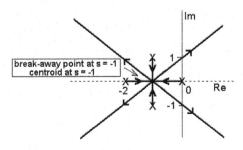

FIGURE 1.46
Exact Root Locus for Example 1.3.19.

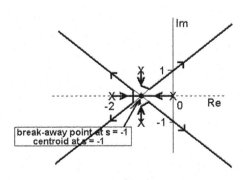

FIGURE 1.47
Computer genreated Root Locus for Example 1.3.19.

The loci are plotted in Figure 1.48. It is left as an exercise to the reader to obtain the break-away/break-in points and the centroid of the asymptotes.

A few typical root loci are given in the following Table 1.5.

At this point, an ardent reader may raise the following questions. For instance, the loci of Examples 1.3.16 and 1.3.17 are straight forward; however, why is a part of the loci of Examples 1.3.18 (and 1.3.20) curved rather than polygonal? We shall provide a rather heuristic answer to this. The loci basically satisfy a certain polynomial equation and they are a collection of continuous branches. In other words, the loci are, in general, smooth curves. For those values of the parameter K for which the coefficients of the polynomial are of the same sign, the roots, if they happen to

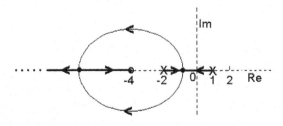

FIGURE 1.48
Root Locus for Example 1.3.20.

TABLE 1.5

Typical Root Loci

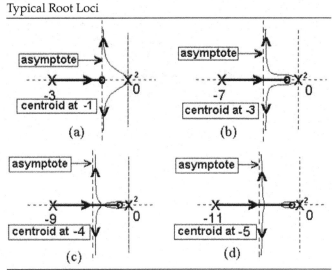

be complex, are conjugates to each other. Hence, we expect symmetry with respect to the real axis. Such portions of the loci, in general, may be expected to be solutions of a (dominant) quadratic polynomial and are elliptical in shape. Similar arguments may be made together with the properties such as the break-away and/or break-in points and the asymptotes. We have already established in property 5 that for large s the loci follow straight lines.

A another question might also arise here. What is the accuracy of root loci plotted using the afore mentioned six properties? Strictly speaking, the answer is "poor." However, with some experience, one might immediately obtain a sketch, albeit rough, of the loci that can be further improved. One can always verify whether every point on the rough sketch satisfies the angle condition. Usually, the error, if found, would be small enough and could be corrected without much difficulty.

1.3.4.2 The Stability Margin

In Examples 1.3.16 to 1.3.18, we see that the loci are completely inside the left half of the s-plane for all $K > 0$. This is to say that we can apply any amount of gain to these systems without affecting their stability. However, in Example 1.3.19, we notice that as K is increased beyond a certain value, two of the branches cross the imaginary axis and enter the right half of the s-plane. In other words, there is an upper limit to the gain K that can be applied to the system. By the same token of reasoning, we see that in Example 1.3.20, there is a lower limit to the gain K. Thus, on the root loci, we can determine the lower and/or the upper limits of the gain K by making use of the magnitude condition evaluated at the crossings of the loci on the imaginary axis. Let us look at the following example.

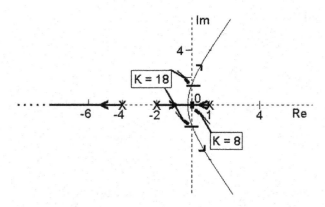

FIGURE 1.49
Root Locus for Example 1.3.21.

Example 1.3.21:

Let

$$P(s) = \frac{1}{(s-1)(s+2)(s+4)}$$

The loci are plotted in Figure 1.49. It can be determined easily that for stability, $8 < K < 18$.

With the six properties of root loci enumerated in this section, sometimes it may not be possible to determine the bounds on K accurately. With little extra effort, it is possible to search a point (or points), graphically using a ruler and a protractor, on the $j\omega$ axis where the angle condition is met. We then apply the magnitude criterion at these point(s). Nevertheless, we may also make use of the Routh-Hurwitz criterion. In fact, this improves the accuracy of the loci.

We close this part with a summary of the properties of Root Loci.

1. If the order of the plant is n, then the root loci consists of n continuous branches. These branches are symmetric with respect to the real axis.
2. The root loci start at the open loop poles for $K = 0$ and terminate at the (open loop) zeros for $K = \infty$.
3. On the real axis, the root loci lie to the left of an odd number of poles and zeros counted together.
4. Any two branches of the root loci meet at a point determined as the solution of the equation

$$\frac{d}{ds}P(s) = 0$$

Such a point is called either a break-away point or a break-in point.
5. For large values of K, the loci follow $n - m$ asymptotic straight lines given in the

point – slope form. The point, called the "centroid," and the slope are given by

$$a = \frac{\Sigma_n \,(poles) - \Sigma_m \,(zeros)}{n - m} \quad \text{and} \quad \phi = \frac{(2q+1)180}{n - m}$$

6. The angle of departure from a complex pole p_r may be found using

$$\theta_{p_r}^{p_r} = \Sigma_{i=1}^m \theta_{z_i}^{p_r} - \Sigma_{j=1\neq r}^n \theta_{p_j}^{p_r} \mp (2q+1)\,180° \quad \forall\, r = 1,\dots,2R$$

Similarly, the angle of arrival $\theta_{z_t}^{z_t}$ may be computed using

$$\theta_{z_t}^{z_t} = \Sigma_{i=1\neq t}^m \theta_{z_i}^{z_t} - \Sigma_{j=1}^n \theta_{p_j}^{z_t} \mp (2q+1)\,180° \quad \forall\, t = 1,\dots,2Q$$

7. Routh-Hurwitz criterion may be used to find the intersection of the loci, if any, on the imaginary axis. This would improve the accuracy of the loci.

1.4 Design of Control Systems

The design problem may be stated as follows:

Does there exist a compensator and a configuration such that given a plant $P(s)$, the overall system response can be forced to meet the desired specifications? If yes, what are they?

We shall make a modest beginning. First, we shall choose the "negative unity feedback" configuration. This is intuitive as was explained earlier in the chapter. Consequently, it remains to choose a suitable compensator $C(s)$ so that the performance specifications are met. This second step basically requires the specifications on the overall transfer function to be translated to those on the loop transfer function $L(s) = P(s)C(s)$.

1.4.1 Development of Classical PID Control

In this part, we will develop the design of classical controllers, both in time domain and in the frequency domain, and converge the design methodology to the well known proportional–integral–derivative (PID) control.

1.4.1.1 Controller Design Using Root Locus

Fortunately, the root locus readily accomplishes the above mentioned translation of specifications. Recall that if the parameter $K = 0$, then we have the open loop poles. However, the moment $K > 0$, we have the feedback systems and obtain the *closed loop poles*. Therefore, it is logical to exploit the properties of the root loci to design

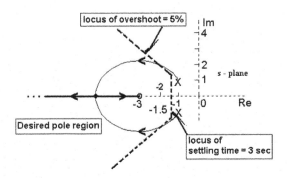

FIGURE 1.50
Feedback control with a controller $C(s)$.

the compensator. The schematic of the control system for a given plant $P(s)$ is shown in Figure 1.50.

In this section, we restrict ourselves to specifications spelt out in time-domain, i.e. specifications on overshoot, settling time, rise time, and steady-state error. As we have seen earlier, the specifications on these attributes may be expressed as a pair of desired (dominant) poles, or more generally, as a desired pole region. It must be remembered, however, that the requirement on stability overrides all other specifications. In other words, the desired poles are expected to lie only in the closed left half of the s-plane.

The magnitude and phase angle conditions may be written as

$$|C(s)| \times |P(s)| = 1$$

$$\angle C(s) + \angle P(s) = \pm (2q + 1) 180^o \qquad (1.81)$$

1.4.1.2 Magnitude Compensation

Recall that we have a standing assumption that the plant $P(s)$ is strictly proper, and the compensator $C(s)$ is proper for well-posedness. Henceforth, we shall examine proper compensators. We begin with the simplest of all compensators—an amplifier, K—called the **constant gain compensator** aka **Proportional Controller**. By definition, $\forall K \in (0, \infty)$, the root loci depict the roots of the characteristic polynomial $F(s, K)$. If we look at this from a design perspective, we may pose the following question:

Does there exist a K given the desired roots, s_d, of $F(s, K)$?

In other words, the problem is to verify whether the root loci pass through the desired poles. If the loci indeed pass through, then the gain K can be computed using the magnitude condition. Let us look at the following example for an illustration.

Example 1.4.1:

Consider the plant given in Example 1.3.17. Suppose that the closed loop poles are required at $-1 \pm j1$. It is clear that the loci pass through these poles. Accordingly, we may compute the required gain on the graph, using the procedure described earlier in property #2, as

$$K \cdot \frac{1}{\sqrt{2}} \cdot \frac{1}{\sqrt{2}} = 1 \quad \text{or} \quad K = 2$$

This simple design procedure may be summarized as follows.

- *Step 1:* From the specifications given, obtain the region/location of the desired poles on the complex s-plane. Make sure that this is in the closed left half of the s-plane.
- *Step 2:* Assume negative unity feedback configuration, and a constant gain compensator.
- *Step 3:* With K as the variable, obtain the loci of the poles of the closed loop system.
- *Step 4:* Apply the magnitude condition and evaluate K at the desired poles s_d.

Example 1.4.2:

Let the plant be

$$P(s) = \frac{s + 3}{s^2 + 2s + 2}$$

Let us assume the following specs to be met:

overshoot $\leq 5\%$

settling time ≤ 3 sec

rise time as small as possible

For these specifications, we may obtain the desired pole region D as shown in Figure 1.51

We have also superimposed the root loci of the plant given in Figure 1.44. Typically, we may choose the dominant poles to be $-2 \pm j2$ on the loci, which are also in D (in fact on the boundary). At this location, we may find the gain to be $K = 2$. Likewise, we may try few other points; a gain of $K = 3$ would fix the closed loop poles at $-2.5 \pm j2.1794$, and $K = 4$ would fix at $-3 \pm j2.236$, and so on.

FIGURE 1.51
Desired Pole region and Root Locus for Example 1.4.2.

Thus, if the desired poles lie on the root loci, a constant gain compensator K is sufficient. However, what if the root loci do not pass through the desired poles? In other words, can we *force* the loci to pass through certain locations in the s-plane of our interest? If yes, how? This we shall address in the next section.

1.4.1.3 Angle Compensation

If the root loci do not pass through the desired pole locations, then it implies that the phase angle condition is not met at the desired poles; the phase angle $\angle P(s)$ measured at these poles could be either *greater than* or *lesser than* 180°. Consequently, we have to choose a compensator, perhaps a higher order one, so that $\angle C(s) = \pm 180° - \angle P(s)$. The answer to this problem lies in making use of properties 3, 4, and 5 of the root loci effectively.

Let us illustrate using an example.

Example 1.4.3:

Let $P(s)$ be

$$P(s) = \frac{1}{s(s+2)}$$

1. If the desired poles are $s_d = -2 \pm j2$, then

$$\angle P(s)|_{s_d} = -135° \pm 90° = -225°$$

This is illustrated in Figure 1.52a.
2. If the desired poles are $s_d = -0.5 \pm j2$, then

$$\angle P(s)|_{s_d} = -104° + -53° = -157°$$

This is illustrated in Figure 1.52b.
Now it is easy to see that, in the former case, $\angle C(s)$ should be +45° and, in the latter case, $\angle C(s)$ should be −23°.

Let this compensator be $C(s) = KC_l(s)$, where K is real and positive, and

$$C_l(s) = \frac{s+Z}{s+P}$$

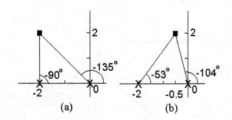

(a)   (b)

FIGURE 1.52
Angle at an arbitrary location.

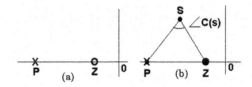

FIGURE 1.53
Lead controller.

a proper rational transfer function with $Z, P \in \Re$.

Case (i): $|Z| < |P|$
The zero at $s = -Z$ and the pole at $s = -P$ are shown in Figure 1.53a.

The angle contributed by $C(s)$ at any point s in the complex s-plane is always *positive*. If S, P, and Z form the vertices of a triangle, we see that from Figure 1.53b,

$$\angle C(s) = \angle PSZ$$

Since this transfer function always contributes to positive phase angle, it is called the **phase lead compensator**.

Example 1.4.4:

In the previous example, we found that the compensator $C_{lead}(s)$ should contribute to an angle of 45° at $-2 \pm j2$. In other words, we need to determine a Z and a P such that $\angle PSZ = 45°$. By using simple geometry, we can find the locations of the zero and the pole as $s = -0.2$, and $s = -2.1$ respectively. We determine the constant K to be equal to 4.21 by applying the magnitude condition. This is shown in Figure 1.54.

Notice that, after the compensation is provided, there is an additional branch. In other words, there are the complex conjugate poles at $-2 \pm j2$ as desired, and in addition, there is another pole at $s = -0.1$. The system is no longer a second order one and we have to verify the effect of the third pole on the transient response. The step response of

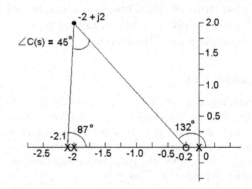

FIGURE 1.54
Lead controller desing for Example 1.4.4.

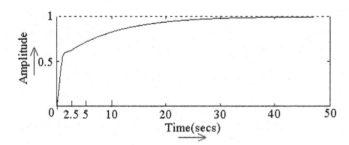

FIGURE 1.55
Step-response for Example 1.4.4.

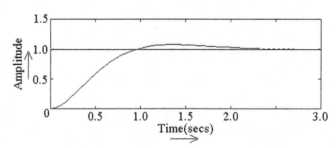

FIGURE 1.56
Step response of Example 1.4.4(b).

this compensated system is plotted in Figure 1.55. Since the plant is of type 1, the steady-state position error is zero. The steady-state velocity error of the uncompensated system is 2, but it is ≈ 5 for the compensated system. Thus, the improvement in the transient response, if any, is at the price of steady-state error.

The basic design requirement is to construct a triangle such that the angle at its apex is equal to the angle deficiency. Accordingly, we have several choices of P and Z of the compensator. From the previous example, we see that the compensator adds another branch to the loci, and this additional pole might play a significant role in the transient response. Indeed, the step response in Figure 1.55 is poor. To preserve the *dominance* of the desired poles so that the specs are met, naturally, we prefer the new pole to be added deeper in the left half plane away from the dominant poles. We modify the design in the previous example as follows.

Example 1.4.4(b):

We shall now fix the zero Z of the compensator at $s = -2.5$. Accordingly, we may find that the pole P of the compensator should be at $s \approx -5.33$. The gain of the compensator is then $A = 10.67$. The closed-loop transfer function is

$$T(s) = \frac{10.67(s+2.5)}{s^3 + 7.33s^2 + 21.33s + 26.675}$$

$$= \frac{10.67(s+2.5)}{(s+3.35)(s+2+j2)(s+2-j2)}$$

The step response is shown in Figure 1.56. Obviously, this design is much better than the earlier one.

Remarks:

1. The design does not end in a unique controller.
2. We may have to try *a lot* of them before we are satisfied.

3. A slight variation in the proposed locations of the poles and zeros of controller may result in significant changes in the loop performance. Generally, this is referred to as **tuning**.
4. An interesting question would be: Can we talk about tuning in more accessible terms such as gain(s) instead of poles and zeros?

Case (ii): $|Z| > |P|$

The angle contributed by $C(s)$ at any point S in the complex s-plane is always *negative*. If S, P, and Z form the vertices of a triangle, we see that in this case

$$\angle C(s) = -\angle PSZ$$

This is depicted in Figure 1.57. Since this transfer function always contributes to negative phase angle, it is called the **phase lag compensator**.

Example 1.4.5:

In Example 1.4.3(2.), we found that the compensator $C_{lag}(s)$ should contribute to an angle of $-23°$ at $-0.5 \pm j2$. We need to determine a Z and a P, such that $\angle PSZ = 23°$. We can easily find the locations of the zero and the pole as $s \approx -1$ and $s = -0.2$, respectively. Again, the coefficient K can be determined to be $K = 5.1$ by applying the magnitude condition. This is shown in Figure 1.58.

Again, in this case, there is a pole at $s = -1.2$ in addition to the desired complex conjugate pair at $s = -0.5 \pm j2$. It is important to verify the effect of this pole on the transient response. The step

FIGURE 1.57
Lag controller.

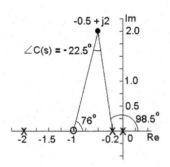

FIGURE 1.58
Lag controller in Example 1.4.5.

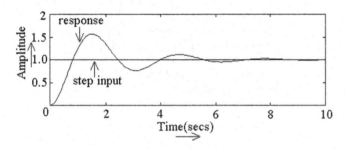

FIGURE 1.59
Step response in Example 1.4.5.

response is plotted in Figure 1.59. Clearly, the step response is poor. The steady-state velocity error in this case is ≈ 0.08. This may be further reduced by shifting the pole of the lag compensator closer to the origin. However, we may also notice that, since the root loci is closer to the imaginary axis, the settling time is larger than that of the plant itself. In other words, we improved the steady-state performance using a lag compensator at the cost of transient performance. We shall come back to this issue a little while later.

The design procedure may be summarized as follows.

- *Step 1:* From the specifications given, obtain the region/location of the desired poles on the complex s-plane.
- *Step 2:* Assume negative unity feedback configuration, and an angle compensator so that the control system is as shown in Figure 6.13.
- *Step 3:* At the desired pole location s_d, find the angle contributed by the plant $P(s)$ and obtain the angle to be contributed by $C(s)$. Call this angle ψ.
- *Step 4:* If ψ is positive then, a lead compensator is required; otherwise, a lag compensator.
- *Step 5:* Construct a triangle PSZ in the s-plane such that the angle at the apex

$$\angle PSZ = |\psi|.$$

The construction is arbitrary and several possibilities exist. However, in what follows, we suggest a more pragmatic method.

- Draw a line SZ (if the choice is a lead compensator; otherwise, draw SP) so that the intersection on the negative real axis closer to origin is at Z (or P).
- Draw another line SP (or SZ if the choice is a lag compensator) so that the triangle is complete with the base PZ on the negative real axis.
- *Step 6:* The intercepts on the negative real axis, viz., Z and P would be the zero and pole respectively of the desired compensator
- $C(s) = KC_l(s)$.
- *Step 7:* The parameter K may now be found using the magnitude condition.

1.4.1.4 Validity of Design

Let us now provide a rigorous justification to several points were deliberately left without discussion.

First, the root locus is a technique in which one of the parameters of a polynomial is varied and the resulting loci of the roots of the polynomial are obtained. If this variable parameter is, for example K, we discussed several nice properties of the loci. We even discussed simple magnitude compensation based on this. If there are two (or more) parameters, as in the case of angle compensation, we can, in principle, obtain a *family of loci* called "root contours" that possess the same properties as before. However, it is easier and more advantageous to apply the angle condition prudently and obtain the loci.

Secondly, certain systems have a dead-time T_d in which case the characteristic equation is of the form

$$1 + G_l(s) \times e^{-T_d s} = 0$$

In such a situation, we make use of Padé's approximation and rewrite $e^{-T_d s}$ as a rational function:

$$e^{-T_d s} \approx \frac{1 - \dfrac{T_d}{2}s}{1 + \dfrac{T_d}{2}s} \tag{1.82}$$

With this approximation, the characteristic equation takes the form

$$1 - G_d(s) = 0 \tag{1.83}$$

The six properties we discussed earlier are applicable with slight modifications to take care of the negative sign. For instance, the magnitude condition remains the same but the phase angle condition says that

$$\angle P(s) = \pm 360°$$

Example 1.4.6:

For the plant in the Example 1.4.3, the CRL is given below in Figure 1.60. The reader may quickly identify that the root loci and the complementary root loci span the entire real axis. CRL is useful in the analysis of systems that may have *positive* feedback.

Thirdly, if to a plant $P(s)$ of order n, a proper transfer function $C(s)$ is cascaded, we notice that the order of the loop transfer function $L(s)$ becomes $n + 1$. Hence, there would be an additional locus. Naturally, we have to exercise care to see that this additional closed loop pole does not affect our specifications.

In the case of lead compensation, since the zero of the compensator is closer to the origin than the pole, the branches on the real axis are altered. A branch starting from one of the open loop poles (either belonging to $P(s)$ or $C(s)$) terminates on the compensator zero at $s = -Z$. A root of the characteristic polynomial exists on this branch in addition to the dominant pair of desired roots. Accordingly, it becomes critical to choose the location of Z. A way out of this is to simply cancel one of the stable plant poles that is *relatively closer* to the origin. However, we should not cancel a pole that is at the origin as it would reduce the type of the plant and affect the steady-state error of the system. If the plant has an open loop pole at origin, Z should be chosen such that the locus starting from the pole at origin and terminating at the zero $s = -Z$ would have negligible effect on the transient response. We may also notice that by virtue of lead compensation, the centroid shifts further to the left on the negative real axis, and consequently, all the loci shift further to the left of the s-plane. As a result, the rise time and the settling time may be expected to decrease, and hence, a considerable amount of improvement in the transient response could be expected. Thus, in general, a phase lead compensator improves the transient response of the system.

We may apply a similar line of reasoning and say that a lag compensator, in general, improves the steady-state response by shifting the loci

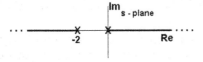

FIGURE 1.60
Complementary Root Loci.

further to the right of the s-plane. We may choose to place the pole of the compensator $s = -P$ close to (or even at) the origin thereby, improving the steady-state error substantially. However, one can expect that this improvement would be at the cost of the settling and/or rise times, and even the stability of the overall system. The zero of the lag compensator at $s = -Z$ may be chosen, to a possible extent, so that there is a pair of complex conjugate poles at least within the desired dominant pole region. This ensures that the transient response is not perturbed drastically while improving the steady-state.

It is worth noting that a lead compensator may not always be the solution to improve the transient performance; it might even improve the steady-state. Similar remarks apply to a lag compensator. It is the open loop poles and zeros, the desired pole location, and the requirement on the steady-state error for the problem at hand that determine which angle (i.e. either lead or lag) compensation to be provided.

On the s-plane we can very clearly see the parameters responsible for transient response. However, the steady-state response can be measured only indirectly, and that too approximately. In general, the steady-state error may be reduced by providing enough gain while taking care to see that the system does not become unstable. If the requirement is on the steady-state performance alone, assuming that the transient response is quite satisfactory, the design of an appropriate lag compensator is more meaningful in frequency domain to be discussed in the next section.

Next, if the design specifications demand an improvement in transient performance as well as the steady-state, a logical way of providing compensation is to first design, for example, a lead section that improves the transient, then a lag section that improves the steady-state of the plant *together with the lead section*. The lead section may be designed using the root loci and the angle compensation discussed until now. The lag section may be designed using the frequency domain techniques taken up next. The design of lead-lag compensators is deferred to the latter part of this chapter.

Next, in the literature as well as in the industry, there exist a few compensators that are extremely popular. They are the Proportional + Derivative (or PD for short), the Proportional + Integral (or PI for short), and the Proportional + Integral + Derivative (or PID) compensators. In general, it is difficult to construct an ideal PD compensator since it results in an improper transfer function $C(s) = K_p + K_d s$. A notable exception, perhaps, is a tachometer. Therefore, this compensation is actually implemented as a special case of the lead compensator with the pole at $s = -P = N \times -Z$, where $3 \le N \le 10$. Similarly, a lag compensator with its pole at the origin would become a PI compensator. It may now be expected that a PID compensator is a special case of the lead-lag compensator. However, the design philosophy here is slightly different and a standard empirical table provided by Ziegler and Nichols in the early 1940s is still in vogue, after nearly eight decades. Discussions on PID compensators may be found in any text book on Process Control. Towards the end of this chapter, we provide a method of designing PID controllers and compare the design with lead-lag controllers.

Last, the design is a trial-and-error procedure that involves a number of trade-offs among the specifications. Therefore, we have to prepare ourselves for several trials before we succeed in designing an acceptable system. It must be remembered throughout that the relationships between the specifications on the peak over-shoot, settling time, and the rise time and the dominant pole locations are not as precise as for quadratic transfer functions with a constant numerator. However, it has become a standard practice to provide the specifications on the transient response in terms of the overshoot and so forth. It is, therefore, important to simulate the resulting system to check whether the specifications are really met. If it does not, the system must be redesigned.

1.4.1.5 Controller Design Using Bode Plots

The response of dynamical systems exhibits transients whenever there is a sudden change in the input excitation. Accordingly, we are interested in the step response characteristics such as overshoot, settling time and the rise time. Moreover, the reference input r may also, in general, be modelled as a step signal (or a linear combination of several step signals). Thus, the specifications in time domain are, generally, the transient response specifications. Having seen controller design to meet the transient specifications, let us now look at the system response at steady-state. We call this response the ac *steady-state* **response**. The name suggests that the input is a sinusoid, such as $\cos(\omega t)$, and we study the response after all the transients have died out. The student might recall, from an earlier course on differential equations, that there would be a transient term in addition to the sinusoidal term in the complete solution of a differential equation to a sinusoidal excitation.

In an earlier part of this chapter, we discussed how to construct Bode plots to convey the frequency response information. Later, we learned how Nyquist criterion can be employed to assess stability of feedback systems. We have also learned about the stability margins—gain margin and phase margin—and how they are better depicted using Bode plots. In what follows, we build

on the above ideas and learn to design controllers that would help us realize our frequency domain specifications. Typically, these are

1. Steady-state error e_{ss}, in the step or ramp response, generally expressed as a percentage
2. Gain margin, typically ≥ 60 dB
3. Phase margin, typically $\geq 60°$
4. Bandwidth (defined below)

1.4.1.6 Definition: Bandwidth

The **bandwidth** is defined as the frequency range in which

$$|L(j\omega)| \geq \frac{1}{\sqrt{2}} \cdot |L(0)| \qquad (1.84)$$

The frequency $\omega = \omega_o$ at which

$$|L(j\omega_o)| = \frac{1}{\sqrt{2}} \cdot |L(0)|$$

is called the **cut-off frequency**.

The bandwidth is related to disturbance rejection.

If the bandwidth is larger, it means that higher frequencies, which might include disturbance/noise, are easily allowed to pass through the system without much attenuation. Therefore, the system shall have poorer disturbance rejection. On the other hand, a system with narrower bandwidth usually exhibits a sluggish response. Intuitively, bandwidth is inversely proportional to time, and hence, larger bandwidth provides faster response. A quick example is the *bandwidth* found in the internet connectivity.

1.4.1.7 The Design Perspective

Let us analyse the situation step by step. The position error is given by:

$$\text{Position Error} = \frac{1}{1 + \lim_{s \to 0} L(s)} \qquad (1.85)$$

If the loop transfer function is type 1, i.e. $L(s)$ has a single pole at origin, the position error is zero, but the velocity error is finite.

Thus, the steady-state error can be reduced by increasing the loop gain, i.e. cascading the loop transfer function with a gain K_e. We shall illustrate this with an example.

Example 1.4.7:

Let

$$L(s) = \frac{s+2}{(s+1)(s+3)}$$

If we assume an amplifier K_e cascaded to the plant/compensator, the modified loop transfer function is then $K_e \cdot L(s)$, and the position error in terms of K_e becomes

$$\text{position error} = \frac{1}{1 + K_e \cdot \lim_{s \to 0} L(s)}$$

$$= \frac{3}{3 + 2K_e}$$

If we desire that

$$\text{position error} \leq 10\%$$

then, the gain K_e has to be chosen so that

$$K_e \geq 13.5$$

Since the loop transfer function is type 0, velocity error is infinite.

Example 1.4.8:

Let

$$L(s) = \frac{s+2}{s(s+1)(s+3)}$$

The position error is 0. The velocity error in terms of K_e is

$$\text{velocity error} = \frac{1}{K_e \cdot \lim_{s \to 0} s \cdot L(s)}$$

$$= \frac{3}{2K_e}$$

If we desire that

$$\text{velocity error} \leq 8\%$$

then the gain K_e has to be chosen so that

$$K_e \geq 18.75$$

Thus, the specification on the steady-state error is met by introducing an amplifier K_e in the loop. However, increasing the gain of the loop transfer function is not without any side effects. Let us look at these now. In the Bode plot, introducing a positive gain K_e shifts the magnitude curve *upwards* by an amount equal to $20\log_{10}K_e$. This, in turn, shifts the gain cross-over frequency further

to the right along the ω axis. Since the loop transfer function is, in general, strictly proper, the phase curve is always monotonically decreasing, and a shift of the cross-over frequency to the right decreases the phase margin. The gain margin, in general, will not be affected since the phase cross-over frequency is high enough and the attenuation magnitude will be sufficiently large. It may be quickly verified that if K_e were less than 1, then the magnitude curve is pushed *downwards*, and the gain cross-over frequency shifts to the left along the ω axis.

In addition to the reduction in phase margin, increasing the loop gain has another, somewhat undesirable, side effect—an increase in bandwidth. As was mentioned earlier, a larger bandwidth may result in a poorer disturbance rejection. We shall discuss more about the bandwidth towards the end of this part.

At this point, the reader should have started appreciating the worth of Bode plots. The Nyquist plots and the Bode plots are equivalent, and theoretically either one can be used in the analysis as well as the design. However, the Bode plots have a distinct advantage that, in addition to being the easiest to draw using the asymptotes, designing a compensator $C(s)$ is simply *adding* the Bode plot of $C(s)$ to that of $P(s)$ until the sum meets the desired specifications. Further, if a system is stable and the model is not explicitly known, then $|L(j\omega)|$ and $\angle L(j\omega)$ may be obtained by measurement. These measurements may then be formatted into a Bode plot, and the transfer function of the system can be obtained, at least approximately. Let us now look at a compensator that improves the stability margins while decreasing the steady-state error. Clearly, a constant gain compensator K_e alone does not serve the purpose. So, let us take up a first order compensator of the form

$$C_{lag}(s) = \frac{P}{Z} \frac{s+Z}{s+P} \quad Z > P(>0) \tag{1.86}$$

Notice that this is the *phase-lag* compensator we have introduced previously. Let us define

$$\alpha = \frac{P}{Z} \quad \text{and} \quad T = \frac{1}{P} \tag{1.87}$$

so that, conforming to the notation for Bode plots, we may rewrite the compensator transfer function as

$$C_{lag}(s) = \frac{1 + \alpha T s}{1 + T s} \tag{1.88}$$

The Bode plot of $C_{lag}(s)$ is shown in Figure 1.61.

The magnitude curve indicates that $C_{lag}(s)$ introduces unity gain until the pole corner frequency and attenuation thereafter. The attenuation is

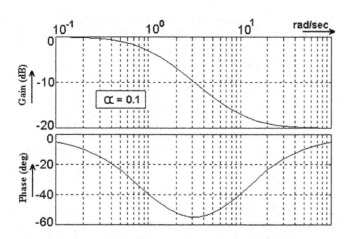

FIGURE 1.61
Bode plot of a typical lag controller.

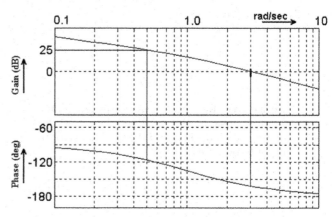

FIGURE 1.62
Illustrating Lag controller design in Example 1.4.9.

$20 \log \alpha$, a constant at all frequencies, beyond the zero corner frequency. The phase curve justifies the name *phase lag* since there is "dip" in the curve between $\omega = \frac{1}{T}$ and $\omega = \frac{1}{\alpha T}$. We particularly notice that the phase is about $-6°$ for $\omega \leq \frac{0.1}{T}$ and for $\omega \geq \frac{10}{\alpha T}$.

As was noted earlier, the phase curve of a strictly proper plant decreases monotonically. Therefore, the phase margin can be improved if we can, somehow, shift the gain cross-over frequency further to the *left* on the ω axis. This would require the magnitude curve to be pushed *downwards* by an appropriate amount. This can be achieved only by introducing *attenuation* in the loop. We shall cleverly employ $C_{lag}(s)$ for this job. For the sake of clarity, we shall use the following example.

Example 1.4.9:

Let

$$P(s) = \frac{1}{s(s+1)}$$

Let us assume that we need

velocity error $\leq 10\%$ and phase margin $\geq 60°$

Then, a gain of $K_e \geq 10$ would meet the specification on velocity error. Let us assume $K_e = 10$. Together with a gain $20 \log K_e = 20$ dB, the Bode plot of the plant is shown in Figure 1.62.

The gain cross-over frequency is to the right of the corner frequency $\omega = 1$ rad/sec, and hence, the phase margin that is available is less than 45). To meet our specification, the gain cross-over frequency has to be at 0.5 rad/sec, and the attenuation needed is about 25 dB. In other words, the magnitude curve has to be pushed down by 25 dB so that the new gain cross-over frequency ω_p^* is 0.5 rad/sec. It appears like an impossibility. While a gain of 20 dB is required to reduce

the steady-state error, an attenuation of 25 dB is required to improve the phase margin. In other words, we have effectively introduced an attenuation of 5dB, which seems to go against the grain. We shall come out of this maze cleverly by noting that $C_{lag}(s)$ provides an attenuation of α at higher frequencies and, hence, equating this to the desired attenuation. In other words, let us put

$$20 \log \alpha = -25 \text{ dB}$$

The negative sign on the right-hand side is required because we are concerned with *pushing the curve down* at the moment. Solving the above equation for α would give

$$\alpha = 0.06$$

Thus, we found the relative locations of the zero and the pole in $C_{lag}(s)$, i.e. if the zero is located at ω_{zero} rad/sec, the pole should be located at $\omega_{pole} = 0.06\omega_{zero}$ rad/sec. Let us now determine ω_{zero}.

It is important to keep in mind that the phase curve of $C_{lag}(s)$ should not disturb the phase curve of the plant, at least, in the vicinity of the new cross-over frequency ω_p^*. Earlier, in Figure 1.61, we have seen that the phase curve of $C_{lag}(s)$ has a dip (below 0)) between ω_{pole} and ω_{zero}. If we can, somehow, shift this dip far to the left of ω_p^*, where the phase margin is quite large, then our problem is solved. We have also noticed, in Figure 1.61, that the phase introduced by $C_{lag}(s)$ is just $-6°$ for $\omega \geq \omega_{zero}$. We shall couple this idea with the fact that the asymptotic phase curve of any transfer function is a straight line between $0.1\omega_c$ and $10\omega_c$, where ω_c is the corner frequency. In our example, the asymptotic phase curve is a horizontal line at $-90°$ until $\omega = 0.1$ rad/sec and is another horizontal line at $-180°$ beyond $\omega = 10$ rad/sec; in between, it is a straight line passing through $-135°$ at the corner frequency 1 rad/sec. In view of this, let us choose

$$\omega_{zero} = 0.1\,\omega_p^*$$

i.e. a decade below the new gain cross-over frequency, so that

$$C_{lag}(s) = 0.06 \cdot \frac{s+0.05}{s+0.003}$$

The resulting Bode plot of the product $C_{lag}(s) \cdot K_e \cdot P(s)$ is shown in Figure 1.63.

The design appears to be tricky since we have provided both a gain and an attenuation. However, let us carry out the following analysis so that the idea becomes clearer. With the lag compensator, the loop transfer function is

$$L(s) = K_e \cdot \alpha \cdot \frac{s+Z}{s+P} \cdot P(s) \qquad (1.89)$$

The steady-state error is

$$\text{velocity error} = \frac{1}{K_e \cdot \alpha \cdot \dfrac{Z}{P}\lim_{s \to 0}P(s)}$$

$$= \frac{1}{K_e \cdot \lim_{s \to 0}P(s)} \qquad (1.90)$$

Clearly, the lag compensator has no effect on the steady-state error specification. The compensator decreases the phase margin at the new cross-over frequency by about 6°. Therefore, it is desirable to search for ω_p^* where the phase margin is 6° in excess. This is allowed since the specification on the phase margin involves "greater than or equal to" inequality.

There is one point worth noting—there is a considerable reduction in the bandwidth. Thus, a lag compensator not only improves the steady-state performance, but also improves the disturbance

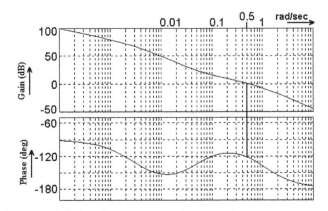

FIGURE 1.63
Bode plot of the lag compensated plant.

rejection. We shall come back to this issue again in the next part of the chapter.

The design procedure may be summarized as follows.

- *Step 1:* From the specification on steady-state error, obtain the gain K_e.
- *Step 2:* Assume negative unity feedback configuration.
- *Step 3:* Obtain the Bode plot of the plant together with the gain K_e. Measure the margins from the plot.
- *Step 4:* Determine the new gain cross-over frequency ω_p^* at which the phase curve has the desired phase margin plus 6°.
- *Step 5:* Measure the attenuation $-a$ dB required to push the magnitude curve down so that the gain cross-over frequency is ω_p^*. This attenuation will be supplied by the phase-lag compensator with α computed as

$$20\log\alpha = -a \qquad (1.91)$$

- *Step 6:* Obtain the location of the compensator zero using the expression

$$\omega_{zero} = 0.1\omega_p^* \qquad (1.92)$$

- *Step 7:* The transfer function of the compensator is then

$$C_{lag}(s) = \alpha \cdot \frac{s+\omega_{zero}}{s+\alpha\omega_{zero}} \qquad (1.93)$$

Example 1.4.10:

Let

$$L(s) = \frac{1}{(s+1)(s+2)}$$

and the specs be

$$\text{position error} \le 10\%$$

$$\text{phase margin} \ge 80°$$

From the spec on position error, we might quickly find the required gain $K_e \ge 18$. Let us fix it at 20. With this gain in the loop, the Bode plot is shown in Figure 1.64.

We notice that the gain cross-over frequency is $\omega_p = 4$ rad/sec; hence, the phase margin is approximately 45°. We have the desired phase margin + the 6° correction at about $\omega_p^* = 1.3$ rad/sec. Accordingly,

$$20\log\alpha = -15\,\text{dB} \quad \text{or} \quad \alpha \approx 0.18$$

and the location of the compensator zero is

$$\omega_{zero} = 0.1 \times 1.3 = 0.13$$

$$\omega_{pole} = 0.18 \times 0.13 = 0.023$$

Thus,

$$C_{lag}(s) = 0.18 \cdot \frac{s + 0.13}{s + 0.023}$$

The Bode plot of the loop-transfer function is given in Figure 1.65.

1.4.1.8 The Lead-Lag Compensator

We have discussed the design of both lead and lag compensators using the root locus. We have prominently noted that a lead compensator improves the transient response, while a lag compensator is likely to spoil the transient performance making the system sluggish. However, we have also seen that the lag compensator improves the steady-state performance and stability margins. Since, at the beginning, we assumed that the frequency response (due to the sinusoids of varying frequencies) is the system's steady-state response after all the transients have decayed, there is no reason discussing the frequency domain design of a lead compensator that would improve the transient response.

We might make the following remarks. The lead compensator increases the bandwidth since it provides a gain in the loop. In other words, we see an intuitive relationship, albeit difficult to establish precisely, between the speed of response and the bandwidth – the faster the system, the larger the bandwidth. Accordingly, we might say that, the lag compensator decreases the speed of response since it reduces the bandwidth. Clearly, the two compensators are complementary. If the design specifications demand an improvement in transient performance as well as the steady-state, a logical way of providing compensation is to first design, for exmaple, a lead controller that improves the transient, then a lag

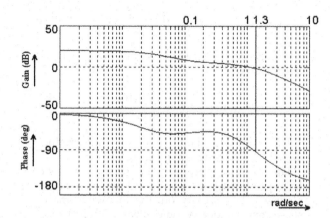

FIGURE 1.65
Bode plot of the lag compensated plant in Example 1.4.10.

controller that improves the steady-state of the plant *together with the lead controller*. The lead controller may be designed using the root loci and the angle compensation, and the lag section may be designed using the frequency domain techniques. Accordingly, the compensation is called **lead-lag controller**. We emphasize that there is always a trade-off between the transient specifications and steady-state specifications, i.e. the given specs have to be relaxed a little so that some uniformity is achieved. Moreover, several iterations may be needed before we freeze a design.

One standard way of designing a lead-lag compensator is to place the product $C_{lead}(s) \times C_{lag}(s)$ in cascade with the plant. This is shown in Figure 1.66.

Notice that there is a single feedback loop. The performance of this compensator becomes clear if we look at its Bode plot. Depending on the choice of the poles and zeros of each of the individual controllers, two possible Bode plots are shown in Figure 1.67a and b.

The Bode plot in Figure 1.67a indicates that the performance of the system at lower frequencies is improved, but the margins are lowered at the higher frequencies. On the contrary, the Bode plot in Figure 1.67b suggests the frequency response is generally improved because at lower frequencies the advantages of a lag compensator may be obtained, and at higher frequencies the advantages of a lead compensator may be obtained. Therefore, we shall select the lead and lag sections as follows.

$$C_{lag}(s) = \frac{s + z_1}{s + p_1} \text{ and } C_{lead}(s) = \frac{s + z_2}{s + p_2} \quad (1.94)$$

with

$$p_1 < z_1 < z_2 < p_2 \quad (1.95)$$

We need to include a gain factor K too.

We study this design procedure in the following example.

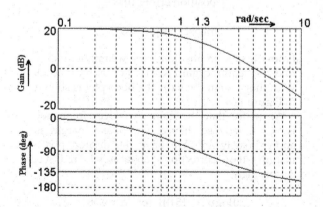

FIGURE 1.64
Illustrating the design in Example 1.4.10.

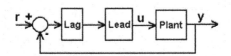

FIGURE 1.66
Conventional lead-lag controller.

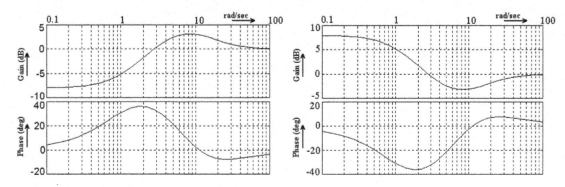

FIGURE 1.67
(a) Bode plot for one choice of poles and zeros. (b) Bode plot for a better choice of poles and zeros.

Example 1.4.11:

Let

$$L(s) = \frac{1}{(s+1)(s+2)}$$

and the specs be

Overshoot $\leq 0.5\%$

Settling time ≤ 2.25 sec

Rise time as small as possible

Position error $\leq 10\%$

Phase margin $\geq 80°$

The dominant pole location corresponding to the transient specifications may be chosen as, for instance, $-2 \pm j1$. Using root locus, the lead compensator that takes care of these transient specifications may be obtained as

$$C_{lead}(s) = 2 \cdot \frac{s+2}{s+3}$$

For the steady-state specifications, we simply use the lag compensator *together with* K_e designed in the earlier example, i.e.

$$C_{lag}(s) = 20 \times 0.18 \times \frac{s+0.13}{s+0.023}$$

The Bode plot of the loop transfer function with gain $K = 7.2$ is given in Figure 1.68. Clearly, the steady-state specifications are met. The Bode plot is almost identical to that of Figure 1.65. This result may be expected since the poles of the plant and the lead compensator are about one decade away from the zero and the pole of the lag compensator.

The root locus of the compensated system (with K as the running parameter between 0 and 8) is given in Figure 1.69. We also notice that there is a pole close to the origin between the pole and the

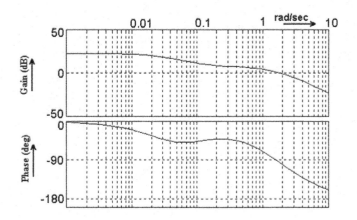

FIGURE 1.68
Bode plot for Example 1.4.11.

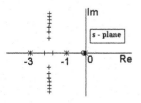

FIGURE 1.69
Root loci for Example 1.4.11.

58 *Control Systems*

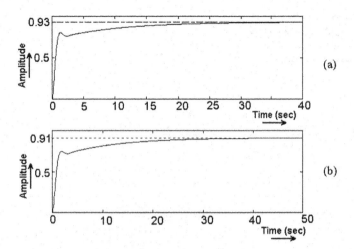

FIGURE 1.70
Step responses for Example 1.4.11.

zero of the lag compensator. The step response of the lead-lag compensated system is shown in Figure 1.70a. For the sake of comparison, we also provide the step response of the lag compensated system in Figure 1.70b. Observe that the step response is not satisfactory owing to larger settling time than desired.

Thus, there is a conflict between the transient specs and steady-state specs. This happens always because of the third pole of the closed loop system near the origin. In fact, there will be two extra poles, since the lead-lag compensator increases the order of the overall system by two. The only way out of this situation is to make a trade-off among the specifications. In other words, we relax the requirements either on the transient response or on the steady-state response or both. For instance, the above design is satisfactory if we relax our requirement on settling time, or phase margin, or both. Thus, several iterations become a rule rather than an exception before arriving at a satisfactory trade-off.

A more logical design of lead-lag compensation would be to make use of *two* feedback loops: one inner loop for, say, the lead compensation and the outer loop for the lag compensation. This is shown in Figure 1.71.

First a lead compensator is designed to improve the transient response. This lead compensator $C_d(s)$ comes along with a negative unity feedback. Then, the steady-state response of the *plant together with the lead compensator*, which may be thought of

as a different plant by itself, is improved by the lag compensator $C_g(s)$, which also comes along with a negative unity feedback. We shall repeat Example 1.4.11 below with this lead-lag compensator.

Example 1.4.11(b):

With the lead compensator

$$C_d(s) = 2\frac{s+2}{s+3}$$

the transfer function of the inner loop is

$$P_1(s) = \frac{2}{s^2 + 4s + 5}$$

We find that a gain $K_e \geq 22.5$ is required to meet the specification on position error. Let us fix $K_e = 25$. The Bode plot of $K_e \cdot P_1(s)$ is given in Figure 1.72.

We find that the phase margin is about 65° at $\omega_p = 3.3$ rad/sec. We may quickly design the lag compensator with K_e as

$$C_g(s) = 25 \cdot 0.16 \cdot \frac{s+0.2}{s+0.032}$$

The step response of the compensated system is given in Figure 1.73a. Clearly, we see an improvement over the previous design. Moreover, we find some hints to improve the design. For instance, if we double the gain in the loop the step response would drastically improve as shown in Figure 1.73b.

However, the Bode plot, given in Figure 1.74, suggests that there is an increase in the bandwidth and a reduction in the phase margin by about 15°. If this relaxation is acceptable, then the design is complete.

Otherwise, we may choose a gain suitably. For instance, if the gain in the loop is chosen to be $K = 14.4$, the step response is shown in Figure 1.75.

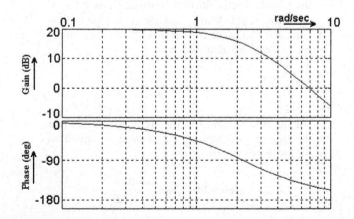

FIGURE 1.72
Bode plot for Example 1.4.11(b).

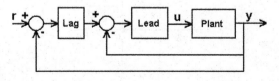

FIGURE 1.71
A novel architecture for lead-lag controller.

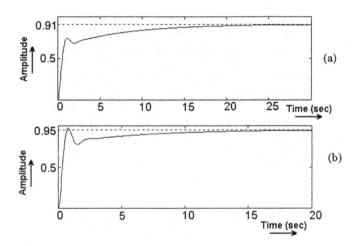

FIGURE 1.73
Step responses for Example 1.4.11(b).

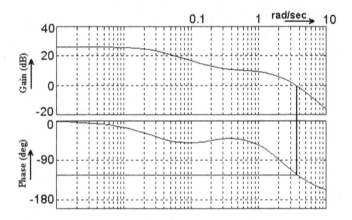

FIGURE 1.74
Bode plot of the plant with the lead-lag controller in Figure 1.71.

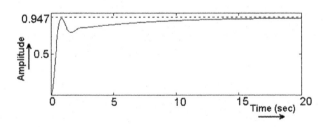

FIGURE 1.75
Step response of the lead-lag controller for Example 1.4.11(b).

We leave it to the reader to verify that the phase margin is 65° approximately. The effect of increasing the gain drastically may be seen on the root loci—the closed loop pole between the lag compensator's pole and zero approaches the zero and, hence, gets canceled. However, the upper limit on the gain would be dictated by the acceptable phase margin.

Although, the idea appears to be simple, there are certain drawbacks. First, the order of the overall system is increased by 2. Hence, retaining the dominance of the closed loop poles for the desired transient performance becomes difficult. Another reason for this is the presence of lag compensator's zero and pole closer, than the dominant poles, to the origin. Accordingly, the resulting performance might be a reasonable trade-off among the desired specifications rather than absolutely achieving them.

In the industry, there exists a controller—the Proportional + Integral + Derivative (or PID) controller—that is extremely popular. We shall briefly look at these controllers now. A PID compensator may be thought of as a special case of the lead-lag compensator. However, the design philosophy here is altogether different and a standard empirical table provided by Ziegler and Nichols in the early 1940s is still in vogue. More detailed discussions on PID compensators may be found in any text book on Process Control.

The transfer function of the PID controller is given by

$$C_{PID}(s) = K_p + \frac{K_i}{s} + K_d s$$

$$= K_p\left(1 + \frac{1}{T_i s} + T_d s\right) \qquad (1.96)$$

The derivative part of the PID controller is the trickiest of all. Since it is not possible to realize an improper transfer function $K_d s$, it is usually taken as an approximation of a proper transfer function with a zero close to origin and a pole far away from the zero. An important issue worth noting in this case is as follows. Suppose the error $r-y$ in the loop has a step change, then the derivative action of PID differentiates this step to generate an impulse. This impulse, which resembles lightning and thunderbolt in practice, might be damaging the plant. Therefore, it is avoided in the feedforward path and is usually preferred in the feedback path as shown in Figure 1.76.

This is similar to our lead-lag compensator in Figure 1.71, with two feedback loops, except that the lead compensation, a general case of PD, is in the feedforward path. However, since we are interested only in the characteristic polynomial,

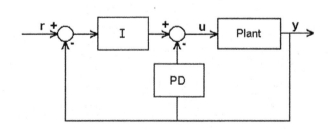

FIGURE 1.76
One of the standard PID controllers.

except for a reduction in the gain, the performance of the overall system does not get altered much.

Typically, a PID controller design requires two parameters to be computed. They are

1. *Ultimate gain K_u,* i.e. the gain at which the system breaks into oscillations, and
2. *Ultimate period T_u,* i.e. the period of induced oscillations at the ultimate gain.

These may be obtained either from the root locus or the Bode plot as follows. On the Bode plot, the ultimate gain is the gain margin at the phase cross-over frequency ω_g, i.e.

$$K_u = 10^{\omega_g/20} \tag{1.97}$$

and accordingly,

$$T_u = \frac{2\pi}{\omega_g} \tag{1.98}$$

On the root locus, the ultimate gain is the gain at which the loci intersect with the imaginary axis at $s = j\omega_u$ and, hence, the ultimate period is $2\pi/\omega_u$.

The following is the Ziegler and Nichols's empirical table of *rules* that is religiously adhered to in the industry, especially in the process industry (Table 1.6).

Since the introduction of Ziegler-Nichols rules in 1942, PID tuning has led to an explosion in research till date. As has been mentioned earlier, tuning in terms of locations of poles and zeros could be laborious; instead, it is convenient if we have explicit parameters like K_p, T_i, and T_d, which affect the pole-zero locations. Consequently, we have a reasonably good estimate about 150 tuning rules for PID controllers. This leads us to a rather natural question: Which one is the best? The answer: It actually depends on the prrocesses, the order, the parameters, the nonlinearity, the uncertainties, and so on; no single tuning rule works well for all systems.

For linear time-invariant lumped (LTIL) systems, the focus of this part of the book, there is no reason to resort to PID controllers. In fact, more *systematic* and mathematically rigorous design techniques are available for a more general class of controllers. Furthermore, it is worth noting that to compute the ultimate gain and the ultimate period

we must have, at least, a third order model for the plant; only then, while the gain is varied, the root loci would follow the asymptotes at ±60°. Then, at some point, there are at least a pair of poles on the imaginary axis, and the closed loop system will exhibit oscillations. In fact, in the industry, these parameters are obtained, rather empirically, using which PID controls are designed; in turn, these controllers are tested on the first order or second order, possibly with time delays, during simulations.

We close this part on classical PID control design with the following implementation issues.

1.4.1.9 PID Implementation

The P-I-D actions may be summarized as follows. The proportional action, P, is fairly intuitive; the integral action, I, drives the position error to zero; and the derivation action, D, attempts to find the trends but unfortunately it is quite sensitive to noise. Accordingly, the implementation of these controllers have attracted quite a good amount of interest, and here we briefly describe a popular method.

1.4.1.10 Reset Windup

The integral action is sometimes counter-intuitive. When a sustained error occurs, the corresponding integral term becomes quite large, and the controller output eventually exceeds saturation limits. Since the actuators, e.g. control valves, have their own limitations, it appears that the controller is demanding more than the actuator is capable of delivering. Consequently, the controller is effectively disabled, but the numerical integration continues to be performed and the controller output keeps on accumulating. This, in turn, results in long transients and large overshoots. This phenomenon is widely known as **reset windup**.

Theoretically, we may avoid the reset windup by simply switching off the integral action whenever the controller output saturates. There are several approaches proposed in practice. One scheme that has become a standard feature in commercial PID controllers is briefly described using the following block diagram [5] in Figure 1.77.

We assume that the actuator is linear over a good enough range of operation and saturates outside the range. More about such behavior in Chapter 3. For now, we simply take

$$u_A(t) = \begin{cases} 0 & u(t) < 0 \\ u(t) & 0 \le u(t) \le 1 \\ 1 & u(t) > 1 \end{cases}$$

Whenever the integral action winds up, the actual control action remains unchanged (e.g., at 1) and can be treated as a reference. We can quickly work out that the controller output may be expressed as

TABLE 1.6

Ziegler-Nichols's Classical PID Rules

Controller	K_p	T_i	T_d
P	$0.5\,K_u$		
PI	$0.4\,K_u$	$0.8\,T_u$	
PID	$0.6\,K_u$	$0.5\,T_u$	$0.12\,T_u$

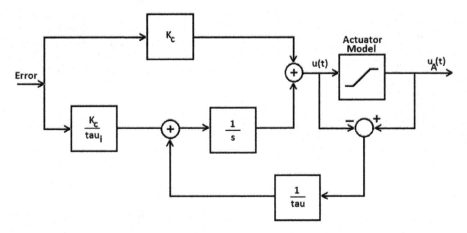

FIGURE 1.77
PID Controller with Reset wind-up. Adapted fromCheng-Ching Yu, Autotuning of PID Controllers, 2/e, Springer Verlag London, 2006.

$$U(s) = \frac{1}{\tau s + 1} U_A(s) + \frac{\tau}{\tau s + 1} E_I(s) \qquad (1.99)$$

With the negative feedback loop around the integrator, with the controller output in the loop, the controller output is driven to the actual control action following first-order dynamics dictated by the constant τ, typically a small value.

1.4.2 Modern Pole-Placement

In this part of the chapter, we address the notion of *controllability* of LTIL systems described in state space and attempt to provide a complete picture of *feedback control*. For the clarity of presentation, we first assume that all the variables needed to implement a feedback control are readily available. Subsequently, we relax this assumption and address related problems.

The principle that the inputs should be computed from the state was enunciated and emphasized by Richard Bellman[9] in the mid-1950s. This is the fundamental idea of control theory. Subsequently, Rudolf Kalman[10] made giant strides and put control theory in a firm algebraic setting where we go from a completely abstract consideration to completely practical results that are, in the end, quite simple. We first

provide a guided tour into this ever exciting concept of controllability.

1.4.2.1 Controllability

Let us first assume the following autonomous system:

$$\dot{x} = Ax \qquad (1.100)$$

where $A \in \Re^{n \times n}$, $x \in \Re^{n \times 1}$. Assume that this system has an arbitrary initial state $x(0) = x_0$. It follows that the system makes a transition to a state $x_1(t_f)$ in a given amount of time $t_f (\geq 0)$ by virtue of the eigenvalues Λ of A.

Example 1.4.12:

Let

$$A = \begin{bmatrix} -1 & 0 \\ 0 & -2 \end{bmatrix} \text{ with } x_0 = \begin{bmatrix} 1 & 1 \end{bmatrix}^T$$

Then, at $t = 1$ sec, we have
$$x_1(1) = \begin{bmatrix} 0.3679 & 0.1353 \end{bmatrix}^T$$

Example 1.4.13:

If the initial state remains same as before, but A is modified to

$$\begin{bmatrix} -3 & 0 \\ 0 & -5 \end{bmatrix}$$

then, at $t = 1$ sec, the state would be

$$x_2 = \begin{bmatrix} 0.0497 & 0.0067 \end{bmatrix}^T$$

[9] Bellman, R.E., and R. Kalaba (eds.) *Selected Papers on Mathematical Trends in Control Theory*, Dover, New York, 1964.
[10] Kalman, R.E., "Mathematical description of linear dynamical systems," *SIAM J. Cont.* vol. 1, pp. 152–192, 1963.

In particular, since $x(t) = e^{At}x_0$ for an autonomous system, given some x_0, $x(t)$ depends on e^{At}, which, in turn, depends on the eigenvalues of A. This idea can be easily extended to non-autonomous systems since, once again, for a fixed initial state x_0 and a fixed input u, $x(t)$ is governed by e^{At} and the eigenvalues of A.

Let us now ask the following question. Is it possible to *force* the system in eqn. 10.1 from its initial state x_0 to an apriori fixed state x_2 (instead of $x_1(t_f)$) in the same amount of time t_f? This is equivalent to asking the existence of a control input u that forces the trajectory $x(t)$ according to our specs.

It is easy to see that the desired state transition takes place if we are allowed to alter the eigenvalues of A to, say, those of another matrix A_1 that has suitable eigenvalues Λ_d for this purpose. To do this, we may need to add a matrix, such as δA to A, so that $A + \delta A$ is *similar* to A_1. The idea should be clear (from Example 1.4.13) that if we imagine diagonal (or, at least, triangular) matrices, then we get a new autonomous system:

$$\dot{x} = (A + \delta A)x$$

$$= Ax + \delta A x \qquad (1.101)$$

The above equation immediately suggests that a linear combination of the state variables needs to be added to the system equation, i.e. we propose that the control $u \in \Re^{p \times 1}$ be expressed as a linear combination of the state variables,

$$u = K x + v \qquad (1.102)$$

where $K \in \Re^{p \times n}$ is the matrix that represents the linear combination, and $v \in \Re^{p \times 1}$ is an arbitrary shift taken into account without loss of generality. The above equation implies a *closed loop* structure, as shown in Figure 1.78.

Since the control input u is obtained by feeding back the state x, we call this the **state feedback control**. We refer to K, hereafter, as the state feedback gain. We emphasize the importance of the input matrix B in forcing the system to attain a different set of eigenvalues. This is easy to observe since the matrix δA in eqn. 1.100 is none other than the product BK. Intuitively, given (A, B), our problem is to find a K such that $A + BK$ has the desired set of eigenvalues Λ_d. In other words, an autonomous system, as in eqn. 1.99, has its own trajectory governed by the eigenvalues of A, and we alter the trajectory by applying an external control signal u via the map B. The autonomous system in eqn. 1.99 becomes the *controlled* system

$$\dot{x} = (A + BK)x + Bv$$

$$y = Cx$$

The following definition summarizes the idea.

1.4.2.2 Definition: Controllability

A system (A, B) is said to be **controllable** if there exists a control $u \in U$, such that the system is *driven* from an arbitrary initial state x_0 to any desired state $x(t_f)$ in a finite amount of time t_f. If the system can be driven to a desired state from any initial condition $x(0)$, and for all times $t \geq 0$, it is said to be **completely controllable**. Here, U is the space of all signals of finite energy on the time interval $[0, t_f]$.

The idea of controllability is not just restricted to LTIL systems.

Let us now address an important question: How do we determine the existence of a control u in eqn. 1.101 that affects the transition $x_0 \to x_2(t_f)$? If it exists, how do we obtain the corresponding matrix K, given the pair (A, B)?

In the rest of the chapter, we refer to the spectrum of a matrix, i.e. the eigenvalues, in terms of its characteristic polynomial. For instance, in the example above, the eigenvalues of A are -1 and -2 that are obtained as the roots of the characteristic polynomial

$$\chi_A = S + 3s + 2$$

Let us first look at the existence conditions. We use the solution of the state equation:

$$x(t_f) = e^{At_f} x_0 + \int_0^{t_f} e^{A(t_f - \tau)} Bu(\tau) d\tau \qquad (1.103)$$

Using the famous Cayley-Hamilton theorem, we may write the following expression:

$$e^{At} = \alpha_0 I + \alpha_1 \frac{At}{1!} + \alpha_2 \frac{A^2 t^2}{2!} + \ldots + \alpha_{n-1} \frac{A^{n-1} t^{n-1}}{(n-1)!}$$

Plugging this into eqn. 1.102, we obtain a linear system of equations:

$$\bar{\gamma} = [B : AB : A^2 B : \cdots : A^{n-1}B] \begin{bmatrix} \bar{\beta}_0 \\ \vdots \\ \bar{\beta}_{n-1} \end{bmatrix} \qquad (1.104)$$

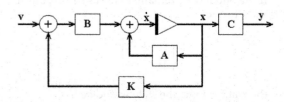

FIGURE 1.78
Standard State Feedback Control System.

where $\bar{\beta}_i$ represents the integral part of eqn. 1.102 and $\bar{\gamma}$ represents the known quantities.

It is easy to see that there exists a control u if and only if

$$\text{rank of } [B : AB : \cdots : A^{n-1}B] = n \qquad (1.105)$$

This result states that every vector $x(t)$ can be expressed as some linear combination of the columns of $[B : \cdots : A^{n-1}B]$, and these columns span the n-dimensional state space. We call the composite matrix in eqn. 1.104 that **controllability matrix** $Q_c \in \Re^{n \times np}$. If the system has only a single input, i.e. $B \in \Re^{n \times 1}$, then the dimension of Q_c is $n \times n$, and each of the n columns have to be independent for the rank to be full. However, if the system has $p(<n)$ *independent* inputs, then each partition in Q_c has p independent columns, and hence, the matrix has np columns out of which n columns are required to be independent. Intuitively, we feel that the greater the number of inputs then the greater the *index* of controllability. We also observe that the number of inputs p need never be greater than the order of the system n. If B has n independent columns, irrespective of A, the first partition itself results in full rank for Q_c, and hence, the system is always controllable. Further, if $p>n$, then from the dimension $n \times p$ of B, we may observe that the extra inputs are merely linear combinations of the n independent inputs. Thus, $p \le n$.

The image of the corresponding map, i.e. the column space of Q_c, is called the **reachable space** of (A, B) and is denoted by R. If Q_c has full rank n, then the column vectors of Q_c span the entire state space X; that is, *every state $x \in X$ can be expressed as a linear combination of the columns of Q_c*, and hence, the system can be steered from any given state x_0 to any arbitrary state $x(t_f)$. However, if Q_c is rank deficient, then only a subspace of the state space is spanned, or equivalently, the system cannot be steered to any arbitrary state but only to a few states. Accordingly, we use the word *reachable* to name the column space of Q_c.

Before we proceed further, we shall look at the following examples. A few important concepts are in order.

Example 1.4.14:

Consider the following mechanical system with two bodies and two forces (Figure 1.79).

Taking $m_1 = 1$, $m_2 = 0.5$, and $k = 1$, we have

$$x = \begin{bmatrix} y_1 \\ \dot{y}_2 \\ y_2 \\ \dot{y}_2 \end{bmatrix}, A = \begin{bmatrix} 0 & 1 & 0 & 0 \\ -1 & 0 & 1 & 0 \\ 0 & 0 & 0 & 1 \\ 2 & 0 & -2 & 0 \end{bmatrix}, B = \begin{bmatrix} 0 & 0 \\ 1 & 0 \\ 0 & 0 \\ 0 & 2 \end{bmatrix}$$

It may be verified that Q_c has rank 4; hence, this system is controllable.

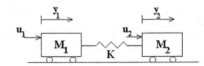

FIGURE 1.79
A controllable system.

Example 1.4.15:

Let us slightly modify the above system to have only one force u as shown in Figure 1.80.

For this case, A is as before, while $u_1 = -u$ and $u_2 = u$. Thus,

$$B = \begin{bmatrix} 0 \\ -1 \\ 0 \\ 2 \end{bmatrix}$$

and it is easy to verify that this system is not controllable.

Example 1.4.16:

Let

$$A = \begin{bmatrix} -3 & 2 & 2 \\ -1 & 0 & 1 \\ -5 & 1 & 4 \end{bmatrix}, B = \begin{bmatrix} -2 & 2 \\ -1 & 1 \\ -3 & 2 \end{bmatrix}$$

For this system,

$$Q_c = [B : AB : A^2B]$$

$$= \begin{bmatrix} -2 & 2 & -2 & 0 & -2 & -2 \\ -1 & 1 & -1 & 0 & -1 & -1 \\ -3 & 2 & -3 & -1 & -3 & -4 \end{bmatrix}$$

Evidently, the rank of Q_c is 2; hence, the system is not controllable. Here, Q_c is effectively the first two columns, which are the only independent columns. This is the basis of the reachable

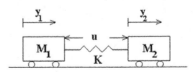

FIGURE 1.80
An uncontrollable system.

subspace R of (A, B). Since the dimension of R is 2, not all states are *reachable*. For instance, the state

$$x_1 = \begin{bmatrix} 1 \\ 1 \\ 0 \end{bmatrix}$$

is not spanned by Q_c; thus, it is not reachable. On the other hand, the state

$$x_2 = \begin{bmatrix} 0 \\ 0 \\ s \end{bmatrix}, \text{ for some } s$$

can be expressed as a linear combination of the two independent columns of Q_c and, hence, is reachable. We may also notice that x_1 is *orthogonal* to any reachable state x_2.

We make the following comments with respect to the *numerical* computation of the rank of Q_c.

- It is easy to form the matrix Q_c by multiplying B with A to get the second partition AB, then multiply AB with A to get the third partition A^2B, then multiply A^2B with A to get the fourth partition, and so on. We need not put effort in computing A^2, A^3, and so on separately.
- *Computing* the rank of a large real matrix numerically is not a trivial task.

Example 1.4.17:

Let us consider the *controllable canonical form* (CCF).

$$\tilde{A} = \begin{bmatrix} 0 & 1 & 0 & \cdots & 0 \\ 0 & 0 & 1 & \cdots & 0 \\ & & & \ddots & \\ 0 & 0 & 0 & \cdots & 1 \\ -a_n & -a_{n-1} & -a_{n-2} & \cdots & -a_1 \end{bmatrix} \quad \tilde{B} = \begin{bmatrix} 0 \\ 0 \\ \vdots \\ 0 \\ 1 \end{bmatrix}$$

There are two things of interest to us:

1. The characteristic polynomial is simply

   $$\chi_{\tilde{A}} : s^n + a_1 s^{n-1} + \cdots + a_n$$

   and,
2. The pair $(\tilde{A}, \tilde{B})$ is always controllable, irrespective of the values of $a_1, \cdots a_n$.

   This canonical form is also called the **controller form** for the following reason. Let us suppose that the eigenvalues

of $\tilde{A}$ have to be displaced so that the resulting (monic) characteristic polynomial is

$$\chi = s^n + h_1 s^{n-1} + \cdots + h_n$$

Corresponding to this characteristic polynomial, the controller form is

$$\tilde{A}_1 = \begin{bmatrix} 0 & 1 & 0 & \cdots & 0 \\ 0 & 0 & 1 & \cdots & 0 \\ & & & \ddots & \\ 0 & 0 & 0 & \cdots & 1 \\ -h_n & -h_{n-1} & -h_{n-2} & \cdots & -h_1 \end{bmatrix} \quad \tilde{B}_1 = \begin{bmatrix} 0 \\ 0 \\ \vdots \\ 0 \\ 1 \end{bmatrix}$$

It is easy to verify that the $\tilde{K}$ matrix for this case is a row vector given by

$$\tilde{K} = [(a_n - h_n) : \cdots : (a_1 - h_1)] \tag{1.106}$$

1.4.2.3 Definition: Similarity

Let (A, B) and (A_1, B_1) be two pairs. Then (A, B) is **similar** to (A_1, B_1), denoted by

$$(A, B) \sim (A_1, B_1)$$

if there exists a nonsingular T such that

$$T^{-1}AT = A_1 \text{ and } T^{-1}B = B_1$$

It is easy to verify the following facts:

1. The similarity relation $\sim$ is an equivalence relation.
2. If $(A, B) \sim (A_1, B_1)$, then (A, B) is controllable if, and only if, (A_1, B_1) is. In other words, a similarity transformation preserves controllability.

Theorem:

The single-input system (A, B) is controllable if, and only if, it is similar to its controller form $(\tilde{A}, \tilde{B})$.

Proof:

For the sake of brevity, we assume $n = 3$ in what follows.

Let A be any arbitrary 3×3 matrix whose characteristic polynomial is

$$\chi_A = s^3 + a_1 s^2 + a_2 s + a_3$$

By the Cayley-Hamilton theorem we have,

$$A^3 + a_1 A^2 + a_2 A + a_3 I = [0]$$

which we would rewrite after post-multiplying by B as follows.

$$A^3 B = -a_1 A^2 B - a_2 AB - a_3 B$$

It is now easy to show that

$$A[B : AB : A^2 B] = [B : AB : A^2 B] \begin{bmatrix} 0 & 0 & -a_3 \\ 1 & 0 & -a_2 \\ 0 & 1 & -a_1 \end{bmatrix}$$

which implies that

$$Q_c^{-1} A Q_c = \begin{bmatrix} 0 & 0 & -a_3 \\ 1 & 0 & -a_2 \\ 0 & 1 & -a_1 \end{bmatrix} \triangleq M \quad (1.107)$$

Further, we may write

$$B = Q_c \begin{bmatrix} 1 \\ 0 \\ 0 \end{bmatrix} \quad (1.108)$$

For the characteristic polynomial χ_A, the controller form is

$$\tilde{A} = \begin{bmatrix} 0 & 1 & 0 \\ 0 & 0 & 1 \\ -a_3 & -a_2 & -a_1 \end{bmatrix} \tilde{B} = \begin{bmatrix} 0 \\ 0 \\ 1 \end{bmatrix}$$

This is the form we wish to transform (A, B) to. Let us define

$$\tilde{Q}_c \triangleq \sim [\tilde{B} : \tilde{A}\tilde{B} : \tilde{A}^2 \tilde{B}]$$

so that we may write

$$\tilde{Q}_c^{-1} \tilde{A} \tilde{Q}_c = \begin{bmatrix} 0 & 0 & -a_3 \\ 1 & 0 & -a_2 \\ 0 & 1 & -a_1 \end{bmatrix} = M (\text{why?}) \quad (1.109)$$

and

$$\tilde{B} = \tilde{Q}_c \begin{bmatrix} 1 \\ 0 \\ 0 \end{bmatrix} \quad (1.110)$$

Thus, we have

$$Q_c^{-1} A Q_c = \tilde{Q}_c^{-1} \tilde{A} \tilde{Q}_c = M$$

and

$$Q_c^{-1} B = \tilde{Q}_c^{-1} \tilde{B} = \begin{bmatrix} 1 \\ 0 \\ 0 \end{bmatrix}$$

Hence, if we define

$$T \triangleq Q_c \tilde{Q}_c^{-1} \quad (1.111)$$

we get the required transformation

$$\tilde{A} = T^{-1} A T \text{ and } \tilde{B} = T^{-1} B \quad (1.112)$$

Thus, any arbitrary pair (A, B) is controllable if, and only if, a non-singular T defined as above exists. It is easy to prove that T is non-singular if, and only if, Q_c is non-singular.

quod erat demonstrandum (QED)

Example 1.4.18:

Let us consider Example 1 again. Let the input matrix be

$$B = \begin{bmatrix} b_1 \\ b_2 \end{bmatrix}$$

If we apply the criterion, we get the condition that $-b_1 b_2 \neq 0$ for the matrix Q_c to have rank 2. In other words, both the rows of B must have non-zero elements. Let us arbitrarily put

$$B = \begin{bmatrix} 1 \\ -1 \end{bmatrix}$$

The matrix A is similar to

$$T^{-1} A T = \tilde{A} = \begin{bmatrix} 0 & 1 \\ -2 & -3 \end{bmatrix} \text{with } T = \begin{bmatrix} 2 & 1 \\ -1 & -1 \end{bmatrix}$$

and

$$T^{-1} B = \tilde{B} = \begin{bmatrix} 0 \\ 1 \end{bmatrix}$$

Since the pair $(\tilde{A}, \tilde{B})$ is always controllable, so is the pair (A, B).

As we have seen earlier, the state feedback control $u = Kx + v$ transforms the system (A, B) to $(A + BK, B)$. The problem of finding the matrix K that contributes to this transformation is called the **pole assignment** or **pole placement** problem (also called spectrum assignment). More precisely,

- Given (A, B), find a K so that the eigenvalues of $A+BK$ are as desired.

  We might like to specify the eigenvalues (i.e. the poles) of $A + BK$ exactly, or we might be satisfied to have them in some *desired pole region* that we have discussed in the context of classical control design for improving the transient or steady-state or both. The great fact is that we can find a K if, and only if, the given pair (A, B) is controllable.

We shall continue with the example above and illustrate the design problem. Suppose we wish to change the pole locations to -3 and -5. The characteristic polynomial for this is

$$\chi = s^2 + 8s + 15$$

And, we may find that

$$\tilde{K} = \begin{bmatrix} -13 & -5 \end{bmatrix}$$

By way of similarity transformation, we have transformed the state x to $z = T^{-1}x$ and the system to the pair $(\tilde{A}, \tilde{B})$. For this we get $u = \tilde{K}z$. When we transform back to (A, B), we get $u = \tilde{K}T^{-1}x$; hence, the matrix K for the original system is

$$K = \tilde{K}T^{-1} = \begin{bmatrix} -8 & -3 \end{bmatrix}$$

We may verify that

$$A + BK = \begin{bmatrix} -9 & -3 \\ 8 & 1 \end{bmatrix}$$

whose eigenvalues are -3 and -5, which can be easily verified.

In this example, we demonstrated how to compute the matrix K via a similarity transformation using a non-singular matrix T. This is summarized below.

1.4.2.4 Algorithm: Pole Assignment – SISO Case

- Step 1: Compute the change of basis matrix T.
- Step 2: Transform the given pair (A, B) into the controller form $(\tilde{A}, \tilde{B})$.
- Step 3: Find $\tilde{K}$.
- Step 4: Obtain the matrix $K = \tilde{K}T^{-1}$.

Before proceeding further, we would make the following observations:

1. Evidently, autonomous systems (with $B = [0]$) are <u>not</u> controllable making the notion of controllability clear.

2. In the single input case, K is *uniquely* determined by the set of desired poles.

Example 1.4.19:

Let

$$A = \begin{bmatrix} 0 & 1 & 0 & 0 \\ 10 & 0 & 0 & 0 \\ 0 & 0 & 0 & 1 \\ 0 & 0 & 20 & 0 \end{bmatrix}, \ B = \begin{bmatrix} 0 \\ -1 \\ 0 \\ -2 \end{bmatrix}$$

Suppose the set of desired poles is

$$\Lambda_d = \{-1, -1, -2 \pm j\}$$

We may find that, for this example,

$$T = \begin{bmatrix} 20 & 0 & -1 & 0 \\ 0 & 20 & 0 & -1 \\ 20 & 0 & -2 & 0 \\ 0 & 20 & 0 & -2 \end{bmatrix}$$

and $K = \begin{bmatrix} -24.5 & -7.4 & 34.25 & 6.7 \end{bmatrix}$

Notice that the given system is not stable. But it is controllable and, in fact, it is *stabilizable*. In other words,

$$(A, B) \text{ controllable} \Rightarrow (A, B) \text{ stabilizable}$$

i.e. controllability is sufficient and, hence, is a stronger property.

The state incorporates *all* the information necessary to determine the control action to be taken. This is because, by definition, the future evolution of a dynamical system is completely determined by its present state and the future inputs. Thus, the state space description of a system is a natural framework for formulating and solving problems in control; the possibility of constructing an arbitrarily good control law is limited only by the controllability property of the system.

Before we proceed to the next section, we shall look at the following examples.

Example 1.4.20:

Suppose we have a scalar system:

$$\dot{y} = -y + u$$

and it is desired to increase the speed of response by a magnitude of 1000 so that the bandwidth of the system becomes 1 kHz. This specification suggests that the gain $K = 999$. However, such a large

---

[11] Imagine the root loci of a third order system with the gain $s = -Z$ tending to infinity.

gain would lead the actuators to saturation and the linear model is no longer valid. In fact, in certain cases, a high gain leads to instability.[11] Therefore, it is natural to constrain the elements of the K matrix; i.e., find a K matrix such that each of its elements lie within certain prespecified bounds. This kind of restriction leads to certain interesting consequences that we shall explore later.

The foregoing theory tells us yet another point. The system designer can actually dictate these properties. If controllability is lacking in the given model, additional control inputs can be provided. In other words, we add more columns to the B matrix. Note that if the number of inputs is equal to the number of states, then the system is necessarily controllable.

2

Nonlinear Systems

In contrast to the previous chapter, wherein we have extensively presented ideas based on the principle of superposition, in this chapter we deal with the analysis and the design of *nonlinear control systems*; i.e., control systems containing at least one nonlinear component and, hence, violating the principle of superposition. In the analysis, a nonlinear closed loop system is assumed to have been designed, and we attempt determine the system's behavior. In the design, we take a nonlinear plant to be controlled and some specifications on the closed loop system behavior, and our task is to construct a controller. This introductory part provides the background for the specific analysis and design methods to be discussed in later chapters.

While linear control appears to be a mature subject with a range of powerful techniques as well as successful industrial applications, there are several reasons why there is a continued interest from the fields of process control, aerospace, biomedical, and so on in the development and applications of nonlinear control.

1. With the key assumption of small range operation for the linear model, a linear controller is likely to fail because the system cannot properly compensate for the nonlinearities. Nonlinear controllers, on the other hand, are expected to directly compensate a wider range of operational swings in the plant.

2. Owing to the discontinuity, simple and essential components, such as switches and relays, in the loop of a system do not allow us to make a linear approximation. Instead of ignoring such effects, for the sake of easier linear controls, we would rather prefer nonlinear techniques.

3. Likewise, we will discard our earlier assumption that the parameters of the system model are reasonably well known. As one learns, there would be several details that are integral to a given physical plant, such as slower time variation (against assumed constant) of the parameters such as ambient air pressure during the flight of an aircraft. Instead of allowing the system behavior to degrade, nonlinearities can be intentionally introduced into the controller part of a control system so that model uncertainties can be tolerated. In general, there are two classes of controllers which find a place in this context—the robust and the adaptive.

4. Lastly, the reader will discover that good nonlinear controllers [11] are likely to be much simpler and more intuitive than their linear counterparts.

It is worth noting that, in the past, nonlinear control methods have been limited due to the lack of software and computing power to obtain quicker solutions, while simultaneously increasing the complexity owing to greater details. However, it is no longer an excuse, and in fact, there is a renewed interest in the research and application of nonlinear control methods.

2.1 Nonlinear Phenomena and Nonlinear Models

In this section we shall briefly encounter nonlinear phenomena using simple examples, tailored to throw a different light on a given differential equation.

Example 2.1.1:

Consider the first-order system [7,8]

$$\dot{x} = -x + x^2 \tag{2.1}$$

with initial condition $x(0) = x_0$.

If x is considered to be small, say of the order of 10^{-2}, the square is even smaller, and often we focus on the linearization

$$\dot{x} = -x \tag{2.2}$$

The solution of this linear equation is $x(t) = x_0 e^{-t}$. Our experience in the previous chapter readily tells us that it is a stable system (with the pole at $s = -1$), no matter what $x(0)$ is, and $x(t)$ exponentially approaches *zero*.

However, the linearized system clearly has a unique equilibrium point at $x = 0$; once $x(t) = 0$, it does not change any further, i.e. $\dot{x} = 0$. In this sense, we call it *equilibrium*. By contrast, the actual response $x(t)$, including the nonlinear part, x^2, can be obtained as

$$x(t) = \frac{x_0 e^{-t}}{1 - x_0 + x_0 e^{-t}}, \quad x(0) = x_0 \tag{2.3}$$

Before looking at the plot of $x(t)$, it would be exciting if the reader carefully pays attention to what this equation says.

- If x_0 is either 0 or 1, $x(t)$ does not change, i.e. $\dot{x} = 0$,
- if $0 < x_0 < 1$ then $x(t) \to 0$, and
- if $x_0 > 1$, the x^2 term in the differential equation dominates, and $x(t) \to \infty$.

The last observation above looks counterintuitive, after all there is a term with an exponential decay, but it may be verified. We place an emphasis on the alacrity of the reader while studying nonlinear phenomena. The response to various initial conditions is given in Figure 2.1.

Coming to notation, we look at the following in this chapter. The nonlinear systems will be represented as

$$\dot{x} = f(x, u, t), \quad t \in \Re \tag{2.4}$$

where $x \in \Re^n$ is the state of the system, $u \in \Re^p$ is the input, and f is a nonlinear function. Recall that we are considering dynamical systems that are modeled by a finite number of coupled, first-order ordinary differential equations, and that the notation above, in terms of vectors and matrices, allows us to represent the system in a compact form.

In some cases, we will look at the properties of the system when f does not depend explicitly on u, i.e. $\dot{x} = f(x, t)$. This is called the unforced response of the system. This does not necessarily mean that the input to the system is zero. Rather, it could be that the input has been specified as a function of time, like $u = u(t)$, or as a given feedback function

of the state, $u = u(x)$, or both. It becomes clearer as we go ahead.

When f does not explicitly depend on t, i.e. $\dot{x} = f(x)$, the system is said to be **autonomous**, or **time-invariant**. An autonomous system is self-governing and is invariant to shifts in the time origin.

Moving back to our Example 2.1, **equilibrium points** are an important class of solutions for a differential equation. They are defined as the points x_e such that:

$$\dot{x} = 0, \quad i.e., \quad f(x) = 0$$

A good place to start the study of an autonomous nonlinear system is by finding its equilibrium points. The system in the example has two equilibrium points, $x = 0$ and $x = 1$, and the qualitative behavior strongly depends on its initial condition.

When we consider the linearized system, stability is seen by noting that for any initial condition the solution always converges to the *only* equilibrium point $x = 0$. However, when we take the actual nonlinear system, the solutions starting with $x_0 < 1$ will indeed converge to the equilibrium point $x = 0$, but those starting with $x_0 > 1$ will go unbounded to infinity. Owing to this, we say that the stability of nonlinear systems is better explained in terms of the stability of the equilibrium points. While at equilibrium, the trajectory does not move, but a little away from it $x(t)$ may either approach the equilibrium or diverge away; moreover, this behavior may depend on initial conditions.

2.1.1 Limit Cycles

Nonlinear systems, interestingly autonomous systems themselves, can display oscillations of fixed amplitude and fixed period without any external excitation. These *self-excited* oscillations are called **limit cycles**, or self-excited oscillations. This important phenomenon can be studied using the following oscillator dynamics, first studied in the 1920s by the Dutch electrical engineer Balthazar van der Pol.

Example 2.1.2:

The second-order nonlinear differential equation, known as the van der Pol equation is given by:

$$m\ddot{x} + 2c(x^2 - 1)\dot{x} + kx = 0 \tag{2.5}$$

where m, c, and k are positive constants. It can be seen as, for instance, a mass-spring-damper system with a position-dependent damping coefficient $2c(x^2 - 1)$.

For large values of x, the damping coefficient is positive and the damper removes energy from the system; hence, the solution $x(t)$ has a

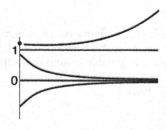

FIGURE 2.1
Trajectories of Example 2.1.1. for various initial conditions.

convergent tendency. Nevertheless, as $x(t)$ becomes smaller (than 1), the damping coefficient is negative and the damper adds energy into the system; hence, the solution $x(t)$ has a divergent tendency. Thus, owing to the nonlinear damping, with $x(t)$, the solution can neither grow unboundedly nor decay to zero. Instead, it displays a sustained oscillation, independent of initial conditions. A typical solution is shown in Figure 2.2. for two different initial conditions; albeit not to scale, the idea is that the amplitude and the frequency of oscillations change with the initial conditions.

While one might think of L-C circuit, i.e., a circuit having only an inductance and a capacitance, with ideal components, for example, as a linear system exhibiting limit cycles, limit cycles in nonlinear systems are different. First, the amplitude of the self-sustained excitation is independent of the initial condition. Secondly, marginally stable linear systems, with the pair of complex conjugate poles on the imaginary axis, are very sensitive to changes in system parameters (with a slight change the system becoming either stable or unstable), whereas limit cycles are not easily affected by parameter changes.

Limit cycles can be found in many areas of engineering. Common examples include aircraft wing fluttering, where a limit cycle caused by the interaction of aerodynamic forces and structural

vibrations is frequently encountered and is sometimes dangerous. As one can see, limit cycles can be undesirable in some cases but desirable in other cases. As a control engineer, we need to know how to eliminate them when they are undesirable and how to nurture them when they are desirable.

2.1.2 Bifurcations

As the parameters of nonlinear dynamic systems are changed, i.e. as the coefficients of the governing differential equation are changed, the solutions of $f(x,t)$ obviously change, and hence, the stability of the equilibrium point; even the number of equilibrium points can change. This phenomenon of *bifurcation*, i.e. quantitative change of parameters leading to qualitative change of system properties, is the topic of bifurcation theory. Values of these parameters at which the qualitative nature of the system's motion changes are known as critical or bifurcation values.

Example 2.1.3:

Let us consider the second-order system described by the so-called undamped Duffing equation

$$\frac{d^2}{dt^2} x = -\alpha x - x^3 \tag{2.6}$$

We can plot the equilibrium points as a function of the parameter α. As α varies from positive to negative, one real equilibrium point at $x = 0$, assuring us stability in some sense, splits into three points

$$x_e = 0, \pm \sqrt{\alpha}$$

as shown in Figure 2.3a. Clearly, the system is unstable with $\alpha < 0$. This represents a qualitative change in the dynamics and, thus, $\alpha = 0$ is a critical bifurcation value. This kind for bifurcation is known as a pitchfork, due to the shape of the equilibrium point plot in Figure 2.3a.

Another kind of bifurcation involves the emergence of the previously discussed limit cycles as parameters are changed. In the Duffing equation, at a certain value of α, it may be shown that the response of the unstable system diverges to a limit cycle. Figure 2.3b depicts the change of typical system state trajectories (states are x and $\dot{x}$) as the parameter α is varied. This type of bifurcation is called a Hopf bifurcation.

2.1.3 Chaos

For any stable linear system, small changes in the initial conditions result in only small changes in the output

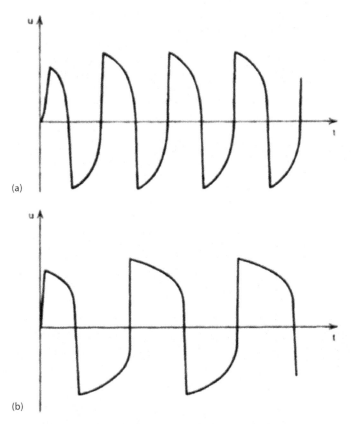

(a)

(b)

FIGURE 2.2
Results of Example 2.1.2. (a) $x(0) = x_1$. (b) $x(0) = x_2 \neq x_1$.

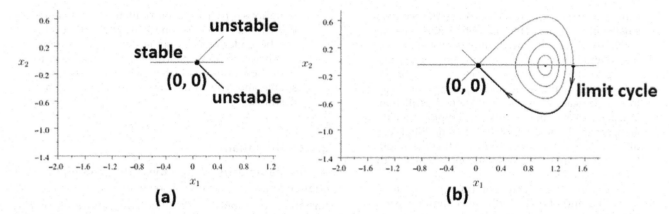

FIGURE 2.3
Results of Example 2.1.3. (a) Pitchfork bifurcation and (b) Hopf bifurcation.

Nonlinear systems, however, can display a phenomenon called **chaos**, by which we mean that the system output is extremely sensitive to the initial conditions. The essential feature of chaos is the *unpredictability* of the system output. What we mean by this is that, even if we have an exact model of a nonlinear system and have an extremely accurate computer, the system's response in the long run still cannot be well predicted.

We emphasize that chaos must be distinguished from random motion. In random motion, the system model, or input, would contain uncertainty, and as a result, the output cannot be predicted exactly (only statistical measures are available). In chaotic motion, on the other hand, the involved problem is deterministic, i.e. there is little uncertainty in system model, input, or initial conditions, but still the output is unpredictable.

Example 2.1.4:

Let us consider the simple nonlinear system [7]

$$\ddot{x} + 0.1\dot{x} + x^5 = 6\sin t \qquad (2.7)$$

which may represent a sinusoidally forced, lightly-damped mechanical structure undergoing large

elastic deflections. Figure 2.4 shows the responses of the system corresponding to two almost identical initial conditions, namely $x(0) = 2$, $\dot{x}(0) = 3$ (thick line), and $x(0) = 2.01$, $\dot{x}(0) = 3.01$ (thin line). The reason for this chaotic behavior is the presence of the nonlinearity in x^5.

Chaotic phenomena can be observed in many physical systems. The most commonly seen physical problem is turbulence in fluid mechanics (such as the swirls of candles or incense sticks, which initially look like bifurcations). Atmospheric dynamics also display clear chaotic behavior, thus making long-term weather prediction impossible. Other systems known to exhibit chaotic vibrations include buckled elastic structures, mechanical systems with backlash, systems with aeroelastic dynamics, and, of course, feedback control devices.

Chaos occurs mostly in strongly nonlinear systems. This implies that, for a given system, if the initial condition or the external input cause the system to operate in a highly nonlinear region, it increases the possibility of generating chaos. In the context of feedback control, it is of course of interest to know when a nonlinear system will get into a chaotic mode (so as to avoid it) and, in case it does, how to recover from it. Such problems are the object of active research.

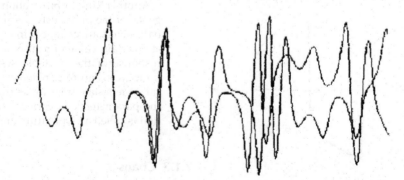

FIGURE 2.4
Results of Example 2.1.4.

Nonlinear systems do exhibit other interesting behaviors, such as jump resonance, subharmonic generation, asynchronous quenching, and frequency-amplitude dependence of free vibrations. However, the above brief introduction to limit cycles, bifurcations, and chaos should give the reader ample evidence that nonlinear systems can have considerably richer and more complex behavior than linear systems, and hence, the reader may look forward to a serious study.

2.2 Fundamental Properties of ODEs [8,9]

In this part, we will solve and analyze first-order differential equations. In an earlier course in mathematics and/or differential equations, the reader would have studied first-order differential equations

$$\frac{dx}{dt} = f(x,t) \tag{2.8}$$

when the differential equation was separable or the differential equation was linear. Recall, for separable differential equations, explicit solutions can be obtained by direct integration. For linear differential equations, the general solution,

$$x(t) = x_p(t) + c x_h(t)$$

is always in the form of a particular solution $x_p(t)$ plus an arbitrary multiple of a homogeneous solution $x_h(t)$. Explicit solutions of first-order linear differential equations can always be obtained using an integrating factor.

Now, we will discuss other properties of first-order differential equations, mostly for differential equations, which are nonlinear. Presently, we will discuss qualitative properties of solutions of first-order differential equations, which are autonomous. Later, we will show that this insight is needed for designing controllers.

First, we will study equilibrium solutions and their stability. Although the autonomous system is separable, we will introduce elementary geometric methods to describe the qualitative behavior of the solution, obtaining different information than that obtained by integration.

2.2.1 Autonomous Systems

We have discussed in previous sections that a specific solution of an autonomous differential equation, $\dot{x} = f(x)$ is called an equilibrium, or steady-state solution, if the solution is constant in time, i.e. $x(t) = a = $ constant. Hence, letting $x = a$ gives $f(a) = 0$. Thus, all roots of the function $f(x)$ are equilibrium solutions. Thinking of

modeling a physical process and assuming we are at an equilibrium, $x = a$ says that everything is in perfect balance (the forces, pressures, balance, and so on) and the process $x(t)$ stays at $x = a$—an equilibrium solution.

Example 2.2.1:

The three equilibria of the system

$$\dot{x} = x - x^3$$

are:

$$x = 0, \text{ and } x = \pm 1$$

2.2.1.1 Stability of Equilibria

Nonetheless, a physical system in perfect balance

1. May be very difficult to achieve, for example a pencil standing on its sharpened tip, or
2. May be easy to maintain, for exmaple a tiny marble at the bottom of a large bowl.

In the fomer instance, any slight perturbation would make the pencil fall down, and in the latter instance, the marble goes back to its original state of rest. Thus, solutions in perfect balance, i.e. equilibrium solutions, may be stable or unstable.

There are more precise mathematical definitions of these ideas, but we will be satisfied with the following definition of stability of an equilibrium:

An equilibrium $x = a$ is stable:

- If all solutions starting near $x = a$ stay in its close vicinity, and
- The closer the solution starts to $x = a$ the closer nearby it stays.

Otherwise the equilibrium is said to be unstable.

To be stable, the solution should stay in the close vicinity of the equilibrium for *all* nearby initial conditions; if the solution does not stay near the equilibrium, for even one nearby initial condition, we would say the equilibrium is unstable. Although these ideas are somewhat vague, we will develop more precise mathematical details shortly.

Sometimes it is helpful to define a stronger sense of stability.

An equilibrium $x = a$ is *asymptotically* stable:

- If $x = a$ is stable, and
- The solution approaches the equilibrium for all nearby initial conditions, i.e. $x(t) \to a$ as $t \to \infty$.

The differential equation states that $\dot{x}$ is a function of x. That relationship can often be *graphed*, possibly easily, since $f(x)$ is given. A typical functional relationship [10] is graphed in Figure 2.5. At time t, the solution $x(t)$ is a point on the x-axis, and as time increases, $x(t)$ moves along the x-axis. We introduce arrows to the right $\rightarrow$ if $x(t)$ is an increasing function of time t, and arrows to the left $\leftarrow$ if $x(t)$ is decreasing. The differential equation easily determines how these arrows should be introduced.

First, we notice that points where the curve intersects the x-axis are the equilibrium points, since $f(x) = 0$ and $\dot{x} = 0$ at these points. Next, in the upper half-plane, where $f(x) > 0$, $\dot{x} > 0$, and it follows that x is an increasing function of time. So the arrows in the graph point to the right, indicating that $x(t)$ moves to the right as time increases. Similarly, where $f(x) < 0$, in the lower half-plane, $\dot{x} < 0$, and the arrows show motion to the left as time increases since there x is a decreasing function of time. These results can be summarized in the **one-dimensional phase line** sketched in Figure 2.5. Not only can we determine graphically where the equilibrium is located, but it is also easy to determine in all cases if the equilibrium is stable or unstable.

- If the arrows indicate that all solutions near an equilibrium move toward the equilibrium as time increases, then the equilibrium is stable, as marked in Figure 2.5.
- If some solutions near an equilibrium move away from that equilibrium as time increases, then that equilibrium is unstable.

In the situation assumed in Figure 2.5, two of the equilibria are clearly stable and one unstable, as marked in the figure. If the slope of the curve is positive when it crosses the x-axis at the equilibrium ($f'(a) > 0$), then the equilibrium is unstable. If the slope of the curve is negative when it crosses the x-axis ($f'(a) < 0$), then the equilibrium is stable.

Thus, the geometric analysis has verified the conclusions of the linear stability analysis in the usual simple cases in

which $f'(a) \neq 0$. In the example graphed in Figure 2.5, we assumed that the three equilibria all satisfied the simple criteria. However, this geometric method can also be used to determine the stability of an equilibrium in the more difficult case in which the linearization does not determine the stability, e.g, $f'(a) = 0$.

Furthermore, we have easily determined the qualitative behavior of the solution to the differential equation corresponding to all initial conditions. We know qualitatively how the solution approaches an equilibrium if the solution starts near an equilibrium, and we even know how the solution behaves when it starts far away from an equilibrium. If there are more than one stable equilibrium, we can determine which initial conditions approach which equilibrium.

We now discuss two illustrative examples.

Example 2.2.2:

Let us consider the first-order nonlinear differential equation [10]

$$\dot{x} = f(x) = x - x^3 = -x(x^2 - 1) = -x(x-1)(x+1)$$

which has three equilibria: $x = 0, +1, -1$. The phase line requires the graph of the cubic $f(x) = x - x^3$ in Figure 2.6a. This graph has two critical points (the reader can easily determine)—one a local maximum and the other a local minimum. The one-dimensional phase line shows that the equilibrium $x = 0$ is clearly unstable since $f'(0) > 1$, while the equilibria $x = \pm 1$ are clearly stable.

We would, in fact, see that we do not need the sketch of $f(x)$ to determine the phase line; all we need is the *sign* of $f(x)$, as shown in Figure 2.6b.

In Figure 2.6, we show the qualitative behavior of x as a function of time for various initial conditions. We begin with the three equilibria. For example, we see that the solution approaches 1 as $t \rightarrow \infty$ for all initial conditions, which satisfy $x(0) > 0$; for initial conditions such that $x(0) < 0$, $x(t) \rightarrow -1$ as $t \rightarrow \infty$; and, for the initial condition $x(0) = 0$, the unstable equilibrium, the solution stays at the equilibrium for all time.

As the reader would have felt, using this method we can only determine the qualitative behavior of these curves. However, the advantage of the present geometric method is the simplicity by which we obtain an understanding of fundamental relationships of equilibrium, stability, and the manner in which the solution depends on the initial conditions. If the differential equation is more complicated, for example, by depending on some parameter, then the qualitative method based on the one-dimensional phase line is particularly effective. If more quantitative details are needed, the one-dimensional phase line can always be supplemented by further analytic work or numerical computations.

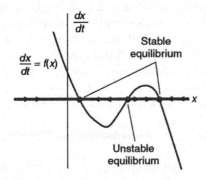

FIGURE 2.5
Building a phase plane.

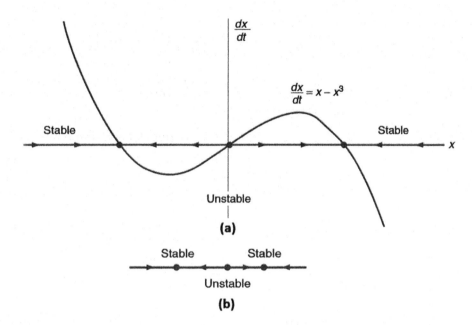

FIGURE 2.6
Phase plane for Example 2.2.2. (a) Graph of the cubic $f(x) = x - x^3$. (b) Sign of $f(x)$.

Example 2.2.3:

Let us consider [10]

$$\dot{x} = x^2$$

The reader would quickly solve this using the method of separation. However, it is interesting to note that the one-dimensional phase line is a quicker and easier method that probably gives a better understanding of some of the behavior of the solutions to this differential equation. In Figure 2.7, $f(x)$ is easily graphed, and we see immediately that $x = 0$ is an unstable equilibrium. If the initial conditions are $x(0) < 0$, then the solution approaches 0 as $t \to \infty$. For these initial conditions, the equilibrium appears stable. However, to be stable, the solutions for *all* nearby initial conditions must approach the equilibrium, not just those initial conditions less than the equilibrium. From Figure 2.7, we readily see that $x(t)$ goes away from zero for initial conditions $x(0) > 0$. Thus, $x = 0$ is an unstable equilibrium.

Furthermore, for these initial conditions, $x(0) > 0$, it is clear that the solution approaches infinity, i.e. $x(t)$ is unbounded. However, from the phase line, we do not know if the solution explodes, i.e. it has an infinite slope in a finite time. We will provide more quantitative details in the next chapter.

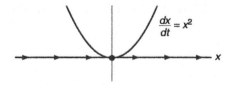

FIGURE 2.7
Stability of the systm in Example 2.2.3.

2.2.2 Non-Autonomous Systems

Linear systems can be classified as either time-varying or time-invariant, depending on whether the coefficients of the differential equation vary with time or not; equivalently, whether the elements of the system matrix A vary with time or not. However, in the more general context of nonlinear systems, these adjectives are traditionally replaced by *autonomous* and *non-autonomous*. As we have seen in the previous part, a nonlinear system is said to be autonomous if $f(x,t)$ does not depend explicitly on time; otherwise, the system is called non-autonomous. Hence, for obvious reasons, linear time-invariant (LTI) systems are autonomous and linear time-varying (LTV) systems are non-autonomous.

Strictly speaking, all physical systems are non-autonomous, because none of their dynamic characteristics is strictly time-invariant. The concept of an autonomous system, therefore, is an idealized notion. In practice, however, system properties often change very slowly, and we can neglect their time variation without causing any practically meaningful error.

It is important to note that for control systems, the above definition is made on the closed loop dynamics. Since a control system is composed of a controller and a plant (including sensor and actuator dynamics), the non-autonomous nature of a control system may be due to a time-variation either in the plant or in the control law. Specifically, a time-invariant plant with dynamics

$$\dot{x} = f(x, u)$$

may lead to a non-autonomous closed loop system if a controller dependent on time t is chosen, i.e. if $u = g(x,t)$. For example, the closed loop system of the simple plant

$$\dot{x} = -x + u$$

can be nonlinear and non-autonomous by choosing u to be nonlinear and time-varying, say, $u = -x^2 \sin t$.

The fundamental difference between autonomous and non-autonomous systems lies in the fact that the state trajectory of an autonomous system is independent of the initial time, while that of a non-autonomous system generally is not. As we will see in the next chapter, this difference requires us to consider the initial time explicitly in defining stability concepts for non-autonomous systems and makes the analysis more difficult than that of autonomous systems. Generally speaking, autonomous systems have relatively simpler properties and their analysis is much easier.

2.2.2.1 Equilibrium Points

For non-autonomous systems, of the form

$$\dot{x} = f(x,t) \tag{2.9}$$

equilibrium points x_e are defined by

$$f(x_e,t) = 0 \ \forall t \geq t_0 \tag{2.10}$$

Note that this equation must be satisfied $\forall t \geq t_0$, implying that the system should be able to stay at the point x_e *all the time*. For instance, one easily sees that the linear time-varying system $\dot{x} = A(t)x$ has a unique equilibrium point at the origin unless $A(t)$ is always singular.

Example 2.2.4:

The system

$$\dot{x} = -\frac{a(t)x}{1+x^2}$$

has an equilibrium point at $x = 0$. However, the same system with an external excitation $b(t)$:

$$\dot{x} = -\frac{a(t)x}{1+x^2} + b(t)$$

with $b(t) \neq 0$, does not have an equilibrium point. It can be regarded as a system under external input or disturbance $b(t)$.

Asymptotic stability of a control system is usually a very important property to be determined. However, the equilibrium points and their stability, just described for both autonomous and non-autonomous systems, are often difficult to apply in

order to assert this property. Nevertheless, a piece of good news is that it is still possible to draw conclusions on asymptotic stability with the help of La Salle's invariant set theorems and other slightly advanced mathematical machinery. We shall explore more about this in the second half of next chapter.

2.2.3 Existence and Uniqueness

While we are interested in the analysis, i.e. obtaining the solutions of the first-order differential equations, it is imperative to question if solutions exist given the initial conditions. To this extent, we need a more precise statement of the mathematical result that guarantees the following.

Picard-Lindelöf Theorem: There exists a unique solution to the differential equation

$$\dot{x} = f(x,t) \tag{2.11}$$

which satisfies given initial conditions

$$x(t_0) = x_0 \tag{2.12}$$

if both the function $f(x,t)$ and its partial derivative $\frac{\partial f}{\partial x}$ are *continuous* functions of t and x at and near the initial point $t = t_0, x = x_0$.

The foundation of the theory of differential equations is the theorem proven by the French mathematician Charles Emile Picard in 1890 and the Finnish mathematician Ernst Leonard Lindelöf in 1894. This therorem guarnatees the existence of solutions for the initial value problem above. There is much more to the deceptively simple appearance of the theorem above, but we'll revisit this in a Chapter 4.

In most cases of practical interest, the continuity conditions are satisfied for most values of (t_0, x_0), so that there exists a unique solution to the initial value problem for most initial conditions.

If the conditions of the basic existence and uniqueness theorem are not met at a given initial condition, then

1. Solutions may not exist at (t_0, x_0), or
2. There may be more than one solution passing through the same point (t_0, x_0), or
3. The solution may not be differentiable at $t = t_0$; for instance, friction.

Example 2.2.5: Local Existence of a Solution

The differential equation

$$\dot{x} = f(x,t) = tx^2$$

satisfies the conditions of the uniqueness theorem for all (t,x) since $f = tx^2$ and $\frac{\partial f}{\partial x} = 2tx$ are continuous functions of t and x. Thus, there should be a unique solution of the initial value problem for any (t_0, x_0).

As a numerical example, it is easy to verify that there should be a solution at the initial condition $(0,2)$. Solving the differential equation by separation gives

$$x(t) = \frac{2}{1-t^2}$$

However, this solution becomes infinite at $t = \pm 1$.

- While the basic existence and uniqueness theorem guarantees a solution of the differential equation, which satisfies the initial condition, it does not say that the solution is defined for all t. It is a local existence theorem, only guaranteeing that the solution exists for some t near the initial t_0. In this example, the solution explodes at $t = 1$.

Example 2.2.6:   Initial Condition with No Solution [10]

The differential equation

$$\dot{x} = f(x,t) = -\frac{1}{t^2}$$

satisfies the basic existence and uniqueness theorem everywhere $t_0 \neq 0$. The general solution may be found to be

$$x(t) = -\frac{1}{t} + c$$

If the initial condition is given at any $t_0 \neq 0$, then the constant c can be determined uniquely by the initial condition. There is a unique solution to the initial value problem in this case, and the solution exists locally, as in the previous example, up to the singularity at $t = 0$. The solutions are shown in Figure 2.8.

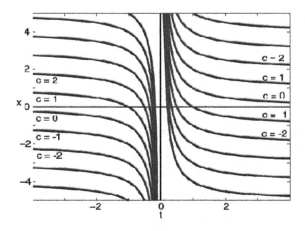

FIGURE 2.8
Example 2.2.6.

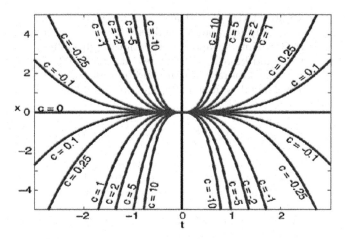

FIGURE 2.9
Solutions are not unique in Example 2.2.7.

Example 2.2.7:   Solutions Not Unique [10]

The differential equation

$$\dot{x} = f(x,t) = \frac{3x}{t}$$

may be easily solved using variable separable method for the general solution

$$x(t) = ct^3$$

Figure 2.9 shows several solutions. The function $f(x,t) = 3x/t$ is continuous everywhere for $t \neq 0$. There do not exist solutions through $(0, x)$ if $x \neq 0$. On the other hand, every solution goes through $(0,0)$, i.e. at the origin we have existence but not uniqueness.

Example 2.2.8:   Continuous but Not Differentiable

Consider the differential equation

$$\dot{x} = f(x,t) = x^{1/3}$$

In this case, while $f(x,t)$ is continuous for all (t,x), the partial derivative is not continuous if $x = 0$. The solutions are

$$x(t) = 0 \quad and \quad x(t) = \left(\frac{3}{2}t + c\right)^{3/2}$$

If the initial condition is given as $x(t_0) = 0$, then there are two solutions, $x = 0$ and

$$x(t) = 0 \quad and \quad x(t) = \left(\frac{3}{2}(t - t_0)\right)^{3/2}$$

To summarize, the existence and uniqueness theorem only guarantees that the solution exists over a certain interval and a precise choice of this interval could be tricky. Let us revisit Example 2.1.3 with an arbitrary initial condition $x(0) = x_0 > 0$. The general solution to the equation would be

$$x(t) = \frac{x_0}{1 - tx_0}$$

which has an interval of existence $t \in (-\infty, 1/x_0)$. Notice that this is an asymmetric interval.

If the reader is interested in the proof of this theorem, a slightly more involved concept of Contraction Maps is needed, which we briefly discuss in the next section.

2.3 Contraction Mapping Theorem [6]

Let us recall that in the state space models of linear systems, particularly the homogeneous system and its solution (cf. 1.2.4), we have

$$x(t) = e^{At} x_0 \qquad (2.13)$$

emphasizing the concept of a map $e^{At} : X \to X$, from a state space X to itself. In the following chapters, we will have many more occasions to use maps in our study of dynamical systems and control. As a prelude, we will first discuss the following theorem, known as the fixed-point theorem due to Stefan Banach (1922).

Before we proceed to the theorem, let us summarize certain mathematical concepts and an appropriate notation.

In any vector space, such as the familiar Cartesian space $\mathfrak{R}^2$ that can be generalized to an n-dimensional $\mathfrak{R}^n$, we have an infinite number of vectors v_i that are generated using a basis of n linearly independent vectors. We will also have the notion of the *size*, called the **norm**, of a given vector v_i, typically the distance from the origin, or more generally, the distance from another vector v_j; for instance, the popular Euclidean norm, also called the 2-norm, is familiar to everyone is:

$$dist\left(v_i, v_j\right) = \sqrt{\left(v_{i1} - v_{j1}\right)^2 + \left(v_{i2} - v_{j2}\right)^2} \quad where \ v_i, v_j \in \mathfrak{R}^2$$

Since the notion of distance is involved, it is typical to define the norm ρ as follows:

$$\rho\left(v_i, v_j\right) \geq 0$$

$$\rho\left(\alpha v_i\right) = |\alpha| \, \rho\left(v_i\right) \qquad \forall \alpha \in F \qquad (2.14)$$

$$\rho\left(v_i + v_j\right) \leq \rho\left(v_i\right) + \rho\left(v_j\right) \qquad (2.15)$$

Such spaces are better known as normed spaces. In the Cartesian space, we are well familiar with the distance, known as the 2-norm. In general, we have a host of norms, including the ∞-norm, also known as sup-norm, which gives us the farthest vector.

We may extend this idea to infinite dimensional function spaces. A function $f : D \to R$ is a map from its domain to its range. In most applications, the domain D is a subset of $\mathfrak{R}^n$ and the range is $\mathfrak{R}^n$; in such a case, the map $f : D \to \mathfrak{R}^n$ is given by n components $f_i(x_1, x_2, \cdots, x_n)$. The set of functions denoted as $C(D)$ consists of those functions on the domain D whose components are continuous. For a function f, the derivative at point x is written as $f'(x) : \mathfrak{R}^n \to \mathfrak{R}^n$; in fact, it is defined to be a linear operator given by the **Jacobian matrix**:

$$f'(x) = \begin{bmatrix} \dfrac{\partial f_1}{\partial x_1} & \dfrac{\partial f_1}{\partial x_2} & \cdots & \dfrac{\partial f_1}{\partial x_n} \\[2ex] \dfrac{\partial f_2}{\partial x_1} & \dfrac{\partial f_2}{\partial x_2} & \cdots & \dfrac{\partial f_2}{\partial x_n} \\[1ex] \vdots & & \ddots & \\[1ex] \dfrac{\partial f_n}{\partial x_1} & \dfrac{\partial f_n}{\partial x_2} & \cdots & \dfrac{\partial f_n}{\partial x_n} \end{bmatrix} \qquad (2.16)$$

A function f is $C^1(D)$—continuously differentiable—if the elements of $f'(x)$ are continuous on the open set D. We also call these *smooth* functions.

Spaces of functions like $C(D)$ and $C^1(D)$ are examples of infinite dimensional (linear) vector spaces. Much of our theoretical analysis will depend upon convergence properties of sequences of functions in some such space. A very broad example is e^{-t} is convergent, while e^t is not; needless to say, we will be interested in converging sequences. We will be further interested in spaces, called the complete normed spaces, where all the sequences converge to an element of the space. For continuous functions, the ∞-norm is defined by

$$\rho(f) = \sup_{x \in D} \| f(x) \| \qquad (2.17)$$

There is so much literature written on this, but in this book, we just extract what is relevant for our study of control systems, provide examples, and bank on the intuition of the reader.

Theorem: Contraction Mapping

Let $T : X \to X$ be a map on a complete normed space. If T is a contraction, i.e. for all $f, g \in X$, there exists a constant $c < 1$, such that

$$\rho(T(f), T(g)) \leq c\rho(f, g) \qquad (2.18)$$

then the map T has a unique equilibrium point $x_e = T(x_e) \in X$.

We will not provide a proof of this theorem in this book but we will supply an example that illustrates the proof.

Example 2.3.1:

Let us consider the space of continuous functions that are periodic with period one: $f(x+1) = f(x)$ and define the map:

$$T(f)(x) = \cos(2\pi x) + \frac{1}{2}f(2x)$$

If we start with $f_0(x) = \sin(2\pi x)$,

$$f_1(x) = \cos(2\pi x) + \frac{1}{2}\sin(4\pi x)$$

$$f_2(x) = \cos(2\pi x) + \frac{1}{2}\cos(4\pi x) + \frac{1}{4}\sin(8\pi x)$$

$$\vdots$$

$$f_j(x) = \sum_{n=0}^{j-1} \frac{\cos(2^{n+1}\pi x)}{2^n} + \frac{1}{2^j}\sin(2^{j+1}\pi x)$$

$$\vdots$$

By the theorem, the equilibrium point is guaranteed to be unique and continuous. We place emphasis on the last word of the previous sentence—continous; hence, the equilibrium point in this example, i.e. the sum of cosines in $f_j(x)$ is not an elementary function, but with some effort, it may be graphed; it may be readily seen that the second sine term of $f_j(x)$ goes to zero.

If the reader needs an even simpler example, then perhaps the simplest example is the well-known *average* operation. Let us recollect how we iteratively determined the root of a given polynomial using bisection method—successive averages contract the interval.

The proof of the Banach theorem may also be obtained iteratively. Also, by using the Picard-Lindelöf existence and uniqueness theorem, it can be proved in a very elegant way. The contraction mapping theorem is very important in studying the stability of nonlinear systems, as well as designing the controllers, as we see in the next two chapters.

3

Nonlinear Stability Analysis

The theory of linear systems is literally linear, though elaborate. Largely the structure is generic with weighted exponentials in the response and excellent graphical tools, as we have seen in Chapter 1. But, the theory of nonlinear systems is not so. The reader would have already gotten the feel of the nonlinearity in the previous chapter; we have worries about the existence of solutions, and we certainly have a heavier dose of mathematics.

The purpose of this chapter is not to scare the reader; in the previous chapter, we provided a glimpse of nonlinear systems, which we will build upon further in this chapter. We will keep this chapter's expository on the analysis part of the nonlinear systems, then take up the design of controllers in the next chapter. In analysis, we need to be extra cautious on the fundamental idea of stability, which was fairly simple in the linear case. We begin this chapter with a classical graphical idea of *looking* at the solutions on a graph. For certain continuity sake, as well as progressively stretching the imagination of the reader, we begin with presenting the ideas using linear systems, then seamlessly move on to nonlinear systems. Later in the chapter, we move on to more analytical techniques.

3.1 Phase Plane Techniques

In this section, we first analyze autonomous, i.e. homogeneous, linear systems of differential equations

$$\dot{x} = Ax, \tag{3.1}$$

with the nonlinear machinery. Accordingly, we restrict ourselves to two dimensions, i.e.

$$\dot{x}_1 = ax_1 + bx_2$$

$$\dot{x}_2 = cx_1 + dx_2 \tag{3.2}$$

Solutions $x_1(t), x_2(t)$ were previously each graphed as functions of time. However, we now introduce the $x_1 - x_2$ plane. Each value of t corresponds to a point (x_1, x_2) in the plane. A solution of the differential equation $x_1(t), x_2(t)$ satisfying a given initial condition now traces out a curve in the $x_1 - x_2$ plane.

Definitions:

This parameterized curve, along with an indication of the direction the solution moves along the curve as time t increases, is called a **trajectory**.

The set of trajectories (corresponding to all initial conditions) in the $x_1 - x_2$-plane, together with an indication of the solutions direction as time increases, is called the **phase plane**.

Usually, only a few representative solutions are drawn. This sketch is sometimes called a phase portrait.

Example 3.1.1:

Consider the simple linear system

$$\dot{x}_1 = x_2$$

$$\dot{x}_2 = -x_1$$

The reader would be tempted to obtain the A matrix:

$$A = \begin{bmatrix} 0 & 1 \\ -1 & 0 \end{bmatrix}$$

and the eigenvalues are purely imaginary, resulting in an oscillatory solution for each of the variables x_1 and x_2.

We will present the same in a newer framework. First, the particular solution associated with the initial conditions $x_1(0) = 1$ and $x_2(0) = 0$ may be found to be

$$x_1(t) = \cos t \quad \text{and} \quad x_2(t) = \sin t$$

in terms of elementary trigonometric functions.

Next, we eliminate t with the simple observation that

$$x_1^2 + x_2^2 = 1$$

which is just a unit circle, with center at the origin of the $x_1 - x_2$-plane—the phase plane.

Next, a little more working allows us to see that the direction of the trajectory, the solution that moves along the curve, as time t increases, is clockwise.

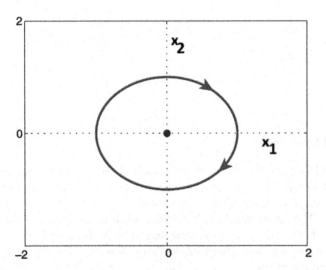

FIGURE 3.1
Unit circle in the phase plane of Example 3.1.1.

All these details are in Figure 3.1. As we move to nonlinear systems in a short while, we will have further discussion on the direction of the trajectories in the phase plane.

Today, it is quite common for phase planes to be readily determined and vividly displayed on computers using software such as MATLAB©. These programs solve the system of differential equations numerically, subject to given initial conditions. It is then easy for the program to take the numerical time-dependent solutions for $x_1(t)$ and $x_2(t)$ and use them directly to graph the phase plane. The time-dependent solutions of the differential equation are the parametric description of the curve shown on the screen, which we call the phase plane. Here, we are not particularly interested in the numerical method used by the program. We recommend a host of available, of course much longer, books on differential equations or specialized books on numerical methods for ordinary differential equations.

To bring in the technicalities of Chapter 2, we notice that the origin $x_1(t) = 0$, $x_2(t) = 0$ is a constant, or equilibrium solution (does not depend on t), of any linear (autonomous) system (3.1). Solutions of the linear system whose initial conditions are at the origin stay at the origin. The phase plane representation for the equilibrium solution is just a point that does not move, which is the equilibrium at the origin. For solutions of the system of differential equations that are not equilibria, the solutions will move in time. Then the trajectory in the phase plane will be curves.

If all solutions of the linear system stay near the equilibrium for all initial conditions near the equilibrium, then we say the equilibrium is stable. If there is *at least* one initial condition for which the solution goes away from the equilibrium, then we say the equilibrium is unstable. We will now spend some time determining conditions for

which the equilibrium (the origin) for the linear system is stable or unstable. In passing, we note that one of the important considerations of autonomous systems is that $\dot{x}$ does not depend explicitly on t. Thus, if two trajectories were to intersect, then at that point there would be two solutions that satisfy the same initial condition, which would violate the uniqueness of solutions. Thus, trajectories cannot intersect other trajectories or cross themselves. This fact will be used repeatedly in what follows.

Example 3.1.2: [10]

For the autonomous system given by:

$$\dot{x} = \begin{bmatrix} -3 & 0 \\ 0 & -1 \end{bmatrix} \begin{bmatrix} x_1 \\ x_2 \end{bmatrix}$$

a. We readily have that $x_1 = x_2 = 0$ is the *only* equilibrium.
b. Figure 3.2 gives an intuitive sketch of the direction field. Since $dx_1/dt = -3x_1$, to the right of the origin, where $x_1 > 0$, the trajectories satisfy $dx_1/dt < 0$, so $x_1(t)$ decreases as time increases and the solution flows to the left. A similar argument explains that, to the left of the origin, the trajectories flow to the right.

Looking at dx_2/dt, we would similarly understand that the trajectories flow downward above the origin and flow upward below the origin.

Thus, we see that *all* the solutions *converge* to the equilibrium at the origin $(0,0)$. Such an equilibrium is called **asymptotically stable**; asymptotic because it takes a while, i.e. $t \to \infty$. Of course, for linear systems, as long as the eigenvalues have

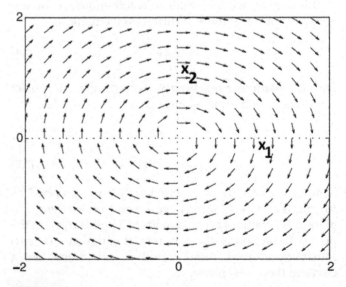

FIGURE 3.2
The direction field in Example 3.1.2.

negative real parts, stability is always asymptotic stability; moreover, the origin is the only stable equilibrium for all stable linear autonomous systems, irrespective of the initial conditions. It turns out very soon that this is the argument where we substantially differ in the case of nonlinear systems.

c. Unlike the previous example where we could rather easily eliminate t, here we may easily understand the nature of the trajectories in the phase plane using the explicit solution of the system:

$$x_1(t) = c_1 e^{-3t}, \quad x_2 = c_2 e^{-t}$$

If $c_2 = 0$, the solution in the phase plane is simple, namely, the line $x_2 = 0$. Since $x_1(t)$ is a decaying exponential, with a time-constant $1/3$, the trajectories along $x_2 = 0$ approach the origin swiftly. More precisely, the trajectories correspond to two rays—one with positive x_1 and one with negative x_1—approaching the origin.

Similarly, the solution corresponding to $c_1 = 0$ may be seen as two rays approaching the equilibrium, in opposite directions, along $x_1 = 0$. Of course, these trajectories move slower, owing to the smaller time-constant, compared to the earlier pair. These are the four straight rays sketched in Figure 3.3. If both $c_1 \neq 0$ and $c_2 \neq 0$, then the picture is more complicated with a cubic polynomial: $\alpha x_1 + \beta x_2^3 = 0$. Nevertheless, all the solutions appropach the equilibrium, asymptotically. This kind of equilibrium is an example of a **stable node**. Naturally, the reader would ask if there exists an unstable node, which we will present next.

Example 3.1.3:

For the following autonomous system,

$$\dot{x} = \begin{bmatrix} 3 & 0 \\ 0 & 1 \end{bmatrix} \begin{bmatrix} x_1 \\ x_2 \end{bmatrix}$$

It is self-explanatory that the trajectories are the same as before, but their directions are reversed in time. Now *all* solutions flow away from the equilibrium, once again, $(0,0)$; the equilibrium is unstable, since the eigenvalues have positive real parts. This kind of equilibrium is called an **unstable node**. All solutions (except the equilibrium itself) fly away, asymptotically towards infinity, from the origin as time increases, as shown in Figure 3.4.

Example 3.1.4:

If the system is

$$\dot{x} = \begin{bmatrix} -3 & 0 \\ 0 & 1 \end{bmatrix} \begin{bmatrix} x_1 \\ x_2 \end{bmatrix}$$

The equilibrium is again the origin, and as can be expected with the experience from the previous examples, the trajectories along $x_2 = 0$ approach the origin, while those along $x_1 = 0$ fly away from the origin. This is shown in Figure 3.5. This kind of equilibrium is an example of a **saddle point**.

We are now adequately prepared for more general definitions of stability, from the point of view of equilibrium points. An equilibrium is defined to be stable if all initial conditions near the equilibrium stay near the equilibrium. Likewise, we have an unstable equilibrium if all initial conditions near the equilibrium go away from the equilibrium. These two cases are depicted as

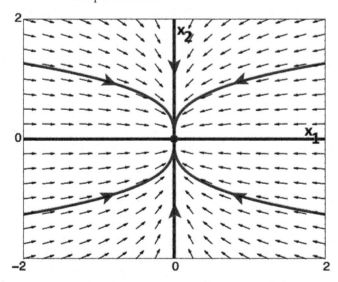

FIGURE 3.3
A stable node.

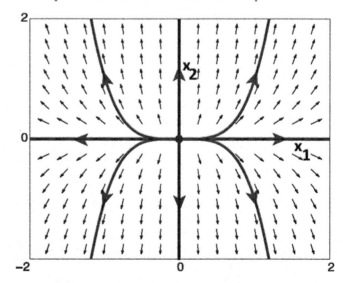

FIGURE 3.4
An unstable node.

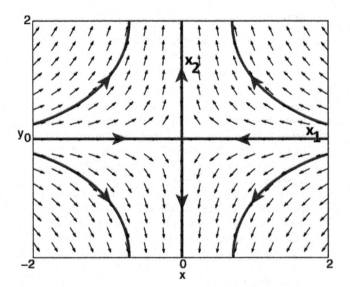

FIGURE 3.5
A saddle point.

stable node and unstable node, respectively, in the phase plane. An equilibrium is a saddle point if some initial conditions near the equilibrium stay closer and others go away. It is customary to treat saddle points as unstable equilibria.

Recall from the first chapter that the solutions of linear systems of differential equations are linked to the nature of the eigenvalues λ_1 and λ_2 of the coefficient matrix A. We will see two more examples, and then we will summarize the results.

Example 3.1.5:

If the system is

$$\dot{x} = \begin{bmatrix} -7 & 6 \\ 6 & 2 \end{bmatrix} \begin{bmatrix} x_1 \\ x_2 \end{bmatrix}$$

Some effort in solving a quadratic characteristic equation $|\lambda I - A| = 0$ gives us the eigenvalues: 5 and −10.

Owing to the opposite signs, the origin is an (unstable) saddle point. We encourage the reader to obtain the solutions $x_1(t)$ and $x_2(t)$ and sketch an indicative phase plane.

Example 3.1.6:

If the system is

$$\dot{x} = \begin{bmatrix} 2 & 1 \\ -1 & 2 \end{bmatrix} \begin{bmatrix} x_1 \\ x_2 \end{bmatrix}$$

the eigenvalues are complex conjugates, $2 \pm j1$, with positive real parts. This is the opposite to our most familiar underdamped case where the sinusoid is progressively damped. Hence, we end up with a progressively exponentially growing sinusoids

for the variables $x_1(t)$ and $x_2(t)$. Eliminating t, we will get an outward growing spiral starting from the initial condition. To determine the direction, we just take one simple non-zero point, say $x_1 = 1$ and $x_2 = 0$, in the plane, in which case the tangent vector is

$$\left[\frac{dx_1}{dt}, \frac{dx_2}{dt} \right]^T = [2, -1]^T$$

which points down and to the right from $(1,0)$. Thus, the *unstable* spiral is clockwise as in Figure 3.6.

In the general case, the spirals may be distorted, but the stability criteria persists. When the eigenvalues $\lambda_i = \alpha \pm j\beta$ are complex numbers, the origin is an equilibrium point that is unstable when $\alpha > 0$ and asymptotically stable when $\alpha < 0$. These equilibria are called foci—**unstable focus** (as in this example) and the **stable focus** (as in the next example), respectively.

Example 3.1.7:

If the system is

$$\dot{x} = \begin{bmatrix} -1 & 4 \\ -4 & -1 \end{bmatrix} \begin{bmatrix} x_1 \\ x_2 \end{bmatrix}$$

the eigenvalues are complex conjugates: $-1 \pm j4$. The result is a stable focus as shown in Figure 3.7.

The reader would now identify that our very first example of this chapter is a special case of this—the eigenvalues are purely imaginary; hence, the phase plane has a well-rounded ellipse, the circle. In general, the phase plane consists of a family of ellipses:

$$x_1^2 + \frac{x_2^2}{\omega^2} = A^2 \tag{3.3}$$

as shown in Figure 3.8.

The trajectories move clockwise since, for example, at $x_1 = 0$, $x_2 = 1$ we see that $dx_1/dt = 1$ so

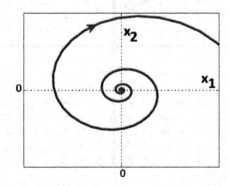

FIGURE 3.6
An unstable focus.

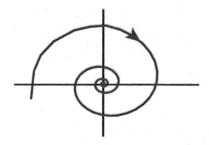

FIGURE 3.7
A stable focus.

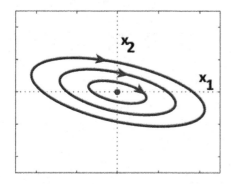

FIGURE 3.8
A stable center.

that x_1 is increasing in time there. The solutions orbit periodically (cyclically) through the same points with the same velocities. We call the equilibrium $(0,0)$ a **center**. It is customary to take centers as stable.

Theorem:

We now summarize the phase plane behavior of a linear system of differential equations and the stability of the equilibrium at the origin:

1. Two positive eigenvalues: unstable node
2. Two negative eigenvalues: stable node
3. One positive and one negative eigenvalue: unstable saddle point
4. Complex eigenvalues (positive real part): unstable focus
5. Complex eigenvalues (negative real part): stable focus
6. Complex eigenvalues (zero real part): stable center

This summary is depicted in Figure 3.9 [10]:

Since we are dealing exclusively with second-order systems, and hence the two-dimensional phase planes, the axes in Figure 3.9 can be seen to be the trace of A, i.e. sum of the eigenvalues, on the horizontal axis, and the determinant, i.e. the product of eigenvalues, on the vertical axis. For instance, when the eigenvalues have

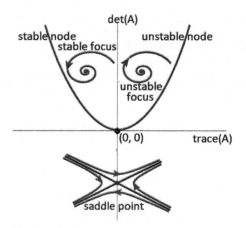

FIGURE 3.9
Summary of phase plane behavior of linear systems.

opposite signs, the determinant is always negative and hence the saddle point is shown in the lower half of the picture; the reader may now easily understand the other stability criteria in the upper half of the picture.

3.1.1 Equilibria of Nonlinear Systems

In this part, we study the nonlinear autonomous system of differential equations:

$$\dot{x}_1 = f(x_1, x_2)$$

$$\dot{x}_2 = g(x_1, x_2) \qquad (3.4)$$

We realize that the equilibrium points satisfy the *pair* of equations:

$$f(x_{1e}, x_{2e}) = 0 \quad \text{and} \quad g(x_{1e}, x_{2e}) = 0$$

where x_{ie} are constants.

Example 3.1.7a:

Let

$$\dot{x}_1 = -x_1 + x_1 x_2$$

$$\dot{x}_2 = -4x_2 + 8x_1 x_2$$

so that we have

$$-x_{1e} + x_{1e} x_{2e} = 0$$

$$-4x_{2e} + 8x_{1e} x_{2e} = 0$$

Notice that these are not a linear pair of equations; in general, they are indeed nonlinear algebraic equations, and we need to use them to determine the equilibria (yes, there is a possibility of multiple solutions to these equations) for a given system of differential equations. As long as we do not go beyond second-order systems, these equations may hopefully be solved.

The first equation implies $x_{1e} = 0$ or $x_{2e} = 1$. Then, using the second equation, we have *two* equilibria:

$$(0,0) \quad \text{and} \quad \left(\frac{1}{2}, 1\right)$$

Once an equilibrium is found, we will develop here some simple ideas that determine whether the equilibrium is stable or unstable. Recall from the previous chapter that we have been interested in determining the behavior of the solutions *near* an equilibrium, with an emphasis on the *neighborhood* of the equilibrium. Suppose that $x_1(t) = x_{1e}, x_2(t) = x_{2e}$ is an equilibrium. The time-tested way of looking at the neighborhood is to make use of a Taylor series, the two-dimensional version of Taylor's approximations learned in calculus, and we get

$$f(x_1, x_2) \approx f(x_{1e}, x_{2e}) + \frac{\partial f}{\partial x_1}\Big|_{(x_{1e}, x_{2e})}(x_1 - x_{1e}) + \frac{\partial f}{\partial x_2}\Big|_{(x_{1e}, x_{2e})}(x_2 - x_{2e})$$

$$g(x_1, x_2) \approx g(x_{1e}, x_{2e}) + \frac{\partial g}{\partial x_1}\Big|_{(x_{1e}, x_{2e})}(x_1 - x_{1e}) + \frac{\partial g}{\partial x_2}\Big|_{(x_{1e}, x_{2e})}(x_2 - x_{2e})$$

$$(3.5)$$

which is only valid as an approximation near the equilibrium. This provides the linearization of a function of two variables. The approximation uses only the linear or first-order terms. Further, since (x_{1e}, x_{2e}) is an equilibrium, it suggests that, near the equilibrium, the solutions of the original nonlinear system of equations can be approximated and resemble those of the following *linearized system*:

$$\dot{z}_1 = \frac{\partial f}{\partial x_1}\Big|_{(x_{1e}, x_{2e})} z_1 + \frac{\partial f}{\partial x_2}\Big|_{(x_{1e}, x_{2e})} z_2$$

$$\dot{z}_2 = \frac{\partial g}{\partial x_1}\Big|_{(x_{1e}, x_{2e})} z_1 + \frac{\partial g}{\partial x_2}\Big|_{(x_{1e}, x_{2e})} z_2 \qquad (3.6)$$

where we have introduced the displacement from the equilibrium:

$$z_1 = x_1 - x_{1e}, \quad z_2 = x_2 - x_{2e}$$

which translates the equilibrium point (x_{1e}, x_{2e}) to the origin. Thus, we end up with

$$\dot{z} = \begin{bmatrix} \dfrac{\partial f}{\partial x_1} & \dfrac{\partial f}{\partial x_2} \\[2ex] \dfrac{\partial g}{\partial x_1} & \dfrac{\partial g}{\partial x_2} \end{bmatrix} z \qquad (3.7)$$

a linear homogeneous system like that discussed earlier in this chapter. We appreciate the matrix notation and define the matrix of first derivatives as the **Jacobian matrix**. This matrix must be evaluated at the equilibrium $x_1 = x_{1e}, x_2 = x_{2e}$.

With this, the phase plane that we discussed earlier with many examples can be extended for nonlinear systems. Some procedures for the phase plane of nonlinear systems will be discussed shortly.

Near an equilibrium, we would expect the phase plane for the nonlinear system to be, more or less, identical to the phase plane of the corresponding linearized system. There are two exceptions, however. One is the case when the eigenvalues of the Jacobian matrix are purely imaginary, and the second is the case when the eigenvalues are complex conjugates with the real part positive. In these two cases, we need to look into the otherwise neglected higher order terms in the Taylor series. We will not explore this in this book.

We recommend the reader to go back a few pages and refer to the theorem and Figure 3.9.

Example 3.1.7b:

Recall that the equilibria are

$$(0,0) \quad \text{and} \quad \left(\frac{1}{2}, 1\right)$$

The Jacobian is

$$\begin{bmatrix} \dfrac{\partial f}{\partial x_1} & \dfrac{\partial f}{\partial x_2} \\[2ex] \dfrac{\partial g}{\partial x_1} & \dfrac{\partial g}{\partial x_2} \end{bmatrix} = \begin{bmatrix} -1 + x_2 & x_1 \\ 8x_2 & -4 + 8x_1 \end{bmatrix}$$

Near the equilibrium $(0,0)$ the matrix is

$$A = \begin{bmatrix} -1 & 0 \\ 0 & -4 \end{bmatrix}$$

Since both the eigenvalues are negative, the equilibrium at origin is stable, and we get a stable node in the phase plane.

For the equilibrium $(1/2, 1)$, the linearization involves the displacement from the equilibrium, so that $z_1 = x_1 - 1/2$ and $z_2 = x_2 - 1$. However, we need not worry about this. We may directly substitute the equilibrium point in the Jacobian to get the eigenvalues to be ± 2, and hence, the equilibrium is a saddle point, an unstable one.

It is important to keep in mind that the theorem describes only the behavior near each equilibrium, as shown in Figure 3.10

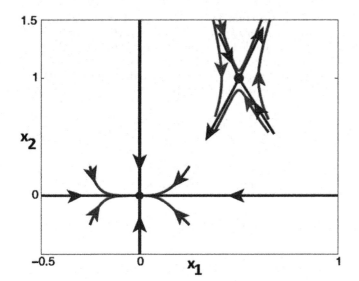

FIGURE 3.10
Phase plane for Example 3.1.7b.

An inquisitive reader might ask if there would be any purpose served by computing the eigenvectors; indeed, it is helpful to compute the eigenvectors to determine the directions as well as the pieces of the trajectories in the phase plane away from the equilibria.

Let us summarize the method for analyzing nonlinear systems:

1. Find the equilibria by solving the non-linear algebraic equations by putting

$$f(.,.) = 0 \quad \text{and} \quad g(.,.) = 0.$$

2. Linearize in the neighborhood of each equilibrium by obtaining the Jacobian matrix.
3. Find the eigenvalues, for each linearization.
4. Roughly sketch the phase plane for the nonlinear system near each equilibrium.
5. Compute the eigenvectors for the real eigenvalues to help determine trajectories in the phase plane away from equilibrium.

More Examples

One of the exciting applications includes developing population models for interacting species [10]; the species can be chemical, economic, or even biological.

Assume that there are two species whose populations at time t are given by $x_1(t), x_2(t)$. Often, it is the growth rate of each population that depends on the other, so that we have

$$\dot{x}_1 = f(x_1, x_2) = x_1 u(x_1, x_2)$$

$$\dot{x}_2 = g(x_1, x_2) = x_2 v(x_1, x_2) \tag{3.8}$$

where the functions $u(.,.)$ and $v(.,.)$ are the growth rates of the species x_1 and x_2, respectively. In passing, we make a remark that merely writing the model as above has ruled out other effects, such as seasonal variations, delay effects, and fluctuating food supply. Nevertheless, models such as these indeed give us an insight into the working of the populations. In all our examples, we will assume $x_1 \geq 0$ and $x_2 \geq 0$.

Example 3.1.8:

In this example, we consider two sets of species competing for the same limited food and shelter. To make a simpler assumption, we assume that the growth rates are

$$u(x_1, x_2) = a - bx_2 - cx_1, \quad \text{and} \quad v(x_1, x_2) = q - rx_1 - sx_2$$

where a, b, c, q, r, s are positive constants, and a and q represent the growth rates if there were no competition. Here, we assume that the growth rates of both species diminishes as either species gets larger due to finite resources for both. These assumptions lead to the nonlinear system

$$\dot{x}_1 = ax_1 - bx_1x_2 - cx_1^2$$

$$\dot{x}_2 = qx_2 - rx_1x_2 - sx_2^2$$

We note that the equilibria are given by

$$x_1 = 0 \quad \text{or} \quad a - bx_2 - cx_1 = 0$$

and

$$x_2 = 0 \quad \text{or} \quad rq - rx_1 - sx_2 = 0$$

so that, with some effort, the four equilibria are

$$(0,0), \left(0, \frac{q}{s}\right), \left(\frac{a}{c}, 0\right), (\alpha, \beta)$$

where α and β are the solutions of

$$cx + by = a, \quad \text{and} \quad rx + sy = q$$

We assume that $cs - br \neq 0$, so that the point (α, β) exists and is unique.

We note that $(0,0)$ is zero population of both species and the next two are the environment's carrying capacity for each species. The fourth equilibrium only corresponds to non-negative populations if both $\alpha \geq 0$ and $\beta \geq 0$.

Suppose $b = q = r = s = 1, c = 3$, and $a = 2$ so that the equilibria are

$$(0,0), (0,1), \left(\frac{2}{3}, 0\right), \left(\frac{1}{2}, \frac{1}{2}\right)$$

We first obtain the Jacobian matrix in general for this example:

$$A = \begin{bmatrix} 2 - x_2 - 6x_1 & -x_1 \\ -x_2 & 1 - x_1 - 2x_2 \end{bmatrix}$$

We then evaluate this matrix at each of the 4 equilibria to get

1. At (0,0), the eigenvalues turn out to be +2 and +1, and the resulting phase plane is an unstable node.
2. At (0,1), the eigenvalues are ±1; hence, there is a saddle point at this equilibrium.
3. At (2/3,0), the eigenvalues of A are −2 and 1/3; hence, there is another saddle point in the phae plane.
4. Finally, at (1/2, 1/2), the eigenvalues are $-1 \pm 1/\sqrt{2}$; hence, the equilibrium point is a stable node.

In Figure 3.11, we present the combined phase planes in the neighborhood of each of the four equilibria.

Other competing species models are considered in the exercises.

Example 3.1.9: Predator-Prey Models

This is yet another interesting application, where we have a prey species x_1, such as rabbits, that is eaten by a predator species x_2, such as lions. If there are no predators, we assume the prey has sufficient sources of food. On the other hand, if there are no prey, we assume the predators would die off. In this situation, the simplest assumptions concerning the growth rates are

$$u(x_1, x_2) = a - bx_2 - cx_1, \quad \text{and} \quad v(x_1, x_2) = -q + rx_1 - sx_2$$

We suggest the reader compare these growth rates to those of the previous example.

Here a,b,c,q,r,s are positive constants, a is the growth rate of prey without predators, and q is the death rate of predators without prey.

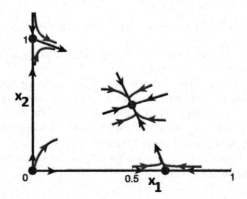

FIGURE 3.11
Phase plane for Example 3.1.8.

The interaction terms are such that increasing the predator x_2 diminishes the growth rate of the prey x_1, while increasing the prey x_1 increases the growth rate of the predator x_2. The growth rates of both species diminish as either species get large due to finite resources for both:

$$\dot{x}_1 = ax_1 - bx_1x_2 - cx_1^2$$

$$\dot{x}_2 = -qx_2 + rx_1x_2 - sx_2^2$$

The four equilibria are given by

$$(0,0), \ \left(0, -\frac{q}{s}\right), \ \left(\frac{a}{c}, 0\right), \ (\alpha, \beta)$$

Here (α, β) is the solution of

$$cx + by = a, \quad \text{and} \quad rx - sy = q$$

and since all the coefficients are assumed to be positive, this equilibrium exists and is unique.

We may readily notice that the equilibrium $(-q/s, 0)$ is not physical and do not consider it anymore. Rest of the equilibria are meaningful.

Let us now assume that $b = q = r = s = 1$, $c = 2$, and $a = 3$, so that the equilibria are

$$(0,0), (0,-1), \left(\frac{3}{2}, 0\right), \left(\frac{4}{3}, \frac{1}{3}\right)$$

Of these, clearly $(0, -1)$ is meaningless.

The Jacobian matrix for this example may be obtained as:

$$A = \begin{bmatrix} 3 - x_2 - 4x_1 & -x_1 \\ x_2 & -1 + x_1 - 2x_2 \end{bmatrix}$$

The reader may verify that

1. The equilibrium at the origin is a saddle point,
2. The equilibrium (3/2, 0) is also a saddle point, but
3. The equilibrium (4/3, 1/3) is a stable node.

While the development of the phase plane could be quite laborious, we show three different ways of development for this example. Figure 3.12 is the raw phase plane based on the linearizations, which are in the immediate neighborhood of the equilibria.

Figure 3.13 shows further development making use of the eigenvectors. For example, for the equilibrium (3/2, 0) the eigenvalues are 1/2 and −3. Corresponding to the eigenvalue 1/2, the eigenvector is $[-3 \ 7]^T$ that has a negative slope, and the direction is away from the equilibrium since the

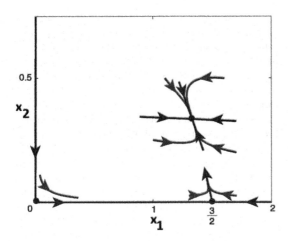

FIGURE 3.12
Coarse phase plane for Example 3.1.9.

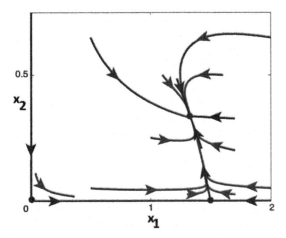

FIGURE 3.13
Finer phase plane for Example 3.1.9.

response has the term $e^{0.5t}$. The eigenvector corresponding to the eigenvalue -3 is the x_1-direction towards the equilibrium point since the response has the term e^{-3t}.

Figure 3.14 shows the phase plane where most initial conditions approach the stable node $(4/3, 1/3)$

3.2 Poincare-Bendixson Theorem

In Chapter 2, using Example 2.2, we became aware of limit cycles; the treatment there was qualitative and very rudimentary. In this part, we will elaborate the same van der Pol's equation and learn about the limit cycles more quantitatively with the background of phase planes.

We will first look at the following system to gain a hands-on experience in the computations.

Example 3.2.1:

Let

$$\dot{x}_1 = x_2 - x_1\left(x_1^2 + x_2^2 - 1\right)$$

$$\dot{x}_2 = -x_1 - x_2\left(x_1^2 + x_2^2 - 1\right)$$

An immediate observation is that there is an equilibrium point at the origin. A little contemplation gives us four more equilibria: $(\pm 1, 0)$ and $(0, \pm 1)$, which are located on the unit circle (centered at the origin). Owing to this observation, let us introduce polar coordinates

$$r^2 = x_1^2 + x_2^2 \quad \text{and} \quad \tan\theta = \frac{x_2}{x_1}$$

so that the differential equations become

$$\dot{r} = -r\left(r^2 - 1\right) \quad \text{and} \quad \dot{\theta} = -1$$

If we have an initial condition on the unit circle, then $\dot{r} = 0$, i.e. the state will circle around the origin with a period of $1/2\pi$. When $r < 1$, then $\dot{r} > 1$, and when $r > 1$, then $\dot{r} < 1$, which means that the state—either from inside or from outside of the unit circle—always tends toward the unit circle.

One may go ahead to obtain the Jacobian and evaluate it at the equilibrium points to confirm that (i) the origin is an unstable node and that (ii) the other 4 are stable nodes.

The foregoing argument about the system's behavior may also be confirmed by obtaining the analytical solution:

$$r(t) = \frac{1}{\left(1 + ce^{-2t}\right)^{1/2}} \quad \text{and} \quad \theta(t) = \theta_0 - 1$$

where

$$c = \frac{1}{r_0^2} - 1$$

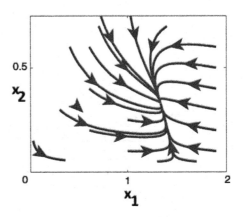

FIGURE 3.14
Complete phase plane for Example 3.1.9.

Observe that wherever the initial r_0 is, $r(t) \rightarrow 1$ asymptotically.

Example 3.2.2:

Now, recall that the van der Pol equation is given as a second-order nonlinear differential equation as

$$m\ddot{x} + 2c(x^2 - 1)\dot{x} + kx = 0$$

We will quickly bring it to our standard form as follows:

$$\dot{x}_1 = x_2$$

$$\dot{x}_2 = -x_1 - \mu(x_1^2 - 1)x_2 \qquad (3.9)$$

where, for simplicity we have taken $m = k = 1$ and $\mu = 2c$. Readily we get an equilibrium point as the origin. Further, the Jacobian is

$$\begin{bmatrix} 0 & 1 \\ -1 & -\mu(x_1^2 - 1) \end{bmatrix}$$

and we notice that the eigenvalues have a positive (non-zero) real part at the origin. Thus, the equilibrium point is an unstable node.

Let us say our initial condition is the origin itself. Since it is an unstable node, the trajectory drifts away from the origin, i.e. either $|x_1|$ or $|x_2|$ or both increase. Consequently, we may expect that at some time $|x_1|$ becomes 1, and hence the eigenvalues of the Jacobian at $(\pm 1, x_2)$ become purely imaginary resulting in an elliptical trajectory; likewise, when $|x_2|$ exceeds 1, the trajectories tend to decay towards the elliptical trajectory. In Figure 3.1.5, we give the phase plane of van der Pol system.

From the foregoing discussion one may ask: What is this unit circle or elliptical trajectory? When does it exist? What are its implications?

The closed curve in the phase portrait is where trajectories inside the curve and those outside the

curve all tend to this curve, while a motion started on this curve will stay on it forever, circling periodically around the origin. This an instance of the so-called "limit cycle" phenomenon. As we have observed earlier, these limit cycles are unique features of nonlinear systems. They are, in fact, isolated closed curves in the phase plane. The trajectory has to be both closed, indicating the periodic nature of the motion, and isolated, indicating the limiting nature of the cycle (with nearby trajectories converging or diverging from it). In other words, there may be several closed curves in the phase portraits of systems, such as the mass-spring-damper, but these are not considered limit cycles unless they are isolated.

Depending on the motion patterns of the trajectories in the vicinity of the limit cycle, one can distinguish three kinds of limit cycles:

1. Stable Limit Cycles: all trajectories in the vicinity of the limit cycle converge to it as $t \rightarrow \infty$
2. Unstable Limit Cycles: all trajectories in the vicinity of the limit cycle diverge to it as $t \rightarrow \infty$
3. Semi-Stable Limit Cycles: some of the trajectories in the vicinity converge to it, while the others diverge from it as $t \rightarrow \infty$

Figure 3.16 [7] depicts the three types of limit cycles.

We would then readily accept that the limit cycle of the van der Pol equation is clearly stable, as well as the limit cycle of Example 3.2.1.

3.2.1 Existence of Limit Cycles

As control engineers, we need to be able to predict the existence of limit cycles in control systems. To this end, we now present three classical theorems.

Poincaré's Index Theorem:

If a limit cycle exists in the second-order autonomous system, then

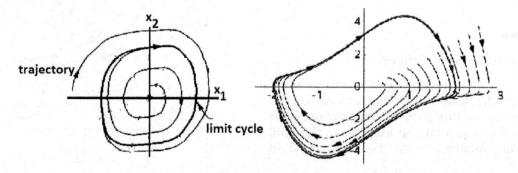

FIGURE 3.15

Phase plane for the van der Pol system. (From Slotine, J.J., and Li, W., *Applied Nonlinear Control*, Englewood Cliffs, NJ, Prentice Hall, 1991.)

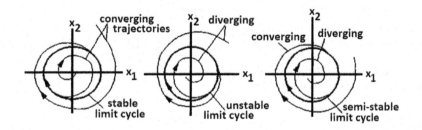

FIGURE 3.16
An illustration of the stable, unstable, and semi-stable limit cycles. (From Slotine, J.J., and Li, W., *Applied Nonlinear Control*, Englewood Cliffs, NJ, Prentice Hall, 1991.)

$$N = S + 1 \qquad (3.10)$$

where N represents the number of nodes, centers, and foci (stable or unstable) enclosed by a limit cycle, and S the number of enclosed saddle points.

The proof of this theorem is too technical and is beyond the scope of this book. However, one might reflect a bit and observe a certain similarity between this result and the Nyquist's criterion for linear system stability.

As an immediate consequence, one might observe that a limit cycle must enclose at least one equilibrium point. Indeed, the limit cycle in Example 3.2.1 and the limit cycle of van der Pol's equation do enclose the origin, an equilibrium point.

At times, negative statements throw more light on a result; a logical statement $A \Rightarrow B$ is equivalent to $\neg B \Rightarrow \neg A$. To this end, we find the next theorem:

Bendixson's Theorem:

In a second-order autonomous system, no limit cycle can exist in a region Ω of the phase plane in which

$$\frac{\partial f}{\partial x_1} + \frac{\partial g}{\partial x_2} \qquad (3.11)$$

does not vanish and does not change sign.

Notice there are three negations. We can afford to prove this theorem by contradiction.

Proof:

From the standard form of the differential equations, it is easy to observe that, by simple division,

$$\frac{dx_1}{dx_2} = \frac{f(x_1, x_2)}{g(x_1, x_2)}$$

from which we can write

$$f dx_2 - g dx_1 = 0 \qquad (3.12)$$

This equation must be satisfied for any system trajectories, including a limit cycle. Thus, assuming there exists a limit cycle, then along the closed curve L of the limit cycle, we have

$$\int_L f dx_2 - g dx_1 = 0 \qquad (3.13)$$

If we borrow Stokes' Theorem from vector calculus, we have

$$\int_L f dx_2 - g dx_1 = \iint \left(\frac{\partial f}{\partial x_1} + \frac{\partial g}{\partial x_2} \right) dx_1 dx_2 = 0 \qquad (3.14)$$

where the integration on the right-hand side is carried out on the area enclosed by the limit cycle.

QED

An alert reader would have noticed that the key expression of this theorem, $(\partial f / \partial x_1 + \partial f / \partial x_2)$, is no other than the trace of the Jacobian matrix. If the trace is zero, i.e. the sum of the eigenvalues is zero, then the equilibrium point is either a saddle point or a center. In particular, if both the principal diagonal elements of the Jacobian are zero, then the eigenvalues are purely imaginary and the equilibrium point is a center; recall that if the center is an isolated curve, it is a limit cycle by definition. Thus, this theorem gives us a sufficient condition for the limit cycles to exist. We may verify this on the two Examples 3.2.1 and 3.2.2. In the former case, the trace never vanishes except on the circle centered at the origin with radius $1/\sqrt{2}$. In the latter case, whenever $x_1 = \pm 1$ the diagonal entries are both 0.

Let us illustrate the result on another example.

Example 3.2.3:

Let us consider the nonlinear system

$$\dot{x}_1 = f(x_1, x_2) = f_1(x_2) + x_1 x_2^2$$

$$\dot{x}_2 = g(x_1, x_2) = g_1(x_1) + x_1^2 x_2$$

Since

$$\frac{\partial f}{\partial x_1} + \frac{\partial g}{\partial x_2} = x_1^2 + x_2^2$$

does not vanish anywhere in the phase plane, except at the origin, Bendixson's theorem guarantees that the system does not have any limit cycles.

The third theorem is concerned with the asymptotic properties of the trajectories of second-order systems.

Poincaré-Bendixson Theorem:

If a trajectory of the second-order autonomous system remains in a finite region Ω, then one of the following is true:

1. The trajectory goes to an equilibrium point
2. The trajectory tends to an asymptotically stable limit cycle
3. The trajectory is itself a limit cycle

While the proof of this theorem is also omitted here, its intuitive basis is easy to see, and can be verified on the previous phase portraits.

To summarize this part, we have presented ideas of stability using a graphical method for an obvious advantage of visual understanding of the global behavior of systems. Nevertheless, the method is mainly limited to second-order systems, and the phase portraits in greater details are obtained only by extensive computations using software. We have also demonstrated the existence of multiple equilibrium points and of limit cycles in the case of nonlinear systems. Some useful classical theorems for the prediction of limit cycles in second-order systems are also presented.

An important result concerning the trajectories of the nonlinear systems is their "looking like" those of the linearized systems. We will take a brief detour to explore this before getting back to the mainstream stability analysis.

3.3 Hartman-Grobman Theorem

We begin with a few definitions and concepts.

Definition:

A linear system is **hyperbolic** if all its eigenvalues have non-zero real parts, i.e. there are no poles on the imaginary axis.

For a nonlinear system, therefore, can we say that an equilibrium point x_e is hyperbolic if none of the eigenvalues of the Jacobian have zero real parts? The Bendixson theorem should have flashed in the mind of the reader. To what extent does the solutions of the nonlinear system look like those of the linearized system at the equilibrium point? And, what does it mean by "looking like?"

Definition:

The map $h : A \to B$ is a **homeomorphism** if it is continuous, bijective, and has a continuous inverse.

Definition:

The map $f : A \to B$ is a **diffeomorphism** if it is a C^1 bijective and has a C^1 continuous inverse.

Example 3.3.1:

The map $f : \Re \to (-1,1)$ defined by $f(x) = \tanh(x)$ is a diffeomorphism. Suppose if we apply this to the trajectory $\phi_t(x) = x_0 e^{-t}$, we get

$$\psi_t(y) = f \circ \phi \circ f^{-1} = \tanh\left(e^{-t} \tanh^{-1}(y_0)\right)$$

whose ordinary differential equation (ODE) is:

$$\dot{y} = g(y) = \frac{d}{dt}\psi_t(y)\mid_{t=0} = \left(y^2 - 1\right)\tanh^{-1}(y)$$

This ODE has only one equilibrium, $y = 0$, in the interval $(-1,1)$. Since $g'(0) = -1$ this trajectory also converges like the original trajectory $\phi_t(x)$.

Definition:

Two trajectories $\phi_t : A \to A$ and $\psi_t : B \to B$ are **conjugate**

$$if \ \exists h : A \to B : \forall t \in \Re \ and \ x \in A, h\left(\phi_t(x)\right) = \psi_t\left(h(x)\right)$$

Example 3.3.2:

The trajectory generated by $\dot{x} = -x$ is $\phi_t(x) = x_0 e^{-t}$. Under the homeomorphism $y = h(x) = x^3$, this is equivalent to the new trajectory

$$\psi_t(y) = \left(x_0 e^{-t}\right)^3 = y_0 e^{-3t}$$

Also, this is the solution of $\dot{y} = -3y$. Consequently, the two ODEs are conjugate.

Conjugacy implies that each trajectory of ϕ corresponds to a trajectory of ψ and vice-versa. Thus, the homeomorphism h provides a one-to-one correspondence between the equilibria of two conjugate trajectories.

In passing, let us note that, as an english word, conjugate means a mathematical value or entity having a reciprocal relation with another.

We now have the following important theorem.

Theorem: Hartman-Grobman

Let x_e be a hyperbolic equilibrium point of a C^1 function $f(x)$ with trajectory $\phi_t(x)$. Then there is a neighborhood N of x_e such that ϕ is conjugate to its linearization on N.

Note that this theorem demands that the equilibrium be hyperbolic. We will discuss some more interesting issues with reference to Example 3.3.2 and close this part.

Example 3.3.3:

First, the trajectories $x_0 e^{-t}$ and $y_0 e^{-3t}$ are conjugate, but x^3 is not a diffeomorphism. Hence, the two trajectories cannot be diffeomorphic.

For two trajectories to be diffeomorphic, quite obviously from the definition, the functions must have been related by the derivative of the conjugacy. Suppose that $\dot{x} = f(x)$ generates the trajectory ϕ and that $\dot{y} = g(y)$ generates ψ. Then,

$$\frac{d}{dt}\psi_t(y) = g\big(\psi_t(y)\big)$$

$$= \frac{d}{dt} h\big(\phi_t(x)\big)$$

$$= h'\big(\phi_t(x)\big)\frac{d}{dt}\phi_t(x) \quad \text{(chain rule)}$$

$$= h'\big(\phi_t(x)\big) f\big(\phi_t(x)\big) \quad (3.15)$$

Simply setting $t = 0$, we get

$$g(y) = g(h(x)) = h'(x)f(x)$$

This process would remind the reader of employing various transformations to solve, for instance, problems in integral calculus. In linear algebra, we have the indispensible similarity transformation, A and $T^{-1}AT$, which preserves the eigenvalues. It may be readily shown that the eigenvalues of the Jacobian matrix can be preserved by a diffeomorphism; refer to Example 3.3.1 again for something that looks like a similarity transformation, albeit in scalars. The problem of solving nonlinear systems of differential equations now, in a way, demands the construction of homeomorphisms and diffeomorphisms.

Example 3.3.4:

Suppose we have the nonlinear system:

$$\dot{x}_1 = x_1$$

$$\dot{x}_2 = -x_2 + x_1^2$$

The equilibrium point is the origin, and the Jacobian at $(0,0)$ is

$$\begin{bmatrix} 1 & 0 \\ 0 & -1 \end{bmatrix}$$

(We warn the reader not to get enticed by trace = 0)

It turns out that the solution of the linearized system is

$$\begin{bmatrix} x \\ y \end{bmatrix} = \begin{bmatrix} e^t & 0 \\ 0 & e^{-t} \end{bmatrix}\begin{bmatrix} x_0 \\ y_0 \end{bmatrix}$$

However, the solution of the nonlinear system (which is relatively easy to obtain analytically) is

$$\begin{bmatrix} x \\ y \end{bmatrix} = \begin{bmatrix} e^t x_0 \\ e^{-t}y_0 + \frac{1}{3}\big(e^{2t} - e^{-t}\big)x_0^2 \end{bmatrix}$$

and the homeomorphism, which is not so intuitive in this case, is

$$H(x_1, x_2) = \big(x_1, x_2 - x_1^2/3\big)$$

What this result says is that the solutions of the llinearized and the original nonlinear ODEs are conjugate with this homeomorphism. As one would have correctly guessed, construction of homeomorphisms and diffeomorphisms is tricky. We will get a better handle on this for a class of nonlinear ODEs in Chapter 4 where we propose to control the nonlinear systems using a technique called feedback linearization.

Lastly, in this example, we leave it to the reader to verify that the nonlinear system has a saddle point at the origin and that there is no limit cycle (despite the trace of the Jacobian being zero).

The rest of this chapter is devoted to a deeper study of stability of nonlinear systems.

3.4 Lyapunov Stability Theory [12,13]

We have already presented the basic idea of stability of nonlinear systems—unlike linear ones, we talk about the stability of the equilibria, which are in general more than one, and we are interested if starting the system in the neighborhood of those points would stay around the point forever. Apart from a mathematical treatment, perhaps, this example would highlight the concept: As an aircraft flies, does the disturbance due to a gust cause a significant deviation in the flight path? An in-depth and general approach for the stability analysis of nonlinear control systems is the theory dating back to the late nineteenth century and contributed by the Russian mathematician Lyapunov.[1] We present this theory in this part of the chapter, and extend its implications in the following parts.

A few simplifying notations are defined at this point. Let B_R denote the spherical region, also known as ball, defined by $\|x\| < R$ in the state space, and S_R the sphere itself, defined by $\|x\| = R$.

Definition: Stability in the sense of Lyapunov

The equilibrium state[2] $x = 0$ is said to be stable if, for any $R > 0$, there exists $\varepsilon > 0$, such that

$$\|x(0)\| < \varepsilon \Rightarrow \|x(t)\| < R \quad \forall t > 0$$

Otherwise, the equilibrium point is unstable.

What this says is simple—for any initial state close (within an epsilon radius) to the equilibrium, the trajectory $x(t > 0)$ can be kept arbitrarily close to the equilibrium, not getting out of the ball of the given radius R.

Conversely, an equilibrium point is unstable if there exists at least one ball B_R, such that for every $\varepsilon > 0$, no matter how small, it is always possible for the system trajectory to start somewhere within the ball B_ε and eventually leave the ball B_R. In passing, we note an important point here. There might exist a $R' \gg R$ such that the size of the trajectory $\|x(t)\| < R'$, but since it is not bounded by the given R, the equilibrium point under investigation is unstable.

As we attempt to expand the notion of stability, we will see that, in many engineering applications, stability in the sense of Lyapunov is not enough. For example, we may want the trajectory to *eventually* get back to the equilibrium.

Definition: Asysmptotic Stability

An equilibrium point is asymptotically stable if it is (i) stable in the sesnse of Lyapunov and if, in addition, (ii) there exists some $\varepsilon > 0$ such that $\|x(0)\| < \varepsilon$ implies that $x(t) \rightarrow 0$ as $t \rightarrow \infty$.

And, one last definition to complete the picture:

---

[1] M.A. Lyapunov, "Probleme general de la stabilite du mouvement," *Annals of Mathematical Studies*, 17, Princeton University Press, 1949.
[2] Hereafter, for notational convenience we assume that the axes are shifted such that the equilibrium point under question is transformed to the origin.

Definition: Asysmptotic Stability in the Large

If asymptotic stability holds for *any* arbitrary initial state, the equilibrium point is said to be asymptotically stable in the large. It is also called globally asymptotically stable.

It may be quickly observed that if a linear system is stable, it is automatically asymptotically stable in the large.

3.4.1 Lyapunov's Direct Method

It is interesting to note that the mathematical theory of Lyapunov is based on a rather physical observation: if the total energy (be it mechanical or electrical) of a system is continuously *dissipated*, then the system, whether linear or nonlinear, must eventually settle down to an equilibrium point. Thus, we may conclude the stability of a system by examining the variation of a single scalar function (of the states x_i), the energy. All we need is to find the most suitable expression for energy, given a set of nonlinear differential equations, and then study its rate of change, with respect to time. Using this approach, we may make some reasonable inferences on the stability of the set of differential equations without using the difficult stability definitions or requiring explicit knowledge of solutions.

3.4.1.1 Positive Definite Lyapunov Functions

The so-called energy function has two properties. The first is a property of the function itself: it is strictly positive unless the state variables x_i are all simultaneously zero. For example,

$$V(x) = \frac{1}{2}\left(x_1^2 + x_2^2\right)$$

The second property is associated with the dynamics: the function is monotonically decreasing when the variables x_i vary according to the given nonlinear differential equations. Continuing with the above example,

$$\frac{d}{dt}V(x) = x_1\dot{x}_1 + x_2\dot{x}_2 = x_1 f(x_1, x_2) + x_2 f(x_1, x_2) < 0$$

In Lyapunov's direct method, the first property is formalized by the notion of positive definite functions, and the second is formalized by the so-called Lyapunov functions.

Definition: Positive Definite Functions

A scalar continuous function $V(x)$ is said to be positive definite

if $V(0) = 0$ and in a ball B_R $x \neq 0 \Rightarrow V(x) > 0$

If $V(0) = 0$ and the above property holds over the entire state-space, i.e. $R \to \infty$, then $V(x)$ is said to be globally positive definite.

Example 3.4.1:

Consider the standard pendulum, hanging from a pivot, set to oscillations. If the length is L, mass is M, the friction at the hinge is b, and the gravity is g, the nonlinear dynamics of the swing θ with respect to the vertical are:

$$ML^2\ddot{\theta} + b\dot{\theta} + MgL\sin\theta = 0$$

which can be put in state space form as

$$\dot{x}_1 = x_2$$

$$\dot{x}_2 = -\left(\frac{g}{L}\sin x_1 + \frac{b}{ML^2}x_2\right)$$

with $\theta = x_1$. The reader would readily identify that the equilibria are $(0,0)$ and $(\pi, 0)$. The mechanical energy—kinetic plus potential—of the pendulum may be obtained as

$$\frac{1}{2}ML^2\dot{x}_1^2 + MgL(1 - \cos x_1)$$

$$= \frac{1}{2}ML^2 x_2^2 + MgL(1 - \cos x_1)$$

which may be verified to satisfy the two properties of the energy function, albeit locally. If we use it as $V(x)$,

$$\dot{V}(x) = ML^2 x_2 \dot{x}_2 + MgL x_2 \cdot \sin x_1 = -bx_2^2 \leq 0$$

asserting Lyapunov's observation that the origin is stable.

A few related concepts can be defined similarly, in a local or global sense.

Definition: Negative Definite Functions

A function $V(x)$ is negative definite if $-V(x)$ is positive definite; $V(x)$ is positive semi-definite if $V(0) = 0$ and $V(x) \geq 0$ for $x \neq 0$; $V(x)$ is negative semi-definite if $-V(x)$ is positive semi-definite. The adjective *semi* reflects the possibility of $V(x)$ becoming 0 even when $x \neq 0$.

A function $V(x)$ that does not fit into any of the above categories is called **indefinite**.

As has been discussed earlier, with x denoting the state of the system, a scalar function $V(x)$ actually represents an implicit function of time t. Assuming that $V(x)$ is differentiable,

$$\dot{V}(x) = \frac{\partial V}{\partial x}\dot{x} = \sum_i \frac{\partial V}{\partial x_i}\dot{x}_i \qquad (3.16)$$

using the product rule of differentiation.

Definition: Lyapunov Function

If, in a ball B_R,

1. The function $V(x)$ is positive definite and has continuous partial derivatives, and
2. If its time derivative along any state trajectory of system is negative semi-definite, i.e.

$$\dot{V}(x) < 0$$

then $V(x)$ is said to be a Lyapunov function for the system.

3.4.1.2 Equilibrium Point Theorems

The relations between Lyapunov functions and the stability of systems are made precise in a number of theorems in Lyapunov's direct method. Such theorems usually have local and global versions. The local versions are concerned with stability properties in the neighborhood of equilibrium point and usually involve a locally positive definite function.

3.4.1.3 Lyapunov Theorem for Local Stability

If, in a ball B_R, there exists a scalar function $V(x)$ with continuous first partial derivatives such that

1. $V(x)$ is positive definite (locally in B_R)
2. $\dot{V}(x)$ is negative semi-definite (locally in B_R)

then the equilibrium point is stable. If, actually, the derivative $\dot{V}(x)$ is locally negative definite in B_R, then the stability is asymptotic.

The proof of this fundamental result is conceptually simple and is typical of many proofs in Lyapunov theory. We omit it here for the brevity of presentation.

In the analysis of a nonlinear system, we need to go through the two steps of choosing a positive definite function and determine its derivative along the path of the nonlinear systems. Example 3.4.1 was done in that spirit. Having found there that $\dot{V}(x) \leq 0$, we concluded that the pendulum is stable; we are not sure of *asymptotic* stability since $\dot{V}(x)$ is negative *semi*-definite. In fact, using physical insight, one easily sees the reason why $\dot{V}(x) < 0$, namely that the damping term absorbs energy. Actually, $\dot{V}$ is precisely the power dissipated in the pendulum.

The following example illustrates the asymptotic stability result.

Example 3.4.2:

Let us consider the nonlinear system:

$$\dot{x}_1 = x_1\left(x_1^2 + x_2^2 - 2\right) - 4x_1x_2^2$$

$$\dot{x}_2 = x_2\left(x_1^2 + x_2^2 - 2\right) + 4x_1x_2^2$$

We observe that the origin $(0,0)$ is an equilibrium point.

Let us *propose* the Lyapunov function to be

$$V(x) = \frac{1}{2}\left(x_1^2 + x_2^2\right)$$

so that its time derivative along the trajectory of the system is

$$\dot{V}(x) = x_1\dot{x}_1 + x_2\dot{x}_2 = \left(x_1^2 + x_2^2\right)\left(x_1^2 + x_2^2 - 2\right)$$

which is negative definite, in the region defined by $x_1^2 + x_2^2 < 2$, i.e. in the two-dimensional ball B_2. Thus, the origin for this system is asymptotically stable.

The above theorem is applicable to local analysis of stability. However, for the asymptotic stability in the large, we need the ball to contain the whole state-space. Not only that, the Lyapunov function has to be $V(x)$ must be radially unbounded, i.e. $V(x) \rightarrow \infty$ as $PxP \rightarrow \infty$ for its suitability.

3.4.1.4 Lyapunov Theorem for Global Stability

Assume there exists a scalar function $V(x)$ with continuous first partial derivatives such that

1. $V(x)$ is positive definite
2. $\dot{V}(x)$ is negative definite
3. $V(x) \rightarrow \infty$ as $\| x \| \rightarrow \infty$

then the equilibrium at the origin is globally asymptotically stable.

The proof is the same as in the local case, with some more work towards the radial unboundedness of V. We do not take it up here.

Example 3.4.3:

Let us consider the first-order nonlinear system:

$$\dot{x} + c(x) = 0$$

where c is any continuous function of the same sign as its scalar argument x, i.e. $xc(x) > 0$ for $x \neq 0$. We may quickly observe that, since c is continuous, it also implies that $c(0) = 0$.

Let us once again *propose* as the Lyapunov function, the square of the distance to the origin:

$$V(x) = \frac{1}{2}x^2$$

This function V is radially unbounded and its derivative is

$$\dot{V} = x\dot{x} = -xc(x)$$

And, it is negative definite as long as $x \neq 0$. Hence, $x = 0$ is a globally asymptotic equilibrium point.

To take up specific cases of the function c,

- $c(x) = x - \sin^2 x$
- $c(x) = x^3$

In the latter case, the linearization gives $\dot{x} = 0$, and we are inconclusive even about the local stability, while the original nonlinear system exhibits strong stability.

Example 3.4.4:

Let us consider the nonlinear system:

$$\dot{x}_1 = -x_1\left(x_1^2 + x_2^2\right) + x_2$$

$$\dot{x}_2 = -x_2\left(x_1^2 + x_2^2\right) - x_1$$

The origin of the state space is an equilibrium point for this system. Let us *propose* the Lyapunov function to be

$$V(x) = \frac{1}{2}\left(x_1^2 + x_2^2\right)$$

so that its time derivative along the trajectory of the system is

$$\dot{V}(x) = -\left(x_1^2 + x_2^2\right)^2 < 0$$

Hence, the origin is a globally asymptotically stable equilibrium point. Moreover, this globalness also implies that the origin is the only equilibrium point of the system.

Remarks:

1. Many Lyapunov functions may exist for the same system.
2. More importantly, for a given system, specific choices of Lyapunov functions may yield more precise results than others. Consider again the pendulum of Example 3.4.1. The function

$$V(x) = \frac{1}{2}\left(\dot{\theta}^2 + \left(\theta + \dot{\theta}\right)^2 + 2(1 - \cos\theta)\right)$$

is also a Lyapunov function for the system, because locally we may find that

$$\dot{V}(x) = -\left(\dot{\theta}^2 + \theta\sin\theta\right) \leq 0$$

However, quite interestingly, we may verify that $\dot{V}$ is actually locally negative definite, and therefore, this modified choice of $V(x)$, without any intuitive physical meaning, allows us to make a wrong conclusion.

3. Along the same lines, it is important to realize that the theorems in Lyapunov analysis are all sufficiency theorems. If for a particular choice of Lyapunov function candidate V, the conditions on $\dot{V}$ are not met, one cannot draw any conclusions on the stability or instability of the system; it only means that we should hunt for a different Lyapunov function candidate. This is the reason why we have emphasized the word "propose" in our earlier examples.

3.4.2 La Salle's Invariant Set Theorems

The above discussion might put us in a somewhat pessimistic situation—the proposed $V(x)$ may not be suitable, and even if it appears to be appropriate, $\dot{V}$ may be appearing to be negative *semi*-definite, guiding us in the wrong direction. But, a piece of good news is that we will still be able to draw conclusions on asymptotic stability, with the help of a set of invariant set theorems attributed to La Salle. The basic idea is that of an invariant set, a generalization of the concept of equilibrium point.

Definition: Invariant Set

A set G is an invariant set for a dynamic system if every system trajectory that starts from a point in G remains in G for all future time.

A little thought on this set tells us that we have a set of points, and if we move the system with an initial condition from the set, the system's state is eventually found in the same set.

Readily, any stable equilibrium point is an invariant set. The set of those points in the close neighborhood, from where the trajectories tend to move towards a stable equilibrium point, is also an invariant set. For an autonomous system, any of the trajectories in state space is an invariant set. Since limit cycles are special cases of system trajectories, i.e. closed curves in the phase plane, they are also invariant sets.

Besides often yielding conclusions on asymptotic stability when $\dot{V}$, the derivative of the Lyapunov function candidate, is only negative semi-definite, the invariant set theorems also allow us to extend the concept of the Lyapunov function so as to describe convergence to dynamic behaviors more general than equilibrium, e.g. convergence to a limit cycle.

The examples earlier have indicated to us that the decrease of a Lyapunov function V has to gradually vanish, i.e. $\dot{V}$ approaches zero. The following theorems assert this intuition.

Local Invariant Set Theorem:

Consider an autonomous system:

$$\dot{x} = f(x)$$

with f continuous, and let $V(x)$ be a scalar function with continuous first partial derivatives. Assume that

1. For some $l > 0$, the region Ω_l, defined by $V(x) < l$, is bounded
2. $\dot{V}(x) \leq 0 \;\forall x \in \Omega_l$

Let R be the set of all points within Ω_l where $\dot{V}(x) = 0$, and M be the largest invariant set in R. Then, every solution $x(t)$ originating in Ω_l tends to M as $t \to \infty$.

Before we demonstrate this in an example, we understand that the word "largest" in the sense that M is the union of all invariant sets (e.g. equilibrium points or limit cycles) within R. In particular, if the set R is itself invariant, then $M = R$. Also note that V, although often still referred to as a Lyapunov function, is not required to be positive definite.

The theorem can be proven in two steps:

1. Showing that $\dot{V}$ goes to zero, and
2. Showing that the state converges to the largest invariant set within the set defined by $\dot{V} = 0$.

The detailed proof is technically involved and is omitted here.

In fact, the asymptotic stability result in the local Lyapunov theorem can be viewed a special case of the above invariant set theorem, where the set M consists only of the origin.

Let us next demonstrate the theorem using the following two examples. The first example shows how to determine a domain of attraction, i.e. the neighborhood of a stable equilibrium point, an issue that has not been specifically addressed before. The next example shows the convergence of system trajectories to a limit cycle.

Example 3.4.5:

Consider again the system in Example 3.4.2. For $l=2$, the region Ω_2, defined by $V(x) = x_1^2 + x_2^2 < 2$, is bounded. The set R is simply the origin 0, the equilibrium point, which is an invariant set. All the conditions of the local invariant set theorem are satisfied, and therefore, any trajectory starting within the circle converges to the origin. Thus, a **domain of attraction** is explicitly determined by the invariant set theorem.

Example 3.4.6:

Consider the following system, which is different from the earlier examples.

$$\dot{x}_1 = x_2 - x_1^7\left(x_1^4 + 2x_2^2 - 10\right)$$

$$\dot{x}_2 = -x_1^3 - 3x_2^5\left(x_1^4 + 2x_2^2 - 10\right)$$

Very quickly we will recognize that the origin is an equilibrium; more about this towards the end of this example.

Let us then look at the set defined by $G(x) = x_1^4 + 2x_2^2 = 10$, essentially the common part of the dynamics. Differentiating this, we get

$$\frac{d}{dt}G(x) = 4x_1^3\dot{x}_1 + 4x_2\dot{x}_2$$

$$= -\left(4x_1^{10} + 12x_2^6\right)\left(x_1^4 + 2x_2^2 - 10\right)$$

$$= -g(x)G(x)$$

and since $\dot{G}(x)$ vanishes on the set $G(x)$, it is an invariant set. Moreover, since $G(x) = 10$ is a closed curve, it represents a limit cycle, along which the state vector moves clockwise.

Next, since G appears in the dynamics, the trajectories on this invariant set may be described by

either $\quad \dot{x}_1 = x_2 \quad$ or $\quad \dot{x}_2 = -x^3$

Next, we will ask the question: Is this limit cycle actually *attractive*, i.e. is it a stable limit cycle?

Let us *propose* the Lyapunov function

$$V(x) = \left(x_1^4 + 2x_2^2 - 10\right)^2$$

It is easy to see that, for any positive l, the region Ω_l is bounded. Using this

$$\dot{V}(x) = -8\left(x_1^{10} + 3x_2^6\right)V(x)$$

is strictly negative, except when

$$x_1^{10} + 3x_2^6 = 0 \quad or \quad x_1^4 + 2x_2^2 = 10$$

i.e. the origin and a limit cycle, both of which are invariant sets. According to the theorem, M is the union of these two sets. Now we need to investigate whether these two are stable.

Suppose we take $l = 10^2$ and look at the region Ω_{100}, such that

$$V(x) = \left(x_1^4 + 2x_2^2 - 10\right)^2 < 100$$

A simple investigation tells us that this region Ω_{100} includes every other point on the limit cycle but it *does not* include the origin. Thus, the set M is just the limit cycle. Therefore, the origin is *not* asymptotically stable as one might have expected in the beginning.

Linearization around the origin gives us $\dot{x}_1 = x_2, \dot{x}_2 = 0$, which is marginally stable, and we are mislead by linearization.

Example 3.4.6 is a representative of an application of the invariant set theorem: conclude asymptotic stability of an equilibrium point for systems with negative semi-definite V. The following corollary of the invariant set theorem is more specifically tailored to such applications.

Corollary:

Assume that in a certain neighborhood Ω of the origin,

1. $V(x)$ is locally positive definite
2. $\dot{V}$ is negative semi-definite
3. The set R defined by $\dot{V}(x) = 0$ contains no trajectories of the autonmous system other than the trivial trajectory $x = 0$

Then, the equilibrium point at the origin is asymptotically stable. Furthermore, the largest connected region of the form Ω_l (defined by $V(x) < l$) within Ω is a domain of attraction of the equilibrium point.

Indeed, the largest invariant set M in R then contains only the equilibrium point at origin. Nevertheless, we need to make a note of the following:

1. The above corollary replaces the negative definiteness condition on $\dot{V}$ in Lyapunov's local asymptotic stability theorem by a negative semi-definiteness condition on $\dot{V}$, combined with a third condition on the trajectories within R.
2. The largest connected region of the form Ω_l within Ω is a domain of attraction of the equilibrium point, but not necessarily the whole domain of attraction because the function V is not unique.
3. The set Ω itself is not necessarily a domain of attraction. Actually, the above theorem does not guarantee that Ω is invariant: some trajectories starting in Ω but outside of the largest Ω_l may actually end up outside Ω.

We now extend the local theorem to a global result to accommodate the radial unboundedness of $V(x)$.

Global Invariant Set Theorem:

Consider the autonomous system with f continuous, and let $V(x)$ be a scalar function with continuous first partial derivatives. Assume that

1. $V(x) \to \infty$ as $\| x \| \to \infty$
2. $\dot{V}(x) \leq 0$ over the whole state space

Let R be the set of all points where $\dot{V}(x) = 0$, and M be the largest invariant set in R. Then all solutions globally asymptotically converge to M as $t \to \infty$.

For instance, the above theorem shows that the limit cycle convergence in the previous example is actually global—all system trajectories converge to the limit cycle.

To demonstrate this theorem, let us look at this example, which we will revisit in the next chapter.

Example 3.4.7:

Consider a second-order system of the form

$$\ddot{x} + b(\dot{x}) + c(x) = 0$$

where b and c are continuous functions verifying the sign conditions

$$\dot{x}b(\dot{x}) > 0 \quad \text{for } \dot{x} \neq 0$$

and

$$xc(x) > 0 \quad \text{for } x \neq 0$$

A positive definite function for this system is the sum of kinetic and potential energies:

$$V(x) = \frac{1}{2}\dot{x}^2 + \int_0^x c(\tau)d\tau$$

not necessarily positive definite.

Next,

$$\dot{V} = \dot{x}\ddot{x} + c(x)\dot{x}$$

$$= -\dot{x}b(\dot{x}) - \dot{x}c(x) + c(x)\dot{x}$$

$$= -\dot{x}b(\dot{x})$$

$$\leq 0$$

Now, let us look at the dynamics from the point of view the assumptions of b and c. The term $xb(s) = 0$ only if only if $x = 0$. Therefore, $\dot{x} = 0$ means that $\dot{x} = -c(x)$, which is non-zero as long as $x \neq 0$. In other words, the only stable equilibrium is the largest invariant set; the origin ($x = 0, \dot{x} = 0$)

The local invariant set theorem indicates that the origin is a locally asymptotically stable point. Furthermore, if the integral in the proposed Lyapunov function is unbounded as $\| x \| \to \infty$, then V is a radially unbounded function and the equilibrium point at the origin is globally asymptotically stable, according to the global invariant set theorem.

We end this discussion with a very profound mathematical logic. As observed earlier, several Lyapunov functions may exist for a given system, and several associated invariant sets may be derived. The system, as everyone would now anticipate, converges to the (necessarily non-empty) intersection of the invariant sets M_i, which may give a more precise result than that obtained from any of the individual Lyapunov functions. One would also expect that the sum of Lyapunov functions for a given system, also a Lyapunov function, whose set R is the intersection of the individual sets R_i.

3.4.3 Krasovskii's Method

In the previous part, we have been, in a way, excessively talking about the proposed Lyapunov functions without actually disclosing how to get one. In this part, we will address the problem of finding Lyapunov functions for general, nonlinear systems. Krasovskii's method suggests a simple form of Lyapunov function.

Krasovskii's Theorem:

Consider the autonomous system $\dot{x} = f(x)$ with the equilibrium point of interest being the origin. Let $A(x)$ denote the Jacobian matrix of the system, i.e.

$$A(x) = \frac{\partial f}{\partial x}$$

If the symmetric matrix

$$F = A + A^T$$

is negative definite in a neighborhood Ω, then the equilibrium point at the origin is asymptotically stable. A Lyapunov function for this system is

$$V(x) = f^T(x)f(x)$$

If Ω is the entire state space along with $V(x) \to \infty$ as $\|x\| \to \infty$, then the equilibrium point is globally asymptotically stable.

A proof of this theorem is relatively simple with the following hint:

$$\dot{V}(x) = \dot{f}^T f + f^T \dot{f} = f^T F f$$

has to be negative definite, given F is negative definite.

Example 3.4.8:

Consider:

$$\dot{x}_1 = -6x_1 + 2x_2$$

$$\dot{x}_2 = 2x_1 - 6x_2 - 2x_2^3$$

from which we have

$$A = \begin{bmatrix} -6 & 2 \\ 2 & -6-6x_2^2 \end{bmatrix} \quad \text{and} \quad F = 2A$$

One would readily agree that F is negative definite over the entire state space, and the origin is asymptotically stable.

The Lyapunov function may be constructed using the theorem as:

$$V(x) = f^T(x)f(x) = \left(-6x_1 + 2x_2\right)^2 + \left(2x_1 - 6x_2 - 2x_2^3\right)^2$$

Since $V(x) \to \infty$ as $\|x\| \to \infty$, the equilibrium state at the origin is actually globally asymptotically stable.

A major drawback of this theorem, while being very attractive, is that the Jacobians of many systems do not satisfy the negative definiteness requirement, and that for systems of high order, it is difficult to check the negative definiteness of the matrix F for all x.

An immediate generalization of Krasovskii's theorem is as follows:

Generalized Krasovskii Theorem:

Consider the autonomous system, with the equilibrium point of interest being the origin, and let $A(x)$ denote the Jacobian matrix of the system. Then, a sufficient condition for the origin to be asymptotically stable is that there exist two symmetric positive definite matrices P and Q, such that $\forall x \neq 0$, the matrix

$$F(x) = A^T P + PA + Q = 0$$

is negative semi-definite in some neighborhood Ω of the origin. The function $V(x) = f^T Pf$ is then a Lyapunov function for the system. If the region Ω is the whole state space and also $V(x) \to \infty$ as $\|x\| \to \infty$, then the system is globally asymptotically stable.

Quite interestlngly, we get a similar matrix equation in A, P, and Q when we apply Lyapunov theory to linear systems.

3.4.4 The Variable Gradient Method

We have a more formal approach to the construction of Lyapunov functions, given the nonlinear dynamics, called the variable gradient method.

To start with, let us note that a scalar function $V(x)$ is related to its gradient ∇V as

$$V(x) = \int_0^x \nabla V dx, \quad \text{where} \quad \nabla V = \left[\frac{\partial f}{\partial x_1} \quad \frac{\partial f}{\partial x_2} \quad \cdots \quad \frac{\partial f}{\partial x_n}\right]^T$$

In order to recover a unique scalar function V, the gradient function has to satisfy the so-called curl conditions

$$\frac{\partial V_i}{\partial x_j} = \frac{\partial V_j}{\partial x_i} \quad i,j,=1 \cdots n$$

The principle of the variable gradient method is to assume a specific form for the gradient ∇V, such as

$$\nabla V_i = \sum_j a_{ij} x_j$$

where the a_{ij} are coefficients to be determined. This leads to the following procedure for seeking a Lyapunov function V:

1. Assume that ∇V is given by the above expression
2. Solve for the coefficients a_{ij} so as to satisfy the curl equations
3. Restrict the coefficients so that $\dot{V}$ is negative semi-definite (at least locally)
4. Compute V from ∇V by integration
5. Check whether V is positive definite

Example 3.4.9:

Let us look at the nonlinear system

$$\dot{x}_1 = -2x_1$$

$$\dot{x}_2 = -2x_2 + 2x_1 x_2^2$$

Let us assume

$$\nabla V_1 = a_{11} x_1 + a_{12} x_2 \quad \text{and} \quad \nabla V_2 = a_{21} x_1 + a_{22} x_2$$

The curl condition gives us

$$\frac{\partial V_1}{\partial x_2} = \frac{\partial V_2}{\partial x_1}$$

which tells us

$$a_{12} + x_2 \frac{\partial a_{12}}{\partial x_2} = a_{21} + x_1 \frac{\partial a_{21}}{\partial x_1}$$

A choice of the coefficients is

$$a_{11} = a_{22} = 1 \quad \text{and} \quad a_{12} + a_{21} = 0$$

which leads to

$$\nabla V_1 = x_1 \quad \text{and} \quad \nabla V_2 = x_2$$

and

$$\dot{V} = \nabla V(x) = -2x_1^2 - 2x_2^2 \left(1 - x_1 x_2\right)$$

Thus, $\dot{V}$ is locally negative definite in the region $\left(1 - x_1 x_2\right) > 0$. Upon integration,

$$V(x) = \frac{1}{2}\left(x_1^2 + x_2^2\right)$$

This is obviously positive definite; therefore, the asymptotic stability is guaranteed.

By taking

$$a_{11} = 1, a_{12} = x_2^2, \ a_{21} = 3x_2^2, \ a_{22} = 3$$

we obtain the positive definite function

$$V(x) = \frac{1}{2}x_1^2 + \frac{3}{2}x_2^2 + x_1 x_2^3$$

which also a Lyapunov function for the system.

We close this part with the following message. We should use as many physical properties as possible in analyzing the behavior of a system. Physical concepts like energy may lead us to some uniquely powerful choices of Lyapunov functions. In the next chapter, we include more about the Lyapunov functions while dealing with controller design.

3.4.5 Stability of Non-Autonomous Systems

We have briefly introduced non-autonomous systems in the previous chapter, 2.2.2. The ideas of stability for such non-autonomous systems are more or less the same as those of autonomous systems, except for the fact that non-autonomous system behavior depends on initial time t_0. In addition, another idea called uniformity becomes necessary.

Recall that for non-autonomous systems of the form:

$$\dot{x} = f(x,t) \tag{3.17}$$

equilibrium points x_e are defined by

$$f\left(x_e, t\right) = 0 \ \forall t \geq t_0 \tag{3.18}$$

Note that this equation must be satisfied $\forall t \geq t_0$.

Once we recognize the significance of the initial conditions in the analysis of nonlinear systems, Lyapunov theory comes in handy. The definition of invariant set is the same for non-autonomous systems as for autonomous systems. However, an important difference is that, unlike in autonomous systems, a system trajectory generally does not belong to an invariant set for a non-autonomous system.

Definition: Stability

The equilibrium point $x = 0$ is stable at t_0 if for any $R > 0$, there exists a positive scalar $\varepsilon(R, t_0)$, such that

$$\| x(t_0) \| < \varepsilon \Rightarrow \| x(t) \| < R \ \forall t \geq t_0$$

Otherwise, the equilibrium point is unstable.

Notice that the starting of the state trajectory is in a ball of radius ε that may depend on the initial time t_0.

Definition: Asymptotic Stability

The equilibrium point $x = 0$ is asymptotically stable at time to if

1. It is stable
2. $\exists \varepsilon(t_0) > 0$ such that $\| x(t_0) \| < \varepsilon(t_0) \Rightarrow \| x(t) \| \Rightarrow 0$ as $t \to \infty$

The size of attractive region and the speed of trajectory convergence may depend on the initial time t_0.

Definition: Global Asymptotic Stability

The equilibrium point $x = 0$ is globally asymptotically stable if $\forall x_0. \ x(f) \to 0$ as $t \to \infty$.

Definition: Local Uniform Stability

The equilibrium point $x = 0$ is locally uniformly stable if the ball radius ε can be chosen independently of t_0.

Once the initial time is removed, we may intuitively understand that we are not bothered by the time variations in stability. In particular, as time progresses, there is a possibility of the system becoming unstable. We shall not bother about such issues. Further, loosely speaking, like with linear time-invariant systems, we do not care at what time the trajectory begins.

Definition: Local Uniform Asymptotic Stability

The equilibrium point $x = 0$ is locally uniformly asymptotically stable if

1. It is uniformly stable
2. There exists a ball of attraction B_{R_0}, whose radius is independent of t_0, such that any system trajectory with initial states in B_{R_0} converges to $x = 0$ uniformly in t_0.

By uniform convergence in terms of t_0, we understand that for all R_1 and R_2 satisfying $0 < R_2 < R_1 < R_o$, $\exists\, T(R_1,R_2) > 0$, such that $\forall t_0 > 0$,

$$\| x(t_0) \| < R_1 \Rightarrow \| x(t_0) \| < R_2 \quad \forall t > t_0 + T(R_1,R_2)$$

i.e. the state trajectory, starting from within a ball B_{R_1}, will converge into a smaller ball B_{R_2} after a time period T, which is independent of t_0.

By definition, uniform asymptotic stability always implies asymptotic stability. However, the converse is not guaranteed.

Example 3.4.10:

Consider the first-order system

$$\dot{x} = -\frac{x}{1+t}$$

This system has the general solution

$$x(t) = \frac{1+t_0}{1+t} x(t_0)$$

We may readily point out that the solution does converge to zero as $t \to \infty$, but not uniformly; it takes progressively longer and longer to get closer to the origin.

The idea of global uniform asymptotic stability can be defined by replacing the ball of attraction B_R; the whole state space.

We now extend the Lyapunov's direct method to non-autonomous systems. When studying non-autonomous systems using Lyapunov's direct method, scalar functions with explicit time-dependence $V(t,x)$ may have to be used. While in autonomous system analysis, time-invariant functions $V(x)$ suffice.

Definition:

A scalar time-varying function $V(x,t)$ is locally positive definite if $V(0,t) = 0$ and there exists a time-invariant positive definite function $V_o(x)$, such that

$$V(x,t) \geq V_o(x) \quad \forall t \geq 0 \tag{3.19}$$

Thus, a time-variant function is locally positive definite if it dominates a time-invariant locally positive definite function. Globally positive definite functions can be defined similarly.

Other related ideas such as negative definite $V(x,t)$, semi-definiteness, and so on, may be readily learned. In addition to these, we also need the idea of decrescent functions.

Definition:

A scalar function $V(x,t)$ is said to be decrescent if $V(0,t) = 0$, and if there exists a time-invariant positive definite function $V_l(x)$, such that

$$V(x,t) < V_l(x) \quad \forall t \geq 0$$

i.e. $V(x,t)$ is decrescent if it is dominated by a time-invariant positive definite function.

The reader may compare the above two definitions to bring more clarity to what follows.

Example 3.4.11:

A simple example of a time-varying positive definite function is

$$V(x,t) = \left(1 + \sin^2 t\right)\left(x_1^2 + x_2^2\right)$$

which is bounded on either side as

$$\left(x_1^2 + x_2^2\right) \leq V(x,t) \leq 2\left(x_1^2 + x_2^2\right)$$

Given a time-varying scalar function $V(x,t)$, its derivative along a system trajectory is

$$\frac{dV}{dt} = \frac{\partial V}{\partial t} + \frac{\partial V}{\partial x} f(x,t)$$

With the above extended background, Lyapunov's direct method for non-autonomous systems can be summarized by the following theorem.

Lyapunov Theorem for Non-autonomous Systems:

If, in a ball B_R around the equilibrium point $x = 0$, there exists a scalar function $V(x,t)$ with continuous partial derivatives such that

1. V is positive definite
2. $\dot{V}$ is negative semi-definite

then the equilibrium point $x = 0$ is **stable in the sense of Lyapunov**. Furthermore, if

3. V is decrescent, then the origin is **uniformly stable**.

If condition 2 is strengthened by requiring that V be negative definite, then the equilibrium point is **uniformly asymptotically stable**.

If the ball B_R is replaced by the whole state space and condition 1, the strengthened condition 2, condition 3, and the condition

4. $V(x,0)$ is radially unbounded

are all satisfied, then the equilibrium point at $x = 0$ is **globally uniformly asymptotically stable**.

Similar to the case of autonomous systems, if (in a certain neighborhood of the equilibrium point) V is positive definite and $\dot{V}$, its derivative along the system trajectories, is negative semi-definite, then V is called a Lyapunov function for the non-autonomous system.

The proof of this important theorem is rather technical, and we avoid it in this book.

Example 3.4.12:

Consider the non-autonomous system defined by [7]

$$\dot{x}_1 = -x_1 - e^{-2t} x_2$$

$$\dot{x}_2 = x_1 - x_2$$

Let us choose the scalar function

$$V(x,t) = \frac{1}{2}\left(x_1^2 + \left(1 - e^{-2t}\right)x_2^2\right)$$

so that

$$\dot{V}(x,t) = -\left(x_1^2 - x_1 x_2 + \left(1 - e^{-2t}\right)x_2^2\right)$$

Some manipulation on this would give us the inequality

$$\dot{V} \leq -\left((x_1 - x_2)^2 + x_1^2 + x_2^2\right)$$

Thus, $\dot{V}$ is negative definite, and therefore, the point $x = 0$ is globally asymptotically stable.

While it is nice to simply add a t and re-define and re-prove the theorems for non-autonomous systems, non-autonmous systems could pose several difficulties, which we demonstrate in the next two examples.

Example 3.4.13: [7]

Let $g(t)$ be a continuously-differentiable function which coincides with the function $e^{-t/2}$ except around some peaks where it reaches the value 1. Specifically, $g^2(t)$ is shown in the Figure 3.17.

There is a peak for each integer value of t. Let us assume that the width of the peak corresponding to abcissa n is assumed to be smaller than $(1/2)^n$. Then,

$$\int_0^\infty g^2(r)dr < \int_0^\infty e^{-r}dr + \sum_{n=1}^\infty \frac{1}{2^n}$$

and therefore, the scalar function

$$V(x,t) = \frac{x^2}{g^2(t)}\left(3 - \int_0^t g^2(r)dr\right)$$

is positive definite.

Let us now use this positive definite function to look into the system given by

$$\dot{x} = \frac{\dot{g}(t)}{g(t)}x(t)$$

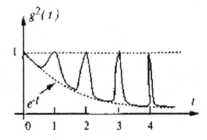

FIGURE 3.17
An illustration of $g^2(t)$ for Example 3.4.13.

We would get

$$\dot{V} = -x^2 < 0$$

But, the solution of the first-order system is

$$x(t) = \frac{g(t)}{g(t_0)} x(t_0)$$

which tells us rather clearly that the origin is *not* an asymptotically stable equilibrium.

Example 3.4.14:

Let us now look at the following variant of an earlier example.

$$\ddot{x} + c(t)\dot{x} + kx = 0$$

where $c(t)$ is a time-varying damping coefficient, and k is the spring constant.

Let us now bring in the following Lyapunov function:

$$V(x,t) = \frac{1}{2}\left((\dot{x} + \alpha x)^2 + b(t)x^2\right)$$

where α is any positive constant smaller than $\sqrt{k}$, and

$$b(t) = k - \alpha^2 + \alpha c(t)$$

$\dot{V}$ can be readily obtained as

$$\dot{V} = (\alpha - c(t))\dot{x}^2 + \frac{\alpha}{2}(\dot{c}(t) - 2k)x^2$$

Thus, for $\dot{V}$ to be negative definite, we need two positive constants

$$c(t) > \alpha > 0 \quad \text{and} \quad \dot{c}(t) \le \beta < 2k$$

Assuming further that $c(t)$ is upper bounded to guarantee the decrescence of V, the above analysis assures that the system is asymptotically stable.

On the other hand, if $c(t) = 2 + e^t$ to keep the time-varying damping greater than a positive constant, the solution would be

$$x(t) = 1 + e^{-t}$$

starting from the initial conditions $(2,-1)$. This solution, apparently, approaches 1 and not the origin.

3.4.6 Instability Theorems

In this part, we will see when a system can be unstable. This perspective of Lyapunov's direct method brings in a lot of clarity in the ideas. While discussing the invariant set theorems earlier, we found that to assess asymptotic stability of the origin, a linearized model might mislead us. In the case of non-autonomous systems, this problem is, most often, even worse. But, the instability theorems, which do not need any linearization, would come handy as we see ahead.

Instability Theorem #1:

If, in a certain neighborhood Ω of the origin, there exists a continuously differentiable, decrescent scalar function $V(x,t)$, such that

1. $V(0,t) = 0 \; \forall t \ge t_0$
2. $V(x,t_0)$ can assume strictly positive values arbitrarily close to the origin
3. $\dot{V}(x,t)$ is positive definite (locally in Ω)

then the equilibrium point 0 at time t_0 is unstable.

Example 3.4.15:

Consider

$$\dot{x}_1 = 2x_2 + x_1\left(x_1^2 + 2x_2^4\right)$$

$$\dot{x}_2 = -2x_1 + x_2\left(x_1^2 + x_2^4\right)$$

Linearizing the system around the origin gives us

$$A = \begin{bmatrix} 0 & 2 \\ -2 & 0 \end{bmatrix}$$

with eigenvalues $\pm j2$. We may not be able to confidently conclude that the system is unstable for sure.

However, if we take the Lyapunov function

$$V(x) = \frac{1}{2}\left(x_1^2 + x_2^2\right)$$

we see that its derivative becomes

$$\left(x_1^2 + x_2^2\right)\left(x_1^2 + x_2^4\right)$$

which is positive definite. Therefore, the system is unstable.

Instability Theorem #2:

If, in a certain neighborhood Ω of the origin, there exists a continuously differentiable, decrescent scalar function $V(x,t)$ satisfying

1. $V(0, t_0) = 0$ and $V(x, t_0)$ can assume strictly positive values arbitrarily close to the origin
2. $\dot{V}(x, t) - \lambda V(x, t) > 0 \ \forall t \geq t_0 \ \forall x \in \Omega$

with λ being a strictly positive constant, then the equilibrium point $x = 0$ at time t_0 is unstable.

Example 3.4.16:

Let

$$\dot{x}_1 = x_1 + 3x_2 \sin^2 x_2 + 5x_1 x_2^2 \sin^2 x_1$$

$$\dot{x}_2 = x_2 + 3x_1 \sin^2 x_2 - 5x_1^2 x_2 \cos^2 x_1$$

and consider the function

$$V(x) = \frac{1}{2}\left(x_1^2 - x_2^2\right)$$

which satisfies condition 1 of the theorem. Its derivative turns out to be

$$\dot{V} = 2V + 5x_1^2 x_2^2$$

which clearly indicates that the equilibrium point at the origin is unstable.

Instability Theorem #3:

Let Ω be a neighborhood of the origin. If there exists a scalar function $V(x, t)$ with continuous first partial derivatives, decrescent in Ω and a region $\Omega_l \in \Omega$, such that

1. V and $\dot{V}$ are positive definite in Ω_l
2. The origin is a boundary point of Ω_l
3. At the boundary points of Ω_l within Ω, $V(x, t) = 0 \ \forall t \geq t_0$

then the equilibrium point 0 at time t_0 is unstable.

Example 3.4.17:

Let

$$\dot{x}_1 = x_1^2 + x_2^3$$

$$\dot{x}_2 = -x_2 + x_1^3$$

Interestingly, the linearization around origin tells us that the eigenvalues are 0 and −1, meaning that the system is stable. However, if we take

$$V = x_1 - \frac{1}{2}x_2^2$$

so that

$$\dot{V} = x_1^2 + x_2^2 + x_2^3 - x_1^3 x_2$$

concluding that the system is actually unstable.

3.4.7 Passivity Framework

We have, so far, presented several examples of nonlinear systems. In most cases, we have had certain linear terms (cf. Example 3.4.15). A question might arise: Is it possible and useful to decompose the system into a linear subsystem and a nonlinear subsystem? The answer is assertive, of course, subject to certain conditions that we lay out in this part.

If the transfer function (or transfer matrix) of the linear subsystem is a so-called **positive real**, then it has important properties that may lead to the generation of a Lyapunov function for the whole system. Such linear systems, called positive linear systems, play a central role in the analysis and design of many nonlinear control problems.

Definition:

A transfer function $H(s)$ is positive real if

$$Re[H(s)] \geq 0 \ \forall \ Re[s] \geq 0$$

It is strictly positive real if $H(s - s_0)$ is positive real for some $s_0 > 0$.

This means that $H(s)$ always has a positive (or zero) real part when s has a positive (or zero) real part.

Theorem:

A transfer function $H(s)$ is positive real if, and only if,

1. $H(s)$ is a stable transfer function, i.e. all the poles are in the open left half of the s-plane.
2. The poles of $H(s)$ on the $j\omega$ axis are simple, and the associated residues are real and non-negative
3. $Re[H(j\omega)] \geq 0$ for any $\omega \geq 0$ such that $j\omega$ is not a pole of $H(s)$

Theorem: Kalman-Yakubovich

Consider a controllable linear time-invariant system

$$\dot{x} = Ax + Bu$$

$$y = Cx$$

The transfer function

$$H(s) = C[sI - A]^{-1}B$$

is strictly positive real if, and only if, there exist positive definite matrices P and Q, such that

$$A^T P + PA = -Q \quad \text{and} \quad C = B^T P$$

Beyond its mathematical statement, the theorem has important physical interpretations and uses in generating Lyapunov functions.

The theorem can be easily extended to positive real systems by simply relaxing Q to be positive semi-definite. The usefulness of this result is that it is applicable to transfer functions containing a pure integrator. This theorem is also referred to as the **positive real lemma**.

3.4.7.1 The Passivity Formalism

Looking at the Lyapunov functions as the generalization of energy, i.e. a scalar, an interesting framework can be built using a combination of systems, analogous to our elementary control systems and block diagram algebra. Intuitively, we expect Lyapunov functions for combinations of systems to be derived by simply adding the Lyapunov functions describing the subsystems. Passivity theory formalizes this intuition and derives simple rules to describe combinations of subsystems or *blocks* expressed in a Lyapunov-like formalism. It also represents an approach to constructing Lyapunov functions, or Lyapunov-like functions, for feedback control purposes.

In the following, we shall more generally consider systems that verify equations of the form

$$\dot{V}(t) = y^T u - g(t)$$

where $V(t)$ and $g(t)$ are scalar functions of time, u is the system input, and y is its output.

Definition:

A system verifying an equation of the above with V lower bounded and $g \geq 0$ is said to be passive (or to be a passive mapping between u and y). Furthermore, a passive system is said to be dissipative if

$$\int_0^\infty y^T(t)u(t)dt \neq 0 \implies \int_0^\infty g(t)dt > 0$$

Example 3.4.18:

Consider the nonlinear mass-spring-damper system

$$m\ddot{y} + y^2\dot{y}^3 + y^7 = u$$

represents a dissipative mapping from external force u (assumed constant) to velocity $\dot{y}$, since

$$\frac{d}{dt}\left(\frac{1}{2}m\dot{y}^2 + \frac{1}{8}y^8\right) = \dot{y}u - y^2\dot{y}^4$$

We may identify here that V is the total energy stored in the system and g is the dissipated power.

Example 3.4.19:

Consider the scalar system

$$\dot{x} + \lambda(t)x = u, \quad y = h(x)$$

where the dynamics are linear time-varying with $\lambda(t) \geq 0$, but the input-output map is an arbitrary nonlinear function $h(x)$, not necessarily continuous, having the same sign as its argument.

If we look at

$$\frac{d}{dt}\int_0^x h(r)dr = h(x)\dot{x} = yu - \lambda(t)h(x)x$$

we notice that the integral is non-negative and $\lambda(t)h(x) \geq 0 \ \forall x$.

Thus, as long as $\lambda(t) \neq 0$, the mapping is dissipative. In terms of state feedback, we may play with the idea of $\lambda(t) = \lambda[x(t)]$, for example,

$$\dot{x} + x^3 = u, \quad y = x - \sin^2 x$$

and verify that the mapping $u \to y$ is dissipative.

Next, we consider another interesting interconnection of subsystems [7] as shown in Figure 3.18.

The forward path is a linear time-invariant system, and the feedback part is a memoryless nonlinearity, i.e. a nonlinear static mapping. The equations of such systems can be written as

$$\dot{x} = Ax - B\phi(y), \quad y = Cx$$

Many systems of practical interest can be represented in this structure. If the feedback path simply contains a constant gain α, then the stability of the whole system, a linear feedback system, can

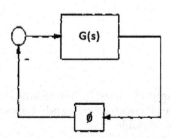

FIGURE 3.18
An interconnection of subsystems. (From Slotine, J.J., and Li, W., *Applied Nonlinear Control*, Englewood Cliffs, NJ, Prentice Hall, 1991.)

be simply determined by examining the eigen-values of the closed loop system matrix $A - \alpha BC$. However, the stability analysis of the whole system with an arbitrary nonlinear feedback function is much more difficult.

In analyzing this kind of system using Lyapunov's direct method, we usually require the nonlinearity to satisfy a so-called sector condition, whose definition is given below.

Definition:

A continuous function ϕ is said to belong to the sector $[k_1,k_2]$, if there exists two non-negative numbers k_1 and k_2, such that

$$y \neq 0 \Rightarrow k_1 \leq \frac{\phi(y)}{y} \leq k_2$$

Geometrically, the nonlinear function always lies between the two straight lines k_1y and k_2y, with two properties: (i) $\phi(0) = 0$ and $y\phi(y) \geq 0$, i.e. the graph of $\phi(y)$ lies in the first and third quadrants.

Assume that the nonlinearity $\phi(y)$ is a function belonging to the sector $[k_1,k_2]$, and that the A matrix of the linear subsystem in the forward path is stable.

Aizerman's Conjecture: [7]

If the matrix $[A - BCk]$ is stable for all values of $k \in [k_1,k_2]$, then the nonlinear system is globally asymptotically stable.

Aizerman's is a very interesting conjecture. If it were true, it would allow us to deduce the stability of a nonlinear system by simply studying the stability of linear systems. However, several counterexamples have shown that this conjecture is false. After Aizerman, many researchers continued to seek conditions that guarantee the stability of the nonlinear system. In Figure 3.18., Popov's criterion imposes additional conditions on the linear subsystem, leading to a sufficient condition for asymptotic stability reminiscent of Nyquist's criterion (a necessary and sufficient condition) in linear system analysis.

A number of versions have been developed for Popov's criterion. The following basic version is fairly simple and useful.

Popov's Criterion

If the system described earlier satisfies the conditions

1. The matrix A has all its eigenvalues strictly in the left half of the s-plane, and the pair $[A, B]$ is controllable

2. The nonlinearity belongs to the sector $[0,k]$
3. There exists a strictly positive number α, such that

$$\forall \omega \geq 0 \quad Re[1 + j\alpha\omega]G(j\omega) + \frac{1}{k} \geq \varepsilon$$

for an arbitrarily small $\varepsilon > 0$

then the point 0 is globally asymptotically stable.

The criterion can be proven constructing a Lyapunov function candidate based on the positive real lemma.

Here are several salient features of Popov's criterion:

1. It is applicable only to autonomous systems.
2. It is restricted to a single memoryless nonlinearity.
3. The stability of the nonlinear system may be determined by examining the frequency-response functions of a linear subsystem, without the need of searching for explicit Lyapunov functions.
4. It gives us only a sufficient condition.

Example 3.4.20:

Let us determine the stability of a nonlinear system, where the linear subsystem is defined by

$$H(s) = \frac{s+3}{s^2 + 7s + 10}$$

and the sector nonlinearity $0 \leq \phi(y) \leq ky$

First, the linear subsystem is strictly stable, because its poles are -2 and -5. It is also controllable, because there is no pole-zero cancellation. Let us now check the Popov inequality.

The frequency response function

$$G(j\omega) = \frac{j\omega + 3}{-\omega^2 + j7\omega + 10}$$

We can build two functions from this:

$$G_1 = \frac{4\omega^2 + 30}{\omega^4 + 29\omega^2 + 100}$$

and

$$G_1 = \frac{-\omega(\omega^2 + 11)}{\omega^4 + 29\omega^2 + 100}$$

so that

$$4\omega^2 + 30\omega + \alpha\omega^2(\omega^2 + 11) + \left(\frac{1}{k} - \varepsilon\right)(\omega^4 + 29\omega^2 + 100) > 0$$

Clearly, any arbitrary pair of strictly positive numbers α and k would satisfy this inequality. Thus, the nonlinear system is globally asymptotically stable.

A more direct generalization of Nyquist's criterion to nonlinear systems is the **circle criterion**, whose basic version can be stated as follows.

Circle Criterion

If the system described earlier satisfies the conditions

1. The matrix A has no eigenvalue on the $j\omega$-axis and has ρ eigenvalues strictly in the right half-plane
2. The nonlinearity ϕ belongs to the sector $[k_1, k_2]$
3. One of the following is true
   a. $0 < k_1 \le k_2$, the Nyquist plot of $G(j\omega)$ does not enter the disk $D(k_1, k_2)$ and encircles it ρ times counterclockwise
   b. $0 = k_1 < k_2$, and the Nyquist plot of $G(j\omega)$ stays in the half-plane $Res >= 1/k_2$
   c. $k_1 < 0 < k_2$, and the Nyquist plot of $G(j\omega)$ stays in the interior of the disk $D(k_1, k_2)$
   d. $k_1 < k_2 < 0$, the Nyquist plot of $-G(j\omega)$ does not enter the disk $D(-k_1, -k_2)$ and encircles it ρ times counterclockwise

then the equilibrium point 0 of the system is globally asymptotically stable.

Figure 3.19 adopted from [7]

Looking at Figure 3.19, we understand that the critical point $(-1/k, 0)$ in Nyquist's criterion is replaced in the circle criterion by the circle, which tends towards the point $-1/k_1$ as k_2 tends to k_1. We also notice that the circle criterion states sufficient but not necessary conditions.

The circle criterion can be extended to non-autonomous systems, but the treatment is beyond the scope of this book.

In the next, and last, part of this chapter, we explore an analysis technique that combines the classical frequency domain methods of linear control with a class of nonlinear systems. The range of this technique is rather limited, but there are two good things we will learn in this last part. First, a good number of practical nonlinearities like relays and hysterisis can be deftly handled. Second, the methodology has a charm of its own, particularly when we learn about its beginnings in the early control engineering in a non-computer era.

3.5 Describing Function Analysis

For some nonlinear systems, with a few reasonable assumptions, the frequency response method, called the describing function method, can be used to approximately analyze and predict nonlinear behavior. The main purpose of this technique is for the prediction of limit cycles in nonlinear systems. This part presents a brief introduction to the describing function analysis of nonlinear systems.

To begin with, we will revisit the classical van der Pol equation

$$\ddot{x} + \mu\left(x^2 - 1\right)\dot{x} + x = 0$$

where μ is a positive constant. We have treated this system using phase plane analysis and Lyapunov analysis. We will uncover the limit cycle in a different way. We first assume that the limit *cycle* exists with an undetermined amplitude and frequency, then work out to either to confim or deny its presence. If it is present, we will then proceed to determine its amplitude and frequency.

Before carrying out this procedure, let us represent the system dynamics in a block diagram form, as shown in Figure 3.20. We write the equation in the following manner:

$$\ddot{x} - \mu\dot{x} + x = -\mu x^2 \dot{x}$$

For the linear part of the equation on the left-hand side, we will use the transfer function representation:

$$\frac{X}{Z} = \frac{\mu}{s^2 - \mu s + 1}$$

and the loop returns $x(t)$ with a negative sign to the nonlinear element.

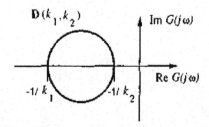

FIGURE 3.19
An illustration of circle criterion. (From Slotine, J.J., and Li, W., *Applied Nonlinear Control*, Englewood Cliffs, NJ, Prentice Hall, 1991.)

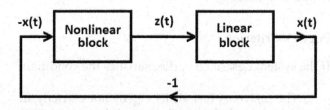

FIGURE 3.20
Block diagram representation for describing function analysis.

Recall the strategy we have in studying the sinusoidal steady-state response of linear systems. The basic premise is that, according to Fourier analysis, any periodic signal can be represented as a sum of sinusoids of different amplitudes and frequencies that are multiples of the fundamental. By the orthogonality principle, when excited by a sinusoid of a certain frequency, only that component of the periodic signal pops up.

Let us then give a signal $x(t) = A \sin \omega t$ to the nonlinear block and work out around the loop.

The output of the nonlinear block is

$$z = -x^2 \dot{x}$$

$$= -A^2 \sin^2 \omega t \cdot A \omega \cos \omega t$$

$$= -\frac{A^3 \omega}{4} \left(\cos(\omega t) - \cos(3\omega t) \right)$$

We will now assume that the linear system in the loop around the nonlinear block works like a low pass filter and removes the third harmonic but retains the fundamental. Accordingly, we consider

$$z = -\frac{A^3 \omega}{4} \cos(\omega t) = \frac{A^2}{4} \frac{d}{dt}(-x(t))$$

Thus, the nonlinear block can be shown to be a differential operator that is a function of the amplitude A of the input sinusoid. We write it as

$$z = N(A)\left[-x\right] \qquad (3.20)$$

Owing to the differentiation, which we take as multiplication by s, or equivalently by $j\omega$ in the linear transfer functions, we write

$$N(A) = j\omega \frac{A^2}{4} \qquad (3.21)$$

in the frequency domain. We call the *quasi-linear* function $N(A)$ the **describing function** of the nonlinearity. In general, the describing function also depends upon the frequency, in which case we write it as $N(A, \omega)$.

If $x(t)$ the sinusoid is assumed to go around the loop exhibiting limit cycle behavior, then

$$1 + j\omega \cdot \frac{A^2}{4} \cdot G(j\omega) = 0$$

at steady state. Plugging in the numbers

$$1 + j\omega \cdot \frac{A^2}{4} \cdot \frac{\mu}{s^2 - \mu s + 1} = 0$$

and solving this we get

$$A = 2, \quad \text{and} \quad \omega = 1$$

i.e. there does exist a limit cycle with an amplitude 2 and frequecy 1. It is interesting to note neither the amplitude nor the frequency obtained above depends on the parameter μ.

In the phase plane, the above approximate analysis suggests that the limit cycle is a circle of radius 2, regardless of the value of μ. However, owing to the assumption that the linear part of the system is a low pass filter with a very sharp cut-off is a bit too far away from the reality - the limit cycle is not a nice circle independent of μ, and as μ grows the nonlinearity becomes more significant.

To summarize, in the above approximate analysis, the critical step is to replace the nonlinear block by the quasi-linear block, which has the frequency response function $N(A, \omega)$. Afterwards, the amplitude and frequency of the limit cycle can be determined from

$$1 + N(A, \omega)G(j\omega) = 0 \qquad (3.22)$$

Before moving on, let us briefly discuss what kind of nonlinear systems it applies to and what kind of information it can provide about nonlinear system behavior. Simply speaking, any system that can be transformed into the configuration in Figure 3.20 can be studied with describing functions. There are at least two important classes of systems in this category. The first important class consists of "almost" linear systems. By "almost" linear systems, we refer to systems that contain hard nonlinearities in the control loop but are otherwise linear. Such systems arise when a control system is designed using linear control but its implementation involves hard nonlinearities, such as motor saturation, actuator or sensor dead-zones, Coulomb friction, or hysteresis in the plant. The second class of systems consists of genuinely nonlinear systems whose dynamic equations can actually be rearranged into the form of Figure 3.20.

3.5.1 Applications of Describing Functions

Prediction of limit cycles is very important, because limit cycles can occur. However, in most control systems, limit cycles are undesirable. This may be due to a number of reasons:

- Limit cycles, as a way of instability, tend to cause poor control accuracy
- The constant oscillation associated with the limit cycles can cause increasing wear, or even mechanical failure, of the control system hardware

- Limit cycling may also cause other undesirable effects, such as passenger discomfort in an aircraft under autopilot

In general, although a precise knowledge of the waveform of a limit cycle is usually not mandatory, the knowledge of the limit cycle's existence, as well as that of its approximate amplitude and frequency, is critical. The describing function method can be used for this purpose. It can also guide the design of compensators so as to avoid limit cycles.

3.5.2 Basic Assumptions

Consider a nonlinear system in the general form of Figure 3.20. In order to develop the basic version of the describing function method, the system has to satisfy the following four conditions:

1. There is only a single nonlinear component
2. The nonlinear component is time-invariant
3. Corresponding to a sinusoidal input $x = \sin \omega t$, only the fundamental component $Z(j\omega)$ in the output $z(t)$ has to be considered
4. The nonlinearity is odd

The second assumption implies that we consider only autonomous nonlinear systems. It is satisfied by many nonlinearities in practice, such as saturation in amplifiers, backlash in gears, Coulomb friction between surfaces, and hysteresis in relays. The reason for this assumption is that the Nyquist criterion, on which the describing function method is largely based, applies only to linear time-invariant systems. The third assumption is often referred to as the filtering hypothesis. The fourth assumption means that the plot of the nonlinearity relation $\phi(x)$ between the input and output of the nonlinear element is symmetric about the origin.

A number of methods are available to determine the describing functions of nonlinear elements in control systems. Convenience and cost in each particular application determine which method should be used. As has been explained in the case of van der Pol oscillator, we obtain an expression for the output of the nonlinear block and express it as a Fourier series, from which we just pick up the fundamental and ignore the rest of the harmonics. We will take up one case and develop the describing function analytically and close the chapter.

Nonlinearities can be either continuous or discontinuous. Because discontinuous nonlinearities cannot

be locally approximated by linear functions, they are called "hard" nonlinearities. Hard nonlinearities are commonly found in control systems, both in small range operation and large range operation. Because of the common occurence of hard nonlinearities, let us briefly discuss the characteristics and effects of **saturation**, which is often found.

When one increases the input to a physical device, the following phenomenon is often observed: when the input is small, its increase leads to a corresponding (often proportional) increase of output; but when the input reaches a certain level, its further increase produces little or no increase of the output. The output simply stays around its maximum value. The device is said to be in saturation when this happens. Simple examples of this are transistor amplifiers and magnetic amplifiers. A saturation nonlinearity is usually caused by limits on component size, properties of materials, and available power. A typical saturation nonlinearity is represented in Figure 3.21.

Consider the input $x(t) = A \sin \omega t$. If $A < a$, then the input remains in the linear range, and therefore, the output is $z(t) = kA \sin \omega t$. Hence, the describing function is simply a constant k.

Now consider the case $A > a$. The input and the output functions are plotted in Figure 3.21. The output is seen to be symmetric over the four quarters of a period. In the first quarter, it can be expressed as

$$z(t)=\begin{cases} kA \sin \omega t & 0 \le \omega t \le \gamma \\ ka & \gamma \le \omega t \le \pi/2 \end{cases} \qquad (3.23)$$

where $\gamma = \sin^{-1}(a/A)$. The odd nature of $z(t)$ implies that the Fourier series of $z(t)$ contains only the "sine" terms, and not any "cosine" terms. The coefficient b_1 of the fundamental may be computed using standard formula as

$$b_1 = \frac{4}{\pi} \int_0^{\pi/2} z(t)\sin(\omega t)\,d(\omega t) \qquad (3.24)$$

which can be shown to evaluate to

$$b_1 = \frac{2kA}{\pi}\left[\gamma + \frac{a}{A}\sqrt{1-\frac{a^2}{A^2}}\right] \qquad (3.25)$$

Since the input is $A \sin \omega t$, and the output (upto the fundamental) is $b_1 \sin \omega t$; therefore, the describing function is

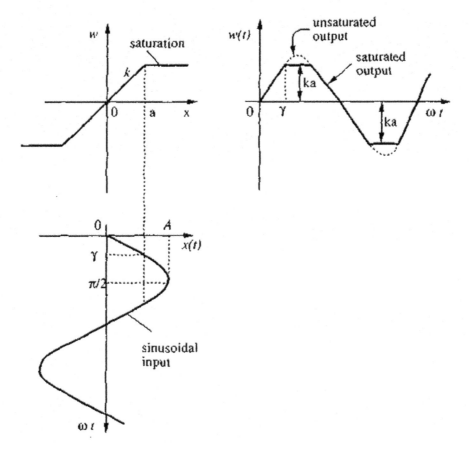

FIGURE 3.21
The working of saturation nonlinearity. (From Slotine, J.J., and Li, W., *Applied Nonlinear Control*, Englewood Cliffs, NJ, Prentice Hall, 1991.)

$$N(A,\omega) = \frac{b_1}{A} = \frac{2k}{\pi}\left[\gamma + \frac{a}{A}\sqrt{1 - \frac{a^2}{A^2}}\right] \quad (3.26)$$

Once again, we notice that this is not explicitly a function of ω. Nevertheless, the reader may find in most standard textbooks other nonlinearities that are functions of ω in addition to A

As a special case, we can see that the describing function for the relay-type (on-off) nonlinearity. This case corresponds to shrinking the linearity range in the saturation function to zero, i.e. $a \to 0$, $k \to \infty$ but the product $ka = M$. By taking the limit, we get

$$N(A,\omega) = \frac{4M}{\pi A} \quad (3.27)$$

We conclude this part with the following comments. Empirical evidence over a long period of time, together with theoretical justification, recommends that the describing function method can effectively solve a large number of practical control problems involving limit cycles. It can be used to approximately analyze and predict the behavior of important classes of nonlinear systems, including systems with hard nonlinearities. The graphical nature and the physically intuitive insights it can provide make it an important tool for practicing engineers. Applications of the describing function method include prediction of limit cycles. Other applications such as predicting subharmonics, jump phenomena, and responses to external sinusoidal inputs, can be found in the literature. However, due to the approximate nature of the technique, it is not surprising that the analysis results are sometimes not very accurate. In the worst cases, a predicted limit cycle does not actually exist, or an existing limit cycle is not predicted.

4

Nonlinear Control Design

In this chapter, we explore the design of controllers for nonlinear systems. At the beginning of the 1980s, generalizations of pole placement and observer design techniques for nonlinear systems were obtained using tools from differential geometry. Since 1990, output feedback algorithms have been given for classes of nonlinear systems. The availability of increasingly more powerful and less expensive microprocessors and the need for better performance stimulated control engineers to design innovative nonlinear control algorithms for advanced applications, such as robotics and aerospace. In these applications, nonlinearities such as Coriolis forces and inertial forces play a significant role and can be exactly modeled using well-known laws of physics. On the basis of a deep understanding of the physics, engineers developed accurate controllers, such as the field oriented control for induction motors and autopilot for helicopters, to meet demanding specifications that otherwise could not be met by linear methods. If we carefully observe the pattern, these algorithms have an interesting innovative feature—using nonlinear changes of state coordinates and of nonlinear state feedback (which aims at nonlinearity cancellation) to make the closed loop system, hopefully, linear in the new coordinates. The background in the Hartman-Grobman theorem and invariant sets has a huge impact in the development of beautiful control system design. As we proceed, we will realize that we are close to the idea of pole placement wherein a linear controllable system is transformed into a system with desired eigenvalues via a state feedback control law.

We devote quite some space to the treatment of the differential geometric technique called **feedback linearization**. We will then conclude the chapter with Backstepping and Sliding Mode Control, which might be considered as special cases of feedback linearization.

To begin with, we consider single input systems of the form

$$\dot{x} = f(x) + g(x)u \quad \text{and} \quad y = \psi(x) \qquad (4.1)$$

in a neighborhood $U_{x_e} \subset \Re^n$ of an equilibrium point x_e corresponding to $u = 0$. The functions f and g are assumed to be *smooth vector fields* on $\Re^n$ with $g(x_e) \neq 0$. By smooth vector fields, we mean that these functions are infinitely differentiable. We proceed ahead to the necessary and sufficient conditions under which the

above system is transformable into a linear controllable system by

1. Nonlinear change of coordinates only or
2. Nonlinear feedback and change of coordinates

For convenience, we will begin with the second problem, known as the state feedback linearization.

4.1 Full-State Linearization

We will first look at a simple example to understand the working of the technique. Subsequently, we present a formal theory.

Example 4.1.1: [14]

Let the system be

$$\dot{x} = ax^3 + u$$

$$y = x$$

Let $u = -kx$, assuming that the nonlinearity can be neglected and that we wish to drive the state to origin asymptotically.

However, the actual system gives

$$\dot{x} = x\left(ax^2 - k\right)$$

and we are not sure of asymptotic stability in the large.

On the other hand, if we have chosen

$$u = -ax^3 - kx$$

the system is globally asymptotically stable.

Example 4.1.2: [14]

Let

$$\dot{x}_1 = ax_1^2 + x_2$$

$$\dot{x}_2 = -2a^2x_1^3 - 2ax_1x_2 + u$$

$$y = x_1$$

We make a *transformation*:

$$z_1 = x_1$$

$$z_2 = ax_1^2 + x_2$$

This results in a linear outer loop control with

$$u = -k_1 z_1 - k_2 z_2$$

$$= -k_1 x_1 - k_2 x_2 - k_2 a x_1^2$$

which assures us global asymptotic stability. Of course, the term ax_1^2 must be exactly known.

There could be counterexamples to defeat us, but we will fortify ourselves with a good theory soon after the following example.

Example 4.1.3: [14]

It may be verified that the system

$$\dot{x}_1 = -x_1^3 + x_2^2 u$$

$$\dot{x}_2 = u$$

$$y = x_1$$

cannot be linearized by state feedback.

Having noticed the transformation in the just posed Example 4.1.2, we have the following questions regarding the importance of transformations. Does there exist one? How do we get the right one? Is it invertible? Moreover, the solution needs n first order differential equations to be solved, where, by solving we mean integration. Can we completely integrate?

Coupling the questions just posed,

1. Can we characterize the transformation?
2. In the process, can we derive a feedback law?

The most familiar idea of geometry, the two-dimensional landscape of plane geometry, was inspired by the ancient vision that the Earth was flat. However, in an era of globalization, Google Earth, and intercontinental air travel, we all need to learn a little about spherical geometry and its modern (roughly 200 years old!) version—differential geometry. At the heart of differential geometry, there are beautiful concepts that can be easily grasped by anyone who has ridden a bicycle, looked at a globe, or stretched a rubber band.

Definition

The **Lie bracket** of f and g is

$$[f,g] = ad_f(g) \overset{\Delta}{=} \frac{\partial g}{\partial x} f - \frac{\partial f}{\partial x} g$$

where $\frac{\partial g}{\partial x}$ and $\frac{\partial f}{\partial x}$ are the Jacobian matrices.

Let us understand this using the following example:

Example 4.1.4: [15]

Let

$$f(x) = \begin{bmatrix} x_2 \\ \sin(x_1) \\ x_3^2 + x_1 \end{bmatrix}, \ g(x) = \begin{bmatrix} 0 \\ x_2^2 \\ 1 \end{bmatrix}$$

The Jacobians are

$$\frac{\partial g}{\partial x} = \begin{bmatrix} 0 & 0 & 0 \\ 0 & 2x_1 & 0 \\ 0 & 0 & 0 \end{bmatrix}, \ \text{and} \ \frac{\partial f}{\partial x} = \begin{bmatrix} 0 & 1 & 0 \\ \cos(x_1) & 0 & 0 \\ 1 & 0 & 2x_3 \end{bmatrix}$$

It may be verified that

$$ad_f(g) = \frac{\partial g}{\partial x} f - \frac{\partial f}{\partial x} g = \begin{bmatrix} -x_2^2 \\ 2x_2 \sin(x_1) \\ -2x_3 \end{bmatrix}$$

We readily observe the following:

1. $[f,g]$ is also a vector, and
2. $[g,f] = -[f,g]$
3. We may now inductively define

$$ad_f^0(g) = g$$

$$ad_f^1(g) = [f,g]$$

$$ad_f^2(g) = [f,[f,g]]$$

$$\vdots$$

$$ad_f^k(g) = [f, ad_f^{k-1}(g)] \qquad (4.2)$$

4. $[f_1 + f_2, g] = [f_1, g] + [f_2, g]$
5. And, with little effort, we may derive the following

$$[[f,g],h] + [[g,h],f] + [[h,f],g] = 0 \qquad (4.3)$$

called the **Jacobi identity**. This is a very important result.

Definitions

Let h be a scalar function of x, i.e. $h: \mathfrak{R}^n \to \mathfrak{R}$. Then,

$$dh \overset{\Delta}{=} \begin{bmatrix} \dfrac{\partial h}{\partial x_1} & \dfrac{\partial h}{\partial x_2} & \cdots & \dfrac{\partial h}{\partial x_n} \end{bmatrix}$$

is a *row vector*. The dual product of the row vector dh and the vector $f\left(= [f_1 \cdots f_n]^T \right)$ is the scalar $< dh, f >$.

This dual product, which appears to be a simple mathematical operation, is better known as the **Lie derivative** of h, with respect to f, and is given by

$$L_f h = \sum_{i=1}^{n} \frac{\partial h}{\partial x_i} f_i(x) \qquad (4.4)$$

The Lie derivative is simply the gradient of h in the direction of $f(x)$. In general, we may recursively write

$$L_f^k h = L_f^{k-1} h \quad k = 1, 2, \cdots, n$$

with $L_f^0 h = h$.

With this Lie derivative, we have another very interesting and important result, which establishes the relationship between the Lie bracket and Lie derivative.

Lemma

Let f and g be vector fields on $\Re^n$, and h the scalar function as before.

$$L_{[f,g]} h = L_f L_g h - L_g L_f h \qquad (4.5)$$

A proof of this is pretty straight with some algebraic manipulations. We prefer to present the proof since it helps the reader acquaint himself with some crucial derivations ahead.

Proof

We first look at the ith component of the vector field $[f, g]$:

$$[f, g]_i = \sum_{j=1}^{n} \frac{\partial g_i}{\partial x_j} f_j - \sum_{j=1}^{n} \frac{\partial f_i}{\partial x_j} g_j$$

so that the left hand side of eqn. 4.5 is

$$L_{[f,g]} h = \sum_{i=1}^{n} \frac{\partial h}{\partial x_i} [f, g]_i$$

$$= \sum_{i=1}^{n} \frac{\partial h}{\partial x_i} \left(\sum_{j=1}^{n} \frac{\partial g_i}{\partial x_j} f_j - \sum_{j=1}^{n} \frac{\partial f_i}{\partial x_j} g_j \right)$$

$$= \sum_{i=1}^{n} \sum_{j=1}^{n} \frac{\partial h}{\partial x_i} \left(\frac{\partial g_i}{\partial x_j} f_j - \frac{\partial f_i}{\partial x_j} g_j \right)$$

Similarly, expanding the right-hand side of eqn. 4.5 can be shown to be equivalent to the above expression.

QED

In addition to the above definitions, we have two more, and we will proceed to an important theorem in differential geometry; the back bone of feedback linearization.

This theorem, known as the Frobenius Theorem, can be thought of as an existence theorem for solutions to certain systems of first order partial differential equations. A rigorous proof of this theorem is beyond the scope of this text, and we suggest some good references in the References for Section 1.

Definition

A linearly independent set of vector fields $\{f_1, \cdots, f_m\} \in \Re^n$ is **completely integrable** *if, and only if*, there are $n - m$ linearly independent functions $h_1, \cdots, h_{n-m}$ satisfying the system of Partial Differential Equation (PDE):

$$L_{f_i} h_j = 0, \quad for \ i=1 \cdots n; j=1 \cdots m \qquad (4.6)$$

Definition

The linearly independent set $\{f_1, \cdots, f_m\}$ is **involutive** *if, and only if*, there exist scalar functions $\alpha_{ijk} : \Re^n \to \Re$, such that

$$[f_i, f_j] = \sum_{k=1}^{m} \alpha_{ijk} f_k \quad \forall i, j, k$$

i.e. the Lie bracket of any pair of vector fields f_i and f_j is a linear combination of the basis fields.

Note that the coefficients in this linear combination are allowed to be smooth functions on $\Re^n$. In the simple case of only two vectors, then involutivity of the set $\{f_1, f_2\}$ defined by eqn. 4.6 is equivalent to the interchangeability of the order of partial derivatives.

The Frobenius Theorem, stated next, gives the conditions for the existence of a solution to the system of PDE in eqn. 4.6.

Frobenius Theorem:

The linearly independent set of vector fields $\{f_1, \cdots, f_m\} \in \Re^n$ is completely integrable *if, and only if*, it is involutive.

Definition

The class of systems

$$\dot{x} = f(x) + g(x)u$$

where f and g are smooth vector fields $\in \Re^n$ with $f(0) = 0$ is said to be feedback linearizable if

1. $\exists$ *a* region $M \in <^n$ (containing the origin)
2. A *diffeomorphism*[1] $T : M \to \Re^n$, and
3. A nonlinear feedback

$$u = \alpha(x) + \beta(x)v \qquad (4.7)$$

with $\beta(x) \neq 0$ on M, such that the transformed variables

$$z = T(x) \qquad (4.8)$$

satisfy

$$\dot{z} = Az + bv \qquad (4.9)$$

where

$$A = \begin{bmatrix} 0 & | & 1 & 0 & 0 & \cdots & 0 \\ 0 & | & 0 & 1 & 0 & \cdots & 0 \\ 0 & | & & \vdots & & & \\ 0 & | & 0 & 0 & 0 & 0 & 1 \\ - & - & - & - & - & - & - \\ 0 & | & 0 & 0 & 0 & 0 & 0 \end{bmatrix}, b = \begin{bmatrix} 0 \\ 0 \\ \vdots \\ 0 \\ - \\ 1 \end{bmatrix} \quad (4.10)$$

called the **Brunovsky form** (multiple integrator system)

When applied to the nonlinear system, the nonlinear transformation eqn. 4.8 and the nonlinear control law eqn. 4.7 result in a linear controllable system eqn. 4.9. The diffeomorphism $T(x)$ can be thought of as a nonlinear change of coordinates in the state space. The idea of feedback linearization is if one first changes to the coordinate system $z = T(x)$, then there exists a nonlinear control law to cancel the nonlinearities in the system. The feedback linearization is said to be global if the region U is all of $\Re^n$.

To determine the sufficient conditions on the vector fields f and g, let us then set $z = T(x)$ and see what conditions the transformation $T(x)$ must satisfy.

Given

$$z = Tx = \begin{bmatrix} T_1 \\ \vdots \\ T_n \end{bmatrix}$$

$$\dot{z} = \frac{\partial T}{\partial x} \dot{x}$$

$$= \frac{\partial T}{\partial x}\left(f(x) + g(x)u\right)$$

$$= Az + bv \qquad (4.11)$$

where $\dfrac{\partial T}{\partial x}$ is the Jacobian matrix.

---

[1] Recall the development of Hartman-Grobman theorem.

Opening up the first equation, i.e.

$$\frac{\partial T}{\partial x}\left(f(x)\right) = Az$$

we get

$$\frac{\partial T_1}{\partial x_1}\dot{x}_1 + \frac{\partial T_1}{\partial x_2}\dot{x}_2 + \cdots + \frac{\partial T_1}{\partial x_n}\dot{x}_n = T_2 \qquad (4.12)$$

More formally:

$$L_f T_1 + L_g T_1 u = T_2 \qquad (4.13)$$

Since we assume that T_i are independent of u while v is not,

$$L_g T_i = 0, \text{ for } i = 1 \cdots n - 1$$

but

$$L_g T_n \neq 0$$

hence, it is not difficult to see that

$$L_f T_i = T_{i+1} \text{ for } i = 1 \cdots n - 1 \qquad (4.14)$$

and

$$L_f T_n + L_g T_n u = v \qquad (4.15)$$

We will make use of the lemma here and bring in some more convenience. The lemma says that

$$L_{[f,g]}h = L_f L_g h - L_g L_f h$$

Replacing h with T_1,

$$L_{[f,g]}T_1 = L_f L_g T_1 - L_g L_f T_1$$

$$= 0 - L_g T_2$$

$$= 0$$

Using eqn. 4.2, we may derive the following:

$$L_{ad_f^k g}T_1 = 0 \text{ for } k = 0, 1 \cdots (n-2) \qquad (4.16)$$

and

$$L_{ad_f^{n-1} g}T_1 \neq 0 \qquad (4.17)$$

This system of PDE serves as the necessary and sufficient condition for the existence of T_1. If we are able to get one such T_1 satisfying eqns. 4.16 and 4.17, the rest of the T_i can be obtained from eqn. 4.14 and the control law from eqn. 4.15 as:

$$u = \frac{1}{L_g T_n}\left(-L_f T_n + v\right)$$

$$= -\frac{\overbrace{\dfrac{L_f T_n}{L_g T_n}}^{\alpha(x)}}{} + \underbrace{\dfrac{1}{L_g T_n}}_{\beta(x)} v$$

$$= \alpha(x) + \beta(x)v \qquad (4.18)$$

We have a system of PDE to solve for T_1, and the solution demands complete integrability. Therefore, bringing in Frobenius theorem, we have that

$$\exists T_1 \Leftrightarrow \left\{g, ad_f(g), ad_f^2(g), \cdots, ad_f^{n-2}(g)\right\} \qquad (4.19)$$

is linearly independent and involutive in M.

Equivalently, we need

$$\text{span}\left\{g, ad_f(g), \cdots, ad_f^{n-1}(g)\right\} = \Re^n \qquad (4.20)$$

One would observe that this sounds quite similar to the familiar linear controllability criterion: rank $[bAb\cdots A^{n-1}b] = n$

To sum up,

Theorem:

The nonlinear system $\dot{x} = f(x) + g(x)u$, with $f(x)$, $g(x)$ smooth vector fields, and $f(0) = 0$ is feedback linearizable *if, and only if,* there exists a region M containing the origin in $\Re^n$ in which the following conditions hold:

1. The vector fields $\{g, ad_f(g), \dots, ad_f^{n-1}(g)\}$ are linearly independent in M
2. The set $\{g, ad_f(g), \dots, ad_f^{n-2}(g)\}$ is involutive in M.

A proof of this has been provided as a derivation ahead of the Theorem. However, interested readers may proceed to formally prove both the necessity and sufficiency. There are three important corollaries of this theorem.

Corollary 1

A *planar* nonlinear system

$$\dot{x} = f(x) + g(x)u \quad x \in \Re^2 \qquad (4.21)$$

is locally state feedback linearizable in a neighborhood of the origin, *if, and only if,* its **linear approximation**

$$\dot{x} = \frac{df}{dx}(0)(x) + g(0)u \overset{\Delta}{=} Ax + bu$$

is controllable.

We will not provide a complete proof but a sketch of the necessity part would be exciting to the reader. The bracket may be found to be

$$ad_f(g)\mid_{x=0} = -\frac{\partial f}{\partial x}g(0) = -Ab$$

since $b = g(0) \neq 0$, b and Ab are linearly independent.

We encourage the reader to prove the sufficiency part and complete the proof.

Corollary 2

If the nonlinear system

$$\dot{x} = f(x) + g(x)u \quad x \in \Re^{n \geq 3} \qquad (4.22)$$

is (locally) state feedback linearizable, then its linear approximation about the origin

$$\dot{x} = \frac{df}{dx}(0)(x) + g(0)u \overset{\Delta}{=} Ax + bu$$

is controllable, i.e.

$$\text{rank}\left[b : Ab : \cdots : A^{n-1}b\right] = n$$

Corollary 3

The system in **triangular form**

$$\dot{x}_i = \phi_i\left(x_1, x_2, \cdots, x_i\right) \quad 1 \leq i \leq n-1$$

$$\dot{x}_n = \phi_n\left(x_1, x_2, \cdots, x_n\right) + u$$

where ϕ_i are smooth functions with $\phi_i(0) = 0$ is locally state feedback linearizable.

A proof of this is not needed if we follow our theorem and observe the pattern in which the linearizing transformations evolve in the present case:

$$z_1 = x_1$$

$$z_2 = x_2 + \phi_1(x_1)$$

$$z_3 = x_3 + \phi_2(x_1, x_2) + \frac{\partial \phi_1}{\partial x_1}\left(x_2 + \phi_1(x_1)\right)$$

$$\vdots$$

$$z_n = x_n + \phi_{n-1}(x_1, \cdots, x_{n-1})$$

$$+ \sum_{i=1}^{n-1} \frac{\partial \phi_i}{\partial x_i}\left(x_{i+1} + \phi_i(x_1, \cdots, x_i)\right)$$

and

$$v=\phi_n\left(x_1,x_2,\cdots,x_n\right)+u+\sum_{i=1}^{n-1}\frac{\partial\phi_i}{\partial x_i}\left(x_{i+1}+\phi_i(x_1,\cdots,x_i)\right)$$

We will revisit the triangular forms while learning the technique called Backstepping later in this chapter.

Thus, the computational procedure is as follows:

Algorithm for controller design

1. Check the existence and find a T_1
2. Using T_1 keep computing T_i, $i = 2\cdots n$
3. With T_n

$$u=\frac{1}{L_gT_n}\left(v-L_fT_n\right)$$

Let us now look at some examples.

Example 4.1.5: [14]

Let us consider

$$\dot{x}_1 = x_2 + x_1^2$$

$$\dot{x}_2 = u$$

Here

$$f(x)=\begin{bmatrix}x_1^2+x_2\\0\end{bmatrix}\text{ and }g(x)=\begin{bmatrix}0\\1\end{bmatrix}$$

$$\frac{\partial f}{\partial x}=\begin{bmatrix}2x_1&1\\0&0\end{bmatrix},\text{ and }\frac{\partial g}{\partial x}=\begin{bmatrix}0&0\\0&0\end{bmatrix}$$

Hence,

$$ad_f(g)=-\begin{bmatrix}1\\0\end{bmatrix}$$

Notice that g and $ad_f(g)$ are linearly independent. We now proceed to find a T_1, such that

$$L_gT_1=0$$

i.e.

$$\begin{bmatrix}\dfrac{\partial T_1}{\partial x_1}&\dfrac{\partial T_1}{\partial x_2}\end{bmatrix}\begin{bmatrix}0\\1\end{bmatrix}=0$$

Thus, we need a T_1 independent of x_2, and hence, let us choose

$$T_1=x_1$$

For instance, another choice could be $T_1=x_1+x_1^3$.

From this $T_1 = x_1$, we may compute T_2 to be

$$T_2=L_fT_1=x_1^2+x_2$$

In terms of the coordinate transformation, we have

$$z=T(x),\text{ i.e., }z_1=x_1,\text{ and }z_2=x_1^2+x_2$$

and the linearized system is

$$\dot{z}_1=z_2,\text{ and }\dot{z}_2=v$$

with

$$u=v-\left(2x_1x_2+2x_1^3\right)$$

We empahsize that the feedback linearizing transformation is not unique. The complete control system is shown in Figure 4.1.

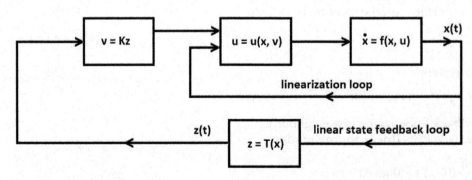

FIGURE 4.1
Feedback linearizing transformation.

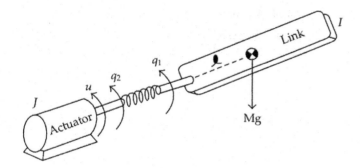

FIGURE 4.2
A robotic single-link arm.

Example 4.1.6: [15]

Consider the single-link arm, coupled through a gear train to a direct current (dc) motor, as shown in Figure 4.2. Let l and M be the length and mass, respectively, of the arm with the moment of inertia I, let the motor exert a torque u and rotate counterclockwise by a position q_2, and let the link, coupled to the motor via a spring with constant k, rotate by a position q_1. We assume that there is no damping.

The equations of motion can be written as

$$I\ddot{q}_1 + Mgl\sin q_1 + k(q_1 - q_2) = 0$$

$$J\ddot{q}_2 + k(q_2 - q_1) = u$$

Let us now define the state variables as:

$$x_1 = q_1, \ x_2 = \dot{q}_1, \ x_3 = q_2, \ \text{and} \ x_4 = \dot{q}_2$$

so that we can first put these dynamics in the standard form:

$$f = \begin{bmatrix} x_2 \\ -\dfrac{Mgl}{I}\sin x_1 - \dfrac{k}{I}(x_1 - x_3) \\ x_4 \\ \dfrac{k}{J}(x_1 - x_3) \end{bmatrix}, \ g = \begin{bmatrix} 0 \\ 0 \\ 0 \\ \dfrac{1}{J} \end{bmatrix}$$

Since $n = 4$, we need the set

$$\{g, ad_f(g), ad_f^2(g), ad_f^3(g)\}$$

to be linearly independent, and the set with the first three terms be involutive.

Indeed, we get

$$\begin{bmatrix} g & ad_f(g) & ad_f^2(g) & ad_f^3(g) \end{bmatrix} = \begin{bmatrix} 0 & 0 & 0 & \dfrac{k}{IJ} \\ 0 & 0 & \dfrac{k}{IJ} & 0 \\ 0 & \dfrac{1}{J} & 0 & -\dfrac{k}{J^2} \\ \dfrac{1}{J} & 0 & -\dfrac{k}{J^2} & 0 \end{bmatrix}$$

which has the rank 4, and that the vector fields are involutive. Hence, the system is feedback linearizable.

Further, the reader may work out and verify that the simplest T_1 is x_1. On this basis, we may build

$$z_1 = T_1 = x_1$$

$$z_2 = L_f T_1 = x_2$$

$$z_3 = L_f T_2 = -\frac{Mgl}{I}\sin x_1 - \frac{k}{I}(x_1 - x_3)$$

$$z_4 = L_f T_3 = -\frac{Mgl}{I}\cos x_1 - \frac{k}{I}(x_2 - x_4)$$

The control law is

$$u = \frac{1}{L_g T_4}(v - L_f T_4)$$

The linear sytem is

$$\dot{z} = Az + bv$$

where

$$A = \begin{bmatrix} 0 & 1 & 0 & 0 \\ 0 & 0 & 1 & 0 \\ 0 & 0 & 0 & 1 \\ 0 & 0 & 0 & 0 \end{bmatrix}, \ b = \begin{bmatrix} 0 \\ 0 \\ 0 \\ 1 \end{bmatrix} \tag{4.23}$$

Next, a few examples demonstrate the corollaries of the theorem.

Example 4.1.7:

The *planar* system

$$\dot{x}_1 = x_2^3, \ \dot{x}_2 = u$$

if linearized about the origin gives us

$$A = \begin{bmatrix} 0 & 0 \\ 0 & 0 \end{bmatrix}, \ B = \begin{bmatrix} 0 \\ 1 \end{bmatrix}$$

And, since this is *not* controllable, the nonlinear system is not state feedback linearizable.

Example 4.1.8:

Consider the system of Example 4.1.5 again:

$$\dot{x}_1 = x_2 + x_1^2$$

$$\dot{x}_2 = u$$

Here,

$$A = \begin{bmatrix} 0 & 1 \\ 0 & 0 \end{bmatrix}, \ B = \begin{bmatrix} 0 \\ 1 \end{bmatrix}$$

and hence, the nonlinear system is state feedback linearizable. Just for the sake of completion, let us take

$$z_1 = T_1 = x_1 + x_1^3$$

different from our earlier choice of $z_1 = x_1$. Building up further,

$$z_2 = L_f T_1 = x_2 + x_1^2 + 3x_1^2 x_2 + 3x_1^4$$

and

$$u = -\frac{2x_1 x_2 + 2x_1^2 + 6x_1 x_2^2 + 18x_1^3 x_2 + 12x_1^5}{3x_1^2 + 1}$$

$$+ \frac{1}{3x_1^2 + 1} v$$

Example 4.1.9:

Consider the following third-order system

$$\dot{x}_1 = x_2 + \frac{1}{3}x_3^3$$

$$\dot{x}_2 = x_3$$

$$\dot{x}_1 = u$$

In this system, we will find that the set

$$\{g, ad_f(g), ad_f^2(g)\}$$

is linearly independent. The first two vectors are not involutive; hence, the nonlinear system is *not* state feedback linearizable. However, if we linearize the system about the origin, we may readily verify that the linear approximation is controllable.

Example 4.1.10:

Consider the system

$$\dot{x}_1 = x_2 + x_1 x_3$$

$$\dot{x}_2 = x_3$$

$$\dot{x}_1 = u$$

This is *not* in the triangular form; nevertheless, it can be shown that the set

$$\{g, ad_f(g), ad_f^2(g)\}$$

is linearly independent, with the first two vectors are involutive. Hence, the nonlinear system is state feedback linearizable.

We encourage the reader to work out and verify that

$$z_1 = x_1 e^{-x_2}$$

$$z_2 = L_f T_1 = x_2 e^{-x_2}$$

$$z_3 = L_f^2 T_1 = x_3 \left(e^{-x_2} - x_2 e^{-x_2} \right)$$

so that

$$u = \frac{1}{L_g L_f^2 T_1} \left(-L_f^3 T_1 + v \right)$$

where

$$L_f^3 T_1 = x_3^2 e^{-x_2} \left(-2 + x_2 \right)$$

and

$$L_g L_f^2 T_1 = e^{-x_2} \left(1 - x_2 \right)$$

We close the disucssion on state feedback linearization with the following observations.

1. Referring to Figure 4.1, the linearization has been achieved by (i) a transformation on the state variables in the inner loop to make the closed loop system linear and (ii) an application of linear state feedback ideas. It is in this sense we call the technique a feedback linearization. Also, this is radically different from the Jacobian based linear-around-equilibrium kind of approximation.
2. The method is not without its challenges. To implement the control law, the new state components in z must be available. If they cannot be measured directly or they do not have any physical interpretation, we need to measure the

original state itself and compute the new state components.

3. We might also address, as in the linear theory, tracking steps and sinusoids. In this case, the desired motion must then be expressed in terms of the new state vector. This could be quite laborious. Instead, if we are able to linearize the input-output map, then we can more easily address the tracking problem. This leads to the idea of input-output linearization, which we deal with in Section 4.2.

4.1.1 Handling Multi-input Systems

The results of the previous section can be generalized [15] to the case of nonlinear systems with multiple inputs:

$$\dot{x} = f(x) + \sum_{i=1}^{p} g_i(x)u_i \overset{\Delta}{=} f(x) + G(x)u \quad x \in \Re^n \quad (4.24)$$

The conditions for feedback linearization of systems with multiple inputs is a bit difficult to state. However, we will first make use of our flexible robot link with the link position, q_1 and motor position q_2 (Example 4.1.6, Figure 4.2), and extend it to have p such links so as to accommodate p inputs. This way we can demonstrate that the conceptual idea is the same as the single-input case.

Consider the equations of motion

$$M(q)\ddot{q} + h(q,\dot{q}) = u \quad (4.25)$$

where $q_{2i}, i = 1, 2, \cdots, p$ are the positions of the p rotors, and $q_{2i-1}, i = 1, 2, \cdots, p$ are the positions of the p links. Accordingly, we define two p-dimensional vectors:

$$q_1 = \begin{bmatrix} q_1 & q_3 & \cdots & q_{2p-1} \end{bmatrix}^T \quad \text{and} \quad q_2 = \begin{bmatrix} q_2 & q_4 & \cdots & q_{2p} \end{bmatrix}^T$$

and assign four p-dimensional state variables $x_1, \cdots, x_4$ as follows:

$$x_1 = q_1, \ x_2 = \dot{q}_1, \ x_3 = q_2, \ and \ x_4 = \dot{q}_2$$

and rewrite the dynamics as follows:

$$\dot{x}_1 = x_2$$

$$\dot{x}_2 = -M(x_1)^{-1}h(x_1, x_2) + M(x_1)^{-1}u \quad (4.26)$$

A state feedback linearizing controller would then look like

$$u = M(x_1)v + h(x_1, x_2) \quad (4.27)$$

so that we get the linearized system as

$$\dot{x}_1 = x_2$$

$$\dot{x}_2 = v \quad (4.28)$$

We advise the reader to take a careful note that eqn. 4.28 is actually a set of p-second order systems of the form

$$\dot{x}_{i1} = x_{i2}$$

$$\dot{x}_{i2} = v_i, \quad i = 1, 2, \cdots, p \quad (4.29)$$

We now state the theorem for multi-input feedback linearization.

Theorem: Multi-Input Feedback Linearization

The nonlinear system in eqn. 4.24 is the state feedback linearizable, i.e. locally transformable in V_0 a neighborhood of the origin contained in M, into a linear controllable system in Brunovsky form by means of

1. A nonsingular state feedback

$$u = K(x) + \beta(x)v \quad \text{with } K(0) = 0$$

   where $K(x)$ is a smooth function from V_0 into $\Re^p$, $\beta(x)$ is a $p \times p$ matrix with smooth entries, nonsingular in V_0.
2. A local diffeomorphism in V_0

$$z = T(x), \quad \text{with } T(0) = 0$$

If, and only if, in M

1. $G_l = \text{Span} \left\{ ad_f^j g_i : 1 \leq i \leq p, 0 \leq j \leq l \right\}$, for $l = 0$ to $p-2$, is involutive and of constant rank, and
2. Rank $G_{p-1} = p$.

The Brunovsky form for the multi-input case has the linear block form:

$$\dot{z} = \begin{bmatrix} 0 & I & 0 & 0 \\ 0 & 0 & I & 0 \\ 0 & 0 & 0 & I \\ 0 & 0 & 0 & 0 \end{bmatrix} z + \begin{bmatrix} 0 \\ 0 \\ 0 \\ I \end{bmatrix} v \quad (4.30)$$

where I is an $p \times p$ identity matrix, 0 is an $p \times p$ zero matrix, $z \in \Re^{4p}$, and $v \in \Re^p$.

We now give a rather technical definition of controllability indices k_i, associated with the multi-input nonlinear system eqn. 4.24.

Definition: Controllability Indices

A set of controllability indices $\{k_1, k_2, \cdots, k_p\}$ *uniquely* associated with systems eqn. 4.24, which satisfy the conditions of the Multi-Input Feedback Linearization theorem, are defined as

$$k_i, 1 \le i \le p = \text{card } \{p_j \ge i : j \ge 0\} \quad (4.31)$$

with

$$p_0 = \text{rank } G_0$$

$$p_1 = \text{rank } G_1 - \text{rank } G_0$$

$$\vdots$$

$$p_{n-1} = \text{rank } G_{n-1} - \text{rank } G_{n-2}$$

and are invariant under feedback transformations.

To understand this idea, one may look at the popular controllability condition for linear systems:

$$\text{controllability} \Leftrightarrow \text{rank} \left[B{:}AB{:}\cdots{:}A^{n-1}B \right] = n$$

If the matrix B has p columns, corresponding to p inputs, the total number of columns of this composite matrix is np, of which it suffices to have n independent columns. While there is no harm having more than n independent columns, we cannot afford to have fewer than n. Since there are n blocks in the composite matrix, we expect there is at least one column in a block, independent of other columns. The controllability indices explore the relation between p and the composite matrix having full rank n. If this idea is somewhat clear, then one can extend it to its counterpart for nonlinear systems:

$$\text{span}\{G, ad_f(G), \cdots, ad_f^{n-1}(G)\} = \Re^n$$

We omit further technical details here, assuring the reader that if the nonlinear multi-input system is state feedback linearizable according to the theorem above, then the Brunovsky form has the linear block form as in eqn. 4.30.

Once this is obtained, we can proceed with the outer loop design; we have also the advantage of n decoupled subsystems.

4.2 Input-Output Linearization

Let us now consider the single-input single-output system:

$$\dot{x} = f(x) + g(x)u \quad \text{and} \quad y = \psi(x) \quad (4.32)$$

with all of our original requirements on smoothness of f, g, and ψ.

The derivative of the output is

$$\dot{y} = \frac{\partial \psi}{\partial x} \dot{x} \overset{\Delta}{=} L_f \psi(x) + L_g \psi(x)u \quad (4.33)$$

If $L_g \psi(x) = 0$, then $\dot{y}$ is independent of u. If we continue to compute the second derivative of y,

$$\ddot{y} = L_f^2 \psi(x) + L_g L_f \psi(x)u \quad (4.34)$$

Also, if $L_g L_f \psi(x) = 0$, then $\ddot{y}$ is independent of u. We repeat this process until we get

$$L_g L_f^{i-1}\psi(x) = 0, i=1,2,\cdots,\rho-1 \quad \text{and} \quad L_g L_f^{\rho-1}\psi(x) \ne 0 \quad (4.35)$$

Thus, u does not appear in the equations of $y, \dot{y}, \cdots, y^{(\rho-1)}$, but appears thereafter:

$$y^{(\rho)} = L_f^\rho \psi(x) + L_g L_f^{\rho-1}\psi(x)u \quad (4.36)$$

from which we get

$$u = \frac{1}{L_g L_f^{\rho-1}\psi(x)}\left(v - L_f^\rho \psi(x)\right) \quad (4.37)$$

which basically reduces the input-output map $y = \psi(x)$ to

$$y^{(\rho)} = v$$

The above control design strategy of first generating a linear input-output relation and then formulating a controller based on linear control is referred to as the input-output linearization approach, and it can be applied to many systems. If we need to differentiate the output of a system ρ times to generate an explicit relationship between the output y and input u, the system is said to have a relative degree ρ.

4.2.1 Definition: Relative Degree

The nonlinear system in eqn. 4.32 is said to have a relative degree ρ, $1 \le \rho \le n$, in the neighborhood containing the origin if

$$L_g L_f^{i-1} \psi(x) = 0, \ i = 1, 2, \cdots, \rho - 1 \ \text{ and } \ L_g L_f^{\rho-1} \psi(x) \ne 0 \tag{4.38}$$

This can be understood intuitively: If it took more than n differentiations, the system would be of an order higher than n; if the control input never appeared, the system would not be controllable.

For a reader with thorough knowledge of classical linear control, this should not be much of a surprise; we are just looking at the pole-zero excess. Recall that for strictly proper transfer functions the number of poles is always greater than that of zeros. Further, we wish to realize the system when the transfer function is a minimal realization, i.e. there is no *pole-zero* cancellation. There are two setbacks if we are tempted to control a linear system using the idea of pole-zero cancellation. First, the structural properties—controllability and observability—are lost. Secondly, if we bring in a zero in the right half of the s-plane to cancel an unstable pole, the system becomes **non-minimum phase,** which has certain obvious disadvantages.

In the present context of feedback linearization, when we attempt to cancel nonlinearities using suitable transformations, we need to be careful about potential pole-zero-like cancellation of the linear systems. Once again, the theory is too technical, but we attempt to provide the basic arguments using appropriate examples.

In Sections 4.1 and 4.2, we have primarily looked at those sysems that are *state feedback linearizable,* i.e. those systems are transformable into linear and controllable systems by state space change of coordinates. We might also expect, if we can transform the nonlinear system into a system with linear input-output map, then such systems are said to be *input-output feedback linearizable.* Before proceeding further, let us see an example.

Example 4.2.1: [14]

Consider the system

$$\dot{x}_1 = x_2 - x_2^2 u$$

$$\dot{x}_2 = u$$

$$y = x_1$$

We suggest to the reader to have a look at the relationship between the input u and the output y.

Using the results of Section 4.1, we may obtain the following transformation:

$$z_1 = x_1 + \frac{1}{3} x_2^3$$

$$z_2 = x_2$$

and the state feedback control is

$$u = k_1 \left(x_1 + \frac{1}{3} x_2^3 \right) + k_2 x_2 + v$$

and the closed loop system in z coordinates is

$$\dot{z}_1 = z_2$$

$$\dot{z}_2 = k_1 z_1 + k_2 z_2 + v$$

$$y = z_1 - \frac{1}{3} z_2^3$$

Glaringly, the input-output map is still nonlinear.

Example 4.2.2:

For the system

$$\dot{x}_1 = x_2 + \theta x_1^2$$

$$\dot{x}_2 = x_3^3 + u$$

$$\dot{x}_3 = x_1 + x_2^3 + x_3^3$$

$$y = x_1$$

there does not exist any transformation, for any constant parameter θ, to make it state feedback linearizable. However, it may be shown that the system is indeed input-output linearizable. For instance, if we propose

$$u = x_3^3 - 2\theta x_1 x_2 - 2\theta^2 x_1^3 + v$$

transforms the system to

$$\dot{x}_1 = z_2$$

$$\dot{z}_2 = v$$

$$\dot{x}_3 = x_1 + x_2^3 + x_3^3$$

$$= x_1 + \left(z_2 - \theta x_1^2 \right)^3 + x_3^3$$

$$y = x_1$$

which is partially linear.

For such systems that are not feedback linearizable, an immediate question is the characterization of the part that can be made linear.

4.2.2 Zero Dynamics and Non-Minimum Phase Systems

Let us consider the following system that has a similar structure to the one in Example 4.3.2:

$$\dot{x}_1 = x_2$$

$$\dot{x}_2 = x_3 + u$$

$$\dot{x}_3 = x_1 + x_2 + x_3$$

$$y = x_1$$

This linear system has the transfer function:

$$Y(s) = \frac{s-1}{s^3 - s^2 - s - 1} U(s)$$

Now, let us assume that we have applied the state feedback

$$u = -x_3 + v$$

so that the closed loop system is

$$\dot{x}_1 = x_2$$

$$\dot{x}_2 = v$$

$$\dot{x}_3 = x_1 + x_2 + x_3$$

$$y = x_1$$

whose transfer function is

$$Y(s) = \frac{s-1}{s^2(s-1)}$$

where there is a pole-zero cancellation; in state space we say the system has become unobservable. The canceled zero is in the right half of the s-plane; a *non-minimum phase zero*. With $x_1 = x_2 = 0$, the unstable dynamics $\dot{x}_3 = x_3$ evolve and are characterized by the pole (eigenvalue), which is equal to the canceled zero. Clearly, this is not desirable for reasons aforementioned.

Let us now look at the **inverse system** for the above. Since $x_1 = y$, $x_2 = \dot{y}$, we have

$$\dot{x}_3 = x_3 + y + \dot{y}$$

$$u = -x_3 + \ddot{y}$$

When this is driven by $y = \dot{y} = \ddot{y} = 0$, then its dynamics coincide with the unstable dynamics of the unobservable system, which caused the pole-zero cancellation.

The dynamics that are made unobservable by the input-output linearizing feedback are called **zero dynamics** and generalize to nonlinear systems the notion of zeros of linear systems. More precisely, they generalize the notion of dynamics evolving on unobservable subspaces. Thus, the existence of unstable zero dynamics implies non-minimum phase systems, and a linearizing state feedback is not feasible.

4.2.2.1 Definition: Partially State Feedback Linearizable

The nonlinear system in eqn. 4.32 is said to be locally, partially state feedback linearizable with index $r \leq n$ if it is locally feedback equivalent to the partially linear system:

$$\dot{\zeta} = \Gamma(\zeta, z), \quad \zeta \in \Re^{n-r}$$

$$\dot{z} = \begin{bmatrix} 0 & 1 & \cdots & 0 \\ 0 & \vdots & \ddots & 0 \\ 0 & 0 & 0 & 1 \\ 0 & 0 & 0 & 0 \end{bmatrix} z + \begin{bmatrix} 0 \\ 0 \\ \vdots \\ 1 \end{bmatrix} v \quad z \in \Re^r \quad (4.39)$$

If a system is partially state feedback linearizable with index $r \leq n$, it is also partially state feedback linearizable with index $r' < r$; if $r = n$, the system is locally state feedback linearizable.

Theorem:

The nonlinear system eqn. 4.32 is locally partially state feedback linearizable with index r if the following conditions hold:

1. The vector fields $\{g, ad_f(g), \ldots, ad_f^{r-1}(g)\}$ are linearly independent in M
2. The set $\{g, ad_f(g), \ldots, ad_f^{r-2}(g)\}$ is involutive in M.

Example 4.2.3:

Consider the system

$$\dot{x}_1 = x_2 + x_4^2 + u$$

$$\dot{x}_2 = x_4^2 + u$$

$$\dot{x}_3 = (x_2 - x_3)^3 + x_4^2 + x_4^3 + u$$

$$\dot{x}_4 = (x_2 - x_3)x_4^2 - x_4$$

We may determine that

$$ad_f(g) \neq 0, \quad ad_f^2(g) = 0$$

and that $r = 2$ since $ad_f(g) \notin$ span $\{g\}$. So, we may find a $T(x_1, x_2)$ such that

$$-\frac{\partial T_1}{\partial x_1} = \gamma \neq 0$$

$$\frac{\partial T_1}{\partial x_1} + \frac{\partial T_1}{\partial x_2} + \frac{\partial T_1}{\partial x_3} = 0$$

One possible solution is $T_1 = x_1 - x_2$ so that $T_2 = L_f T_1 = x_2$

The other two coordinates $\zeta_1(x)$ and $\zeta_2(x)$ must be

$$\zeta_i(0) = 0, \quad L_{\zeta_i(x)}(g) = 0, \quad i = 1, 2$$

i.e.

$$\frac{\partial \zeta_i}{\partial x_1} + \frac{\partial \zeta_i}{\partial x_2} + \frac{\partial \zeta_i}{\partial x_3} = 0 \quad i = 1, 2$$

Further, we also need to satisfy the rank condition:

$$\text{rank} \begin{bmatrix} d\zeta_1 \\ d\zeta_2 \\ dT_1 \\ dT_2 \end{bmatrix} = 4$$

One possibility is $\zeta_1 = x_3 - x_2$, and $\zeta_2 = x_4$

The final change of coordinates is a linear system as we can observe from the above. In z-coordinates, we have

$$\dot{z}_1 = z_2$$

$$\dot{z}_2 = z_4^2 + u$$

$$\dot{z}_3 = -z_3^3 + z_4^3$$

$$\dot{z}_4 = -z_3 z_4^2 - z_4$$

which by applying the feedback

$$u = v - z_4^2$$

becomes a partially linear system of the form given in eqn. 4.39.

4.3 Stabilization

In this section, let us investigate what happens if there are uncertainties or computational errors, which do not allow us to cancel the nonlinearities exactly. Let us

then consider the state feedback linearizable system with single-input:

$$\dot{z}_i = z_{i+1} \quad i = 1, 2, \cdots, n - 1$$

$$\dot{z}_n = v = \beta^{-1}(x)(u - \alpha x)$$

(after appropriate coordinate transformation).

Assuming that exact nonlinear cancellation is not possible, let us take that the control law is

$$u = \hat{\alpha}(x) + \hat{\beta}(x)v \tag{4.40}$$

where $\hat{\alpha}$ and $\hat{\beta}$ are computed versions of α and β, respectively. There might be several reasons behind this, such as computational errors, uncertainty, and even our own intention to simplify the control to facilitate online computation. If we now substitute this approximation in u and write the linearized system in matrix form, we get

$$\dot{z} = Az + b(v + \eta(z, v)) \tag{4.41}$$

where A and b conform to the Brunovsky form, and

$$\eta(z, v) = ((\beta^{-1}\hat{\beta} - 1)v + \beta^{-1}(\hat{\alpha} - \alpha)) \tag{4.42}$$

is the description of the uncertainty. Now we appreciate that the system is nonlinear as long as $\eta \neq 0$. Moreover, the nonlinearity is $b\eta$, which is in the range space of the input. This is popularly known as the **matching condition**. This turns out to be a very good idea as other nonlinear control strategies, such as backstepping and sliding mode control (discussed later in the chapter) stem from this formulation.

For the time being, we will write the control law v as a sum of two parts: $v + \Delta v$. The first part $v = Kz + v_d$ may be designed to *stabilize* the approximated linear part of the system, with $\eta = 0$, via standard pole placement technique. The second part Δv then helps us take care of the performance, such as tracking a reference trajectory v_d. Hence, Δv may be viewed as an additional inner control loop for performance. However, this is only a convenient perspective and, as we proceed further, we do not really need this Δv.

The outer loop, though as we argued in the previous paragraph could be a simple pole placement problem, has to guarantee stability in the face of uncertainty. Therefore, we choose a Lyapunov-based method to accomplish this task. Since the uncertainty $\eta(z, v)$ is unknown, it may be possible for us to get a known function $\rho(z, t)$, such that

$$\|\eta(z, v)\| \leq \rho(z, t) \tag{4.43}$$

The reader must have observed that the new function $\rho(.,.)$ is chosen deliberately to be independent of v. In practice, additional statistics, e.g. frequency response, of the uncertainty may be needed to choose the right $\rho(.,.)$. If this $\rho(.,.)$ bounds the uncertainty allowing us to have everything in eqn. 4.41 known, the outer loop control v may be obtained as follows.

First, it is convenient for us to assume that all the eigenvalues of the matrix A in eqn. 4.41 have negative real parts; this is not a very demanding assumption since $\eta(.,.)$ is supposed to have contained all such uncertain information, including the instability of A and then has been enveloped by $\rho(.,.)$. If A is indeed stable, then Lyapunov's stability theory applied to linear systems allows us to compute a unique symmetric positive definite matrix P satisfying the Lyapunov equation:

$$A^T P + PA + Q = 0 \qquad (4.44)$$

where Q is an arbitrary positive definite matrix, say the identity matrix. With this matrix P, the control law may be computed as

$$v = \begin{cases} -\rho(z,t) \cdot \dfrac{b^T Pz}{\| b^T Pz \|} & if \quad \| b^T Pz \| \neq 0 \\[3mm] 0 & if \quad \| b^T Pz \| = 0 \end{cases} \qquad (4.45)$$

Theorem:

Consider the Lyapunov function $V(z) = z^T Pz$. Then, with the control law in eqn. 4.45, $\dot{V} < 0$ along the solutions of the system eqn. 4.41.

We may approximate the control law in eqn. 4.45 to a continuous control law, with an $\varepsilon > 0$ as follows:

$$v = \begin{cases} -\rho(z,t) \cdot \dfrac{b^T Pz}{\| b^T Pz \|} & if \quad \| b^T Pz \| \geq \varepsilon \\[3mm] -\dfrac{\rho(z,t)}{\| b^T Pz \|} \cdot b^T Pz & if \quad \| b^T Pz \| < \varepsilon \end{cases} \qquad (4.46)$$

in which case we have the following theorem.

Theorem:

The solution $z(t)$ of the system eqn. 4.41 with initial condition $z(t_0)$ is uniformly ultimately bounded using the control law as seen in eqn. 4.46.

Example 4.3.1: [15]

Consider the single link robotic arm, with direct coupling for simplicity, with the equation of motion:

$$I\ddot{\theta} + Mgl \sin\theta = u$$

Let us choose the control law as

$$u \approx \hat{I}(v + \Delta v) + \widehat{Mgl}\sin\theta$$

If we wish to track a trajectory θ_d, in terms of the tracking error $e_1 = \theta - \theta_d$ and $e_2 = \dot{e}_1$,

$$\dot{e}_1 = e_2$$

$$\dot{e}_2 = \frac{1}{I}u - \frac{Mgl}{I}\sin\theta - \ddot{\theta}_d$$

$$= v + \underbrace{\left(\frac{\hat{I}}{I} - 1\right)v + \frac{\Delta Mgl}{I}\sin\theta - \ddot{\theta}_d}_{\eta(z,v)\,cf.\,eqn.4.42}$$

Since we assume that A in eqn. 4.41 has all its eigenvalues in the left half of the s-plane, let

$$A = \begin{bmatrix} 0 & 1 \\ -100 & -20 \end{bmatrix} \quad with \quad b = \begin{bmatrix} 0 \\ 1 \end{bmatrix}$$

so that we have the error dynamics:

$$\dot{e} = Ae + b(v + \eta)$$

Let us also have the worst case bounds

$$5 \leq I \leq 10 \quad and \quad 5 \leq Mgl \leq 10$$

and choosing

$$\hat{I} = \hat{Mgl} = 5$$

we will have the estimates

$$\left(\frac{\hat{I}}{I} - 1\right) \leq \frac{1}{2} = \alpha \quad and \quad \left| \frac{\Delta Mgl}{I}\sin\theta \right| \leq \frac{2}{5}$$

and hence,

$$\| \eta \| \leq \alpha(\| v \| + \rho) + \frac{2}{5} = \rho$$

from which we get

$$\rho = \| v \| + \frac{4}{5}$$

With $Q = I$, solving the Lyapunov equation for P, we get

$$P = \begin{bmatrix} 2.625 & 0.005 \\ 0.005 & 0.2525 \end{bmatrix}$$

and we choose v according to eqn. 4.45.

Next, we will look at a special case of feedback linearization, where the system of equations has a nice triangular structure.

4.4 Backstepping Control

This part briefly discusses the fundamentals of a technique called **the recursive backstepping control** [16], related to feedback linearization. The idea of backstepping is suitable for strict-feedback systems, which are also known as *lower triangular*. An example of a strict-feedback systems is:

$$\dot{x}_i = \phi_i\left(x_1, x_2, \cdots, x_i\right) \quad 1 \le i \le n-1$$

$$\dot{x}_n = \phi_n\left(x_1, x_2, \cdots, x_n\right) + u \tag{4.47}$$

where ϕ_i are smooth functions with $\phi_i(0) = 0$. Recall that, from corollary 3 of this chapter, such systems are state feedback linearizable.

A typical feedback linearization approach, in most cases, leads to the cancellation of useful nonlinearities. The backstepping design exhibits more flexibility compared to feedback linearization since it does not require that the resulting input-output dynamics be linear. Cancellation of potentially useful nonlinearities can be avoided resulting in less complex controllers. The main idea is to use some of the state variables of eqn. 4.47 as "virtual controls" and, depending on the dynamics of each state, design intermediate control laws. For a straight picture in the mind, we encourage the reader to think of the first canonical form in linear state space theory where the input to an integrator is the output of the previous integrator and so on. The difference with triangular systems is that, in addition to an input from the previous stage output, there shall also be an output feedback from the current stage.

We will briefly illustrate the idea using a linear time-invariant system, with the following transfer function written in the factored form:

$$\frac{y(s)}{u(s)} = K \cdot \frac{s+z_1}{s+p_1} \cdot \frac{s+z_2}{s+p_2} \cdot \frac{s+z_3}{s+p_3}$$

$$= K \cdot \frac{y_1(s)}{u_1(s)} \cdot \frac{y_2(s)}{u_2(s)} \cdot \frac{y_3(s)}{u_3(s)} \tag{4.48}$$

Notice that $u_1(s)=y_2(s)$, $u_2(s)=y_3(s)$, and $u_3(s)=u(s)$.

Each of the factors

$$\frac{y_i(s)}{u_i(s)} = \frac{s+z_i}{s+p_i} \quad i=1,2,3$$

has the following state model:

$$x_i = y_i - u_i$$

$$or \quad y_i = x_i + u_i$$

$$and \quad \dot{x}_i = -p_i y_i + z_i u_i \tag{4.49}$$

The schematic for the third order system may be obtained as shown in Figure 4.3 by *cascading* the individual schematics.

The state model from this schematic is

$$\Sigma^p(A,B,C,D):$$

$$\begin{bmatrix} \dot{x}_1 \\ \dot{x}_2 \\ \dot{x}_3 \end{bmatrix} = \begin{bmatrix} -p_1 & z_1-p_1 & z_1-p_1 \\ 0 & -p_2 & z_2-p_2 \\ 0 & 0 & -p_3 \end{bmatrix} \begin{bmatrix} x_1 \\ x_2 \\ x_3 \end{bmatrix} + K \cdot \begin{bmatrix} z_1-p_1 \\ z_2-p_2 \\ z_3-p_3 \end{bmatrix} u$$

$$y = \begin{bmatrix} 1 & 1 & 1 \end{bmatrix} x + \begin{bmatrix} K \end{bmatrix} u \tag{4.50}$$

Notice that the system matrix is upper triangular. The eigenvalues, i.e. the diagonal elements, are once again the poles of the system.

Example 4.4.1:

Let the linear system be

$$y^{(3)} + 6y^{(2)} + 11y^{(1)} + 6y = 7u$$

The schematic is shown in Figure 4.4. From this, we would get the following matrices:

$$A = \begin{bmatrix} -1 & 1 & 0 \\ 0 & -2 & 1 \\ 0 & 0 & -3 \end{bmatrix} \quad B = \begin{bmatrix} 0 \\ 0 \\ 1 \end{bmatrix}$$

$$C = \begin{bmatrix} 1 & 0 & 0 \end{bmatrix} \quad D = \begin{bmatrix} 0 \end{bmatrix}$$

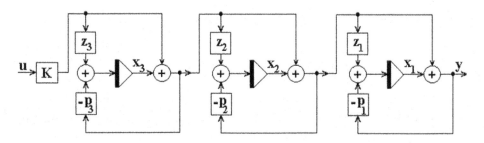

FIGURE 4.3
Schematic of a third order system.

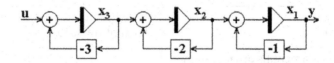

FIGURE 4.4
Schematic for Example 4.4.1.

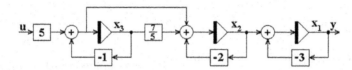

FIGURE 4.5
An augmented schematic for Example 4.4.1.

If the linear time-invariant lumped (LTIL) system is

$$y^{(3)} + 6y^{(2)} + 11y^{(1)} + 6y = 5u^{(1)} + 7u$$

The schematic is shown in Figure 4.5 and the model is

$$A = \begin{bmatrix} -3 & 1 & 0 \\ 0 & -2 & \dfrac{2}{5} \\ 0 & 0 & -1 \end{bmatrix} \quad B = \begin{bmatrix} 0 \\ 1 \\ 1 \end{bmatrix}$$

$$C = \begin{bmatrix} 1 & 0 & 0 \end{bmatrix} \quad D = \begin{bmatrix} 0 \end{bmatrix}$$

The backstepping design is a recursive procedure where a Lyapunov function is derived for the entire system. The recursive procedure can be easily expanded from the nominal case of a system augmented by an integrator. This is also referred to as integrator backstepping. Based on the design principles of the integrator backstepping, the control design can be easily expanded for the case of strict-feedback systems given in eqn. 4.47.

An illustration of the backstepping procedure based on the generic formulation of the strict feedback systems is given in eqn. 4.47 and would result in the derivation of tedious recursive formulas that are difficult to follow. Instead, we explain the idea using an example.

Example 4.4.2:

Let us consider a simple third order strict feedback system:

$$\dot{x}_1 = f_1(x_1) + x_2$$

$$\dot{x}_2 = f_2(x_2) + x_3$$

$$\dot{x}_3 = u$$

The objective is to design a state feedback control law, such that $x_i \to 0$ as $t \to 0$. The idea is to use the state variable x_2 as an input for stabilizing x_1.

Consider the Lyapunov function

$$V_1 = \frac{1}{2} x_1^2 \tag{4.51}$$

so that

$$\dot{V}_1 = x_1 \left(f_1(x_1) + x_2 \right) \tag{4.52}$$

The objective of this step is to find a virtual control law $\phi_2(x_1)$ with $\phi_2(0) = 0$, such that when $x_2 = \phi_2(x_1)$ then $\dot{V}_1(x_1) \le \hat{a} - W_1(x_1)$ where W_1 is a positive definite function for every $x_1 \in \Re$.

An obvious choice would be to remove the effect of the function $f_1(x_1)$ and inject a stabilizing feedback term. Thus, we pick:

$$\phi_2(x_1) = -f_1(x_1) - k_1 x_1 \tag{4.53}$$

where k_1 is a positive gain. This choice readily yields $\dot{V}_1 = -k_1 x_1^2 < 0$.

Let us denote the error $e_2 = x_2 - \phi_2(x_1)$. Using the new coordinate e_2, the given system can be written as:

$$\dot{x}_1 = -k_1 x_1 + e_2$$

$$\dot{e}_2 = -\dot{\phi}_2(x_1) + f_2(x_1, e_2) + x_3$$

$$\dot{x}_3 = u$$

Notice that the implementation of the derivative $\dot{\phi}_2(x_1)$ does not require a differentiator since

$$\dot{\phi}_2(x_1) = \frac{\partial \phi_2}{\partial x_1} \left(f_1(x_1) + x_2 \right) \tag{4.54}$$

Now, let

$$V_2(x_1,x_2)=\frac{1}{2}e_2^2 \tag{4.55}$$

The goal of this second design step is to determine a virtual control $\phi_3(x_1,e_2)$ with $\phi_3(0,0)=0$, such that when $x_3=\phi_3(x_1,e_2)$ then $\dot{V}_2(x_1,e_2)\leq\hat{a}-W_2(x_1,e_2)$ where W_2 is a positive definite function for every $x_1,e_2\in\mathfrak{R}$.

Consequently,

$$\dot{V}_2=-k_1x_1^2+e_2\left(x_1-\dot{\phi}_2(x_1)+f_2(x_1,e_2)+\phi_3(x_1,e_2)\right) \tag{4.56}$$

An obvious choice would be:

$$\phi_3=-x_1+\dot{\phi}_2-f_2(x_1,e_2)-k_2e_2 \tag{4.57}$$

where k_2 is a positive constant. Using the change of variables $e_3=x_3-\phi_3(x_1,e_2)$, the system dynamics become:

$$\dot{x}_1=-k_1x_1+e_2$$

$$\dot{e}_2=-x_1-k_2e_2+e_3$$

$$\dot{x}_3=-\dot{\phi}_3(x_1,e_2)+u$$

Once again, we need not compute the derivative $\dot{\phi}_3$.

Now that the pattern is clear, with

$$V_3=V_2+\frac{1}{2}e_3^2 \tag{4.58}$$

as a candidate Lyapunov function, the choice of u is:

$$u=-e_2+\dot{\phi}_3(x_1,e_2)-k_3e_3 \tag{4.59}$$

where k_3 is a positive constant. This choice yields:

$$\dot{V}_3=-k_1e_1^2-k_2e_2^2-k_3e_3^2 \tag{4.60}$$

Therefore, the origin of the error system is globally asymptotically stable. Since $\phi_2(0)=\phi_3(0,0)=0$, then $x_i\to0$ as $t\to0$.

The final system dynamics are

$$\begin{bmatrix}\dot{x}_1\\\dot{e}_1\\\dot{e}_2\end{bmatrix}=\begin{bmatrix}-k_1&1&0\\-1&-k_2&1\\0&-1&-k_3\end{bmatrix}\begin{bmatrix}x_1\\e_1\\e_2\end{bmatrix} \tag{4.61}$$

As one would readily notice, the system matrix is composed by the sum of a negative diagonal and a skew-symmetric matrix. This is a typical structural pattern when the backstepping design is based on a sequential construction of the Lyapunov functions. Notice the similarity with the linear case given at the beginning of this part.

The key feature of the backstepping methodology is the fact that it provides significant design freedom. The choice of the virtual controls ϕ_2 and ϕ_3 and the control input u is not unique. For example, an alternative could have been

$$\phi_2(x_1)=-f_1(x_1)-k_1x_1$$

$$\phi_3(x_1,e_2)=\dot{\phi}_2(x_1)-f_2(x_2)-k_2e_2$$

$$u=\dot{\phi}_3(x_1,e_2)-k_3e_3 \tag{4.62}$$

and the stabilization of the same system could have been achieved with a much simpler design. Apparently, the backstepping methodology is a powerful design tool for the development of simplistic controllers for practical nonlinear systems.

An interesting case study in aerospace is presented in Chapter 14. For now, let us illustrate the idea behind the design with the following numerical examples.

Example 4.4.3:

In this example, we will see the second order system:

$$\dot{x}_1=x_1^2-x_1^3+x_2$$

$$\dot{x}_2=u$$

In the first equation, we will consider x_2 as the input and design the feedback control:

$$x_2=\phi_2(x_1)=-x_1^2-x_1$$

to obtain

$$\dot{x}_1=-x_1^3-x_1$$

and propose the Lyapunov function:

$$V_1(x_1)=\frac{1}{2}x_1^2$$

It is easy to see that

$$\dot{V}_1=-x_1^4-x_1^2\leq-x_1^2\ \forall x_1\in\mathfrak{R}$$

and hence, the origin is asymptotically stable.

Let us backstep with the change of variables:

$$e_2=x_2-\phi_2(x_1)=x_2+x_1+x_1^2$$

so that the dynamics of the system become

$$\dot{x}_1=-x_1^3-x_1+e_2$$

$$\dot{e}_2=u+(1+2x_1)(-x_1^3-x_1+e_2)$$

Let us now propose

$$V_2(x_1,x_2)=V_1+\frac{1}{2}e_2^2$$

A little exercise gives us

$$\dot{V}_2 = x_1\dot{x}_1 + e_2\dot{e}_2$$

$$= x_1\left(-x_1^3 - x_1 + e_2\right)$$

$$+ e_2\left(u + (1+2x_1)\left(-x_1^3 - x_1 + e_2\right)\right)$$

The choice of u is clear:

$$u = -x_1 - (1+2x_1)\left(-x_1^3 - x_1 - e_2\right)$$

so that

$$\dot{V}_2 = -x_1^4 - x_1^2 - e_2^2$$

And, the origin is globally asymptotically stable.

Example 4.4.4:

Let us now take up a third order system:

$$\dot{x}_1 = x_1^2 - x_1^3 + x_2$$

$$\dot{x}_2 = x_3$$

$$\dot{x}_3 = u$$

Compared to Example 4.4.3, we have one more integrator and we will quickly look at the results of this example as an extension of the results of the previous example. However, we will approach this (for learning sake) example in a direction opposite to that in Example 4.4.2.

Backstepping once, the second order system is

$$\dot{x}_1 = x_1^2 - x_1^3 + x_2$$

$$\dot{x}_2 = x_3$$

with x_3 as input can be made asymptotically stable by the choice

$$x_3 \overset{\Delta}{=} \phi_3(x_1, x_2)$$

$$= -x_1 - (1+2x_1)\left(x_1^2 - x_1^3 + x_2\right) - \left(x_2 + x_1 + x_1^2\right)$$

with the Lyapunov function

$$V_2(x_1, x_2) = \frac{1}{2}x_1^2 + \frac{1}{2}\left(x_2 + x_1 + x_1^2\right)^2$$

Backstepping once again, we will change the variables as

$$e_3 = x_3 - \phi_3(x_1, x_2)$$

to rewrite the dynamics as

$$\dot{x}_1 = x_1^2 - x_1^3 + x_2$$

$$\dot{x}_2 = \phi_3(x_1, x_2) + e_3$$

$$\dot{e}_3 = u - \frac{\partial\phi_3}{\partial x_1}\left(x_1^2 - x_1^3 + x_2\right) - \frac{\partial\phi_3}{\partial x_2}\left(\phi_3 + e_3\right)$$

We now routinely propose the Lyapunov function:

$$V_3 = V_2 + \frac{1}{2}e_3^2$$

We now leave it to the reader to verify that

1. The choice of control is

$$u = -\frac{\partial V_2}{\partial x_2} + \frac{\partial\phi_3}{\partial x_1}\left(x_1^2 - x_1^3 + x_2\right) + \frac{\partial\phi_3}{\partial x_2}\left(\phi_3 + e_3\right) - e_3$$

which yields

$$\dot{V}_3 = -x_1^2 - x_1^4 - \left(x_2 + x_1 + x_1^2\right)^2 - e_3^2$$

2. And that the origin is globally asymptotically stable.

4.5 Sliding Mode Control

In this last part of the chapter, we will briefly explore a design technique that has, in all likelihood, attracted a vast majority of engineers as well as researchers for more than five decades now. Historically, there were other terms, particularly Variable Structure Systems (VSS), which have been used to refer to this technique, but Sliding Mode Control has more or less became universally accepted. There were reports of its application to power systems in Japan in as early as the 1970s. Subsequently, there have been applications in robotics and power electronics among many. As the reader will see, this technique is a very interesting combination of both intuition—as intuitive as a thermostat in homes—and a rigorous nonlinear system theory—we will bring in the ideas of Lyapunov theory, feedback linearization, and relative degree.

We will begin with an example [15] that throws some light on the theoretical ideas.

Example 4.5.1:

Consider the second order autonomous system:

$$\dot{x}_1 = x_2, \quad \dot{x}_2 = -\mu_1 x_1$$

Our earlier experience with the phase plane would immediately suggest that

1. If $\mu_1 > 0$, the eigenvalues $\pm j\sqrt{\mu_1}$ are purely imaginary and we get a center.
2. If $\mu_2 > 0$ is added to μ_1, still we get a center, but the magnitude of the eigenvalues is larger.

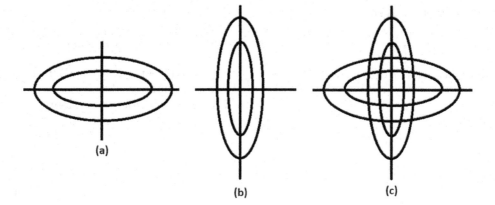

FIGURE 4.6
Illustration of sliding modes. (Adapted from Spong, M.W. et al., *Robot Modelling and Control*, John Wiley & Sons, Hoboken, NJ, 2005.)

Typical trajectories are shown in Figure 4.6. Figure 4.6a is the case of μ_1 alone, Figure 4.6b when μ_2 is added, and Figure 4.6c is a superposition of the trajectories in both the above cases.

Before we comment further on the figures, let us see the following *logical* system:

$$\dot{x}_1 = x_2, \quad \dot{x}_2 = -\left(\mu_1 + \alpha\mu_2\right)x_1$$

where $\alpha = 0$ corresponds to the system with eigenvalues $\pm j\sqrt{\mu_1}$, Figure 4.6a, and $\alpha = 1$ describes the modified system.

If we now consider α as a control parameter and implement the system as

$$\alpha = \begin{cases} 1 & if \quad x_1 x_2 > 0 \\ \\ 0 & if \quad x_1 x_2 < 0 \end{cases}$$

From Figure 4.3c, we may understand that whenever both the trajectories are in first and third quadrants α will be 1 and it is 0; otherwise, this works like *switching* the thermostat on and off, and the trajectories jump from bigger ellipses to smaller ones once in every quadrant, eventually spiral towards the equilibrium point, the origin. In other words, by switiching, we made a marginally stable system an asymptotically stable one. Notice that it is not only the dynamics that are important, but the switching logic also.

To implement the switching logic appropriately, we need to first define a quantity—the sliding surface—in the state space. Secondly, once a trajectory reaches this surface we need to switch the control input so that the trajectory, in some sense, gets trapped on the surface and slides towards the origin, possibly in finite time. Since the design of sliding surface is in our hands, we should be able to handle the uncertainties as well.

To intuitively understand the theory ahead, suppose we need the state x to follow a specific trajectory, say, x_d. If the error is $e = x - x_d$, then what else choice than the linear error dynamics:

$$\dot{e} + \tau e = 0$$

would assure us asymptotic stability? Accordingly, our chose of sliding surface would be

$$S = \dot{e} + \tau e$$

When we deal with the state vector having more than one component, S is more appropriately called as a *hypersurface*.

Taking the derivative of S we get

$$\dot{S} = \ddot{e} + \tau\dot{e} = \ddot{x} + \tau\dot{e} - \ddot{x}_d \qquad (4.63)$$

In $\ddot{x}$, we would usually (many examples in the previous parts of this chapter) come across the control input u; hence, we have systems with relative degree 1. We will be able to *choose* the control signal u to cancel the nonlinearities and ensure stability of the error dynamics. It is easy to see that for scalar second order systems with $x_2 = \dot{x}_1$, the sliding surface in the phase plane is a straight line in second and fourth quadrants (when $x_1 x_2 < 0$) passing through the origin.

Let us now go ahead with the theory briefly. Suppose we have the following nonlinear system

$$\dot{x} = f(x) = \begin{cases} f^+(x) & if \quad S(x) > 0 \\ \\ f^-(x) & if \quad S(x) < 0 \end{cases} \qquad (4.64)$$

We assume that the vector fields $f^\pm(x)$ are well defined.

Since the system is not defined for $S = 0$, we shall take

$$\lim_{x \to 0^-} f(x) = f^-(x)$$

$$\lim_{x \to 0^+} f(x) = f^+(x)$$

Taking the derivative of S, we get

$$\dot{S}=L_f S \tag{4.65}$$

To attract the trajectories in the vicinity of S towards the surface, we may find that it is sufficient to have

$$\lim_{S\to 0^-}\dot{S}>0 \quad and \quad \lim_{S\to 0^+}\dot{S}<0 \tag{4.66}$$

These two can be combined into one equation:

$$\lim_{S\to 0}S\dot{S}<0 \tag{4.67}$$

Also, our ideas of Lyapunov theory tell us that the most appropriate Lyapunov function is

$$V(x)=\frac{1}{2}S^2(x) \tag{4.68}$$

Still, we have certain difficulty in defining the trajectory on the surface when $S=0$. To overcome this, we first consider that the logic will ideally switch infinitely between $f^{\pm}(x)$ on the surface. Secondly, the system should respond as if it is being governed by

$$\dot{x}=f_0(x)=p_1 f^+(x)+p_2 f^-(x), \quad p_1+p_2=1 \tag{4.69}$$

In order for the system to remain on the sliding surface, the vector field $f_0(x)$ must be tangential to the surface. Accordingly,

Theorem:

Defining

$$f_0^+(x)=\lim_{s\to 0}L_{f^+}S, \quad and \quad f_0^-(x)=\lim_{s\to 0}L_{f^-}S \tag{4.70}$$

then, if a trajectory exists on the surface $S=0$, we have

$$f_0^-(x)-f_0^+(x)>0 \tag{4.71}$$

and

$$p_1(x)=\frac{f_0^-(x)}{f_0^-(x)-f_0^+(x)} \tag{4.72}$$

A proof of this theorem directly follows the definition of Lie derivatives and is left as an exercise to the reader.

The original solution to the problem by Fillipov is difficult to implement as a controller. Instead, we obtain the control signal, called the equivalent control, as follows.

The two conditions:

$$S=0 \quad and \quad \dot{S}=0 \tag{4.73}$$

basically tell us that the trajectory is on the surface and that it does not leave the surface.

If we now consider the system:

$$\dot{x}=f(x)+g(x)u \tag{4.74}$$

then

$$\dot{S}=L_{f+gu}S=L_f S+L_g Su=0 \tag{4.75}$$

from which we get the *equivalent control* as

$$u_{eq}=-\frac{L_f S}{L_g S} \tag{4.76}$$

With this as the control, the dynamics *on* the sliding surface are

$$\dot{x}=f(x)-g(x)\frac{L_f S}{L_g S} \tag{4.77}$$

The equivalent control is simple depending on the given system's parameters. Nevertheless, parametric uncertainty would certainly influence the control. Therefore, we do not *calculate* this equivalent control straightaway, but rather arrive at it "on an average" approach.

We will define the switching control

$$u=\begin{cases} -u^+ & if \quad S>0 \\ \\ -u^- & if \quad S<0 \end{cases} \tag{4.78}$$

Just for convenience, let us use the symbol u alone in the following. From eqns. 4.72 and 4.73, we have

$$\dot{S}=L_g S(u-u_{eq}) \tag{4.79}$$

Therefore, from the sliding mode conditions in eqn. 4.63, a sliding mode exists using the contol law of eqn. 4.75 *if, and only if,*

$$u^-<u_{eq}<u^+ \tag{4.80}$$

These bounds give us a lot of freedom in the choice of the control signal, while sliding. This should remind the reader of employing a bisection method to solve for the roots of a polynomial by initially guessing two of them on either side of the actual root. It is in this sense we talk of arriving at the surface on an average.

One popular way of implementing this control is as

$$u = -K \mid \hat{u}_{eq} \mid sgn(S) \qquad (4.81)$$

where $\hat{u}_{eq}$ is an estimate of the equivalent control, K is a gain, and $sgn(S)$ is the signum function:

$$sgn(S) = \begin{cases} 1 & if \quad S > 0 \\ -1 & if \quad S < 0 \end{cases} \qquad (4.82)$$

Example 4.5.2:

Consider the single link robotic arm of example 4.3.1 once again:

$$I\ddot{\theta} + Mgl\sin\theta = u$$

and choose the inner loop control law as

$$u \approx \hat{I}v + Mgl\sin\theta$$

so that

$$\ddot{\theta} = \frac{\hat{I}}{I}v + \frac{\Delta Mgl}{I}\sin\theta$$

Let us assume that the desired trajectory θ_d is a constant so that the sliding surface is

$$S = \dot{\theta} + \tau(\theta - \theta_d)$$

Then

$$\dot{S} = \ddot{\theta} + \tau\dot{\theta}$$

$$= \tau S - \tau^2(\theta - \theta_d) + \frac{\hat{I}}{I}v + \frac{\Delta Mgl}{I}\sin\theta$$

We can now identify the equivalent control as

$$v_{eq} = \frac{\hat{I}}{I}\tau^2(\theta - \theta_d) - \frac{\Delta Mgl}{I}\sin\theta$$

As before, let us assume that

$$5 < I, Mgl < 10$$

and we take $\hat{I} = 5 = \widehat{Mgl}$. With these estimates we get

$$\mid v_{eq} \mid \leq 2\tau^2 \mid (\theta - \theta_d) \mid + \frac{2}{5}$$

Therefore, we choose

$$\mid \hat{v}_{eq} \mid = 2\tau^2 \mid (\theta - \theta_d) \mid + \frac{2}{5}$$

and

$$v_{eq} = -K \mid \hat{u}_{eq} \mid sgn(S)$$

In practice, the small but non-zero switching delay causes what is known as **chattering**—the trajectory slightly overshoots the surface each time the control is switched. This phenomenon is both useful, as in modern power electronic circuits, as well as undesirable, as in hydraulic actuators. We recommend the reader to some good references in the References for Section 1.

Finally, to ensure asymptotic stability in practice, it is sufficient to show that

$$\dot{S} < -\varepsilon S, \quad for \; some \; \varepsilon > 0$$

4.6 Chapter Summary

Most mechanical systems, robotic examples in this chapter and aerospace examples in Chapter 14, are governed by second order differential equations and evolve on nonlinear manifolds. At a superficial level, with more linear theory behind, typical control techniques for these systems rely on the introduction of local coordinates so as to set up a smooth one-to-one correspondence between the actual state space and Cartesian space within an admissible range. Although this converts control on a manifold to control in Cartesian space, the engineer would have enjoyed the intuition. The issue of singularity exists for any choice of local coordinates on certain nonlinear manifolds, such as the special orthogonal group SO(3). We have purposefully avoided the kinematics, inverse kinematics, and the possibility of singularities in this text, while placing emphasis on touching upon the controller desgin. Interested readers may look into books [17–24] on dynamics and control.

Geometric control is developed to circumvent this dilemma of singularity in mechanical systems. Contrary to controllers that are based on Cartesian spaces, geometric controllers treat the configuration spaces of mechanical systems as manifolds and develop feedback laws based on differential geometry. The idea of feedback linearization is fairly generic, and accordingly, we have devoted good amount of space on that. Subsequently, we have introduced backstepping control design methodology, as well as the sliding mode control, that enable the construction of a Lyapunov candidate function iteratively.

In Chapter 14, we present a case where a hybrid controller—intelligent and nonlinear—has been designed for an interesting aerospace application.

Appendix IA: Performance Error Criteria

In any control system, it is important to monitor and evaluate the performance of the control design, for which some criteria are specified in terms of transient-responses to specific inputs; like step, ramp, or parabolic. Also, these criteria are used for the design of the control systems [AIA.1], wherein these criteria are to be minimized with respect to the parameters that are sought in order to obtain best design. The discrete time versions of some of these and other performance indices are presented in Appendix IIB.

IA.1 Performance Indexes

A performance index (PI) indicates the "goodness" of (closed loop) control system's performance. In control analysis and design, the parameters and state variables/control inputs should be chosen so that a PI is either minimum or maximum, depending upon how the index is defined. So, the optimal values of these parameters depend on the chosen PI, which must be a function of the parameter of the control system and should be easily computable. The error PIs are integrals of some functions or the weighted functions of the error in the system, say a deviation of the actual output from the desired output that is also called as (closed) loop error. Some important error performance indexes are discussed next.

IA.1.1 Integral Square Error Criterion

The integral square error (ISE) is specified as

$$ISE = \int_0^\infty e^2(t)dt \qquad (IA.1)$$

In (IA.1), the error is defined as difference of the desired output and the actual output:

$$e(t) = y_d(t) - y(t); \ \lim_{e \to \infty} e(t) = 0; \ \text{OR}$$
$$e(t) = y_d(\infty) - y(t) \qquad (IA.2)$$

The actual output could be a measured output (or true output in certain filtering problems). The optimal system would be such that the *ISE* is minimized. Interestingly, this PI is suitable for deterministic as well as stochastic control systems. This PI weighs large error heavily, and small error lightly. Due to this, a system designed by this criterion will have a fast decrease in a large initial error, and the response would be fast, yet oscillatory, indicating a poor relative stability. The significance of

this PI criterion is that, for example, the minimization of the PI results in the minimization of power consumption for a spacecraft system.

IA.1.2 Integral-of-Time-Multiplied Square-Error Criterion

This *ITSE* PI is defined as

$$ITSE = \int_0^\infty t \cdot e^2(t)dt \qquad (IA.3)$$

Due to the explicit appearance of time *t* in (IA.3), a large initial error is weighted lightly and the errors occurring at later stages are weighted heavily.

IA.1.3 Integral Absolute Square-Error Criterion

This *IAE* criterion is defined as

$$IAE = \int_0^\infty |e(t)| dt \qquad (IA.4)$$

With this criterion, the highly overdamped/underdamped systems cannot be made optimum, it will yield a design of the control system that has reasonable damping, and satisfactory transient response characteristics. The minimization of *IAE* is directly related to the minimization of the fuel consumption, for example of a spacecraft system.

IA.1.4 Integral-of-Time-Multiplied Absolute-Error Criterion

This *ITAE* PI is defined as

$$IAE = \int_0^\infty t \cdot |e(t)| dt \qquad (IA.5)$$

A large initial error is weighted lightly and the errors occurring at later stages are weighted heavily. The design with this criterion would yield a system with a small overshoot, and the oscillations would be well damped.

Reference

AIA.1 Ogata, K. *Model Control Engineering.* Prentice Hall of India, New Delhi, India, 1985.

Appendix IB: Table of Laplace Transforms

The Laplace Transform (LT) is defined as $Y(s) = \int_0^\infty e^{-st} y(t) dt$, $y(t)$ is a continuous time signal. The inverse LT is defined as $y(t) = \frac{1}{j2\pi} \int_{c-j\infty}^{c+j\infty} e^{st} Y(s) ds$, here, s is a complex operator (frequency) [AIB.1–AIB.4]. These are used for transfer functions-based analysis, design and synthesis of linear control systems, mostly for LTI systems. For nonlinear systems, the equivalent approach is that of describing function analysis (Chapter 3, and Appendix IC). For discrete time control systems, the pulse TFs are used (Chapter 9). (See www.crcpress.com/9780815346302).

References

AIB.1 Jan Tum. *Engineering Mathematics Handbook*, McGraw-Hill, New York, 1979.

AIB.2 Oberhettinger and L. Badii. *Table of Laplace Transforms*, Springer-Verlag, New York, 1972.

AIB.3 M. Abramowitz and I. Stegun (Editors). *Handbook of Mathematical Functions with Formulas, Graphs, and Mathematical Tables*, National Bureau of Standards, Washington, DC, 1964.

AIB.4 Tom Irvine. Table of Laplace Transforrms. September 20, 2011. www.vibrationsdata.com/Laplace_Transorms.doc; accessed Nov 2018.

Appendix IC: Describing Function Analysis for a Model with a Saturation Nonlinearity

Here, we present an example of a sinusoidal-input describing function analysis (SIDFA) for a model with a saturation nonlinearity [AIC.1]. The DFA is a widely-known technique to study frequency response of nonlinear systems, and it is an extension of linear frequency response analysis; very early analysis of the technique of DF can be found in [AIC.2]. When a sinusoidal input (SI) signal is applied to a nonlinear element, one can represent this element by a function, known as SIDF that depends not only on frequency, but also on input amplitude (Chapter 3). To use SIDFA, the model should satisfy the conditions: (i) nonlinearity should be time-invariant, (ii) nonlinearity should not generate any subharmonic as a response to the SI, and (iii) the system filters out the super-harmonics generated by the nonlinearity. The results and plots of this example can be generated by running the script "sidfasatnl.m" in MATLAB. When the script runs, it will open the Simulink model with a saturation nonlinearity, which has the SIDF as

$$N_X(\gamma) = -1, \ if \ \gamma \leq 1$$

$$N_X(\gamma) = \frac{2}{\pi}(\sin^{-1}(\gamma) + \{\gamma\sqrt{(1-\gamma^2)}\}), \ \ if -1 < \gamma < 1 \quad (IC.1)$$

$$N_X(\gamma) = 1, \ if \ \gamma \geq 1$$

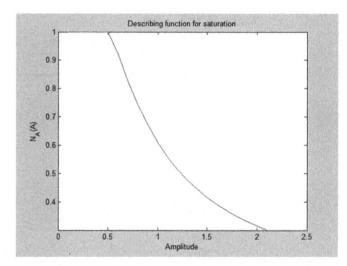

FIGURE AIC.1
Sinusoidal DF of the saturation nonlinearity (x-axis: Amplitude of A = 0.1 to 2.1).

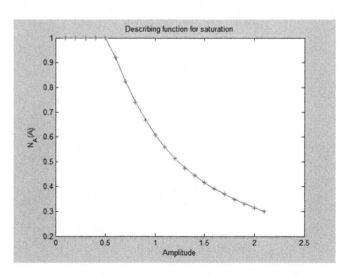

FIGURE AIC.2
SIDF for the saturation nonlinearity with "frestimate."

In (IC.1), $\gamma = 0.5 / X$ and the limits are $+0.5$, and -0.5, and X is the amplitude of the sinusoidal input signal to the nonlinearity. A plot of this SIDF is shown in Figure AIC.1. One can compute the DF for saturation nonlinearity using "frestimate" over the same set of amplitudes for a fixed frequency of 5 rad/sec. It is sufficient to run the analysis at a single frequency (since this DF does not depend on frequency). One can run a loop over all amplitudes to create a "sinestream" input with the (fixed) frequency and given amplitude, then run "frestimate" using this input signal at each iteration, the resulting plot is shown in Figure AIC.2. Now, one can run a CLDFA over a frequency range, first by analytically computing the frequency response from reference to output using DFs. First compute the amplitude of the input signal "nonlinear_input" for the nonlinearity, given the reference amplitude and frequency. The input amplitude for the nonlinearity is not necessarily equal to reference amplitude. Then, compute the analytical frequency response of the closed loop from reference to output with DF for each amplitude and store it in an "frd"-array. Obtain frequency response for the closed loop from reference to input using "frestimate" in a similar way to the DFA (of the saturation nonlinearity). Then, plot analytically calculated closed loop (Bode-) magnitude along with the one from "frestimate," see Figure AIC.3. Some nonlinearities and their DFs are presented in Table AIC.1 [AIC.2].

TABLE AIC.1

Some Nonlinearities with Their Describing Functions

Nonlinearity [a]	Describing Function
	$$N = k_2 + \frac{2(k_1 - k_2)}{\pi}\left(\sin^{-1}\left(\frac{S}{X}\right) + \frac{S}{X}\sqrt{1 - \frac{S^2}{X^2}} \right); X \geq S$$
If $k = 0$, this nonlinearity is an on-off nonlinearity.	$$N = k + \frac{4M}{\pi X};$$ If $k = 0$, we have the DF for the on-off nonlinearity $$N = \frac{4M}{\pi X}$$
	$$N = k - \frac{2k}{\pi}\left(\sin^{-1}\left(\frac{\Delta}{X}\right) + \frac{(4 - 2k)\Delta}{\pi X}\sqrt{1 - \frac{\Delta^2}{X^2}} \right); X \geq \Delta$$
	For the dead-zone nonlinearity we have the DF as $$N = k - \frac{2k}{\pi}\left[\sin^{-1}\left(\frac{\Delta}{X}\right) + \frac{\Delta}{X}\sqrt{1 - \frac{\Delta^2}{X^2}} \right]$$
	The DF for the on-off nonlinearity is $$N = \frac{4M}{\pi X} \angle -\sin^{-1}(h / X)$$

[a] X-axis is x, and Y-axis is y.

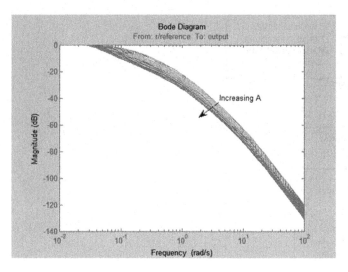

FIGURE AIC.3
The closed loop (Bode-) magnitude from the reference to output.

References

AIC.1 Anon. https://in.mathworks.com/help/slcontrol/ examples/describing-function-analysis-of-nonlinear-simulink-models.html, accessed November 2018.

AIC.2 Ogata, K. *Model Control Engineering*. Prentice Hall of India, New Delhi, India, 1985.

Appendix ID: An Iterative Control Design Process

Here, a comprehensive design procedure for a flight control system is briefly discussed; however, it is applicable to many other mechanical/mechatronic and electrical engineering systems [AID.1]. Such a design is multi-disciplinary in nature, e.g. in case of an aircraft, the aerodynamic, structural aspects, propulsive features, and control functions need to be considered; the modern flight controllers may excite structural modes of the aircraft. These modes would interact with the control-actuator dynamics. Also, because of the increasing need to integrate flight controls with engine controls, the interactions between the aerodynamics, propulsive, and structural modes/forces should be investigated and taken into account during the design cycles, the latter leading to the need of an IFPC (integrated flight propulsion control) design process.

Design and development of a FBW/FCS (fly-by-wire/flight control system, whereby actually a digital computer houses the control laws and flies the aircraft) and auto-pilot presuppose good understanding of the dynamics of the aircraft and easy availability of the mathematical models of these dynamics over the entire flight envelope. Such information is available in the form of aero-database for the aircraft and is in the form of look-up tables. These tables are constructed based on the extensive wind tunnel tests carried out on scaled (-down) models of the aircraft well before (and even during) the design/development program. This database would be in terms of aerodynamic coefficients as function of several independent variables (Mach no., AOA, control surface deflection, store configurations, and any such parameter) that would have effect on these coefficients. At a given flight condition (altitude, Mach no.) and the specification of other relevant parameters, one would be able to determine the values of these aerodynamic coefficients. These coefficients are used in the EOM (equations of motion) to obtain the flight responses of the aircraft. A thorough scrutiny of this aero-database is made, and the so-called flight mechanics parameters are computed, which are in terms of aerodynamic (or stability and control) derivatives (and other compound derivatives). Thus, dimensional linear mathematical models of the aircraft (flight dynamics) are obtained at several flight conditions and configurations. These models and the design specifications (how the controlled aircraft/plant should perform) form a starting point for the design of control laws for the given aircraft.

The design specifications could be: (a) frequency domain parameters and/or (b) time domain parameters. These performance indices collected as a vector is to be optimized (in the optimal control method), and the structures of controller blocks (controller transfer functions/filters) are specified. At times, some criteria would be of conflicting nature, and hence, some relaxation/compromise would be required in the performance of the designed system, such a procedure is well suited to be attempted using the H-Infinity/robust design by using some weighing functions/filters. Also, certain constraints on the controller gains (actual control gains and time constants) could be specified so that these gains are not unrealizable (this can be carried out using constraints in an optimization based/robust design process). In general, such a design process is the multiple-input multiple-output (MIMO) pareto-optimal or conditionally optimal control law design procedure.

In such an interactive, the control design engineer also simultaneously looks at the dynamic responses of the designed CLCS for checking the limits/shapes of the responses to guide her in the design process. The entire design process can be almost fully automated thereby

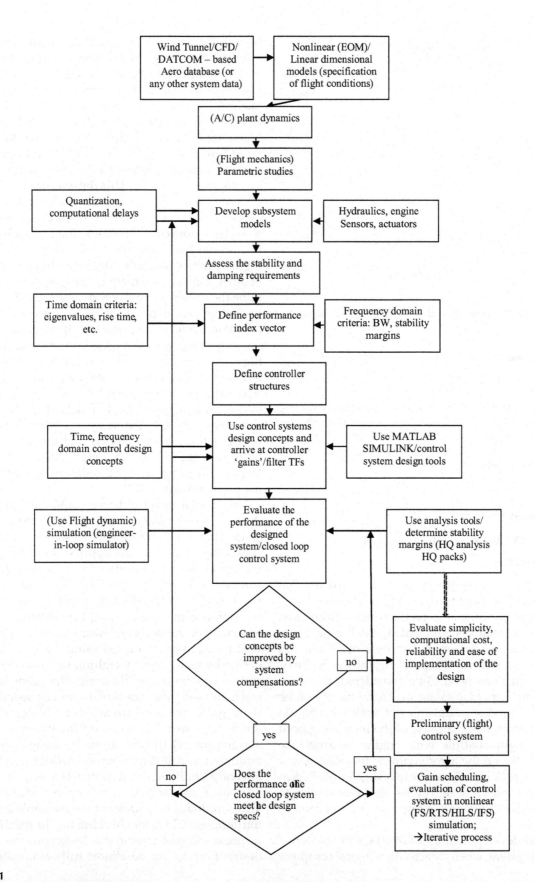

FIGURE AID.1
A typical iterative process for design of a (flight) controller and evaluation by simulation/analysis. (Adapted from With kind permission from CRC Press: *Flight Mechanics Modelling and Analysis*, Boca Raton, FL, 2009, Raol, J.R. and Singh, J.)

freeing the designer from the tedious design cycles/iterations; leading to a very practical design procedure for determination of (flight) control laws and can form a large part of the rapid prototyping formulation and computational paradigms.

In general, the procedure centers around: (a) availability of all the subsystem models that form the entire (aircraft) close loop system: actuators, sensors, anti-aliasing filters, ADC/DACs, quantization errors and computational delay; (b) controller structures; and (c) closed loop system performance criteria: stability margins and conventional control system criteria. The computational delays can be modeled by a second order lag TF; a first order TF of the form (K/(K + s)) would suffice as a model for the actuator; and the proposed controller structure/s could be in state space or TF forms.

Subsequently, the handling qualities (of the aircraft: pilot-aircraft interactions) are evaluated and full nonlinear (flight) simulation (FS) is carried out; which is used in the flight control design cycle iterations. Here, an engineer (-in-the loop) would fly the simulator and perform several different types of maneuvers and failure modes and give the assessment of the performance of the CLCS (aircraft and controllers). A comprehensive and iterative procedure for design of a (flight) control system is shown in Figure AID.1.

Reference

AID.1 Raol, J. R., and J. Singh. *Flight Mechanics Modelling and Analysis*, CRC Press, Boca Raton, FL, 2009.

Exercises for Section I

1. Design a phase lead compensator for a plant given by

$$P(s) = \frac{2}{s(s+1)(s+5)}$$

such that the following specifications are met: (a) overshoot $\leq$ 4%, (b) settling time $\leq$ 4.5 sec, and (c) as small rise time as possible. What is the steady-state velocity error?

2. Design a phase lag compensator for a plant given by

$$P(s) = \frac{1}{s(s+10)}$$

such that the settling time is smaller than 3 sec, the overshoot is smaller than 10%, and the velocity error is zero.

3. Consider the plant transfer function

$$P(s) = \frac{300}{s(s+0.225)(s+3.997)(s+179.8)}$$

Design a suitable compensator so that (i) phase margin is at least 55° and (ii) the gain cross-over frequency is not less than that of the uncompensated plant.

4. Use the first instability theorem to show the instability of the verticalup position of a pendulum.

5. Consider a nonlinear system whose output $w(t)$ is related to the input $u(t)$ by an odd function, of the form

$$w(t) = F(u(t)) = -F(-u(t))$$

Derive the following very simple approximate formula for the describing function $N(A)$:

$$N(A) = \frac{2}{3A}\left[F(A) + F\left(A/2\right)\right]$$

6. Using Poincaré-Bendixson theorem, show that the following system has at least one periodic solution:

$$\dot{x} = -x - y + x\left(x^2 + 2y^2\right)$$

$$\dot{y} = x - y + y\left(x^2 + 2y^2\right)$$

7. Show that the system:

$$\dot{x}_1 = x_2 + e^{-x_2}x_3$$

$$\dot{x}_2 = x_3$$

$$\dot{x}_3 = u$$

is state linearizable and determine the corresponding linearizing transformation.

8. Design a state feedback that makes the origin globally asymptotically stable for the system

$$\dot{x}_1 = x_1 x_2^2$$

$$\dot{x}_2 = x_2^2 + x_3$$

$$\dot{x}_3 = u$$

9. Consider the system: $\dot{x}_1 = x_1 x_2$

$$\dot{x}_2 = x_1 + u$$

a. Design a continuous globally stabilizing state feedback controller using the sliding mode control technique.

b. Is the origin globally stabilizable via feedback linearization?

10. Use backstepping to design a state feedback controller to globally stabilize the system

$$\dot{x}_1 = x_2 + a + \left(x_1 - a^{1/3}\right)^3$$

$$\dot{x}_2 = u$$

where a is a known constant.

References for Section I

1. Franco, S. *Design with Operational Amplifiers and Analog Integrated Circuits.* 3rd ed., Tata McGraw-Hill, New Delhi, India, 2002.

2. Brogan, W. L. *Modern Control Theory.* 3rd ed., Prentice Hall, Englewood Cliffs, NJ, 1990.

3. Chen, C. T. *Analog and Digital Control System Design: Transfer-Function, State-Space, and Algebraic Methods.* Oxford University Press, New York, 2006.

4. Chen, C. T. *Linear System Theory.* 3rd ed., Oxford University Press, New York, 1998.

5. Yu, C.-C. *Autotuning of PID Controllers: A Relay Feedback Approach.* 2/e, Springer Verlag, London, UK, 2006.

6. Vidyasagar, M. *Nonlinear Systems.* SIAM Classics Edition, Philadelphia, PA, 2002.

7. Slotine, J. J., and Li, W. *Applied Nonlinear Control.* Prentice Hall, Englewood Cliffs, NJ, 1991.

8. Strogatz, S. H. *Nonlinear Dynamics and Chaos, with Applications to Physics, Biology, Chemistry, and Engineering.* 2/e, Westview Press, 2015.

9. May, R. L. Simple mathematical models with very complicated dynamics. *Nature,* Vol. 261, pp. 459–467, 1976.

10. Campbell, S. L., and Haberman, R. *Introduction to Differential Equations with Dynamical Systems.* Princeton University Press, Princeton, NJ, 2008.

11. Meiss, J. D. *Differential Dynamical Systems.* Revised Edition, SIAM, Philadelphia, PA, 2017.

12. Hinrichsen, D., and Pritchard, A. P. *Mathematical System Theory I: Modelling, State Space Analysis, Stability and Robustness.* Springer, New York, 2011.

13. Sastry, S. *Nonlinear Systems: Analysis, Stability, and Control.* Springer Science & Business Media, New York, Vol. 10, 2013.

14. Marino, R., and Tomei, P. *Nonlinear Control Design: Geometric, Adaptive, Robust.* Prentice Hall International, Upper Saddle River, NJ, 1995.

15. Spong, M. W., Hutchinson, S., and Vidyasagar, M. *Robot Modelling and Control.* John Wiley & Sons, New York, 2005.

16. Khalil, H. K. *Nonlinear Systems.* 3/e, Prentice Hall, Upper Saddle River, NJ, 2002.

17. Krstic, M., Kokotovic, P. V., and Kanellakopoulos, I. *Nonlinear and Adaptive Control Design.* John Wiley & Sons, New York, 1995.

18. Bullo, F. *Geometric Control of Mechanical Systems.* Springer Science & Business Media, New York, Vol. 49, 2005.

19. Bullo, F., and Murray, R. M. Tracking for fully actuated mechanical systems: A Geometric framework. *Automatica,* Vol. 35, No. 1, pp. 17–34, 1999.

20. Guillemin, V. and Pollack, A. Differential topology. *Am. Math. Soc.,* Vol. 370, 2010.

21. Kelly, S. D., and Murray, R. M. Geometric phases and robotic locomotion. *J. Robotic Syst.,* Vol. 12, No. 6, pp. 417–431, 1995.

22. Lee, T, Leok, M., and McClamroch, N. H. Nonlinear robust tracking control of a quadrotor uav on SO (3). *Asian J. Control,* Vol. 15, No. 2, pp. 391–408, 2013.

23. Liu, Y. and Yu, H. A survey of underactuated mechanical systems. *Control Theory A., IET,* Vol. 7, No. 7, pp. 921–935, 2013.

24. Spong, M. W. Partial feedback linearization of underactuated mechanical systems. *Proceedings of the IEEE/RSJ/GI International Conference on, 'Intelligent Robots and Systems Advanced Robotic Systems and the Real World', IROS'94,* Vol. 1. IEEE, pp. 314–321, 1994.

Section II

Optimal and H-Infinity Control

Often, it is sufficient to design a controller for a dynamic system that achieves an acceptable closed loop control behavior, and in such cases, the classical PI tuning approach would be considered adequate. However, such approaches would not be sufficient when the demands are on better efficiency and performance, of course, the fundamental requirement of overall closed loop stability should be assured; the latter is a must, since when any new and more control transfer functions (and feedback loops) are incorporated in the control system, the system's dimension (and perhaps, even nonlinearity, due to switching, if required to be incorporated) would increase, and it would be prone to instability. Hence, to have just minimal/sufficient control quality and elimination of the steady-state offsets is not enough; the latter could have been achieved with the classical PID approaches. Interestingly, the availability of inexpensive computing power (in terms of memory, speed, throughput, and hardware/software/information/network redundancies) has made the use of more complex and sophisticated controllers feasible for several practical applications. In this section, starting with calculus of variation, we study the so-called modern control approaches: H_2-optimal, LQG (linear quadratic Gaussian) control, and robust H_∞ (H-Infinity/HI) control, including model predictive control (MPC).

Remaining part of the modern control, the digital and adaptive approaches are discussed in Section III, and the ultra-modern (!) approaches are dealt with in Section IV.

A few applications areas of the modern control are: (a) chemical process control systems: distillation, chemical reactors, cements kilns, pH control, and basis weight control in paper machines, and so on. Chemical processes are typically subject to random disturbances and large modeling uncertainties and the objective is to control a quality variable whose variations should be made as small as possible in-spite of disturbances/model uncertainties; (b) mechanical systems: active suspension systems, elevator control for a smooth and efficient ride in spite of varying loads, positioning systems in ships offshore operations, magnetic bearings to achieve very high rotation speeds and to avoid vibrations, and engine controls; (c) active vibration control in high rise buildings in earthquake-prone areas, these structures have an infinite number of vibrating modes and resonance frequencies with the need of multivariable control methods; (d) active control of acoustic noise for which anti-noise is applied to suppress acoustic noise: active noise suppression in headsets and the interior of luxury cars and airplanes; and (e) aerospace and robotic systems.

5

Optimization-Extremization of Cost Function

The optimal control theory and methods are considered very important parts of modern control theory and practice. In fact, the advent of the optimal control principles and approaches themselves, are where from the modern control theory came to be known as such. Later on, the digital and adaptive control joined the optimal control, and these could be called as very important additional ingredients of the modern control theory. The basic aspect of all these being the state-space representation of the dynamic systems. A big part of optimal control is developed in continuous time domain. The optimal includes, in its definition, either a maximum or a minimum (value) of the so-called cost function. This cost function is found in many practical cases: maximum benefit-returns/profits, maximum power transfer from one system/circuit to another, minimum power/energy consumption for carrying out a certain task/function, minimum cost (in terms of budget or effort involved) or minimum time taken to travel between two points, and maximum efficiency. Thus, in optimal control, in most cases the minimization (in a few cases, the maximization) of the appropriately chosen cost function is important. This function is also often named the loss function, as is obvious from the foregoing qualitative definitions (of course, the maximization is handled with a negative sign to the cost function). In fact, a specific cost function that is function of the system's state variables and the control input is chosen, and the minimization of this function (some error-criterion) is sought with respect to the independent variable, e.g. control input. The resultant set of equations is solved to explicitly determine the formula for the control input, which is also called control law (in the sense of a rule, i.e. how to generate the control input). This would lead to the desired optimal control input within the constraints that might have been imposed in/on the states of the dynamic system as well as the control input's magnitude or the maximum energy. Thus, the optimal control is considered as a standard approach for solving several dynamic optimization problems; invariably, often these are represented in continuous time framework [1]. The optimal control can be considered as a generalization of the calculus of variations—the mathematics of how the variations, in some important state variable/s and/or input variable/s of the appropriate mathematical model (since a real system is appropriately represented by its mathematical model) of the dynamic system take place. These variations should be such that the chosen cost function is minimized, or maximized, based on the type of the involved cost. In a direct sense, optimal control deals with an optimization problem in which one is seeking an optimal control law/rule (in many cases in the form of feedback transfer function/s; or state space models) and/or an optimal control input that can be used to excite a dynamic system to transfer its states from one point to another as time evolves in a certain specific time (-period), invoking here, the condition of controllability of the dynamic system. Much of the development in this chapter is based on [1,2].

5.1 Optimal Control Theory: An Economic Interpretation

For example, the problem posed is: (a) a firm wants to maximize its total profits over a certain duration of time; (b) let it have a stock of $x(t)$, which is the capital amount, on a certain date, t, inherited by the firm from its past actions/performance; and (c) given $x(t)$, the firm is free to make decision $u(t)$, which might be, output, price, or anything else [1]. Now, assume that with $x(t)$ and $u(t)$, the firm derives a flow of benefits/unit time (per unit time), denoted as

$$\phi = \phi(x(t), u(t), t) \tag{5.1}$$

Consider the time t and let $\Phi(x(t), \tilde{u}, t)$ denote the flow of profit over the time interval $[t, T]$:

$$\Phi(x(t), \tilde{u}, t) = \int_t^T \phi(x(\tau), u(\tau), \tau) d\tau \tag{5.2}$$

In (5.2), $\tilde{u}$ is the entire time trajectory of the decision variable $u(.)$ from time t (say a certain date) up to the final time/date T. The time trajectory of $x(.)$ will depend on past values of x and the decision $u(.)$ that the firm makes. This is the constraint in the optimization problem written as

$$\dot{x}(t) = f(x(t), u(t), t) \tag{5.3}$$

The interpretation of (5.3) is that the mathematical model of a dynamic system, or a dynamic problem, must be adhered to, i.e. even after the control law is synthesized and used in the control system, and the constraint (5.3) should be satisfied. Then, in the optimal control problem, one has to choose the entire time trajectory $\tilde{u}$ to maximize (5.2) (here it is a profit), subject to the constraint (5.3), which is the equation of motion that specifies the trajectory $x(t)$.

5.1.1 Solution for the Optimal Path

The optimization problem posed by (5.2) and (5.3) is in general a difficult one, since one is seeking an optimal trajectory $\tilde{u}$ and not an optimal value of $u(.)$. The latter can be obtained by using standard calculus, i.e. the minimization of the cost function wrt to $u(t)$; the point estimate. Then, the idea is to transform the problem into another problem from which we need to determine only one number or a few numbers. We have the optimal control (capital) problem posed in (5.2). This problem, i.e. (5.2) is partitioned on/for: (a) a short interval of some delta length of time that begins at time t; and (b) the remaining one from t+delta to final time T. The assumption is that the small delta is short enough so that the firm would not change $u(t)$ during this short interval. Then, (5.2) can be written as

$$\Phi(x(t),\tilde{u},t) = \phi(x,u(t),t)\Delta + \int_{t+\Delta}^{T} \phi(x(\tau),u(\tau),\tau)d\tau \quad (5.4)$$

In (5.4), the first term is the contribution from the delta interval. Here, u is kept constant, and the second term gives the influence of the $u(t)$, starting from $u(t+\Delta)$, on the $x(t)$. We can now use (5.1) and (5.2) to rewrite (5.4) as

$$\Phi(x(t),\tilde{u},t) = \phi(x,u(t),t)\Delta + \Phi(x(t+\Delta),\tilde{u},t+\Delta) \quad (5.5)$$

Next, denote the time trajectory that maximizes Φ, as $\hat{x}$, and $\hat{\Phi}$ as the maximum value

$$\hat{\Phi}(x(t),t) = \max_{\tilde{u}} \Phi(x(t),\tilde{u},t) \quad (5.6)$$

Choose a policy: (i) for the delta interval, $u(t)$ is arbitrary, and (ii) from $(t+\Delta)$, it is an optimal trajectory $\hat{u}$, then we have

$$\tilde{\Phi}(x(t),u,t) = \phi(x,u(t),t)\Delta + \hat{\Phi}(x(t+\Delta),t+\Delta) \quad (5.7)$$

In (5.7), the first term represents the benefit/s obtained from the decision $u(t)$ in the delta interval, and the second term is the maximum benefit/s that can be obtained over the interval $[t+\Delta,T]$, provided that $u(t+\Delta)$ is determined by $x(t)$ and $u(t)$. Also, $u(t)$ does not appear in this term since the term is optimal itself.

The main aspect of (5.7) is to determine the value of $u(t)$ that would maximize $\tilde{\Phi}(x(t),u,t)$. The simple approach would be to use (ordinary) calculus and take partial derivatives of (5.7), with respect to $u(t)$, and set the resultant expressions equal to zero

$$\frac{\partial \tilde{\Phi}}{\partial u(t)} = \Delta \frac{\partial \phi}{\partial u(t)} + \frac{\partial \hat{\Phi}}{\partial x(t+\Delta)} \frac{\partial x(t+\Delta)}{\partial u(t)} = 0 \quad (5.8)$$

From (5.8), we get the following two considerations

1. Since the delta interval is small,

$$x(t+\Delta) \approx x(t) + \dot{x}(t)\Delta$$
$$= x(t) + f(x(t),u(t),t)\Delta \quad (5.9)$$

$$\frac{\partial x(t+\Delta)}{\partial u(t)} = \frac{\partial f(x(t),u(t),t)}{u(t)}\Delta \quad (5.10)$$

2. $\dfrac{\partial \hat{\Phi}(x(t+\Delta),t+\Delta)}{\partial x(t+\Delta)}$ is the increment that results from a small increase in $x(t+\Delta)$, signifying the marginal value of the capital, which is now denoted as $\lambda(t+\Delta)$.

Then, we obtain the following from (5.8) by using (5.10), and inference in (2):

$$\Delta \frac{\partial \phi}{\partial u(t)} + \lambda(t+\Delta)\frac{\partial f}{\partial u(t)}\Delta = 0 \quad (5.11)$$

Since, the Δ interval is small, we can delete it from (5.11) as a common term and, letting it go to zero, in the parenthesis we obtain

$$\frac{\partial \phi(x(t),u(t),t)}{\partial u(t)} + \lambda(t)\frac{\partial f(x(t),u(t),t)}{\partial u(t)} = 0 \quad (5.12)$$

Equation (5.12) is the first necessary condition to be solved for an optimal control problem. The first term of (5.12) is the marginal short-term/run effect due to the change in the decision $u(t)$ along the optimal path/trajectory. The second term of (5.12) is the effect on the capital stock, due to a change in u times the marginal value of the capital stock. So, the essence of this first necessary condition is that the marginal immediate gain balances the marginal long-term/run cost; and the variable $u(t)$ is chosen in this sense optimally. However, (5.12) also has the variable $\lambda(t)$ that is the marginal value of the capital and this is yet as such unknown. So, now we presume that $u(t)$ is chosen so that the first condition is satisfied. Then, we can write (5.7) as

$$\hat{\Phi}(x(t),t) = \phi(x,u(t),t)\Delta + \hat{\Phi}(x(t+\Delta),t+\Delta) \quad (5.13)$$

Differentiate (5.13) with respect to $x(t)$ to obtain

$$\frac{\partial \hat{\Phi}(x(t),t)}{\partial x(t)} = \lambda(t) = \Delta \frac{\partial \phi}{\partial x(t)} + \frac{\partial x(t+\Delta)}{\partial x(t)} \lambda(t+\Delta) \quad (5.14)$$

$$\lambda(t) = \Delta \frac{\partial \phi}{\partial x(t)} + \left[\frac{\partial \{x(t) + \dot{x}(t)\Delta\}}{\partial x(t)} \right] [\lambda(t) + \dot{\lambda}(t)\Delta] \quad (5.15)$$

$$= \Delta \frac{\partial \phi}{\partial x(t)} + \left[1 + \frac{f}{\partial x(t)}\Delta \right] [\lambda(t) + \dot{\lambda}(t)\Delta] \quad (5.16)$$

The expansion of the second term in (5.16) would yield four terms and, canceling out the one common term from both the sides of (5.16), results in the following

$$0 = \Delta \frac{\partial \phi}{\partial x(t)} + \dot{\lambda}(t)\Delta + \lambda(t)\frac{f}{\partial x(t)}\Delta + \dot{\lambda}(t)\frac{f}{\partial x(t)}\Delta^2 \quad (5.17)$$

Rearranging (5.17) and dividing the entire Equation (5.17) by the delta term, and letting the delta tend to zero, one gets the second necessary condition of the optimal control problem as

$$-\dot{\lambda}(t) = \frac{\partial \phi(x(t),u(t),t)}{\partial x(t)} + \lambda(t)\frac{f(x(t),u(t),t)}{\partial x(t)} \quad (5.18)$$

The third necessary condition is obviously (5.3), and these three conditions are grouped as follows:

$$\frac{\partial \phi(x(t),u(t),t)}{\partial u(t)} + \lambda(t)\frac{\partial f(x(t),u(t),t)}{\partial u(t)} = 0$$

$$-\dot{\lambda}(t) = \frac{\partial \phi(x(t),u(t),t)}{\partial x(t)} + \lambda(t)\frac{f(x(t),u(t),t)}{\partial x(t)}$$

$$\dot{x}(t) = f(x(t),u(t),t) \quad (5.19)$$

5.1.2 The Hamiltonian

The three necessary conditions (5.19) are the basis of the Pontryagin Maximum Principle (PMP) in optimal control and can be obtained by putting the optimization problem in a standard form. The problem is rewritten as to determine

$$\max_{u(t)} \int_t^T \phi(x(\tau),u(\tau),\tau)d\tau \quad (5.20)$$

subject to the constraint (5.3); then, use the so-called Hamiltonian function

$$H = \phi(x(t),u(t),t) + \lambda(t)f(x(t),u(t),t) \quad (5.21)$$

In (5.21), the Hamiltonian (function) is defined as such. Then the three conditions can be obtained from the following:

$$\frac{\partial H}{\partial u} = 0 \quad (5.22)$$

$$\frac{\partial H}{\partial x} = -\dot{\lambda} \quad (5.23)$$

$$\frac{\partial H}{\partial \lambda} = \dot{x} \quad (5.24)$$

In real control applications, the $u(t)$ is a control variable, i.e. control input, and our aim there is to optimize an objective function by changing the state variable $x(t)$ of the state space dynamic system. The variable $\lambda(t)$ is known by various names: auxiliary variable, co-state, or even as Langrage multiplier (even it is called an adjoin(t) variable). Thus, the three Equations in (5.19), or equivalently (5.21–5.24) collectively, determine the time trajectories of the state variable $x(.)$, the control variable $u(.)$, and the co-state $\lambda(.)$, of which the two equations are the differential equations that require specification of boundary conditions, which might occur differently for the different variable. That is to say, the boundary condition of one variable, $x(t)$ will be known at the initial time, whereas for the co-state it might be known at the final time T; leading to the case of the two-point boundary value problem (TPBVP). The boundary condition/s themselves are known as transversality conditions. This problem generally requires multiple solutions back and forth and can be handled by the so-called multiple shooting methods, or the method of invariant embedding.

5.2 Calculus of Variation

In calculus of variation we encounter the problem of maximization and minimization. We consider that the optimization problem is valid for real valued functions [2].

5.2.1 Sufficient Conditions

Let $f{:}X \to R$ be a real valued function with domain X. Then, the maximum and minimum points are defined; the value $f(x_0)$ is called the maximum value of $f(.)$

$$f(x_0) \geq f(x); \forall x \in X \quad (5.25)$$

In (5.25), $x_0 \in X$ is called the maximum point/value for the function f and it is also called the solution of the function that is dependent of/from x. The minimum value $f(x_0)$ of $f(.)$ is

$$f(x_0) \leq f(x); \forall x \in X \qquad (5.26)$$

In (5.26), $x_0 \in X$ is called the minimum point/value for the function f and it is also called the solution of the function that is dependent of/on x.

5.2.1.1 Weierstrass Result

Let f be well defined function as earlier. The function f will have a maximum/minimum for the sufficient conditions:

1. $f : S \rightarrow R$ is a continuous function
2. $S \subset R$ is a bound(-ed) and closed/compact set of R.

Interestingly enough, these conditions are sufficient and not necessary; i.e. even if one or both the conditions are not obeyed, there still could be maximum or minimum of the function f.

5.2.2 Necessary Conditions

We also need to have conditions when the function f is differentiable and also if it is a function of several variables. Let us have $f : S \rightarrow R$ as a differentiable function. Also, $x_0 \in X$ is an internal point/value of S. This x_0 could be either a minimum or maximum point of the function f. Then, we have the following result

$$\frac{df}{dx} = 0 \qquad (5.27)$$

The condition of (5.27) is necessary, but not sufficient. The point x_0 is called a stationary point if (5.27) is satisfied. The sufficient condition is given as, the x_0 is a minimum point if

$$\frac{d^2 f}{dx^2}(x_0) > 0 \; ; f''(x_0) > 0 \qquad (5.28)$$

The x_0 is a maximum point if

$$\frac{d^2 f}{dx^2}(x_0) < 0 \; ; f''(x_0) < 0 \qquad (5.29)$$

Often one uses f' and f'' for representing the differentiation in (5.27) and (5.28) respectively. Let us have f as a real valued function of n variables defined on the real space R of dimension n. It is also assumed that f admits

its partial derivatives with respect to the components of x. Then x_0 is either a minimum or maximum point of the function f, if

$$\frac{\partial f}{\partial x_1} = 0; \; \frac{\partial f}{\partial x_2} = 0; ... \frac{\partial f}{\partial x_n} = 0; \text{ for all at } x = x_0 \quad (5.30)$$

Let us define a matrix of the 2nd order partials as

$$H = \begin{bmatrix} \dfrac{\partial^2 f}{\partial x_1 \partial x_1} & \dfrac{\partial^2 f}{\partial x_1 \partial x_2} & \cdots & \dfrac{\partial^2 f}{\partial x_1 \partial x_n} \\[3mm] \dfrac{\partial^2 f}{\partial x_2 \partial x_1} & \dfrac{\partial^2 f}{\partial x_2^2} & \cdots & \dfrac{\partial^2 f}{\partial x_2 \partial x_n} \\[3mm] \cdots & \cdots & \cdots & \cdots \\[3mm] \dfrac{\partial^2 f}{\partial x_n \partial x_1} & \dfrac{\partial^2 f}{\partial x_n \partial x_2} & \cdots & \dfrac{\partial^2 f}{\partial^2 x_n^2} \end{bmatrix} \qquad (5.31)$$

Then x_0 is a minimum point of the function f, if the matrix H is positive definite, and it is maximum point of the function f, if the matrix H is negative definite, the matrix H is known as Hessian matrix, and its elements as Hessians.

5.3 Euler-Lagrange Equation

The Euler-Lagrange (EL) equation gives the necessary condition for extremum. Let $y(x)$ be an optimizing/extremizing function for the problem

$$J(y) = \int_{x_1}^{x_2} F(x, y, y') dx \qquad (5.32)$$

Minimize/maximize the cost function $J(y)$, (5.32) with the end point conditions $y(x_1) = y_1; y(x_2) = y_2$; then, $y(x)$ satisfies the following EL (Euler-Lagrange) equation

$$\frac{d}{dx}\left(\frac{\partial F}{\partial y'} \right) - \frac{\partial F}{\partial y} = 0 \qquad (5.33)$$

This is known as the boundary value problem (BVP). The proof of (5.33) is presented next. If $y(x)$ is an extremizing function of the cost function $J(y)$, then we have $Y = y(x) + \varepsilon \eta(x)$ as a variation of $y(x)$ and $Y' = y'(x) + \varepsilon \eta'(x)$. Here, ε is a small (positive) constant, $\eta(x_1) = \eta(x_2) = 0$ (local end conditions), and $\eta(.)$ is a continuously differential function. Then, the trajectory/path Y induces the following cost function variation in (5.32)

$$J(\varepsilon) = \int_{x_1}^{x_2} F(x, y(x) + \varepsilon\eta(x), y'(x) + \varepsilon\eta'(x))dx \quad (5.34)$$

$$J(\varepsilon) = \int_{x_1}^{x_2} F(x, Y(x), Y'(x))dx \quad (5.35)$$

For (5.35), $y(x)$ is an extremizing function; hence, $J(\varepsilon)$ will have extremum when $\varepsilon = 0$. Then, by the (standard/chain rule) calculus, we have the following

$$\frac{dJ}{d\varepsilon} = 0; \quad at \, \varepsilon = 0;$$

$$\frac{dJ}{d\varepsilon} = \int_{x_1}^{x_2} \left[\frac{\partial F}{\partial x}\frac{\partial x}{\partial \varepsilon} + \frac{\partial F}{\partial Y}\frac{\partial Y}{\partial \varepsilon} + \frac{\partial F}{\partial Y'}\frac{\partial Y'}{\partial \varepsilon} \right]dx \quad (5.36)$$

Using the full expressions for Y and Y' in (5.36), evaluating their partials with respect to ε, and since x does not depend on ε, we obtain the following

$$\frac{dJ}{d\varepsilon} = \int_{x_1}^{x_2} \left[\frac{\partial F}{\partial Y}\eta(x) + \frac{\partial F}{\partial Y'}\eta'(x) \right]dx \quad (5.37)$$

Next, the second term of (5.37) is expanded using the "integrating it by parts" as

$$\int_{x_1}^{x_2} \frac{\partial F}{\partial Y'}\eta'(x)dx = \frac{\partial F}{\partial Y'}\{\eta(x_2) - \eta(x_1)\} - \int_{x_1}^{x_2} \frac{d}{dx}\left(\frac{\partial F}{\partial Y'}\right)\eta(x)dx \quad (5.38)$$

Using $\eta(x_1) = \eta(x_2) = 0$ in (5.38), we obtain

$$\int_{x_1}^{x_2} \frac{\partial F}{\partial Y'}\eta'(x)dx = -\int_{x_1}^{x_2} \frac{d}{dx}\left(\frac{\partial F}{\partial Y'}\right)\eta(x)dx \quad (5.39)$$

Substituting (5.39) in (5.37), we obtain

$$\frac{dJ}{d\varepsilon} = \int_{x_1}^{x_2} \left[\frac{\partial F}{\partial Y} - \frac{d}{dx}\left(\frac{\partial F}{\partial Y'}\right) \right]\eta(x)dx = 0; \quad with \, \varepsilon \to 0. \quad (5.40)$$

Since $Y(x) = y(x)$ and $Y'(x) = y'(x)$ at $\varepsilon = 0$, we get

$$\frac{dJ}{d\varepsilon} = \int_{x_1}^{x_2} \left[\frac{\partial F}{\partial y} - \frac{d}{dx}\left(\frac{\partial F}{\partial y'}\right) \right]\eta(x)dx = 0 \quad (5.41)$$

Since the common variable $\eta(x)$ is an arbitrary function, the bracketed term can be separately written as

$$\frac{d}{dx}\left(\frac{\partial F}{\partial y'}\right) - \frac{\partial F}{\partial y} = 0 \quad (5.42)$$

Another form of the EL equation is derived next. Let $y(x)$ be an extremizing function of $J(y)$. Then, we have the following for $F = F(x, y, y')$ by applying the chain rule of derivatives

$$\frac{d}{dx}(F) = \frac{dF(x,y,y')}{dx} = \frac{\partial F}{\partial x}\frac{dx}{dx} + \frac{\partial F}{\partial y}\frac{dy}{dx} + \frac{\partial F}{\partial y'}\frac{dy'}{dx} \quad (5.43)$$

$$= \frac{\partial F}{\partial x} + \frac{\partial F}{\partial y}y' + \frac{\partial F}{\partial y'}y'' \quad (5.44)$$

Also, we have the following expressions:

$$\frac{d}{dx}\left(y'\frac{\partial F}{\partial y'}\right) = y'\frac{d}{dx}\left(\frac{\partial F}{\partial y'}\right) + \frac{\partial F}{\partial y'}y'' \quad (5.45)$$

Subtract (5.45) from (5.44) to get

$$\frac{dF}{dx} - \frac{d}{dx}\left(y'\frac{\partial F}{\partial y'}\right) = \frac{\partial F}{\partial x} + \frac{\partial F}{\partial y}y' - \frac{d}{dx}\left(\frac{\partial F}{\partial y'}\right)y' \quad (5.46)$$

$$\frac{d}{dx}\left(F - y'\frac{\partial F}{\partial y'}\right) - \frac{\partial F}{\partial x} = \left[\frac{\partial F}{\partial y} - \frac{d}{dx}\left(\frac{\partial F}{\partial y'}\right)\right]y' \quad (5.47)$$

Using (5.42), the EL equation derived earlier, in (5.47) we obtain

$$\frac{d}{dx}\left(F - y'\frac{\partial F}{\partial y'}\right) - \frac{\partial F}{\partial x} = 0 \quad (5.48)$$

5.4 Constraint Optimization Problem

In certain problems of calculus of variations, while considering the extremization or optimization of a given cost functional $J(y)$; along with the conditions $y(x_1) = y_1$, and $y(x_2) = y_2$, one also might need that the extremizing function has to satisfy an additional constraint [2]. This would arise due to the physical constraint of the problem at hand. So, the general problem of optimization with a constraint can be represented as: maximize/minimize the function $J(y)$, (5.32) subject to the constraint specified as

$$\int_{x_1}^{x_2} G(x, y, y')dx = L = \text{Constant} \quad (5.49)$$

The end conditions are the same as for the problem of (5.32). The problem of the constraint optimization can be converted into that of the unconstraint one by the Lagrange multiplier (LM) technique. In this case, a new functional is defined as

$$H(x,y,y') = F(x,y,y') + \lambda G(x,y,y') \qquad (5.50)$$

Then, the new functional to be optimized is

$$J(y) = \int_{x_1}^{x_2} H(x,y,y')dx \qquad (5.51)$$

We see that (5.51) is now a problem similar to the one posed as (5.32) and can be easily handled by using the EL equation specified in terms of H instead of F, with the same end conditions:

$$\frac{\partial H}{\partial y} - \frac{d}{dx}\left(\frac{\partial H}{\partial y'}\right) = 0 \qquad (5.52)$$

5.5 Problems with More Variables

In this section, we consider three very important problems that occur in many practical situations.

5.5.1 With Higher Order Derivatives

Optimize

$$J(y) = \int_{x_1}^{x_2} F(x,y,y',y'')dx \qquad (5.53)$$

The end conditions are: $y(x_1) = y_1$; $y(x_2) = y_2$; $y'(x_1) = y_1'$; $y'(x_2) = y_2'$. The problem can be solved by using the following Euler-Poisson equation

$$\frac{\partial F}{\partial y} - \frac{d}{dx}\left(\frac{\partial F}{\partial y'}\right) + \frac{d^2}{dx^2}\left(\frac{\partial F}{\partial y''}\right) = 0 \qquad (5.54)$$

5.5.2 With Several Unknown Functions

Optimize

$$J(u,v) = \int_{x_1}^{x_2} F(x,u,v,u',v'')dx \qquad (5.55)$$

The end conditions are: $u(x_1) = u_1$; $u(x_2) = u_2$; $v(x_1) = v_1$; $v(x_2) = v_2$. The appropriate EL equations are

$$\frac{\partial F}{\partial u} - \frac{d}{dx}\left(\frac{\partial F}{\partial u'}\right) = 0$$

$$\frac{\partial F}{\partial v} - \frac{d}{dx}\left(\frac{\partial F}{\partial v'}\right) = 0 \qquad (5.56)$$

5.5.3 With More Independent Variables

Let z be the dependent variable, and x and y be the independent variables; then, optimize

$$J(z) = \int_{x_1}^{x_2} F(x,y,z,z_x,z_y)dydx \qquad (5.57)$$

This problem can be solved by using the EL equation

$$\frac{\partial F}{\partial z} - \frac{d}{dx}\left(\frac{\partial F}{\partial z_x}\right) - \frac{d}{dy}\left(\frac{\partial F}{\partial z_y}\right) = 0 \qquad (5.58)$$

5.6 Variational Aspects

The variational aspects are, in fact, analogous to derivatives in the calculus. When a function changes its value from $y(x)$ to $y(x + \Delta x)$, the rate of change defines the derivative $y'(x)$. Whereas, in the variational calculus, the function $y(x)$ is changed to a new function $y(x) + \varepsilon\, \eta(x)$; here, ε is a small constant. In this case, $\eta(x)$ is a continuous and differentiable function. The change $\varepsilon\, \eta(x)$ in $y(x)$ as a function is called the variation of y and is denoted by δy. Hence, the change in F due to variation in x is given as

$$\Delta F = F(x, y + \varepsilon\eta, y' + \varepsilon\eta') - F(x,y,y') \qquad (5.59)$$

Expand the terms on the RHS of (5.59) in Taylor's series to obtain

$$\Delta F = F(x,y,y') + \left(\frac{\partial F}{\partial y}\eta + \frac{\partial F}{\partial y'}\eta'\right)\varepsilon$$

$$+ \left[\frac{\partial^2 F}{\partial y^2}\eta^2 + 2\frac{\partial^2 F}{\partial y \delta y'}\eta\eta' + \frac{\partial^2 F}{\partial y'^2}(\eta')^2\right]\frac{\varepsilon^2}{2!} \qquad (5.60)$$

$$+ \text{higher order terms of}(\varepsilon) - F(x,y,y')$$

Simplify (5.60) to obtain

$$= \frac{\partial F}{\partial y}\delta y + \frac{\partial F}{\partial y'}\delta y'$$

$$+ \frac{1}{2!}\left[\frac{\partial^2 F}{\partial y^2}(\delta y)^2 + 2\frac{\partial^2 F}{\partial y \delta y'}\delta y \delta y' + \frac{\partial^2 F}{\partial y'^2}(\delta y')^2\right] \qquad (5.61)$$

$$+ \text{higher order terms of}(\varepsilon)$$

From (5.61), we obtain the following two variations

$$\delta F = \frac{\partial F}{\partial y}\delta y + \frac{\partial F}{\partial y'}\delta y'$$

$$\delta^2 F = \frac{1}{2}\left[\frac{\partial^2 F}{\partial y^2}(\delta y)^2 + 2\frac{\partial^2 F}{\partial y \delta y'}\delta y \delta y' + \frac{\partial^2 F}{\partial y'^2}(\delta y')^2\right] \quad (5.62)$$

The first Equation of (5.62) is the first (order) variation, and the second Equation of (5.62) is the second (order) variation of function *F*. We now derive the result of variation of the cost function *J*(*y*);

$$\text{if } J(y) = \int_{x_1}^{x_2} F(x,y,y')dx \quad (5.63)$$

then, determine the variation of *I*, δJ.

$$\delta J = \delta\left(\int_{x_1}^{x_2} F(x,y,y')dx\right)$$

$$= \int_{x_1}^{x_2} \delta F(x,y,y')dx \quad (5.64)$$

$$= \int_{x_1}^{x_2}\left[\frac{\partial F}{\partial y}\delta y + \frac{\partial F}{\partial y'}\delta y'\right]dx$$

Another result is derived as follows. Optimize

$$J(y) = \int_{x_1}^{x_2} F(x,y,y')dx; \quad y(x_1) = y_1; \; y(x_2) = y_2 \quad (5.65)$$

Then, if *y*(*x*) is an extremizing function for (5.65), then the first variation/s are $\delta J(y) = 0$. The first variation is given as

$$\delta J = \int_{x_1}^{x_2}\left[\frac{\partial F}{\partial y}\delta y + \frac{\partial F}{\partial y'}\delta y'\right]dx$$

$$= \int_{x_1}^{x_2}\left[\frac{\partial F}{\partial y}\delta y + \frac{\partial F}{\partial y'}\frac{d}{dx}(\delta y)\right]dx \quad (5.66)$$

Integrating by parts the second term in (5.66), we get

$$\int \frac{\partial F}{\partial y'}\frac{d}{dx}(\delta y)dx = \frac{\partial F}{\partial y'}(\delta y)\Big|_{x_1}^{x_2}\{=0, \text{ from } x_1 \text{ to } x_2\}$$

$$-\int_{x_1}^{x_2}\frac{d}{dx}\left(\frac{\partial F}{\partial y'}\right)\delta y dx \quad (5.67)$$

Substitute (5.67) into (5.66) to obtain:

$$\delta J = \int_{x_1}^{x_2}\left[\frac{\partial F}{\partial y} - \frac{d}{dx}\left(\frac{\partial F}{\partial y'}\right)\right]\delta y dx \quad (5.68)$$

Substitute (5.42), the EL Equation in (5.68) to obtain the desired result

$$\delta J = \int_{x_1}^{x_2}[0]\delta y dx; \text{ if } y(x) \text{ is an extremizing function.} \quad (5.69)$$

$$= 0$$

5.7 Conversion of BVP to Variational Problem

As we have seen in Section 5.5 that for all the variational problems, there is a corresponding Euler-Lagrange equation, which is inherently a boundary value problem (BVP). In a BVP, we specify boundary conditions for the two variables, *x*(.), the state variables/vector and $\lambda(t)$, the co-state variable/vector: *x*(0)=?, *x*(*T*)=?, $\lambda(0) = ?, \lambda(T) = ?$; and this will depend on the kind of problem that is being solved. This is an interesting problem in itself, and has some methods for solution that can be used in the context of estimation of states of nonlinear dynamic systems, both continuous and discrete time, in deterministic setting. The main idea is that the TPBVP is converted into an initial value problem and this yields estimator equations, similar to Kalman filter. The state estimator equation can be solved by the numerical integration method, and the state error Gramian matrix equation (similar to the matrix Riccati differential equation) is solved by the transition matrix method. We can find a corresponding variational problem if a BVP is given. We study the reduction of the following BVP

$$y'' - y + x = 0; \quad y(0) = y(1) = 0 \quad (5.70)$$

Given the BVP problem as in (5.70), convert it to a variational problem. Multiply both the sides of (5.70) with $\delta\lambda$ to obtain

$$y''\delta y - y\delta y + x\delta y = 0 \quad (5.71)$$

Integrate (5.71) over the interval [0, 1] to obtain

$$\int_0^1 y''\delta y dx - \int_0^1 y\delta y dx + \int_0^1 x\delta y dx = 0 \quad (5.72)$$

Integrating the first term of (5.72) by parts, we obtain

$$y'\delta y\{0 \text{ to } 1\} - \int_0^1 y'\delta y' dx - \int_0^1 y\delta y dx + \int_0^1 x\delta y dx = 0 \quad (5.73)$$

Now, using the properties of the variational calculus in (5.73), we obtain the following expressions:

$$-\int_0^1 \frac{1}{2}\delta y'^2\, dx - \int_0^1 \frac{1}{2}\delta y^2\, dx + \int_0^1 \delta(xy)dx = 0 \qquad (5.74)$$

$$\int_0^1 \delta\left[-\frac{1}{2}y'^2 - \frac{1}{2}y^2 + xy\right]dx = 0 \qquad (5.75)$$

$$\delta\left(\int_0^1 \left[y'^2 + y^2 - 2xy\right]dx\right) = 0 \qquad (5.76)$$

From (5.76), we deduce that it is of the form

$$\delta J(y) = 0 \qquad (5.77)$$

Thus, the corresponding variational problem is given as

$$J(y) = \int_0^1 (y'^2 + y^2 - 2xy)dx; \quad y(0) = 0; y(1) = 0 \quad (5.78)$$

5.7.1 Solution of a Variational Problem Using a Direct Method

One can use the Rayleigh Ritz approach to find an approximate solution to a variational problem. Obtain the minimum of the following cost functional

$$J(y) = \int_{x_1}^{x_2} F(x,y,y')dx; \quad y(x_1) = y_1; y(x_2) = y_2 \quad (5.79)$$

Let $y(x) \in C[x_1, x_2]$ be the solution to the variational problem of (5.79). Also, $C(.,.)$ is a set of all continuously differentiable functions on the interval $[x_1, x_2]$. Also, we have the bases functions for the infinite dimensional vector space $C[x_1, x_2]$. For making a solution practical, let $\bar{y}(x)$ be an approximation of y and is given by

$$\bar{y}(x) = \sum_{i=0}^{n} c_i\phi_i(x) \qquad (5.80)$$

The bases functions should be taken to satisfy the boundary conditions $\bar{y}(x_1) = y_1; \bar{y}(x_2) = y_2$. Then, the problem of (5.79) is converted to the equivalent problem as follows: minimize

$$J(y) = \int_{x_1}^{x_2} F\left\{x, \sum_{i=0}^{\infty} c_i\phi_i(x), \sum_{i=0}^{\infty} c_i\phi_i'(x)\right\}dx; \qquad (5.81)$$

$$y(x_1) = y_1; y(x_2) = y_2$$

One can find an approximate solution $\bar{y}$ by using the following cost functional

$$J(\bar{y}) = \int_{x_1}^{x_2} F\left\{x, \sum_{i=0}^{n} c_i\phi_i(x), \sum_{i=0}^{n} c_i\phi_i'(x)\right\}dx; \qquad (5.82)$$

$$\bar{y}(x_1) = y_1; \bar{y}(x_2) = y_2$$

So, the idea is to minimize (5.82) subject to the known bases functions, since these are chosen a priori. Hence, in (5.82), we have the unknown coefficients, $c(i = 1$ to $n)$, then the problem reduces to

$$\min J(\bar{y}) = \min_{c_1, c_2, \dots, c_n} J(c_1, c_2, \dots, c_n) \qquad (5.83)$$

Then, we have

$$\frac{\partial J(.)}{\partial c_i} = 0; \quad i = 0, 1, 2, \dots, n \qquad (5.84)$$

The solution of the set of the resultant Equation in (5.84) will give the required coefficients that can be used in (5.80) to obtain an approximate solution to the minimization problem.

5.8 General Variational Approach

Often, the sufficient conditions for an optimal control problem to have a solution, how much these might be reassuring, are not so useful in finding the solutions; hence, we need conditions that any optimal control must necessarily satisfy; for several optimal control problems, such conditions point to a small subset of controls—even a single control [3]. In our optimal control, the simplest problem involves a free value of the state variables:

$$\text{minimize} \int_{t_0}^{t_f} F(x(t), u(t), t)dt; \text{ with constraint} \\ \dot{x}(t) = f(x(t), u(t), t); x(t_0) = x_0 \qquad (5.85)$$

In (5.85), x is $n \times 1$ state vector, and u is a control function/variable that serves as an input signal to open/closed loop control system (CLCS). For an optimal control problem, a variation of the state trajectory x cannot be explicitly related to a variation of the control input signal $u(.)$, because the state and control vectors are implicitly related by a nonlinear differential equation, which is also regarded as a constraint—the second

Equation in (5.85). Hence, we use a one-parameter family of comparison trajectories as

$$u(t) + \varepsilon\eta(t); \ \eta \in C[t_0, t_f]; \ with \|\eta\|_\infty \le d \quad (5.86)$$

5.8.1 First Order Necessary Conditions

Let us consider the problem to minimize the following functional

$$J(\mathrm{u}) = \int_{t_0}^{t_f} F(x(t), u(t), t) dt;$$
$$\text{Constraint } \dot{x}(t) = f(x(t), u(t), t); x(t_0) = x_0 \quad (5.87)$$

Let $u \in C[t_0, t_f]$; with fixed end points $t_0 < t_f$. The functions F and f are continuous in (x, u, t). Also, these functions have continuous first (order) partial derivatives with respect to x and u for all $\{x(.), u(.), t\}$ within the domain specified by the end conditions and vector spaces of x and u [3]. Let $\hat{u}(.) \in C[t_0, t_f]$ be a local minimizer for the problem, and $\hat{x}(.)$ be the corresponding state vector response. There is a function $\hat{\lambda}(.)$ such that the triple $\{\hat{x}(.), \hat{\lambda}(.), \hat{u}(.)\}$ satisfies the following system of equations

$$\dot{x}(t) = f(x(t), u(t), t); \qquad x(t_0) = x_0 \quad (5.88)$$

$$\dot{\lambda}(t) = -F_x(x(t), u(t), t) - f_x^T(x(t), u(t), t)\lambda(t); \ \lambda(t_f) = 0 \quad (5.89)$$

$$0 = F_u(x(t), u(t), t) + f_u^T(x(t), u(t), t)\lambda(t); \ t_0 \le t \le t_f \quad (5.90)$$

The Equations (5.88–5.90) are the EL equations; (5.89) is the adjoint/co-state equation. In order to establish the result, we consider that a one-parameter family of controls exists:

$$\tilde{u}(\varepsilon, t) = u(t) + \varepsilon\eta(t);$$
$$\eta(t) = \in C[t_0, t_f], \text{a fixed funtion}; \varepsilon, \text{a scalar parameter} \quad (5.91)$$

Because of the continuity and differentiability of f, there exists $\bar{\varepsilon} > 0$, such that the response $y(\varepsilon, t) \in C[t_0, t_f]$ associated with $\tilde{u}(\varepsilon, t)$ via (5.88) exists, is unique, and differentiable with respect to $\varepsilon > 0; \forall \varepsilon, t$. Then, $\varepsilon = 0$ would provide the optimal response:

$$y(t) = y(0, t) \equiv \hat{x}(t); t_0 \le t \le t_f \quad (5.92)$$

Since, the control $\tilde{u}(\varepsilon, t)$ is admissible and its associated response is $y(\varepsilon, t)$; the following expressions can be formulated [3]:

$$J(\tilde{u}(\varepsilon, .)) =$$
$$\int_{t_0}^{t_f} [F(y(\varepsilon, t), \tilde{u}(\varepsilon, t)) + \lambda^T(t)\{f(y(\varepsilon, t), \tilde{u}(\varepsilon, t)) - \dot{y}(\varepsilon, t)\}] dt \quad (5.93)$$

$$J(\tilde{u}(\varepsilon, .)) = \int_{t_0}^{t_f} [F(y(\varepsilon, t), \tilde{u}(\varepsilon, t))$$
$$+ \lambda^T(t)f(y(\varepsilon, t), \tilde{u}(\varepsilon, t)) - \lambda^T(t)\dot{y}(\varepsilon, t)] dt \quad (5.94)$$

Integrate the last term in (5.94) by parts, and substitute the time interval's limits to obtain

$$J(\tilde{u}(\varepsilon, .)) =$$
$$\int_{t_0}^{t_f} [F(y(\varepsilon, t), \tilde{u}(\varepsilon, t), t) + \lambda^T(t)f(y(\varepsilon, t), \tilde{u}(\varepsilon, t), t) + \dot{\lambda}^T(t)y(\varepsilon, t)] dt$$
$$- \lambda^T(t_f)y(\varepsilon, t_f) + \lambda^T(t_0)y(\varepsilon, t_0) \quad (5.95)$$

By differentiating (5.95), we get

$$\frac{\partial}{\partial \varepsilon} J(\tilde{u}(\varepsilon, .)) =$$
$$\int_{t_0}^{t_f} [F_u(y(\varepsilon, t), \tilde{u}(\varepsilon, t), t) + f_u^T(y(\varepsilon, t), \tilde{u}(\varepsilon, t), t)\lambda(t)]^T \eta(t) dt$$
$$+ \int_{t_0}^{t_f} [F_x(y(\varepsilon, t), \tilde{u}(\varepsilon, t), t) + f_x^T(y(\varepsilon, t), \tilde{u}(\varepsilon, t), t)\lambda(t)$$
$$+ \dot{\lambda}(t)]^T y_\varepsilon(\varepsilon, t) dt$$
$$- \lambda^T(t_f)y_\varepsilon(\varepsilon, t_f) + \lambda^T(t_0)y_\varepsilon(\varepsilon, t_0) \quad (5.96)$$

Evaluate the limit of (5.96) as $\varepsilon \to 0$, and since the last term in (5.96) is 0; to obtain

$$\delta J(\hat{u}, \eta) = \int_{t_0}^{t_f} [F_u(\hat{x}(t), \hat{u}(t), t) + f_u^T(\hat{x}(t), \hat{u}(t), t)\lambda(t)]^T \eta(t) dt$$
$$+ \int_{t_0}^{t_f} [F_x(\hat{x}(t), \hat{u}(t), t) + f_x^T(\hat{x}(t), \hat{u}(t), t)\lambda(t) + \dot{\lambda}(t)]^T y_\varepsilon(0, t) dt$$
$$- \lambda^T(t_f)y_\varepsilon(0, t_f) \quad (5.97)$$

Since, $\hat{u}$ is a local minimizer, we get

$$0 = \int_{t_0}^{t_f} \{[\hat{F}_x + \hat{f}_x^T\lambda(t) + \dot{\lambda}(t)]^T y_\varepsilon(0, t) + [\hat{F}_u + \hat{f}_u^T\lambda(t)]^T \eta(t)\} dt$$
$$- \lambda^T(t_f)y_\varepsilon(0, t_f) \quad (5.98)$$

We can set the terms in the brackets to zero, assuming that the individual terms $y_\varepsilon(0,t)$ and $\eta(t)$ are non-zer, and the terminal condition $\lambda(t_f) = 0$ to obtain

$$\dot{\lambda}(t) = -\hat{F}_x - \hat{f}_x^T \lambda(t) \qquad (5.99)$$

$$\hat{F}_{u_i} + \hat{f}_{u_i}^T \hat{\lambda}(t) = 0; \quad i = 1, \ldots, n_x \qquad (5.100)$$

As is seen from (5.88) and (5.89), the boundary conditions (for the state and the co-state variables) are split, yielding a TPBVP. In (5.90), if we have $f(x(t), u(t), t) = u(t)$, then (5.90) becomes

$$\dot{\lambda}(t) = -F_u(\hat{x}(t), \hat{u}(t), t) \qquad (5.101)$$

Using (5.101) in (5.89), one obtains

$$\frac{d}{dt}\{F_u(\hat{x}(t), \hat{u}(t), t)\} = F_x(\hat{x}(t), \hat{u}(t), t) \qquad (5.102)$$

$$\frac{d}{dt}\{F_u(\hat{x}(t), \dot{\hat{x}}(t), t)\} = F_x(\hat{x}(t), \dot{\hat{x}}(t), t); \qquad (5.103)$$

$$\text{with } F_u(\hat{x}(t), \dot{\hat{x}}(t), t) = 0, t = t_f$$

Equation (5.103) is the Euler equation.

5.8.2 Mangasarian Sufficient Conditions

The essence of minimization of (5.85) and finding an appropriate control $u(.)$ signifies that

$$J(\hat{u}) \leq J(u); \quad \forall u(.) \qquad (5.104)$$

The inequality (5.103) indicates that one is seeking global minimum of $J(.)$. Let us consider the problem of minimization as in (5.87), with $F(.,.,.)$, and $f(.,.,.)$ as the continuous functions of their arguments, and have continuous partial derivatives of first order with respect to x and u; and these functions are convex in x and u. Let us now assume that the state and co-state variables $\hat{x}, \hat{u}, \hat{\lambda}$ satisfy the EL equations. Also, assume that $\hat{\lambda}(t) \geq 0$, then $\hat{u}(t)$ is a global minimizing control $u(t)$ for the problem of (5.87). Let us form the following difference for any $x(t), u(t)$ using (5.87) [3]:

$$J(u) - J(\hat{u}) = \int_{t_0}^{t_f} [F(x(t), u(t), t) - F(\hat{x}(t), \hat{u}(t), t)]dt \qquad (5.105)$$

$$\geq \int_{t_0}^{t_f} [\hat{F}_x^T(\hat{x}(t), \hat{u}(t), t)\{x(t) - \hat{x}(t)\} - \hat{F}_u^T(\hat{x}(t), \hat{u}(t), t)\{u(t) - \hat{u}(t)\}]dt \qquad (5.106)$$

Using the compact notation, one can write (5.106) as

$$\geq \int_{t_0}^{t_f} [\hat{F}_x^T(.)\{x(t) - \hat{x}(t)\} - \hat{F}_u^T(.)\{u(t) - \hat{u}(t)\}]dt \qquad (5.107)$$

Using the EL conditions/equations (5.99) and (5.107) in (5.107), we obtain

$$J(u) - J(\hat{u}) \geq \int_{t_0}^{t_f} (-[\hat{f}_x^T \hat{\lambda}(t) + \dot{\hat{\lambda}}(t)]^T \{x(t) - \hat{x}(t)\} - [\hat{f}_u^T \hat{\lambda}(t)]^T \{u(t) - \hat{u}(t)\})dt \qquad (5.108)$$

$$J(u) - J(\hat{u}) \geq \int_{t_0}^{t_f} -([\hat{f}_x^T \hat{\lambda}(t)]^T \{x(t) - \hat{x}(t)\} - \dot{\hat{\lambda}}(t)^T \{x(t) - \hat{x}(t)\} - [\hat{f}_u^T \hat{\lambda}(t)]^T \{u(t) - \hat{u}(t)\})dt \qquad (5.109)$$

Next, integrate by parts the term associated with $\dot{\lambda}(t)$ and rearrange, then one obtains

$$J(u) - J(\hat{u}) \geq \int_{t_0}^{t_f} \hat{\lambda}^T(t)[(f - \hat{f}) - \hat{f}_x\{x(t) - \hat{x}(t)\} - \hat{f}_u\{u(t) - \hat{u}(t)\}]dt$$
$$-\hat{\lambda}^T(t_f)\{x(t_f) - \hat{x}(t_f)\} + \hat{\lambda}^T(t_0)\{x(t_0) - \hat{x}(t_0)\} \qquad (5.110)$$

For the optimizing variables $x(t)$ and $u(t)$, the terms on the RHS of (5.109) disappear, and hence, we obtain

$$J(u) \geq J(\hat{u}) \qquad (5.111)$$

5.8.3 Interpretation of the Co-State Variables

It is well-known that one can incorporate a constraint, say of the dynamics of the system, into the cost function for optimization by using the LM. As such, this artifice does not alter the cost function, since the additional term is anyway zero. But, when the resultant equations are solved, they not only give the optimal control strategy, but are such that the optimal trajectory, both $x(t)$ and $u(t)$, would satisfy the original constraint! This LM is a sensitive coefficient/variable of the cost function with respect to a change in the particular constraint [3]. Again, the optimal problem of (5.87) is considered to establish the interpretation of the co-state/adjoint variables. Let $J(x_0, t_0)$ be the minimum value of the cost function $J(u)$ for the chosen initial conditions, and the corresponding control strategy is $\hat{u}(t_0, t_f)$, control response is $\hat{x}(t_0, t_f)$, and the adjoint trajectory is $\hat{\lambda}(t_0, t_f)$. Next, we consider a small alteration in the initial state as $x_0 + \varepsilon$; let the unique optimal control be $\upsilon(\varepsilon,t)$ for the altered initial condition/optimal problem, and let the related optimal response be given as $y(t,\varepsilon)$. The equation for this response is given as

$$\dot{y}(\varepsilon,t) = f(y(\varepsilon,t),\upsilon(\varepsilon,t),t); \ y(\varepsilon,t_0) = x_0 + \varepsilon \quad (5.112)$$

Then, we have $\upsilon(\varepsilon = 0,t) = \hat{u}(t); \ y(\varepsilon = 0,t) = \hat{x}(t)$. It is also assumed that the variables $\upsilon(\varepsilon,t)$ and $y(\varepsilon,t)$ are continuously differentiable with respect to ε. Next, form the combined cost functional as [3]:

$$J(\upsilon(\varepsilon,t_0),t_0) =$$

$$\int_{t_0}^{t_f} [F(y(\varepsilon,t),\upsilon(\varepsilon,t),t) + \hat{\lambda}^T(t)\{f(y(\varepsilon,t),\upsilon(\varepsilon,t),t) - \dot{y}(\varepsilon,t)\}]dt \quad (5.113)$$

Differentiate (5.113) with respect to ε to obtain

$$\frac{\partial}{\partial \varepsilon} J(\upsilon(\varepsilon,t_0),t_0) =$$

$$\int_{t_0}^{t_f} [F_u(y(\varepsilon,t),\upsilon(\varepsilon,t),t) + \lambda^T(t)f_u(y(\varepsilon,t),(\varepsilon,t),t)]^T \upsilon_\varepsilon(\varepsilon,t)dt$$

$$+ [F_x(y(\varepsilon,t),\upsilon(\varepsilon,t),t) + \lambda^T(t)f_x(y(\varepsilon,t),\upsilon(\varepsilon,t),t)$$

$$+ \dot{\lambda}(t)]^T y_\varepsilon(\varepsilon,t)dt - \lambda^T(t_f)y_\varepsilon(\varepsilon,t_f) + \lambda^T(t_0)y_\varepsilon(\varepsilon,t_0) \quad (5.114)$$

Take the limit of (5.114) as $\varepsilon \to 0$ to obtain

$$\frac{\partial}{\partial \varepsilon} J(\upsilon(\varepsilon,t_0),t_0)_{\varepsilon=0} =$$

$$\int_{t_0}^{t_f} [F_u(\hat{x}(t),\hat{u}(t),t) + \hat{\lambda}^T(t)f_u(\hat{x}(t),\hat{u}(t),t)]^T \upsilon_\varepsilon(0,t)dt \quad (5.115)$$

$$+ [F_x(\hat{x}(t),\hat{u}(t),t) + \hat{\lambda}^T(t)f_x(\hat{x}(t),\hat{u}(t),t) + \dot{\hat{\lambda}}(t)]^T y_\varepsilon(0,t)dt$$

$$- \hat{\lambda}^T(t_f)y_\varepsilon(0,t_f) + \hat{\lambda}^T(t_0)y_\varepsilon(0,t_0)$$

We note that the triple $(\hat{x},\hat{u},\hat{\lambda})$ satisfies the EL Equations (5.89) and (5.90); hence, by using these in (5.115) and since $y(\varepsilon,t_0) = x_0 + \varepsilon; \ y(0,t_0) = x(t_0) = x_0 \Rightarrow y_\varepsilon(0,t_0) = I$ in (5.115), one obtains

$$\hat{\lambda}(t_0) = \frac{\partial}{\partial \varepsilon} J(\upsilon(\varepsilon,t_0),t_0)|_{\varepsilon=0} = J_x(x_0,t_0) \quad (5.116)$$

We see from (5.116) that the LHS is the adjoint variable/co-state at the initial time, $t(0)$, and comparing it with the RHS, one can infer that the adjoint variable can be interpreted as the sensitivity of the cost functional to a change in the initial condition $x(0)$.

5.8.4 Principle of Optimality

We need to also interpret the adjoint variable at the end condition. For this we need to establish the result on the principle of optimality [3]. Again, start with (5.87), with the usual notion of the optimal control input and the optimal response trajectories: $(\hat{u}(t),\hat{x}(t)); [t_0,t_f]$. Then, the principle of optimality states that for any time $t_1 \in [t_0,t_f]$, the control strategy $\hat{u}(t); t_1 \le t \le t_f$ is also an optimal control strategy for the following optimization problem

minimize $J(u) =$

$$\int_{t_1}^{t_f} F(x(t),u(t),t)dt; \ \text{Constraint } \dot{x}(t) = f(x(t),u(t),t); x(t_1) = x_1 \quad (5.117)$$

Let us have the following as the minimum of the cost function

$$J(x_0,t_0) = \int_{t_0}^{t_1} F(x(t),u(t),t)dt + \int_{t_1}^{t_f} F(x(t),u(t),t)dt \quad (5.118)$$

We use the method of contradiction to establish the result of the principle of optimality; i.e. $\hat{u}(t); t_1 \le t \le t_f$ is not the optimal control strategy for the optimal control problem of (5.117). This means that there is a control strategy $\tilde{u}(t); t_1 \le t \le t_f$ that will give the following functional form

$$\int_{t_1}^{t_f} F(\tilde{x}(t),\tilde{u}(t),t)dt < \int_{t_1}^{t_f} F(\hat{x}(t),\hat{u}(t),t)dt \quad (5.119)$$

Then, substitute for the RHS in (5.119), from (5.118), we get

$$\int_{t_1}^{t_f} F(\tilde{x}(t),\tilde{u}(t),t)dt < J(x_0,t_0) - \int_{t_0}^{t_1} F(\hat{x}(t),\hat{u}(t),t)dt \quad (5.120)$$

Now, shift the second term of the RHS of (5.120) to the LHS to obtain the following

$$\int_{t_0}^{t_1} F(\hat{x}(t),\hat{u}(t),t)dt + \int_{t_1}^{t_f} F(\tilde{x}(t),\tilde{u}(t),t)dt < J(x_0,t_0) \quad (5.121)$$

From (5.118), it is seen that the entire control strategy $\hat{u}(t); t_0 \le t \le t_f$ is optimal by assumption and this is contradicted by (5.121). Thus, it is established that the part of the control $u(t); t_1 \le t \le t$ should also be optimal, i.e. $\hat{u}(t); t_1 \le t \le t$, and hence, the principle of optimality is established.

Now, the arguments and the development of Section 5.8.3 can be used along with the principle of optimality to establish the interpretation of the adjoint variable at the final time $\lambda(t_f)$ to obtain the following

$$\hat{\lambda}(t_1) = J_x(\hat{x}(t_1), t_1) \qquad (5.122)$$

In (5.122), the time t_1 is arbitrary, one gets

$$\hat{\lambda}(t) = J_x(\hat{x}(t), t); \quad t_0 \le t \le t_f \qquad (5.123)$$

From (5.123), one can see that for an exogenous small perturbation in the state variable at time t (RHS) and for the optimally modified control input, thereafter, the optimal cost (5.123) changes at the rate $\lambda(t)$(LHS); thus, the latter is a small/marginal valuation in the optimal control (problem) of the state variable x at time t. The optimal cost value would remain unchanged in case of a perturbation at terminal time t_f; since, $\hat{\lambda}(t_f) = J_x(\hat{x}(t_f), t_f) = 0$, see (5.115).

5.8.5 General Terminal Constraints

In the previous sections, we have attended to the variational/optimal control problem with fixed initial time and terminal time; and free terminal state $x(t_f)$. However, some optimal control problems need different terminal conditions: a) the terminal time is free, i.e. to traverse in a minimum time; (b) the state variables at the final time are either fixed or restricted to lie on a smooth manifold. These conditions can be handled by the artifice: (i) specify end-point constraint, and/or (ii) add a terminal cost. The later problem is called the Bolza problem [3]. A free terminal time can be considered as an additional variable in the main optimization problem.

5.8.5.1 Necessary Conditions for Equality Terminal Constraints

Let us consider the optimal problem to

$$\text{minimize: } J(u, t_f) = \int_{t_0}^{t_f} F(x(t), u(t), t)dt + \phi(x(t_f), t_f);$$

with Constraints $\dot{x}(t) = f(x(t), u(t), t); x(t_0) = x_0; \qquad (5.124)$

$$\Psi_k(u, t_f) = \psi_k(x(t_f), t_f) = 0; \quad k = 1, ..., n_\psi$$

The functions ϕ and ψ are continuous and have continuous first order partial derivatives with respect to x and t. Assume that $(\hat{u}, \hat{t}_f)$ are a local minimizer for the optimal problem and terminal time T is free. Also, $\hat{x}$ is the corresponding state response. Next, the differential of the constraint variable $\Psi_k(\hat{u}, t_f)$ has the following conditions

$$\begin{bmatrix} \delta \Psi_1(\hat{u}, \hat{t}_f; \eta_1, \tau_1) ... \delta \Psi_1(\hat{u}, \hat{t}_f; \eta_{n_\psi}, \tau_{n_\psi}) \\ \\ ... \\ \\ \delta \Psi_{n_\psi}(\hat{u}, \hat{t}_f; \eta_1, \tau_1) ... \delta \Psi_{n_\psi}(\hat{u}, \hat{t}_f; \eta_{n_\psi}, \tau_{n_\psi}) \end{bmatrix} \ne 0 \qquad (5.125)$$

In (5.125), n_ψ are independent directions: $\eta_1, \tau_1, ..., \eta_{n_\psi}, \tau_{n_\psi}$. Then, for $\hat{\lambda}$, $(\hat{u}, \hat{x}, \hat{\lambda}, \hat{\upsilon}, \hat{t}_f)$ satisfies the following EL equations in the Hamiltonian

$$\dot{x}(t) = H_\lambda(x(t), u(t), \lambda(t), t); \quad x(t_0) = x_0 \qquad (5.126)$$

$$\dot{\lambda}(t) = -H_x(x(t), u(t), \lambda(t), t); \quad \lambda(t_f) = \Phi_x(x(t_f), t_f) \qquad (5.127)$$

$$0 = H_u(x(t), u(t), \lambda(t), t) \qquad (5.128)$$

Additional conditions to be admitted are at the terminal time

$$\Phi_t(x(t_f), t_f) + H(x(t_f), u(t_f), \lambda(t_f), t_f) = 0$$
$$\psi(x(t_f), t_f) = 0 \qquad (5.129)$$

In (5.129), we have the Hamiltonians as

$$\Phi = \phi + \upsilon^T \psi \text{ and } H = F + \lambda^T f \qquad (5.130)$$

Let us consider again, a one-parameter family $\tilde{u}(\varepsilon, t) = \hat{u}(t) + \varepsilon \eta(t)$; and let $y(\varepsilon, t)$ be the response corresponding to $\tilde{u}(\varepsilon, t)$; i.e. $y(0, t) \equiv \hat{x}(t)$. The Gateaux variation at $(\hat{u}, \hat{t}_f)$ in any direction (η, t) of the cost functional $J(.)$ subject to the initial value problem is given as

$$\delta J(\hat{u}, \hat{t}_f; \eta, \tau) = \int_{t_0}^{\hat{t}_f} [F_u(\hat{x}(t), \hat{u}(t), t) + f_u^T(\hat{x}(t_f), \hat{u}(t), t)\lambda^{(0)}(t)]^T \eta(t)dt$$

$$+ [\phi_t(\hat{x}(\hat{t}_f), \hat{t}_f) + F(\hat{x}(\hat{t}_f), \hat{u}(\hat{t}_f), \hat{t}_f) + f^T(\hat{x}(\hat{t}_f), \hat{u}(\hat{t}_f), \hat{t}_f)\lambda^{(0)}(t)]\tau \qquad (5.131)$$

The adjoint variables in (5.131) are evaluated as

$$\dot{\lambda}^{(0)}(t) = -F_x(\hat{x}(t), \hat{u}(t), t) - f_x^T(\hat{x}(t), \hat{u}(t), t)\lambda^{(0)}(t);$$
$$\lambda^{(0)}(\hat{t}_f) = \phi_x(\hat{x}(\hat{t}_f), \hat{t}_f). \qquad (5.132)$$

Similarly, the Gateaux variations for the functional (5.124) and (5.125) is given by

$$\delta\Psi_k(\hat{u},\hat{t}_f;\eta,\tau) = \int_{t_0}^{\hat{t}_f}[f_u^T(\hat{x}(t_f),\hat{u}(t),t)\lambda^{(k)}(t)]^T\eta(t)dt \tag{5.133}$$

$$+[(\psi_k)_t(\hat{x}(\hat{t}_f),\hat{t}_f)+f^T(\hat{x}(\hat{t}_f),\hat{u}(\hat{t}_f),\hat{t}_f)\lambda^{(k)}(t)]\tau$$

The adjoint variables in (5.133) are evaluated as

$$\dot{\lambda}^{(k)}(t) = -f_x^T(\hat{x}(t),\hat{u}(t),t)\lambda^{(k)}(t);$$

$$\lambda^{(k)}(\hat{t}_f) = (\psi_k)_x(\hat{x}(\hat{t}_f),\hat{t}_f); \ k=1,...,n_\psi \tag{5.134}$$

Now, there exists a vector $\hat{\upsilon}$ such that the following is true

$$0 = \delta\left(J+\sum_{k=1}^{n_x}\hat{\upsilon}_k\Psi_x(\hat{u},\hat{t}_f,\eta,\tau)\right) \tag{5.135}$$

$$0 = \int_{t_0}^{\hat{t}_f}H_u^T(\hat{x}(t),\hat{u}(t),\hat{\lambda}(t),t)\eta(t)dt$$

$$+[\Phi_t(\hat{x}(\hat{t}_f),\hat{t}_f)+H(\hat{x}(\hat{t}_f),\hat{u}(\hat{t}_f),\hat{\lambda}(\hat{t}_f),\hat{t}_f)]\tau \tag{5.136}$$

In (5.136), we have the following expressions

$$\Phi = \phi + \hat{\upsilon}^T\psi$$

$$\hat{\lambda} = \lambda^{(0)} + \sum_{k=1}^{n_x}\hat{\upsilon}_k\lambda^{(k)} \tag{5.137}$$

$$H = F + \hat{\lambda}^T f$$

If we take $\tau = 0$, and focus attention on $\eta(t)$ so that

$$\eta_i(t) = H_{u_i}(\hat{x}(t),\hat{u}(t),\hat{\lambda}(t),t); \ \eta_j(t) = 0 \text{ for } j \neq i \tag{5.138}$$

then, one gets the necessary condition of optimality

$$H_{u_i}(\hat{x}(t),\hat{u}(t),\hat{\lambda}(t),t) = 0; \ i=1,...,n_x \tag{5.139}$$

One can choose $\eta(t) = 0$, to get

$$\Phi_t(\hat{x}(\hat{t}_f),\hat{t}_f)+H(\hat{x}(\hat{t}_f),\hat{u}(\hat{t}_f)\hat{\lambda}(\hat{t}_f),\hat{t}_f) = 0 \tag{5.140}$$

Because the adjoint differential equations are linear, the adjoint co-states should satisfy the following equations and the conditions

$$\dot{\hat{\lambda}}(t) = -H_x(\hat{x}(t),\hat{u}(t),\hat{\lambda}(t),t); \ \hat{\lambda}(\hat{t}_f) = \Phi_x(\hat{x}(\hat{t}_f),\hat{t}_f) \tag{5.141}$$

The optimality conditions of (5.126–8.128) to obtain the optimal values of $x(t)$, $u(t)$, and the co-state for the optimal control problem of (5.124) constitute the TPBV problem, which can be solved by using the multiple shooting procedure or the method of invariant embedding (IE), as stated earlier.

Appendix 5A: Conditions in Calculus of Variation

Some important results that have been used for the theoretical development of Chapter 5 are collected here [1–3].

5A.1 Fundamental Lemma

Let $f(x)$ be a continuous time function defined on the interval $[a, b]$. If the following is true for every function, $g(x) \in C(a,b)$, such that $g(a) = g(b) = 0$;

$$\int_a^b f(x)g(x)dx = 0 \tag{5A.1}$$

Then $f(x) \equiv 0; \ \forall x \in [a,b]$.

Let $f(x) \neq 0$ for some $c \in (a,b)$. Also, assume that $f(c) > 0$. Since f is continuous, we also have $f(x) > 0$ for the interval $[x_1,x_2] \subset [a,b]$ that contains the point c. Let us specify the follow function [2]:

$$g(x) = \begin{cases} (x-x_1)(x_2-x) & \text{for } x\in[x_1,x_2] \\ 0 & \text{outside } [x_1,x_2] \end{cases} \tag{5A.2}$$

In (5A.2), $(x-x_1)(x_2-x)$ is positive for $x \in (x_1,x_2)$. Expand (5A.1) as follows

$$\int_a^b f(x)g(x)dx = \int_a^{x_1}f(x)g(x)dx + \int_{x_1}^{x_2}f(x)g(x)dx + \int_{x_2}^b f(x)g(x)dx$$

$$\tag{5A.3}$$

Since $g(x) = 0$, outside the interval $[x_1, x_2]$, (5A.3) simplifies to

$$\int_a^b f(x)g(x)dx = \int_{x_1}^{x_2} f(x)g(x)dx \qquad (5A.4)$$

Substitute (5A.2) in (5A.4) to obtain

$$\int_a^b f(x)g(x)dx = \int_{x_1}^{x_2} f(x)(x - x_1)(x_2 - x)dx \qquad (5A.5)$$

Since $g(x)$ is positive for the interval that is specified in (5A.5)

$$\int_a^b f(x)g(x)dx > 0 \qquad (5A.6)$$

This is contradicted from what is given in the Lemma; hence, $f(x) \equiv 0$.

5A.2 Variational Properties

Some variational properties [2] that look almost the same as the classical derivative rules are:

1. $\delta(F_1 \pm F_2) = \delta F_1 \pm \delta F_2$ $\qquad$ (5A.7)
2. $\delta(F_1 F_2) = (\delta F_1)F_2 \pm F_1(\delta F_2)$ $\qquad$ (5A.8)
3. $\delta\left(\dfrac{F_1}{F_2}\right) = \dfrac{F_2 \delta F_1 - F_1 \delta F_2}{F_2^2}; \ F_2 \neq 0$ $\qquad$ (5A.9)
4. $\delta(F^n) = nF^{n-1}\delta F$ $\qquad$ (5A.10)

5A.3 Hamilton's Principle

Let K be the kinetic energy and P be the potential energy of a particle in motion; and $L = K - P$ be the kinetic potential or the Lagrangian function [2]. Let the action integral be given as

$$A = \int_{t_1}^{t_2} L\,dt \qquad (5A.11)$$

As per the principle of least action, we have

$$\delta A = 0 \qquad (5A.12)$$

Then

$$\delta\left(\int_{t_1}^{t_2} L\,dt\right) = \delta\left(\int_{t_1}^{t_2} (K - P)\,dt\right) = 0 \qquad (5A.13)$$

From (5A.13), we deduce that the motion of the particle is such that the integral of the difference between the kinetic and potential energies of the particle is stationary for the true path. The integral is minimum over a sufficiently small interval. Thus, Nature tends to equalize the kinetic and potential energies over the motion trajectory; hence, (5A.12) is true along the trajectory. It would be rather better to say that, what we observe in the nature of the balance of the kinetic and potential energies, in most or all the cases, (there might be some cases where some complex laws might be working, which we do not know as yet), can be described by (5A.12 and 5A.13).

5A.4 Legendre Test for Extremum

For the functional $J(y)$ and y being an optimizer of $J(y)$, the following are valid [2], which again look similar to the familiar conditions on Hessian matrix for the extrema:

1. $\delta J(y) = 0$; Euler-Lagrange Equation $\qquad$ (5A.14)
2. $\delta^2 J(y) > 0 \rightarrow y$ is a minimizing variable $\qquad$ (5A.15)
3. $\delta^2 J(y) < 0 \rightarrow y$ is a maximizing variable $\qquad$ (5A.16)

5A.5 Sufficiency of Necessary Conditions

Let $f(x, u, t)$ and $g(x, u, t)$ be differentiable functions [1] for:

$$\max_u = \int_{t_0}^{t_2} f(x, u, t)\,dt \qquad (5A.17)$$

This is subject to

$$\dot{x} = g(x, u, t); \ x(t_0) = x_0 \qquad (5A.18)$$

If f and g are jointly concave in x, u, and $\lambda \geq 0, \forall t$, then, the necessary conditions are also sufficient.

5A.6 Piecewise Continuous Extremals

In certain cases, it may be necessary to find extremals in the large class of piecewise continuous controls [3]. Discontinuous controls would give rise to discontinuities in the slope of the response, $x(t)$, and then these are referred to as corner points. Hence, certain conditions should hold at the corner points. Let us consider the optimal control problem of (5.85), for which we have the solution $\hat{u} \in C[t_0, t_f]$; $C \rightarrow$ Continuous vector space, and the associated response is $\hat{x}(.)$, and the co-state response is $\hat{\lambda}(.)$. Then, at each corner point $\tau \in (t_0, t_f)$ of the control solution $u(t)$, we should have the following

$$\hat{x}(\tau^-) = \hat{x}(\tau^+)$$
$$\hat{\lambda}(\tau^-) = \hat{\lambda}(\tau^+)$$
(5A.19)

$$H(\tau^-, \hat{x}(\tau), \hat{u}(\tau^-), \hat{\lambda}(\tau)) = H(\tau^+, \hat{x}(\tau), \hat{u}(\tau^+), \hat{\lambda}(\tau))$$
(5A.20)

In (5A.19), τ^-, τ^+ denote the time just before and just after the corner, and the values of the variables are signified accordingly, on the left, and right limits.

5A.7 Reachability Condition

We consider the problem of (5.124), Section 5.8.5.1, with fixed terminal time t_f and a single terminal state constraint [3]:

$$\Psi(u) = x_k(t_f) - x_{f_k} = 0; \quad \text{for some } k \in \{1, ..., n_x\}$$
(5A.21)

The Gateaux variation of the function in (5A.21) at u^* in any direction $\eta \in [t_0, t_f]^{n_u}$ is given as

$$\delta \Psi_u(\hat{u}; \eta) = \int_{t_0}^{t_f} \left[f_u^T(\hat{x}(t_f), \hat{u}(t), t) \lambda^{(k)}(t) \right]^T \eta(t) dt$$
(5A.22)

We have $\dot{\lambda}^{(k)}(u) = -f_x^T(\hat{x}(t_f), \hat{u}(t), t) \lambda^{(k)}(t)$, $t_0 \leq t \leq t_f$ with terminal conditions as

$$\lambda_k^{(k)}(t_f) = 1; \ \lambda_i^{(k)}(t_f) = 0, \text{ for } i \neq k$$
(5A.23)

Now, set $\eta(t) = f_u^T(\hat{x}(t_f), \hat{u}(t), t) \lambda^{(k)}(t)$ in (5A.22) to obtain the following

$$\int_{t_0}^{t_f} (\lambda^{(k)}(t))^T f_u(\hat{x}(t_f), \hat{u}(t), t) f_u^T(\hat{x}(t_f), \hat{u}(t), t) \lambda^{(k)}(t) dt \neq 0$$
(5A.24)

The condition in (5A.24) is regarded as a reachability condition for dynamic system. It means that if the condition does not hold true, then it may not be possible to find a control law $u(.)$, such that the terminal condition $x_k(t_f) = x_{f_k}$ is satisfied at the final time. The reachability is defined as: a state x_f is said to be reachable at time t_f if for some finite $t_0 < t_f$ there exists an input $u(t)$, $t_0 \leq t \leq t_f$, then that transfers the state $x(t)$ from the origin at t_0, to x_f at time t_f. In general for the considered optimal control problems, with the fixed terminal time t_f and η_ψ terminal constraints $\psi_j(x(t_f)) = 0$, the sufficient condition becomes rank$\{\Psi(\lambda, f)\} = \eta_\psi$, and the matrix is defined as

$$\Psi_{ik}(\lambda, f) = \int_{t_0}^{t_f} (\lambda^{(i)}(t))^T f_u(\hat{x}(t_f), \hat{u}(t), t) f_u^T(\hat{x}(t_f), \hat{u}(t), t) \lambda^{(k)}(t) dt,$$
$$1 \leq i, k \leq \eta_\psi$$
(5A.25)

The adjoint variables are $\lambda^{(j)}(t), t_0 \leq t \leq t_f$.

5A.8 Summary of Necessary Conditions

For the problem of (5.124), we have the summary of necessary conditions [3]:

1. The Euler-Lagrange equations are

$$\dot{x} = H_\lambda \qquad t_0 \leq t \leq t_f$$
$$\dot{\lambda} = -H_x$$
(5A.26)
$$H_u = 0; \quad H = F + \lambda^T f$$

2. Legendre-Clebsch condition is

$$H_{uu} \text{ is semi positive definite, } t_0 \leq t \leq t_f$$
(5A.27)

3. Transversal conditions (based on (5.129), (5.130), and (5.141) are

$$[H + \phi_t + \upsilon^T \psi_t]_{t_f} = 0, \text{ if } t_f \text{ is free}$$
$$[\lambda - \phi_x + \upsilon^T \psi_x]_{t_f} = 0; \quad [\psi]_{t_f} = 0,$$
(5A.28)

and ψ satisfy a regularity condition.

5A.9 Inequality Terminal Constraints

If the considered optimal control problem of (5.124) has the inequality state constraints of the type $\psi_k(x(t_f), t_f) \leq 0$, instead of the equality constraint as in (5.124), the conditions of (5.124–5.128) and the first equation of (5.129) still remain as the necessary ones; however, the constraint condition of the second equation of (5.129) is replaced by the following ones [3]

$$\psi(x(t_f), t_f) \leq 0; \quad \upsilon \geq 0; \quad \upsilon^T \psi(x(t_f), t_f) = 0 \quad (5A.29)$$

6

Optimal Control

Optimal control refers to the class of problems in which one needs to minimize (sometimes maximize) a function of state variables and input variables; this function is often called a functional and, in most cases, it is nonlinear in nature. These variables are the state, or phase, variables and the control (input) variables. In general, the time-evolution (or -propagation) of the states is governed by the control input and the dynamics of the underlying system; the governing mechanism is either a set of differential or difference equations. Interestingly, due to some natural physical limits in the dynamics of the real systems of which these equations are the mathematical models, the control as well as the state variables would also have some constraints, for example, the limit on their maximum amplitude (as we have seen in Chapter 5). One important constraint is that while one should and can find an optimal control variable $u(t)$ for a given system, the same input should be such that the original dynamic system's differential/difference equation should be satisfied with this new optimal signal $u(t)$ as the control input. Another important requirement is that the optimal control input, when applied to a dynamic system, especially if this system is already working in a closed loop, should not render the overall closed loop system unstable. Also, there would be some limit on the amplitude of the optimal control input, due to the fact that large input/s, when applied, can damage the physical system. Many problems in optimal control are non-classical, due to the requirement of constraints on state and control variables. Thus, the problem in/of optimal control is to determine the control signal/s that will make a dynamic system to satisfy the physical constraints and minimize (or maximize) some appropriately chosen performance criterion, which is based on the functional already mentioned in the foregoing [3]. Often, the resultant set of equations would be nonlinear and might require (linearization or) numerical methods to solve these equations to obtain the optimal control and state variables' trajectories. Theory of optimal control has evolved as an extension of the theory and methods of calculus of variation/s [3]. The development of this chapter is based on [3,4]. In most cases, we consider real vector spaces, and their dimensions and meanings will be clear from the underlying states, outputs, and control variables.

6.1 Optimal Control Problem

In the problem of optimal control, certain very important considerations are: (a) existence of the solution, (b) the solution or procedure of the optimal control should satisfy certain constraints, and (c) satisfaction of certain necessary/sufficient conditions. In most cases the optimal control is based on the dynamic systems; hence, it is necessary to define such a system.

6.1.1 Dynamic System and Performance Criterion

In general, one would use a mathematical model of a real dynamic system. This is often and mostly represented in the form of differential (and for discrete time systems, difference) equations. This system of equations when solved analytically or numerically would produce adequate responses of the state of the systems, for the obtained/used optimal control input. Such a mathematical model of a dynamic system is given as

$$\dot{x}(t) = f(x(t), u(t), t); \quad x(t_0) = x_0 \quad (6.1)$$

In (6.1), $x(t)$ is a state space vector, and $u(t)$ is a control input vector, both of appropriate dimensions, and f is a vector valued nonlinear function. In the optimal control problem, it is necessary to assume that this function is continuous in its arguments and is continuously differentiable (at least with respect to the state variable $x(t)$). This means that the first derivative of f wrt x is not infinite. There will be controllers in a control system, and these controllers will receive optimal control input/s $u(t)$, in turn, and the response of the entire closed loop dynamic system will depend on the responses of the controller. Even if one gets an optimal control, its admissibility might be limited due to the physical constraint/s naturally present in the overall dynamic system. This might be due to the facts: quantity of fuel could be limited, temperature should not exceed certain limits, too much current in a motor cannot be allowed to flow, rise in a voltage should be limited, and many such or similar parameters in electro-mechanical-chemical dynamic systems [3,4]. This requires that the numerical/physical constraints should be put on the dynamic behavior of the system, as well as on the control input $u(.)$. This is done by incorporating the mathematical constraints on u: $|u(k)| \leq U_{\max}$; $k = 1,...,n_u$. Next, one needs to define or specify a performance criterion that can be used to evaluate quantitatively—the performance of the dynamic system (6.1). Such a criterion is called a cost functional and is specified as

$$J(u(t)) = \int_{t_0}^{t_f} F(x(t),u(t),t)dt \qquad (6.2)$$

The idea is to minimize (or maximize, in some cases, depending upon the definition of the cost functional and the problem at hand) the cost functional (6.2) with respect to $u(t)$ in order to obtain the optimal control $u(t)$. This cost functional is said to be in the Lagrange form, and the function $F(x,u,t)$ is called Lagrangian. The Lagrangian should be continuous and differential with respect to $x(t)$ (and $u(t)$). The initial or final time may be fixed or a free variable. Some other objective functions can also be defined as [3]

$$\text{\textbf{\textit{Mayor form:}}} \ J(u) = \varphi(x(t_0),t_0,x(t_f),t_f) \qquad (6.3)$$

In (6.3), the function is a real valued continuous function and its first level differentiation should exist.

$$\begin{aligned} \text{\textbf{\textit{Bolza form:}}} \quad & J(u(t)) = \int_{t_0}^{t_f} F(x(t),u(t),t)dt \\ & +\varphi(x(t_0),t_0,x(t_f),t_f) \end{aligned} \qquad (6.4)$$

It can be seen that the form in (6.4) is a sum of Lagrange and Mayor forms.

6.1.2 Physical Constraints

As said earlier, in an optimal control problem one needs to use certain constraints on state variables, $x(t)$, and control input $u(t)$. These constraints are [3]: (i) point constraints, (ii) path constraints, (iii) isoperimetric constraints; these constraints could be of equality or inequality type.

6.1.2.1 Point Constraints

The terminal constraints are the point constraints. One constraint is

$$\psi(x(t_f),t_f) \leq 0 \qquad (6.5)$$

The interpretation of (6.5) is that the system's response is forced to be of a certain value at the final/terminal time and signifies that the final response is stabilized.

Sometimes, one may want to put a restriction on the way a process (dynamic state) changes over from one steady state to a new steady state by the following constraint

$$\psi'(x(t_f),t_f) = 0 \qquad (6.6)$$

6.1.2.2 Isoperimetric Constraints

This constraint is related to the integral of a given functional over certain time interval

$$\int_{t_0}^{t_f} h(x(t),u(t),t)dt \leq C \qquad (6.7)$$

The constraint (6.7) can be transformed to the point constraint/s by using the same procedure of transforming the Lagrange form to the Mayor form as in Section 6.1.1. The interpretation of (6.7) is that the total energy/work/power (or whatever is represented) over a fixed interval should be limited and upper bounded by a constant C.

6.1.2.3 Path Constraints

The idea in the path constraint is to restrict the range of values that a function of the state and control variables can take and is defined as

$$\phi(x(t),u(t),t) \leq 0; \quad \forall t \in [t_0,t_f] \qquad (6.8)$$

A more specific path constraint can be stated as

$$x_k(t) \leq x; \text{may be for a given } u \in U; \forall t \in [t_0,t_f]; \ k = 1,...,n_x \qquad (6.9)$$

Based on the foregoing definitions/specifications of the constraints, one can say that the pair of $x(t)$ and $u(t)$ that satisfies one or more constraint (as required) is called an admissible pair and results in the feasible control $u(t)$.

6.1.3 Optimality Criteria

The optimal control problem now can be stated as: find an admissible control $u(t)$ that satisfies one or more of the physical constraints as specified (deemed necessary) and mathematically represented as in Section 6.1.2, such that the cost functional $J(u(t))$ has a minimum value; when one has obtained this optimality, the control input $u(t)$ is called the optimal control input $\hat{u}(t)$, or optimal control law [3]:

$$J(\hat{u}\ (t)) \leq J(u(t)); \quad \forall u \in U[t_0,t_f] \qquad (6.10)$$

Also, a description of the local minimum of $J(.)$ can be specified as

$$\exists \delta > 0 \text{ such that } J(\hat{u}(t)) \le J(u(t)); \quad \forall u \in U_\delta(\hat{u}) \cap U[t_0, t_f] \tag{6.11}$$

This requires a specification of a norm

$$\|u\|_\infty = \sup_{t \in \bigcup_{k=0}^{N}(\tau_k, \tau_{k+1})} \|u(t)\|; \quad t = \tau_0 < \tau_1 < \dots < \tau_N < \tau_{N=1} = t_f \tag{6.12}$$

In (6.12), the controls could be piecewise continuous functions. The norm in (6.12) is supported by Banach space: $\hat{C}[t_0, t_f]^{n_u}; \|.\|_\infty$. In case of the assumption of the control being differentiable between two consecutive discontinuities $[\tau_0, \tau_f]$, $k = 0, \dots, N$, the norm is given as

$$\|u\|_{1,\infty} = \sup_{t \in \bigcup_{k=0}^{N}(\tau_k, \tau_{k+1})} \|u(t)\| + \sup_{t \in \bigcup_{k=0}^{N}(\tau_k, \tau_{k+1})} \|\dot{u}(t)\| \tag{6.13}$$

6.1.4 Open Loop and Closed Loop Optimal Control

In an optimal control problem, one needs to synthesize an optimal control input; within the physical constraints of states $x(t)$ and the control input $u(t)$ itself. A closed loop optimal control law (or optimal feedback control) $U(t)$ is obtained if one can find a functional of the following form [3]

$$\hat{u}(t) = U(x(t), t); \quad \forall t \in [t_0, t_f] \tag{6.14}$$

The question of (existence of) synthesizing a control law is difficult, however, for simple linear systems, it is rather easy; for the latter an optimal control can be found in the following form

$$\hat{u}(t) = -K(t)x(t) \tag{6.15}$$

In (6.15), $K(.)$ is a feedback gain. This is a state-feedback control law, which means that some (or all) of the state components are fed-back and the control input $u(t)$ is synthesized. This presupposes the availability of some (or all) of the state vector components, which may not be true in general. Therefore, an observer (or state estimator) that determines the states of the given dynamic system that is to be controlled is required. This observer determines the states of the system from its (input and) output measurements. The observer uses the same system dynamics in its structure and has its own feedback mechanism. For the design of optimal control of the type (6.15), the use of some estimation technique is necessary [5]. The optimal control is said to be in open loop form if the control law is determined as function of time for a specified initial state

$$\hat{u}(t) = U(x(t_0), t); \quad \forall t \in [t_0, t_f] \tag{6.16}$$

From (6.16), one can see that the open loop optimal control is optimal only for a given initial state. So, if an optimal control law (means rule or procedure to synthesize) is known, then one can determine an optimal input trajectory from any initial state of the system. Although, in most cases, when the closed loop control/s are applied, the knowledge of an open loop optimal control law for a given dynamic system would provide considerable insight on the aspect of improvement of the system's operation. It also gives an idea of the benefit obtained upon optimization. Also, one should consider the aspect of existence of an optimal control. The control system response $x(t)$ should satisfy a bound

$$\|x(u(.), x_0, t)\| \le \alpha; \quad \forall t \ge t_0 \tag{6.17}$$

The bound in (6.17) must be true for every feasible control strategy.

6.2 Maximum Principle

In Chapter 5 and its Appendix, we have dealt with the problem of calculus of variation on which optimal control theory has considerable bearing. We also studied the first order conditions that every optimal control should necessarily satisfy, provided that no path restriction is placed on the control $u(t)$ or the state variables $x(t)$. In this Chapter, we discuss more general necessary conditions of optimality for the problems having path constraints; these conditions are known as the Pontryagin Maximum Principle (PMP).

6.2.1 Hamiltonian Dynamics

Here, the problem of calculus of variation is revisited and linked to the problem of optimal control via definition and specification of the Hamiltonian dynamics. The idea is to find a trajectory $\hat{x}(.) : [0, T] \to R$, in an n-dimensional real space, so that the following functional is minimized

$$J(x(.)) = \int_0^T F(x(t), \dot{x}(t)) dt; \quad x(0) = x_0, \ x(T) = x_f \tag{6.18}$$

Let F be written as $F = F(x, v)$. Here, x can be considered as some position and, hence, v as some corresponding velocity. Then, the partial derivative of F is denoted as:

$$\frac{\partial F}{\partial x_j} = F_{x_j}; \ \frac{\partial F}{\partial v_j} = F_{v_j}; \ j = 1, \dots, n;$$

$$\nabla_x F = \nabla_x \{F(x,v)\}$$

$$= (F_{x_1},...,F_{x_n}); \nabla_v F = \nabla_v \{F(x,v)\} = (F_{v_1},...,F_{v_n}) \quad (6.19)$$

Then, the trajectory, $x(.)$, that solves the calculus of the variational problem satisfies/solves the following Euler-Lagrange (EL) differential equations [4]:

$$\frac{d}{dt}[\nabla_v F(\hat{x}(t),\dot{\hat{x}}(t)] = [\nabla_x F(\hat{x}(t),\dot{\hat{x}}(t)];$$

$$\frac{d}{dt}[\nabla_{v_j} F(\hat{x}(t),\dot{\hat{x}}(t)] = [\nabla_{x_j} F(\hat{x}(t),\dot{\hat{x}}(t)]; \quad (6.20)$$

$$j = 1,...,n$$

The result of (6.20) is established next. Let us chose any smooth trajectory $y = y[0,T]$ in the same n-dimensional real space so that it satisfies $y(0) = y(T) = 0$. The variation in the cost functional (6.18) is obtained as

$$j(\tau) := J[\hat{x}(.) + \tau y(.)] \quad (6.21)$$

From (6.21), it is noticed that since $\hat{x}(t)$ is the minimizing trajectory, i.e. $j(.)$ has a minimum at $\tau = 0$, one obtains

$$j(\tau) \geq J[\hat{x}(.)] = j(0); \quad j'(0) = 0 \quad (6.22)$$

From (6.18), one has

$$j(\tau) = \int_0^T F(\hat{x}(t) + \tau y(t), \dot{\hat{x}}(t) + \tau \dot{y}(t))dt \quad (6.23)$$

By differentiating both the sides of (6.23), with respect to τ, one gets

$$j'(\tau) = \int_0^T \left[\begin{array}{c} \sum_{j=1}^n F_{x_j}\{\hat{x}(t) + \tau y(t), \dot{\hat{x}}(t) + \tau \dot{y}(t)\} y_j(t) \\ + \sum_{j=1}^n F_{v_j}\{\hat{x}(t) + \tau y(t), \dot{\hat{x}}(t) + \tau \dot{y}(t)\} \dot{y}_j(t) \end{array} \right] dt \quad (6.24)$$

Specify $\tau = 0$ to obtain the following simplification

$$j'(0) = \sum_{j=1}^n \int_0^T \left[F_{x_j}\{\hat{x}(t), \dot{\hat{x}}(t)\} y_j(t) + F_{v_j}\{\hat{x}(t), \dot{\hat{x}}(t)\} \dot{y}_j(t) \right] dt = 0;$$

$$y(0) = y(T) = 0$$

$$(6.25)$$

Now, let $1 \leq k \leq n$; choose $y_j(t) \equiv 0, j \neq k$; $y_k(t) = \psi(t)$, $\psi(.)$ is an arbitrary function. $\quad (6.26)$

Next, use (6.26) to obtain

$$\int_0^T [F_{x_k}\{\hat{x}(t),\dot{\hat{x}}(t)\}\psi(t) + F_{v_k}\{\hat{x}(t),\dot{\hat{x}}(t)\}\dot{\psi}_k(t)]dt = 0 \quad (6.27)$$

Integrate the second term of (6.27) by parts, and realizing that $\psi(0) = \psi(T) = 0$, one obtains

$$\int_0^T [F_{x_k}(\hat{x}(t),\dot{\hat{x}}(t)) - \frac{d}{dt}\{F_{v_k}(\hat{x}(t),\dot{\hat{x}}(t))\}]\psi(t)dt = 0 \quad (6.28)$$

Since ψ is an arbitrary function/variable, (6.28) can be written as

$$F_{x_k}(\hat{x}(t),\dot{\hat{x}}(t)) - \frac{d}{dt}\{F_{v_k}(\hat{x}(t),\dot{\hat{x}}(t))\} = 0 \quad (6.29)$$

The result of (6.29) is true for all times from 0 to T, the end points inclusive. This establishes the result in (6.20). Let us now define a new variable called as the generalized momentum

$$\lambda(t) := \nabla_v\{F(x(t),v(t))\} = \nabla_v\{F(x(t),\dot{x}(t))\}; \quad t = 0,...,T;$$
$$\text{In short } \lambda = \nabla_v F(x,v) \quad (6.30)$$

Equation (6.30) can be solved for $v(t)$ in terms of $x(t)$ and $\lambda(t)$; i.e. $v = v(x, \lambda(t))$. We define the Hamiltonian dynamics as

$$H(x,\lambda) = \lambda \cdot v(x,\lambda) - F(x,v(x,\lambda)) \quad (6.31)$$

The partial derivatives of H (6.31) are given as

$$\frac{\partial H}{\partial x_j} = H_{x_j}; \quad \frac{\partial H}{\partial \lambda_j} = H_{\lambda_j}; \quad j = 1,...,n \quad (6.32)$$

$$\nabla_x H = (H_{x_1},...,H_{x_n}); \quad \nabla_v F = (H_{\lambda_1},...,H_{\lambda_n}) \quad (6.33)$$

If $x(t)$ satisfies the EL equations (6.20), then with (6.30) and (6.31), the $\{x(t),\lambda(t)\}$ satisfies and solves the Hamiltonian dynamics as follows:

$$\dot{x}(t) = \nabla_\lambda\{H(x(t),\lambda(t))\}$$

$$\dot{\lambda}(t) = -\nabla_x\{H(x(t),\lambda(t))\} \quad (6.34)$$

Using (6.30) and (6.31), we obtain the following

$$\nabla_x H(x,\lambda) = \lambda \nabla_x v - \nabla_x F(x,v(x,\lambda)) - \nabla_v F(x,v(x,\lambda))\nabla_x v$$

$$= -\nabla_x F(x,v(x,\lambda)) \quad (6.35)$$

The final term in (6.35) is obtained because $\lambda = \nabla_v F(.,.)$, and hence, the first and the last terms in (6.35) cancel out. Since $\lambda(t) = \nabla_v F(x(t),\dot{x}(t))$ and because $\dot{x}(t) = v(x(t),\lambda(t))$, the EL equations imply from (6.29),

$$\dot{\lambda}(t) = \nabla_x F(x(t), \dot{x}(t)) \tag{6.36}$$

$$= \nabla_x F(x(t), v(x(t), \lambda(t))) \tag{6.37}$$

Then, from (6.35), one obtains

$$\dot{\lambda}(t) = -\nabla_x H(x(t), \lambda(t)) \tag{6.38}$$

Also, we have from (6.31)

$$\nabla_\lambda H(x, \lambda) = v(x, \lambda) + \lambda \cdot \nabla_\lambda v - \nabla_v F \cdot \nabla_\lambda v = v(x, \lambda) \tag{6.39}$$

Again, using (6.30) in (6.39), and since the last two terms cancel out, one obtains

$$\nabla_\lambda H(x, \lambda) = v(x, \lambda); \nabla_\lambda H(x(t), \lambda(t)) = v(x(t), \lambda(t)) \tag{6.40}$$

Using (6.30) and realizing that $\dot{x}(t) = v(x(t), \lambda(t))$, one gets from (6.40)

$$\dot{x}(t) = \nabla_\lambda H(x(t), \lambda(t)) \tag{6.41}$$

We can now obtain the total time derivative of the Hamiltonian as follows:

$$\frac{d}{dt}\{H(x(t), \lambda(t))\} = \nabla_x H \dot{x}(t) + \nabla_\kappa H \dot{\lambda}(t)$$

$$= \nabla_x H \cdot \nabla_\lambda H + \nabla_\lambda H \cdot (-\nabla_x H) \tag{6.42}$$

$$= 0; \frac{dH}{dt} = 0$$

Equation (6.42) implies that the Hamiltonian is constant because its derivative is zero.

6.2.2 Pontryagin Maximum Principle

In the optimal control problem, the Pontryagin Maximum Principle (PMP) is a very important aspect that ascertains a feasible existence of an optimal control law/input/signal, $u(t)$, such that a function called co-state/adjoint $\lambda(t)$ exists and it satisfies a certain PMP. This function/variable is also called a Lagrange multiplier that allows one to add a constraint in the cost functional, and this constraint is that the optimal state trajectory $x(t)$ must satisfy its own differential equation (6.1), which is excited by the optimal control input $u(t)$; so the triple is well connected: $\{x(t), u(t), \lambda(t)\}$.

6.2.2.1 Fixed Time, Free Endpoint Problem

Let us define the control law as $u(.) \in U$ that satisfies the dynamic system

$$\dot{x}(t) = f(x(t), u(t)); \quad x(0) = x_0; t \geq 0 \tag{6.43}$$

The payoff (POF)/cost functional is given as

$$J(u(.)) = \int_0^T \varphi(x(t), u(t))dt + g(x(T)) \tag{6.44}$$

In (6.44), $\varphi(.,.)$, the integrand is the running POF, and g is the terminal POF; and these are specified. The optimal control problem is to find an optimal control input/ law $\hat{u}(t)$ such that $J(.)$ in (6.44) is maximized

$$J(\hat{u}(.)) = \max_{u(.) \in U} J(u(t)) \tag{6.45}$$

The PMP assures the existence of the two variables/ functions, the state trajectory (or called as state time history) and the ACOLM (adjoint/co-state/Lagrange multiplier), such that these variables satisfy some appropriate Hamiltonian dynamics (HD); similar to (6.38), and (6.41); i.e. (6.34); where for the Hamiltonian is defined in (6.31). However, since now (6.45) involves as such the triple $\{x(t), u(t), \lambda(t)\}$; $or \{\hat{x}(t), \hat{u}(t), \hat{\lambda}(t)\}$, the control theoretic Hamiltonian is defined as

$$H(x, \lambda, u) = f(x, u)\lambda + \varphi(x, u) \tag{6.46}$$

Pontryagin Maximum Principle (PMP) [4]: If it is assumed that $\hat{u}(t)$ is the optimal for the HD and (6.45) the maximum POF, and if $\hat{x}(t)$ is the corresponding trajectory, then there exists ACOLM such that the following equations are satisfied

$$\dot{\hat{x}}(t) = \nabla_\lambda \{H(\hat{x}(t), \hat{\lambda}(t), \hat{u}(t))\}$$

$$\dot{\hat{y}}(t) = -\nabla_x \{H(\hat{x}(t), \hat{\lambda}(t), \hat{u}(t))\} \tag{6.47}$$

$$H(\hat{x}(t), \hat{\lambda}(t), \hat{u}(t)) = \max_{u \in U} H(\hat{x}(t), \hat{\lambda}(t), u(t)); 0 \leq t \leq T \tag{6.48}$$

In (6.47), the first equation is often known as ODE (ordinary differential equation), the second equation is known as the ADJ (adjoint equation, also a differential equation), and (6.48) is often known as the maximum principle, PMP (or maximization principle, MP). The following mapping is also constant

$$t \mapsto H(\hat{x}(t), \hat{\lambda}(t), \hat{u}(t)) \tag{6.49}$$

The transversality condition is given as

$$\hat{\lambda}(T) = \nabla\{g(\hat{x}(T)\} \tag{6.50}$$

In (6.50), g is some given function and is specified by the following region

$$R := \{x \in \mathbb{R}^n / g(x) \leq 0\} \tag{6.51}$$

6.2.2.2 Free Time, Fixed Endpoint Problem

Let the control be $u(.) \in U$, and it satisfies the dynamic system of (6.43). Next, it is assumed that the target/end point is given as x_f, and the POF functional cost is given as

$$J(u(.)) = \int_0^\tau \varphi(x(t), u(t)) dt \qquad (6.52)$$

In (6.44), $\varphi(.,.)$, the integrand is the given running POF; and $\tau = \tau[u(.)] \le \infty$. The control problem is to find an optimal control input/law $\hat{u}(t)$, such that $J(.)$ in (6.52) is maximized

$$J(\hat{u}(.)) = \max_{u(.) \in U} J(u(t)) \qquad (6.53)$$

The PMP, as defined earlier, assures the existence of the two variables/functions—the state trajectory and the ACOLM, such that these variables satisfy some appropriate Hamiltonian dynamics (HD) for the Hamiltonian defined as in (6.31).

Pontryagin Maximum Principle (PMP) [4]: If it is assumed that $\hat{u}(t)$ is the optimal for the HD, and (6.53) the maximum POF, and if $\hat{x}(t)$ is the corresponding trajectory, then there exists ACOLM such that the following equations are satisfied

$$\dot{\hat{x}}(t) = \nabla_\lambda \{H(\hat{x}(t), \hat{\lambda}(t), \hat{u}(t))\}$$
$$\dot{\hat{\lambda}}(t) = -\nabla_x \{H(\hat{x}(t), \hat{\lambda}(t), \hat{u}(t))\} \qquad (6.54)$$

$$H(\hat{x}(t), \hat{\lambda}(t), \hat{u}(t)) = \max_{u \in U} H(\hat{x}(t), \hat{\lambda}(t), u(t)); \; 0 \le t \le \tau^* \quad (6.55)$$

Also, we have

$$H(\hat{x}(t), \hat{\lambda}(t), \hat{u}(t)) \equiv 0 \; 0 \le t \le \tau^* \qquad (6.56)$$

In (6.55), τ^* is the first time the state trajectory $\hat{x}(t)$ hits the target point x_f.

6.2.3 Maximum Principle with Transversality Conditions

The considered system dynamics are

$$\dot{x}(t) = f(x(t), u(t)); \; t > 0 \qquad (6.57)$$

For this problem, we consider that the initial and the final positions are constraint to lie within a given set $x_0, x_f \subset \mathbb{R}$; it is considered that the initial and end trajectories are smooth surfaces in the defined real space. The idea is to maximize again the POF functional, with $\tau = \tau[u(.)]$ as the first time one hits x_f

$$J(u(.)) = \int_0^\tau \varphi(x(t), u(t)) dt \qquad (6.58)$$

For this problem, T_0, T_f denote the tangent planes to X_0 at x_0 and X_f at x_f, respectively.

Pontryagin Maximum Principle (PMP) [4]: Let $\hat{u}(t)$ be the optimal control input and $\hat{x}(t)$ be the corresponding trajectory that solves the optimal control problem with $x_0 = \hat{x}(0)$; *and* $x_f = \hat{x}(\tau^*)$. In such a case, there exists ACOLM $\hat{\lambda}(.): [0, \tau^*]$ that satisfies the ODE and ADJ equations (6.47) and maximum principle (PMP) equation (6.48). Also, we have the following transversality conditions (TC)

$$\hat{\lambda}(\tau^*) \perp T_f; \; \hat{\lambda}(0) \perp T_0 \qquad (6.59)$$

If one has the following POF functional

$$J(u(.)) = \int_0^T \varphi(x(t), u(t)) dt + g(x(T)); \; T > 0 \text{ fixed} \quad (6.60)$$

Then, from TC we have $\hat{\lambda}(T) = \nabla g(\hat{x}(T))$.

6.2.4 Maximum Principle with State Constraints

We again consider the following ODE

$$\dot{x}(t) = f(x(t), u(t)); \; x(0) = x_0 \qquad (6.61)$$

The POF functional is (6.58), with $\tau = \tau[u(.)]$ as the first time one hits that $x(\tau) = x_f$; the fixed end point problem. The state constraint is as in (6.51). A new variable/function is introduced as

$$c(x, u) := \nabla g(x) \cdot f(x, u) \qquad (6.62)$$

Pontryagin Maximum Principle (PMP) [4]: Assume that $\hat{u}(.), \hat{x}(.)$ solve the control theoretic optimal problem of (6.61) and (6.58), and also $\hat{x}(.) \in \partial R$, $s_0 \le t \le s_1$, then there exists an ACOLM $\hat{\lambda}(.): [s_0, s_1] \to \mathbb{R}$ (n-dimensional vector space), such that ODE (6.61) is satisfied. Also, there exists another function/variable $\hat{\lambda}_1(.): [s_0, s_1] \to \mathbb{R}$, such that the following new ADJ1 and PMP1 are satisfied

$$\dot{\hat{\lambda}}(t) = -\nabla_x \{H(\hat{x}(t), \hat{\lambda}(t), \hat{u}(t))\} + \lambda_1 \nabla_x \{c(\hat{x}(t), \hat{u}(t))\}; \text{ ADJ1} \quad (6.63)$$

$$H(\hat{x}(t), \hat{\lambda}(t), \hat{u}(t)) = \max_{u \in U} \{H(\hat{x}(t), \hat{\lambda}(t), u) / c(\hat{x}(t), u) = 0\}; \quad (6.64)$$
$$\text{PMP1}$$

Let us have the form of u as

$$U = \{u \in \mathbb{R}^m / g_1(u) \le 0, ..., g_s(u) \le 0\} \qquad (6.65)$$

In (6.65), the functions $g(.)$ are given. Using the Lagrange multipliers, one can obtain PMP2 from PMP1 as

$$\nabla_u \{H(\hat{x}(t), \hat{\lambda}(t), \hat{u}(t))\} = \lambda_1(t) \nabla_u \{c(\hat{x}(t), \hat{u}(t))\}$$

$$+ \sum_{j=1}^{s} \hat{\mu}_j(t) \nabla_u g_j(\hat{x}(t)) \quad \text{PMP2} \tag{6.66}$$

If $\hat{x}(t)$ lies in the domain R, for $0 \le t < s_0$, then the ordinary PMP holds true.

6.3 Dynamic Programming

In dynamic programming (DP), a particular control problem is generalized. Then this larger problem is solved, and this solution is specialized to our original problem, i.e. the DP problem is finally solved [4]. Let us consider the following dynamic system

$$\dot{x}(s) = f(x(s), u(s)); \quad x(0) = x_0; \, 0 < s < T \tag{6.67}$$

The POF function is given correspondingly as

$$J(u(.)) = \int_0^T \varphi(x(s), u(s))ds + g(x(T)) \tag{6.68}$$

This problem is embedded in a larger class of problem (compare with [6.67]):

$$\dot{x}(s) = f(x(s), u(s)); \quad t < s < T$$
$$x(t) = x \tag{6.69}$$

The corresponding POF function (compare with [6.68]) for any x and t, is given as

$$J_{x,t}(u(.)) = \int_t^T \varphi(x(s), u(s))ds + g(x(T)); \, x \in \mathbb{R}, \, 0 \le t \le T \tag{6.70}$$

We define a value function $v(x,t)$ that is the greatest POF possible, if one starts at x at time t. This is shown as

$$v(x,t) := \sup_{u(.) \in U} J_{x,t}(u(.)); \quad x \in \mathbb{R}, \, 0 \le t \le T \tag{6.71}$$

Then, we have

$$v(x,T) = g(x); \quad x \in \mathbb{R}, n - \text{dimensional space} \tag{6.72}$$

Thus, our aim is now to show that the value function, $v(x,t)$, satisfies a certain nonlinear partial differential

equation (PDE). The given value function is a continuous function of the variables x and t. Then, the function $v(.,.)$ solves the following nonlinear PDE with the terminal condition given in (6.72).

Hamilton-Jacobi-Bellman (HJB) equation [4]:

$$v_t(x,t) + \max_{u(.) \in U} \{f(x,u) \cdot \nabla_x v(x,t)$$
$$+ \varphi(x,u)\} = 0; \quad x \in \mathbb{R}, \, 0 \le t \le T \tag{6.73}$$

The HJB is rewritten as

$$v_t(x,t) + H(x, \nabla_x v) = 0$$

$$H(x,\lambda) = \max_{u(.) \in U} H(x,\lambda,u);$$

$$= \max_{u(.) \in U} \{f(x,u) \cdot \lambda + \varphi(x,u)\}; \tag{6.74}$$

$$x \in \mathbb{R}, \, 0 \le t \le T$$

In (6.74), H is the Hamiltonian for the PDE (6.73). Let us use the constant control $u(.) = u$; for times $t \le s \le t + \tau$, letting the dynamics reach at the point $x(t+\tau)$, here $t + \tau < T$. Next, at time $t + \tau$, we switch to an optimal control and use it for the remaining times $(t + \tau) \le s < T$. The ODE, the dynamics of our systems are

$$\dot{x}(s) = f(x(s), u); \, t \le s \le t + \tau; \quad x(t) = x \tag{6.75}$$

Then, the POF for this is given as

$$\int_t^{t+\tau} \varphi(x(s), u)ds \tag{6.76}$$

The POF incurred from time $(t + \tau)$ to $(T) \rightarrow v(x(t+\tau), t+\tau)$, then the total payoff is given as

$$\int_t^{t+\tau} \varphi(x(s), u)ds + v(x(t+\tau), t+\tau) \tag{6.77}$$

However, the greatest possible POF, if one starts from (x,t) is $v(x,t)$; hence,

$$v(x,t) \ge \int_t^{t+\tau} \varphi(x(s), u)ds + v(x(t+\tau), t+\tau) \tag{6.78}$$

Rearrange (6.78) and then divide by the step size to obtain

$$\frac{v(x(t+\tau), t+\tau) - v(x,t)}{\tau} + \frac{1}{\tau}\int_t^{t+\tau} \varphi(x(s), u)ds \le 0 \tag{6.79}$$

Let the step size $\to 0$ to obtain

$$v_t(x,t) + \nabla_x\{v(x(t),t)\} \cdot \dot{x} + \varphi(x(t),u) \le 0 \qquad (6.80)$$

Using (6.75) in (6.80), one obtains

$$v_t(x,t) + f(x,u) \cdot \nabla_x v(x,t) + \varphi(x,u) \le 0 \qquad (6.81)$$

The inequality (6.81) holds for all the control parameters; hence,

$$\max_{u \in U}\{v_t(x,t) + f(x,u) \cdot \nabla_x v(x,t) + \varphi(x,u)\} \le 0 \qquad (6.82)$$

It now remains to show that the maximum in (6.82) is equal to zero. Assume that $u(.)$ and $x(.)$ are optimal for the foregoing problem. Then, the POF is

$$\int_t^{t+\tau} \varphi(\hat{x}(s),\hat{u}(s))ds \qquad (6.83)$$

The balance POF is given as: $v(\hat{x}(t+\tau),t+\tau)$; hence, the total POF is

$$v(x,t) = \int_t^{t+\tau} \varphi(\hat{x}(s),\hat{u}(s))ds + v(\hat{x}(t+\tau),t+\tau) \qquad (6.84)$$

Rearrange (6.84) and divide it by the step size to obtain

$$\frac{v(\hat{x}(t+\tau),t+\tau) - v(x,t)}{\tau} + \frac{1}{\tau}\int_t^{t+\tau} \varphi(\hat{x}(s),\hat{u}(s))ds = 0 \qquad (6.85)$$

Let the step size tend to zero; and $\hat{u}(t) = \hat{u}$, then one obtains the following

$$v_t(x,t) + \nabla_x\{v(x,t)\} \cdot \dot{x} + \varphi(x,\hat{u}) = 0 \qquad (6.86)$$

Using the dynamics, (6.75) in (6.86), one gets

$$v_t(x,t) + f(x,\hat{u}) \cdot \nabla_x v(x,t) + \varphi(x,\hat{u}) = 0 \qquad (6.87)$$

Thus, (6.87) establishes that for some $\hat{u} \in U$ the equation (6.73) holds and, hence, the HJB is established.

6.3.1 Dynamic Programming Method

The steps in the application of the DP method are [4]:

1. Solve the HJB equation and compute the value of the function v.
2. Use this value function and the HJB PDE to design optimal control $\hat{u}(.)$ as follows:

a. Define for each point x and each time $0 \le t \le T$, $u(x,t) = u \in U$, to be a parameter value where the maximum in HJB is obtained; i.e. we select $u(x,t)$ such that

$$v_t(x,t) + f(x,u(x,t)) \cdot \nabla_x v(x,t) + \varphi(x,u(x,t)) = 0 \qquad (6.88)$$

b. Solve the ODE assuming that $u(.,t)$ is fairly and adequately regular

$$\dot{\hat{x}}(s) = f(\hat{x}(s),\hat{u}(x(s),s)); \ t \le s \le T$$
$$x(t) = x \qquad (6.89)$$

3. Define the (feedback) control (from (i)):

$$\hat{u}(s) := u(\hat{x}(s),s) \qquad (6.90)$$

6.3.2 Verification of Optimality

It is necessary to establish that the control obtained by the DP procedure in Section 6.3.1 is optimal. From (6.68), we have

$$J_{x,t}(\hat{u}(.)) = \int_t^T \varphi(\hat{x}(s),\hat{u}(s))ds + g(\hat{x}(T)) \qquad (6.91)$$

From (6.87) we have

$$\varphi(\hat{x}(s),\hat{u}(s)) = -[v_t(\hat{x}(s),s) + f(\hat{x}(s),\hat{u}(s)) \cdot \nabla_x v(\hat{x}(s),s)] \qquad (6.92)$$

Substituting (6.92) in (6.91), one gets

$$J_{x,t}(\hat{u}(.)) = \int_t^T -[v_t(\hat{x}(s),s) + f(\hat{x}(s),\hat{u}(s)) \cdot \nabla_x v(\hat{x}(s),s)]ds + g(\hat{x}(T)) \qquad (6.93)$$

$$J_{x,t}(\hat{u}(.)) = \int_t^T -[v_t(\hat{x}(s),s) + \nabla_x v(\hat{x}(s),s) \cdot \dot{\hat{x}}(s)]ds + g(\hat{x}(T)) \qquad (6.94)$$

$$J_{x,t}(\hat{u}(.)) = \int_t^T -\frac{d}{ds}\{v(\hat{x}(s),s)\}ds + g(\hat{x}(T)) \qquad (6.95)$$

$$J_{x,t}(\hat{u}(.)) = -v(\hat{x}(T),T) + v(\hat{x}(t),t) + g(\hat{x}(T)) \qquad (6.96)$$

$$J_{x,t}(\hat{u}(.)) = -g(\hat{x}(T) + v(\hat{x}(t),t) + g(\hat{x}(T)) \qquad (6.97)$$

$$J_{x,t}(\hat{u}(.)) = v(x(t),t) = \sup_{u(.)\in U} J_{x,t}(u(.)) \qquad (6.98)$$

6.3.3 Dynamic Programming and Pontryagin Maximum Principle

Let us consider the initial value problem for the Hamiltonian–Jacobi (HJ) equation

$$v_t(x,t) + H(x, \nabla_x v(x,t)) = 0; \quad x \in \mathbb{R}, 0 < t < T; \quad v(x,0) = g(x) \tag{6.99}$$

We need to find a trajectory $x(.)$ along which one can compute $v(x,t)$. Let us have the following notations:

$$x(t) = \begin{bmatrix} x^1(t) \\ . \\ . \\ . \\ x^n(t) \end{bmatrix}; \quad \lambda(t) = \nabla_x \{v(x(t),t)\} = \begin{bmatrix} \lambda^1(t) \\ . \\ . \\ . \\ \lambda^n(t) \end{bmatrix} \tag{6.100}$$

6.3.3.1 Characteristic Equations

Since we have $\lambda^k(t) = v_{x_k}(x(t),t)$, one gets

$$\dot{\lambda}^k(t) = v_{x_k t}(x(t),t) + \sum_{i=1}^{n} v_{x_k x_i}(x(t),t) \cdot \dot{x}^i \tag{6.101}$$

Assume that the HJ equation is solved by v, then differentiate this PDE (6.99) with respect to x_k to obtain

$$v_{t x_k}(x,t) = -H_{x_k}(x, \nabla v(x,t)) - \sum_{i=1}^{n} H_{\lambda_i}(x, \nabla v(x,t)) v_{x_k x_i}(x,t) \tag{6.102}$$

Now with $x = x(t)$, substitute (6.102) in (6.101) to obtain

$$\dot{\lambda}^k(t) = -H_{x_k}(x(t), \nabla v(x(t),t)) + \sum_{i=1}^{n} [\dot{x}^i(t) - H_{\lambda_i}(x, \nabla v(x,t))] v_{x_k x_i}(x,t) \tag{6.103}$$

Select $x(.)$ as

$$\dot{x}^i(t) = H_{\lambda_i}(x, \lambda(t)); \quad \lambda(t) = \nabla v(x,t); 1 \le i \le n \tag{6.104}$$

Substituting (6.104) in (6.103), one gets

$$\dot{\lambda}^k(t) = -H_{x_k}(x(t), \lambda(t)); \quad 1 \le k \le n \tag{6.105}$$

Equations (6.104) and (6.105) are the Hamilton's (H) equations:

$$\dot{x}(t) = \nabla_\lambda \{H(x(t), \lambda(t))\}$$
$$\dot{\lambda}(t) = -\nabla_x \{H(x(t), \lambda(t))\} \tag{6.106}$$

If we solve this Hamilton's equation, then it obtains the solution to HJ equation (6.99) and satisfies the initial conditions $u = g$, at $t = 0$. We set $\lambda^0 = \nabla g(x^0)$, then Hamilton's H equations are solved with $x(0) = x^0$, $\lambda(0) = \lambda^0$. Then, we can have the following

$$\frac{d}{dt} v(x(t),t) = v_t(x(t),t) + \nabla_x v(x(t),t) \cdot \dot{x}(t) \tag{6.107}$$

In (6.107), one can use (6.99) to obtain

$$\frac{d}{dt} v(x(t),t) = -H\{\nabla_x v(x(t),t), x(t)\} + \nabla_x v(x(t),t) \cdot \nabla_\lambda H(x(t), \lambda(t)) \tag{6.108}$$

$$= -H(x(t), \lambda(t)) + \lambda(t) \cdot \nabla_\lambda H(x(t), \lambda(t)) \tag{6.109}$$

We also have: $v(x(0),0) = v(x^0,0) = g(x^0)$. Integrating (6.109), one gets

$$v(x(t),t) = \int_0^t \{-H + \nabla_\lambda H \cdot \lambda(t)\} ds + g(x^0) \tag{6.110}$$

When we computed $x(.)$ and $\lambda(.)$, (6.109) gives the solution for $v(x(.),.)$.

6.3.3.2 Relation between Dynamic Programming and the Maximum Principle

We start again with (6.75), the control theory problem, ODE

$$\dot{x}(s) = f(x(s), u); \quad t \le s \le T; \quad x(t) = x \tag{6.111}$$

The POF functional (POF) from (6.68) is given as

$$J_{x,t}(u(.)) = \int_t^T \varphi(x(s), u(s)) ds + g(x(T)) \tag{6.112}$$

The value function (6.98) is

$$v(x(t),t) = \sup_{u(.) \in U} J_{x,t}(u(.)) \tag{6.113}$$

Then, it is established that the ACOLM variable in the Pontryagin's maximum principle (PMP) is the gradient in x of the value function:

$$\hat{\lambda}(s) = \nabla_x v(\hat{x}(s), s); \quad t \le s \le T \tag{6.114}$$

For (6.114), it is assumed that $\hat{u}(.), \hat{x}(.)$ solve the control problem, ODE (6.111), with the POF, (6.112); also, it is assumed, in (6.114) that $v(.)$ twice differentiable continuous function. From (6.108) and (6.109), we have $\hat{\lambda}(t) = \nabla_x v(\hat{x}(t), t)$; this satisfies the ADJ equation and the maximum principle equation of the PMP problem. Next, one has [4]

$$\dot{\hat{\lambda}}^i(t) = \frac{d}{dt} v_{x_i}(\hat{x}(t), t) = v_{x_i t}(\hat{x}(t), t)$$
$$+ \sum_{j=1}^n v_{x_i x_j}(\hat{x}(t), t) \dot{\hat{x}}^j(t) \qquad (6.115)$$

We also know that the $v(.,.)$ solves the following equation

$$v_t(\hat{x}, t) + \max_{u \in U}\{f(\hat{x}, u) \cdot \nabla_x v(\hat{x}, t) + \varphi(\hat{x}, u)\} = 0 \qquad (6.116)$$

Now apply the optimal control to obtain

$$v_t(\hat{x}(t), t) + f(\hat{x}(t), \hat{u}(t)) \cdot \nabla_x v(\hat{x}(t), t) + \varphi(\hat{x}(t), \hat{u}(t)) = 0 \quad (6.117)$$

Next, let us freeze time t and define the following function

$$h(\hat{x}) = v_t(\hat{x}, t) + f(\hat{x}, \hat{u}) \cdot \nabla_x v(\hat{x}, t) + \varphi(\hat{x}, \hat{u}) \leq 0 \qquad (6.118)$$

The function $h(.)$ has a maximum at the point $\hat{x} = \hat{x}(t)$, and hence, for $i = 1, \ldots, n$, one has

$$0 = h_{x_i}(\hat{x}(t)) = v_{t x_i}(\hat{x}(t), t) + f_{x_i}(\hat{x}(t), \hat{u}(t)) \cdot \nabla_x v(\hat{x}(t), t)$$
$$+ f(\hat{x}(t), \hat{u}(t)) \cdot \nabla_x v_{x_i}(\hat{x}(t), t) + \varphi_{x_i}(\hat{x}(t), \hat{u}(t)) \qquad (6.119)$$

Now, from (6.115), we have the following

$$\dot{\hat{\lambda}}^i(t) = v_{x_i t}(\hat{x}(t), t) + \sum_{j=1}^n v_{x_i x_j}(\hat{x}(t), t) f_j \qquad (6.120)$$

$$= v_{x_i t}(\hat{x}(t), t) + f \cdot \nabla_x v_{x_i}(\hat{x}(t), t) \qquad (6.121)$$

Now, substituting for RHS in (6.121) from (6.119), one obtains

$$= -f_{x_i}(\hat{x}(t), \hat{u}(t)) \cdot \nabla_x v(\hat{x}(t), t) - \varphi_{x_i}(\hat{x}(t), \hat{u}(t)) \qquad (6.122)$$

Substitute for the ACOLM variable in (6.122) to get the following, in short hand notation

$$\dot{\hat{\lambda}}(t) = -f_{x_i}(\hat{x}(t), \hat{u}(t)) \cdot \hat{\lambda}(t) - \varphi_{x_i}(\hat{x}(t), \hat{u}(t))$$
$$\qquad (6.123)$$
$$= -(\nabla_x f)\hat{\lambda}(t) - \nabla_x(\varphi)$$

We also have from (6.46)

$$H = f \cdot \lambda + \varphi; \nabla_x H = (\nabla_x f) \cdot \lambda + \nabla_x \varphi \qquad (6.124)$$

Comparing (6.124) with (6.123), we get the ACOLM equation, the seond equation of (6.54), i.e. the adjoint equation:

$$\dot{\hat{\lambda}}(t) = -\nabla_x\{H(x(t), \hat{\lambda}(t))\} \qquad (6.125)$$

Next, we must establish the maximum condition, using the HJB equation (6.73)

$$v_t(x(t), t) + \max_{u(.) \in U}\{f(x(t), u) \cdot \nabla_x v(x(t), t) + \varphi(x(t), a)\} = 0$$
$$\qquad (6.126)$$

Since, the maximum occurs for $u = \hat{u}(t)$, one has

$$\max_{u(.) \in U}\{H(\hat{x}(t), \hat{\lambda}(t), u\} = H(\hat{x}(t), \hat{\lambda}(t), \hat{u}(t)) \qquad (6.127)$$

Thus, (6.127) establishes the maximum principle. The development of the present section allows us to relook at the transversality conditions.

The free endpoint case:

The condition is

$$\hat{\lambda}(T) = \nabla g(\hat{x}(T)) \qquad (6.128)$$

This is for the POF functional

$$\int_t^T \varphi(x(s), u(s))ds + g(x(T)) \qquad (6.129)$$

We also note that $\hat{\lambda}(s) = -\nabla v(\hat{x}(s), s)$; $v(x, t) = g(x)$; thus,

$$\hat{\lambda}(T) = \nabla_x v(\hat{x}(T), T) = \nabla g(\hat{x}(T)) \qquad (6.130)$$

Constrained initial and target cases:

For this case, we had earlier the following transversality conditions

$$\hat{\lambda}(0) \perp T_0; \ \hat{\lambda}(\tau^*) \perp T_f \qquad (6.131)$$

In (6.131), (τ^*) is the first time that the optimal trajectory touches the target at X_f. The value function for this case is v

$$v(x) = \sup_{u(.)} J_x(u(.)) \qquad (6.132)$$

This case has the constraint as $x_0 \in X_0$; $x_f \in X_f$, start and end points, respectively, and $v(.)$ will be constant on both the sets. Since ∇v is perpendicular to any level surface, it is perpendicular to both ∂X_0; ∂X_f; thus, we get

$$\hat{\lambda}(t) = \nabla v(\hat{x}(t)) \tag{6.133}$$

$$\hat{\lambda} \perp \partial X_0, t = 0; \quad \hat{\lambda} \perp \partial X_f, t = \tau^* \tag{6.134}$$

6.4 Differential Games

This would need a model for a two-person, zero-sum (TPZS) differential game. The main idea is that two players control the dynamics of an evolving system. In this case, one player tries to maximize a payoff functional and the other tries to minimize; this functional depends upon the trajectory of the given dynamic system [4]. The problem is quite involved since each player's control decision/s would depend on the other player's "just" previous action. Let us consider the control problem: given sets U and V as real control domains of appropriate dimensions for control functions $u(.)$ and $v(.)$ for the time $0 \le t \le T$, for the players 1 and 2, respectively; we consider that the sets and control functions are measureable. The corresponding dynamics are

$$\dot{x}(s) = f(x(s), u(s), v(s)); \quad x(t) = x; \quad t \le s \le T \tag{6.135}$$

The POF functional for the game is given as

$$J_{x,t}(u(.), v(.)) = \int_t^T \varphi(x(s), u(s), v(s))ds + g(x(T)) \tag{6.136}$$

Player P1 would like to maximize the payoff functional with the control $u(.)$, and player P2 would like to minimize the functional (6.136) with her/his control $v(.)$ leading to a two-person zero-sum differential game. What is important to note is that at each time instant, neither player knows the other player's intended future move. The P1 will select in advance her responses to all the possible controls that could be selected by her opponent; thus, P1 does not select her control in advance. We now define the mappings and the strategies as follows:

A mapping $\Phi : V(t) \to U(t)$ is P1's strategy for t to T

$$v(\tau) \equiv \hat{v}(\tau); \quad \Phi\{v(\tau)\} \equiv \Phi\{\hat{v}(\tau)\}; t \le \tau \le s \tag{6.137}$$

We can think of the function $\Phi(v)$ as the response of P1 to the selection of $v(.)$ by P2. This also signifies that P1 cannot foresee the future. Similarly, P2 has the following strategy:

A mapping $\Psi : U(t) \to V(t)$ is P2's strategy for t to T

$$u(\tau) \equiv \hat{u}(\tau); \quad \Phi\{u(\tau)\} \equiv \Phi\{\hat{u}(\tau)\}; \ t \le \tau \le s \tag{6.138}$$

Thus, the two strategies are:

$U(t)$: = Strategies for P1 starting at time t.

$V(t)$: = Strategies for P2 starting at time t.

The suitable value functions are given as,

$$p(x,t) := \inf_{\Psi \in V(t)} \sup_{u(.) \in U(t)} J_{x,t}[u(.), \Psi\{u\}(.)] \tag{6.139}$$

$$q(x,t) := \sup_{\Phi \in U(t)} \inf_{v(.) \in V(t)} J_{x,t}[v(.), \Phi\{v\}(.)] \tag{6.140}$$

The lower (p) and the upper (q) value functions are given, respectively, in (6.139) and (6.140). The game strategy is: one player announces her strategy in response to the other's choice of the control, the other player chooses the control; the player who plays second, the one who chooses the strategy, has an advantage. Also, the following is always true

$$p(x,t) \le q(x,t) \tag{6.141}$$

6.4.1 Isaacs's Equations and Maximum Principle/ Dynamic Programming in Games

We assume that $q(.,.)$ and $p(.,.)$ are continuously differentiable. The $q(.,.)$ solves the upper Isaacs's equation

$$q_t + \min_{v \in V(t)} \max_{u(.) \in U(t)} \{f(x,u,v) \cdot \nabla_x q(x,t)$$
$$+ \varphi(x,u,v)\} = 0; \ q(x,T) = g(x) \tag{6.142}$$

Then, $p(.,.)$ solves the lower Isaacs' equation

$$p_t + \max_{u \in U(t)} \min_{v(.) \in V(t)} \{f(x,u,v) \cdot \nabla_x p(x,t)$$
$$+ \varphi(x,u,v)\} = 0; \ p(x,T) = g(x) \tag{6.143}$$

The Isaacs's equations (for the differential games) are similar to the HJB equation (of the DP method) in the two-person, zero-sum game (control theoretic game) and can be written as:

For upper Hamiltonian (H)

$$q_t + H^+(x, \nabla_x(q)) = 0 \tag{6.144}$$

Upper H-PDE: $H^+(x,\lambda) = \min_{v \in V(t)} \max_{u(.) \in U(t)} \{f(x,u,v) \cdot \lambda$

$$+ \varphi(x,u,v)\} \tag{6.145}$$

For lower Hamiltonian

$$p_t + H^-(x, \nabla_x(p)) = 0 \tag{6.146}$$

Lower H-PDE: $H^-(x,\lambda) = \max_{u \in U(t)} \min_{v(.) \in V(t)} \{f(x,u,v) \cdot \lambda$

$$+\varphi(x,u,v)\} \tag{6.147}$$

In general, we have the following hold true and the value functions $q \neq p$

$$\max_{u \in U(t)} \min_{v(.) \in V(t)} \{f(x,u,v) \cdot \lambda + \varphi(x,u,v)\} < \min_{v \in V(t)} \max_{u(.) \in U(t)} \{f(x,u,v) \cdot \lambda$$

$$+\varphi(x,u,v)\} \tag{6.148}$$

$$H^-(x,\lambda) < H^+(x,\lambda) \tag{6.149}$$

The situation in the differential game is that the two players take turns exerting their controls over some short intervals; hence, it is a disadvantage to go first, because the other player then would know what control strategy is selected. The value function $q(.,.)$ represents a sort of "infinitesimal" version of this situation, for which P1 has the advantage; whereas, the value function $p(.,.)$ models the reverse case, for which P2 has the advantage. For all x, λ one can say that the game satisfies the min-max condition, i.e. Isaacs's condition, if the following holds

$$\max_{u \in U(t)} \min_{v(.) \in V(t)} \{f(x,u,v) \cdot \lambda + \varphi(x,u,v)\} = \min_{v \in V(t)} \max_{u(.) \in U(t)} \{f(x,u,v) \cdot \lambda$$

$$+\varphi(x,u,v)\} \tag{6.150}$$

As is done in the case of the DP, if (6.150) holds, one can solve the Isaacs's equation for $q \equiv p$. Then, one can design the optimal controls for players P1 and P2, in principle, though. In case $\hat{u}(.), \hat{v}(.)$ are optimal control laws/signals, then $\{\hat{u}(.), \hat{v}(.)\}$ is a saddle point for the payoff functional $J(.,.)$

$$J_{x,t}(u(.), \hat{v}(.)) \le J_{x,t}(\hat{u}(.), \hat{v}(.)) \le J_{x,t}(\hat{u}(.), v(.)),$$

for all controls, $u(.), v(.)$. $\tag{6.151}$

Interestingly, tP1 will select, $\hat{u}(.)$, because she is afraid that P2 will play $\hat{v}(.)$; and P2 will select $\hat{v}(.)$, because she is afraid that P1 will play $\hat{u}(.)$. Now, for the PMP, i.e. the maximum principle, it is assumed that for (6.150) the minimax condition holds, and the control laws $\{\hat{u}(.), \hat{v}(.)\}$ are designed from the foregoing theory; the solution of the ODE (6.135) is denoted as $\hat{x}(.)$, corresponding to

the control laws, $\hat{u}(.), \hat{v}(.)$; then the following ACOLM expressions are valid

$$\hat{\lambda}(t) := \nabla_x q\{\hat{x}(t),t\} = \nabla_x p\{\hat{x}(t),t)\};$$

$$\dot{\hat{\lambda}}(t) = -\nabla_x[H(\hat{x}(t), \hat{\lambda}(t), \hat{u}(t), \hat{v}(t)] \tag{6.152}$$

Then, the game-theory Hamiltonian is given as

$$H(x,\lambda,u,v) := f(x,u,v) \cdot \lambda + \varphi(x,u,v) \tag{6.153}$$

6.5 Dynamic Programming in Stochastic Setting

We discuss dynamic programming method from a stochastic (theory) point-of-view [4]. An almost self-contained treatment of stochastic processes and stochastic calculus is given in [5, Appendix B]. However, a very brief treatment is given in Appendix B of the present book. A continuous time stochastic differential equation (SDE) is given as

$$\dot{x}(t) = f(x(t)) + w(t); \ t > 0; \quad x(0) = x_0 \tag{6.154}$$

In (6.154), $w(.)$ is white noise. Though, we know that the solution to (6.154) is a random stochastic process, $x(.)$, i.e. the state variable is a stochastic process (the solution, $x(.)$ is a collection of several sample paths of a stochastic process), the notation is kept the same as for the deterministic state variable for simplicity. Also, the solution $x(t)$ of (6.154) contains, or carries, the probabilistic information as to the likelihoods of the various paths; this is the main difference between stochastic and deterministic differential equations. Then the controlled, i.e. with the control input signal, SDE is given as

$$\dot{x}(s) = f(x(s), u(s)) + w(s); \quad x(t) = x_0; \ t \le s \le T \tag{6.155}$$

In (6.155), the "white noise" $w(.)$ is formally interpreted as $\frac{dW}{dt}$, here, W is a Brownian motion (which is often denoted as $B(.)$); this Brownian motion is also called "Wiener process," although the latter is some limit of the former [5,Appendix B]; and $W(.)$ is a process with independent increments. Representation of SDE, as in (6.154), and (6.155), is a formal (from an engineering point-of-view), but the proper SDE is given as

$$dx(s) = f(x(s), u(s))ds + \sigma dw(s); \ t \le s \le T \tag{6.156}$$

$$x(t) = x$$

The solution of (6.156) is given as

$$x(\tau) = x + \left[\int_t^T f(x(s), u(s))ds] + \sigma[w(\tau) - w(t)] \right]; \ t \le \tau \le T$$

(6.157)

The expected POF is given as

$$J_{x,t}[(u(.)] := E\left\{ \int_t^T \varphi(x(s), u(s))ds + g(x(T)) \right\}$$

(6.158)

The value function is given as

$$v(x,t) = \sup_{u(.) \in U} J_{x,t}[(u(.)]$$

(6.159)

To use the method of DP, one needs to find a PDE that would be satisfied by $v(.,.)$. Then one can use this PDE to design an optimal control $\hat{u}(.)$. Let $u(.)$ be any control and used for times $t \le s \le t+h$, $h > 0$; subsequently, an optimal control law is used. Then one has the following inequality [4]

$$v(x,t) \ge E\left\{ \int_t^{t+h} \varphi(x(s), u(s))ds + v(x(t+h), t+h) \right\}$$

(6.160)

The inequality of (6.160) will be an equality for the optimal control law: $u(.) = \hat{u}(.)$. Note here the introduction of the expectation operator, E in (6.160), due to the reason that we are dealing with the random/stochastic processes. For an arbitrary control input $u(.)$, (6.160) becomes

$$0 \ge E\left\{ \int_t^{t+h} \varphi(x(s), u(s))ds + v(x(t+h), t+h) - v(x,t) \right\}$$

(6.161)

$$0 \ge E\left\{ \int_t^{t+h} \varphi(.)ds \right\} + E\{v(x(t+h), t+h) - v(x,t)\}$$

(6.162)

Now, one can use the following Ito formula

$$dv(x(s),s) = v_t(x(s),s)ds + \sum_{i=1}^n v_{x_i}(x(s),s)dx^i(s)$$

$$+ \frac{1}{2} \sum_{i,j=1}^n v_{x_i x_j}(x(s),s)dx^i(s)dx^j(s)$$

(6.163)

$$dv(x(s),s) = v_t \cdot ds + \nabla_x v(f \cdot ds + \sigma \cdot dw(s)) + \frac{\sigma^2}{2} \sum_{i=1}^n \frac{\partial^2}{\partial x_i^2}(v)ds$$

(6.164)

From (6.164), one can deduce the following by introducing the integrals (equivalent to introducing the summation signs and may be not necessarily integrating!) without affecting the equality

$$v(x(t+h), t+h) - v(x(t), t) = \int_t^{t+h} (v_t + \nabla_x v \cdot f + \frac{\sigma^2}{2} \Delta\{v\})ds$$

$$+ \int_t^{t+h} \sigma \nabla_x v \cdot dw(s)$$

(6.165)

In (6.165), $\Delta\{v\} = \sum_{i=1}^n \frac{\partial}{\partial x_i^2}(v)$.

Now, take the expected value of (6.165) to get

$$E\{v(x(t+h), t+h) - v(x(t), t)\} = E\{ \int_t^{t+h} (v_t + \nabla_x v \cdot f + \frac{\sigma^2}{2} \nabla\{v\})ds\}$$

(6.166)

Substitute (6.166) in (6.162) to obtain

$$0 \ge E\{ \int_t^{t+h} [\varphi + v_t + \nabla_x v \cdot f + \frac{\sigma^2}{2} \Delta\{v\}]ds\}$$

(6.167)

From (6.167), one gets the following by assigning $x(t) = x$ and $U(t) = u$

$$0 \ge \varphi(x,u) + v_t(x,t) + f(x,a) \cdot \nabla_x v(x,t) + \frac{\sigma^2}{2} \Delta v(x,t) \quad (6.168)$$

For the optimal control law, (6.168) becomes equality

$$\max_{u \in U} \left\{ \varphi(x,u) + v_t(x,t) + f(x,u) \cdot \nabla_x v(x,t) + \frac{\sigma^2}{2} \Delta v(x,t) \right\} = 0$$

(6.169)

In simple form, (6.169) is given as

$$\max_{u \in U} \left\{ v_t + f \cdot \nabla_x v + \frac{\sigma^2}{2} \Delta v + \varphi \right\} = 0$$

(6.170)

The stochastic HJB is given as

$$v_t(x,t) + \frac{\sigma^2}{2} \Delta v(x,t) + \max_{u \in U}\{f(x,u) \cdot \nabla_x v(x,t)$$

$$+ \varphi(x,u)\} = 0; \ v(x,T) = g(x)$$

(6.171)

Once, the HJB equation (6.171) is solved, then we know $v(.,.)$. Then, compute at each point (x,t) a value of u (of set U) for which $\{f(x,u) \cdot \nabla_x v(x,t) + \varphi(x,u)\}$ attains its maximum, so we get $\hat{u}(t)$. Then solve the following

$$d\hat{x}(s) = f\{\hat{x}(s), \hat{u}(\hat{x}(s),s)\}ds + \sigma dw(s); \ \hat{x}(t) = x \quad (6.172)$$

6.6 Linear Quadratic Optimal Regulator for Time-Varying Systems

We can now focus on a special case when the system dynamics are linear and the cost is quadratic; i.e. the integrand in the cost functional is quadratic, in eqn. (6.2); hence, the linear quadratic regulator (LQR) problem [2,3,6,7]. This additional specification of structure for the dynamic system makes the optimal control problem more tractable. This might help us go further and develop a more complete understanding of optimal solutions. Let us consider the problem of reaching to a terminal state $x(t_f) \approx 0; (t_f$ is given) from an initial state $x(t_0) \neq 0$ for the following linear time varying (LTV) system [3]

$$\dot{x}(t) = A(t)x(t) + B(t)u(t); \quad x(t(0)) = x(t_0) = x_0 \quad (6.173)$$

For (6.172), it is assumed that the trajectories $x(t)$ and $u(t)$ are within some 'acceptable' limits. The optimal control problem can be specified for the following cost functional:

$$J(u) = \int_{t_0}^{t_f} \frac{1}{2}[u^T(t)Q(t)u(t) + x^T(t)R(t)x(t)]dt + \frac{1}{2}x^T(t_f)P_f x(t_f) \quad (6.174)$$

In (6.174), the matrices $R(.)$ and $P_f(.)$ are symmetric and positive semi-definite; and matrix $Q(.)$ is symmetric positive definite (PD); $Q(.)$ should be strictly positive definite, since its inverse would be required. Keeping in tune with the theory of deterministic/robust estimation, these matrices can/should be called Gramians; as such, these are weighting matrices as well as the normalizing matrices. In a normal course, for example, a "covariance" matrix (it could be a state-error covariance matrix) will appear in the "denominator" of the term $x^T(t)x(t)$; equivalently, then an inverse of this "covariance" matrix will appear in the "numerator" of term $x^T(t)x(t)$; hence, in the present case, the matrix $R(.)$ may be called "information" Gramian (since "information" is inverse of "covariance;" whereas, in the Kalman filter theory $P(.)$ denotes the covariance matrix of state errors; then $Y(.) = P^{-1}(.)$ is called the information matrix; also, in estimation/filtering theory the measurement covariance matrix is denoted as R, and the process noise covariance matrix is denoted as Q!). However, for our purpose, matrix $Q(.)$ can be called information Gramian. Now, one can use the necessary conditions and apply these to the optimal control problem of (6.174) with the constrain (6.173), thereby satisfying the EL equations (Chapter 5)

$$\dot{x}(t) = H_\lambda(x(t), u(t), \lambda(t), t); \quad x(t_0) = x_0 \quad (6.175)$$

$$\dot{\lambda}(t) = -H_x(x(t), u(t), \lambda(t), t); \quad \lambda(t_f) = P_f x(t_f) \quad (6.176)$$

$$0 = H_u(x(t), u(t), \lambda(t), t) \quad (6.177)$$

In (6.175), the Hamiltonian is given as

$$H(x, u, \lambda, t) = \frac{1}{2}u^T Q(t)u + \frac{1}{2}x^T R(t)x + \lambda^T[A(t)x + B(t)u] \quad (6.178)$$

The last term in (6.178) arises due to the fact that we have used the system's dynamics (6.173) in the cost functional as a constraint via Lagrange multiplier (LM), which is termed as a co-state or even an adjoint variable. The Hamiltonian (6.178) is quadratic and, hence, twice differentiable. The LQR problem could have been also treated by using the calculus of variation in Chapter 5. The solution of (6.175) through (6.177) will give the solution of the optimal control problem. Thus, using (6.178) in (6.177), one gets

$$\hat{u}(t) = -Q^{-1}(t)B^T(t)\hat{\lambda}(t) \quad (6.179)$$

Again, using (6.178) in (6.175) and (6.176), one gets

$$\begin{bmatrix} \dot{\hat{x}}(t) \\ \dot{\hat{\lambda}}(t) \end{bmatrix} = \begin{bmatrix} A(t) & -B(t)Q(t)^{-1}B^T(t) \\ -R(t) & -A^T(t) \end{bmatrix} \begin{bmatrix} \hat{x}(t) \\ \hat{\lambda}(t) \end{bmatrix}; \quad (6.180)$$

$$\hat{x}(t_0) = x_0; \quad \hat{\lambda}(t_f) = P_f \hat{x}(t_f)$$

In (6.180), we have the TPBVP and can solve it by the sweep method [3]. If we can determine the missing initial condition $\lambda(t_0)$, then (6.180) can be integrated forward in time, regarding it as an initial value problem. To do this, the coefficients of the terminal condition in (6.180) are swept backward to the initial time; hence, we have $\hat{\lambda}(t_0) = P(t_0)\hat{x}(t_0)$. Then, substitute $\hat{\lambda}(t) = P(t)\hat{x}(t)$ for intermediate times in (6.180) to obtain the following matrix Riccati equation

$$\dot{P}(t) = -P(t)A(t) - A^T(t)P(t) + P(t)B(t)Q^{-1}(t)B^T(t)P(t)$$

$$-R(t); P(t_f) = P_f$$

$$(6.181)$$

Now, in order to obtain the initial condition for the co-state variable, (6.181) is integrated backward (from the final time to the initial time) to obtain

$$\hat{\lambda}(t_0) = P(t_0)\hat{x}(t_0) \quad (6.182)$$

Since $\hat{\lambda}(t_0)$ is known from (6.182), one can find $\hat{\lambda}(t)$ and $\hat{x}(t)$ by integrating (6.180) forward and its initial conditions from (6.182). Now, since the entire trajectories of $P(t)$ and $\hat{x}(t)$ are known, the optimal feedback control law can be determined from

$$\hat{u}(t) = -\{Q^{-1}(t)B^T(t)P(t)\}\hat{x}(t) \qquad (6.183)$$

From (6.183), one can easily see that the optimal control law/rule/input signal is essentially the linear state-feedback control law. This necessitates that one must know the entire state vector in order to realize the optimal control. If the state vector is not available, then one must use state estimation or observer methods [5]. Also, even if the dynamic system is time-invariant, the feedback gain in (6.183) is time-varying, since the matrix $P(t)$ is time-varying due to (6.181). One can use the steady-state solution, P of the Riccati equation, if the dynamics (6.173) are time-invariant; however, this will not work well, if the dynamics are time-varying. In that case, one needs to solve the Riccati equation recursively along with the solutions of other equations. The Riccati equation can be solved by using the transition method [5].

6.6.1 Riccati Equation

We start with (6.182) for general time t

$$\hat{\lambda}(t) = P(t)\hat{x}(t) \qquad (6.184)$$

Next differentiate (6.184) to obtain

$$\dot{\hat{\lambda}}(t) = \dot{P}(t)\hat{x}(t) + P(t)\dot{\hat{x}}(t) \qquad (6.185)$$

Substitute (6.180) and (6.184) in (6.185) to obtain the following

$$-R(t)\hat{x} - A(t)P(t)\hat{x}(t) = \dot{P}(t)\hat{x}(t) + P(t)[A(t)\hat{x}(t) \\ -B(t)Q^{-1}(t)B^T(t)P(t)\hat{x}(t)] \qquad (6.186)$$

Since $x(.)$ is optimal and occurs in all the terms of (6.186), it can be considered as an arbitrary variable and eliminated to get

$$\dot{P}(t) = -R(t) - A(t)P(t) - P(t)A(t) + P(t)B(t)Q^{-1}(t)B^T(t)P(t) \qquad (6.187)$$

Then by rearranging (6.187), one gets the matrix Riccati equation as (6.181). We can now sum up the results of application of the necessary conditions of the maximum principle to the LQR problem: the unique optimal control law is given by (6.183), which needs the solution of the Riccati equation (6.187).

6.6.2 LQ Optimal Regulator for Mixed State and Control Terms

Let the cost functional be specified as

$$J(u) = \int_{t_o}^{t_f} \frac{1}{2} \left\{ \begin{bmatrix} u(t) \\ x(t) \end{bmatrix}^T \begin{bmatrix} Q & S \\ S^T & R \end{bmatrix} \begin{bmatrix} u(t) \\ x(t) \end{bmatrix} \right\} dt + \frac{1}{2} x^T(t_f)P_f x(t_f) \qquad (6.188)$$

The matrix Riccati equation for this problem is given as

$$\dot{P}(t) = -P(t)[A(t) - B(t)Q^{-1}(t)S^T(t)] - [A(t) \\ -B(t)Q^{-1}(t)S^T(t)]^T P(t) \\ +\{P(t)B(t)Q^{-1}(t)B^T(t)P(t)\} + S(t)Q^{-1}(t)S^T(t) \\ -R(t); \quad P(t_f) = P_f \qquad (6.189)$$

The state and the ACOLM equations are

$$\begin{bmatrix} \dot{\hat{x}}(t) \\ \dot{\hat{\lambda}}(t) \end{bmatrix} = \begin{bmatrix} A - BQ^{-1}S^T & -BQ^{-1}B^T \\ -R(t) + SQ^{-1}S^T & -(A - BQS^T)^T \end{bmatrix} \begin{bmatrix} \hat{x}(t) \\ \hat{\lambda}(t) \end{bmatrix}; \\ \hat{x}(t_0) = x_0; \hat{\lambda}(t_f) = P_f \hat{x}(t_f) \qquad (6.190)$$

The corresponding control law is given as

$$\hat{u}(t) = -Q^{-1}(t)[S^T(t) + B^T(t)P(t)]\hat{x}(t) \qquad (6.191)$$

6.7 Controller Synthesis

In this section, we study the linear quadratic optimal control in discrete time setting: the basic algorithm, offset free tracking, regulation, and nominal stability [6,7].

6.7.1 Dynamic Models

We first consider a discrete time stochastic state space model

$$x(k+1) = \varphi x(k) + Bu(k) + w(k) \qquad (6.192)$$
$$z(k) = Hx(k) + v(k)$$

The process $w(.)$ and $v(.)$ are zero mean white noise sequences with their statistics as:

$$Q = E\{w(k)w^T(k)\}; S = E\{w(k)v^T(k)\}; R = E\{v(k)v^T(k)\} \qquad (6.193)$$

174

Control Systems

The noise covariance matrices Q, S, and R are assumed as per the standard literature on stochastic linear system theory, and $w(k)$ and $v(k)$ are sequences of the unmeasured state and measurement disturbances (often regarded as state/process noise and observation noise). Models such as in (6.192) with (6.193) can be obtained from: (a) linearization of a grey box (nonlinear) model or (b) state realization of a time-series model that has been developed from input-output (I/O) data of a dynamic system that is to be controlled; its control law is to be determined by using optimal control approach [6]. The linearization of the first principles/grey box model would yield the following model:

$$x(k+1) = \varphi x(k) + Bu(k) + B_d d(k)$$
$$z(k) = Hx(k) + v(k)$$
(6.194)

The model in (6.194) is obtained for the neighborhood of an operating point and starts from a nonlinear model. On the other hand, it could be a grey box model, and $d(.)$ denotes the unmeasured disturbances that could be fluctuations in feed concentration, feed flows, system temperature, etc.; this disturbance is considered as a piecewise constant function. In case $d(.)$ is a zero mean white noise sequence with a known covariance matrix, then we have, in equivalence to (6.192), the following

$$w(k) = B_d d(k); \; E\{w(k)\} = 0;$$
$$Cov(w(k)) = Q = E\{w(k)w^T(k)\} = B_d Q_d B_d^T$$
(6.195)

The state and measurement noise processes can be considered as uncorrelated, and hence, the matrix S is null; equal to 0. Also, one may have the following as a measurement model

$$z_c(k) = H_c x(k); \text{ with } z_c(k) \neq z(k), \text{ in general}$$
(6.196)

Alternatively, one can use an innovations form of the state space mode as

$$x(k+1) = \varphi x(k) + Bu(k) + Ke(k)$$
$$z(k) = Hx(k) + e(k)$$
(6.197)

Models such as in (6.197) might have been obtained from time series models like: ARX (autoregressive exogenous input), ARMAX (autoregressive moving average exogenous input), BJ (Box-Jenkins model) models; these models might have been identified/determined primarily from the I/O data of the dynamic system under consideration. In (6.197), $e(.)$ is a zero mean Gaussian white noise sequence. In (6.197), K represents

a steady-state Kalman gain matrix/vector. The model in (6.197) can also be rewritten as

$$x(k+1) = \varphi x(k) + Bu(k) + w(k)$$
$$z(k) = Hx(k) + v(k)$$
(6.198)

In (6.198), we have $w(.)$ and $v(.)$ as the zero mean white noise sequences with the following statistics

$$Q = E\{w(k)w^T(k)\} = KR_e K^T$$
$$S = E\{w(k)v^T(k)\} = KR_e$$
$$R = E\{v(k)v^T(k)\} = R_e$$
(6.199)

6.7.2 Quadratic Optimal Control

The models discussed in Section 6.7.1 can be utilized for the design of a state feedback controller, and the following steps are involved in the process [6]:

1. Obtain the solution of the regulator/control problem by assuming that the complete state vector, i.e. all the state variables/components are available for the feedback.
2. Design a state estimator and implement the control law/controller, now since the full state vector is available, since this is the state-feedback controller.

6.7.2.1 Linear Quadratic Optimal State Regulator

For this regulator (-control), it is required to take the system state from non-zero initial state to zero (initial-) state (the origin of the state space). The desired control law is specified as

$$u(k) = -Fx(k)$$
(6.200)

The expected transition to the zero initial state should happen in an optimal manner. We consider only the deterministic model for the design purpose

$$x(k+1) = \varphi x(k) + Bu(k)$$
$$z(k) = Hx(k)$$
(6.201)

The regulator design is considered as an optimization problem, and we need to determine the control sequence, $u(0)$, $u(1)$,...,$u(N-1)$, such that this sequence takes the dynamic system to the origin in the optimal way; the main aim is to move the present state to the origin as soon as possible. This requirement dictates the inclusion of a penalty term in the cost function for the optimization as follows

$$\|x(k)\|_2^2 = [(x(k)-0)^T(x(k)-0)] = x^T(k)x(k) \qquad (6.202)$$

In (6.202), the distance from the origin is penalized. Also, since the individual components of the state vector $x(.)$ could have different levels of magnitudes, it would be necessary to use the weighting factor/matrix (weightage) as follows

$$\|x(k)\|_{W_x,2}^2 = x^T(k)W_x x(k) \qquad (6.203)$$

In (6.203), $W(x,.)$ is positive definite matrix, and can be considered as a normalizing matrix. Also, one should specify a similar mechanism for the components of the control vector, since different inputs should have different cost-penalty values associated with them; hence, the following term should be incorporated in the general cost function

$$\|u(k)\|_{W_u,2}^2 = u^T(k)W_u u(k) \qquad (6.204)$$

In (6.204), $W(u,.)$ is a positive definite matrix and can be considered a normalizing matrix. Now, with the individual terms of the cost function defined in (6.203) and (6.204), we can have the total cost functional as

$$J = \sum_{k=0}^{N-1} [x^T(k)W_x x(k) + u^T(k)W_u u(k)] + x^T(N)W_N x(N) \qquad (6.205)$$

In (6.205), N is the final/terminal time. The problem is considered a minimization of the cost functional (6.205), with respect to the control values: $\{u(0), u(1),\ldots, u(N-1)\}$, with appropriate weighting matrices; $W(x,.)$, $W(u,.)$, and $W(N,.)$, which can be assumed to be symmetric and positive definite matrices. This optimization problem is solved by the method of dynamic programming: obtain the solution of the problem at time-instant k, with the presumption that the problem up to the time $k-1$ has been already solved optimally. In the present case, we begin at $k = N$ and proceed backward in time; and define $P(N) = W_N$ and the cost functional J as [6]:

$$J(k) = \min_{u(k),\ldots,u(N-1)} \left\{ \sum_{i=k}^{N-1} [x^T(i)W_x x(i) + u^T(i)W_u u(i)] + x^T(N)W_N x(N) \right\} \qquad (6.206)$$

From (6.206), we have

$$J(N) = x^T(N)W_N x(N) \qquad (6.207)$$

From (6.206) and (6.207), we get the following very easily for $k = N-1$

$$J(N-1) = \min_{u(N-1)} \{x^T(N-1)W_x x(N-1) + u^T(N-1)W_u u(N-1) + J(N)\} \qquad (6.208)$$

Next, using (6.201) in (6.207), one gets

$$J(N) = x^T(N)W_N x(N)$$

$$= [\varphi x(N-1) + Bu(N-1)]^T P(N)[\varphi x(N-1) + Bu(N-1)] \qquad (6.209)$$

Now, substitute (6.209) in (6.208) to obtain the following

$$J(N-1) = \min_{u(N-1)} \{x^T(N-1)W_x x(N-1) + u^T(N-1)W_u u(N-1) + [\varphi x(N-1) + Bu(N-1)]^T P(N)[\varphi x(N-1) + Bu(N-1)]\} \qquad (6.210)$$

Combining certain terms in (6.210) and just rearranging, one gets

$$J(N-1) = \min_{u(N-1)} \{x^T(N-1)[W_x + \varphi^T P(N)\varphi]x(N-1) + x^T(N-1)\varphi^T P(N)Bu(N-1) + u^T(N-1)B^T P(N)\varphi x(N-1) + u^T(N-1)[B^T P(N)B + W_u]u(N-1)\} \qquad (6.211)$$

Now, substitute the following short notations in (6.211)

$$a = B^T P(N)\varphi x(N-1); \; A = [B^T P(N)B + W_u] \qquad (6.212)$$

We obtain the following from (6.211) and (6.212)

$$J(u) = a^T u + u^T Au + u^T a \qquad (6.213)$$

Since the first term in (6.211) does not depend on $u(.)$, it is sufficient to consider the minimization of the $J(u)$ as in (6.213). Next, we complete the squares in (6.213) as follows

$$J(u) = a^T u + u^T Au + u^T a + a^T A^{-1}a - a^T A^{-1}a \qquad (6.214)$$

$$J(u) = (u + A^{-1}a)^T A(u + A^{-1}a) - a^T A^{-1}a \qquad (6.215)$$

The matrix A is positive definite. From (6.215), we can see that the first term is non-negative; hence, the minimum is obtained as

$$u = -A^{-1}a \qquad (6.216)$$

Then, the minimum value of $J(u)$ is given as

$$J(u)_{\min} = -a^T A^{-1}a \qquad (6.217)$$

Thus, after substituting back the terms from (6.212) into (6.216) and (6.217), we obtain the following optimal solution

$$u(N-1) = -[B^T P(N)B + W_u]^{-1} B^T P(N)\varphi x(N-1);$$
$$= -F(N-1)x(N-1) \qquad (6.218)$$

$$F(N-1) = [B^T P(N)B + W_u]^{-1} B^T P(N)\varphi \qquad (6.219)$$

The $F(.)$ is the feedback gain for the control law of (6.218), which is clearly the state-feedback law. Utilizing the substitutions from (6.212) in (6.217), one gets

$$J(N-1) = x^T(N-1)P(N-1)x(N-1) \qquad (6.220)$$

In (6.220), the Riccati type equation is given as

$$P(N-1) = \varphi^T P(N)\varphi + W_x - F^T(N-1)[B^T P(N)B + W_u]F(N-1) \qquad (6.221)$$

If we use the similar procedure as done to obtain (6.218) again for time $N-2$, we have to start with a similar equation as (6.208)

$$J(N-2) = \min_{u(N-2)} \{ x^T(N-2)W_x x(N-2) $$
$$+ u^T(N-2)W_u u(N-2) + J(N-1) \} \qquad (6.222)$$

Since the problem in (6.222) is similar to the one in (6.208), one can obtain the solution for the problem of (6.222) from the solutions of (6.219) and (6.221) as follows

$$F(k) = [B^T P(k+1)B + W_u]^{-1} B^T P(k+1)\varphi \qquad (6.223)$$

$$P(k) = [\varphi - BF(k)]^T P(k+1)[\varphi - BF(k)] + W_x + F^T(k)W_u F(k) \qquad (6.224)$$

Equation (6.224) is the discrete time Riccati equation. If the horizon N tends to be very large, then the matrix P ($N \rightarrow$ large) reaches a steady state and one can write the algebraic Riccati equation (ARE) as

$$P = [\varphi - BF]^T P[\varphi - BF] + W_x + F^T W_u F \qquad (6.225)$$

The steady-state gain and the control law are given as

$$F = [B^T PB + W_u]^{-1} B^T P\varphi; \ u(k) = -Fx(k) \qquad (6.226)$$

6.7.2.2 *Linear Quadratic Optimal Output Regulator*

When one is interested in controlling certain outputs of a dynamic system and there is no direct physical meaning of the states, then it is convenient to define the regulatory control problem in terms of measured

outputs. In this case, the objective function (6.205) can be written as [6]

$$J = E\left\{ \sum_{k=0}^{N-1} [z^T(k)W_z z(k) + u^T(k)W_u u(k)] + z^T(N)W_{zN} z(N) \right\} \qquad (6.227)$$

Using (6.201) in (6.227), one can rewrite it as

$$J = E\left\{ \sum_{k=0}^{N-1} [x^T(k)[H^T W_z H]z(k) + u^T(k)W_u u(k)] + x^T(N)[H^T W_{zN} H]x(N) \right\} \qquad (6.228)$$

Next, use the following in (6.228)

$$W_x = [H^T W_z H]; \quad W_N = [H^T W_{zN} H] \qquad (6.229)$$

Then, we have the following cost functional

$$J = E\left\{ \sum_{k=0}^{N-1} [x^T(k)W_x z(k) + u^T(k)W_u u(k)] + x^T(N)W_N x(N) \right\} \qquad (6.230)$$

The cost functional (6.230) is similar to (6.205); hence, the solution of the output regulator problem can be handled by using the Riccati equation developed in Section 6.7.2.1.

6.7.3 Stability of the Linear Quadratic Controller/Regulator

We have derived the optimal controller in the foregoing sections for a discrete time system, i.e. the LQ regulator/s, the stability of which can now be established. We consider the dynamic system as in (6.192)

$$x(k+1) = \varphi x(k) + Bu(k) \qquad (6.231)$$

The point is that when we use the optimal control law to move the current state of the dynamic system to the origin, we need to assure that the overall control system is stable; otherwise, the control law does not do any good. It is assumed that the cost function is given as (6.205). It is also assumed that the Riccati equation is (6.225) and the steady-state matrix P is available. Also, the steady-state feedback control strategy from (6.226) is given as (with its associated steady-state feedback gain):

$$u(k) = -[B^T PB + W_u]^{-1} B^T P\varphi x(k) \qquad (6.232)$$

Since $u(k) = -Fx(k)$ and utilizing this in (6.231), one gets

$$x(k+1) = (\varphi - BF)x(k) \qquad (6.233)$$

Thus, we need to see that the closed loop control system as in (6.233) is asymptotically stable. This is because an additional term has now been added to the system matrix. In order to derive the stability results, we need to first define an appropriate Lyapunov energy (LE) functional as [6]

$$V((x(k)) = x^T(k)Px(k) \qquad (6.234)$$

The LE functional in (6.234) is positive definite (PD), since P is as such PD and a symmetric matrix, and the form of $V(.)$ is a quadratic function. The idea is to establish that the time derivative of $V(.)$, or its differential, should be negative definite in order that the closed loop control is stable. Thereby, meaning that the energy is eventually dissipated in the control system and attains a steady state, after the optimal control law has been operative and the system has attained a new state. The differential is given as

$$\Delta V((x(k)) = x^T(k+1)Px(k+1) - x^T(k)Px(k) \qquad (6.235)$$

Substitute (6.233) in (6.235) to obtain the following

$$\Delta V((x(k)) = x^T(k)(\varphi - BF)^T P(\varphi - BF)x(k) - x^T(k)Px(k) \qquad (6.236)$$

$$\Delta V((x(k)) = x^T(k)[(\varphi - BF)^T P(\varphi - BF) - P]x(k) \qquad (6.237)$$

Use (6.225) in (6.237) to obtain the following

$$\Delta V((x(k)) = -x^T(k)[W_x + F^T W_u F]x(k) \qquad (6.238)$$

In (6.238), the bracketed term is PD, and hence, the differential of the LE functional as in (6.238) is negative definite. This establishes that the closed loop control system with the use of the optimal control regulator law is asymptotically stable. From (6.233) and (6.238), we can deduce that all the poles of $(\varphi - BF)$ are strictly inside a unit circle and the LQ optimal controller guarantees the performance and the asymptotic stability under the nominal conditions.

6.7.4 Linear Quadratic Gaussian (LQG) Control

We first need to consider the state estimation problem. We might not have all the components of the state vector available for the design of the optimal controller, since it depends on the state vector as in (6.226). Also, since the state estimation would be done by using noisy measurements, one would need an estimator based on stochastic estimation theory, such as the Kalman filter.

6.7.4.1 State Estimation and LQ Controller

We can use the system model as in (6.192). The optimal state predictor can be given, with the measurement prediction defined, as

$$e(k) = z(k) - H\hat{x}(k/k-1) \qquad (6.239)$$

$$\hat{x}(k+1/k) = \varphi\hat{x}(k/k-1) + B(k)u(k) + K(k)e(k) \qquad (6.240)$$

In (6.239), $e(k)$ is the residual in state estimation, and in (6.240), K is the gain matrix/vector. The gain can be obtained by the solution of the steady-state Riccati equation:

$$P = \varphi P \varphi^T + Q - K(HPH^T + R)K^T \qquad (6.241)$$

$$K = [\varphi PH^T + S][HPH^T + R]^{-1} \qquad (6.242)$$

The matrices, Q, S, and R are defined in (6.193), and we use the same symbol for the solution of the Riccati equation, (6.241) as in the earlier development of the control law; hence, here P denotes the steady-state covariance matrix of the state errors due to using a predictor-estimator. Equations (6.239) through (6.242) are in fact the equations of a steady-state Kalman filter. In fact, first (6.241) is solved for P, then this matrix is used in (6.242) to obtain the filter gain that is also known as the Kalman gain, which is then used in (6.240), along with the computation of $e(.)$ by (6.239). Thus, the predicted state is used in (6.226) to obtain the optimal control law:

$$u(k) = -F\hat{x}(k/k-1) \qquad (6.243)$$

The variable $e(.)$, (6.239), the residual is also called innovations process, which is zero mean Gaussian white noise sequence with its covariance matrix as $R_e = [HPH^T + R]$. We can interpret (6.243) as follows, let $u(k)$ be given as

$$u(k) = -F\hat{x}(k/k) = -F[x(k) - e(k/k-1)] \qquad (6.244)$$

$$u(k) = -Fx(k) + Fe(k/k-1) \qquad (6.245)$$

Comparing (6.245) with (6.226), we can now say that the optimal control law is modified in order to account for the mismatch between the true states (which are anyway not known) and the estimates of the states by incorporating the Kalman state predictor in the design of the overall LQ controller, which is now called the LQG controller, and the Gaussianess coming from the assumption of the process noise as a Gaussian stochastic process, since the system is linear as defined earlier, the state process $x(.)$ is also Gaussian. Thus, the optimal combination of the LQ controller and the Kalman filter is known as the LQG controller, here Q signifying the quadratic criterion, i.e. the cost function.

6.7.4.2 Separation Principle and Nominal Closed Loop Stability

We have seen in Section 6.7.4.1 that the LQG controller is a combination of the LQ controller and the (linear) Kalman state-predictor (which is based on the assumption that the stochastic processes $w(.)$, and $x(.)$ are Gaussian) The latter is also known as the observer, in the sense that the linear Kalman filter is the best linear unbiased estimator (BLUE); hence, the BLUE state-observer. Thus, now we have two dynamic systems working in bootstrap manner, i.e. hand-in-hand: the control input $u(.)$ is based on the Kalman state-predictor ($u = -Fx$). This input then generates the new state and new predicted measurements, which are then used in computing the innovations sequence $e(.)$. This, in turn, is used to update the state estimation in the Kalman filter and again is used to compute the new/next value of the control input signal $u(.)$. Interestingly, this mechanism works well due to the separation principle that holds true for this joint dynamics: (i) the original system and control law incorporation and (ii) the Kalman filter (as a dynamic) system. We present this principle next [6]. We consider the dynamic system as in (6.192) and the corresponding observer, i.e. the state-predictor

$$\hat{x}(k+1/k) = \varphi\hat{x}(k/k-1) + Bu(k) + L[z(k) - H\hat{x}(k/k-1)]$$
$$(6.246)$$

The state-error in estimation is defined as

$$e(k/k-1) = x(k) - \hat{x}(k/k-1) \qquad (6.247)$$

By using the dynamics of the system, (6.192) and the state predictor (6.246) one can obtain the following observer (i.e. Kalman filter) error-dynamics

$$e(k+1/k) = [\varphi - FH]e(k/k-1) + w(k) - Lv(k) \qquad (6.248)$$

Now, we can combine the system dynamics and the error dynamics (6.248) to obtain the joint dynamics for the closed loop dynamics

$$\begin{bmatrix} x(k+1) \\ e(k+1/k) \end{bmatrix} = \begin{bmatrix} [\varphi - BF] & BF \\ [0] & [\varphi - LH] \end{bmatrix} \begin{bmatrix} x(k) \\ e(k/k-1) \end{bmatrix}$$
$$+ \begin{bmatrix} I \\ I \end{bmatrix} w(k) + \begin{bmatrix} [0] \\ -L \end{bmatrix} v(k)$$
$$(6.249)$$

The transition matrix of the closed loop dynamics is given as follows

$$\varphi_c = \begin{bmatrix} [\varphi - BF] & BF \\ [0] & [\varphi - LH] \end{bmatrix} \qquad (6.250)$$

The determinant of the combined dynamics of (6.250) is

$$\det(\lambda I - \varphi_c) = \det(\lambda I - (\varphi - BF)) \cdot \det(\lambda I - (\varphi - LH)) \qquad (6.251)$$

It is obvious from (6.251) that the combined dynamics are determined by the dynamics of the controller (with feedback gain F) as well as the dynamics of the Kalman-filter/observer (with feedback gain L). If both are indeed separated from each other, the latter means that if the controller and the observer are designed individually as stable systems, then the combined closed loop system is also stable. This is the essence of the separation principle. This also implies that if the underlying governing dynamic system is observable and controllable, the closed loop control system is guaranteed to be stable; of course, subject to the proper choice of the tuning matrices as specified in the cost function of the control problem and the cost function of the Kalman filter. It may be noted that the separation principle as specified in (6.251) is valid even if the feedback gain matrix F is obtained by using the approach of pole placement, and the Luenberger observer is used that is also designed by pole placement.

6.7.5 Tracking and Regulation with Quadratic Optimal Controller

The LQ regulator (LQR) design presented in Section 6.7.2.1 can generate an offset if: (a) the unmeasured disturbances are non-stationary, or (b) there is mismatch between the plant's real dynamics and its mathematical model. These discrepancies can be greatly reduced if an integral action in the control input signal is introduced (see Section 6.1) to deal with the model plant mismatch and reject the drifting unmeasured disturbances. Also, the regulator designed only solves the limited problem of moving the system from any initial state to the origin; however, if it is desired to move the system from any initial condition to an arbitrary set-point, the state feedback control laws needs to be modified. The problem of regulation to handle such issues is solved by modifying the regulatory control law as follows:

$$u(k) - u_s = -F[x(k) - x_s] \qquad (6.252)$$

$$u(k) = u_s - F[x(k) - x_s] \qquad (6.253)$$

In (6.253), x_s is the steady-state final position/state that corresponds to the set point, r, and u_s is the steady-state control input that is required for the system to reach this steady state of the target. This is equivalent to the change of the origin. Here, F is the steady-state gain.

6.7.5.1 Transformation of the Model for Output Regulation and Tracking

Let us consider the problem of designing a LQ controller for the following system

$$x(k+1) = \varphi x(k) + Bu(k) + B_d d$$
$$z(k) = Hx(k) + H_d v_d \tag{6.254}$$

In (6.254), the $d(.)$ is the input disturbance, and $v(d,.)$ is the output disturbance. It is assumed that these disturbance vectors and the associated gain factors/matrices are known. It is aimed to control this system at an arbitrarily specified set point nonzero point r, i.e. $z(k) = r$ as the time index $k \to$ to infinity. One can easily see that the steady-state behavior of the dynamic system (6.254) is given by

$$x_s = \varphi x_s + Bu_s + B_d d \tag{6.255}$$

From (6.256), the steady-state value of $x(.)$ can be found as

$$x_s = [I - \varphi]^{-1}[Bu_s + B_d d] \tag{6.256}$$

Since one needs $z(k) = r$, as $k \to$ infinity, so at the steady state we get from the output equation (6.254)

$$r = Hx_s + H_d v_d \tag{6.257}$$

Substituting in (6.257) for steady-state x from (6.256), one gets

$$r = H[I - \varphi]^{-1}[Bu_s + B_d d] + H_d v_d \tag{6.258}$$

Assign the following as the steady-state gains

$$K_u = H[I - \varphi]^{-1}B; \; K_d = H[I - \varphi]^{-1}B_d \tag{6.259}$$

Substituting (6.259) in (6.258), one gets

$$r = K_u u_s + K_d d + H_d v_d \tag{6.260}$$

If the controller gain is invertible, one gets

$$u_s = K_u^{-1}[r - K_d d - H_d v_d] \tag{6.261}$$

Next, substitute (6.261) in (6.256) to obtain the steady-state x as

$$x_s = [I - \varphi]^{-1}Bu_s + [I - \varphi]^{-1}B_d d \tag{6.262}$$

$$x_s = [I - \varphi]^{-1}B\{K_u^{-1}[r - K_d d - H_d v_d]\} + [I - \varphi]^{-1}B_d d \tag{6.263}$$

$$x_s = [I - \varphi]^{-1}\{BK_u^{-1}[r - K_d d - H_d v_d] + B_d d\} \tag{6.264}$$

Now, subtract (6.255) from (6.254) to obtain the difference in the state as well as in the output

$$x(k+1) - x_s = \varphi(x(k) - x_s) + B(u(k) - u_s) \tag{6.265}$$

$$z(k) - r = H(x(k) - x_s) \tag{6.266}$$

From (6.265) and (6.266), we can obtain the transformed system in the perturbation variables

$$\Delta x(k+1) = \varphi \Delta x(k) + B\Delta u(k)$$
$$\Delta z(k) = \Delta Hx(k) \tag{6.267}$$

Using the dynamic model of the differential states as in (6.267), one can develop a LQ controller and obtain the following control law

$$\Delta u(k) = -F\Delta x(k) \tag{6.268}$$

In (2.263), F is the steady-state gain for the new controller for the transformed dynamics. This control law regulates the transformed system at the origin, $\Delta x(k) = 0$; this is equivalent to obtaining the offset free action for the original system for the preassigned set point. This also includes the behavior of the controller for the given level of the deterministic disturbance. The LQ controller that handles the servo together with the regulatory problems can also be given as

$$u(k) = u_s - F\Delta x(k) = u_s - F(x(k) - x_s) \tag{6.269}$$

If d and v_d are slowly varying with time and/or the set point r is varying with time, then one can use the time varying target steady-state and time varying target steady-state inputs

$$u_s(k) = K_u^{-1}[r(k) - K_d d(k) - H_d v_d(k)] \tag{6.270}$$

$$x_s(k) = [I - \varphi]^{-1}\{BK_u^{-1}[r(k) - K_d d(k) - H_d v_d(k)] + B_d d(k)\} \tag{6.271}$$

Then the control law action is modified as

$$u(k) = u_s(k) - F(x(k) - x_s(k)) \tag{6.272}$$

In (6.271), it can be safely assumed that the unmeasured disturbances remain constant at future times:

$$d(k+j+1) = d(k+j); \quad v_d(k+j+1) = v_d(k+j); j = 0,1,2,...,$$

(6.273)

The foregoing modification in the control law is useful in the face of drifts in the measured disturbances and slowly time varying set point.

6.7.5.2 Unmeasured Disturbances and Model Mismatch

It is assumed that a Kalman predictor is given for the nominal model

$$e(k) = z(k) - H\hat{x}(k/k-1) \qquad (6.274)$$

$$\hat{x}(k+1/k) = \varphi\hat{x}(k/k-1) + Bu(k) + Le(k) \qquad (6.275)$$

If there is no mismatch in the model and plant and/or there is no drift in the unmeasured disturbances, then the innovations sequence $e(.)$ is a zero mean Gaussian process; however, that is not so in the following cases:

1. Plant-model mismatch (MPM): The plant dynamics evolve according to the following model

$$\tilde{x}(k+1) = \bar{\varphi}\tilde{x}(k) + \bar{B}u(k) + w(k)$$

$$z(k) = \bar{H}\tilde{x}(k) + v$$

(6.276)

2. In (6.276), the plant's matrices are different from the ones used in the math model (6.192).

3. Plant-model structural mismatch: Often one uses lower order approximations of the plant dynamics for controller synthesis in the desired range. This means that the system could be evolving as per the following dynamics

$$X(k+1) = \tilde{\varphi}X(k) + \tilde{B}u(k) + \tilde{w}(k)$$

$$z(k) = \tilde{H}X(k) + v$$

(6.277)

4. In this case, the synthesis might have been done using (6.192). Since there are structural differences between (6.192) and (6.277), the dimensions of the plant state vector X and the model state vector x are different. Hence, the numerical values of the coefficients entering in (6.277) as well the dimensions of the matrices in (6.277) are different from those of (6.192).

5. Unmeasured drifting/colored disturbances: In this case, the plant dynamics are affected by some unknown drifting/colored disturbance

$$\tilde{x}(k+1) = \varphi\tilde{x}(k) + Bu(k) + B_d d(k) + w(k)$$

$$z(k) = H\tilde{x}(k) + v$$

(6.278)

6. In (6.278), $d(.)$ is an auto-correlated stochastic process that has not been accounted in the model.

7. Nonlinear plant dynamics: The actual plant/system is nonlinear, but the dynamics used for the design of the controller are linearized local models. This would lead to the errors that are called the model errors and will further lead to errors in the overall controlled system.

6.7.5.3 Innovations Bias Approach

In this method, the filtered innovations signal $e(.)$ is used as a proxy/alternative for the unmeasured disturbances:

$$e(k) = z(k) - \hat{z}(k/k-1) \qquad (6.279)$$

In the presence of MPM, the innovations sequence $e(.)$ becomes a colored signal and has significant power at the low frequencies, which are relevant to control region of the closed loop system. The low-frequency drifting mean of the sequence $\{e(.)\}$ can be estimated using the following filter

$$\hat{e}(k) = \varphi_e\hat{e}(k-1) + (I - \varphi_e)e(k) \qquad (6.280)$$

In (6.280), $\varphi_e = diag\{\alpha_1\ \alpha_2...\alpha_r\}$ and its elements are tuning parameters; $0 \le \alpha_i < 1$. The choice of these parameters govern the regulatory behavior of the LQG controller and provide some robustness against the MPM; the usual range is: 0.8 to 0.99. This filtered signal is used in place of the low-frequency unmeasured disturbances/model plant mismatch. Thus, we have the following assumptions after (6.254):

1. Unmeasured disturbance in the state dynamics: $d(k) \equiv \hat{e}(k)$, $B_d \equiv L$
2. Unmeasured disturbance in output: $v_d(k) \equiv \hat{e}(k)$, $H_d \equiv I_r$

Thus, due to the foregoing equivalence, the control law synthesis equations in (6.270) through (6.272) can be used. Also, it is assumed that the disturbances remain almost time-invariant:

$$d(k+j+1) \equiv d(k+1); \quad d(k) = \hat{e}(k); \quad j = 1,2,..., \quad (6.281)$$

$$v_d(k+j+1) \equiv v_d(k+1); \quad v_d(k) = \hat{e}(k); \quad j = 1,2,..., \quad (6.282)$$

The LQG control law is now modified as follows:

$$u(k) = u_s(k) - F[\hat{x}(k/k-1) - x_s(k)] \qquad (6.283)$$

In (6.283), the $x(s,.)$ is computed using

$$x_s(k) = \varphi x_s + B u_s + L \hat{e}(k) \qquad (6.284)$$

$$r(k) = H x_s(k) + \hat{e}(k) \qquad (6.285)$$

If the transition matrix has no poles on the unit circle, (6.283) and (6.284) reduce to

$$u_s(k) = K_u^{-1}[r(k) - K_e \hat{e}(k)] \qquad (6.286)$$

$$x_s(k) = [I - \varphi]^{-1}\{B u_s(k) + L \hat{e}(k)\} \qquad (6.287)$$

In (6.286), the gains are given as

$$K_u = H(I - \varphi)^{-1} B; \quad K_e = H(I - \varphi)^{-1} L + I \qquad (6.288)$$

6.7.5.4 State Augmentation Approach

In this method, the disturbances are considered as additional states and are estimated along with the plant states. Reconsider the models of (6.192), which then are augmented with additional states as

$$x(k+1) = \varphi x(k) + B u(k) + B_d d(k) + w(k)$$

$$d(k+1) = d(k) + w_d(k)$$

$$z(k) = H x(k) + H_d v_d(k) + v(k)$$

$$v_d(k+1) = v_d(k) + w_{vd}(k)$$

$$\qquad (6.289)$$

The first and the fourth equations of the set (6.289) are the models of the disturbances that are considered as remaining as constant, but are to be estimated as the additional states. The noise processes, $w(d,.)$ and $w(v_d,.)$, are zero mean white noise sequences with covariance matrices as: Q_d, and Q_{vd}, respectively. The model coefficients of $B(d,.)$ and $H(d,.)$ are treated as tuning parameters as is the coefficients of the respective covariance matrices; this is done to attain the desired disturbance rejection characteristics for the closed loop system. Since the additional states should be observable, the total number of the extra states should not exceed the number of measurements. The choices that can be made for the additional states and their coupling matrices are:

Output bias formulation

The drifting disturbance is considered to cause a bias in the measured outputs:

$$B_d = 0; \quad Q_d = 0$$

$$H_d = I; \quad Q_{vd} = \sigma^2 I$$

$$\qquad (6.290)$$

Input bias formulation

The coefficients of $d(.)$ are viewed as bias in r-manipulated variables. If the number of manipulated inputs equals the number of measured outputs, $r = m$, then one can choose

$$B_d = B; \quad Q_d = \sigma^2 I$$

$$H_d = 0; \quad Q_{vd} = 0$$

$$\qquad (6.291)$$

If the manipulated inputs are more than the number of measurements, then r linearly independent columns of B can be chosen as B_d.

Disturbance bias formulation

If the state space model is obtained from first principles, one can choose

$$Q_d = \sigma^2 I \qquad (6.292)$$

This is for the number of disturbance variables $(d) = r$. Now, one can get the augmented set of equations

$$x_a(k+1) = \varphi_a x_a(k) + B_a u(k) + w_a(k)$$

$$z(k) = H_a x_a(k) + v(k)$$

$$\qquad (6.293)$$

In (6.293), we have

$$x_a(k) = \begin{bmatrix} x(k) \\ d(k) \\ v_d(k) \end{bmatrix}; \quad w_a(k) = \begin{bmatrix} w(k) \\ w_d(k) \\ w_{vd}(k) \end{bmatrix}$$

$$\varphi_a = \begin{bmatrix} \varphi & B_d & 0 \\ 0 & I_d & 0 \\ 0 & 0 & I_{vd} \end{bmatrix}; \quad B_a = \begin{bmatrix} B \\ 0 \end{bmatrix}; \quad H_a = [H \ 0 \ H_d]$$

$$\qquad (6.294)$$

Also, the covariance matrices are given as

$$Q_a = \begin{bmatrix} Q & 0 & 0 \\ 0 & Q_d & 0 \\ 0 & 0 & Q_{vd} \end{bmatrix}; \quad S_a = \begin{bmatrix} S \\ 0 \end{bmatrix}; \quad R_a = [R] \qquad (6.295)$$

The KF predictor is then given as

$$e_a(k) = z(k) - H_a \hat{x}_a(k/k-1) \qquad (6.296)$$

$$\hat{x}_a(k+1/k) = \varphi_a \hat{x}_a(k/k-1) + B_a u(k-1) + L e_a(k) \qquad (6.297)$$

The steady-state gain is given as

$$L_a = [\varphi_a P_a H_a^T + S_a][H_a P H_a^T + R]^{-1} \qquad (6.298)$$

The matrix Riccati equation is given as

$$P_a = \varphi_a P_a \varphi_a^T + Q_a - L_a [H_a P H_a^T + R] L_a^T \qquad (6.299)$$

If the original model (6.197) is used in Kalman predictor is observable and stable (with no integrating modes), then the augmented state space model will be observable/detectable. The control law can be designed as

$$u_s(k) = K_u^{-1}[r(k) - K_d \hat{d}(k/k-1) - H_d \hat{v}_d(k/k-1)] \qquad (6.300)$$

$$x_s(k) = [I - \varphi]^{-1}[BK_s^{-1}\{r(k) - H_d \hat{v}_d(k/k-1)\} \\ + (B_d - BK_u^{-1}K_d)\hat{d}(k/k-1)] \qquad (6.301)$$

When $m = r$, the two special cases of the quadratic optimal tracking control emerge:

1. Output bias formulation: Here, $B_d = 0$ and $H_d = I$; hence, $K_d = 0$

$$u_s(k) = K_u^{-1}[r(k) - \hat{v}_d(k/k-1)] \qquad (6.302)$$

$$x_s(k) = [I - \varphi]^{-1} BK_u^{-1}\{r(k) - \hat{v}_d(k/k-1)\} \qquad (6.303)$$

2. Input bias formulation: Here, $B_d = B$ and $H_d = 0$; then, $K_d = K_u$

$$u_s(k) = K_u^{-1}r(k) - \hat{d}(k/k-1) \qquad (6.304)$$

$$x_s(k) = [I - \varphi]^{-1} BK_u^{-1}r(k) \qquad (6.305)$$

If the number of the manipulated inputs (m) is not the same as the number of controlled outputs r, the matrix K_u can be replaced by its pseudo-inverse.

6.8 Pole Placement Design Method

Let us consider the following linear dynamic system in the continuous time domain

$$\dot{x} = Ax + Bu; \quad y = Cx \qquad (6.306)$$

The idea is to stabilize the system in (6.306) or improve its transient response by using the full state feedback. The state feedback is given as

$$U = -Fx \qquad (6.307)$$

With the state feedback chosen as in (6.307), the open loop system (6.306) becomes the following closed loop system

$$\dot{x} = (A - BF)x; \quad y = Cx \qquad (6.308)$$

Hence, the idea now translates to stabilize the system (6.308) so that all the closed loop eigenvalues are placed in the left half of the complex s-plane (see Section 6.1). This approach of pole placement [8]. The theoretical result is that if the system dynamic matrix pair (A, B) is controllable, i.e. if the dynamic system is controllable, then there exists a feedback matrix F, such that the closed loop dynamic system (6.308) is stabilized and its eigenvalues can be placed in any arbitrary locations. However, the practical design engineer has to look into implementation aspects very carefully, even though the freedom of choice is given by the analytical result, which is established here for the single-input single-output (SISO) system. If the pair of the SISO system is controllable, then the original system is transformed into the phase variable canonical form. This means that there exists a nonsingular transformation [8]

$$x = Pz \qquad (6.309)$$

such that the new system in its canonical form is given as

$$\dot{z} = \begin{bmatrix} 0 & 1 & 0 & \dots & 0 \\ 0 & 0 & 1 & \dots & 0 \\ : & : & : & \dots & : \\ 0 & 0 & 0 & \dots & 1 \\ -a_0 & -a_1 & -a_2 & \dots & -a_{n-1} \end{bmatrix} z + \begin{bmatrix} 0 \\ 0 \\ : \\ 0 \\ 1 \end{bmatrix} u \qquad (6.310)$$

In (6.310), the coefficients $a(.)$ form the characteristic polynomial of the matrix A and is given in terms of the characteristic equation using the complex frequency $s = \sigma + j\omega$:

$$\det(\lambda I - A) = \lambda^n + a_{n-1}\lambda^{n-1} + a_{n-2}\lambda^{n-2} + \dots + a_1\lambda + a_0 \\ = s^n + a_{n-1}s^{n-1} + a_{n-2}s^{n-2} + \dots + a_1 s + a_0 \qquad (6.311)$$

Then, for the SISO the state feedback control input, $u(.)$ is given as

$$u(z) = -f_1 z_1 - f_2 z_2 - \dots - f_n z_n = -f_c z \qquad (6.312)$$

After closing the feedback loop, by using (6.312) in (6.308) the following closed loop system, we get

$$\dot{z} = \begin{bmatrix} 0 & 1 & 0 & \cdots & 0 \\ 0 & 0 & 1 & \cdots & 0 \\ \vdots & \vdots & \vdots & \cdots & \vdots \\ 0 & 0 & 0 & \cdots & 1 \\ -(a_0 + f_1) & -(a_1 + f_2) & -(a_2 + f_3) & \cdots & -(a_{n-1} + f_n) \end{bmatrix} z; \dot{z} = A_z z \tag{6.313}$$

The closed loop eigenvalues would be given by

$$\det(\lambda_c I - A_z) = \lambda_c^{n} + (a_{n-1} + f_n)\lambda_c^{n-1} + (a_{n-2} + f_{n-1})\lambda_c^{n-2} \tag{6.314}$$
$$+ \ldots + (a_1 + f_2)\lambda_c + (a_0 + f_1)$$

In terms of the complex frequency, (6.314) is written as

$$\Delta^d(\lambda) = (\lambda - \lambda_1^d)(\lambda - \lambda_2^d)\ldots(\lambda - \lambda_n^d) \tag{6.315}$$
$$= s^n + a_{n-1}^d s^{n-1} + a_{n-2}^d s^{n-2} + \ldots + a_1^d s + a_0^d$$

It is noted that the closed loop matrix in (6.313) contains in its last term the coefficients of the characteristic polynomial of the original system, (6.310); hence, the required feedback gains $f(.)$ can be obtained from the following formulae:

$$a_0 + f_1 = a_0^d \Rightarrow f_1 = a_0^d - a_0;$$
$$a_1 + f_2 = a_1^d \Rightarrow f_2 = a_1^d - a_1; \tag{6.316}$$
$$\ldots$$
$$a_{n-1} + f_n = a_{n-1}^d \Rightarrow f_n = a_{n-1}^d - a_{n-1}$$

6.9 Eigenstructure Assignment

The LQG and LQR designs in control are examples of the H_2 synthesis procedure. The aim of the H_2 optimal problem is to find a controller that minimizes a quadratic performance index, i.e. the H_2 norm of the system. It offers a method of combining the design criteria of quadratic performance and disturbance rejection, or at least its attenuation; however, such a controller design method cannot necessarily guarantee that the closed loop system would have good transient responses; the latter are determined mainly by the locations of the systems' eigenvalues. As an important design method associated with eigenvalues

and eigenvectors in control theory, eigenstructure assignment has attracted much attention of many researchers; one such approach parameterizes all the solutions to the problem [9]. This parametric approach presents complete, explicit, and parametric expressions of all the feedback gain matrices and the closed loop eigenvector matrices. It also offers design degrees of freedom, which can be further utilized to satisfy some additional performances like robustness. Here, the purpose is to design a state feedback controller so that the closed loop system would have (i) the desired poles that would lie in some desired domains and (ii) the disturbance attenuation performance. The parametric solution for state feedback eigenstructure assignment is utilized, the disturbance attenuation index is parameterized, and the H_2-optimal control (problem) with regional pole assignment is changed into a minimization problem with certain constraints: so it is H_2-optimal control problem with regional pole assignment via state feedback.

6.9.1 Problem Statement

Let us consider the following time-invariant continuous time dynamic system

$$\dot{x} = Ax + Bu + Gw; \quad y = Cx + Du \tag{6.317}$$

In (6.317), w is an exogenous input and other variables have appropriate dimensions and usual meanings. It is assumed that the matrix pair (A,B) is controllable, i.e. in fact, the system (6.317) is controllable. We use the state feedback concept for obtaining the control law:

$$u = Fx \tag{6.318}$$

In some cases, $u = -Fx$ is used but it does not alter the development of the theory. The modified system, i.e. the closed loop control system, is now given as

$$\dot{x} = (A + BF)x + Gw; \quad y = (C + DF)x \tag{6.319}$$

It is assumed that the eigenvalues of (6.319) are distinct and self-conjugate. Let λ_i $(i = 1,2,\ldots,n)$ be the eigenvalues of the closed loop system (6.319), and v_i $(i = 1,2,\ldots,n)$ be the corresponding eigenvectors. Then, the following is the valid eigenstructure assignment problem

$$(A + BF)v_i = \lambda_i v_i, \quad (i = 1,2,\ldots,n) \tag{6.320}$$

Now, the closed loop transfer function, from w to y, using (6.319) is given as

$$\text{TF}(F) = (C + DF)(sI_{n \times n} - A - BF)G \tag{6.321}$$

In (6.321), we use s as a complex frequency $(s = \sigma + j\omega)$. Where we want to emphasize specifically the importance of the closed loop eigenvalues, we would use: λ_i $(i = 1,2,...,n)$, since both being complex numbers and the eigenvalues would, in general, be also complex numbers. Now, let P be the positive semi-definite as the solution of the following algebraic matrix equation

$$(A + BF)^T P + P(A + BF) + (C + DF)^T (C + DF) = 0 \quad (6.322)$$

Then, it can be shown that

$$\|\text{TF}(F)\| = \sqrt{\text{trace}(G^T PG)} \quad (6.323)$$

Given the dynamic system as in (6.317) and the assumption that this system is controllable, then the objective is to find a state feedback controller such that the minimization of the trace of transfer function (TF) (between output y and exogenous input w) is achieved

$$\min_F \|\text{TF}(F)\| \quad (6.324)$$

Additionally, the following conditions should be satisfied

1. Equation (6.320) holds, and $\det(V) \neq 0$; and
2. λ_i $(i = 1,2,...,n)$, i.e. the closed loop eigenvalues lie in a stable region. Here, V is the matrix that has the eigenvectors as its columns, corresponding to each eigenvalue.

6.9.2 Closed Loop Eigenstructure Assignment

Let the eigenvalues and eigenvectors matrices be specified as

$$\Lambda = diag\{\lambda_1 \lambda_2 ... \lambda_3\}; \; V = [v_1 v_2 ... v_3] \quad (6.325)$$

With (6.325), (6.320) becomes

$$AV + BFV = V\Lambda \, ; \; W = FV; \; AV + BW = V\Lambda \quad (6.326)$$

Since the matrix/vector pair (A,B) is controllable, by using a series of elementary transformations, one can obtain the following with s being a complex frequency:

$$P(s)[A - sI_n \; B]Q(s) = [0 \; I_n] \quad (6.327)$$

Next, we partition $Q(s)$ as

$$Q(s) = \begin{bmatrix} Q_{11}(s) & Q_{12}(s) \\ Q_{21}(s) & Q_{22}(s) \end{bmatrix} \quad (6.328)$$

Given the controllable pair (A,B) and the associated assumptions and conditions defined in the foregoing, it can be stated that the parametric expressions of the feedback gain matrix can be given as

$$F = WV^{-1} \quad (6.329)$$

In (6.329), we have

$$\begin{aligned} V &= \begin{bmatrix} v_1 v_2 ... v_n \end{bmatrix}; \; v_i = Q(s_i)f_i \, ; \\ W &= \begin{bmatrix} w_1 w_2 ... w_n \end{bmatrix}; \; w_i = Q_{21}(s_i)f_i \end{aligned} \quad (6.330)$$

In (6.330), $(w(.)$ are *not* the components of the exogenous input as in [6.317]). Here, f_i $(i = 1,2,...,n)$ are a set of free parametric vectors and satisfy the following conditions:

Condition 1: $s_i = \bar{s}_j \Leftrightarrow f_i = \bar{f}_j \, ; i, j = 1,2,...,n$ (6.331)

Condition 2: $\det(V) \neq 0$

The parametric vectors and the closed loop eigenvalues can be regarded as the design freedom for the synthesis of the control laws. We now need to obtain the parametric solution to (6.322). Given all the conditions and assumptions in the foregoing, the solution to P can be given as

$$P = -V^{-T} \left[\frac{(v_i^T C^T + w_i^T D^T)(Cv_j + Dw_j)}{s_i + s_j} \right]_{n \times n} V^{-1} \quad (6.332)$$

For (6.332), all the components and variables have been defined in the foregoing description. This result is established next. Noting (6.326), we get

$$A + BF = V\Lambda V^{-1} \quad (6.333)$$

Now substitute (6.333) into (6.322) to obtain

$$(V\Lambda V^{-1})^T P + P(V\Lambda V^{-1}) = -(C + DF)^T (C + DF) \quad (6.334)$$

Denote

$$\tilde{P} = V^T PV; \; \text{or} \; P = V^{-T}\tilde{P}V^{-1} \quad (6.335)$$

Substituting (6.335) in (6.334), one gets after simplification

$$\Lambda\tilde{P} + \tilde{P}\Lambda = -V^T (C + DF)^T (C + DF)V \quad (6.336)$$

Next, denote $\tilde{P} = [\tilde{p}_{ij}]_{n \times n}$; $\tilde{p}_{ij} = \tilde{p}_{ji}$, then (6.336) becomes

$$\tilde{p}_{ij} = -\frac{(v_i^T C^T + w_i^T D^T)(Cv_j + Dw_j)}{s_i + s_j} \quad (6.337)$$

Using (6.337) in (6.335), we see that (6.332) holds. Now, we can set

$$P = P(s_i, f_j); \ i = 1, 2, ..., n \quad (6.338)$$

The equation (6.338) signifies that the matrix P in (6.323) is parametrized by s_i, and f_i; these parameters are determined by the following minimization proposition

$$\min_{\{s_i\}, \{f_i\}} \sqrt{\text{trace}(G^T P(s_i, f_j)G)} \quad (6.339)$$

The foregoing result (6.339) is subject to the condition in (6.331). Let us denote the real eigenvalues of s_i by

Condition 3: $a_i \leq \sigma_i \leq b_i$; $i = 1, 2, ..., n$,

with $a(.)$, and $b(.)$ as some specified real numbers, then (6.339) becomes, with the $h(.)$ as the corresponding numbers for $f(.)$:

$$\min_{\{\sigma_i\} \subseteq R, \{h_i \subseteq R\}} \sqrt{\text{trace}(G^T P(\sigma_i, h_j)G)} \quad (6.340)$$

The algorithm for the synthesis of the control law is now given as:

1. Compute a pair of unimodular matrices $P(s)$, $Q(s)$ to satisfy (6.327), and partition $Q(s)$ as per (6.328).

2. Specify and set the parametric free vectors $f(.)$ and compute the parametric expressions of matrices V and W, as per (6.330).

3. Using (6.332), compute the parametric expressions of P.

4. Determine s_i and f_i satisfying the conditions in (6.331) by solving the minimization problem, as in (6.339) or (6.340).

5. Compute the gain matrix F as in (6.329) and obtain the matrices V and W.

The optimization problem can be solved by the following procedure: Let the following cost functional be defined as

$$J_1 = \sqrt{\text{trace}(G^T P(s_i, f_j)G)} \quad (6.341)$$

$$J_2 = \sqrt{\text{trace}(G^T P(\sigma_i, h_j)G)} \quad (6.342)$$

Then, we have the following necessary conditions for the optimization

$$\frac{\partial J_1}{\partial s_i} = 0; \ \frac{\partial J_1}{\partial f_i} = 0; \ \frac{\partial J_2}{\partial \sigma_i} = 0; \ \frac{\partial J_2}{\partial h_i} = 0; \ i = 1, 2, ..., n. \quad (6.343)$$

The partials in (6.343) satisfy the conditions 1, 2, and 3, already specified earlier.

6.10 Minimum-Time and Minimum-Fuel Trajectory Optimization

Derivation of the exact solution for the weighed fuel and time optimal control problem is very interesting, since it incorporates, in the limit, the time-optimal and fuel-optimal cases. In this section, we address the problem of designing a fuel-time optimal (FTO) controller for a two-mass-spring (TMS) system representative of a dynamic system with non-collocated sensor-actuator pair [10]. This problem is representative of many flexible structures whose response is characterized by one rigid body mode and one vibrational mode. The design of a FTO controller is posed as the design of a time-delay (TD) filter that generates a bang-off-bang (BOB) profile when it is subject to a step input; wherein the magnitudes of the time-delayed signals are constrained to satisfy the saturation limits of the actuator. In fact, the closed form solutions of the times delays as functions of the weighting parameter and the net rigid body motion are derived for the large a weighting parameter case. This optimal control problem leads to an integer programming problem that would uniquely determine the optimal control trajectory. It is also shown that, for a cost function with a very small weighting parameter associated with the fuel, the optimal control profile is a sequence of pulses characterized by six switches. When the weighting parameter increases, the optimal control trajectory/profile approaches the well-known structure of a fuel optimal control profile that is characterized by two switches.

6.10.1 Problem Definition

Let the dynamic system be described as

$$\dot{x} = Ax + Bu \quad (6.344)$$

In (6.344), we have the constraint on control input (i.e. the fuel) as

$$-1 \leq u \leq 1 \tag{6.345}$$

Then, the FTO control problem leads to the following cost function for the optimization

$$J = \int_0^{t_f} (dt + \alpha |u| dt) = \int_0^{t_f} (1 + \alpha |u|) dt \tag{6.346}$$

The problem is to determine u that minimizes the cost function J, as in (6.346). It can be easily seen that the cost function is a weighted combination of the maneuver time (dt) and the fuel consumed. From the development of the earlier sections of this chapter, we can easily define the Hamiltonian as follows

$$H(x, \lambda, u) = 1 + \alpha |u| + \lambda^T (Ax + Bu) \tag{6.347}$$

The necessary conditions for the optimality are

$$\dot{x} = Ax + Bu \, ;$$

$$\dot{\lambda} = -A^T \lambda \tag{6.348}$$

$$u = -dez(\frac{B^T \lambda}{\alpha}); \; \forall t \in [0, t_f]$$

The conditions in (6.348) are a result of the Pontryagin's minimum principle, and the following Hamiltonian equation holds

$$H(x, \lambda, u) = 1 + \alpha |u| + \lambda^T (Ax + Bu) = 0;$$

$$x(0) = 0; \text{and } x(t_f) = \text{specified} \tag{6.349}$$

The dead-zone function *dez* is prescribed as

$$\phi = dez(\psi) \Rightarrow \begin{cases} \phi = 0 & if \; |\psi| < 1 \\ \phi = sgn(\psi) & if \; |\psi| > 1 \\ 0 \leq \phi \leq 0 & if \; \psi = 1 \\ -1 \leq \phi \leq 0 & if \; \psi = -1 \end{cases} \tag{6.350}$$

The two-mass spring (TMS) system is given as

$$m\ddot{x}_1 + k(x_1 - x_2) = u$$

$$m\ddot{x}_2 + k(x_1 - x_2) = 0 \tag{6.351}$$

The eigenvector matrix Φ is used to decouple the equations of motion into the following form

$$\ddot{\theta} = \phi_0 u \tag{6.352}$$

$$\ddot{q} + \omega^2 q = \phi_1 u$$

$$\begin{bmatrix} \phi_0 \\ \phi_1 \end{bmatrix} = \Phi^T \begin{bmatrix} 0 \\ 1 \end{bmatrix} \tag{6.353}$$

In (6.353), the first equation is the rigid body mode and the second equation is the vibratory mode that is characterized by frequency ω. The specified FTO control problem is next and solved by the design of the transfer function of a TD filter—a set of the zeros (of) that are constrained to cancel the poles of the system. Also, some more constraints on actuator limits and boundary conditions are incorporated in the design process. For simplicity, we assumed that $m = 1$ and $k = 1$ in (6.351).

6.10.2 Parameterization of the Control Problem

Now, with the knowledge that the FTO control profile is BOB, we construct a TD filter whose output is BOB when it is subject to a step input [10]. The amplitudes of the TD signals are constrained by the bounds on the control and that requires the determination of the TDs to completely define the TD filter. To do this, one can formulate a constrained parameter optimization problem, where the parameters to be determined are the TDs. To eliminate any residual vibration of the system, the complex poles located at $\pm j\omega$ should be canceled by a complex conjugate pair of zeros of the TD filter, and this is a first constraint. The boundary condition constraint is satisfied by integrating the rigid-body equation of motion, as is the normal constraint on the dynamics of the system. It has been established that the TF of a TD filter that generates a control profile (that is antisymmetric about its mid-maneuver time) has a minimum of two zeros at the origin of the complex plane, which facilitate the cancelation of the rigid-body poles of the system. This fact is utilized to reduce the dimension of the parameter search space. The re-orientation of the problem is considered with the following boundary conditions:

$$x_1(0) = x_2(0) = 0, \; x_1(t_f) = x_2(t_f) = \frac{\theta_f}{2\phi_0} \tag{6.354}$$

$$\dot{x}_1(0) = \dot{x}_2(0) = 0, \; \dot{x}_1(t_f) = \dot{x}_2(t_f) = 0$$

The conditions in (6.354) correspond to the following conditions

$$\theta(0) = q(0) = 0, \; \theta(t_f) = \theta_f, \; q(t_f) = 0 \tag{6.355}$$

$$\dot{\theta}(0) = \dot{q}(0) = 0, \; \dot{\theta}(t_f) = \dot{q}(t_f) = 0$$

The two-switch control profile for the FTO problem is parameterized in the frequency domain as

$$u = (\frac{1}{s})(1 - e^{-s(T_2 - T_1)} - e^{-s(T_2 + T_1)} + e^{-2sT_2}) \qquad (6.356)$$

In (6.356), T_2 is mid-maneuver time and $T_2 - T_1$ is the first switch time. The TF of the TD filter that generates the BOB control profile is given as

$$G(s) = (1 - e^{-s(T_2 - T_1)} - e^{-s(T_2 + T_1)} + e^{-2sT_2}) \qquad (6.357)$$

From (6.352) and (6.356), we have

$$s^2 \theta(s) = \left(\frac{\phi_0}{s}\right)(1 - e^{-s(T_2 - T_1)} - e^{-s(T_2 + T_1)} + e^{-2sT_2}) \qquad (6.358)$$

The inverse Laplace transform of (6.358) yields

$$\theta(t) = \phi_0 \begin{pmatrix} \frac{1}{2}t^2 - \mathrm{H}[t - (T_2 - T_1)]\left\{\frac{1}{2}[t - (T_2 - T_1)]^2\right\} \\ -\mathrm{H}[t - (T_2 + T_1)]\left\{\frac{1}{2}[t - (T_2 + T_1)]^2\right\} \\ -\mathrm{H}[t - 2T_2]\left\{\frac{1}{2}[t - 2T_2^2]\right\} \end{pmatrix} \qquad (6.359)$$

In (6.359), $H(t-T_i)$ is the Heaviside unit function. Since, the final state value of $\theta(2T_2) = \theta_f$; we have

$$\theta_f = \phi_0(T_2^2 - T_1^2); \quad \frac{\theta_f}{\phi_0} = (T_2 - T_1)(T_2 + T_1) \qquad (6.360)$$

In (6.357), we substitute $s = j\omega$ to obtain; equating the real and imaginary parts to zero:

$$1 - \cos[\omega(T_2 - T_1)] - \cos[\omega(T_2 + T_1)] + \cos(2\omega T_2) = 0$$
$$-\sin[\omega(T_2 - T_1)] - \sin[\omega(T_2 + T_1)] + \sin(2\omega T_2) = 0 \qquad (6.361)$$

Simplifying (6.361), we obtain

$$2\cos(\omega T_2)[-\cos(\omega T_1) + \cos(\omega T_2)] = 0$$
$$2\sin(\omega T_2)[-\cos(\omega T_1) + \cos(\omega T_2)] = 0 \qquad (6.362)$$

From (6.363), one gets the following constraint

$$-\cos(\omega T_1) + \cos(\omega T_2) = 0 \qquad (6.363)$$

From (6.364), we get

$$\omega T_2 = \pm \omega T_1 + 2n\pi; \quad n = \pm 1, \pm 2, \dots, \infty \qquad (6.364)$$

Substitute (6.365) in (6.360) to obtain

$$T_1 = \begin{cases} \dfrac{\omega \theta_f}{4n\pi\phi_0} - \dfrac{n\pi}{\omega} & for\ \omega T_2 = \omega T_1 + 2n\pi \\[3mm] \dfrac{n\pi}{\omega} - \dfrac{\omega \theta_f}{4n\pi\phi_0} & for\ \omega T_2 = -\omega T_1 + 2n\pi \end{cases} \qquad (6.365)$$

The cost function to be minimized is

$$J = \int_0^{2T_2} (1 + \alpha|u|)dt \qquad (6.366)$$

The cost function (6.367) is also written as

$$J = 2T_2 + 2\alpha(T_2 - T_1) \qquad (6.367)$$

$$J = \begin{cases} \dfrac{\omega \theta_f}{2n\pi\phi_0} + (2 + 4\alpha)\dfrac{n\pi}{\omega} & for\ \omega T_2 = \omega T_1 + 2n\pi \\[3mm] (1 + 2\alpha)\dfrac{\omega \theta_f}{4n\pi\phi_0} + \dfrac{2n\pi}{\omega} & for\ \omega T_2 = -\omega T_1 + 2n\pi \end{cases} \qquad (6.368)$$

The cost function in (6.368) is now minimized with respect to n assuming it is continuous

$$\frac{dJ}{dn} = 0 \qquad (6.369)$$

We obtain the following expressions for n

$$n^2 = \begin{cases} \left(\dfrac{\omega}{\pi}\right)^2 \dfrac{\theta_f}{(2 + 4\alpha)\phi_0} & for\ \omega T_2 = \omega T_1 + 2n\pi \\[3mm] \left(\dfrac{\omega}{\pi}\right)^2 \dfrac{(1 + 2\alpha)\theta_f}{2\phi_0} & for\ \omega T_2 = -\omega T_1 + 2n\pi \end{cases} \qquad (6.370)$$

The optimal control profile leads to the displacement of the rigid-body mode at mid-maneuver time, which is half the total maneuver time, and the displacement of the vibratory mode, which is zero at the mid-maneuver time. Now, we can arrive at a closed form solution of the FTO control profile for a rigid body that has the same structure as the TMS system. Consider the dynamic system as in (6.352) and the input as in (6.356). The boundary condition leads to

$$T_2^2 = \left(\frac{\theta_f}{\phi_0}\right) + T_1^2 \qquad (6.371)$$

Substitute (6.371) in (6.366) to obtain

$$J = 2(\alpha + 1)\sqrt{\frac{\theta_f}{\phi_0} + T_1^2} - 2\alpha T_1 \qquad (6.372)$$

The T_1 that optimizes the cost function is obtained from

$$\frac{dJ}{dT_1} = 0 \qquad (6.373)$$

From (6.372) and (6.373), we obtain

$$T_1^2 = \frac{\alpha^2 \theta_f}{(1+2\alpha)\phi_0} \qquad (6.374)$$

$$J = 2\sqrt{\theta_f(1+2\alpha)/\phi_0} \qquad (6.375)$$

Because n is a constant for a given range of α; the T_1 and T_2 are constant in the same interval; signifying that the optimum control is identical for different values of α: (i) for $\alpha = 0$, corresponding to the time optimal case, the first switch time coincides with the mid-maneuver time for the rigid-body case; and (ii) implies a control law/trajectory that is bang-bang. From (6.370), one sees two possible solutions that would be a function of θ_f and α: (i) for small θ_f, the solution for the range $\omega T_2 = -\omega T_1 + 2n\pi$ is optimum; however, (ii) for the large values of θ_f, $\omega T_2 = \omega T_1 + 2n\pi$ points to the optimum solution.

6.10.3 Control Profile for Small α

It has been shown that the control profile for $\alpha = 0$ (the time optimal control) has a three-switch bang-bang structure, which is antisymmetric about its mid-maneuver time; this control profile is quite different from that of one parametrized to optimize the weighted FTO cost function. This would indicate that there should exist a control structure that, in the limit, can generate the time optimal control profile. The antisymmetric FTO control for small α is given as

$$u = \left(\frac{1}{s}\right)(1 - e^{-s(T_4-T_3)} - e^{-s(T_4-T_2)}$$
$$+ e^{-s(T_4-T_1)} - e^{-s(T_4+T_3)} - e^{-s(T_4+T_2)} \qquad (6.376)$$
$$+ e^{-s(T_4+T_1)} + e^{-2sT_4})$$

In (6.376), T_4 is the mid-maneuver time, and the constraint for the parameter optimization problem is given as

$$\theta_f = \phi_0(T_4^2 - T_3^2 + T_2^2 - T_1^2) \qquad (6.377)$$

The condition (6.377) is derived from the rigid-body boundary condition. Also, we have

$$-\cos(\omega T_3) + \cos(\omega T_2) - \cos(\omega T_1) + \cos(\omega T_4) = 0 \qquad (6.378)$$

The condition (6.378) forces the TF of the TD filter to cancel the frequency corresponding to the vibratory mode of the dynamic system. Next, the parameter optimization problem is formulated to obtain the solution for the TDs. Then, the cost function to be minimized is subject to the conditions in (6.377) and (6.378), given as:

$$J = \int_0^{T_f} (1 + \alpha|u|)dt = 2T_4 + 2\alpha[(T_4 - T_3) + (T_2 - T_1)] \qquad (6.379)$$

Because the constraints are nonlinear and the number of parameters is more than the number of constraints, multiple solutions might exist. In order to prove optimality of the profile, the one suggested by the switching function should coincide with that predicted by the parameter optimization; to determine the switching function, one needs to solve for the initial co-states, for which we have

$$\lambda(t) = e^{(-A^T t)}\lambda(0) \qquad (6.380)$$

We also know that the switching function has to satisfy the following equation

$$\begin{bmatrix} B^T \exp[-A^T(T_4 - T_3)] \\ B^T \exp[-A^T(T_4 - T_2)] \\ B^T \exp[-A^T(T_4 + T_2)] \\ B^T \exp[-A^T(T_4 + T_3)] \end{bmatrix} \lambda(0) = \begin{bmatrix} -\alpha \\ \alpha \\ -\alpha \\ \alpha \end{bmatrix} \qquad (6.381)$$

We solve (6.381) for $\lambda(0)$ and substitute in

$$u = -\text{dez}[B^T \exp(-A^T t)\lambda(0)/\alpha] \qquad (6.382)$$

6.10.4 Determination of Critical α

We consider (6.344) and (6.345) for the TMS system. The matrices A and B are given as

$$A = \begin{bmatrix} 0 & 1 & 0 & 0 \\ 0 & 0 & 0 & 0 \\ 0 & 0 & 0 & 1 \\ 0 & 0 & -\omega^2 & 0 \end{bmatrix}; B = \begin{bmatrix} 0 \\ \phi_0 \\ 0 \\ \phi_1 \end{bmatrix} \qquad (6.383)$$

The optimal control is given as

$$u(t) = -\text{dez}[B^T \lambda(t)/\alpha]; \quad \lambda(t) = \exp(-A^T t)\lambda(0) \qquad (6.384)$$

The value of α_{cr}, i.e. "alpha" critical corresponds to the α, when the switching function $B^T \lambda(t)$ equals to a value of α and $-\alpha$ at times t_{cr} and $T - t_{cr}$, respectively, as seen below:

$$B^T \lambda(t_{cr}) = B^T \exp(-A^T t_{cr})\lambda(0) = \alpha_{cr}$$
$$B^T \lambda(T_f - t_{cr}) = B^T \exp(-A^T(T_f - t_{cr}))\lambda(0) = -\alpha_{cr} \qquad (6.385)$$

This should ensure that the switching function touches the lines parallel to the abscissa with ordinate intercepts of $\pm\alpha_{cr}$, tangentially, requiring that the slope of $B^T\lambda(t)$ at the critical times should be equal to zero:

$$\frac{d(B^T\lambda(t))}{dt}\Big/_{\{at\,t=t_{cr}\}} = -B^T A^T \exp(-A^T t_{cr})\lambda(0) = 0$$

$$\frac{d(B^T\lambda(t))}{dt}\Big/_{\{at\,t=T_f-t_{cr}\}} = -B^T A^T \exp(-A^T(T_f-t_{cr}))\lambda(0) = 0 \tag{6.386}$$

Write (6.385) and (6.386) together to get

$$\begin{bmatrix} B^T \exp(-A^T t_{cr}) \\ B^T \exp(-A^T(T_f-t_{cr})) \\ -B^T A^T \exp(-A^T t_{cr}) \\ -B^T A^T \exp(-A^T(T_f-t_{cr})) \end{bmatrix}\lambda(0) = \begin{bmatrix} \alpha_{cr} \\ -\alpha_{cr} \\ 0 \\ 0 \end{bmatrix} \tag{6.387}$$

Solving (6.387), one arrives at the values of t_{cr} and α_{cr}. Next, to determine co-state at time $t=0$; we have

$$B^T \exp(-A^T(T_2-T_1))\lambda(0) = -\alpha$$

$$B^T \exp(-A^T(T_2+T_1))\lambda(0) = \alpha \tag{6.388}$$

The switching function evaluated at time T_2 gives (from Hamiltonian evaluated at time = 0):

$$B^T \exp(-A^T T_2)\lambda(0) = 0; \text{ and } B^T\lambda(0) = -1-\alpha \tag{6.389}$$

It has been observed that the switch times T_1 and T_2 remain constant for small α, i.e. for the range of α greater than α_{cr} and less that $\alpha = 9.87$ (for maneuver of $\theta_f = 1$). The equations (6.388) and (6.389) can be written together as

$$\begin{bmatrix} B^T \exp(-A^T(T_2-T_1)) \\ B^T \exp(-A^T(T_2+T_1)) \\ B^T \exp(-A^T T_2) \\ B^T \end{bmatrix}\lambda(0) = \begin{bmatrix} -\alpha \\ \alpha \\ 0 \\ -1-\alpha \end{bmatrix} \Rightarrow P\lambda(0) = \begin{bmatrix} -\alpha \\ \alpha \\ 0 \\ -1-\alpha \end{bmatrix} \tag{6.390}$$

Now, combining the equations (6.387) and (6.390) appropriately, we get

$$\begin{bmatrix} B^T \exp(-A^T t_{cr})P^{-1} \\ B^T \exp(-A^T(T_f-t_{cr}))P^{-1} \\ -B^T A^T \exp(-A^T t_{cr})P^{-1} \\ -B^T A^T \exp(-A^T(T_f-t_{cr}))P^{-1} \end{bmatrix} \begin{bmatrix} -\alpha_{cr} \\ \alpha_{cr} \\ 0 \\ -1-\alpha_{cr} \end{bmatrix} = \begin{bmatrix} \alpha_{cr} \\ -\alpha_{cr} \\ 0 \\ 0 \end{bmatrix} \tag{6.391}$$

From (6.391), we have $\alpha_{cr} = 0.6824$ for of $\theta_f = 1$. When the masses were fixed, the frequency of the vibratory mode was found to be proportional to the stiffness [10]. It was revealed that α_{cr} tends to zero as the stiffness increases, then increases rapidly as the stiffness tends to zero. It was also found that the transition switch time was between the first switch time and the mid-maneuver time, which asymptotically approach each other making intuitive sense, since the system is approaching a rigid body as the stiffness increases. Also, with the fact that α_{cr} tends to zero as the stiffness increases, the control should resemble a bang-bang controller for a rigid body.

Appendix 6A: More Topics of Optimal Control

6A.1 Bang-Bang Principle

A control $u(.)$ (of the set U) is called bang-bang (BB) if for each $t \geq 0$ and each index, $i = 1,...,m$, we have [4]:

$$|u^i(t)| = 1, \text{ and}$$

$$u(t) = \begin{bmatrix} u^1(t) \\ : \\ u^m(t) \end{bmatrix} \tag{6A.1}$$

Let the standard dynamic system be given as

$$\dot{x}(t) = Ax(t) + Bu(t), \; x(0) = x_0 \tag{6A.2}$$

Then, there exists a control signal $u(.)$, called "bang-bang control" that takes the system's state from $x(0)$ (of some numerical value, x_0) to the "0" (i.e. of zero value) at time t (>0), i.e. $x(t) = 0$. We need to specify the conditions: (i) a set S is convex if $ax + (1-a)\hat{x} \in S$, $\forall x, \hat{x}$; and $\forall a$, the real numbers; $0 \leq a \leq 1$, and (ii) a point $y \in S$ is called extreme if there do not exist points $x, \hat{x} \in$, and $0 < a < 1$ such that $y = ax + (1-a)\hat{x}$, and (iii) then the set S has at least one extreme point. In our case of the optimal control, the set S is the set of controls $(u(.))$ that take the system's state $x(0)$ (some value) to $x(t) = 0$. We consider the dynamic system as in (6A.2). Then, we have the set $S = \{u(.) \in U; \text{and } u(.) \text{ takes } x(0) \text{ to } 0 \text{ at time } t\}$. Since $x(0)$ is finite, the set S is non-zero. Now, since the set S is convex, then $u(.) \in S$, if, and only if, the following is true (from the solution of the inhomogeneous equation (6A.2), for $x(t) = 0$):

$$x(t) = 0 \Rightarrow e^{As}x_0 + e^{As}\int_0^t e^{-As}Bu(s)ds \qquad (6A.3)$$

From (6A.3), we get

$$x_0 = -\int_0^t e^{-As}Bu(s)ds; \text{ and } x_0 = -\int_0^t e^{-As}B\hat{u}(s)ds$$

$$;\text{hence, } ax + (1-a)\hat{x} \in S$$

$$x_0 = -\int_0^t e^{-As}B\Big[au(s) + (1-a)\hat{u}(s)\Big]ds$$

$$(6A.4)$$

By the definition of weak convergence, the following is also implied

$$x_0 = -\int_0^t e^{-As}Bu_{n_k}(s)ds \Rightarrow -\int_0^t e^{-As}Bu(s)ds; \text{ if } n_k \to \infty \quad (6A.5)$$

Next, one needs to show that for almost all times $0 \le s \le t$ and for each $i = 1,\ldots, m$, we have the bang-bang control:

$$\left|\hat{u}^i(s)\right| = 1 \qquad (6A.6)$$

Start with the contradiction that (6A.6) is not a BB control, which means that there exists an index $i \in \{1,\ldots,m\}$ and that $\left|\hat{u}^i(s)\right| < 1$ for a set of $T_s \subset [0,t]$, *for* $s \in T_s$; or we can say

$$\left|\hat{u}^i(s)\right| \le 1-\varepsilon, \text{ and } \left|T_\varepsilon\right| > 0, \text{ for } s \in T_\varepsilon; \text{ and } \varepsilon > 0, \ T_\varepsilon \subseteq T_s$$

$$(6A.7)$$

Next, define

$$I_{T_\varepsilon}(v(.)) = \int_{T_\varepsilon} e^{-As}Bv(s)ds; \text{ for } v(.) = \{0,\ldots,v(.),\ldots,0\}^T \quad (6A.8)$$

We can choose any real-valued function $v(.) \ne 0$, and we have $I_{T_\varepsilon}(v(.)) = 0$, $|v(.)| \le 1$ $\qquad$ (6A.9)

Let us define the following

$$u_1(.) = \hat{u}(.) + \varepsilon v(.); \ u_2(.) = \hat{u}(.) - \varepsilon v(.); \ v \text{ is the zero off the set}$$

$$(6A.10)$$

Based on the foregoing, it is claimed that $u_1(.), u_2(.) \in S$, and this can be seen based on the following development

$$-\int_0^t e^{-As}Bu_1(s)ds = -\int_0^t e^{-As}B\hat{u}(s)ds - \varepsilon \int_0^t e^{-As}Bv(s)ds$$

$$(6A.11)$$

$$= x_0 - \varepsilon \int_0^t e^{-As}Bv(s)ds = x_0$$

In (6A.11), we have used (6A.8) and (6A.9). We also have the following equivalence

$$u_1(.) = \hat{u}(.), \text{ for}(s \notin T_\varepsilon); \ u_1(.) = \hat{u}(.) + \varepsilon v(.), \text{ for}(s \in T_\varepsilon)$$

$$(6A.12)$$

However, on the set T_ε one has $\left|\hat{u}^i(s)\right| \le 1-\varepsilon$, and hence

$$\left|u_1(s)\right| \le \left|\hat{u}(s)\right| + \varepsilon\left|v(s)\right| \le 1-\varepsilon+\varepsilon = 1 \qquad (6A.13)$$

Similarly, for $u_2(s)$; hence, $u_1(.), u_2(.) \in S$. Also, we have

$$u_1 = \hat{u}(.) + \varepsilon v, \text{ for } u_1 \ne \hat{u}; \ u_2 = \hat{u} - \varepsilon v, \text{ for } u_2 \ne \hat{u} \quad (6A.14)$$

However, we have

$$\frac{1}{2}u_1 + \frac{1}{2}u_2 = \hat{u} \qquad (6A.15)$$

The result in (6A.15) is a contradiction, since $\hat{u}$ is an extremal point in the set S (of $u(.)$).

6A.2 Proof of Pontryagin Maximum Principle

Here, we discuss various proofs of the results PMP that were used in the main Chapter 6 on optimal control [4]. This is based on the control theory of Hamiltonian H and the co-state; the latter is called ADCOLM.

6A.2.1 Informal Approach

Here, we study the variations for the free endpoint problem. The linear dynamics of the time-variant system are given as

$$\dot{y}(t) = Fy(t), \ 0 \le t \le T; \ y(0) = y_0 \qquad (6A.16)$$

The adjoint system is given as

$$\dot{\lambda}(t) = -F^T\lambda(t), \ 0 \le t \le T; \ \lambda(T) = \lambda_0 \qquad (6A.17)$$

By introducing the adjoint dynamics we achieve the following

$$\frac{d(\lambda.y)}{dt} = \dot{\lambda}y + \lambda\dot{y} = -(F^T\lambda)y + \lambda(FY) \Rightarrow 0 \quad (6A.18)$$

The identity (6A.18) implies that over a time $y(t).\lambda(t)$ is constant and we also have $y(T)\lambda_0 = y_0\lambda(0)$; this will be useful later on. The variation in control is introduced by considering the basic free endpoint control problem as defined next with the dynamics:

$$\dot{x}(t) = f(x(t), u(t)); \ x(0) = x_0; \ t \ge 0 \qquad (6A.19)$$

The POF is given as

$$J(u(.)) = \int_0^T \varphi(x(t), u(t))dt + g(x(T)) \qquad (6A.20)$$

For the sake of simplicity, the distinction between the symbols of the states and control input, and their respective estimatees is avoided time being. Let $\varepsilon > 0$, and the variation be

$$u_\varepsilon(t) = u(t) + \varepsilon v(t), \quad 0 \le t \le T \qquad (6A.21)$$

In (6A.21), $v(.)$ is an acceptable variation and it is assumed to exist. We have the following solution

$$x_\varepsilon(t) = x(t) + \varepsilon y(t) + o(\varepsilon) \quad 0 \le t \le T \qquad (6A.22)$$

For (6A.22) and (6A.16), one can also write as follows

$$\dot{y}(t) = f_x(x, u)y + f_a(x, u)v; \quad 0 \le t \le T;$$
$$y(0) = y_0; \; F(t) = f_x(x(t), u(t)) \qquad (6A.23)$$

In (6A.23), $f_x(x(t), u(t)) \equiv \nabla_x f$. Now, the variation in the POF can be incorporated as follows. Since the control $u(.)$ maximizes the POF, we have:

$$\frac{dJ(u_\varepsilon(.))}{d\varepsilon} \Big|_{\varepsilon=0} \le 0 \qquad (6A.24)$$

Insert (6A.21), (6A.22) in (6A.24), and evaluating we get

$$\frac{dJ(u_\varepsilon(.))}{d\varepsilon} \Big|_{\varepsilon=0} = \int_0^T \left[\nabla_x \varphi(x, u)y + \nabla_a \varphi(x, u)v \right] ds + \nabla g(x(T))y(T)$$
$$(6A.25)$$

We need to enforce the terminal condition

$$\lambda(T) = \nabla g(x(T)); \; \nabla g(x(T))y(T) = \lambda(T)y(T) \quad (6A.26)$$

We assume that (6A.17) can be written as

$$\dot{\lambda}(t) = -\nabla_x f \lambda - \nabla_x \varphi; \quad 0 \le t \le T$$
$$\lambda(T) = \nabla g(x(T)) \qquad (6A.27)$$
$$\dot{\lambda}(t) = -F^T \lambda - \nabla_x \varphi$$

$$\frac{d(\lambda.y)}{dt} = \dot{\lambda}y + \lambda\dot{y} = -(F^T \lambda + \nabla_x \varphi)y + \lambda(Fy + \nabla_a fv) \qquad (6A.28)$$
$$= -\nabla_x(\varphi)y + \lambda\nabla_a(f)v$$

Integrate (6A.28) to obtain, with $y(0) = 0$:

$$\nabla g(x(T))y = \int_0^T \left[\lambda\nabla_a(f)v - \nabla_x(\varphi)y \right] ds \qquad (6A.29)$$

Use (6A.29) in (6A.25) to obtain

$$\int_0^T \left[p\nabla_a(f) + \nabla_a(\varphi) \right] vds = \frac{dJ(u_\varepsilon(.))}{d\varepsilon} \Big|_{\varepsilon=0} \le 0 \quad (6A.30)$$

With (6A.30), the variation in the POF is written in the terms of the control variation $v(.)$. Now, we rewrite (6A.30) in terms of the estimatee variable as follows

$$\int_0^T \left[\hat{p}\nabla_a(f(\hat{x}, \hat{u})) + \nabla_a(\varphi(\hat{x}, \hat{u})) \right] vds \le 0 \qquad (6A.31)$$

From (6A.31) we can see that we have the Hamiltonian term included in it

$$\nabla_a H(\hat{x}, \hat{\lambda}, \hat{a}); \text{ where } H(x, \lambda, a) = f(x, a)\lambda + \varphi(x, u) \quad (6A.32)$$

From the foregoing development we can see that the variational approach has led to the control theoretic Hamiltonian.

6A.2.2 Control Variation

Let us consider the system of (6A.19) again here, and fix a time $s > 0$ and a control parameter value as $a \in U$; select $\varepsilon > 0$, so small that $0 < s - \varepsilon < s$ and define the modified control

$$u_\varepsilon(t) = \begin{cases} a & \text{if } s - \varepsilon < t < s \\ u(t) & \text{otherwise} \end{cases} \qquad (6A.33)$$

We have a corresponding dynamic system as

$$\dot{x}_\varepsilon(t) = f(x_\varepsilon(t), u_\varepsilon(t)); \; x_\varepsilon(0) = x_0; \, t > 0;$$

Also, we have

$$A(t) = \nabla_x \{ f(x(t), u(t)) \} \qquad (6A.34)$$

$$\dot{y}_\varepsilon(t) = f(y_\varepsilon(t), u(t)), \quad t \ge 0$$
$$y_\varepsilon(0) = x_0 + \varepsilon y_0 + o(\varepsilon) \qquad (6A.35)$$

$$y_\varepsilon(t) = x(t) + \varepsilon y(t) + o(\varepsilon) \text{ as } \varepsilon \to 0 \qquad (6A.36)$$

We then also have

$$\dot{y}(t) = A(t)y(t), \; y(0) = y_0, \quad t \ge 0 \qquad (6A.37)$$

$$x_\varepsilon(t) = x(t) + \varepsilon y(t) + o(\varepsilon) \text{ as } \varepsilon \to 0 \qquad (6A.38)$$

$$\dot{y}(t) = A(t)y(t), \quad y(s) = y_s, \quad t \geq s$$

$$y_s(0) = f(x(s),a) - f(x(s),u(s)), \quad y(t) = Y(t,s)y_0, \quad t \geq s$$
$$(6A.39)$$

We have $x_\varepsilon(t) = x(t), 0 \leq t \leq s - \varepsilon$, and for $s - \varepsilon \leq t \leq s$, we have the following

$$x_\varepsilon(t) - x(t) = \int_{s-\varepsilon}^{t} f(x(\tau),a) - f(x(\tau),u(\tau))d\tau + o(\varepsilon) \quad (6A.40)$$

Specifically, one would have

$$x_\varepsilon(s) - x(s) = [f(x(s),a) - f(x(s),u(s))]\varepsilon + o(\varepsilon) \quad (6A.41)$$

For the time interval $(s,\infty), x(.)$ and $x_\varepsilon(.)$, solve the same ODE but with different initial conditions specified by

$$x_\varepsilon(s) = x(s) + \varepsilon y_s + o(\varepsilon), \quad x_\varepsilon(t) = x(t) + \varepsilon y(t) + o(\varepsilon), t \geq s$$
$$(6A.42)$$

6A.2.3 Free Endpoint, and No Running Cost

We again consider the same dynamic system. We also have the POF as $J(u(.)) = g(x(T))$, which is to be maximized. The Hamiltonian is

$$H(x,\lambda,a) = f(x,a)\lambda \quad (6A.43)$$

We need to determine the co-state vector so that the following hold true

$$\dot{\hat{\lambda}}(t) = -\nabla_x H(\hat{x}(t),\hat{\lambda}(t),\hat{u}(t)), \quad 0 \leq t \leq T$$
$$(6A.44)$$
$$H(\hat{x}(t),\hat{\lambda}(t),\hat{u}(t)) = \max_{a \in U} H(\hat{x}(t),\hat{\lambda}(t),a)$$

The co-state equations are

$$\dot{\lambda}(t) = -F^T(t)\lambda(t); \quad 0 \leq t \leq T$$
$$(6A.45)$$
$$\lambda(T) = \nabla g(x(T))$$

For further development, we avoid the "cap/hat" notation for the optimal trajectories of the states, control input, and the co-state variables. The variation of the terminal POF is obtained as follows:

$$\frac{dJ(u_\varepsilon(.))|_{\varepsilon=0}}{d\varepsilon} = \lambda(s)[f(x(s),a) - f(x(s),u(s))] \quad (6A.46)$$

$$J(u_\varepsilon(.)) = g(x_\varepsilon(T)) = g(x(T)) + \varepsilon y(T) + o(\varepsilon) \quad (6A.47)$$

$$\frac{dJ(u_\varepsilon(.))|_{\varepsilon=0}}{d\varepsilon} = \nabla g(x(T))y(T) \quad (6A.48)$$

We also have

$$\frac{d(\lambda(t)y(t)}{dt} = \dot{\lambda}(t)y(t) + \lambda(t)\dot{y}(t)$$

$$= -F^T(t)\lambda(t)y(t) + \lambda(t)A(t)y(t) = 0; \text{ Since } F \equiv A$$
$$(6A.49)$$

Hence, we obtain

$$\nabla g(x(T))y(T) = \lambda(T)y(T) = \lambda(s)y(s) = \lambda(s)y^s \quad (6A.50)$$

Also, we have

$$y^s = f(x(s),a) - f(x(s),u(s)) \quad (6A.51)$$

Using (6A.50) and (6A.48), the equation (6A.46) is established. The PMP is established as follows:

Let $0 < s < T, a \in U$. Since $J(u_\varepsilon(.))|_{\varepsilon=0}$ has maximum at $\varepsilon = 0; 0 \leq \varepsilon \leq 1$, we have

$$0 \geq \frac{dJ(u_\varepsilon(.))}{d\varepsilon} = \hat{\lambda}(s)[f(\hat{x}(s),a) - f(\hat{x}(s),\hat{u}(s))]$$

$$H(\hat{x}(s),\hat{\lambda}(s),a) = f(\hat{x}(s),a)\hat{\lambda}(s)$$

$$\leq f(\hat{x}(s),\hat{u}(s))\hat{\lambda}(s) = H(\hat{x}(s),\hat{\lambda}(s),\hat{u}(s))$$
$$(6A.52)$$

The problem of the free endpoint with running cost can be handled by incorporating an extra variable for the running part/term of the POF

$$\dot{x}^{n+1}(t) = \varphi(x(t),u(t)), \quad x^{n+1}(0) = 0$$
$$(6A.53)$$
$$\bar{f}(\bar{x},a) = \begin{bmatrix} f(x,a) \\ \varphi(x,a) \end{bmatrix}$$

This would transform the problem to the free endpoint with no running cost

$$\dot{\bar{x}}(t) = \bar{f}(\bar{x}(t),u(t)), \quad \bar{x}(t) = \begin{bmatrix} x(t) \\ x^{n+1}(t) \end{bmatrix}$$

$$\bar{x}(0) = \bar{x}_0 \qquad ; \bar{H}(\bar{x},\bar{\lambda},a) = \bar{f}(\bar{x},a)\bar{\lambda}$$

$$\bar{J}(u(.)) = \bar{g}(\bar{x}(T))$$

$$(6A.54)$$

The problem of (6A.54) is similar to the one in 6A.2.3, and the same procedure can be used to solve the problem of free endpoint with running cost. The problem of fixed endpoint can also be handled by the procedure described in the foregoing [4].

The global optimality of the optimal control law obtained by the PMP and the global existence of the solution of the Riccati differential equation are discussed in [7], wherein also, the problem of infinite horizon LQR is discussed.

7

Model Predictive Control

The schemes related to model predictive control (MPC) were developed in late 1970s and became known as dynamic matrix control (DMC) or model algorithmic control (MAC). These have found applications in the process industry, and with the availability of fast computers/microprocessors, MPC has found several applications in robotics, automobiles, nuclear, and aerospace industries. MPC schemes can handle multivariable interactions and operating constraints in a proper manner and are formulated as a constraint optimization problem, which can be solved online repeatedly by carrying out model-based forecasting/prediction over a sliding window of time. The MPC facilitates optimal control of systems with unequal number of inputs and measured outputs [6]. The linear MPC is related to the classical state space linear quadratic optimal control (LQOC). The dynamic models are the key bases for the MPC methods: (i) finite impulse response, (ii) step response, (iii) linear models, (iv) nonlinear models, (v) black box, and/or (vi) mechanistic models. The MPC schemes can be viewed as modified versions of LQ (LQG) formulation and can deal with the operating constraints in a proper manner [6]. An MPC is based on: (a) given reasonable accurate mathematical model of the dynamic system, (b) study of the consequences of the current and future (manipulated) input moves on the future plant (system's) behavior (keeping in mind the possible exceedance of the constraint in future), and (c) study of these consequences that can be predicted online and used while deciding the input moves in optimal manner. One needs a powerful computer to implement an MPC scheme, since, to facilitate computational tractability, online prediction is performed over a sliding window [6]. Much of the development of this chapter is based on [6,11].

7.1 Model-Based Prediction of Future Behavior

One of the major aspects of MPC formulation is the state estimator that is used to carry out online/real time prediction. Before that, we need to specify the dynamics of the system for which control law is to be determined:

$$x(k+1) = \varphi x(k) + Bu(k) + w(k)$$
$$z(k) = Hx(k) + v(k) \tag{7.1}$$

For the prediction of the current states, we need to use a Kalman predictor for the model (7.1)

$$\hat{x}(k/k-1) = \varphi \hat{x}(k-1/k-2) + Bu(k-1) + Le(k-1) \tag{7.2}$$

The corresponding innovations sequence is given as

$$e(k-1) = z(k-1) - H\hat{x}(k-1/k-2) \tag{7.3}$$

The predictor (7.2) is used at each sampling instant k for predicting future behavior of the system (7.1); its states for a finite future time horizon of length p (i.e. the prediction horizon) starting from the current time instant k for the chosen p future manipulated control input (MCI) values

$$\{u(k/k), u(k+1/k), \dots, u(k+p-1k)\} \tag{7.4}$$

In the absence of model plant/system mismatch (MPM; unmeasured drifting disturbance [UDD], Chapter 6), the innovations sequence $e(k)$ is a white noise sequence; hence, its expected values of the future innovations are zero: $E\{e(k+i)\} = 0$; $i = 1, 2, \dots$. In the absence of MPM/UDD, an observer (an estimator) can be used for forecasting the future behavior as

$$\hat{x}(k+1/k) = \varphi \hat{x}(k/k-1) + Bu(k/k) + Le(k) \tag{7.5}$$

$$\hat{x}(k+j+1/k) = \varphi \hat{x}(k+j/k) + Bu(k+j/k); \\ j = 1, \dots, p-1 \tag{7.6}$$

Such an ideal situation, i.e. the absence of MPM, generally does not exist in reality, and the innovations sequence would be a colored noise (with some dynamics of the noise process itself); the latter signifies existence of MPM/UDD. It then becomes necessary to compensate the future predictions for the MPM/UDD for which two approaches are available: (a) innovations bias approach and (b) state augmentation approach.

7.2 Innovations Bias Approach

One can use the filtered innovations for the unknown component and utilize it to correct the future predictions by employing unmeasured disturbance models (Chapter 6). At the beginning of the prediction, one uses the prediction from (7.2) as

$$\hat{x}(k) = \hat{x}(k / k - 1) \tag{7.7}$$

Then, the MPM that are compensated for the future predictions are generated as

1. At future time instant $(k + 1)$

$$\hat{x}(k+1) = \varphi \hat{x}(k) + Bu(k / k) + Le_f(k) \tag{7.8}$$

$$\hat{z}(k+1 / k) = H\hat{x}(k+1) + e_f(k) \tag{7.9}$$

2. At future time instant $(k + 2)$

$$\hat{x}(k+2) = \varphi \hat{x}(k+1) + Bu(k+1/k) + Le_f(k)$$

$$= \varphi^2 \hat{x}(k) + \varphi Bu(k / k) + \varphi Bu(k+1/k) \tag{7.10}$$

$$+ (\varphi + I)Le_f(k)$$

$$\hat{z}(k+2 / k) = H\hat{x}(k+2) + e_f(k) \tag{7.11}$$

3. For general future time instants $(k + j)$

$$\hat{x}(k+j) = \varphi \hat{x}(k+j-1) + Bu(k+j-1/k) + Le_f(k)$$

$$= \varphi^j \hat{x}(k) + \varphi^{j-1}Bu(k / k) + \varphi^{j-2}Bu(k+1/k) \tag{7.12}$$

$$+ ... + Bu(k+j-1) + (\varphi^{j-1} + \varphi^{j-2} + ... + I)Le_f(k)$$

$$\hat{z}(k+j / k) = H\hat{x}(k+j) + e_f(k) \tag{7.13}$$

Let us now examine the output prediction equation at the j-th time instant

$$\hat{z}(k+j / k) = H\varphi^j \hat{x}(k / k - 1)$$

$$+ \{H\varphi^{j-1}Bu(k / k) + H\varphi^{j-2}Bu(k+1/k) + ...$$

$$+ HBu(k+j-1/k)\} \tag{7.14}$$

$$+ [H(\varphi^{j-1} + \varphi^{j-2} + ... +)L + I]e_f(k)$$

The future prediction (7.14) has three aspects: (a) $H\varphi^j \hat{x}(k / k - 1)$ is the effect of the current state estimate over the (future) dynamics, (b) the term in $\{...\}$ is the quantification of the effects of the current and future inputs over the predictions, and (c) the term [...] is the effect of MPM over the predictions. It must be noted here, from (7.14), that at a given instant k the first and the third terms do not change. Thus, one can choose only the current and future MCI moves: $\{u(k / k), ..., u(k + p - 1 / k)\}$ to affect the future dynamic behavior of the considered system. Certain observations are in order here. For a stable open loop system, the predictions can be carried out using the open loop observer

$$\hat{x}(k / k - 1) = \varphi \hat{x}(k - 1 / k - 2) + Bu(k - 1) \tag{7.15}$$

Then, the model predictions are performed as

$$\hat{x}(k + j) = \varphi \hat{x}(k + j - 1) + Bu(k + j - 1 / k) \tag{7.16}$$

$$\hat{z}(k + j / k) = H\hat{x}(k + j) + e_f(k) \tag{7.17}$$

$$\hat{x}(k) = \hat{x}(k / k - 1) \tag{7.18}$$

For $j = 1,2,...,p$, $e_f(k)$ represents the residuals

$$e(k) = z(k) - H\hat{x}(k / k - 1) \tag{7.19}$$

The residuals in (7.18) are obtained by using the filter (Chapter 6, eqn. [6.275]), thereby the predictions are qualitatively similar to those by the DMC. In earlier times, the initial MPC formulations, DMC or MAC, utilized finite impulse response (FIR), or step response, models for modeling and predictions of the dynamic responses/systems. Interestingly, the predictions from using the state space model (7.2) are not different from those resulting from a FIR model, as seen from the following development. Let us start with the state space models

$$x(k + 1) = \varphi x(k) + Bu(k)$$
$$z(k) = Hx(k) \tag{7.20}$$

Let the input be given at $k = 0$ (with zero order hold) as an impulse function

$$u(0) = [1\ 1\ 1\ ...\ 1]^T; u(k) = 0; \text{ for } k > 0; x(0) = 0. \tag{7.21}$$

Then, the output of the system (7.20) is given as

$$z(k) = H_I(k)u(0); \quad H_I(k) = H\varphi^{k-1}B, \text{ for } k > 0 \tag{7.22}$$

In (7.22), $H_I(.)$, $k = 1,2,...$, are the coefficients of the impulse response (IR). Let $L = [0]$, then the output prediction (7.14) is given as

$$\hat{z}(k + j / k) = H\varphi^j \hat{x}(k / k - 1) + e_f(k)$$

$$+ \{H_I(j)u(k / k) + H_I(j - 1)u(k + 1 / k) + ... \tag{7.23}$$

$$+ H_I(1)u(k + j - 1 / k)\}$$

Thus, the prediction obtained by (7.22) is similar to the prediction equation used in some industrial MPC formulations.

7.3 State Augmentation Approach

Another approach to handle the MPM is to supplement the state space model with artificial input and/or output disturbance variables; the latter are treated as integrated white noise sequences, as given by equations (6.284) (Chapter 6). The extended model can be used for developing a Kalman (filter) predictor of the form

$$\hat{x}_a(k/k-1) = \varphi_a \hat{x}(k-1/k-2) + B_a u(k-1) + L_a e_a(k-1) \quad (7.24)$$

$$e_a(k) = z(k) - H \hat{x}_a(k/k-2) \quad (7.25)$$

In (7.25), $L(a,.)$ is the steady state Kalman gain and is obtained by solving the associated steady state Riccati equation. The observer (i.e. Kalman filter) accounts for MPM via explicit estimation of drifting disturbance terms, and it can be expected that the sequence $\{e(a,.)\}$ is zero mean white noise process. The optimal predictions of the states, based on the supplemented model, are now generated by

$$\hat{x}_a(k+1/k) = \varphi_a \hat{x}(k/k-1) + B_a u(k/k) + L_a e_a(k) \quad (7.26)$$

$$\hat{x}_a(k+j+1/k) = \varphi_a \hat{x}(k+j/k-1) + B_a u(k+j/k);$$
$$\text{for } j = 0,1,\dots,p-1 \quad (7.27)$$

$$\hat{z}(k+j/k) = H_a \hat{x}_a(k+j/k); \quad \text{for } j = 1,\dots,p \quad (7.28)$$

A modification to (7.26) through (7.28) can be made by restricting the future degrees of freedom to q future MCI values

$$\{u(k/k), u(k+1/k),\dots,u(k+q-1/k)\} \quad (7.29)$$

Then, one can impose the constraints on the remaining of the future MCI values

$$u(k+q/k) = u(k+q+1/k) = ,\dots, = u(k+p-1/k)$$
$$= u(k+q-1/k) \quad (7.30)$$

In (7.30), q is called the control horizon. In fact, the q degrees of freedom over the future are mapped over the horizon using the concept of input blocking:

$$u(k+j/k) = u(k/k); \quad \text{for } j = m_0 + 1,\dots,m_1 - 1$$

$$u(k+j/k) = u(k+m_1/k); \quad \text{for } j = m_1 + 1,\dots,m_2 - 1$$

$$\dots$$

$$u(k+j/k) = u(k+m_i/k); \quad \text{for } j = m_i + 1,\dots,m_{i+1} - 1 \quad (7.31)$$

$$\dots$$

$$u(k+j/k) = u(k+m_{q-1}/k); \quad \text{for } j = m_{q-1} + 1,\dots,m_q - 1$$

In (7.31), the constants $m(j)$ are selected such that

$$m_0 = 0 < m_1 < m_2 < \dots < m_{q-1} < m_q = p \quad (7.32)$$

7.4 Conventional Formulation of MPC

Sometimes a mathematical model used for developing an industrial MPC strategy is obtained from the operating data, and the states might not have any physical meaning; thus, MPC is formulated as an output control scheme. For this, only q future MCI values, equal to the control horizon, are to be specified/determined. Assume that, at the k-th instant, the sequence $\{z_r(k+j/k); j = 1,2,\dots,p\}$ denotes the future desired set point trajectory, then given the set point trajectory, the MPC at the same instant is defined as a constraint optimization problem. In this case, the future MCI values $u(k/k)$, $u(k+1/k)$,... ,$u(k+q-1/k)$ are determined by minimizing the objective function

$$J = \varepsilon(k+p/k)^T W_\infty \varepsilon(k+p/k)$$

$$+ \sum_{j=1}^{p} \varepsilon(k+j/k)^T W_e \varepsilon(k+j/k) \quad (7.33)$$

$$+ \sum_{j=0}^{q-1} \Delta u(k+j/k)^T W_{\Delta u} \Delta u(k+j/k)$$

In (7.33), we have the predicted error control as

$$\varepsilon(k+j/k) = z_r(k+j/k) - \hat{z}(k+j/k); \quad j = 1,2,\dots,p \quad (7.34)$$

The changes in the MCI values are given as

$$\Delta u(k+j/k) = u(k+j/k) - u(k+j-1/k); j = 1,\dots,q-1$$
$$\Delta u(k/k) = u(k/k) - u(k-1) \quad (7.35)$$

In (7.33), the various weights are: W_e is the positive definite error weighting matrix, $W_{\Delta u}$ is the positive semi-definite weight matrix, and W_∞ is the terminal weight matrix. The foregoing minimization problem is subject to the following constraints:

1. The MP equations defined in the previous section
2. MCI constraints

$$u(k+q/k) = u(k+q+1/k) = ,\dots, = u(k+p-1/k)$$
$$= u(k+q-1/k)$$
$$u^L \le u(k+j/k) \le u^H; \quad \Delta u^L \le \Delta u(k+j/k) \le \Delta u^H; \quad (7.36)$$
$$j = 0,1,2,\dots,q-1$$

3. Output inequality constraint

$$z^L \leq \hat{z}_c(k+j/k) \leq z^H; \quad j = p_1, p_1+1, ..., p \qquad (7.37)$$

In (7.37), p_1 is the constraint horizon. The resulting constraint optimization problem is solved by a standard nonlinear programming method, i.e. sequential quadratic programming (SQP), then the controller is implemented in a moving horizon setup. Then, after solving the optimization problem, the first move is implemented on the system/plant

$$u(k) = u_{opt}(k/k) \qquad (7.38)$$

Then, the optimization problem is reformulated at the next sampling instant based on the updated information from the system. A typical MPC scheme is shown in Figure 7.1. Figure 7.2 shows the moving window

formulation for the same [6]. For input blocking, the constraints are modified:

$$\Delta u(k+m_j/k) = u(k+m_j/k) - u(k+m_{j-1}/k); j = 1,2,...,q-1$$

$$\Delta u(k+m_0/k) = u(k/k) - u(k-1)$$

$$u^L \leq u(k+m_j/k) \leq u^H;$$

$$\Delta u^L \leq \Delta u(k+m_j/k) \leq \Delta u^H; j = 0,1,2,...,q-1$$

$$(7.39)$$

If the number of controlled outputs exceeds the number of MCI values, then one can specify set points only for the number of outputs equal to or less than the number of MCI values; then, for the remaining controlled outputs, one can define only domains in which these outputs should lie by imposing bounds on the predicted outputs—referred to as domain control variables in MPC. If the number of MCI values (m) exceeds the number of controlled outputs (MCO) (r), then it is possible to specify ($m-r$) inputs via optimization, leaving only m DoF to MPC.

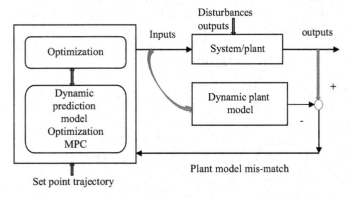

FIGURE 7.1
A model predictive scheme. (Modified from Patwardhan, S.C., A Gentle Introduction to Model Predictive Control (MPC) Formulations based on Discrete Linear State Space Models, https://nptel.ac.in/courses/103101003/downloads/lecture-notes/LQG_MPC_Notes.pdf, accessed December 2017.)

7.5 Tuning Parameters

The predictive control solutions would need the tuning of several parameters:

1. Prediction horizon (PZ p) and control horizon (CH q): The closed loop stability and the desired closed loop performance are obtained

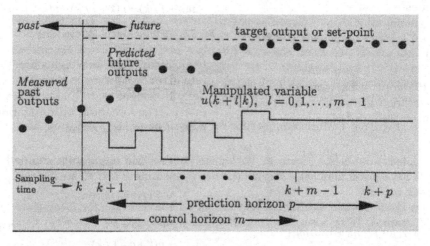

FIGURE 7.2
Moving window formulation. (Adapted from Patwardhan, S.C., A Gentle Introduction to Model Predictive Control (MPC) Formulations based on Discrete Linear State Space Models, https://nptel.ac.in/courses/103101003/downloads/lecture-notes/LQG_MPC_Notes.pdf, accessed December 2017.)

by adequately selecting three parameters, p and q; here, PZ is selected close to the open loop settling time, and the CH is selected significantly smaller (i.e. 1 to 5). In this moving horizon formulation (MHF), the constrained problem has to be solved online, in real time, at each sampling instant, and time required for solving the optimization problem is a concern in/for an MPC. However, the choice of relatively smaller CH would reduce the dimension of the optimization at each sampling instant and thereby the time for online computations.

2. Weighting matrices: The matrices W_e and $W_{\Delta u}$ can be diagonal and used to specify relative importance of elements of control error vectors and elements of MCI values.

3. Future set point trajectory: Along with the future output trajectory, at each instant, a filtered future set point trajectory is generated using a reference system:

$$x_r(k+j+1/k) = \varphi_r x_r(k+j/k) + [I - \varphi_r][r(k) - z(k)];$$

$$x_r(k/k) = 0.$$

$$z_r(k+j+1/k) = z(k) + x_r(k+j+1/k); j = 0, 1, ..., p-1$$

$$\varphi_r = diag\{\gamma_1 \ \gamma_2 ... \ \lambda_r\}; \ 0 \le \gamma_i < 1; i = 1, 2, ..., r$$

$$\tag{7.40}$$

In (7.40), γ_i are the tuning parameters, and $r(k)$ is the set point vector. The coefficient matrices of the reference system are regarded as the tuning parameters that are selected to achieve the desired closed loop performance (γ_i from 0.8 to 0.99). To ensure the free servo responses for step changes in the set point, these matrices should be selected so that the steady state (SS) gain of the reference system is identity:

$$H_{rf}(I - \varphi_r)^{-1} B_r = I \tag{7.41}$$

In most cases, the reference system is such that its transfer function (TF) matrix is diagonal with unit gain first (or higher) order low pass filters on the main diagonal.

4. Robustness against MPM: When the innovation bias formulation for generating the output predictions is used, then, the elements of matrix φ_e in (6.275) can be chosen to incorporate robustness against MPM (α_i is from 0.8 to 0.99). In the state augmentation formulation, the selection of covariance matrices in (6.285)

decides the quality of the estimates of the artificially incorporated disturbance variables and thereby influences the quality of the regulatory responses.

7.6 Unconstrained MPC

The connection between the conventional MPC and state feedback controller (SFC) can be seen by deriving unconstrained MPC control law using innovation bias formulation and assuming the CH being equal to the PH; in the next two sections we can show how to modify the formulation for $(q < p)$ and input blocking. Let the future input vector $U_p(k)$ and the predicted output vector $\hat{Z}(k)$ over the future horizon be given as

$$U_p(k) = [u^T(k/k) \ u^T(k+1/k) ... \ u^T(k+p-1/k)]^T$$

$$U_s(k) = [u_s^T(k) \ u_s^T(k) ... \ u_s^T(k)]^T \tag{7.42}$$

$$\hat{Z}(k) = [\hat{z}^T(k+1/k) \ \hat{z}^T(k+2/k) ... \ z^T(k+p/k)]^T$$

The prediction model is expressed as

$$\hat{z}(k+j/k) = H\varphi^j \hat{x}(k/k-1)$$
$$+ \{H\varphi^{j-1} bu(k/k) + H\varphi^{j-2} Bu(k+1/k) + ...$$
$$+ HBu(k+j-1/k)\}$$
$$+ [H(\varphi^{j-1} + \varphi^{j-2} + ... + I)L + I]e_f(k); \ j = 1, 2, ..., p$$
$$\tag{7.43}$$

This model can also be expressed as a single vector equation as

$$\hat{Z}(k) = S_x \hat{x}(k/k-1) + S_u U_p(k) + S_e e_f(k) \tag{7.44}$$

In (7.44), we have

$$S_x = \begin{bmatrix} H\varphi \\ H\varphi^2 \\ ... \\ H\varphi^p \end{bmatrix}; \ S_e = \begin{bmatrix} HL + I_r \\ H(\varphi + I_n)L + I_r \\ ... \\ H(\varphi^{p-1} + \varphi^{p-2} + ... + I_n)L + I_r \end{bmatrix} \tag{7.45}$$

$$S_u = \begin{bmatrix} HB & 0 & 0...0 \\ H\varphi B & HB & 0...0 \\ ... & ... & ......0 \\ H\varphi^{p-1}B & H\varphi^{p-2}B & ......HB \end{bmatrix} \tag{7.46}$$

$$= \begin{bmatrix} H(1) & 0 & 0...0 \\ H(2) & H(1) & 0...0 \\ ... & ... & ......0 \\ H(p) & H(p-1) & ......H(1) \end{bmatrix}$$

In (7.46), the $H(.)$ represents the impulse response coefficients, and $S(u)$ is the dynamic matrix of the system. To implement the input blocking constraints, $S(u)$ matrix is required to be altered for which let us consider the following matrix

$$I_{m_i} = \begin{bmatrix} I_m \\ I_m \\ ... \\ I_m \end{bmatrix}_{(m_i - m_{i-1} + 1)m \times m} \tag{7.47}$$

Define a matrix as follows

$$\psi_{pq} = \text{block diagonal} \begin{bmatrix} I_{m_1} & I_{m_2} & ... & I_{m_q} \end{bmatrix}_{(pm \times pm)} \tag{7.48}$$

Next, the vector for future q manipulated input moves is defined

$$U_q(k) = [u^T(k/k) \, u^T(k+m_1/k) \, ... \, u^T(k+m_{q-1}/k)]^T \tag{7.49}$$

Then, the vector $U(p)$ is given as (to verify)

$$U_p(k) = \psi_{pq}U_q(k) \tag{7.50}$$

$$U_p(k) = [u^T(k/k)...u^T(k/k)...u^T(k+m_{q-1}/k)...u^T \\ (k+m_{q-1}/k)]^T \tag{7.51}$$

Then, the vector form of the predictor (7.42) is given as

$$\hat{Z}(k) = S_x \hat{x}(k/k-1) + S_u U_q(k) + S_e e_f(k) \tag{7.52}$$

$$S_u = \begin{bmatrix} HB & 0 & 0...0 \\ H\varphi B & HB & 0...0 \\ ... & ... & ......0 \\ H\varphi^{p-1}B & H\varphi^{p-2}B & ......HB \end{bmatrix} \psi_{pq} \tag{7.53}$$

Now, let us define the future reference trajectory vector $Z(r,k)$ as

$$Z_r(k) = [z_r{}^T(k+1/k) \, z_r{}^T(k+2/k) \, ... \, z_r{}^T(k+p/k)]^T \tag{7.54}$$

Hence, the predicted control error at time k is given as

$$\varepsilon(k) = Z_r(k) - \hat{Z}(k) \tag{7.55}$$

The unconstraint MPC problem can now be represented as

$$\min_{U_q(k)} \{\varepsilon^T(k)W_E\varepsilon(k) + \Delta U_p^T(k)W_{\Delta U}\Delta U_p(k) + [U_q(k) \\ - U_s(k)]^T W_U [U_q(k) - U_s(k)]\} \tag{7.56}$$

The weighting matrices in (7.56) are defined as

$$W_E = \text{block diag}\{w_e \, w_e \, ... \, w_e\}_{(pn \times pn)}$$

$$W_{\Delta U} = \text{block diag}\{w_{\Delta u} \, w_{\Delta u} \, ... \, w_{\Delta u}\}_{(qm \times qm)} \tag{7.57}$$

$$W_U = \text{block diag}\{w_u \, w_u \, ... \, w_u\}_{(qm \times qm)}$$

The other variables are defined as

$$\Delta U_q(k) = \begin{bmatrix} u(k/k) - u(k-1) \\ u(k+m_1/k) - u(k/k) \\ ... \\ u(k+m_{q-1}/k) - u(k+m_{q-2}/k) \end{bmatrix} = \psi U_q(k) - \psi_0 u(k-1) \tag{7.58}$$

$$\psi = \begin{bmatrix} I_m & 0 & 0 & 0 \\ -I_m & I_m & 0 & 0 \\ ... & ... & ... & ... \\ 0 & ... & -I_m & I_m \end{bmatrix}_{(qm \times qm)} ; \psi_0 = \begin{bmatrix} I_m \\ 0 \\ ... \\ 0 \end{bmatrix}_{(qm \times m)} ;$$

$$I_m \text{ is } m \times m \text{ matrix} \tag{7.59}$$

The first term in the objective function (7.56) is as follows

$$\varepsilon^T(k)W_E\varepsilon(k) = [Z_r(k) - \hat{Z}(k)]^T W_E [Z_r(k) - \hat{Z}(k)] \tag{7.60}$$

Define the estimation error compensated set point trajectory as

$$\xi(k) = [Z_r(k) - S_x \hat{x}(k\,/\,k-1) - S_e e_f(k)] \quad (7.61)$$

Then, we have

$$Z_r(k) - \hat{Z}(k) = \xi(k) - S_u U_p(k)$$

$$\varepsilon^T(k) W_E \varepsilon(k) = \xi^T(k) W_E \xi(k) + U_q^T(k) S_u^T W_E U_q(k) \quad (7.62)$$

$$-2 U_q^T(k) S_u^T W_E \xi(k)$$

Similarly we have

$$\Delta U_q^T(k) W_{U\Delta} \Delta U_q(k) = [\psi U_p(k) - \psi_0 u(k-1)]^T W_{\Delta U}[\psi U_p(k)$$
$$-\psi_0 u(k-1)]$$
$$= U_q^T(k)[\psi^T W_{\Delta U} \psi] U_q(k)$$
$$-2 U_q^T(k)[\psi^T W_{\Delta U} \psi_0] u(k-1)$$
$$+u^T(k-1)[\psi_0^T W_{\Delta U} \psi_0] u(k-1)$$
$$\quad (7.63)$$

Also, we have

$$[U_q(k) - U_s(k)]^T W_U[U_q(k) - U_s(k)] =$$

$$U_q^T(k) W_U U_q(k) - 2 U_q^T(k) W_U U_s(k) \quad (7.64)$$

$$+U_s^T(k) W_U U_s(k)$$

Next, define the following matrix

$$\psi_u = \begin{bmatrix} I_m \\ I_m \\ \cdots \\ I_m \end{bmatrix}_{(qm\times m)} ; \quad U_s(k) = \psi_u u_s(k) \quad (7.65)$$

The term in (7.65) consists of $I(m)$ stacked q times, then (7.64) reduces to

$$[U_q(k) - U_s(k)]^T W_U[U_q(k) - U_s(k)] =$$

$$U_q^T(k) W_U U_q(k) - 2 U_q^T(k) W_U \psi_u u_s(k) \quad (7.66)$$

$$+u_s^T(k) \psi_u^T W_U \psi_u u_s(k)$$

Next, using the foregoing equations, the unconstraint optimization problem (7.56) can be recast in the form of the minimization of the quadratic objective function

$$\min_{U_q(k)} \left\{ \frac{1}{2} U_q^T(k) H_c U_q(k) + F_c^T(k) U_q(k) \right\} \quad (7.67)$$

In (7.67) we have

$$H_c = 2(S_u^T W_E S_u + \psi^T W_{\Delta U} \psi + W_U) \quad (7.68)$$

$$F_c(k) = -2[(S_u^T W_E)\xi(k) + (\psi^T W_{\Delta U} \psi_0) u(k-1) \quad (7.69)$$
$$+ W_U \psi_u u_s(k)]$$

The unconstraint problem of (7.67) can be solved to obtain the closed form of the control law, the least squares solution of which is given as

$$[U_q(k)]_{opt} = -H_c^{-1} F_c(k) \quad (7.70)$$

The optimal control law is given as

$$u_{opt}(k\,/\,k) = \psi_0^T [U_q(k)]_{opt} = -\psi_0^T H_c^{-1} F_c(k) \quad (7.71)$$

By doing some algebraic manipulations, the control law can be rearranged as

$$u_{opt}(k\,/\,k) = -K_x \hat{x}(k\,/\,k-1) + K_u u(k-1) + K_{us} u_s(k) \quad (7.72)$$
$$+ K_e e_f(k) + K_r Z_r(k)$$

In (7.72), K(x/u/us/e/r) are the feedback gains that can be obtained from (7.68), (7.69), and (7.71). It is obvious from (7.72) that the unconstraint MPC law can be viewed as a type of state feedback controller.

7.7 Quadratic Programming (QP) Formulation of MPC

The conventional MPC formulation of (7.56) with (7.52) is represented with constraints on the MCI values and predicted outputs as

$$Z^L \le \hat{Z}(k) \le Z^H ; \quad \psi_u u_L \le U_q(k) \le \psi_u u_H ; \quad (7.73)$$
$$\psi_u \Delta u_L \le \Delta U_q(k) \le \psi_u \Delta u_H$$

The formulation of (7.56), (7.52), and (7.73) has a quadratic objective function and linear constraints. This formulation can be transformed into an equivalent QP formulation, (7.67), and subject to

$$A U_q(k) \le B \quad (7.74)$$

with the following specifications

$$A = \begin{bmatrix} I_{qm} \\ -I_{qm} \\ \psi \\ -\psi \cdots \\ S_u \\ -S_u \end{bmatrix} ; \quad B = \begin{bmatrix} \psi_u u_H \\ -\psi_u u_L \\ \psi_u \Delta u_H + \psi_0 u(k-1) \\ -\psi_u \Delta u_L - \psi_0 u(k-1) \\ Z^H(k) - S_x \hat{x}(k\,/\,k-1) - S_e e_f(k) \\ -Z^L(k) + S_x \hat{x}(k\,/\,k-1) + S_e e_f(k) \end{bmatrix} \quad (7.75)$$

The variables in (7.73) through (7.75) have appropriate dimensions that follow from Section 7.5.

7.8 State-Space Formulation of the MPC

This is based on the input bias approach, and the state predictions are constructed as

$$\tilde{x}(k+j) = \varphi\tilde{x}(k+j-1) + Bu(k+j-1/k) + Le_f(k)$$
$$\tilde{x}(k) = \hat{x}(k/k-1); \quad j = 1,2,...,p \tag{7.76}$$

In (7.76), $\hat{x}(k/k-1)$ is generated using a suitable predictor/estimator. One needs to obtain the target states and the inputs: $x_s(k)$ and $u_s(k)$. Here, target state computation is carried out by solving a constrained minimization problem; if the number of controlled outputs is more or equal to the number of MCI values, the target state can be obtained by solving the optimization problem

$$\min_{u_s(k)}\{(r(k)-z_s(k))^T W_e(r(k)-z_s(k))\} \tag{7.77}$$

The conditions to be met for (7.77) are

$$[I-\varphi]x_s(k) = Bu_s(k) + Le_f(k) \tag{7.78}$$

$$z_s(k) = Hx_s(k) + e_f(k); \quad u_L \le u_s(k) \le u_H \tag{7.79}$$

In case the number of the MCI values is more than the number of the controlled outputs, then the time varying target state can be computed by

$$\min_{u_s(k)} u_s^T(k)W_U u_s(k) \tag{7.80}$$

The conditions to be met for (7.80) are

$$[I-\varphi]x_s(k) = Bu_s(k) + Le_f(k)$$
$$r(k) = Hx_s(k) + e_f(k); \quad u_L \le u_s(k) \le u_H \tag{7.81}$$

Now, given the target states, the MPC problem at the sampling instant k is defined as a constraint optimization problem, where by the future MCI values $u(k+1/k),...$ can be determined by minimizing the objective function J

$$J = [\tilde{x}(k+p)-x_s(k)]^T W_\infty[\tilde{x}(k+p)-x_s(k)]$$
$$+ \sum_{j=1}^{p-1}[\tilde{x}(k+j)-x_s(k)]^T W_x[\tilde{x}(k+j)-x_s(k)] \tag{7.82}$$
$$+ \sum_{j=0}^{p-1}[u(k+j/k)-u_s(k)]^T W_u[u(k+j/k)-u_s(k)]$$

In (7.82), $W(..)$ are the respective positive definite weighting matrices. The terminal state weighting matrix in the first term of (7.82) can be determined by solving discrete Lyapunov equation (when the poles of the transition matrix are within the unit circle):

$$W_\infty = W_x + \varphi^T W_\infty\varphi \tag{7.83}$$

Next, the optimization problem has the following constraints

1. The model prediction equations
2. MCI values' constraints

$$u(k+q/k) = u(k+q+1/k) = ... = u(k+p-1/k)$$
$$= u(k+q-1/k) \tag{7.84}$$

$$u^L \le u(k+j/k) \le u^H; \tag{7.85}$$
$$\Delta u_L \le \Delta u(k+j/k) \le \Delta u_H; \quad j=0,1,2,...,q-1$$

3. For the linearized version of a mechanistic model (for the MPC), the constraints on the predicted states are also used

$$x_L \le \tilde{x}(k+j) \le x_H; \quad j=p_1,p_1+1,...,p;$$
$$p_1 \text{ is the constraint horizon.} \tag{7.86}$$

7.9 Stability

The MPC scheme is based on minimization of a performance measure, and the resulting formulation should guarantee stable closed loop behavior. We examine nominal stability of a deterministic version of MPC with the assumptions:

1. No model plant mismatch, or the unmeasured disturbances are absent, and the internal model (observer) and the plant/system evolve according to

$$x(k+1) = \varphi x(k) + Bu(k) \tag{7.87}$$

2. The true states are accurately measurable, and it is desired to control the dynamic system at the origin.

The MPC scheme of Section 7.7 can now be examined as the minimization of the following cost function

$$J(x(k), U_p(k)) = x^T(k+p)W_\infty x(k+p)$$

$$+ \sum_{j=1}^{p-1} x^T(k+j)W_x x(k+j) + \sum_{j=0}^{p-1} u^T(k+j/k)W_u u(k+j/k)$$

$$(7.88)$$

In (7.88), the MPC is represented as the minimization of a quadratic loss function for the sake of simplicity and to facilitate connection with LQOC. However, the MPC objective function can also be reformulated using any suitable positive function of the vectors involved; i.e., formulate the MPC in terms of a generalized loss function as

$$J(x(k), U_p(k)) = \Phi\{x(k+p)\} + \sum_{j=1}^{p} \Psi\{x(k+j), u(k+j/k)\}$$

$$(7.89)$$

The functions in (7.89) are positive functions of their arguments; hence $J(.,.) \to \infty$, as $\|x(k)\| \to \infty$. Thus, the cost function $J(.,.)$ is itself a Lyapunov function. Here, it is desired to minimize $J(x(.), U(.,.))$ subject to the condition that $U_p(k) \in S_U$, S_U is the close and bounded set. Also, we have the constraint on the moving horizon optimization problem: $x(k+p) = 0$. Now, let the optimal solution to the optimization problem at time $t = k$ be given as

$$\hat{U}_p(k) = [\hat{u}^T(k/k)\, \hat{u}^T(k+1/k)\, ... \, \hat{u}^T(k+p-1/k)]^T \quad (7.90)$$

In the MPC scheme, we set, $u(k) = \hat{u}(k/k)$ and the problem is solved over the window $\{k+1, k+p+1\}$; $\hat{u}(k/k)$ is a function of $x(k)$, and is written as $\hat{u}(k/k) = F(x(k))$. Then, we have

$$x(k+1) = \varphi x(k) + B\hat{u}(k/k) = \varphi x(k) + BF(x(k)) \quad (7.91)$$

In (7.91), $x(k+p) = 0$ is the terminal state from the use of $\hat{U}_p(k)$ on the system of (7.87). We have

$$\hat{U}_p(k+1) = [\hat{u}^T(k+1/k+1)\, \hat{u}^T(k+p-1/k+1)\, ...$$
$$\hat{u}^T(k+p/k+1)]^T \quad (7.92)$$

In (7.92), the optimal solution of the MPC is represented over the window $\{k+1, k+p+1\}$, then one would like to examine the differential

$$\Delta J\{x, U_p\} = J\{x(k+1), \hat{U}_p(k+1)\} - J\{x(k), \hat{U}_p(k)\} \quad (7.93)$$

A non-optimal, yet, feasible solution to the problem over the chosen window is obtained as

$$U_p(k+1) = [\hat{u}^T(k+1/k)\, \hat{u}^T(k+2/k)\, ...$$
$$\hat{u}^T(k+p-1/k)\, 0^T]^T \quad (7.94)$$

The following inequality holds for the feasible solution of (7.94)

$$J\{x(k+1), \hat{U}_p(k+1)\} \le J\{x(k+1), U_p(k+1)\} \quad (7.95)$$

From (7.93) and (7.95), the following is obtained

$$J\{x(k+1), \hat{U}_p(k+1)\} - J\{x(k), \hat{U}_p(k)\}$$
$$\le J\{x(k+1), U_p(k+1)\} - J\{x(k), \hat{U}_p(k)\} \quad (7.96)$$

With $x(k+p) = 0$ by assumption, then with the application of the last element of $U_p(k+1)$, i.e. 0, on (7.87), results in $x(k+p+1) = 0$; hence, $\Psi(x(k+p+1), 0) = 0$. Now, since $\hat{U}_p(k)$ and $U_p(k+1)$ share $\{p-1\}$ sub-vectors, we obtain the following

$$J\{x(k+1), U_p(k+1)\} - J\{x(k), \hat{U}_p(k)\}$$

$$= \sum_{j=2}^{p} \Psi\{x(k+j), \hat{u}(k+j/k)\} - \sum_{j=1}^{p} \Psi\{x(k+j), \hat{u}(k+j/k)\}$$

$$(7.97)$$

$$= -\Psi\{x(k), F(x(k))\} < 0; \text{ if } x(k) \ne 0 \quad (7.98)$$

Thus, we have the following final condition for the stability of the controller based on MPC scheme

$$\Delta J\{x, U_p\} = J\{x(k+1), \hat{U}_p(k+1)\} - J\{x(k), \hat{U}_p(k)\} \le$$
$$-\Psi\{x(k), F(x(k))\} < 0; \text{ if } x(k) \ne 0 \quad (7.99)$$

The inequality in (7.99) states that the nominal closed loop system with the controller designed by the MPC is globally asymptotically stable. It is also interesting to note that the MPC cost function can be used to construct a Lyapunov function. Additionally, under the nominal conditions, the MPC controller guarantees the optimal performance as well as guarantees the global asymptotic stability.

Appendix 7A: Certain Basic Aspects of MPC and Illustrative Examples

In fact, the MPC is a regulatory control that uses an explicit dynamic model of the responses of the process (i.e. system/plant) variables (to the changes in the manipulated variables, MCI, i.e.

manipulated control inputs) to compute the control "actions," also known as "moves" [12]. Thus, in MPC the control actions are intended to guide, or even "force" these process variables to follow a pre-specified trajectory to the target from the current operating point. So, the basic control action is on the current observations and the future predictions; it means that the future values of the output variables are predicted using a dynamic model (of the system or the process/plant that is to be controlled) and the current measurements. In the MPC, the predictions are made for more than one-time steps ahead, unlike the time (step-) delay compensation methods. The previously known dynamic matrix control (DMC) is now the MPC. The optimal controller is based on minimizing error from the set point, and the basic version uses linear models, of course several possible system's models are in use. The MCI variables $u(.)$ are computed so that they minimize an objective function J. Here, this J is a criterion of the sum of the squares of the errors between predicted future outputs and specified reference trajectory. As we have seen in Chapter 7, the corrections for unmeasured disturbances, and model errors (that are also called deterministic discrepancies) are handled as required; and the inequality/equality constraints and the measured disturbances are incorporated in the computations of the control law. The MPC can work in single step/multi-step versions, and also, for the multivariable and feedforward controls. The predictive control (PC) is used in the circumstances when: (a) processes are difficult to control with standard PID strategies, (b) there is a lot of dynamic interaction amongst controls-more than one manipulated (control) variable (MCVs) will have a significant effect on an important process variable, and (c) there are constraints on process and manipulated variables for normal control behavior. The computed manipulated variables are realized as set point for lower level control loops. Thus, overall objectives are: (i) curtail the violations of I/O constraints, (ii) drive some output variables to their optimal set points, at the same time keeping other outputs within given ranges, (iii) prevent excessive excursions of the input, especially control variables, i.e. the control constraints, and (iv) if a sensor or actuator is not available, control as much of the process as feasible. The MPC evolved as the techniques developed by industry in the form of: (a) dynamic matrix control (DMC) and (b) model algorithmic control (MAC). There have been numerous applications of MPC since 1980.

7A.1 Computations Involved in MPC

Step-1: At the k-th sampling instant, the values of the MCVs, u, at the next M sampling instants, $\{u(k), u(k+1),\ldots, u(k+M-1)\}$ are computed; see Figure 7A.1. The set of control actions is computed so as to minimize the predicted deviations from the reference trajectory over the next p (P) sampling instants (P is the prediction horizon) at the same time satisfying the specified

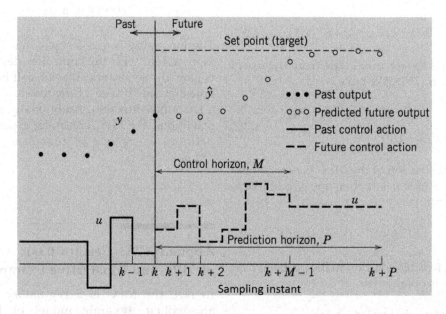

FIGURE 7A.1
The concept of MPC. (Adapted From Anon. Model predictive control-1, PPT slides;*utw10249.utweb.utexas.edu/edgar_group/che391/.../ChE%20391%20MPC(1).ppt, accessed December 2017.*)

constraints. This linear quadratic programming problem is solved at each sampling instant.

Step-2: Then the first control action $u(k)$, is implemented.

Step-3: At the next sampling instant $k+1$, the M-step control policy (M is the control horizon) is re-computed for the next m sampling instants $k+1$ to $k+m$, and the first control action $u(k+1)$ is implemented.

Step-4: Then Steps 1 and 2 are repeated for subsequent sampling instants.

7A.2 Dynamic Models Used for MPC

These could be: (i) physical or empirical, and (ii) linear or nonlinear. The linear models used are step response, TFs, and state space models. In the case of DTS (discrete time system), the model is finite impulse response (FIR) type, which is also the convolution between the impulse functions of the system, and the control inputs, and is represented as

$$y(k) = \sum_{i=1}^{N} h_{k-i} u(k-i) \quad (7A.1)$$

The variables in (7A.1) have usual meanings and dimensions (Chapter 9). The incremental change in $u(k)$ is given as $\Delta u(k) = u(k) - u(k-1)$; the response $y(.)$ to a unit step change in u at $t = 0$, $\Delta u(0) = 1$, is depicted in Figure 7A.2. For Figure 7A.2, we have the following notations and descriptions:

$$\{S_i\} \triangleq \text{step resposne coefficients}$$
$$\{h_i\} \triangleq \text{impulse resposne coefficients} \quad (7A.2)$$

Also, we have

$$h_i = S_i - S_{i-1}$$
$$y(1) = y(0) + S_1 \Delta u(0)$$
$$y(2) = y(0) + S_2 \Delta u(0); \{\Delta u(0)=1 \text{ for unit step change at } t=0\}$$
$$\vdots$$
$$y(n) = y(0) + S_n \Delta u(0)$$

$$(7A.3)$$

If the step change Δu_1 occurs at $t = \Delta t$, then we have the following description

$$y(2) = y(0) + S_1 \Delta u(1)$$
$$y(3) = y(0) + S_2 \Delta u(1)$$
$$\vdots \quad\quad (7A.4)$$
$$y(n) = y(0) + S_{n-1} \Delta u(1)$$

From Figure 7A.2, we have the following description, if the step changes in $u(.)$ occur at both $t = 0$, $\Delta u(0)$ and $t = \Delta t$, Δu_1 (by using the principle of superposition for linear systems):

$$y(1) = y(0) + S_1 \Delta u(0)$$
$$y(2) = y(0) + S_2 \Delta u(0) + S_1 \Delta u(1)$$
$$y(3) = y(0) + S_3 \Delta u(0) + S_2 \Delta u(1) \quad (7A.5)$$
$$\vdots$$
$$y(n) = y(0) + S_n \Delta u(0) + S_{n-1} \Delta u(1)$$

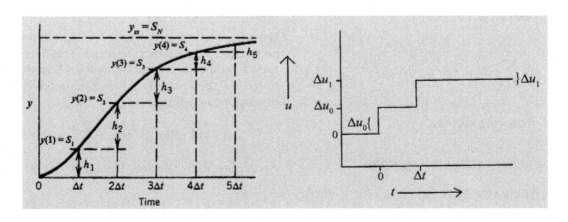

FIGURE 7A.2
A typical unit step response (left graph) and the unit steps (in incremental/s, right graph). (Adopted from Anon. Model predictive control-1, PPT slides; *utw10249.utweb.utexas. Edu/edgar_group/che391/.../ChE%20391%20MPC(1).ppt, accessed December 2017.*)

7A.3 Predictions for MPC

For the step response models, the output is given as

$$y(k+1) = y(0) + \sum_{i=1}^{N-1} S_i \Delta u(k-i+1) + S_N \Delta u(k-N+1); \tag{7A.6}$$

N = the model horizon

The one-step-ahead prediction for the SISO model can be obtained from (7A.6) by assuming $y(0) = 0$:

$$\hat{y}(k+1) = \sum_{i=1}^{N-1} S_i \Delta u(k-i+1) + S_N \Delta u(k-N+1) \tag{7A.7}$$

By expanding (7A.7), one gets the individual contributions

$$\hat{y}(k+1) = \underbrace{S_1 \Delta u(k)}_{\substack{\text{effect of current} \\ \text{control action}}} + \underbrace{\sum_{i=2}^{N-1} S_i \Delta u(k-i+1) + S_N \Delta u(k-N+1)}_{\text{effect of past control actions}}$$

$$\tag{7A.8}$$

In the similar way, the j-th step ahead prediction is given as

$$\hat{y}(k+j) = \underbrace{\sum_{i=1}^{j} S_i \Delta u(k+j-i)}_{\substack{\text{effects of current and} \\ \text{future control actions}}} + \underbrace{\sum_{i=j+1}^{N-1} S_i \Delta u(k+j-i) + S_N \Delta u(k+j-N)}_{\text{effect of past control actions}}$$

$$\tag{7A.9}$$

Now, we define the predicted (unforced) response as

$$y_0(k+j) \triangleq \sum_{i=j+1}^{N-1} S_i \Delta u(k+j-i) + S_N \Delta u(k+j-N) \tag{7A.10}$$

Using (7A.10), one can write (7A.9) as

$$\hat{y}(k+j) = \sum_{i=1}^{j} S_i \Delta u(k+j-i) + \hat{y}_0(k+j) \tag{7A.11}$$

We define the following vectors:

$$\hat{Y}(k+1) \triangleq col\{\hat{y}(k+1)\,\hat{y}(k+2)...\hat{y}(k+P)\}$$

$$\hat{Y}_0(k+1) \triangleq col\{\hat{y}_0(k+1)\,\hat{y}_0(k+2)...\hat{y}_0(k+P)\} \tag{7A.12}$$

$$\Delta U(k) \triangleq col\{\Delta u(k)\,\Delta u(k+1)...\Delta u(k+M-1)\}$$

Using (7A.12), (7A.11) can be written as the dynamic matrix model:

$$\hat{Y}(k+1) = S\Delta U(k) + \hat{Y}_0(k+1) \tag{7A.13}$$

$$S \triangleq \begin{bmatrix} S_1 & 0 & ... & 0 \\ S_2 & S_1 & 0 & : \\ : & : & ... & 0 \\ S_M & S_{M-1} & ... & S_1 \\ S_{M+1} & S_M & ... & S_2 \\ : & : & ... & : \\ S_P & S_{P-1} & ... & S_{P-M+1} \end{bmatrix}_{P \times M} \tag{7A.14}$$

The bias correction can be done for unmeasured disturbance as

$$\hat{y}(k+j) \triangleq \hat{y}(k+j) + [y(k) - \hat{y}(k)] \tag{7A.15}$$

Using (7A.15), one obtains the following

$$\hat{Y}(k+1) = S\Delta U(k) + \hat{Y}_0(k+1) + [y(k) - \hat{y}(k)] \tag{7A.16}$$

The column vector is given as:

$$\hat{Y}(k+1) \triangleq col\{\hat{y}(k+1)\,\hat{y}(k+2)...\hat{y}(k+P)\} \tag{7A.17}$$

7A.4 MPC Control

If a reference trajectory is used, the desired set point can be reached in a gradual manner. This reference trajectory Y_r is specified in many ways. Over the prediction horizon P the vector Y_r (mP vector, and m is the number of outputs) is defines as

$$Y_r(k+1) \triangleq col\{y_r(k+1)\,y_r(k+2)...y_r(k+P)\} \tag{7A.18}$$

An exponential trajectory from $y(k)$ to $y_{sp}(k)$ is given as

$$y_{i,r}(k+j) = (a_i)^j y_i(k) + [1-(a_i)^j]y_{i,sp}(k); \\ i=1, 2, ..., m; j = 1, 2, ..., P. \tag{7A.19}$$

The control computation is based on the minimization of the predicted deviations from the reference trajectory. The prediction error is given as, using the corrected prediction, (7A.17):

$$\hat{E}(k+1) \triangleq Y_r(k+1) - \hat{Y}(k+1) \quad (7A.20)$$

The prediction unforced error is defined as

$$\hat{E}_0(k+1) \triangleq Y_r(k+1) - \hat{Y}_0(k+1) \quad (7A.21)$$

The aim of the control computations is to determine/obtain the control policy/law for the next M time intervals $\Delta U(k)$ (rM dimensional vector) as specified in (7A.12), the third expression, which is determined to minimize: (i) prediction errors over the prediction horizon, P, and (ii) the size of the control action over the control horizon M. The performance index is specified as

$$\min_{\Delta U(k)} J = \hat{E}^T(k+1)R\hat{E}(k+1) + \Delta U^T(k)Q\Delta U(k) \quad (7A.22)$$

In (7A.22), Q is a positive semi-definite matrix, and R is a positive definite matrix (here, Q and R are like information matrices, and in fact, inverses of the respective matrices that occur in the theory of KFs. Ironically, the symbols used are the same!) and can be considered as the respective weighing matrices or the Gramians, and often are chosen as the diagonal matrices with the elements as positive constants, and are chosen to weigh the most important I/Os, and these elements can be considered as the tuning parameters for obtaining the optimal control law. The MPC law for the unconstrained case that minimizes the objective function, (7A.22) can be determined analytically (with S defined as in (7A.14) as:

$$\Delta U(k) = (S^T RS + Q)^{-1} S^T R\hat{E}_0(k+1);$$
$$\Delta U(k) = K\hat{E}_0(k+1); K = (S^T RS + Q)^{-1} S^T R \quad (7A.23)$$

In (7A.23), K is the $(rM \times mP)$ controller gain matrix that can be computed off-line, if the dynamic matrix S, and the weighting matrices, Q, and R are considered to be constant, and if these are time dependent, the gain has to be computed online. In the receding horizon control problem, only the first step of the M-step control law is implemented as

$$\Delta u(k) = K_1 \hat{E}_0(k+1); K_1 = \dim(r \times mP) \quad (7A.24)$$

Thus, the control gain is specified as the first r rows of the gain matric K. The design of MPC law requires to select a number of design parameters: (i) N, model horizon; (ii) Δt, sampling period; (iii) P, prediction horizon (number of predictions); (iv) M, control horizon (number of control actions); (v) R, the weighting matrix for predicted errors $(R > 0)$; and (vi) Q, the weighting matrix for control moves $(Q \geq 0)$. The N and Δt are selected so that $N \Delta t \geq$ open loop settling time, typical values of N being $30 \leq N \leq 120$. For the prediction horizon, P, increasing its value results in less aggressive control action, so set $P = N + M$. Increasing M, the control horizon makes the controller more aggressive and increases computational effort, typically $5 \leq M \leq 20$. The weighting matrices R and Q are chosen to be the diagonal matrices with largest elements corresponding to most important variables.

For predictions used in MPC, some transformations might be required and are briefly given here [11].

Example 7A.1: From FIR to the state space model. A FIR model is given as

$$y(k) = h_1 u(k-1) + h_2 u(k-2) + h_3 u(k-3) \quad (7A.25)$$

The model in (7A.25) is written as a state space model:

$$\begin{bmatrix} x_1(k+1) \\ x_2(k+2) \\ x_3(k+3) \end{bmatrix} = \begin{bmatrix} 0 & 1 & 0 \\ 0 & 0 & 1 \\ 0 & 0 & 0 \end{bmatrix} \begin{bmatrix} x_1(k) \\ x_2(k) \\ x_3(k) \end{bmatrix} + \begin{bmatrix} h_1 \\ h_2 \\ h_3 \end{bmatrix} u(k) \quad (7A.26)$$

$$y(k) = \begin{bmatrix} 1 & 0 & 0 \end{bmatrix} \begin{bmatrix} x_1(k) \\ x_2(k) \\ x_3(k) \end{bmatrix} \quad (7A.27)$$

Example 7A.2: Transformation of a polynomial model (PM) to state space model (SSM). The PM is given as

$$y(k) = a_1 y(k-1) + b_0 u(k-1) + b_1 u(k-2); a_1 = 0.8,$$
$$b_0 = 0.4, b_1 = 0.6 \quad (7A.28)$$

With the prediction horizon chosen as $P = 3$; we obtain the following equations from (7A.28):

$$y(k+1) = y(k+1)$$
$$y(k+2) = y(k+2) \quad (7A.29)$$
$$y(k+3) = a_1 y(k+2) + b_0 u(k+2) + b_1 u(k+1)$$

One can express (7A.29) in a compact form of a SSM

$$\begin{bmatrix} y(k+1) \\ y(k+2) \\ y(k+3) \end{bmatrix} = \begin{bmatrix} 0 & 1 & 0 \\ 0 & 0 & 1 \\ 0 & 0 & a_1 \end{bmatrix} \begin{bmatrix} y(k) \\ y(k+1) \\ y(k+2) \end{bmatrix} + \begin{bmatrix} 0 & 0 & 0 \\ 0 & 0 & 0 \\ 0 & b_1 & b_0 \end{bmatrix} \begin{bmatrix} u(k) \\ u(k+1) \\ u(k+2) \end{bmatrix} \quad (7A.30)$$

The prediction model is given as

$$\begin{bmatrix} y(k+1) \\ y(k+2) \\ y(k+3) \end{bmatrix} = Y(k) + \begin{bmatrix} 0.4 & 0 & 0 \\ 1.32 & 0.4 & 0 \\ 2.056 & 1.32 & 0.4 \end{bmatrix} \begin{bmatrix} u(k) - u(k-1) \\ u(k+1) - u(k) \\ u(k+2) - u(k+1) \end{bmatrix} \quad (7A.31)$$

$$Y(k) = \begin{bmatrix} y(k-2) \\ y(k-1) \\ y(k) \end{bmatrix} + \begin{bmatrix} 0 & 1 & 1.8 \\ 0 & 0 & 2.44 \\ 0 & 0 & 1.952 \end{bmatrix} \begin{bmatrix} y(k-2) - y(k-3) \\ y(k-1) - y(k-2) \\ y(k) - y(k-1) \end{bmatrix}$$

$$+ \begin{bmatrix} 0 & 0.6 \\ 0 & 1.08 \\ 0 & 1.464 \end{bmatrix} \begin{bmatrix} u(k-2) - u(k-3) \\ u(k-1) - u(k-2) \end{bmatrix} \quad (7A.32)$$

Example 7A.3: The Kuhn-Tucker conditions for QP problem. Let the QP be given as to minimize the following cost function

$$J(x) = x^T P x + 2f^T x; \; P \text{ is a weighting matrix} \quad (7A.33)$$

The constrains are given as

$$A_1 x - b_1 = 0; \; A_2 x - b_2 \le 0 \quad (7A.34)$$

The Lagrange function (not the LM, but, the total amended cost function $J_a(x)$) is given as

$$L_a(x) = x^T P x + 2f^T x + \lambda^T (A_1 x - b_1) + \mu^T (A_2 x - b_2)$$

The first order Kuhn-Tucker conditions for the minimum are given as

$$\frac{\partial L_a(x)}{\partial x} = 2Px + 2f + A_1^T \lambda + A_2^T \mu = 0$$

$$\frac{\partial L_a(x)}{\partial \lambda} = A_1 x - b_1 = 0 \quad (7A.35)$$

$$\mu^T \frac{\partial L_a(x)}{\partial \lambda} = \mu^T (A_2 x - b_2) = 0$$

Let the cost function be given as

$$J = \frac{1}{2}(u-1)^2 = \frac{1}{2}u^2 - u + \frac{1}{2}; \; u \le \frac{3}{4} \quad (7A.36)$$

This QP has the following values:
$J_0 = 1/2; \; P = 1/2; \; f = -1/2; \; A = 1; \text{ and } b = 3/4$. We can easily see from (7A.36) that the minimum of the unconstrained problem is $u = 1$. Since the constraint on u is given, the minimization of the augmented cost function with the LM would give $u^* = 3/4$. Formally we can proceed as follows:

$$L_a(u, \lambda) = \frac{1}{2}u^2 - u + \lambda \left(u - \frac{3}{4} \right) \quad (7A.37)$$

Next, using the first order necessary conditions for minimization, we get

$$\frac{\partial L_a(u, \lambda)}{\partial u} = u - 1 + \lambda = 0$$

$$\frac{\partial L_a(u, \lambda)}{\partial \lambda} = u - 3/4 = 0 \quad (7A.38)$$

Solving, (7A.39), we obtain, $u = 3/4$; and $\lambda = 1/4$.

Appendix 7B: MPC Illustrative Examples

Some MATLAB-based examples on model predictive control [A7B.1,A7B.2] are given here. Where feasible the results and the plots for this Appendix are generated by using the MATLAB scripts assembled from [A7B.1,A7B.2], (and re-run by the first author of the current volume).

Example 7B.1:

To simulate a MPC under a mismatch between the predictive plant model and the actual plant. The predictive plant model has 2 manipulated variables (MV), 2 unmeasured input disturbances (UD), and 2 measured outputs (MO) [A7B.1]. The actual plant has different dynamics. The results can be obtained by running the MATLAB script **mpcpmm.m**. This defines the parameters of the nominal plant/system on which the MPC is based, and the systems from MV to MO and UD to MO are identical. The inputs 1 and 2 are the manipulated variables (MV), and 3 and 4 are the unmeasured disturbance-variables (UD). The parameters of the actual/real plant in CL with the MPC controller are defined, and the CL response is simulated for which the reference trajectories and unmeasured disturbances entering the plant are defined. Then MPC simulation object is created. The CL MPC simulation with MM and unforeseen/unmeasured disturbance inputs is run. The results are presented in Figures 7B.1 and 7B.2 from which it is seen that the CL tracking performance is satisfactory in the presence of unmeasured disturbances.

Example 7B.2:

The idea is to create and test a model predictive controller. Create a state space model of the system and set some of the optional model properties [A7B.2]. The results can be obtained by running the MATLAB script **mpcontroller.m**, and are shown in Figures 7B.3 through 7B.9.

This program creates an MPC with a control interval or sampling interval of 1 sec., and can display the controller properties in the command window. Also, one can view and modify the controller properties, display a list of the controller properties and their current values; these might be different for your controller, since it depends on when the controller was created. For the editable properties of an MPC controller, use "mpcprops;" use "."/dot notation to modify these properties. Many of the controller properties are structures containing additional fields; can use "."/dot notation to view/modify these field values, e.g. one can set the measurement units for the controller output variables; the "OutputUnit" property is for

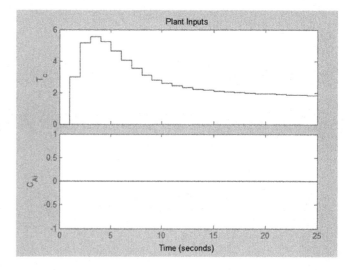

FIGURE 7B.3
Plant inputs (Example 7B.2).

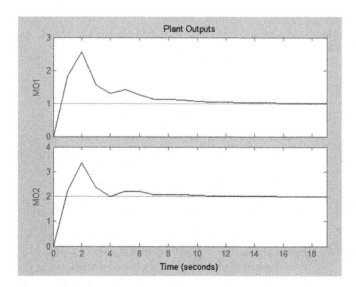

FIGURE 7B.1
Plant inputs for an MPC controller (Example 7B.1).

FIGURE 7B.2
Plant outputs for an MPC controller (Example 7B.2).

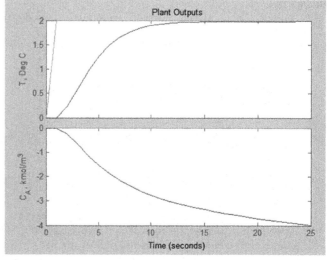

FIGURE 7B.4
Plant outputs (Example 7B.2).

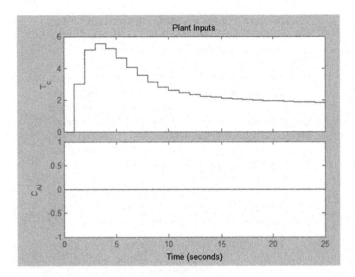

FIGURE 7B.5
Plant inputs (Example 7B.2).

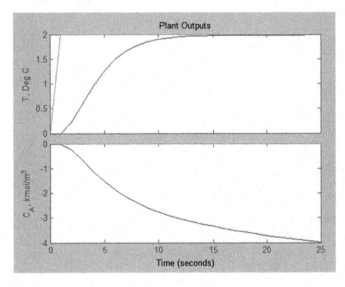

FIGURE 7B.6
Plant outputs (manipulated variable constraints turned-off) (Example 7B.2).

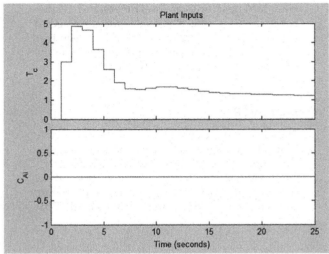

FIGURE 7B.7
Plant inputs (Example 7B.2).

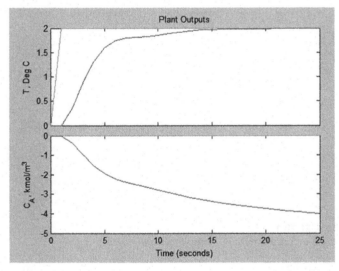

FIGURE 7B.8
Plant outputs (plant/model mismatch; 50% larger gains than in the model used) (Example 7B.2).

display purposes only and is optional. By default, the controller has no constraints on manipulated variables and output variables; and one can view and modify these constraints using "dot" notation; e.g. set constraints for the controller manipulated variable. Also, one can view/modify the controller tuning weights; e.g. modify the weights for the manipulated variable rate and the output variables. You can also define time-varying constraints and weights over the prediction horizon, which shifts at each time step. Time-varying constraints have a nonlinear effect when they are active; e.g. to force the manipulated variable to change more slowly towards the end of the prediction horizon, Generate a report on potential run-time stability and performance issues. Use the "sim" function to run a linear simulation of

the system; simulate the CL response of "MPCobj" for 26 control intervals. Specify "setpoints" of 2 and 0 for the reactor temperature and the residual concentration respectively. The "setpoint" for the residual concentration is ignored because the tuning weight for the second output is zero. You can modify the simulation options using "mpc-simopt;" run a simulation with the manipulated variable constraints turned off; the first move of the manipulated variable now exceeds the specified 3-unit rate constraint, see Figure 7B.6. You can also perform a simulation with a plant/model mismatch; define a plant with 50% larger gains than those in the model used by the controller; the plant/model mismatch degrades controller performance slightly, see Figure 7B.8. Degradation can be severe and should be tested on a case-by-case

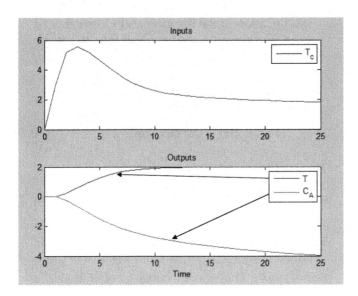

FIGURE 7B.9

Plant I/Os (manipulated variables and both output variables in the same plot) (Example 7B.2).

basis. Other options are: (i) the addition of a specified noise sequence to the manipulated variables or measured outputs, (ii) open loop simulations, and (iii) a look-ahead option for better "setpoint" tracking or measured disturbance rejection. Store the simulation results in the MATLAB Workspace. This syntax suppresses automatic plotting and returns the simulation results. You can use the results for other tasks, including custom plotting; plot the manipulated variable and both output variables in the same plot, see Figure 7B.9. Restore the setting: mpcverbosity(old _ status).

Example 7B.3:

Nonlinear MPC for a boiler unit. In some cases, a fault tolerance control (FTC) system may be more effective for providing the back-up than hardware redundancy. The cost of production losses and unnecessary energy consumption due to a fault can be large [A7B.3]. An objective of FTC system is to maintain the performance of the system closer to the desirable one, and to preserve the stability in the presence of faults. In an active method for FTC, a fault diagnosis unit is necessary to provide the information about the fault, its localization and size. It requires a reconfigurable controller and the mechanism for the same. The MPC is an active method for FTC, and the representation of fault and control objectives is relatively simple, and current control signal is obtained online, by solving a finite horizon optimal control problem. The actual control is derived by taking into considerations the internal changes and interactions in the controlled process. MPC is a specific control strategy that uses a math model of the plant to derive the control signal, $u(t)$ by minimizing a cost function over a finite

receding horizon. The math model predicts future plant outputs (based on past/current outputs) as well as future control signals, which are computed by minimizing the cost function taking into account constraints. For MPC design, two elements are required: (a) a math model of the plant working in normal conditions and (b) the robust optimization procedure. For the FTC system, the information provided by the FDI block (fault detection and identification; localization and size of faults) is used during the optimization in order to redefine the constraints imposed on the state and inputs.

Optimization criterion: In a GPC (generalized predictive control), at each iteration a cost function is minimized:

$$J = \sum_{i=N_1}^{N_2} (r(k+i) - \hat{y}(k+i))^2 + \lambda \sum_{i=1}^{N_u} (\Delta u(k+i-1))^2;$$
$$u(k) = [u(k), ..., u(k+N_u-1)]^T \tag{7B.1}$$

The constraints are

$$\Delta u(k+i) = 0, \ N_u \le i \le N_2 - 1 \tag{7B.2}$$

In (7B.1), we have: $r(.)$ as the future reference signal, $\hat{y}(.)$ as the prediction of future outputs, $\Delta u(k+i-1) = u(k+i-1) - u(k+i-2)$, N_1 the minimum prediction horizon, N_2 the prediction horizon, N_u the control horizon, and λ is a tuning/weighing factor to penalize the changes in $u(.)$. In case of a nonlinear system, the optimization problem has to be solved at each sample time to obtain a sequence of future controls $u(.)$, wherefore the first element is taken to control the plant. The distinct features of MPC are: (i) the receding horizon; the control signal is derived to achieve the desired behaviour, in the subsequent N_2 time steps; and (ii) the control horizon $N_u < N_2$; only the first N_u future controls are determined; and from that point the control is assumed to be constant. To reduce the computational burden, N_u is kept very small; say to 1 or 2.

Neural modeling: In the cost function, $\hat{y}(k+i)$ is the minimum variance i-step ahead prediction that can be obtained by: (i) instantaneous linearization of a nonlinear model of the plant, here, a unique solution of (7B.1) exists and the future control signals are calculated directly; however, the instantaneous linearization has a limited validity in certain regimes of the operating range; and (ii) successive recursion of a one-step ahead nonlinear model, here, the control system uses a nonlinear model of the plant, and the minimization of (7B.1) has to be carried out by an iterative procedure. This kind of control system is called Nonlinear Predictive Control (NPC). The one-step ahead prediction is obtained as:

$$\hat{y}(k+1) = f(y(k), ..., y(k-n+1), u(k), ..., u(k-m+1)) \quad (7B.3)$$

In (7B.3), n and m are the numbers denoting past outputs and inputs. The i-step ahead prediction is obtained as;

$$\hat{y}(k+i) = f(y(k+i-1), ..., y(k+i-n), u(k+i-1), ..., u(k+i-m))$$
$$(7B.4)$$

It is assumed that the observations of $y(.)$ are available up to k. Hence, for the future outputs only the predicted values can be used:

$$y(k+i) = \hat{y}(k+i), \ \forall i > 1 \quad (7B.5)$$

A nonlinear function f can be realized by an ANN, since it would provide a very good math model. It has an important property due to the use of a sigmodial activation function, and any continuous nonlinear relation can be approximated with arbitrary accuracy. In order to design the i-step ahead predictor, one needs to feed the network with past predicted outputs (7B.4 and 7B.5); hence, the recurrent neural network (RNN) can be used. However, in the case when $i > n$, the neural network structure is in between the FFNN and RNN, as it uses both measured past outputs (up to time k) and predicted outputs (for time instants $k + 1$, ..., $k + i$).

Updating of the control law: When a nonlinear FFNN/RNN is used as a plant model, output predictions are nonlinear in the control inputs, and we have a complex nonlinear programming problem. This has to be solved in real-time, and the optimization procedure should assure fast convergence/numerical robustness. The control law update at each sample time i is given as:

$$u(i+1) = u(i) + \eta(i)h(i) \quad (7B.6)$$

In (7B.6), $u(i)$ is the current iterate of the future control inputs, $\eta(i)$ is the step size, and $h(i)$ is the search direction. A second optimization algorithm seems to be a reasonable choice, since other algorithms are too slow, or very complex. The Levenberg-Marquardt (LM) algorithm does not guarantee rapid convergence. Hence, a

modification of the Newton and LM method is used. The search direction $h(i)$ at the i-th sample time is derived as:

$$h(i) = -\tilde{H}^{-1}g(i); \ \bar{H} = H(i) + \mu(i)I; H(i) = \frac{\partial J^2}{\partial u^2}|_{u=u(i)};$$

$$H \text{ is the Hessian.} \quad (7B.7)$$

The gradient is calculated as

$$g(i) = \frac{\partial J}{\partial u}|_{u=u(i)} \quad (7B.8)$$

The parameter $\mu(i)$ in (7B.8) should be selected/adjusted so that the Hessian is positive definite.

Boiler unit: The system consists of a boiler, a storage tank, control valve with a positioner, a pump, and transducers to measure process variables. The boiler introduces strong nonlinearity into the static characteristic of the system. The specification of process variables is shown in Table A7B.1.

The aim of the control system is to keep the required level of the water in the boiler, and three reference signals were used: (i) the constant value $r(k) = 0.25$; (ii) the sinusoidal signal given as

$$r(k) = \begin{cases} 0, & for\ k < 120 \\ \frac{1}{20}\sin(\frac{2\pi k}{600}) + \frac{1}{4}, & otherwise \end{cases}; \quad (7B.9)$$

and (iii) random steps with levels from the interval (0, 0.5); each step lasts for 240 sec. The boiler unit-control system was implemented in MATLAB/Simulink [A7B.3], with $T = 0.05$. The specifications of faults are given in Table A7B.2, and the faults in different parts of the installation are considered: sensor, actuator and component faults, and the scenarios being additive as well multiplicative faults.

TABLE A7B.1

Specifications of Plant Variables

Variable	Specification	Range
CV	Control value	0%–100%
dP	Pressure difference on valve V_1	0–275 kPa
P	Pressure before valve V_1	0–500 kPa
F_1	Flow (electromagnetic flowmeter)	0–5 m³/h
F_2	Flow (vortex flowmeter)	0–5 m³/h
L	Water level in boiler	0–0.5 m

TABLE A7B.2

Specifications of Faults

Fault	Description	Type
f_1	Leakage from the boiler	Additive (−0.05)
f_2	Outflow checking	Partly closed (50%)
f_3	Loss of internal pipe diameter	Partly closed (50%)
f_4	Leakage form pipe	Additive (−1)
f_5	Level transducer failure	Additive (−0.05)
f_6	Positioner fault	Multiplicative (0.7)
f_7	Valve head or servo-motor fault	Multiplicative (0.8)
f_8	Pump productivity reduction	Multiplicative (0.8)

TABLE A7B.3

Fault-Tolerance Results

Fault	$\|e_l(k)\|^2$	SEE	Δe
no	0.038	0.0014	–
f_1	0.038	0.0014	0
f_2	0.0733	0.0054	0.0352
f_3	12.112	146.691	12.074
f_4	0.0843	0.007	0.046
f_5	2.2123	4.894	2.174
f_6	0.069	0.0048	0.031
f_7	0.057	0.0053	0.019
f_8	0.0395	0.0016	0.0014

Experimental details:

Training data: The data were collected in the OL control: (i) random steps with levels from the interval (0,100), each step lasts 240 sec., (ii) pseudorandom binary signal (PRBS) simulated as a sum of five sinusoidal signals with large amplitude (0, 100), and (iii) the sum of sinusoidal signals of 0.00066, 0.0008, 0.0016, 0.002, and 0.005 Hz, with the amplitude of 0.05 for each signal, with the mean value of 0.25 [A7B.3]. After the spectrum analysis of these signals and with respect to (wrt) the boiler dynamics, the random step sequence of 50,000 samples was selected for training the chosen RNN.

Plant modeling: Two NN models were tried: NNARX (Neural Network Auto Regressive with eXogenous input), and the NNOE (Neural Networks Output Error) model. The model I/O were the control value (*CV*) and the level in the boiler (*L*). Both models had the same number of hidden neurons (5) and the same number of I/O delays ($m = 2$, $n = 2$). However, the final model of the system was designed by means of NNOE. The final settings are: one input, one output, five hidden neurons with hyperbolic tangent activation function, one linear output neuron, the number of past inputs $m = 2$ and the number of past outputs $n = 2$. The modeling was carried out in OL control in normal operating conditions, with the random step function as input, containing 50,000 samples, and the training was carried out off-line for 100 steps with the LM algorithm. The quality of the obtained model was checked using a testing set containing 300,000 samples. The model output duplicated the behaviour of the plant pretty well [A7B.3]. Also, the one step ahead prediction of the neural model fed with the sum of sinusoids, showed that the model followed the plant output almost immediately, proving good generalization of the model.

Control: For the design of MPC for the boiler unit using a neural model, the values of control parameters chosen are: (i) the prediction horizon N_1 for the controlled plant is the one-step ahead predictor; hence, it was taken as 1; (ii) the prediction horizon N_2 of the plant was taken as 15; (iii) the control horizon N_u was set to 2; (iv) the penalty factor in the cost function was taken as 0.0003; (v) the maximum number of iterations taken as 5; and (vi) constraints: the control signal should have a value from the interval [0, 100]. When the output of the plant reached the reference after some decreasing oscillations, the control became a sinusoidal-type signal, the optimization procedure ensured the signal close to the sinusoidal, and the tracking the reference was carried out with a high quality. Also, every time the reference changed, the controller adapted the control signal to the changes. In general, the quality of the proposed MPC was much better than in the case of the classical PID design [A7B.3].

Fault tolerance: This was investigated introducing several faults as in Table A7B.2 and observing the control quality; each fault was inducted at the 3000th second as a permanent fault. Fault tolerant control results are given in Table A7B.4.

In Table A7B.3, we have: $\Delta e = \left\|e_l^F(k)\right\|^2 - \left\|e_l^N(k)\right\|^2$; F denotes a faulty condition and N the normal condition for the tracking error, which is defined as $e_l(k) = L(k) - r(k)$. As seen from this table, all faults were properly compensated except the fault f_3. In most cases, the norm of the tracking error increases, a bit

TABLE A7B.4

Performance of FTC for Fault Detection and Compensation

Fault	Reference Signal: (7B.9)		Reference Signal: $r(k) = 0.25$	
	t_d in sec.	t_c in sec.	t_d in sec.	t_c in sec.
f_1	ND	–	ND	–
f_2	5.9	0	1.15	0
f_3	19.45	NC	3.55	NC
f_4	12.35	0	9.1	0
f_5	0.05	154.75	0.05	64.55
f_6	69.65	0	7.7	0
f_7	42.15	0	5.7	0
f_8	33.85	0	7.7	0

compared to the normal operation conditions; this indicates that the MPC works pretty well and the fault effect is not observable. The highest difference between norms of the tracking error is for the fault f_5, this is a sensor fault of an additive type. It is seems that the fault f_1 had a negligible effect on the control system; the leakage from the boiler was small.

Fault compensation: The tolerance region and interval to evaluate the fault compensation was specified as $[0.95r(k), 1.05r(k)]$. The fault compensation time was calculated as a difference between the time in which the system response is inside the interval and the time of fault occurrence. The fault compensation times are given in the third column of Table A7B.4 for the reference signal (7B.9), and the fifth column for the $r(k) = 0.25$; t_c is fault compensation time in sec., and NC (not compensated) denotes that a fault was not compensated.

Fault detection: The automatic control system can hide faults from being observed, Table A7B.4; $t_c = 0$; hence, to diagnose a system condition, a fault detection and isolation algorithm is required. The already designed one-step ahead predictor can be used to construct model-based fault detection system; the

difference between the system output and the one-step ahead predictor gives a residual signal. Then, to carry out fault detection, a decision making technique based on constant thresholds can be used: if the residual is smaller than the threshold, the plant is normal, otherwise it is faulty. Two thresholds can be used to detect both increasing and decreasing trends in the residual: $T_u = t_a v + m$, $T_l = t_a v - m$; where T_u and T_l denote the upper and lower threshold values; t_a is $N(0, 1)$ (Gaussian pdf) tabulated value assigned to 1-α confidence interval, v is standard deviation of the residual and m is mean value of the residual. The results of fault detection, t_d are presented in the second and fourth columns of Table A7B.4; Not Detected (ND) denotes that a fault was not detected by the fault diagnosis block. The detection time is a period of time needed for the detection of a fault measured from the fault start-up time to a permanent, true decision about the fault. A fault is declared if the residual permanently exceeds the uncertainty region.

References

A7B.1 Anon. Simulating model predictive controller with plant model mismatch. https://in. mathworks.com/help/mpc/examples/simulating-model-predictive-controller-with-plant-model-mismatch.html, accessed August 2018.

A7B.2 Anon. Design MPC controller at the command line. http://in.mathworks.com/help/mpc/gs/design-controller-using-the-command-line.html, accessed August 2018.

A7B.3 Patan, K., and Korbicz, J. Nonlinear model predictive control of a boiler unit: A fault tolerant control study. *Int. J. Appl. Math. Comput. Sci.*, AMCS-Vol. 22, No. 1, pp. 225–237, 2012; http://matwbn.icm.edu.pl/ksiazki/amc/amc22/amc22117.pdf, accessed December 2018.

8

Robust Control

As we have seen in Section I of this book, a PI (Proportional-Integral) controller provides an example of how feedback is applied to reduce the uncertainty of low frequency disturbances; thus, a PI controller can be adjusted, or tuned, to eliminate steady-state (SS) offsets for unknown load changes using a very simple model of the plant dynamics. A zero SS offset is obtained even though the magnitude of the constant disturbance might be completely unknown. Also, feedback provides the only way to stabilize unstable plant dynamics. It might be true that, if one has a perfect mathematical model (MM) of the plant and selects the input in such a way that the states/outputs of the model remain bounded, then the same input might be thought to stabilize the real plant. However, if the plant is (even slightly) unstable, then very small discrepancies between plant and the mathematical model and in the initial state will cause the states/outputs of the MM and plant to diverge, and the stabilization of the system without some feedback would not be possible. As we have seen in Chapters 5 through 7, the disturbance attenuation and related problems are dealt with by optimal control methods, where some measure (loss, cost, or payoff functional) of the magnitude of the output (state or output errors) is minimized subject to appropriate assumptions on the disturbances. A linear quadratic Gaussian (LQG) control problem is formulated, where the sum of the output variances (energies) is minimized with the assumption that the disturbances are characterized as stochastic processes. The theory of LQG control was developed during the 1960s. The cost function used in LQG theory is a particular norm, i.e. the H_2 norm of the closed loop transfer function (TF). The LQG control problem, therefore, minimizes this H_2 norm; hence, this is also known as H_2 optimal control. Also, the H_2 norm is a measure of the mean of the square of the gain taken over all frequencies (see Appendix IIB), [13]. As we have also seen in Section I, robustness (then understood in the sense of the assurance of the performance of the closed loop control system against some unknown disturbance) against model uncertainties is classically handled by phase and gain margins to ensure closed loop stability despite the presence of modeling errors. The classical methods, however, cannot be easily generalized to multivariable plants and they do not handle the problem of simultaneously achieving good performance (against disturbances) and robustness (against model uncertainties).

The modern approach to design controllers that are robust against model uncertainties is provided by the H-Infinity (HI) control theory (based on H_∞ norm) and was developed largely during the 1980s [13]. A good and realistic controller, actually the system with controller, should have both robustness against model uncertainty, characterized by an HI norm bound, and good disturbance rejection properties, characterized by an H_2- or HI-norm. Often, a controller design based on optimal (LQG or H_2) and robust HI control is considered as a standard procedure for control systems when a high demand is imposed on control quality. Some fairly good software is available in MATLAB toolboxes: basic control system analysis and design, signal processing, system identification, optimization, and mu-synthesis. It is natural to work in continuous time domain, since the controlled processes mostly operate in continuous time; however, for most applications, the controllers would be normally implemented digitally and would involve a discretization of the continuous time controller. A practical way would be to apply discrete time or sampled data control theory directly to arrive at the control laws. The standard controller design is restricted to linear or weakly nonlinear processes, and the design of optimal and robust controllers for nonlinear systems is relatively a complicated problem. Some simpler suboptimal methods can be used to modify the linear optimal and robust control methods to nonlinear dynamic plant, which is described as a linear parameter varying (LPV) system, and an optimal robust controller is determined using the estimated maximum rate of change of the system's parameters. This would lead to robust gain-scheduled controllers. Interestingly enough, very popular control methods in the industry—fuzzy control and neural control—do not explicitly address many fundamental control issues in a quantitative way. However, these methods would often work in simple applications and appeal to practicing engineers with a non-control background. Also, the functions of these method can easily be explained to plant/process operators and other users. Yet, the ability to analyze a particular method quantitatively cannot be considered a prerequisite for its successful application. We present the approach of the intelligent control, based on the soft computing paradigms: artificial neural networks (ANNs), fuzzy logic (FL), and genetic algorithms (GAs) in Section IV of this book.

8.1 Robust Control of Uncertain Plants

Let us consider a dynamic system with G denoting the plant described by

$$z = Gu; \quad z(t) = Gu(t) \tag{8.1}$$

In (8.1), $u, u(t)$ is a manipulated/control input (MCI), and $z, z(t)$ is the controlled output variable, see Figure 8.1; $y(t)$ is the measured output, $v(t)$ is the disturbance signal, G is the plant dynamics, and K is the (feedback) controller, traditionally $H(s)$ is used. The operator type (in the form of the symbol) mapping G represents a dynamic system, which relates the input signal $u(t)$ and the output signal $z(t)$, and is described by a differential equation for continuous time systems. The purpose is to manipulate input $u(t)$ so that the output $z(t)$ follows, or is at least close to, a given reference signal $r(t)$. This is represented as

$$Gu = r \tag{8.2}$$

In (8.2), $r(t)$ is the desired reference for $z(t)$. One could always solve (8.2) approximately, if G is known, by inverse of G, or if the inverse does not exist, by the pseudoinverse. However, such a simplistic procedure generally does not work; the main reason is that there are certain uncertainties present in reality: (a) unknown disturbances (UD) that affect the process dynamics, $G(s)$ and/or (b) incomplete knowledge about the plant dynamics; the model uncertainties could be present, $G(s)$ is known only approximately. The uncertainties can be compensated by using the observations y from the process and making the MCI $u(t)$ a function of $y(t)$. This is the basic concept of the feedback control, as we have seen in Section I of this book: the feedback is a means to reduce the effect of uncertainty on the plant's performance; the use of feedback makes it possible to manage with a not-so-accurate process/plant model. Because of this simple property of feedback, control engineers work with simple and often linear models, even though the models might be originally complex. For developing the robust approach for control, it is assumed that the plant dynamics are described as

$$G = G_0 + \Delta G; \quad \text{(often the last } G \text{ is omitted)} \tag{8.3}$$

In (8.3), the actual plant (model) is considered as the sum of the nominal plant (dynamics) and the model of the uncertainty: the nominal plant is often linear dynamics, and the plant uncertainty, unknown but assumed to belong to a kind of uncertainty set, and the latter might be nonlinear. Thus, the normal control system is called robust if it is stable despite the presence of uncertainty in the plant dynamics. Also, the robustness applies to the achievable performance despite the presence of the plant uncertainty. So, in general, a control system should be robustly stable and should have robust performance, i.e. the error performance of the control system should be acceptable despite the presence of the plant uncertainty within a certain specified domain. Interestingly, the condition that the closed loop control system (CLCS), with robustness taken into account, is stable for all possible uncertainties in the specified domain, is equivalent to a bound on the HI norm of the closed loop transfer function (CLTF).

8.1.1 Robust Stability and HI Norm

For a stable and scalar TF $G(s)$, the HI norm is defined as

$$\|G\|_\infty = \max_\omega |G(j\omega)| \tag{8.4}$$

In (8.4), the HI norm is the maximum, the largest gain of the system taken over all the admissible frequencies. The HI norm can be related to the robust stability. For uncertain plant, we have

$$z = T_{zw}w; \quad T_{zw} = (I - KG_0)^{-1}K \tag{8.5}$$

We know from the conventional control theory (Section I of this book) that the closed loop feedback control is stable if

$$|\Delta(j\omega)T_{z\omega}(j\omega)| < 1, \text{ for all } \omega \tag{8.6}$$

Also, equivalently

$$|T_{z\omega}(j\omega)| < 1/l(\omega), \text{ for all } \omega \tag{8.7}$$

The eqn. (8.7) implies that the closed loop system is stable for all the uncertainties that satisfy the bound

$$|\Delta(j\omega)| \le l(\omega), \text{ for all } \omega \tag{8.8}$$

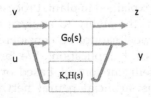

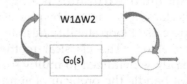

FIGURE 8.1
Basic feedback control system and the uncertain plant.

It can be shown that the condition (8.7) is also necessary for robust stability. Thus, if (8.7) does not hold, then there exists an uncertainty that satisfies (8.8) and makes the closed loop unstable. The connection to HI norm is obtained by observing that the condition (8.7) is equivalent to

$$\left\| IT_{zw} \right\|_{\infty} < 1 \qquad (8.9)$$

Thus, the CLCS with model uncertainty is robustly stable with respect to the uncertainty class specified by (8.8), if, and only if, the CLTF satisfies the HI norm bound of (8.9), compare this with (8.4) qualitatively.

8.1.2 Disturbance Rejection and Loop-Shaping Using HI Control

Although the robustness of the designed control system to model-uncertainties is very important drive for the HI control approach/action (HICA), the latter can also be applied to the disturbance rejection problem—the cost function takes the form of a worst-case cost. The HICA is also suited for loop shaping; in several control problems, the overall closed loop behavior can be specified by requiring that various CLTFs satisfy given bounds. For example, one could require that the TF, $T_{zv}(s)$ from the disturbance v to the output z for the system in Figure 8.1, satisfies the frequency domain bound

$$\left| T_{zv}(j\omega) \right| < \frac{1}{\left| W(j\omega) \right|}, \ \forall \omega \qquad (8.10)$$

The bound (8.10) can be restated in terms of the HI norm

$$\left\| W(j\omega)T_{zv}(j\omega) \right\|_{\infty} < 1; \qquad (8.11)$$

The loop-shaping problems with the design specifications (8.10) can be solved efficiently using HICA; however, what makes it better suited for the robust stability problem and for loop-shaping than the H_2 (LQ) control problem, is the fact that the HI norm bound (8.11) guarantees (8.10) for all frequencies. Ironically, the H_2 norm only gives an average measure taken over all frequencies, and therefore, it does not guarantee that the frequency domain bound would hold at all frequencies. As a precursor to the HI control, the H_2 control is considered.

8.2 H_2 Optimal Control

In this section, we consider the problem of optimal control, which is also conventionally known as H_2 optimal control problem/approach/action (H_2CA).

The setting is as per Figure 8.1, and the partition of G is introduced as

$$\begin{bmatrix} z \\ y \end{bmatrix} = \begin{bmatrix} G_{11} & G_{12} \\ G_{21} & G_{22} \end{bmatrix} \begin{bmatrix} v \\ u \end{bmatrix} \qquad (8.12)$$

The CLCS/CLTF is given as

$$Z = F(G, K)v \qquad (8.13)$$

$$F(G, K) = G_{11} + G_{12}(I - KG_{22})^{-1}KG_{21} \qquad (8.14)$$

The H_2 optimal control (H_2OC/H_2CA) problem is to determine a casual controller K that would stabilize the plant dynamics G and minimize the cost function

$$J(K) = \left\| F(G, K) \right\|_2^2 \qquad (8.15)$$

In (8.15), the $\left\| F(G, K) \right\|_2$ is the H_2 norm (Appendix B). The H_2OC is very conveniently solved in the time domain. We consider that the TF, G can be given by the following state-space representation

$$\dot{x}(t) = Ax(t) + B_1 v(t) + B_2 u(t)$$

$$z(t) = C_1 x(t) + D_{12} u(t) \qquad (8.16)$$

$$y(t) = C_2 x(t) + D_{21} v(t)$$

In the control literature, the state-space representation of $G(s)$ is written in equivalent forms as

$$G(s) = C(sI - A)^{-1}B + D$$

$$G(s) = \begin{bmatrix} A & B \\ C & D \end{bmatrix} = \begin{bmatrix} A & B_1 & B_2 \\ \hline C_1 & D_{11} & D_{12} \\ C_2 & D_{21} & D_{22} \end{bmatrix} \qquad (8.17)$$

In (8.17), the diagonal elements of the matrix D are zeros. The direct feedback from v to z in (8.16) is zero to obtain the finite H_2 norm for the closed loop system. Also, the direct feedback from u to y is zero, since the physical systems have a zero gain at the infinite frequency. Certain assumptions are made for the theory of the H_2OC/ H_2CA (might be applicable to HICA also) studied in this chapter [13]:

a1) The pair (A, B_2) is stabilizable; a2) $D_{12}^T D_{12}$ is invertibe, a3) $D_{12}^T C_1 = 0$, a4) The pair (C_1, A) has no unobservable modes on the imaginary axis; and b1) The pair (C_2, A) is detectable, b2) $D_{21}^T D_{21}$ is invertibe, b3) $D_{21} B_1^T = 0$, and b4). The pair (A, B_1) has no uncontrollable modes on the imaginary axis.

The first four assumptions are related to the state feedback control problem, and the next four assumptions are related to the state estimation problem. The assumptions a1) and b1) are required for a stabilizing controller $u = Ky$ to exist. The assumptions a2), a3), b2), and b3) may be relaxed (not very restrictive) and would help obtain convenient simplification of the solution. The assumptions a4) and b4) are required for the (various) Riccati equations that characterize the optimal controller to have stabilizing solutions. These assumptions can be relaxed, but then the solution of the optimal control (OC) problem should be characterized in terms of the linear matrix inequalities (LMIs). Now, let the time domain cost function of (8.15) be given as

$$J(K) = \sum_{k=1}^{m} \left[\int_0^\infty z^T(t)z(t)dt : v = e_k\delta(t) \right] \quad (8.18)$$

In the time domain, the H_2 norm is interpreted by observing that a function with constant Laplace Transform is the Dirac's δ function with the following property

$$\delta(t) = \begin{cases} \infty \ if\ t = 0 \\ 0 \ if\ t \neq 0 \end{cases}; \quad \int_{-\infty}^\infty \delta(t)dt = 1 \quad (8.19)$$

Alternatively, the linear quadratic optimal control (LQOC) cost function is defined as

$$J_{LQ}(K) = \sum_{k=1}^{m} \left[\int_0^\infty [x^T(t)Rx(t) + u^T(t)Qu(t)]dt : v = e_k\delta(t) \right] \quad (8.20)$$

In fact, R can be easily replaced by P, and Q by R to match the notations in some literature on estimation, or filtering; however, we have in (8.20) R as a symmetric positive definite matrix and Q is a symmetric positive semidefinite matrix, and these can be considered as appropriately chosen weighting matrices. The cost functions (8.18) and (8.20) are equivalent. The H_2OC is solved in two stages: (a) an optimal state feedback law is constructed, and (b) an optimal estimator is determined.

8.2.1 The Optimal State Feedback Problem

Let us consider the dynamics in (8.16) and that the assumptions a1) to a4) hold. Also, the control signal $u(t)$ can have the present and past values of the state, $x(\tau), \tau \leq t$. Then, the cost function (8.18) is minimized by the state feedback controller

$$u(t) = K_{opt}x(t); \quad K_{opt} = -(D_{12}^T D_{12})^{-1}B_2^T P \quad (8.21)$$

In (8.21), P is the unique symmetric (semi) positive definite solution to the algebraic Riccati equation

$$A^T P + PA - PB_2(D_{12}^T D_{12})^{-1}B_2^T P + C_1^T C_1 = 0 \quad (8.22)$$

For (8.22), the following matrix has all its eigenvalues with negative real parts (i.e. this matrix is stable)

$$A + B_2 K_{opt} \quad (8.23)$$

Also, the minimum value attained by the cost function (8.20), with the control law specified in (8.21), is given by

$$\min_{K_x} J(K_x) = tr\{B_1^T P B_1\} \quad (8.24)$$

Often in the literature, matrix S is used instead of P, and subsequent with some special conditions, the matrix S is replaced by P; here, we continue to use matrix P throughout. Now, consider the integral in (8.18), which can be partitioned as

$$\int_0^\infty z^T(t)z(t)dt = \int_0^{0^+} z^T(t)z(t)dt + \int_{0^+}^\infty z^T(t)z(t)dt \quad (8.25)$$

Also, with $x(0) = 0$ and $v(t) = e_k\delta(t)$, one can have the following

$$x(0^+) = \int_0^{0^+} e^{A(0^+ - \tau)}[B_1 e_k \delta(\tau) + B_2 u(\tau)]d\tau = B_1 e_k \quad (8.26)$$

The first integral in (8.25) is zero. The second integral can be expanded as

$$\int_{0^+}^\infty z^T(t)z(t)dt = \int_{0^+}^\infty [C_1 x(t) + D_{12}u(t)]^T[C_1 x(t) + D_{12}u(t)]dt \quad (8.27)$$

$$= \int_{0^+}^\infty \left\{ [C_1 x(t) + D_{12}u(t)]^T[C_1 x(t) + D_{12}u(t)] + \frac{dx^T(t)Px(t)}{dt} \right\}dt - x^T(\infty)Px(\infty) + x^T(0^+)Px(0^+) \quad (8.28)$$

$$= \int_{0^+}^\infty [u(t) - u^0(t)]^T D_{12}^T D_{12}[u(t) - u^0(t)]dt + x^T(0^+)Px(0^+) \quad (8.29)$$

The last three terms in (8.28), in fact, cancel out, but these are introduced to bring in the effect of matrix P. In (8.29), we have $x(0^+)$ denoting the state immediately after the impulse input at time $t = 0$.

$$u^0(t) = K_{opt}x(t) \quad (8.30)$$

It is also assumed that $x(\infty) = 0$ for the sake of stability. Now, the cost function can be expressed as

$$
\begin{aligned}
J(K) &= \sum_{k=1}^{m} \left[\int_{0}^{\infty} z^{T}(t) z(t) dt : v = e_k \delta(t) \right] \\
&= \sum_{k=1}^{m} \left[\int_{0}^{\infty} [u(t) - u^{0}(t)]^{T} D_{12}^{T} D_{12} [u(t) \right. \\
&\qquad \left. - u^{0}(t)] dt + x^{T}(0^{+}) P x(0^{+}) : v = e_k \delta(t) \right] \\
&= \sum_{k=1}^{m} \left[\int_{0}^{\infty} [u(t) - u^{0}(t)]^{T} D_{12}^{T} D_{12} [u(t) - u^{0}(t)] dt \right. \\
&\qquad \left. + e_k^{T} B_1^{T} P B_1 e_k : v = e_k \delta(t) \right]
\end{aligned}
\tag{8.31}
$$

In (8.31), we have

$$
\begin{aligned}
\sum_{k=1}^{m} e_k^{T} B_1^{T} P B_1 e_k &= \sum_{k=1}^{m} Tr(B_1^{T} P B_1 e_k e_k^{T}) \\
&= Tr(B_1^{T} P B_1 \sum_{k=1}^{m} e_k e_k^{T}) = Tr(B_1^{T} P B_1)
\end{aligned}
\tag{8.32}
$$

With (8.32), (8.31) becomes

$$
J(K) = \sum_{k=1}^{m} \left[\int_{0}^{\infty} [u(t) - u^{0}(t)]^{T} D_{12}^{T} D_{12} [u(t) - u^{0}(t)] dt : v \right. \\
\left. = e_k \delta(t)] + Tr(B_1^{T} P B_1) \right]
\tag{8.33}
$$

The integrals in (8.33) are non-negative and equal to zero when $u(t) = u^{0}(t)$, and the cost function in (8.33) is minimized by the state feedback (8.21). This minimum cost is given by (8.24). The optimal controller (8.21) does not depend on the disturbance v, because the controller is independent of B1; this matrix only enters the minimum cost (8.24). The control law, as in (8.30), is optimal for all the initial states $x(0^{+})$, and that the solution of the H_2 optimal control problem can be interpreted (in a worst-case setting) as the controller that would minimize the worst-case cost function

$$
\begin{aligned}
J_{worst}(K) &:= \max_{x(0)} \left\{ \|z\|_2^2 : x^{T}(0) x(0) \le 1 \right\} \\
&= \max_{x(0)} \left\{ \int_{0}^{\infty} z^{T}(t) z(t) dt : x^{T}(0) x(0) \le 1 \right\}
\end{aligned}
\tag{8.34}
$$

This result of the optimal state feedback control indicates how one can obtain an optimal output feedback control law: $\hat{u}(s) = K(s) \hat{y}(s)$. In the case of the optimal controller, for (8.31), the integral in the expansion cannot be made

equal to zero, if the entire state vector $x(t)$ is not available to the controller; however, in the output feedback controller, the integral can be made as small as feasible by utilizing the optimal state estimate $\hat{x}(t)$.

8.2.2 The Optimal State Estimation Problem

Let us consider the following dynamic equations

$$
\dot{x}(t) = A x(t) + B_1 v(t)
$$

$$
z(t) = C_1 x(t)
\tag{8.35}
$$

$$
y(t) = C_2 x(t) + D_{21} v(t)
$$

As can be seen from (8.35), that explicitly $u(.)$ does not appear for the estimation problem; since it is assumed known, or it will come from the designed optimal control law, $u(t) = K_{opt} x(t)$, (8.21); for which, of course, the estimation of $x(.)$ is required. It is also assumed that $v(.)$ is a disturbance that has a unit (covariance) dispersion. In fact, it is not really necessary to assume that these are stochastic disturbances (at least for HI/robust estimation problem). When a situation like this is considered, then the dispersion matrices can be called the covariance matrices; however, for the deterministic and robust HI-based estimation theory, these dispersions are called Gramians. In the case of the application of HI theory for estimation/filtering, the variables are considered as "generalized random" variables. However, in the case of the H_2 estimation theory, we can consider the noise processes as deterministic disturbances, if not the stochastic variables, and their corresponding dispersion like matrices as Gramians, or weighting matrices. It would be better to define such dispersions or the energy functions for the sake of uniformity of notations and definitions: (a) for the stochastic case as the covariance/dispersion/uncertainty matrices (for the continuous time systems, the noise process appearing in the system dynamic equation is said to have certain intensity or the spectral density matrix (Q), and if the noise appears in the (algebraic) discrete time equation, it is said to have the covariance matrix (R); (b) for the noise (and even the other processes) in the case of HI theory, being called as the generalized stochastic variables, the strengths of these uncertainties are called the Gramians; and (c) for the disturbances appearing in the theory of H_2 estimation, the strengths may be called as the covariance-Gramians, or the dispersion-Gramians. Next, we consider the stable causal state estimators F, such that the state estimate $\hat{x}(.)$ is determined from the measured output y according to $\hat{x}(s) = F(s) y(s)$. For the H_2-optimal estimation, the quadratic H_2-type cost is minimized:

$$
J(F) = \sum_{k=1}^{m} \left[\int_{0}^{\infty} [x(t) - \hat{x}(t)]^{T} C_1^{T} C_1 [x(t) - \hat{x}(t)] dt : v = e_k \delta(t) \right]
\tag{8.36}
$$

From (8.36) and (8.35), one can see that the cost function is based on the residuals (output errors) of the estimator and the difference between the actual and the predicted measurements. The H_2 estimator problem is treated in great detail in [5]. The solution to this estimation problem is given as

The state estimator/estimatee formula:

$$\dot{\hat{x}}(t) = A\hat{x}(t) + L(y(t) - C_2\hat{x}(t)) \tag{8.37}$$

In (8.37), the noise terms $v(.)$ do not appear since the stochastic noise is not known, only its statistics are assumed to be known. The gain L, which is often known as the Kalman gain in the theory of linear filtering/estimation, is given as

The estimator gain:

$$L = PC_2^T(D_{21}D_{21}^T)^{-1} \tag{8.38}$$

The matrix Riccati equation:

$$AP + PA^T - PC_2^T(D_{21}D_{21}^T)C_2P + B_1B_1^T = 0 \tag{8.39}$$

In fact, (8.39) is a SS equation, since it is equated to zero. In many situations, the SS solution is required, or even might be sufficient; thus, in the gain equation (8.38), this steady solution, P, from (8.39) is used. The matrix P is the unique symmetric positive (semi-) definite solution to the Riccati equation, and in the filtering theory, it is known as the state error covariance matrix. Whereas, in the H_2 theory, it can be called the covariance-Gramian of the state errors, or simply a weighting matrix. One interesting and simple method for the solution of (8.39) is given in [5] and is known as the transition matrix method. The condition for (8.39) is that the following matrix is stable

$$A - LC_2 \tag{8.40}$$

All the eigenvalues of (8.40) then have negative real parts. The minimum value of the cost function (8.36) is

$$\min_F J(F) = Tr\{C_1PC_1^T\} \tag{8.41}$$

This minimum cost (8.41) is obtained by using (8.37) in the cost function (8.36), in turn the formula for L and P are utilized. It is now obvious that the covariance-Gramian of the noise affecting the dynamic system (8.35) is considered unity, as can be seen from equations (8.38), and (8.39), since the process noise covariance matrix Q (or intensity matrix) and the measurement noise covariance matrix R (here, the Q and R are defined in the context of the conventional Kalman filter theory) do not explicitly appear in

these equations; like these appear in the linear filtering theory. In fact, this H_2 estimator is known as the Kalman-Bucy (KB) filter in the literature on linear filtering theory; where we can say that the KB filter is in the stochastic frame work, and the present H_2 estimator is in the deterministic setting. It can be further observed that the solution (8.38) and (8.39) is dual to the solution (8.21) and (8.22) of the H_2 optimal control problem. This can be seen by using the following substitutions:

$$A \to A^T; C_1 \to B_1^T; B_2 \to C_2^T; D_{12} \to D_{21}^T \tag{8.42}$$

The minimum costs are also dual with the substitution $B_1 \to C_1^T$. Generally, the KB filter has been considered in the stochastic setting. The cost function (8.36) is equal to the H_2 norm of the TF from the disturbance v to the estimation error $C_1(x(t) - \hat{x}(t))$. In fact, if the disturbance v is considered as the stochastic process and white Gaussian noise, then the cost function is represented as

$$J(F) = \lim_{t_f \to \infty} \left[E\left\{ \frac{1}{t_f} \int_0^\infty \left[x(t) - \hat{x}(t) \right]^T C_1^T C_1 [x(t) - \hat{x}(t)] dt \} \right] \right] \tag{8.43}$$

In (8.43), E is the mathematical expectation that is required due to the stochastic nature of the disturbance v. Various linear filtering algorithms, based on the Kalman filtering theory, are treated in a recent book [5]. In such a treatment, first the disturbance v appearing in (8.35) is treated as the process noise w appearing in the state equation, and the v measurement noise appearing in the measurement equation for $y(.)$; these noise processes are considered as independent from each other for the sake of simplicity. Subsequently, these noise processes are considered as correlated, and/or colored, noise processes and the filtering algorithms derived [5]. From the foregoing, we can easily see that the H_2 estimator is the deterministic equivalent (or interpretation) of the stochastic KB filtering theory, the former, i.e. (8.36), being the estimator that minimizes the pointwise worst-case error, here F could be $F(G, K)$, or only K:

$$J_{e,worst}(F) = \max_v \left\{ \left\| x(t) - \hat{x}(t) \right\|^2 : \|v\|_2 \le 1 \right\}$$

$$= \max_v \left\{ [x(t) - \hat{x}(t)]^T [x(t) - \hat{x}(t)] : \int_{-\infty}^t v^T(\tau)v(\tau)d\tau \le 1 \right\}$$

8.2.3 The Optimal Output Feedback Problem

Since now that the solution to the optimal state feedback control is available, we can represent the cost function of (8.33) by using (8.30) as

$$J(K) = \sum_{k=1}^{m} \left[\int_0^\infty [u(t) - K_{opt}x(t)]^T D_{12}^T D_{12}[u(t) - K_{opt}x(t)]dt : \right.$$

$$\left. v = e_k\delta(t)] + Tr(B_1^T PB_1) \right]$$

(8.44)

The best we can do is to have the estimate of $x(t)$ and determine the control law signal $u(t)$ as

$$u(t) = K_{opt}\hat{x}(t)$$

(8.45)

$$J(K) = \sum_{k=1}^{m} \left[\int_0^\infty [\hat{x}(t) - x(t)]^T K_{opt}^T D_{12}^T D_{12} K_{opt}[\hat{x}(t) - x(t)]dt : \right.$$

$$\left. v = e_k\delta(t)] + Tr(B_1^T PB_1) \right)$$

(8.46)

The restatement of the H_2 optimal control is given as follows. For the dynamic system (8.16), if all the assumptions a1–a4 and b1–b4 hold, then the controller $u = Ky$ that minimizes the H_2 cost functional, (8.18) is given with (8.45) as

$$\dot{\hat{x}}(t) = (A + B_2K_{opt})\hat{x}(t) + L(y(t) - C_2\hat{x}(t))$$

(8.47)

The optimal gain L in the observer-type/estimator (8.47) is given as (8.38), with its associated weighting matrix P, i.e. the covariance-Gramian (in H_2 theory), as discussed and defined in Section 8.22, or the state-error covariance matrix, as in Kalman-Bucy theory. The optimal controller gain is given by (8.21), with its associated weighting matrix P (often S is used to avoid any confusion in eqn. [8.22] and the optimal controller-related equations) as the solution of (8.2.2), the Riccati equation. For the sake of simplicity and brevity, as well as to imply the connection between H2CA (in fact H_2 estimation) and the KB filtering, the covariance-Gramian is abbreviated as Gramiance; implying first we signify the Gramian (matrix) and then covariance (matrix), since the H_2 estimator is a special case of the KB filter. Also, the minimum cost achieved by the optimal controller is given as

$$\min_K J(K) = Tr(D_{12}K_{opt}PK_{opt}^T D_{12}^T) + Tr(B_1^T SB_1)$$

(8.48)

The optimal controller (8.45) is a combination of an H_2 optimal state estimator and an H_2 optimal state feedback of the thus estimated states. The CLCS with the optimal controller is a stable system, which can be easily seen by the following reformulation

$$\begin{bmatrix} \dot{x} \\ \dot{\hat{x}} \end{bmatrix} = \begin{bmatrix} A + B_2K_{opt} & -B_2K_{opt} \\ 0 & A - LC_2 \end{bmatrix} \begin{bmatrix} x \\ \hat{x} \end{bmatrix} + \begin{bmatrix} B_1 \\ B_1 - LD_{21} \end{bmatrix} v$$

(8.49)

It can be easily seen from (8.49) that if the diagonal matrices are stable, then the closed loop controller/controlled system is also stable. The former follows from the fundamental requirement of the individual systems being controllable and observable. Also, a very interesting feature of this is that the optimal state estimator and the state feedback (controller) can be calculated independently; the property being known as the separation principle that is established in Chapter 6. The results related to the optimal controller, (8.47), are classical in the theory of linear optimal control, in which case we assume that the disturbances are stochastic white processes with a Gaussian PDF/pdf (probability density function), and then the cost function is given as

$$J_{LQG}(K) = \lim_{t_f \to \infty} E\left\{ \frac{1}{t_f} \int_0^{t_f} [x^T(t)Rx(t) + u^T(t)Qu(t)]dt \right\}$$

(8.50)

The problem in (8.50) is known as the LQG control problem. The H_2 optimal control problem solves a well-defined optimal control problem specified by the quadratic cost function, as we have seen in the foregoing. However, the practical applications require that the cost function defined commensurate with the intended control purpose and its objectives. This might not be so straightforward, and next, we discuss two such situations.

8.2.4 H_2 Optimal Control against General Deterministic Inputs

For the deterministic interpretation of the H_2, in (8.18), it is not realistic to assume the impulse functions, as in certain cases, a step disturbance might be more appropriate as a disturbance. Let us consider the dynamic system (8.16) so that it is driven by an input v, which has a rational Laplace transform, and this type of input in time domain is characterized by the state space equations as follows:

$$\dot{x}_v(t) = A_v x_v(t); \quad x(0) = b;$$

$$v(t) = C_v x_v(t); \quad \text{OR}$$

$$\dot{x}_v(t) = A_v x_v(t) + bw(t); \quad w(t) = \delta(t);$$

$$v(t) = C_v x_v(t)$$

(8.51)

If we augment the system dynamics equations (8.16) with (8.51), the problem is transformed to the standard form with an impulse input. However, for some inputs like step, ramp, and sinusoid-type disturbances, the model of (8.51) could be unstable. The problem is that since the disturbance dynamics are is not affected by the controller, the augmented plant will not be stabilizable. Also, the corresponding Riccati equation would not have a stabilizable solution either. For the step input, the problem can be handled in the following way. Let $v(t)$ be the step disturbance

$$v(t) = \begin{cases} 0, & t < 0 \\ v_{step} & t \geq 0 \end{cases}; \quad \dot{v}(t) = v_{step}\,\delta(t) \qquad (8.52)$$

The model in (8.52) is not stable. It can be made stable by using $\dot{v}(t) = -av(t) + v_{step}\,\delta(t)$ for $a \gg 0$. Another approach is to eliminate the unstabilizable mode. We can differentiate the system of equations (8.16) to obtain the following

$$\ddot{x}(t) = A\dot{x}(t) + B_1\dot{v}(t) + B_2\dot{u}(t)$$
$$\dot{y}(t) = C_2\dot{x}(t) + D_{21}\dot{v}(t) \qquad (8.53)$$

If the input $u(.)$ is weighted in the cost, a SS offset will result, so the input variation $\dot{u}(t)$ is given the weightage. The controller output z is now defined as

$$z(t) = C_1 x(t) + D_{12}\dot{u}(t) \qquad (8.54)$$

Next, the composite system model can be written as

$$\begin{bmatrix} \ddot{x}(t) \\ \dot{z}_x(t) \\ \dot{y}(t) \end{bmatrix} = \begin{bmatrix} A & 0 & 0 \\ C_1 & 0 & 0 \\ C_2 & 0 & 0 \end{bmatrix}\begin{bmatrix} \dot{x}(t) \\ z_x(t) \\ y(t) \end{bmatrix} + \begin{bmatrix} B_1 \\ 0 \\ D_{21} \end{bmatrix}\dot{v}(t) + \begin{bmatrix} B_2 \\ 0 \\ 0 \end{bmatrix}\dot{u}(t) \qquad (8.55)$$

$$z(t) = \begin{bmatrix} 0 & I & 0 \end{bmatrix}\begin{bmatrix} \dot{x}(t) \\ z_x(t) \\ y(t) \end{bmatrix} + \begin{bmatrix} D_{12} \end{bmatrix}\dot{u}(t) \qquad (8.56)$$

$$y(t) = \begin{bmatrix} 0 & 0 & I \end{bmatrix}\begin{bmatrix} \dot{x}(t) \\ z_x(t) \\ y(t) \end{bmatrix} \qquad (8.57)$$

The assumption b2, as in section (8.2), does not hold true. With the impulse disturbance input, (8.52), the formulation of (8.55) through (8.57) is in the standard form. If a noise term is added to $y(t)$, (8.57) this is valid, as is often the case, then the assumption b2 can be satisfied,

$$y(t) = \begin{bmatrix} 0 & 0 & I \end{bmatrix}\begin{bmatrix} \dot{x}(t) \\ z_x(t) \\ y(t) \end{bmatrix} + v_{meas}(t) \qquad (8.58)$$

8.2.5 Weighting Matrices in H_2 Optimal Control

The choice of the weighting matrices Q and R in (8.50) are very important. Also, the cost as it appears in the cost function (8.50) sounds a bit artificial, since it often does not correspond physically to any practical problem. However, if one considers these as the two components; the variances (or energies) of the state variables and the inputs, then it would sound meaningful: (i) in the former case, if the states $x(.)$ are the quality variables like concentrations, then the variance/energy is a measure of the variation in the quality, and (ii) in the latter case, the control effort required is included in the cost function. Thus, the overall quadratic cost function reflects that all the individual costs are weighted properly and are to be kept as small as possible. Since all the individual costs cannot be made arbitrarily small concurrently, the problem can be solved as a multi-objective optimization problem, wherein and where for the set of the optimal points is characterized by the pareto-optimal or non-inferior set/points. The pareto-optimality condition states that a pareto-optimal controller is simply the one in/for which, no individual cost can be improved without making at least one of the other costs a bit higher. This means, overall, some trade-off is required and seems implicit in such problems, which are also called conditionally optimal. Thus, in the pareto-optimal control problem, the linear combination of the individual costs is minimized, with a suitable choice of the weighting factors/matrices as

$$J_{LQG}(K) = \sum_{i=1}^{n} q_i J_i(K) + \sum_{i=1}^{m} r_i J_{j+1}(K) \qquad (8.59)$$

$$J_{LQG}(K) = E\left\{ \frac{1}{t_f}\int_0^{t_f}[x^T(t)Rx(t) + u^T(t)Qu(t)]dt \right\} \qquad (8.60)$$

In (8.60), $Q = \text{diag}\{q_1, q_2, ..., q_n\}$; $R = \text{diag}\{r_1, r_2, ..., r_m\}$; the $q(.)$, and $r(.)$ are positive constants. The individual cost functions are given as

$$J_i(K) = \lim_{t_f \to \infty} E\left\{ \frac{1}{t_f}\int_0^{t_f} x_i(t)dt \right\}; \ i = 1, 2, ..., n$$

$$J_{n+1}(K) = \lim_{t_f \to \infty} E\left\{ \frac{1}{t_f}\int_0^{t_f} u_i(t)dt \right\}; \ i = 1, 2, ..., m \qquad (8.61)$$

Thus, the costs in (8.61) are the variances of the state variables and the variances of the inputs, respectively. In the deterministic H_2 problem, these variance matrices are called Gramians. Thus, very interestingly, the multi-objective optimization problem is reformulated as a Pareto-optimal control problem, which in turn, reduces to the appropriate choices of the weighting matrices so that the individual costs have the satisfactory values as acceptable by the controller designer based on the given problem at hand and other criteria dictated by the same problem for the good performance of the overall combined closed loop control system. Each individual criterion can be relaxed, if the cost of other components is acceptable; this is often done by doing simulations and checking the time histories of the predicted states and the measurements, along with the control signal $u(t)$ to see its form and magnitude for their acceptability. Thus, if K_o is the pareto-optimal controller, then there is no other controller K for which the individual costs are lesser than the individual costs of the pareto-optimal solution. Thus, if one cost $J_j(K)$ is the primary cost, then other costs should be restricted, and one can solve this as a constraint minimization problem

$$\text{Minimize } J_j(K); \ J_i(K) \le c^2, \ i = 1, 2, \ldots, n+m; \ i \ne j$$

$$(8.62)$$

8.3 H$_\infty$ Control

In the HICA, the performance measure to be minimized is H$_\infty$-norm (HI norm) of the CLTF:

$$J_\infty(K) = \|F(G, K)\|_\infty \qquad (8.63)$$

In (8.63), the function $F(G, K)$ is the TF of the CLCS:

$$z = F(G, K)v \qquad (8.64)$$

This TF is given as

$$F(G, K) = G_{11} + G_{12}(I - KG_{22})KG_{21} \qquad (8.65)$$

As done in the previous sections, the HICA problem is very conveniently solved in the time domain using the usual state space modeling approach; thus, the plant G is represented as in (8.16) with $x(0) = 0$. The assumptions made are the same as the ones made earlier in the case of the H$_2$ optimal control problem, Section 8.2. The straightforward minimization of the cost $J_\infty(K)$ is a very difficult problem, so it is aimed at achieving a stabilizing controller, such that the H$_\infty$-norm bound is obeyed as follows

$$J_\infty(K) < \gamma; \ \gamma > 0 \qquad (8.66)$$

One can easily see that we have to achieve the condition of (8.66) and iteratively tune the resulting control-generating algorithm by varying the values of γ. Interestingly, this factor then turns out to be a tuning parameter, as well as the upper bound on the cost function. From this point of view, the H$_\infty$ control can achieve the robust controller but not really the optimal control; hence, it results in the sub-optimal control solution. The H$_\infty$ performance measure needs to be now characterized in terms of the worst-case gain, in terms of L$_2$-norm

$$J_\infty(K) = \sup \left\{ \frac{\|z\|_2}{\|v\|_2}; \ v \ne 0 \right\} \qquad (8.67)$$

Equivalently, one can write the performance bound as

$$\frac{\|z\|_2}{\|v\|_2} < \gamma; \ J(v, u) = \left\{ \|z\|_2^2 - \gamma^2 \|v\|_2^2 \right\} < 0; \ \forall v \ne 0 \qquad (8.68)$$

The time domain formula for the HI cost function inequality is given as follows

$$J(v, u) = \int_0^\infty [z^T(t)z(t) - \gamma^2 v^T(t)v(t)]dt < 0; \ \forall v \ne 0 \qquad (8.69)$$

Thus, the problem of determining the (sub-optimal and yet the robust) controller $u = Ky$ that satisfies the inequality in (8.69) is stated as

$$\max_{v \ne 0} \{ \min_{u=Ky} J(v, u) \} < 0; \ \forall v \ne 0 \qquad (8.70)$$

As we have seen in Chapter 6, this min-max problem is like a dynamic game problem, in which the first player v tries to make the cost as large as possible, while the second player tries to make $L(v, u) < 0$; inspite of the action of v. This game problem can be called the linear quadratic game problem, since the system is linear, and the cost function is still quadratic. The solution can be obtained as: (i) an H$_\infty$ (optimal) state feedback control problem and associated transformations, and (ii) an H$_\infty$ optimal state estimation problem.

8.3.1 H$_\infty$ Optimal State Feedback Control

We have the proposition that: (a) the dynamic system is as in (8.16), (b) the assumptions a1–a4 hold true, and (c) the controller, i.e. the control law/signal $u(t)$ has the present and past values of the state of the system available; $u(\tau), \ \tau \le t$. In this case, there exists a state feedback controller, such that $J_\infty(K) < \gamma; \ \gamma > 0$, if, and only if, there exists a positive (semi-) definite solution to the algebraic Riccati equation

$$A^T S + SA - SB_2(D_{12}^T D_{12})^{-1} B_2^T S + \gamma^{-2} SB_1 B_1^T S + C_1^T C_1 = 0 \tag{8.71}$$

Also, the following matrix has all its eigenvalues as negative real parts

$$A - B_2(D_{12}^T D_{12})^{-1} B_2^T S + \gamma^{-2} B_1 B_1^T S \tag{8.72}$$

With the foregoing true, the controller is given as

$$u(t) = K_\infty x(t); \quad K_\infty = -(D_{12}^T D_{12})^{-1} B_2^T S \tag{8.73}$$

The state feedback controller can be obtained in the similar manner as the H_2 optimal controller has been obtained in the first place.

$$\int_0^\infty [z^T(t)z(t) - \gamma^2 v^T(t)v(t)]dt$$

$$= \int_0^\infty \{[C_1 x(t) + D_{12} u(t)]^T [C_1 x(t) + D_{12} u(t)] - \gamma^2 v^T(t)v(t)\}dt$$

$$= \int_0^\infty \left\{ [C_1 x(t) + D_{12} u(t)]^T [C_1 x(t) + D_{12} u(t)] - \gamma^2 v^T(t)v(t) \right.$$

$$\left. + \frac{dx^T(t)Sx(t)}{dt} \right\} dt - x^T(\infty)Sx(\infty) + x^T(0)Sx(0)$$

$$= \int_0^\infty \{[u(t) - u^0(t)]^T D_{12}^T D_{12}[u(t) - u^0(t)]$$

$$- \gamma^2 [v(t) - v^0(t)]^T [v(t) - v^0(t)]\}dt \tag{8.74}$$

In (8.74), we have the following formulas

$$u^0(t) = -(D_{12}^T D_{12})^{-1} B_2^T Sx(t) = K_\infty x(t)$$

$$v^0(t) = \gamma^{-2} B_1^T Sx(t) \tag{8.75}$$

For the foregoing, we have assumed $x(0)$ and $x(\infty) = 0$ for the sake of stability. From the expansion in (8.74), one can see that using $u(t) = u^0(t)$, the inequality $J(v,u) \le 0$ holds true for all v; and the maximum value of $J(v, u)$ is zero and is obtained if, and only if, $v(t) = v^0(t)$ for all t. However, since $x(0) = 0$, which implies that $x(t) = 0$ and $v(t) = 0$ for all $t \ge 0$; hence, the inequality (8.69) holds true for all $v \ne 0$. Further, one can show that if a state feedback controller, which obtains the bound in (8.69)

exists, then Riccati equation (8.71) has a symmetric positive definite (or semi-definite) solution so that (8.72) is a stable matrix. The H_∞ state feedback law can be obtained in a nearly similar manner as the H_2 optimal controller. For getting the H_∞ (optimal) robust output feedback controller, one can use the expansion (8.74) to transform the problem to an equivalent H_∞ estimation problem (the terms optimal or robust are avoided for brevity). In this case, we can define the following variables

$$\tilde{z}(t) = (D_{12}^T D_{12})^{1/2} [u(t) - u^0(t)]$$

$$= (D_{12}^T D_{12})^{1/2} u(t) + (D_{12}^T D_{12})^{-1/2} B_2^T Sx(t) \tag{8.76}$$

$$\tilde{v}(t) = v(t) - v^0(t) = v(t) - \gamma^{-2} B_1^T Sx(t) \tag{8.77}$$

Further, we have the following equivalence

$$J(v,u) = \int_0^\infty [z^T(t)z(t) - \gamma^2 v^T(t)v(t)]dt$$

$$= \int_0^\infty [\tilde{z}^T(t)\tilde{z}(t) - \gamma^2 \tilde{v}^T(t)\tilde{v}(t)]dt \tag{8.78}$$

Equivalently, we have

$$J(v,u) = \left\{ \|z\|_2^2 - \gamma^2 \|v\|_2^2 \right\} = \left\{ \|\tilde{z}\|_2^2 - \gamma^2 \|\tilde{v}\|_2^2 \right\} \tag{8.79}$$

The inequality (8.69) holds true for all $v \ne 0$, if, and only if, the following is true

$$\tilde{J}(\tilde{v},u) = \left\{ \|\tilde{z}\|_2^2 - \gamma^2 \|\tilde{v}\|_2^2 \right\} < 0; \forall \tilde{v} \ne 0 \tag{8.80}$$

The problem can be restated in terms of the new signals appearing in (8.80) as follows by substituting for these signals in the equations of the dynamic system (8.16)

$$\dot{x}(t) = (A + \gamma^{-2} B_1 B_1^T S)x(t) + B_1 \tilde{v}(t) + B_2 u(t); \quad x(0) = 0$$

$$\tilde{z}(t) = (D_{12}^T D_{12})^{-1/2} B_2^T Sx(t) + (D_{12}^T D_{12})^{1/2} u(t) \tag{8.81}$$

$$y(t) = (C_2 + \gamma^{-2} D_{21} B_1^T S)x(t) + D_{21} \tilde{v}(t)$$

Equivalently (8.81) can be written as follows with the appropriate terms defined:

$$\dot{x}(t) = \tilde{A}x(t) + \tilde{B}_1 \tilde{v}(t) + \tilde{B}_2 u(t); \quad x(0) = 0$$

$$\tilde{z}(t) = \tilde{C}_1 x(t) + \tilde{D}_{12} u(t) \tag{8.82}$$

$$y(t) = \tilde{C}_2 x(t) + \tilde{D}_{21} \tilde{v}(t)$$

In the state feedback controller situation, the output $\tilde{z}$ can be made zero by the controller $u(t) = K_\infty x(t)$. In the case of the output feedback, the controller can be based on the estimate of $x(t)$, or the output $\tilde{z}$; hence, we need an H_∞ optimal estimation algorithm.

8.3.2 H_∞ Optimal State Estimation

Let us consider the state space dynamic system of (8.35) with $x(0) = 0$. Next, we consider the stable causal estimators F of the output z, based on the measured output y, such that $\hat{z}(s) = F(s)y(s)$. For the H_∞ optimal (in fact robust and suboptimal) estimation problem, one can define the H_∞ norm of the TF from the disturbance v to the estimation error $z - \hat{z}$ as follows

$$J_{e,\infty}(F) = \sup\left\{\frac{\left\|z - \hat{z}\right\|_2}{\left\|v\right\|_2}; v \neq 0\right\} \quad (8.83)$$

Similar to the state feedback controller problem, one needs to seek the conditions for the existence of an estimator that obtains the bound as

$$J_{e,\infty}(F) < \gamma;$$
$$J_e(v, \hat{z}) = \{\left\|z - \hat{z}\right\|_2^2 - \gamma^2 \left\|v\right\|_2^2\} < 0; \ \forall v \neq 0 \quad (8.84)$$

The solution to this H_∞ estimation problem has a similar structure as the H_2 estimation. Consider the proposition: (a) the dynamic system as in (8.35) and (b) the assumptions b1–b4 hold true. Then, there exists a stable estimator F that obtains the H_∞ norm bound as in (8.84), if, and only if, there exists a symmetric positive definite or semi-definite solution P to the following Riccati equation

$$AP + PA^T - PC_2^T(D_{21}D_{21}^T)^{-1}C_2P + \gamma^{-2}PC_1^TC_1P + B_1B_1^T = 0 \quad (8.85)$$

With the following matrix that has all its eigenvalues as negative real parts

$$A - PC_2^T(D_{21}D_{21}^T)^{-1}C_2 + \gamma^{-2}PC_1^TC_1 \quad (8.86)$$

The corresponding estimator that would satisfy the bound in (8.84) is given as

$$\dot{\hat{x}}(t) = A\hat{x}(t) + L_\infty[y(t) - C_2\hat{x}(t)]; \ \hat{x}(0) = 0$$
$$\hat{z}(t) = C_1\hat{x}(t) \quad (8.87)$$

The filter gain is given as

$$L_\infty = PC_2^T(D_{21}D_{21}^T)^{-1} \quad (8.88)$$

In the similar way, as in the case of the H_2 optimal estimator, the H_∞ optimal estimator is also dual to the H_∞ optimal state feedback controller. The H_∞ optimal estimation is fully dealt with in the recent book [5, Chapter 3].

8.3.3 H_∞ Optimal Output Feedback Problem

The H_∞ optimal estimate of the output $\hat{z}$ of the transformed system can be written as

$$\hat{\tilde{z}}(t) = \tilde{C}_1\hat{x}(t) + \tilde{D}_{12}u(t) \quad (8.89)$$

Since, matrix $\tilde{D}_{12} = D_{12}^TD_{12}$ is invertible, we obtain the following formulas

$$u(t) = -\tilde{D}_{12}^{-1}\tilde{C}_1\hat{x}(t) = -(D_{12}^TD_{12})^{-1}B_2^TP\hat{x}(t) = K_\infty\hat{x}(t) \quad (8.90)$$

This controller is utilized to obtain the output of the system as $\tilde{z}(t) = \tilde{C}_1(x(t) - \hat{x}(t))$:

$$\tilde{J}(\tilde{v}, u) = \left\|C_1(x - \hat{x})\right\|_2^2 - \gamma^2\left\|\tilde{v}\right\|_2^2 \quad (8.91)$$

Hence, the inequality $\tilde{J}_e(\tilde{v}, u) < 0$ holds true $\forall v \neq 0$, if, and only if, the estimate $\hat{\tilde{z}}$ satisfies:

$$J_e(\tilde{v}, \hat{\tilde{z}}) = \left\|(z - \hat{z})\right\|_2^2 - \gamma^2\left\|\tilde{v}\right\|_2^2 < 0; \ \tilde{v} \neq 0 \quad (8.92)$$

Thus, the H_∞ optimal output controller is characterized for the dynamics considered as in (8.16), and the assumptions a1–a4 and b1–b4 holding true, as $u = Ky$, and that it achieves the H_∞ norm bound if, and only if, the following conditions are satisfied:

i. There exists a symmetric positive or semi-definite solution S to the Riccati equation (8.22), such that the matrix (8.23) is stable.

ii. There exists a symmetric positive or semi-definite solution Z to the following Riccati equation that is associated with the system (8.81) and (8.82), and the estimation performance bound as in (8.92)

$$\tilde{A}Z + Z\tilde{A}^T - Z\tilde{C}_2^T(\tilde{D}_{21}\tilde{D}_{21}^T)^{-1}\tilde{C}_2Z$$
$$+ \gamma^{-2}Z\tilde{C}_1^T\tilde{C}_1Z + \tilde{B}_1\tilde{B}_1^T = 0 \quad (8.93)$$

Such that the following matrix is stable

$$\tilde{A} - Z\tilde{C}_2^T(\tilde{D}_{21}\tilde{D}_{21}^T)^{-1}\tilde{C}_2 + \gamma^{-2}Z\tilde{C}_1^T\tilde{C}_1 \quad (8.94)$$

When the conditions (i) and (ii) are satisfied, a controller that obtains the performance bound (the H_∞ norm) is given by the following equations

$$\dot{\hat{x}}(t) = \tilde{A}\hat{x}(t) + \tilde{B}_2 u(t) + L_Z[y(t) - \tilde{C}_2\hat{x}(t)] \qquad (8.95)$$

$$u(t) = K_\infty \hat{x}(t); \quad K_\infty = -(D_{12}^T D_{12})^{-1} B_2^T S; \quad L_Z = Z\tilde{C}_2^T (\tilde{D}_{21}\tilde{D}_{21}^T)^{-1} \qquad (8.96)$$

8.3.4 The Relation between *S*, *P* and *Z*

If there is a symmetric positive (semi-)definite solution S to the corresponding Riccati equation of the H_∞ optimal state feedback control problem, then there is a symmetric positive (semi-) definite solution Z to the corresponding Riccati equation of the H_∞ optimal output feedback problem, if, and only if, (i) there exists positive (semi-)definite solution P to the corresponding Riccati equation of the H_∞ optimal state estimation problem, and (ii) the maximum eigenvalue has the condition $\lambda(SP) < \gamma^2$. Also, when these conditions hold true, then the Z is given as

$$Z = P(I - \gamma^{-2}SP)^{-1}$$

The foregoing controller in Section 8.3.3 is the central controller. As the limit $\gamma \to \infty$, this controller approaches the H_2 optimal controller. In order to see that the closed loop norm is as small as possible, we should iterate on γ and check if there exists a controller that obtains the performance bound as specified as the cost function with a sufficient degree of accuracy. The minimum achievable norm is denoted as

$$\gamma_{\text{inf}} = \inf\left\{\|F(G, K)\|_\infty : u = Ky, K \text{ is stabilizing gain}\right\} \quad (8.97)$$

8.4 Robust Stability and H_∞ Norm

It is important here to make it clear that the robustness and the H_∞ control problem are intimately connected, in that we are seeking control solutions and algorithms that would be robust against model uncertainties. The condition that a control system is robustly stable in spite of a certain kind of uncertainties in a model can be stated quantitatively in terms of an H_∞ norm bound, and that this bound should be satisfied by the designed control

system. We can consider the plant dynamics and its characterization as

$$y = Gu;$$

$$G = G_0 + \Delta_a \qquad (8.98)$$

In (8.98), it is presumed that the nominal plant is affected by the additive unknown uncertainty; the basic plant is assumed to be linear and finite dimensional. However, it is assumed that an upper bound on its magnitude can be determined as a function of frequency. Specifically, we can have a filter, $W(.)$, that a SISO system can have

$$|\Delta_a(j\omega)| \le |W(j\omega)|; \quad \forall \omega \qquad (8.99)$$

Thus, the filter-based uncertainty bound characterizes the maximum uncertainties related to the chosen model as a function of frequencies. This filter can be determined from the empirical data; for example, if the model of the plant is determined from the frequency response experiments, then one can determine the maximum error of the frequency response from the empirically available data. The plant characterization is stated as

$$G = G_0 + W\Delta; \quad |\Delta(j\omega)| \le 1; \quad \forall \omega \qquad (8.100)$$

In (8.100), it is assumed that the uncertainty is bounded uniformly at all frequencies and it is incorporated in the filter W, which is then known as the uncertainty weighting filter, or the uncertainty weight. For multiple-input multiple-output (MIMO) plant, the uncertainty magnitude is bounded by its induced matrix norm, in terms of the maximum singular value:

$$\bar{\sigma}(\Delta_a(j\omega)) \le |W(j\omega)|, \quad \forall \omega; \quad \|\Delta(j\omega)\| = \bar{\sigma}(\Delta(j\omega)) \qquad (8.101)$$

Then, the plant is characterized as

$$G = G_0 + W\Delta; \quad \bar{\sigma}(\Delta(j\omega)) \le 1; \quad \forall \omega; \text{ OR}$$

$$G = G_0 + W\Delta; \quad \|\Delta\|_\infty \le 1 \qquad (8.102)$$

In (8.102), the definition of HI norm is used. The uncertainty description can be extended to output multiplicative uncertainty:

$$G = (I + \Delta_m)G_0; \quad \bar{\sigma}(\Delta_m(j\omega)) \le |W(j\omega)|; \quad \forall \omega$$

$$G = (I + \Delta W)G_0 = G_0 + \Delta W G_0; \quad \|\Delta\|_\infty \le 1 \qquad (8.103)$$

One can characterize the input uncertainty in the similar way. Figure 8.1 is given as

$$G = G_0 + W_1 \Delta W_2; \quad \|\Delta_\infty\| \le 1$$

For a control system, one should see that the CLCS is stable for all possible G; especially for all norm bounded Δ such that $\|\Delta_\infty\| \le 1$. Such a control system is robustly stable, and the controller gain K is robustly stabilizing; the system has the robust stability performance. Thus, the determination of a controller for a situation, as in (8.103) with the associated bound, is the robust stabilization problem. This problem can also be expressed in the terms of an H_∞ optimal control problem (HIOC); in the equivalent form as, see Figure 8.2:

$$P = \begin{bmatrix} P_{11} & P_{12} \\ P_{21} & P_{22} \end{bmatrix} = \begin{bmatrix} 0 & W_2 \\ W_1 & G_0 \end{bmatrix} \qquad (8.104)$$

We can obtain the following TF between output y and the input u for the system

$$y = [P_{22} + P_{21}\Delta(I - P_{11}\Delta)^{-1} P_{12}]u$$

$$= [G_0 + W_1 \Delta (I - 0 \cdot \Delta)^{-1} W_2] u \qquad (8.105)$$

$$= [G_0 + W_1 \Delta W_2] u$$

It can be seen that the uncertainty block appears in a feedback loop, and the classical stability analysis can be applied to arrive at the stability conditions; the system equations are

$$z = (I - F\Delta)^{-1} Fv; \quad F = F(P, K), \text{for } \Delta = 0 \qquad (8.106)$$

In (8.106), $F = F(P, K)$ is the closed loop transfer (function) from v to z for the nominal system, i.e. with no uncertainty, then we get

$$F(P, K) = P_{11} + P_{12}(I - KP_{22})^{-1} KP_{21} \qquad (8.107)$$

As per the SISO control theory, the feedback loop is stable if $F(P, K)$ is stable and the loop TF $F\Delta$ has the magnitude less than 1. This translates into the following requirement for the uncertain system

$$|F(j\omega)\Delta(j\omega)| < 1, \quad \forall \omega \qquad (8.108)$$

The condition in (8.108) implies the following at the cross over frequency ω_c where $F\Delta$ has phase $-\pi$:

$$|F(j\omega_c)\Delta(j\omega_c)| < 1; \quad \text{at } \omega = \omega_c \qquad (8.109)$$

The foregoing result is known as the small gain theorem. We assume the disturbance, as in (8.100), is norm bounded, the robust stability is assured if the closed loop TF, F satisfies $|F(j\omega)| < 1$; at $\omega = \omega_c$, and equivalently the HI norm bound $\|F\|_\infty < 1$ holds true. It so happens that the assumption on the uncertainty block Δ, $\|\Delta\|_\infty \le 1$, gives sufficient freedom so that a norm bounded uncertainty that destabilizes the system can always be found if $|F(j\omega)| > 1$ for some ω, i.e. $\|F\|_\infty \ge 1$, so this is a sufficient and necessary condition for the robust stability. In case of a MIMO system, one can use the following development. The TF of (8.106) can be written as

$$(I - F\Delta)^{-1} F = \frac{1}{\det(I - F\Delta)} adj(I - \Delta F) F \qquad (8.110)$$

We assume that F and the uncertainty block are stable, then the stability of (8.110) is determined by the zeroes of $\det(I - F\Delta)$. Since the system is stable for sufficiently small magnitude of the uncertainly block, the robust stability is equivalent to the condition as follows

$$\det(I - F(j\omega)\Delta(j\omega)) \ne 0; \ \forall \omega$$
$$\text{and } \Delta; \text{such that} \|\Delta\|_\infty \le 1 \qquad (8.111)$$

Thus, the considered dynamic system is stable for all uncertainties that satisfy the norm bound, as in (8.111), if, and only if, the nominal closed loop TF $F = F(P, K)$ is stable and $\|F\|_\infty < 1$. An interesting interpretation is that, the "if" part, $\|F\|_\infty < 1 \rightarrow$ robust stability, follows from the small gain theorem, since the HI norm bound guarantees $|F(j\omega)\Delta(j\omega)| < 1$, for SISO systems, and $\bar{\sigma}(F(j\omega)\Delta(j\omega)) < 1$, for MIMO systems. And, the "only if" part, robust stability $\rightarrow \|F\|_\infty < 1$, follows from the fact that the uncertainty can have the arbitrary phase; hence, if $\|F\|_\infty \ge 1$, it follows that a norm-bound uncertainty $\|\Delta\|_\infty \le 1$ can always be found that destabilizes the system. Often, one might determine the largest norm

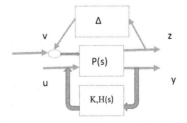
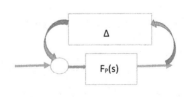

FIGURE 8.2
Representations of uncertain control systems.

bound uncertainty that can be permitted or tolerated so that it is still possible to determine a controller that gives robust stability. The robust stability result can be extended to nonlinear and time varying uncertainties. Let the uncertainty satisfy the norm bound $\|\Delta\| \leq 1$, with the induced L_2 norm

$$\|\Delta\| = \sup\left\{\frac{\|\Delta v\|_2}{\|v\|_2} : v \neq 0\right\} \tag{8.112}$$

The L_2 induced norm coincides with the definition of HI norm from for the linear time-invariant systems; however, the uncertainty function could be nonlinear and/or time varying. Then, it can be shown that the considered dynamic system is stable for all such uncertainties that satisfy the L_2 induced norm bound, if, and only if, the nominal closed loop TF, $F = F(P, K)$ is stable and satisfies the HI norm bound, $\|F\|_\infty < 1$. Thus, interestingly, the condition/s for robust stability against nonlinear, time varying uncertainties is the same as the condition for robust stability against linear time-invariant uncertainties. The robust stability condition provides a quantitative way of treating plant nonlinearities as uncertainties about a nominal linear plant model; this approach is only feasible for not too strong nonlinear plants, where the norm bound of the uncertainty, which captures the nonlinearities, is not very large. The foregoing robust stability condition gives a quantitative characterization of robust stability in terms of the HI norm; and earlier of the HICA, the solution of the robust stabilization problem was not known, although the condition of stability of the closed loop system was known. Theoretically, though important, the robust stabilization problem is of limited practical interest since: (i) many chemical process control systems are open loop stable, the maximally robustly stabilizing control strategy is then to operate the plant in open loop, because any feedback controller gives a reduction in robust stability for an open loop stable plant; and (ii) even for an unstable uncertain plant, a controller that is only designed for robust stability will in general be conservative (somewhat restrictive); in practice, the robust stability condition is useful as a robustness bound used in combination with other performance/design criteria.

8.5 Structured Uncertainties and Structured Singular Values

In Section 8.4, we considered the uncertainties with only norm boundedness without imposing any structure on the uncertainties block. Yet, in many cases, we might know more about the uncertainties related to the plant dynamics, in which case, the robust stability criteria would result in overly conservative/restrictive control design. In this section, we study a structured uncertainty that signifies structured singular values (SSV) and the control design based on these. Consider a system with single input and two outputs (SIMO) described by $y = Gu$,

$$G = \begin{bmatrix} G_1 \\ G_2 \end{bmatrix} \tag{8.113}$$

In (8.113), the plants are described by

$$\begin{aligned} G_1 &= G_{0,1} + W_{01}\Delta_1; \quad \|\Delta_1\|_\infty \leq 1 \\ G_2 &= G_{0,2} + W_{02}\Delta_2; \quad \|\Delta_2\|_\infty \leq 1 \end{aligned} \tag{8.114}$$

The plant now is represented as

$$G = \begin{bmatrix} G_{0,1} \\ G_{0,2} \end{bmatrix} + \begin{bmatrix} W_{01}\Delta_1 \\ W_{02}\Delta_2 \end{bmatrix} = \begin{bmatrix} G_{0,1} \\ G_{0,2} \end{bmatrix} + \begin{bmatrix} W_{01} & 0 \\ 0 & W_{02} \end{bmatrix}\begin{bmatrix} \Delta_1 & 0 \\ 0 & \Delta_2 \end{bmatrix}\begin{bmatrix} I \\ I \end{bmatrix} \tag{8.115}$$

$$= G_0 + W_1 \begin{bmatrix} \Delta_1 & 0 \\ 0 & \Delta_2 \end{bmatrix} W_2 \tag{8.116}$$

The plant characterization with the uncertainty is of the same form as earlier one, but now the uncertainty has a structure as in (8.116)

$$\Delta = \begin{bmatrix} \Delta_1 & 0 \\ 0 & \Delta_2 \end{bmatrix}, \|\Delta_i\|_\infty \leq 1, i = 1, 2. \tag{8.117}$$

So, we can see that the present uncertainty structure is a subset of the unstructured norm bounded uncertainty studied in Section 8.4. The general structured norm bounded uncertainty with norm threshold, $\delta > 0$ (the radius of uncertainty), can be defined as

$$\begin{aligned} \Delta_s(\delta) = \{\Delta = \text{block diag}(\Delta_1, ..., \Delta_s), \\ \Delta_i \in H_\infty^{\bar{n} \times \bar{n}}, \|\Delta_i\|_\infty \leq \delta\} \end{aligned} \tag{8.118}$$

In (8.118), $H(.)$ are the set of stable TFs with HI norm. The uncertainty is characterized by the parameters: (i) s number of blocks, (ii) the $r(i)$-dimensional uncertainty blocks, and (iii) "radius of uncertainty;" the latter is the maximum norm of the uncertainty blocks.

The structure of $\Delta \in \Delta_s(\delta)$ implies $\|\Delta\|_\infty \leq \delta$. An uncertain plant with the structured uncertainty is represented as

$$G = G_0 + W_1 \Delta W_2, \quad \Delta \in \Delta_s(\delta) \qquad (8.119)$$

In (8.119), the distinction between the set of the uncertainties and its member is avoided. The robust stabilization problem for dynamic plants with structured uncertainties is specified as: The plant is as in (8.119) with the uncertainty as in (8.118). Then, the aim is to determine a controller so that the considered control system is stable for all the uncertainties as specified in (8.119). Interestingly, the closed loop plant F is stable for all the uncertainties specified in (8.119), if, and only if

$$\det(I - F(j\omega)\Delta(j\omega)) \neq 0; \ \forall \ \omega \ \text{and} \ \Delta \in \Delta_s(\delta) \qquad (8.120)$$

The condition of (8.120) cannot be reduced to a simple condition on the HI norm of the closed loop TF, F. However, based on (8.120), a quantity $\mu(F(j\omega))$ can be defined. Let us have the smallest value of the uncertainty magnitude δ, such that there exists $\Delta \in \Delta_s(\delta)$ that destabilizes the system

$$\begin{aligned} \delta_{\min}(F(j\omega)) = \min\{&\delta : \det(I - F(j\omega)\Delta(j\omega)) \\ &= 0 \ \text{for some} \ \Delta \in \Delta_s(\delta)\} \end{aligned} \qquad (8.121)$$

The uncertainty plant is stable for all structured uncertainties as in (8.121), with the specified uncertainty radius, if, and only if

$$\delta < \delta_{\min}(F(j\omega)), \ \forall \omega \qquad (8.122)$$

The robust stability for the set of structured uncertainties is now characterized as

$$\mu(F(j\omega)) = \delta_{\min}(F(jw))^{-1} \qquad (8.123)$$

In this case, the uncertain plant is stable for all the uncertainties specified in (8.120), for the specified radius, if, and only if, the following holds

$$\mu(F(j\omega)) < \delta^{-1}, \ \forall \omega \qquad (8.124)$$

The quantity in (8.124) is known as the structured singular value (SSV) of F, and it depends on F itself and the structure of the uncertainty set, as in (8.118). The SSV can be treated as a generalization of the maximum singular value $\bar{\sigma}$; it degenerates to $\bar{\sigma}$ itself, if there is only one uncertainty block: $\mu(F) = \bar{\sigma}(F)$, if $s = 1$. Thus, the robust stabilization problem for systems with the structured uncertainties is to determine a stabilizing controller (control law) so that the SSV of the closed loop system satisfies the following condition

$$\sup_\omega \mu(F(j\omega)) < \delta^{-1} \qquad (8.125)$$

This problem is known as the SSV, or μ-synthesis. It so happens that to compute SSV is difficult; hence, a more tractable approach is to determine/compute an upper bound of SSV. Let us define the following matrix

$$D_s = \{D = \text{diag}(d_1 I_{r_1}, \ldots, d_s I_{r_s})\} \qquad (8.126)$$

In (8.126), the $d(.)$ are the complex valued scalar constants. Then we have

$$\mu(F) \leq \bar{\sigma}(DFD^{-1}), \ D \in D_s \qquad (8.127)$$

Then, the upper bound on SSV is given as

$$\mu(F(j\omega)) \leq \inf_{D(j\omega) \in D_s} \bar{\sigma}(D(j\omega)F(j\omega)D(j\omega)^{-1}) \qquad (8.128)$$

In case of s being equal to or less than 3, then we have the following equality

$$\mu(F(j\omega)) = \inf_{D(j\omega) \in D_s} \bar{\sigma}(D(j\omega)F(j\omega)D(j\omega)^{-1}); \ s \leq 3 \qquad (8.129)$$

The numerical studies indicate that when $s > 3$, the upper bound is nearly within 15% larger than the SSV [13]. This is based on the difference between the two bounds on μ, and the upper bound provides a useful and not overly conservative approximation of μ. Let $D(.)$ (8.126) be a stable TF with the structure, as specified therein

$$D(j\omega) = \text{diag}\{d_1(j\omega)I_{r_1}, \ldots, d_s(j\omega)I_{r_s}\} \in D_s, \ \forall \omega \qquad (8.130)$$

In (8.130), the $d(.)$ are the stable scalar TFs. Then, we obtain the following from (8.127) and the HI norm

$$\begin{aligned} \sup_\omega \mu(F(j\omega) &\leq \sup_\omega \bar{\sigma}[(D(j\omega)F(j\omega)D(j\omega)^{-1}] \\ &= \|DFD^{-1}\|_\infty, \ D(j\omega) \in D_s \end{aligned} \qquad (8.131)$$

Then, with the robust stability condition and the structured uncertainties, the existence of the filter D with $D(j\omega) \in D_s$, such that

$$\|DFD^{-1}\|_\infty < \delta^{-1} \qquad (8.132)$$

signifies that the TF, F is robustly stable for the structured uncertainties (8.118). Unless s is less than or equal to 3, the existence of $D(j\omega) \in D_s$, such that the condition in (8.134) holds is not a necessary condition for robust stability. Next, the robust stabilization problem for plants with structured uncertainties can be studied. Because of the fact that (8.132) implies robust stability, a natural approach to design a controller which obtains robust stability for structured uncertainties involves the construction of a controller K and a filter $D \in D_s$, such that the CLTF DFD^{-1} satisfies the HI norm bound (8.132); D is a scaling filter, it scales the components of the I/O vectors of the closed loop systems. An approximate robust stabilization problem for plants with structured uncertainties (or the DK optimization problem) is: for the considered plant, find a stabilizing controller K, and $D(.)$, such that the scaled closed loop TF, $DF(P, K)D^{-1}$ satisfies the HI norm bound as in (8.132); and if the norm bound is attained, the closed loop system is robustly stable with respect to the structured uncertainty $\Delta_s(\delta)$. The $F(P, K)$ is the CLTF of the plant P (with the controller K), and $DF(P, K)D^{-1}$ is the CLTF from the signal $v(D)$ to the signal $z(D)$ of the following system

$$\begin{bmatrix} z_D \\ y \end{bmatrix} = \begin{bmatrix} D & 0 \\ 0 & I \end{bmatrix} P \begin{bmatrix} D^{-1} & 0 \\ 0 & I \end{bmatrix} \begin{bmatrix} v_D \\ u \end{bmatrix}; \ u = Ky \quad (8.133)$$

With the uncertainty as in (8.119) and the augmented plant specified by (8.104), we have:

$$\begin{bmatrix} D & 0 \\ 0 & I \end{bmatrix} P \begin{bmatrix} D^{-1} & 0 \\ 0 & I \end{bmatrix} = \begin{bmatrix} 0 & DW_2 \\ W_1D^{-1} & G_0 \end{bmatrix} \quad (8.134)$$

Since there is no explicit method for the simultaneous solution to obtain D and K, such that the condition in (8.132) holds true, the boot-strap procedure is adopted to reduce the HI cost. This is known as the DK iteration.

It would also be of interest to study the case of the structured uncertainties when these are nonlinear and/or time varying. Let the system with the uncertainty be as in (8.118), then the system is robustly stable with respect to the structured nonlinear and/or time varying uncertainties with an L_2 induced norm less than or equal to δ, i.e. $\|\Delta\| \le \delta$, if, and only if, the nominal closed loop TF, $F = F(P, K)$ is stable and there exists a real valued (frequency independent) diagonal matrix with the structure as in (8.126), $D \in D_s$, D being real, such that $\|DFD^{-1}\|_\infty < \delta^{-1}$. This is a sufficient and necessary robust stability condition; however, it is more restrictive (on the choice of D) than that of the unstructured uncertainty. Also, the DK iteration problem is easier. Also, the system is robustly stable with respect to the structured (and arbitrarily slowly time varying linear) uncertainties with an L_2 induced norm less than or equal to δ, i.e. $\|\Delta\| \le \delta$, if, and only if, the nominal closed loop TF, $F = F(P, K)$ is stable and there exists a scaling filter $D(j\omega) \in D_s$, such that $\|DFD^{-1}\|_\infty < \delta^{-1}$.

8.6 Robust Performance Problem

We need to address the problem of robust stability as well as robust performance, the former has been studied in Section 8.5. Here, we study the performance issue. Let us take the CLTF:

$$z_P = F_P(P, K, \Delta)v_P \quad (8.135)$$

The idea is to minimize a performance measure $J(F_P(P, K, \Delta))$. Normally, the cost function $J(F_P)$ signifies the HI (or the square of H_2) norm. In the present case, the actual value of the cost (function) depends on the unknown uncertainty. In any case, the controller must be minimally robustly stabilizing; (i) the robust performance problem and (ii) nominal performance optimization subject to robust stability are important. In the former, the worst cost obtained in the assumed uncertainty set is minimized and stated as: determine a robustly stabilizing controller for the uncertain plant that minimizes the worst-case cost $J_{worst}(P, K)$ specified as, see Figure 8.2

$$J_{worst}(P, K) = \sup_\Delta \{J(F_P(P, K, \Delta)) : \|\Delta\|_\infty \le \delta\} \quad (8.136)$$

for unstructured uncertainty, and

$$J_{worst}(P, K) = \sup_\Delta \{J(F_P(P, K, \Delta)) : \Delta \in \Delta_s(\delta)\} \quad (8.137)$$

for structured uncertainty. Thus, a controller designed for the robust performance attains a given performance bound, such that the $J_{worst}(P, K) < \gamma$, for all the norm bounded uncertainties. The HI performance problem is equivalent to a μ–synthesis problem. The robust H_2 performance problem has to deal with: (i) H_2 norm for performance optimization and (ii) the HI norm for robustness; signifying a need of the mixed H_2/HI optimal/robust control problem. The nominal performance problem (with robust stability) is to determine a robustly stabilizing controller for the uncertain plant that minimizes the nominal cost $J(F_P(P, K, 0))$ defined for the nominal plant with no uncertainty. Explicitly, the uncertainty is not taken into account but it is implicit in the requirement of robust stability. Although the nominal performance might be good, there is no assurance that the performance for nonzero uncertainty would be acceptable.

8.6.1 The Robust HI Performance Problem

The idea is to determine a robustly stabilizing controller for the uncertain plant that achieves the robust HI performance bound

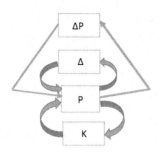

FIGURE 8.3
Schematic for robust performance.

$$\sup_{\Delta}\{\|F_P(P,K,\Delta)\|_{\infty} : \Delta \in \Delta_s(\delta)\} < \delta^{-1} \quad (8.138)$$

We know from Section 8.4 that the considered plant satisfies the HI norm bound

$$\|F_P\|_{\infty} < \delta^{-1} \quad (8.139)$$

If, and only if, the uncertain plant is robustly stable with respect to norm bound (unstructured) uncertainties Δ_P that satisfy the norm bound as $\|\Delta_P\|_{\infty} \leq \delta$. Thus, the HI performance problem is equivalent to a robust stability problem, in which case the uncertainty Δ_P is considered as a fictitious uncertainty associated with the HI norm bound as in (8.139). Now, the application of the HI performance and robust stability to the closed loop system (Figure 8.3) indicates that the control system satisfies the HI norm bound

$$\|F_P(P,K,\Delta)\|_{\infty} < \delta^{-1} \quad (8.140)$$

If, and only if, the system is robustly stable with respect to norm bound (unstructured) uncertainty Δ_P that satisfies the norm bound $\|\Delta_P\|_{\infty} \leq \delta$. Thus, the plant obtains the robust HI performance bound (8.138), or equivalently, satisfies the HI norm bound (8.140) for all $\forall \Delta \in \Delta_s(\delta)$, if, and only if, the system is robustly stable with respect to all norm bounded Δ_P that satisfies $\|\Delta_P\|_{\infty} \leq \delta$ and $\forall \Delta \in \Delta_s(\delta)$. However, this is equivalent to robust stability with respect to the set $\Delta_{P,s}(\delta)$ of uncertainties defined as

$$\Delta_{P,s}(\delta) = \{\tilde{\Delta} = \text{block diag}(\Delta, \Delta_P),$$
$$\Delta \in \Delta_s, \Delta_P \in H_{\infty}^{r_p \times r_p}, \|\Delta_P\|_{\infty} \leq \delta\}; \Delta_P \text{ is } r_p \times r_p; \quad (8.141)$$

Thus, the robust HI performance problem is equivalent to a robust stability with respect to the structured uncertainty that belongs to $\Delta_{P,s}(\delta)$, obtained by extending the original uncertainty $\Delta_s(\delta)$ with a performance-related uncertainty block Δ_P. Thus, the HI robust performance problem is equivalent to the $\Delta_{P,s}(\delta)$ obtained by extending the original uncertainty $\Delta_s(\delta)$ with a

performance-related uncertainty block Δ_P. In summary, we have the following statement. Let us specify the closed loop TF, $F = F(P, K)$ from $[v_\Delta^T \ v_P^T]^T$ to $[z_\Delta^T \ z_P^T]^T$

$$\begin{bmatrix} z_\Delta \\ z_P \end{bmatrix} = F(P,K) \begin{bmatrix} v_\Delta \\ v_P \end{bmatrix} \quad (8.142)$$

The considered system, with the uncertainty Δ in $\Delta_s(\delta)$, is robustly stable and obtains the robust HI performance bound as in (8.138), if, and only if, the CLTF, $F = F(P,K)$, (8.142) is stable and

$$\sup_{\omega} \mu_P(F(j\omega)) < \delta^{-1} \quad (8.143)$$

In (8.143), the SSV corresponds to the structure of the extended uncertainty $\Delta_{P,s}(\delta)$ as in (8.141). The problem of determination of a controller that obtains HI performance is tackled by following the procedures for the robust stabilization problem.

8.6.2 The Robust H_2 Performance Problem

In this case, the objective is to determine a controller that obtains the robust H_2 performance for the uncertain plant. The main aim is to determine a robustly stabilizing controller for the uncertain plant that minimizes the worst-case H_2 cost

$$J_{2,worst}(P,K) = \sup_{\Delta} \left\{ \|F_P(P,K,\Delta)\|_2^2 : \|\Delta\|_{\infty} \leq \delta \right\} \quad (8.144)$$

for the unstructured uncertainties, and

$$J_{2,worst}(P,K) = \sup_{\Delta} \left\{ \|F_P(P,K,\Delta)\|_2^2 : \Delta \in \Delta_s(\delta) \right\} \quad (8.145)$$

for the structured uncertainties. This problem is a combination of a performance-related H_2 cost and a robustness-related HI cost. The problem is a bit difficult to solve, and hence, an upper bound on the worst-case H_2 cost, (8.145) is introduced, as is implied by the condition as in (8.144)

$$\|v_\Delta\|_2^2 \leq \delta^2 \|z_\Delta\|_2^2 \quad (8.146)$$

In a similar way, a structured uncertainty signifies quadratic inequalities for the various components of v_Δ, and z_Δ, and an upper bound on the worst H_2 cost can be obtained by using a H_2 cost, (with a quadratic constraint) for the considered plant

$$J_{rob}(P,K) = \sup_{v_\Delta =} \left\{ \sum_{k=1}^{m} \left[\int_0^{\infty} z_P^T(t) z_P(t) dt : v_P = e_k \delta(t) \right] \right\} \quad (8.147)$$

This is subject to the following constraint

$$\sum_{k=1}^{m}\left[\int_{0}^{\infty}[z_\Delta^T(t)z_\Delta(t)-\delta^{-2}v_\Delta^T(t)z_\Delta(t)]dt:v_P=e_k\delta(t)\right]\geq0 \quad (8.148)$$

The constraint in (7.146) is called an integral quadratic constraint, and the cost (8.147) is evaluated by introducing the associated Lagrange multiplier, λ (in the literature it is often used without squaring it, and it is called LM or co-state, or even the adjoint variable):

$$J_{rob}(P,K,\lambda)=\sup_{v_\Delta=}\left\{\sum_{k=1}^{m}\left[\int_{0}^{\infty}\left[z_P^T(t)z_P(t)+\lambda^2(z_\Delta^T(t)z_\Delta(t)\right.\right.\right.$$
$$(8.149)$$
$$\left.\left.\left.-\delta^{-2}v_\Delta^T(t)v_\Delta(t))\right]dt:v_P=e_k\delta(t)\right]\right\}$$

It can be shown that, if the HI norm from v_Δ to z_Δ is less than δ^{-1}, then λ can be chosen such that the cost in (8.147) is bounded, and

$$J_{2,worst}(P,K)\leq J_{rob}(P,K),$$
$$\text{such that } (8.146)\leq J_{rob}(P,K,\lambda),\forall\lambda>0 \quad (8.150)$$

Thus, the robust performance problem is to determine a constant λ and a stabilizing controller for the considered plant, such that the cost in (8.149) is minimized. A linear time-invariant uncertainty implies that the following is true

$$\left\|v_\Delta(j\omega)\right\|^2\leq\delta^2\left\|z_\Delta(j\omega)\right\|^2,\ \forall\omega \quad (8.151)$$

The inequality of (8.151) can be used for constructing a tighter upper bound on the worst-case H_2 cost, and the associated constraint cost can be evaluated in terms of Lagrange function (LF) that is frequency dependent LM, $\lambda(j\omega)$. The scalar λ that minimizes $J_{rob}(P,K,\lambda)$ can be found by some direct search method. For the problem of determining a controller that minimizes (8.149), with a fixed λ, a new variable is chosen as

$$z=\begin{bmatrix}\lambda z_\Delta\\z_P\end{bmatrix} \quad (8.152)$$

Also, consider the following plant

$$\dot{x}(t)=Ax(t)+B_0v_P(t)+B_1v_\Delta(t)+B_2u(t),\ x(0)=0$$
$$z(t)=C_1x(t)+D_{12}u(t) \quad (8.153)$$
$$y(t)=C_2x(t)+D_{20}v_P(t)+D_{21}v_\Delta(t)$$

Then, the cost function, (8.149) becomes

$$J_{mxd}(P,K)$$
$$=\sup_{v_\Delta=}\left\{\sum_{k=1}^{m}\left[\int_{0}^{\infty}\left[z^T(t)z\ (t)-c^2v_\Delta^T(t)v_\Delta(t)\right]dt:v_P=e_k\delta(t)\right]\right\}$$
$$(8.154)$$

In (8.154), $c=\lambda/\delta$. The mixed H_2HI problem is stated as: For the considered system, assume that the conditions/assumptions a1–a14 are valid, then there exists a state feedback controller, such that the mixed H_2H_∞ cost $J_{mxd}(P,K)$ is bounded, if, and only if, there also exists a positive (semi-) definite solution to the following algebraic Riccati equation, (8.71):

$$A^TS+SA-SB_2(D_{12}^TD_{12})^{-1}B_2^TS$$
$$+\gamma^{-2}SB_1B_1^TS+C_1^TC_1=0 \quad (8.155)$$

With the following matrix also being stable, (8.72)

$$A-B_2(D_{12}^TD_{12})^{-1}B_2^TS+\gamma^{-2}B_1B_1^TS \quad (8.156)$$

When the forgoing conditions are satisfied, then the state feedback controller that minimizes the cost $J_{mxd}(P,K)$ is given by

$$u(t)=K_{mxd}x(t) \quad (8.157)$$

In (8.157), the controller gain is given as

$$K_{mxd}=-(D_{12}^TD_{12})^{-1}B_2^TS \quad (8.158)$$

The results of (8.157) and (8.158) can be obtained by following the procedures similar to the optimal H_2 and HI problems, for which the following expansion is useful, (8.74):

$$\int_{0}^{\infty}[z^T(t)z(t)-\gamma^2v_\Delta^T(t)v_\Delta(t)]dt=$$
$$\int_{0^+}^{\infty}\left\{[u(t)-u^0(t)]^TD_{12}^TD_{12}[u(t)-u^0(t)]\right. \quad (8.159)$$
$$\left.-\gamma^2[v_\Delta(t)-v_\Delta^0(t)]^T[v_\Delta(t)-v_\Delta^0(t)]\right\}dt$$

In (8.159), we have, see (8.75)

$$u^0(t)=K_{mxd}x(t),\text{ and }v_\Delta^0(t)=\gamma^2B_1^TSx(t) \quad (8.160)$$

We see from (8.159) that the mixed H_2HI optimal output feedback controller can be constructed in terms of the optimal state feedback law and an optimal estimator for the transformed system:

$$\dot{x}(t) = \tilde{A}x(t) + B_0 v_P(t) + \tilde{B}_1 \tilde{v}_\Delta(t) + \tilde{B}_2 u(t)$$

$$\tilde{z}(t) = \tilde{C}_1 x(t) + \tilde{D}_{12} u(t) \qquad (8.161)$$

$$y(t) = \tilde{C}_2 x(t) + D_{20} v_P(t) + \tilde{D}_{21} \tilde{v}_\Delta(t)$$

In (8.161), we have the following equivalence

$$\tilde{z}(t) = (D_{12}^T D_{12})^{1/2} [u(t) - u^0(t)]; \quad \tilde{v}_\Delta(t) = v_\Delta(t) - v_\Delta^0(t) \qquad (8.162)$$

The matrices in (8.161) are as in (8.81) and (8.82). Similar to the optimal H_2 and HI problems, the expansion of (8.159) degenerates to the optimal output feedback control problem for the mixed H_2HI problem for the system of (8.161). The mixed H_2HI estimation problem is to determine a stable estimator $\hat{\tilde{z}} = FY$ that minimizes the H_2HI cost

$$J_{mxd,e}(F) = \sup_{\tilde{v}_\Delta} \left\{ \sum_{k=1}^{\infty} \left[\int_0^\infty \left[\tilde{z}(t) - \hat{\tilde{z}}(t) \right]^T [\tilde{z}(t) - \hat{\tilde{z}}(t)] \right. \right.$$
$$\left. \left. - \gamma^2 \tilde{v}_\Delta^T(t) \tilde{v}_\Delta(t) \right] dt : v_P = e_k \delta(t) \right] \right\} \qquad (8.163)$$

This problem does not lead to a closed form solution. However, the following result can be used. Let us consider the system as in (8.161). There exists a stable estimator $\hat{\tilde{z}} = FY$ that obtains a bounded cost $J_{mxd, e}(F)$ if, and only if, there also exists a matrix L, such that all the eigenvalues of the matrix

$$\tilde{A} + L\tilde{C}_2 \qquad (8.164)$$

have negative real parts and the following algebraic Riccati equation

$$(\tilde{A} + L\tilde{C}_2)^T P + P(\tilde{A} + L\tilde{C}_2) + \gamma^{-2}$$
$$P(\tilde{B}_1 + L\tilde{D}_{21})(\tilde{B}_1 + L\tilde{D}_{21})^T P + \tilde{C}_1^T \tilde{C}_1 = 0 \qquad (8.165)$$

has a symmetric positive (semi-) definite solution, such that the following matrix is stable

$$\tilde{A} + L\tilde{C}_2 + \gamma^{-2}(\tilde{B}_1 + L\tilde{D}_{21})(\tilde{B}_1 + L\tilde{D}_{21})^T P \qquad (8.166)$$

When the foregoing conditions are satisfied, then the estimator F_L defined as

$$\dot{\hat{x}}(t) = \tilde{A}\hat{x}(t) - L[y(t) - \tilde{C}_2 \tilde{x}(t)]; \quad \hat{x}(0) = 0$$
$$\hat{\tilde{z}}(t) = \tilde{C}_1 \hat{x}(t) \qquad (8.167)$$

obtains the bounded cost

$$J_{mxd,e}(F_L) = Tr\{(B_0 + LD_{20})P^T(B_0 + LD_{20})\} \qquad (8.168)$$

The filter structure (8.167) is optimal

$$\inf_F J_{mxd,e}(F) = \inf_L J_{mxd,e}(F_L) \qquad (8.169)$$

The infimum on the left of (8.169) is wrt all the stable estimators. If we note the state error as $\tilde{x} = x - \hat{x}$, then it is given as

$$\dot{\tilde{x}}(t) = (\tilde{A} + L\tilde{C}_2)\tilde{x}(t) + (B_0 + LD_{20})v_P(t) + (\tilde{B}_1 + L\tilde{D}_{21})\tilde{v}_\Delta(t)$$
$$\tilde{z}(t) - \hat{\tilde{z}}(t) = \tilde{C}_1 \tilde{x}(t) \qquad (8.170)$$

The mixed H_2HI optimal output feedback controller problem is now stated as follows: Consider the system of (8.153) and the assumptions a1–a4 hold good. Then, there exists a controller $u = Ky$ that obtains a bounded cost $J_{mxd}(P, K)$ if, and only if, the conditions of the previous two results are satisfied. When these conditions are satisfied, the controller, $u = K_L y$ defined by

$$\dot{\hat{x}}(t) = \tilde{A}\hat{x}(t) + \tilde{B}_2 u(t) - L[y(t) - \tilde{C}_2 \tilde{x}(t)]$$
$$u(t) = K_{mxd}\hat{x}(t) \qquad (8.171)$$

obtains the following bounded cost

$$J_{mxd}(P, K_L) = Tr\{B_0^T S B_0\} + Tr\{(B_0 + LD_{20})^T P(B_0 + LD_{20})\} \qquad (8.172)$$

The controller structure of (8.171) is optimal

$$\inf_K J_{mxd}(P, K) = \inf_L J_{mxd}(P, K_L) \qquad (8.173)$$

8.7 Design Aspects

In the theory of controller design, the controller synthesis is the final stage involving the selection of the: (i) cost function, (ii) controlled output z (or y as the case may be) that is to be used in the synthesis step, (iii) design parameters like weights for the I/O in the chosen cost function, and (iv) weights that depend on the frequencies of the operating range of the dynamic system that is to be controlled.

Realistic magnitudes and the structure of/for the uncertainties related to the plant are also important. Often, the control objectives are stated in a qualitative manner: the requirement of fast response, small overshoot, and adequate robustness; hence, it is necessary to translate these objectives into some quantitative synthesis problem.

8.7.1 Some Considerations

Let us consider the general control system, with y (this might not be the measured output) and u as the I/O control signal, and w and v as the process and measurement disturbances, respectively. These need not necessarily be the random noise processes, see Figure 8.4. The reference signal is r. Often the signals are meant to be in the Laplace domain, and yet, we use the lowercase letters. First, we have the following relation between y, r, w, and v for the closed loop system

$$y = (I+GK)^{-1}GKr + (I+GK)^{-1}G_w w - (I+GK)^{-1}GKv$$
(8.174)

The sensitivity (S) and the complementary sensitive (T) functions are given as

$$S = (I+GK)^{-1}; \; T = (I+GK)^{-1}GK$$
(8.175)

With S and T, (8.174) is rewritten as

$$y = Tr + SG_w w - Tv$$
(8.176)

It is easy to see, by directly adding (8.175), that the S and T are related as

$$S + T = I$$
(8.177)

We see that the control input is given as

$$u = K(r - y_m); \; u = KSr - KSG_w w - KSv$$
(8.178)

In the considered control system, the aim is to see that the output y follows the reference signal r and, hence, to make $e = y-r$ as small as possible. For the closed loop control system, we have

$$e = -Sr + SG_w w - Tv$$
(8.179)

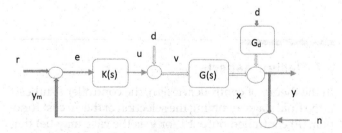

FIGURE 8.4
General control system (often $C(s)$ or $H(s)$ are used for $K(s)$; and for $G(s)$, $P(s)$ is used; also, the disturbance is added between u and v and in that case n shifts to d and G_d is identity).

The sensitive S gives the reduction in sensitivity in the output y that is obtainable by the use of feedback. Without feedback for the open loop, we have

$$y = GKr + G_w w$$
(8.180)

Then, the response from the reference signal r to the disturbance signal w with feedback is obtained by multiplying (8.180) by S; actually S equals the relative sensitivity dT/T of the CLTF (8.176) to the relative plant model error dG/G at a given frequency

$$S = \frac{dT/T}{dG/G}$$
(8.181)

The relation of (8.181) can be obtained by differentiating (8.175) wrt G. From (8.179), one can see that a good tracking of signal r can be achieved, i.e. smaller and smaller e, if the S is quite small. However, the suppression of the measurement noise v requires that T should be quite small. Interestingly, due to (8.177), both sensitivities cannot be made small simultaneously, this requires a trade-off: good tracking of the reference signal and lower sensitivity to the measurement noise. Interestingly enough, the reference tracking is a low-frequency activity, and the noise being high-frequency activity. One can keep the sensitivity function S smaller at the low frequencies, and the complementary sensitivity function T smaller at the higher frequencies. This is the performance requirement, however, the robustness is also important, and the sensitivity S is also associated with this aspect. In the scalar case, the distance from a point on the Nyquist plot of the loop TF, $G(j\omega)K(j\omega)$ to the point $(-1,0)$ is $|1+G(j\omega)K(j\omega)| = 1/S(j\omega)$; thus, the distance of the Nyquist plot to the point $(-1,0)$ is greater than $1/M_S$ if, and only if

$$|S(j\omega)| < M_S, \; \forall \omega$$
(8.182)

Thus, the classical/conventional stability margins impose a bound (in the form of a norm) on the sensitivity function S. Interestingly enough, the robustness aspects do not only lead to a bound on the sensitivity function S, but it happens that the complementary sensitivity function T is also associated with the robustness. For an uncertain plant model, the real plant is described by the nominal model G and a multiplicative uncertainty as follows

$$G_\Delta = (1+\Delta_M)G$$
(8.183)

In (8.183), the Δ_M is the relative uncertainty that is assumed bounded as

$$|\Delta_M(j\omega)| \leq l(\omega)$$
(8.184)

In fact, this uncertainty is unknown as such. We have by the simple manipulations

$$1 + G_\Delta K = 1 + (1 + \Delta_M)GK = (1 + GK)(1 + \Delta_M T) \quad (8.185)$$

In fact, (8.185) is the denominator of the CLTF, and for the stability, the following must hold

$$1 + G_\Delta K \neq 0 \quad (8.186)$$

By using (8.185), (8.186) must be true for all Δ_M, such that (8.184) is true if, and only if

$$|T(j\omega)| < 1/l(\omega); \quad \forall \omega \quad (8.187)$$

From (8.187), we see that the requirement of the robustness bound imposes a bound on T. The conditions specified for these sensitivities for a certain requirement of performance and robustness can be used directly for controller design; particularly, in various loop shaping procedures: the controller is designed by shaping the magnitudes $|S(j\omega)|$ and $|T(j\omega)|$ as functions of frequency so that performance and robustness criteria are satisfied. These procedures are: (i) the classical ones for SISO systems, and (ii) for HI optimal (robust) control problem.

8.7.2 Basic Performance Limitations

There are some performance situations that apply to a stable control system: (i) for the considered control system, it is theoretically not possible to obtain an arbitrary disturbance reduction at all frequencies (even though the system might be accurately known; (ii) this leads to certain limits on the attainable performance; and (iii) the latter are related to some basic properties that any stable control system TF (CSTF) must satisfy. In this direction, we have the relationship for a minimum phase CSTF, known as Bode's gain-phase relationship (BGPR): we have the $G(s)$ as a proper stable TF with no zeros on the right half plane (RHP) and no time delays, and the also $G(0) > 0$. Then, there is a unique relationship between the magnitude and the phase of the TF:

$$\arg(G(j\omega_0)) = \frac{1}{\pi}\int_{-\infty}^{\infty}\frac{d\log\{|G(j\omega)|\}}{d\log(\omega)}\log\left\{\left|\frac{\omega+\omega_0}{\omega-\omega_0}\right|\right\}\frac{d\omega}{\omega} \quad (8.188)$$

With (8.188), one gets the minimum (possible) phase lag that any system with the given magnitude response can have; such systems are called the minimum phase systems. The non-minimum phase systems have more phase lags and are the ones that have RHP zeros and/or time delays. The factor $\log\left\{\left|\frac{\omega+\omega_0}{\omega-\omega_0}\right|\right\}$ is infinite at $\omega = \omega_0$ and finite otherwise; hence, $\arg(G(j\omega_0))$ is primarily determined by $\frac{d\log\{|G(j\omega)|\}}{d\log(\omega)}$ at $\omega = \omega_0$. Then, using

using the fact that $\int_{-\infty}^{\infty}\log\left\{\left|\frac{\omega+\omega_0}{\omega-\omega_0}\right|\right\}\frac{d\omega}{\omega} = \frac{\pi^2}{2}$, one gets the approximate formula

$$\arg(G(j\omega_0)) \approx \frac{\pi}{2}\left(\frac{d\log\{|G(j\omega)|\}}{d\log(\omega)}\right)|_{\omega=\omega_0} \; rad. \quad (8.189)$$

From (8.189), one can see that a slope of −20 dB/decade in the gain at the vicinity of ω_0, i.e. $\left(\frac{d\log\{|G(j\omega)|\}}{d\log(\omega)}\right) = -1$, would mean the phase angle is approximately $-\pi/2$ rad; and a slope of −40 dB/decade would mean phase of $-\pi$ rad. Thus, the classical design concept is that to ensure closed loop stability, the slope of the loop TF, $\{G(.)K(.)\}$ should be in the range −20 to −30 (not −40) dB/decade at the gain cross over point, where the magnitude of $\{G(.)K(.)\}$ is equal to 1. This would mean that the phase lag is less than −180 deg. The BGPR and related results have significant implications for the attainable control system performance; this is because the BGPR holds for the loop TF $L(s) = G(s)K(s)$. A small value of the sensitivity S implies that the loop TF is large; however, one knows from classical SISO control theory that for stability, magnitude of $L(.) < 1$ should hold at the phase crossover frequency ω_p, at which the phase lag is −180 deg., i.e. $\arg(L(j\omega_c)) = -180$ deg. The relationship (8.188) and (8.189) for the loop TF $L(.)$ gives the minimum phase for any given $|L(j\omega)|$. Thus, there are limitations on the gain-functions $|L(j\omega)|$, and $|S(j\omega)|$, so that the closed loop system is stable. That the $S(.)\sim I$ (identity) should hold at high frequencies is implied by robustness considerations. Let us look at the multiplicative uncertainty (8.183) and the stability condition (8.187). We cannot determine the phase shift or the relative magnitude of $G(.)$ accurately at high frequencies; thus, we have $G(.)\to 0$, and, due to large uncertainty in the phase and relative magnitude, the bound (8.184) on the relative uncertainty will be large, $l(\omega) >> 1$; and by (8.187), robust stability implies the condition $|L(j\omega)| << 1$; and by (8.177), $S(.) \sim I$ at high frequencies. In fact, we need a good reference signal tracking and disturbance rejection at low frequencies; and $S(j\omega)$ should be small for small ω. On the other hand, we know that all physical dynamical systems attenuate high frequencies, so that $G(j\omega)\to 0$ as $\omega\to\infty$; hence, $S(j\omega)\to I$ as $\omega\to\infty$. From the foregoing, it follows that the sensitivity S cannot be made small at all frequencies simultaneously; so, make $S(j\omega)$ small in a given frequency band. However, if we try to make $S(j\omega)$ small in some frequency band, its value at some other frequencies would necessarily increase. Despite this, the attainable average performance can be expressed quantitatively. There is another classical result for SISO systems called the Bode's sensitivity integral (BSI); Let $L(s)$ have: (i) no RHP zeros, (ii) at least two more poles than zeros, and (iii) N_p number of RHP poles, p_i; then, for the closed loop stability, S must satisfy the following condition

$$\int_0^\infty \log\{|S(j\omega)|\}d\omega = \pi \sum_{i=1}^{N_p} \text{Re}(p_i) \qquad (8.190)$$

In (8.190), Re(.) denotes the real part of the argument; and for the stable plant $N_p = 0$ and (8.190) reduces to

$$\int_0^\infty \log\{|S(j\omega)|\}d\omega = 0 \qquad (8.191)$$

From (8.191), one can interpret that when $\log |S|$ is negative, i.e. $|S| < 1$, it is the area of the sensitivity reduction, and when $\log |S|$ is positive, i.e. $|S| > 1$, it is the area of the sensitivity increase; so balancing out the benefits and the cost of feedback. For a MIMO system, the condition (8.190) is stated as

$$\int_0^\infty \log\{|\det(S(j\omega))|\}d\omega = \pi \sum_{i=1}^{N_p} \text{Re}(p_i) \qquad (8.192)$$

Now S is a matrix one uses the determinant in (8.192). By (8.190), the obtainable performance is degraded when RHP has unstable poles, since $\text{Re}(p_i) > 0$. Also, there is a corresponding BSI relation for the non-minimum phase loop TFs with RHP zeros, and for this case, the performance is degraded because the area of the sensitivity increases when $\log |S|$ is positive. From the BSI, we infer that one should concentrate the control effort to a frequency range: $0 \le \omega \le \omega_1$. As the frequency range $\omega_1 \le \omega \le \omega_\infty$ (that contributes to the positive part of (8.190)) is infinite, $|S(j\omega)|$ may be made arbitrarily close to one at all ω. This leads to the result of obtainable performance for the minimum phase systems. Let $G(s)$ have no RHP zeros, and let $\omega_1 > 0$; then, for any $\varepsilon > 0, \delta > 0$, there exists a stabilizing K, such that

$$|S(j\omega)| < \varepsilon, \forall |\omega| \le \omega_1; \ |S(j\omega)| < 1 + \delta, \forall \omega \qquad (8.193)$$

The result of (8.193) applies to the case when the only requirement on $S(.)$ is that it should be small in some frequency range. It is also required that the loop TF should fall off sufficiently at high frequencies $|L(j\omega)| < \frac{1}{\omega^m}$ for some $m > 0$. The BSI (8.190) signifies that, if $|S|$ is made small in a frequency range $0 \le \omega \le \omega_1$, then $|S|$ must necessarily be large at some other frequencies. Performance limitation in non-minimum systems is stated next. For a non-minimum system, define

$$M = \max_{|\omega \le \omega_1|} |S(j\omega)| \qquad (8.194)$$

Now, assume that $G(s)$ has at least one RHP zero, $G(s_0) = 0$, for some s_0 with $\text{Re}(s_0) > 0$; then, for a positive number $c > 0$, that depends only on ω_1 and G, for every stabilizing K we have

$$M|S(j\omega)|^c \ge 1; \ \text{for some} \ \omega > \omega_1 \qquad (8.195)$$

The result in (8.195) signifies that as $M \to 0$ for a non-minimum phase system, $|S(j\omega)|$ increases without bound at some other frequencies.

8.7.3 Application of H$_\infty$ Optimal Control to Loop Shaping

In the foregoing section/s, we have seen that the control objectives can be stated in terms of the magnitudes of CLTF, $S(.)$, and $T(.)$. Interestingly, these types of the problems can be solved by HI optimal (robust) synthesis approaches (HICA). Let have the considered control system and the error signal as $e = y-r$, (8.179), and the control input signal as in (8.178). Also, there are bounds on: (i) $S(.)$; (ii) $T(.)$, (8.177); and (iii) the control signal, i.e. on KS, (8.178). Ironically, the design specifications normally imply that S, T, and KS should be small in different frequency ranges: S should be small at low frequency and T and KS should be small at higher frequencies. This requirement is expressed as follows

$$J_1 = \begin{vmatrix} W_S S \\ W_T T \\ W_u KS \end{vmatrix} \qquad (8.196)$$

Obviously, in (8.196), there are weighting factors associated with the functions, S, T, and KS. Then, the design specifications can be met by selecting these weights properly and minimizing the cost function J_1; the problem is a mixed sensitivity type. In the conventional sense, the design can be carried out using the H$_2$ cost function

$$J_2(Ji) = \|J_1\|_2^2 = \frac{1}{2\pi}\int_{-\infty}^\infty \begin{vmatrix} W_S S \\ W_T T \\ W_u KS \end{vmatrix}^T (-j\omega) \begin{vmatrix} W_S S \\ W_T T \\ W_u KS \end{vmatrix}(j\omega)d\omega \qquad (8.197)$$

$$= \|W_S S\|_2^2 + \|W_T T\|_2^2 + \|W_u KS\|_2^2$$

For (8.197), even though the overall cost function could be made smaller, there is no assurance that S, T, and KS would be automatically small at all frequencies in the range of interest. By using the HICA, one can put different and specified bounds on $S(.)$, $T(.)$, and $K(.)S(.)$ at all the frequencies. This is illustrated in the case of the SISO control system. The bound in (8.196) implies

$$\|J_1\|_\infty < \gamma; \|W_S S\|_\infty < \gamma; \|W_T T\|_\infty < \gamma; \|W_u KS\|_\infty < \gamma \qquad (8.198)$$

$$\left|W_S(j\omega)S(j\omega)\right| < \gamma \ ; \ \left|W_T(j\omega)T(j\omega)\right| < \gamma \ ;$$

$$\left|W_u(j\omega)K(j\omega)S(j\omega)\right| < \gamma \ ; \ \forall \omega \tag{8.199}$$

$$\left|S(j\omega)\right| < \gamma / \left|W_S(j\omega)\right| ;$$

$$\left|T(j\omega)\right| < \gamma / \left|W_T(j\omega)\right| ; \tag{8.200}$$

$$\left|K(j\omega)S(j\omega)\right| < \gamma / \left|W_u(j\omega)\right| ; \ \forall \omega$$

Often, the frequency bands are non-overlapping, where the different TFs should be constrained. At low frequencies, a small sensitivity function is required for good performance; hence, we can have $\left|W_S(j\omega)\right| \gg \left|W_T(j\omega)\right|$; and $\left|W_S(j\omega)\right| \gg \left|W_u(j\omega)\right|$ for $|\omega| < \omega_1$. We can have the maximum singular value $\bar{\sigma}\{Ji(j\omega) \approx \left|W_S(j\omega)S(j\omega)\right|$. In HI controller, if γ is not too much larger than γ_{inf}, the CLTF is close to the HI norm bound, at least at low frequencies. That is, we have $\bar{\sigma}\{Ji(j\omega) \approx \gamma$ and

$$\left|S(j\omega)\right| \approx \gamma / \left|W_S(j\omega)\right|, \ |\omega| < \omega_1 \tag{8.201}$$

From the foregoing discussion, one can see that the weight filter W_S determines the shape of the magnitude of $S(.)$ in the low frequency range; similarly, for other TFs in their frequency ranges. Thus, by (8.201), HICA can be utilized in loop shaping procedures, where by the magnitudes $|S(.)|$ can be modulated by the choice of the weight filters. Thus, to solve the HICA problem that attains the bound (8.198) on the TF matrix J_1, one can specify a state space model that has the CLTF matrix J_1. In this case, we see that the signal y_m is given by

$$y_m = Tr + SG_w w + Sv \tag{8.202}$$

We have the following, with $w = 0$ and $r = 0$:

$$\begin{vmatrix} y_m \\ y \\ u \end{vmatrix} = \begin{vmatrix} S \\ -T \\ -KS \end{vmatrix} v \tag{8.203}$$

Then, a state-space representation with the CLTF matrix J_1 can be constructed in a simple way. These $W(.)$ filters should be chosen large on frequency bands to constrain the magnitudes of the associated CLTFs and small at other bands; these are typically low-pass (LP), high-pass (HP), or band-pass (BP) filters. Let us specify the requirement to construct a weight filter $W(.)$, such that: (i) $1/\left|W(j\omega)\right| = A$ at small ω, (ii) $1/\left|W(j\omega)\right| = B$ at large ω, and (iii) cross over is at $\sim \omega_0$. Hence, we have $A > 1$ and $B < 1$ (HP, high pass filter) or $A < 1$, and $B > 1$ (LP, low pass filter) that obtains this condition of the filter and it is given by the following weight TF

$$W(s) = \frac{1}{B} \frac{s + B\omega_0}{s + A\omega} \tag{8.204}$$

The crossing (crossover) is made steeper by selecting an n-th order filter

$$W_n(s) = \frac{1}{B} \frac{(s + B^{1/n}\omega_0)^n}{(s + A^{1/n}\omega_0)^n} \tag{8.205}$$

If required, a band-pass filter can be constructed by combining appropriately an LP filter and a HP filter, whose passbands would then overlap. In many situations, a requirement would be that the controller should eliminate any SS offsets in the error signal e; i.e. $S(0) = 0$ would hold, this is attained by allowing the filter W_S having the factor $1/\text{sec}$; this is the classical requirement that the SS offset can be eliminated if the controller has an integrator, $1/\text{sec}$. The state corresponding to the unstable factor $1/\text{sec}$ would not be observable from the measured output (available to the controller); hence, the factor $1/(s + \varepsilon)$ is used instead, where $\varepsilon > 0$ is a small constant. The selection of proper weights in HICA might require several design iterations. It would be very advisable to use some rules of thumb (similar to those used in tuning of PID controllers, and Ziegler-Nichols). The frequency regions are specified with respect to ω_π, at which the phase equals -180 deg., $\arg(G(j\omega_\pi)) = -\pi$. The three ranges can be chosen: (i) a medium frequency (*MF*) range close to ω_π, (ii) low frequency (*LF*) range, and (iii) a high frequency (*HF*) range; and usually the width of the MF range is taken to be one decade:

   i. *LF*: $\omega < 0.3\omega_\pi$
   ii. *MF*: $0.3\omega_\pi < \omega < 3\omega_\pi$
   iii. *HF*: $\omega > 3\omega_\pi$

With the foregoing discussion, one can see that different specifications for the closed loop system can be given for the different frequency regions: (a) in the *LF* case, the aim is to make the closed loop gain (from the process disturbance w, to the tracking error $e = y-r$) small, and to eliminate steady state offsets. So, let $G_w = G$, the disturbance enters at the input and the CLTF from w to e is $S(s)G(s)$, then the *LF* cost is taken as

$$J_{LF} = \left\|W_{LF}SG\right\|_\infty \tag{8.206}$$

In (8.206), $W_{LF}(s) = 1/(s + \varepsilon)$; (b) for the *MF*, one has to ensure acceptable stability margins. We know that $||S||_\infty$ is the inverse of the smallest distance of the Nyquist plot of the LTF from the $(-1,0)$ point, and hence, one can use the following cost

$$\left\|S\right\|_\infty < M_S \tag{8.207}$$

The suggested values for the $M(.)$ are in the band of 1.4 to 2.0; the good one being 1.7. Also, since $T(.)$ is linked to robustness, the following bound can be used

$$\|T\|_\infty < M_T \tag{8.208}$$

In (8.208), one can select $M_T < M_S$; and a good choice is $M_T = (M_S/3)$. This bound is called the "peak M value" in the classical controller design, because a peak in $|T(.)|$ gives rise to oscillations at the corresponding frequency after a step change is made in the set point; (c) in the case of the *HF* region, the requirement is to keep the control signal limited. Thus, the magnitude of the CLTF, $K(s)S(.)$ should be kept small; hence, we have the cost function as

$$J_{HF} = \|W_{HF}KS\|_\infty \tag{8.209}$$

In (8.209), the filter $W_{HF}(s)$ is a *HP* filter with the pass-band in the *HF* region and making J_{HF} small would ensure that the high frequency robustness condition would hold true. Thus, the controller design focuses on determining a controller that makes the costs J_{LF} and J_{HF} as small as possible and ensures that the *MF* bounds are also satisfied. So, the main aspect is to find proper weights in the cost function (8.196), like using $W_S = W_{LF}G$ to capture the *LF* cost J_{LF}, then solving the associated HI control problem to obtain the performance bound as in (8.198). Here, such a trade-off is still involved for the relative magnitudes of the *LF* and *HF* frequency costs, and hence, a compromise for these two costs, J_{LF} and J_{HF}, to be sufficiently small has to me made.

Example 8.1:

Small gain theorem [14]. Let $G(s)$ and $H(s)$ (often $K(s)$ is used) be the plant and the feedback controller TFs, see Figure 8.5 for various equivalent representations of both. These are real-rational, proper, and stable TFs. The TF $G(s)$ or $H(s)$ (or both) are assumed to be strictly proper and are zero at $s = \infty$. The feedback system is assumed to be internally stable. Mostly there are four TFs from two inputs to two outputs individually, as seen from Figure 8.5. The TF from input u_1 to the loop error is $E(s)/U_1(s) = 1/(1-G(s)H(s))$. From the Nyquist criterion, one would gather that the feedback system is internally stable if, and only if, the Nyquist plot $GH(s)$ does not pass through or encircle the point $s = 1$. Thus, a sufficient condition for internal stability is the small gain condition $||GH(s)||_\infty < 1$. This idea is now extended to the problem of robust stabilization. Next, we have the plant with TF as $G(s) + \Delta G(s)$, and $H(s)$. Here $G(s)$ is the nominal plant, and the $\Delta G(s)$ is an unknown perturbation. The latter could be due to unmodeled dynamics or variation/s of the parameter/s of the basic and nominal plant. We assume that these three TFs are real-rational, $G(s)$ and $\Delta G(s)$ are strictly proper and stable, and $H(s)$ is proper. We also assume that the feedback system is internally stable for $\Delta G(s) = 0$. The question is how much large $|\Delta G(s)|$ can be tolerated so that the internal stability is still maintained. If we use the frequency responses of these TFs, then we can obtain the ideas of the magnitudes and phase at several frequencies and get an upper bound for the disturbance plant $|\Delta G(s)|$ at several values of frequency ω. Let us define the bound on this perturbed plant as

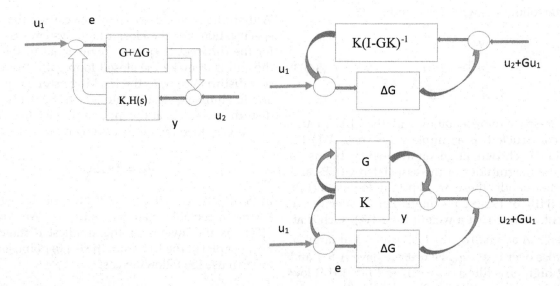

FIGURE 8.5
Feedback systems with various (equivalent) representations of the perturbed plant and controller.

$$|\Delta G(j\omega)| < |\gamma(j\omega)| \; \forall 0 \leq \omega \leq \infty$$

$$\left\| \gamma^{-1}\Delta G \right\|_\infty < 1 \tag{8.210}$$

In (8.210), γ is a radius of the function belonging to HI norm. From (8.210), the question arises, how large $|\gamma|$ can be so that the internal stability is maintained. We assume that the nominal feedback system, $H(s)(1-P(s)H(s))^{-1} \in \gamma(s)$, $\gamma \in H_\infty$ norm. Based on the small gain theorem, we say that the perturbed system will be internally stable if the following holds

$$\left\| \Delta GH(1-GH)^{-1} \right\|_\infty < 1 \tag{8.211}$$

Combining (8.210) with (8.211), we obtain the sufficient condition as

$$\left\| \gamma H(1-GH)^{-1} \right\|_\infty \leq 1 \tag{8.212}$$

To obtain (8.212), we have used the following sub-multiplicative property of the HI (H-Infinity) norm:

$$\left\| \gamma G \right\|_\infty \leq \left\| \gamma \right\|_\infty \left\| G \right\|_\infty \tag{8.213}$$

Thus, an HI norm bound on a weighted CLTF, i.e. condition (8.212) is sufficient for robust stability.

Example 8.2:

Choosing weight filter [15]. Let the nominal and the perturbed plants be given as

$$G_0(s) = \frac{1}{s^2}; \;\; G(s) = \frac{1}{s^2}e^{-\tau s}; \; \tau \in [0, 0.1] \tag{8.214}$$

In (8.214), time delay is given to the extent of the range specified. Then, the uncertain plant $G(s)$ is specified as

$$G \in G_\Delta = \{(1 + W\Delta)G_0 \; \|\Delta\|_\infty \leq 1\} \tag{8.215}$$

The weight $W(.)$ should be chosen, such that

$$\left| \frac{G(j\omega)}{G_0(j\omega)} - 1 \right| \leq |W(j\omega)|; \; \forall \omega \in R \tag{8.216}$$

Based on (8.216), the weight can be chosen using the series approximation of the exponential function as follows

$$|W(j\omega)| \geq \left| \frac{\frac{1}{s^2}e^{-j\tau\omega}}{\frac{1}{s^2}} - 1 \right| \Rightarrow W(s) = \frac{0.21s}{(1+0.1s)} \tag{8.217}$$

Example 8.3:

Use of Hardy space H$_2$ [14]. This space consists of all complex valued functions $G(s)$ that are analytic in the open LHS plane and satisfy

$$\left[\sup_{\xi > 0} \left(\frac{1}{2\pi} \right)^{-1} \int_{-\infty}^{\infty} |G(\sigma + j\omega)|^2 \, d\omega \right]^{1/2} < \infty \tag{8.218}$$

In (8.219), the LH side is the H$_2$ norm of G, $||G||_2$. Presently, we study the real function belonging to RH_2 if, and only if, it is sable and strictly proper; and R (like "γ") is a radius of the function belonging to H$_2$ norm. For the function $G(s)$, it can be seen that its H$_2$ norm can be obtained by integrating over the imaginary axis

$$\|G\|_2 = \left[\left(\frac{1}{2\pi} \right)^{-1} \int_{-\infty}^{\infty} |G(j\omega)|^2 \, d\omega \right]^{1/2} \tag{8.219}$$

Now, take a one-sided $x(t)$, zero for $t < 1$, with its Laplace transform (LT) belonging to RH_2, then the Plancherel's theorem gives

$$\int_{-\infty}^{\infty} x(t)^2 \, dt = \|X(s)\|_2^2 \tag{8.220}$$

The RH side of (8.220) is interpreted physically as the energy of the signal $x(t)$. Let us now consider a system with TF $G(s)$ in RH_∞, with its I/O signal as $x(t)$ and $y(t)$. It can be seen that if

$$X(s) \in RH_2, \text{ and } \|X(s)\|_2 = 1, \text{ then}$$

$$Y(s) \in RH_2 \text{ and } \|Y(s)\|_2 \leq \|G\|_\infty \tag{8.221}$$

The HI norm of the TF provides a bound on the system gain

$$\sup\{\|Y(s)\|_2 : X(s) \in RH_2, \|X(s)\|_2 = 1\} \tag{8.222}$$

It means that HI norm of the TF equals the system's gain.

If $G \in H_\infty$, and $x \in H_2$, then $Gx \in H_2$ and

$$\|G\|_\infty = \sup\{\|Gx\|_2 : x \in H_2, \|x\|_2 = 1\} \tag{8.223}$$

We assume that the main input $u_1 = 0$ and another input u_2 is a disturbance signal that is affecting the output of the plant $G(s)$. The aim is to attenuate the effect of this disturbance on the output y_2 in a properly defined manner. The TF from u_2 to y_2 is the sensitivity function

$$S = \frac{1}{(1 - G(s)H(s))} \qquad (8.224)$$

It is assumed that the disturbance input u_2 is the function as follows

$$\{u_2 : u_2 = Wx \text{ for some } x \in H_2, \|x\|_2 \leq 1\} \qquad (8.225)$$

In (8.225), the weight is W, $W^{-1} \in H_\infty$. The class of the disturbance signal consists of all u_2 in H_2, such that

$$\left\| W^{-1}y_2 \right\|_2 \leq 1 \qquad (8.226)$$

We presume that the boundary values $y_2(j\omega)$ and $W(j\omega)$ are well defined, then (8.226) is interpreted as a constraint on the weighted energy of y_2; i.e. the energy spectrum $|y_2(j\omega)|^2$ is weighed by the factor $|W(j\omega)|^{-2}$. If $|W(j\omega)|$ is relatively large on a certain frequency band and small otherwise, then (8.226) will generate a class of signals that have their energy concentrated on that band. Thus, for disturbance rejection, one should minimize the energy of u_2 for the worst-case y_2 in the class (8.225); or equivalently, by (8.223), minimize $\|WS\|_\infty$, the HI norm of the weighted sensitivity function. In a synthesis problem, $G(s)$ and W would be given and $H(s)$ would be chosen to minimize $\|WS\|_\infty$ with the added constraint of the requirement of the internal stability. In practice, W is used as a design parameter and adjusted to shape the magnitude Bode plot of S.

Appendix 8A: Basic Aspects of Robustness and Robust Control with Examples

In this appendix, we discuss some basic aspects of control system related to uncertainty and the requirement of robustness and robust control and present several examples [A8A.1–A8A.3] on robust stability and robust performance.

8A.1 Need of Robustness

In the usual state space model of a dynamic system, for example in (8.16), one can describe model uncertainties as variations in the element of the system coefficient matrixes. Yet, this is very restricted class of perturbations and do not cover all the neglected dynamics or small time delays. The latter can be easily described in frequency domain [A8A.1]. In fact, very strong robustness could be well established when all the states were measured, and in

this case, a phase margin of 60° and infinite gain margin would be available; this would indicate a very good robustness. In fact, nice robustness properties of systems with the state feedback do not hold so good for the systems with output feedback; this can be recovered by full state feedback by using very fast observers; the latter led to a design method called loop transfer recovery (LTR). Interestingly, the formal requirement on the system to be controlled in the state space theory is that the system be observable and controllable, and ironically, there is no consideration of right hand s-plane (RHS) poles and zeros or time delays; hence, it is necessary to investigate the aspect of robustness of the control system design and to make appropriate modifications, if feasible in the otherwise good (classical) design approaches, to obtain good robustness.

Example 8A.1:

A fast system with a low BW.
Consider a dynamic system

$$\dot{x} = \begin{bmatrix} -1 & 1 \\ 0 & 0 \end{bmatrix} x + \begin{bmatrix} a \\ 1 \end{bmatrix} u; \quad y = \begin{bmatrix} 1 & 0 \end{bmatrix} x \qquad (8A.1)$$

In (8A.1), if $a \neq 1$, then the system is controllable; so, let us have $a = -10$. The system is of second order, and one state variable is measured directly. The system can be controlled by using an observer of first degree, with a state feedback. We assume that the closed loop system is given as

$$(s + \alpha\omega_0)(s^2 + \omega_0 s + \omega_0^2) \qquad (8A.2)$$

The TF of the system/plant, (8A.1) is

$$G(s) = \frac{as + 1}{s(s + 1)} \qquad (8A.3)$$

In order to obtain a fast closed loop response, one can choose $\omega_0 = 10$, and $\alpha = 1$. Then, we obtain the controller TF as

$$C(s) = \frac{s_0 s + s_1}{s + b} \qquad (8A.4)$$

In (8A.4), we have $b = 9274.5$, $s_0 = 925.5$, and $s_1 = 1000$. The loop TF is

$$G(s)C(s) = \frac{as + 1}{s(s + 1)} \cdot \frac{s_0 s + s_1}{s + b}$$
$$= -9255 \frac{(s - 0.1)(s + 1.0805)}{s(s + 1)(s + 9274)} \qquad (8A.5)$$

From (8A.5), we can see that the plant pole at $s = -1$ is nearly canceled by the controller zero. The Bode diagram; obtained by *bode(GC,{1e-2,1e4})*, *grid,*

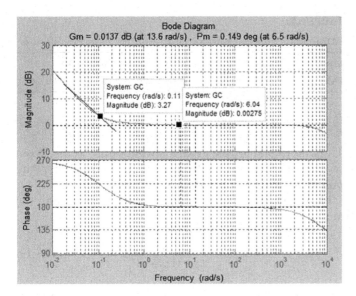

FIGURE 8A.1
Open loop TF gain and phase of $C(s)G(s)$ (Example 8A.1).

margin(GC); of the (open) loop TF, (8A.5) is shown in Figure 8A.1, from which it can be seen that the stability margin is very poor, and the slope of the magnitude curve at the cross over frequency is very small, again indicating that the system has poor robustness. The maximum sensitivities, eqn. (8.175), are max(S) = 678 and max(T) = 677; this also shows that the system is very sensitive. The system was controllable, and a compensator was provided, but the closed loop control system is very sensitive and, hence, not robust.

We see that the state space theory, as such is an elegant way to approach a control problem (Chapters 5 through 8, 10 and 11), has a nice aspect that naturally deals with multi-variable (MIMO) systems. Also, it has given important concepts such as observability and reachability and given several design methods: (a) as linear quadratic control (LQG), (b) LTR, and (c) pole placement (PP), and has also introduced powerful computational methods based on numerical linear algebra. A limitation is that the robustness is not dealt with properly. This means that it is possible to formulate control problems, as we have seen in Example 8A.1, which give solutions that have very poor robustness properties. It is easy to avoid the difficulties when we are aware of them, simply by evaluating the robustness of a design and to reduce the requirements until a suitable compromise is reached.

8A.2 Basic Limitations

In this section, we study certain fundamental limitations in the design of a (robust) control system that would be useful to determine the performance that can be achieved without sacrificing robustness. We can study the limitations that would arise from poles and zeros in/on the RHS plane and the presence of time delay/s. Let a system with the TF $G(s)$ be factored as the minimum phase and the non-minimum phase TFs

$$G(s) = G_{mp}(s)G_{nmp}(s) \qquad (8A.6)$$

The factorization is normalized so that $|G_{nmp}(j\omega)| = 1$ and the sign is chosen so that $G_{nmp}(s)$ has negative phase. The achievable BW is characterized by the gain cross over frequency ω_g.

8A.2.1 The Crossover Frequency Inequality

The (open) loop TF is $G(s)C(s) = L(s)$, and for the required PM we have

$$\arg L(j\omega_g) = \arg G_{nmp}(j\omega_g) + \arg G_{mp}(j\omega_g)$$
$$+ \arg C(j\omega_g) \geq -\pi + \phi_m \qquad (8A.7)$$

Now, assume that the controller is chosen such that the loop TF $G_{mp}(s)C(s)$ is equal to Bode's ideal loop TF, i.e. $L(s) = \left(\frac{s}{\omega_g}\right)^n$, then we have

$$\arg G_{mp}(j\omega_g) + \arg C(j\omega_g) = n\frac{\pi}{2} \qquad (8A.8)$$

In (8A.8), n is the slope of the loop TF at the crossover frequency, and the equation is also a good approximation for other controllers because the magnitude curve is typically close to a straight line at the crossover frequency; and n is the slope n_g at the crossover frequency. From the Bode's relation, the phase is $n_g\frac{\pi}{2}$. Thus, from (8A.7) and (8A.8) the crossover frequency inequality is

$$\arg G_{nmp}(j\omega_g) \geq -\left(\pi - \phi_m + n_g\frac{\pi}{2}\right) = -(\theta) \qquad (8A.9)$$

Thus, (8A.9) gives the limitations due to the non-minimum phase factors. One can plot the LHS of (8A.9) and determine when this inequality holds true.

Example 8A.2:

A rule of thumb. Let us make some design choice: with a PM of 45° and a slope of −1/2, then we have from (8A.9):

$$\arg G_{nmp}(j\omega_g) \geq -\frac{\pi}{2} \qquad (8A.10)$$

From (8A.10), we get a simple rule that the phase lag of the minimum phase components should be less than 90°, at the gain crossover frequency.

8A.2.2 A Zero in/on the RHS Plane

Let us study the system with one zero in the RHS plane. The non-minimum phase part of the TF is

$$G_{nmp}(s) = \frac{z-s}{z+s}, \text{ here } z \text{ is a zero} \qquad (8A.11)$$

The G_{nmp} should be chosen to have unit gain and negative phase. Thus, we have

$$\arg G_{nmp}(j\omega_g) = -2\arctan\left(\frac{\omega}{z}\right) \qquad (8A.12)$$

Using the crossover inequality (8A.9) and the rule of thumb (8A.10), we obtain the following relations from (8A.12):

$$\frac{\omega_g}{2} \leq \tan(\theta/2), \ \omega_g < z \qquad (8A.13)$$

Thus, the RHS plane zero gives an upper bound to the achievable bandwidth (BW), and the BW decreases with decreasing frequency of the zero, making it more difficult to control systems with slow zeros.

8A.2.3 Time Delays

In such cases, the TF has an essential singularity at infinity. The non-minimum phase part of the TF of the plant is

$$G_{nmp}(s) = e^{-sT}; \ \arg G_{nmp}(j\omega) = -\omega T \qquad (8A.14)$$

From (8A.9) and (8A.10), we have

$$\omega_g T \leq \pi - \phi_m + n_g \frac{\pi}{2} = \theta; \ \omega_g T \leq \frac{\pi}{2} = 1.57 \qquad (8A.15)$$

The existing of a time delay in the system gives an upper bound on the achievable BW.

8A.2.4 Pole/Zero in the RHS Plane

First, let us study a system with one pole on RHS plane:

$$G_{nmp}(s) = \frac{s+p}{s-p}, \text{ here } p \text{ is a pole; } p > 0. \qquad (8A.16)$$

The TF (8A.16) is normalized so it has a unity gain and −ve phase. Then, we have

$$\arg G_{nmp}(j\omega) = -2\arctan\left(\frac{p}{\omega}\right) \qquad (8A.17)$$

Using the crossover frequency inequality and the rule of thumb, we have

TABLE 8A.1

Attainable Phase Margin for $ng = -1/2$, for Zero-Pole Ratios

z/p	2	2.24	3.86	5	5.83	8.68	10	20
PM	−6.0	0	30	38.6	45	60	64.8	84.6

$$\omega_g \geq p \qquad (8A.18)$$

Thus, an unstable pole gives a lower bound on the crossover frequency, i.e. with the systems with the RHS plane poles, the BW must be adequately large.

Now let us consider a system with RHS plane pole and zero:

$$G_{nmp}(s) = \frac{(z-s)}{(z+s)}\frac{(s+p)}{(s-p)}, \ z > p \qquad (8A.19)$$

Using (8A.9) and (8A.10), we have the following expressions

$$\arg G_{nmp}(j\omega) = -2\arctan\left(\frac{\omega}{z}\right) - 2\arctan\left(\frac{p}{\omega}\right)$$
$$= -2\arctan\left(\frac{\omega/z + p/\omega}{1 - p/z}\right) \qquad (8A.20)$$

$$z \geq 25.3p$$

In Table 8A.1, phase margins as a function of the ratio of z/p for PM as 45°, and $n = -1/2$ are given. The PM that can be attained for a given ratio p/z is

$$\phi_m < \pi + n_g \frac{\pi}{2} - 4\arctan\sqrt{\frac{p}{z}} \qquad (8A.21)$$

From Table 8A.1, it is seen that when the unstable zero is faster than the unstable pole, i.e. $z > p$, the ratio z/p should be sufficiently large so that a desired PM is obtained.

Example 8A.3:

The case of X-29 (experimental US) aircraft. For this case, one of the design criterion was that the PM should be greater than 45° for all flight conditions. At one of the flight conditions, the math model has the non-minimum component:

$$G_{nmp}(s) = \frac{s-26}{s-6} \qquad (8A.22)$$

Since $z/p = 4.33$, it follows from Table 8A.1 and 8A.21 that the achievable PMs for $n_g = -0.5$ and $n_g = -1$ are 32.3° and −12.6°, respectively. Any design method could not have achieved the desired PM of 45°, at this particular flight condition. When the unstable zero is slower than the unstable pole,

TABLE 8A.2

Various Limits for the Design

Maximum Sensitivities	A RHS Plane Zero z $\omega_g/z \leq$	A RHS Plane Pole p $p/\omega_g \geq$	A Time Delay T $\omega_g T \leq$	A RHS Plane Pole-Zero Pair with $z > p$ $z/p \geq$	A RHS Plane p and a Time Delay T $pT \leq$
MS, MT < 2	0.5	2	0.7	6.5	0.16
MS, MT < 1.4	0.2	5	0.37	14.4	0.05

the crossover frequency inequality cannot be satisfied unless PM < 0 and $n_g > 0$.

8A.2.5 A Pole in the RHS Plane and a Time Delay

The TF in this case is

$$G_{nmp}(s) = \frac{s+p}{s-p}e^{-sT} \qquad (8A.23)$$

The condition of (8A.9) gives

$$2\arctan\left(\frac{\omega_g}{p}\right) - \omega_g T \geq \phi_m - n_g \frac{\pi}{2} \qquad (8A.24)$$

If $pT < 2$, then the LH side of (8A.24) has its smallest value for $\omega_g/p = \sqrt{\frac{2}{pT}-1}$; then with this and using the thumb rule, we get for PM = 45° and $n_g = -0.5$: $pT \leq 0.326$

Example 8A.4:

Balancing a pole.
We consider balancing of an inverted pendulum. A pendulum of length l has a RHS plane pole $\sqrt{\frac{g}{l}}$, and assuming that the neural lag of a human is 0.07 sec., we get $\left(\sqrt{\frac{g}{l}}\right) 0.07 < 0.326$, hence $l > 0.45$; this indicates that a human with a lag of 0.07 sec., can balance a pendulum with length of 0.5 m. To balance a pendulum with length 0.1 m, the time delay must be less than 0.03 sec. With a video rate of 20 Hz (video cameras used as angle sensors), it follows from that, the shortest pendulum that can be balanced with PM = 45°, and $n_g = -0.5$ can be $l = 0.23$ m.

From the foregoing and other related studies, the limits for various combination/s of zero, pole and a time delay for two maximum sensitivities are given in Table 8A.2.

8A.3 Comparison of Two Systems

If a system structure with 2-DoF is used, the problems of set-point response can be dealt with separately, and we can focus on robustness and attenuation of disturbances. The system has two inputs the measurement noise n and

the load disturbance d. The problem is to design a controller with the properties: (i) insensitive to changes in the process properties, (ii) ability to reduce the effects of the load disturbance d, and (iii) does not inject too much measurement noise into the system. Stability is a primary robustness requirement. Because the stability is based on the loop TF only, there may be cancellations of poles and zeros in the plant and the controller. This does not pose any problems if the canceled factors are stable. The results will, however, be strongly misleading if the canceled factors are unstable because there will be internal signals in the system that will diverge.

Example 8A.5:

Let the plant and the compensator/controller be given as

$$G(s) = 1/(s-1) \text{ and } C(s) = (s-1)/s \qquad (8A.25)$$

From (8A.25), one can see that the (open) loop TF = $G(S)C(s) = 1/s$ and the system appears stable. However, the TF from the disturbance d to the output is, see Figure 8.5 (Chapter 8; first LH side; $u_1 = d$; and y; K or H = C; and with only G):

$$G_{yd}(s) = \frac{s}{(s+1)(s-1)} \qquad (8A.26)$$

From (8A.26), it can be seen that the system is not stable, since the load disturbance will make the output diverge.

The problem of example 8A.5 is conventionally resolved by a rule that the cancellation of unstable poles is not permissible, and this is ensured in the computation that the division by factors having roots in the RHS plane is not permitted. This problem can also be handled directly by introducing a stability concept. It follows that the closed loop control system (CLCS) is completely characterized by the overall transfer functions, see Figure 8.4, with only four independent TFs given as

$$G_{xr} = \frac{GC}{1+GC} = T(s); \ G_{xd} = \frac{G}{1+GC};$$

$$G_{yn} = \frac{1}{1+GC} = S(s); \ G_{ur} = \frac{C}{1+GC}$$

$$G_O(s) = \begin{bmatrix} \dfrac{1}{1+GC} & -\dfrac{C}{1+GC} \\[2ex] \dfrac{G}{1+GC} & -\dfrac{GC}{1+GC} \end{bmatrix} \qquad (8A.27)$$

In (8A.27), S is the sensitivity function and T is the complementary sensitivity function and $T(s) + S(s) = 1$. We can say that the system is stable if all these TFs in (8A.27) are stable, which is called the internal stability. Let the TFs of the plant and the controller be represented by

$$G(s) = \frac{B_p}{A_p}; \; C(s) = \frac{B_c}{A_c} \qquad (8A.28)$$

Then we have (8A.27) in terms of (8A.28) as

$$G_O(s) = \begin{bmatrix} \dfrac{A_c A_p}{A_c A_p + B_c B_p} & -\dfrac{A_p B_c}{A_c A_p + B_c B_p} \\[2ex] \dfrac{A_c B_p}{A_c A_p + B_c B_p} & -\dfrac{B_p B_c}{A_c A_p + B_c B_p} \end{bmatrix} \qquad (8A.29)$$

From (8A.29), we can see that the stability criterion is that the common denominator in (8A.29) has all its roots in the left hand s-plane (LHS). This is an internal stability, and say that the system (G, C) is stable.

Example 8A.6:

We now apply the results of (8A.29) to example 8A.5 and obtain the following $G_o(s)$:

$$G_O(s) = \begin{bmatrix} \dfrac{s}{s+1} & -\dfrac{s-1}{s+1} \\[2ex] \dfrac{s}{(s-1)(s+1)} & -\dfrac{1}{s+1} \end{bmatrix} \qquad (8A.30)$$

From (8A.30), the characteristic polynomial is $= (s - 1)(s + 1)$ and this shows that there is a root in the RHS plane.

A basic problem in robustness is to determine when two systems are close. For feedback control, it would be natural to claim that two systems are close if they have similar behavior under a given feedback; however, if the two systems have similar open loop characteristics that does not mean that they will behave similarly under feedback.

Example 8A.7:

Comparison of two systems with similar open loops and different closed loops. Let these systems' TFs be given as

$$G_1(s) = \frac{1000}{(s+1)}, \; G_2(s) = \frac{1000a^2}{(s+1)(s+a)^2} \qquad (8A.31)$$

These systems have very similar open loop step responses for large values of a as can be seen from Figure 8A.2. The close loop TFs with unity feedback are given as for $a = 100$.

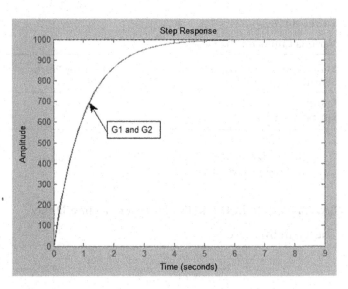

FIGURE 8A.2
Two systems with similar open loop responses (Example 8A.7).

$$G_{1cl}(s) = \frac{1000}{(s+1001)}, \qquad (8A.32)$$

$$G_{2cl}(s) = \frac{10^7}{(s-287)(s^2 + 86s + 34879)}$$

From (8A.33), we can see that the CLTFs of the very similar open loop systems are quite different, since the second one is unstable, as can also be seen from Figure 8A.3.

Now, consider the comparison of two systems with different open loop responses and similar closed loop responses. Let these systems' TFs be given as

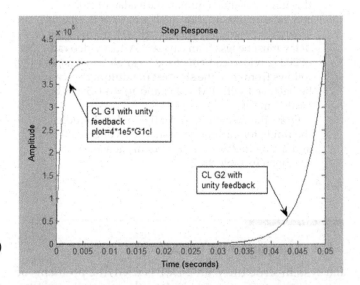

FIGURE 8A.3
The CLTF (G_1, G_2 with feedback) responses are different for the similar plant responses (Example 8A.7).

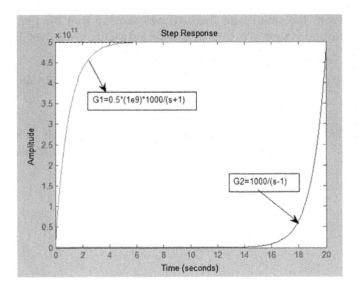

FIGURE 8A.4
Responses (that are different) of the plants only, G_1, and G_2 (Example 8A.7).

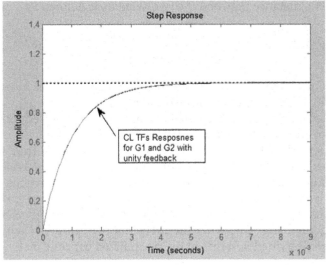

FIGURE 8A.5
Similar CL responses for the plants that have different open loop responses (Example 8A.7).

$$G_1(s) = \frac{1000}{(s+1)}, \; G_2(s) = \frac{1000}{(s-1)} \qquad (8A.33)$$

From (8A.33), we see that the second system is open loop unstable, and their step responses are shown in Figure 8A.4. The CLTFs for a unity feedback are given as

$$G_{1cl}(s) = \frac{1000}{(s+1001)}, \; G_{2cl}(s) = \frac{1000}{(s+999)} \qquad (8A.34)$$

It is apparent from (8A.34) that, though the two open loop TFs are quite different, their CLTFs are very close, which is seen from Figure 8A.5 also.

8A.3.1 Coprime Factorization

An approach is to compare the outputs when the inputs are restricted to the class of inputs that give bounded outputs. Let the process/plant be described by the rational transfer function where $A(s)$ and $B(s)$ are polynomials, and let there be a stable polynomial $C(s)$ whose degree is not smaller than the degrees of $A(s)$ and $B(s)$:

$$G(s) = \frac{B(s)}{A(s)}, \; G(s) = \frac{B(s)/C(s)}{A(s)/C(s)} = \frac{N(s)}{D(s)} \qquad (8A.35)$$

Two rational functions $N(s)$ and $D(s)$ are called coprime if there exist rational functions X and Y that satisfy the following equation

$$XN(s) + YD(s) = 1 \qquad (8A.36)$$

The condition is that $N(s)$ and $D(s)$ do not have any common factors. The functions $N(s)$ and $D(s)$ can be chosen such that

$$NN^* + DD^* = 1, \; N^*(s) = N(-s) \qquad (8A.37)$$

A factorization of $G(s)$ where $N(s)$ and $D(s)$ satisfy (8A.37) is called a normalized coprime factorization of $G(s)$, then the polynomials $A(s)$ and $B(s)$ do not have common factors. A Vinnicombe metric is appropriate for comparing two systems with feedback. Let the two systems be given as

$$G_1(s) = \frac{N_1}{D_1}, \; G_2(s) = \frac{N_2}{D_2} \qquad (8A.38)$$

It is required that the following is true if we want to compare these two systems

$$\frac{1}{2\pi} \Delta\{\arg_{nq}(N_1 N_2^* + D_1 D_2^*)\} = 0 \qquad (8A.39)$$

In (8A.39), nq is the Nyquist counter. In the polynomial form of (8A.35), it is given as

$$\frac{1}{2\pi} \Delta\{\arg_{nq}(B_1 B_2^* + A_1 A_2^*)\} = \deg A_2 \qquad (8A.40)$$

The distance between the two systems is defined as

$$\delta_v(G_1,G_2) = \sup_{\omega} \frac{|G_1(j\omega) - G_2(j\omega)|}{\sqrt{(1+|G_1(j\omega)|^2)(1+|G_2(j\omega)|^2)}} \quad (8A.41)$$

The metric in (8A.41) is Vinnicombe metric and we have $|\delta_v(G_1,G_2)| \le 1$, which is also called v-gap metric.

Classical sensitivity results are obtained based on additive perturbations (Chapter 8). The system G is perturbed to $G = G + \Delta G$, where $G(s)$ is a stable transfer function. These types of perturbations are not well suited to deal with feedback systems. A more sophisticated way to describe perturbations is required for this. The development of the metrics for systems gives a good insight into what should be done. Uncertainty will be described in terms of the normalized coprime factorization of a system, Figure 8A.6a:

$$G(s) + \Delta G(s) = \frac{N + \Delta N}{D + \Delta D} \quad (8A.42)$$

In (8A.42), N and D is a normalized coprime factorization of G. Also, the perturbing TFs are stable and proper TFs. Let z be the intermediate output, then we have

$$z = \frac{D^{-1}}{1+GC} w_1 - \frac{D^{-1}C}{1+GC} w_2$$

$$= D^{-1}\left(\frac{1}{1+GC} \quad -\frac{C}{1+GC}\right)\binom{w_1}{w_2} \quad (8A.43)$$

Also we have

$$\binom{w_1}{w_2} = \binom{\Delta D}{\Delta N} z \quad (8A.44)$$

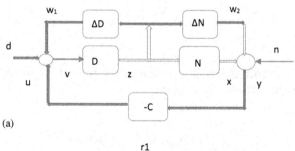

(a)

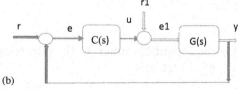

(b)

FIGURE 8A.6
(a) A plant with coprime factors uncertainty and controller; (b) 2-input 2-output system for study of internal stability.

From the small gain theorem, we can say that the perturbed system will be stable of the loop gain is less than 1:

$$\left\|\binom{\Delta D}{\Delta N}\right\|_\infty \left\|D^{-1}\left(\frac{1}{1+GC} \quad -\frac{C}{1+GC}\right)\right\|_\infty < 1 \quad (8A.45)$$

The condition in (8A.45) can be satisfied if we use the fact that N and D is a normalized coprime factorization:

$$\left\|D^{-1}\left(\frac{1}{1+GC}\binom{w_1}{w_2} - \frac{C}{1+GC}\right)\right\|_\infty$$

$$= \left\|\binom{D}{N}D^{-1}\left(\frac{1}{1+GC} \quad -\frac{C}{1+GC}\right)\right\|_\infty$$

$$\left\|\binom{I}{G}\left(\frac{1}{1+GC} \quad -\frac{C}{1+GC}\right)\right\|_\infty \quad (8A.46)$$

$$= \left\|\begin{bmatrix} \dfrac{1}{1+GC} & -\dfrac{C}{1+GC} \\ \dfrac{G}{1+GC} & -\dfrac{GC}{1+GC} \end{bmatrix}\right\| = \left\|G_0(G,C)\right\|_\infty$$

If we have

$$\gamma(G,C) = \sup_{\omega}\left\|G_0(G(j\omega),C(j\omega)\right\|_\infty \quad (8A.47)$$

Then, the closed loop system is stable for all normalized coprime factorization ΔN, ΔD such that

$$\left\|\binom{\Delta D}{\Delta N}\right\|_\infty < 1/\gamma \quad (8A.48)$$

8A.4 H-Infinity Loop Shaping

The goal of H-Infinity control system (HICS) is to design control systems that are largely insensitive to model uncertainty. One can see from (8A.47) and (8A.48) that this can be accomplished by finding a controller $C(s)$ that gives a stable CLCS and minimizes the H_∞ norm of the TF $G_0(G, C)$ given by (8A.46). Also, from (8A.48) one can see that such a design admits the largest deviation of the normalized coprime deviations. One can observe that the TF G_o also describes the signal transmission from the output disturbances, and a robust controller obtained will also attenuate the disturbances very well. In the design procedure for H_∞ loop shaping, a frequency weighting W as a design parameter is introduced; and H_∞ control problem for the process $G_w = GW$ is then solved giving the controller C_w. One obtains a controller that has high gain at specified frequency ranges and high frequency roll off. A generalization of the classical stability margin can also be used. For a CLCS consisting of the process G and the controller C, we define the generalized stability margin (GSM) as

$$b(G,C) = \begin{cases} 1/\gamma & if\ (G,C)\ is\ stable \\ 0 & otherwise \end{cases} \qquad (8A.49)$$

The margin "0" means the system is unstable, and if it is close to 1, then it is a good margin of stability. Reasonable values of the margin are in the range of 1/3 to 1/5. The H_∞ loop shaping would give a controller that maximizes the stability margin

$$b_{opt} = \sup_C b(G,C) \qquad (8A.50)$$

Some interesting theorems that relate model uncertainty to robustness are known as Vinnicombe theorems.

Proposition 1: Consider a nominal plant G, the controller C, and a parameter β. Then, the controller C stabilizes all plants G_1 such that $\delta_v(G,G_1) \le \beta$ if, and only if, $b(G_1,C) > \beta$. A controller C designed for the plant G with a generalized stability margin greater than parameter β will stabilize all the plants G_1 in a δ_v domain of G provide that $\delta_v(G,G_1) \le \beta$.

Proposition 2: Given a nominal plant G, a perturbed plant G_1 and a number $\beta < b_{opt}(G,C)$, then (P_1, C) is stable for all compensators C, such that $b(G_1, C) > \beta$ if, and only if, $\delta_v(G,G_1) \le \beta$. The classical robustness criteria like maxima of the sensitivity and complementary sensitivity depend only on the loop TF $G(s)C(s)$. However, the generalized stability margin $b(G, C)$ and the generalized sensitivity $\gamma(G,C)$ depend on both $G(s)$ and $C(s)$. Interestingly, if $G(s)$ is multiplied with a constant and the controller TF is divided by the same constant, then $G(s)C(s)$ would remain the same, but the generalized stability margin and the generalized sensitivity would change. If the disturbances affecting the system have equal weights, then this is reasonable, provided we have sufficient information about

these disturbances. We can have the following modified TFs: $P_w = PW$, and $C_w = CW^{-1}$, then we have the same loop TF. However, the generalized sensitivity becomes

$$\gamma(G_w,C_w) = \sup_\omega \frac{\sqrt{(1+|C(j\omega)W^{-1}(j\omega)|^2)(1+|G(j\omega)W(j\omega)|^2)}}{|1+G(j\omega)C(j\omega)|} \qquad (8A.51)$$

This sensitivity is minimized for $W = \sqrt{|C|/|G|}$. We can use the weighted sensitivity function to obtain an interesting connection between H_∞ loop shaping and classical robustness theory. The parameter (8A.47) is a natural generalization of the maxima, M_S and M_T of the sensitivity and the complementary sensitivity functions, S and T, respectively. Let the combined sensitivity be M requiring that both M_S and M_T should at most be equal to M, which implies that the Nyquist curve of the loop TF is outside a circle with diameter on

$$\left(-\frac{M+1}{M-1}, -\frac{M-1}{M+1} \right) \qquad (8A.52)$$

We then have the following inequalities

$$2M-1 < \gamma < 2M; \quad \frac{\gamma}{2} < M < \frac{\gamma+1}{2} \qquad (8A.53)$$

The GSM b is also a natural generalization of the classical stability margin A_m. The normal stability margin takes values between 1 and ∞ while the GSM takes values between zero and one; for compatibility, the classical stability margin should be redefined as the distance between the critical point and the intersection of the Nyquist curve with the negative real axis; and we get:

$$A_m^* = 1 - \frac{1}{A_m} \qquad (8A.54)$$

Example 8A.8:

Internal stability. Let us consider the arrangement of the control system as in Figure 8A.6b, with the controller $C(s)$, then the plant $G(s)$, and then the unity feedback from the output of the plant to the input to the controller [A8A.2]. Let the TFs be given as $G(s) = 8(s + 5)/(s(s - 2))$ and $C(s) = 0.5(s - 2)/(s + 4)$. The normal CLTF for this Figure 8A.6b (also known as the complementary sensitivity function $T(s)$) is given as

$$G_o(s) = G_{cl}(s) = T(s) = \frac{C(s)G(s)}{1+C(s)G(s)} \qquad (8A.55)$$

From Figure 8A.6, we have the following TF:

$$\begin{bmatrix} E(s) \\ E_1(s) \end{bmatrix} = \begin{bmatrix} \dfrac{1}{1+C(s)G(s)} & \dfrac{-G(s)}{1+C(s)G(s)} \\ \dfrac{C(s)}{1+C(s)G(s)} & \dfrac{1}{1+C(s)G(s)} \end{bmatrix} \begin{bmatrix} R(s) \\ R_1(s) \end{bmatrix} \qquad (8A.56)$$

Using given TFs in (8A.55), we get

$$T(s) = \frac{4(s+5)(s-2)}{(s-2)(s+4-j2)(s+4+j2)} = \frac{4(s+5)}{(s+4-j2)(s+4+j2)} \qquad (8A.57)$$

The system of (8A.57) is stable. However, let us now get the TF(1, 2) from (8A.56), i.e.

$$TF_{1,2}(s) = G_{o12}(s) = \frac{-G(s)}{1+C(s)G(s)}$$
$$= \frac{-8(s+4)(s+5)}{(s-2)(s+4-j2)(s+4+j2)} \qquad (8A.58)$$

From (8A.58), we see that the internal TF, i.e. $G_{o12}(s)$ is not stable, and hence, although the $T(s)$ is stable, the overall system, i.e. the closed loop (CL) system is internally not stable. This means that any disturbance being injected into the system from $r_1(t)$ will produce an unbounded response at $e(t)$. That unstable response will contain (or is due to) an exponential of the form e^{2t} due to the pole at $s = 2$ (on the RHS plane and, hence, unacceptable situation). The compensator zero at $s = 2$ makes the compensator unresponsive to that unstable exponential; the zero is at the one location guaranteed to make the overall system unstable. The signal injected at $r_1(t)$ will propagate through the unstable plant to reach $e(t)$, but then it will not be processed by the compensator to complete the loop. The control signal $u(t)$ will have no knowledge of the instability, so the unstable signal e^{2t} will remain at the output $e(t)$. If we have the following compensator then we have the following TFs $T(s)$, and $TF_{12}(s)$:

$$C(s) = \frac{0.75(s+0.5)}{(s+6)} \qquad (8A.59)$$

$$T(s)$$
$$= \frac{6(s+5)(s+0.5)}{(s+7.4521)(s+1.2739-j0.6244)(s+1.2739+j0.6244)}$$
$$TF_{12}(s) = G_{o12}(s)$$
$$= \frac{-8(s+6)(s+5)}{(s+7.4521)(s+1.2739-j0.6244)(s+1.2739+j0.6244)} \qquad (8A.60)$$

We see that the CL system is internally stable.

Example 8A.9:

Robust stability, robust performance, and sensitivity bounds [A8A.2]. The TF of a nominal plant along with the TF of one extreme system is given: these define a family of plants modeled by a multiplicative perturbation to the nominal system. The extreme plant is far away from the nominal plant in terms of the frequency domain magnitude of the perturbation at each frequency, as is feasible. The aim is to design a compensator that provides both robust stability and performance for this family: robust stability means that the compensator provides internal CL stability for each member of the family, and robust performance means that each CL system satisfies a given frequency domain constraint on the sensitivity function $S(j\omega)$. The results for this example can be generated by running the script "**robustperformance.m**." The nominal and extreme plants are given as:

$$G(s) = \frac{9}{s(s+2)}; \ G_e(s) = \frac{5}{s(s+1)} \qquad (8A.61)$$

A lag-lead compensator is designed for the nominal system to place the dominant CL poles at $s = -4 \pm j2$ and to make the steady state (SS) error for a ramp input equal to $e_{ss} = 0.05$. A compensator that achieves these objectives is given, along with the OLTFs for the nominal and the one extreme plant as:

$$C(s) = \frac{2.7745(s+0.04)(s+2.2654)}{(s+6.4075e-3)(s+8.8284)};$$

$$L(s) = C(s)G(s) = \frac{24.971(s+0.04)(s+2.2654)}{s(s+6.4075e-3)(s+2)(s+8.8284)}$$

$$L_e(s) = C(s)G_e(s) = \frac{13.873(s+0.04)(s+2.2654)}{s(s+6.4075e-3)(s+1)(s+8.8284)} \qquad (8A.62)$$

The compensator $C(s)$ is placed in series with the system, $G(s)$ and $G_e(s)$, which are to be compensated; the nominal and extreme loop gains (TFs) are given in (8A.62). The performance can be defined by the specifications imposed on the frequency domain magnitude of the sensitivity function TF S, i.e. on

$$|S(j\omega)| = |1 + L(j\omega)|^{-1} \qquad (8A.63)$$

Nominal performance (NP) is defined as

$$|S(j\omega)| < \frac{1}{|W_P(j\omega)|} \Rightarrow |W_P(j\omega)S(j\omega)| < 1,$$

$$\forall \omega; \ W_P = \frac{0.5s+0.8}{s} \qquad (8A.64)$$

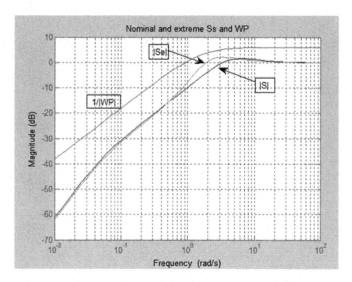

FIGURE 8A.7
The sensitivity magnitudes for the nominal, $|S|$ and extreme $|S_e|$ plants, and the magnitude of $1/W_P(s)$ (Example 8A.9).

In (8A.64), $W_P(s)$ is a specified weighting function. The time domain specifications, like overshoot and settling time, or frequency domain specifications like phase margin and BW can be achieved by proper choice of this function. For robust performance (RP), each member of the plant family must satisfy an expression like in (8A.64). Figure 8A.7 shows the sensitivity magnitudes for the nominal and extreme plants and the magnitude of $1/W_P(s)$, and it is seen that the two plants satisfy the performance requirement; say nominal performance (NP).

However, we need to study the stability and then the RP. Nominal stability (NS) is defined by the nominal CLCS being stable; that is, all the CL poles of the compensator-nominal plant combination (Loop TF) must be in the open left-half of the complex s-plane (LHS-plane). With the assumption of an unstructured multiplicative uncertainty model, robust stability (RS) is defined by the following constraint on the magnitude of the nominal complementary sensitivity function $|T(j\omega)|$:

$$|T(j\omega)| < \frac{1}{|W_I(j\omega)|} \Rightarrow$$

$$|W_I(j\omega)T(j\omega)| < 1, \ \forall \omega; \ W_I = \frac{0.4444(s+0.25)}{(s+1)} \quad (8A.65)$$

The nominal complementary sensitivity function TF, which is also the CLTF of the overall system, is given by

$$T(s) = \frac{L(s)}{1+L(s)} \quad (8A.66)$$

The $W_I(s)$ is the uncertainty weighting function that over-bounds the magnitude of the maximum uncertainty as follows:

$$L_I(\omega) = \max_{L_i \in L_P} \left[\frac{L_i(j\omega) - L(j\omega)}{L(j\omega)} \right]$$

$$= \left[\frac{L_e(j\omega) - L(j\omega)}{L(j\omega)} \right] = \left[\frac{G_e(j\omega) - G(j\omega)}{G(j\omega)} \right] \quad (8A.67)$$

$$L_I(s) = \left[\frac{G_e(s) - G(s)}{G(s)} \right] = \frac{-0.4444(s - 0.25)}{(s+1)}$$

The set $L_P(s)$ represents the loop gain (TF) for all the members of the family of plants. Since we are interested in only the magnitude for over-bounding, the $W_I(s)$ is chosen as in (8A.65), as a stable and minimum phase TF. Figure 8A.8 shows nominal $T(s)$ and the bound, and since the $T(.)$ magnitude is less than the bound for frequencies, the family is robustly stable.

For the robust performance, the domain/range of uncertainty must not encircle or intersect the circle for performance at any frequency and, therefore, $|1+L(j\omega)|$ must be greater than the sum of the radii of the circles representing uncertainty and performance:

$$|W_P(j\omega)| + |W_I(j\omega)L(j\omega)| < |1 + L(j\omega)|$$

$$\frac{|W_P(j\omega)|}{|1+L(j\omega)|} + \frac{|W_I(j\omega)L(j\omega)|}{|1+L(j\omega)|} < 1 \quad (8A.68)$$

$$|W_P(j\omega)S(j)| + |W_I(j\omega)T(j\omega)| < 1$$

From the foregoing expressions, one can infer that the necessary conditions for RP are NP and RS. As seen from (8A.68), the sum should be less

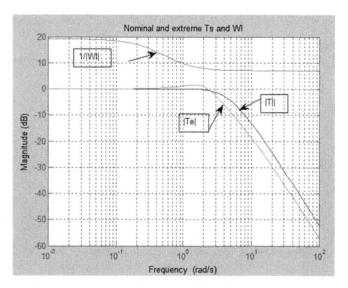

FIGURE 8A.8
Nominal $T(s)$ ($|T|$, $|Te|$) and the bound (Example 8A.9).

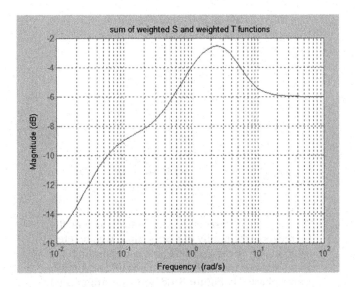

FIGURE 8A.9
Sum of the weighted sensitivity and complementary sensitivity functions (Example 8A.9).

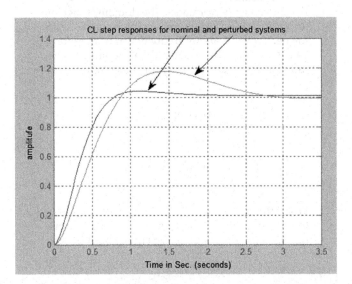

FIGURE 8A.10
The CL step responses for the nominal and the perturbed systems (Example 8A.9).

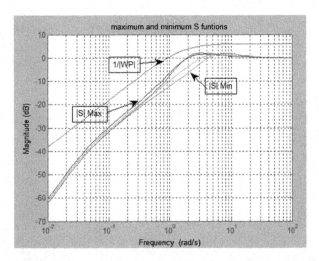

FIGURE 8A.11
The extremal magnitudes for the sensitivity function S for any plant model (Example 8A.9).

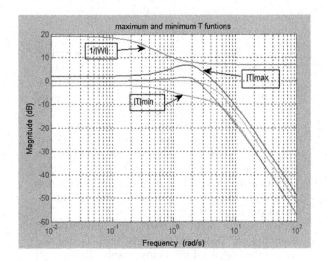

FIGURE 8A.12
The extremal magnitudes for the complementary sensitivity function T for any plant model (Example 8A.9).

than 1 for RP. As can be seen from Figure 8A.9, the peak value is less than 1 (0 dB), and this family of plants has robust performance.

Figure 8A.10 shows the CL step responses for the nominal and the perturbed systems, and it is seen that the nominal system clearly has better performance in terms of overshoot, and it also has a shorter settling time that is a benefit. The performance of the perturbed (extreme) system also satisfies the specification on the sensitivity function, so this must also be considered acceptable. Figures 8A.11 and 8A.12 show the maximum and minimum magnitudes for both the sensitivity functions for any plant model allowed in the family of systems defined by $W_I(s)$; and these given by

$$|S(j\omega)|_{\max} = \frac{1}{|1+L(j\omega)|-|W_I(j)L(j\omega)|};$$

$$|S(j\omega)|_{\min} = \frac{1}{|1+L(j\omega)|+|W_I(j)L(j\omega)|}$$

$$|T(j\omega)|_{\max} = \frac{L(j\omega)+|W_I(j)L(j\omega)|}{|1+L(j\omega)|-|W_I(j)L(j\omega)|}; \qquad (8A.69)$$

$$|T(j\omega)|_{\min} = \frac{L(j\omega)-|W_I(j)L(j\omega)|}{|1+L(j\omega)|+|W_I(j)L(j\omega)|}$$

Although, obtaining the Bode-magnitude is not exactly the same as the magnitudes specified in (8A.69), the Bode-magnitudes give a good picture of the achievable performance. Knowing that the family of systems is robustly stable (from $||W_IT||_\infty < 1$), the fact that the maximum sensitivity magnitude for all systems in the family is below $1/|W_P(j\omega)|$ at all frequencies indicates that the family also has robust performance. Thus, if all plants in the family are stable, and the worst-case system satisfies the performance criterion, then all systems in the family satisfy the performance criterion and robust performance is achieved.

Example 8A.10:

Loop shaping of high level mobility assessment tool (HiMAT) pitch axis controller [A8A.3]. The example shows how to design a MIMO controller by shaping the gain of an OL response across frequency. The idea is to choose a suitable desired/target loop shape (TLS) and use the "loopsyn" function to compute a multivariable controller that would optimally match the desired loop shape. Loop shaping is a trade-off/balance between two conflicting objectives: (i) to maximize the OL gain to get the best possible performance, but (ii) for robustness, the gain needs to be dropped below 0 dB where the model accuracy is poor and high gain might induce instabilities. This requires a good model where performance is needed (typically at low frequencies), and sufficient roll-off at higher frequencies, so the gain would start dropping, since in this range the model would be often poor. The frequency ω_g where the gain crosses the 0 dB line, is the (gain) crossover frequency and

indicates the transition between performance and robustness requirements. Some guidelines for choosing a TLS, i.e. G_d are: (i) for stability robustness, the TLS should have a gain of less than 0 dB at high frequencies, where the phase error may approach 180°; hence, toward the instability, (ii) for performance, the G_d should have high gain where we want good control accuracy and good disturbance rejection, and (iii) for **crossover and roll-off,** G_d should cross the 0 dB line between these two frequency regions, and roll off with a slope of −20 to −40 dB/decade beyond the crossover frequency ω_c.

The simplest target loop shape is $G_d(s) = \omega_c/s$, the crossover ω_c is the reciprocal of the rise-time of the desired step response. The math model of NASA's HiMAT aircraft is specified as: a six-state model (6 DoF) of the longitudinal dynamics of the HiMAT aircraft trimmed at 25,000 ft. and 0.9 Mach; the aircraft dynamics are unstable, with two right-half plane phugoid modes. The math model has two control inputs: (i) elevon deflection, and (ii) canard deflection. It has two measure outputs/observables: (i) angle of attack, α, and (ii) pitch angle, θ. This math model is unreliable beyond 100 rad/sec, the fuselage bending modes and other uncertain factors induce deviations between the model and the true aircraft dynamics of as much as 20 dB (or 1000%) beyond frequency 100 rad/sec. The results for this example can be generated by running the

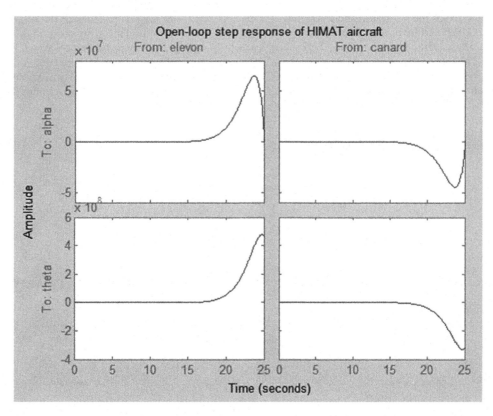

FIGURE 8A.13
Step responses of the aircraft without a controller (Example 8A.10).

script **"loopshaping."** From Figure 8A.13, we see the unstable responses and very high peaks and troughs. The task is to control α and θ by issuing appropriate elevon and canard commands and want minimum spill-over between channels—that is, a command in α should have minimum effect on θ and vice versa. Loop (loop gain) shaping design steps for designing a controller with "loopsyn" are:

Step 1: Study the original plant dynamics and responses.
Step 2: Specify the desired loop shape (TLS) G_d.
Step 3: Use "loopsyn" to compute the optimal loop shaping controller.
Step 4: Analyze the shaped-loop L, and the sensitivities: closed loop T, and sensitivity S.
Step 5: Verify the CLCS responses to see if the specifications are met.

The description of each step and the results generated is briefly given next:

Step 1: Analyze and study the plant dynamics: The aircraft model G has two unstable RHS plane phugoid-modes/poles: [−5.6757, 0.6898±j0.2488, −0.2578, −30.00, −30.00]; and one zero, which is in the LHS plane: −0.0210. In Figure 8A.14, the "sigma" plots are shown to see the minimum and maximum I/O gain as a function of frequency.
Step 2: Choose the TLS G_d: The TLS $G_d(s) = 8/s$ that corresponds to a rise time of $1/8 = 0.125$ sec., is specified, and the plot is shown in Figure 8A.15.

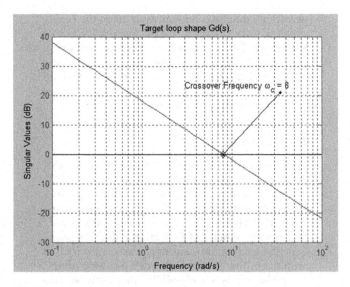

FIGURE 8A.15
The desired/target loop shape (Example 8A.10).

Step 3: Use "loopsyn" to compute the optimal loop shaping controller: Design an HI controller K, such that the gains of the OL response $G(s) * K(s)$ match the TLS $G_d(s)$ as closely as possible, while stabilizing the aircraft dynamics also. The value γ, GAM = 1.6445 indicates that the TLS shape was met within ±4.3 dB (20 * log10(GAM) = 4.32). In Figure 8A.16, the singular values of the OL, $L = G * K$ with the TLS G_d are compared.
Step 4: Analyze the shaped-loop L, CL T (complementary sensitivity function for stability), and sensitivity S: Various singular values' plots are shown in Figure 8A.17.

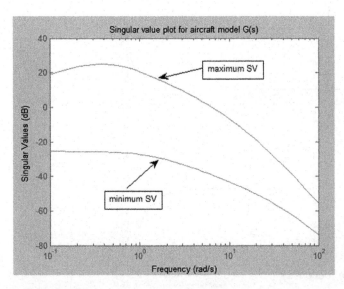

FIGURE 8A.14
The "sigma" (SVD) plots of the min and max I/O gain of the plant (Example 8A.10).

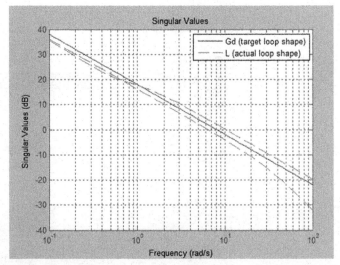

FIGURE 8A.16
The singular values of the OL (loop TF) $L = G * K$ with the TLS G_d (Example 8A.10).

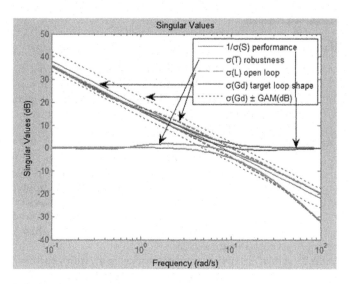

FIGURE 8A.17
CL *T* (complementary sensitivity function for stability) and sensitivity *S*: various singular values (Example 8A.10).

Step 5: Verify the CL responses to meet the specifications: The step responses of the CLTF *T* are plotted in Figure 8A.18. The design looks good if the α and θ controls are fairly decoupled, overshoot is less the 15%, and peak time is 0.5 sec.

Now, we can attempt to simplify the controller and see if the results are acceptable. The designed (2 × 2) controller *K* has fairly high order, it has 16 states. We can use model reduction algorithms to simplify the controller and still retaining its performance characteristics. We compute the Hankel singular values of *K* to understand how many controller states effectively contribute to the control law: the Hankel singular values (HSV) measure the relative energy of each state in a balanced realization of *K*; as noticed from Figure 8A.19, the HSV for the 10th state is four orders of magnitude smaller than for the 9th state, so one can try computing a 9th-order approximation of *K*, the simplified controller *K_r*

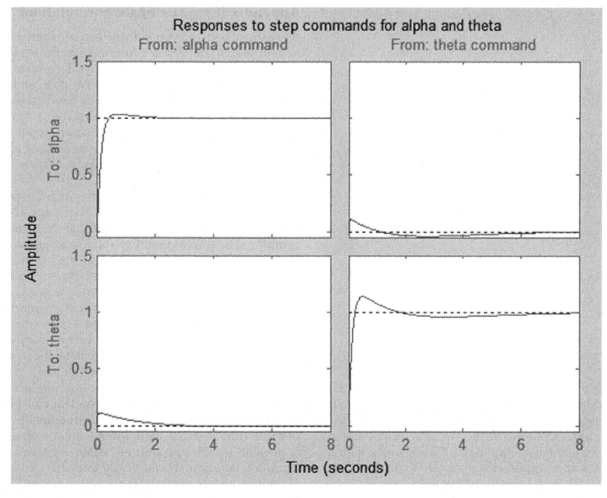

FIGURE 8A.18
The step responses of the CLTFs, *T* after the design (Example 8A.10).

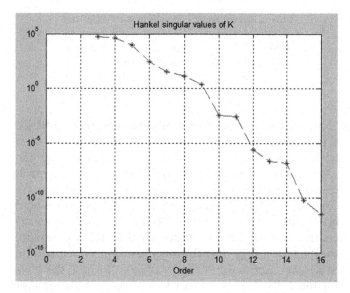

FIGURE 8A.19

The Hankel singular values of the original controller (Example 8A.10).

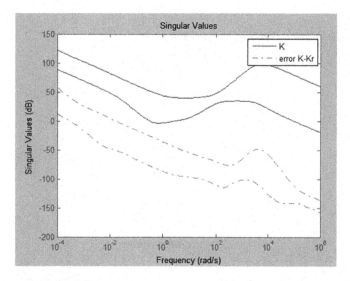

FIGURE 8A.20

The approximation error $K–K_r$ across frequency with the gains of K (Example 8A.10).

has order 9 compared to 16 for the initial controller K. In Figure 8A.20, the approximation error $K–K_r$ across frequency with the gains of K is compared, and it is seen that the approximation error is relatively small; hence, we can now compare the CL responses with the original, K and simplified K_r, as in Figure 8A.21, and it is seen that, the two responses are almost the same; thus, the reduced ninth order controller K_r can be a good replacement of K for practical purposes.

References

A8A.1 Åström, K. J. Model uncertainty and robust control. Deptartment of Automatic Control, Lund University, Lund, Sweden. https://www.researchgate.net/publication/228602986_Model_Uncertainty_and_Robust_Control, accessed March 2018.

A8A.2 Anon. http://teal.gmu.edu/~gbeale/ece_421/examples_421.html; Lecture notes by Prof. Guy Beale for presentation in ECE 720, Multivariable and Robust Control, in the Electrical and Computer Engineering Department, George Mason University, Fairfax, VA, USA. Additional notes can be found at: http://teal.gmu.edu/~gbeale/examples.html, accessed August 2018.

A8A.3 Anon. https://in.mathworks.com/help/robust/examples/loop-shaping-of-himat-pitch-axis-controller.html, accessed August 2018.

Appendix IIA: State Space Formulations

As we have seen in Section I, one can use for the design of controllers, the transfer functions of the systems to be controlled. The TFs of the systems provide final/overall relation between output variable and input variable; and yet the system might have other internal variables that would provide some insight into the behavior of the system. State space or the state-variable representation takes into account of all such internal variables of the dynamic system. Interestingly, the controller design using root locus or frequency domain method are limited to LTI systems, especially SISO systems, because for MIMO systems the controller design using classical methods becomes very complex. The development of state space approach of system modeling, analysis, control, and design then formed a basis of modern control theory, including adaptive and computer-based control systems. State space models are basically time domain models, in which the dynamics of the systems are characterized by some variables called the state variables, and these along with the input represent the state of a system at a given time. Thus, in the so-called modern control/system theory, the dynamic systems are described in the state space forms. These models can be used to represent linear, nonlinear, continuous time and discrete time systems with almost equal ease and are applicable to SISO as well as MIMO systems. These formulations are mathematically tractable for varieties of control related problems: (i) optimization, (ii) system identification, (iii) state/parameter estimation, (iv) simulation, (v) control analysis/design-cum-synthesis,

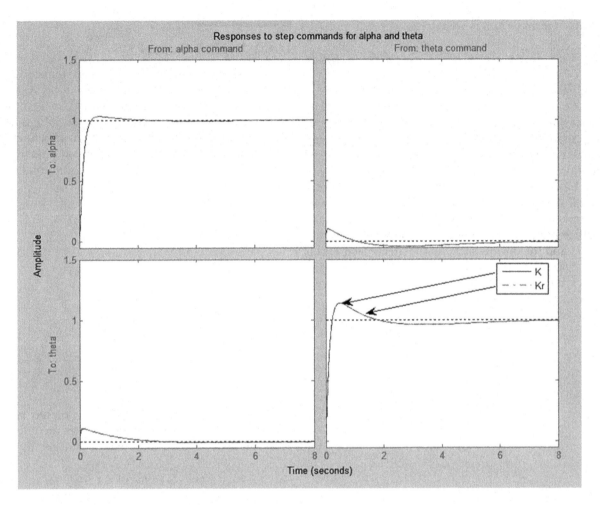

FIGURE 8A.21
The CL responses with the original, K and simplified K_r (Example 8A.10).

and (vi) forecasting/prediction. The TF models can be obtained easily from the state space models and they will be unique, because TF is input/output behavior of a system. However, the state space model from the TF model may not be unique, since inherently the states are not necessarily unique. This means that internal states could be defined in several ways, but the system's TF from those state space models would be unique. The state of a system at any time t is a minimum set of values $x_1, ..., x_n$ that along with the input to the system for all time t, $t \geq t_s$, is sufficient to determine the behavior of the system for all (future) $t \geq t_s$. In order to specify the solution to an n-th order differential (and difference) equation completely, we must prescribe n initial conditions at the initial time t_0 and the forcing function for all (future times) $t \geq t_0$ of interest: there are n quantities required to establish the system "state" at t_0. But t_0 can be any time of interest, so one can see that n variables are required to establish the state of the dynamic system at any given time.

IIA.1 Continuous Time System

Let a second order differential equation be given as:

$$\frac{d^2 z(t)}{dt^2} + a_1 \frac{dz(t)}{dt} + a_0 z(t) = u(t) \qquad \text{(IIA.1)}$$

Let $z(t)$ be the output; and $z(t) = x_1$ and $\dot{z}(t) = x_2$ be the two now defined states of the system. Then using this assignment in (IIA.1) we have

$$\begin{aligned} \dot{x}_1 &= \dot{z}(t) = x_2 \\ \dot{x}_2 &= \ddot{z}(t) = u(t) - a_1 x_2 - a_0 x_1 \end{aligned} \qquad \text{(IIA.2)}$$

In vector/matrix form (IIA.2) is written as

$$\begin{bmatrix} \dot{x}_1 \\ \dot{x}_2 \end{bmatrix} = \begin{bmatrix} 0 & 1 \\ -a_0 & -a_1 \end{bmatrix} \begin{bmatrix} x_1 \\ x_2 \end{bmatrix} + \begin{bmatrix} 0 \\ 1 \end{bmatrix} u(t); \; z(t) = \begin{bmatrix} 1 & 0 \end{bmatrix} \begin{bmatrix} x_1 \\ x_2 \end{bmatrix} \qquad \text{(IIA.3)}$$

In compact form (IIA.3) is written as

$$\dot{x} = Ax + Bu(t) \; ; \quad z = Cx \qquad \text{(IIA.4)}$$

The equivalence between (IIA.3) and (IIA.4) is very obvious. One can easily see that any number of state variables (x) and the observables (z) can be added and the systems of very higher order (degrees of freedom) can be described by the state-space representation. Also, the equations of the systems with higher degree of freedom can be easily solved using the algorithms on modern digital computers. The part of (IIA.4) without the $Bu(t)$ term is called the homogeneous system. The system coefficient matrix is A, the input coefficient matrix/vector is B, and the output coefficient matrix/vector is C, which is also called the measurement model or matrix. The TF of the system of (IIA.4) can be easily derived using the Laplace transform method:

$$
\begin{aligned}
G(s) &= \frac{y(s)}{u(s)} = C(sI - A)^{-1}B + D \\[2mm]
&= C\frac{adj(sI - A)}{|sI - A|}B + D \\[2mm]
&= \frac{C \cdot adj(sI - A) \cdot B + D \cdot |sI - A|}{|sI - A|} \\[2mm]
&= \frac{N(s)}{|sI - A|}
\end{aligned}
\qquad \text{(IIA.5)}
$$

In (IIA.5), we have used the lowercase notation for the Laplace transformed variables and s is the complex frequency. In (IIA.5), the denominator of the TF $G(s)$ is obviously the characteristic polynomial and when equated to zero is called the characteristic equation. The roots of this characteristic equations are the zeros of the same; however, since the characteristic polynomial occurs in the denominator of (IIA.5) (hence of the TF $G(s)$), these roots/zeros of the characteristic equation are the poles of the TF $G(s)$, and these determine the fundamental and the characteristic behavior of the dynamic system of which $G(s)$ it the TF.

The solution of the matrix differential equation (IIA.4) is given as

$$x(t) = e^{A(t-t_0)}x(t_0) + \int_{t_0}^{t} e^{A(t-\tau)}Bu(t)d\tau \qquad \text{(IIA.6)}$$

In (IIA.6), the exponential term is known as the system's transition matrix, $e^{At} = \varphi(t)$, since it is mainly responsible for taking the system state from one time to another time under the influence of the input $u(t)$.

IIA.2 Digital Control System

Here, we have two distinctions, the systems: (i) resulting from the sampling of the continuous time system's output at discrete instants of time; called the sampled data systems; (ii) that are inherently discrete where the states of the system are defined only at the discrete instants of time. The state equations of sampled data system are obtained as follows. Let the basic continuous time system be given as

$$
\begin{aligned}
\dot{x}(t) &= Ax(t) + Bu(t) \\
y(t) &= Cx(t) + Du(t)
\end{aligned}
\qquad \text{(IIA.7)}
$$

The solution of (IIA.7) is given as

$$x(t) = \varphi(t - t_0)x(t_0) + \int_{t_0}^{t} \varphi(t - \tau)Bu(t)d\tau \qquad \text{(IIA.8)}$$

Now, let the system of (IIA.7) be sampled with time period of T, then assuming that inputs are constants in between the two sampling instants, we get

$$u(t) = u(kT) \; for \; kT \le \tau \le (k+1)T \qquad \text{(IIA.9)}$$

From (IIA.8) and (IIA.9) we have the following

$$x(t) = \varphi(t - kT)x(kT) + \int_{kT}^{t} \varphi(t - \tau)Bu(kT)d\tau \quad \text{(IIA.10)}$$

Then we have

$$x(t) = \varphi(t - kT)x(kT) + \phi(t - kT)u(kT); \; t_0 = kT \qquad \text{(IIA.11)}$$

The equivalence between the second terms in (IIA.10) and (IIA.11) is obvious. Finally, with $t = (k + 1)T$ in (IIA.11) we have

$$
\begin{aligned}
&x((k+1)T)) = \varphi(T)x(kT) + \phi(T)u(kT); \\[2mm]
&\varphi(T) = e^{AT}; \phi(T) = \int_{kT}^{(k+1)T} \varphi((k+1)T - \tau)Bd\tau
\end{aligned}
\qquad \text{(IIA.12)}
$$

The first equations of (IIA.12) and (IIA.7) have similar forms. The sampled data part of the observation equation of (IIA.12) can be easily obtained by the process of sampling since this equation is algebraic in nature:

$$x(k+1) = \varphi(1)x(k) + \phi(1)u(k); \quad \text{when } T = 1$$

$$y(k) = Cx(k) + Du(k) \tag{IIA.13}$$

The state space equations of an inherently discrete system are given as

$$x(k+1) = \varphi(k+1,k)x(k) + B_d u(k)$$

$$y(k) = Hx(k) \tag{IIA.14}$$

The discrete time approximation of a continuous time system can be obtained as follows. Let the continuous time system be given as in (IIA.7), then using the forward difference method of approximation of the derivative of x at $t = kT$, we obtain

$$\dot{x}(t)\,|_{t=kT} = \frac{1}{T}[x((k+1) - x(kT)]$$

$$\Rightarrow \frac{1}{T}[x((k+1) - x(kT)] = Ax(kT) + Bu(kT)$$

$$y(kT) = Cx(kT) + Du(kT) \tag{IIA.15}$$

Now, rearrange the first equation in (IIA.15) to obtain

$$x((k+1)T) = (I + AT)x(kT) + BTu(kT) \tag{IIA.16}$$

The equation (IIA.16) can be written as the first equation of (IIA.14) with obvious equivalence, and when $T = 1$, we have

$$x(k+1) = (I + A)x(k) + Bu(k) \tag{IIA.17}$$

Thus, in an overall sense the state space equations of a system obtained from the sampling process of the corresponding continuous time system or a digital control system can be written as:

$$x(k+1) = \varphi_{k,k+1}x(k) + B_d u(k)$$

$$y(k) = Hx(k) + D_d u(k) \tag{IIA.18}$$

Form (IIA.12) and (IIA.17) we see that the first order (series) approximation of the transition matrix e^{AT} is given as $I + AT$.

Between the continuous time systems and the discrete time/digital systems the matrices/vectors $H = C$, and $D_d = D$, because the observation equations are algebraic.

IIA.3 Several State-Space Representations

Since the state-space representation is not unique, we consider four major representations here. These forms would be useful in optimization, control design, and parameter estimation methods.

IIA.3.1 Physical Representation

For example, a state space model of an aircraft (or a space vehicle, car, etc.) has the states as actual response variables like position, velocity, accelerations, Euler angles, flow angles and so on, and we can see that in such a model the states have obvious physical meaning and they are measurable (or computable from measurable) quantities. The corresponding state equations result from physical laws governing the dynamic process, e.g. aircraft equations of motion based on Newtonian mechanics, and satellite trajectory dynamics based on Kepler' laws of orbital mechanics. These physical variables are desirable for applications where feedback is needed, since they are directly measurable, the use of actual variables depends on specific system.

Example IIA.1:

Let the model of system be

$$\dot{x} = \begin{bmatrix} -1.43 & (40-1.48) \\ 0.216 & -3.7 \end{bmatrix} x + \begin{bmatrix} -6.3 \\ -12.8 \end{bmatrix} u(t); \tag{IIA.19}$$

Here, $C = I_2 \times_2$ and $D = [0]$. The eigenvalues of the system (i.e. of the coefficient matrix A) are $eig(A) = 2.725, -4.995$ (obtained by using MATLAB): one pole of the system is on the RHS plane (of the complex s plane) and indicates that the system is unstable. If we change the sign of the term $A(2,1)$ of the A matrix, i.e. use -0.216, then the eigenvalues are: $eig(A) = -1.135 \pm 1.319$; the system has become stable, because the real part of the complex roots is negative. The system has damped oscillations: the damping ratio is 0.652 (damp(A)), and the natural frequency of the dynamic system is 1.74 rad/sec. In fact (IIA.19) describes the short period motion of a light transport aircraft. The elements of the A matrix are directly related to the dimensional aerodynamic derivatives [16]; and the term $A(2,1)$ is the change in the pitching moment due to a small change in the vertical speed of the aircraft, and for the static stability of the aircraft, this term should be negative; the physical states of the system are: vertical speed (w m/sec) and the pitch rate (q rad/sec) and have obvious physical meaning. Also, the elements of the matrices A and B have the physical meaning. The actual equations for this example of short period dynamics of the aircraft are:

$$\dot{w} = -1.43w + 38.52q - 6.3\delta_e$$

$$\dot{q} = -0.216w - 3.7q - 12.8\delta_e \tag{IIA.20}$$

We can see from (IIA.20) that the variables w and q inherently provide some feedback to state variable in general and signify internally connected control system characteristics of a dynamic system.

IIA.3.2 Controllable Canonical Form

The state variables are the outputs of the integrators (in the classical sense of analog computers). The inputs to these integrators are differential states ($\dot{x}$, etc.), and can be easily obtained from a given TF of the system by inspection:

$$G(s) = \frac{c_m s^m + c_{m-1} s^{m-1} + \dots + c_1 s + c_0}{s^n + a_{n-1} s^{n-1} + \dots + a_1 s + a_0} \quad \text{(IIA.21)}$$

$$\dot{x} = \begin{bmatrix} 0 & & & \\ 0 & & I & \\ \dots & & & \\ 0 & 0 & \dots & 0 \\ -a_0 & -a_1 & \dots & -a_{n-1} \end{bmatrix} x(t) + \begin{bmatrix} 0 \\ 0 \\ \dots \\ 1 \end{bmatrix} u(t); \ x(t_0) = x_0$$

$$\text{(IIA.22)}$$

$$y(t) = [c_0 \ c_1 \ \dots \ c_m] \ x(t) \quad \text{(IIA.23)}$$

The last row of matrix A of (IIA.22) is the denominator coefficients of the TF, $G(s)$, (IIA.21) in the reverse order. This form is so called because this system (model) is controllable as can be seen from the following example. The discrete time systems have similar controller canonical form as in (IIA.22), with $G(z)$ given in the form of (IIA.21).

Example IIA.2:

Let the TF $G(s)$ be

$$\frac{2s^2 + 4s + 7}{s^3 + 4s^2 + 6s + 4} \quad \text{(IIA.24)}$$

Let $y(s)/u(s) = (2s^2 + 4s + 7) \dfrac{1}{s^3 + 4s^2 + 6s + 4}$;
$y(s)/u(s) = (2s^2 + 4s + 7) \ v(s)/u(s)$

$y(s) = (2s^2 + 4s + 7)v(s)$; and equivalently, $y(t) = 2\ddot{v} + 4\dot{v} + 7v$; next define: $v = x_1, \dot{v} = x_2 \ and \ \ddot{v} = x_3$ to obtain

$$y(t) = 2x_3 + 4x_2 + 7x_1; \text{ and } \dot{x}_1 = x_2 \text{ and } \dot{x}_2 = x_3 \quad \text{(IIA.25)}$$

Also, $v(s) = \dfrac{u(s)}{s^3 + 4s^2 + 6s + 4}$ \quad (IIA.26)

$$\dddot{v} = u(t) - 4\ddot{v} - 6\dot{v} - 4v \quad \text{(IIA.27)}$$

$$\dot{x}_3 = u(t) - 4x_3 - 6x_2 - 4x_1 \quad \text{(IIA.28)}$$

In the compact form we have

$$\dot{x} = \begin{bmatrix} 0 & 1 & 0 \\ 0 & 0 & 1 \\ -4 & -6 & -4 \end{bmatrix} x + \begin{bmatrix} 0 \\ 0 \\ 1 \end{bmatrix} u(t); \text{ and } y = [7 \ 4 \ 2] \ x$$

$$\text{(IIA.29)}$$

With $x = \begin{bmatrix} x_1 \ x_2 \ x_3 \ x_4 \end{bmatrix}$, It can be seen that the numerator coefficients of the TF, (IIA.24), $G(s)$ appear in the y-equation, (IIA.25), i.e. the observation model and the denominator coefficients appear in the last row of the matrix A, (IIA.29). The denominator of the TF mainly governs the dynamics of the system, and the numerator mainly shapes the output response; thus, the matrix A signifies the dynamics of the system and the vector C the output of the system. We also see that $eig(A)$ and the roots of the denominator of TF have the same numerical values: there is no cancellation of any pole with zero of the TF; zeros=roots([2 4 7]); verifiable using MATLAB control system toolbox.

The controllable matrix $C_0 = \begin{bmatrix} 0 & 0 & 1 \\ 0 & 1 & -4 \\ 1 & -4 & 10 \end{bmatrix}$ is

obtained from (IIA.29) as $C_0 = \begin{bmatrix} B & AB & A^2 B \dots \end{bmatrix}$.
For the system to be controllable, the rank of C_0 should be n (the dimension of the state vector), i.e. rank$(C_0) = 3$, and this state space form/system is controllable; hence, the name controller canonical form.

IIA.3.3 Observable Canonical Form

This form is given as

$$\dot{x}(t) = \begin{bmatrix} 0 & 0 \dots 0 & -a_0 \\ 1 & 0 \dots 0 & -a_1 \\ 0 & 1 \dots 0 & -a_2 \\ . & & \\ . & & \\ . & & \\ . & & \\ 0 & 0 \dots 1 & -a_n \end{bmatrix} x(t) + \begin{bmatrix} b_0 \\ b_1 \\ . \\ . \\ . \\ b_{n-1} \end{bmatrix} u(t); \quad x(t_0) = x_0$$

$$\text{(IIA.30)}$$

$$z(t) = [0 \ 0, \dots, 1] \ x(t)$$

It is so called since this system (model) is observable as can be seen from the following example.

Example IIA.3:

We now obtain the observable canonical form of the TF of (IIA.24), Example IIA.2. The TF, $G(s)$ can be given as

$$G(s) = C(sI - A)^{-1}B \qquad \text{(IIA.31)}$$

Since the TF is a scalar, we have

$$G(s) = \{C(sI - A)^{-1}B\}^T; \; G(s) = B^T(sI - A^T)^{-1}C^T \qquad \text{(IIA.32)}$$

$$G(s) = C_o(sI - A^T)^{-1}B_o \qquad \text{(IIA.33)}$$

Thus, if we use the A, B, C of the solution of Example IIA.2, their "transpose" form we obtain the observable canonical form:

$$\dot{x} = \begin{bmatrix} 0 & 0 & -4 \\ 1 & 0 & -6 \\ 0 & 1 & -4 \end{bmatrix} x + \begin{bmatrix} 7 \\ 4 \\ 2 \end{bmatrix} u(t) ; \text{ and } y = [0\ 0\ 1]\, x \qquad \text{(IIA.34)}$$

The observability matrix of this system is obtained as $O_b = \text{obsv}(A, C)$ (from MATLAB):

$$O_b = \begin{bmatrix} 0 & 0 & 1 \\ 0 & 1 & 4 \\ 1 & 4 & 10 \end{bmatrix} \qquad \text{(IIA.35)}$$

As the rank $(O_b) = 3$, the system (IIA.34) is observable and, the name, the observable canonical form. The discrete time systems have similar observable canonical form as in (IIA.30), with $G(z)$ given in the form of (IIA.21).

IIA.3.4 Diagonal Canonical Form

This form provides decoupled system modes, the matrix A is diagonal with entries as the eigenvalues of the system:

$$\dot{x}(t) = \begin{bmatrix} \lambda_1 & 0...0 \\ 0 & \lambda_2... \\ 0 & & \\ .. & & \\ . & ....\lambda_n \end{bmatrix} x(t) + \begin{bmatrix} 1 \\ 1 \\ . \\ . \\ 1 \end{bmatrix} u(t); \qquad \text{(IIA.36)}$$

$$y(t) = \begin{bmatrix} c_1 & c_2 ... c_n \end{bmatrix} x(t) ; \; x(t_0) = x_0 \qquad \text{(IIA.37)}$$

If the eigenvalues of A are distinct, then the diagonal form is possible. We see from (IIA.36) that every state is

independently controlled by the input $u(t)$, and it does not depend on the other states.

Example IIA.4:

We obtain the diagonal canonical form of the TF

$$G(s) = \frac{3s^2 + 14s + 14}{s^3 + 7s^2 + 14s + 8} = \frac{y(s)}{u(s)} \qquad \text{(IIA.38)}$$

The TF (IIA.38) is expanded into its partial fractions:

$$y(s) = u(s)/(s+1) + u(s)/(s+2) + u(s)/(s+4);$$

$$y(t) = [1\ 1\ 1] \begin{bmatrix} x_1 \\ x_2 \\ x_3 \end{bmatrix} \qquad \text{(IIA.39)}$$

Since $x_1(s) = u(s)/(s+1)$, we have

$$\begin{matrix} \dot{x}_1 = -1x_1 + u \\ \dot{x}_2 = -2x_2 + u \\ \dot{x}_3 = -4x_3 + u; \end{matrix} \quad \dot{x} = \begin{bmatrix} -1 & 0 & 0 \\ 0 & -2 & 0 \\ 0 & 0 & -4 \end{bmatrix} x + \begin{bmatrix} 1 \\ 1 \\ 1 \end{bmatrix} u(t) \qquad \text{(IIA.40)}$$

It is seen from the diagonal form (IIA.40) that each state is independently controlled by input; and the eigenvalues: eig(A) are −1, −2, −4 and are the diagonal elements of the matrix A. A state space form can be directly obtained from the given TF, (IIA.38) by using the MATLAB function [A, B, C, D] = tf2ss(num, den); i.e. [A, B, C, D] = tf2ss([3 14 14], [1 7 14 8]) as follows:

$$A = \begin{bmatrix} -7 & -14 & -8 \\ 1 & 0 & 0 \\ 0 & 0 & 1 \end{bmatrix} ; B = [1, 0, 0]; C = [3\ 14\ 14]; D = [0] \qquad \text{(IIA.41)}$$

We see from (IIA.41) that the obtained matrix A is not the same as the one in the diagonal form, (IIA.40); and this shows that for the same TF we get different state space forms indicating the non-uniqueness of the state-space representations. Yet, we can see that the eigenvalues of this new matrix A are identical to the original matrix: eig(A_{new}) = −1,−2,−4; and can also quickly verify that the TF obtained by using the function [num, den] = ss2tf(A, B, C, D, 1) for both the state-space representations are the same as in (IIA.38); the input/output representation of a linear dynamic system is unique as should be the case.

Interestingly enough, if the matrix A has repeated eigenvalues, it cannot be expressed in the proper diagonal form; however, it can be represented in a Jordan canonical form that is nearly a diagonal form of the matrix. If the system has the eigenvalues as: $\lambda_1, \lambda_1, \lambda_2, \lambda_3$, then the Jordan canonical form is given as

$$A = \begin{bmatrix} \lambda_1 & 1 & 0 & 0 \\ 0 & \lambda_1 & 0 & 0 \\ 0 & 0 & \lambda_2 & 0 \\ 0 & 0 & 0 & \lambda_3 \end{bmatrix} \quad \text{(IIA.42)}$$

In the Jordan canonical form: (i) the diagonal elements of the matrix A are the eigenvalues of the matrix A itself, (ii) the elements below the principal diagonal are zero, and (iii) some of the elements of just above the principal diagonal are one. The Jordan matrix can be divided into a number of blocks along the diagonal and are called the Jordan blocks. For example, for a DTS, one Jordan block is given as

$$A = \begin{bmatrix} z_1 & 1 & 0 & 0 \\ 0 & z_1 & 1 & 0 \\ 0 & 0 & z_1 & 1 \\ 0 & 0 & 0 & z_1 \end{bmatrix} \quad \text{(IIA.43)}$$

Example IIA.5:

We obtain the three canonical forms for the DTTF

$$\frac{y(z)}{u(z)} = G(z) = \frac{0.17z + 0.04}{z^2 - 1.1z + 0.248} \quad \text{(IIA.44)}$$

The controller canonical form can be directly obtained from the TF by observation:

$$\varphi_c = \begin{bmatrix} 0 & 1 \\ -0.24 & 1.1 \end{bmatrix}; B_{dc} = \begin{bmatrix} 0 \\ 1 \end{bmatrix}; H_c = \begin{bmatrix} 0.04 & 0.17 \end{bmatrix}; D_c = [0]$$

$$\text{(IIA.45)}$$

The observable canonical form can be obtained by observing that for this form we have:

$$\varphi_o = \varphi_c^T, B_{do} = H_c^T, \text{and } H_0 = B_{dc}^T$$

$$\varphi_o = \begin{bmatrix} 0 & -0.24 \\ 1 & 1.1 \end{bmatrix}; B_{do} = \begin{bmatrix} 0.04 \\ 0.17 \end{bmatrix}; H_o = \begin{bmatrix} 0 & 1 \end{bmatrix}; D_o = [0]$$

$$\text{(IIA.46)}$$

To obtain the diagonal state space form from (IIA.45), the TF is expanded into its partial fractions as:

$$G(z) = \frac{0.17z + 0.04}{z^2 - 1.1z + 0.248} = \frac{0.352}{z - 0.8} + \frac{-0.182}{z - 0.3} \quad \text{(IIA.47)}$$

The diagonal form of the state-space representation is now

$$\varphi_d = \begin{bmatrix} 0.8 & 0 \\ 0 & 0.3 \end{bmatrix}; B_{dd} = \begin{bmatrix} 1 \\ 1 \end{bmatrix}; H_d = \begin{bmatrix} 0.352 & -0.182 \end{bmatrix}; D_d = [0]$$

$$\text{(IIA.48)}$$

Interestingly, in the literature, the diagonal form is often called the Jordan canonical form; however, we will reserve this name (Jordan) when the system matrix has some eigenvalues as repeats.

IIA.4 Eigenvalues and Eigenvector

The eigenvalues of a system are the characteristics values of matrix A. Let

$$Ax = \lambda\, x \quad \text{(IIA.49)}$$

The operation, (IIA.49) means that a matrix operation on vector x simply upgrades the vector x by scalar λ, here, x is known as the eigen vector, and it cannot be [0], if the eigenvector of A is multiplied by a scalar, then the scaled vector is also the eigenvector of A. If the eigenvalues are distinct, then they can be solved directly from (IIA.49), in this case, the eigen vectors are linearly independent. We formulate the eigenvalues/eigenvector problem as:

$$(\lambda\, x - Ax) = 0 \quad \Rightarrow \quad (\lambda I - A)\, x = 0 \rightarrow |\lambda\, I - A| = 0 \quad \text{(IIA.50)}$$

For a system of (IIA.50) to have a solution, we should have $|\lambda\, I - A| = 0$ and λ_i are the so-called eigenvalues of the matrix A. If λ_i are distinct, then $A = T\Lambda T^{-1}$ and Λ, the diagonal matrix has its elements as eigenvalues. The T is the modal matrix with its columns as eigenvectors corresponding to each eigenvalue. A real symmetric matrix has distinct eigenvalues.

$$\text{Also } \lambda(A) = \frac{1}{\lambda(A^{-1})} \quad \text{(IIA.51)}$$

In case of (IIA.51), the matrix inverse should exist. Also, the eigenvalues of A and A^T are the same when the matrix A is a real matrix, if the matrix is real and symmetric, then all its eigenvalues are real (means not complex numbers). Now, let us consider the CLTF as:

$$\frac{y(s)}{u(s)} = \frac{G(s)}{1 + G(s)H(s)} \quad \text{(IIA.52)}$$

In (IIA.52), $GH(s) + 1 = 0$ is the characteristic equation, and its roots are the poles of the CLTF. We also have

$$\dot{x}(t) = Ax(t) + Bu(t)$$
$$y(t) = Cx(t) \tag{IIA.53}$$

Then taking Laplace transform of (IIA.53), with zero initial conditions, we get

$$sx(s) = Ax(s) + Bu(s); \quad y(s) = Cx(s) \tag{IIA.54}$$

By rearranging (IIA.55) we obtain

$$\frac{y(s)}{u(s)} = C(sI - A)^{-1}B = \frac{C \cdot adj(sI - A) \cdot B}{|sI - A|} \tag{IIA.55}$$

Comparing (IIA.50) and (IIA.55), we see the following similarities:

$$|\lambda I - A| = 0 \text{ and } |sI - A| = 0 \tag{IIA.56}$$

The second equality, (IIA.56) gives solution for s and they are the poles of the system $\frac{y(s)}{u(s)}$, we also obtain the poles of the system from $GH(s) + 1 = 0$. Due to the first equality, we say that the system has eigenvalues; and due to the similarity of the first equation with the second equation (except that the λ is replaced by s in the second equality), we say that "eigenvalues" are the same as the "poles" [17]; except that there could be cancellation of some "poles" due to "zeros" of $\frac{G(s)}{1+G(s)H(s)}$. In general, a system will have more eigenvalues than poles, this means that all the poles are eigenvalues, but all eigenvalues are not the poles. However, for a system with minimal realization (and minimum phase systems) poles and eigenvalues are the same, because there are no excess eigenvalues. For MIMO systems, there are specialized definitions for zeros (and poles). Eigenvalues are very useful in control theory; however, they have certain limitations when smallness or largeness of a matrix is defined, also when the matrix A has eigen-modes that are mixed; some are slow and some other are very fast. These limitations are partially overcome it the concept of singular values is used. For a DTS, the characteristic equation is given as

$$|zI - \varphi| = 0 \tag{IIA.57}$$

The roots of (IIA.57) are the eigenvalues of the system matrix. If matrices A and $\bar{A}$ are square, then they are called similar if

$$AP = P\bar{A}; \quad \bar{A} = P^{-1}AP$$
$$A = P\bar{A}P^{-1} \tag{IIA.58}$$

The matrix P has to be non-singular and it is called the similarity transformation matrix. The eigenvalues of this square matrix do not change due to the similarity transformation. One can formulate the matrix P with its columns as the eigenvectors associated with each eigenvalue:

$$P = [v_1 \ v_2 \ ... \ _nv]; \ A[v_1 \ v_2 \ ... \ _nv] = [v_1 \ v_2 \ ... \ _nv]\begin{bmatrix} \lambda_1 & 0 & ...0 \\ 0 & \lambda_2 & ...0 \\ ... & & \\ 0 & 0 & ...\lambda_n \end{bmatrix};$$
$$\tag{IIA.59}$$

The matrix P is invertible, and from (IIA.59) we see that by this transformation we get the diagonal form of the matrix A. This also signifies that the original discrete model can be converted into the one with its coefficient matrix as diagonal:

$$x(k+1) = \varphi x(k) + B_d u(k); \quad x(k) = Pz(k)$$
$$z(k+1) = P^{-1}\varphi Px(k) + P^{-1}B_d u(k); \ with \ P^{-1}\varphi P = \Lambda \tag{IIA.60}$$

The second equation in (IIA.60) is the diagonalization of the first equation. Interestingly, we need the transition matrix for the discrete time system as in (IIA.60), which can be obtained by: (i) using inverse Laplace transform, $\varphi = e^{At} = L^{-1}\{(sI - A)^{-1}\}$, (ii) since,

$$P^{-1}AP = \Lambda; \ \varphi = e^{\Lambda t} = \begin{bmatrix} e^{\lambda_1 t} & 0 & ...0 \\ 0 & e^{\lambda_2 t} & ...0 \\ ... & & \\ 0 & 0 & ...e^{\lambda_n t} \end{bmatrix}, \text{ or (iii) using Caley}$$

Hamilton theorem.

Appendix IIB: Signal and System Norms and Performance Indices

In several applications of signal processing, estimation-cum-filtering and control, it is necessary to quantify the magnitude (and/or strength or power) of signals, which are represented by vectors or matrices. Such a quantification is needed in order to assess the performance and robustness of the control systems, especially in/ for the design process of the controller; that is to define the performance criteria that are used in optimization procedure to arrive at the optimal solution of the filtering/estimation or the control problem. This quantification is obtained and specified by defining a set of norms that could also be used for assessing parameter and state estimation errors, and residuals/innovations in filtering applications, and control system loop errors. Also, such norms are useful in design of robust control laws and to specify the limits on the tolerable uncertainties in the plant and/or the deterministic discrepancies, or the model error. Most of these norms evolved from the optimal and robust control fields; however, these norms are the formal ways of specification of some of the classical criteria to obtain good control system designs; one can say that the latter are the special cases of the former. The optimal and robust designs still should meet the classical criteria and specifications (at the minimal/base level), which are based on the intuition/experience of the design engineers and practical realities of the workable control systems. In most practical control applications, it also becomes necessary to evaluate the robustness of LTI systems and this is accomplished by defining certain system norms [13,15]. As we have seen in Chapters 5 through 8, and then in Chapters 9 and 10, the optimality of any control law, in most of cases, is defined, specified, or signified in terms of some criterion, which is either to be minimized, maximized, and hence, optimized (or extremized) with respect to the control input (variable) $u(t)$, or $u(k)$. Such criteria are called performance indices (PIs). We also briefly discuss these PIs [18].

IIB.1 Spaces, Operators and Norms

Often, in the study of robustness, and linear estimation-cum-filtering theory, one looks upon the class of linear systems as a linear space. The norms and inner products are utilized as a way to measure distance between vector spaces, and these are useful in defining the closeness of vectors and systems in vector/matric spaces. In case of control system analysis and design, in most cases, the concept is that the feedback is required if there is some uncertainty in the system, and in some of its subsystems/components, a feedback is also required to stabilize an otherwise unstable or poorly damped system, and to improve the error performance of the basic system, i.e. some error criterion should be minimized. The main idea of coping with the uncertainty is to live with it, meaning going for a robust control, despite the presence of this uncertainty (Chapter 8), or try to reduce an uncertainty, meaning going for an adaptive control (Chapter 11). Early theoretical ideas for (robust) control centered around developments due to Nyquist, Bode, and Horowitz: (i) Nyquist curve, (ii) Bode diagram, Bode's relations, Bode's integrals, and Bode's ideal loop TF, and (iii) templates and quantitative feedback theory (QFT). Many useful concepts depended on: (a) state of a dynamic system under study, or for which robust

control is needed, (b) observability, reachability, and (c) Kalman filter and principle of separation of control design and estimator design for state feedback control. The uncertainty was regarded as "parameter errors" or additive disturbances. In case of the multivariable dynamic systems singular values (which are, for a non-square matrix A, the square roots of the eigenvalues of the matrix A^TA) turn out to be very important. The development of the H-Infinity (HI) theory laid the emphasis on uncertainty, and structured/unstructured uncertainty could be dealt with. So, the main idea in robust control is to design a controller C (TF $C(s)$), for a given mathematical model, M that represents a real/true process/plant/dynamic system P with TF $G(s)$, so that this plant P behaves well. It is known that the M might not be a precise math model of the plant P; and even if $M = P$, there would be errors due to implementation of the controller. So, the robustness philosophy is that the controller C is robust if the following holds true:

$$P \approx M, \text{and} \, C_r \approx C \Rightarrow (P, C_r) \approx (M, C) \qquad \text{(IIB.1)}$$

The problem of checking and asserting (IIB.1) is that of the analysis and the problem of finding the controller is called the synthesis. Fundamentally vector/matrix spaces and norms are very useful in the study of robust control, and in general for control system analysis/design and synthesis. Consider a set $X = \{x\}$; if the following operations on the elements of the set, then X is called a linear space:

$$1. x_1 + x_2 = x_2 + x_1$$

$$2. (x_1 + x_2) + x_3 = x_1 + (x_2 + x_3)$$

$$3. \exists 0 \in X \text{ such that } x + 0 = x, \, \forall x \in X$$

$$4. \forall x \in X, \exists (-) \in X \text{ such that } x + (-x) = 0$$

$$5. (\lambda_1 + \lambda_2)x = \lambda_1 x + \lambda_2 x \qquad \text{(IIB.2)}$$

$$6. \lambda(x_1 + x_2) = \lambda x_1 + \lambda x_2$$

$$7. \lambda_1(\lambda_2 x) = (\lambda_1 \lambda_2)x$$

$$8. 1x = x$$

A linear space is called a normed (space) if every vector $x \in X$ has an associated real number $||x||$, i.e. its length, and it is called the norm of the vector x, and has the following properties:

$$1. \| x \| \geq 0 \text{ and } \| x \| = 0 \Leftrightarrow x = 0$$

$$2. \| \lambda x \| = | \lambda | \cdot \| x \| \qquad \text{(IIB.3)}$$

$$3. \| x_1 + x_2 \| \leq \| x_1 \| + \| x_2 \|$$

From (IIB.2) and (IIB.3), we can say that $x_1 \approx x_2$, if $\| x_2 - x_1 \|$ is small. Let us denote the set of all linear systems as L, then we should support it with a norm. A choice should provide understanding and permit the computational analysis and synthesis. Thus, we can consider a linear system as an operator from the input space U to the output space Y, and if U and Y are the normed linear spaces, then the following system norm is said to be the induced norm by the signal norms on U and Y:

$$\| G \| = \sup_{\|u\| U \leq 1} \| Gu \|_Y \qquad \text{(IIB.4)}$$

An inner product is a functional with the properties

$$1. \langle x, x \rangle \geq 0, \text{ and } \langle x, x \rangle = 0, iff\ x = 0$$

$$2. \langle x_1, x_2 \rangle = \overline{\langle x_2, x_1 \rangle}$$

$$3. \langle x_1 + x_2, x_3 \rangle = \langle x_1, x_3 \rangle + \langle x_2, x_3 \rangle \qquad \text{(IIB.5)}$$

$$4. \langle \lambda x_1, x_2 \rangle = \lambda \langle x_1, x_2 \rangle$$

If there is an inner product on X, then the norm can be defined as

$$\| x \| = \sqrt{\langle x, x \rangle} \qquad \text{(IIB.6)}$$

A complete normed linear space is called Banach space. A Banach space with inner product and the norm is called Hilbert space. Completeness means that there are no holes/gaps in the vector space, and it is very important property; people deal with real numbers rather than with rational numbers because the space with the latter is not the complete space. Existence of the inner product gives an additional nice property of the corresponding norm, which makes the space be very similar to $R_{(n \times n)}$ – dimensional real space, this property is

$$\| x_1 + x_2 \|^2 + \| x_1 - x_2 \|^2 = 2(\| x_1 \|^2 + \| x_2 \|^2) \quad \text{(IIB.7)}$$

The property in (IIB.7) simplifies considerably the optimization in Hilbert spaces.

Example IIB.1:

L_2 space. Consider the linear space of all the matrix-valued function on R (the real space):

$$L_2(\mathbb{R}) = \left\{ F \int_{\mathbb{R}} : tr[F^*(t)F(t)]dt < \infty \right\} \qquad \text{(IIB.8)}$$

The space in (IIB.8) is the Hilbert space with the inner product given by

$$(F, G)_2 = \int_{\mathbb{R}} tr[F^*(t)G(t)]dt \qquad \text{(IIB.9)}$$

Example IIB.2:

L_∞ space. Consider the linear space of all the matrix-valued function on R (the real space):

$$L_\infty(\mathbb{R}) = \{ F : ess \sup \sigma_{max}[F(t)] < +\infty \} \qquad \text{(IIB.10)}$$

The space in (8A.10) is a Banach space with

$$\| F \|_\infty = ess \sup_{t \in \mathbb{R}} \sigma_{max}[F(t)] \qquad \text{(IIB.11)}$$

Now we focus on the choice of U and Y as L_2 space. One simple choice of the input and output spaces is L_2 (R, $n \times n$ real space) mainly because it is the Hilbert space. In this case the linear system G is a linear operator on L_2; $G: L_2 \rightarrow L_2$ and the norm of the linear system is L_2-induced norm $\| G \| = \sup_{\|u\|_2 \leq 1} \| Gu \|_2 = \| G(j\omega) \|_\infty$; here $G(s)$ is the TF of LTI system.

Now, we discuss the stability that is an important problem in control system analysis and design. This motivates the introduction of Hardy spaces: H_2 and H_∞. Define for $p = 2$ and $p = \infty$:

$$H_p = \{ F \in L_p(j\mathbb{R}) : F \text{ is analytic in the RHP} \} \qquad \text{(IIB.12)}$$

$$\| F \|_{H_p} = \sup_{\sigma > 0} \| F(\sigma + j\omega) \|_{L_p}$$

If the plant/process/system G is stable, rational and strictly proper, then

$$\| G \|_p := \| G(j\omega) \|_{L_p} = \| G \|_{H_p} \qquad \text{(IIB.13)}$$

In (IIB.13) $\| G \|_2$ is finite iff G is strictly proper. Let $G(s) = C(sI - A)^{-1}B$ be the TF of an LTI dynamic system, and let the coefficient matrix of the system A be stable matrix, then we have

$$\| G \|_2^2 = tr(B^*QB) = tr(CPC^*) \qquad \text{(IIB.14)}$$

In (IIB.14), the vector/matrix C (for the continuous time system) is the measurement model of the system, and B is the input coefficient matrix; C_o and O_b are respectively the controllability and observability Gramians, and we have the following result:

$$AC_o + C_oA^* + BB^* = 0$$
$$A^*O_b + O_bA + C^*C = 0 \qquad \text{(IIB.15)}$$

The TF $G(s)$ is the Laplace transform of the impulse response:

$$g(t) = \begin{cases} Ce^{At}B, & t \le 0 \\ 0, & t < 0 \end{cases} \qquad \text{(IIB.16)}$$

By using the Parseval's formula we have

$$\|G\|_2^2 = \frac{1}{2\pi}\int_{-\infty}^{\infty} tr\{G(j\omega)^* G(j\omega)\}d\omega = \int_{0}^{\infty} tr\{g(t)^* g(t)\}dt$$

$$= \int_{0}^{\infty} tr\{B^* e^{A^* t} C^* C e^{At} B\}dt = tr(B^* QB)$$

$$O_b = \int_{0}^{\infty} e^{A^* t} C^* C e^{At} dt$$

$$\text{(IIB.17)}$$

One can consider the L_∞ and H_∞ norms. For real and rational plants $||G||_\infty < +\infty$ if $G(s)$ is proper. When we include the stability requirement then L_2 and L_∞, give H_2 and H_∞.

IIB.2 Signal Norms

L_2 norm of a scalar valued signal $x(t)$ in time domain is defined as

$$x_2 = \left(\int_{0}^{\infty} x(t)^2 dt\right)^{1/2} \quad \text{for all } t \ge 0. \qquad \text{(IIB.18)}$$

For the Laplace transformed signal, $X(s)$, the L_2 norm in frequency domain is defined as

$$X_2 = \left(\frac{1}{2\pi}\int_{-\infty}^{\infty} |X(j\omega)|^2 d\omega\right)^{1/2} \qquad \text{(IIB.19)}$$

Using Parseval's theorem one can see that $x_2 = X_2$. The square of the L_2 norm, namely x_2^2 is proportional to the energy content in the signal $x(t)$. Likewise, the L_2 norm of a vector valued signal $x(t) = [x_1(t),\dots,x_n(t)]^T$ is given as:

$$x_2 = \left(\sum_{i=0}^{n} x_2^2\right)^{1/2} = \left(\int_{0}^{\infty}\sum_{i=0}^{n} x_i(t)^2 dt\right)^{1/2} = \left(\int_{0}^{\infty} x(t)^T x(t)dt\right)^{1/2}$$

$$\text{(IIB.20)}$$

The L_p norm is defined to generalise the measure in many practical situations wherein we require computation of: (i) the norm of the absolute value of a signal, which could be defined using L_1 norm, (ii) the L_∞ norm, which represents the maximum value or least upper bound of the absolute value of a signal, and (iii) the L_2 norm the square, which represents the power/variance/energy (time factor involved in power) in the signal

$$Lp\,norm x_p = \left(\int_{0}^{\infty} |x(t)|^p dt\right)^{1/p}, p \ge 1 \qquad \text{(IIB.21)}$$

$$L1\,norm x_1 = \int_{0}^{\infty} |x(t)|dt \qquad \text{(IIB.22)}$$

For the signals when the maximum exists, $L_\infty\,norm$ is defined as

$$x_\infty = \frac{\max}{t}|x(t)| \qquad \text{(IIB.23)}$$

For the signals when clear maximum does not exists, it is defined as the least upper bound of the absolute signal value

$$x_\infty = \frac{\sup}{t}|x(t)| \qquad \text{(IIB.24)}$$

The L in L_p means that the signal should be Lebesgue (L) integrable for the norm to exist. So, we ascertain here that the L_2 is a Hilbert space, and L_∞ is the Banach space.

IIB.3 System Norms

These norms consider the transfer functions of the dynamic systems. The H_2 (Hardy space, H) norm for a stable SISO system with TF $G(s)$ is defined as

$$G_2 = \left(\frac{1}{2\pi}\int_{-\infty}^{\infty} |G(j\omega)|^2 d\omega\right)^{1/2} \qquad \text{(IIB.25)}$$

For a multivariable system with TF matrix $G(s) = [g_{kl}(s)]$, the H_2 norm becomes

$$G_2 = \left(\sum_{kl} g_{kl2}^2\right)^{1/2} = \left(\frac{1}{2\pi}\int_{-\infty}^{\infty} |g_{kl}(j\omega)|^2 d\omega\right)^{1/2} \qquad \text{(IIB.26)}$$

It is expressed as

$$G_2 = \left(\frac{1}{2\pi} \int_{-\infty}^{\infty} \sum_{kl} g_{kl}(-j\omega) g_{kl}(j\omega) d\omega \right)^{1/2}$$

$$= \left(\frac{1}{2\pi} \int_{-\infty}^{\infty} tr \left[G(-j\omega)^T G(j\omega) \right] d\omega \right)^{1/2} \tag{IIB.27}$$

In terms of a trace of a matrix A:

$$tr\left[A^T A \right] = \sum_{k=1}^{n} \sum_{l=1}^{m} a_{kl}^2 = a_{11}^2 + \ldots + a_{kl}^2 \tag{IIB.27}$$

The H_2 norm can be computed in the time domain using the state-space representation. Consider the system described by the state space model

$$\dot{x}(t) = Ax(t) + Bu(t); \quad y(t) = Cx(t) \tag{IIB.28}$$

The TF for this system is given by

$$G(s) = C(sI - A)^{-1} B \tag{IIB.29}$$

The output $y(t)$ obtained by integrating (IIB.28) is given by

$$y(t) = Ce^{At} x(0) + \int_0^t h(t-\mu) u(\mu) d\mu \tag{IIB.30}$$

In (IIB.30), $h(t-\mu)$ is the impulse response of the system defined as

$$H(\tau) = \begin{cases} He^{At}B & if \tau \geq 0 \\ 0 & if \tau < 0 \end{cases} \tag{IIB.31}$$

And we have

$$G(s) = \int_0^{\infty} h(\tau) e^{-s\tau} d\tau \tag{IIB.32}$$

Using Parseval's theorem we have

$$G_2 = H_2 = \left(\int_0^{\infty} tr \left| h(t)^T h(t) \right| dt \right)^{1/2} \tag{IIB.33}$$

For H_2 to be stable, all the eigenvalues of A must be in the LHS plane. The equation (IIB.33) can be written as

$$G_2^2 = H_2^2 = tr \left[C \int_0^{\infty} e^{At} BB^T e^{A^T t} dt C^T \right] = tr \left[CPC^T \right] \tag{IIB.34}$$

where

$$P = \int_0^{\infty} e^{At} BB^T e^{A^T t} dt \tag{IIB.35}$$

is the unique solution of the linear matrix Lyapunov equation

$$AP + PA^T + BB^T = 0 \tag{IIB.36}$$

It can be shown that G_2 and H_2 represent the average system gain computed over all frequencies. For stochastic systems, when the input is a white noise signal with unit covariance matrix, it can be proven that the sum of the stationary variances of the outputs is given by the square of the H_2 norm of the system transfer function.

The H-Infinity (HI), H_{∞} norm is a measure of the worst-case system gain of the TF/System. For a SISO system with transfer function $G(s)$, the H_{∞} norm is defined by

$$G_{\infty} = \frac{max}{\omega} \left| G(j\omega) \right| - \text{if the maximum exists} \tag{IIB.37}$$

$$G_{\infty} = \frac{sup}{\omega} \left| G(j\omega) \right| - \text{if the maximum does not exist} \tag{IIB.38}$$

In (IIB.38), $\left| G(j\omega) \right|$ represents the TF of the system and is the factor by which the amplitude of a sinusoid with angular frequency ω is multiplied, and H_{∞} norm is a measure of the largest factor by which any sinusoid is amplified by the system.

If the L_2 (Hilbert space) norm of an input signal $u(t)$ with Laplace transform $U(s)$ is bounded, the system output $Z(s) = G(s)U(s)$ has L_2 norm, which is also bounded and it can be shown that

$$G_{\infty} \geq \frac{GU_2}{U_2} \text{ for all } U \neq 0 \tag{IIB.39}$$

Also, H_{∞} (Hardy space) norm is given by

$$G_{\infty} = sup \left\{ \frac{GU_2}{U_2} \right\} : U \neq 0 \tag{IIB.40}$$

In (IIB.40), U can be selected such that its Laplace transform is concentrated to a frequency range where $\left| G(j\omega) \right|$ is arbitrarily close to G_{∞} and with U $U(j\omega) = 0$ elsewhere. The H_{∞} norm gives the maximum factor by which the L_2 norm of any input is amplified by the system, and hence, it can be regarded as the gain of the system. For MIMO systems, the SISO gain $\left| G(j\omega) \right|$ at a given frequency is generalized to the multivariable case by intoducing the maximum gain of $G(j\omega)$ at the frequency ω

$$G(j\omega) = \frac{max}{U}\left\{\frac{G(j\omega)U}{U} : U \neq 0, U \in C^m\right\}$$

$$\text{or}\quad G(j\omega) = \frac{max}{U}\left\{G(j\omega)U : U = 1, U \in C^m\right\}$$
$$\text{(IIB.41)}$$

In (IIB.41), $U = [U_1, \ldots, U_m]^T \in C^m$ is the complex-valued vector with the Euclidean norm

$$U = \left(|U_1|^2 + \ldots + |U_m|^2\right)^{\frac{1}{2}} \qquad \text{(IIB.42)}$$

For SISO systems, the H_∞ norm of the TF matirx $G(s)$ is defined by

$$G_\infty = \frac{\sup}{\omega} G(j\omega) \quad \text{where } G(j\omega) \text{ is as in (IIB.43)}.$$

In terms of the maximum singular value (SVD) of the matrix $G(j\omega)$, the H_∞ norm is expressed as

$$G_\infty = \frac{\sup}{\omega}\sigma(G(j\omega)) \qquad \text{(IIB.44)}$$

IIB.4 Performance Indices

In certain cases these PIs would contain more than one term or index, depending on the kind of the control problem that is being handled [18]. In general, and in many cases, the criterion is chosen such that the numerical value of the criterion decreases as the "quality" of the control law is enhanced, in fact if the index decreases, then the control law tends to be optimal and even robust in most cases. Such indices are the performance measures of the overall control law design, analysis and synthesis, without which the entire gamut is subjective and not objective. Let us consider a DTS:

$$x(k+1) = \varphi x(k) + Bu(k),\ y(k) = Hx(k);\ x(k_0) = x_0 \quad \text{(IIB.45)}$$

Our aim is to see that over a fixed interval $[T_0, T_f]$, the components of state variables/vector or/and the control input $u(.)$ are as small as possible. We have in fact several PIs available at our disposal.

$$\text{PI1:}\quad J1 = \sum_{k=T_0}^{T_f} x^T(k)x(k) \qquad \text{(IIB.46)}$$

When $J1$ is very small, it means that the (H_2) norm of x is very small too.

$$\text{PI2:}\quad J2 = \sum_{k=T_0}^{T_f} y^T(k)y(k) = \sum_{k=T_0}^{T_f} x^T(k)H^THx(k)$$
$$\text{(IIB.47)}$$
$$= \sum_{k=T_0}^{T_f} x^T(k)Rx(k);\ R \text{ is a symmetric matrix}$$

In (IIB.47), the matrix R can be viewed as a weighting matrix for optimal control problems, covariance matrix for the stochastic optimization, control and estimation problems, and Gramian matrix for the deterministic control and estimation problems, especially for robust control and estimation problems. Interestingly their roles are almost similar in each kind of problem, they provide some weightage to individual (error) terms, and happen to be tuning parameters.

$$\text{PI3, PI4:}\quad J3 = \sum_{k=T_0}^{T_f} u^T(k)u(k)$$
$$\text{(IIB.48)}$$
$$J4 = \sum_{k=T_0}^{T_f} u^T(k)Qu(k);\ Q \text{ is a symmetric}$$
$$\text{positive definite matrix}$$

In (IIB.48), the aim is to control the system such that the input $u(.)$ is not very large in its magnitude. The role of matrix Q is similar as discussed for R.

$$\text{PI5, PI6:}\quad J5 = aJ1 + (1-a)J3$$
$$= \sum_{k=T_0}^{T_f} [ax^T(k)x(k) + (1-a)u^T(k)u(k)$$
$$\text{(IIB.49)}$$
$$J6 = \sum_{k=T_0}^{T_f} [x^T(k)Rx(k) + u^T(k)Qu(k)]$$

The PI $J5$ is needed because one cannot minimize $J1$ and $J3$ simultaneously, because the minimization of the $J1$ requires large control inputs; however, $J3$ requires that the $u(.)$ should be as small as possible. Thus, $J5$ is a compromise between $J1$ and $J3$, and its generalization is $J6$, and the latter is called the quadratic performance index, and is very commonly used in optimal control problems.

$$\text{PI7:}\quad J7 = x^T(T_f)Q_f x(T_f) \qquad \text{(IIB.50)}$$

By using (IIB.50), we want the final state of the system is very close to zero or a constant.

$$\text{PI8:}\quad J8 = \frac{1}{2}x^T(T_f)Q_f x(T_f) + \frac{1}{2}\sum_{k=T_0}^{T_f} x^T(k)Rx(k) + u^T(k)Qu(k)]$$
$$\text{(IIB.51)}$$

The aim of the index PI8 is very obvious, it combines the requirements of PI6 and PI7: we want small states, the control input not to be very large, and the final states close to zero (or some constant) as much as possible. In case, it is desired to track a desired trajectory overall the entire interval, then in (IIB.51), the state variable $x(.)$ is replaced by the error between the desired trajectory and the actual state trajectory: $x_d(k) - x(k)$, leading to PI9:

PI9:
$$J8 = \frac{1}{2}[(x_d(T_f) - x(T_f))^T Q_f (x_d(T_f) - x(T_f))]$$
$$+ \frac{1}{2} \sum_{k=T_0}^{T_f} [(x_d(k) - x(k))^T R(x_d(k) - x(k)) + u^T(k)Qu(k)] \quad \text{(IIB.52)}$$

PI10:
$$J10 = \sum_{k=T_0}^{\infty} [x^T(k)Rx(k) + u^T(k)Qu(k)] \quad \text{(IIB.53)}$$

In (IIB.53), the infinite time (infinite horizon) problem is represented.

Appendix IIC: Illustrative Examples

In this appendix, we present examples related to some of the topics discussed in Chapters 5 through 8, and most of the examples and their solutions are based on [1–3].

Example IIC.1:

Maximize the given function and obtain the optimal trajectory and control law:

Function: $x(t) + u(t)$; $t_0 = 0$ to $t_f = 1$;

i.e. $\max_{u(t)} \int_0^1 (x(t) + u(t))dt$; constraints   (IIC.1)

$$\dot{x}(t) = 1 - u^2(t); x(0) = 1$$

The Hamiltonian (cost) is given as:
$$H = x(t) + u(t) + \lambda(t)(1 - u^2(t)) \quad \text{(IIC.2)}$$

The COV (calculus of variance) conditions give

$$\frac{\partial H}{\partial u} = 1 - 2\lambda(t)u(t) = 0; \frac{\partial H}{\partial x} = 1 = -\dot{\lambda}(t); \frac{\partial H}{\partial \lambda} = 1 - u^2(t) = \dot{x}(t) \quad \text{(IIC.3)}$$

Using one of these conditions we get

$$\lambda(t) = \lambda(0) + \int_0^t \dot{\lambda}(\tau)d\tau; \dot{\lambda}(t) = -1$$

$$\lambda(t) = \lambda(0) - t; at\ t = 1, \lambda(1) = 0; \lambda(1) = \lambda(0) - 1 = 0; \lambda(0) = 1 \quad \text{(IIC.4)}$$

Hence, we get the co-state variable as $\lambda(t) = 1 - t$, and putting this in the first of equation (IIC.2), we obtain the control law as

$$u(t) = \frac{1}{2(1-t)} \quad \text{(IIC.5)}$$

Using (IIC.5) in the state constraint equation and integrating it we get the state trajectory:

$$x(t) = x(0) + \int_0^t x(\tau)d\tau = 1 + \int_0^t 1 - \frac{1}{4(1-\tau)^2}d\tau = 1 + t - \frac{1}{4(1-t)} + \frac{1}{4} \quad \text{(IIC.6)}$$

$$x(t) = t - \frac{1}{4(1-t)} + \frac{5}{4} \quad \text{(IIC.7)}$$

Thus, the optimal control law and the associated state trajectory are given by (IIC.5), and (IIC.7).

Example IIC.2:

Optimize the cost function

$$J(y) = \int_{x_1}^{x_2} (1 + (y')^2)dx; y(x_1) = y(x_2) = 0 \quad \text{(IIC.8)}$$

We have the internal function as $F = 1 + (y')^2$; and $\frac{\partial F}{\partial y'} = 2y'$, and the Euler equation is

$$\frac{dF(2y')}{dx} = 0 \Rightarrow 2y'' = 0 \quad \text{(IIC.9)}$$

$$y'(x) = a; y(x) = ax + b; \text{ hence, } a = \frac{y_1 - y_2}{x_1 - x_2}$$

Example IIC.3:

Find an approximate solution to the BVP: $y'' - y + x = 0$; $y(0) = y(1) = 0$. Use the Rayleigh-Ritz approach. We have the cost function as: $J(y) = \int_0^1 (2xy - y^2 - (y')^2)dx$; Let $\bar{y}(x) = c_0 + c_1 x + c_2 x^2$ be an approximate solution. After the application of the both conditions, we obtain the following

$$\bar{y}(0) = 0 \Rightarrow c_0 = 0; \bar{y}(1) = 0 \Rightarrow c_1 + c_2 = 0, c_1 = -c_2 \quad \text{(IIC.10)}$$
$$\bar{y}(x) = c_1 x(1 - x)$$

$$J(c_1) = \int_0^1 (2x\bar{y} - \bar{y}^2 - (\bar{y}')^2)dx = \frac{1}{6}c_1 - \frac{11}{30}c_1^2 \quad \text{(IIC.11)}$$

$$\frac{dJ(c_1)}{dc_1} = 0 \Rightarrow c_1 = \frac{5}{22}$$

Thus, the approximate and the exact solutions are:

$$\bar{y}(x) = \frac{5}{22}x(1 - x); y(x) = x - \frac{e^x - e^{-x}}{e^x - e^{-1}} \quad \text{(IIC.12)}$$

Example IIC.4:

Consider the optimal control: minimize

$$J(u) = \int_0^1 \frac{1}{2} u^2(t) dt; \text{ constraints} \tag{IIC.13}$$
$$\dot{x}(t) = u(t) - x(t); \ x(0) = 1, x(1) = 0.$$

We will use the reachability condition (Appendix 5A), from which we obtain

$$\dot{\lambda}^{(1)}(t) = \lambda^{(1)}(t), \ \lambda^{(1)}(1) = 1, \Rightarrow \lambda^{(1)}(t) = e^{t-1}$$
$$\int_0^1 \left(\lambda^{(j)}\right)^T f_u f_u^T \lambda^{(j)} dt = \int_0^1 e^{2t-2} dt = \frac{1-e^2}{2} \neq 0 \tag{IIC.14}$$

The Hamiltonian function is given as

$$H(x, u, \lambda) = \frac{1}{2} u^2 + \lambda(u - x) \tag{IIC.15}$$

The EL equations are

$$\dot{x}(t) = H_\lambda = u(t) - x(t); \ x(0) = 1$$
$$\dot{\lambda}(t) = -H_x = \lambda(t); \qquad \hat{\lambda}(1) = \upsilon \tag{IIC.16}$$
$$H_u = 0 = u(t) + \lambda(t)$$

From the optimality condition we have the solutions for control input and co-state variable as

$$\hat{\lambda}(t) = \hat{\upsilon} e^{t-1}; \ \hat{u}(t) = -\hat{\upsilon} e^{t-1}; \text{ also } H_{uu} = 1 > 0, \text{ for } 0 \le t \le 1 \tag{IIC.17}$$

Finally we have the following optimal trajectory solutions

$$\dot{\hat{x}}(t) = -\hat{\upsilon} e^{t-1} - x(t); \ x(0) = 1; \ \hat{u}(t) = -\hat{\upsilon} e^{t-1}$$
$$\hat{x}(t) = e^{-t}[1 + \frac{\hat{\upsilon}}{2e} - \frac{\hat{\upsilon}}{2} e^{2t-1}] = e^{-1} - \frac{\hat{\upsilon}}{e} \sinh(t) \tag{IIC.18}$$
$$\hat{\lambda}(t) = \hat{\upsilon} e^{t-1}$$

The terminal condition is

$$\hat{\upsilon} = \frac{2}{e - e^{-1}} = \frac{1}{\sinh(1)} \tag{IIC.19}$$

Example IIC.5:

Illustrates the control problem formulation of driving a car in a straight line, and certain aspects of existence of the optimal solution, whether it exist or not. Let $p(t)$ denote position. The car is assumed to have a unit mass, the control input is $u(t) \ge 0$, accelerating; $u(t) \le 0$, decelerating. We have 2-DoF dynamics, $x(t) = \{p(t), \dot{p}(t)\}$; position, velocity:

$$\dot{x}(t) = \begin{bmatrix} 0 & 1 \\ 0 & 0 \end{bmatrix} x(t) + \begin{bmatrix} 0 \\ 1 \end{bmatrix} u(t); \ x(t) = \begin{bmatrix} p_0 \\ 1 \end{bmatrix} u(t) \tag{IIC.20}$$

The car starts from rest, and the control problem is to bring it to rest: p_f; thus, we have the terminal conditions:

$$x(t_0) = \begin{bmatrix} p_0 \\ 0 \end{bmatrix}; \ x(t_f) - \begin{bmatrix} p_f \\ 0 \end{bmatrix} = 0 \tag{IIC.21}$$

Also, there is constraint on the use of the fuel gas

$$\int_{t_0}^{t_f} [k_1 u(t) + k_2 x_2(t)] dt \le G \tag{IIC.22}$$

In (IIC.22), $u(t)$ is in fact the accelerating force (control input), and $x_2(t)$ is the speed of the car. The PI, the cost function is chosen such that the aim is to reach destination as early as possible:

$$J = t_f - t_0 = \int_{t_0}^{t_f} dt \tag{IIC.23}$$

Next, consider the car reaching with the minimal amount of fuel spent:

$$\min_{u(.), t_f} J(u, t_f) = \int_{t_0}^{t_f} [u(t)]^2 dt; \ x(t_f) - p_f = 0; \ 0 \le u(t) \le 1, \forall t \tag{IIC.24}$$

Since the state trajectories are continuous, we have the final position greater than the initial position; hence, $u(t) = 0$ as the minimum fuel is not feasible, and $J(u) > 0$ for every feasible control input. Assume that the permissible controls are: $u^k(t) = 1/k; \ t \ge t_0, k \ge 1$, i.e. as the time elapses, as the car moves forward, less and less control input (i.e. less and less fuel) is required, so we have

$$x_1(t) = \frac{1}{2k} (t - t_0)^2 + p_0(t - t_0)$$
$$x_2(t) = \frac{1}{k} (t - t_0)^2 + p_0$$
$$t_f^k = t_0 + 4k \left(\sqrt{p_0^2 + \frac{2}{k} p_f} - p_0 \right); \text{ the destination reaching time} \tag{IIC.25}$$

Also, we have

$$J(u^k) = \int_{t_0}^{t_f^k} \frac{1}{k^2} dt = \frac{4}{k} \left(\sqrt{p_0^2 + \frac{2}{k} p_f} - p_0 \right) \to 0, \text{ as } k \to +\infty \tag{IIC.26}$$

From (IIC.26) we see that the inf{J(u)} = 0, and the problem does not have a real minimum. Let us consider another scenario:

$$J(u) = \int_0^1 \sqrt{x^2(t) + u^2(t)}\,dt; \text{ constraint}: x(1) = 1 \quad \text{(IIC.27)}$$

$$\dot{x}(t) = u(t); \, x(0) = 0$$

In (IIC.27) the aim is to minimize functional $J(u)$, and the optimal control can be rewritten as

$$J(x) = \int_0^1 \sqrt{x^2(t) + \dot{x}^2(t)}\,dt \quad \text{(IIC.28)}$$

However, with the finite time horizon, and bounded system response for each feasible control, an optimal control law may not exist, (IIC.26), (and for (IIC.28), because the corresponding variational problem/s also would not have the global optimal [3]. In order that the optimal control exists for these problems some additional conditions should be met: (a) the controls should satisfy a Lipschitz condition, or (b) the control should be piecewise constant with a few points of discontinuity.

Example IIC.6:

Let us have a control problem defined as

$$J(u) = \int_0^1 \frac{1}{2} u^2(t)\,dt; \text{ constraints}: \quad \text{(IIC.29)}$$

$$\dot{x}(t) = u(t) - x(t); \, x(0) = 1, x(1) = 0$$

The constrain on the control is $-0.6 \le u(t) \le 0$; $t \in [0,1]$. The Hamiltonian is given as

$$H(x, u, \lambda) = \frac{1}{2} \lambda_0 u^2 + \lambda(u - x) \quad \text{(IIC.30)}$$

The co-states should satisfy the following conditions

$$\dot{\lambda}_0(t) = -H_c = 0 \quad \Rightarrow \hat{\lambda}_0(t) = K_0; \; K_0 = 1(chosen) \quad \text{(IIC.31)}$$

$$\dot{\lambda}(t) = -H_x = \lambda(t) \Rightarrow \hat{\lambda}(t) = Ke^t$$

From PMP, we get the following solution for the optimal control problem

$$\hat{u}(t) = \begin{cases} 0 & if\,\hat{\lambda}(t) \le 0 \\ -0.6 & if\,\hat{\lambda}(t) \ge 0.6 \\ -\hat{\lambda}(t) = -Ke^t & otherwise \end{cases} \quad \text{(IIC.32)}$$

Note that $K \le 0 \Rightarrow \hat{\lambda}(t) \le 0$, and $\hat{u}(t) = 0$, $0 \le t \le 1$. This control results in an infeasible response $\hat{x}(1) \ne 1$, hence $K > 0$; i.e. every optimal control is a piecewise continuous function that takes the values $-Ke^t$ or -0.6, and has at most 1 corner point:

$$\hat{u}(t) = \hat{u}_{(1)}(t) = -Ke^t, 0 \le t \le \hat{t}_s;$$
$$\hat{u}(t) = \hat{u}_{(2)}(t) = -0.6, \hat{t}_s < t \le 1 \quad \text{(IIC.33)}$$

In (IIC.33), $t(s) \rightarrow$ optimal switching. There must be a corner point because $\hat{u}(t) = -0.6, 0 \le t \le 1$ results in an infeasible response. We have the following solutions for the state trajectories:

1. For $0 \le t \le \hat{t}_s$: $\hat{x}_{(1)}(t) = C_1 e^{-t} 1 \left(-\frac{K}{2C_1} e^{2t} \right)$

$$H(\hat{u}_{(1)}(t), \hat{x}_{(1)}(t), \hat{\lambda}^*(t)) = -KC_1 \quad \text{(IIC.34)}$$

2. For $\hat{t}_s < t \le 1$: $\hat{x}_{(2)}(t) = C_2 e^{-t} - 0.6$

$$H(\hat{u}_{(2)}(t), \hat{x}_{(2)}(t), \hat{\lambda}^*(t)) = -KC_2 + \frac{(-0.6)^2}{2} \quad \text{(IIC.35)}$$

Since the arc $\hat{x}_{(1)}$ starts at $t = 0$ with $\hat{x}_{(1)}(0) = 1$ and the arc $\hat{x}_{(2)}$ ends at $t = 1$ with $\hat{x}_{(2)}(1) = 0$, the constants of integration C_1 and C_2 are:

$$C_1 = 1 + \frac{K}{2}; \; C_2 = 0.6e \quad \text{(IIC.36)}$$

Since the Hamiltonians (IIC.34) and (IIC.35) should remain the same, then by substituting the values of the constants from (IIC.36) and equating these Hamiltonians we obtain

$$K = -(1 - 0.6e) - \sqrt{(1 - 0.6e)^2 - (0.6)^2} \approx 0.436 \quad \text{(IIC.37)}$$

By equating the state trajectories from (IIC.34) and (IIC.34) at switching time, we obtain its value as 0.320.

Example IIC.7:

Bang-bang control: The dynamic system is given as

$$\dot{x}_1(t) = x_2(t); \, \dot{x}_2(t) = u(t), \; -1 \le u(t) \le 1$$

$$\hat{u}(t) = \begin{cases} 1 & if\,\hat{\lambda}_2(t) < 0 \\ -1 & if\,\hat{\lambda}_2(t) > 0 \end{cases} \quad \text{(IIC.38)}$$

The co-states are

$$\dot{\hat{\lambda}}_1(t) = 0; \, \dot{\hat{\lambda}}_2(t) = -\hat{\lambda}_1(t);$$

$$\hat{\lambda}_1(t) = A_1 \quad \text{(IIC.39)}$$

$$\hat{\lambda}_2(t) = -A_1(t) + A_2$$

The switching function is a linear function of time: $\hat{\lambda}^T(t)g = -A_1t + A_2$ and every optimal control, for $t_0 \le t \le \hat{t}_f$ is a piecewise constant function that takes the values ±, and has two intervals on which it is constant:

1. For the time interval when $\hat{u}(t) = 1$

$$\hat{x}_2(t) = t + K_1, \quad \hat{x}_1(t) = \frac{t^2}{2} + K_2 t + K_1$$

$$= \frac{1}{2}(t + K_2)^2 + \left(K_1 - \frac{K_2^2}{2}\right), \quad \text{(IIC.40)}$$

$$\hat{x}_1(t) = \frac{1}{2}(\hat{x}_2(t))^2 + K; \quad K = K_1 - \frac{1}{2}K_2^2 \quad \text{(IIC.41)}$$

The portion of the optimal response for which $u(t) = 1$, is an arc of a parabola along which the phase points move upwards.

2. For the time interval when $\hat{u}(t) = -1$

$$\hat{x}_2(t) = -t + K_1', \quad \hat{x}_1(t)$$

$$= \frac{t^2}{2} + K_2' t + K_1' \quad \text{(IIC.42)}$$

$$= -\frac{1}{2}(t + K_2')^2 + \left(K_1' - \frac{K_2'^2}{2}\right)$$

$$\hat{x}_1(t) = -\frac{1}{2}(\hat{x}_2(t))^2 + K'$$

The portion of the optimal response for which $u(t) = -1$, is an arc of a parabola along which the phase points move downwards. If the optimal control input is initially 1, and afterwards −1, the response consists of two adjoining parabolic arcs, and the second arc lies on that parabola (IIC.42) that passes through the origin: $\hat{x}_1(t) = -\frac{1}{2}(\hat{x}_2(t))^2$, however, if the optimal control input is initially −1, and afterwards 1, the second arc lies on that parabola (IIC.41) that passes through the origin: $\hat{x}_1(t) = \frac{1}{2}(\hat{x}_2(t))^2$. Thus, the optimal feedback control law can be written as

$$\hat{u}(t) = \begin{cases} 1 & if \ \hat{x}_2^2 \ sign \ x_2 < -2\hat{x}_1, \ \text{OR} \ \hat{x}_2^2 \ sign \ x_2 = -2\hat{x}_1, \ \hat{x}_1 > 0 \\ -1 & if \ \hat{x}_2^2 \ sign \ x_2 > -2\hat{x}_1, \ \text{OR} \ \hat{x}_2^2 \ sign \ x_2 = -2\hat{x}_1, \ \hat{x}_1 < 0 \end{cases}$$

(IIC.43)

Example IIC.8:

Linear quadratic regulator design. Let us take the same example as Example IIC.5 with the following system matrix/vector:

$$A = \begin{bmatrix} 0 & 1 \\ 0 & 0 \end{bmatrix}, \quad B = \begin{bmatrix} 0 \\ 1 \end{bmatrix}; \quad R = eye(2); Q = eye(1). \quad \text{(IIC.44)}$$

We can use the MATLAB function: [K, P, Ec] = lqr(A, B, R, Q, S), which computes the optimal gain matrix K, the solution of the matrix Riccati equation, P, and the closed loop eigenvalues, $E_c = eig(A-B*K)$. The cost function minimized (subject to the dynamics of the system IIC.18, IIC.44) is

$$J = \int (x'Rx + u'Qu + 2 \times x'Su)dt \quad \text{(IIC.45)}$$

The results are: K = [1.00 1.73]; P = [1.73 1.00;1.00 1.73]; E_c = −0.8660 ± 0.5000i. The plot of these eigenvalues is shown in Figure IIC.1.

Example IIC.9:

Problem with mixed inequality constraints:

$$\text{Minimize: } J(u) = \int_0^1 u(t)dt; \quad \dot{x}(t) = -u(t), \ x(0) = -1 \quad \text{(IIC.46)}$$

$$u(t) \le 0, \ x(t) - u(t) \le 0, \ 0 \le t \le 1$$

In (IIC.46), u is piecewise continuous function, and the constraints are of the mixed type; hence

$$x(t) \le u(t) \le 0, \ 0 \le t \le 1 \quad \text{(IIC.47)}$$

The Hamiltonian and the Lagrangian functions are given as

$$H(x,u,\lambda) = u(1-\lambda);$$

$$L(x,u,\lambda,\mu) = H(x,u,\lambda) + \mu_1(x-u) + \mu_2 u \quad \text{(IIC.48)}$$

$$= (1 - \lambda - \mu_1 + \mu_2)u + \mu_1 x$$

In turn, we have the following solutions for optimal state trajectory and the control

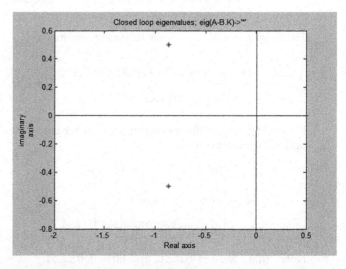

FIGURE IIC.1
Closed loop eigenvalues (Example IIC.8).

$$\hat{u}(t) = \hat{x}(t);\ \hat{x}(t) = -e^{-t};\ \ \hat{\mu}_2(t) = 0$$

$$0 = 1 - \hat{\lambda}(t) - \hat{\mu}_1(t) + \hat{\mu}_2(t) = 1 - \hat{\lambda}(t) - \hat{\mu}_1(t) \quad \text{(IIC.49)}$$

$$\hat{\mu}_1(t) = 1 - \hat{\lambda}(t)$$

The adjoint equation and the solution are

$$\dot{\hat{\lambda}}(t) = -\hat{\mu}_1(t) = \hat{\lambda}(t) - 1;\ \hat{\lambda}(1) = 0 \quad \text{(IIC.50)}$$

$$\hat{\lambda}(t) = 1 - e^{t-1} < 1;\ \hat{\mu}_1(t) = e^{t-1} > 0$$

The Hamiltonian function is constant:

$$H(\hat{x}, \hat{u}, \hat{\lambda}) = \hat{u}(t)(1 - \hat{\lambda}(t)) = -e^{-1}$$

$$\hat{u}(t)\begin{cases} 0 & \text{if } \hat{\lambda}(t) > 1 \\ \hat{x}(t) & \text{if } \hat{\lambda}(t) < 1 \end{cases} \quad \text{(IIC.51)}$$

Example IIC.10:

A simple LQR problem: Let us have the linear system with associated cost functionals:

$$\dot{x}(t) = x(t) + u(t),\ x(0) = x_0$$

$$\int_0^T \{x^2(t) + u^2(t)\}dt,\ \text{the quadratic}$$
$$\text{cost function}$$

$$J(u(.)) = -\int_0^T \{x^2(t) + u^2(t)\}dt,\ \text{the pay}$$
$$\text{off functional}(POF) \quad \text{(IIC.52)}$$

The PMP problem: Since our problem is a scalar one, from (IIC.52) we get

$$f(x, u) = x + a;\ g = 0;\ \text{and}$$

$$\varphi(x, u) = -x^2 - u^2 \quad \text{(IIC.53)}$$

The Hamiltonian is given as

$$H(x, u, \lambda) = \lambda f + \varphi = \lambda(x + u) - (x^2 + u^2) \quad \text{(IIC.54)}$$

The condition of maximum is

$$H(\hat{x}(t), \hat{u}(t), \hat{\lambda}(t)) = \max_u \{-(x^2(t) + u^2(t)) + \lambda(t)(x(t) + u(t))\} \quad \text{(IIC.55)}$$

We get the formula for optimal u, the state constraint, and the adjoint as follows:

$$H_u(x, u, \lambda) = \lambda - 2u = 0 \Rightarrow \hat{u}(t) = \frac{\hat{\lambda}(t)}{2}$$

$$\dot{\hat{x}}(t) = \hat{x}(t) + \frac{\hat{\lambda}(t)}{2} \quad \text{(IIC.56)}$$

$$\dot{\hat{\lambda}}(t) = -H_x = 2\hat{x} - \hat{\lambda}(t);\quad \lambda(T) = 0$$

From (IIC.52) and (IIC.56), we obtain the composite system (A_c) from using the PMP

$$\begin{bmatrix} \dot{\hat{x}}(t) \\ \dot{\hat{\lambda}}(t) \end{bmatrix} = \begin{bmatrix} 1 & 0.5 \\ 2 & -1 \end{bmatrix}\begin{bmatrix} \hat{x}(t) \\ \hat{\lambda}(t) \end{bmatrix} \Rightarrow \begin{bmatrix} \hat{x}(t) \\ \hat{\lambda}(t) \end{bmatrix} = e^{A_c t}\begin{bmatrix} \hat{x}_0 \\ \hat{\lambda}_0 \end{bmatrix} \quad \text{(IIC.57)}$$

Now, we need to obtain the optimal feedback control using the following rule

$$\hat{u}(t) = K(t)\hat{x}(t) = \frac{\hat{\lambda}(t)}{2} \Rightarrow K(t) = \frac{\hat{\lambda}(t)}{2\hat{x}(t)} \quad \text{(IIC.58)}$$

Now, let us define a new variable in (IIC.58)

$$p(t) = \frac{\hat{\lambda}(t)}{\hat{x}(t)};\ K(t) = \frac{p(t)}{2} \quad \text{(IIC.59)}$$

Now, from the definition in (IIC.59), one can formulate a differential equation as follows:

$$\dot{p}(t) = \frac{\dot{\hat{\lambda}}(t)}{\hat{x}(t)} - \frac{\hat{\lambda}(t)\dot{\hat{x}}(t)}{\hat{x}^2(t)} \quad \text{(IIC.60)}$$

Substitute the dynamics from (IIC.57) in (IIC.60) to obtain the following after simplification

$$\dot{p}(t) = 2 - 2p(t) - \frac{p^2(t)}{2};\ \lambda(T) = 0 \Rightarrow p(T) = 0 \quad \text{(IIC.61)}$$

Thus, (IIC.61) is a nonlinear first order ordinary differential equation and it happens to be the familiar Riccati equation, the solution of which will provide us with the adjoint variable, (IIC.58) and (IIC.59), and the solution for the state feedback gain K:

$$\hat{u}(t) = \frac{1}{2}p(t)\hat{x}(t) \quad \text{(IIC.62)}$$

Example IIC.11:

The motion of a rocket railroad car is described as

$$\begin{bmatrix} \dot{x}_1(t) \\ \dot{x}_2(t) \end{bmatrix} = \begin{bmatrix} 0 & 1 \\ 0 & 0 \end{bmatrix}\begin{bmatrix} x_1(t) \\ x_2(t) \end{bmatrix} + \begin{bmatrix} 0 \\ 1 \end{bmatrix}u(t),\ |u| \le 1 \quad \text{(IIC.63)}$$

$$J(u(.)) = -\int_0^\tau 1\,dt = -\tau = -\text{time to reach}(0,0) \quad \text{(IIC.64)}$$

In order to use the dynamic programming (DP) approach, $v(x_1, x_2)$ is defined as minus the least time it takes to get to the origin $(0, 0)$, if we start at the point (x_1, x_2). Next, the HJB equation is obtained. Since $v(.,.)$ does not depend on t, one obtains

$$\max_{u \in U}\{f \cdot \nabla_x v + \varphi\} = 0; \quad U = [-1,1], \quad f = \begin{pmatrix} x_2 \\ u \end{pmatrix}, \varphi = -1 \quad \text{(IIC.65)}$$

The PDE (partial differential equation) and the HJB equation are respectively obtained as

$$\max_{|u| \le 1}\{x_2 v_{x_1} + u v_{x_2} - 1\} = 0$$

$$x_2 v_{x_1} + |v_{x_2}| - 1 = 0; \quad v(0,0) = 0 \quad \text{(IIC.66)}$$

Now, it is required to establish that $v(.,.)$ satisfies the HJB equation. Let is define the regions as

$$R_1 := \{(x_1, x_2) \mid x_1 \ge -\frac{1}{2} x_2 \mid x_2 \mid\};$$

$$R_2 := \{(x_1, x_2) \mid x_1 \le -\frac{1}{2} x_2 \mid x_2 \mid\} \quad \text{(IIC.67)}$$

Finally, we get

$$v(x) = \begin{cases} -x_2 - 2(x_1 + \frac{1}{2}x_2^2)^{1/2} & in\ R_1 \\ x_2 - 2(-x_1 + \frac{1}{2}x_2^2)^{1/2} & in\ R_2 \end{cases} \quad \text{(IIC.68)}$$

In region R_1, we have

$$v_2(x) = -1 - (x_1 + \frac{1}{2}x_2^2)^{-1/2} x_2;$$

$$v_1(x) = -(x_1 + \frac{1}{2}x_2^2)^{-1/2} \quad \text{(IIC.69)}$$

Substituting (IIC.69) in (IIC.66), one obtains

$$x_2 v_{x_1} + |v_{x_2}| - 1 = 0 \quad \text{(IIC.70)}$$

This establishes that HJB equation holds true in region R_1 (and R_2). Then, the optimal control is given as

$$\hat{u}(t) = \text{sgn}(v_{x_2}) \quad \text{(IIC.71)}$$

Example IIC.12:

DC Motor control design; comparison of three methods: (i) feedforward command, (ii) integral feedback control, and (iii) LQR-linear quadratic regulator [AIIC.1]. In an armature-controlled DC motor, the applied voltage V_a controls the angular velocity ω of the shaft.

The idea is to reduce the sensitivity of the angular velocity to load variations, i.e. the changes in the torque opposed by the motor load. The torque T_d models the load disturbances, and we should minimize the speed variation induced by such disturbances. The results for this example can be obtained by running the script **dcmotorcontrol.m,** which has the physical constants of the motor already entered. The state space model of the DC motor with two inputs: V_a, T_d, and one output ω is obtained. The response of the angular velocity to a step change in the applied voltage is shown in Figure IIC.2, which also shows the settling time. Now, one can use simple feedforward structure to command the angular velocity to a given reference ω_{ref} value, i.e. this reference command is modified with a feedforward gain K_{ff}, and then acts as a voltage command to the DC motor whose output, the angular velocity is monitored, and T_d is also directly affecting the DC motor as an additional input. This gain is set ($K_{ff} = 4.1$) as a reciprocal of the DC gain from V_a to the speed. The load disturbance is set to $T_d = -0.1$ Nm between $t = 5$ and $t = 10$ secs., and a reference command of 1 unit is applied. The step response for this case is shown in Figure IIC.3, and it is seen that the feedforward design does not handle the load disturbance properly. Next, we can try the feedback around the DC motor by feeding back the angular velocity to the reference command and using the integral action control ($C(s) = K/s$) in the forward loop before the

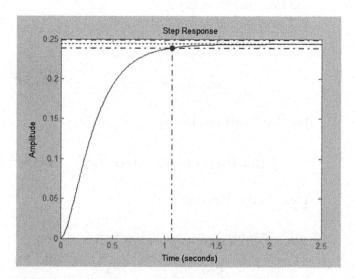

FIGURE IIC.2
Response of the angular velocity to a step change in the applied voltage (Example IIC.12).

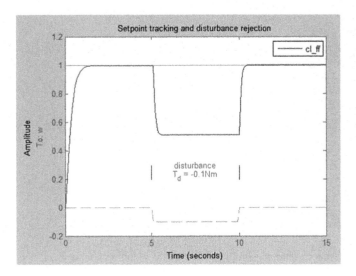

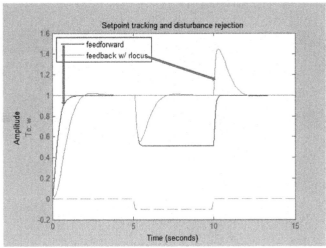

FIGURE IIC.3
Step response from feedforward design (Example IIC.12).

FIGURE IIC.5
Comparison of two designs: feedforward (FF), and feedback (FB) (Example IIC.12).

motor, and this will generate a new velocity V_a. We can use the root locus to determine the gain K, the plot is shown for the $1/s * G_p(s)$; i.e. the controller and the DC motor model as the (open) loop TF, see Figure IIC.4, where we can see that the reasonable gain is $K = 5$. We can now compare the two designs as in Figure IIC.5, and see that the integral action design (with root locus) is better at the rejecting the load disturbances. Next, we try the LQR based design for DC motor control. There is an integral action in the forward loop to the gain K_{lqr} that goes to compute the voltage signal V_a, and there are two feedback loops: (i) inner loop, from the motor output, i.e. angular velocity to the K_{lqr}, i.e. the state feedback (an optimal control action), and (ii) the outer loop, from the motor output to the reference

signal; then the loop error signal acts on the integral controller, and the torque disturbance acts as earlier. The LQR uses the state vector, the armature current i and the motor speed as the state feedback to synthesize the motor driving voltage, input V_a. This voltage is given by the formula

$$V_a = K_1\omega + K_2\frac{\omega}{s} + K_3 i \qquad (IIC.72)$$

We use the cost function to penalize the large integral errors, and for better disturbance rejection

$$J = \int_0^\infty \left\{ 20\left(\frac{\omega(t)}{t}\right)^2 + \omega^2(t) + 0.01V_a^2(t) \right\} dt \qquad (IIC.73)$$

The LQR design is done by using "lqry" function. The CL Bode plot for the three designs is shown in Figure IIC.6, whereon each design is identified. Now, the three designs are compared in Figure IIC.7, from which it is clear that the LQR design performs better than the other two designs.

Example IIC.13:

Thickness control of a steel beam. To design a MIMO LQG regulator to control the horizontal and vertical thickness of a steel beam in a hot steel rolling mill [AIIC.2], in which the process of shaping a beam of hot steel is carried out by compressing it with the rolling cylinders, which are the two pairs (one per axis) positioned by hydraulic actuators. The gap between the two cylinders is called the roll gap. The aim is to maintain the x and y thickness within the specified tolerances. This thickness variations arise from

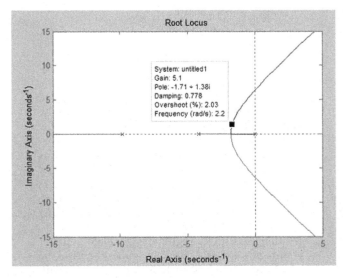

FIGURE IIC.4
The root locus to determine the gain K, for the feedback design (Example IIC.12).

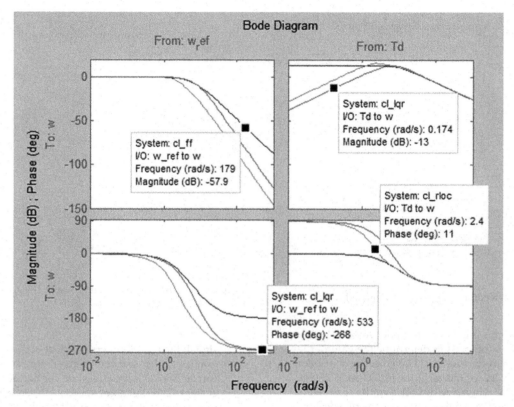

FIGURE IIC.6
The CL Bode plot for the three designs (Example IIC.12).

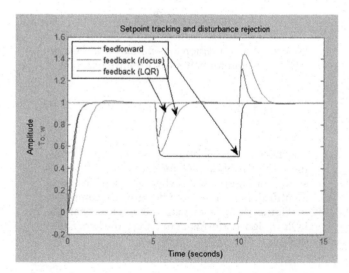

FIGURE IIC.7
Comparison of three designs: step responses (Example IIC.12).

the variations in the thickness and hardness of the incoming beam (that is regarded as an input disturbance) and the eccentricities of the rolling cylinders. An open loop model for the x (and y) axis is built up. The eccentricity disturbance is modelled as a white noise w_e driving a band-pass filter F_e. The input thickness disturbance is modelled as a white noise w_i driving a low-pass filter F_i. Feedback control is necessary to counter

such disturbances. Because the roll gap δ cannot be measured close to the stand, the rolling force f is used for feedback. The open loop models are specified as

$$F_{ex}(s) = \frac{3 \times 10^4 s}{s^2 + 0.125s + 36}; \quad F_{ix}(s) = \frac{10^4}{s + 0.05}$$

$$H_x(s) = \frac{2.4 \times 10^8}{s^2 + 72s + (90)^2}$$

(IIC.74)

In (IIC.74), $F_e(s)$ is an eccentricity filter (in x-axis), $F_i(s)$ is the input disturbance filter, $H(s)$ is the actuator model, and $g = 10^{-6}$ is the gap-to-force gain. The TFs are constructed from u, w_e, w_i, to f_1 and f_2. The results for this example are obtained by running the script **tcsteelbeamlgq.m**. Now, the frequency responses from the normalized disturbances w_e and w_i to the outputs are obtained as in Figure IIC.8, from which the peak at 6 rad/sec due to the periodic eccentricity disturbance is apparent. An LQG regulator design to attenuate the thickness variations due to the eccentricity and input thickness disturbances, w_e and w_i, is attempted. LQG regulators generate actuator commands $u = -Kx_e$, where x_e is an estimate of the plant states. This estimate is derived from available measurements of the rolling force f using the Kalman filter. Thus, the LQG-regulator acts a feedback to the overall system by obtaining a

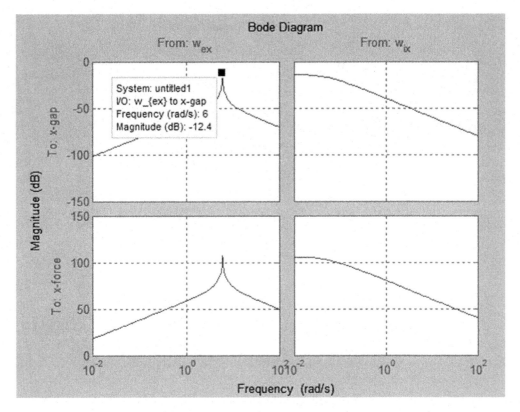

FIGURE IIC.8
Frequency responses from the normalized disturbances w_e and w_i to the outputs (Example IIC.13).

control signal u (from the force via LQG-regulator gain, by using "lqry"), i.e. state feedback gain K is obtained; the latter is chosen to minimize the cost function

$$J(u) = \int_0^\infty (\delta^2(t) + \beta u^2(t)) dt \qquad (IIC.75)$$

In (IIC.75), the weighting factor is used to balance the performance and the control effort, and it is chosen as $\beta = 1e-4$. The feedback gain obtained is $K = [0.0621\ 0.1315\ 0.0222\ -0.0008\ -0.0074]$. Next, the Kalman filter is designed with the measurement noise covariance set to 1e4 to limit the gain at high frequencies. Now, the "lqgreg" is used to assemble the LQG regulator Regx from K_x and E_x, the estimates of the states of the system. The Bode plot of the LQG regulator is shown in Figure IIC.9. The regulator loop is now closed and the results of comparison of the open loop and closed loop responses to eccentricity and input thickness disturbances are shown in Figure IIC.10, from which it can be seen that ~20 dB attenuation of disturbance effects is achieved. The disturbance induced thickness variations with and without the LGQ regulator is shown in Figure IIC.11, and we see the improvement with the LQG design. Next, the LQG regulator can be designed for the

y-axis using the same procedure that used for the x-axis. One can assume that the two axes are decoupled, and use these two regulators independently to control the process, but, this will be valid if the actual coupling in the process is fairly week. However, in reality these processes have a

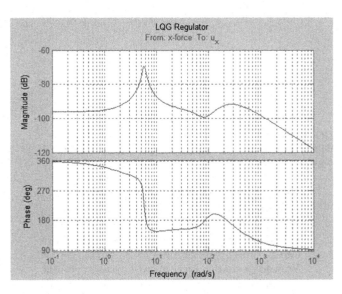

FIGURE IIC.9
The Bode plot of the LQG regulator (Example IIC.13).

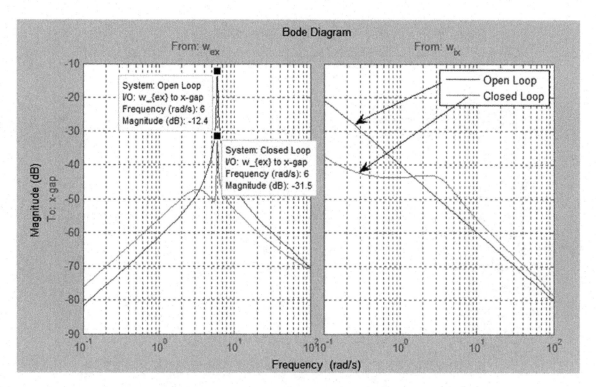

FIGURE IIC.10

Comparison of the OL and CL responses to eccentricity and input thickness disturbances (Example IIC.13).

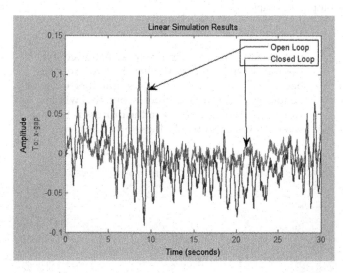

FIGURE IIC.11

Disturbance induced thickness variations with and without the LGQ regulator (Example IIC.13).

cross-coupling between axes, since an increase in force along x-axis compresses the material and caused a relative decrease in force along the y-axis. The cross-coupling effects are taken into account by assuming $g_{xy} = 0.1$, and $g_{yx} = 0.4$. Now, to study the effect of cross-coupling on the decoupled SISO loops, construct the two-axis model, and close the x-axis, and y-axis loops using the previously designed LQG regulators.

Next, simulate the x and y thickness gaps for the two-axis model, see Figure IIC.12, from which it is seen that the x-axis disturbance is still large. The treating of axes now can be done jointly by using MIMO design to handle the cross-coupling effects. This MIMO design consists of a single regulator that uses both force measurements f_x and f_y to compute the actuator commands u_x and u_y. We can use the same procedure as used for SISO designs: (i) compute the feedback gain, (ii) then compute the estimator, and (iii) assemble these two components using "lqgreg." Then, close the MIMO design loop. Simulate the x and y axis gaps for the two-axis model, see the results in Figure IIC.13, from which it can be seen that the MIMO design shows no performance loss in the x axis. The improvement can be seen by comparing the principal gains of the CL responses from input disturbances to thickness gaps x-gap, and y-gap, and it is seen from Figure IIC.14 that the MIMO regulator does a better job at keeping the gain equally low in all directions.

Example IIC.14:

Pressure regulation in a drum boiler. Operating point search function is used for model linearization, then state observer and LQR design are done for a drum boiler model [AIIC.3]. The control problem is to regulate boiler pressure in the presence of random heat variations from the

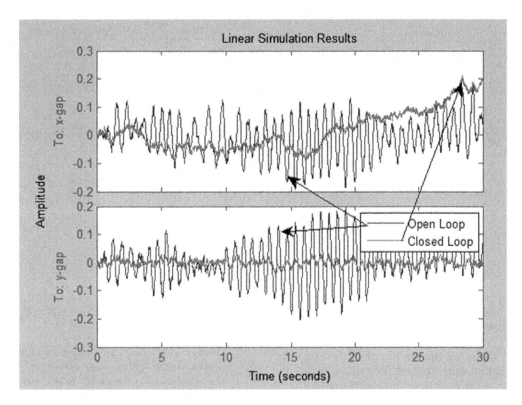

FIGURE IIC.12
Simulation of the x and y thickness gaps for the two-axis model (Example IIC.13).

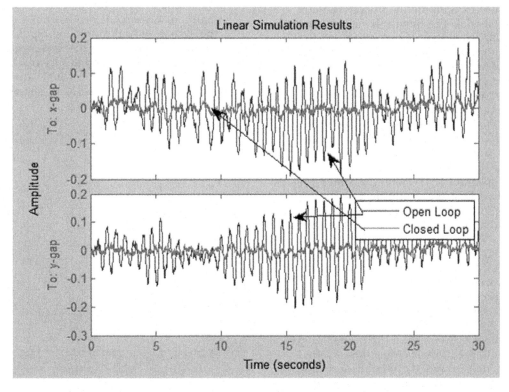

FIGURE IIC.13
Simulation of the x and y axis gaps for the two-axis model with the MIMO design loop closed (Example IIC.13).

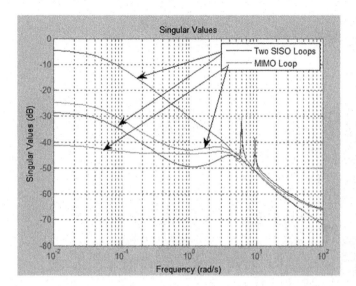

FIGURE IIC.14
Principal gains (SVD, singular values) of the CL responses from input disturbances to thickness gaps x-gap, and y-gap (Example IIC.13).

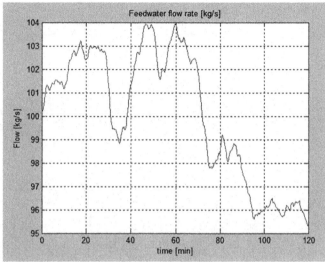

FIGURE IIC.15
The feed water actuation signal in kg/s (Example IIC.14).

furnace by adjusting the feed water flow rate and the nominal heat applied. For this example, 95% of the random heat variations are less than 50% of the nominal heating value, which is usual for a furnace-fired boiler. The results for this example can be generated by using the script **rpdrumboiler.m.** The running will open the Simulink model. The boiler control model's preload function would initialize the controller sizes, because to compute the operating point and linear model, the Simulink model must be executable. Note that u_0 and y_0 are set after computing the operating point and are initially set to zero. The observer and regulator are computed during the controller design step and are initially set to zero. Next, task is to find a nominal operating point and linearize the model for which the model's initial states are defined in the Simulink model. Using these state values, one would find the SS operating point using the "findop" function. Now, create an operating point specification where the states are known; let's adjust the operating point specification to indicate that the inputs must be computed that are lower bounded. Then, add an output specification to the operating point specification; this is needed to ensure that the output operating point is computed during the solution process. Next, compute the operating point and generate a report. Before linearizing the model around this point, specify the I/O signals for the linear model; specify the input points for linearization, specify the open loop output points for linearization; thus, we find a linear model around the chosen operating point. Using the "mineral" function, make sure that the model is a minimum realization, (there are no pole/zero cancellations). Using this

linear model, one can design an LQR regulator and Kalman filter/state observer. Determine the controller offsets to make sure that the controller is operating around the chosen linearization point by retrieving the computed operating point. Now design the regulator using the "lqry" function; a tight regulation of the output is required while input variation should be limited. The Kalman state observer is designed using "Kalman" function; here the main noise source is process noise, and enters the system only through one input; hence, the form of G and H. For the designed controller the process inputs and outputs are shown next. Figure IIC.15 shows the feed water actuation signal in kg/s. Figure IIC.16 shows the heat

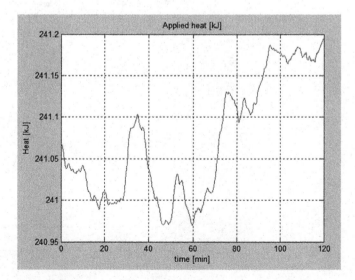

FIGURE IIC.16
The heat actuation signal in KJ (Kilo-Jules) (Example IIC.14).

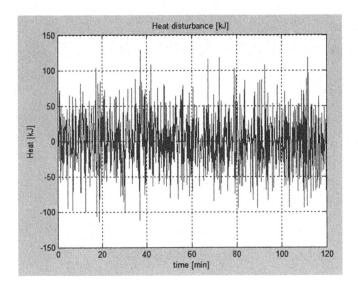

FIGURE IIC.17
The heat disturbance in KJ (Example IIC.14).

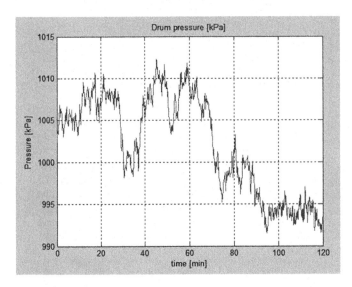

FIGURE IIC.18
the corresponding drum pressure in KPa (Kilo-Pascals) (Example IIC.14).

actuation signal in KJ (Kilo-Jules). Figure IIC.17 shows the heat disturbance in KJ. Note that the disturbance varies by as much as 50% of the nominal heat value. Figure IIC.18 shows the corresponding drum pressure in KPa (Kilo-Pascals), the pressure varies by ~1% of the nominal value even though the disturbance is relatively large.

Example IIC.15:

To design a robust controller for an active suspension system, i.e. a quarter-car suspension model. Conventional passive suspensions use a spring

and damper between the car body and wheel assembly [AIIC.4]. The spring-damper characteristics are selected to emphasize one of several conflicting objectives such as passenger comfort, road handling, and suspension deflection. Active suspensions allow the designer to balance these objectives using a feedback-controller hydraulic actuator between the chassis and wheel assembly. This example uses a quarter-car model of the active suspension system. The mass m_b (in kilograms) represents the car chassis (body) and the mass m_w (in kilograms) represents the wheel assembly. The spring ks and damper b_s represent the passive spring and shock absorber placed between the car body and the wheel assembly. The spring k_l models the compressibility of the pneumatic tire. The variables x_b, x_w, and r (all in meters) are the body travel, wheel travel, and road disturbance, respectively. The force f_s (in Kilo-Newtons) applied between the body and wheel assembly is controlled by feedback and represents the active component of the suspension system. The linearized state space model for the quarter-car is given as

$$\dot{x}_1 = x_2$$
$$\dot{x}_2 = -(1 / m_b)[k_s(x_1 - x_3) + b_s(x_2 - x_4) - 10^3 f_s]$$
$$\dot{x}_3 = x_4$$
$$\dot{x}_4 = -(1 / m_w)[k_s(x_1 - x_3) + b_s(x_2 - x_4)$$
$$- k_l(x_3 - r) - 10^3 f_s]$$

(IIC.76)

The states in (IIC.76) are: $(x_1, x_2, x_3, x_4) = (x_b, \dot{x}_b, x_w, \dot{x}_w)$. This state model is created in MATLAB and the TFs from the actuator to body travel and acceleration are obtained. The results for this example can be generated using the script "**rcactivesuspension**." As can be seen from Figure IIC.19, the TFs from the actuator force to the body travel/acceleration, and to the suspension deflection have zeros on the imaginary axis:$-0.0000 + j56.2731$; $-0.0000 - j56.2731$ (this is called the tire-hop frequency); and $0.0000 + j22.9734$; $0.0000 - j22.9734$ (this is called rattle space frequency) respectively. Road disturbances influence the motion of the car and suspension. Passenger comfort is associated with small body acceleration. The allowable suspension travel is constrained by limits on the actuator displacement. The Bode plots are shown in Figure IIC.19; the open loop gain from road disturbance and actuator force to body acceleration and suspension displacement. Because of the imaginary-axis zeros, feedback control cannot improve the response from road disturbance r to body acceleration a_b at the tire-hop frequency, and from r to suspension deflection s_d at the rattlespace frequency. Moreover, because of the relationship $x_w = x_b - s_d$ and the fact that the wheel position x_w roughly follows "r" at low frequency (less than 5 rad/sec), there is an inherent trade-off between passenger comfort and

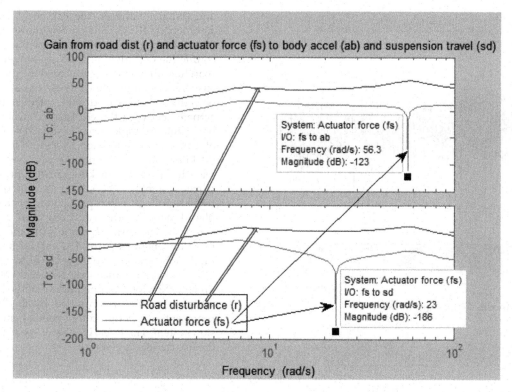

FIGURE IIC.19
The TFs from the actuator force to the body travel/acceleration, and to the suspension deflection have zeros on the imaginary axis (Example IIC.15).

suspension deflection: any reduction of body travel at low frequency will result in an increase of suspension deflection. Next, the uncertainty analysis is carried out. The hydraulic actuator used for active suspension control is connected between the body mass m_b and the wheel assembly mass m_w. The nominal actuator dynamics are represented by the first-order transfer function $1/(1+s/60)$ with a maximum displacement of 0.05 m. This nominal model only approximates the physical actuator dynamics. We can use a family of actuator models to account for modelling errors and variability in the actuator and quarter-car models. This family consists of a nominal model with a frequency-dependent amount of uncertainty. At low frequency, below 3 rad/sec, the model can vary up to 40% from its nominal value. Around 3 rad/sec, the percentage variation starts to increase. The uncertainty crosses 100% at 15 rad/sec and reaches 2000% at approximately 1000 rad/sec. The weighting function W_{unc} is used to modulate the amount of uncertainty with frequency. The result "Act" is an uncertain state space model of the actuator. Figure IIC.20 shows the Bode response of 20 sample values of "Act" and compares with the nominal value. Now, for the design setup, the main control objectives are formulated in terms of passenger comfort and road handling, which relate to body acceleration a_b and suspension travel s_C. Other factors that influence the control design include the characteristics of the road disturbance, the quality of the sensor measurements for feedback, and the limits on the

available control force. To use HI synthesis algorithms, we must express these objectives as a single cost function to be minimized. The feedback controller uses measurements y_1, y_2 of the suspension travel s_d and body acceleration a_b to compute the control signal u driving the hydraulic actuator. There are external sources of disturbance: (i) the road disturbance r, modelled as a normalized signal d_1 shaped by a weighting function W_{road}, to model broadband

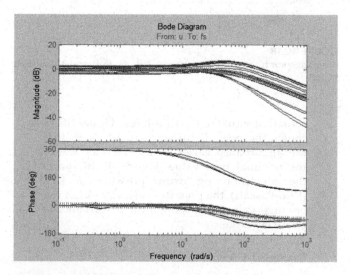

FIGURE IIC.20
Bode response of 20 sample values of "Act" and comparison with the nominal value (Example IIC.15).

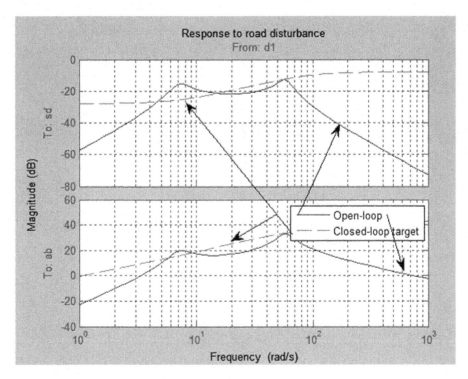

FIGURE IIC.21
The magnitude responses of the TFs to road disturbance (Example IIC.15).

road deflections of magnitude seven centimeters, we use the constant weight $W_{road} = 0.07$; and (ii) sensor noise on both measurements, modelled as normalized signals d_2 and d_3 shaped by weighting functions W_{d2} (=0.01) and W_{d3} (0.5) to model broadband sensor noise of intensity 0.01 and 0.5, respectively. In a more realistic design, these weights would be frequency dependent to model the noise spectrum of the displacement and acceleration sensors. The control objectives can be reinterpreted as a *disturbance rejection* goal: minimize the impact of the disturbances (d_1, d_2, d_3) on a weighted combination of control effort u, suspension travel s_d, and body acceleration a_b. When using the HI norm (peak gain) to measure "impact," this amounts to designing a controller that minimizes the HI norm from disturbance inputs (d_1, d_2, d_3) to error signals (e_1, e_2, e_3). Create the weighting functions and label their I/O channels to facilitate interconnection. Use a high-pass filter for W_{act} to penalize high-frequency content of the control signal and, thus, limit the control bandwidth. Specify closed loop targets for the gain from road disturbance r to suspension deflection s_d (handling) and body acceleration a_b (comfort). Because of the actuator uncertainty and imaginary-axis zeros, we only seek to attenuate disturbances below 10 rad/sec. The magnitude responses of the TFs to road disturbance are shown in Figure IIC.21. The corresponding performance weights W_{sd}, W_{ab} are the reciprocals of these comfort and handling targets. To investigate the trade-off between passenger comfort and road handling, construct three sets of weights $(\beta W_{sd}, (1-\beta)W_{ab})$

corresponding to three different trade-offs: comfort ($\beta = 0.01$), balanced ($\beta = 0.5$), and handling ($\beta = 0.99$). Finally, use connect to construct a model "qcaric;" which is an array of three models, one for each design point β, and it is an uncertain model since it contains the uncertain actuator model "Act." One can now use the "hinfdyn" to compute an HI controller for each value of the blending factor β, the values of γ are: 0.9410, 0.6724, and 0.8877. The three controllers achieve closed loop HI norms of 0.94, 0.67 and 0.89, respectively. From the corresponding closed loop models and the gains from road disturbance to x_b, s_d, a_b for the passive and active suspensions, Figure IIC.22, one can see that all three controllers reduce suspension deflection and body acceleration below the rattlespace frequency, 23 rad/sec. Next, we can do the time domain evaluation. To further evaluate the three designs, perform time-domain simulations using a road disturbance signal $r(t)$ representing a road bump of height 5 cm; the responses are shown in Figures IIC.23 and IIC.24. Once can observe that the body acceleration is smallest for the controller emphasizing passenger comfort and largest for the controller emphasizing suspension deflection. The "balanced" design achieves a good compromise between body acceleration and suspension deflection. Next, we can try the robust mu design. So far we have designed HI controllers that meet the performance objectives for the *nominal* actuator model, however, this model is only an approximation of the true actuator and one needs to make sure that the controller performance is maintained in the face of

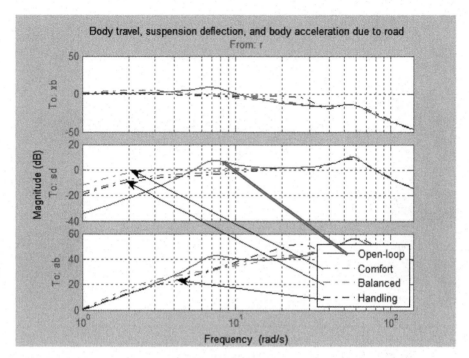

FIGURE IIC.22
Closed loop models and the gains from road disturbance to x_b, s_d, a_b for the passive and active suspensions (Example IIC.15).

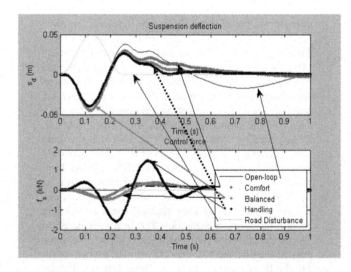

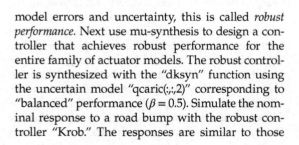

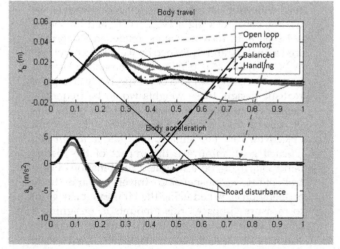

FIGURE IIC.23
Time-domain simulations using a road disturbance signal $r(t)$ representing a road bump of height 5 cm (Example IIC.15).

FIGURE IIC.24
The time-domain simulations using a road disturbance signal $r(t)$ representing a road bump of height 5 cm. (Example IIC.15).

model errors and uncertainty, this is called *robust performance*. Next use mu-synthesis to design a controller that achieves robust performance for the entire family of actuator models. The robust controller is synthesized with the "dksyn" function using the uncertain model "qcaric(:,:,2)" corresponding to "balanced" performance ($\beta = 0.5$). Simulate the nominal response to a road bump with the robust controller "Krob." The responses are similar to those

obtained with the "balanced" HI controller as can be seen from Figure IIC.25. Next simulate the response to a road bump for 100 actuator models randomly selected from the uncertain model set Act. The robust controller "Krob" reduces variability due to model uncertainty and delivers more consistent performance, as is seen from Figures IIC.26 and IIC.27. Now we try controller simplification. The robust controller "Krob" has eleven states. It is often the case

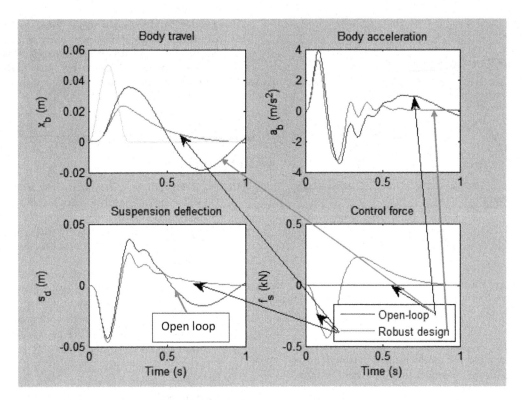

FIGURE IIC.25
Simulation of the nominal responses to a road bump with the robust controller "Krob" (Example IIC.15).

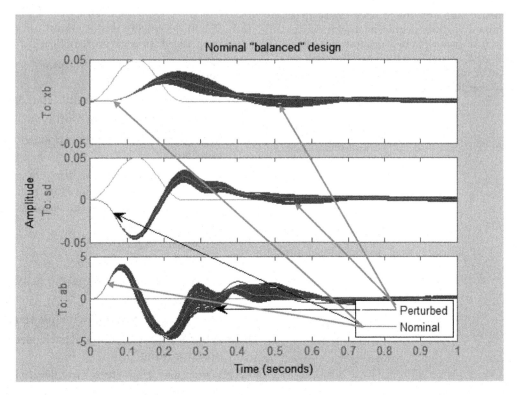

FIGURE IIC.26
Simulated responses to a road bump for 100 actuator models randomly selected from the uncertain model set Act, nominal balanced design (Example IIC.15).

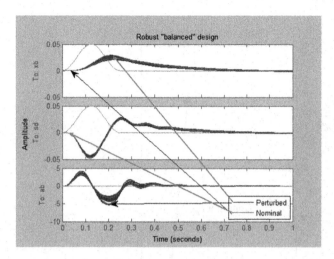

FIGURE IIC.27

Simulated responses to a road bump for 100 actuator models randomly selected from the uncertain model set Act, robust balanced design (Example IIC.15).

that controllers synthesized with "dksyn" have high order. One can use the model reduction functions to find a lower-order controller that achieves the same level of robust performance. Use "reduce" to generate approximations of various orders. Next use "robustperf" to compute the robust performance margin for each reduced-order approximation. The performance goals are met when the closed-loop gain is less than $\gamma = 1$. The robust performance margin measures the amount of uncertainty that can be sustained without degrading performance (exceeding $\gamma = 1$). A margin of 1 or more indicates that we can sustain 100% of the specified uncertainty. As can be seen from Figure IIC.28, the robust performance margin is well below 1 for controllers

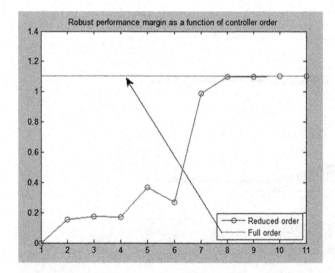

FIGURE IIC.28

Robust performance margin is well below 1 for controllers of order 7 and lower (Example IIC.15).

of order 7 and lower, however there is no significant difference in performance margin between the eighth and eleventh order controllers, so one can safely replace "Krob" by its eighth order approximation.

References

AIIC.1 Anon. https://in.mathworks.com/help/control/ examples/dc-motor-control.html# d119e8126, accessed August 2018.

AIIC.2 Anon. https://in.mathworks.com/help/control/ examples/thickness-control-for-a-steel-beam.html, accessed August 2018.

AIIC.3 Anon. https://in.mathworks.com/help/slcontrol/ examples/regulating-pressure-in-a-drum-boiler. html, accessed Aug 2018.

AIIC.4 Anon. http://in.mathworks.com/help/robust/ examples/robust-control-of-an-active-suspension. html, accessed August 2018.

Exercises for Section II

II.1 What is a functional?

II.2 What is an extremizing function and extremum?

II.3 When do we say that a minimum or maximum is attained for a given function?

II.4 What is a two-point boundary value problem (TPBVP)?

II.5 What is the Lagrange multiplier (LM) technique?

II.6 How is the LM approach utilized?

II.7. What is the result of the application of the LM method?

II.8 In some equations for the cost function (in Chapter 6), some weighting matrices like Q and R appear, what is the significance of these matrices?

II.9 What is an LQG problem?

II.10 What is an LQR problem?

II.11 What are the major ingredients of a maximum principle problem?

II.12 How is the maximum principle stated?

II.13 What is the basic technical trick used in Dynamic programming?

II.14 What are the steps involved in Dynamic programming?

II.15 What are two approaches in solving a DP problem?

II.16 What do the dynamic and programming signify in the DP problem?

II.17 What is the two-person differential game?

II.18 What is model predictive control (MPC) in the context of optimal control?

II.19 What does an MPC algorithm consist of?

II.20 What is the strategy in MPC?

II.21 What are the prediction models that can be used in MPC?

II.22 Which algorithms cannot be used in MPC to control unstable systems or the systems with integrator?

II.23 What are important differences between the MPC and the LQ/LQG methods?

II.24 In which way does the MPC differs from the PID controller?

II.25 In a TITO (two input-two output) system, what is the internal stability?

II.26 What is the exponential stability?

II.27 Is it always safe to cancel the poles and zeros in a TF, if they are (~) equal?

II.28 Does the robust stably assure robust performance?

II.29 When is a CLCS called well-posed?

II.30 What is the main difference between the optimal and suboptimal control problems?

II.31 What is a Hamiltonian matrix?

II.32 What is a good performance design's requirement, say for a TITO system?

II.33 What is small-gain theorem?

II.34 Why do we often use system norms?

II.35 What is the ∞-norm for an LTI stable system?

II.36 What is linear matrix inequality?

II.37 Is $AX + XA^T < 0$ LMI?

II.38 What does LMI describe or signify?

II.39 What is the bounded real lemma?

II.40 Minimize

$$\int_{0}^{t_f=1} P(t)dt; \dot{T} = P - T; 0 \le P \le P_{\max}; T(0) = 0, T(1) = 1,$$

here, H is the heat effect and T is the temperature.

II.41 Let the model of a dynamic system be given as

$$x(k+1) = \varphi_k x(k) + Bu(k); \ y(k) = Hx(k)$$

Assume that the given system is unstable and its responses are increasing without bound. Reformulate the equations by using an appropriate transformation method on the unstable data.

II.42 The state space model of the short period dynamics of a transport aircraft are given as:

$$\dot{w} = Z_w w + (u_0 + Z_q)q + Z_{\delta_e}\delta_e;$$

$$\dot{q} = M_w w + M_q q + M_{\delta_e}\delta_e$$

Assume there is instability in pitch dynamics, and hence, the M_w has positive numerical value. Stabilize the system with appropriate feedback from the vertical speed (an aircraft state variable). Give new state space equations.

Does the solution of this problem indicate any relationship between the state feedback and the pole placement/assignment?

References for Section II

1. Thompson, P. Lecture notes on optimal control. Carnegie Mellon University, USA, January 2003. http://www.andrew.cmu.edu/course/88-737/optimal_control/optimal_control.pdf, accessed November 2017.

2. George, R. K. Calculus of variations, Lecture notes, IIST. https://www.iist.ac.in/sites/ default/files/people /COV Main.pdf, *accessed November 2017.*

3. Chachuat, B. *Nonlinear and Dynamic Optimization: From Theory to Practice.* Automatic Control Laboratory, EPFL, Switzerland, 2007. http://la.epfl.ch/files/content/sites/la/files/shared/import/migration/IC_32/ic-32_lectures-17-28.pdf, accessed November 2017.

4. Evans, L. C. *An Introduction to Mathematical Optimal Control Theory*, Ver. 0.2. https://math.berkeley.edu/~evans/control.course.pdf, accessed November 2017.

5. Raol, J. R., Gopalratnam, G. and Twala, B. *Nonlinear Filtering: Concepts and Engineering Applications.* Taylor & Francis Group, Boca Raton, FL, 2017.

6. Patwardhan, S. C. *A Gentle Introduction to Model Predictive Control (MPC) Formulations based on Discrete Linear State Space Models.* https://nptel.ac.in/courses/103101003/downloads/lecture-notes/LQG_MPC_Notes.pdf, accessed December 2017.

7. Liberzon, D. Calculus of Variations and Optimal Control Theory-A Concise Introduction. pp. 1–200, 2011. http://liberzon.csl.illinois.edu/teaching/cvoc.pdf, accessed November 2017.

8. Anon. Pole placement design technique. http://www.ece.rutgers.edu/~gajic/psfiles/poleplacement.pdf, accessed December 2017.

9. Wang, G.-S., Liang, B., and Duan, G.-R. H_2-optimal control with regional pole assignment via state feedback. *IJCAS*, Vol. 4, No. 5, pp. 653–659, 2006.

10. Singh, T. Fuel/time optimal control of the benchmark problem. *J. Guid. Control Dyn.*, AIAA, USA. Vol. 18, No. 6, 1995.

11. Ruscio, D. Di. Model predictive control and optimization. Lectures notes Model predictive control, System and Control Engineering, Department of Technology, Telemark University College, 2010.

12. Anon. Model predictive control-1, PPT slides; utw10249. utweb.utexas.edu/edgar_group/che391/.../ChE%20391%20 MPC(1).ppt, accessed December 2017.

13. Toivonen, H. T. Robust Control Methods, Process Control Laboratory, Abo Akademi University, Turku (Abo.) Finland, 1998. http://users.abo.fi/htoivone/courses/ robust/ rob1.pdf; rob1-rob8, accessed January 2018.

14. Francis B. A. *Lectures Notes in Control and Information Sciences*. Springer-Verlag, Berlin, Germany, 1987. http:// individual.utoronto.ca/brucefrancis1/second_level/ Papers/H_infinity.pdf, accessed January 2018.

15. Rantzer, A., and Khong, S. Z. http://www.control.lth.se/ Education/Doctorate Program /robust-control/robust-control-2015.html, accessed February 2018.

16. Raol, J. R., and J. Singh. *Flight Mechanics Modelling and Analysis*. Taylor & Francis Group, Boca Raton, FL, 2008.

17. Raol, J. R., Girija, G., and Singh, J. *Modelling and Parameter Estimation of Dynamical Systems*. IEE/IET Control Series Vol. 65, IEE/IET Society, London, UK, 2004.

18. Kar, I., and Somanath, M. Introduction to Digital Control. Lecture Notes; Modules 1 to 11. Department of Electronics and Electrical Engineering, Indian Institute of Technology (IIT) Guwahati, Guwahati, Assam, India. https://nptel.ac.in/courses/108103008/PDF/module1/ m1_lec1.pdf, accessed November 2017.

Section III

Digital and Adaptive Control

In this section, we will deal with digital and/or sampled data and adaptive control systems. The digital signals are the sampled signals of the respective continuous time signals. The magnitudes of these discrete time signals are then represented by the digital real numbers. These digital data are directly amenable as the data to be used in digital computers for further analyses. Also, correspondingly, the continuous time systems are to be represented in equivalent sampled data systems. These systems are then called digital (control) systems, and they can be easily coded in a digital computer. So, the digital/sampled data (signals) and these sampled-data control systems can be easily handled in a digital computer, since the solutions of the equations of the control systems, their digital controllers, and the corresponding estimators/filters (for state feedback control) can be obtained by/in recursive formulas; the complex and very involved methods like those required for the continuous time systems are not required here in most cases. Starting from the initial conditions, one can, in a recursive fashion, compute the states and control signals required for realization of the algorithms in a digital computer. Almost every technique developed for continuous time systems are now implemented in digital form, either by converting these continuous time systems/solutions to discrete time systems/algorithms or by finding out directly the similar solutions/algorithms for discrete time systems from first principles only.

The main aim in an adaptive control is to arrive at a set of techniques for automatic adjustment of the controllers in real time (e.g., while aircraft is flying, an industrial plant/process is running, a robot is fining its optimal path) in order to achieve or to maintain a desired level of performance of the control system (say autopilot) when certain parameters of the plant/dynamic (aircraft) model change in time and/or the disturbances that affect the dynamics are unknown (turbulence). Application of adaptive control technology has been made very easy with the advent of high-speed computers, large bandwidth (communications) networks, and associated digital processing technology.

9

Discrete Time Control Systems

Many basic concepts of linear control systems, which we studied in Section I, are also applicable to the sample-data and digital control systems (SDDCS); however, there are certain differences due to the very nature of the sampling process and digitization of the continuous time control systems. This calls for a separate treatment that will be covered in this chapter and the next chapter. In fact, the continuous time algorithms for control systems/controllers/estimators can be converted in the discrete (or digital) form so that their implementation on a digital computer is relatively easy; however, in most cases, it would be necessary to convert not only these algorithms, but also all the sub-systems into discrete time systems/algorithms. From the signal processing point of view, it has been found that [1]: (i) the sampled signals are easier to transmit and, at the receiving end, the signals can be regenerated or recovered without much, or with little, transmission error; (ii) the sampled signals can more easily be coded, i.e. use of cryptology, to encode the signal with some secrete code, that would be known to the concerned people; and (iii) the sampled signals can be modulated and multiplexed with relative ease; thus, the introduction of a sampler in a control system improves the dynamic behavior of the control (loop), in an overall sense with the merits: (i) better sensitivity behavior, (ii) better reliability, (iii) drift is absent, (iv) noise can be reduced, (v) overall lesser weight (control/filter, hardware [HW] boxes are avoided), and space is saved, (vi) the HW cost is reduced, and (vii) the HW redundancy shifts to software [SW] and/or the code redundancy; the latter is also called information redundancy; in fact SW burden might increase, by way of accounting for extra SW and/or algorithmic redundancy in the digital computer, but basically the reliability of the overall digital control system is enhanced. If needed, one can use HW/SW hybrid redundancy to derive merits from both. Some more characteristic of the SDDCS from the control point of view are [1]: (i) one can have multiple use of the (expensive) equipment, like a digital computer, data transfer channels, measuring instruments and sensors, by way of multiplexing several tasks needed for the digital control and other monitoring tasks; (ii) the data required for control and estimation are available at particular time instants only, if these are needed for future computation, then one has to store the past data, which is lot easier if a lot of memory in the computer is available; (iii) these data then can be

modified only at these or other particular time instants only, in a known specific way to improve predictive capability, say by fusing the data from several sources/sensors; (iv) these data are naturally discrete, due to the introduction of the sampler in the appropriate channel, like from an optical position sensor, or a stepping motor, in some latter cases the data come as discrete pulses only; (v) the digital computer would produce a new control signal every sampling time or instant, but due to the finite word length of the computer (arithmetic, like 16, 32, or 64 bits), only a finite number of the levels is possible; for an 8 bit machine there are only $2^n = 256$ different levels; (vi) interestingly, while the discretization in time is linear, the one in space is not, and hence, the effect is accounted for by introducing noise terms in the state space equations of the discrete time/digital dynamic systems; and (vii) the state change can occur at any time and take any value, however, in a finite time interval, the states of the dynamic systems would change only a finite number of times, this situation is important for sequencing control in/for assembly line manufacturing processes, and robot controls requiring good control strategies, for the latter one can use simulation and optimization via nonlinear programming. In general, we know more about how to control linear systems, (some portion of Sections I and II) than nonlinear ones. Hence, we often avoid nonlinear control operations and compensate the inherent nonlinearities of these systems and apply linear or nonlinear approximations approaches for the design and control of the systems. Much of the study and discussions in this chapter are based on [1,2].

Some very important developmental aspects of the digital control theory have been and are [2]: (i) the sampling theorem, since most computer-controlled systems operate at discrete instants of time only, it is important to know the situations for which a signal can be retrieved, back in nearly the continuous form, from its sampled values, in fact, Nyquist studied this issue and Claude Shannon provided a technically sound solution that is known as Shannon's sampling theorem; (ii) the difference equations are key ingredients of the theory of SDDCS that is closely related to numerical analysis; (iii) difference equations are used in place of the differential equations; (iv) derivatives and integrals are evaluated/computed approximately with differences, e.g. finite differences, and sums; (v) Z-transform is used in the place of Laplace transform; and (vi), in the state space theory, the discrete time representation of state models are obtained only at sampling points. In the very early development of digital control systems, an analog system (that did not contain a digital device. e.g. computer), in which some of the signals were sampled, was referred to as a sampled data system (SDS); however, with the arrival of the digital computer, the term discrete time system, in which all its signals are in a digital coded form, started being recognized as digital control systems (DCSs) [2]. Yet, many practical systems are of a hybrid nature: contain both analog and digital components/devices and these can be referred to as SDDCS, or simply as DC systems (DCSs). The use of computers for implementing controllers, in fact their programs/algorithms, has several merits [2]: (i) it is easy to implement very complicated and sophisticated algorithms/programs, (ii) it is easy to include logic and nonlinear functions, since these need to be programmed only, and (iii) the controllers can be made reconfigurable, which is very useful in the domain of fault tolerance and adaptive control applications. There are systems that are inherently sampled (data) systems and are natural descriptions for many phenomena, mainly because, in some such cases, sampling occurs naturally due to the nature of measurement/sensing system, whereas in some other cases, it occurs because information/data are transmitted in pulsed form. The theory and techniques of the sampled data systems has applications in/to: (i) communication using radar—when a radar antenna rotates, information-data about range and direction are naturally obtained once per revolution of the antenna; (ii) economic systems—accounting procedures in economic systems are generally tied to the calendar, and information about important variables is accumulated only at certain times, e.g. daily, weekly, monthly, quarterly, or yearly, even, if the transactions occur at any point of time; and (iii) biological processes in the evolutionary systems, since the signal transmission in the nervous system occurs in pulsed form and, hence, are inherently sampled.

9.1 Representation of Discrete Time System

A digital control system (DCS) and its mathematical model (MM) can be viewed from different perspectives including [2]: (i) control algorithms, (ii) computer program/software (SW), (iii) conversion from analog to digital, and back to continuous time, and (iv) system performance, in terms of stability and error performance. The most important aspect in SDDCS is the sampling process; in continuous time control systems, all the system variables are continuous signals, in the latter case, whether the system is linear or nonlinear, all the variables are

continuously present and, therefore, known (available) at all times. This is also true in the case of the digital control system, but now the values are available only at certain prespecified instants of time, called sampling instants. In a DCS, a control (law) algorithm is implemented in a digital computer: the error signal, the difference between the reference and the output signals, is discretized and fed to the computer by using an A/D (analog to digital) converter. The controller output (from the control algorithm that has been processed in the digital computer) is again a discrete time signal, which is applied to the plant after using a D/A (digital to analog) converter, see Figure 9.1. In a DCS, the control and communication signals are generated by converting the continuous time signals into sequence/s of numbers at discrete time intervals. Various fundamental aspects of the sampling process and related analyses are presented in Appendix 9A. One would like to expect that a DCS behaves like a continuous time system if the sampling period is reasonably and sufficiently small, which is true only under feasible assumptions and situations. Very fast sampling (very small sampling interval/time T) would generate a huge amount of data and, if these data are to be stored for some future computations and use, then huge memory with involved data logging capability would be required.

A simple way to obtain a digital control algorithm is by writing the continuous time control law as a differential equation and approximating the derivatives by finite differences and integrations by summations; however, this will work (reasonably well) when the sampling period (T) is very small, but various parameters, like over-shoot and settling time, will be slightly higher than those of the continuous time control system (CCS) [2]. One can discretize a continuous time PD (proportional plus derivative) controller as

$$u(t) = K_p e(t) + K_d \frac{de(t)}{dt} \Rightarrow u(kT)$$

$$= K_p e(kT) + K_d \frac{[e(kT) - e((k-1)T)]}{T} \tag{9.1}$$

In (9.1), k is the discrete time instant/s and T is the discrete time step, or the sampling period or interval. Interestingly enough, one can obtain control strategies with different behaviors, like deadbeat control/response (DBCR) with/in computer control/

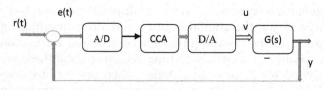

FIGURE 9.1
General digital control system (CCA: Computer-Control Algorithm).

digital control that are not possible with a continuous time control. One important aspect of the SDDCS is that of the aliasing: stable continuous time linear systems (CTLS) have property that the steady-state (SS) response to sinusoidal excitations is sinusoidal with same frequency as that of the input signal; but, DCSs behave in a more complicated way because sampling will create signals with new frequencies. This phenomenon is called aliasing, or frequency folding, which is an effect of the sampling that causes different signals to become indistinguishable. The effect of aliasing is that the signal reconstructed from the samples may become different than the original continuous signal. This can greatly degrade the performance of the SDDCS.

9.1.1 Numerical Differentiation

Continuous time systems and related formulas/algorithms can be converted into discrete time systems and corresponding formulas by using difference equations. Let us consider the linear time-invariant (LTI) systems. Mainly two approaches are used: (i) backward difference and (ii) forward difference.

i. Backward difference

   a. First order: Continuous time signal is $u(t) = \dot{e}(t)$, then discrete time signal is represented as

$$u(kT) = \frac{e(kT) - e((k-1)T)}{T} \qquad (9.2)$$

   b. Second order: Continuous time signal is $u(t) = \ddot{e}(t)$, then the discrete time signal is represented

$$u(kT) = \frac{\dot{e}(kT) - \dot{e}((k-1)T)}{T} \qquad (9.3)$$

$$u(kT) = \frac{e(kT) - e((k-1)T) - e((k-1)T) + e((k-2)T)}{T^2} \qquad (9.4)$$

$$= \frac{e(kT) - 2e((k-1)T) + e((k-2)T)}{T^2} \qquad (9.5)$$

ii. Forward difference

   a. First order: Continuous time signal is $u(t) = \dot{e}(t)$, then discrete time signal is represented as

$$u(kT) = \frac{e((k+1)T) - e(kT)}{T} \qquad (9.6)$$

   b. Second order: Continuous time signal is $u(t) = \ddot{e}(t)$, then the discrete time signal is represented

$$u(kT) = \frac{\dot{e}((k+1)T) - \dot{e}(kT)}{T} \qquad (9.7)$$

$$= \frac{e((k+2)T) - 2e((k+1)T) + e(kT)}{T^2} \qquad (9.8)$$

9.1.2 Numerical Integration

These methods depend on the approximation of the instantaneous continuous time signal; the integral function can be approximated by a number of rectangular pulses and the area under the curve is represented by summation of the areas of all the small rectangles: assume the input signal as

$$u(t) = \int_0^t e(\tau)d\tau \; ; u(NT) = \int_0^{NT} e(\tau)d\tau \cong \sum_{k=0}^{N-1} e(kT)\Delta T$$

$$= \sum_{k=0}^{N-1} e(kT)T \qquad (9.9)$$

In (9.9), $k = 0,1,2,\ldots,N-1$; $N > 0$; and $\Delta t = T$. From (9.9) we can obtain the following

$$u((N-1)T) = \int_0^{(N-1)T} e(\tau)d\tau \cong \sum_{k=0}^{N-2} e(kT)T \qquad (9.10)$$

Thus, we get the following formulas

$$u(NT) - u((N-1)T) = Te((N-1)T)$$

$$u(NT) = u((N-1)T) + Te((N-1)T) \qquad (9.11)$$

Interestingly, the formula (9.11) is the recursion for the backward rectangular integration. The forward integration formulas are given as

$$u(NT) = \sum_{k=1}^{N} e(kT)T$$

$$= u((N-1)T) + Te(NT) \qquad (9.12)$$

One can use the polygonal or trapezoidal form of approximations instead of rectangular shapes for numerical integration.

9.1.3 Difference Equations

A general linear difference equation of an n-th order causal LTI-SISO system is

$$y((k+n)T) + a_1 y((k+n-1)T)$$

$$+ a_2 y((k+n-2)T) + \ldots + a_n y(kT) \qquad (9.13)$$

$$= b_0 u((k+m)T) + b_1 u((k+m-1)T) + \ldots + b_m u(kT)$$

In (9.13), $y(.)$ and $u(.)$ are the output and the input of the system, and in order to avoid the anticipatory or the non-causal behavior, we should have $m \leq n$.

9.2 Modeling of the Sampling Process

The sampling process in a SDDCS models either the sample and hold operation (SAH), or that the signal is inherently digitally coded. In case of the sampler representing the SAH and analog to digital (A/D) operations, it might involve some errors due to (time-)delays, finite sampling duration, and quantization; however, if the sampler is used for digitally coded data, then the model will be simpler [2]. The two popular sampling operations are: (i) single rate, or periodic sampling, and (ii) multi-rate sampling.

9.2.1 Finite Pulse Width Sampler

A sampler converts a continuous time signal into a pulse modulated or discrete signal; the most common being in SAH operation the pulse amplitude modulation (PAM), see Figure 9.2. The pulse duration is p_s. and sampling period/interval is T sec. The uniform rate sampler satisfies the principle of superposition and is a linear device/process. In Figure 9.2, $p(t)$ is a unit pulse sequence and has a period of T:

$$p(t) = \sum_{k=-\infty}^{\infty} [u_s(t-kT) - u_s(t-kT-p)] \quad (9.14)$$

In (9.14), $u(.)$ is the unit step function. If we assume that the leading edge of the pulse at $t = 0$ coincides with $t = 0$, then we can write the function $f(..)$ as

$$f_p^*(t) = f(t) \sum_{k=-\infty}^{\infty} [u_s(t-kT) - u_s(t-kT-p)] \quad (9.15)$$

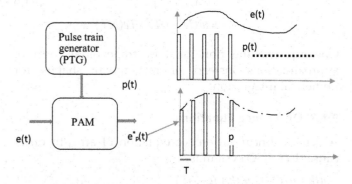

FIGURE 9.2
Operation of a finite pulse width sampler (PAM-pulse amplitude modulation) T = sampling period is from start of one pulse to the start of the next pulse; p = width of the pulse itself.

The frequency domain characteristic is developed next. Here, $p(t)$ is a periodic function and it can be represented by a Fourier series

$$p(t) = \sum_{n=-\infty}^{\infty} C_n e^{jn\omega_s t}; \; \omega_s = \frac{2\pi}{T}, \text{the sampling frequency} \quad (9.16)$$

In (9.16), the $C(.)$ are the complex Fourier series coefficients given as

$$C_n = \frac{1}{T} \int_0^T p(t) e^{-jn\omega_s t} dt \quad (9.17)$$

Because $p(t) = 1$ for $0 \leq t \leq p$ and zero for the rest of the period, one gets

$$C_n = \frac{1}{T} \int_0^T e^{-jn\omega_s t} dt = [\frac{1}{-jn\omega_s T} e^{-jn\omega_s t}]_0^p$$

$$= \frac{[1 - e^{-jn\omega_s p}]}{jn\omega_s T} \quad (9.18)$$

The coefficients $C(.)$ are simplified as

$$C_n = \frac{e^{-jn\omega_s p/2}(e^{jn\omega_s p/2} - e^{-jn\omega_s p/2})}{jn\omega_s T}$$

$$= \frac{2je^{-jn\omega_s p/2} \sin(n\omega_s p/2)}{jn\omega_s T} \quad (9.19)$$

$$= \frac{p}{T} \frac{\sin(n\omega_s p/2)}{n\omega_s p/2} e^{-jn\omega_s p/2}$$

Since, the function in (9.15) is periodic, it can be written as

$$f_p^*(t) = \sum_{n=-\infty}^{\infty} C_n f(t) e^{jn\omega_s t} \quad (9.20)$$

The Fourier transformed function is given as

$$F_p^*(j\omega) = F\{f_p^*\}; \quad F \text{ is the Fourier transform}$$

$$= \int_{-\infty}^{\infty} f_p^*(t) e^{-j\omega t} dt \quad (9.21)$$

We can now use the complex shifting theorem to obtain

$$F\{e^{jn\omega_s t} f(t)\} = F(j\omega - jn\omega_s)$$

$$F_p^*(j\omega) = \sum_{n=-\infty}^{\infty} C_n F(j\omega - jn\omega_s) \quad (9.22)$$

Also, since n is from $-\infty$ to ∞, (9.22) can also be written as

$$F_p^*(j\omega) = \sum_{n=-\infty}^{\infty} C_n F(j\omega + jn\omega_s);$$

$$C_0 = \lim_{n \to 0} C_n = \frac{p}{T}$$

$$F_p^*(j\omega)\,|_{n=0} = C_0 F(j\omega) = \frac{p}{T} F(j\omega) \qquad (9.23)$$

One can see that in (9.23), the frequency contents of the original signal $f(t)$ are still present in the sampler output, except that the amplitude is multiplied by a factor p/T. For $n \neq 0$, $C(.)$ is a complex quantity, with the magnitude as

$$|C_n| = \frac{p}{T} \left| \frac{\sin(n\omega_s p / 2)}{n\omega_s p / 2} \right| \qquad (9.24)$$

The magnitude of $F_p^*(j\omega)$ is given as

$$|F_p^*(j\omega)| = \left| \sum_{n=-\infty}^{\infty} C_n F(j\omega + jn\omega_s) \right|$$

$$\leq \sum_{n=-\infty}^{\infty} |C_n| \, |F(j\omega + jn\omega_s)| \qquad (9.25)$$

In fact, the sampling process would retain the fundamental frequency, yet the sampler output would contain some harmonic components

$$F(j\omega + jn\omega_s), \text{ for } n = \pm 1, \pm 2, \dots \qquad (9.26)$$

Shannon's sampling theorem states that "if a signal contains no frequency higher than ω_c rad/sec., then it is completely characterized by the values of the signal measured at the instants of time separated by $T/(\pi\omega_c)$ sec." [2] In order to avoid aliasing, the sampling frequency rate should be reasonably greater than the Nyquist rate, which is twice the highest frequency component of the original continuous time signal; if the sampling rate is less than twice the input frequency, the output frequency will be different from the input, which is known as aliasing. Then, the output frequency is called alias (-ed) frequency, and the period is called alias (-ed) period. The overlapping of the high frequency components with the fundamental component in the frequency spectrum is sometimes also referred to as frequency-folding, and the frequency $w_s/2$ is often known as folding frequency; w_c is called Nyquist frequency. Thus, a low sampling rate/frequency would normally have an adverse effect on the closed loop control system (CLCS) stability. Thus, often one might have to select a sampling frequency much higher than the theoretical minimum. Possibly, one should sample the original signal at 4–5 times higher than the minimum requirement.

9.2.2 An Approximation of the Finite Pulse Width Sampling

The Laplace transform (LT) of $f_p^*(t)$ can be written as

$$F_p^*(s) = \sum_{n=-\infty}^{\infty} \frac{1 - e^{-jn\omega_s p}}{jn\omega_s T} F(s + jn\omega_s) \qquad (9.27)$$

If the sampling duration $p \ll T$ and $\ll T_c$ (the smallest time constant of the signal $f(t)$), then the sampler output can be approximated by a sequence of rectangular pulses, because during the sampling duration, the variation of $f(t)$ will be less significant. Hence, for $k = 0,1,2,\dots, f_p^*(t)$ is expressed as an infinite series

$$f_p^*(t) = \sum_{k=0}^{\infty} f(kT)[u_s(t - kT) - u_s(t - kT - p)] \qquad (9.28)$$

And the LT of (9.28) is

$$F_p^*(s) = \sum_{k=0}^{\infty} f(kT)\left[\frac{1 - e^{-ps}}{s}\right] e^{-kTs} \qquad (9.29)$$

Because p is very small, e^{-ps} is approximated by the Taylor series to obtain

$$1 - e^{-ps} = 1 - [1 - ps + \frac{(ps)^2}{2!} + \dots +]$$

$$\cong ps \qquad (9.30)$$

Then, we have using (9.30) in (9.29)

$$F_p^*(s) \cong p \sum_{k=0}^{\infty} f(kT) e^{-kTs} \qquad (9.31)$$

And in the time domain, one gets

$$f_p^*(t) = p \sum_{k=0}^{\infty} f(kT) \delta(t - kT) \qquad (9.32)$$

In (9.32), $\delta(t)$ is the unit impulse function; the finite pulse width sampler is viewed as an impulse modulator, or an ideal sampler, in series with an attenuator with magnitude p.

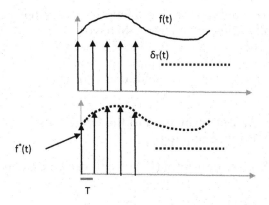

FIGURE 9.3
Ideal sample operation; $p = 0$.

9.2.3 Ideal Sampler

Here, the carrier signal is a train of unit impulse, as shown in Figure 9.3; the sampling duration p approaches 0, signifying that its operation is instantaneous, and output is given as

$$f^*(t) = \sum_{k=0}^{\infty} f(kT)\delta(t - kT) \qquad (9.33)$$

$$F^*(s) = \sum_{k=0}^{\infty} f(kT)e^{-kTs} \qquad (9.34)$$

The output of a hold device (of SAH) will be the same regardless of the nature of the sampler (the attenuation factor p is dropped, in that case); the sampling process is approximated by an ideal sampler or impulse modulator.

9.3 Reconstruction of the Data

We have to convert the outputs of a DCS into analog/continuous signals, because most of the control systems have analog controlled processes that are inherently driven by analog inputs. The high-frequency components of $f(t)$ should be removed before applying to analog devices, and a low-pass filter, or a data reconstruction device, is required to be used for this operation. The hold operation is used for reconstruction: given a sequence of numbers, $f(0), f(T), f(2T), \cdots, f(kt), \cdots$, a continuous time signal $f(t)$ is to be constructed from the information contained in the given sequence; this data reconstruction process is then an extrapolation process, because the continuous time signal $f(t)$ has to be obtained based on the information available at past sampling instants of the discrete time signal, i.e. the digital samples. For the latter, one needs to determine

the original signal $f(t)$, between two consecutive sampling instants kT and $(k + 1)T$, based on the values of $f(t)$ at previous instants of kT, i.e. $(k - 1)T$, $(k - 2)T$, $\cdots$ 0. For this purpose, a power series expansion can be used:

$$f_k(t) = f(kT) + f^{(1)}(kT)(t - kT) + \frac{f^{(2)}(kT)}{2!}(t - kT)^2 + \ldots$$

$$f_k(t) = f(t) \ for \ kT \le t \le (k+1)T; \qquad (9.35)$$

$$f^{(n)}(kT) = \frac{d^n f(t)}{dt^n}; \ n = 1, 2, \ldots$$

The derivatives of $f(t)$ must be estimated from the value of $f(kT)$ as follows

$$f^{(1)}(kT) \cong \frac{1}{T}[f(kT) - f((k-1)T]$$

$$f^{(2)}(kT) \cong \frac{1}{T}[f^{(1)}(kT) - f^{(1)}((k-1)T] \qquad (9.36)$$

$$f^{(1)}((k-1)T) \cong \frac{1}{T}[f((k-1)T) - f^{(1)}((k-2)T]$$

9.3.1 Zero Order Hold

If we need to estimate the higher order derivatives, for example in (9.35), we need to use a larger number of the delayed pulses. Using these higher order derivatives might enhance the accuracy of the reconstructed continuous time signal; however, this might cause a degradation of the stability of the CLCS because of the time delay/s in the system. Also, the use of higher order extrapolation would need more complex circuitry and, hence, higher HW costs. To avoid this, one uses only the first term in the power series to approximate $f(t)$ (during the time interval $kT \le t < (k+1)T$) and, hence, a zero order hold (ZOH) or zero order extrapolator (that holds the value of $f(kT)$ for $kT \le t < (k+1)T$ until the next sample $f((k + 1)T)$ arrives) is used, see Figure 9.4. The accuracy of the ZOH depends on the sampling rate: when $T \to 0$, the output of the ZOH approaches the continuous time signal, and the ZOH is a linear device/artifice. The impulse response of the ZOH is written as

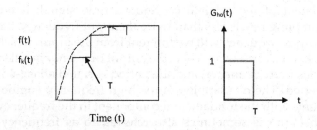

FIGURE 9.4
Zero order hold (ZOH) operation (left) (-- Original signal; - Reconstructed signal) and Impulse response of the ZOH (IRZ) (right)

$$g_{h_0}(t) = u_s(t) - u_s(t-T)$$

$$G_{h_0}(s) = \frac{1-e^{-Ts}}{s}; \ G_{h_0}(j\omega) = \frac{1-e^{-j\omega T}}{j\omega} \quad (9.37)$$

$$= T\frac{\sin(T/2)}{\omega T/2}e^{-j\omega T/2}$$

$$|G_{h_0}(j\omega)| = \frac{2\pi}{\omega_s}|\frac{\sin(\pi\omega/\omega_s)}{\pi\omega/\omega_s}| \text{ is the magnitude,}$$

$$\text{since } T = \frac{2\pi}{\omega_s} \ \angle\{(G_{h_0}(j\omega)\} \quad (9.38)$$

$$= \sin(\pi\omega/\omega_s) - \pi\omega/\omega_s \ rad$$

9.3.2 First Order Hold

In the first order hold (FOH), the first two terms of the power series are used for extrapolation of $f(t)$, over the same time interval, $kT \le t < (k+1)T$:

$$f_k(t) = f(kT) + f^{(1)}(kT)(t-kT);$$

$$f^{(1)}(kT) = \frac{1}{T}[f(kT) - f((k-1)T]$$

$$f_k(t) = f(kT) + \frac{1}{T}[f(kT) - f((k-1)T](t-kT) \quad (9.39)$$

One can then obtain the impulse response of the FOH by applying a unit impulse at t = 0, and the corresponding output by setting k = 0,1,2, ... :

$$f_0(t) = f(0) + \frac{1}{T}[f(0) - f(-1)T]t;$$

$$f_{h_1}(t) = 1 + \frac{t}{T}; \ \ 0 \le t < T \ for \ k = 0; \ 0 \le t < T;$$

$$f(0) = 1, \text{ and } f(-T) = 0.$$

Next we have

$$f_1(t) = f(T) + \frac{1}{T}[f(T) - f(0)](t-T); T \le t < 2T;$$

$$\text{since } f(T) = 0, \text{ and } f(0) = 1, \text{ we have}$$

$$f_{h_1}(t) = 1 - \frac{t}{T}; f_{h_1}(t) = 0, for \ t \ge 2T, \text{ since } f(t) = 0 \ for \ t \ge 2T$$

Finally, we get the impulse response as

$$g_{h_1}(t) = \left(1+\frac{t}{T}\right)u_s(t) + \left(1-\frac{t}{T}\right)u_s(t-T)$$

$$-\left(1+\frac{t}{T}\right)u_s(t-T) - \left(1-\frac{t}{T}\right)u_s(t-2T)$$

$$=\left(1+\frac{t}{T}\right)u_s(t) - 2\frac{t}{T}u_s(t-T) - \left(1-\frac{t}{T}\right)u_s(t-2T) \quad (9.40)$$

It can be verified, from (9.40), when $0 \le t < T$, only the first term produces a nonzero value as $(1 + t/T)$; similarly, when $T \le t < 2T$, the first two terms produce nonzero values as $(1 - t/T)$; and when $t \ge 2T$, all the three terms would produce nonzero values as 0. The TF of the FOH is given as

$$G_{h_1}(s) = \frac{1+Ts}{T}\left[\frac{1-e^{-Ts}}{s}\right]^2;$$

$$G_{h_1}(j\omega) = \frac{1+j\omega T}{T}\left[\frac{1-e^{-j\omega T}}{j\omega}\right]^2 \quad (9.41)$$

$$|G_{h_1}(j\omega)| = \left|\frac{1+j\omega T}{T}\right||G_{h_0}(j\omega)|^2$$

$$= \frac{2\pi}{\omega_s}\left(\sqrt{1+\frac{4\pi^2\omega^2}{\omega_s^2}}\right)|\frac{\sin(\pi\omega/\omega_s)}{\pi\omega/\omega_s}|^2 \quad (9.42)$$

$$\text{is the magnitude, since } T = \frac{2\pi}{\omega_s}$$

$$\angle\{(G_{h_1}(j\omega)\} = \tan^{-1}(2\pi\omega/\omega_s) - 2\pi\omega/\omega_s \ rad$$

9.4 Pulse Transfer Function

As we have seen in section I of this book, in/for the s-domain, the TF of a (SISO) LTI continuous time system is specified as

$$G(s) = \frac{Y(s)}{R(s)} \quad (9.43)$$

In (9.43), $Y(s)$ is the LT of the output $y(t)$, and $R(s)$ is the LT of the reference input signal $r(t)$; assuming the initial conditions are zero. For discrete time systems, equivalently, one can use the pulse transfer function (PTF) that relates the z transform (or Z transform [ZT]) of the output at the sampling instants to the ZT of the sampled input. In a usual manner, one would write the output of the system as $Y(s) = G(s)R^*(s)$; however, the TF of such a system is difficult to handle, since the DCS contains a mixture of analogue and digital components/devices (see Figure 9.5a), for which the system's characteristics are expressed by a TF that relates $r(t)^*$ to $y^*(t)$, pseudo sampler output:

$$Y^*(s) = \sum_{k=0}^{\infty}y(kT)e^{-kTs} \quad (9.44)$$

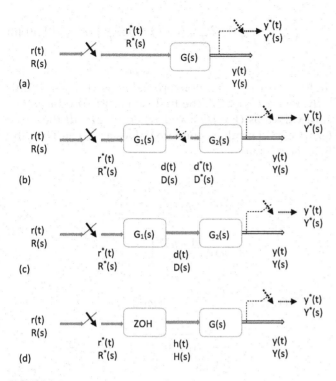

FIGURE 9.5

(a) A system subjected to a sampling input. (b) Discrete data system (DDS) with cascaded elements separated by a sample. (c) Discrete data system (DDS) with cascaded elements, Not separated by a sample. (d) System with sample and hold process.

Since the output is periodic, $Y^*(s) = \dfrac{1}{T} \sum\limits_{n=-\infty}^{\infty} Y(s + jn\omega_s)$; $y(0) = 0$. Similarly, we have for $R(s)$

$$R^*(s) = \frac{1}{T} \sum_{n=-\infty}^{\infty} R(s + jn\omega_s)$$

Also, using the expression for $Y^*(s)$, we get

$$Y^*(s) = \frac{1}{T} \sum_{n=-\infty}^{\infty} R^*(s + jn\omega_s)G(s + jn\omega_s)$$

Since R(s) is periodic, we have

$$Y^*(s) = \frac{1}{T} \sum_{n=-\infty}^{\infty} R^*(s)G(s + jn\omega_s)$$

$$\text{(9.45)}$$

$$= R^*(s)\frac{1}{T} \sum_{n=-\infty}^{\infty} G(s + jn\omega_s)$$

Now, we define $G^*(s) = \dfrac{1}{T} \sum\limits_{n=-\infty}^{\infty} G(s + jn\omega_s)$, then we have

$$Y^*(s) = R^*(s)G^*(s) \text{ and } G^*(s) = \frac{Y^*(s)}{R^*(s)} \quad (9.46)$$

The TF in (9.46) is known as the PTF. It is also known as the discrete time TF (DTF). Now, in (9.44), we substitute $z = e^{Ts}$ to obtain the z-TF of G(s) as

$$G(z) = \frac{Y(z)}{R(z)}, \text{ and } G(z) = \sum_{k=0}^{\infty} g(kT)z^{-k} \quad (9.47)$$

In (9.47), $g(kT)$ represents the sequence of the impulse response $g(t)$ of the system's TF $G(s)$, also known as impulse sequence. The PTF/ZTF characterizes the DDS responses only at the sampling instants. The output information between the sampling instants is not captured.

Pulse transfer function of discrete data systems separated by a sampler

The block diagram for this case is shown in Figure 9.5b. The I/O relations of the two systems are given by

$$D(z) = G_1(z)R(z)$$

$$Y(z) = G_2(z)D(z)$$

$$Y(z) = G_2(z)G_1(z)R(z)$$

Pulse transfer function of discrete data systems, Not separated by a sampler

The block diagram for this case is shown in Figure 9.5c. The I/O relations of the two systems are given by:

$$Y(s) = G_1(s)G_2(s)R^*(s)$$

The output of the fictitious sampler is

$$Y(s) = Z[G_1(s)G_2(s)]R(z)$$

The ZT of the product of two TFs is denoted as

$$Z[G_1(s)G_2(s)] = G_1G_2(z) = G_2G_1(z)$$

The final output is given as

$$Y(s) = G_1G_2(z)R(z)$$

9.4.1 Pulse Transfer Function of the ZOH

The TF of the ZOH is given as

$$G_{h_0}(s) = \frac{1 - e^{-Ts}}{s}; PTF: G_{h_0}(z) = Z\left\{\frac{1 - e^{-Ts}}{s}\right\}$$

$$\text{(9.48)}$$

$$= (1 - z^{-1})Z\left\{\frac{1}{s}\right\} = (1 - z^{-1})\frac{z}{z-1} = 1$$

The result of (9.48) is so because the ZOH simply holds the discrete time signal for one sampling period; hence, the ZT of ZOH would give its original signal. In a usual situation, a SAH device precedes a system, as shown in Figure 9.5d. The TF of the system with ZOH preceding the system can be obtained as follows: first the ZT of the output is obtained as

$$Y(z) = Z\{G_{h_0}(s)G(s)\}R(z)$$

$$= Z\left\{\frac{1-e^{-Ts}}{s}G(s)\right\}R(z) \qquad (9.49)$$

$$= (1-z^{-1})Z\left\{\frac{G(s)}{s}\right\}R(z)$$

When the sampling frequency reaches to infinity, a DTS can be regarded as a continuous time system (CTS). Yet, this does not imply that if the signal $r(t)$ is sampled by an ideal sampler, then $r^*(t)$ can be reversed back to $r(t)$ by setting the sampling time T to zero, because this will put all the samples together. However, if the output (of the sampler) is passed through a hold device, then if we set the sampling time T to zero, the signal $r(t)$ can be retrieved. Thus, $\lim_{T->0} H(s) = R(s)$.

9.4.2 Pulse Transfer Function of a Closed Loop System

One needs the TF of the CLCS; in which case, the output of the sampler is regarded as the input to the dynamic system $G(s)$, and the input to the sampler is regarded as another output. The sample is in the forward path of the closed loop system, see Figure 9.6a. Hence, we obtain

$$E(s) = R(s) - G(s)H(s)E^*(s); E(s) = R(s) - H(s)G(s)E^*(s) \quad (9.50)$$

$$Y(s) = G(s)E^*(s); \text{ Thus, } E(s) = R(s) - H(s)Y(s) \quad (9.51)$$

For SISO systems, both the expressions in (9.50) are equivalent. Even for the SISO, we cannot write down (9.51) properly if we use the first expression of (9.50). The PTF of (9.50) gives

$$E^*(s) = R^*(s) - H^*(s)G(s)E^*(s);$$

$$E^*(s) = \frac{R^*(s)}{1+H^*(s)G(s)}$$

$$= \frac{R^*(s)}{1+G(s)H^*(s)}; \text{ for SISO Systems} \qquad (9.52)$$

The magnitude of $GH^*(s)$ is given as

$$GH^*(s) = |G(s)H(s)|^* = \frac{1}{T}\sum_{n=-\infty}^{\infty} G(s+jn\omega_s)H(s+jn\omega_s) \quad (9.53)$$

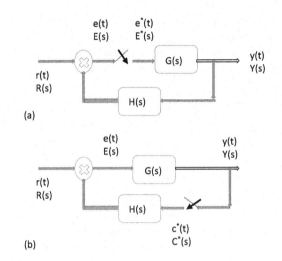

(a)

(b)

FIGURE 9.6
(a) A closed loop system with a sampler in the forward path.
(b) A closed loop system with a sampler in the feedback path.

From (9.51) and (9.52), we can write

$$Y(s) = G(s)E^*(s) = \frac{G(s)R^*(s)}{1+G(s)H^*(s)} \qquad (9.54)$$

The pulse transformation of (9.54) gives

$$Y^*(s) = |G(s)E^*(s)|^* = G^*(s)E^*(s) = \frac{G^*(s)R^*(s)}{1+G(s)H^*(s)} \qquad (9.55)$$

$$\frac{Y^*(s)}{R^*(s)} = \frac{G^*(s)}{1+G(s)H^*(s)}; \quad \frac{Y(z)}{R(z)} = \frac{G(z)}{1+G(z)H(z)};$$

$$GH(z) = Z\{G(s)H(s)\}$$

Consider the CLCS with a sampler in the feedback path, see Figure 9.6b, then we have

$$E(s) = R(s) - H(s)Y^*(s);$$

$$Y(s) = G(s)E(s) = G(s)R(s) - G(s)H(s)Y^*(s) \qquad (9.56)$$

The pulse transformation of (9.56) gives

$$E^*(s) = R^*(s) - H^*(s)Y^*(s);$$

$$Y^*(s) = G(s)R^*(s) - G(s)H^*(s)Y^*(s); \qquad (9.57)$$

$$GR^*(s) = |G(s)R(s)|^* ; GH^*(s) = |G(s)H(s)|^*$$

Also, we have from (9.57)

$$Y^*(s) = \frac{GR^*(s)}{1+GH^*(s)} \Rightarrow Y(z) = \frac{GR(z)}{1+GH(z)} \qquad (9.58)$$

Interestingly enough, we cannot define the I/O TF of this system by $\frac{C^*(s)}{R^*(s)}$ Or $\frac{C(z)}{R(z)}$, because $r(t)$ is not sampled and $r^*(t)$ does not exist. Using (9.56) and (9.58), we can obtain the continuous time data output (Ys) as

$$Y(s) = G(s)R(s) - \frac{G(s)H(s)}{1+GH^*(s)}GR^*(s) \qquad (9.59)$$

9.4.3 Characteristics Equation

As we have seen in Section I of this book, the characteristic equation plays a very important role in the study of linear systems, especially for the CLCSs. The n-th order difference equation represents the n-th order LTI discrete time system

$$y(k+n)+a_{n-1}y(k+n-1)+...+a_1y(k+1)+a_0y(k)$$
$$= b_m r(k+m)+b_{m-1}r(k+m-1)+...+b_0r(k) \qquad (9.60)$$

The ZT of (9.60) is given as

$$\frac{Y(z)}{R(z)} = \frac{b_m z^m + b_{m-1}z^{m-1}+...+b_0}{z^n + a_{n-1}z^{n-1}+...a_1z+a_0} \qquad (9.61)$$

If we equate the denominator of (9.61) to 0/zero, we get the so-called characteristic equation

$$\text{denominator}\{Y(z)/R(z)\} = z^n + a_{n-1}z^{n-1}+...a_1z+a_0 = 0 \qquad (9.62)$$

The roots (zeros) of (9.62) give the poles of the system $G(z)$ and define the characteristic dynamics of the system. In a causal system, the output does not precede the input, i.e. the output depends only on the past values of the input, and not the future values, to maintain the cause-effect principle. The TF of such a system is physically realizable, and the system can be realized/synthesized by using the physical elements/components. For a causal discrete data system (DDS), the power series of its TF should not contain any positive power in z since this indicates prediction, and for TF (9.61), the conditions to be satisfied are: proper TF → $m = n$; strictly proper TF → $n > m$.

9.5 Stability Analysis in z-Plane

There are two major stability aspects for LTI systems: (a) BIBO stability or zero state stability and (b) internal stability of zero input stability. The BIBO is the bounded input bounded output stability, and for the relaxed LTI system

(initial conditions are zero), it is BIBO stable if for every bounded input, the output is bounded. The stability of the $Y(z)/R(z)$ TF can be determined by using the location of the closed loop poles in the z-plane, and these are the root of the characteristic equation, (9.61) and (9.62):

$$\frac{Y(z)}{R(z)} = \frac{G(z)}{1+GH(z)}; \quad 1+GH(z)=0 \qquad (9.63)$$

If the closed loop poles, or the roots (zeros) of the characteristic equation, lie within the unit circle in the z-plane, the system is stable; otherwise, it is unstable. Also, if a simple pole lies at $|z| = 1$, the system is marginally stable (or marginally unstable!). If a pair of complex conjugate poles lie on the $|z| = 1$, the system is marginally stable (or marginally unstable!), multiple poles on the unit circle make the system unstable. There are three stability tests that can be applied directly to the characteristic equation without solving for the roots: (i) Schur-Cohn stability test, (ii) Jury stability test, and (iii) Routh stability coupled with bi-linear transformation. The other stability test is the Lyapunov stability analysis, which is applicable for state space system models.

9.5.1 Jury Stability Test

Let the characteristic equation be given as

$$P(z) = a_0 z^n + a_1 z^{n-1}+...+a_{n-1}z+a_n \qquad (9.64)$$

When $a_0 > 0$, form the Jury table as

Row	z^0	z^1	z^2	z^3	z^4	...	z^n
1	a_n	a_{n-1}	a_{n-2}	...	...	...	a_0
2	a_0	a_1	a_2	...	...	...	a_n
3	b_{n-1}	b_{n-2}	...	...	b_0		
4	b_0	b_1	...	...	b_{n-1}		
5	c_{n-2}	c_{n-3}	...	c_0			
6	c_0	c_1	...	c_{n-2}			
.	...	...	...	...			
.	...	...	...	...			
.	...	...	...	...			
2n−3	q_2	q_1	q_0				

In the table, entries are generated as follows

$$b_k = \begin{vmatrix} a_n & a_{n-1-k} \\ a_0 & a_{k+1} \end{vmatrix}; k=0,1,2,3,...,n-1;$$

$$c_k = \begin{vmatrix} b_{n-1} & b_{n-2-k} \\ b_0 & b_{k+1} \end{vmatrix}; k=0,1,2,3,...,n-2; q_k = \begin{vmatrix} c_3 & c_{2-k} \\ c_0 & c_{k+1} \end{vmatrix} \qquad (9.65)$$

The system will be stable under the following conditions

(i) $|a_n| < a_0$; (ii) $P(z)|_{z=1} > 0$; (iii) $P(z)|_{z=-1} > 0$ for "even" n, and $P(z)|_{z=-1} < 0$ for "odd;" and (iv) $|b_{n-1}| > |b_0|$; $|c_{n-2}| > |c_0|$; ... $|q_n| > |q_0|$

9.5.2 Singular Cases

If some or all the elements of a row in the Jury table are zero, the tabulation would end up, prematurely, leading to a singular case. This is avoided by expanding or contracting the unit circle by a very small amount ε, which is equivalent to moving the roots of $P(z)$ off the unit circle; the transformation being: $z_1 = (1 + \varepsilon)z$. If ε is positive, the unit circle is expanded, and if it is negative the unit circle is contracted. The difference between the number of zeros found inside or outside the unit circle, when the unit circle is expanded or contracted, is the number of zeros on the unit circle. Since $(1 + \varepsilon)^n z^n = (1 + n\varepsilon)z^n$ for both positive and negative ε, the transformation requires the coefficient of the z^n term to be multiplied by $(1 + n\varepsilon)$.

9.5.3 Bilinear Transformation and Routh Stability Criterion

Bilinear transformation with Routh stability criterion is also a useful method for stability analysis of DTSs. The bilinear transformation (BLT) is given as

$$z = \frac{aw + b}{cw + d}; \text{ if } a = b = c = 1; \text{ and } d = -1, \text{ then}$$

$$z = \frac{w + 1}{w - 1}; w = \frac{z + 1}{z - 1} \tag{9.66}$$

The BLT (9.66) maps the inside of the unit circle of the z-plane into the left half (LH) of the w-plane. Let w be represented as $w = \alpha + j\beta$, then the inside of the unit circle of the z-plane can be represented as

$$|z| = \left| \frac{w + 1}{w - 1} \right| = \left| \frac{\alpha + j\beta + 1}{\alpha + j\beta - 1} \right| < 1 \tag{9.67}$$

$$\Rightarrow \frac{(\alpha + 1)^2 + \beta^2}{(\alpha - 1)^2 + \beta^2} < 1 \Rightarrow (\alpha + 1)^2 + \beta^2 < (\alpha - 1)^2 + \beta^2 \tag{9.68}$$

$$\Rightarrow \alpha < 0$$

From (9.68), it can be observed that the inside of the unit circle of the z-plane maps into the LH of the w-plane, and the outside of the unit circle of the z-plane maps into the right half (RH) of the w-plane. For the stability analysis, one transforms the characteristic equation in z to that in w as, say $Q(w) = 0$. After this transformation, the Routh stability criterion that is used for continuous time systems is directly used for equation $Q(w) = 0$.

9.5.4 Singular Cases

In the Routh array, the tabulation may end: (a) if the first element in any row is zero, or (b) all the elements in a single row are zero. In case of (a), one replaces zero by a small number ε and proceeds with the tabulation, and the stability is checked for the limiting case. The second case indicates pairs of: (a) real roots with opposite signs, (b) imaginary roots, and (c) complex conjugate roots which are equidistant from the origin. If a row of all zeros occurs, an auxiliary equation $A(w) = 0$ is formed by using the coefficients of the row just above the row of all zeros. Interestingly, the roots of the auxiliary equation are also the roots of the characteristic equation; then, the tabulation is continued by replacing the row of zeros by the coefficients of $\{dA(w)/dw\}$. If we look at the relation between w-plane and z-plane, when an all zero row occurs, then two situations could happen: (a) pairs of real roots in the z-plane that are inverse of each other and (b) pairs of roots on the unit circle simultaneously.

9.6 Time Responses of Discrete Time Systems

For a control system, the basic requirements are: (a) absolute stability, (b) a good relative stability, and (c) steady-state accuracy, the steady-state (SS) error is zero. The transient response of a control system corresponds to the system's closed loop poles and steady-state response corresponds to the (input) excitation poles or poles of the input signal/function.

9.6.1 Transient Response Specifications and Steady-State Error

For most practical and man-made control systems, the desired performance characteristics are specified in terms of certain time domain quantities, called (performance) specifications. For control system's basic analysis, a step input signal is widely used, since it is easy to generate and signifies a drastic change that provides useful information on both transient and steady-state responses of the control system. The transient response of a system depends on the initial conditions, and it is usual practice to first study the control system when it is initially at rest. A block diagram of a closed loop digital system is shown in Figure 9.7a. A transient response of a DCS is characterized by: (a) rise time (t_r) is the time required for the system's (unit step) response to rise from 0% to 100% of its final value in case of a underdamped system, or 10%–90% of its final value in case of an overdamped system; (b) delay time (t_d) is the required for a unit step response to reach 50% of its final value; (c) peak

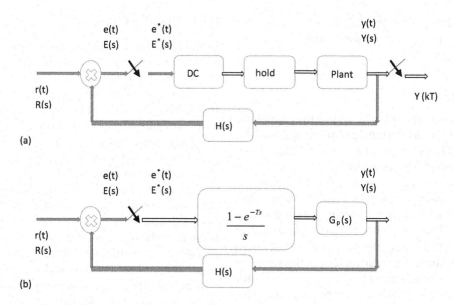

FIGURE 9.7
(a) Closed loop digital control system (DC-digital controller). (b) Closed loop system with TF of ZOH.

time (t_p) at which maximum peak of the output response occurs; (d) peak overshoot (M_p) is the difference between the maximum peak and the steady state value of the unit step response; and (e) settling time (t_s) is time required for the unit step response to reach and stay within 2% (or 5%) of its steady state value. As the output response is discrete, the computed performance measures (numerical values) might be slightly different from the actual values: the response output might be a maximum value y_{max}, (of the original analog/continuous time output), whereas the maximum value of the discrete output could be y^*_{max}, which is always less than or equal to y_{max}. Of course, if the sampling period is quite small compared to the time periods of the oscillations of the response (i.e. 1/frequency), then this difference will be very small.

The steady-state behavior of a stable control system is measured by the steady error due to step, ramp, or parabolic inputs depending on the system type. One can consider the steady-state error (SSE) at the sampling instants for the following system, see Figure 9.7b:

$$E(s) = R(s) - H(s)Y(s) \qquad (9.69)$$

Applying the final value theorem (Section I), we get

$$\lim_{k \to \infty} e(kT) = \lim_{z \to 1}[(1 - z^{-1})E(z)] \qquad (9.70)$$

$$G(z) = (1 - z^{-1})Z\left\{\frac{G_p(s)}{s}\right\}; GH(z) = (1 - z^{-1})Z\left\{\frac{G_p(s)H(s)}{s}\right\};$$

$$(9.71)$$

$$\frac{Y(z)}{R(z)} = \frac{G(z)}{1 + GH(z)}; E(z) = R(z) - GH(s)E(s); \qquad (9.72)$$

$$E(z) = \frac{1}{1 + GH(z)}R(z) \qquad (9.73)$$

$$e_{ss} = \lim_{k \to \infty} e(kT) = \lim_{z \to 1}[(1 - z^{-1})\frac{1}{1 + GH(z)}R(z)] \quad (9.74)$$

From (9.74), one can see that the SSE of a system with feedback depends on the input signal $r(t)$; $R(z)$ and the loop TF $GH(z)$.

9.6.2 Type-n Discrete Time Systems

Control systems having a finite nonzero SSE with a step input (i.e. zero order polynomial) are called type-0 systems, and for such systems, the position error constant is defined as outlined below:

$$e_{ss} = \lim_{z \to 1}\left[(1 - z^{-1})\frac{1}{1 + GH(z)}\frac{1}{(1 - z^{-1})}\right];$$

$$\text{since } R(z) = \frac{1}{(1 - z^{-1})} \qquad (9.75)$$

for $r(t)$ = unit step input, $u_s(t)$

$$e_{ss} = \lim_{z \to 1}\left[\frac{1}{1 + GH(z)}\right] = \frac{1}{1 + K_p}; \qquad (9.76)$$

In (9.76), $K_p = \lim_{z \to 1}\{GH(z)\}$ is the so called the position error constant (PEC).

Control systems with a finite nonzero SSE with a ramp input (i.e. first order polynomial) are called type-1 systems. The velocity error constant for a system is defined as outlined below:

$$e_{ss} = \lim_{z \to 1}[(1-z^{-1})\frac{1}{1+GH(z)}R(z)];$$

$$\text{since } R(z) = \frac{Tz}{(z-1)^2} = \frac{Tz^{-1}}{(1-z^{-1})^2}$$

$$\text{for } r(t) = \text{unit ramp input}, u_r(t) \tag{9.77}$$

$$e_{ss} = \lim_{z \to 1}[\frac{T}{(z-1)GH(z)}] = \frac{1}{K_v}$$

In (9.77), the velocity error constant (VEC) is given as

$$K_v = \frac{1}{T}\lim_{z \to 1}\{(z-1)GH(z)\} \tag{9.78}$$

Control systems with a finite nonzero SSE with a parabolic input (i.e. second order polynomial) are called type-2 systems, and the acceleration error constant for such a system is defined as outline below:

$$R(z) = \frac{T^2 z(z+1)}{2(z-1)^3} = \frac{T^2(1+z^{-1})z^{-1}}{2(1-z^{-1})^3};$$

$$\text{for } r(t) = \text{a parabolic input}, u_a(t)$$

$$e_{ss} = \frac{T^2}{2}\lim_{z \to 1}\left\{\frac{(z+1)}{(z-1)^2(1+GH(z))}\right\} \tag{9.79}$$

$$= \frac{1}{\lim_{z \to 1}\left\{\frac{1}{T^2}(z-1)^2 GH(z)\right\}} = \frac{1}{K_a}$$

In (9.79), K_a is the acceleration error constant (AEC). Table 9.1 collects these constants for comparison:

9.6.3 Study of a Second Order Control System

The study of a second order control system is very important, since many higher order systems can be approximated by a second order math model. If these higher order poles are located such that their contributions to transient response are negligible, and the dominant response is due to the second order system:

TABLE 9.1

Control Systems Error Constants

System Type	Step Input	Ramp Input	Parabolic Input
Type-0	$\frac{1}{1+K_p}$	∞	∞
Type-1	0	$\frac{1}{K_v}$	∞
Type-2	0	0	$\frac{1}{K_a}$

$$G(s) = \frac{\omega_n^2}{s(s+2\xi\omega_n)};$$

$$\text{Closed loop } G_c(s) = \frac{\omega_n^2}{s^2 + 2\xi\omega_n s + \omega_n^2}$$

$$\xi = \text{dampling ratio}; \tag{9.80}$$

$$\omega_n = \text{natural undamped frequency}$$

$$\text{Roots: } -\xi\omega_n \pm j\omega_n\sqrt{1-\xi^2}$$

Let us consider a space vehicle as a continuous time control system with an objective to control the attitude in one dimension, and the vehicle is considered as a rigid body, and the position $y(t)$ and the velocity $v(t)$ are fed back, see Figure 9.8. The open loop transfer function (OLTF) can be obtained as [2]:

$$G(s) = \frac{Y(s)}{E(s)} = KK_P \frac{1}{K_R + J_v s}\frac{1}{s} = \frac{KK_P}{s(K_R + J_v s)} \tag{9.81}$$

The CLTF is obtained as

$$G_c(s) = \frac{G(s)}{1+G(s)} = \frac{KK_P}{J_v s^2 + K_R s + KK_P} \tag{9.82}$$

In (9.82), we have K_p, the position sensor gain $= 1.65 \times 10^6$; K_R, the rate sensor gain $= 3.71 \times 10^5$;

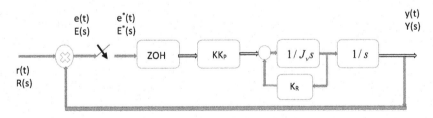

FIGURE 9.8
Discrete form of space vehicle attitude control.

K = amplifier gain; and J_v = moment of inertia. Then, the $G(s)$ is given as

$$G(s) = \frac{39.45K}{s(s+8.87)}; \quad \frac{Y(s)}{R(s)} = \frac{39.45K}{s^2 + 8.87s + 39.45K} \quad (9.83)$$

The characteristic equation $s^2 + 8.87s + 39.45k = 0$; $\omega_n = \sqrt{39.45K}$ rad/sec.; $\xi = \frac{8.87}{2\omega_n}$. This continuous time system will always be stable, if all four constants in (9.81) are positive. Now, the same system is considered with the sample data features and with all the constants, as those in the continuous time system. Then, we have the TF as follows:

$$G(s) = \frac{Y(s)}{E^*(s)} = G_{ho}(s)G_p(s) = \frac{1 - e^{-Ts}}{s} \cdot \frac{KK_p/J_v}{s(s + K_R/J_v)} \quad (9.84)$$

$$G(z) = (1 - z^{-1})\frac{KK_p}{K_R}Z\left\{\frac{1}{s^2} - \frac{J_v}{K_R s} + \frac{J_v}{K_R(s + K_R/J_v)}\right\}$$

$$= (1 - z^{-1})\frac{KK_p}{K_R}\left\{\frac{Tz}{(z-1)^2} - \frac{J_v z}{K_R(z-1)} + \frac{J_v z}{K_R(z - e^{-K_R T/J_v})}\right\}$$

$$= \frac{KK_p}{K_R^2}\left[\frac{(TK_R - J_v + J_v e^{-K_R T/J_v})z - (TK_R + J_v)e^{-K_R T/J_v} + J_v}{(z-1)(z - e^{-K_R T/J_v})}\right]$$

$$(9.85)$$

The characteristic equation for the CLDCS is obtained as: $z^2 + \alpha_1 z + \alpha_0 = 0$ with

$$\begin{aligned}\alpha_1 &= f_1(K, K_p, K_R, J_v) \\ \alpha_0 &= f_0(K, K_p, K_R, J_v)\end{aligned} \quad (9.86)$$

Using the known constants in (9.86), we obtain the following

$$\alpha_1 = 0.000012K(3.71 \times 10^5 T - 41822 + 41822 e^{-8.87T})$$

$$-1 - e^{-8.87T}$$

$$\alpha_0 = e^{-8.87T} + 0.0000K[41822$$

$$-(3.71 \times 10^5 T + 41822)e^{-8.87T}] \quad (9.87)$$

For the CLDCS to be stable, we should have: (i) $|\alpha_0| < 1$; (ii) $P(1) = 1 + \alpha_1 + \alpha_0 > 0 = 1 - e^{-8.87T} > 0$, which is always satisfied when T is positive, which is so; and (iii) $P(-1) = 1 - \alpha_1 + \alpha_0 > 0$. From studies of continuous time systems (Section I), we know that increasing the value of

K generally reduces the damping ratio, increases peak overshoot and bandwidth, and decreases the SSE, if it is finite and nonzero.

9.6.4 Correlation between Time Response and Root Locations in s- and z-Planes

For continuous time control systems, the correlation between root locations in the s-plane and time responses is very well placed: (i) a root on negative real axis of s-plane produces an output exponentially decaying with time, (ii) complex conjugate pole pairs in negative s-plane produce damped oscillations, (iii) the conjugate poles on imaginary axis produce undamped oscillations, and (iv) complex conjugate pole pairs in positive s-plane produce growing oscillations. The DCS should be studied carefully due to the sampling operation involved; e.g. if the sampling operation is not performed according to the sampling rule, the folding effect (aliasing) would change the true response of the system: if the system is subject to sampling with frequency $\omega_s < 2\omega_1$, it will generate an infinite number of poles in the s-plane at $s = \sigma_1 \pm j\omega_1 + jn\omega_s$, for $n = \pm 1, \pm 2, \ldots$, and these poles will fold back into the primary strip where $-\omega_s/2 < \omega < \omega_s/2$; the net effect would be equivalent to having a system with poles at $s = \sigma_1 \pm j(\omega_s - \omega_1)$.

9.6.5 Dominant Closed Loop Pole Pairs

Similar to the case of the s-plane, some of the roots in the z-plane would have more effects on the control system's response than the others. In the s-plane, the roots that are closest to $j\omega$ axis in the left plane (have smaller relative magnitudes) and are called the dominant roots because the corresponding time response would have the slowest decay; whereas, the roots that are far away from $j\omega$ axis correspond to fast decaying response, see Figure 9.9. In the z-plane: (i) dominant roots are those that are inside and closest to the unit circle; whereas, the less significant region is near the origin (the roots near the origin are less significant from the maximum overshoot and damping point of view), (ii) the negative real axis is generally avoided since the corresponding time response is oscillatory in nature with alternate signs. However, these roots should not be completely ignored, since the excess number of poles over zeros has a delay effect in the initial region of the time response: adding a pole at z = 0 would not affect the maximum overshoot or damping, but the time response would have an additional delay of one sampling period. Thus, a proper way of simplifying a higher order system in the z-domain is to replace the poles near origin by poles at z = 0, since the poles at z = 0 would correspond to pure time delays.

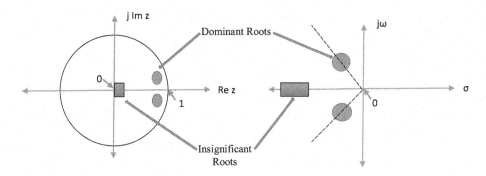

FIGURE 9.9
Pole-zero map of a typical 2nd order system (left: z-plane; right: s-plane).

Appendix 9A: Some Basic Results of Sampled Data and Digital Control Systems

Basic aspects of sampled data/digital control systems are briefly presented here to support the material of Chapter 9. These will also be useful in understanding the material of Chapter 10; [1, 2].

9A.1 Types of Sampled Data/Digital Control Systems

Analog-to-digital (A/D) converter/device/circuit transforms an analog (continuous time) signal to digital (discrete time signal); whereas, digital-to-analog (D/A) converter performs a reverse operation. These operations/devices/circuits are used in the DCS. There are mainly three types of the digital control systems wherein: (i) the continuous time error signal is sampled and held by a ZOH, the output of which is then fed to the continuous controller that in turn produces the control signal $u(t)$ and drives the continuous system under consideration, this output is fed back to the reference signal which is continuous; (ii) the error signal is sampled and it is followed by ZOH, the output of which is fed to an analog to digital converter (A/D) which in turn feeds to a digital controller, the output of this controller is then converted to an analog signal by using D/A, this signal then is followed by sample and ZOH device which then drives the continuous plant, this output is fed back to the reference signal which is continuous; and (iii) the reference/command input, which is discrete/digital signal is given to the digital controller itself, and the output of this controller is converted by using D/A converter which is followed by sample and ZOH device which then drives the continuous plant, the output of which is then followed by a sample and ZOH device, and this signal is converted by A/D converter to the discrete/digital signal and fed back to the reference signal. In general, we have

three types of sampling processes: (a) pulse amplitude modulation (PAM), (b) pulse width modulation (PWM), and (c) pulse frequency modulation (PFM). An A/D converter in itself is a closed loop control circuit: the error signal is quantized and then fed to a control logic, the output of which is then converted to an analog signal by using D/A device/circuit and this output is then fed back to the reference analog input. With an n-bit converter, $2n$ different numbers can be represented. The sampling period is limited as

$$T_{\min} \leq T \leq T_{\max} \qquad (9A.1)$$

The minimum limit is dictated by the technology of the converters/computer speed, and the upper limit is dictated by the requirement of the no-loss of the information in the signals/systems.

9A.2 Z-Transform

The analysis and design methods for the control systems are based on either transform approaches or modern (optimal) techniques, the latter are based on the state space models of the systems. For the continuous time systems, the Laplace transform method is popular, and this can be used for digital control systems; however, it poses problems since the transform contains a term e^{Ts}. Denote the output of an ideal sampler as $f^*(t)$

$$L\{f^*(t)\} = F^*(s) = \sum_{k=0}^{\infty} f(kT)e^{-kTs} \qquad (9A.2)$$

The exponential term in (9A.2) is not a rational function of s (here, s is a complex number and is complex frequency). Hence, it is preferable to use the following artefact

$$z = e^{Ts} \Rightarrow s = \frac{1}{T}\ln(z); \ s = \sigma + j\omega$$
$$(9A.3)$$
$$\mathrm{Re}(z) = e^{\sigma T}\cos(\omega T); \ \mathrm{Im}(z) = e^{\sigma T}\sin(\omega T)$$

With (9A.3) and (9A.2), the Z-transform (z-transform, ZT) is obtained as

$$F(z) = F^* \left\{ s = \frac{1}{T} \ln(z) \right\} = \sum_{k=0}^{\infty} f(kT)z^{-k} \qquad (9A.4)$$

If $f(t)$ is Laplace transformable, then it will also have the ZT:

$$L\{f(t)\} = \int_0^{\infty} f(t)e^{-sT}dt; \quad Z\{f(t)\} = F(z) = \sum_{k=0}^{\infty} f(kT)z^{-k} \quad (9A.5)$$

The ZT is a powerful transformation technique for analysis and design of DTSs. If only the sampled values of a signal $x(t)$, i.e. $x(0)$, $x(T)$, $x(2T)$, ..., are considered, then:

$$X(z) = Z\{x(t)\} = Z\{x(kT)\} = \sum_{k=0}^{\infty} x(kT)z^{-k} \qquad (9A.6)$$

For a sequence of numbers, we have

$$X(z) = Z\{x(k)\} = \sum_{k=0}^{\infty} x(k)z^{-k} \qquad (9A.7)$$

In one-sided transforms as in (9A.6) and (9A.7), the values $x(t)$, $x(k) = 0$ for $t, k < 0$. In two-sided transforms we use the ranges as: $-\infty < t < \infty$, or $k = 0, \pm 1, \pm 2, ...$:

$$X(z) = Z\{x(kT)\} = \sum_{k=-\infty}^{\infty} x(kT)z^{-k};$$

$$X(z) = Z\{x(k)\} = \sum_{k=-\infty}^{\infty} x(k)z^{-k} \qquad (9A.8)$$

The one sided ZT has a closed form solution in its domain/range of convergence (DOC); whenever $X(z)$, an infinite series in z^{-1}, converges outside the circle $|z| = R$ (the radius of absolute convergence), it is not required every time to specify the values of z over which $X(z)$ is convergent:

$$|z| > R \to \text{convergent; and } |z| < R \to \text{divergent} \quad (9A.9)$$

The ZTs of some basic functions are given as follows:

Unit step function/sequence:

$$u_s(t) = 1, \text{for } t \geq 0; \qquad u_s(k) = 1, \text{for } k = 0,1,2...$$
$$= 0, \text{for } t < 0; \qquad \qquad = 0, \text{for } k < 0$$

$$U_s(z) = \sum_{k=0}^{\infty} u_s(kT)z^{-k}$$

$$= \sum_{k=0}^{\infty} z^{-k} = 1 + z^{-1} + z^{-2} + z^{-3} + ... + \qquad (9A.10)$$

$$= \frac{1}{1-z^{-1}} = \frac{z}{z-1};$$

The series converges if $|z| > 1$.

Unit ramp function:

$$u_r(t) = 1, \text{for } t \geq 0;$$
$$= 0, \text{for } t < 0; \qquad (9A.11)$$

$$U_r(z) = \frac{Tz}{(z-1)^2} = T\frac{z^{-1}}{(1-z^{-1})^2}; \quad \text{The } DOC\,|\,z\,|> 1.$$

For a polynomial function:

$$x(k) = a^k$$

$$X(z) = \frac{1}{1-az^{-1}} = \frac{z}{z-a}; \quad \text{The } DOC\,|\,z\,|> a. \qquad (9A.12)$$

For an exponential function:

$$x(t) = e^{-at}, \text{for } t \geq 0$$
$$= 0, \quad \text{for } t < 0$$
$$x(kT) = e^{-akT}, \ k = 0,1,2,... \qquad (9A.13)$$

$$X(z) = \frac{1}{1-e^{-aT}z^{-1}} = \frac{z}{z-e^{-aT}}$$

Some properties of the ZT are as follows:

1. Multiplication: $Z\{ax(k)\} = aX(z)$; $Z\{a^k x(k)\} = X(a^{-1}z)$.
2. Linearity:
   If $x(k) = af(k) \pm bg(k)$, then $X(z) = aF(z) \pm bG(z)$.
3. Real shifting:

   $$Z\{x(t-nT)\} = z^{-n}X(z); \ Z\{x(t+nT)\}$$

   $$= z^n \left[X(z) - \sum_{k=0}^{n-1} x(kT)z^{-k} \right].$$

4. Complex shifting: $Z\{e^{\pm at}x(t)\} = X(ze^{\mp aT})$.
5. Initial value theorem: $x(0) = \lim_{z \to 1} X(z)$.
6. Final value theorem: $\lim_{k \to \infty} x(k) = \lim_{z \to 1}[(1-z^{-1})X(z)]$.

The inverse ZT is not necessarily similar to the inverse Laplace transform, since the latter is not equal to $f(t)$ but it is equal to $f(kT)$ at the sampling instants only. The inverse ZT formula is given as

$$f(kT) = Z^{-1}\{F(z)\} = \frac{1}{2\pi j} \oint_{\Gamma} F(z)z^{k-1}dz \qquad (9A.14)$$

Other important properties of the ZT are as follows:

1. Partial differentiation: $Z\left\{ \frac{\partial}{\partial a}[f(t,a)] \right\} = \frac{\partial}{\partial a}F(z,a)$.

2. Real value theorem: If $f_1(t)$ and $f_2(t)$ have the ZT as $F_1(z)$, and $F_2(z)$ and $f_1(t) = 0 = f_2(t)$ for $t < 0$; then

$$F_1(z)F_2(z) = Z\left[\sum_{n=0}^{k} f_1(nT)f_2(kT - nT)\right].$$

3. Complex convolution:

$$Z\{f_1(t)f_2(t)\} = \frac{1}{2\pi j}\oint_{\Gamma} \frac{F_1(\xi)F_2(z\xi^{-1})}{\xi} d\xi.$$

In (9A.14), Γ is the circle/closed path in z-plane that lies in the domain

$$\sigma_1 < |\xi| < \frac{|z|}{\sigma_2}; \ \sigma_1 \text{ and}$$

σ_2: radii of convergence of $F_1(\xi)$ and $F_1(\xi)$.

Certain limitations of the ZT are: (a) the sample is assumed to be ideal, ZT represents the function, $f(t)$ only at the sampling instants; (b) it is not unique; (c) its accuracy depends on the magnitude of the sampling frequency relative to the highest frequency component in $f(t)$; and (d) a good approximation of $f(t)$ can be obtained from $f(kT)$, the inverse ZT of $F(z)$ by connecting $f(kT)$ with a smooth curve. Some aspects of the representations of the DCS and their relations are presented in Table 9A.1.

The regular ZT gives information about the continuous time signal in the SDCS at the sampling instants only. In order to have a smoother correspondence between $y(t)$ and $y^*(t)$, one needs a ZOH following the sample (in a modified ZT). A way to obtain the

information (continuously) a delay is added to the CTS time system: $u(t) \rightarrow$ sampler $\rightarrow u^*(t) \rightarrow G(s) \rightarrow y(t) \rightarrow$ Time delay $\rightarrow y(t-\Delta t) \rightarrow$ sampler $\rightarrow y^*(t, \Delta t)$.

9A.3 Multirate Sampling

The multirate sampling (MRS) is useful for several reasons: (i) where the individual subsystems might have a different bandwidth, and (ii) multi-layered control systems need simple and fast algorithms for the innermost loops, and somewhat complicated and slow algorithms for the outer most loops. It is assumed that the various sampling operations happen synchronously. For analysis purposes, the MRS is reduced to a single rate sampling (SRS) using a reference sampler, for the latter the slowest sampler is taken as the reference sampler. In this case, each sampled signal is represented by adding a delay compared to the reference signal. Then, these delays are added to the TFs to the left and right, and we obtain one reduced system that would have the sampler synchronized with the reference signal. However, we have now a complicated equivalent one system with an SRS. This technique is closely related to modified ZT.

9A.4 Relation Between s-Plane and z-Plane

For the CTCS, we know that the left half s-plane (LHS) is a stable region and the right half s-plane is the unstable region. Also, for the relative stability the LHS plane is segmented where the poles of the control loop TF should be located. The TF $F^*(s)$ has an infinite number of poles that are located with the intervals of $\pm m\omega_s; \ m = 0, 1.2,\dots$ in the s-plane. The primary strip

TABLE 9A.1

Some Representations

Representation	Mathematical	Physical
1. $g(t) \rightarrow g^*(t)$	Evaluate $g(t)$ at kT: 0, T, 2T,... $\rightarrow g(0), g(T), g(2T)$...	$g(t) \rightarrow$ PAM $\rightarrow g^*(t)$
2. $g^*(t) \rightarrow g(t)$	$g^*(t) \rightarrow$ FFT $\rightarrow G^*(jw) \rightarrow$ LPF $\rightarrow G(jw) \rightarrow$ FFT$^{-1} \rightarrow g(t)$	$g^*(t) \rightarrow$ LPF $\rightarrow g(t)$
3. $g(t) \rightarrow g_h(t)$	Same as in (1)	$g(t) \rightarrow$ SAH $\rightarrow g_h(t)$
4. $g_h(t) \rightarrow g(t)$	Same as in (2)	$g_h(t) \rightarrow$ PAM $\rightarrow$ LPF $\rightarrow g(t)$
5. $G^*(s) \rightarrow G(z)$	Replace e^{-Ts} by z^{-1}	–
6. $G(z) \rightarrow G^*(s)$	Replace z^{-1} by e^{-Ts}	–
7. $G(s) \rightarrow G(z)$	(i) use formula; (ii) expand $G(s)$ by partial fractions, and then each term is mapped.	
8. $G(z) \rightarrow G(s)$	Obtain $\{G^\wedge(z) = G(z)/z\} \rightarrow$ decompose by partial fractions; then $G(z) = zG^\wedge(z) \rightarrow$ map back.	
9. $g^*(t) \rightarrow G(z)$	$g^*(t) = \{g_1, g_2, g_3,\dots\}; \ G(z) = \sum_{k=0}^{\infty} g_k z^{-k}$	
10. $G(z) \rightarrow g^*(t)$	$G(z) = \dfrac{b_n z^n + b_{n-1}z^{n-1} + \dots + b_0}{z^n + a_{n-1}z^{n-1} + \dots + a_0}$; equate it to $\sum_{k=0}^{\infty} g_k z^{-k}$; expand and compare the coefficients: $g_k = b_{n-k} - a_{n-1}g_{k-1} - a_{n-2}g_{k-2} - \dots - a_0 g_{k-n}; \ b_k, g_k = 0; \forall k < 0.$	

Note: w-frequency; h-ZOH; SAH-sample and hold.

(and all the complementary strips) of the s-plane will be mapped into a unit circle in the z-plane and will be centred at the origin:

$$e^{(s+jm\omega_s)T} = e^{sT}e^{j2\pi m} = e^{sT} = z \qquad (9A.15)$$

Some important points to note are: (i) all the points in the LHS plane correspond to points inside the unit circle, (ii) all the points in the RHS plane correspond to points outside the unit circle, (iii) points on the $j\omega$ axis in the s-plane correspond to points on the unit circle $|z| = 1$ in the z-plane: $s = j\omega$; $z = e^{Ts} = e^{j\omega T} \rightarrow$ magnitude = 1, (iv) constant damping loci: the real part σ of a pole $s = \sigma + j\omega$, of a TF s-domain determines the damping factor which represents the rate of rise or decay of time response of the system. The large real part σ represents the small time constant; thus, a faster decay or rise and vice versa, (v) the loci in the LHS plane (vertical line parallel to $j\omega$ axis) denote positive damping since the system is stable; the loci in the RHS plane denote negative damping, (vi) constant damping loci in the z-plane are concentric circles with the centre at $z = 0$, (vi) negative damping loci map to circles with radii > 1 and positive damping loci map to circles with radii < 1, (vii) the constant frequency loci are the horizontal lines in the s-plane, and are parallel to the real axis; corresponding ZT: $z = e^{Ts} = e^{j\omega T}$, when ω = constant, it is a straight line from the origin at an angle of $\theta = \omega T$ rad, measured from positive real axis, and (viii) constant damping ratio loci:

$$s = \varsigma\omega_n \pm j\omega_n\sqrt{1-\varsigma^2};$$

$$\omega_n \text{ is natural undamped frequency}$$

$$= -\frac{\omega}{\sqrt{1-\varsigma^2}}\varsigma \pm j\omega \qquad (9A.16)$$

$$= -\omega\tan\beta \pm j\omega; \quad \beta = \sin^{-1}\{\varsigma\}$$

The ZT is given as

$$z = e^{T(-\omega\tan\beta+j\omega)} = e^{-2\pi\omega\tan\beta/\omega_s} \angle(2\pi\omega/\omega_s) \qquad (9A.17)$$

In the z-plane, the constant damping locus will be a logarithmic spiral, except for $\varsigma = 0$ or $\beta = 0°$, and $\varsigma = 1$ or $\beta = 90°$.

9A.5 Several Stability Aspects of Discrete and Sample Data Systems

When a CTS is sampled, we get $G(s) \rightarrow G^*(s)$, and the resultant system is yet the "continuous" system. So, the features of the CTS still hold true here: $G^*(s)$ is stable if and only if all "poles" in the s-plane are in the LHS.

Next, we use the variable $z = e^{Ts}$, then we have the following equivalences:

$$s = \sigma + j\omega \rightarrow z = e^{T(\sigma+j\omega)} = e^{T\sigma}e^{j\omega T} = |z| \angle z;$$

$$\text{Thus, we have } |z| = e^{T\sigma} \text{ and } \angle z = T\omega \qquad (9A.18)$$

From (9A.18), we can see that the stability is assured if $\sigma < 0$, i.e. $|z| < 1$; the system is marginally stable if $\sigma = 0$, i.e. $|z| = 1$; and this is the imaginary axis in the s-plane. Interestingly enough for the Lyapunov stability, a single pole can be allowed on the unit circle, in the z-plane, but the multiple poles on the unit circle would render the system unstable/instable. As in the CTS, a DTS is BIBO stable if all the poles of $G(z)$ lie inside the unit circle: $|z_i| < 1$. A DTS is Laypunov stable (i.e. the stability with respect to the initial conditions) if all the eigenvalues of the system coefficient matrix lie inside or on the unit circle: $|\lambda_i| \leq 1$ and if anyone is on the unit circle, then these must be single and not multiple. Hence, for a DTS to be stable it must be BIBO and Lyapunov stable. Let us have the DTS as

$$x(k+1) = \varphi x(k) + B_d u(k); \quad y(k) = Hx(k); \quad x(0) = x_0 \quad (9A.19)$$

We have several aspects of system's responses and stability related to (9A.19): (a) the output response of the system only due to the input $u(.)$, is called zero state response, (b) if the response is due only to the initial states, then it is called zero input response, (c) for any bounded input $u(k)$, if the output $y(k)$ is also bounded, then it is called BIBO stability, (d) if for any bounded input $u(k)$, the states are also bounded, then it is called BIBS (bounded input bounded states stability), (e) L_2 norm stability (boundedness):

$$\|x(k)\|_2 = \left[\sum_{i=1}^{n} x_i^2(k)\right]^{1/2} < M; \quad \forall k, M \text{ is finite,} \qquad (9A.20)$$

and (f) if the zero-input response of the system with finite initial condition is bounded, and reaches zero as the time index $k \rightarrow \infty$, then the system is considered to be internally stable; this condition can be stated as

$$|y(k)| \leq M < \infty; \quad \lim_{k\to\infty}|y(k)| = 0 \qquad (9A.21)$$

The conditions just stated in the foregoing are also the requirements of asymptotic stability. Ultimately, to ensure all the stability conditions for an LTI DTS, the overriding condition is that the roots of the characteristic equation should be inside the unit circle in the z-plane.

A general DTS is described, without any input $u(k)$ as

$$x(k+1) = f(x(k)) \qquad (9A.22)$$

For such a system, at the equilibrium point/state the dynamics of the system are zero

$$f(x(k)) = 0 \qquad (9A.23)$$

The condition of (9A.23) signifies that the states, $x(k)$ and the solution of (9A.23) will remain at this point forever once the states have reached there, and this implies that the equilibrium point/s are the solutions of autonomous system, (9A.22), (9A.23); however, interestingly enough, the nonlinear DTS can have multiple equilibrium points and the stability of these systems are with respect to such points. For the LTI systems

$$x(k+1) = \varphi x(k) \Rightarrow 0 \qquad (9A.24)$$

In case of (9A.24), the equilibrium point is $x(k) = 0$, since the transition matrix is finite.

The stability in the sense of Lyapunov is stated as: the equilibrium point $x = 0$ of (9A.22), is stable if, for each small and positive ε, there exist, $\delta = \delta(\varepsilon, k_0) > 0$, such that the following condition holds true

$$\|x(k_0)\| < \delta \Rightarrow \|x(k)\| < \varepsilon; \ \forall k \geq k_0 \geq 0 \qquad (9A.25)$$

The equilibrium point $x = 0$ of (9A.22) is asymptotically stable if it is stable and there is a positive constant $c = c(k_0)$ such that

$$x(k) \to 0 \ as \ k \to \infty, \ \forall \|x(k_0)\| < c \qquad (9A.26)$$

The equilibrium point $x = 0$ of (9A.22) is uniformly stable if, for each $\varepsilon > 0$, there exist, a $\delta = \delta(\varepsilon) > 0$, independent of k_0, such that, see (9A.25):

$$\|x(k_0)\| < \delta \Rightarrow \|x(k)\| < \varepsilon; \ \forall k \geq k_0 \geq 0 \qquad (9A.27)$$

The equilibrium point $x = 0$ of (9A.22) is uniformly asymptotically stable if it is uniformly stable and there is a positive constant c, independent of k_0, such that the following is true, see (9A.26):

$$x(k) \to 0 \ as \ k \to \infty, \ \forall \|x(k_0)\| < c \ (\text{uniformly in } k_0) \qquad (9A.28)$$

The equilibrium point $x = 0$ of (9A.22) is globally uniformly asymptotically stable, if it is uniformly asymptotically stable for such δ, when $\delta(\varepsilon)$ can be chosen such that the following is true

$$\lim_{\varepsilon \to \infty} \delta(\varepsilon) = \infty \qquad (9A.29)$$

The equilibrium point $x = 0$ of (9A.22) is exponentially stable, if there exist constants c, γ, λ such that

$$\|x(k)\| \leq \gamma \|x(k_0)\| e^{-\lambda(k-k_0)}; \ \forall \|x(k_0)\| < c \qquad (9A.30)$$

The equilibrium point $x = 0$ of (9A.22) is globally exponentially stable, if it is exponentially stable for any initial state $x(k_0)$.

9A.6 Lyapunov Stability Results

The two methods are: (i) Lyapunov's first, or indirect, method and (ii) Lyapunov's second or direct method (LSM).

First method: expanding the system of (9A.22) in the Taylor series around the point x_e, and neglecting the higher order terms we get:

$$\Delta x(k+1) = F \Delta x(k); \quad F = \frac{\partial f}{\partial x}|_{x=x_e} \qquad (9A.31)$$

The nonlinear system (9A.22) is, at the least, asymptotically stable around x_e, if the linear system of (9A.31) is stable, i.e. if all the eigenvalues of F are inside the unit circle. The linearization would work well in many cases, but if the nonlinearity is hard, then it might not work. However, if some eigenvalues are on the unit circle, then the equilibrium might be (marginally) stable or even unstable.

Second method: The method is a generalization of the Lagrange's concept of stability of minimum potential energy, i.e. any dynamic system, under favorable situations, could reach to the state of lowest potential/static energy, if it contains energy-dissipating sub-systems. Let the Lyapunov normalized energy functional (LNEF) $V(x)$ be considered with: (i) $V(0) = 0$, i.e. it is zero at the equilibrium; (ii) $V(x) > 0$; $x \neq 0$, i.e. it is positive definite everywhere; and (iii) $\Delta V(x) < 0$ along the trajectories of our system, it is continuously decreasing as the equilibrium is being reached. If these three conditions hold true, then the origin of the given DTS is asymptotically stable. Let us consider the method for a linear or linearized system of (9A.22), i.e. (9A.31), in the terms of state variables that are small perturbations from the equilibrium point/s:

$$x(k+1) = Fx(k) \qquad (9A.32)$$

Lyapunov energy functional, with Y as the symmetric positive definite matrix, is chosen as

$$V(x(k)) = x^T(k)Yx(k) \qquad (9A.33)$$

We can interpret (9A.33) as the normalized
Lyapunov energy functional, by presuming that
the small perturbations (from the equilibrium
point) are like some "state errors" and their
outer product is a dispersion, or is a Gramian
(the latter term being used for the deterministic
systems and variables), then the matrix Y (the
symbol used in the filtering theory) is called the
information Gramian. The dispersion, covariance Gramian can be termed as the Gramianance
to maintain the connectivity with its stochastic
companion which is called covariance matrix.
The covariance matrix's units might match with
the unit of (electrical) power (it is proportional
to the squared voltage or current), but the term
energy sounds better, and the Lyapunov energy
functional as stated in (9A.33) is called the normalized energy functional. Also, for simplicity,
the matrix is taken as constant. Then we have the
differential of the functional, and using (9A.33):

$$\Delta V(x(k)) = V(x(k+1)) - V(x(k))$$
$$= x^T(k+1)Yx(k+1) - x^T(k)Yx(k) \quad (9A.34)$$

Now, substitute (9A.32) in (9A.34) and simplify
to obtain

$$\Delta V(x) = x^T F^T PFx - x^T Yx$$
$$= x^T(F^T YF - Y)x \quad (9A.35)$$
$$= -x^T Y_e x$$

In (9A.35), we have

$$F^T YF - Y = -Y_e \quad (9A.36)$$

If Y_e is positive definite matrix, then the system
is asymptotically stable, using the concept of
the LSM. Alternatively, if we choose $Y_e = I$ and
solve (9A.36), called Lyapunov's matrix equation,
for Y from $F^T YF - Y = -I$ and see if Y is positive definite, then the system is asymptotically
stable. The symbol Y, as the information matrix-
Gramian, is deliberately used here to indicate the
connection of the LNEF with the information concepts. Accordingly, the LNEF can then be called
the Lyapunov normalized energy Gramiance.

10

Design of Discrete Time Control Systems

The main aim for a control system is to design and synthesize a controller that gives, as its output, a control law $u(t)$ (input signal to the actual system to be controlled), either in the forward or the feedback path, so that the closed loop system (if it is unstable) is made stable and has some desired stability and error performance [2]. The classical and conventional approaches for the design of control law for continuous time control systems (CTCS) are: (i) root locus and (ii) frequency domain methods, presented in Section 10.1.

10.1 Design Based on Root Locus Method

In general, we are interested in studying: (i) the effect of system gain, and/or sampling period, on the absolute and relative stability and (ii) the transient response characteristics of the CLCS. The root locus method for CTCS can be extended to a discrete time control system (DTCS) with minor changes. This is because the characteristic equations, for both the type of systems, are of the same form. A characteristic equation would be of the following form

$$1+G(z)H(z) = 0; \text{ or } 1+GH(z) = 0;$$
$$1 + G(s)H(s) = 0 \text{ for the CTCS} \quad (10.1)$$

In general, it can be written as

$$1+L(z) = 0 \quad (10.2)$$

In (10.2), $L(.)$ is known as the loop TF, $GH(z)$ (like $GH(s)$), and we can rewrite (10.2) as

$$L(z) = -1 \quad (10.3)$$

Because $L(.)$ is a complex function, we can write it as:
magnitude criterion: $|L(z)| = 1$; phase criterion:

$$\angle L(z) = \pm 180°(2k+1), k = 0, 1, 2, ... \quad (10.4)$$

The values of z that satisfy both these criteria in (10.4) are the root of the characteristic equation (or the closed loop TF poles). The characteristic equation (10.2) is rearranged as

$$1+K\frac{(z+z_1)(z+z_2)...(z+z_m)}{(z+p_1)(z+p_2)...(z+p_n)} = 0 \quad (10.5)$$

In (10.5), z_i and p_i are the zeros and poles of the (open) loop TF (OLTF).

10.1.1 Rules for Construction of the Root Locus

These rules for digital systems are same as that for a CTCS [2]: (i) the root locus (RL) is symmetric about real axis, and the number of root locus branches (RLB) is equal to the number of open loop poles; (ii) the RLBs start from the open loop poles at gain $K = 0$ and end at the open loop zeros at $K = \infty$; when there are no open loop zeros, the RL tends to ∞ when $K \rightarrow \infty$. The number of branches that tend to ∞ is equal to the difference between the number of poles and number of zeros; (iii) any portion of the real axis will be a part of the RL, if the number of poles plus number of zeros to the right of that portion is odd; (iv) if there are n open loop poles and m open loop zeros, then $n - m$ RL branches tend to ∞ along the straight line asymptotes drawn from a single point $s = \sigma$, which is called centroid of the loci:

$$\sigma = \frac{\left(\sum \begin{array}{c} \text{real parts of the} \\ \text{open loop poles} \end{array}\right) - \left(\sum \begin{array}{c} \text{real parts of the} \\ \text{open loop zeros} \end{array}\right)}{n-m} \quad (10.6)$$

Angle of asymptotes

$$\phi_q = \frac{180°(2q+1)}{n-m}; q = 0, 1, ..., n-m-1; \quad (10.7)$$

(v) breakaway (break-in) points or the points of multiple roots are the solution of

$$\frac{dK}{dz} = 0 \quad (10.8)$$

In (10.8), K is expressed as a function of z from the characteristic equation, that is a necessary, but not the sufficient condition, and if the solutions lie on the root locus should be checked; (vi) the intersection, if any, of the RL with the unit circle can be determined from the

Routh array; (vii) the angle of departure from a complex open loop pole is given by

$$\phi_p = 180° + \phi \qquad (10.9)$$

In (10.9), ϕ is the net angle contribution of all other open loop poles and zeros to that pole.

$$\phi = \sum_i \psi_i - \sum_{j \neq p} \gamma_j \qquad (10.10)$$

In (10.10), ψ_i are the angles contributed by zeros and γ_j are the angles contributed by the poles; (viii) the angle of arrival at a complex zero is given by

$$\phi_z = 180° - \phi; \qquad (10.11)$$

and (iv) the gain at any point z on the root locus is given by

$$K = \frac{\prod_{j=1}^{n} |z_0 + p_j|}{\prod_{i=1}^{m} |z_0 + z_i|} \qquad (10.12)$$

10.1.2 Root Locus of a Digital Control System

Let us first consider the effect of controller gain K and the sampling time/interval T on the relative stability of the closed loop system, say for $T = 0.5$ sec.; see Figure 9.7a (and assume $H(s) = 1$, DC is the controller TF $G^*_D(s)$, and ZOH) [2]; then, we have the TF as

$$Z[G_{ho}(s)G_p(s)] = Z\left[\frac{1 - e^{-Ts}}{s} \frac{1}{s + 1}\right]$$

$$= (1 - z^{-1})Z\left[\frac{1}{s(s + 1)}\right]$$

$$= \frac{z - 1}{z}\left[\frac{z}{z - 1} - \frac{z}{z - e^{-T}}\right]$$

$$= \frac{1 - e^{-T}}{z - e^{-T}}; \text{ use partial fractions}$$

For an integral controller, $G_D(z) = \frac{Kz}{z-1}$, we obtain $G(z)$ as

$$G(z) = G_D(z)G_h G_p(z)$$

$$= \frac{Kz}{z - 1}\frac{1 - e^{-T}}{z - e^{-T}}$$

Then, the characteristic equation is:

$$1 + G(z) = 0 \Rightarrow 1 + \frac{Kz(1 - e^{-T})}{(z - 1)(z - e^{-T})} = 0;$$
$$\qquad (10.13)$$
$$\text{With } T = 0.5 \text{ sec., } L(z) = \frac{0.3935Kz}{(z - 1)(z - 0.6065)}$$

$L(z)$ has poles at $z = 1$, and $z = 0.6065$, and one zero at $z = 0$. The breakaway/break-in points are ($dK/dz = 0$):

$$\frac{dK}{dz} = -\frac{z^2 - 0.6065}{0.3935z^2} = 0; \text{ since } K = -\frac{(z - 1)(z - 0.6065)}{0.3935z} \qquad (10.14)$$

This gives $z^2 = 0.6065 \rightarrow z_1 = 0.7788$, $z_2 = -0.7788$. The critical value of K is (from the magnitude criterion):

$$\left|\frac{0.3935z}{(z - 1)(z - 0.6065)}\right| = 1/K, \text{ with } z = -1,$$
$$\qquad (10.15)$$
$$\left|\frac{-0.3935}{(-2)(-1.6065)}\right| = 1/K \Rightarrow K = 8.165$$

The root locus of the system for $K = 0$ to $K = 10$, can be obtained as: (i) the two RLBs would start from two open loop poles at $K = 0$; (ii) if K is increased, one branch will go towards the zero, and the other one would tend to infinity. Thus, the stable range of K is $0 < K < 8.165$. If $T = 1$ sec., then we have $G(z) = \frac{0.6321Kz}{(z-1)(z-0.3679)}$. Breakaway/break-in points are: $z^2 = 0.3679 \rightarrow z_1 = 0.6065$, $z_2 = -0.6065$. The critical gain $K_c = 4.328$. For this case, the radius of the inside circle decreases, and the maximum value of stable K also decreases to $K = 4.328$. Similarly, if $T = 2$ sec., then we have $G(z) = \frac{0.8647Kz}{(z-1)(z-0.1353)}$. In this case, the critical gain is reduced to $K_c = 2.626$.

10.1.3 Effect of Sampling Period T

From the example of Section 10.1.2, it is seen that large T, i.e. the sampling period, worsens the relative stability of the considered control system. One can sample, eight to ten times during a cycle of the damped sinusoidal oscillation of the output, i.e. if it is underdamped. The smaller sampling period allows the critical gain to be larger (maximum allowable gain can be made larger by increasing sampling frequency/rate). Also, the damping ratio decreases with the decrease in T. However, the damping ratio of the closed loop poles of a DTCS indicates the relative stability only if the sampling frequency is sufficiently high (8 to 10 times); otherwise, the prediction of overshoot from the damping ratio will be much higher, and the actual overshoot will be larger than the predicted one. Let us have $K = 2$. We obtain the SSE as:

1. $T = 0.5$ sec., $G(z) = \frac{0.787z}{(z-1)(z-0.6065)}$, this being the second order system, the velocity error constant will be a finite quantity:

$$K_v = \lim_{z \to 1} \frac{(1-z^{-1})G(z)}{T} = 4; \ e_{ss} = 0.25 \qquad (10.16)$$

2. $T = 1$ sec., $G(z) = \frac{1.2642z}{(z-1)(z-0.3679)}$,

$$K_v = \lim_{z \to 1} \frac{(1-z^{-1})G(z)}{T} = 2; \ e_{ss} = 0.5 \qquad (10.17)$$

3. $T = 2$ sec., $G(z) = \frac{1.7294z}{(z-1)(z-0.1353)}$,

$$K_v = \lim_{z \to 1} \frac{(1-z^{-1})G(z)}{T} = 1; \ e_{ss} = 1 \qquad (10.18)$$

We see from the foregoing example that when the sampling period is increased (the sampling frequency is decreased), the SSE increases.

10.1.4 Design Procedure

The controller design for a continuous time system using root locus is based on the approximation that the closed loop system has a complex conjugate pole pair, since such a pair would dominate the system behavior [2]. Similarly, for a DTCS, the controller can be designed based on the features of a dominant pole pair. For CTCS (Section 10.1), there are PI, PD, and PID controllers. The PI (the proportional + Integral control actions) controller is generally used for improving the system's SS performance; and the PD (proportional + derivative control actions) controller is used for improving the relative stability (i.e. increasing the damping effect) or the transient response of the CLCS. Also, a phase lead compensator would improve the dynamic performance and a lag compensator would improve the SS response. The usual approach in the design/synthesis of controllers in s- or z-planes is to cancel the undesired poles or zeros of the plant (system) TF by the zeros and poles of the (designed) controller. Also, new poles and zeros can be added in some locations if desired. In case the undesired poles are near the $j\omega$ axis, an inexact cancellation might lead to a marginally stable/unstable CLCS. Hence, canceling an unstable pole is always risky, and hence should be avoided. Let us have a compensator as $K(z+a)/(z+b)$; and if the zero lies on the RH side of the pole, it will be a lead compensator. The design procedure is to [2]: (i) compute

the desired closed loop pole pairs based on the specified design criteria; (ii) map the s-domain poles to z-domain (poles); (iii) check if the sampling frequency is 8–10 times the desired damped frequency of oscillation; (iv) compute the angle contributions of all open loop poles and zeros to the desired closed loop pole; (v) compute the required contribution by the controller transfer function to satisfy angle criterion; (vi) place the controller zero in a suitable location and calculate the required angle contribution of the controller pole; (vii) compute the location of the controller pole to provide the required angle; and (viii) determine the gain K from the magnitude criterion.

10.2 Frequency Domain Analysis

A sinusoidal input to a stable LTI system would produce its response as a sinusoidal output of the same frequency but with different magnitude and phase (this is because of the gain and phase contributed by the dynamic system itself), however, this is not so for the nonlinear systems, and we know from the study in Section 10.1, that the variation of the magnitude and phase of the output of the system with input frequency/ies is known as frequency response of the system. Experimental results can be used to construct frequency response/s, even if the plant's (system's) mathematical model is not known. Analysis of DCSs in frequency domain can be carried out by using the techniques that are used for CTCSs. Very popular representations/models in frequency domain are: (i) Nyquist plot and (ii) Bode diagram, as we have seen in Section 10.1.

10.2.1 Nyquist Plot

The Nyquist plot of a TF, which is usually the loop transfer function (LTF), $GH(z)$, is a mapping of Nyquist contour in z-plane onto $GH(z)$ plane that is in polar coordinates; it is also known as a polar plot (similar to the continuous time case). Absolute and relative stabilities of a CLCS can be determined from the Nyquist plot using Nyquist stability criterion, i.e. by using only the LTF. For an LTF, $GH(z)$ of a DCS, the polar plot of $GH(z)$ is obtained by setting $z = e^{j\omega T}$, and varying ω from 0 to ∞. The CLTF of a SISO DCS is specified as

$$C(z) = G_c(z) = \frac{G(z)}{1 + GH(z)} \qquad (10.19)$$

As we know, the stability of the system depends on the roots (or zeros) of the characteristic equation (10.1) or the poles of the CLTF (system) (10.19), and all these roots should lie inside of the unit circle for the CLCS to be stable. The Nyquist path in the z-plane that encloses the exterior of the unit circle is defined; here, the region to the left of a closed path is considered to be enclosed by that path, when the direction of the path is taken anti-clockwise; then one maps the Nyquist path in z-plane onto the $GH(z)$ plane that yields Nyquist plot of $GH(z)$. Then, the stability of the CLCS is investigated by studying the behavior of Nyquist plot with respect to the critical point $(-1, j0)$ in the $GH(z)$ plane. Two Nyquist paths are defined [2]: (i) path one, z_1 that does not enclose the poles on the unit circle, (ii) the other path, z_2 that encloses the poles on the unit circle. Define the parameters as: (a) Z_{-1} = number of zeros of $1 + GH(z)$ outside the unit circle in the z-plane; (b) P_{-1} = number of poles of $1 + GH(z)$ outside the unit circle in the z-plane; (c) P_0 = number of poles of $GH(z)$ (same as number of poles of $1 + GH(z)$) that are on the unit circle; (d) N_1 = number of times the $(-1, j0)$ point is encircled by the Nyquist plot of $GH(z)$ corresponding to z_1; (e) N_2 = number of times the $(-1, j0)$ point is encircled by the Nyquist plot of $GH(z)$ corresponding to z_2; and (f) as per the principle of argument (from the complex variable theory) we have the following condition

$$N_1 = Z_{-1} - P_{-1} \qquad (10.20)$$

Denote the angle traversed by the phasor drawn from $(-1, j0)$ point to the Nyquist plot of $GH(z)$ as ω varies from $\omega_s/2$ to 0, on the unit circle of z_1 excluding the small indentations, by ϕ. It can be shown that

$$\phi = (Z_{-1} - P_{-1} - 0.5P_0)180° \qquad (10.21)$$

For the closed loop digital control system to be stable, Z_{-1} should be equal to zero; thus, the Nyquist criterion for stability of the closed loop digital control systems is

$$\phi = -(P_{-1} + 0.5P_0)180 \qquad (10.22)$$

For the CL digital control system to be stable, the angle traversed by the phasor drawn to the $GH(z)$ plot from $(-1, j0)$ point, as ω varies from $\omega_s/2$ to 0, must satisfy equation (10.22).

10.2.2 Bode Plot, and Gain and Phase Margins

As we have seen in Section 10.1, a Bode plot, often known as a Bode diagram, is a graphical tool for drawing the frequency response of a control system. In fact, it is a diagram represented by two plots: (i) the magnitude of a TF v/s frequency and (ii) the phase of the TF v/s frequency; generally, this frequency is in rad/sec. (or rad/s); the magnitude is in dB; and the frequency is plotted in log scale. One important merit is that, in the s-domain the magnitude curve can be approximated by straight lines, and hence, the magnitude plot can be sketched without exact computation of the TF. However,

this feature is not available for the Bode diagram in the z-domain. To have this feature, one uses bilinear transformation to transform unit circle of the z-plane into the imaginary axis of another complex plane, the w-plane:

$$w = \frac{1}{T}\ln(z); \quad w = \frac{2}{T}\frac{(z-1)}{(z+1)} \Rightarrow z = \frac{1+wT/2}{1-wT/2} \qquad (10.23)$$

One can then transform a given z-domain TF by using the bilinear transformation as in (10.23), i.e. from $GH(z)$ to $GH(w)$ and then construct the Bode plot for the $GH(w)$. Similar to the case of a continuous time system (CTS), the gain margin and phase margin are the measures of relative stability of a control system, and we need to first define phase and gain cross over frequencies. Gain margin (GM) is the safety-allowance factor by which the open loop gain of a system can be increased before the control system becomes unstable:

$$GM = 20\log_{10}\left|\frac{1}{GH(e^{i\omega_p T})}\right| dB \qquad (10.24)$$

In (10.24), ω_p is the phase crossover frequency, where the phase of the loop TF $GH(e^{i\omega T})$ is 180°. Similarly, the Phase margin (PM) is

$$PM = 180° + \angle GH(e^{i\omega_g T}) \qquad (10.25)$$

In (10.25), ω_g is the gain crossover frequency, where the loop gain magnitude (amplitude ratio of the system's TF) is 1.

10.3 Compensator Design

A compensator, or so-called controller TF (that is also called a filter in forward path or the feedback path, or in both), is added to a basic to improve its steady state as well as dynamic responses/performance. Often the Bode plot is used since two crucial design specifications, (i) phase margin and (ii) gain crossover frequency, are visible from the Bode plot along with gain margin. Hence, it is important to note that: (a) low frequency asymptote of the amplitude ratio plot is indicative of one of the error constants K_p, K_v, K_a depending on the type $(-n)$ of the system under consideration, and (b) specifications on the transient response can be stated in terms of phase margin (PM), gain margin (GM), gain crossover frequency, and bandwidth.

10.3.1 Phase Lead, Phase Lag, and Lag-Lead Compensators

We have important points to note [2]: (a) phase lead compensation is used to improve stability margins, because it increases system bandwidth; (b) phase lag compensation reduces the system gain at high frequencies without reducing low frequency gain; the total gain (low frequency gain) can be increased to improve the steady state accuracy, high frequency noise can also be attenuated, but stability margin and bandwidth would reduce; and (c) using a lag-lead compensator, where a lag compensator is cascaded with a lead compensator, both steady state and transient responses can be improved. It would be easy to use bilinear transformation, since it transfers the loop TF of the z-plane to that in the w-plane. Also, qualitatively, the w-plane is similar to the s-plane because the design technique used in the s-plane is employed to design a controller in w-plane. Once the design is completed, the controller of the z-plane can be determined by using the inverse transformation from the w-plane to z-plane.

10.3.2 Compensator Design Using Bode Plot

It is interesting to note here that the design methods (Section 10.1) that are used for CTCS can be used for the design of controllers for DTCS also; this is made feasible and easier because of the bilinear transformation of the TF from the z-plane to w-plane.

10.3.2.1 Phase Lead Compensator

In the frequency response of a simple PD controller (Section 10.1), the magnitude of the compensator continuously grows with the increase in frequency; this is not desirable because it would amplify any high frequency noise, if present, in any real system. In the case of lead compensator, a first order pole is added in the denominator of the PD controller at frequencies well above the corner frequency of the PD controller. A lead compensator is

$$C_{lead}(s) = K \frac{\tau s + 1}{\alpha \tau s + 1}, \alpha < 1 \qquad (10.26)$$

The magnitude and phase angle of the lead compensator (i.e. lead providing TF) are

$$K \frac{\sqrt{1 + \omega^2 \tau^2}}{\sqrt{1 + \alpha^2 \omega^2 \tau^2}} \qquad (10.27)$$

$$\phi = \tan^{-1}(\omega \tau) - \tan^{-1}(\alpha \omega \tau) \qquad (10.28)$$

Thus, the lead compensator TF would provide a good amount of the phase (in fact it will provide a phase lead) with much less amplitude (magnitude) at high frequencies. The frequency where the maximum phase occurs and the magnitude of $C(s)$ at this frequency are given by $\omega_{max} = \frac{1}{\sqrt{\alpha}\tau}$; $\sin(\phi_{max}) = \frac{1-\alpha}{1-\alpha} \Rightarrow \alpha + \left(\frac{1-\sin(\phi_{max})}{1+\sin(\phi_{max})}\right)$; $C_{lead}(s) = \frac{K}{\sqrt{\alpha}}$.

10.3.2.2 Phase Lag Compensator

A lag compensator provides more gain, (of α) at low frequencies, thus decreasing the steady state error, and it does not change the transient response significantly and also would have the system with sufficient phase margin. The TF of the lag compensator is given as

$$C_{lag}(s) = \alpha \frac{\tau s + 1}{\alpha \tau s + 1}, \alpha > 1 \qquad (10.29)$$

The final compensator is given as $C_{com}(s) = K C_{lag}(s)$; $s \rightarrow 0, C_{lag}(s) \rightarrow \alpha; s \rightarrow \infty, C_{lag}(s) \rightarrow 1$.

Because, the lag compensator provides the maximum lag near the two corner frequencies, the PM is maintained if the zero of the compensator is chosen, such that $\omega = 1/\tau$ is much lower than the gain crossover frequency of the uncompensated system; τ is accordingly chosen.

10.3.2.3 Lag-Lead Compensator

If the use of a lead or lag compensator (TF) in a control system does not meet the specified design criteria, a lag-lead compensator is used; in which the lag part precedes the lead part:

$$C_{lela}(s) = K \frac{1 + \tau_1 s}{1 + \alpha_1 \tau_1 s} \frac{1 + \tau_2 s}{1 + \alpha_2 \tau_2 s}, \alpha_1 > 1, \alpha_2 < 1 \quad (10.30)$$

As can be seen from (10.30), the corner frequencies are: $\frac{1}{\alpha_1 \tau_1}, \frac{1}{\tau_1}, \frac{1}{\alpha_2 \tau_2}, \frac{1}{\tau_2}$. One can use the guidelines: (a) one should first check the PM and BW (bandwidth) of the uncompensated system with adjustable gain K; (b) if the BW is smaller than the acceptable BW one should use lead compensator, and if the BW is large (broader), lead compensator would not be useful since it would provide high frequency magnification; for this a lag compensator may be used, provided the open loop system is stable; and (c) if the lag compensator yields a too low BW (low speed of response), a lag-lead compensator may be used.

10.4 Design with Deadbeat Response

The deadbeat response design is applicable to DCS only. Let us consider an all-digital control system (ADCS) with the TFs of the plant and the cascade controller (see Figure 10.1) as

$$G_p(z) = \frac{z + 0.6}{3z^2 - z - 1}; D_c(z) = \frac{3z^2 - z - 1}{(z-1)(z+0.6)} \qquad (10.31)$$

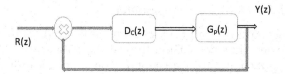

FIGURE 10.1
All digital control system ($D_c(z)$-digital controller).

Then, the loop TF (like $G(s)H(s)$) is given as

$$G(z) = D_c(z)G_p(z) = \frac{1}{z-1} \tag{10.32}$$

The closed loop TF is given as

$$\frac{G(z)}{1+G(z)} = \frac{1}{z} \tag{10.33}$$

For a step input to this ADCS, the output is

$$Y(z) = \frac{1}{z} \frac{z}{(z-1)} = z^{-1} + z^{-2} + \dots \tag{10.34}$$

We can see from (10.34) that $y(k)$ reaches the desired steady-state value 1, in one sampling period, without any overshoot, and stays there for ever. This type of response is known as deadbeat response (DBR); if $G_p(z)$ were a result of sampling a CTS, the $D_c(z)$ does not guarantee that no ripples would occur between two sampling instants in constant output $y(t)$.

10.4.1 DBR Design of a System When the Poles and Zeros Are in the Unit Circle

The design (i.e. the design requirements or specifications) criteria are: (i) the system must not have an SSE at sampling instants, (ii) the final output should be attained in finite and minimum time, and (iii) the controller should be causal, i.e. physically realizable. The CLTF is taken as;

$$C(z) = \frac{Y(z)}{R(z)} = \frac{D_c(z)G_p(z)}{1+D_c(z)G_p(z)} \tag{10.35}$$

$$D_c(z) = \frac{1}{G_p(z)} \frac{C(z)}{1-C(z)} \tag{10.36}$$

The error TF is given as

$$E(z) = R(z) - Y(z) = \frac{R(z)}{1+D_c(z)G_p(z)} \tag{10.37}$$

Let the reference input signal TF be

$$R(z) = \frac{A(z)}{(1-z^{-1})^N}; N \text{ is positive integer} \tag{10.38}$$

In (10.38), $A(z)$ is the polynomial in z^{-1} with no zeros at $z = 1$; for unit step input $A(z) = 1$, $N = 1$; and for unit ramp input $A(z) = Tz^{-1}$, $N = 2$. To obtain zero SSE, we have

$$\lim_{k\to\infty}\left\{e(kT)\right\} = \lim_{z\to1}\left\{(1-z^{-1})E(z)\right\} \tag{10.39}$$

$$= \lim_{z\to1}\left\{\frac{(1-z^{-1})A(z)(1-C(z))}{(1-z^{-1})^N}\right\} = 0 \tag{10.40}$$

The requirement for the zero SSE is that $1-C(z)$ should contain $(1-z^{-1})^N$ as a factor,

$$1 - C(z) = (1-z^{-1})^N F(z)$$

$$C(z) = 1 - (1-z^{-1})^N F(z) = \frac{Q(z)}{z^p}; \ p > N \tag{10.41}$$

In (10.41), $F(z)$ is polynomial in z^{-1}, and $Q(z)$ is a polynomial in z. We also have $E(z) = A(z)F(z)$, and since $A(z)$ and $F(z)$ are polynomials of z^{-1}, $E(z)$ will have a finite number of terms in the inverse power (series) of z, and the error will go to zero in a finite number of sampling periods.

10.4.1.1 Physical Realizability of the Controller $D_c(z)$

The requirement of physical realizability needs some constraints on the form of the CLTF $C(z)$. Let the plant dynamics TF be given as

$$G_p(z) = g_n z^{-n} + g_{n+1} z^{-n-1} + \dots;$$
$$C(z) = m_k z^{-k} + m_{k+1} z^{-k-1} + \dots \tag{10.42}$$

In (10.42), n and k are the excess poles over zeros of the two TFs, implying

$$D_c(z) = d_{k-n} z^{-(k-n)} + d_{k-n+1} z^{-(k-n+1)} + \dots \tag{10.43}$$

In order that the compensator $D_c(z)$ is realizable, $k \geq n$, the additional poles over the zeros for $C(z)$ must be at least equal to the additional poles over zeros of $G_p(z)$. Thus, if $G_p(z)$ does not have poles or zeros outside the unit circle, then $C(z)$ should have the forms: (i) for step input, $R(z) = \frac{z}{z-1}$, $C(z) = \frac{1}{z^n}$; and (ii) for ramp input, $C(z) = \frac{(n+1)z-n}{z^{n+1}}$.

DBR design depends on the cancellation of poles and zeros of the plant TF; if the poles are on or outside the unit circle, then any imperfect cancellation may lead to instability. Thus, for practical reasons, one should avoid the canceling of these poles and use a different design approach.

10.4.2 DBR When Some of the Poles and Zeros Are on or outside the Unit Circle

Let the system TF be given as

$$G_p(z) = \frac{\prod_{i=1}^{m}(1-z_i z^{-1})}{\prod_{j=1}^{n}(1-p_j z^{-1})} B(z) \qquad (10.44)$$

In (10.44), m and n are the number of zeros and poles on or outside the unit circle, and $B(z)$ is a rational TF in z^{-1} with poles and zeros inside the unit circle. With this, the compensator TF is given as

$$D_C(z) = \frac{\prod_{j=1}^{n}(1-p_j z^{-1})}{\prod_{i=1}^{m}(1-z_i z^{-1})} \frac{C(z)}{B(z)(1-C(z))}; \qquad (10.45)$$

$C(z)$ is the closed loop TF

Since the cancellation of the poles and zeros that are on or outside the unit circle should not be done by the compensator/controller $D_c(z)$, one should choose the $C(z)$ such that these cancel out. Then, $C(z)$ should contain the factors such as

$$\prod_{i=1}^{m}(1-z_i z^{-1}) \qquad (10.46)$$

The following factors should be contained in $(1-C(z))$

$$\prod_{j=1}^{n}(1-p_j z^{-1}) \qquad (10.47)$$

Thus, we would have

$$C(z) = \prod_{i=1}^{m}(1-z_i z^{-1})(c_k z^{-k}+c_{k+1}z^{-k-1)}+...);$$
$$\qquad (10.48)$$
$$1-C(z) = \prod_{i=1}^{n}(1-p_j z^{-1})(1-z^{-1})^l(1+a_1 z^{-1}+a_2 z^{-2}+...)$$

In (10.48), l (el) equals either the order of the poles of $R(z)$ or the order of poles of $G_p(z)$ at $z = 1$, whichever is greater. The indicated truncation in (10.48) depends on: (i) the order of poles of $C(z)$ and $(1-C(z))$ must be equal, and (ii) total number of unknowns must be equal to the order of $C(z)$ to be able to solve independently.

10.4.3 Sampled Data Control Systems with DBR

For a CTCS, the output $y(t)$ is a function of time t and the DBR design, based on cancellation of stable poles and zeros, may lead to inter sampling ripples in the output; this is because, the process/plant zeros are canceled by controller poles, and the continuous dynamics are excited by the input and are not affected by feedback. In case of a DBR-design for a SDCS, with the plant TF $G_{h0}G_p(z^{-1})$ having at least one zero, it is not advisable to cancel the zeros (whether they are inside or outside the unit circle). If

$$G_{h0}G_p(z^{-1}) = \frac{Q(z^{-1})}{P(z^{-1})} \qquad (10.49)$$

Then, the digital controller for ripple free DBR to step input is given as

$$D_c(z) = \frac{P(z^{-1})}{Q(1)-Q(z^{-1})} \qquad (10.50)$$

The ripple free DBR design can be done using a similar approach, as discussed in the foregoing sections, but with an added constraint that will increase the response time of the system.

10.5 State Feedback Controller

Hither to, the design of the controller for control system was based on the TFs of the considered dynamic systems/plants/processes; these methods are applicable only for linear time-invariant (LTI) and initially relaxed systems and to the linearized (nonlinear) plants. Next, we discuss the state variable/space methods of designing controllers, which can be easily applied to MIMO systems. Let us start with the state space model of a SISO discrete time system (DTS):

$$x(k+1) = \varphi x(k) + Bu(k)$$
$$y(k) = Hx(k) \qquad (10.51)$$

The variables in (10.51) have usual meanings and dimensions, as per section 10.2. The main idea of the state feedback is to feedback some or all the states (variables/state vector) to the input side (meaning these states are added to the control input that is already used for controlling the same system) to place the closed loop poles at the desired locations (in the s-plane or z-plane). There are two important design problems:

1. In regulation, the states should approach zero starting from any arbitrary initial state; here, the internal stability of the system with desired transient is attained,

$$u(k) = -Kx(k) \tag{10.52}$$

2. In tracking, the output has to track a reference (input) signal $r(k)$,

$$u(k) = -Kx(k) + K_r r(k)$$

Now, if we use (10.52) in (10.51), we get

$$x(k+1) = (\varphi - BK)x(k) \tag{10.53}$$

The problem of regulator design is theoretically solved when one can determine appropriate gain K, such that the eigenvalues of $(\varphi - BK)$ are within the unit circle (for a SDCS/DCS). For this, a necessary and sufficient condition for arbitrary pole placement is that the pair (φ, B) must be controllable. Also, we assume that all the states are available for the feedback.

10.5.1 Designing K by Transforming the State Model into Controllable Canonical Form

Let U_c be the controllability matrix and consider a transformation matrix T as

$$T = U_c W;$$

$$W = \begin{bmatrix} a_{n-1} & a_{n-2} & \cdots & a_1 & 1 \\ a_{n-2} & a_{n-3} & \cdots & 1 & 0 \\ : & : & : & : & : \\ a_1 & 1 & \cdots & 0 & 0 \\ 1 & 0 & \cdots & 0 & 0 \end{bmatrix} \tag{10.54}$$

In (10.54), the $a(.)$ are the coefficients of the characteristic polynomial

$$|zI - \varphi| = z^n + a_1 z^{n-1} + \ldots + a_{n-1}z + a_n \tag{10.55}$$

Now, define a new state vector $x = T\bar{x}$ to transform the original system into the controllable canonical form

$$\bar{x}(k+1) = \bar{\varphi}\bar{x}(k) + \bar{B}u(k) \tag{10.56}$$

One gets from (10.55) and (10.56) the following equivalence

$$\bar{\varphi} = T^{-1}\varphi T = \begin{bmatrix} 0 & 1 & 0 & \cdots & 0 \\ 0 & 0 & 1 & \cdots & 0 \\ : & : & : & \cdots & : \\ 0 & 0 & 0 & \cdots & 1 \\ -a_n & -a_{n-1} & -a_{n-2} & \cdots & -a_1 \end{bmatrix}; \tag{10.57}$$

$$\bar{B} = T^{-1}B = \begin{bmatrix} 0 \\ 0 \\ : \\ 1 \end{bmatrix}$$

In the design procedure, first one can find the gain for the new system, i.e. $\bar{K}$, such that the feedback $u(k) = -\bar{K}\bar{x}(k)$ would place the poles in the desired/specified locations. Since the eigenvalues remain unaffected under the similarity transformation, $u(k) = -\bar{K}T^{-1}x(k)$ will also place the poles of the original control system in the desired locations. If the poles are placed at $z_1, z_2, \ldots, z_n$, the desired characteristic equation is

$$(z - z_1)(z - z_2)\ldots(z - z_n) = 0$$
$$z^n + \alpha_1 z^{n-1} + \ldots + \alpha_{n-1}z + \alpha_n = 0 \tag{10.58}$$

Since, the new pair $(\bar{\varphi}, \bar{B})$ is in the controllable-companion form, one gets

$$\bar{\varphi} - \bar{B}\bar{K} = \begin{bmatrix} 0 & 1 & 0 & \cdots & 0 \\ 0 & 0 & 1 & \cdots & 0 \\ : & : & : & \cdots & : \\ 0 & 0 & 0 & \cdots & 1 \\ -(a_n - \bar{k}_1) & -(a_{n-1} - \bar{k}_2) & -(a_{n-2} - \bar{k}_3) & \cdots & -(a_1 - \bar{k}_n) \end{bmatrix} \tag{10.59}$$

The characteristic equation for both the original and the new canonical form is expressed as

$$|zI - \varphi| = |zI - \bar{\varphi}| = z^n + a_1 z^{n-1} + \ldots + a_{n-1}z + a_n = 0 \tag{10.60}$$

Then, the characteristic equation of the CLCS (from the CLTF) with $u(k) = -\bar{K}\bar{x}(k)$ can be given as

$$z^n + (a_1 + \bar{k}_n)z^{n-1} + (a_2 + \bar{k}_{n-1})z^{n-2} + \ldots + (a_n + \bar{k}_1) = 0 \tag{10.61}$$

We now compare (10.61) with (10.58) to obtain the following

$$\bar{k}_n = \alpha_1 - a_1, \ \bar{k}_{n-1} = \alpha_2 - a_2, \ \bar{k}_1 = \alpha_n - a_n \qquad (10.62)$$

Once we compute the matrix T, we can find the actual gain factor $K = \bar{K}T^{-1}$, the gain factor being $\bar{K} = [\bar{k}_1, \bar{k}_2, ..., \bar{k}_n]$.

10.5.2 Designing *K* by Ackermann's Formula

We consider the SISO dynamic system as in (10.51), the control law as in (10.52), the CLCS would be as in (10.53), and the desired characteristic equation is as in (10.58); the latter is obtained from

$$|zI - \varphi + BK| = |zI - \hat{\varphi}| = 0 \qquad (10.63)$$

We can use Cayley-Hamilton as

$$\hat{\varphi}^n + \alpha_1 \hat{\varphi}^{n-1} + ... + \alpha_{n-1} \hat{\varphi} + \alpha_n I = 0 \qquad (10.64)$$

Let us take the design case with $n = 3$, then we obtain the following for various powers in/of (10.64)

$$\hat{\varphi} = \varphi - BK$$

$$\hat{\varphi}^2 = (\varphi - BK)^2 = \varphi^2 - \varphi BK - BK\varphi - BKBK = \varphi^2 - \varphi BK - BK\hat{\varphi}$$

$$\hat{\varphi}^3 = (\varphi - BK)^3 = \varphi^3 - \varphi^2 BK - \varphi BK\hat{\varphi} - BK\hat{\varphi}^2 \qquad (10.65)$$

We can now substitute the corresponding terms from (10.65) into (10.64) to obtain:

$$\hat{\varphi}^3 + \alpha_1 \hat{\varphi}^2 + \alpha_2 \hat{\varphi} + \alpha_3 I = 0$$

$$\alpha_3 I + \alpha_2 (\varphi - BK) + \alpha_1 (\varphi^2 - \varphi BK - BK\hat{\varphi})$$

$$+ \varphi^3 - \varphi^2 BK - \varphi BK\hat{\varphi} - BK\hat{\varphi}^2 = 0 \qquad (10.66)$$

$$\alpha_3 I + \alpha_2 \varphi + \alpha_1 \varphi^2 + \varphi^3 - \alpha_2 BK - \alpha_1 \varphi BK - \alpha_1 BK\hat{\varphi}$$

$$- \varphi^2 BK - ABK\hat{\varphi} - BK\hat{\varphi}^2 = 0$$

Comparing (10.64) and the last equation (the third one) of (10.6), we obtain the following equation

$$\phi(\varphi) = B(\alpha_2 K + \alpha_1 K\hat{\varphi} + K\hat{\varphi}^2) + \varphi B(\alpha_1 K + K\hat{\varphi}) + \varphi^2 BK$$

$$= \begin{bmatrix} B & \varphi B & \varphi^2 B \end{bmatrix} \begin{bmatrix} \alpha_2 K + \alpha_1 K\hat{\varphi} + K\hat{\varphi}^2 \\ \alpha_1 K + K\hat{\varphi} \\ K \end{bmatrix} \qquad (10.67)$$

$$= U_c \begin{bmatrix} \alpha_2 K + \alpha_1 K\hat{\varphi} + K\hat{\varphi}^2 \\ \alpha_1 K + K\hat{\varphi} \\ K \end{bmatrix}$$

In (10.67), $\phi(.)$ is the closed loop characteristic polynomial, and because the controllable matrix is nonsingular, we have the following

$$U_c^{-1}\phi(\varphi) = \begin{bmatrix} \alpha_2 K + \alpha_1 K\hat{\varphi} + K\hat{\varphi}^2 \\ \alpha_1 K + K\hat{\varphi} \\ K \end{bmatrix} \qquad (10.68)$$

$$\begin{bmatrix} 0 & 0 & 1 \end{bmatrix} U_c^{-1}\phi(\varphi) = \begin{bmatrix} 0 & 0 & 1 \end{bmatrix} \begin{bmatrix} \alpha_2 K + \alpha_1 K\hat{\varphi} + K\hat{\varphi}^2 \\ \alpha_1 K + K\hat{\varphi} \\ K \end{bmatrix} = K$$

One can easily ascertain that (10.68) is valid for any order, n

$$K = \begin{bmatrix} 0 & 0...1 \end{bmatrix} U_c^{-1}\phi(\varphi); \ U_c = [B \ \varphi B \ \varphi^2 B \ ... \ \varphi^{n-1}B] \qquad (10.69)$$

In the literature, (10.69) is known as the Ackermann's formula.

10.5.3 Set Point Tracking

We have seen in the previous section that one can do the state feedback design for the controller and place the poles so that the closed loop system is stable; however, this does not guarantee good performance in tracking. Let us again consider the system as in (10.51), Figure 10.2, and the control law as

$$u(k) = -Kx(k) + K_r r(k) \qquad (10.70)$$

The closed loop dynamics, after combining (10.51) and (10.70), are

$$x(k+1) = (\varphi - BK)x(k) + BK_r r(k) \qquad (10.71)$$

At the steady state, we have $x = x_{ss}$, $y = Hx_{ss} = r$, and $u = u_{ss}$; and the states and the output do not change with time in this SS condition, one can write

$$x_{ss} = \varphi x_{ss} + Bu_{ss} \qquad (10.72)$$

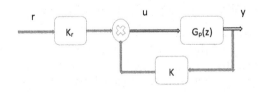

FIGURE 10.2
State feedback with feedforward gain for set point tracking.

Then, we have the following new variables

$$\tilde{u}(k) = u(k) - u_{ss}, \quad \tilde{x}(k) = x(k) - x_{ss}, \quad \tilde{y}(k) = y(k) - r \quad (10.73)$$

In the DTS form, we get the following

$$\tilde{x}(k+1) = \varphi x(k) + Bu(k) - \varphi x_{ss} - Bu_{ss}$$

$$= \varphi \tilde{x}(k) + B\tilde{u}(k) \quad (10.74)$$

$$\tilde{y}(k) = H\tilde{x}(k)$$

For a designed stable control $\tilde{u}(k) = -K\tilde{x}(k)$ in the new/transformed domain, the state variables/vector in the shifted domain would be driven to zero:

$$\tilde{x}(k) \to 0 \Rightarrow x(k) \to x_{ss} \Rightarrow y(k) = r \quad (10.75)$$

So, the set point tracking is converted into a simple regulator problem:

$$u(k) - u_{ss} = -Kx(k) + Kx_{ss}$$

$$\Rightarrow u(k) = -Kx(k) + u_{ss} + Kx_{ss} \quad (10.76)$$

Also, we have the following expressions

$$x_{ss} = \varphi x_{ss} + Bu_{ss}$$

$$0 = (\varphi - I)x_{ss} + Bu_{ss}$$

$$= (\varphi - I)x_{ss} + BKx_{ss} - BKx_{ss} + Bu_{ss} \quad (10.77)$$

$$= (\varphi - BK - I)x_{ss} + B(u_{ss} + Kx_{ss})$$

$$\Rightarrow x_{ss} = -(\varphi - BK - I)^{-1}B(u_{ss} + Kx_{ss})$$

$$Hx_{ss} = -H(\varphi - BK - I)^{-1}B(u_{ss} + Kx_{ss}) = r; u_{ss} + Kx_{ss} = K_r r \quad (10.78)$$

Comparing the two equations in (10.78), we obtain the following

$$(K_r)^{-1} = -H(\varphi - BK - I)^{-1}B \quad (10.79)$$

Then, the control input is obtained as

$$u(k) = -Kx(k) + K_r r \quad (10.80)$$

10.5.4 State Feedback with Integral Control

Actual computation of feed forward gain requires exact knowledge of the system parameters, and any change in these parameters will affect the SSE; hence, the solution to this problem is to feedback the states as well as the integral of the output error, which will eventually make the actual output follow the desired one. One can

do this by introducing the integrator, that is to augment the integral state v with the plant state vector x, since v integrates the difference between the output $y(k)$ and reference r. By using a difference equation, we have

$$v(k) = v(k-1) + y(k) - r \quad (10.81)$$

$$v(k+1) = v(k) + y(k+1) - r$$

$$= v(k) + H(\varphi x(k) + Bu(k)) - r \quad (10.82)$$

Combining the system's equations with (10.82), we get

$$\begin{bmatrix} x(k+1) \\ v(k+1) \end{bmatrix} = \begin{bmatrix} \varphi & 0 \\ H\varphi & 1 \end{bmatrix} \begin{bmatrix} x(k) \\ v(k) \end{bmatrix} + \begin{bmatrix} B \\ HB \end{bmatrix} u(k) + \begin{bmatrix} 0 \\ -1 \end{bmatrix} r \quad (10.83)$$

Since the reference input r is constant and if the system is stable, then $x(k+1) = x(k)$ and $v(k+1) = v(k)$ in the steady-state condition, so we have the following from (10.83)

$$\begin{bmatrix} 0 \\ -1 \end{bmatrix} r = \begin{bmatrix} x_{ss} \\ v_{ss} \end{bmatrix} - \begin{bmatrix} \varphi & 0 \\ H\varphi & 1 \end{bmatrix} \begin{bmatrix} x_{ss} \\ v_{ss} \end{bmatrix} - \begin{bmatrix} B \\ HB \end{bmatrix} u_{ss} \quad (10.84)$$

With the new augmented state (-error) vectors: $\tilde{x}_a = [x - x_{ss} \quad v - v_{ss}]^T$; $\tilde{u} = u - u_{ss}$, we have

$$\tilde{x}_a(k+1) = \bar{\varphi}\tilde{x}_a(k) + \bar{B}\tilde{u}(k); \bar{\varphi} = \begin{bmatrix} \varphi & 0 \\ H\varphi & 1 \end{bmatrix}, \bar{B} = \begin{bmatrix} B \\ H\varphi \end{bmatrix} \quad (10.85)$$

The original control problem is now transformed to a standard regulator problem; and one needs to design

$$\tilde{u}(k) = -K\tilde{x}_a(k); K = [K_p K_i]; K_i \text{ is the integral gain} \quad (10.86)$$

Then, we have the steady-state terms as follows:

$$u - u_{ss} = -[K_p \ K_i] \begin{bmatrix} x - x_{ss} \\ v - v_{ss} \end{bmatrix} = -K_p(x - x_{ss}) - K_i(v - v_{ss}) \quad (10.87)$$

Balancing the steady-state terms, we get

$$u(k) = -K_p x(k) - K_i v(k) \quad (10.88)$$

And, at the steady state, we obtain

$$\tilde{x}_a(k+1) - \tilde{x}_a(k) = 0$$

$$v(k+1) - v(k) = 0 \quad (10.89)$$

One can also have $y(k) - r = 0$ at steady state, this $y(k)$ follows r.

10.6 State Observers

Realization of a state feedback controller is possible only when all the state variables/vector are (directly) measurable. However, due to unavailability of these states directly (from the observations), in several realistic situations, one cannot meet the requirement of precise computation of the control signal $u(.)$, $\{u(k) = -K(x(k)\}$, even if the feedback gain K is computed accurately. In practice, only a subset of state variables, or their combinations, may be available from the observations. Sometimes only output y is available; hence, one needs an estimator or state observer that would estimate all the state variables. For the purpose of obtaining the state estimation, one uses: (i) full order state observer, if the state observer estimates all the state variables, (ii) reduced order state observer, if fewer than n states of the system are available, or (iii) minimum order state observer, if the order of the observer is minimum possible. The state observers, linear filtering, and nonlinear filtering were described in a great detail recently in [3].

10.6.1 Full Order Observers

Let the dynamic system be described by (10.51). The main assumption is that the pair (ϕ, H) is observable. We need to have the estimate of the states so that we can use these in the state feedback controller, $u(.) = -Kx(.)$. In the observer design, we need to use: (i) the input/output of the system and (ii) the observer structure that itself is the same as, or almost similar to, the original plant dynamics. So, this will give us a mechanism to determine/estimate/observe the states of the dynamic system for which we need the state feedback controller.

10.6.1.1 Open Loop Estimator

The dynamics of the open loop estimator are described as

$$\hat{x}(k+1) = \varphi \hat{x}(k) + Bu(k)$$
$$\hat{y}(k) = H\hat{x}(k) \tag{10.90}$$

From (10.51) and (10.90), we have the state estimation error as $\tilde{x} = \hat{x}(k) - x(k)$, and the error dynamics for the observer are defined as

$$\tilde{x}(k+1) = \hat{x}(k+1) - x(k+1)$$

$$= \varphi \hat{x}(k) + Bu(k) - \varphi x(k) - Bu(k)$$
$$\tag{10.91}$$

$$= \varphi \tilde{x}(k)$$

$$\tilde{x}(0) = \hat{x}(0) - x(0)$$

If the eigenvalues of the transition matrix are inside the unit circle, then the state error in (10.91) will tend to zero; however, one does not have any control over this convergence rate. If the eigenvalues of the transition matrix are outside the unit circle, then the state error would diverge from zero; hence, the open loop estimator/observer is not very useful.

10.6.1.2 Luenberger State Observer

A practical observer (estimator) is the one that has some feedback in (10.90) that can control the dynamics of the observer itself, and the most suitable is the output of the same dynamic system for which we are trying to design a state feedback controller. Hence, the structure of the observer (Figure 10.3a) is:

$$\hat{x}(k+1) = \varphi \hat{x}(k) + Bu(k) + L(y(k) - \hat{y}(k))$$
$$\hat{y}(k) = H\hat{x}(k) \tag{10.92}$$

By subtracting (10.51) the original plant's dynamics from (10.92), one obtains the observer's state error dynamics

$$\tilde{x}(k) = \hat{x}(k) - x(k); \quad \tilde{x}(0) = \hat{x}(0) - x(0)$$

$$\hat{x}(k+1) - x(k+1) = \varphi(\hat{x}(k) - x(k)) + LH(x(k) - \hat{x}(k)) \tag{10.93}$$

$$\tilde{x}(k+1) = (\varphi - LH)\tilde{x}(k)$$

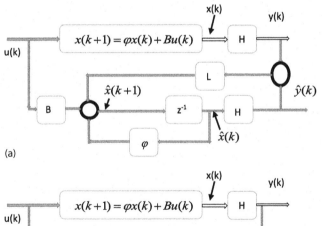

(a)

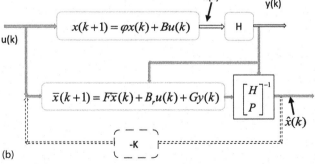

(b)

FIGURE 10.3
(a) The Luenberger observer (Full order) (b) The reduced order observer and controller (with state feedback, SFB, -K).

One can easily see from (10.93) that if the eigenvalues of $(\varphi - LH)$ are made to lie inside of the unit circle of the z-plane, then the observer state error would tend to zero asymptotically; this requires a proper choice of the gain L, because it is the only factor unknown, and for $y(.)$, one uses the measured output (from the appropriate sensors) of the dynamic system. Hence, the main design aspect is to place the poles of $|\varphi - LH|$ such that the convergence of the observer is assured, and a necessary and sufficient condition for arbitrary pole placement is that the pair (φ, C) should be observable, and from the theorem of duality, one knows that the pair (φ^T, H^T) should be controllable. One should note that the eigenvalues of $(\varphi^T - H^T L^T)$ are same as that of $(\varphi - HC)$; hence, the determination of the observer gain problem is same as a (fictitious/hypothetical) pole placement problem for the new system $\bar{x}(k+1) = \varphi^T \bar{x}(k) + H^T \bar{u}(k)$, using a control law $\bar{u}(k) = -L^T \bar{x}(k)$.

10.6.1.3 Controller with Observer

We have the observer dynamics as

$$\hat{x}(k+1) = \varphi \hat{x}(k) + Bu(k) + LH(x(k) - \hat{x}(k)) \quad (10.94)$$

Now, we combine (10.94) with the system dynamics (10.51) to obtain

$$\begin{bmatrix} x(k+1) \\ \hat{x}(k+1) \end{bmatrix} = \begin{bmatrix} \varphi & 0 \\ LH & \varphi - LH \end{bmatrix} \begin{bmatrix} x(k) \\ \hat{x}(k) \end{bmatrix} + \begin{bmatrix} B \\ B \end{bmatrix} u(k)$$

$$y(k) = [H \ 0] \begin{bmatrix} x(k) \\ \hat{x}(k) \end{bmatrix} \quad (10.95)$$

Since the actual states are not available, the control input is taken to depend as

$$u(k) = -K\hat{x}(k) \quad (10.96)$$

Substituting the control law (10.96) in (10.95), we obtain

$$\begin{bmatrix} x(k+1) \\ x(k+1) \end{bmatrix} = \begin{bmatrix} \varphi & -BK \\ LH & \varphi - LH - BK \end{bmatrix} \begin{bmatrix} x(k) \\ \hat{x}(k) \end{bmatrix}$$

$$y(k) = [H \ 0] \begin{bmatrix} x(k) \\ \hat{x}(k) \end{bmatrix} \quad (10.97)$$

The observer error dynamics are given as the third equation of (10.93). We now augment the system dynamics (10.51), with the state error formula, and the observer error dynamics (10.93):

$$\begin{bmatrix} x(k+1) \\ \tilde{x}(k+1) \end{bmatrix} = \begin{bmatrix} \varphi - BK & -BK \\ 0 & \varphi - LH \end{bmatrix} \begin{bmatrix} x(k) \\ \tilde{x}(k) \end{bmatrix}$$

$$y(k) = [H \ 0] \begin{bmatrix} x(k) \\ \tilde{x}(k) \end{bmatrix} \quad (10.98)$$

From (10.98), we can infer that the design of the control law to obtain K can be accomplished independently of the design of the observer to obtain L, and this is known as the separation principle, which is apparent from the form of the combined system matrix of (10.98), since here, the combined system eigenvalues are the eigenvalues of the systems of $\varphi - BK$, and $\varphi - LH$, the controller and the observer closed loop systems, respectively.

10.6.2 Reduced Order Observers

An observer that estimates a fewer number of the states than n states, is called the reduced order observer. We consider the same system as in (10.51), and that the output vector is of dimension p, with $p < n$, we can use p outputs to determine the p states of the state vector and design an observer/estimator of order $n-p$ to estimate the rest of the states. If the rank $(H) = p$, then $y(k) = Hx(k)$ can be used to solve for p of the $x(.)$ in terms of $y(.)$ and the remaining $n-p$ state variables of $x(.)$ need be actually estimated. The dynamics of the new observer are (see Figure 10.3b):

$$\bar{x}(k+1) = F\bar{x}(k) + B_r u(k) + Gy(k) \quad (10.99)$$

Use the transformation P such that

$$\bar{x} = Px \quad (10.100)$$

Using (10.100), we can transform the original plant (10.51) to the following

$$Px(k+1) = P\varphi x(k) + PBu(k) \quad (10.101)$$

Subtracting (10.101) from (10.99) and utilizing (10.100) also in (10.99), we get the following

$$(P\varphi - FP - GH)x(k) + (PB - B_r)u(k) = 0 \quad (10.102)$$

The expression (10.102) will hold true for all k and arbitrary input $u(k)$ for the conditions:

$$(P\varphi - FP - GH) = 0 \ \Rightarrow P\varphi - FP = GH; \text{ and } PB = B_r \quad (10.103)$$

If $\bar{x} \neq Px$, yet if (10.103) holds true, then we can write (10.99) and (10.101), and using (10.103) the following

$$\bar{x}(k+1) - Px(k+1) = F\bar{x}(k) - P\varphi x(k) + GHx(k)$$

$$= F\bar{x}(k) - FPx(k) \quad (10.104)$$

$$= F(\bar{x}(k) - Px(k))$$

If the eigenvalues of F are inside of the unit circle, then we can write

$$\bar{x}(k) \to Px(k) \ as, \ k \to \infty \qquad (10.105)$$

If P is an identity matrix, and $G = L$, then we have

$$\varphi - LH = F, \ B_r = B \qquad (10.106)$$

From the foregoing, we can see that the resulting observer is the Luenberger full order observer, and (10.106) can be solved for L such that F has its eigenvalues at the prescribed locations. Since, we have the following

$$y(k) = Hx(k), \ and \ \bar{x}(k) = Px(k), \ as \ k \to \infty$$

$$\begin{bmatrix} y \\ \bar{x} \end{bmatrix} = \begin{bmatrix} H \\ P \end{bmatrix} x \qquad (10.107)$$

The estimated state vector can be obtained from (10.107) as

$$\hat{x} = \begin{bmatrix} H \\ P \end{bmatrix}^{-1} \begin{bmatrix} y \\ \bar{x} \end{bmatrix} \qquad (10.108)$$

In (10.108), the rank of $\begin{bmatrix} H \\ P \end{bmatrix}$ should be equal to n, for it to be invertible; and $P\varphi - FP = GH$ can be uniquely solved if no eigenvalues of F is an eigenvalue of φ.

10.6.3 Controller with Reduced Order Observer

Let us consider the observer dynamics (10.99) again and the control law as (10.96), see the Figure 10.3b (with the addition of the dotted lines, and then the independent input is assumed to be absent). The state estimates are obtained from (10.108). Now, assume that the matrix in (10.108) is given as

$$\begin{bmatrix} H \\ P \end{bmatrix}^{-1} = [Q_1 \ Q_2] \qquad (10.109)$$

Then, we have the control law as

$$u(k) = -K[Q_1 \ Q_2] \begin{bmatrix} y(k) \\ \bar{x}(k) \end{bmatrix} = -KQ_1 y(k) - KQ_2 \bar{x}(k) \qquad (10.110)$$

Now, combine the observer (10.99), with the system dynamics (also use the second equation of [10.103]), to obtain the composite dynamics as

$$\begin{bmatrix} x(k+1) \\ \bar{x}(k+1) \end{bmatrix} = \begin{bmatrix} \varphi & 0 \\ GH & F \end{bmatrix} \begin{bmatrix} x(k) \\ \bar{x}(k) \end{bmatrix} + \begin{bmatrix} P \\ PB \end{bmatrix} (-KQ_1 y(k) - KQ_2 \bar{x}(k))$$

$$= \begin{bmatrix} \varphi - BKQ_1 H & -BKQ_2 \\ GH - PBKQ_1 H & D - PBKQ_2 \end{bmatrix} \begin{bmatrix} x(k) \\ \bar{x}(k) \end{bmatrix} \qquad (10.111)$$

$$y(k) = [H \ 0] \begin{bmatrix} x(k) \\ \bar{x}(k) \end{bmatrix}$$

$$\begin{bmatrix} x(k) \\ \bar{x}(k) - Px(k) \end{bmatrix} = \begin{bmatrix} I_n & 0 \\ -P & I_{n-p} \end{bmatrix} \begin{bmatrix} x(k) \\ \bar{x}(k) \end{bmatrix}; \ \begin{bmatrix} I_n & 0 \\ -P & I_{n-p} \end{bmatrix}^{-1} = \begin{bmatrix} I_n & 0 \\ P & I_{n-p} \end{bmatrix} \qquad (10.112)$$

$$x(k+1) = (\varphi - BKQ_1 H)x(k) - BKQ_2 \bar{x}(k))$$

$$= (\varphi - BKQ_1 H)x(k) - BKQ_2(\bar{x}(k) - Px(k)) - BKQ_2 Px(k)$$

$$= (\varphi - BKQ_1 H - BKQ_2 P)x(k) - BKQ_2(\bar{x}(k) - Px(k)) \qquad (10.113)$$

Also, we have the following using the second equation of (10.103)

$$\bar{x}(k+1) - Px(k+1)$$

$$= F\bar{x}(k) + B_r u(k) + GHx(k) - P\varphi x(k) - PBu(k) \qquad (10.114)$$

$$= (FP - P\varphi + GH)x(k) + F(\bar{x}(k) - Px(k))$$

Then, we get

$$\begin{bmatrix} x(k+1) \\ \bar{x}(k+1) - Px(k+1) \end{bmatrix}$$

$$= \begin{bmatrix} \varphi - BKQ_1 H - BKQ_2 P & -BKQ_2 \\ FP - P\varphi + GH & F \end{bmatrix} \begin{bmatrix} x(k) \\ \bar{x}(k) - Px(k) \end{bmatrix} \qquad (10.115)$$

Using (10.103) and $Q_2 P = I_n - Q_1 H$, we obtain

$$\begin{bmatrix} x(k+1) \\ \bar{x}(k+1) - Px(k+1) \end{bmatrix} = \begin{bmatrix} \varphi - BK & -BKQ_2 \\ 0 & F \end{bmatrix} \begin{bmatrix} x(k) \\ \bar{x}(k) - Px(k) \end{bmatrix} \qquad (10.116)$$

Now, we can say from (10.116) that if the matrices $(\varphi - BK)$ and F have their eigenvalues inside the unit circle, then $x(k) \to 0$ and $\bar{x}(k) \to Px(k)$ (meaning these states converge); and the eigenvalues of the composite matrix in (10.116) are the same as the eigenvalues of the block diagonal matrices individually. Hence, the controller and the observer can be designed separately, as per the separation principle, and this principle is also valid for the reduced order observer. However, certain points should be noted: (i) the principle assumes that the state observer uses an exact dynamics of the plant, but the model dynamics are often not very accurate; (ii) the data known about the real process/states might be complicated to be simply used in the observer; (iii) separation principle is not always sufficient, since the robustness of the observer must also be assured; (iv) the controller gain K should such that the designed $u(.)$ is limited in its magnitude because of hardware limitation; because the large control input might excite the system's nonlinear behavior; and (v) the observer dynamics should be much faster than the controller dynamics.

10.6.4 Deadbeat Control by State Feedback and Deadbeat Observer

Let us consider the same system as in (10.51) and the control law as $u(k) = -Kx(k)$, then the closed loop system has the dynamics as

$$x(k+1) = (\varphi - BK)x(k) \tag{10.117}$$

The desired characteristic equation is given as in (10.58), (the second equation), then we select K, such that the coefficients of $|zI - (\varphi - BK)|$ match with those of the desired characteristic equation. If we consider the desired characteristic equation as

$$z^n = 0; \quad \text{with } \alpha_1 = \alpha_2 = ... = \alpha_n = 0. \tag{10.118}$$

By the Cayley-Hamilton theorem, we have

$$(\varphi - BK)^n = 0 \tag{10.119}$$

Thus, we get

$$x(k) = (\varphi - BK)^n x(0) = 0, \text{ for } k \geq n \tag{10.120}$$

Any initial state $x(0)$ is driven to the equilibrium state $x = 0$ at most n steps. The control law that obtains all the poles to origin can be viewed as a deadbeat control; also, when all observer poles are at zero, it is a deadbeat observer (DBO). In the DBR response, settling time depends on the sampling period, for a very small T, settling time is very small, but the control signal becomes very high. The designer should make a trade-off between the two.

10.6.5 Incomplete State Feedback

In some controller designs, we can use only partial state feedback or the output feedback for the reasons of economy. We consider the same dynamic system as in (10.51), and the control law is given as $u(k) = -Gx(k)$.

We assume that the state $x_i(k)$ are not available for the feedback, where $i \rightarrow [1,n]$, then corresponding columns of G are zero; $G = WG^*$; the columns of G^* that correspond to the zero columns of G must be equal to zero. Our aim is to choose W such that (φ, BW) is controllable with partial state feedback; G^* for single input case is related to the desired closed loop poles, the system parameters, and W. One can have

$$G^* = -[\Delta_{01} \Delta_{02} ... \Delta_{0n}]K^{-1} \tag{10.121}$$

In (10.121), $\Delta_{0i} = \Delta_0(\lambda_i)$, where $\Delta_0(.)$ represents the open loop characteristic equation and λ_i, the i-th desired close loop eigenvalue. Also, we have $K = [k_1 \ k_2 \ ...k_n]$, and $k_i = k(\lambda_i), k(z) = adj(zI - \varphi)BW$. Some constraints on the desired close loop eigenvalues, when one or more columns of G^* are forced to be zero, are required.

10.6.6 Output Feedback Design

The output feedback control law uses the system outputs since these are always measurable (as the sensors' outputs) and available for feedback. We consider the same system as earlier (10.51), and the control input as $u(k) = -Gy(k)$; then, choose G such that the eigenvalues of the CLDCS are at the desired locations. However, since $p \leq m \leq n$, all the eigenvalues cannot be arbitrarily specified, and this depends on the ranks of H and B. For the SISO case, we have the following by combining the system's dynamics with the control law

$$x(k+1) = (\varphi - BGH)x(k) \tag{10.122}$$

In (10.122), the GH can be obtained by using the pole placement technique if the pair (ϕ, B) is controllable, and since H is not a square matrix, G cannot be obtained from $GH = K$. There are p gain elements in G, but only r of them are free as independent parameters, where r is the rank of H and $r \leq p$; if

$$H = \begin{bmatrix} 1 & 0 & 0 \\ 0 & 1 & 2 \\ 0 & 0 & 0 \end{bmatrix}; \text{ and } G = [g_1 \ g_2 \ g_3]; GH = [g_1 \ g_2 \ 2g_2] \tag{10.123}$$

From (10.123), one can see that only two elements of GH are independent; hence, only two of the total eigenvalues can be placed arbitrarily. For an SISO system, if the rank of H is equal to the order of the system, n, then the output feedback is equivalent to the complete state feedback. Assume the order $n = 3$, then the gain K is designed as $[k_1 \ k_2 \ k_3]$, and

$$H = \begin{bmatrix} 1 & 0 & 1 \\ 0 & 1 & 1 \\ 0 & 0 & 1 \end{bmatrix} \qquad (10.124)$$

Also, rank of H is 3, and we have G as in (10.123); hence

$$GH = [g_1 \ g_2 \ g_1 + g_2 + g_3] \qquad (10.125)$$

The $g(.)$ can be solved as
$$g_1 = k_1, \ g_2 = k_2, \ g_3 = k_3 - g_1 - g_2 \qquad (10.126)$$

For a multi-input system, let $B^* = BW$ and let $G = WG^*$, where $G^* = [g_1^* \ g_2^* \ g_3^* ... g_p^*]$, then we have

$$BGH = BWG^*H = B^*G^*H \qquad (10.127)$$

The characteristic equation of the CLCS is

$$|zI - (\varphi - BGH)| = |zI - (\varphi - B^*G^*H)| \qquad (10.128)$$

The G^*H is determined by using the pole placement technique, the solution for the gain depends on the ranks of H and B: if the rank of H is greater than or equal to the rank of B, the elements of W can be arbitrarily chosen if the pair (φ, B) is controllable; and if (rank of B) > (rank of H), this arbitrary assignment of all the elements of W is not feasible.

10.7 Optimal Control

We studied and discussed, in the foregoing sections, several designed controllers based on certain specifications; however, the aspect of optimality of the controller was not formally brought into the design procedure; of course, the optimal control was introduced in Chapter 6. In optimal control theory, one uses a performance criterion (or index, PI, Appendix IIB) that is either maximized or minimized, and this index is a measure of the effectiveness of the controller. In this context, especially for the techniques based on the calculus of variation (COV, Chapter 5), the Euler-Lagrange (EL) equation is very popular in the context of minimization or maximization of a cost (loss) functional; the latter is a mapping or transformation that depends on one or more functions and can be numerically evaluated in most cases of the optimization problems.

10.7.1 Discrete Euler-Lagrange Equation

In most cases of the optimal controller problems for SDCS/DCS/DTCS (as in the case of the continuous

time control systems, CTCS), the goal is to maximize or minimize (i.e. to optimize/extremize) a certain PI or functional

$$J = \sum_{k=0}^{N-1} F(x(k+1), x(k), u(k), k) \qquad (10.129)$$

In (10.129), F is scalar functional that is differentiable with respect to its arguments. The optimization of the cost function (10.129) is subject to the constraint

$$x(k+1) = f(x(k), u(k), k) \qquad (10.130)$$

In fact, (10.130) is usually a state space equation, or math model, of a DCS. According to the approaches studied in Chapters 5 and 6, the problem of minimizing the cost functional, with its equality constraints, is solved by adjoining this constraint (taking it concurrently) to the function to be optimized. Define $\lambda(k+1)$ as the Lagrange multiplier, and since the constraint (10.130) will be adjoined with the functional F in the cost (criterion/ performance index) (10.129), via this multiplier, it is also known as the co-factor, or co-state, and the resulting equation of this multiplier will be called an adjoin (in fact, adjoint) or co-state equation. The augmented/ adjoined cost is given as

$$J_a = \sum_{k=0}^{N-1} \begin{array}{l} F(x(k+1), x(k), u(k), k) \\ + \lambda^T(k+1)[x(k+1) - f(x(k), u(k), k)] \end{array} \qquad (10.131)$$

From the theory of COV in Chapter 5, we know that the minimization of $J_a(u(k))$ with constraint is just the same as the minimization of J_a without any constraint; however, the artifice of the LM allows the constraint to be included in the original cost function without affecting the cost function at all, and still respecting the dynamics of the system. Now, the variables $x^*(k)$, $x^*(k+1)$, $u^*(k)$, and $\lambda^*(k+1)$ represent the optimal trajectories, and one has the normal state/co-state variables and the input as

$$x(k) = x^*(k) + \varepsilon\eta(k)$$

$$x(k+1) = x^*(k+1) + \varepsilon\eta(k+1)$$

$$u(k) = u^*(k) + \delta\mu(k) \qquad (10.132)$$

$$\lambda(k+1) = \lambda^*(k+1) + \gamma v(k+1)$$

In (10.132), $\eta(k), \mu(k), v(k)$, as arbitrary variables, and $\varepsilon, \delta, \gamma$, as small positive constants, depict the deviations of the optimal trajectories from the normal trajectories. Now, using (10.132) into the cost function (10.131), we obtain

$$J_a = \sum_{k=0}^{N-1} F(x^*(k+1) + \varepsilon\eta(k+1), x^*(k) + \varepsilon\eta(k), u^*(k) + \delta\mu(k), k)$$

$$+ (\lambda^*(k+1) + \gamma\nu(k+1))^T [(x^*(k+1) + \varepsilon\eta(k+1))$$

$$- f(x^*(k) + \varepsilon\eta(k), u^*(k) + \delta\mu(k), k)]$$

$$\tag{10.133}$$

Equivalently, we have the cost functional in short notation as

$$J_a = \sum_{k=0}^{N-1} F_a(x(k+1), x(k), u(k), \lambda(k+1), k) \tag{10.134}$$

Using the Taylor series around the optimal trajectories (*s), we obtain from (10.133) and (10.134):

$$F_a(x(k+1), x(k), u(k), \lambda(k+1), k)$$

$$= F_a(x^*(k+1), x^*(k), u^*(k), \lambda^*(k+1), k)$$

$$+ \varepsilon \left[\frac{\partial F_a^*(k)}{\partial x^*(k)}\right]^T \eta(k) + \varepsilon \left[\frac{\partial F_a^*(k)}{\partial x^*(k+1)}\right]^T \eta(k+1) + \delta \left[\frac{\partial F_a^*(k)}{\partial u^*(k)}\right]^T \mu(k)$$

$$+ \gamma \left[\frac{\partial F_a^*(k)}{\partial \lambda^*(k+1)}\right]^T \nu(k+1) + \text{higher order terms}$$

$$\tag{10.135}$$

In (10.135), we have

$$F_a^*(k) = F_a(x^*(k+1), x^*(k), u^*(k), \lambda^*(k+1), k) \tag{10.136}$$

The necessary conditions for the optimization are

$$\frac{\partial J_a}{\partial \varepsilon}\Big|_{\varepsilon=\delta=\gamma=0} = 0; \quad \frac{\partial J_a}{\partial \delta}\Big|_{\varepsilon=\delta=\gamma=0} = 0; \quad \frac{\partial J_a}{\partial \gamma}\Big|_{\varepsilon=\delta=\gamma=0} = 0 \tag{10.137}$$

Now, we substitute the expansion of (10.135) into the cost function (10.134), apply the conditions of (10.137), and note that (10.136) will be zero at the optimal condition, then we obtain the following expressions

$$\sum_{k=0}^{N-1} \left\{ \left[\frac{\partial F_a^*(k)}{\partial x^*(k)}\right]^T \eta(k) + \left[\frac{\partial F_a^*(k)}{\partial x^*(k+1)}\right]^T \eta(k+1) \right\} = 0$$

$$\sum_{k=0}^{N-1} \left\{ \left[\frac{\partial F_a^*(k)}{\partial u^*(k)}\right]^T \mu(k) \right\} = 0 \tag{10.138}$$

$$\sum_{k=0}^{N-1} \left\{ \left[\frac{\partial F_a^*(k)}{\partial \lambda^*(k+1)}\right]^T \nu(k+1) \right\} = 0$$

The first equation of (10.138) can be written as

$$\sum_{k=0}^{N-1} \left\{ \left[\frac{\partial F_a^*(k)}{\partial x^*(k)}\right]^T \eta(k) \right\} = -\sum_{k=1}^{N} \left\{ \left[\frac{\partial F_a^*(k-1)}{\partial x^*(k)}\right]^T \eta(k) \right\}$$

$$= -\sum_{k=0}^{N-1} \left\{ \left[\frac{\partial F_a^*(k-1)}{\partial x^*(k)}\right]^T \eta(k) \right\} + \left\{ \left[\frac{\partial F_a^*(k-1)}{\partial x^*(k)}\right]^T \eta(k) \right\}\Big|_{k=0}$$

$$\tag{10.139}$$

$$- \left\{ \left[\frac{\partial F_a^*(k-1)}{\partial x^*(k)}\right]^T \eta(k) \right\}\Big|_{k=N}$$

In (10.139), we have the following from (10.136)

$$F_a^*(k-1) = F_a(x^*(k-1), x^*(k), u^*(k-1), \lambda^*(k), k-1) \tag{10.140}$$

Now, let us rearrange the terms of (10.139) as follows

$$\sum_{k=0}^{N-1} \left\{ \left[\frac{\partial F_a^*(k)}{\partial x^*(k)}\right]^T \eta(k) \right\} + \sum_{k=0}^{N-1} \left\{ \left[\frac{\partial F_a^*(k-1)}{\partial x^*(k)}\right]^T \eta(k) \right\}$$

$$- \left\{ \left[\frac{\partial F_a^*(k-1)}{\partial x^*(k)}\right]^T \eta(k) \right\}\Big|_{k=0} + \left\{ \left[\frac{\partial F_a^*(k-1)}{\partial x^*(k)}\right]^T \eta(k) \right\}\Big|_{k=N} = 0;$$

$$\sum_{k=0}^{N-1} \left\{ \left[\frac{\partial F_a^*(k)}{\partial x^*(k)}\right]^T + \left[\frac{\partial F_a^*(k-1)}{\partial x^*(k)}\right]^T \right\} \eta(k)$$

$$+ \left\{ \left[\frac{\partial F_a^*(k-1)}{\partial x^*(k)}\right]^T \eta(k) \right\}\Big|_{k=0}^{k=N} = 0$$

$$\tag{10.141}$$

According to the fundamental lemma of COV, the second equation in (10.141) is satisfied for any arbitrary $\eta(k)$, only if its two components are individually zero:

$$\left[\frac{\partial F_a^*(k)}{\partial x^*(k)}\right]^T + \left[\frac{\partial F_a^*(k-1)}{\partial x^*(k)}\right]^T = 0$$

$$\tag{10.142}$$

$$\left\{ \left[\frac{\partial F_a^*(k-1)}{\partial x^*(k)}\right]^T \eta(k) \right\}\Big|_{k=0}^{k=N} = 0$$

The first equation of (10.142) is known as the discrete Euler-Lagrange (DEL) equation and the second is known as transversality condition, that is the boundary condition required to solve the first one. Thus, the DEL equation is a necessary condition that must be satisfied for the composite cost function to be an extremal, as is the case with the continuous time case of the calculus of variation. From the second and third equations of (10.138), we obtain other conditions for arbitrary $\mu(k), \nu(k+1)$

$$\left[\frac{\partial F_a^*(k)}{\partial u_j^*(k)}\right]^T = 0, \; j = 1, 2, ..., m$$

$$\left[\frac{\partial F_a^*(k)}{\partial \lambda_i^*(k+1)}\right]^T = 0, \; i = 1, 2, ..., n \quad (10.143)$$

Interestingly, the second equation of (10.143) leads to

$$x^*(k+1) = f(x^*(k), u^*(k), k) \quad (10.144)$$

The equation (10.144) means that the original state equation (10.130) of the dynamic system should satisfy the optimal trajectory. The firstequation of (10.143) gives the optimal control $u^*(k)$ in terms of $\lambda^*(k+1)$. In many situations, the initial state $x(0)$ is given, then $\eta(0) = 0$, since $x(0)$ is fixed, and the transversality condition reduces to

$$\left\{\left[\frac{\partial F_a^*(k-1)}{\partial x^*(k)}\right]^T \eta(k)\right\}|_{k=N} = 0 \quad (10.145)$$

The optimal control problems are then classified based on the transversality conditions: (a) if $x(N)$ is given and fixed, the problem is known as fixed-endpoint design, for this ($x(N)$ =fixed, $\eta(N) = 0$) problems, no transversality condition is required to solve; and (b) if $x(N)$ is free, the problem is called a free endpoint design, for this the transversality condition is given as

$$\left[\frac{\partial F_a^*(k-1)}{\partial x^*(k)}\right]^T |_{k=N} = 0 \quad (10.146)$$

10.7.2 Linear Quadratic Regulator

Let us consider the LTI dynamic system as in (10.51), with the initial condition as $x(k_0) = x_0$. Our aim is to obtain a stabilizing linear state feedback controller $u(k) = -Kx(k)$ that will minimize the quadratic PI:

$$J = \sum_{k=0}^{\infty} \{x^T(k)Rx(k) + u^T(k)Qu(k)\} \quad (10.147)$$

In (10.147), $R = R^T \geq 0; Q = Q^T \geq 0$. The controller is denoted as u^*. We assume that such a controller exists such that the following CLCS is asymptotically stable

$$x(k+1) = (\varphi - BK)x(k) \quad (10.148)$$

In this case, there exists a Lyapunov function

$$V(x(k)) = x^T(k)Px(k) \quad (10.149)$$

for the CLCS for which the following forward difference is negative definite, since the Lyapunov function (10.149) is positive definite (also, because of the conditions on the weighting matrices R, and Q):

$$\Delta V(x(k)) = V(x(k+1)) - V(x(k)) \quad (10.150)$$

If the controller u^* is optimal, then the following holds true

$$\min_u\{\Delta V(x(k)) + x^T(k)Rx(k) + u^T(k)Qu(k)\} = 0 \quad (10.151)$$

Now, we have to find an appropriate Lyapunov function that can be utilized for constructing the optimal controller. We have to find the u^* that minimizes the function, see (10.151):

$$F = F(u(k)) = \Delta V(x(k)) + x^T(k)Rx(k) + u^T(k)Qu(k) \quad (10.152)$$

Now, substitute (10.150) in (10.152) to obtain

$F(u(k))$

$= x^T(k+1)Px(k+1) - x^T(k)Px(k) + x^T(k)Rx(k) + u^T(k)Qu(k)$

$= (\varphi x(k) + Bu(k))^T P(\varphi x(k) + Bu(k)) - x^T(k)Px(k)$

$\quad + x^T(k)Rx(k) + u^T(k)Qu(k)$

$$(10.153)$$

Take the partial derivative of $F(.)$ and (10.153) with respect to $u(k)$ to obtain the following

$$\frac{\partial F(u(k))}{\partial u(k)} = 2(\varphi x(k) + Bu(k))^T PB + 2u^T(k)Q = 0$$

$$= 2x^T(k)\varphi^T PB + 2u^T(k)(B^T PB + Q) = 0 \quad (10.154)$$

The matrix $(B^T PB + Q)$ is positive definite (and hence it is invertible), and we get from (10.154) the control law as follows

$$u^*(k) = -(B^T PB + Q)^{-1}B^T P\varphi x(k) = -Kx(k) \quad (10.155)$$

$$u^*(k) = -S^{-1}B^T P\varphi x(k) = -Kx(k) \quad (10.156)$$

Next, one can check the second order sufficient condition for the minimization

$$\frac{\partial^2 F(u(k))}{\partial u^2(k)} = \frac{\partial}{\partial u(k)}[2x^T(k)\varphi^T PB + 2u^T(k)(B^T PB + Q)] > 0$$

$$= 2(B^T PB + Q) > 0 \quad (10.157)$$

Again, since the matrix $(B^TPB + Q)$ is positive definite, the condition in (10.157) holds true, and u^* is the optimal control law. Now, for the optimal controller, we need to determine suitable matrix P (also called Lyapunov matrix), since that is the only unknown in (10.155); of course, the weights Q and R need also to be chosen appropriately. We have the CLCS after the incorporation of the control law (10.155) in (10.148)

$$x(k+1) = (\varphi - BS^{-1}B^TP\varphi)x(k) \qquad (10.158)$$

Since, the controller should satisfy (10.151), utilizing (10.149) and (10.150), we have, after introducing the optimal control law:

$$x^T(k+1)Px(k+1) - x^T(k)Px(k)$$
$$+ x^T(k)Rx(k) + u^{*T}(k)Qu^*(k) = 0 \qquad (10.159)$$

Now, we put the expression (10.158) and the control law formulae (10.156) in (10.159) to obtain

$$x^T(k)(\varphi - BS^{-1}B^TP\varphi)^T P(\varphi - BS^{-1}B^TP\varphi)x(k) - x^T(k)Px(k)$$

$$+ x^T(k)Rx(k) + x^T(k)\varphi^TPBS^{-1}QS^{-1}B^TP\varphi x(k) = 0 \qquad (10.160)$$

The expansion and simplification of (10.160) yields the following

$$x^T(k)(\varphi^TP\varphi - P + R - \varphi^TPBS^{-1}B^TP\varphi)x(k) = 0 \qquad (10.161)$$

Since (10.161) should hold true for any $x(.)$, we obtain the following discrete algebraic Riccati equation

$$\varphi^TP\varphi - P + R - \varphi^TPBS^{-1}B^TP\varphi = 0 \qquad (10.162)$$

Since only P is unknown, the solution of (10.162) can be utilized in (10.156) to obtain the optimal control law that will minimize the given quadratic performance index (10.147).

11

Adaptive Control

Adaptive control is needed for a variety of reasons as highlighted in the introduction of the book, which also traces its history in brief. In many cases, the topic of system identification/parameter estimation, continuous and discrete time both, is covered as a part of the adaptive control, since, in many adaptive schemes, system identification-cum-parameter estimation (SIPE) would be required explicitly (for indirect adaptive control, EIAC) or implicitly (for direct adaptive control, IDAC); so, we cover some important methods for these in Appendix 11A. Much of the discussion in the present chapter is based on the material from [4–12]. Adaptive control can be viewed as a special type of nonlinear feedback control, since when the adaption/adaptive action is on, the linear control system, as a whole (in overall sense), behaves like a nonlinear system. In an adaptive control, in most cases, the states of the systems are such that some change slowly with time and the latter can be regarded as parameters also.

In an adaptive control, there is a feature for automatic adjustment of the controllers in real time to achieve and maintain a desired level of performance of the control system. This is especially needed when the parameters of the plant's model are unknown and/or change with respect to time. Additionally, there are unknown but deterministic (i.e. not necessarily stochastic) disturbances affecting the dynamics. Various aspects of adaptive control (and techniques) are [4]: (i) some high performance control systems would require precise tuning of the controller's parameters that are changing, but the plant (disturbance) model parameters may be unknown or time varying, (ii) these methods provide a logical and systematic approach for automatic online/ real time (OLRT) tuning of controller's parameters, and (iii) the methods can be viewed as approximations of some nonlinear stochastic control problems that are difficult to solve in practice. The basic idea is that one starts with an ordinary feedback control loop that is already, say stabilizing the plant/process, and with a regulator with some adjustable parameters. Then, the next thing to do is to find or determine a method of changing some or all of these regulator's parameters in response to changes in the plant and the disturbance dynamics. The latter situations could arise due to the wear and tear of the components of the sub-systems; hence, their properties would vary as time elapses, and the numerical values of these properties used in the design of the

basic control are no longer valid and we need to upgrade these values. Also, there are plants that have some inherent variations of some such parameters depending upon their operational ranges/domains, e.g. a fighter aircraft (and missile) flight mechanics/dynamics would vary from one flight condition to another when there is a change in the altitude and/or Mach number (vehicle's speed relative the speed of sound).

In a basic conventional control (control system, CS), we have two main blocks: (i) plant/system/process and (ii) controller that is used to stabilize the plant dynamics and/or improve its (error) performance. However, for the design of the controller, we need to know the mathematical model (MM) of the real/actual plant, since the plant is directly not known, except mainly its I/O signals and some overall information that it is affected by some unknown disturbances and/or random noises. For the design of the controller and for this plant, we need to know its mathematical model, which is, in general, an offline activity. Once an adequate MM is known, the well-known control design procedures are used to arrive at a suitable controller to meet the specified goals; stability and performance. These procedures are discussed in Section 11.1, as the conventional/classical control design methods, and Chapters 5 through 10 as the modern control approaches. Hence, the plant modeling and the controller design tasks are done, mainly offline, and constitute the principles of controller design, see Figure 11.1a. Subsequently, the performance is verified by analytical means for the linear control systems and by extensive simulations (Monte Carlo runs) for the nonlinear control systems. In contrast to the conventional procedure of the controller design, for the adaptive controller, the controller is made adjustable (this was fixed in the conventional design), see Figure 11.1b. These required adjustments are carried out by the controller parameters that are obtained as the outputs from the adaptation scheme (AS); and this AS is governed by the I/O of the plant that is operating in the real time. This is a gross AS, however, for more details, specifically five blocks in a typical AS can be specified as: (i) plant, (ii) performance measurements, (iii) comparison decision, (iv) adaptive mechanism, and (v) adjustable controller; see Figure 11.2. The decision block presupposes that some reference signal has been used, in some cases, the reference model (RM) is used; and the latter will lead to what is called a MRAC (model reference

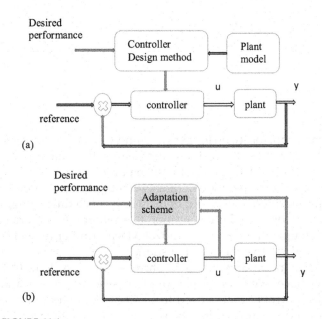

FIGURE 11.1
(a) Conventional model based control design and (b) An adaptive control system.

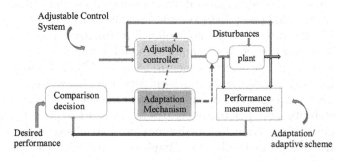

FIGURE 11.2
An adaptive control system with adaptation scheme.

present; and (iv) the main unit in CFC is the controller, and in ACS, it is the adaptive mechanism. Thus, in CFC, the aim is the elimination of the effect of disturbances upon the controlled variables, and in ACS, it is to eliminate the effect of parameter disturbances (because of the variations in the parameters of the plant) upon the performance of the control system. Thus, any ACS would look like being a hierarchical system: (i) conventional (or similar) feedback control system and then (ii) supported/augmented/monitored by a suitable adaptation loop. Hence, there is a basic/fundamental concept in/for ACS: for any possible values of plant (disturbance) model parameters, there is a controller with a fixed structure and complexity such that the specified performance/s can be achieved with appropriate values of the (adjustable) controller parameters. The task of the adaptation loop is to solely search for the good values of these controller parameters. Interestingly, the AC techniques/mechanisms can improve the performance of a robust control (already synthesized by some appropriate robust optimization/control techniques, Chapter 8) system by expanding the range of uncertainty for which the performance specification can be attained and by better tuning of the nominal controller (already synthesized by other methods of Sections 11.1 and 11.2). Thus, in enhancing the performance of the existing robust controller (Chapter 8) a suitable adaptive mechanism can be used. In an adaptive robust control, the existing robust control is made adaptive by adding suitable adaptive mechanism; whereas in robust adaptive control, the existing adaptive control is made more robust by making adaptive mechanism more reliable. In general, at the higher level, we might want robust adaptation of a robust controller.

adaptive control, since a mathematical model-reference is used). The scheme is known as direct one, since the system identification/parameter estimation procedure is not explicitly used to determine the changes of the plant parameters and for control law adjustment.

Certain main aspects and differences between the conventional feedback control (CFC) and adaptive control system/scheme (ACS) are: (i) in CFC, the aim is to monitor the controlled variables according to some performance index (PI criterion), in this case, the parameters are known; in ACS, the performance of the control system is monitored for unknown and (time) varying parameters; (ii) in CFC, the emphasis is on the controlled variables (measured by some sensors), whereas in ACS, it is on the index of performance (which is measured and also specified); (iii) in CFC, there is reference input and comparison block, and in ACS, there is "desired" PI to be achieved and the comparison decision block is

11.1 Direct and Indirect Adaptive Control Methods

There are basically three schemes for adaptive mechanisms: (i) open loop adaptive control (OAC), often known as gain scheduling system (GSS), Figure 11.3; (ii) indirect adaptive control (IAC) based on EIAC plant model estimation (PME) Figure 11.4a and b; and (iii) direct adaptive control (model reference adaptive control, MRAC, IDAC), Figure 11.5. It is interesting to note that this MRAC/IDAC resembles the plant model (parameter) estimation (PME/PPE) scheme of Figure 11.4b. In most cases, the idea in adaptive control is to estimate parameters of the uncertain plant/controller online, while using the measured signals, $u(.)$ and $y(.)$, of the system. Then, these estimated parameters are used in the computation of the control input. Thus, the adaptive controller is a

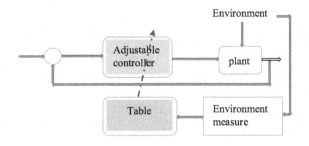

FIGURE 11.3
Open loop adaptive control.

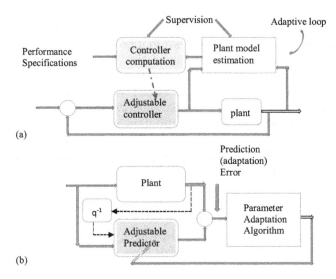

FIGURE 11.4
(a) EIAC-Explicit indirect adaptive control and (b) PME scheme for EIAC-Explicit indirect adaptive control.

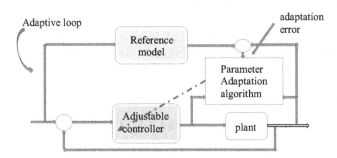

FIGURE 11.5
IDAC-Implicit direct adaptive control (MRAC).

dynamic system with online parameter estimation in most cases, and being adaptive, the ACS then becomes inherently nonlinear. Hence, the analysis and design of such systems would use Lyapunov stability theory [5], Appendix C.

In OAC, there are nearly six blocks, see Figure 11.3: (i) plant, (ii) adjustable controller, (iii) environment, (iv) environment measurements, (v) environment-plant

relationship, and (vi) adaptation mechanism. In OAC, there is an underlying assumption that there is a known and a solid/fixed relationship between some measurable variables that characterize the environment, and the plant model parameters. This relationship is captured in the form of tables (or some appropriate formulas), and the controller gains are adjusted based on these values. So, if there is some change is an environment (-al parameter/s), then the corresponding and an appropriate value of the gain is chosen from the table, and then the existing gain of the controller (or its TF) is changed to this new value. This scheme presupposes that a lot of analysis must have been done offline for the plant and the controller, and the gain scheduling tables should have been prepared well in advance. Also, the entire scheme should have been tried out for its performance by carrying out several offline simulation exercises. In many cases, the tables are converted into formulas that are obtained by the method of curve fitting, i.e. least squares method (LS), between the gain/s and the relevant environmental parameter/s that are expected to change. So, when a particular parameter changes, then correspondingly, the gain is computed by using the formula, and this gain is used in the controller.

In EIAC, the blocks are, Figure 11.4a: (i) plant, (ii) adjustable controller, (iii) PME, and (iv) controller computation; the last two blocks are supervised. The PME, also called plant parameter estimation (PPE), consists of three blocks, Figure 11.4b: (a) plant, (b) parameter adaptation algorithm (PAA), and (c) adjustable predictor. In the EIAC, the plant parameters are estimated, and the controller parameters are computed; hence, the technique depends on the convergence of the estimated parameters to their true (but unknown values) [5]. The adaptive loop consists of the plant model estimation block and the controller computation block. This loop is also the supervision loop of the changes in the plant parameters. The estimation program/block (PME) has its inputs from the I/O of the plant. The controller computation has inputs from the estimation block and the performance specifications. The controller computation block then feeds to the adjustable controller. For the plant model estimation, one can use real time SIPE method (explicitly), which has good properties of convergence and accuracy. This is also called PAA, Figure 11.4b.

In EIAC, as we know, the parameters and/or state variables (of the plant that might be changing, and are required to be adapted) are computed, in fact, estimated (explicitly) using an RLS-PAA, or any suitable recursive method that is relatively fast, accurate, and stable. These estimated parameters are then used to adjust the parameters of the controller. The controller parameters are computed using the certainty equivalence principle. EIAC is used with plants that have non-minimum phase zeros; plant unstable zeros are not canceled. Interestingly, such control schemes are referred to as self-tuning regulators (STR). Thus, in EIAC, a model of the plant is identified online and is used for choosing the parameters of the controller; whereas, in the IDAC (MRAC) scheme, a model of the desired behavior is explicitly used. In EIAC, the (system) identification error is used for updating the control parameters, whereas for IDAC (MRAC), for the same purpose, the control error is used. For the EIAC, if the adaptive observer/estimator/filtering algorithm converges, then the adaptive controller should also converge. The stability problems of both direct and indirect control can be handled simultaneously, and stable adaptive control can be achieved using either the control error for IDAC or the identification error for EIAC [6].

In IDAC (MRAC), there are four blocks, Figure 11.5: (i) plant, (ii) explicit reference model, (iii) PAA, and (iv) adjustable controller. In some cases, the controller is designed using the plant model, then the reference model is used, and the output of which is also fed to the controller design for adjustment. In actual IDAC, there is no direct plant parameter estimation done, but the controller parameters, i.e. the gains, are estimated and adjusted directly (with the results form PAA). Actually, for DAC, a reference model is used in parallel to the controller-plant cascade arrangement. This reference model specifies the idea, or desired response, to the external command and is a guiding factor in the AC. The idea is to force the plant output to follow the reference model output. The reference model continually (or continuously) provides the desired output. The difference between the reference output and the plant output is fed to the PAA, which also receives the output from the adjustable controller, and in turn, the PAA feeds the controller with the adjusted parameters/gains. Thus, a PAA is required in both the schemes: IAC and DAC systems; in the former, it is a part of the recursive SIPE technique (as an explicit technique), and in the latter, it is implicit (for estimation) but directly used for the adjustable controller.

So, in general, the plant has a known structure but its parameters are unknown. In fact, these or some of these might be known, but as the dynamics of the plant change (based on its operative environment or regime, e.g. aircraft flying in different flight conditions of altitude and speed, the Mach number), there will be changes in the parameters and these changes need to be determined online by using some real time algorithm, as is required in the EIAC scheme. In the IDAC, the changes in the plant would reflect in the changes in the plant output that then mismatches with the reference model's output. This difference is used to determine an adaptation law, which in turn, tunes the controller. So, here the explicit

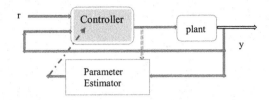

FIGURE 11.6
STC/R- self tuning controller/regulator.

SIPE of the plant parameters is not required but is used implicitly. However, the adaptation law needs to be determined online/real time. The controller is parametrized in both the schemes and it provides the tracking.

There is another scheme of the AC, known as the self-tuning controller/regulator (STC/R), shown in Figure 11.6, where the controller is directly affected by the estimator (of the plant parameters) and it performs a simultaneous parameter estimation and control. In STC, the controller parameters are computed from the estimates of the plant parameters as if these parameters were the true values.

Since, in both the schemes EIAC and IDAC, the PAA is a commonly used block, many books on adaptive control discuss several real time parameter estimation schemes in main sections; however, we briefly discuss these in the Appendix 11A. For PAA, a general approach is: new parameter (vector of plant or controller) = old parameter (vector of plant or controller) + gain factor/matrix × prediction/estimation error. The gain is dependent on the I/O data of the system of which the parameters are to be estimated. The error is between the measured output and the previously predicted output. One example of such a scheme is a recursive least squares (RLS) method, which is reasonably accurate and has a relatively fast convergence, as well as being relatively less sensitive to the spread of the eigenvalues in the basic dynamic system.

11.1.1 Adaptive Control and Adaptive Regulation

In order to obtain IDAC, one can express the performance error in terms of the difference between the parameters of an unknown optimal controller and those of the adjustable controller, then reparametrize the EIAC scheme (if feasible), such that the adaptive predictor will provide directly the estimated parameters of the controller; however, the number of cases for which an IDAC scheme can be developed is limited. In an adaptive domain, the activities are [4]: (i) plant parameters are estimated every sampling time; (ii) controller parameters can be updated every sampling time (instant), which might be difficult (due to a finite delay involved in the process of estimation) or only every *n* samples (where n is reasonably small), this can be done if the plant parameters vary slowly with time; and (iii) for adaptation to work, sufficient excitation is required, which has always been the necessary requirement or condition for SIPE process. For self-tuning regime: (i) the PAA with decreasing adaptation gain can be used, the parameters are assumed to be unknown but constant; (ii) the controller parameters are either updated at every sampling time or kept constant during the parameter estimation process; and (iii) an external excitation is applied during tuning or plant SIPE process. If the controller parameters are kept constant during the SIPE operation, then it is "auto-tuning," and for the EIAC this corresponds to "plant identification in closed loop operation (PICL) and controller redesign situation." The PICL using appropriate algorithms, e.g. RLS method, would provide better models for design, and this, with the redesign of the controller, is a powerful (auto-) tuning procedure. For PICL and controller redesign (CRD), first the controller is kept constant and a new model is identified such that estimation error is reduced; and then for controller redesign, a new controller is determined, such that the error is reduced. An important distinction between adaptive control (EIAC) and adaptive regulation (AR) can now be made: (i) in AC, the plant model is unknown and it would be time varying, and the disturbance model is known and constant; and (ii), in AR, the plant model is known and constant, and the disturbance model is unknown and time varying. The combined AC-AR (i.e. ACR) is a very difficult problem, since it is not easy to distinguish in the performance (prediction) error, what comes from the plant model error, and what comes from the disturbance model error. In all these cases, the "internal model principle" has to be used.

In AC, the focus is on adaptation with respect to plant model variation/s, and the aim is of monitoring tracking/disturbance attenuation performance; the idea is that in the presence of changing parameter/s (i.e. even if the parameter/s have changed), the performance has to be maintained at the same level as it was if the parameter/s had not changed. In AC, the model of the disturbance is assumed to be known and constant, only a level of attenuation in a frequency band is needed; however, for the known DC disturbance, this is not feasible, since the integrator might be used. Also, no effort is made to concurrently estimate the model of the disturbance. In AR, the focus is on adaptation with respect to the disturbance model parameter/s' variation/s, and the aim is of suppressing the effect of the unknown disturbance that is characterized by a rational power spectrum. Since, in AR, the plant model is assumed to be known (by a priori system identification) and almost constant, any small plant parameter/s' variation/s are handled by a robust control design. Here, also no effort is made to concurrently estimate the plant model.

Primary or preliminary design process for an adaptive control has these steps; (i) analysis of the system under control exercise, (ii) plant identification in various regimes of operation, (iii) to ensure the availability of models for various regimes of operation, since this would allow design of sensibly/reasonably robust controllers that in turn would assure satisfactory performance (in the domain of the parameter space around each of the identified models), (iv) to ascertain the current region of the operation of the system, so the appropriate controller can be utilized, (v) EIAC cannot detect this fast enough but can make a fine tuning over a certain time/period, (vi) when the parameters change rapidly, the adaptation-transients in EIAC may be unacceptable; however, these transients can be improved by utilizing the available information.

In the supervisory control (SC), there are several models of the plant (for its different operational regions) and for each model, there is a controller, connected in switching modes. The supervisor checks what plant-model (combination) error is minimum, and then it will switch to the controller associated with the selected model. The SC can provide a very fast decision, if there are only a few models used, but it cannot do fine tuning. In the case of the AC with multiple models, there are some fixed models and one adaptive model; then the supervisor selects the best fixed model, and then the adaptive model is selected; in the scheme, these fixed models are for improving the adaptation transients, and the adaptive plant model (estimation) is used for performance improvement.

11.2 Gain Scheduling

Sometimes it is possible to determine some auxiliary variables that correlate reasonably well with the changes in the plant dynamics; then one can change the parameters of the regulator as functions of these variables. Since the approach was originally used for considering the changes in the plant/system gains only, it is known as the gain scheduling (GS) method. Actually, the concept of GS (gain scheduling) was developed for flight control systems [7], since the Mach number and the dynamic pressure are measured by air-data systems, then used as the GS variables to change the gains of the control TFs; the latter are the results of the control law synthesis, mainly based then on the classical control design methods. The main aspect in GS is to choose the appropriate gain scheduling variable/s (GSVs): for example, in process control, the production rate can be taken, since the time constants and time delays are often inversely proportional to this rate. Then the regulator parameters are determined at a number of operating conditions using some appropriate design technique. Subsequently, the stability and performance of the designed GS-based control system are evaluated by computer simulations at these operating points as well as other intermediate points, and the transitions between different operating conditions are considered. One can obtain the GS by using the normalized dimension-free parameters, such that the normalized model is not dependent on these operating conditions. In this case, the auxiliary measurements are used along with the plant measurements to calculate the normalized measurement variables; the normalized control variable is computed and retransformed before it is applied to the plant for control purposes [7]. The GS is an Open Loop (OL) compensation, if there is an incorrect schedule, there is no feedback for the compensation in the control system; so, it is viewed as a feedback control system in which the feedback gains are adjusted by the feed forward compensation. However, in the GS, the parameters can be changed quickly in response to the changes in the plant/process, since the corresponding scheduling variable/s and the related gains are already available. The GS is not an adaptive control strategy in the strict sense. Yet, it can be used to handle the plant parameter variations in (e.g. the flight) control system and, in general, a very useful approach to reduce the effects of the parameter variations; hence, it is mostly studied in the adaptive control literature.

The technique of GS can be viewed as switching or blending [8] of: (i) gain values of controllers, or models, or (ii) complete controllers or model dynamics according to different operating conditions (or according to pre-set times), and that the GS may involve continuous/discrete scheduling of controllers/model dynamics. In general, GS covers the attenuation of: (a) nonlinear dynamics over a range of operations, (b) environmental time-variations, and (c) parameter variations and uncertainties; the main point here is to reduce or eliminate the effects of these disturbing factors, the nonlinearities, environment, and variations, which would otherwise degrade the performance of the CLCS. The technique of (offline) linearization of dynamics of a nonlinear system and the design of the corresponding controllers using linearized models is the conventional GS approach, where the scheduling variable is used that corresponds to a design (gains) related to linearized models. In online GS, the gains are scheduled such that the scheduling variable (SV) is time dependent and signifies the nearby operating condition. The aim is to extend a single linearization-based control design over an entire operating regime. Because of the scheduling process, and introduction of such a SV, the stability aspects need to be ascertained. There are three GS approaches: (i) classical

or conventional, (ii) linear fractional transformation (LFT), and (iii) fuzzy logic (FL)-based.

An important aspect in linear controller design involves the robustness and performance of the CL system in the presence of: (i) dynamical uncertainty, which corresponds to neglected plant dynamics (e.g. high frequency behavior, nonlinearities), and (ii) constant parametric uncertainty, which results either from inaccurate knowledge of the physical parameters or from variations of these during operation. To handle the former, one can use robust H-Infinity (HI) theory, and for the latter, μ-synthesis and Lyapunov-based techniques can be used.

Linear Parameter Varying (LPV) plants are: (i) LTI plants subject to a time varying parametric uncertainty, as GSV(t) or (ii) the models of LTV plants, either derived as a nonlinear model or a set of linear models describing a nonlinear plant. When GSV(t) can be measured, it can be used in an appropriate control strategy. Assuming that the uncertainties are constant or only slowly varying, one would get LTI robust control techniques in case of (i). In case of large parameter variations, a single (robust) controller could be very restrictive and the plant-stabilization by a single LTI controller would not be feasible. If the varying parameter GSV(t) can be measured during operation, GS would prove to be less restrictive. For GSV(t), one needs to know its range of operation and its rate of variation. A GS controller is a function of the GSV(t); hence, it can adjust itself in accordance to the changes in the plant dynamics and is usually less restrictive compared to a robust one. The parameter GSV(t) can be divided as: (i) measured in real time and (ii) uncertain. The controller adapts to changes in the plant dynamics due to the GS that is a measured part and is robust to changes in the uncertain part. The GSV(t) can be treated as a time varying uncertainty; it can then represent a synthesis of a robust HI performance criterion with time varying uncertainties. Recent LPV and LFT control analysis/synthesis techniques use: (i) parameter varying Lyapunov functions and (ii) scaled small gain theorems.

11.2.1 Classical GS

In general, there is no guarantee on the robustness, performance, and nominal stability of the closed loop for this GS design. The steps involved in the classical GS approach are:

Step 1: The LTI approximations of a nonlinear plant at constant operating points are computed. These are parameterized by constant values of (some convenient plant) variables or exogenous GSVs. Then, the implementation of the controller uses the measureable GSV. There are other approaches to derive a parameter-dependent model: (i) when zero error equilibrium points are not present, off-equilibrium, or velocity-based linearizations can be used, (ii) the quasi-LPV in which the plant dynamics are formulated to distinguish nonlinearities as time-varying parameters; the latter are used as GSVs, and (iii) direct LPV modeling, based on a linear plant incorporating time-varying parameters.

Step 2: The LTI controllers corresponding to the already derived set of local LTI models are designed at each operating point, and the set of controllers is parameterized by GSV. Although, the GSV is time-varying, the classical design methods are based on fixed GSV. Although, to enable subsequent scheduling, the set of LTI controllers might require fixed-structure controller designs, however: (i) in case a direct derivation of a LPV controller for a corresponding LPV plant model is feasible, subsequent scheduling and interpolation becomes superfluous, and (ii) when discrete or hybrid scheduling is required, the set of controller designs are not necessarily required to be fixed-structured.

Step 3: Next, the GSV implemented. At each operating point, the scheduled controller has to linearize to the corresponding linear controller design and provide a constant control value yielding zero error at these points. As in Step 2, in case of direct scheduling, this step is not needed. In case of discrete scheduling, the implementation of the LTI controllers involves the design of a scheduled selection procedure. The latter is applied to the set of LTI controllers, rather than the design of a family of scheduled controllers.

Step 4: The performance assessment of the control system is carried out either analytically or by extensive simulations.

11.2.2 LPV and LFT Synthesis

These yield direct synthesis of a controller utilizing L$_2$ norm-based methods, with assurance of the robustness, performance, and nominal stability of the overall GS design. This has the following steps:

Step 1: This is the classical approach. The LTI approximations of a nonlinear plant at equilibria, parameterized by constant values of convenient plant variables, GSVs is computed. Then, the implementation of the controller requires, GSV = GSV(t) to be a measurable variable. Also, the LFT description serves as a basis for LFT controller synthesis.

Step 2: LPV and LFT synthesis yield a GS controller. The performance can be assured a priori as the GSV(t) instead of its fixed value GSV, here only continuous GS is considered.

11.2.3 Fuzzy Logic-Based Gain Scheduling (FGS)

The fuzzy modeling considers the transient dynamics of the nonlinear model instead of local linearizations. The steps are:

Step 1: A set of local LTI models and corresponding LTI controllers are to be designed. The attention is on the domains of the envelope of operating conditions.

Step 2: For fuzzy modeling, the weighting functions are designed, corresponding to the selected domains, and the weights are utilized for blending the local models. Based on a specific approximation accuracy of the fuzzy model, the required number of local models are obtained.

Step 3: The set of local controllers is blended using the weighting functions. This blending obtains the scheduling of the controller outputs. Hence, the set of LTI controllers does not need to have fixed structure/dimension.

Step 4: The performance is established by simulations. Here, the global and local specifications are checked from simulations, since the characteristic of the fuzzy models cannot be related to the dynamics of the set of local models.

11.3 Parameter Dependent Plant Models

Let us describe such plant as

$$\dot{x}(t) = f(x(t), u(t), w(t))$$

$$z(t) = g(x(t), u(t), w(t)) \qquad (11.1)$$

$$y = h(x(t), w(t))$$

In (11.1), $z(.)$ is the error signal to be controlled, $y(.)$ is a measured output that is available to the controller (a penalized variable, a tracking commands, some state variables, or a scheduling parameter that accounts for the nonlinearities), and $w(.)$ is the external inputs, like a reference command, disturbance/noise. The linear approximation is parameterized by the GSV, θ, and it yields a family of linear plant models:

$$\dot{x}(t) = A(\theta)x + B_1(\theta)w + B_2(\theta)u$$

$$z(t) = C_1(\theta)x + D_{11}(\theta)w + D_{12}(\theta)u \tag{11.2}$$

$$y = C_2(\theta)x + D_{21}(\theta)w$$

A true LPV model is given with GSV(t), $\theta(t)$:

$$\dot{x}(t) = A(\theta(t))x + B_1(\theta(t))w + B_2(\theta(t))u$$

$$z(t) = C_1(\theta(t))x + D_{11}(\theta(t))w + D_{12}(\theta(t))u \tag{11.3}$$

$$y = C_2(\theta(t))x(t) + D_{21}(\theta(t))w$$

The GS based on (11.3) would be called linearization-based GS. Here, the variations in state, and I/O are neglected.

11.3.1 Linearization Based GS

Assume that $f(x,u,w)$ is continuously differentiable, then we have the conventional linearization:

$$\dot{x} = f_0 + \nabla_x f(x - x_0) + \nabla_u f(u - u_0)$$

$$+ \nabla_w f(w - w_0) + HOT$$

$$z = g_0 + \nabla_x g(x - x_0) + \nabla_u g(u - u_0)$$

$$+ \nabla_w g(w - w_0) + HOT \tag{11.4}$$

$$y = h_0 + \nabla_x h(x - x_0) + \nabla_w h(w - w_0)$$

$$+ HOT$$

The higher order terms (HOT) are neglected. By coordinate transformation, we obtain

$$x = x_0 + \tilde{x}; \ u = u_0 + \tilde{u};$$

$$w = w_0 + \tilde{w}; \tag{11.5}$$

$$y = y_0 + \tilde{y}; \ z = z_0 + \tilde{z}$$

Substituting (11.5) in (11.4) and comparing with (11.2), we obtain the linearization family:

$$\dot{\tilde{x}} = A\tilde{x} + B_1\tilde{w} + B_2\tilde{u}$$

$$\tilde{z} = C_1\tilde{x} + D_{11}\tilde{w} + D_{12}\tilde{u} \tag{11.6}$$

$$\tilde{y} = C_2\tilde{x} + D_{21}\tilde{w}$$

Now, the GS variable is introduced with equilibrium conditions as

$$0 = f(x_0(\theta), u_0(\theta), w_0(\theta))$$

$$z_0(t) = g(x_0(\theta), u_0(\theta), w_0(\theta)) \tag{11.7}$$

$$y_0 = h(x_0(\theta), w_0(\theta))$$

Combining the concept of (11.7) with (11.6), we obtain

$$\dot{\tilde{x}} = A(\theta)\tilde{x} + B_1(\theta)\tilde{w} + B_2(\theta)\tilde{u}$$

$$\tilde{z} = C_1(\theta)\tilde{x} + D_{11}(\theta)\tilde{w} + D_{12}(\theta)\tilde{u} \tag{11.8}$$

$$\tilde{y} = C_2(\theta)\tilde{x} + D_{21}(\theta)\tilde{w}$$

The system matrices of (11.8) are now parametrized. We finally obtain the models for the set point design:

$$\dot{\tilde{x}} = A(\theta(t))\tilde{x} + B_1(\theta(t))\tilde{w} + B_2(\theta(t))\tilde{u}$$

$$\tilde{z} = C_1(\theta(t))\tilde{x} + D_{11}(\theta(t))\tilde{w} + D_{12}(\theta(t))\tilde{u} \tag{11.9}$$

$$\tilde{y} = C_2(\theta(t))\tilde{x} + D_{21}(\theta(t))\tilde{w}$$

11.3.2 Off Equilibrium Linearizations

In this approach, the nonlinearities are explicitly introduced as follows

$$\dot{x}(t) = Ax + B_1w + B_2u + f(\theta)$$

$$z(t) = C_1x + D_{11}w + D_{12}u + g(\theta) \tag{11.10}$$

$$y = C_2x + D_{21}w + h(\theta)$$

Corresponding velocity based linearizations are given as

$$\dot{x}_0 = \zeta_0$$

$$\dot{\zeta}_0 = \left(A + \frac{\partial f}{\partial t}(\theta_0)\right)\zeta_0 + \left(B_1 + \frac{\partial f}{\partial t}(\theta_0)\right)\dot{w}_0 + \left(B_2 + \frac{\partial f}{\partial t}(\theta_0)\right)\dot{u}_0$$

$$\dot{z}_0 = \left(C_1 + \frac{\partial g}{\partial t}(\theta_0)\right)\zeta_0 + \left(D_{11}\frac{\partial g}{\partial t}(\theta_0)\right)\dot{w}_0 + \left(D_{12} + \frac{\partial g}{\partial t}(\theta_0)\right)\dot{u}_0$$

$$\dot{y}_0 = \left(C_2 + \frac{\partial h}{\partial t}(\theta_0)\right)\zeta_0 + \left(D_{21} + \frac{\partial h}{\partial t}(\theta_0)\right)\dot{w}_0$$

$$\tag{11.11}$$

Now, since there is no restriction on the equilibrium point, one can have the following velocity-based linearizations for the original nonlinear model (11.1):

$$\dot{x} = \zeta$$

$$\dot{\zeta} = \left(A + \frac{\partial f}{\partial t}(\theta)\right)\zeta + \left(B_1 + \frac{\partial f}{\partial t}(\theta)\right)\dot{w} + \left(B_2 + \frac{\partial f}{\partial t}(\theta)\right)\dot{u}$$

$$\dot{z} = \left(C_1 + \frac{\partial g}{\partial t}(\theta)\right)\zeta + \left(D_{11}\frac{\partial g}{\partial t}(\theta)\right)\dot{w} + \left(D_{12} + \frac{\partial g}{\partial t}(\theta)\right)\dot{u} \quad (11.12)$$

$$\dot{y} = \left(C_2 + \frac{\partial h}{\partial t}(\theta)\right)\zeta + \left(D_{21} + \frac{\partial h}{\partial t}(\theta)\right)\dot{w}$$

These velocity-based linearizations can be adopted for GS controller design.

11.3.3 Quasi LPV Method

In this, nonlinear terms are masked by including them in the GSV, and it requires a transformation. The model (11.1) is rewritten as

$$\begin{bmatrix} \dot{y} \\ \dot{x}' \end{bmatrix} = f(y) + A'(y)\begin{bmatrix} y \\ x' \end{bmatrix} + B_1'(y)w' + B_2'(y)u' \quad (11.13)$$

In (11.13), the variable y is regarded as a GSV. The equilibrium family parameterized by y and the related expressions are obtained as:

$$0 = f(y) + A'(y)\begin{bmatrix} y \\ x_0'(y) \end{bmatrix} + B_1'(y)w_0'(y) + B_2'(y)u_0'(y) \quad (11.14)$$

$$f(y) + \begin{bmatrix} A'^{(11)}(y) \\ A'^{(21)}(y) \end{bmatrix} y = -\begin{bmatrix} A'^{(12)}(y) \\ A'^{(22)}(y) \end{bmatrix} x_0'(y)$$
$$\quad (11.15)$$
$$-B_1'(y)w_0'(y) - B_2'(y)u_0'(y)$$

$$\begin{bmatrix} \dot{y} \\ \dot{x}' \end{bmatrix} = -\begin{bmatrix} A'^{(12)}(y) \\ A'^{(22)}(y) \end{bmatrix} x_0'(y) - B_1'(y)w_0'(y) - B_2'(y)u_0'(y)$$

$$+ \begin{bmatrix} A'^{(12)}(y) \\ A'^{(22)}(y) \end{bmatrix} x' + B_1'(y)w' + B_2'(y)u'$$
$$\quad (11.16)$$

$$= \begin{bmatrix} A'^{(12)}(y) \\ A'^{(22)}(y) \end{bmatrix}(x' - x_0'(y)) + B_1'(y)(w' - w_0'(y))$$

$$+ B_2'(y)(u' - u_0'(y))$$

Next, we apply the change of variables

$$x = \begin{bmatrix} y \\ x' - x_0'(y) \end{bmatrix}; \; w = w' - w_0'(y); \; u = u' - u_0'(y); \quad (11.17)$$

$$\dot{x} = \begin{bmatrix} \dot{y} \\ \dot{x}' - \dot{x}_0'(y) \end{bmatrix} = \begin{bmatrix} \dot{y} \\ \dot{x}' - \nabla_y\{x_0'(y)\}\dot{y} \end{bmatrix} \quad (11.18)$$

We then have

$$\dot{x} = \begin{bmatrix} 0 & A'^{(12)}(y) \\ 0 & A'^{(22)}(y) \end{bmatrix} x + \begin{bmatrix} 0 \\ -\nabla_y\{x_0'(y)\}\dot{y} \end{bmatrix} + B_1'(y)w + B_2'(y)u$$
$$\quad (11.19)$$
$$= A(y)x + B_1(y)w + B_2(y)u$$

In (11.19), we have the following equivalences

$$A(y) = \begin{bmatrix} 0 & A'^{(12)}(y) \\ 0 & A'^{(22)}(y) - \nabla_y\{x_0'(y)\}A'^{(12)}(y) \end{bmatrix}$$

$$B_1(y) = \begin{bmatrix} B_1'^{(1)}(y) \\ B_1'^{(2)}(y) - \nabla_y\{x_0'(y)\}B_1'^{(1)}(y) \end{bmatrix} \quad (11.20)$$

$$B_2(y) = \begin{bmatrix} B_2'^{(1)}(y) \\ B_2'^{(2)}(y) - \nabla_y\{x_0'(y)\}B_2'^{(1)}(y) \end{bmatrix}$$

Subsequently, the controller synthesis can be based on the GSV as $y = y(t)$ and should be used if it is really required.

11.3.4 Linear Fractional Transformation

In certain cases, the parameter dependency (of LPV or quasi-LPV modeling) is modeled as a LFT, which represents uncertainty in models as the: (i) known represented by M, and (ii) unknown represented by Δ (the possible values of which are bounded), in a feedback like connection, Figure 11.7:

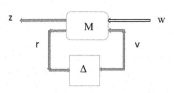

FIGURE 11.7
LFT-Uncertainty modeling.

$$M = \begin{bmatrix} M^{(11)} & M^{(12)} \\ M^{(21)} & M^{(22)} \end{bmatrix} \qquad (11.21)$$

The lower and upper LFTs are then obtained as

$$z = F_L(M,\Delta)w = (M^{(11)} + M^{(12)}\Delta(I - M^{(22)}\Delta)^{-1}M^{(21)})w$$

$$F_U(M,\Delta) = (M^{(22)} + M^{(21)}\Delta(I - M^{(11)}\Delta)^{-1}M^{(12)}) \qquad (11.22)$$

If the matrices of (11.4) depend on the unknown, i.e. GSV, an LFT on a corresponding diagonal, Δ(GSV) is activated with $\Delta = diag(\theta_1 I_n, ..., \theta_p I_n)$ for θ with p components and M of the order n.

11.4 Classical Gain Scheduling

The task is to design a set of LTI controllers $\Lambda(\theta)$ called parameter dependent controllers.

11.4.1 LTI Design

The parameterized family $\Lambda(\theta)$ of linear controllers is signified as

$$\dot{\xi}(t) = A_c(\theta)\xi(t) + B_c(\theta)\tilde{z}(t)$$

$$\tilde{u}(t) = C_c(\theta)\xi(t) + D_c(\theta)\tilde{z}(t) \qquad (11.23)$$

In (11.23), $\tilde{z}(t)$ is the error signal obtained from $z(.)$ (it could be $z(t) = r(t) - y(t)$, where $r(.)$ is the reference signal), and $y(t)$ is the usual output signal. The standard loop shaping, pole placement, or HI and mu-synthesis approaches can be employed for the LTI controller synthesis based on the models of (11.23).

11.4.2 GS Controller Design

For actual implementation, the GSV(t) is replaced by the measured variable $\theta(t)$ to obtain the scheduling controller:

$$\dot{\xi}(t) = f_c(\xi(t), \tilde{z}(t), \tilde{w}(t), \theta(t))$$

$$\tilde{u}(t) = h_c(\xi(t), \tilde{z}(t), \tilde{w}(t), \theta(t)) \qquad (11.24)$$

The details of the scheduling/interpolation method depends on the structure of the set of LTI controllers: (i) fixed structured LTI controllers with only one or a few varying gains for changing plant dynamics, here classical linear control methods are applied to design the

set of LTI controllers; and (ii) operating conditions with respect to a set of LTI controllers with varying structure, here multivariable state space design procedures could be used. For discrete scheduling, the implementation of the LTI controllers involves the design of a scheduled selection procedure or switching.

11.4.2.1 Linearization Scheduling

The aim of linearization scheduling is that the equilibrium set of the controller matches the equilibrium set of the plant. This lead to the requirement of

$$0 = f_c(\xi_e(\theta), z_e(\theta), w_e(\theta))$$

$$u_e(t) = h_c(\xi_e(\theta), z_e(\theta), w_e(\theta)) \qquad (11.25)$$

$$\nabla_\xi \{f_c(\xi_e(\theta), z_e(\theta), w_e(\theta))\} = A_c(\theta)$$

$$\nabla_{\tilde{z}} \{f_c(\xi_e(\theta), z_e(\theta), w_e(\theta))\} = B_c(\theta)$$

$$\nabla_\xi \{h_c(\xi_e(\theta), z_e(\theta), w_e(\theta))\} = C_c(\theta) \qquad (11.26)$$

$$\nabla_{\tilde{z}} \{h_c(\xi_e(\theta), z_e(\theta), w_e(\theta))\} = D_c(\theta)$$

The general controller family is given as the GS controller

$$\dot{\xi}(t) = A_c(\theta(t))(\xi(t) - \xi_e(\theta(t))) + B_c(\theta(t))(z(t)$$
$$- z_e(\theta(t)))$$

$$u(t) = C_c(\theta(t))(\xi(t) - \xi_e(\theta(t))) + D_c(\theta(t))(z(t)$$
$$- z_e(\theta(t))) \qquad (11.27)$$

11.4.2.2 Interpolation Methods

For isolated point designs, corresponding interpolation of the controller (coefficients/gains) is required for which: (i) ad hoc interpolation of local point design controllers is utilized to arrive at a GS LPV controller, and (ii) theoretically justified methods can be used. Ad hoc interpolation methods require a set of local LTI controller designs with fixed structure and dimension; all controllers should have the same number of poles and zeros. One can interpolate point design HI controllers by means of interpolating the corresponding poles, zeros, and gains. A set of LTI-HI controllers using Riccati equations can be designed wherein linear interpolation of the Riccati equations yields a family of parameter dependent controllers. In case of theoretically justified interpolation technique, the resulting LPV controller $\Lambda(\theta)$ is stability preserving if: (i) the coefficients of $\Lambda(\theta)$ are continuous functions of θ, (ii) $\Lambda(\theta) = \Lambda(\theta_i)$, *for* $\theta = \theta_i$, and (iii) $\Lambda(\theta)$ stabilizes the models (11.8) for all θ; the method then involves a form of controller blending and yields switching between controllers with assured smoothness. There are various approaches: (i) a method to linearly interpolate the controller gains of a state feedback controller based on eigenvalue placement; (ii) an observer based state feedback and linearly interpolating the observer gains; (iii) a TF interpolation method; (iv) a generalized approach for state space controllers as the application of HI and μ-synthesis, and (v) interval mathematics to determine regions for which closed loop poles remain within specified bounds.

11.4.2.3 Velocity Based Scheduling

The velocity-based linearization set (VLS) may be adopted as a suitable basis for subsequent GS controller design for which: (i) classical GS, (ii) LPV synthesis, and (iii) fuzzy-like blending methods may be employed. VLS has several constant operating points as linearizations are assigned to every operating point and not only to equilibrium operating points; hence, VLS broadens the application range of classical GS. The LTI controller design is applied to a grid of operating points. Also, one can use a blended multiple model representation, which has similarity to Takagi-Sugeno or fuzzy modeling. The VLS is approximated by a small number of local models; and the blended local models directly equal the velocity-based linearization model at the corresponding operating point; the blended model is approximated by the weighted combination of the local model solutions. The actual implementation requires derivative of y, and the corresponding differentiator/integrator may be absorbed in the controller. The performance is established using small gain or Lyapunov theories. So, conservative results might be obtained, since, except in special cases, only sufficient conditions are known for the stability of nonlinear systems; however, the resulting nonlinear controller is valid throughout the operating envelope of the plant.

11.4.3 Hidden Coupling Terms

Since the scheduling variable $\theta(t)$ is time varying and the controller design is based on corresponding fixed values, this would introduce so called hidden coupling terms, which represent additional feedback terms and the local stability and performance of the resulting closed loop would differ from the original LTI designs. The idea is then that the linearized closed loop system at any operating point (parametrized by fixed values of the GSV) should accurately match the interconnection of the corresponding linearized plant and the linear

controller design at that operating condition. If the following conditions hold true, then the hidden terms would not be present:

$$\nabla_\theta\{f_c(\xi_0(\theta), z_0(\theta), w_0(\theta))\} = 0$$
$$\nabla_\theta\{h_c(\xi_0(\theta), z_0(\theta), w_0(\theta))\} = 0 \tag{11.28}$$

11.4.4 Stability Properties

Stability/performance of the CLCS, incorporating the original nonlinear plant and the parameter dependent controller, is important for the application of GS in case of the use of: (i) the equilibrium-operating-point-based controller designs (that disregard off-equilibrium operating points) and (ii) a fixed-valued GSV θ in place of (that neglect transient dynamics). The use of any approximations might affect the performance. In the vicinity of equilibrium points, the deviations would be small if: (i) the initial conditions lie in the neighborhood of an equilibrium point, i.e.

$$\left\| \begin{pmatrix} x_i(\theta(t)) \\ \xi_i(\theta(t)) \end{pmatrix} - \begin{pmatrix} x_e(\theta(t)) \\ \xi_e(\theta(t)) \end{pmatrix} \right\| < \gamma, \ \gamma > 0;$$

(ii) if at all, there is only slow variation present in the GSV, $\|\dot{\theta}(t)\| < \mu$, $\mu > 0$; thus, the permissible inputs and initial conditions should remain quite close to equilibrium points; and (iii) the linearization of the nonlinear plant equals the local point designs; then the stability of the nonlinear system may be related to the stability of a set of LTI models representing the nonlinear system.

11.5 LPV Controller Synthesis

The LPV control synthesis indicates controller synthesis based on either LPV or LFT models: (i) it yields parameter dependent controllers with a priori guaranteed stability and performance properties, (ii) uses the real time information of the parameter variation, (iii) focuses on an LPV model rather than a nonlinear model of the plant, (iv) has a theoretical foundation guaranteeing a priori stability and performance for all GSV(t) given a corresponding range and rate of variation, (v) the controller design is global with respect to the parameterized operating envelope (classical GS techniques focus on local system properties), (vi) a controller is synthesized directly, rather than from a set of local linear controllers, (vii) focuses on appropriate problem formulation rather than the actual controller design, (viii) a conservatism has to be introduced to arrive at a feasible and convex problem, and (ix) a predefined controller design synthesis has to be adopted; as the latter point already indicates, LPV synthesis constitutes a specific performance evaluation framework, whereas classical GS provides an open framework.

LPV synthesis utilizes the induced L_2-norm as a performance measure, that is directly related to linear HI techniques; hence, when applied to a time-invariant system, it is equivalent to a standard HI approach. LPV control synthesis is categorized into techniques utilizing: (i) Lyapunov based approach; (ii) the specific structure of systems with LFT parameter dependence, also determined as LFT approaches; and (iii) a combination of the two preceding points, as 'mixed' LPV-LFT approaches.

11.5.1 LPV Controller Synthesis Set Up

In this case, the GSV is regarded as $GSV(t) = GSV(w(t))$, (as an exogenous variable), and when $\theta(t)$ is an endogeneous variable, we have $\theta(t) = \theta(x(t))$, the latter is called quasi-LPV. A general LPV controller is specified as

$$\dot{\xi}(t) = A_c(\theta(t))\xi(t) + B_c(\theta(t))y(t)$$
$$u(t) = C_c(\theta(t))\xi(t) + D_c(\theta(t))y(t) \tag{11.29}$$

If we consider the rate of $GSV(t)$ as also a GSV, then we have

$$\dot{\xi}(t) = A_c(\theta(t),\dot{\theta}(t))\xi(t) + B_c(\theta(t),\dot{\theta}(t))y(t)$$
$$u(t) = C_c(\theta(t),\dot{\theta}(t))\xi(t) + D_c(\theta(t),\dot{\theta}(t))y(t) \tag{11.30}$$

Now, the corresponding CL system is

$$\begin{bmatrix} \dot{\zeta}(t) \\ z(t) \end{bmatrix} = \begin{bmatrix} A(\theta(t)) & B(\theta(t)) \\ C(\theta(t)) & D(\theta(t)) \end{bmatrix} \begin{bmatrix} \zeta(t) \\ w(t) \end{bmatrix}; \quad \zeta = [x,\xi]^T$$

$$\begin{bmatrix} A & B \\ C & D \end{bmatrix} = \begin{bmatrix} A + B_2 D_c C_2 & BC_c & B_1 + B_2 D_c D_{21} \\ \hline B_c C_2 & A_c & B_c D_{21} \\ C_1 + D_{12} D_c C_2 & D_{12} C_c & D_{11} + D_{12} D_c D_{21} \end{bmatrix} \tag{11.31}$$

11.5.1.1 Stability and Performance Analysis

For the LTI systems the asymptotic stability is characterized by (LMIs, linear model inequalities) as

$$A^T P + PA < 0; \ P > 0, \text{positive defintie matrix} \tag{11.32}$$

In (11.32), A is the CLS matrix of (11.31). For LPV dependent CLCSs we have

$$A^T(\theta(t))P + PA(\theta(t)) < 0 \tag{11.33}$$

A time varying system is stable if: i) the corresponding time-invariant system is stable, and ii) if at all, there is only slow variation present in the GSV, $\|\dot{\theta}(t)\| < \mu, \ \mu > 0$. In particular Lyapunov normalized energy functional can be used, and one can establish that the closed loop system with LPV GS controller is asymptotically stable, and derive the necessary conditions for the same.

The performance analysis of the CLS and LPV controller is related to the L_2 induced norm of the signal $w(t)$ to error signal $z(t)$, i.e. induced L_2 gain

$$\sup_{w(t) \neq 0} \frac{\|z(t)\|_2}{\|w(t)\|_2} < \gamma \tag{11.34}$$

The maximum of the error gain should be bounded as in (11.34).

11.5.2 Lyapunov Based LPV Control Synthesis

Often the synthesis is based on quadratic Lyapunov functions, and continuous time state feedback controller is designed using $V(x) = x^T P x$, here x is the CLS state. Also, the parameter dependent Lyapunov functions can be used; a gridding approach can be used by computing solution to these fixed points, this would require interpolation of intermediate points, and one can have a dense grid, and large LMIs, this would lead to the larger optimization problem. Another approach is to use affine or polytopic parameter dependencies, $V(p)$, of the plant and the controller: $A(\theta(t)) = A_0 + \theta(t)A_1$. This does not require gridding, and all the parameters can be vary independently, and would lead to a conservative overestimation.

11.5.3 LFT Synthesis

Utilizing LFT parameter dependency of LPV systems allows application of a generalized HI synthesis (with the optimal small gain theorem), and these parameter variations are temporarily regarded as unknown perturbations. Multipliers-cum-scaling variables (MSVs) describing the nature of the unknown $\theta(t)$ are introduced to decrease the conservatism of the LPV methods using a constant, common quadratic Lyapunov function. An LFT model is a special case of a LPV model, that is transformed using an upper LFT, resulting in constant known part M, and a corresponding time varying unknown part $\Delta(\theta(t))$. Similarly, transformation of the LPV controller using a lower LFT results in a constant known part K, and a corresponding unknown part $\Delta_K(\theta(t))$ that is probably a nonlinear controller scheduling variable. The CL interconnects T for the channel $w \rightarrow z$, Figure 11.7, is given as

$$T = F_L\{F_U(M, \Delta(\theta(t))), F_L(K, \Delta_K(\theta(t)))\}$$

$$T = F_U\{F_L(M, K), \Delta\}; \Delta = diag\{\Delta(\theta(t)), \Delta_K(\theta(t))\}$$ (11.35)

The CLCS is defined as

$$\begin{bmatrix} \dot{\zeta}(t) \\ z_u(t) \\ z(t) \end{bmatrix} = \begin{bmatrix} A & B_u & B_p \\ C_u & D_u & D_{up} \\ C_p & D_{pu} & D_p \end{bmatrix} \begin{bmatrix} \zeta(t) \\ w_u(t) \\ w(t) \end{bmatrix}; z_u(t) = \Delta w_u(t)$$ (11.36)

With $\Delta(\theta(t))$ is block-diagonal with diagonal terms $|\Delta_i(\theta(t))| \lessapprox 1$, an LTI controller K is determined to render the augmented system M, so that robust quadratic performance is achieved for the performance channel $w \rightarrow z$. Then, the corresponding parameter variations are incorporated in the controller again, yielding an LPV controller and with LFT parameter dependence. Consider the closed loop LFT interconnection system for which the corresponding analysis involves three main conditions regarding the CL model T: (i) well-posedness, i.e. the system should have a unique solution for all initial conditions corresponding to the augmented system M and all inputs $w(t)$ to the system; (ii) exponential stability of the CLCS (analyzed via the small-gain theory), i.e. a CLCS is stable provided that the loop gain <1; and (iii) quadratic performance, i.e. asymptotic convergence of the channel $w(t) \rightarrow z(t)$, or L_2-performance, i.e., the loop gain $< \gamma$. Combination of all the three conditions yields a set of LMIs, which is directly obtained from the general LPV synthesis constraints.

11.5.4 Mixed LPV-LFT Approaches

The infinite dimensional LFT/LMI is replaced by single LMI using an additional multiplier, and parameter dependent Lyapunov function $P(p)$ is introduced, this approach is called extended Kalman-Yakubovich-Popov (KYP) lemma. In an another approach the infinite dimensional LPV/LMI is replaced by single LMI via quadratic LFT Lyapunov functions and full block multipliers, and this will result in general parameter dependent controller.

11.6 Fuzzy Logic-Based Gain Scheduling

Here, the blending is done using fuzzy modeling (FM), i.e. the blending of a set of models describing some nonlinear dynamics via point designs; and

examples of multiple model representations are: (i) Takagi-Sugeno (TS fuzzy, TSF) models, (ii) local model networks, and (iii) Polytopic Linear Models (PLMs). A set of N_m local, indexed models parametrized by θ^i, cover the system's equilibrium operating points (or a set of constant operating points that span the nonlinear, changing system dynamics). These points corresponding to the indexed models are defined by $\Re^0(\theta^i) = \{x_0(\theta^i), w_0(\theta^i), u_0(\theta^i), z_0(\theta^i), y_0(\theta^i)\}$. The nonlinear dynamics are approximated by a weighted ($\alpha(.)$, weighting factors) sum of the individual models to obtain the fuzzy model:

$$\dot{x}(t) = \sum_{i=1}^{N_m} \alpha^i(\theta_\alpha)\{A(\theta^i)\tilde{x}^i + B_1(\theta^i)\tilde{w}^i + B_2(\theta^i)\tilde{u}^i\}$$

$$z(t) = \sum_{i=1}^{N_m} \alpha^i(\theta_\alpha)\{C_1(\theta^i)\tilde{x}^i + D_{11}(\theta^i)\tilde{w}^i + D_{12}(\theta^i)\tilde{u}^i\}$$ (11.37)

$$y(t) = \sum_{i=1}^{N_m} \alpha^i(\theta_\alpha)\{C_2(\theta^i)\tilde{x}^i + D_{21}(\theta^i)\tilde{w}^i\}$$

Equations in (11.37) can be written after the coordinate transformation (11.5) as

$$\dot{x}(t) = \sum_{i=1}^{N_m} \alpha^i(\theta_\alpha)\{A(\theta^i)x + B_1(\theta^i)w + B_2(\theta^i)u + f_0(\theta^i)\}$$

$$z(t) = \sum_{i=1}^{N_m} \alpha^i(\theta_\alpha)\{C_1(\theta^i)x + D_{11}(\theta^i)w + D_{12}(\theta^i)u + g_0(\theta^i)\}$$

$$y(t) = \sum_{i=1}^{N_m} \alpha^i(\theta_\alpha)\{C_2(\theta^i)x + D_{21}(\theta^i)w + h_0(\theta^i)\}$$

(11.38)

With the following expressions for the nonlinear terms

$$f_0(\theta^i) = -A(\theta^i)x_0(\theta^i) - B_1(\theta^i)w_0(\theta^i) - B_2(\theta^i)u_0(\theta^i)$$

$$g_0(\theta^i) = z_0(\theta^i) - C_1(\theta^i)x_0(\theta^i) - D_{11}(\theta^i)w_0(\theta^i) - D_{12}(\theta^i)u_0(\theta^i)$$

$$h_0(\theta^i) = y_0(\theta^i) - C_2(\theta^i)x_0(\theta^i) - D_{21}(\theta^i)w_0(\theta^i)$$

(11.39)

From (11.38) and (11.39) we see that the model is a time varying, weighted combination of linear and/or affine state space systems. With FM the nonlinear blended model dynamics cannot be related to the original set of the local point design models; because of the presence of cross terms involving derivatives of the weighing factors/functions. This aspect has to be taken into account by the controller design; because these weights are dependent on the operating points. The blending using $\alpha^i(\theta_\alpha)$ is essential to the FGS approach. The weighting functions are normalized: $\sum_{i=1}^{N_m} \alpha^i(\theta_\alpha) = 1$. The functions are parameterized such that the local models switch over into each other smoothly. A distinction can be made between the case of local models describing the original, nonlinear plant dynamics at: (i) distinct operating points and (ii) operating regions, validity is in an extended operating region and the mixing is restricted to small transitions at the boundaries between these regions. The main merit of FM is that the original nonlinear model may be approximated arbitrarily close by a fuzzy model; which is a global approximation of the nonlinear model, and takes into account transient dynamics. The approximation accuracy result is stated as follows. Let us consider the nonlinear dynamics as

$$\dot{x} = f(x, w, u); \text{ with}$$
$$f(x, w, u) = F^T(x, w, u) + f_{nl}(x, w, u) \quad (11.40)$$

The distance between the nonlinear dynamics $f(x,w,u)$ and its nonlinear FM approximation is given as

$$d(f, \hat{f}) = \sup_{(x,w,u)} \left\| f(x, w, u) - \hat{f}(x, w, u) \right\|_2 \quad (11.41)$$

If N_m local models are utilized with N_m given as follows, then the $d(f, \hat{f}) \le \varepsilon$ is satisfied:

$$N_m = \prod_{i=1}^{n_{nl}} \text{ceil}\left(\frac{\beta_i}{2\sqrt{2}\varepsilon} \sqrt{\lambda_H n_{nl} \sqrt{n}} \right) \quad (11.42)$$

In (11.42), we have n_{nl} as the dimension of the nonlinear terms, $\beta(.)$ as the range of the nonlinear terms, $n = \dim (F)$, i.e. the order of the model, and $\lambda_H = \max(\lambda_{H_j})$ as the maximum eigenvalue of all the eigenvalues of the Hessian matrices that are from the nonlinear terms.

For FGS, the FM based on a set of local, indexed models (i) (parameterized by the equilibrium/constant operating points) covering the nonlinear changing system dynamics are utilized as a basis and then appropriately blended (by using local controller design techniques, or LMI based LPV methods). The parallel distributed control (PDC) synthesis involves the blending of a set of LTI controllers using the same weighting factors as used for the modeling to yield a nonlinear GS controller. The controller outputs are blended/scheduled rather than scheduling based on the controller coefficients. In PDC a feedback (FB) and a feedforward (FF) part are used in a like manner:

1. The FB part directly comes from the local controller designs

$$\Lambda_{FB}(\theta_\alpha) = \sum_{i=1}^{N_m} \alpha^i(\theta_\alpha)\Lambda(\theta^i) \quad (11.43)$$

The FB part holds for the parametrized linear part, the offsets are not compensated for; then the FF part is determined;

2. This FF part is not strictly so, since it depends on $\theta_\alpha(x, u, w)$ which may have measured states. The FF compensator is given as

$$\Lambda_{FF}(\theta_\alpha) = -\{B(\theta_\alpha)^T B(\theta_\alpha)\}^{-1} B(\theta_\alpha)^T \sum_{i=1}^{N_m} \alpha^i(\theta_\alpha)f_0(\theta^i)$$

$$B(\theta_\alpha) = \sum_{i=1}^{N_m} \alpha^i(\theta_\alpha)B_2(\theta^i) \quad (11.44)$$

The GS controller is given as $\Lambda(\theta_\alpha) = \Lambda_{FB}(\theta_\alpha) + \Lambda_{FF}(\theta_\alpha)$, $\theta_\alpha = \theta_\alpha(x, u, w)$, the latter being the GS vector.

11.7 Self-Tuning Control

The self-tuning regulator (STR) has: (i) the inner loop of the process and usual linear feedback regulator, and (ii) the outer loop that consists of a recursive parameter estimator and a design computation [9], and this adjusts the parameters of the regulator. It may be necessary to use some perturbation signal/s in order to obtain good estimates of the required parameters, as is almost always the case for SIPE methods. A STRC (self-tuning regulator/controller) has two coupled algorithms for: (i) the online estimation of the parameters of an assumed model, and (ii) evaluating the control action from a suitable control law design procedure [7]. The online solution to the design problem for a system with known parameters is represented by the regulator design (box). The latter aspect is called the underlying design problem, and is associated with most ACS, and this would give the characteristics of the system under the ideal conditions when the parameters of the systems are known exactly. The STR has many extensions: (i) self-tuners based on gain/amplitude and phase margins, (ii) pole-placement self-tuners, (iii) minimum variance self-tuners, and (iv) LQG based self-tuners. The self-tuners use recursive parameter estimation scheme/s [9]: (i) stochastic approximation, (ii) LS method, (iii) extended and generalized LS methods, (iv) instrumental variables method, (v) extended Kalman filter, and (vi) the maximum likelihood method. There are two main self-tuners: (i) explicit STR that is based on estimation of an explicit process/plant model, and (ii) implicit STR which is based on the estimation of the implicit plant model; in this case, the process is re-parametrized so that it is expressed in terms of the regulator parameters, and hence, the design computations are eliminated.

In order to fully utilize the STC concept, the models are required to be repeatedly updated as the system is driven over the operational domain of interest. If the operating domain is small, then a local linear model with fixed parameters may be sufficient, however, if the operational domain is enlarged, the assumptions on local linearity for the system to be controlled may not remain valid [10], and the overall CL performance will reduce due to the increase in the mismatch between the system and model. As we have seen in the previous sections, alternative approaches of gain scheduling (i.e. look-up tables and multiple switched- or blended model) can be utilized. The STC, wherein the model parameters are continually updated (as the operational domain is scanned) is in effect an infinite model approach, and here, as the system and/or subsystem components (in fact their parameters) change over time, so do the resulting models. In this updating process, the estimated values at each time step are assumed to be correct. Further it is possible, using the same measured I/O data, to detect the onset of a faulty condition. Such a concept permits: (i) the determination of thresholds, adherence to certain inequalities allowing the implementation of adaptive control via STC, and (ii) a fault detection/tolerant control scheme. There are the indirect (or explicit) and direct (i.e. implicit) STC schemes: (i) in an indirect direct approach, the control law is obtained from the estimated model parameters like EIAC scheme, and (ii) in the direct approach, the control law is estimated from the I/O data along with the estimated model parameters, the latter being implicit within the scheme, like IDAC/MRAC. For both the former schemes we have: (i) nondual STC, the control action performs the role of control signal only; and (ii) dual STC, the control action is ideal for control as well as for the estimation. When a linear STC is applied to nonlinear systems, the self-tuning principle (that holds when estimated model parameters converge to steady values) may not remain valid; hence, a nonlinear (here, bilinear) STC might be required. The STC model structures that can be used for parameter estimation and some simple methods for the latter are described in Appendix 11A.

11.7.1 Minimum Variance Regulator/Controller

The aim is to minimize the variance (MV) of the system output $y(t)$ via an optimal control input $u(k)$. The assumptions are [7]: (i) the system to be controlled

is linear, (ii) the cost function is quadratic, and (iii) the noise affecting the system is Gaussian. In the regulator problem the desired output or the set point $r(k) = 0$. The MV regulator cost function is defined as

$$J = E\{y^2(k+d)\}; d \text{ is the time delay, often } d = 1. \quad (11.45)$$

The objective is to determine the optimum input $u(k)$ that minimizes the cost function of (11.45). The output is written in two parts as

$$y(k+1) = \hat{y}(k+1/k) + e(k+1) \quad (11.46)$$

Then we have (see Appendix 11A for the structure of the original model used):

$$\hat{y}(k+1/k) = -a_1\hat{y}(k) - a_2\hat{y}(k-1) + b_0u(k) + bu_1(k-1) \quad (11.47)$$

Substitute (11.47) in (11.46), which is then inserted in (11.45), and after the simplification one gets

$$J = E\{\hat{y}(k+1/k)\}^2 + \sigma_e^2$$
$$= [-a_1\hat{y}(k) - a_2\hat{y}(k-1) + b_0u(k) + b_1u(k-1)]^2 + \sigma_e^2 \quad (11.48)$$

Minimization of (11.48) with respect to $u(k)$ and equating the resultant expression to zero we obtain the formula for the optimal control

$$\hat{u}(k) = \frac{1}{b_0}[a_1\hat{y}(k) + a_2\hat{y}(k-1) - b_1\hat{u}(k-1)] \quad (11.49)$$

In the case of the MV controller the aim is to track a reference signal $r(k)$, and this scheme is also referred to as servo controller. The cost function is defined as

$$J = E\{y(k+d) - r(k)\}^2 \quad (11.50)$$

Following the same procedure as done for the MV regulator we obtain the following expression for the optimal control input

$$\hat{u}(k) = \frac{1}{b_0}[a_1\hat{y}(k) + a_2\hat{y}(k-1) - b_1\hat{u}(k-1) + r(k)] \quad (11.51)$$

A property of the MV controller is illustrated as follows. Let us have the ARX (which can also be called as the LS model as discussed in Appendix 11A) model structure as

$$\frac{Y(z)}{U(z)} = \frac{z(b_0 + b_1z)}{1 + a_1z + a_2z^2};$$
$$z = q^{-1}, \text{unit delay shift operator} \quad (11.52)$$

Then, the MV control input TF is given from (11.49) as

$$\frac{U(z)}{Y(z)} = \frac{a_1 + a_2 z}{b_0 + b_1 z} \qquad (11.53)$$

It can be seen from (11.53) that the denominator consists of the numerator of (11.52). Thus, if the system's zeros, of (11.52) are outside the unit circle, then the controller, (11.53) is unstable, because its poles are also outside of the unit circle. In this case, the system is non-minimum phase. The CLTF is given as

$$\frac{Y(z)}{R(z)} = \frac{a_1 z^{-1} + a_2}{z^{-2}}; \ q = z^{-1} \qquad (11.54)$$

In (11.54), the CL poles lie at the origin and the response of such a system is as fast as possible, and it is referred to as "deadbeat" response. This would need large demand on the control action $u(k)$. In order to balance the tracking performance and the cost of control one can use the weighted performance index:

$$J = E\{[Py(k+d) - Rr(k)]^2 + [Qu(k)]^2\} \qquad (11.55)$$

If $Q > 0$, the control input is penalized and, hence, constrained. However, the CL poles would be displaced away from the origin and response would not be a "deadbeat." By following the same procedure as for the regulator and controller, we get the solution for the control law/input for the generalized MV controller, from (11.51) as

$$\hat{u}(k) = \frac{Pb_0[P\{a_1\hat{y}(k) + a_2\hat{y}(k-1) - b_1\hat{u}(k-1)\} + Rr(k)]}{Pb_0^2 + Q^2} \qquad (11.56)$$

The general GMV controller input is given from the predictor

$$\hat{y}(k+j/k) = A_j(z)\hat{y}(k) + B_j(z)\hat{u}(k) \qquad (11.57)$$

The choice of the cost weighting parameters is crucial and application specific. The P, R and Q parameters can be chosen either by an operator or adaptively (with some initial *a priori* values) within a STC framework: (i) $P = R = 1$ and $Q = 0$ results in MV control, (ii) $P = R = 1$ and varying $Q > 0$ achieves a trade-off between tracking ability and reduction of control effort. By over constraining the control effort, $u(k)$ (e.g. energy) the GMV controller may not achieve the set point and the SS errors would occur; this problem is overcome by $P = 1$ and $R > 1$, which results in a new fictitious set point aim. Note that the importance of tracking ability versus reducing control cost is governed by the ratio P/Q. The SS offset problems can also be overcome by using

IGMV (incremental) control, where the inherent integral action would guarantee type-1 performance [7].

11.7.2 Pole Placement Control

This is the problem of self-tuning pole placement/assignment control wherein the aim is to match the CL (transient) behavior of a feedback system to a desired/user prescribed specifications/form. This is an eigenvalue assignment problem, the effect of PPC (pole placement control) is that of relocation of the CL poles of the system [10]. The system is represented by the following system model (see Appendix 11A):

$$A(z)y(k) = z^d B(z)u(k) \qquad (11.58)$$

The control law is defined as

$$B(z)u(k) = A(z)y(k) + Cr(k) \qquad (11.59)$$

The controller polynomials are expressed in the form of their coefficients

$$A(z) = \alpha_0 + \alpha_1 z + \alpha_2 z^2 + \ldots + \alpha_{n_\alpha} z^{n_\alpha}; \ \alpha_0 = 1 \qquad (11.60)$$

$$B(z) = \beta_0 + \beta_1 z + \beta_2 z^2 + \ldots + \beta_{n_\beta} z^{n_\beta}; \ \beta_0 \neq 0;$$

$$n_\alpha = n_a + d - 1; \ n_\beta = n_\beta - 1 \qquad (11.61)$$

The CL TF is given by

$$\frac{Y(z)}{R(z)} = \left[\frac{CB(z)}{B(z)A(z)} z^d\right]\left[1 - z^d \frac{A(z)B(z)}{B(z)A(z)} z^d\right]^{-1} \qquad (11.62)$$

$$= \frac{z^d B(z)C}{B(z)A(z) - z^d A(z)B(z)} \qquad (11.63)$$

The design aim is to assign the CL poles to a desired/specified location, and this is being done by equating the characteristic equation, i.e. the denominator of the CLTF to a user specified/designed polynomial $D(z)$:

$$B(z)A(z) - z^d A(z)B(z) = D(z);$$

$$D(z) = d_0 + d_1 z + d_2 z^2 + \ldots + d_{n_d} z^{n_d}; \ d_0 = 1 \qquad (11.64)$$

In (11.64), $D(z)$ is the desired CL characteristic polynomial. It is possible that the SS gain will be affected, hence, one can choose the gain C, such that the SS gain = 1, i.e. using the final value theorem:

$$\text{SS gain} = \left[z^d \frac{B(z)C}{D(z)}\right]_{z=1} = \frac{B(1)C}{D(1)} = 1 \qquad (11.65)$$

Thus, we have the gain $C = D(1)/B(1)$, which cancels the offset due to $D(z)$ on the CL SS gain, so that, if there is no model mismatch, the SS output matches the reference signal $r(t)$. Thus, a gain compensated PPC would be able to achieve the transient response and desired SS gain.

11.7.3 A Bilinear Approach

The bilinear systems are a small/important sub-class of nonlinear systems; and many real-life processes can be appropriately described using bilinear models that are signified by linear behavior in both state and control when considered separately, however, the nonlinearity comes about as a product of system state and control variables. Such bilinear processes may be found in engineering, ecology, medicine and socioeconomics. The potential merits are: (i) improved control/efficiency compared to (linear) linearizations models, (ii) reduced wastage (from implementation point-of-view), (iii) increased profitability, and (iv) improved product quality. A general single-input single-output (SISO) can be modeled using a nonlinear ARMAX (NARMAX) model representation as:

$$y(k) = \sum_{i=1}^{n_a} -a_i y(k-i) + \sum_{i=0}^{n_b} b_i u(k-d-i)$$

$$+ \sum_{i=0}^{n_b} \sum_{j=1}^{n_a} \gamma_{i,j} y(k-i-d)u(k-i-j-d+1) + e(k) \tag{11.66}$$

The control law is based on the bilinear model of (11.66). One can consider the problem with a quasi-linear model so that the bilinear coefficients are combined with either the $a(.)$ or $b(.)$, the coefficients of the original linear model of the plant; the combined parameters are:

$$\tilde{a}_i(k) = a_i(k) - u(k-d-i)\gamma(i-1); \; OR$$

$$\tilde{b}_i(k) = b_i(k) + y(k-i)\gamma(i) \tag{11.67}$$

As a result of utilizing the bilinear (bi-linearized) model for the purpose of predicting the system output, the prediction error would decrease. However, since the combined coefficients are inherently time varying and input/output dependent, these are required to be updated at each time step.

11.8 Adaptive Pole Placement

In pole placement (pole assignment), the aim is to place the poles of the CLTF in reasonable positions. As we have seen the process of a self-tuning regular would contain three/four blocks, (see Figure 11.6): (i) plant, (ii) parameter estimation, (iii) computations, and (iv) the controller, the latter receives the outputs form the computation

block, which in fact, would obtain the controller parameters based on the inputs from the estimation block. Here, the system identification/parameter estimation (SIPE) is explicitly done, and the controller parameters are computed based on these results. The scheme is also called EIAC. For the SIPE, one can use RLS as suggested earlier, then the task is of pole placement for the controller design. The plant/process dynamics could be some electrical/electronic circuits, actuators in a mechanical control systems/aircraft control, and sensors; it is important to note that these processes might be slow/fast and even (inherently) unstable. The certainty equivalence principle (i.e. the estimated plant parameters of the model are regarded as the true parameters) is used for pole placement, tracking control, and deterministic disturbance rejection controller synthesis methodology: (i) estimate the plant parameters using say RLS-PAA (parameter adaptation algorithm), and (ii) re-compute the controller parameters at every sampling instant by assuming that the last/latest plant parameters are the real/true parameters. The idea of the adaptive pole placement (APP) is based on the servo following [11]. The plant and the control dynamics are represented as

$$Ay(k) = Bu(k-1) + Ce(k) \; ; \; B_c u(k) = C_c r(k) - A_c y(k) \tag{11.68}$$

In (11.68), we have the following expansions for the polynomials, $B_c(z)$, $C_c(z)$ and $A_c(z)$;

$$B_c(z) = 1 + \beta_1 z + ... + \beta_{n_f} z^{n_f}$$

$$A_c(z) = \alpha_0 + \alpha_1 z + ... + \alpha_{n_g} z^{n_g} \tag{11.69}$$

$$C_c(z) = \gamma_0 + \lambda_1 z + ... + \gamma_{n_h} z^{n_h}$$

The closed loop description of the system is given as

$$[AB_c + zBA_c]y(k) = zBC_c r(k) + CB_c e(k) \tag{11.70}$$

The dynamics of the closed loop are governed by the denominator in (11.70), and we can specify the stable poles inside unit circle:

$$D(z) = 1 + d_1 z + ... + d_n z^n \tag{11.71}$$

The basic idea of the APP control is: (i) the requirement is specified for the closed loop poles in $D(z)$, (ii) the plant is given I/O data with the underlying dynamics governed by B/A, (11.68), and (iii) the control designer chooses the controller dynamics $B_c(z)$, $C_c(z)$, and $A_c(z)$, such that the desired closed loop response is obtained. Thus, the pole assignment is done as

$$AB_c + zBA_c = DC \tag{11.72}$$

One can get many solutions of (11.72). In order to obtain the unique solution, one can choose:

$$n_f = n_b; \; n_g = n_a - 1; \; n_d \le n_a + n_b - n_c \qquad (11.73)$$

We can also have: (i) $C = 1$, (ii) A, B must be coprime (no common factors), and (iii) C that has been canceled must be stable. For the real time pole placement, self-tuning one can follow the procedure: (i) at each sampling instant k, collect the new system's output $y(k)$, (ii) update the model parameters' estimates A^*, and B^* using RLS, (iii) synthesize the control polynomials B_c^*, A_c^* using (11.72) based on the desired closed loop response D, (iv) apply the control $B_c^* u(k) = C_c r(k) - A_c^* y(k)$, and (v) repeat from (i).

11.9 Model Reference Adaptive Control/Systems (MRACS)

As we know, in IDAC (MRACS), the controller parameters are updated directly using a RLS- PAA. It is used with plants that do not have non-minimum phase zeros; i.e. they have minimum phase zeros; the plant zeros are canceled. In this ADC/S, the specifications are given in terms of a reference model that suggests how the plant output ideally should respond to the command signal. The reference model is the part of the complete control system. The MRAC can be thought of having two loops [9]: (i) the inner loop of an ordinary control loop consisting of the plant and the regulator, and (ii) the outer loop that adjusts the parameters of the inner loop, in such a way that the error between the reference model and the plant output becomes as small as possible; thus, the outer loop is also a regulator loop. The main point is to determine the adjustment mechanism so that a stable system with zero error is obtained.

In the initial stages, the MIT-rule was used for the MRAC schemes:

$$\frac{d\theta}{dt} = ke \frac{\partial e}{\partial \theta}; \; e \text{ is the model error,}$$

$$\theta \text{ the regulator parameters} \qquad (11.74)$$

Thus, the adjustment rule depends on the sensitivity derivatives of the error with respect to the adjustable parameters. The sensitivities can be obtained as outputs of a linear system driven by the plant I/Os, and the k determines the adaptation rate. An adjustment as follows is typical for many ACSs:

$$\theta(t) = -k \int e(s) \frac{\partial e(s)}{\partial \theta} ds \qquad (11.75)$$

The adaptation rule (11.75) has three aspects: (i) a linear filter that computes the sensitivity derivatives (from using the plant's I/Os), ii) a multiplier, and (iii) an integrator. The MIT-rule will work well if the rate of adjustment k is small. Interestingly, the MRAC scheme is called a direct scheme, since the regulator parameters are updated directly; however, in some MRACSs, these parameters are updated indirectly. One can see that the MRACS and STR are closely related: both have the two feedback loops; the inner loop is an ordinary feedback loop with a plant and a regulator, and the regulator has adjustable parameters that are dictated by the outer loop; and these adjustments are based on the feedback from the plant I/Os. The methods for the design of the inner loop and the methods used for adjustment of the parameters in the outer loop might be different. The direct MRACS is closely related to the implicit STR, and the indirect MRACS to the explicit STR.

11.9.1 MRAC Design of First Order System

Let the system dynamics be given as

$$\dot{x} = ax + b(u - f(x)) \qquad (11.76)$$

In (11.76), the constants a and b are unknown parameters. The uncertain nonlinear function is given as

$$f(x) = \sum_{i=1}^{N} \theta_i \varphi_i(x) = \theta^T \Phi(x) \qquad (11.77)$$

In (11.77), we have the nonlinear function expressed in terms of the constant and yet unknown parameters and the basic function. Let us choose the stable reference model as

$$\dot{x}_m = a_m x_m + b_m r; \; a_m < 0 \qquad (11.78)$$

In (11.78), r is the reference input signal. The control goal is to determine the control input signal $u(.)$ such that the following is true:

$$\lim_{t \to \infty} \{x(t) - x_m(t)\} = 0 \qquad (11.79)$$

We consider the basic state feedback strategy, and hence, we have the control law/feedback as

$$u = \hat{k}_x x + \hat{k}_r r + \hat{\theta}^T \Phi(x) \qquad (11.80)$$

Thus, from (11.76), (11.77), and (11.80), we see that we need to estimated $N + 2$ parameters online. Substituting (11.80) in (11.76), we obtain the closed loop system as [5]:

$$\dot{x} = ax + b(\{\hat{k}_x x + \hat{k}_r r + \hat{\theta}^T \Phi(x)\} - f(x))$$
$$= (a + b\,\hat{k}_x)x + b\{\,\hat{k}_r r + (\hat{\theta} - \theta)^T \Phi(x)\} \tag{11.81}$$

The desired dynamics are as given in (11.78). There exist the ideal gains (k_x, k_r) such that the following holds

$$a + bk_x = a_m$$
$$bk_r = b_m \tag{11.82}$$

Interestingly, the knowledge of the idea gains is not required, but their existence is needed. Now, the following equivalence is established

$$a + b\hat{k}_x - a_m = a + b\hat{k}_x - a - bk_x = b\Delta k_x$$
$$b\hat{k}_r - b_m = b\hat{k}_r - bk_r = b(\hat{k}_r - k_r) = b\Delta k_r \tag{11.83}$$

Now, as we know from (11.79), the tracking error is given as

$$e(t) = x(t) - x_m(t) \tag{11.84}$$

From (11.78) and (11.81) and adding and subtracting one term, we obtain the following for the error dynamics

$$\dot{e}(t) = \dot{x}(t) - \dot{x}_m(t) = (a + b\hat{k}_x)x$$
$$+ b\{\hat{k}_r r + (\hat{\theta} - \theta)^T \Phi(x)\} - a_m x_m \tag{11.85}$$
$$- b_m r + a_m x - a_m x$$

Using (11.82) and (11.83) in (11.85) and simplifying, we obtain

$$\dot{e}(t) = a_m e + b\{\Delta k_x x + \Delta k_r r + \Delta \theta^T \Phi(x)\} \tag{11.86}$$

Next, we can choose the following Lyapunov (semi-normalized) function (LF)

$$V(e(t), \Delta k_x, \Delta k_r, \Delta \theta(t)) = e^2 + |b|(\gamma_x^{-1}\Delta k_x^2 + \gamma_r^{-1}\Delta k_r^2$$
$$+ \Delta \theta^T \Gamma_\theta^{-1} \Delta \theta) \tag{11.87}$$

The γ constants in (11.87) are strictly positive, and the γ matrix is symmetric and positive definite. The time derivative of the LF is given as

$$\dot{V}(e(t), \Delta k_x, \Delta k_r, \Delta \theta(t)) = 2e\dot{e} + 2|b|(\gamma_x^{-1}\Delta k_x \dot{\hat{k}}_x$$
$$+ \gamma_r^{-1}\Delta k_r \dot{\hat{k}}_r + \Delta \theta^T \Gamma_\theta^{-1} \dot{\hat{\theta}}) \tag{11.88}$$

Substituting the error dynamics (11.86) in (11.88) and combining and simplifying, we obtain the following

$$\dot{V}(.,.,.,.) = 2a_m e^2 + 2|b|\Delta k_x (xe\,\mathrm{sgn}(b) + \gamma_x^{-1}\dot{\hat{k}}_x)$$
$$+ 2|b|\Delta k_r (re\,\mathrm{sgn}(b) + \gamma_r^{-1}\dot{\hat{k}}_r) \tag{11.89}$$
$$+ 2|b|\Delta \theta^T (\Phi(x)e\,\mathrm{sgn}(b) + \Gamma_\theta^{-1}\dot{\hat{\theta}})$$

Since a_m is negative, we see from (11.89) that the derivative of the LF will be (semi-)negative definite if the following control gains are chosen

$$\dot{\hat{k}}_x = -\gamma_x xe\,\mathrm{sgn}(b)$$
$$\dot{\hat{k}}_r = -\gamma_r re\,\mathrm{sgn}(b)$$
$$\dot{\hat{\theta}} = -\Gamma_\theta \Phi(x)e\,\mathrm{sgn}(b); \dot{V}(.,.,.,.) = 2a_m e^2(t) \le 0 \tag{11.90}$$

With the gains as in (11.90) and the adaptive laws (11.80) by online parameter updates, the time derivative of the LF would decrease along the error dynamics trajectories, and hence, a stable adaptive control system is obtained. From the foregoing results, we see that since the LF is PD (semi-definite) and its derivative is negative semi-definite, all the parameter estimation errors are bounded. Also, since the unknown parameters are constant, all the estimated parameters are bounded; and since the reference input $r(t)$ is bounded, the reference model trajectories are bounded. Also, since $x = e + x_m$, the following are also bounded: (i) x, (ii) the adaptive control feedback u, (iii) error dynamics, and (iv) Hessain of V; $\ddot{V} = 4a_m e(t)\dot{e}(t)$. Using Barbalat's Lemma, one can infer that the time derivative of LF is a uniformly continuous function of time. Since we have (11.90) using Lyapunov-like Lemma, $\lim_{t\to\infty} \dot{V}(x,t) = 0$ we see that $\lim_{t\to\infty} e(t) = 0$; hence, we conclude that the asymptotic tracking has been achieved: $x(t) \to x(t)_m$, $as\ t \to \infty$ and all the signals in the closed loop system are bounded.

11.9.2 Adaptive Dynamic Inversion (ADI) Control

Let the system dynamics be given as

$$\dot{x} = ax + bu + f(x) \tag{11.91}$$

Or

$$\dot{x} = \hat{a}x + \hat{b}u + \hat{f}(x) - (\hat{a} - a)x - (\hat{b} - b)u - (\hat{f}(x) - f(x))$$

In (11.91), the parameters a and b are unknown but constant, and uncertain nonlinear function is the same as in (11.77). The reference model is also as in (11.78) with the same control goal as in (11.79). The function estimation error is given as

$$\Delta f(x) = \hat{f}(x) - f(x) = (\hat{\theta} - \theta)^T \Phi(x) = \Delta \theta \Phi(x) \tag{11.92}$$

The parameters to be estimated online and the parameter estimation errors are now obvious from (11.91). The ADI control feedback law, considering (11.91) and (11.78) as

$$u = \frac{1}{\hat{b}}\{(a_m - \hat{a})x + b_m r\} - \hat{\theta}^T \Phi(x) \quad (11.93)$$

It should be noted that the estimatee of b should not be zero. The CLS system is given as

$$\dot{x} = a_m x + b_m r - (\Delta a)x - (\Delta b)u - (\Delta \theta)\Phi(x) \quad (11.94)$$

As before, the tracking error dynamics are

$$\dot{e}(t) = a_m e - \Delta a x - \Delta b u - \Delta \theta \, \Phi(x) \quad (11.95)$$

The semi-normalized LF is given as [5]:

$$V(e(t), \Delta a, \Delta b, \Delta \theta(t)) = e^2 + \gamma_a^{-1} \Delta a^2 + \gamma_b^{-1} \Delta b^2 \\ + \Delta \theta^T \Gamma_\theta^{-1} \Delta \theta \quad (11.96)$$

Following the procedure of the Section 11.9.1 and adjusting (11.89) and (11.90) for the present case, we get the following time derivative of the LF (11.96) and the control laws:

$$\dot{V}(e, \Delta a, \Delta b, \Delta \theta) = 2a_m e^2 + \Delta a(-xe + \gamma_a^{-1} \dot{\hat{a}})$$

$$+ \Delta b(-ue + \gamma_b^{-1} \dot{\hat{b}}) \quad (11.97)$$

$$+ \Delta \theta^T (-\Phi(x)e + \Gamma_\theta^{-1} \dot{\hat{\theta}})$$

$$\dot{\hat{a}} = \gamma_a x e$$

$$\dot{\hat{b}} = \gamma_b u e \quad (11.98)$$

$$\dot{\hat{\theta}} = \Gamma_\theta \Phi(x)e; \, \dot{V}(.,.,.,.) = 2a_m e^2(t) \le 0$$

Using Barbalat's Lemma and Lyapunov-like Lemma, we have $\lim_{t\to\infty} \dot{V}(x,t) = \lim_{t\to\infty}[2a_m e^2(t)] = 0$, and we see that $\lim_{t\to\infty} e(t) = 0$; hence, we conclude that the asymptotic tracking has been achieved: $x(t) \to x(t)_m$, $as \, t \to \infty$ and all the signals in the closed loop system are bounded.

11.9.3 Parameter Convergence and Comparison

Interestingly, the convergence of the adaptively online estimated parameters to their true and unknown values depends on the reference signal $r(t)$. We see from (11.86) that the error dynamics represents a stable filter

$$\dot{e}(t) = a_m e + b\{\Delta k_x x + \Delta k_r r + \Delta \theta^T \Phi(x)\}$$

$$= a_m e + b(\text{input}) \quad (11.99)$$

Since the filter input signal is uniformly continuous and the tracking error asymptotically converges to zero as time tends to be very large, we have the following condition

$$\Delta k_x x + \Delta k_r r + \Delta \theta^T \Phi(x) \cong 0; \, (x \; r \; \Phi^T(x)) \begin{bmatrix} \Delta k_x \\ \Delta k_r \\ \Delta \theta^T \end{bmatrix} \cong 0 \quad (11.100)$$

If $r(t)$ is such that $v(t) = (x \; r \; \Phi^T(x))^T$ satisfies the so-called persistent excitation conditions, then the adaptive parameter convergence will take place:

$$\int_t^{t+T} v^T(\tau)v(\tau)d\tau > \alpha I_{N+2}, \, I \text{ is the identity matrix.} \quad (11.101)$$

It is implied by (11.101) that there is sufficient excitation for the online adaptive parameter estimation, and the parameter errors would converge to zero.

In case of ADI, the knowledge of sgn(b) is not needed. The adaptive laws in both the procedures ADI and MRAC are similar. Both the methods yield the asymptotic tracking; however, the ADI need that the estimatee of b should not cross zero. The regression vector in (11.80) must have bounded components, since this is needed for the proof of the stability.

11.9.4 MRAC for n-th Order System

Let the system dynamics be given as

$$\dot{x} = Ax + B\Lambda(u - f(x)); \, \Lambda = \text{diag}\{\lambda_1...\lambda_m\} \text{ constant}$$

$$\text{unknown matrices} \quad (11.102)$$

In (11.102), B is a known constant matrix. The nonlinear function with constant unknown parameters and the known basis function is given as

$$f(x) = \Theta^T \Phi(x) \quad (11.103)$$

The stable reference model is selected as

$$\dot{x}_m = A_m x_m + B_m r \quad (11.104)$$

The control goal is to determine control input signal $u(t)$, such that

$$\lim_{t\to\infty} \|x(t) - x_m(t)\| = 0 \quad (11.105)$$

Let us select the control law as the state feedback and reference input feedback with obvious parameters to be estimated [5]:

$$u = \hat{K}_x^T x + \hat{K}_r^T r + \hat{\Theta}^T \Phi(x) \qquad (11.106)$$

The closed loop dynamics are given as

$$\dot{x} = (A + B\Lambda \hat{K}_x^T)x + B\Lambda(\hat{K}_r^T r + (\hat{\Theta} - \Theta)^T \Phi(x)) \qquad (11.107)$$

From the model matching requirement, we obtain

$$A + B\Lambda K_x^T = A_m; \quad B\Lambda K_r^T = B_m \qquad (11.108)$$

Finally, we have

$$A + B\Lambda \hat{K}_x^T - A_m = B\Lambda \Delta K_x^T$$
$$B\Lambda \hat{K}_r^T - B_m = B\Lambda \Delta K_r^T \qquad (11.109)$$

Since we know the tracking error, we obtain the error dynamics as

$$\dot{e}(t) = \dot{x}(t) - \dot{x}_m(t) = A_m e + B\Lambda(\Delta K_x^T x + K_r^T r$$
$$+ \Delta\Theta^T \Phi(x)) \qquad (11.110)$$

The normalized LF is selected as

$$V(e, \Delta K_x, \Delta K_r, \Delta\Theta) = e^T P e + \text{trace}\{\Delta K_x^T \Gamma_x^{-1} \Delta K_x |\Lambda|\}$$
$$+ \text{trace}\{\Delta K_r^T \Gamma_r^{-1} \Delta K_r |\Lambda|\} \qquad (11.111)$$
$$+ \text{trace}\{\Delta\Theta^T \Gamma_\Theta^{-1} \Phi(x) |\Lambda|\}$$

The weighting matrices in (11.111) are symmetric and positive definite, and P is unique symmetric PD matrix that is solution of the $PA_m + A_m^T P = -Q$, and here Q is any symmetric PD matrix. Following the procedure of the Section 11.9.1, we obtain the following results for the adaptive control law

$$\dot{\hat{K}}_x = -\Gamma_x x e^T P B \, \text{sgn}(\Lambda)$$

$$\dot{\hat{K}}_r = -\Gamma_r r e^T P B \, \text{sgn}(\Lambda)$$

$$\dot{\hat{\Theta}} = -\Gamma_\theta \Phi(x) e^T P B \, \text{sgn}(\Lambda);$$

$$\dot{V}(e(t), \Delta K_x, \Delta K_r, \Delta\Theta(t)) = -e^T(t) Q e(t) \le 0 \qquad (11.112)$$

Using Barbalat's Lemma and Lyapunov-like Lemma, we have $\lim_{t\to\infty} \dot{V}(x,t) = 0$, and we see that $\lim_{t\to\infty} \|e(t)\| = 0$; hence, we conclude that the asymptotic tracking has been achieved: $x(t) \to x(t)_m$, as $t \to \infty$ and all the signals in the closed loop system are bounded.

11.9.5 Robustness of Adaptive Control

Since the adaptive controllers are designed to control real physical systems, we have certain aspects to consider; non-parametric uncertainties may lead to performance degradation and/or instability due: (i) low frequency un-modeled dynamics, structural vibrations and Coulomb friction; (ii) measurement noise; and (iii) computation round-off errors and sampling delays; hence, one needs to enforce robustness of MRAC. When $r(t)$ is persistently exciting the system, MRACSs are robust with respect to non-parametric uncertainties. If $r(t)$ is not persistently exciting, small uncertainties may lead to drift of the estimated parameters (an might even diverge), and the adaptation has difficulty in distinguishing the parameter information and the noise. The parameter drift in MRAC is caused because of the presence of measurement noise and some disturbances, though the drift does not affect the tracking accuracy until the instability occurs; however, it might lead to sudden failure. The remedy is to turn off the adaptation process for "small" tracking errors: let the gains be zeros when $\|e(t)\| \le \varepsilon$, otherwise use the normally estimated gains; a dead-zone effect is used. As a result, a bounded tracking is achieved.

The uncertainty in the mass in an acceleration equation is a parametric uncertainty; however, the neglected motor dynamics, measurement noise, and sensor dynamics are non-parametric uncertainties. In function approximation (11.77), both kinds of uncertainties can occur:

$$\hat{f}(x) = \sum_{i=1}^{N} \theta_i \varphi_i(x) + \varepsilon(x) \qquad (11.113)$$

The non-parametric uncertainty can be handled by dead-zone modification, and for parametric uncertainty, one needs to use certain bases functions that can approximate a large class of functions with a given tolerance: (i) Fourier series, (ii) splines, (iii) polynomials, and (iv) artificial neural networks; sigmoidal and radial basis function (RBF).

11.10 A Comprehensive Example

A design problem for systems with known parameters is considered. Then different adaptive control laws are discussed [9], and a pole placement design is considered as the underlying design study. This will bring out the similarities and differences between self-tuners and MRACS.

11.10.1 The Underlying Design Problem for Known Systems

Let us take a discrete time SISO system

$$Ay = Bu \qquad (11.114)$$

In (11.114), y is the output signal, and A and B denote prime polynomials in the form of forward shift operator as we have seen in Section 11.7.1. The aim is to determine a regulator such that the correspondence between the command signal and the desired output signal is given as with appropriately defined polynomials

$$A_m y_m = B_m u_c \qquad (11.115)$$

A control law that combines the feedback and feedforward control effects is given as

$$Ru = Tu_c - Sy \qquad (11.116)$$

In (11.116), R, T, and S are appropriate polynomials. Using (11.114) and (11.116), one gets the following for the CLS

$$(AR + BS)y = BTu_c \qquad (11.117)$$

The process zeroes are given by the polynomial B set equal to 0; and are the CL zeros unless these are canceled by the corresponding CL poles. Because the unstable or poorly damped zeros cannot be canceled, the polynomial B is factored as

$$B = B^+ B^- \qquad (11.118)$$

In (11.118), the factored polynomial $B(+)$ contains the factors that can be canceled and the remaining factors are in $B(-)$; thus, the zeros of $B(+)$ must be stable and well damped; and the $B(+)$ should be monic, so that the factorization is unique. In (11.117), we see that the CL system is $AR + BS$, and this combined polynomial can be designed to assign: (i) the canceled plant zeros, (ii) the desired model poles, and (iii) the desired observer poles. The latter suggests the use of the observer polynomial A_o, and it is dictated by the state space solution to the problem that is a combination of the state feedback and an observer. Thus, the CL combined polynomial (11.117) is now given in terms of these three factor-polynomials:

$$AR + BS = B^+ A_m A_o \qquad (11.119)$$

Since we know that $B(+)$ divides the polynomial B, it follows that $B(+)$ also divides the polynomial R:

$$R = B^+ R_1 \qquad (11.120)$$

Substituting (11.120) in (11.119), we obtain

$$A B^+ R_1 + B^+ B^- S = B^+ A_m A_o \qquad (11.121)$$

Simplifying (11.121), one obtains the following

$$AR_1 + B^- S = A_m A_o \qquad (11.122)$$

Comparing (11.115), (11.117), and (11.122), one gets the following so that the correspondence between the command signal and the plant output is the desired relation as in (11.115):

$$B_m = B^- B_m^+; \text{ and } T = A_o B_m^+ \qquad (11.123)$$

From (11.123), one can ascertain that the polynomial $B(-)$ should be a polynomial factor of $B(m)$, otherwise the design problem is not solved. The following conditions are required to be satisfied if the solution of (11.119) is required to exist and that give a causal control law:

1. The feedback TF S/R is causal if

$$\deg(S) \le \deg(R) \qquad (11.124)$$

There is always a solution with

$$\deg(S) \le \deg(A) - 1 \qquad (11.125)$$

It also follows from (11.119) that

$$\deg(R) = \deg(A_o) + \deg(A_m) + \deg(B^+) - \deg(A) \qquad (11.126)$$

2. The feedback TF S/R is causal and is also guaranteed by

$$\deg(A_o) \ge 2\deg(A) - \deg(A_m) - \deg(B^+) - 1 \qquad (11.127)$$

3. The feed forward TF T/R is causal and is implied by

$$\deg(A_m) - \deg(B_m) \ge \deg(A) - \deg(B) \qquad (11.128)$$

The pole placement control design (PPP) is solved by using (11.122) to obtain R_1 and S; this will give the selected poles $A(m)$ and $A(0)$ with A and B given. The desired feedback is given by (11.116), and R and T are obtained from (11.120) and (11.123). There might be several solutions to the Diophantine equation (11.119) that satisfy the causality conditions (11.127) and (11.128); however, all these solutions would give the same CLTF; the difference solutions would give different responses to the disturbances and measurement errors.

From foregoing development, it follows that the control law can be expressed as

$$u = G_m G_p^{-1} u_c - \frac{S}{R}(y - y_m)$$

$$G_p = \frac{B}{A}, \; G_m = \frac{B_m}{A_m}, \; y_m = G_m u_c$$

(11.129)

From (11.129), one can observe that the PPP can be interpreted as the model following control. This also would establish the relation between the STR and MRACS, yet the form of (11.116) would be preferable for realizations of the control laws.

11.10.2 Parameter Estimation

In order that the control law of (11.116) is realized, we need to know the parameters of the model and this is achieved by using a suitable online real time parameter estimation method. A recursive method can be employed

$$\theta(k) = \theta(k-1) + K(k)\varphi(k)\varepsilon(k)$$

(11.130)

In (11.130), $K(.)$ is the gain factor/matrix, and can also be considered as a tuning parameter, if it is constant, then it gives the updating formula similar to the MIT-rule. Basically, the recursion (11.130) minimizes the sum of squares of the prediction errors. For the LS method, the prediction error is obtained as

$$\varepsilon(k) = z^{\deg(A)}\{A(z^{-1})y(k) - B(z^{-1})u(k)\}$$

$$= y(k) - \varphi^T(k)\theta(k-1)$$

(11.131)

In (11.131), we have the regression vector/matrix that contains an arrangement of the I/O data, i.e. (the delayed values of) the input and output signals from the plant whose parameters are to be estimated:

$$\varphi(k) = [-y(k-1), \dots, -y(k - \deg(A))$$

$$u(k-d), \dots, u(k - \deg(A)] \;;$$

$$d = \deg(A) - \deg(B).$$

(1.132)

The quantity $K(.)$ is chosen variously as follows:

1. Based on a recursive solution of a set of linear equations, it is chosen as a scalar

$$K(k) = \frac{1}{\varphi^T(k)\varphi(k)}$$

(11.133)

2. For the stochastic approximation method, it is chosen as (a scalar)

$$K(k) = \left[\sum_{i=1}^{k} \varphi^T(i)\varphi(i) \right]^{-1}$$

(11.134)

3. For RLS, it is taken as

$$K(k) = \left[\sum_{i=1}^{k} \varphi(i)\varphi^T(i) \right]^{-1}$$

(11.135)

The quality of the estimates depends on the method and the kind of disturbances affecting the system. The estimates converge in a finite number of steps for the case when there are no disturbances (for $K(.)$ as in [11.135]). The ones with a constant K converge exponentially when there is persistent excitation. In case if the output, y has an additive random variables/signals, the algorithm should use $K(.)$, such that it goes to zero for increasing k, then the estimates would converge to the correct values (for $K(.)$ as in [11.134], [11.135]); these have the decreasing gains, but not useful if the process parameters are changing; for the latter one can use (11.133), or (11.135) can be replaced by the following formula

$$K(k) = \left[\sum_{i=1}^{k} \lambda^{k-i} \varphi(i)\varphi^T(i) \right]^{-1}$$

(11.136)

In (11.136), $0 \le \lambda \le 1$ is a forgetting factor or a discounting factor, and this choice of $K(.)$ corresponds to the LS estimator with an exponential discounting of the past data.

11.10.3 An Explicit Self-Tuner

An explicit self-tuner (EST) based on the PP design can be expressed as follows:

Algorithm 1

Step 1. Estimate the coefficients of the polynomials A and B recursively using the recursion (11.130) and the gain as in any one from (11.133) through (11.136).

Step 2. Substitute A and B by the estimates from the Step 1. Solve (11.122), and obtain R_1, and compute R using (11.120), and T from (11.123).

Step 3. Compute the control signal using (11.116).

Then, repeat steps 1 to 3 at each sampling instant (i.e. within each sampling period). This EST sounds more like an indirect adaptive control method (IAC).

11.10.4 An Implicit Self-Tuner

For this implicit self-tuner (IST), the design computations are eliminated, and the regulator parameters are updated directly; thus, this IST sounds more like direct adaptive control (DAC) method, as is the case with MRAC. In fact, the MRAC is often called an implicit adaptive control scheme. Firs,t let us note the following from Section 11.10.1:

$$A_m A_o y = AR_1 y + B^- Sy = BR_1 u + B^- Sy = B^- \{Ru + Sy\} \quad (11.137)$$

The model (11.137) is interpreted as a process model, and is parametrized by $B(-)$, R and S, the estimation of these gives the regulator parameters directly. In a special case of the minimum phase systems, $B(-) = b_0$, the IST algorithm can be expressed as follows:

Algorithm 2

Step 1. Estimate the coefficients of R and S of (11.137) recursively using (11.130) with the prediction error as

$$\varepsilon(k) = z^{\deg(A_o A_m)}[A_o A_m(z^{-1})y(k)$$

$$- b_0\{R(z^{-1})u(k) + S(z)y(k)\}] \quad (11.138)$$

$$= z^{\deg(A_o A_m)} A_o A_m(z^{-1})y(k) - \varphi^T(k)\theta(k-1)$$

In (11.138), we have the regression vector/matrix as

$$\varphi(k) = [-y(k-d),...,-y(k-d-\deg(S))$$

$$b_0 u(k-d),...,b_0 u(k-d-\deg(R))] \quad (11.139)$$

Step 1. For estimation, one can use either one from (11.133) through (11.136).

Step 2. Compute the control signal using (11.116) by substituting R and S with the estimates obtained on Step 1.

Then, repeat steps 1 to 2 at each sampling instant (i.e. within each sampling period).

11.10.5 Other Implicit Self-Tuners

The algorithm 2 is based on the re-parametrization of the process model (11.114), and one can easily see that (11.137) has more parameters than the original process model, and the model obtained is not linear in the parameters. Another possibility of parametrization is as follows:

$$A_m A_o y = B^- \{Ru + Sy\} = \Re u + \Im y; \Re = B^- R; \Im = B^- S \quad (11.140)$$

The estimated polynomials will have the common factor $B(-)$, (11.140), and this represents the unstable modes. It is necessary to cancel the common factor before computing the control law, in order to avoid the cancellation of such unstable modes. Then we obtain the following algorithm.

Algorithm 3

Step 1. Estimate the coefficients of the modified polynomials in (11.140).

Step 2. Cancel the factors in $\Re$, and $\Im$ to obtain R and S.

Step 3. Compute the control signal using (11.116) by substituting R and S with the estimates obtained on Step 2.

Then, repeat steps 1 to 3 at each sampling instant (i.e. within each sampling period).

This algorithm avoids a nonlinear estimation problem, but estimates more parameters than the ones in the algorithm 2, since the parameters of the polynomial $B(-)$ are estimates two times. There are other possibilities, like if $B(+)$ is constant, then one can do the following:

Let the model be written as

$$Az = u, \text{ and } y = Bz \quad (11.141)$$

If the polynomials A and B are coprime, then there exist two polynomials U and V, such that

$$UA + VB = 1 \quad (11.142)$$

It follows that

$$A_o A_m z = A_o A_m(UA + VB)z = (RA + SB)z \quad (11.143)$$

Then, from (11.141), one obtains

$$A_o A_m Uu + A_o A_m Vy - Ru - Sy = 0 \quad (11.144)$$

$$U(A_o A_m u) + V(A_o A_m y) - Ru - Sy = 0 \quad (11.145)$$

Thus, we obtain the equation (11.145) as linear in parameters. Thus, an algorithm similar to Algorithm 3 can be devised based on (11.145).

The example presented in the foregoing allows us to see a unification of many different types of control algorithmic schemes. The simple self-tuner based on the LS method (and some minimum variance control scheme/s) is the special case of Algorithm 2 with $A_o A_m = z^m$ and $B^- = b_0$. There are several algorithms in the literature [9], that are special cases of Algorithm 2 and Algorithm 1.

11.11 Stability, Convergence, and Robustness Aspects

The CLSs with the AC mechanisms become nonlinear in an overall sense of the dynamic systems [9], and the analysis becomes difficult if there are stochastic disturbances, The stability analysis, convergence of the algorithms, and error performance are main aspects for the ACSs. Also, these analyses can answer if the control structures/laws proposed in Section 11.11 are reasonable.

11.11.1 Stability

As we know, the stability is a basic requirement of any (or many) control systems. For nonlinear differential equations, the stability pertains to the stability of a particular solution, because one solution might be stable and the other unstable. As we have seen, for example in Appendix C, the stability theories of Lyapunov (and Popov) can be and have been applied for ACSs. In case of MRACSs, one desires to construct adjustable mechanisms that give stable solutions. Many analyses can be applied for ascertaining the stability of the MRACSs based on the fact that the CLS can be represented as if it is composed of a linear system and a nonlinear passive system. Then, if the linear system is strictly positive real, then using the so-called passivity theorem, one can see that the error e goes to zero. To obtain the desired setting, it is necessary to parametrize the model so that it is linear in parameters (LIP), of course, we will get only limited algorithms to our disposal:

$$y(k) = \varphi^T(t)\theta \qquad (11.146)$$

A problem with an output feedback is that it is not possible to obtain the desired setting by filtering the model error. One can add the additional signal/s, i.e. the prediction error will thus be the augmented error. Also, in order to have established stability (proofs), the regression vector should be bounded. If the regression vector contains the plant I/O signals (both), then assurance of boundedness is not easy. If the system has only a variable gain, then the regression vector contains only the command signal and this is taken and assured to be bounded. Also, normally the prediction error would go to zero; however, this will not happen unless the matrix $\sum \varphi\varphi^T/k$ is always larger than a PD matrix.

Also, when the TF $G(s)$ is not positive real, then the prediction error signal is the augmented error.

Let the system be controlled by Algorithm 2 with $A_oA_m = z^m$ and $B^- = b_0$ and PE approach using (11.130), and the following assumptions:

A1 The pole excess d is known

A2 The estimated model is at least of the same order as the plant

A3 The polynomial B has all zeros inside of the unit circle/disc ($z < 1$)

Then, signal u and y are bounded and $y(k)$ approaches the command signal $y_c(k)$ as time tends to infinity.

The foregoing result is a simple and rigorous stability proof for a reasonable adaptive control design problem. However, certain assumptions as stated are restrictive. A1 means that the time delay for discrete time systems is known accurately. For the continuous time systems, it means that the slope of the high frequency asymptote of the Bode plot is known, then it is possible to design a robust high gain regulator for the design problem. The assumption is more restrictive, because it implies that the estimated model should be at least as complex as is the true system; however, the actual system might be nonlinear with distributed parameters, and yet, all the control systems are designed using the simplified mathematical models, which neglect the high frequency dynamics. Thus, the design method should be able to cope with the model uncertainties. The parameters estimated and obtained for lower order models depend on the frequency content of the input signal, and in an AC it is important that the parameters are not updated unless the input signal has sufficient energy at these relevant frequencies. For assumption A3, the model should be LIP, and this is possible if $B(-) = b_0$. This implies that the underlying design method presupposes the cancellation of all process zeros; however, such a design will not work even for the systems with known constant parameters if the system has an unstable inverse. The foregoing results are valid if there are no disturbances, otherwise additional assumptions are necessary: (i) to put bound on the parameter estimates a priori (by introducing a saturation in the estimator), and (ii) to introduce a dead zone in the estimator that keeps the estimates constant if the residuals are small. The foregoing additional assumptions also hold good for continuous time systems.

11.11.2 Convergence

For an adaptive control system, the convergence analysis is mainly related to the online real time estimation of the model parameters, e.g. (11.130). The convergence conditions, points, and rates are important for the ACSs. In an AC case, the plant input is generated by the feedback, and hence, the excitation for the estimation process depends on the plant disturbances; in some cases, this excitation might not be persistently exciting, and the latter is a necessary condition for the convergence of the parameter estimates. Also, due to the feedback, the input is correlated with the disturbances, and the regression vector (11.130) would depend on the past estimates; hence, it is not a simple equation.

For the convergence analysis, it is assumed that the system is driven by disturbances, and hence, the methods from ergodic and martingale theories can be applied to establish the convergence for some ACSs.

11.11.2.1 Martingale Theory

The convergence condition is simply that $K(k) \to 0$ as $k \to$ infinity (based on martingale convergence theorem). The extension to adaptive system is limited to the model

$$Ay = Bu + e \quad (11.147)$$

In (11.147), the noise process is assumed to be white. The approach is Bayesian meaning the parameters are assumed to be random variables, one cannot say much about the convergence of a particular (values of) parameter. Another approach is based on modified stochastic approximation estimation and minimum variance control. This corresponds to the special case of Algorithm 2. A system model is chosen as

$$Ay = Bu + Ce \quad (11.148)$$

Application of the martingale convergence theorem has the following result: let the plant (11.148) be controlled by the Algorithm 2, with

$$A_o A_m = z^m \text{ and } B^- = b_0 \quad (11.149)$$

and $d = 1$ about a modified stochastic approximation identification (method), i.e. (11.130) with $K(k) = a_0/k$, and the assumption of Section 11.11.1, A1–A3, and assume that the function

$$G(z) = C(z) - a_0/2 \quad (11.150)$$

is strictly positive real, then inputs and outputs are mean square bounded, and the adaptive regulator converges to the minimum variance regulator for the system (11.148).

11.11.2.2 Averaging Methods

The control law algorithms in Section 11.10 are based on the assumption that the parameters change slower than the change of the state variable of the dynamic system under control action. The state variables of the overall CL control system are separated in to two groups: (i) the parameters and the estimator's tuning gain $K(.)$ change slowly, and (ii) the input and output signals of the system change somewhat faster. In such a situation,

the parameters can be described by some approximation of $K\varphi\varepsilon$ in (11.130). One can replace this by its mean value, and this method can be used to determine possible equilibrium points for such parameters. For the simple self-tuner that is based on minimum variance control and LS method or stochastic approximation, it is shown [9] that the equilibria are characterized by

$$E\{\varphi\varepsilon\} = 0 \quad (11.151)$$

Since (11.151) contains the correlations of the I/O signals, it implies the following

$$\lim_{k \to \infty} \frac{1}{k} \sum_{i=1}^{k} y(i+j)y(i) = 0; \ j = d, d+1, ..., d + \deg(S) \quad (11.152)$$

$$\lim_{k \to \infty} \frac{1}{k} \sum_{i=1}^{k} y(i+j)u(i) = 0; \ j = d, d+1, ..., d + \deg(R)$$

and characterizes the possible equilibrium even if the plant is nonlinear or of higher order. It has also been shown in the literature [4] that even when C is different from 1 in (11.149), the minimum variance control is an equilibrium and the LS estimates are biased. Also, since the gain $K(.)$ goes to zero as time goes to infinity, the separation of fast and slow modes will improve as time increases. Also, it has been shown that asymptotically as time goes to infinity, the estimates are described by ordinary differential equation. Let us take the algorithm based on the stochastic approximation (11.130) with $K(.)$ given by (11.134). Let us have the transformed equation as

$$j(k) = c \sum_{i=1}^{k} \| K(i) \| \quad (11.153)$$

Then, the estimates would be approximately described by the solutions to the ordinary differential equation

$$\frac{d\theta}{dj} = f(\theta); \ f(\theta) = E\{\varphi\varepsilon\} \quad (11.154)$$

The estimates converge to the solution of (11.154) as time increases. Due to this, the approach is also referred to as "ODE-approach." The method is useful for determination of possible equilibrium points and their local stability and also for determination of convergence rates.

However, the method is based on the assumption that the signals are bounded. A similar analysis of the self-tuner based on minimum variance control and LS estimation obtained the condition that the following function is strictly positive real:

$$G(z) = 1/C(z) - 1/2 \quad (11.155)$$

When averaging technique is applied to the tracking problem, the parameters would not converge because the gain $K(k)$ does not go to zero.

11.12 Use of the Stochastic Control Theory

The regulators like STR and MRAC are based on the heuristic arguments. Similar regulators can be obtained based on the unified theoretical framework, say by using nonlinear stochastic control theory. The plant and its environment are described by a stochastic model, and the criterion is formulated as to minimize the expected value of a loss function, and this is a scalar function of the stated and the control signals. With the assumption that a solution exists, a functional equation, called Bellman equation (Chapter 6), for the optimal loss function can be derived using the dynamic programming, which can be solved for simple cases. Such an adaptive regulator would consist of three blocks: (i) plant, (ii) computation of the hyper state based on the I/O of the plant, and (iii) nonlinear function that is driven by the hyper state and the commanded control input, this block then outputs a signal that controls the plant. The controller is composed of two parts: (i) the estimator that generates the conditional PDF of the states from the measurements, (this distribution is the hyper state of the problem), and (ii) the feedback regulator that is nonlinear function which maps the hyper state into the space of the control variables. The STR can be regarded as an approximation where the conditional PDF is replaced by a distribution with all the mass at the conditional mean value. In the hyper state adaptive regulator (HSAR), there is no distinction between the parameters and the other state variables, the regulator handles very rapid parameter variations. The control attempts to drive the output to its desired value, but this will also introduce perturbations when the parameters are uncertain. This will improve the estimates and the future control actions, and the optimal control gives the correct balance between maintaining good control and small estimation errors.

11.13 Uses of Adaptive Control Approaches

Some ways of using the adaptive control techniques are discussed next.

11.13.1 Auto-Tuning

It is possible to tune the regulators with three/four parameters by hand if there is not much interaction between adjustments of different parameters. For more complex regulators, it is necessary to have suitable tuning tools. Classically, the tuning of more complex regulators has used the route of modeling or identification and regulator design. The STR and MRAC are the constant gain feedback controls when the estimated parameters are constant. The adaptive loop can be used as a tuner for a control loop. By simply switching on the adaptive loop, the perturbation signals can be added, and the adaptive regulator is run until the satisfactory performance is obtained. Then, the adaptation loop is discontinued and the system is left running with fixed regulator parameters. Auto-tuning can be considered as a convenient way to incorporate automatic modeling and design into a regulator. Auto-tuning is useful for design methods like feed forward, which critically depend on good models. It is useful to combine auto-tuning with diagnostic tools for checking the performance of the control loops. For the minimum variance control, the performance evaluation can be done by monitoring the covariances of the inputs and outputs. In a custom package, one tuning algorithm can serve many loops. Auto-tuning can be included in single-loop regulators, with three conditions of the mode switch: manual, automatic, tuning. The tuning algorithm would represent a major part of the software of a single loop regulator.

11.13.2 Automatic Construction of Gain Schedulers and Adaptive Regulators

The adaptive control loop can also be used for gain scheduling. The parameters obtained when the system is running in one operating condition are stored in a table. The GS is obtained when the plant has operated at a range of operating conditions that covers the operating range. The GS can be also combined with the adaptation: a GS can be used to quickly get the parameters into the correct domain/range and the adaptation can be used for fine tuning.

The adaptive techniques can be used for adaptive control of systems with time varying parameters. The adaptive regulators may also have parameters that must be chosen. The regulators without any externally adjusted parameters can be designed for specific applications where the aim of control can be stated a priori. One can introduce dials that give the desired properties of the CLS, and are called performance related dials. It is possible to have a regulator with one dial that is labeled with the desired CL bandwidth, another with the weighting between state deviation and control action in an LQG based regulator, and yet another could be a dial labeled with the phase margin of the amplitude margin. There are many cases when a constant gain feedback can do as well as an adaptive regulator. The aircrafts, in many cases, could be easily be controlled with conventional methods. There are many cases where a constant gain feedback can do well with considerable variations in system dynamics, and some design techniques for constant gain feedback can cope with considerable gain variations.

11.13.3 Practical Aspects and Applications

The system obtained when a PID regulator is connected to a linear system is well understood and can be analysed with great precision; however, there are several aspects that need attention: (i) hand/automatic transfer, (ii) bumpless parameter changes, (iii) reset windup and (iv) nonlinear output. Certain PID regulators can be connected via logical selectors. Then, the overall systems become nonlinear and the linear analysis would be of limited use. For adaptive control, the situation is similar, the windup can occur in the inner and outer loops.

11.13.3.1 Parameter Tracking

Since the ability to track the variations in the plant dynamics is needed, the performance of parameter estimator is important. The main aspect for the SIPE is that the input signal to the plant (and hence, for SIPE) must be persistently exciting or sufficiently rich. In an adaptive system, the input signal is generated by the feedback, and there is no assurance that the plan is properly excited. A good regulation might give a poor excitation, and hence, some extra perturbation signals might be required. This is the aspect of dual control theory. If the perturbation signal is not introduced, then when the system is poorly excited, the parameter updated should be turned off. Also, to track the parameter variation, it is necessary to discount old/past data. The data should not be discounted too fast, else the estimates will be uncertain, and not too slow, else the tracking of rapid parameter variations will not be good. One can use the exponential discounting of the data, as is done for LS with gain as in (11.136). The forgetting factor is given as

$$\lambda = 1 - \frac{\Delta t}{\Delta t_e} \; ; \; \Delta t - \text{the sampling period;}$$

(11.156)

Δt_e – the time constant of exponential discounting

The time constant of the exponential discounting can be chosen as a trade-off between the precision of the estimates and the ability to track the parameter variations.

11.13.3.2 Estimator Windup and Bursts

When the major excitation is caused by set point changes, there might be long periods of no excitation at all; and when the set point is kept constant, the relevant data will be discounted even if no new data are obtained. Let us take the following formula for the updating of the gain factor

$$K(k) = [\varphi(k)\varphi^T(k) + \lambda K^{-1}(k-1)]^{-1}$$

(11.157)

If the plant is poorly excited, the I/O data signals are small, and consequently the regression vector is also small, and in the limit case, this vector goes to zero and $K(.)$ grows exponentially. The latter behavior makes the estimator unstable; the small residuals induce large changes in the parameters and the CLS might become unstable. This will interestingly and ironically result in the plant being well excited, and the parameter estimates will quickly obtain good values. This phenomenon is that of periods of good regulation followed by bursts, the latter behavior is due to the fact that the estimator with exponential forgetting factor is an OL unstable system, lending the name of estimator windup, a similar thing happens as the integer windup in simple regulators. The phenomenon of the estimator windup occurs because of the poor excitation and discounting of the old data, also suggesting the remedy for the problem: (i) the condition of the persistent excitation can be monitored and perturbation can be introduced if required, (ii) the covariance windup can be avoided by discounting the old data only when there is poor excitation, and (iii) update the estimates only when the amplitude of the regression vector $|\varphi|, or |K\varphi|$ is larger than a given quantity. One can use a varying forgetting factor.

11.13.3.3 Robustness

It is important for the adaptive control to be robust, e.g. with respect to additive and multiplicative uncertainties. Various methods for design of robust control have been discussed in Chapter 8. There could be instabilities in the adaptive control (laws, algorithms) due to small disturbances, high gains, high frequency operations, fast adaptation, and time varying parameters of the system/model [12]. Sliding mode control is one of the approaches as an alternative to the robust adaptive control. In a conventional approach, one would design an adaptive controller and include robustness later. However, one can design a robust controller first to guarantee transient performance to some error bound and include parameter adaptation later using an appropriate method. Certain work on nonlinear controller design also led to the design of adaptive nonlinear observers. Adaptive sliding mode control (ASMC) is a variable structure control method that specifies a manifold or surface along with the system that will operate or "slide." When the performance deviates from the manifold, the controller provides an input in the direction back towards the manifold to force the system back to the desired output; in fact, the sliding mode adds some robustness to the system. In general, we need the overall robustness of adaptive controllers to external disturbances and un-modeled system dynamics. It must be ensured that the regulator works well under extreme conditions. One can limit the control signal and its rate of change, or one can use robust estimation scheme (Chapter 8). One can use the following estimator

$$\theta(k) = \theta(k-1) + K(k)\varphi(k)f\{\varepsilon(k)\} \qquad (11.158)$$

In (11.158), f is a saturation function or a similar non-linearity, and it is determined from the PDF of the disturbances.

11.13.3.4 Numerics and Coding

Numerical problems should not occur in any operating mode/s. If the models are over parametrized, then the estimation methods are poorly conditioned. The LS method is poorly conditioned for high SNRs. The DC level also should be removed from the signals, and use square root algorithm/s instead of the ordinary conventional algorithms. Many control design methods are poorly conditioned for some model parameters, and singularities are associated with any loss of observability or controllability of the system, in fact of the model. In such cases, the parameter estimator algorithm might give any value, and it is important to be safe against the difficulties that might arise in explicit or indirect schemes.

The adaptive regulators are easily implemented using microprocessor/microcomputers. A simple STR can be coded in less than 100 lines in a high-level computer programming language. The adaptive algorithms are at least an order of magnitude more complex than fixed gain regulators.

11.13.3.5 Integral Action

In a control system, the control action should maintain small errors in spite of large low-frequency disturbances. For the fixed gain regulators, this is achieved by the integral action that gives a high loop gain at low frequencies. In ACSs, similar actions can be introduced automatically when needed. This can be also added via some parameter estimation schemes, or one can use special regulator structure/s.

11.13.3.6 Supervisory Loops

As we have seen, the STR/MRACSs have some hierarchical systems of levels: (i) a lower level of ordinary feedback loop with the plant and the regulator, and (ii) the adaptation loop that adjusts the parameters, a higher level loop. In STR/MRACSs, one would need other loops for model order (updating), forgetting factor, and the sampling period. Thus, a third layer can be added to the existing hierarchical levels to set these parameters. The forgetting factor can be determined by monitoring the excitation process. Also, by storing the plant I/O data signals, it is possible to estimate the models with different sampling periods, different model orders, and different structures.

11.13.3.7 Applications

1. Laboratory experiments. The aim has been to understand the algorithms and to investigate many aspects to complement the theoretical developments.

2. Industrial feasibility studies. These have been carried out in/for: (i) autopilots for aircrafts, missiles, and ships, (ii) cement mills, glass furnaces, heat exchangers, heating and ventilation, titanium oxide kiln, and chemical reactors, (iii) diesel engines, motor drivers, rolling mills, and ore crushers, and (iv) optical telescopes, paper machines, pH-control, power systems, and digesters. Many such applications and associated feasibility studies have shown that the adaptive control was very useful. Although in some cases, the benefits were marginal.

3. Signal processing applications. The algorithm studies in this chapter have been used in the communication field, and almost parallel development of AC algorithms has happened. In the case of signal processing, the main use has been of adaptive filtering. Other uses are and have been in/for: (i) adaptive noise cancellation and (ii) adaptive echo cancellation (based on VLSI technology).

4. Industrial products. The GS is being used for the design of autopilots for high performance aircrafts and in the process industry. It has been found very easy to implement the GS using the modern hardware for distributed process control. There are products based on STR and MRACSs as standalone systems and as software (in dedicated) packages.

Appendix 11A: STC Model Structures for System Identification and Parameter Estimation

We discuss several mathematical model structures and forms that would be useful for system identification and parameter estimation in self-tuning control (STC), explicit indirect adaptive control (EIDAC), and IDAC schemes. The so-called auto regressive with exogenous/external explicit inputs (ARX) model with additive disturbance as a white Gaussian noise with zero mean and known covariance is a relatively simple model preferred for the purpose. An extension of ARX is the auto regressive moving average with exogenous inputs (ARMAX) model structure, where the noise is modeled as the output of a moving average process. However, we introduce here various model structures [13] that might be useful for the indirect (explicit) STC/adaptive control, i.e. for explicit indirect adaptive control (EIAC); and some of these can be used for implicit direct adaptive control (IDAC), i.e. for model reference adaptive control (MRAC). In some special cases, it might be useful to employ the state space mathematical models in/for system identification-parameter estimation (SIPE) for the parameter adaptation algorithm (PAA), which are discussed in Appendix IIA.

11A.1 Time Series Models

These are very simple and useful models for SIPE, wherein the problem of determining coefficients of a differential or difference equation is addressed, which in turn, are fitted to the empirical data; often one obtains coefficients of a transfer function model of a system from its I/O data. The main aim of time series modeling is the use of the model for prediction of the future behavior of the system or phenomena; and may also be used in model predictive control (MPC). For the use in STC, the parameters of the models are estimated using the I/O data obtained from the plant that is operating within the closed loop. We will generally deal with DTSs. Though, many phenomena occurring in nature are of continuous type and can be described by CTS, the theory of the DT modeling is very handy, and the estimation algorithms can be easily implemented using a digital computer and for adaptive control (AC). A general linear stochastic discrete time model is given as:

$$x(k+1) = \Phi_k x(k) + Bu(k) + w(k)$$
$$z(k) = Hx(k) + Du(k) + v(k) \tag{11A.1}$$

The time series-canonical form known as Astrom's model is given as

$$A(q^{-1})z(k) = B(q^{-1})u(k) + C(q^{-1})e(k) \tag{11A.2}$$

In (11.A.2), A, B, and C are polynomials in $q^{-1}(=z)$, which is a shift operator (with one sampling interval) defined as

$$q^{-n}z(k) = z(k-n) \tag{11A.3}$$

The SISO expanded form is:

$$z(k) + a_1 z(k-1) + \ldots$$
$$+ a_n z(k-n) = b_0 u(k) + b_1 u(k-1) + \ldots \tag{11A.4}$$
$$+ b_m u(k-m) + e(k) + c_1 e(k-1) + \ldots$$
$$+ c_p e(k-p)$$

In (11A.4), z is the discrete measurement sequence, u is the input sequence, and e is the random noise/error sequence. We have the following equivalence with (11A.2):

$$A(q^{-1}) = 1 + a_1 q^{-1} + \ldots + a_n q^{-n}$$
$$B(q^{-1}) = b_0 + b_1 q^{-1} + \ldots + b_n q^{-m} \tag{11A.5}$$
$$C(q^{-1}) = 1 + c_1 q^{-1} + \ldots + c_n q^{-p}$$

In (11A.5), a_i, b_i and c_i are the coefficients of the polynomial to be estimated and the plant parameters of significance that are required to be ascertained in order to capture the changing dynamics of the plant for AC. We also assume here that the noise processes w and v are uncorrelated and white. Also, the time series (I/O data) are assumed to be stationary, in the sense that the first and second order (and higher) statistics are not dependant on time t explicitly; for mildly non-stationary time series, the appropriate models can be fitted to the segments of such time-series. We assume that the inputs would excite the modes of the system, i.e. they contain sufficient frequencies to excite the dynamic modes of the system, the case of the persistent excitation. This will assure that in the output, there is sufficient effect of the modes of the plant and the information, so that from I/O time-series data, one can accurately estimate the characteristics of the process/plant/system:

1. *Astrom's Model*: Given input (u)/output (z) data, the parameters can be estimated by some iterative process, e.g. ML (maximum likelihood) method. The transfer function form is given by, Figure 11A.1:

$$z = \frac{B(q^{-1})}{A(q^{-1})} u + \frac{C(q^{-1})}{A(q^{-1})} e \tag{11A.6}$$

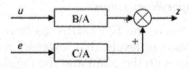

FIGURE 11 A.1
Astrom's Model.

This model can be used to fit time-series data, which can be considered to be arising out of a system with a control input u and a random excitation.

2. *Autoregressive (AR) Model*: If we use $b_i = 0$ and $c_i = 0$ in the Astrom's model, we get

$$z(k) = -a_1 z(k-1) - \ldots - a_n z(k-n) + e(k) \quad (11\text{A}.7)$$

The transfer function form can be easily obtained as:

$$z = \frac{1}{A(q^{-1})} e \quad (11\text{A}.8)$$

Here, the output process $z(k)$ depends on its previous values (autoregressive) and it is excited by the random signal e. It is assumed that the parameters a_i are constants such that the process z is stationary.

The form $1/A(q^{-1})$ is an operator (Figure 11A.2), which transforms the process e into the process z. The polynomial A determines the characteristics of the output signal z and the model is the "all poles" model, since the roots of $A(q^{-1}) = 0$ are the poles of the TF model. The input process e is inaccessible and immeasurable. The parameters of A are estimated by using the least squares (LS) method.

3. *Moving Average (MA) Model*: For $a_i = 0$ and $b_i = 0$ in the Astrom's model, we have

$$z(k) = e(k) + c_1 e(k-1) + \ldots + c_p e(k-p) \quad (11\text{A}.9)$$

The z is now a linear combination of the past and present values of the inaccessible random input process e.

The roots of $C(q^{-1}) = 0$ are the zeros of the model, Figure 11A.3. The process z is called the MA process and is always stationary since $A(q^{-1}) = 1$; the output signal does not regress over its past values.

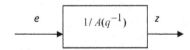

FIGURE 11 A.2
AR Model.

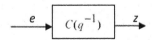

FIGURE 11 A.3
MA Model.

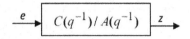

FIGURE 11 A.4
ARMA Model.

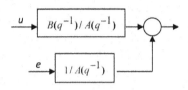

FIGURE 11 A.5
LS Model.

4. *Autoregressive Moving Average (ARMA) Model*: If $b_i = 0$ in the Astrom's model, we obtain an ARMA model since it contains both AR and MA parts (Figure 11A.4):

$$z(k) + a_1 z(k-1) + \ldots + a_n z(k-n) = \\ e(k) + c_1 e(k-1) + \ldots + c_p e(k-p) \quad (11\text{A}.10)$$

$$z = \frac{C(q^{-1})}{A(q^{-1})} e \quad (11\text{A}.11)$$

The model is a zero/pole type model and has the structure of the input/output (O/I) model. More complex time-series can be accurately modeled using this model.

5. *Least Squares Model*: With $c_i = 0$ in the Astrom's model, we get

$$z(k) + a_1 z(k-1) + \ldots + a_n z(k-n) = b_0 u(k) + b_1 u(k-1) \\ + \ldots + b_m u(k-m) \quad (11\text{A}.12) \\ + e(k)$$

Control input u is present. The model is so called (LS), since its parameters can be easily estimated by LS method. The transfer function form is, Figure 11A.5:

$$z = \frac{B(q^{-1})}{A(q^{-1})} u + \frac{1}{A(q^{-1})} e \quad (11\text{A}.13)$$

It has an AR model for the noise part and the O/I model for the signal part. Determination of $B(q^{-1}) / A(q^{-1})$ gives the TF model of the system.

11A.2 Model Identification

The estimation of the parameters of MA and ARMA models can be done using the ML approach, since the unknown parameters appear in MA part, which represents itself as unknown time series e; however, parameters of AR and LS models can be estimated using the LS method. However, actually for online real time (OLRT) estimation, as is deemed necessary for STC and adaptive control in general, one needs to resort to some computationally simpler, efficient, and numerically stable/accurate estimation methods. The stability of the estimator is very important, since any divergence of the parameter estimation algorithm cannot be tolerated while the algorithm is working in the overall (closed loop) control and the controlled system is operating in real time. However, here the batch method is presented for the sake of introduction of parameter estimation. Assumption of the identifiability of the coefficients of the postulated models is pre-supposed (Appendix A). Let the LS model be given as in (11A.12) and (11A.13). We define the equation error as (Figure 11A.6):

$$e(k) = A(q^{-1})z(k) - B(q^{-1})u(k)$$

$$r(k) = \hat{A}(q^{-1})z(k) - \hat{B}(q^{-1})u(k) \qquad (11A.14)$$

In the compact form we have

$$z = H\beta + e, \ z = \{z(n+1), z(n+2), \dots z(n+N)\}^T, \text{ where}$$

$$H = \begin{bmatrix} -z(n) & -z(n-1) - z(1) & u(n) & u(n-1) \ u(1) \\ -z(n+1) & -z(n) & -z(2) & u(n+1) & u(n) \ u(2) \\ \vdots & & \vdots & & \vdots \\ -z(N+n-1) & \cdots & -z(n) \ u(N+n-1) & \dots & u(N) \end{bmatrix} \qquad (11A.15)$$

N = number of total data used; $m = n$ and $b_0 = 0$; for example, let $n = 2$ and $m = 1$, then

$$e(k) = z(k) + a_1 z(k-1) + a_2 z(k-2) - b_0 u(k) - b_1 u(k-1) \quad (11A.16)$$

$$z(k) = \begin{bmatrix} -z(k-1) - z(k-2) \end{bmatrix} \begin{bmatrix} a_1 \\ a_2 \end{bmatrix}$$
$$+ \begin{bmatrix} u(k) \ u(k-1) \end{bmatrix} \begin{bmatrix} b_0 \\ b_2 \end{bmatrix} + e(k) \qquad (11A.17)$$

$$z(k+1) = \begin{bmatrix} -z(k) - z(k-1) \end{bmatrix} \begin{bmatrix} a_1 \\ a_2 \end{bmatrix}$$
$$+ \begin{bmatrix} u(k+1) \ u(k) \end{bmatrix} \begin{bmatrix} b_0 \\ b_2 \end{bmatrix} + e(k+1) \qquad (11A.18)$$

The foregoing leads to the following formulation for the LS system identification/parameter estimation:

$$z = H\beta + e$$

$$\hat{\beta} = \{\hat{a}_1, \hat{a}_2, \dots, \hat{a}_n : \hat{b}_1, \hat{b}_2, \dots, \hat{b}_m\} = (H^T H)^{-1} H^T z \qquad (11A.19)$$

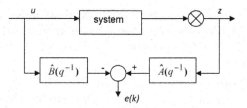

FIGURE 11 A.6
Equation error (EE) formulation.

The important aspect in time series modeling is that of selection of model structure (AR, MA, ARMA, or LS) and the number of coefficients required for fitting this model to the time series I/O data. It is better to auto-regressive (AR) or a least squares (LS) model structure. The problem of model order determination is to assign a model dimension so that it adequately represents the unknown system: involves selecting a model structure and just a sufficient amount/degree of complexity. A model structure can be ascertained based on the knowledge of the physics of the system; and if physics is not well understood, then a black-box approach can be used. However, in many situations, some knowledge about the system or the process is always available; and so is the case for the use of such models for PAA in EIAC, IDAC.

11A.3 Least Squares Methods

We discuss the least squares/equation error methods (LS/EEM) for parameter estimation. The measurement equation/model is given as:

$$z = H\beta + v, \quad y = H\beta \qquad (11A.20)$$

In (11A.20), y is (mx1) vector of true outputs, z is (mx1) vector that denotes the measurements (affected by noise) of the unknown parameters (through H), β is (nx1) vector of the unknown parameters, and v is the measurement noise/errors, which are assumed to be zero mean and Gaussian. These noisy measurements are utilized in the estimation procedure/algorithm/software to improve upon the initial guesstimate of the parameters that characterize the signal or system. One of the objectives of the estimator is to produce the estimates of the signal (i.e. the predicted signal using the estimated parameters) with errors much less than the noise affecting the signal. For this, the signal and the noise should have significantly differing characteristics, e.g. different frequency spectra and widely differing statistical properties (true signal is deterministic, and the noise is random). This means that the signal is characterized by a structure or a mathematical model (like $H\beta$) and the noise (v), often or usually, is assumed as zero mean and white process. In most cases, the measurement noise is also considered Gaussian. The LS method is a special case of the well-known ML estimation method for linear systems with Gaussian noise. For this method, it is assumed that the system parameters do not rapidly change with time, (the plant or the process is stationary); the plant is assumed quasi-stationary during the measurement period. This should not be confused with the requirement of non-steady I/O data over the period for which the data is collected for parameter estimation; i.e. during the measurement period there should be some activity. We choose an estimator of β that minimizes the sum of the squares of the error

$$ J \cong \frac{1}{2}\sum_{k=1}^{N} v_k^2 = \frac{1}{2}(z-H\beta)^T(z-H\beta) \quad (11A.21) $$

Here, J is a cost function, v the residual errors at time k (index), and T stands for the vector/matrix transposition. The minimization of J w.r.t. β yields

$$ \partial J/\partial\beta = -(z-H\hat\beta_{LS})^T H = 0 \text{ OR} $$
$$ H^T(z-H\hat\beta_{LS}) = 0 \quad (11A.22) $$

$$ H^T z - (H^T H)\hat\beta_{LS} = 0 \quad \text{or} \quad \hat\beta_{LS} = (H^T H)^{-1} H^T z \quad (11A.23) $$

Since the matrix H and the vector z are known quantities, $\hat\beta_{LS}$, the LS estimate of β, can be readily obtained.

Equation error method (EEM) is based on the principle of LS and minimizes a quadratic cost function of the error in the (state) equations; and the states, their derivatives, and control inputs are required to be accurately measured. The EEM is relatively fast, simple, and applicable to linear as well as linear-in-parameter systems. The system is described by the state equation

$$ \dot x = Ax + Bu \quad \text{with} \quad x(0) = x_0 \quad (11A.24) $$

The equation error is written as

$$ e(k) = \dot x_m - A x_m - B u_m \quad (11A.25) $$

Here, xm is the measured state. Parameter estimates are obtained by minimizing the equation error w.r.t. β. The error equation is written as

$$ e(k) = \dot x_m - A_a x_{am} \quad (11A.26) $$

Also, $A_a = [A \quad B]$ and $x_{am} = \left[\frac{x_m}{u_m}\right]$. The cost function is given by

$$ J(\beta) = \frac{1}{2}\sum_{k=1}^{N} \left[\dot x_m(k) - A_a x_{am}(k)\right]^T \left[\dot x_m(k) - A_a x_{am}(k)\right] \quad (11A.27) $$

The estimates are given as:

$$ \hat A_a = \dot x_m(x_{am}^T)(x_{am}\,x_{am}^T)^{-1} \quad (11A.28) $$

The EEM requires accurate measurements of the states and their derivatives. It can be applied to unstable systems because it does not involve any numerical integration of the dynamic system that would otherwise cause divergence. Also, the utilization of measured states and state-derivatives in the algorithm enables estimation of the parameters of even an unstable system directly. However, if the measurements are noisy, the method will give biased estimates. We would like to mention here that the EEM formulation can be programmed in the structure of recurrent neural network as well.

11A.4 Real Time Parameter Estimation

Kalman filter being a recursive algorithm is more suitable for real time estimation of parameters. For OLRT estimation problem, several aspects are important: (i) the recursive estimation algorithm should be robust, (ii) it should converge to an estimate close to the true value, (iii) computational requirements should be low or very low, and (iv) the algorithm should be numerically reliable and stable. If the plant dynamics are changing, as is the case for AC, there is always some activity that is useful for parameter estimation.

11A.4.1 Kalman Filter-UD Filter

The Kalman filter in its factorization form of UD filtering algorithm is a feasible approach for such a purpose (U is unit upper triangular matrix with its diagonal with unitary values, and D is the diagonal matrix). It is computationally efficient, numerically reliable and stable, and algebraically equivalent to KF. For parameter estimation, it can be used in the extended Kalman filter/UD filter mode (EKF/EUD). For recursive least squares (RLS) method, factorization scheme can also be used, in fact here the measurement update part of the KF/EKF is in fact RLS method, and the state equation has the identity transition matrix. The reconfigured state equations are given as

$$\beta(k) = \beta(k-1) + w(k); \quad Q(k) = E\{w(k)w^T(k)\}$$
$$z(k) = H(k)\beta(k) + v(k): \ R(k) = E\{v(k)v^T(k)\} \tag{11A.29}$$

The reconfigured parameter estimator in the recursive mode for use in STC/IAC is given as

$$\tilde{\beta}(k+1) = \hat{\beta}(k); \tilde{P}(k+1) = \hat{P}(k) + Q \text{ (a priori parameter}$$
estimation and covariance) $\tag{11A.30}$

$$r(k+1) = z(k+1) - H \ \tilde{x}(k+1); K = \tilde{P}H^T(H\tilde{P}H^T + R)^{-1}$$
(residuals; Kalman gain) $\tag{11A.31}$

$$\hat{\beta}(k+1) = \tilde{\beta}(k+1) + K \ r(k+1); \hat{P} = (I - KH)\tilde{P} \text{ (a posteriori}$$
estimation) $\tag{11A.32}$

11A.4.2 Parameter Estimation Using Recurrent Neural Networks

The artificial neural networks (ANNs) provide new paradigms to handle the problem of parameter estimation with potential application to OLRT estimation, especially for STC/IAC. In fact, the recurrent neural networks (RNNs) are easily amenable to such possibilities due to their special structure, i.e. these RNNs are feed forward neural networks (FFNNs) with feedback feature (Figure 11A.7) and with their massively parallel processing capacity, can be easily adapted to OLRT parameter estimation.

The Hopfield neural network (HNN) has a number of mutually interconnected information processing units called neurons, and the outputs of the network are nonlinear function of the states of the network. The dynamic representation of the network is given as in Figure 11A.8:

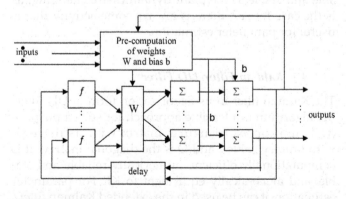

FIGURE 11 A.7
An RNN structure: Feedforward ANN with feedback. (Adapted from Raol, J.R. et al., *Modelling and Parameter Estimation of Dynamic Systems*, IET/IEE Control Series Vol. 65, IET/IEE Society, London, UK, 2004.)

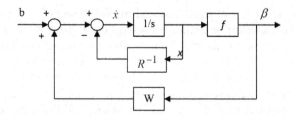

FIGURE 11 A.8
HNN-RNN structure-Dynamics. (Adapted from: Raol, J.R. et al., *Modelling and Parameter Estimation of Dynamic Systems*, IET/IEE Control Series Vol. 65, IET/IEE Society, London, UK, 2004.)

$$\dot{x}_i(t) = -x_i(t)R^{-1} + \sum_{j=1}^{n} w_{ij}\beta_j(t) + b_i; j = 1,...,n \tag{11A.33}$$

Here, x is the internal state of the neurons, β the output state, $\beta_j(t) = f(x_j(t))$, w_{ij} are the neuron weights, b the bias input to the neurons, and f the sigmoid nonlinearity. R is the neuron impedance and n is the dimension of the neuronal state. One can write (11A.33) as

$$\dot{x}(t) = -x(t)R^{-1} + W\{f(x(t)\} + b \tag{11A.34}$$

The RNN (11A.34) obtains a simple system: neuron as a transducer of input to output and a smooth sigmoidal response up to a maximum level of output, and feedback nature of connections.

Consider the dynamic system for which the parameters, the elements of matrices A, and B are to be estimated:

$$\dot{x} = Ax + Bu; x(0) = x_0 \tag{11A.35}$$

For parameter estimation, $\beta = \{A, B\}$ represents the parameter vector to be estimated and n is the number of parameters to be estimated. We can consider that the dynamics are affected by the nonlinear function f, i.e. $\beta_i = f(x_i)$, and the cost function is given as

$$E(\beta) = \frac{1}{2}\sum_{k=1}^{N} e^T(k) \ e(k) = \frac{1}{2}\sum_{k=1}^{N}(\dot{x} - Ax - Bu)^T \ (\dot{x} - Ax - Bu) \tag{11A.36}$$

Here, $e(k)$ is the equation error:

$$e = \dot{x} - Ax - Bu \tag{11A.37}$$

By appropriate optimization, we have:

$$\frac{d\beta}{dt} = -\frac{\partial E(\beta)}{\partial \beta} = -\frac{1}{2}\frac{\partial\left\{\sum_{k=1}^{N} e^T(k) \ e(k)\right\}}{\partial \beta} \tag{11A.38}$$

Since β as a parameter vector contains the elements of A and B, we can obtain the following expressions for $\partial E / \partial A$ and $\partial E / \partial B$ for A and B vectors, with $\sum(.) = \sum_{k=1}^{N}(.)$

$$\frac{\partial E}{\partial A} = \sum(\dot{x} - Ax - Bu)(-x^T) = A\sum x\,x^T + B\sum u\,x^T - \sum \dot{x}\,x^T$$

$$\frac{\partial E}{\partial B} = \sum(\dot{x} - Ax - Bu)(-u) = A\sum x\,u + B\sum u^2 - \sum \dot{x}\,u$$

$$(11A.39)$$

Expanding we get, for example for $A(2,2)$ and $B(2,1)$:

$$\begin{bmatrix} \dfrac{\partial E}{\partial a_{11}} & \dfrac{\partial E}{\partial a_{12}} \\[2mm] \dfrac{\partial E}{\partial a_{21}} & \dfrac{\partial E}{\partial a_{22}} \end{bmatrix} = \begin{bmatrix} a_{11} & a_{12} \\ a_{21} & a_{22} \end{bmatrix} \begin{bmatrix} \sum x_1^2 & \sum x_1 x_2 \\ \sum x_2 x_1 & \sum x_2^2 \end{bmatrix} +$$

$$\begin{bmatrix} b_1 \\ b_2 \end{bmatrix} \begin{bmatrix} \sum ux_1 & \sum ux_2 \end{bmatrix} - \begin{bmatrix} \sum \dot{x}_1 x_1 & \sum \dot{x}_1 x_2 \\ \sum \dot{x}_2 x_1 & \sum \dot{x}_2 x_2 \end{bmatrix}$$

$$(11A.40)$$

Simplifying (11A.40), we get:

$$\frac{\partial E}{\partial a_{11}} = a_{11}\sum x_1^2 + a_{12}\sum x_2 x_1 + b_1 \sum x_1 u - \sum \dot{x}_1 x_1$$

$$\frac{\partial E}{\partial a_{12}} = a_{11}\sum x_1 x_2 + a_{12}\sum x_2^2 + b_1 \sum ux_2 - \sum \dot{x}_1 x_2$$

$$(11A.41)$$

$$\frac{\partial E}{\partial a_{21}} = a_{21}\sum x_1^2 + a_{22}\sum x_2 x_1 + b_2 \sum ux_1 - \sum \dot{x}_2 x_1$$

$$\frac{\partial E}{\partial a_{22}} = a_{21}\sum x_1 x_2 + a_{22}\sum x_2^2 + b_2 \sum ux_2 - \sum \dot{x}_2 x_2$$

In addition, we have:

$$\frac{\partial E}{\partial b_1} = a_{11}\sum x_1 u + a_{12}\sum x_2 u + b_1 \sum u^2 - \sum \dot{x}_1 u$$

$$(11A.42)$$

$$\frac{\partial E}{\partial b_2} = a_{21}\sum x_1 u + a_{22}\sum x_2 u + b_2 \sum u^2 - \sum \dot{x}_2 u$$

Next, assuming that the impedance R is very high, we describe the dynamics of RNN-S as:

$$\dot{x}_i = \sum_{j=1}^{n} w_{ij}\beta_j + b_i \qquad (11A.43)$$

We also have $E = -\frac{1}{2}\sum_i \sum_j W_{ij}\beta_i \beta_j - \sum_i b_i \beta_i$ as the energy landscape of the RNN, then, we get

$$\frac{\partial E}{\partial \beta_i} = -\sum_{j=1}^{n} w_{ij}\beta_j - b_i \qquad (11A.44)$$

$$\text{or } \frac{\partial E}{\partial \beta_i} = -\left[\sum_{j=1}^{n} w_{ij}\beta_j + b_i\right] = -\dot{x}_i \qquad (11A.45)$$

$$\text{or } \dot{x}_i = -\frac{\partial E}{\partial \beta_i}$$

Since $\beta_i = f(x_i)$, $\dot{x}_i = (f^{-1})'\,\dot{\beta}_i \qquad (11A.46)$

Thus, $(f^{-1})'\,\dot{\beta}_i = -\dfrac{\partial E}{\partial \beta_i}$; here, $'$ denotes derivative w.r.t β.

Hence, $\dot{\beta}_i = -\dfrac{1}{(f^{-1})'(\beta_i)}\,\dfrac{\partial E}{\partial \beta_i}$

$$= \frac{1}{(f^{-1})'(\beta_i)}\left[\sum_{j=1}^{n} w_{ij}\beta_j + b_i\right] \qquad (11A.47)$$

The expressions for the weight matrix W and the bias vector b are:

$$W = -\begin{bmatrix} \sum x_1^2 & \sum x_2 x_1 & 0 & 0 & \sum ux_1 & 0 \\ \sum x_1 x_2 & \sum x_2^2 & 0 & 0 & \sum ux_2 & 0 \\ 0 & 0 & \sum x_1^2 & \sum x_2 x_1 & 0 & \sum ux_1 \\ 0 & 0 & \sum x_1 x_2 & \sum x_2^2 & 0 & \sum ux_2 \\ \sum x_1 u & \sum x_2 u & 0 & 0 & \sum u^2 & 0 \\ 0 & 0 & \sum x_1 u & \sum x_2 u & 0 & \sum u^2 \end{bmatrix}$$

$$(11A.48)$$

$$b = -\begin{bmatrix} \sum \dot{x}_1 x_1 \\ \sum \dot{x}_1 x_2 \\ \sum \dot{x}_2 x_1 \\ \sum \dot{x}_2 x_2 \\ \sum \dot{x}_1 u \\ \sum \dot{x}_2 u \end{bmatrix} \qquad (11A.49)$$

The PE algorithm is given as:

1. Compute W matrix, (11A.48) since, the measurements of x, $\dot{x}$ and u are available (equation error formulation) for a certain time interval T.

2. Compute bias vector in the similar way using (11A.49).

3. Choose the initial values of β_i randomly for (11A.47).

4. Solve the following differential equation: since $\beta_i = f(x_i)$ and the sigmoid nonlinearity f is known, by differentiating it and simplifying (11A.47), we get

$$\frac{d\beta_i}{dt} = \frac{\lambda \ (\rho^2 - \beta_i^2)}{2\rho} \left[\sum_{j=1}^{n} w_{ij}\beta_j + b_i \right];$$

$$f(x_i) = \rho \left(\frac{1 - e^{-\lambda x_i}}{1 + e^{-\lambda x_i}} \right) \qquad (11A.50)$$

Integration of (11A.50) yields the solution to PE problem posed in the structure of HNN-EEM. For good convergence of the estimates to the true parameters, the proper tuning of λ and ρ is very essential; λ is chosen small, i.e. less than 1.0; ρ is chosen such that when x_i (of RNN) approaches $\pm\infty$, the function f, approaches $\pm\rho$. The foregoing scheme is termed as non-recursive, since the required computation of elements of W and b is performed by considering all the data. The equation (11A.50) can be discretized to obtain the estimates by recursion:

$$\beta_i(k+1) = \beta_i(k) + \frac{\lambda(\rho^2 - \beta_i^2(k))}{2\rho} \left[\sum_{j=1}^{n} w_{ij}\beta_j(k) + b_j \right] \qquad (11A.51)$$

11A.4.3 Recursive Information Processing Scheme

In this scheme, the data are processed in a sequential manner for the real time parameter estimation as required for the adaptive control, it is feasible to use this scheme for OLRT application. The data $x, \dot{x}$ and u are processed as soon as they are available to obtain the elements of W and b without waiting to receive the complete set of the data. Thus, the scheme uses the current data ($x, \dot{x}$ and u in cumulative manner). It is not necessary to store the previous data until the estimation process is completed. This is because the previous data have been already incorporated in the computation of W and b. However, in the start W and b are based on partial information. The recursions are given as:

Step 1: Choose initial values of β randomly

Step 2: Compute W and b based on currently available data (at time index k)

$$W(k) = \frac{k-1}{k} \left[W(k-1) - \frac{1}{k-1} P(k) \, \Delta t \right]$$

$$\qquad (11A.52)$$

$$b(k) = \frac{k-1}{k} \left[b(k-1) + \frac{1}{k-1} Q(k) \, \Delta t \right]$$

With $W(1) = -W_w(1) \, \Delta t$ and $b(1) = -b_b(1) \, \Delta t$

Step 3: The parameters are computed one-time step ahead using the following equation

$$\beta_i(k+1) = \beta_i(k)$$

$$+ \frac{\lambda \ (\rho^2 - \beta_i^2(k))}{2\rho} \left[\sum_{j=1}^{n} w_{ij}(k)\beta_j(k) + b_i(k) \right] \qquad (11A.53)$$

Step 4: Recursively cycle through steps 2, 3 until convergence is reached or no more data are available.

It can be readily seen that the scheme has the following recursive form for information processing:

$$I_{W,b}(k) = h\big(I_{W,b}(k-1), x(k), \dot{x}(k), u(k) \big) \qquad (11A.54)$$

In the foregoing, W_w and b_b are essentially the correlation elements computed by using $x, \dot{x}, u$ etc. Here, h is some functional relationship between present and past information. Thus, the utilization of data, computation of W and b and the solution of (11A.53) for the estimation of parameters are carried out in a recursive manner within the Hopfield neural network structure/equation error formulation.

11A.4.4 Implementation Aspects of Real Time Estimation Algorithms

With the availability of microprocessors/fast computers, the real time implementation of the estimation algorithm has become greatly feasible and viable. Some aspects need to be kept in mind for real time implementation:

1. Reliable and stable algorithms should be used; the UD filter is one such algorithm.

2. The algorithm structure should be as simple as possible. The system models should not be complex, otherwise, they will put heavy burden on computation. Uncertainties in the model will cause additional errors in the estimation.

3. As much knowledge as possible on the system and data should be used in filter design (tuning, etc.), based on the previous experiments.

4. Necessary noise characterization modules be included or used.

5. It may be necessary to split the data processing tasks and program on two or more individual (parallel) processors, which can have interprocessor communication links for transfer of data or results of state/parameter estimation. This calls for use of multi-programming concepts.

11A.5 Relationship Between EIAC and IDAC Schemes

As we have seen there is one important difference between EIAC, and IDAC: in an EIAC scheme, one adaptation mechanism tunes the parameters of an adjustable predictor and then these parameters are used to compute the controller parameters; whereas in IDAC, the parameters of the controller are directly estimated/adapted by the adaptation mechanism. However, by an appropriate parametrization of the adjustable predictor, i.e. by re-parametrization, the parameter adaptation algorithm (PAA) of the EIAC can be used directly to estimate the parameters of the controller, and this will be an IDAC scheme [4]. In such a case the other adaptation mechanism, i.e. the design block, is eliminated and we get a direct adaptive control scheme (DACS)/MRAC: the output of the adjustable predictor, the parameters of which are known at each sampling instant, would behave as the output of a reference model; the latter is also known as "implicit" MRAC for which there are four blocks; (i) plant, (ii) re-parametrized adjustable predictor, (iii) adaptation mechanism, and (iv) adjustable controller. The re-parametrization is illustrated as follows; let a discrete time plant math model be given as

$$y(k+1) = -a_1 y(k) + u(k) \qquad (11A.55)$$

The parameter in (11A.55) is unknown. The objective is to find $u(k)$ such that

$$y(k+1) = -c_1 y(k) \qquad (11A.56)$$

The parameter in (11A.56) is a desired CL pole. The appropriate control law when the plant parameter is known, is given as

$$u(k) = -r_0 y(k); \quad r_0 = c_1 - a_1 \qquad (11A.57)$$

One can rewrite (11A.55) as

$$y(k+1) = -c_1 y(k) + r_0 y(k) + u(k) \qquad (11A.58)$$

In case of (11A.58), the estimation of the unknown parameter r_0 will directly give the parameter of the controller. One can use the adjustable predictor

$$\hat{y}(k+1) = -c_1 y(k) + \hat{r}_0 y(k) + u(k) \qquad (11A.59)$$

And a controller as $u(k) = -\hat{r}_0 y(k)$, we get

$$\hat{y}(k+1) = -c_1 y(k) \qquad (11A.60)$$

Hence, we obtain the desired output at the instant $k + 1$, the output of the implicit reference model from the combination of the predictor and the controller. Interestingly, even in the minimum variance self-tuning control (including the generalized) schemes, though presented as EIAC schemes, estimation of the controller parameters is directly carried out; hence, these can be classified as the IDAC schemes.

Appendix 11B: Sliding Mode Control

A sliding model control (SMC) provides robustness against parameter variations and disturbances, and it is a special class of the variable structure systems (VSSs). The main aspect of SMC is that in a (small) neighborhood of a specified switching manifold (usually some curved surface), the velocity vector of the controlled state trajectories always points toward this manifold. This motion is induced by disruptive/discontinuous control actions in the form of switching control operations. For an ideal sliding mode (SM), the system state should satisfy the dynamic equation that governs the sliding mode for all the time. This would require infinite switching operations to ensure the SM. Several interesting problems of SMC are: (i) chattering, (ii) the effects of unmodeled dynamics, (iii) handling of disturbances/uncertainties, (iv) adaptive learning, and (v) enhancing of robustness [14]. One important aim of the recent SMC research is to make it more intelligent, and this leads to the introduction of intelligent agents, via soft computing (SC) into SMC paradigms.

11B.1 Basic Aspects of SMC

The SMC addresses the specific problem of a special class of VSSs that are the control systems involving discontinuous control actions. The design procedure has two aspects: (i) reaching phase, where the system state is driven from any initial state to reach the switching manifold/s in finite time, and (ii) SM phase, where the system is goaded into the SM on the switching manifold/s; in this case the latter become the attractor. These two phases lead to the main design steps: (i) switching manifold selection, a set is selected with specified/desirable dynamic characteristics, e.g. a set of linear hyperplanes; and (ii) a discontinuous control (design) strategy is utilized to ensure the finite time reachability of the switching manifolds, this controller may be either local or global, depending upon specific control requirements. A MIMO SMC system considered is given by

$$\dot{x} = f(x,t) + B(x,t)u + g(x,t) \qquad (11B.1)$$

The switching manifold can be denoted as $s(x) = 0$, $s = (s_1,\dots,s_m)^T$, the m-dim vector selected based on the desired dynamic properties. Then, SMC is characterized by the control structure defined by

$$u_i = \begin{cases} u_i^+(x), & for\ s_i(x) > 0 \\ u_i^-(x), & for\ s_i(x) < 0 \end{cases}; i = 1,2,\dots,m. \quad (11B.2)$$

In SMC theory, when the sliding mode occurs, an equivalent control is induced

$$\dot{s} = \frac{\partial s}{\partial x}\dot{x} = \frac{\partial s}{\partial x}\{f(x,t) + B(x,t)u + g(x,t)\} \qquad (11B.3)$$

In (11B.3), it is assumed that $\frac{\partial s}{\partial x}\{B(x,t)\}$ is non-singular. In the SM, there exists a virtual control signal u (equivalent) that drives the system dynamics $\dot{s} = 0$ to yield the following control formula from (11B.3):

$$u_{eq} = -\left(\frac{\partial s}{\partial x}B(x,t)\right)^{-1}\left\{\frac{\partial s}{\partial x}\big(f(x,t) + g(x,t)\big)\right\} \qquad (11B.4)$$

In the SM, if the $g(x,t)$ satisfied the matching condition, $g(x,t) = B(x,t)u_g$; the motions is governed by

$$\dot{x} = \left[I - B(x,t)\left(\frac{\partial s}{\partial x}B(x,t)\right)^{-1}\frac{\partial s}{\partial x}\right]f(x,t) \qquad (11B.5)$$

Equation (11B.5) is the invariant property of the SMC, and it is immune to the external force $g(x,t)$, since this term does not appear in (11B.5).

11B.2 Design Approach for SMC

The approach is based on the Lyapunov stability theory, and the following LF is chosen

$$V = \frac{1}{2}s^T s \qquad (11B.6)$$

Then, the design task is to find a suitable discontinuous control signal such that the time derivative of the

LF is negative definite in the neighborhood of the equilibrium. In case some learning and adaptation is to be incorporated into the SMC then one can use the LF as

$$V = \frac{1}{2}s^T Q s + \frac{1}{2}z^T R z \qquad (11B.7)$$

In (11B.7), $z = 0$ denotes a desirable outcome and Q, R, are symmetric and nonnegative definite weighting/normalizing matrices. If system parameter learning is required then, one can use (11B.7); thus, $\dot{V} < 0$ in the neighborhood of $V = 0$ would lead to the convergence of the system states during which time, certain parameters also would be learnt. However, some parameters might not converge to their true values, and z may not converge to zero. The typical SMC strategies for MIMO dynamic systems are: (i) equivalent control based SMC:

$$u = u_{eq} + u_s \qquad (11B.8)$$

In (11B.8), the $u(s)$ is a switching control part that may have two types of switching: (a) $u_{s_i} = -\alpha_i(x)\mathrm{sgn}(s_i)$ for $\alpha_i(x) > 0$, and (b) $u_{s_i} = -\beta_i(x)(s_i/\|s_i\|)$, $\beta > 0$; (ii) Bang-bang type SMC, i.e. the direct switching

$$u_{s_i} = -M\,\mathrm{sgn}(s_i) \qquad (11B.9)$$

In (11B.9) $M > 0$, sufficiently large to supress all the bounded uncertainties and unstructured dynamics; (iii) enforce

$$\dot{s} = -R\,\mathrm{sgn}(s) - K\sigma(s) \qquad (11B.10)$$

This is used to realize the finite time reachability, where $R > 0$, and $K > 0$ are the diagonal matrices, and $s_i\sigma(s_i) < 0$; and (iv) embedding adaptive estimation and learning of parameters or uncertainties in items (i) to (iii); which largely follows the methodologies of adaptive control and iterative learning control mechanism. Other methods are: (a) in the terminal sliding modes, a fractional power is introduced to improve the SMC convergence speed and precision; (b) the second order SMC that has the merit of reducing chattering, and it does not require derivative information for the control; and (c) the higher order SMC uses the sequential integration to smoothen the chattering. In case of the linear time-invariant systems, the model used is

$$\dot{x} = Ax + Bu + g(x,t) \qquad (11B.11)$$

For this case, $s(x)$ is a set of switching hyperplanes defined as $s(x) = Cx = 0$, C is constant matrix to be determined. For the original control u, $u_{eq} = -(CB)^{-1}CAx$, such that under the control, the dynamics in the SM becomes

$$\dot{x} = [I - B(CB)^{-1}C]Ax = A_{eq}x \qquad (11B.12)$$

During the sliding phase, the system dynamics become of $n-m$ dim., due to the constraint $Cx = 0$, confined to $\dot{s}(x) = 0$. The matrix A_{eq} is a projection operator along the range space of B onto the null space of C; i.e. $A_{eq} B = 0$, $A_{eq} x = x$, subject to $Cx = 0$. Here, C can be designed such that the $n-m$ eigenvalues of A_{eq} are allocated in the LHS of the complex plane and the remaining m eigenvalues are zero.

11B.3 Certain Aspects in SMC

These are:

1. Chattering: SMC would suffer from the chattering, which is a motion oscillating around the predefined switching manifold/s; for this two causes are: (i) the presence of parasitic dynamics in series with the control systems due to which a small-amplitude high-frequency oscillation would occur; the parasitic dynamics signify the fast actuator and sensor dynamics that are normally neglected during the control design; and (ii) the switching non-idealities can cause such oscillations: small time delays due to sampling, ZOH, and/or execution time required for computation of $u(.)$, and transmission delays in interconnected control systems. The approaches to soften the chattering are: (a) the boundary layer control in which the sign function is replaced

$$\operatorname{sgn}_b(s_i) = \operatorname{sgn}(s_i); if\, |s_i| > \varepsilon_i, \varepsilon_i > 0, \text{ and}$$

$$\operatorname{sgn}_b(s_i) = \rho(s_i), if\, |s_i| \le \varepsilon_i,\ \varepsilon_i \text{ is small and}$$

$$\rho(s_i) \text{ is a function of } s_i$$

which could be a constant or a function of states as well; and (b) the sign function is replaced by a continuous approximation $\operatorname{sgn}_a(s_i) = (s_i / (|s_i| + \varepsilon))$, ε is small postive number. This gives rise to a high-gain control when the states are in the vicinity of the switching manifold.

2. Unmatched uncertainties: if the matching condition, (11B.5) is not satisfied, the SM motion is dependent on the uncertainties g that is not desirable because the condition for the well-known robustness of SMC does not hold.

3. Unmodeled dynamics: (i) these might contain high-frequency oscillatory dynamics that may be excited by the high-frequency control switching of SMC; (ii) the unmodeled dynamics can refer to the presence of parasitic dynamics in series with the control system, and (iii) if these cannot be modeled, then they lumped with uncertainties and a sufficiently large switching control has to be imposed. In the complex systems' environment, one needs to design an effective SMC without knowing the dynamics of the entire system.

4. Interconnections between chattering, unmatched uncertainties, and unmodeled dynamics: for the SMC system with the equivalent control $u_s = -\alpha \operatorname{sgn}(s)$: (i) if we know g completely, the switching magnitude α does not need to be large to ensure that the switching manifold $s = 0$ is reached in finite time. Of course, one has to increase the control magnitude α to shorten the reaching time; (ii) in the absence of the knowledge of g, provided it is bounded, one can always find a large enough α to overcome the detrimental effect of g. This might result in chattering if the switching device is not ideal; the same effect of the unmodeled dynamics, if we have no knowledge of it at all; and (iii) if we partially know the uncertainties or the unmodeled dynamics, this can help reduce the requirement for large switching magnitude α, and thereby reducing the tendency of larger chattering.

5. Sliding, non-sliding, and zeno motions: due to switching-type controls, each with its own characteristics. In the SM, a subsystem becomes an attractor sucking in all system states. In non-sliding motion, the switching manifolds (representing a subsystem that are not attractive and switchings that occur) correspond to the crossings of the manifolds. The zeno motion refers to infinite switching resulting in finite reaching time. Ironically, a theoretically global SMC strategy does not exist, it is always confined within a limited domain in the state space. Assume the upper $u(+)$ and the lower $u(-)$ bounds of control magnitudes, then the attraction region of the sliding motion is confined within $u(-) < u_{eq} < u(+)$. In order to be able to utilize the invariant properties of SMC, finite time reachability of the switching manifolds must be realized. Often, the reachability condition is loosely defined as only being $\dot{V} < 0$, which then would result in asymptotical stability, in such a case, the switching manifolds would never reach in finite time; the system does not exhibit the invariant properties strictly. In reality, the induced motion could be a mixed type: (i) in one region of the state space, the system exhibits a SM, and (ii) in another region, it exhibits a non-sliding motion.

Appendix 11C: Difficulties with Adaptive Control and Their Resolutions

As we have seen in Chapter 11, the adaptive control is a very appealing technology. However, it is fraught with some problems. Mainly these were normally reflected in unexpected instabilities, with the adaptive control mechanism being operational in a closed loop: (i) the MIT rule, an intuition based gradient descent algorithm works in most cases, but in some cases had an unpredictable performance; (ii) a phenomenon of temporary instability, bursting, in algorithmic implementation (1970s); (iii) the Rohrs' counterexample (i.e. the adaptive control laws, at earlier times could not be used with confidence in practice, because the unmodeled dynamics could be excited and might result in an unstable control system); (iv) iterative controller redesign and system identification-parameter estimation (SIPE), an intuitively appealing approach (for updating controllers) can lead to unstable performance; and (v) multi-model adaptive control (MMAC), another intuitively appealing approach with the potential to include nonlinear systems, can lead to unstable performance [15]. A large portion of AC theory presumes the unknown-ness of the plant, to some degree at least, together with a performance index (PI). The PI should be minimized, however, due to unknownness of the plant, it can never be minimized.

11C.1 Adaptive Control

In the context of an adaptive control, the plant is initially unknown (or partially known), or it may be slowly varying. Because in certain cases, a single controller will not be adequate, it would be required to utilize some learning capability in/for the controller. A typical non-adaptive controller maps the error signal r–y (reference signal-output) into the plant input signal u (control law) in a causal time-invariant manner as we have seen in Chapters in Section II; whereas in an adaptive control some elements of the various systems' matrices are adjusted. Interestingly, in the AC three time scales occur associated with/for: (i) the plant dynamics/controller systems, (ii) identifying the plant (by SIPE block/PAA), and (iii) plant parameter variations. The SIPE time scale needs to be faster than the plant variation time scale, else identification cannot keep up with the requirement of adaptation. It is difficult to develop good adaptive controllers that would identify (and adjust the controllers) at a time scale comparable with that of the closed loop (CL) dynamics: interaction of the two processes could occur and cause instability. For the case of a sudden damage to/of a plant (where controller reduction must occur very fast to avoid catastrophe), designing an adaptive control would be challenging.

11C.2 The MIT Rule

The MIT rule is a scalar parameter adjustment law/rule/formula for the adaptive control of a linear system that is modeled as a cascade of a linear stable plant and a single unknown gain. The rule is a gradient descent procedure and seeks the minimum of an integral-squared PI. For the control of aircraft dynamics, a single unknown parameter was related to dynamic pressure, and the rule was used for adaptation. Performance happened to be unpredictable, the instability was a result of interaction of the adaptive loop dynamics with the plant dynamics, a phenomenon that could occur, if the time scales are comparable, not otherwise. The MIT rule scheme block is: (i) from reference input $r(t)$ to (a) $K_c(t) \rightarrow K_p G(s) \rightarrow y_p(t)$, and (b) $K_m G(s) \rightarrow y_m(t)$; and (ii) then the difference between these two channels $e(t) = y_p(t) - y_m(t)$, is minimized by adjusting the gain K_c. For the plant, K_p is unknown, and $G(s)$ is a known stable TF; K_m is a known gain with the same sign as K_p: $K_p K_c = K_m$, then zero error will result and we can get $K_p = K_m / K_c$; so, if the error is nonzero then we adjust K_c to go to zero; this adjustment is done by the steepest descent:

$$\dot{K}_c = -g \frac{\partial}{\partial K_c} \left\{ \frac{1}{2} e^2(t) \right\}; \ g \text{ is a positive gain} \quad (11C.1)$$

The result (11C.1) is from the optimization theory: the time rate of the parameter (to be adjusted) is equal to negative of the gradient of the cost function with respect to the same parameter. From (11C.1), we obtain the expression for tuning K_c,

$$\dot{K}_c = -g e(t) \frac{\partial e(t)}{\partial K_c} = -g(y_p(t) - y_m(t)) \frac{\partial (y_p(t) - y_m(t))}{\partial K_c}$$

$$= -g(y_p(t) - y_m(t)) \frac{\partial (r(t) K_c(t) K_p G(s) - r(t) K_m G(s))}{\partial K_c}$$

$$= -g(y_p(t) - y_m(t)) K_p G(s) r(t)$$

$$\quad (11C.2)$$

We assume that $r(t) = R$ a constant input, then simplifying we get

$$\dot{K}_c = -g \{ R K_c(t) K_p G(s) - R K_m G(s) \} K_p G(s) R$$

$$= -g \{ K_c(t) K_p - K_m \} K_p G^2(s) R^2 \quad (11C.3)$$

If the convergence has reached, $K_p K_c = K_m$, then the rate of the adjustment of the gain $K_c(t)$ is zero. Also, we obtain the characteristic equation for K_c:

$$sK_c + g\{K_c(t)K_p - K_m\}K_pG^2(s)R^2 = 0$$

$$s + g\{K_p - K_m / K_c\}K_pG^2(s)R^2 = 0 \qquad (11C.4)$$

In a simple form, (11C.4) can be written

$$s + gKG^2(s)R^2 = 0 \qquad (11C.5)$$

If in (11C.5), we have $G(s) = 1/(s + 1)$, then

$$s + gKR^2 \frac{1}{(s+1)^2} = 0 \Rightarrow s^3 + 2s^2 + s + gKR^2 = 0 \qquad (11C.6)$$

From (11C.6), when the convergence of the gain K_c approaches K_m/K_p; $K{\to}0$, the characteristic equation has a zero $s = 0$, signifying neutral stability, which occurs WHEN the adaptive rule itself is converging! In case we have $G(s) = (s{-}1)/(s + 1)$, i.e. the plant has a RHS plane zero, then we have the new characteristic equation as

$$s + gKR^2 \frac{(s-1)^2}{(s+1)^2} = 0 \Rightarrow s^3 + 2s^2 + s + (s^2 - 2s + 1)gKR^2 = 0$$
$$(11C.7)$$

$$\Rightarrow s^3 + (2 + gKR^2)s^2 + (1 - 2gKR^2)s + gKR^2 = 0 \qquad (11C.8)$$

From (11C.8), we can see that if the gR^2 is very high then, the equation is simplified to

$$s^3 + gKR^2s^2 - 2gKR^2s + gKR^2 = 0 \Rightarrow s^2 - 2s + 1$$
$$= 0 \Rightarrow (s-1)(s-1) = 0 \qquad (11C.9)$$

The characteristic equation (11C.9) indicates definite cause for the instability. If the plant model $G_m(s)$ for the Y_m loop is not the same as the plant $G(s)$, then we have the following expression from (11C.2):

$$\dot{K}_c = -g(y_p(t) - y_m(t)) \frac{\partial(r(t)K_c(t)K_pG(s) - r(t)K_mG_m(s))}{\partial K_c}$$

$$= -g(y_p(t) - y_m(t))K_pG(s)r(t)$$

$$(11C.10)$$

However, from (11C.3), we obtain the following characteristic equation

$$\dot{K}_c = -g\{RK_c(t)K_pG(s) - RK_mG_m(s)\}K_pG(s)R$$

$$= -g\{RK_c(t)K_pG(s)K_pG(s)R\} + g\{RK_mG_m(s)K_pG(s)R\}$$

$$= -gK_c(t)[K_pG(s)R]^2 + g[K_mG_m(s)R][K_pG(s)R]$$

$$(11C.11)$$

The characteristic equation from (11C.11) is obtained as

$$s + g[K_pG(s)R]^2 - g[(K_m / K_c(t))G_m(s)R][K_pG(s)R] \qquad (11C.12)$$

$$s + \{[K_pG(s)]^2 - (K_m / K_c(t))K_pG_m(s)G(s)\}gR^2 \qquad (11C.13)$$

From (11C.11) and (11C.13), we see that when the adjustment rule converges, the results would be mostly the same as when the plant TF and the model are the same, if $G_m(s) = G(s)$:

$$s + gK_p^2R^2\{G^2(s) - G_m(s)G(s)\} = 0 \qquad (11C.14)$$

Interestingly, in both the cases, the characteristic equation would have a zero at $s = 0$, signifying the neutral stability, near to the convergence of the adjustment mechanism. In case the $G(s)$ has an RHS plane zero and the model nearly the same or differing somewhat from the plant, the results would be similar to the one obtained in (11C.9). Using the theory of averaging, one can see from (11C.11) that if the average value of the following is positive, then the stability can be assured.

$$gK_c(t)[K_pG(s)R]^2 \qquad (11C.15)$$

From the foregoing, it is clear that the adaptation rule itself introduces its own dynamics, and the separation of the time scales would help towards the resolution of the problem of instability; the major aspect is that the rule is for the adjustment of the gain that would tend to make the two outputs nearly the same and that the error would converge to zero, and this aspect per se means that the gain K_c is not necessarily for making the unstable system stable, or correct the system with overly RHS plane zero/s; the gain is to adjust the control law if the parameters of the plant change. Hence, the basic stability should have been addressed in the first place; because the gain K_c is for estimating the parameter of the plant, and adjusting the control law, to compensate for the change in the system, while operating at some points in its domain of functioning.

11C.3 Bursting

For an adaptive controller for set-point tracking, the plant is described as

$$\dot{x} + ax = bu \qquad (11C.16)$$

In (11C.16), a and b are unknown. The normal "identifier" in the adaptive control block (Chapter 11) uses the measurements of u and x to determine the constants b and a. If u is a nonzero constant (for set-point control), the identifier can identify the plant's DC gain, the ratio b/c, but not b and a separately. The identifier is ignorant of the constant type of the input, and that it cannot identify the parameters separately; it just runs an algorithm driven by u and x. A typical identifier, by the way, has the form

$$\begin{bmatrix} \dot{\hat{b}} \\ \dot{\hat{a}} \end{bmatrix} = \begin{bmatrix} \text{time function derived from } u, x \\ \text{another time function derived from } u, x \end{bmatrix} \quad (11C.17)$$

The error obeys

$$\frac{d}{dt}\begin{bmatrix} b-\hat{b} \\ a-\hat{a} \end{bmatrix} = -\begin{bmatrix} u \\ x \end{bmatrix}\begin{bmatrix} u & x \end{bmatrix}\begin{bmatrix} b-\hat{b} \\ a-\hat{a} \end{bmatrix} \quad (11C.18)$$

If u and x are both constant, then we have, after simplification $(d(.)/dt \to 0)$, and from the expansion of the right hand side expression of (11C.18), and since $dx/dt = 0$ in (11C.16), we get:

$$u(\hat{b}-b) + x(\hat{a}-a) \to 0;\ ax = bu;\ \frac{\hat{b}}{\hat{a}} \to \frac{b}{a} \quad (11C.19)$$

Since the u and x are constants, the actual time histories followed by $\hat{b}$, and $\hat{a}$ will depend on the initial conditions, drift, and noise, and yet the controller will use the estimates to implement a control law. If a gain K is used in the controller, for a high design of BW: $d = a + bK$, then we have $\hat{K} = (d-\hat{a})/\hat{b}$, and the actual CL pole will be at $-a - b\hat{K}$, and if $\frac{\hat{b}}{\hat{a}} = \frac{b}{a}$, this pole will be at $-\frac{bd}{\hat{b}}$. Since the estimates of b and a can move around, and if randomly, then in many cases, we might have $-\frac{bd}{\hat{b}} > 0$; and this will cause the instability. Interestingly enough, this instability would cause the signals to be richer, and the identification would improve, which in turn will provide a stabilizing controller. The latter action recovers the process, until the time the drifting signals induce again instability in the closed loop. A sufficient and virtually necessary condition, the persistence of excitation is that the following is true for all the constants that are positive, for parameter estimation error to tend to zero:

$$\alpha_1 I < \int_s^{s+T} \begin{bmatrix} u(\sigma) \\ x(\sigma) \end{bmatrix}\begin{bmatrix} u(\sigma) & x(\sigma) \end{bmatrix}d\sigma < \alpha_2 I \quad (11C.20)$$

If there are $m + n$ coefficients (the plant with $m-1$ finite zeros, and n poles) to be identified, then an external input (complex sinusoid) signal should excite $m + n$ distinct frequencies; then, the condition (11C.20) is fulfilled, and the identifier can learn each unknown parameter. Even, a broadband noise signal would be a rich enough input signal. Thus, in an essence, a set-point control and adaptation are incompatible. For practical purposes, one can turn off the adaptation when the input is constant or superimposing a small and yet rich excitation on the external set-point signal to prevent bursting.

11C.4 The Rohrs' Counterexample

The possible unstable behavior of adaptive systems could be due to: (i) unmodeled dynamics at sufficiently high frequencies (and modeling these dynamics might not be useful), and (ii) the plant is affected by unknown disturbances (e.g., 60 Hz hum), however small. Typical problems were/are: (a) the reference signal $r(t)$ is a step, which is not persistently exciting, and the instability/bursting; and/or (b) a sinusoidal excitation with the frequency intrusion into the region of unmodeled (unmodel-able) dynamics; the number of parameters in the model was/is too small to capture all the dynamics present in the true plant.

11C.5 Iterative Control and Identification

In an iterative identification and controller redesign; the processes of the identification and control are separated; and a performance index has to be minimized: (i) identify the plant with the current controller (using some OLRT RLS method), then (ii) redefine the controller using the newly identified model (with the minimization of PI), and (iii) implement the new controller for plant. If the plant is unknown but constant, the convergence would occur and the controller would settle down; however, if the plant is slowly varying, this might not happen; if the modeling is based on only fewer parameters than needed, the controller might settle to an incorrect value (PI is not minimized), or the signal would not remain bounded. The controller change is determined by the current plant model, but here the new controller is attached to the actual plant rather than the model. Consider two TFs: a model (plant) transfer function $G_m(s) = 1/(s+1)$ and a plant (model) transfer function $G_p(s) = e^{(-0.01s)}(0.1s + 1)/(s + 1)$. These two TFs, as such, are almost equal, Figure 11C.1: the open loop step responses (without controller) are nearly identical. Either one TF can serve as the plant and the model to the

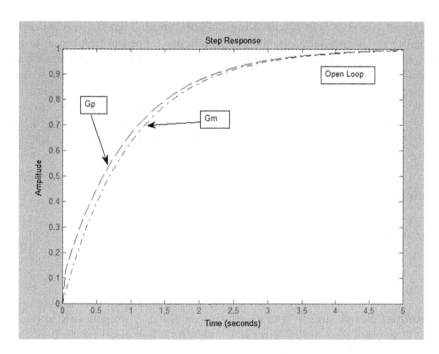

FIGURE 11 C.1
Open loop system TFs' responses without any controller.

other one, since if $G_p(s)$ is the plant, then $G_m(s)$ can be (an approximate) math model of the plant, and vice versa. Figure 11C.2 depicts the closed loop step responses with two different constant gain controllers: G_p with $K_1 = 1$, and $K_2 = 7$; and G_m with $K_1 = 1$, and $K_2 = 7$; thus, with gain 7, the closed loop responses are very different, and for the $G_p(s)$, it is even oscillatory during the initial period;

despite the fact that the two original TFs are almost the same. Thus, a model may be good for the plant with one controller, but not for all controllers: if it is used as a basis for controller redesign, then, a new controller (while fine with the model) could destabilise the plant. In an iterative big controller change design, one can go from the current controller "in the direction of" (but not all the

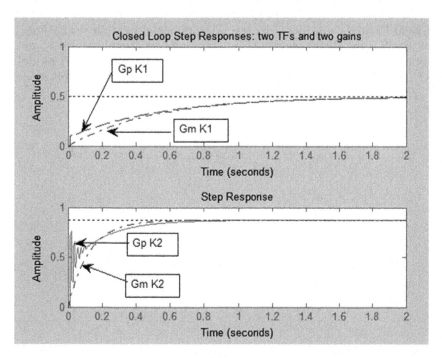

FIGURE 11 C.2
The CL responses with two different constant gain controllers: G_p with $K_1 = 1$, and $K_2 = 7$; and G_m with $K_1 = 1$, and $K_2 = 7$.

way to) the newly designed controller. In an iterative identification and controller redesign approach; at each controller redesign step, one can expand the CLBW by a small amount, which is appropriate with making a safe and small controller change, out until a BW adequate to the design objective is attained.

11C.6 Multiple Model Adaptive Control

Take a plant $G(b)$ where b is the vector of unknown plant parameters. It would be good to learn b-vector from the observations:

$$\dot{\hat{b}} = f(\hat{b}, \text{measurements}) \qquad (11C.21)$$

Suppose b lies in a bounded closed simply-connected region B with $G_t(s)$, the true plant. For multiple model adaptive control (MMAC): let a set of values $b_1, b_2,...,b_N$, in the region (B), with associated plants $G_1, G_2,..., G_N$ be chosen. Then, design N controllers C_i such that C_i gives good performance with G_i. Also, for the plants "near" G_i with one controller connected, run an algorithm that estimates at any time instant the particular representative model from $G_1, G_2,...,G_N$, call it G_l, which is the best model to explain the measurements of the I/Os of G_t. Then connect up C_l.

The issues are: (i) how many plants, G_i should be chosen, how does one choose a good set of plants $G_1, G_2,...,G_N$, and how to ensure the controllers C_i will give good performance for plants near P_i?; ii) if controller C_J is connected, and it turns out that P_I is the best explainer of G_t (a good model), there is no assurance that after switching in of C_I to replace C_J, G_I will continue to be a good model of G_t. Instead, G_K for some other index K might be

a good model. For the second problem, one can use safe switching. For the first issue, one sequentially picks G_1, $G_2,...,G_N$, chooses b_1 and, thus, G_1 arbitrarily, and designs C_1; determine an open ball B_1 around G_1 for which C_1 constitutes a satisfactory design. Then an approach is to choose G to be in the ball if, and only if, the v-gap metric (distance between G and G_1) is below the generalized stability margin (GSM):

$$\delta_v(G, G_1) < 0.3\beta_{G_1, C_1} \qquad (11C.22)$$

$$\beta_{G_1, C_1} = \| T(G_1, C_1) \|_\infty^{-1}; \ T_{G_1, C_1} = \begin{bmatrix} G_1 \\ I \end{bmatrix} [I - C_1 G_1]^{-1} [-C_1 \ I]$$

$$(11C.23)$$

Then, b_2 near the outer limit of B_1, and design C_2 using $G_2 = G(b_2)$; determine an open ball B_2 around G_2 for which C_2 is a good controller; choose b_3 near the outer limit of $B_1 \cup B_2$, and so on. Of course, the b_i must be drawn from B, and the procedure terminates at some finite N (by the Heine-Borel Theorem). However, for nonlinear plants, it is difficult to generalise v-gap metric, and generalized stability margin, which are necessary in the linear system case for a quantitative version of the in-principle algorithm.

Keeping adaptation and plant dynamics time scales separate would reduce the likelihood of problems. One should not use more parameters than one needs for modeling purposes. To learn the plant (parameters), one needs satisfactory experimental conditions, i.e. persistently rich excitations. A sudden controller change can cause instability, even if the new controller is defined with a good model. For MMAC, one can pick representative models systematically, at least for the linear cases.

12

Computer-Controlled Systems

In past decades, there has been a paradigm shift in methodology of measurements/sensing, interpretation of data/signals, and implementation of control systems/ strategies, mainly due to continuous technological innovations supported by fast and accurate digital technology and components, such as analog-to-digital converters (ADC), transducers, microprocessors/ microcontrollers and, hence, integrated digital computer technology, all resulting into computer (-based) control and measurement systems [16]. Also, this has been supported by the associated advancements in communication technology that has replaced natural scale-up procedures of manual monitoring and control by highly advanced automated systems. Today, computers and the associated software (SW) have become very powerful in terms of speed, throughput, and memory, and the application of computer-controlled system is justified automatically because of its low cost, its capabilities to handle large, complex industrial processes, and it use in many aerospace control applications. The computers: (a) ensure the repeatability/ precision and accuracy of the controlled processes, (b) permit flexibility to modify the sequencing and control procedures, and (c) provide increased understanding of the behavior of the processes.

12.1 Computers in Measurement and Control

The industrial revolution of some past decades was, in fact, propelled by the then concurrently developing digital computers (based technology). This has led to the rapid development of (computer-) machine-based controls (e.g. autopilots for aircrafts, and remotely operateable control systems, RC controls), measuring systems, and smart sensors. The use of computers (i.e. digital computing elements/logics for setting up the control functions) for measurements and controls, in real time, has been for both feedback and feed-forward loops in a control system. The first digital computer, specifically for control in a real time application, was for airborne (military) operation and used to provide an automatic

flight and weapons control system (in 1954). The computer control systems were used in textile and chemical industries during the early 1960s as supervisory systems. The logged data were used for SS optimization computations to determine the set-points for standard analog controllers. These computer-based systems did not control directly the movement of the valves/final control elements. The first direct-digital control (DDC) computer system for process monitoring and control was Ferranti Argus 200; with 120 control loops and 256 measurement inputs. The main aim of computer-based measurement and control (CBMC) is to acquire the information from field devices/sensors (input) and compute a logical decision to manipulate the material and energy flow of given process in a desired way to obtained optimal output. The expectations from a process computer are primarily in terms of response time, computing power, flexibility, and fault tolerance, which are needed to be precise and reliable; also, the control has to function in real time. Digital computer control applications are: (i) passive type involves only acquisition of process data (data acquisition/data logging), and (ii) active type involves acquisition and manipulation of data and uses it for process control in real time. The passive application is mainly for monitoring, alarming, and data reduction: the smart devices (sensors, transmitters, actuators, final control element/s) with embedded computer help to receive real time process information and automatic transmission for further processing by the process control computer: video display terminals supervise the entire plant from a control room; a few keyboards and screens are used; the plant managers/ engineers have comprehensive information concerning the status of plant operations to aid effective operation. The major active form of application of computers is in process control, plant optimization (optimal control, Chapter 6), robotics, and aerospace vehicle control systems: process control computers implement sophisticated mathematical models and automatically tune the controller parameters for best operating performance. The expert systems and advanced control techniques, such as model predictive control (MPC) (Chapter 7), are being applied with the help of computers for optimization of the process operation.

12.2 Components in Computer-Based Measurement and Control System (CMCS)

The components in CMCS are: (i) measurement/data gathering/accumulation (as inputs to the systems to be controlled), (ii) data conversion/scaling/checking/formatting, (iii) visual display (iv) electronic/digital computations-comparators (for limiting and alarm raising), (v) events, sequence and trends-monitoring, and data logging (for storage of the response data if required for off-line analyses), and (vi) integrated control actions. In most CMCS, the controlled variable, the process output, is measured as a continuous (analog) electrical signal and converted into a discrete time signal using an ADC (Chapter 9). This digital signal is fed back to a (digital) comparator and compared with the discrete form of the set point (the desired value), by the digital computer; this produces an error signal. An appropriate computer program (control algorithm representing the controller; Chapters 10 and 11) is executed to yield a discrete controller output. The latter is then converted into a continuous electrical signal using a digital to analog converter (DAC) and is fed to the final control element. This control strategy is repeated at some predetermined frequency (time division multiplexed with other control loops/activities) to achieve the closed loop computer control of the process.

12.3 Architectures

The three mainly used architectures of computer-aided industrial process/plant are discussed next.

12.3.1 Centralized Computer Control System

The centralized computer (CC)-based control system has usually (and not always) a large computer with huge space and magnetic core memory, wired-in arithmetic and gate logics; and, of course, it has high cost, and hence, every possible control function and supervisory/DDC were/are incorporated in this single computer system. Such a mainframe system was also required to provide communication systems for bringing in the (field) signals to the computer location; and output control signals to the field devices; valves, motors, actuators, and so on). Electrical noise problems for large distance communication of signals has been a major issue: process interruptions and complete stoppage of plant/process.

12.3.2 Distributed Computer Control Systems (DDCS)

With the advent of microprocessors/microcomputers (of course decades ago), distributed computer control architecture/distributed digital control (DDCS/DDC) have become very popular; the total task is divided up and spread/distributed across several computers. Also, the computing power required to control several (distributed) processes is also distributed and emphasis is put to locations where major (control) activity takes place. Such a system is flexible and has a good hardware/software (HW/SW) redundancy in case of a failure of a few systems. The sub-system computers are connected by a single high-speed data link for communication between each of the microprocessor-based systems with centralized operator station.

12.3.3 Hierarchical Computer Control Systems

In the hierarchal system, the upper level computers depend on lower level devices for process data, and the lower level systems depend upon the higher level systems for sophisticated control functions, see Figure 12.1. At Level 0, the field devices for the measurement of process parameters would forward the measured data to Level 1 for process control functions, here the tasks of process monitoring/control, inter process operation, and system coupling are performed. At Level 2, supervisory is implemented: data collection/logging and process optimization; the mathematical models of the process take care of overall process optimization. At Level 3, the tasks are: plant resource allocation, production planning/maintenance scheduling, and production accounting; the information is communicated

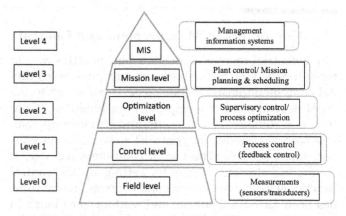

FIGURE 12.1
Hierarchical computer-based process control system. (Adapted from Anon, Introduction to Computer-Based Control Systems, IDC Technologies Pty Ltd., West Perth, Western Australia, pp. 1–24, 2012, www.idc-online.com, accessed October, 2018.)

down to the supervisory Level 2 computers. At level 2, the computed set-points for various parameters to meet the production schedule/mission goals are passed on to the process controllers at Level 1. The computers at Level 1 take requisite control actions to maintain the process/mission conditions based on set points. The monitoring/recording of all the plant are done by computers at Level 3: plant/process parameters, alarm conditions, production/mission and quality issues; and selectively, the data are transferred to the mission Level 4 for management information (MIS) purposes. The hierarchical systems involve a type of a distributed network, and hence, most systems may be termed as a mix of hierarchical and DCCS.

12.3.4 Tasks of Computer Control Systems and Interfaces

The computing systems are concerned with many variables operating over a wide range of process dynamics and tasks: the process/control algorithms require the development of many complex functions that work on several final control elements. The tasks carried out by each level in the automation hierarchy are:

1. Field Level (Level-0): Measurement of process parameters and signal conditioning.
2. Control Level (Level 1): Maintain direct control of the plant units, detect emergency condition and take appropriate action, undertake system coordination/reporting-transmission to higher level computer (Level 2), and take up reliability assurance activities by performing diagnostics on the various control equipment.
3. Supervisory Level (Level 2): Optimize the processes under its control, coordinate the plant operation, collect/maintain process/production database, and communicate with the higher and lower level computer systems.
4. Plant Level (Level 3): Undertake the mission planning/scheduling, diagnose the various control equipment to help in detecting the fault, and keep the standby system live.
5. Management level (Level 4): Manage the tasks.

12.3.4.1 HMI-Human Machine Interface

At this level, technically, the human and the machine interact during a given task by touch, sight, sound, heat transference, or any other physical or cognitive function. The operator needs to provide minimal input

to achieve the desired output, and the machine minimizes undesired outputs to the human HMI for process monitoring/control. The standard SW packages may have the features: (a) mimic diagram of plant/process overview, (b) alarm information/status of large areas of the plant, (c) multiple area displays presenting information on the control system, and (d) loop displays for information of particular control loop/s. HMI devices are: (i) display unit/s (CRT), (ii) keyboard/s, (iii) input unit/s, (iv) printing unit/s, (v) control panel/desks, mimic board/panel, and (vi) recorders. The computer system designer should consider the human as an integral and indispensable part in the design of MMI (man-machine interface) at each stage.

12.3.4.2 Hardware for Computer-Based Process/Plant Control System

Most such systems have the standard hardware, as in the case of any computer; the four basic components of a general computer are: (i) central processing unit (CPU), (ii) memory, (iii) I/O, and (iv) bus interface. The programmer writes a set of instructions (programs/SW) to make these physical devices to perform specified functions. The hardware components work as per instructions from software, because without software (programming) the hardware simply cannot work.

12.3.4.3 Interfacing Computer System with Plant

In a computer-based process control system, it is required to convert the process parameters (the physical quantities) from the analog to digital pattern. All the sensors/instruments should have standard interface compatible to the host computer to achieve this conversion. The field instruments/sensors' measured data could have the categories: (i) analog quantities; continuous variables; (ii) digital quantities, binary or binary coded decimal (BCD), (iii) pulses or pulse rates, and (iv) telemetry. There are a variety of interface cards: (a) analog interfaces, (b) digital interfaces, (c) pulse interfaces, (d) real time clock, and (e) standard (bus) interfaces.

12.4 Smart Sensor Systems

A smart sensor/device is the combination of a sensing element with processing capabilities provided by a microprocessor and are the basic sensing elements with embedded computational intelligence [17], Figure 12.2. A complete self-contained sensor system has the

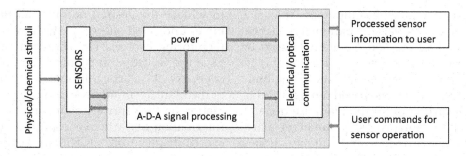

FIGURE 12.2
A smart sensor (system). (Adapted from Hunter, G.W. et al., Smart sensor systems. The Electro-chemical Society Interface, pp. 29–34, 2010, https://www.electrochem.org /dl/interface/wtr/wtr10/wtr10_p029-034.pdf, accessed October 2018.)

capabilities of logging, data processing, self-contained power, and transmitting/displaying informative data to the user. The key idea in a smart sensor is integration of silicon microprocessors with sensor technology to provide customized outputs and significantly improve sensor system performance/capability: signal detection from discrete sensing elements, signal processing, data validation/interpretation, signal transmission/display, and multiple sensors. Some or most important functions as in any computer-based controlled system could and would be part of the smart sensor. The embedded intelligence will continuously monitor the discrete sensor elements, validate the engineering data, and periodically verify sensor calibration and health. The smart sensor system can optimize the performance of the individual sensors and lead to a better understanding of the measurement data and the environment; the presence of the microprocessor-based sensor allows the design to meet the needs of a wide range of different applications. The new generation smart sensors can be networked to have the capability of individual network self-identification and of the smart sensor system as a whole as necessary; the information can be shared in a rapid, reliable, and efficient manner with on-board communications capability in the sensor and helps in making concrete decisions.

12.4.1 Components of Smart Sensor Systems

These are basic usual sensors, power/battery, communication, and signal processing capability provided by a microprocessor [17]. The microfabrication approaches have revolutionized the microprocessor technology and micro-electro-mechanical systems/sensors (MEMS) sensor elements leading to the development of smart sensor systems.

Low-powered sensor elements

Microfabrication methods make it possible to build very small, low-power sensors; a micro-hotplate; these have extremely low-power consumption, on the order of 4 mW continuous and <4 µW, when operated on a duty cycle to read every millisecond. The microfabrication, compatible with complementary metal-oxide semiconductors (CMOS) process, makes integration of the electronic interface for the sensor feasible on a single substrate.

Battery or Energy Harvesting

A smart sensor system will have certain energy considerations: (i) to support and operate all components and the sensors, (ii) small scale energy systems consider batteries and energy harvesting options, (iii) primary and rechargeable (or secondary) batteries will be important, (iv) in energy harvesting, energy is derived from an external source, captured, and stored, (v) piezoelectric crystals/fibres, thermoelectric generators, solar cells, electrostatic, and magnetic energy capture devices are used for local power needs. A piezoelectric energy system produces a small voltage when physically deformed. Thermoelectric generators with the junctions of two dissimilar materials produce a small voltage due to a thermal gradient.

Wireless Communication

The smart sensor systems would have small wireless telemetric interfaces and a multichannel wireless telemetric microsystem with four input channels and a calibration channel. This system has a package size of approximately 1 cm × 1 cm × 0.5 cm (and a small watch battery) with a total weight of 1.0 gram, for which a monolithic integrated circuit (IC) chip (a low-power BiCMOS signal processor chip of 2 mm × 2 mm in size) is fabricated. The chip can also amplify, filter, and time-division multiplex the signals that are, in turn, transmitted via an RF link to an external radio receiver. Ultra-low-power wireless systems are also being designed for interfacing with biological systems for sensory data acquisition/transmission.

Smart Sensor System

A typical smart sensor contains an analog circuit for power management, control, and interface to the digital world. In the latter, sensor inputs are processed to reduce noise and are integrated with other sensory inputs for compensation, redundancy, and reliability improvements, and the data can be fed back to sensors or fed forward to appropriate system controls. The increased complexity is achieved with less expense than that of larger, hand-assembled systems.

12.5 Control System Software and Hardware

The computer code from the control laws is usually less than 25% of the total SW for the entire control system, for which a building block approach for embedded control system development would be a fast/efficient SW design process. The modern control systems consist of: (i) the plant to be controlled, (ii) the control SW running on the control computer(s), (iii) the specific interface hardware, and (iv) command interfaces, safety measures, and data integrity checks [18]. Use of the plug-and-play capabilities of the blocks allows efficient design processes, and concurrent engineering is efficiently supported by the simulatability of the design process.

12.5.1 Embedded Control Systems

In embedded control systems (ECS), the dynamic behavior of the appliance is essential for the functionality of the embedded system; the central control loop is hard real time. In embedded/modern control systems, the ingredients are: (i) SW-principal functions are user interfacing, data processing and appliance control, the communication latency and its jitter should be small compared to the sampling time; (ii) HW-the computer and the I/O interfacing, the specific processors are ASICs, DSPs, MCUs, and also sensors/actuators get integrated with the processor on one chip, also used are FPGAs for flexibility and upgrading; and (iii) appliance/plant/process, the machine part of the ECS, e.g. a robot including its actuators/motors/sensors. In fact, the ECS embodies a closed loop control system (CLCS), where the control loop is spread over the EC (embedded computer) and plant, wherein the time constants of the system dictate the timing constraints of the SW, and for its optimality, the complete ECS should be considered. For the embedded data systems (EDS), computational latency and jitter should be small compared to the reaction time of a user of cellular phones and other telecom systems.

12.5.2 Building Blocks

THE Building blocks for the complex systems should have the features of: (i) overview of the system description, (ii) high reusability of the blocks to allow for competitive fast development, and (iii) simulatability of the total description. To accomplish these, one can use an object-oriented approach for all the parts of the ECS, since it allows for hierarchy and encapsulation. The object-oriented approaches for ECS are: (a) compositional programming techniques for the SW parts using content security policy/protocol (CSP)-based channels for information exchange between processes; (b) very high speed integrated circuit (VHSIC) hardware description language (VHDL) for the specific I/O HW parts, which remain configurable when using FPGA's; and (c) bond graphs for the appliance to be controlled. These three description methods support hierarchy and encapsulation, and the combination of the methods is possible.

12.5.2.1 Software and Hardware Building Blocks

To describe the SW, one uses data flow diagrams (DFD) as directed graphs: the vertices denote the processes, and the edges denote the communication of data, such a DFG depicts the structure of the SW and allows for hierarchy. For the data communication, one uses channels that are simply synchronisation primitives that provide communication between concurrent or distributive processes; the channels control synchronisation and scheduling of processes. Channels are one-way, fully synchronised, and basically unbuffered. Scheduling is no longer a part of the operating system but is hidden in the channels. A process (this is not a plant now) is a group of tasks and need not necessarily be sequential. Processes may run in parallel or in some sequence. CSP specifies fundamental control-flow constructs that describe the sequence of executing processes: PAR (parallel), SEQ (sequential), or ALT. At the ALT, alternative, construct, each process is preceded by a so-called guard determining whether the guarded process will be executed. Analysing the CSP description of the software part of an ECS allows for formal checking on deadlock, starvation and life-lock allowing to verify the SW before it is tested on the real appliance. In addition to Java, which serves as a design pattern, implementations in C++ and C can also be developed.

One can use VHDL descriptions of the computer HW. Either realisation can be done in specific circuits (ASIC) or using FPGA chips.

12.5.2.2 Appliance/System Building Blocks

For modeling the appliance/system, one can use Bond Graphs, showing the relevant dynamic behavior; vertices are the sub-models and the edges, the ideal exchange of energy. Bond Graphs are physical-domain independent due to the fact that physical concepts are analogous for the different physical domains: mechanical, electrical, hydraulic system parts are all modeled with the same graphs. Most SW packages supporting bond graphs have model libraries available. The models are written as directed graphs: parts are interconnected by bonds, for exchange of energy, the flow of which is described as the product of two variables (effort and flow). The interfaces of bond-graph sub-models consist of so-called ports, consisting of two variables, whose product is the power; (i) a pair can be specified as voltage and current, force and velocity; and (ii) the sub-model equations are specified as real equalities, differential equations are generated after model processing, where the port variables obtain a computational direction (input and output) and the equations are rewritten to assignment statements.

interfacing, reaction to external commands (from the operator/connected systems) is specified; and (iv) effects due to non-idealness of computer HW/SW are added, and effects of computational latency and accuracy are checked. The impact of scheduling and/or algorithm optimisation techniques on the behavior of the ECS can be checked by extensive simulations. A top–down decomposition may be applied: (i) define the global architecture of the system, (ii) those control algorithms in which problems are expected may be developed, and (iii) the parts of the controller can be developed incrementally and combined to obtain the description of the total controller. In the realization step, simulation plays a relevant role when the design project is implemented in a concurrent engineering fashion: the available part of an ECS is tested together with the other parts that are still the simulated models; the verification process is a hardware-in-the-loop simulation (HILS).

12.6 ECS-Implementation

Different parts of an ECS can be developed separately, provided that the overall model is competent for testing, the development design process is concurrent engineering activity, Figure 12.3. The design trajectory of ECS is: (i) the dynamic behavior of the system is object–oriented-ly modeled, using bond graphs as a main modeling paradigm; (ii) using the model, the control laws are designed; (iii) ECS implementation is done by transforming the control laws to efficient concurrent algorithms (i.e. computer code) via a stepwise refinement process, after each step, the results are verified by simulation; and (iv) realisation of the ECS is done on a stepwise sequence. The stepwise refinement procedure for the embedded SW consists of: (i) control laws only, for the ideal situation; (ii) non-ideal components are modeled more precisely by considering their relevant dynamic effects; (iii) safety, and command

12.7 Aspects of Implementation of a Digital Controller

Most control systems are based on computer control. A computer-controlled system (CCS) has at least five blocks/operations: (i) AD converter/conversion, (ii) algorithm, (iii) DA converter/conversion, (iv) the plant to be controlled, and (v) the clock that synchronises the events; except the plant the four blocks/operations are performed or are part of the overall computer system [19]. Interestingly enough, it should be noted that the control signal is kept constant between the conversions; in fact, the system runs in open loop during the sampling interval. The CCS has continuous and discrete time signals, and hence can be called as sampled data system (SDS). Due to the mixture of signals, the events are of interest only at the sampling instants, and hence such systems are called discrete time systems (DTS)

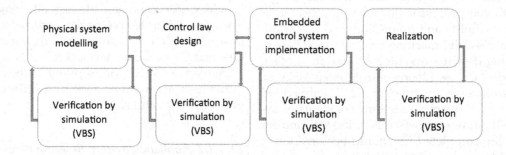

FIGURE 12.3
ECS design trajectory flow diagram: development, implementation and simulation. (Adapted from Broenink, J.F. and Hilderink, G.H., Building blocks for control system software, *Proceedings of the 3rd Workshop on European Scientific and Industrial Collaboration WESIC2001*, Enschede, the Netherlands, (Eds.), J. V. Amerongen, B. Jonker, and P. Regtien, pp. 329–338, 2001.)

which deal with sequences of numbers, and hence are studied using difference equations. The development of CS (CCS) is spread over six periods: (i) pioneering (1955), (ii) direct digital control, DDC (1962), (iii) minicomputers (1967), (iv) microcomputer (1972), (v) general use of digital control (1980), and (v) distributed control (1990). The merits of CCS are: (a) no problem of accuracy and drift, (b) sophisticated computations can be done, (c) can easily include logic and nonlinear functions, and (d) the tables can be used to store the data for future use. The CCSs become time varying due to inclusion of the clock. In most cases, periodic sampling is used, but one can use different sampling rates for different (local) loops in the system, i.e. use the multi-rate sampling. A digital filter itself is a simple example of the CCS. A first order lag compensator is approximated by a first order difference equation. The resultant sampled system with A–D→computer→D–A conversions and the clock does not remain time-invariant, the response depends on the time when the step input occurs. The response of this CCS to an external input will depend on how the external event is synchronized with the internal clock of the CCS. Because of the periodic sampling, CCS will result in CLCS that are linear periodic. Under reasonable conditions, a CCS will behave as a continuous time system if the sampling period is sufficiently small. In general, it is simple to obtain an algorithm for computer control by writing continuous time control law as a differential equation and by approximating the derivatives with finite differences. In the case of a CCS, the overshoot and the settling time would be a little larger than the continuous counterpart of the control. However, by using the periodic nature of the control action, one can obtain control actions with better performances, often the response time can be reduced by a factor of 2. A deadbeat control response (DBR) occurs only in a digital control system (DCS), here the system is at rest when the desired position is reached. In CCS, the sampling creates signals with new frequencies, lower as well as higher than the excitation frequency. This phenomenon is known as aliasing. All signal components with frequencies higher than the Nyquist frequency should be removed before a signal is sampled, so that these components do not alias or fold back (frequency folding) to the frequency of interest, and the filters to do this are called anti-aliasing filters.

12.7.1 Representations and Realizations of the Digital Controller

One can obtain different forms of the control laws based on the design methods that are studied in Chapter 5 through 11. Some algorithms are in the form of continuous time systems, and some are of the discrete time systems, the former can be easily converted into the latter forms (Chapters 9 and 10): (i) a state representation with an explicit observer (based on the pole placement by state feedback), (ii) a general state representation of a discrete time dynamic system (the states may not have a physical interpretation), and (iii) a general input-output (I/O) representation in the polynomial form (design based on SISO system). The different realizations can be obtained based on the choice of different state variables, and all are equivalent from an input-output point of view (assuming the transformations are performed with infinite precision); otherwise, the quantization and round-off would introduce nonlinearities, and the controller (control algorithm) would be sensitive to these errors. Several aspects of controller implementation are discussed next.

12.7.1.1 Pre-Filtering and Computational Delays

As suggested earlier, an anti-aliasing (pre-filter) filtering might be required, and there is always a time delay (of one sampling interval) due to computations, and these would introduce additional dynamics in the CCS, and one must account for these dynamics while designing the control system. One can use faster sampling (to avoid aliasing) and generate more data, then use a digital filter to thin down the data. The computational delay is accounted by including a time delay of say τ in the plant/process model. One can analyze the CLCS for effect of this delay, say using root locus.

12.7.1.2 Nonlinear Actuators

Nonlinear actuators must be considered at the design stage by using the design theory that takes the nonlinearity into account (e.g. the optimal control), because the design based on the assumption of the linearity of the plant model will not be valid, since the saturation, hysteresis, and backlash might cause problems during the start-up and shutdown during the control system operations. The optimal control law might be complex, and one should use some heuristic method.

12.7.1.3 Antiwindup with an Explicit Observer

If the output of the nonlinearity is measured, the state estimate and the state of the controller will be correct. If the output is not measured, it can be estimated if the nonlinear characteristics are known. If the system transition matrix has eigenvalues outside the unit circle and if the control variable saturates, windup would occur; the state and the control signal will continue to grow, even though the plant output is restricted because of the saturation. Often, one introduces a special tracking mode that makes the state of the controller correspond to the input-output sequence. In such a design, the tracking problem is formulated as an explicit observer problem.

Let the controller be specified as

$$x(k+1) = \varphi x(k) + Gy(k)$$
$$u(k) = Hx(k) + Dy(k) \tag{12.1}$$

We assume that the output is at its limit and the control error is $y(.)$. Let (12.1) be written as

$$x(k+1) = \varphi x(k) + Gy(k) + K[u(k) - Hx(k) - Dy(k)]$$
$$= (\varphi - KH)x(k) + (G - KD)y(k) + Ku(k) \tag{12.2}$$
$$= \varphi_0 x(k) + G_0 y(k) + Ku(k)$$

For the observable system, the matrix K can be chosen, so that $\varphi_0 = (\varphi - KH)$ has the prescribed eigenvalues inside the unit circle. Now, the control law can be chosen as

$$x(k+1) = \varphi_0 x(k) + G_0 y(k) + Ku(k)$$
$$u(k) = Hx(k) + Dy(k)$$

$$sat\{u\} = \begin{cases} u_{low} & u \le u_{low} \\ u & u_{low} < u < u_{high} \\ u_{high} & u \ge u_{high} \end{cases} \tag{12.3}$$

The values of the thresholds (i.e. the $sat(.)$ function) in (12.3) are chosen to correspond to the actuator limits. Now, let the controller be given in the form of an input-output model:

$$R(q)u(k) = T(q)u_c(k) - S(q)y(k); \ q$$
$$= z^{-1}, \text{a shift operator} \tag{12.4}$$

The controller (12.4) is rewritten in the form of a dynamic system with the observer driven by the inputs: u_c, y, and u:

$$A_{aw}u = Tu_c - Sy + (A_{aw} - R)u \tag{12.5}$$

In (12.5), A_{aw} is the desired characteristic polynomial. The controller with the antiwindup compensation is

$$A_{aw}v = Tu_c - Sy + (A_{aw} - R)u$$
$$u = sat\{u\} \tag{12.6}$$

If the controller saturates, then it is interpreted as an observer with its dynamics given by the polynomial A_{aw}.

12.7.2 Operational and Numerical Aspects

The operator may change the factors of the proportional–integral–derivative (PID) control: (i) gain, (ii) integration time, and (iii) the derivative time. The controller can be run as automatic or in a manual mode. The switching should be bumpless transfer/transition, for which a tracking mode could be introduced, which can be viewed as a realization of an observer. To run it in a tracking mode, let $u_{low} = u_{high} = u_{manual}$. The state of the controller will be at rest due to internal feedback in the controller, and the $sat\{.\}$ will give a bumpless transfer. It is conventional to initialize a controller, e.g. PI (proportional-integral), by operating it in a manual mode until the plant output comes close to its desired value. If the controller has an observer, then the controller state is initialized by keeping $u(k)$ fixed to allow the observer settle down. If the controller has an antiwindup, then it is run in a manual mode for the period of the observer's settling. For bumpless parameters, changes required in the design-cum-operation process can store the past input-output data and run an observer when the controller parameters are changed.

The aspects of accuracy/precision and fixed point, or floating point, arithmetic need to be studied to understand their effects on the CLCS. There is an involved interaction of the feedback algorithm and the sampling rate. The main errors can occur due to: quantizations in A–D/D–A converters and parameters, roundoff, and overflow/underflow in arithmetic computations. For the converters, the accuracy of 10 bits is at least needed; higher the bit accuracy, the greater the resolution accuracy, e.g. 14 bits has 0.006% resolution accuracy. Since the digital algorithms are implemented on micro/mini computers with word lengths of 8, 16, or 32 bits, then accuracy can be lost in numerical computation, say the product of two 32-bit numbers will be only so long, and not higher. If we need better accuracy, then we need to devise numerically stable and accurate algorithms, say square root type for maintaining the accuracy, or use double precision computations and then rounding to single precision. Often DSP (digital signal processing) units and VLSI (very large scale integrated) circuits can be used, and custom-built systems are used. The roundoff and quantization could lead to limit cycles, since the then the overall system becomes nonlinear. These effects can be modeled by a linear model with additive or multiplicative deterministic or stochastic disturbances. If there is a nonlinearity in a system, then one can use describing function (DF) method, like Laplace transform method used in case of analysis of linear systems. The transmission of a sinusoidal signal with amplitude A through a nonlinearity can be characterized by DF. The nonlinear analysis can be done by using

Nyquist criterion with the critical point −1, replaced by −1/$G_n(A)$; where $G_n(.)$ is a DF of the nonlinearity, which is amplitude dependent. In a simple CLCS, this approach would predict a limit cycle if

$$G_p(e^{i\omega T}) = -\frac{1}{G_n(A)}; \ T = \text{sampling period} \quad (12.7)$$

One can see that the critical point is now a line segment, and the analysis would thus predict the oscillations due to quantization, if the Nyquist curve of the loop gain interacts with the line segment. Accuracy of the converters can have a greater effect on the performance of a digital controller/algorithm.

12.7.3 Realization of Digital Controllers

The roundoff errors in computation of a control law, i.e. the algorithm can also cause quantization effects. The realization of a controller, say control filter TF, can be done in several ways: (i) direct, (ii) companion, (iii) serial-Jordan, (iv) parallel-diagonal, (v) ladder, and (vi) δ-operator.

If we use finite precision for representation of the coefficients in a TF (or the controller), say its characteristic function, then this will affect the locations of poles of the CLCS, and thereby disturb the stability of the digital control system. For a linear filter, we have

$$\begin{aligned} P(z, a_i) &= z^n + a_1 z^{n-1} + ... + a_n \\ &= (z - p_1)...(z - p_n) = f(z, a_i) \end{aligned} \quad (12.8)$$

By equating (12.8) to zero, then expanding it in the Taylor's series, and retaining only the first order terms, we obtain

$$\delta_{p_k} \approx -\frac{\delta P / \delta a_i}{\delta P / \delta z} |_{z=p_k} \cdot \delta a_i; \text{ and we have } \frac{\delta P}{\delta a_i} |_{z=p_k}$$

$$= p_k^{n-i}, \ \frac{\delta P}{\delta z} |_{z=p_k} = \prod_{j \neq k} (p_k - p_j) \quad (12.9)$$

The change in the value of the pole, i.e. its location due to a change in the coefficient of the characteristic equation, is then given as

$$\delta_{p_k} \approx -\frac{p_k^{n-1}}{\prod_{j \neq k} (p_k - p_j)} \cdot \delta a_i \quad (12.10)$$

One can see that, if the poles are close, then the denominator would be small, and there will be a larger change in the pole location, i.e. the system is affected greatly by a small change in the coefficients.

Let us consider some forms for the controller realizations. Let the controller TF be given as

$$y(k) = G_c(z)u(k) = \frac{b_0 + b_1 z + ... + b_m z^m}{1 + a_1 z + a_2 z^2 + ... + a_n z^n} u(k) \quad (12.11)$$

12.7.3.1 Direct/Companion Forms

The TF (12.11) is put in the direct form as

$$y(k) = \sum_{i=0}^{m} b_i u(k - i) - \sum_{i=1}^{n} a_i u(k - i) \quad (12.12)$$

For this direct form, (12.12) one needs to store $n + m$ variables, and it is also not a minimal realization, whereas the controller and observable canonical forms have only n states. In the direct form, the variables are the delayed form of the I/O signals. Since all these forms have the coefficients in their realizations, the same as the ones in the characteristic polynomial, the errors in these coefficients would affect the accuracy of the realizations.

12.7.3.2 Well-Conditioned Form

The control algorithm can be represented in the form of model, by a combination of first- and second-order systems, and the complex poles are represented using real variables:

$$z_i(k+1) = \lambda_i z_i(k) - \beta_i y(k); \ i = 1,...,n_r \quad (12.13)$$

$$v_i(k+1) = \begin{bmatrix} \sigma_i & \omega_i \\ -\omega_i & \sigma_i \end{bmatrix} v_i(k) + \begin{bmatrix} \gamma_{i1} \\ \gamma_{i2} \end{bmatrix} y(k); \ i = 1,...,n_c$$

$$u(k) = Dy(k) + \sum_{i=1}^{n_r} \gamma_i z_i(k) + \sum_{i=1}^{n_c} \mu_i^T v_i(k)$$

12.7.3.3 Ladder Form

One can use the continuous fraction expansion of the PTF (pulse TF) to avoid the coefficient sensitivity:

$$G_c(z) = \alpha_0 + \cfrac{1}{\beta_1 z + \cfrac{1}{\alpha_1 + \cfrac{1}{\beta_2 z + \cfrac{1}{\ddots}}}} \quad (12.14)$$

$$\cfrac{}{\alpha_{n-1} + \cfrac{1}{\beta_n z + \cfrac{1}{\alpha_n}}}$$

One can use z^{-1} form also. Let the controller TF be given as

$$G_c(z) = \frac{B(z)}{A(z)}; \ \deg\{A(z)\} = \deg\{B(z)\} = n \qquad (12.15)$$

The coefficients of the TF polynomials can be computed in the following manner:

$$\alpha_0 = B(z) div A(z); A_1(z) = A(z); \ and \ B_1(z) \\ = B(z) mod A(z) \qquad (12.16)$$

The computation is repeated for $i = 1$ to n:

$$\beta_i = A_i \, div \, z B_i; \qquad A_{i+1} = A_i \, mod \, z B_i \\ \alpha_i = B_i \, div \, A_{i-1}; \qquad B_{i+1} = B_i \, mod \, A_{i+1} \qquad (12.17)$$

The ladder NW can be expressed in the state space form as:

$$\beta_1 x_1(k+1) = \frac{1}{\alpha_1}(x_2(k) - x_1(k)) + u(k)$$

$$\beta_2 x_2(k+1) = \frac{1}{\alpha_1}(x_1(k) - x_2(k)) + \frac{1}{\alpha_2}(x_3(k) - x_2(k))$$

$$\vdots$$

$$\beta_i x_i(k+1) = \frac{1}{\alpha_{i-1}}(x_{i-1}(k) - x_i(k)) + \frac{1}{\alpha_i}(x_{i+1}(k) - x_1(k)) \quad (12.18)$$

$$\vdots$$

$$\beta_n x_n(k+1) = \frac{1}{\alpha_{n-1}}(x_{n-1}(k) - x_n(k)) + \frac{1}{\alpha_n} x_n(k)$$

$$y(k) = x_1(k) + \alpha_0 u(k)$$

12.7.3.4 Short-Sampling-Interval Modification and δ-Operator Form

If the sampling period is small, the transition matrix would be closer to the unity matrix, and the matrix G will be proportional to the sampling period:

$$x(k+1) = x(k) + (\varphi - I)x(k) + Gy(k) \qquad (12.19)$$

In (12.20), the matrix $(\varphi - I)$ will also be proportional to the sampling period T, the numerical representation would require fewer decimals, and the second and third terms are corrections that will be small if the period T is small.

In the δ-operator form, the following development with the system (12.20) is written as:

$$x(k+1) = x(k) + T(\bar{\varphi}x(k) + \bar{G}y(k)) \qquad (12.20)$$

Here, equivalence between (12.19) and (12.20) is obvious. With the δ operator, we obtain the following expressions:

$$\delta x(k) = \bar{\varphi}x(k) + \bar{G}y(k); \ \delta = \frac{q-1}{T} \qquad (12.21)$$

Then, we have the PTF in the δ operator form is given as

$$G_c(q) = \frac{B(q)}{A(q)} = \frac{B(\delta T + 1)}{A(\delta T + 1)} = \frac{\bar{B}(\delta)}{\bar{A}(\delta)} = \bar{G}_c(\delta); \ q = z^{-1} \qquad (12.22)$$

The δ operator has the property like the forward difference approximation of the operator d/dt:

$$\delta f(kT) = \frac{f(kT + T) - f(kT)}{T} \qquad (12.23)$$

Due to this property, the δ operator is closer to the continuous time domain rather than the shift operator. The stability domain is a circle with radius $1/T$, the origin as $-1/T$, and as $T \to 0$ (reasonably very small), then the stability region becomes the left half plane. In the limit as $T \to 0$, we have

$$\lim_{T \to 0} \bar{G}(\delta) = G(\delta) \Rightarrow G(s) \qquad (12.24)$$

The δ operator can be interpreted as a shift of origin and scaling, and the δ form can obtain better numerical properties than the shift form. A controller can be represented as

$$\delta x(kT) = \bar{\varphi}x(kT) + \bar{G}y(kT) = d(kT)$$

$$u(kT) = Hx(kT) + Dy(kT) \qquad (12.25)$$

$$x(kT) = \delta^{-1}\{d(kT)\} = x(kT - T) + Td(kT)$$

Since in (12.25), the second term is much smaller than the first one, it is necessary to represent the first term with enough word length. The sampling period also has some influence on the conditioning of the controller algorithm/computations, and a rapid sampling (higher sampling frequency) requires a high precision in the coefficients. Since the sampling period occurs in the denominator in some computations, being smaller would introduce larger errors.

12.7.4 Programming

Most discrete time (DT) controllers are implemented in a real time (RT) operating system, and in some systems, the different parts of the algorithms are distributed among different processors. The controller script/code is programmed in C or C++. Ada (developed by the U.S.

Department of Defence for computer-control applications) was the first language designed and developed for such programming. A code for the operator communication is larger than the control-only code. One can map several control loops as concurrent activities that are running in parallel, using special purpose software-real-time (SWRT) operating systems that can schedule tasks without making a strict sequential program. Here, the notions of process and task are basic concepts and represent activities that may be thought of as running in parallel in time. The processes may be scheduled to run periodically or in response to events like interrupts or completion of other tasks. In the case of shared variables/resources, it is necessary to ensure that the system does not enter the deadlock, where both the processes are waiting for each other. When there is an interaction with an operator, the control is considered as a foreground task/process, and the operator communication is called the background task/process. The real time operating systems are large programs written in assembly code and can handle more complex situations.

One can use special methods (DDC-packages, direct digital control) to program control systems that contain a large number of identical control loops: (i) read analog inputs and store in a table, (ii) convert signals to engineering units and store results in a table, (iii) apply the control algorithm sequentially to all values in the table using controller parameters stored in some table, and (iv) carry out D–A conversion to all variables stored in the output table. The control algorithms are of the PID type, and the modules for gain scheduling, logic, supervision, and adaptation are also available. Since all programming is reduced to entering the appropriate data in the tables, these are called table-driven. DDC-packages can contain modules for start-up, shutdown, and alarm handling.

Appendix III: Illustrative Examples

Most of the examples and their solutions are based on [1], [2], and [20] (however, some MATLAB scripts for generating the associated results, plots, and figures are reassembled/rewritten by the first author; and where applicable, these scripts are provided in the present books' URL, CRC Press).

Example III.1

Determine the stability of the given closed loop system, Figure III.1 (m3_lec1). The open loop TF is given as

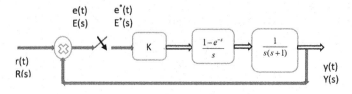

FIGURE III.1
A closed loop system for Example III.1.

$$G(z) = Z\left\{\frac{1-e^{-s}}{s} \cdot \frac{1}{s(s+1)}\right\} = (1-z^{-1})Z\left\{\frac{1}{s^2(s+1)}\right\};$$

since we have $H(s) = 1$, (III.1)

$$
\begin{aligned}
\frac{Y(z)}{E(z)} &= \frac{z-1}{z}\left\{\frac{z}{(z-1)^2} - \frac{(1-e^{-1})z}{(z-1)(z-e^{-1})}\right\} \\
&= \frac{(z-e^{-1})-(1-e^{-1})(z-1)}{(z-1)(z-e^{-1})} \\
&= \frac{z-0.368-0.632z+0.632}{(z-1)(z-0.368)} \\
&= \frac{0.368z+0.264}{(z-1)(z-0.368)}
\end{aligned}
$$

(III.2)

Since the characteristic equation is $\rightarrow 1+G(z) = 0$, we obtain

$$(z-1)(z-0.368)+0.368z+0.264 = 0 \rightarrow$$
$$z_{1,2} = 0.5 \pm 0.618j$$

(III.3)

Since the magnitudes of the two roots is $|z_1| = |z_2| < 1$, the closed loop system is stable.

Example III.2

Jury stability test: The characteristic equation is given as

$$P(z) = 1+G(z) = z^4 - 1.2z^3 + 0.07z^2 + 0.3z - 0.08 = 0 \quad \text{(III.4)}$$

The stability conditions are:

$$|a_n| = |a_4| = 0.08 < a_0$$

$$= 1 \Rightarrow \text{the first condition is satisfied}$$

$$P(1) = 1 - 1.2 + 0.07 + 0.3 - 0.08$$

$$= 0.09 > 0 \Rightarrow \text{the 2nd condition is satisfied}$$

$$P(-1) = 1 + 1.2 + 0.07 - 0.3 - 0.08$$

$$= 1.89 > 0 \Rightarrow \text{the 3rd condition is satisfied}$$

(III.5)

From the Jury's table we have the following details

$$b_3 = \begin{vmatrix} a_n & a_0 \\ a_0 & a_n \end{vmatrix} = 0.0064 - 1 = -0.9936$$

$$b_2 = \begin{vmatrix} a_n & a_1 \\ a_0 & a_3 \end{vmatrix} = -0.08 \times 0.3 + 1.2 = 1.176 \tag{III.6}$$

Other elements are computed in the similar manner. We have the following values:

$b_1 = -0.0756, \ b_0 = -0.204, \ c_2 = 0.946,$

$c_1 = -1.184, \ c_0 = 0.315 \ |b_3| = 0.9936 > |b_0| = 0.204; \tag{III.7}$

$|c_2| = 0.946 > |c_0| = 0.315$

Since all the criteria are satisfied, the closed loop system is stable.

Example III.3

Determine the value of the gain K, Figure III.1, for which the system is stable. We have

$$G(z) = \frac{K(0.368z + 0.264)}{(z-1)(z-0.368)}; \ \frac{Y(z)}{R(z)}$$

$$= \frac{K(0.368z + 0.264)}{z^2 + (0.368K - 1.368)z + 0.368 + 0.264K} \tag{III.8}$$

Setting the characteristic equation, the denominator of the 2nd TF of (III.8), one gets

$|a_2| < a_0, \ P(1) > 0; \ P(-1) > 0, n = 2, \text{even number.}$

1. $|0.368 + 0.264K| < 1 \Rightarrow 2.39 > K$

2. $P(1) = 1 + (0.368K - 1.368) + 0.368 + 0.264K$

$= 0.632K > 0 \Rightarrow K > 0$ $\tag{III.9}$

3. $P(-1) = 1 - (0.368K - 1.368) + 0.368 + 0.264K$

$= 2.736 - 0.104K > 0 \Rightarrow 26.31 > K$

From (III.9), the stabilizing range for the open loop gain K is found to be $0 < K < 2.39$.

Example III.4

Determine the value of the gain K for which the closed loop system of Figure III.2 is stable using bilinear transformation. The open loop TF is given as

$$G(z) = \frac{K(0.084z^2 + 0.17z + 0.019)}{(z^3 - 1.5z^2 + 0.553z - 0.05)}$$

$$1 + G(z) = z^3 + (0.084K - 1.5)z^2 \tag{III.10}$$

$$+ (0.17K + 0.553)z + (0.019K - 0.05) = 0$$

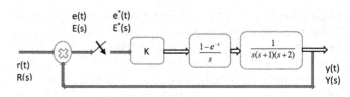

FIGURE III.2
The closed loop system for Example III.4.

To transform the characteristic equation $\tag{IIIA.10}$ by using the bilinear transformation, we get

$$Q(w) = (0.003 + 0.27K)w^3 + (1.1 - 0.11K)w^2$$

$$+ (3.8 - 0.27)w + (3.1 + 0.07K) = 0 \tag{III.11}$$

The Routh table is formed as follows

w^3	$0.003 + 0.27K$	$3.8 - 0.27K$
w^2	$1.1 - 0.11K$	$3.1 + 0.07K$
w^1	$\dfrac{0.01K^2 - 1.15K + 4.17}{1.1 - 0.11K}$	
w^0	$3.1 + 0.07K$	

The closed loop system will be stable if all the entries in the first column have the same sign; hence, the for stability we have the following conditions

$$0.003 + 0.27K > 0 \Rightarrow K > -0.011$$

$$1.1 - 0.11K > 0 \Rightarrow K > 10$$

$$0.01K^2 - 1.55K + 4.17 > 0 \tag{III.12}$$

$$\Rightarrow K < 2.735 \ or \ K > 152.47$$

$$3.1 + 0.07K > 0 \Rightarrow K > -44.3$$

From (III.12), the range for the gain K for the stability $-0.011 < K < 2.74$, because this is the overriding condition.

Example III.5

Singular case: Let the characteristic equation be given as

$$P(z) = 1 + G(z) = z^3 - 1.7z^2 - z + 0.8 = 0 \tag{III.13}$$

Now, transform $P(z)$ to $Q(w)$ to obtain

$$Q(w) = 0.9w^3 + 0.1w^2 - 8.1w - 0.9 = 0 \tag{III.14}$$

The Routh table is formed as

w^3	0.9	-8.1
w^2	0.1	-0.9
w^1	0	0

The auxiliary equation is formed by using the coefficients of w^2 row: $A(w) = 0.1w^2 - 0.9 = 0$, and then taking the derivative we obtain $\frac{dA(w)}{dw} = 0.2w$. Then, the Routh table is continued further

w^3	0.9	−8.1
w^2	0.1	−0.9
w^1	0.2	0
w^0	−0.9	

Since there is one sign change in the first row, one of the roots that lies in w-plane, is on the right hand s-plane (RHS) of the w-plane, implying that one root lies outside of the z-plane. Directly solving the characteristic equation (III.13), one obtains the roots: $z = 0.5$, −0.8, and 2; thus, the third root is greater than 1, hence, it lies outside the unit circle of the z-plane.

Example III.6: (m4_lec1 figure)

The steady state errors for unit step, ramp, and parabolic inputs are determined for the closed loop system of Figure III.3. The (open) loop TF is given as

$$G(s) = \frac{Y(s)}{E^*(s)} = G_{ho}(s)G_p(s)$$

$$= \frac{1-e^{-Ts}}{s} \cdot \frac{1000/10}{s(s+500/10)}$$

(III.15)

The Z-transform is given as

$$G(z) = 2(1-z^{-1})Z\left\{\frac{1}{s^2} - \frac{10}{500s} + \frac{10}{500(s+5000)}\right\}$$

$$= 2(1-z^{-1})\left[\frac{Tz}{(z-1)^2} - \frac{10z}{500(z-1)} + \frac{10z}{500(z-e^{-50T})}\right]$$

$$= \frac{1}{250}\left[\frac{(500T-10+10e^{-50T})z - (500T+10)e^{-50T}+10}{(z-1)(z-e^{-50T})}\right]$$

(III.16)

Steady state errors are: for step input→ 0, for ramp input→0.5, and for parabolic input→∞.

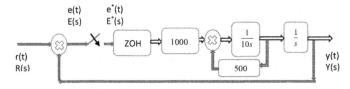

FIGURE III.3
The closed loop system for Example III.6.

Example III.7

(Section 10.1.2) (m5_lec1) Digital control system-root locus diagram for the system of Figure III.4. The details of this example are presented in Section 10.1.2. Figure III.5 shows the root locus of the LTF (OLTF) system for $T = 1$. Two root locus branches start from two open loop poles at $K = 0$. If we further increase K, one branch will go towards the zero and the other one will tend to infinity. The dotted circle represents the unit circle. Thus, the stable range of K is $0 < K < 8.165$. For $T = 1$ sec., we have the following (open) loop TF:

$$\frac{(0.6321z)}{(z^2 - 1.368z + 0.368)}$$

The breakaway/break-in points are: $z^2 = 0.368\rightarrow$ $z_1 = 0.6065$ and $z_2 = -0.6065$; critical gain $K_c = 4.328$. One can verify that when $T = 0.5$ sec.,

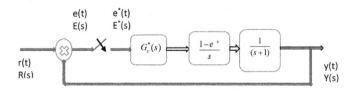

FIGURE III.4
The closed loop system for Example III.7.

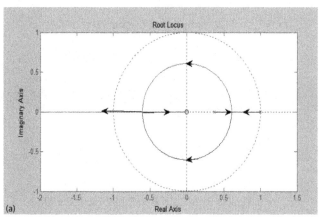

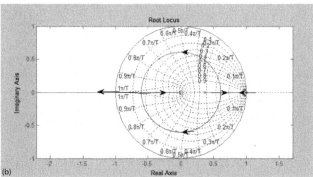

FIGURE III.5
(a) Root locus for $T = 1$ sec., No grid; Example III.7 and (b) Root locus for $T = 1$ sec., with grid; Example III.7.

then the inside circle would increase and the critical gain would be equal to 8.165. One can use the following commands in the MATLAB command window to generate the root locus as shown in Figure III.5a: $G = $ tf([0 0.6321 0], [1 −1.368 0.368],−1), and rlocus(G). If the grid is required on the plot, one can use: zgrid, axis equal, see Figure III.5b.

Example III.8

Controller design. The aim is to design a digital controller for the system of Figure III.4 (now with different plant dynamics), such that the dominant closed loop poles have damping ration = 0.5, settling time of 2 sec. for 2% tolerance band, and $T = 0.2$ sec. The dominant pole pair in continuous time is

$$-\xi\omega_n \pm j\omega_n\sqrt{1-\xi^2}\,; \xi - \text{damping ratio,} \qquad \text{(III.17)}$$

ω_n natural undamped frequency rad/sec.

Since the settling time is given as 2 sec., we have other parameters as

$$t_s = \frac{4}{\xi\omega_n} = 2 \to \omega_n = 4\,\text{rad/sec};$$

$$\omega_d = 4\sqrt{1-\xi^2} = 3.46, \text{damped frequency rad/sec;} \quad \text{(III.18)}$$

$$\omega_s = \frac{2\pi}{T} = 31.4; \text{samplinmg frequency rad/sec.}$$

Since 31.4/3.46 = 9.07, one gets 9 samples per cycle of the damped oscillation. The closed loop poles in s-plane are: $s_{1,2} = -\xi\omega_n \pm j\omega_n\sqrt{1-\xi^2} = -2 \pm j3.46$; the same in the z-plane are

$$z_{1,2} = \exp(T(-2 \pm j3.46)); |z| = e^{-T\xi\omega_n}$$

$$= \exp(0.4) = 0.67\angle z = T\omega_d = 0.2 \times 3.464$$

$$= 0.69\,\text{rad} = 39.69°; z_{1,2} = 0.67\angle 39.7 \cong 0.52 + j0.43$$

$$\text{(III.19)}$$

The $z(1,2)$ are the desired poles of the closed loop digital control system. The (open) loop TF is given as

$$G(z) = Z\left\{\frac{1-e^{-Ts}}{s} \cdot \frac{1}{s(s+1)}\right\} = (1-z^{-1})Z\left\{\frac{1}{s^2(s+1)}\right\}$$

$$\cong \frac{0.02(z+0.93)}{(z-1)(z-0.82)} = \frac{0.02z+0.0186}{z^2-1.82z+0.82}$$

$$\text{(III.20)}$$

The root locus of the uncompensated system (III.2) is depicted in Figure III.6 and is obtained by $G = $ tf([0 0.02 0.0186], [1 −1.82 0.82],−1), rlocus(G).

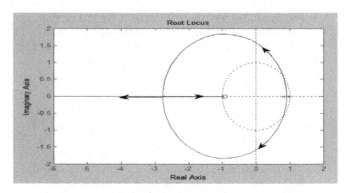

FIGURE III.6
Root locus of the uncompensated system; Example III.8.

From the poles, $z_1 = 1$, $z_2 = 0.82$, and the zero $z = -0.93$; and the desired poles $z_{1,2} = 0.52 \pm j0.43$, one can calculate the phase angle contribution of the two poles and the zero with respect to the location of the desired pole as (Figure III.7, by "using pzplot(G)"):

$$\theta = \theta_1 \text{ (by the zero)} - \theta_2 \text{ (by pole 1)} - \theta_3 \text{ (by pole 2)};$$

$$= 16.5° - 124.9° - 138.1° = -246.5°$$

$$\text{(III.21)}$$

Using the angle criterion, a point will lie on the root locus if the total angle contribution at that point is ±180°. Thus, we see that the angle deficiency is −246.5−180 = −66.5°. Thus, the controller pulse TF must provide an angle of 66.5°. Hence, one needs a Lead Compensator:

$$G_D(z) = K\frac{z+a}{z+b} \qquad \text{(III.22)}$$

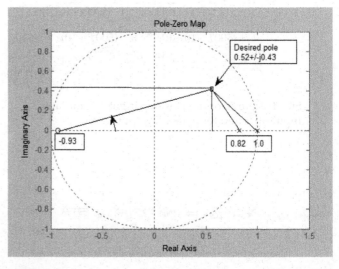

FIGURE III.7
Pole zero map for computation of the angel contributions; Example III.8.

We can place the controller zero at $z = 0.82$ to cancel the pole at $z = 0.82$; then the controller pole should provide an angle of $124.9 - 66.5 = 58.4°$. To satisfy the angle criterion we have

$$\tan^{-1}\left\{\frac{0.43}{0.52 - |b|}\right\} = 58.4° \Rightarrow \frac{0.43}{0.52 - |b|} \quad (\text{III.23})$$

$$= \tan(58.4°) = 1.625 |b| = 0.253$$

As the required angle is greater than $\tan^{-1}(0.43/0.52) = 39.6°$., the pole must lie on the right half of the unit circle, and be must be negative, $b = 0.253$. Hence, the controller TF is

$$G_D(z) = K\frac{z - 0.82}{z - 0.253} \quad (\text{III.24})$$

The (open) loop TF is given (using (III.24) and (III.2)) as

$$G(z) = K\frac{(z - 0.82)}{(z - 0.253)}\frac{(0.02z + 0.0186)}{(z^2 - 1.82z + 0.82)}$$

$$= K\frac{0.02z^2 - 0.0022z - 0.01525)}{z^3 - 2.073z^2 + 1.2805z - 0.2075)} \quad (\text{III.25})$$

Figure III.7 shows the pole-zero map for computation of the angel contributions. Then, the root locus of the compensated system is obtained using $G = $ tf([0.02 0.0022 −0.01525], [1 −2.073 1.281 −0.2075],−1), rlocus(G), which is shown in Figure III.8. One can see that for the compensated

system, the stable region of K is much larger than the uncompensated system as is seen from Figures III.7, and III.8. Now, from (III.20), and (III.24) we obtain the following using the magnitude criterion

$$\left|\frac{0.02K(z + 0.93)}{(z - 1)(z - 0.253)}\right|\{at\, z = 0.52 + j0.43\} = 1; \quad K = 10.75$$

$$(\text{III.26})$$

The controller is given as $G_D(z) = 10.75\dfrac{(z - 0.82)}{(z - 0.253)}$.

Example III.9

Controller design using MATLAB. The aim is to design a digital controller for the system of Figure III.4 (now with different plant dynamics) as (with $T = 0.1$ sec.)

$$G_p(s) = \frac{10}{(s + 1)(s + 2)}$$

$$G(z) = Z\left\{\frac{1 - e^{-0.1s}}{s}\cdot\frac{10}{(s + 1)(s + 2)}\right\}$$

$$= (1 - z^{-1})Z\left\{\frac{10}{s(s + 1)(s + 2)}\right\} \quad (\text{III.27})$$

$$\approx \frac{0.04528(z + 0.9048)}{(z - 0.9048)(z - 0.8187)}$$

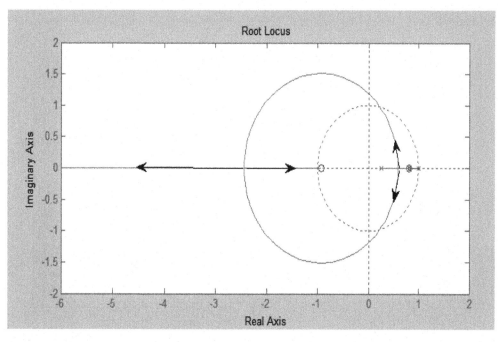

FIGURE III.8
Root locus of the compensated system, Example III.8.

Use the following script in MATLAB command window to obtain (open) loop TF, and root locus:

```
>>s = tf(s);
>>G_p = 10/((s + 1)×(s + 2));
>>G_hG_p = c2d(G_p,0.1,zoh), rlocus(G_hG_p)
```

$$GhGp = \frac{0.04528z + 0.04097}{z^2 - 1.724z + 0.7408} \qquad (III.28)$$

The pole-zero map of the uncompensated system obtained by using pzplot(GhGp) is shown in Figure III.9. The root locus of the uncompensated system is shown in Figure III.10. One of the design criteria is to have a zero SS error for a unit step input, and for this a PI controller is required that would have the TF in z-domain (with backward rectangular integration) as

$$G_D(s) = K_p + \frac{K_i T}{(z-1)} = \frac{K_p z - (K_p - K_i T)}{(z-1)} \qquad (III.29)$$

The integration constant K is decided based on the velocity error constant

$$k_v(s) = \frac{1}{T} \lim_{z \to 1} (z-1) G_D(z) G_{ho}(z) G_p(z)$$
$$= 5K_i \ge 5 \Rightarrow K_i \ge 1 \qquad (III.30)$$

We take $K_i = 1$, and the characteristic equation of the system is obtained as

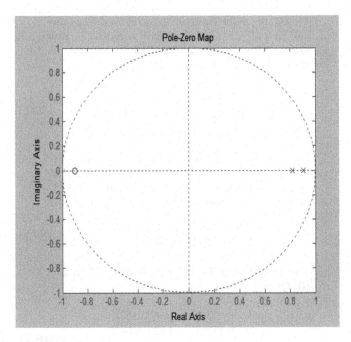

FIGURE III.9
Pole zero map of the uncompensated system (Example III.9).

$$1 + G_D(z) G_{ho}(z) G_p(z) = 0$$

$$1 + \frac{K_p z - (K_p - 0.1)}{(z-1)} \frac{0.04528(z + 0.9048)}{(z - 0.9048)(z - 0.8187)} = 0$$

$$1 + \frac{0.04528 K_p (z-1)(z + 0.9048)}{z^3 - 2.724z^2 + 2.469z - 0.7367} = 0$$

$$(z-1)(z - 0.9048)(z - 0.8187)$$

$$+ 0.004528(z + 0.9048) + 0.04528 K_p (z-1)(z + 0.9048) = 0$$

$$(III.31)$$

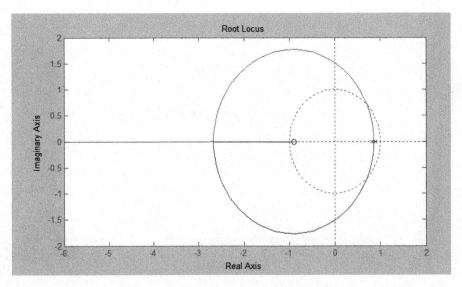

FIGURE III.10
Root locus of the uncompensated system (Example III.9).

We can obtain the root locus of the compensated system with K_p as the variable parameter:

```
>> z = tf('z',0.1);
>>
Gcomp = 0.04528 × (z-1) × (z + 0.9048)/
(z³-2.724 × z² + 2.469 × z-0.7367);
>> zero(Gcomp);
>> pole(Gcomp);
>> rlocus(Gcomp)
```

$$\text{Gcomp} = \frac{0.04528z^2 - 0.004311z - 0.04097}{z^3 - 2.724z^2 + 2.469z - 0.7367};$$

zeros are $= 1, -0.9048$; poles are 1.0114

$$+/-0.1663j, 0.7013$$

The root locus of the PI compensated control system is shown in Figure III.11. The system is stable for a very small range of K_p. This stable portion of the root locus is expanded and shown in Figure III.12, from which it is seen that the stable range of K_p is: $0.239 < K_p < 6.31$. The best achievable overshoot is 45.5%, for $K_p = 1$, which is very high for any practical system. To improve the relative stability, we need to introduce the derivative action; the PID controller TF in z-domain is given as

$$G_D(z) = \frac{(K_pT + K_d)z^2 + (K_iT^2 - K_pT - 2K_d)z + K_d}{Tz(z-1)} \quad \text{(III.32)}$$

We need to satisfy the earlier velocity constant. If we assume 15% overshoot (for damping ratio = 0.5), and 2 sec. settling time (for natural frequency = 4 rad/sec), the desired poles can be calculated as $s_{1,2} = -\xi\omega_n \pm j\omega_n\sqrt{1-\xi^2} = -2 \pm j3.46$, and the CL poles in z-plane are

$$z_{1,2} = \exp(T(-2 \pm j3.46)) \cong 0.77 \pm 0.28$$

The pole-zero map of (i) the original digital system, (ii) the PID controller only including its poles, and (iii) the desired poles is depicted in Figure III.13. The following commands are used in the command window (for the controller $G_D(z)$TF):

```
>>G_d = 1/((z)×(z-1))
```

And since the G_hG_p (z) is known form (III.27), one can use the following

```
>> pzplot(G_d × G_hG_p)
```

Compare Figures III.7 and III.13 for the pole/zero configurations, and except the new (to be designed) PID controller (with poles at $z = 0$, and $z = 1$ and the new desired poles $z_{1,2} = 0.77 \pm 0.28$, both are same. The total angle contribution from the left most zero, and up to the right most poles is = $(9.5-20-99.9-115.7-129.4) = -355.5°$. Angle deficiency is $-355.5 + 180 = -175.5°$. Thus, the two zeros of the PID must provide an angle of 175.5°, then we can place two zeros at the same location, the required angle by the individual zero is 87.75°,

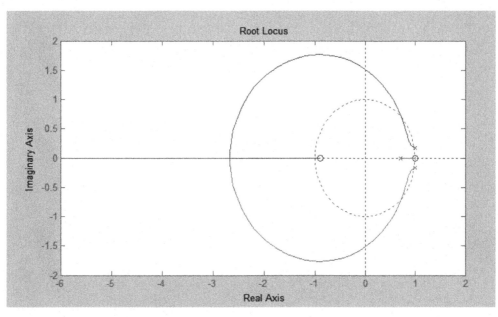

FIGURE III.11
Root locus of the compensated, PI controller, system (Example III.9).

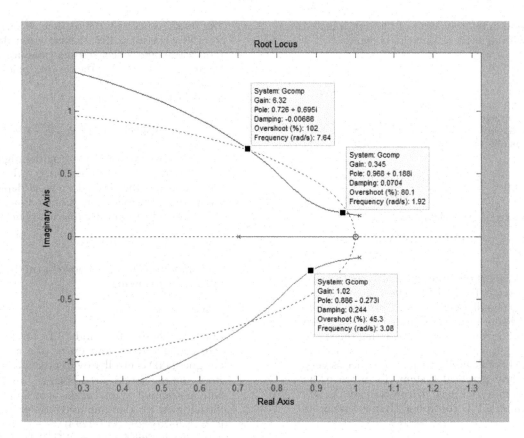

FIGURE III.12
Root locus with PI controller (zoomed) (Example III.9).

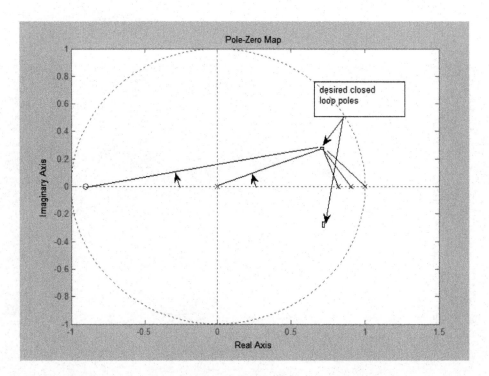

FIGURE III.13
Pole zero map for computation of the angel contributions from the plant and the PID (only poles) controller; Example III.9.

the zeros must lie of the left of the desired closed loop pole. Hence, we can compute the magnitude of this zero from

$$\tan^{-1}\left\{\frac{0.28}{0.77-z_{pid}}\right\}=87.75° \Rightarrow \frac{0.28}{0.77-z_{pid}}$$

$$=\tan(87.75°)=25.45$$

$$z_{pid}=0.759$$

Thus, the new PID controller is

$$G_D(z)=K\frac{(z-0.759)^2}{z(z-1)} \qquad (III.33)$$

The root locus plot of the system with the new PID controller, (III.33) is obtained using the following commands, and is shown in Figure III.14:

```
>>G_d = (z-0.759)²/((z) × (z-1))
>>rlocus(G_d × G_hG_p)
```

From the figure the desired pole corresponds to $K = 4.33$ (in the Figure it is approximately shown, because of the coarser movement of the data curser). The new PID controller is then given as

$$G_D(z)=4.33\frac{z^2-1.518z+0.5761}{z(z-1)} \qquad (III.34)$$

Comparing the $G_D(z)$ of (III.34) with the general PID controller form we can obtain the following constants

$$K_d/T=0.5761\times4.33 \rightarrow K_d=0.2495$$

$$K_p+K_d/T=4.33 \rightarrow K_p=1.835$$

$$K_iT-K_p-2K_d/T=-1.518\times4.33 \rightarrow K_i=2.521$$

The K_i satisfies imposed constraint on it.

Example III.10

Compensator design using Bode plot. Let us consider the continuous time plant $G(s)$, and $H(s)$ for which a lead compensator is required to be designed

$$G(s)=\frac{1}{s(s+1)}, H(s)=1 \qquad (III.35)$$

The criterion is that the SS error for a ramp input is ≤ 1 and the phase margin (PM) is at least 45°. For the closed loop system. For this the lead compensator is given in equation (10.26). The steady state error for the ramp input is obtained as

$$\frac{1}{\lim\limits_{s\to0}\{sC(s)g(S)\}}=\frac{1}{C(s)}=1/K \qquad (III.36)$$

$$=0.1 \Rightarrow K=10.$$

The gain cross over frequency of the uncompensated system with gain K is found as follows:

$$G(j\omega)=\frac{1}{j\omega(j\omega+1)}; \text{Mag.}$$

$$=\frac{1}{\omega\sqrt{1+\omega^2}}; \text{Phase}=-90°-\tan^{-1}(\omega) \qquad (III.37)$$

$$\frac{1}{\omega_g\sqrt{1+\omega_g^2}}=1 \Rightarrow \omega_g=3.1$$

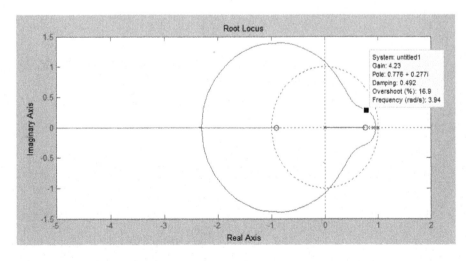

FIGURE III.14
Root locus of the compensated (with PID controller), system (Example III.9).

The phase of the uncompensated system at the gain cross over frequency (of 3.1 rad/sec.) is $-90° - \tan^{-1}(3.1) = -162°$, and the PM of this system is 18°; thus, the additional phase (lead required from the lead compensator) is $45 - 18 = 27°$. at gain cross over frequency. By adding a compensator (and maintaining same low frequency gain) would increase the crossover frequency, the actual PM will deviate from the desired one and a safety margin of 5 to 10° is added. Thus, the phase lead to be provided by the lead compensator is $27 + 10°$, and for this we have

$$\phi_{max} = 37° \Rightarrow \alpha = \left(\frac{1 - \sin(\phi_{max})}{1 + \sin(\phi_{max})} \right) = 0.25 \qquad (III.38)$$

For determining the time constant in the lead compensator, locate the frequency at which the uncompensated system had a logarithmic magnitude of $-20 \log_{10} \left\{ \frac{1}{\sqrt{\alpha}} \right\}$. Selecting this frequency as the new gain crossover frequency (the compensator) provides a gain of $20 \log_{10} \left\{ \frac{1}{\sqrt{\alpha}} \right\}$ at ω_{max}; and we have $\omega_{max} = \omega_{gnew} = \frac{1}{\tau \sqrt{\alpha}}, \omega_{max} = 4.41 \text{rad/sec.} \Rightarrow \tau = 0.4535$ Then the lead compensator is

$$C_{lead}(s) = 10 \frac{(0.4535s + 1)}{(0.1134s + 1)} \qquad (III.39)$$

The Bode plot of the compensated system is shown in Figure III.15, from which one can see

that the actual PM is ~49.6°, and the GM and PM are obtained using

```
>> bode(G_c*Gs)
>> [Gm, Pm, Wcg, Wcp] = margin(Gc*Gs).

[Gm, Pm, Wcg, Wcp]=[Inf,49.615,
Inf,4.4193].
```

Example III.11

Use of the bilinear transformation and its Bode diagram. Let us consider the $GH(z)$ as

$$GH(z) = \frac{0.095z}{(z-1)(z-0.9)}; T = 0.1 \text{sec.}, use\, z = \frac{1 + wT/2}{1 - wT/2};$$

$$GH(w) = \frac{10.02(1 - 0.0025w^2)}{w(1 + 1.0026w)}; like\, s = j\omega_s, w = j\omega_\omega$$

$$= \frac{10.02(1 - 0.05w)(1 + 0.05w)}{w(1 + 1.0026w)}$$

$$= \frac{10.02(1 - 0.05j\omega_w)(1 + 0.05j\omega_w)}{j\omega_w(1 + 1.0026j\omega_w)}$$

The corner frequencies are $1/1.0026 = 0.997$ rad/sec., and $1/0.05 = 20$ rad/sec. The Bode diagram of the $GH(w)$, with bilinear transformation is shown in Figure III.16. Let us take now the system

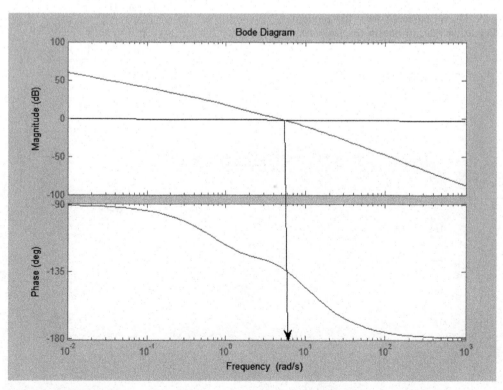

FIGURE III.15
Bode diagram of the compensated system (Example III.10).

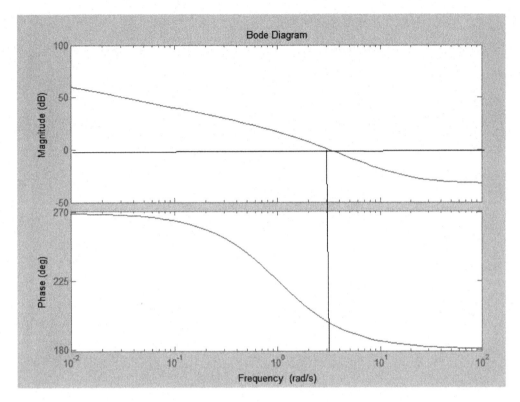

FIGURE III.16
Bode diagram of the $GH(w)$, with bilinear transformation (Example III.11).

(of Example III.10), and with sample data control system with $T = 0.2$ sec.:

$$G_z(z) = (1 - z^{-1})Z\left\{\frac{1}{s^2(s+1)}\right\}$$

$$= \frac{0.0187z + 0.0175}{z^2 - 1.8187z + 0.8187};$$

(III.40)

Now, using the bilinear transformation we get

$$z = \frac{1 + wT/2}{1 - wT/2} = \frac{1 + 0.1w}{1 - 0.1w}$$

$$G_w(w) = \frac{(1 + w/300)(1 - w/10)}{w(w+1)}$$

(III.41)

The phase lead compensator should be designed such that the PM of the compensated system is at least 50° with $K_v = 2$. The compensator in w-plane is in the following form

$$C(w) = K\frac{1 + w\tau}{1 + \alpha w\tau}; \quad 0 < \alpha < 1$$

(III.42)

The design steps are: (i) K has to be found out from the K_v requirement, (ii) compute the gain crossover frequency and the PM of the uncompensated system with K in the system, (iii) at this gain crossover frequency determine the additional phase lead required, and add the safety margin, and determine the constant α from the required

maximum phase angle, (iv) determine the new gain crossover frequency (as done in the Example III.10), where the maximum phase lead should occur, and (v) calculate τ and K from the following

$$\omega_{gnew} = \omega_{max} = \frac{1}{\tau\sqrt{\alpha}}$$

$$; K_v = \lim_{w \to 0}\{wC(w)G_w(w)\} = 2 \Rightarrow K = 2$$

(III.43)

The $G(z)$ in the w-plane/domain is now obtained as

$$G_w(w) = \frac{(1 + w/300)(1 - w/10)}{w(w+1)}$$

$$= \frac{-w^2 - 290w + 3000}{3000w^2 + 3000w}$$

(III.44)

The Bode diagram of the uncompensated system using the TF of (III.44) is obtained using the following commands, and is shown in Figure III.17:

```
>>s=tf('s'),
>>Gwuc=(-s^2-290*s+3000)/(3000*s^2 +
3000*s)
>>bode(Gwuc)
```

The maximum phase is chosen as 30°, and hence, we have

$$\phi_{max} = 30° \Rightarrow \alpha = \left(\frac{1 - \sin(\phi_{max})}{1 + \sin(\phi_{max})}\right) = 0.33$$

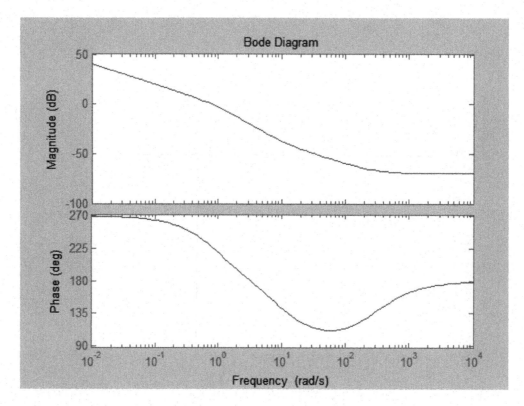

FIGURE III.17
Bode diagram of the uncompensated system (in w-plane/Example III.11).

From the Bode diagram (III.17) one can find that the following parameters hold

$$\omega_{max} = \omega_{gnew} = 1.75\,rad\,/\,Sec.$$

$$= \frac{1}{\tau\sqrt{\alpha}} \Rightarrow \tau = \frac{1}{1.75\sqrt{0.33}} = 0.99$$

The controller is then obtained as

$$C(w) = 2\frac{(1+0.99w)}{(1+0.327W)} \qquad (III.45)$$

The TF of the compensated system is obtained as

```
>>Gwc=(2*(1 + 0.99*s)/(1 + 0.327*s))*
(-s^2-290*s+3000)/(3000*s*(s+1))

  -1.98 s^3 - 576.2 s^2 + 5360 s + 6000
 ------------------------------------------
        981 s^3 + 3981 s^2 + 3000 s

>>bode(Gwc)
>>margin(Gwc)
```

The bode plot of the compensated system is shown in Figure III.18. One can transform the controller to the z-plane:

$$C(z) = 2\frac{(2.5527z - 2.0843)}{(z - 0.5316)} \qquad (III.46)$$

Example III.12

Lag compensator design. First we do it in the s-plane, then in the w-plane. Before that the form and the bode plot of a lag compensator itself is obtained

$$C_{lag}(s) = \alpha\frac{(\tau s + 1)}{(\alpha\tau s + 1)},\ \alpha > 1; for\,s \rightarrow 0,$$

$$C_{lag}(s) \rightarrow \alpha; for\,s \rightarrow \infty, C_{lag}(s) \rightarrow 1 \qquad (III.47)$$

The idea in using a lag compensator is to provide and additional gain of α in the low frequency region and to have the system with sufficient phase margin. The Bode diagram (frequency response) of the lag compensator TF with $\alpha = 4$ and $\tau = 3$ is depicted in Figure III.19.

```
>> Cs=4*(1 + 3*s)/(1 + 12*s)
>>bode(Cs)
>>margin(Cs)
```

Since the lag compensator $C(s)$ provides the maximum lag near the two corner frequencies, in order to maintain the PM of the system, the zero of the compensator should be chosen so that $\omega = 1/\tau$ is much lower than the gain crossover frequency of the uncompensated system; and in general τ is selected such that $1/\tau$ is at least one decade below the gain crossover frequency of the

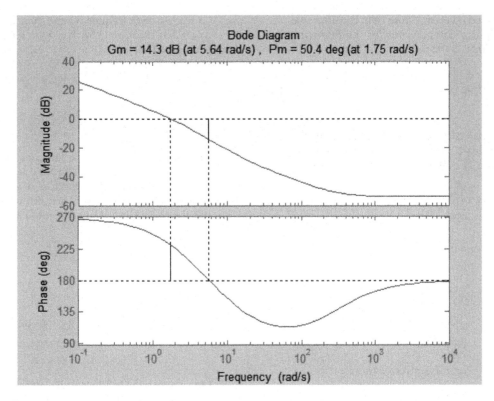

FIGURE III.18
Bode diagram of the compensated system (in w-plane/Example III.11).

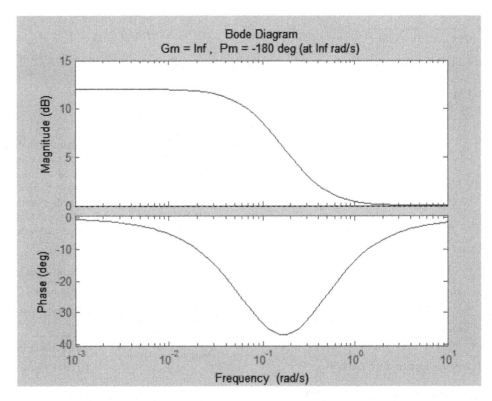

FIGURE III.19
Bode diagram of the lag compensator itself (s-plane/Example III.12).

uncompensated system. Now, we consider the design of the lag compensator for the following control system

$$G(s) = \frac{1}{(s+1)(0.5s+1)} \qquad \text{(III.48)}$$

The lag compensator is to be designed such that the PM is at least 50°, and the SS error to a unit step input is ≤ 0.1. The overall compensator is

$$C_{lag}(s) = K\alpha \frac{(\tau s + 1)}{(\alpha \tau s + 1)}; \text{ for } s \to 0, C_{lag}(s) \to K\alpha \qquad \text{(III.49)}$$

The SS error for a unit step input is

$$\frac{1}{1 + \lim\limits_{s \to 0} C(s)G(s)} = \frac{1}{1 + C(0)}$$

$$= \frac{1}{1 + K\alpha} = 0.1 \Rightarrow K\alpha = 9 \qquad \text{(III.50)}$$

The modified original control system is

$$G_m(s) = \frac{K}{(s+1)(0.5s+1)}; G_m(j\omega)$$

$$= \frac{K}{(j\omega + 1)(0.5j\omega + 1)}; \qquad \text{(III.51)}$$

$$Mag. = \frac{K}{(\sqrt{1+\omega^2})(\sqrt{1+0.25\omega^2})};$$

$$Phase = -\tan^{-1}\omega - \tan^{-1}0.5\omega$$

The required PM = 50°, and this is achieved only by selecting K. With the safety margin the required PM = 55°. This yields

$$180° - \tan^{-1}\omega_g - \tan^{-1}0.5\omega_g = 55°;$$

$$\tan^{-1}\frac{\omega_g + 0.5\omega_g}{1 - 0.5\omega_g^2} = 125°$$

$$0.715\omega_g^2 - 1.5\omega_g - 1.43 = 0 \Rightarrow \omega_g$$

$$= 2.8 \, rad / \sec.$$

For this to be the gin crossover frequency of the modified system the magnitude should be 1:

$$Mag. = \frac{K}{(\sqrt{1+\omega_g^2})(\sqrt{1+0.25\omega_g^2})}$$

$$= 1 \Rightarrow K = 5.1, and \alpha = 9 / K = 1.76$$

Since the PM is already attained with the gain K, we can place $\omega = 1/\tau$ so that it does not affect the PM much, then we place $1/\tau$ one decade below the

gain crossover frequency, and hence, we obtain $1/\tau = 2.8/10$, or, $\tau = 3.57$. The overall compensator is obtained as

$$C(s) = 9\frac{3.57s+1}{6.3s+1} \qquad \text{(III.52)}$$

```
>> Gsc=9*((1 + 3.57*s)/(1 + 6.3*s))*
(1/((s+1)*(0.5*s+1)))
```

The Bode diagram of the lag-compensated system is given in Figure III.20. Now let the original system, (III.48) be described as a sampled data control system with $T = 0.1$ sec., for which one can use the MATAB to obtain the plant TF as follows;

```
>> s=tf('s')
>> Gs=1/((s+1)*(0.5*s+1))
>> Gz=c2d(Gs,0.1,'zoh')
```

We would get the G(z) as

```
    (0.009056 z + 0.008194)
    ---------------------------
    (z^2 - 1.724 z + 0.7408)
```

One can use the bilinear transformation to obtain the TF in the w-plane:

$$z = \frac{1+wT/2}{1-wT/2} = \frac{1+0.05w}{1-0.05w} \qquad \text{(III.53)}$$

```
>> aug=[0.1,1];
>> gws = bilin(ss(Gz),-1,'S_Tust',aug)
>> Gw=tf(gws)
```

$$G(w) = \frac{(-0.0002488 \ w^2 - 0.09461ws + 1.992)}{(w^2 + 2.993 \ w + 1.992)}$$

```
>>bode(Gw)
>>margin(Gw)
```

The Bode plot of the uncompensated system in depicted in Figure III.21. Now, we need to design the compensator so that PM of the compensated system is at least 50°, and the SS error to a unit step input is ≤ 1, and the lag compensator's TF is as in (III.49). We have $C_{lag}(0) = K\alpha$, and $G_w(0) = 1$, $K\alpha = 9$ for SS $= 0.1$. The modified system is obtained approximately from G(w) as

$$G_m(w) \cong \frac{-0.00025K(w-20)(w+400)}{(w+1)(w+2)} \qquad \text{(III.54)}$$

The magnitude and phase of (III.54) are given as

$$Mag.(G_m(w)) = \frac{0.00025K(\sqrt{w^2+400})(\sqrt{w^2+160000})}{(\sqrt{1+\omega^2})(\sqrt{4+\omega^2})};$$

$$Phase(G_m(w)) = -\tan^{-1}\omega - \tan^{-1}0.5\omega$$

$$-\tan^{-1}0.05w + \tan^{-1}0.0025\,\omega$$

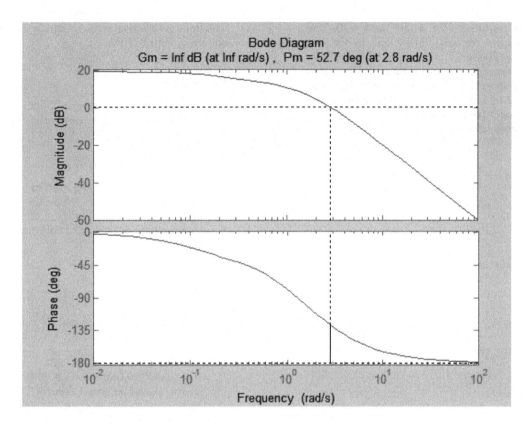

FIGURE III.20
Bode diagram of the lag-compensated system (s-plane/Example III.12).

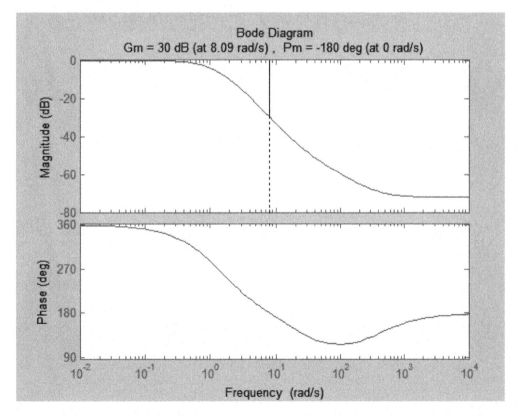

FIGURE III.21
Bode diagram of the uncompensated system (w-plane/Example III.12).

Since the required PM $= 50°$ and with the safety margin we get PM $= 55°$; hence, we get $180° + \text{Phase}(G_m(w)) = 55°$, which yields $\omega_g = 2.44$ rad/sec. Since the magnitude at this frequency should be 1, we obtain $K = 4.13$, and $\alpha = 9/K = 2.18$; placing $1/\tau$ one decade below the gain crossover frequency, we get $1/\tau = 2.44/10 \rightarrow \tau = 4.1$. Hence, the compensator/controller TF is

$$C_{lag}(w) = 9\frac{1+4.1w}{1+8.9w}; \; w = 20\frac{z-1}{z+1};$$

$$C_{lag}(z) = 9\frac{83z-81}{179z-177}$$

(III.55)

We can obtain the TF of the compensated system as

```
>>Gwc=9*((1 + 4.1*s)/(1 + 8.9*s))*Gw
```

$$\text{Gc(w)} = \frac{(-0.009179\ s^3 - 3.493\ s^2 + 72.64\ s + 17.93)}{(8.9\ s^3 + 27.63\ s^2 + 20.72\ s + 1.992)}$$

```
>>bode(Gwc)
>>margin(Gwc)
```

The Bode plot of the compensated system is shown in Figure III.22.

Example III.13

Lag-lead compensator design. The design procedure is first illustrated for the continuous TF

$$G(s) = \frac{1}{s(1+0.1s)(0.2s+1)}$$

(III.56)

The lag-lead controller is required to be designed such that the PM is at least $45°$ at the gain crossover frequency around 10 rad/sec, and the velocity error constant K_v is 30. The compensator TF is given as

$$C_{lale}(s) = K\frac{(\tau_1 s + 1)}{(\alpha_1\tau_1 s + 1)}\frac{(\tau_2 s + 1)}{(\alpha_2\tau_2 s + 1)};$$

$$\alpha_1, \text{and}\, \alpha_2 > 1; \text{when}\, s \rightarrow 0, C_{ll}(s) \rightarrow K$$

(III.57)

From (III.56), and (III.57), we have $K_v = \lim_{s\to 0} sG(s)C(s) = C(0) = 30 \Rightarrow K = 30$. The Bode plot of the uncompensated control TF is shown

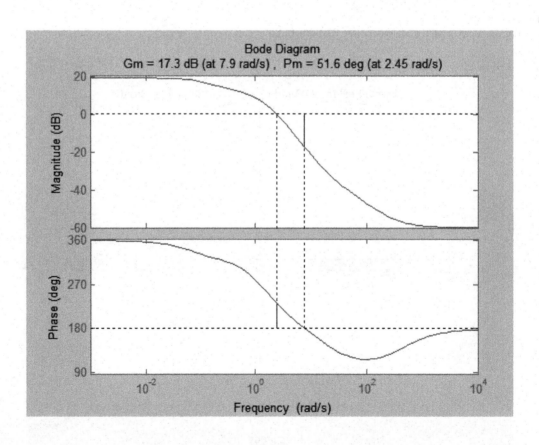

FIGURE III.22
Bode diagram of the compensated system (w-plane/Example III.12).

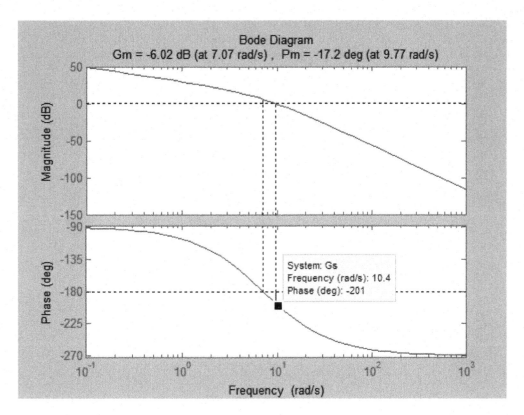

FIGURE III.23
Bode diagram of the uncompensated system (s-plane/Example III.13).

if Figure III.23, from which we observe that the PM is 17.2°, at 9.77 rad/sec. Since the PM of the system is negative, we use a lead compensator.

However, it is known that this will increase the gain crossover frequency to maintain the low frequency gain. As a result the gain crossover frequency of the system cascaded with a lead compensator is likely to be much above the specified one, and hear the gain crossover frequency of the uncompensated system is with K is already 9.77 rad/sec., and this requires the use of a lag-lead compensator. So, first we would use the lead compensator. One can deduce, from Figure III.23 that the phase angle of the system is −198° at 10 rad/sec. Since the new crossover frequency should be 10 rad/sec, the required additional phase at this frequency is 45−(180−198)=63°, to maintain the specified PM, and with a small safety margin we have

$$\alpha_2 = \left(\frac{1 - \sin(65°)}{1 + \sin(65°)} \right) = 0.05; \; and$$

$$10 = \frac{1}{\tau_2 \sqrt{\alpha_2}} \Rightarrow \tau_2 = 0.45;$$

$$C_{lead}(s) = \frac{(0.45 \; s \; + \; 1)}{(0.0225 \; s \; + \; 1)}$$

$$Glc(s) = \frac{(13.5 \; s \; + \; 30)}{(0.00045 \; s^4 + 0.02675 \; s^3 + 0.3225 \; s^2 + s)}$$

(III.58)

As can be seen from Figure III.24, with only the lead compensated control system, the gain crossover frequency is increased where the phase characteristics are different than the desired one; the gain crossover frequency is 23.2 rad/sec., and at 10 rad/sec., the phase angle is −134°, and the gain is 12.6 dB. We need to make this at the actual gain crossover frequency, and lag compensator need to be cascaded to provide an attenuation of −12.6 dB at high frequencies. Thus, we have $20\log_{10}\alpha_1 = 12.6 \Rightarrow \alpha_1 = 4.27$. The $1/\tau_1$ should be placed much below the new gain crossover frequency to retain the desired PM, and hence, 0.25 gives τ_1 as 4. Thus, the overall lag-lead compensator is obtained as

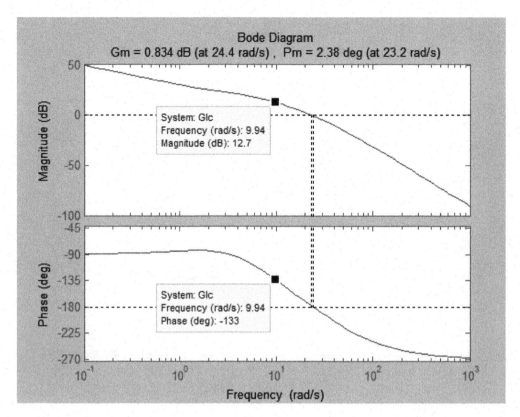

FIGURE III.24
Bode diagram of the only lead-compensated system (s-plane/Example III.13).

$$C_{lale}(s) = 30 \frac{(4s+1)}{(17.08s+1)} \frac{(0.45s+1)}{(0.0225s+1)} \qquad \text{(III.59)}$$

```
                 (54 s^2 + 133.5 s + 30)
G_lag-lead C(s) = ---------------------------------          (III.60)
                 (0.007686 s^5 +
                 0.4573 s^4 + 5.535 s^3 +
                 17.4 s^2 + s)
```

The Bode plot of the lag-lead compensated control system is shown in Figure III.25, and one can see that the desired characteristics are met. Now, for the same original control system we illustrate the design procedure considering it as a sampled data control system with $T = 0.1$ sec. We can use the following MATLAB commands to obtain the w-plane TF

```
>>s=tf('s')
>>Gc=1/(s*(1 + 0.1*s)*(1 + 0.2*s))
>>Gz=c2d(Gc,0.1,'zoh')
```

```
        (0.005824 z^2 + 0.01629 z + 0.002753)
G(z) = ---------------------------------------
        (z^3 - 1.974 z^2 + 1.198 z - 0.2231)
```
                                                        (III.61)

Now, we use the bilinear transformation to obtain the plant TF in the w-plane as in Example III.12:

```
        (0.001756 w^3 - 0.06306 w^2 - 1.705 w + 45.27)
G(w) = -----------------------------------------------
        (w^3 + 14.14 w^2 + 45.27 w + 2.815e-13)
```
                                                        (III.62)

The Bode plot of the uncompensated original system with $K = 30$, in w-plane is shown in Figure III.26, and since PM of 45° is needed, and the system has the phase lag of 139 °., we need a large lead = (180–139) + 45 = 86°. The single lead compensator cannot provide such a phase. We can lower the need of the PM to 20° at the gain cross over of 10 rad/sec., the required additional phase at this frequency to maintain PM of 20° is 20–(180–221) = 61°, and with a small safety margin we set

$$\alpha_2 = \left(\frac{1-\sin(66°)}{1+\sin(66°)}\right) = 0.045; \text{ and}$$

$$10 = \frac{1}{\tau_2\sqrt{\alpha_2}} \Rightarrow \tau_2 = 0.47$$
                                                        (III.63)

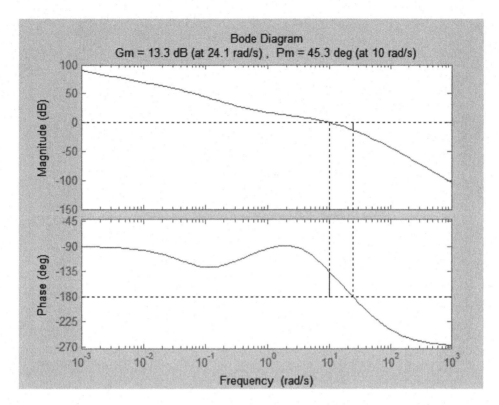

FIGURE III.25
Bode plot of the only lag & lead-compensated system (s-plane/Example III.13).

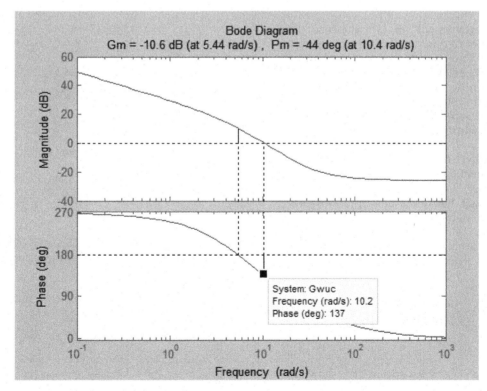

FIGURE III.26
Bode plot of the uncompensated system (w-plane/Example III.13).

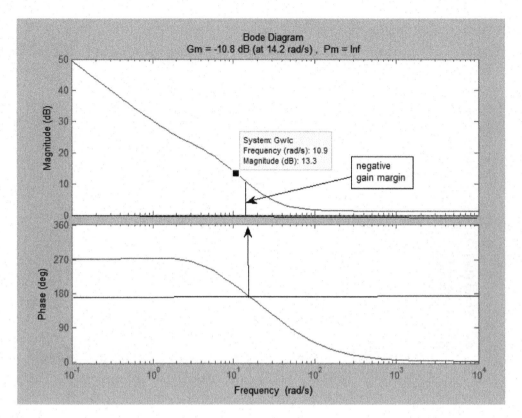

FIGURE III.27
Bode plot of the only lead-compensated system (w-plane/Example III.13).

This compensator will in fact increase the gain crossover frequency where the phase will be different than the desired one, as can be seen from Figure III.27. We can also see that the gain margin is negative of this only-lead compensated system. This requires the use of a lag compensator to lower the magnitude at 10 rad/sec., which is 14.2 dB. To make this as the actual

Gain crossover frequency, lag TF should provide an attenuation of −14.2 dB. at high frequencies; $20\log_{10}\alpha_1 = 14.2 \Rightarrow \alpha_1 = 5.11$. The $1/\tau_1$ should be placed much below the new gain crossover frequency to retain the desired PM, and hence, 1 (10/10) gives τ_1 as 1. Thus, the oveall new lag-lead compensator is obtained as

$$C_{lale}(w) = 30\frac{(w+1)}{(5.11w+1)}\frac{(0.47w+1)}{(0.02115w+1)} \qquad \text{(III.64)}$$

The lag-lead compensated system TF is shown in Figure III.28.

$$G_{\text{lag-leadlc}}(w) = \frac{(0.02476\ s^5 - 0.8118\ s^4 - 26.76\ s^3 + 561.3\ s^2 + 1945\ s + 1358)}{(0.1081\ s^5 + 6.659\ s^4 + 78.45\ s^3 + 246.4\ s^2 + 45.27\ s + 2.815e\text{-}13)}$$

The controller in the z-plane is obtained as

$$C_{lale}(z) = 30\frac{(0.2035z - 0.1841)}{(z - 0.9806)}\frac{(7.309z - 5.903)}{(z + 0.4055)}$$

$$\text{(III.65)}$$

The compensator of (III.65) can be obtained from (III.64) by using the bilinear transform as

```
>> aug=[0.1,1];
>> Cws = bilin(ss(Cw),1,'S_Tust',aug)
>> Gz=tf(Cws)
```

$$G_{\text{lag-leadc}}(z) = \frac{(44.62\ z^2 - 76.4\ z + 32.6)}{(z^2 - 0.5751\ z - 0.3976)}$$

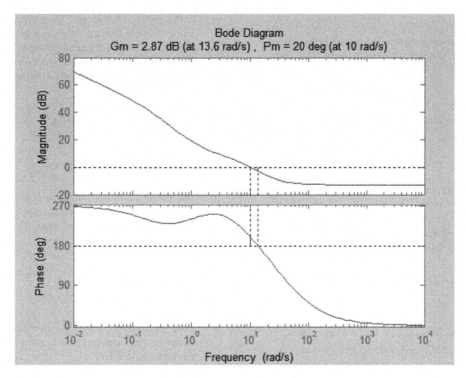

FIGURE III.28
Bode plot of the lag-lead compensated system (w-plane/Example III.13).

Example III.14

Deadbeat response design for the plant TF given as

$$G_p(z) = \frac{z+0.6}{3z^2 - z - 1} \Rightarrow D_c(z) = \frac{1}{G_p(z)} \frac{C(z)}{1 - C(z)}$$

$$D_c(z) = \frac{3z^2 - z - 1}{(z+0.6)} \frac{1/z}{1 - 1/z} = \frac{3z^2 - z - 1}{(z+0.6)(z-1)}$$

(III.66)

So, when the input is a unit step function, and the CLTF is $C(z) = 1/z$; the output will follow the input after one sampling instant as given by equation (10.34). When the input is a ramp function, and the closed loop response is

$$C(z) = \frac{2z - 1}{z^2} \Rightarrow D_c(z) = \frac{(3z^2 - z - 1)}{(z+0.6)} \frac{(2z-1)/z^2}{1 - (2z-1)/z^2}$$

$$D_c(z) = \frac{(3z^2 - z - 1)(2z - 1)}{(z^2 - 2z + 1)(z + 0.6)}$$

(III.67)

The output for the ramp input is now $Y(z) = \frac{T(z-0.5)}{z(z-1)^2} = T(2z^{-2} + 3z^{-3} + ...)$, which follows

the input after 2 sampling periods, as can be seen from the Figure III.29.

For yet another plant as

$$G_p(z) = \frac{0.05(z+0.5)}{(z-0.9)(z-0.8)(z-0.3)} \Rightarrow$$

$$D_c(z) = \frac{20(z-0.9)(z-0.8)(z-0.3)}{(z+0.5)(z+1)(z-1)}$$

(III.68)

$$Y(z) = C(z)R(z) = \frac{1}{z^2} \frac{z}{z-1} = \frac{1}{z^2 - z} = z^{-2} + z^{-3} + ...$$

FIGURE III.29
Deadbeat response to a ramp input (to the digital control system); Example III.14.

Example III.15

Deadbeat response design-when some poles and zeros are on or outside the unit circle. Let the plant TF be given as

$$G_p(z) = \frac{0.01(z+0.2)(z+2.8)}{z(z-1)(z-0.4)(z-0.8)}$$

$$= \frac{0.01z^{-2}(1+0.2z^{-1})(1+2.8z^{-1})}{(1-z^{-1})(1-0.4z^{-1})(1-0.8z^{-1})} \quad (III.69)$$

The $G_p(z)$ has a zero at $z = -2.8$, and a pole at $z = 1$; so $C(z)$ should contain the term $1 + 2.8z^{-1}$, and $(1-C(z))$ should contain $1-z^{-1}$. Also, plant TF has two more poles than zeros, and this implies that the CLTF has the following form

$$C(z) = (1+2.8z^{-1})c_2 z^{-2};$$

$$1-C(z) = (1-z^{-1})(1+a_1 z^{-1}+a_2 z^{-2}) \quad (III.70)$$

Since the minimal order of the CLTF is 3, we need to determine 3 unknowns; hence combining the two equations we obtain the following equivalences:

$$a_1 = 1; \ a_1 - a_2 = c_2; \ a_2 = 2.8c_2 \Rightarrow c_2 = 0.26; \ a_2 = 0.73$$

$$C(z) = 0.26z^{-2}(1+2.8z^{-1}); \quad (III.71)$$

$$1-C(z) = (1-z^{-1})(1+z^{-1}+0.73z^{-2})$$

We now substitute the expressions of (III.71) in the controller TF we obtain

$$D_c(z) = \frac{(1-z^{-1})}{(1+2.8z^{-1})} \frac{(1+2.8z^{-1})0.26z^{-2}}{(1-z^{-1})(1+z^{-1}+0.73z^{-2})} \frac{(1-0.4z^{-1})(1-0.8z^{-1})}{0.01z^{-2}(1+0.2z^{-1})}$$

$$= \frac{0.26z^{-2}(1-0.4z^{-1})(1-0.8z^{-1})}{0.01z^{-2}(1+0.2z^{-1})(1+z^{-1}+0.73z^{-2})} \quad (III.72)$$

$$= \frac{26z(z-0.4)(z-0.8)}{(z+0.2)(z^2+z+0.73)}$$

For a unit step input the output response is now obtained as

$$Y(z) = C(z)R(z) = \frac{0.26(z+2.8)}{z^2(z-1)} = 0.26z^{-2}+z^{-3}+... \quad (III.73)$$

From (III.73), it can be seen that the poles on or outside the unit circle are still present in the expression of the output $Y(z)$. The output tracks the unit step perfectly after 3 sampling periods.

If the plant TF did not have any poles or zeros on or outside the unit circle, it would have taken only two sampling periods to track the step input, when the plant TF has two more poles than zeros. Now let us consider another plant TF

$$G_p(z) = \frac{0.0004(z+0.2)(z+2.8)}{(z-1)^2(z-0.28)}$$

$$= \frac{0.0004z^{-1}(1+0.2z^{-1})(1+2.8z^{-1})}{(1-z^{-1})^2(1-0.28z^{-1})} \quad (III.74)$$

The plant TF has an outside zero at $z = -2.8$, and two poles at $z = 1$, and the number of poles exceed the number of zeros by one. The CLTF should contain the terms as follows

$$C(z) = (1+2.8z^{-1})(m_1 z^{-1}+m_2 z^{-2});$$

$$1-C(z) = (1-z^{-1})^2(1+a_1 z^{-1}) \quad (III.75)$$

Then, combining the two equations of (III.75), and equating the like powers of z^{-1}, we finally obtain the following CLTF

$$C(z) = (0.72z^{-1}-0.457z^{-2})(1+2.8z^{-1});$$

$$1-C(z) = (1-z^{-1})^2(1+1.28z^{-1}) \quad (III.76)$$

Substituting the expressions of (III.76) into the form of the controller and simplifying we obtain

$$D_c(z) = \frac{1800(z-0.635)(z-0.28)}{(z+0.2)(z+1.28)} \quad (III.77)$$

We have the CLTF output response for the unit step as

$$Y(z) = C(z)R(z) = \frac{z(0.72z^2+1.56z-1.28)}{z^3(z-1)}$$

$$= 0.72z^{-1}+2.28z^{-2}+z^{-3}+z^{-4}+... \quad (III.78)$$

In this case the output tracks the unit step input perfectly after 3 sampling periods (starts tracking at the third period only), there is some overshoot, this is because the digital plant is of type 2, and a deadbeat response cannot be obtained for a step input without an overshoot.

Example III.16

Sampled data control system with deadbeat response. Let a sampled data plant be given as

$$G_p(s) = \frac{2}{s(s+2)}; \ G_{ho}G_p(z) = \frac{0.01(z+0.9)}{(z-1)(z-0.8)} \quad (III.79)$$

The normal design procedure would lead to the following

$$C(z) = z^{-1}; \quad 1 - C(z) = 1 - z^{-1}; \quad \text{and the controller is}$$

$$D_c(z) = \frac{100(z - 0.8)}{(z + 0.9)} \tag{III.80}$$

For the unit step response the output response is obtained as

$$Y(z) = C(z)R(z) = \frac{1}{z - 1} = z^{-1} + z^{-2} + \dots \tag{III.81}$$

From (III.81) it is implied that the output response is deadbeat only at the sampling instants. However, the actual output $y(t)$ has inner sampling ripples, and the system takes forever to reach the SS. The necessary and sufficient condition for the output $y(t)$ to track a unit step input in a finite time is naturally the following

$$y(NT) = 1; \quad \frac{dy(t)}{dt} = 0 \tag{III.82}$$

Let the derivative be represented by another variable, say $w(t)$, then

$$w(t) = \frac{dy(t)}{dt}; \quad W(z) = \frac{D_c(z)(1 - z^{-1})Z\{G_p(s)\}}{1 + D_c(z)(1 - z^{-1})Z\{\frac{G_p(s)}{s}\}} R(z)$$

$$= \frac{A_1(z - 1)}{z(z - 0.9)} R(z) \tag{III.83}$$

Since the poles of the $W(z)/R(z)$ are not all at $z = 0$, the unit step response of $W(z)$ will not go to zero in a finite time. Applying the condition that zero of $G_{ho}G_p(z)$ at $z = -0.9$ should not be cancelled by the controller TF ($D_c(z)$), we obtain

$$C(z) = (1 + 0.9z^{-1})m_1 z^{-1};$$

$$1 - C(z) = (1 - z^{-1})(1 + a_1 z^{-1}), \Rightarrow m_1 \tag{III.84}$$

$$= 0.53, \text{ and } a_1 = 0.47$$

The values in (III.84) yield the following CLTF and the controller TF as

$$C(z) = \frac{0.53(z + 0.9)}{z^2}; \quad D_c(z) = \frac{A_2(z - 0.8)}{z + 0.47}$$

$$Y(z) = C(z)R(z) = A_3 z^{-1} + z^{-2} + z^{-3} + z^{-4} + \dots \tag{III.85}$$

From (III.85), it can be seen that the deadbeat response reaches the SS after two sampling periods; since $W(z) = 2z^{-1}$, and that to without inter sample ripples. Now consider another plant TF such that

$$G_{ho}G_p(z) = \frac{0.01(z + 0.2)(z + 2.8)}{z(z - 1)(z - 0.4)(z - 0.8)} \tag{III.86}$$

Using the condition that we do not want to cancel the two zeros of the TF (III.86), then we have the CLTF as

$$C(z) = (1 + 0.2z^{-1})(1 + 2.8z^{-1})m_1 z^{-2};$$

$$1 - C(z) = (1 - z^{-1})(1 + a_1 z^{-1} + a_2 z^{-2} + a_3 z^{-3}) \tag{III.87}$$

We need to keep the points in mind: (i) $C(z)$ should contain all the zeros of $G_{ho}G_p(z)$, (ii) the number of poles over zeros of $C(z)$ should be at least equal to that of $G_{ho}G_p(z)$, i.e. 2 here, (iii) $1 - C(z)$ must include the term $(1 - z^{-1})$, and iv) the orders of $C(z)$ and $1 - C(z)$ should be same and equal to the number of unknown coefficients. Using the procedure of the previous example, we obtain the following TFs:

$$C(z) = \frac{0.219z^2 + 0.657z + 0.123}{z^4}; \text{since } m_1$$

$$= 0.219, a_1 = 1, a_2 = 0.781, a_3 = 0.123 \tag{III.88}$$

$$D_c(z) = \frac{21.9z(z - 0.4)(z - 0.8)}{z^3 + z^2 + 0.781z + 0.123}$$

The output for a unit step input is obtained as

$$Y(z) = C(z)R(z) = \frac{0.219z^2 + 0.657z + 0.123}{z^3(z - 1)} \tag{III.89}$$

$$= 0.219z^{-2} + 0.876z^{-3} + z^{-4} + z^{-5} + \dots$$

From (III.89) one can see that the output reaches the SS in four sampling instants. For sampled data control system, the deadbeat response $y(t)$ reaches the SS after three sampling periods and the inter sample ripples occur; however, after four sampling instants these disappear. The z-transform of $w(t)$ (i.e. the derivative of $y(t)$) is obtained as $W(z) = A_1 z^{-2} + A_2 z^{-3}$, and we see that the derivative of $y(t)$ is zero for $kT \geq 4T$; thus, the step response reaches the SS in four sampling instants and without inter sample ripples.

Example III.17

Nyquist plot is obtained for the following digital control system with $T = 0.1$:

$$GH(z) = \frac{0.095Kz}{(z - 1)(z - 0.9)} \tag{III.90}$$

With $K = 40$ the plot is shown in Figure III.30. The $GH(z)$ has one pole on the unit circle and

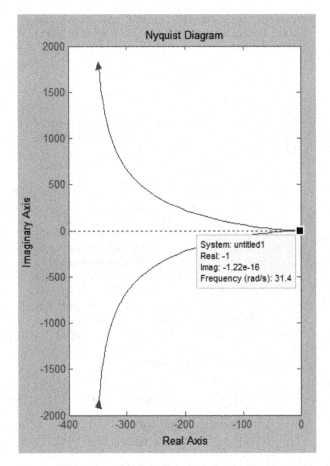

FIGURE III.30
Nyquist plot for the digital control system (Example III.17).

does not have any pole outside the unit circle, $P_{-1} = 0$, and $P_0 = 1$. Nyquist path has a small indentation at $z = 1$ on the unit circle. The plot intersects at the negative real axis at $-0.025 \times 40 = 1$, obtained as >>nyquistplot (40×GHz,{0.2,35}) when $\omega = \omega_s / 2 = 31.4\,\text{rad/sec}$. Using equation (10.21), we can calculate the angle traversed by the phasor drawn from $(-1, j0)$ point to the Nyquist plot of $GH(z)$ as ω varies from $\omega_s / 2$ to 0, on the unit circle:

$\phi = -(0 + 0.5 \times 1)180° = -90°$. For this angle to be $-90°$, the $(-1, j0)$ point should be located at the left of $-0.025K$ point; thus, for the stability, $-1 < -0.025K$, $K < 40$. If $K > 40$, the $(-1, j0)$ point will be at the right of $-0.25K$ point making the angle $= 90°$. If this angle is $+90°$, then we have from (10.21): $Z_{-1} = \frac{\phi}{180°} + 0 + 0.5 = 1$, so for $K > 40$, one of the CL poles will be outside the unit circle. If K is negative, we can still use the same Nyquist plot but refer $(+1, j0)$ point as the critical point. The angle ϕ still equals $+90°$ and the system is unstable. Hence, the stable range of K is between 0 and 40.

Example III.18

State feedback control design. Consider the following state space system and we have to design the controller, i.e. determine the gain factor/matrix K, such that the CL poles are located at 0.5, 0.6, and 0.7:

$$x(k+1) = \begin{bmatrix} 0 & 1 & 0 \\ 0 & 0 & 1 \\ -1 & -2 & -3 \end{bmatrix} x(k) + \begin{bmatrix} 0 \\ 0 \\ 1 \end{bmatrix} u(k) \qquad \text{(III.91)}$$

The controllability matrix is found by $U_c = \text{ctrb}(a, b); \text{rank}(U_c)$, and is obtained as

$$U_c = \begin{bmatrix} 0 & 0 & 0 \\ 0 & 1 & -3 \\ 1 & -3 & 7 \end{bmatrix},$$

and the rank of this matrix is 3; hence, the basic system is controllable. Based on the system given in (III.91), the open loop characteristic equation is obtained as $z^3 + 3z^2 + 2z + 1 = 0$ and the desired characteristic equation is given as $(z - 0.5)(z - 0.6)(z - 0.7) = 0 \Rightarrow z^3 - 1.8z^2 + 1.07z - 0.21 = 0$. Since the given open loop system is in the controllable canonical form, $T = I$, and the K can be easily formulated as $K = [(\alpha_3 - a_3)\ (\alpha_2 - a_2)\ (\alpha_1 - a_1)]$. However, we use the MATLAB function "place" as Kp = place(a,b,Dp); with Dp = [-0.5 -0.6 -0.7] that contains the desired closed loop poles, to obtain the gain K as [-0.79 -0.93 -1.2]. Now, using the Ackermann's formula the feedback gain can be obtained by the procedure described in Section 10.5.2, and alternatively by using the MATLAB function: Ka = acker(a,b,Dp), and we obtain the feedback gain again as Ka = [-0.79 -0.93 -1.20]. The closed loop system matrix and the closed loop eigenvalues can be obtained as: CLm = a-b × K; and eig(CLm)→[-0.5 -0.6 -0.7] as the desired closed loop poles. If we also use the pole placement function of the MATLAB, Kp = place(a,b,Dp), then we get the same gain values, closed loop system matrix, and the closed loop poles, as obtained by using the Ackemann's script.

Example III.19

Feedforward gain design. Let the system be given as

$$x(k+1) = \begin{bmatrix} 0 & 1 \\ -0.47 & -1.47 \end{bmatrix} x(k) + \begin{bmatrix} 0 \\ 1 \end{bmatrix} u(k); y(k) \qquad \text{(III.92)}$$

$$= [0.007\ 0.009] x(k)$$

We need to design a state feedback controller such that the output follows a step input with the desired CL poles at 0.5, and 0.6, this gives the desired characteristic equation as $z^2 - 1.1z + 0.3 = 0$ and we have the OL characteristic equation as $z^2 - 1.47z + 0.47 = 0$. Then, the controller can be designed using: $u(k) = -Kx(k) + K_r r(k)$. The system is in controllable form, see (III.92), and the controllable matrix is given as $U_c = [0\ 1;1\ 1.47]$ (U_c = ctrb(a,b)), which is non-singular, the state feedback gain be obtained as $K = [0.3-0.47\ -1.1 + 1.47] = [-0.17\ 0.37]$. The other gain can be obtained as $(K_r)^{-1} = -H(\varphi - BK - I)^{-1}B \rightarrow 0.08$ and this yields the final state feedback controller formula, i.e. $u(k) = -[0.17\ 0.37]\ x(k) + 12.5\ r(k)$.

Example III.20

State feedback with integral action. Let the plant be described as

$$G(s) = \frac{1}{s(s+2)}; T = 0.1 \sec. \qquad (III.93)$$

The idea is to design a state feedback with an integral action such that the output follows a step input. The CL continuous poles of the system be located as $-1 +/- j1$ and -5. The plant is discretized as

$$GH(z) = Z\left\{\left(\frac{1-e^{-Ts}}{s}\right)\left(\frac{1}{s(s+2)}\right)\right\} = \frac{0.005z + 0.004}{z^2 - 1.8z + 0.82}$$

$$(III.94)$$

From (III.94) we obtain the model of the plant in the form of discrete state space from

$$x(k+1) = \begin{bmatrix} 0 & 1 \\ -0.82 & 1.82 \end{bmatrix} x(k) + \begin{bmatrix} 0 \\ 1 \end{bmatrix} u(k); \qquad (III.95)$$

$$y(k) = [0.004\ 0.005]x(k)$$

Now, the system's state vector is augmented with the integral state $v(k) = v(k-1) + y(k)-r$; thus, we have the following system

$$\begin{bmatrix} x(k+1) \\ x(k+1) \\ v(k+1) \end{bmatrix} = \begin{bmatrix} 0 & 1 & 0 \\ -0.82 & 1.82 & 0 \\ -0.0041 & 0.0132 & 1 \end{bmatrix} x(k) + \begin{bmatrix} 0 \\ 1 \\ 0.005 \end{bmatrix} u(k) + \begin{bmatrix} 0 \\ 1 \\ -1 \end{bmatrix} r$$

$$(III.96)$$

Following the procedure outlined in Section (10.5.4), and using the equivalence specified there, we obtain the control law as $u(k) = -[-0.328\ 0.416]x(k)-0.889v(k)$. If we use the Ackerman's and

pole placement script of the MATLAB as done for the example III.18 (both Ackermann's and pole placement scripts), we obtain the gain as $K = [-0.3257\ 0.4169\ 0.6255]$, and the closed loop eigenvalues as: $[0.6000 + 0.0000i; 0.9000 + 0.1000i; 0.9000 - 0.1000i]$.

Example III.21

The observer design. Let the discrete time system be given as

$$x(k+1) = \begin{bmatrix} 0 & 1 \\ 20 & 0 \end{bmatrix} x(k) + \begin{bmatrix} 0 \\ 1 \end{bmatrix} u(k); y(k) = [1\ 0]x(k) \qquad (III.97)$$

The observability matrix is obtained as U_o = obsv(a_d, c_d) = $[1\ 0;0\ 1]$, and the system is observable. The observer is given as $\hat{x}(k+1) = \varphi\hat{x}(k) + Bu(k) + LH(x(k) - \hat{x}(k))$, and the observer gain L is designed such that the observer poles are at 0.2, and 0.3. We use Koa=acker(ad',cd', $[0.2\ 0.3]$) to obtain the gain as $L = [-0.5000\ 20.0600]$.

Example III.22

Reduced order observer. Let the system be given as in (III.97), and F for the estimator is taken as 0.5, and we have $P\varphi - FP = GH$, and assume that $P = [p_1\ p_2]$, then substituting these values in equation (10.103) we obtain

$$[p_1\ p_2]\begin{bmatrix} 0 & 1 \\ 20 & 0 \end{bmatrix} - 0.5[p_1\ p_2] = GH \qquad (III.98)$$

$$[20p_2 - 0.5p_1\ p_1 - 0.5p_2] = GH$$

If we take $G = 20$, then $GH = [20\ 0]$. Then, from (III.98) we obtain $P = [0.51\ 1.01]$, and finally we obtain $B_r = PB = 1.01$.

Example III.23

Incomplete state feedback. Let the system be given as

$$x(k+1) = \begin{bmatrix} 0 & 1 \\ -1 & -2 \end{bmatrix} x(k) + \begin{bmatrix} 0 \\ 1 \end{bmatrix} u(k) \qquad (III.99)$$

The state feedback is defined as (see Section 10.6.5) $u(k) = -Gx(k)$, and $G = [g_1\ g_2]$. It is assumed that the state x_2 is not available for the feedback. Then, the characteristic equation of the CL system is obtained as

$$|zI - \varphi + BG| = z^2 + 2z + (1 + g_1) = 0 \qquad (III.100)$$

Because we have only one parameter g_1 to design, the two eigenvalues cannot be chosen arbitrarily chosen simultaneously. We rewrite (III.100) as follows

$$1 + \frac{g_1}{z^2 + 2z + 1} = 0 \qquad (\text{III.101})$$

The variation of the CL poles with respect to the parameter g_1 can be seen in the root locus plot as in Figure III.31. By noting the directions of the arrows, one can see that for the positive values of g_1, both the roots are outside the unit

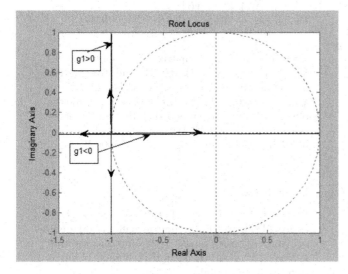

FIGURE III.31
Root locus with variation in g_1; Example III.23.

circle, and for the negative values of g_1, one root is outside and one root is inside the unit circle. It is concluded that when x_1 is available, then the system cannot be stabilized for any value of g_1. If we consider that now only x_2 is available for the feedback, then $G = [0 \ g_2]$, and the characteristic equation is obtained as

$$1 + \frac{g_2 z}{z^2 + 2z + 1} = 0 \qquad (\text{III.102})$$

The root locus for this case, (III.102) is shown in Figure III.32. The system is not stabilizable.

Example III.24

Output feedback. The dynamic system considered is given as

$$x(k+1) = \begin{bmatrix} 0 & 1 & 0 \\ 0 & 0 & 1 \\ -1 & 0 & 0 \end{bmatrix} x(k) + \begin{bmatrix} 0 & 1 \\ 1 & 0 \\ 0 & 0 \end{bmatrix} u(k); H = \begin{bmatrix} 1 & 0 & 0 \\ 1 & 1 & 0 \end{bmatrix};$$

and the control law is $u(k) = -Gy(k)$ $\qquad (\text{III.103})$

Since the matrices H and B have each rank 2, it is feasible to place minimum 2 eigenvalues at desired location, say at 0.1, and 0.2. The characteristic equation of the original given control system is ($|zI - \varphi| = z^3 + 1$), and

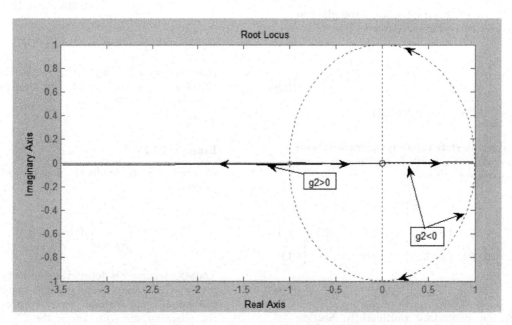

FIGURE III.32
Root locus with variation in g_2; Example III.23.

$$B^* = BW = \begin{bmatrix} 0 & 1 \\ 1 & 0 \\ 0 & 0 \end{bmatrix} \begin{bmatrix} w_1 \\ w_2 \end{bmatrix} = \begin{bmatrix} w_2 \\ w_1 \\ 0 \end{bmatrix} \tag{III.104}$$

The controllability matrix is given as

$$U_c^* = \begin{bmatrix} w_2 & w_1 & 0 \\ w_1 & 0 & -w_2 \\ 0 & -w_2 & -w_1 \end{bmatrix} \tag{III.105}$$

This matrix will be non-singular if $w_1^3 - w_2^3 \neq 0$. Also, we have

$$G^* = [g_1^* \ g_2^*] \Rightarrow G^* H = [g_1^* + g_2^* \ g_2^* \ 0]$$

The closed loop characteristic equation is $\phi(z) = z^3 + \alpha_3 z^2 + \alpha_2 z + \alpha_1 = 0$. By using the Caley-Hamilton principle, the system coefficient matrix satisfies its own characteristic equation, we obtain the following

$$G^* H = [0 \ 0 \ 1] U_c^{*-1} \phi(\varphi);$$

$$\begin{bmatrix} g_1^* + g_2^* \\ g_2^* \\ 0 \end{bmatrix} = \frac{1}{(w_1^3 - w_2^3)} \begin{bmatrix} -\alpha_3 w_2^2 + \alpha_2 w_1^2 - (\alpha_1 - 1)w_1 w_2 \\ \alpha_3 w_1^2 - \alpha_2 w_1 w_2 + (\alpha_1 - 1)w_2^2 \\ -\alpha_3 w_1 w_2 + \alpha_2 w_2^2 - (\alpha_1 - 1)w_1^2 \end{bmatrix}$$

$$\tag{III.106}$$

From (III.106), we see that the last row gives a constraint equation

$$-\alpha_3 w_1 w_2 + \alpha_2 w_2^2 - (\alpha_1 - 1)w_1^2 = 0 \tag{III.107}$$

Since two of the three eigenvalues can be arbitrarily placed, w_1 and w_2 can be arbitrary, provided the condition $w_1^2 \neq w_2^3$ is satisfied, but, should be selected such that the third eigenvalue is stable; for example, the necessary condition for the closed loop system to be stable is $|\alpha| < 1$, and to satisfy this, w_2 cannot be equal to zero. Next, for $z = 0.1$ and 0.2 to be the roots of the characteristic equation

$$z^3 + \alpha_3 z^2 + \alpha_2 z + \alpha_1 = 0 \tag{III.108}$$

the following equations must be satisfied

$$\alpha_1 + 0.001 + \alpha_3 0.01 + 0.1 \alpha_2 = 0$$
$$\alpha_1 + 0.008 + \alpha_3 0.04 + 0.2 \alpha_2 = 0 \tag{III.109}$$

Solving (III.107) and (III.109) we get

$$\alpha_1 = \frac{0.02 w_1^2 + 0.0004 w_2^2 + 0.006 w_1 w_2}{0.3 w_2^2 + w_1 w_2 + 0.02 w_1^2}$$

$$\alpha_2 = \frac{-0.2996 w_1^2 - 0.07 w_1 w_2}{0.3 w_2^2 + w_1 w_2 + 0.02 w_1^2}$$

$$\alpha_3 = \frac{0.994 w_1^2 - 0.07 w_2^2}{0.3 w_2^2 + w_1 w_2 + 0.02 w_1^2} \tag{III.110}$$

If $w_1 = 0$, and $w_2 = 1$; we have $\alpha_3 = -0.23333$, $\alpha_2 = 0$, and $\alpha_1 = 0.00133$; and with these values the roots are $z_1 = 0.1, z_2 = 0.2, z_3 = -0.0667$; the third pole is within the unit circle, and hence, the closed loop system is stable. Using these values of w_1 and w_2 and corresponding values of the coefficients of the characteristic equation in $G^* C$, we obtain

$$[g_1^* + g_2^* \ g_2^* \ 0] = \begin{bmatrix} \alpha_3 \\ -\alpha_1 + 1 \\ \alpha_2 \end{bmatrix}^T = \begin{bmatrix} -0.23333 \\ 0.99867 \\ 0 \end{bmatrix}^T \tag{III.111}$$

The feedback gain is given as $G^* = [-1.232 \ 0.99867]$; and

$$G = WG^* = \begin{bmatrix} 0 \\ 1 \end{bmatrix} [-1.232 \ 0.99867]$$

$$= \begin{bmatrix} 0 & 0 \\ -1.232 & 0.99867 \end{bmatrix} \tag{III.112}$$

Example III.25

Linear quadratic regulator (LQG). Let the linear system be described as

$$x(k+1) = \begin{bmatrix} 0 & 1 \\ 0.5 & 0.8 \end{bmatrix} x(k) + \begin{bmatrix} 0 \\ 1 \end{bmatrix} u(k); x_0 = \begin{bmatrix} 1 \\ 1 \end{bmatrix}; y(k) = [1 \ 0]x(k)$$

$$\tag{III.113}$$

We need to design an optimal controller to minimize the PI:

$$J = \sum_{k=0}^{\infty} \{x_1^2 + x_1 x_2 + x_2^2 + 0.1 u^2\} \tag{III.114}$$

The PI is now written using (10.147)

$$J = \sum_{k=0}^{\infty} \{x^T(k)\begin{bmatrix} 1 & 0.5 \\ 0.5 & 1 \end{bmatrix} x(k) + 0.1u^2\}; R = \begin{bmatrix} 1 & 0.5 \\ 0.5 & 1 \end{bmatrix}; Q = 0.1$$

(III.115)

Let us choose $P = \begin{bmatrix} p_1 & p_2 \\ p_2 & p_3 \end{bmatrix}$. Now substituting the known values of the matrices and variables from (III.113), and (III.115) in (10.162) and simplifying, we obtain finally (with $S = 0.1 + p_3$):

$$0.25p_3 - p_1 + 1 - \frac{0.25p_3^2}{(0.1 + p_3)} = 0$$

$$0.4p_3 - 0.5p_2 + 0.5 - \frac{0.5p_2p_3 + 0.4p_3^2}{(0.1 + p_3)} = 0 \qquad \text{(III.116)}$$

$$p + 1.6p_2 - 0.36p_3 + 1 - \frac{p_2^2 + 1.6p_2p_3 + 0.64p_3^2}{(0.1 + p_3)} = 0$$

The solution of (III.116) gives the following values for the matrix P: $p_1 = 1.0238$, $p_2 = 0.5513$, and $p_3 = 1.9811$, and the optimal control law is obtained as using (10.156) as

$$u^*(k) = -0.476x_1(x) - 1.0265x_2(k) \qquad \text{(III.117)}$$

The optimal cost is given as

$$J_{opt} = x_0^T P x_0 = \begin{bmatrix} 1 & 1 \end{bmatrix}\begin{bmatrix} 1.0238 & 0.5513 \\ 0.5513 & 1.9811 \end{bmatrix}\begin{bmatrix} 1 \\ 1 \end{bmatrix} = 4.1075 \qquad \text{(III.118)}$$

Example III.26

Design of a digital controller for the system that has a continuous controller.

The MATLAB script is given as **ctdtcontroller.m** for the $G(s) = 1/[(s-1)(s-2)]$ as the OLTF, and first the controller is designed using the pole placement, for which the $K = [5\ 0]$, the required closed loop poles are at $1+/-j1$, and with this gain the requirement was met, and the time response of the closed loop system to a unit step is shown in Figure III.33. The responses of the continuous time and discrete time closed loop system

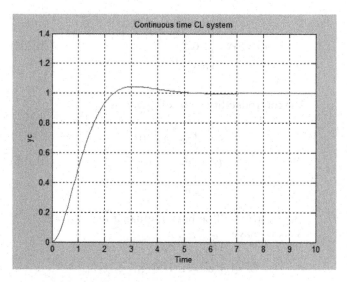

FIGURE III.33
Output response of the continuous time closed loop system; Example III.26.

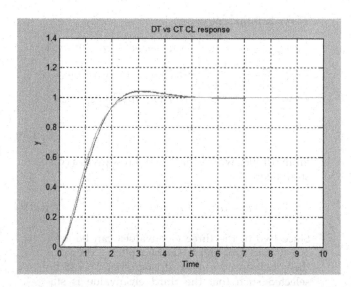

FIGURE III.34
Output response of the CT vs. DT closed loop systems; Example III.26.

(with different sampling time, see the MATLAB script, but the controller gain as used in the CT system) are shown in Figure III.34, from which it is seen that the DT responses grossly match. Now, the controller is redesigned (by adjusting the value of K), and the responses of the CT and DT closed loop systems are shown in Figure III.35.

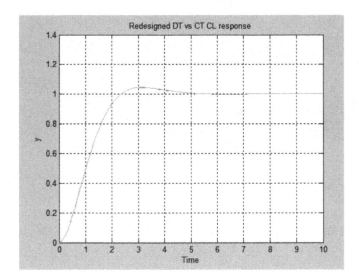

FIGURE III.35
Output response of the CT vs. DT closed loop systems (redesigned); Example III.26.

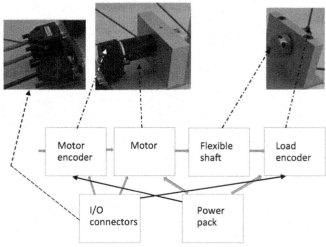

FIGURE III.36
Block diagram of a motion control system, Example III.27. (Adapted/ modified from https://www.mathworks.com/help/control/examples/ tuning-of-a-digital-motion-ontrol-system.html.)

Example III.27

Tuning of digital motion control system. The example shows how to use the Control System Toolbox (MATLAB™) to tune a digital motion system that consists of [20]: (i) power amplifier, (ii) I/O connectors, (iii) motion encoder, (iv) motor, (v) flexible shaft, and (vi) load encoder; and such a sub-system/device, Figure III.36 [20], could be a part of some production machine and is intended to move some load (a gripper, a tool, or a nozzle) from one angular position to another and back again. This task is a part of the production cycle that has to be completed to create each item or batch of items, for which a digital controller must be tuned to maximize the production speed of the machine to obtain good accuracy and product quality. In order to do perform the control-tuning task, the control system is modelled in Simulink using a 4th-order model of the

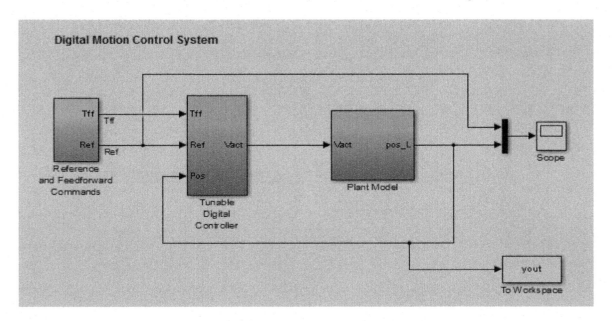

FIGURE III.37
Motion control system in Simulink (Source: generated by running the MATLAB script: dtmotioncontroller, m, by the first author of the current book); Example III.27.

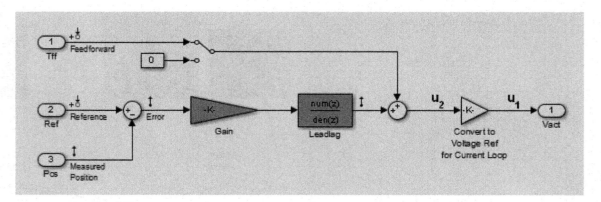

FIGURE III.38
Digital controller in Simulink (Source: generated by running the MATLAB script: dtmotioncontroller.m, by the first author of the current book); Example III.27.

inertia and flexible shaft, Figures III.37 and III.38 (run the MATLAB script: **dtmotioncontroller.m** provided with the present CRC book, for Section III): The digital controller, Figure III.38 has a gain in series with a lead-lag controller. A 0.5 sec. response time to a step command in angular position with minimum overshoot is desired that corresponds to a target bandwidth of approximately 5 rad/sec. The frequency response of the system is shown in Figure III.39. The "looptune" can be conveniently used to tune a fixed-structure

compensator, for this, first instantiate the "slTuner" interface to acquire the control structure from Simulink. Use the "looptune" to tune the compensator parameters for the desired gain crossover frequency 5 rad/sec. The following output would be obtained:

```
Final: Peak gain = 0.975, Iterations = 21
Achieved target gain value TargetGain=1.
Block 1: rct_dmc/Tunable Digital
Controller/Gain =
d =
```

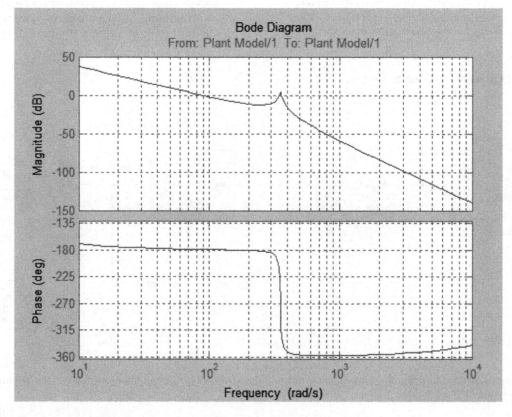

FIGURE III.39
Frequency response of the plant; Example III.27.

```
u1
y1 2.753e-06
Name: Gain
Static gain.
-----------------------------------
Block 2: rct_dmc/Tunable Digital
Controller/Leadlag =
30.54 s + 59.01
---------------
s + 18.94
Name: Leadlag
```

Continuous-time transfer function.

The comparison of the CL responses of the (original and tuned) systems is shown in Figure III.40; and the tuned response satisfies the response time requirement. However, the simulation is done using a CT lead/lag compensator ("looptune" operates in continuous time); hence, we need to further validate the design in Simulink using a digital implementation of the lead/lag compensator, for which one can use "writeBlockValue" to apply the tuned values to the Simulink model and automatically discretize the lead/lag compensator to the rate specified in Simulink. The response of the CT plant with the digital controller is obtained as seen in Figure III.41, and the simulated responses closely match. Now, we tune an additional Notch filter, since the plant has a flexible mode near 350 rad/sec. We will try to increase the control BW from 5 to 50 rad/sec. Since the plant has a resonance near 350 rad/sec, the lead/lag compensator is not sufficient to obtain the adequate stability margins and a small (smaller) overshoot. Hence, a notch filter is required to be added after the lead/lag filter. The modified control architecture is tuned by an "s1Tuner" instance with the three blocks. The notch filter TF is taken as

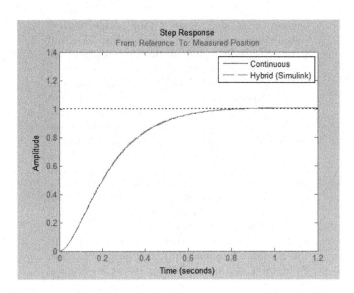

FIGURE III.41
Step response comparison of the plant (with digital controller); Example III.27.

$$N_f(s) = \frac{s^2 + 2\varsigma_1\omega_n s + \omega_n^2}{s^2 + 2\varsigma_2\omega_n s + \omega_n^2} \quad \text{(III.119)}$$

The coefficients in (III.119) are real parameters, and are specified to create a parametric model of the $N_f(s)$ TF. This model is associated with the "Notch" block in the Simulink model; since the control system is tuned in CT, we can use a CT parameterization of the notch filter, even though the "Notch" block it-self is discrete. Then, use "looptune" to tune the "Gain," "lead-lag," and "Notch" blocks together with a 50 rad/sec desired crossover frequency. Also, to eliminate residual oscillations form the resonance of the

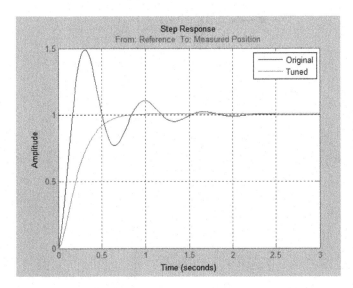

FIGURE III.40
Step response comparison of the plant; Example III.27.

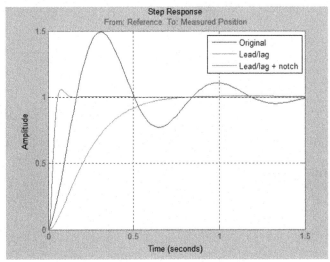

FIGURE III.42
CL responses with the previous designs; Example III.27.

system, specify a desired loop shape with a −40 dB/decade roll off past 50 rad/sec. The final gain is close to 1, and all the requirements are met. The comparison of the CL responses is shown in Figure III.42. Also, the Bode diagrams of the total compensator, the OL response, and the system are shown in Figure III.43, from which it can be seen that the system's resonance has been notched out. Now, use "write-BlockValue" to discretize the tuned lead/lag and notch filters and write their values back to Simulink. Compare the MATLAB and Simulink responses, as shown in Figure III.44, from which we can see that the Simulink shows small residual oscillations. It might be due to the dis-cretization, since the notch frequency is close to the Nyquist frequency pi/0.002 = 1570 rad/sec. By default, the notch filter is discretized using the ZOH method. One can compare this with the Tustin method pre-warped at the notch frequency. The Bode diagrams are shown in Figure III.45. The Tustin filter seems better. Now,

we can use the Tustin, for which the desired rate conversion method for the notch filter is speci-fied. The "writeBlockValue" now uses Tustin pre-warped at the notch frequency to discretize the notch filter and write it back to Simulink, and this gets rid of the oscillations as can be seen from Figure III.46. One can tune the controller directly in discrete time to avoid discretization problems with the notch filter, which specifies that the Simulink model should be linearized and tuned at the controller sample time of 0.002 sec to prevent high-gain control and saturations, a requirement that limits the gain from refer-ence to control signal (output of Notch block) is added. Then, retune the controller at the speci-fied sampling rate and verify the tuned open- and closed loop responses, which are shown in Figure III.47. Also, the Bode diagrams are shown in Figure III.48, and the results are similar to the ones obtained when the controller was tuned in the continuous time. Further, the digital con-troller is validated against the continuous time

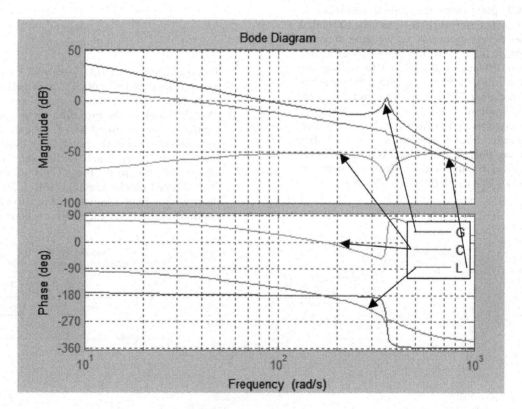

FIGURE III.43
Bode diagram; Example III.27.

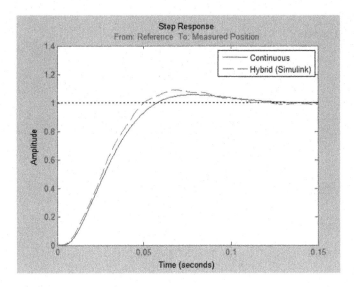

FIGURE III.44
CL responses with the previous designs; Example III.27.

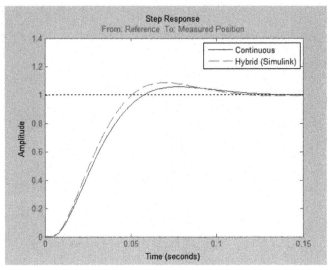

FIGURE III.46
Step responses show that the small oscillations are removed with Tustin; Example III.27.

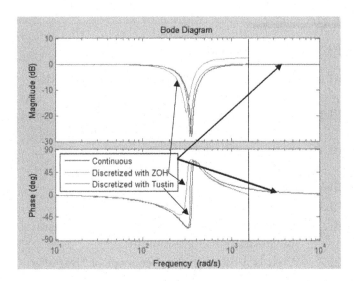

FIGURE III.45
Bode diagrams of the Notch filters; Example III.27.

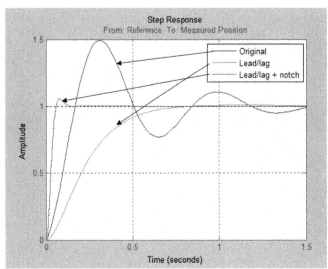

FIGURE III.47
Step responses of the OL and CL systems; Example III.27.

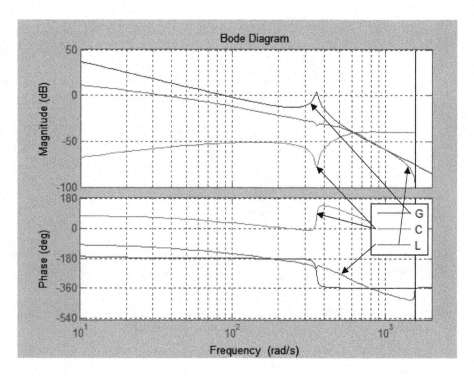

FIGURE III.48
Bode diagrams with the re-tuned controller; Example III.27.

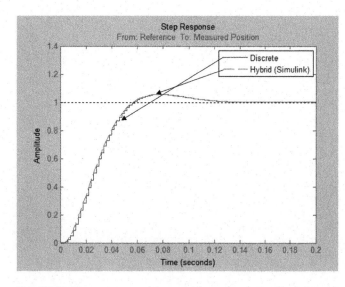

FIGURE III.49
Final design step responses; Example III.27.

system in Simulink, and the step responses are compared in Figure III.49, and we see that the hybrid response matches closely to its discrete time approximation, and the design is complete.

Exercises for Section III

III.1   What are the important parts/blocks of a discrete time control system?

III.2   In an inherently discrete time system, what happens in between discrete time instances?

III.3   In which way does the output of the discrete time controller depends on the error input $e(k)$?

III.4   What is Euler transformation?

III.5   What is Tustin transformation?

III.6   What are the equivalents of Euler and Tustin transformations in z-domain?

III.7   What is a mathematical concept of "impulse sampling?"

III.8   How can the sampling be interpreted?

III.9   How can zero order hold (ZOH) be interpreted?

III.10   What is the differentiation property of the z-transform?

III.11   What role does a z-(Z) transform plays in the analysis of discrete time system?

III.12   If the continuous time coefficient matrix A is in the diagonal form, will its exponential also form the diagonal form?

III.13   If A is in the controller canonical form, will its exponential form also? How does one restore its canonical form?

III.14   What are the dominant poles in a discrete time system?

III.15   If in a continuous time system a ZOH is added, does its stability behavior remain the same?

III.16   What happens to the steady state error in the case of III.15, above?

III.17   What are the main features/effects of the lag and lead compensators?

III.18   When should primarily the lag compensator be used?

III.19   When should primarily the lead compensator be used?

III.20   When can the poles of a closed loop system be selected freely?

III.21   Can the uncontrollable poles be affected by the state feedback?

III.22   How many solutions of a matrix Riccati equation lead to a stable feedback?

III.23   Why will, in principle, a (state) observer work?

III.24   Why is the feedback itself used in the design of an observer?

III.25   How can you make the continuous system faster, and what/how will its response be?

III.26   Is such a behavior practically achievable, as in III.25?

III.27   If a system posed in III.25 is sampled, what will happen to the poles?

III.28   Can you express the condition of III.26 mathematically?

III.29   What then will be a result of such a condition as in III.8, in terms of the response of the discrete time system?

III.30   What is the main issue in achieving a deadbeat response?

III.31   Do the zeros change due to the state feedback design?

III.32   Can observer poles be chosen freely? Does this apply to the controller poles?

III.33   Can the observer and the controller poles be chosen independent of each other?

III.34   What is the lowest possible order of the observer?

III.35   What is the importance of the causality and physical realizability?

III.36   For a discrete time causal system, why does the power series of its TF should not contain any positive power in z?

III.37   If the lag compensator results in a too low BW, what should be done?

III.38   What is a necessary and sufficient condition for arbitrary pole placement?

III.39   What is the basic assumption in the separation principle?

III.40   Why are observer dynamics faster than the controller dynamics?

III.41   Can the deadbeat control be achieved by the state feedback and the observer?

III.42   What is the deadbeat observer?

III.43   In an adaptive control, in principle, what is very important?

III.44   What is the main task of the adaptive mechanism in the adaptive control?

III.45   Why, in addition to robust control, might adaptive control be required?

III.46   How can the robust and adaptive controls be used in conjunction with one another?

III.47   What is the sufficient condition for parameter convergence in an adaptive control scheme?

III.48   Under what circumstances is the robustness of an adaptive controller needed?

III.49   What steps are followed for embedded control system design?

III.50   What are the constituents of a smart sensor?

III.51   What is the main idea in/of a smart sensor?

References for Section III

1. Cellier, F. E. Lectures notes on digital control, University of Arizona, USA, 2015. https:// www.inf.ethz.ch/ personal/cellier/Lect/DC/Lect_dc_index.html. (dc_1 to dc_29), accessed December 2017.

2. Kar, I., and Somanath, M. Introduction to Digital Control. Lecture Notes; Modules 1 to 11. Dept. of Electronics and Electrical Engg., Indian Institute of Technology (IIT) Guwahati, Guwahati, Assam, India. https://nptel. ac.in/ courses/108103008/PDF/module1/m1_lec1.pdf; (https://nptel.ac.in/downloads/108103008/), Accessed November 2017.

3. Raol, J. R., Gopalratnam, G., and B. Twala. *Nonlinear Filtering: Concepts and Engineering Applications.* Taylor & Francis Group, Boca Raton, FL, 2017.

4. Landau, I. D., Lozano, R., M'Saad, M., and Karimi, A. Adaptive control, Chapter 1: Introduction to Adaptive Control, PPT slides; http://www.landauadaptivecontrol.org/Slides %20Ch1.pdf; Algorithms, Analysis, and Applications.http://www.springer.com/978-0-85729-663-4, Springer, Berlin, Germany, 2011, accessed December February 2017.

5. Lavretsky, E. Adaptive control: Introduction, overview, and applications. PPT slides, Robust and Adaptive Control Workshop. http://web.iitd.ac.in/~sbhasin/docs/IEEE_WorkShop_Slides_Lavretsky_Adaptive_Control.pdf, accessed July 2017.

6. Narendra, K. S., and Valavani, L.S. Direct and indirect adaptive control. AD-A058235, Dept. of Engineering and Applied Science, Yale University, New Haven, CT, 1978.

7. Ioannou, A., and Sun J. *Robust Adaptive Control.* Prentice Hall, New York, 1996; http://www-bcf.usc.edu/~ioannou/RobustAdaptiveBook95pdf/Robust_Adaptive _ Control.pdf, accessed January 2018.

8. Naus, G. J. L. Gain scheduling: Robust design and automated tuning of automotive controllers. University of Technology, Eindhoven, Dept. of Mechanical Engineering., Division of Dynamical Systems Design, Control System Technology group. 2009.

9. Astrom, K. J. Theory and applications of adaptive control-A survey. *Automatica,* (IFAC, Pergamon Press), Vol. 19, pp. 471–486, 1983.

10. Bunham, K. J., and Larkowski, T. Self-tuning and adaptive control, Advanced Informatics and Control, Wroclaw University of Technology, Poland. Wroclaw 2011.

11. Anon. Adaptive pole placement (or pole-assignment). PPTs; pp. 1–13. http://www. personal.reading.ac.uk/~sis01xh/teaching/ComputerControl/adaslide4.pdf, accessed September 2018.

12. Black, W. S., Haghi, P., and Ariyur, K. B. Adaptive systems: History, techniques, problems, and perspectives. Systems (ISSN 2079-8954), 2, 606-660, 2014. www.mdpi.com/journal/systems, accessed December 2017.

13. Raol, J. R., Girija, G., and J. Singh. *Modelling and Parameter Estimation of Dynamic Systems,* IET/IEE Control Series Vol. 65, IET/IEE Society, London, UK, 2004.

14. Yu, X., and Kaynak, O. Sliding-mode control with soft computing: A survey. *IEEE Trans. On Industrial Electronics* (USA), Vol. 56, No. 9, pp. 3275–3285, 2009.

15. Anderson, B. D. O. Failures of adaptive control theory and their resolutions. Communications in Information and Systems; International Press. Vol. 5, N0. 1, pp. 1-20, 2005; users.cecs.anu.edu.au/~briandoa/pubs/hidden/R412AN798.pdf, accessed August 2018.

16. Anon. Introduction to Computer Based Control Systems. IDC Technologies Pty Ltd., West Perth, Western Australia, 2012, pp. 1–24, www.idc-online.com, accessed October 2018.

17. Hunter, G W., Stetter, J. R., Hesketh, P. J., and Liu, C.-C. Smart sensor systems. The Electro-chemical Society Interface, pp. 29–34, winter, 2010. https://www.electrochem.org /dl/interface/wtr/wtr10/wtr10_p029-034.pdf, accessed October 2018.

18. Broenink, J. F., and Hilderink, G. H. Building blocks for control system software. *Proceedings of the 3rd Workshop on European Scientific and Industrial Collaboration WESIC2001,* Enschede, the Netherlands, (Eds.) J. V. Amerongen, B. Jonker, and P. Regtien, pp. 329–338, June 2001.

19. Astrom, K. J., and Wittenmark, B. *Computer Controlled Systems-Theory and Design* (3rd Ed.), Tsingua University Press, Prentice Hall, Upper Saddle River, NJ, 1997.

20. Anon. https://www.mathworks.com/help/control/examples/tuning-of-a-digital-motion-control-system.html, accessed August 2018.

Section IV

AI-Based Control

To give a historical flavor, we might recall that the approximation properties of artificial neural networks by Cybenko in 1989 and by Park and Sandberg in 1991 are basic to their applications in feedback control. So, it is easier for us to think of Artificial Intelligence as belonging to biologists, sociologists, neuroscience, and computer science until 1990. Nevertheless, it is the control community that is the pioneer to embrace the advances in artificial neural networks (ANNs), fuzzy logic (FL), and genetic algorithms (GAs) and attempted to solve optimization problems that were otherwise thought to be highly demanding in terms of computational power due to complexity. Evidently, since the mid-1990s, there has been a lot of interest in proving control problems, such as static state feedback control problem and the output feedback control problem, to be computationally intractable and, henceforth, providing a rigorous framework called randomized algorithms. Simultaneously, there have been very encouraging results, particularly in the applications sector, obtained via the so-called intelligent control techniques. Thereafter, there are traces of active collaboration between the two schools as well.

13

Introduction to AI-Based Control

13.1 Motivation for Computational Intelligence in Control

Computational intelligence refers to the theory, design, application, and development of paradigms that are deeply inspired by nature—both biological and physical sciences. Looking back for the past three to four decades, we find three strong pillars—artificial neural networks, fuzzy systems, and genetic algorithms—keeping computational intelligence aloft. Over the last few years, there has been an explosion of research on deep learning, in particular deep convolutional neural networks. Computational intelligence plays a major role in developing successful intelligent control systems. In this chapter, we will briefly explore each of these pillars at the conceptual level [1–13]. The next three chapters contain several success stories of each of these techniques.

13.2 Artificial Neural Networks

Inspired by the human brain and its intelligence, artificial neural networks (ANNs) emerged as one of the most exciting paradigms. For instance, given a straight line $ax + by + c = 0$, we understand that there would be several points in the plane that are on either side of the line. A curious mind would then ask a question in the opposite direction: given two sets of points, can we *fit* a line in between? If we dub this question into a method, we get ANNs. A neural network *learns* from observational data, figuring out its own solution to the problem at hand.

Automatically learning from data sounds promising. However, until 2006 we didn't know how to train neural networks to surpass more traditional approaches, except for a few specialized problems. What changed in 2006 was the discovery of techniques for learning in so-called deep neural networks. These techniques are now known as deep learning. They've been developed further, and today deep neural networks and deep learning achieve outstanding performance on many important problems in computer vision, speech recognition, and natural language processing, not to discount any major control problem. They're being deployed on a large scale by companies such as Google, Microsoft, and Facebook.

The purpose of this section of the book is to provide certain central concepts of neural networks, including modern techniques for deep learning. Our intent is to essay the durable, lasting insights underlying how neural networks work. The mathematical requirements to read this part of the book are modest—elementary algebra and plots of functions. Only occasionally do we need more advanced mathematics: multivariable calculus and linear algebra. In a subsequent chapter, we apply these techniques to the solutions of problems in control engineering.

13.2.1 An Intuitive Introduction

Consider the following sequence of handwritten digits: 1 3 5 7 2 4 6 9 8, which most of us would recognize in various disguises. For instance, consider the CAPTCHA we encounter in online transactions, such as the one in Figure 13.1:

This is considered a trivial task for humans. Technically speaking, in each hemisphere of our brain, humans have a primary visual cortex, known as V_1, which contains 140 million neurons with tens of billions of connections between them. Following this, we have an entire series of visual cortices—V_2 to V_5—involved in progressively more complex image processing. This is just one instance of what we may call a supercomputer, tuned by evolution over hundreds of millions of years and superbly adapted to understand the visual world. We said it is trivial to recognize handwritten digits just to bring forth two issues: (i) nearly all that work is done unconsciously and (ii) hence we fail to appreciate how tough a problem our visual systems solve.

On the other hand, attempting to teach a machine to recognize a CAPTCHA, by way of a computer program, we begin to see the magnitude of the task— what description of, say, 4 when programmed will allow the machine to recognize it, in contrast to the handwriting of our kindergarten teacher we have learned from?

FIGURE 13.1

Artificial neural networks approach the problem in a different way. The idea is to take a large enough number of handwritten digits, known as training examples, and then devise a method that helps the ANN learn from those training examples. On a related note, one might explore on the internet to find that a computer program that would successfully implement the recognition of handwritten digits, with about 90% accuracy, is hardly 70–80 lines long, and that it does not require any special libraries. Once we are able to, somehow, improve the accuracy to much higher levels, we would then be tempted to use them at banks to securely process financial instruments.

To motivate the reader on contemporary research, a case in point is autonomous driving. To think conventionally, the technology has too many variables to address—if it's rainy, if it's dark, if a child runs into the street, and so on. No matter how many lines of code are written, we still hesitate to put self-driving vehicles on the road. To think unconventionally, we are more comfortable: vehicles enabled with artificial intelligence are able to watch humans drive to figure out a lot more. In other words, autonomous driving, while far from perfect, as well as scary, can now be thought of as getting an answer to the question: What would a good human driver do? By extension, a top management team can review the organizational workflows and break them into tasks that have a significant prediction component. For example, banks can learn from voluminous old records to predict and detect fraud.

We shall now move on developing many key ideas about neural networks, including two important types of artificial neurons—the perceptron and the sigmoidal neuron—and the standard learning algorithm for neural networks. Throughout, the emphasis is on why things are done the way they are, which is needed to understand what deep learning is and why it matters.

13.2.2 Perceptrons

Perceptrons were developed in the 1950s and 1960s by Frank Rosenblatt, inspired by earlier collaborative work (in the 1940s) by Warren McCulloch and Walter Pitts, a mathematician and a biologist. A perceptron takes several binary inputs, $x_1, x_2, \ldots$, and produces a single binary output, as shown in Figure 13.2:

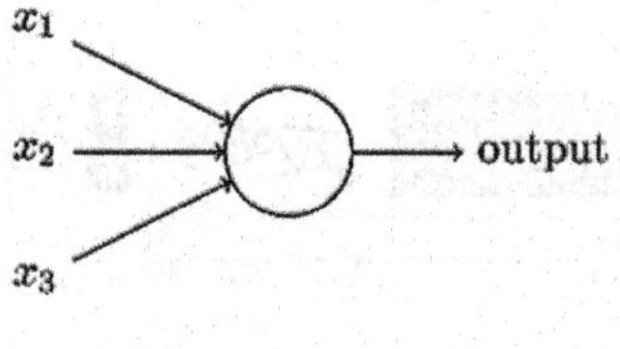

Rosenblatt proposed a simple rule to compute the output. He introduced weights, $w_i, i = 1, 2, \ldots$, real numbers expressing the importance of the respective inputs to the output. The neuron's output, 0 or 1, is determined by whether the weighted sum $\sum_i w_i x_i$ is less than or greater than some threshold value, better known as **bias**, θ. Just like the weights, the bias is a real number, which is a parameter of the neuron. To put it in more precise algebraic terms:

$$\text{Output} = \begin{cases} 0 & \text{if } \sum_i w_i x_i + \theta \le 0 \\ 1 & \text{if } \sum_i w_i x_i + \theta > 0 \end{cases} \tag{13.1}$$

In two dimensions, the reader would have immediately recognized the equation of a straight line, which we have already mentioned in the opening paragraph. What shall we call the polynomial in higher dimensions? And, in terms of functions, does this remind one of the famous Fourier series?

That's the basic mathematical model. A way we can think about the perceptron is that it's a little device that tells what can be done by weighing up evidence. By varying the weights and the bias, we can get different models of decision making.

Obviously, a perceptron by itself cannot be a complete model of human decision making. Perhaps a complex network of perceptrons, as in Figure 13.3, could make quite subtle decisions:

In this network, the first column of perceptrons—we'll call it the first *layer*—is making three very simple decisions by weighing the input evidence. Each of those perceptrons in the second layer is making a decision by weighing up the results from the first layer, i.e. a perceptron in the second layer makes a compound decision at a more complex and more abstract level than perceptrons in the first layer. And, even more complex decisions can be made by the perceptron in the third layer. In this way, a many-layer network of perceptrons can engage in sophisticated decision making.

x_1
x_2 → output
x_3

FIGURE 13.2

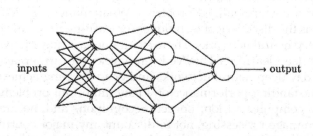

inputs → output

FIGURE 13.3

We draw the attention of the reader to Figure 13.3. While we define perceptrons as having a single output, in the network above, the perceptrons appear to have multiple outputs. In fact, they're still single output. The multiple output arrows are merely a useful way of indicating that the output from a perceptron is being fed as the input to several other perceptrons in the following layer.

Let's simplify the way we describe perceptrons by simply noting that a linear combination such as $\sum_i w_i x$ is actually the dot product (aka inner product) of two vectors **w** and **x** whose components are w_i and x_i, respectively. We'll see later that this description leads to further notational simplifications.

A more popular way a set of perceptrons can be used is to compute the elementary logical functions, such as AND, OR, and NAND. For example, suppose we have a perceptron with two inputs, each with weight 2, and a bias of 3, i.e. the governing polynomial is

$$-2x_1 - 2x_2 + 3 \tag{13.2}$$

Then it is easy to see that input $x_1 = x_2 = 0$ produces output 1, since $(-2) \times 0 + (-2) \times 0 + 3 = 3 > 0$. By the same token of reasoning, we may verify that the inputs $x_1 = 0, x_2 = 1$ and $x_1 = 1, x_2 = 0$ produce output 1. But, the input $x_1 = 1, x_2 = 1$ produces output 0 since $(-2) \times 1 + (-2) \times 1 + 3 = -1 < 0$. Thus, with the set of weights and the bias chosen, the perceptron implements a NAND gate. Moreover, the NAND gate is a universal gate and we can build any logical computation out of NAND gates. For example, we can use NAND gates to build a circuit that adds two bits, x_1 and x_2. This requires computing the bitwise sum, as well as a carry bit that is set to 1 when both x_1 and x_2 are 1. To get an equivalent network of perceptrons, we simply replace all the NAND gates by perceptrons with two inputs, each with weight 2, and a bias of 3. The resulting network is shown in Figure 13.4.

A comparison of the two networks above should clarify many minor issues in the symbols we prefer to use for the inputs, the perceptrons, and the outputs.

Thus, the adder example demonstrates how a network of perceptrons can be used to simulate a circuit containing many NAND gates. And because NAND gates are universal for computation, it follows that perceptrons are also universal for computation.

This idea that the perceptrons could compute everything is both reassuring and disappointing. It's reassuring because it tells us that networks of perceptrons can be as powerful as any other computing device. But it's also disappointing because it makes it seem as though perceptrons merely replace a NAND gate. But, there is much more than this perspective. We might contemplate *learning algorithms*, which can automatically tune the weights and biases of a network of perceptrons. This tuning happens in response to external stimuli, without direct intervention by a programmer. These learning algorithms enable us to use perceptrons in a way that is radically different than conventional logic gates. Instead of explicitly laying out a circuit of NAND and other gates, our neural networks can simply learn to solve problems, sometimes problems where it would be extremely difficult to directly design a conventional circuit.

MIT's Marvin Minsky and Seymour Papert then put a damper on this research in their 1969 book, "Perceptrons: An Introduction to Computational Geometry," by showing mathematically that the perceptron could only perform very basic tasks. Their book also discussed the difficulty of training multilayer neural networks.

13.2.3 Sigmoidal Neurons

Suppose we have a network of perceptrons that we'd like to use to learn to solve some problem. To continue with our handwritten digit recognition, let the inputs to the network be the raw pixel data from a scanned, handwritten image of a digit. And we wish the network to learn the weights w_i and biases θ_i so that the output from the network correctly classifies the digit. We will be able to detect whether some sort of learning has happened or not by employing a simple test. Suppose we make a small change in some weight

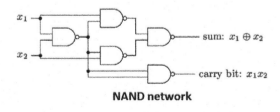

NAND network

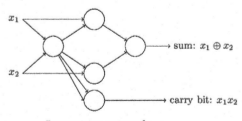

Perceptron network

FIGURE 13.4

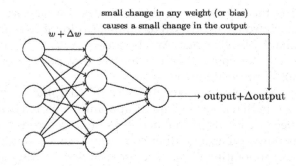

FIGURE 13.5

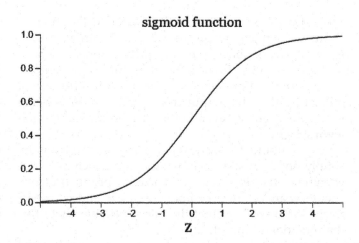

FIGURE 13.6

(or bias) in the network. This should, invariably, cause only a small corresponding change in the output from the network. Some tolerance is permissible, but a 6 cannot be read as 9.

Schematically, we would put our idea into the following Figure 13.5:

At first, we may be a bit suspicious—is this simple network going to recognize handwritten digits? We'll present the following arguments to clear this.

If it were true that a small change in a weight (or bias) causes only a small change in output, then this differential relationship should give us some pointer towards minimizing the changes in the output. For instance, suppose the network was mistakenly classifying an image as "6" while it must be "9." We would perhaps attempt to figure out how to make a small change in the weights and biases so the network gets a little closer to classifying the image as indeed "9." And then we would iterate on this to produce better and better output. With this convergence to the correct output, we would say that the network is be learning.

Once we are talking about infinitesimal changes, so to say, we cannot have non-smooth functions, like the 0–1 pulses. In fact, it may be readily shown that a "small" change in the weights or bias of any single perceptron in the network can cause the output of that perceptron to completely flip, say from 0 to 1. And, this flip may lead to a very complicated change in the working of the rest of the network. Further, this makes it difficult to see how to gradually modify the weights and biases so that the network converges to the desired behavior.

We can overcome this problem by introducing a new type of artificial neuron called a sigmoidal neuron. Sigmoidal neurons are similar to perceptrons, but modified so that small changes in their weights and bias cause only a small change in their output, i.e. the underlying function is a smooth one. A sigmoid is shown in Figure 13.6.

This function is defined as follows:

$$\sigma(z) = \frac{1}{1 + e^{-z}} \quad (13.3)$$

Therefore, instead of being just 0 or 1, as in the case of perceptrons, the inputs can take on any values between 0 and 1, for instance, 0.638887. But, just like a perceptron, the sigmoidal neuron has weights w_i and an overall bias, θ. The output is now

$$\sigma(w \cdot x + \theta) \quad (13.4)$$

The function σ is also known as the logistic function, and the neurons are also known as logistic neurons. However, in this book, we prefer the term sigmoidal neurons.

Indeed, it's the smoothness of the σ function that is the crucial fact, not its detailed form. Calculus tells us that:

$$\Delta \text{output} = \sum \frac{\partial \text{output}}{\partial w_i} \Delta w_i + \frac{\partial \text{output}}{\partial \theta} \Delta \theta \quad (13.5)$$

To take this expression easy, Δ output is a linear function of the changes Δw_i and $\Delta \theta$. It is this linearity that makes it easy to choose small changes in the weights and biases to achieve any desired small change in the output.

More than the shape and/or size of the sigmoid function, a genuine question is in terms of the meaning of the output from a sigmoidal neuron—it can have as output any real number between 0 and 1. We modify the output, in such cases, either by rounding or by quantizing as demanded by the application.

13.2.4 The Architecture of Neural Networks

Suppose we have the network in Figure 13.7:

As mentioned earlier, the leftmost layer in this network is called the **input layer**, and the neurons within the layer are called input neurons. The rightmost or **output layer** contains the output neurons, or, as in this case, a single output neuron. The middle layer is called a **hidden layer**, since the neurons in this layer are neither inputs nor outputs.

The network above has just a single hidden layer, but some networks have multiple hidden layers. For example, the four-layer network in Figure 13.8 has two hidden layers:

Somewhat confusingly, and for historical reasons, such multiple layer networks are sometimes called multilayer perceptrons or MLPs, despite being made up of sigmoid neurons, not perceptrons.

The design of the input and output layers in a network is often straightforward. For example, suppose we are attempting to determine whether a handwritten image depicts a "9" or not. A natural way to design the network is to encode the intensities of

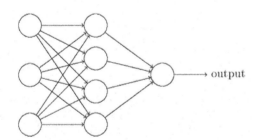

FIGURE 13.7

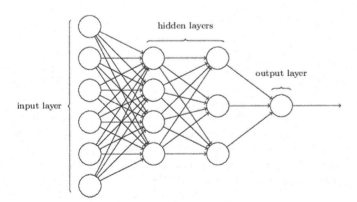

FIGURE 13.8

the image pixels into the input neurons. If the image is, say, a 64 by 64 grayscale image, then we'd have $64 \times 64 = 4096$, or simply 4k input neurons, with the intensities scaled appropriately between 0 and 1. The output layer will contain just a single neuron, with output values of less than 0.5 indicating "input image is not a 9" and values greater than 0.5 indicating "input image is a 9."

However, it is quite an art to design the hidden layers. In particular, it's not possible to sum up the design process for the hidden layers with a few simple rules of thumb. Instead, neural network researchers have developed many design heuristics for the hidden layers, which help people get the behavior they want out of their neural networks. For example, such heuristics can be used to help determine how to trade off the number of hidden layers against the time required to train the network. The reader would have rightly asked the following question: Are we just talking about the number of neurons in one hidden layer between the input and the output? Or, a larger number of hidden layers? We will have this in mind while going through the following arguments in this part of the book.

Example 13.2.1: A Simple Network to Classify Handwritten Digits

Having defined neural networks, let's return to handwriting recognition. We can split the problem of recognizing handwritten digits into two sub-problems. First, a way of breaking an image containing many digits into a sequence of separate images, each containing a single digit. We humans solve this *segmentation* problem with ease, but it could be a challenge for a computer program to correctly break up the image. Once the image has been segmented, the program then needs to classify each individual digit. We'll focus on writing a program to solve the second problem, classifying individual digits. We do this because it turns out that the segmentation problem is not so difficult to solve, once you have a good way of classifying individual digits, and there are many approaches to solving the segmentation problem.

To recognize individual digits, we will use a three-layer neural network in Figure 13.9:

The input layer of the network contains neurons encoding the values of the input pixels. As discussed in the previous section, our training data for the network will consist of many 28 by 28 pixel images of scanned handwritten digits, and so the input layer contains $28 \times 28 = 784$. For clarity, we have not shown as many neurons in the picture above. The input pixels are grayscale, with a value of 0.0 representing white, a value of 1.0 representing black, and in between values representing gradually darkening shades of gray.

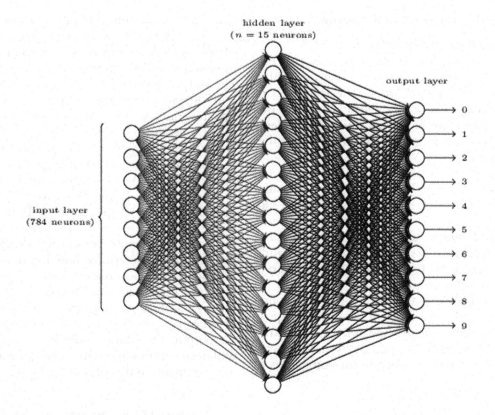

hidden layer
($n = 15$ neurons)

output layer

input layer
(784 neurons)

0
1
2
3
4
5
6
7
8
9

FIGURE 13.9

The second layer of the network is one hidden layer. The example shown illustrates a small hidden layer containing just 15 neurons.

The output layer of the network contains 10 neurons. If the first neuron fires, i.e. has an output ≈ 1, then that will indicate that the network thinks the digit is a 0. If the second neuron fires then that will indicate that the network thinks the digit is a 1. And so on. A little more precisely, we number the output neurons from 0 through 9 and figure out which neuron has the highest activation value.

13.2.5 Learning with Gradient Descent

Now that we have a design for our neural network, how can it learn to recognize digits? A version of one of the methods encountered in optimization theory, the gradient descent may be developed for the following training/learning of the neural networks with sigmoidal neurons. The **error back propagation algorithm** (BPA), also known as the generalized δ rule, provides a way to calculate the gradient of the error function efficiently using the chain rule of differentiation. The error after initial computation in the forward pass—from the input layer through the hidden layer to the output layer—is propagated backward, justifying the name given to the algorithm. This algorithm is fairly general,

but first we will attempt to understand its working on a multilayer feedforward network.

The first thing we need is a data set to learn from—a so-called training data set, a huge one, represented as a vector $x_t \in \Re^n, t = 1 \dots N$. For example, we may use the well-known MNIST data set[1], which contains N = tens of thousands of scanned images of handwritten digits together with their correct classifications, the desired output vector $y_t \in \Re^m, t = 1 \dots N$ for the input vector x. Together, we call this as a set of exemplars. The output of the jth neuron in the output layer is designated as $\hat{y}_j \in \Re^m$. Intuitively, our algorithm's goal is to minimize the error between y_j and $\hat{y}_j$ for all the outputs, at least asymptotically, over the entire data set.

Let us assume there are h hidden layers between the input layer and the output layer. We shall designate the weights from the ith neuron, in one layer, to the jth neuron in the following layer as w_{ij}. To be a little more precise, we shall employ a superscript (l) with $w_{ij}^{(l)}$ to denote the fact that the layer containing the jth neuron is l layers ahead of the output layer; thus, $l = 0$ denotes the output layer, and, say, $l = h + 3$ is the second hidden layer after the input. For convenience, we will omit the subsscript on the output layer.

---

[1] a modified subset of two data sets collected by NIST, the United States National Institute of Standards and Technology

Let m be the number of output neurons. If $y_j(t)$ is the desired output from the jth neuron and $\hat{y}_j(t)$ is the actual output, corresponding to the tth input vector $x(t)=\left[x_1(t)\ldots x_n(t)\right]^T$, we define the sum of squares of the error over all output unitis for the tth exemplar by:

$$E(t)=\frac{1}{2}\sum_{i=1}^{m}\left(\hat{y}_i(t)-y_i(t)\right)^2 \in \Re \qquad (13.6)$$

and the total accumulated error over the entire set of N exemplars by

$$E=\sum_{t=1}^{N}E(t) \in \Re \qquad (13.7)$$

Computing the scalar $E(t)$ in eqn.13.6 is called a *forward pass*.

Referring to Figure 13.24 and eqn.13.5, we have

$$\frac{\partial E(t)}{\partial \hat{y}_j}=\hat{y}_j-y_j \qquad (13.8)$$

Following the notation we have previously used, the net input to neuron j in the output layer is of the form:

$$s_j=\sum_i y_i^{(1)}w_{ij}+\theta_j \qquad (13.9)$$

Before we proceed, let us quickly observe that

$$\frac{\partial s_j}{\partial y_i^{(1)}}=w_{ij}$$

This will be useful after a few more equations.

Since we have employed the sigmoid, which has all the nice properties required, we now readily obtain

$$\frac{\partial E(t)}{\partial s_j}=\frac{\partial E(t)}{\partial \hat{y}_j}\times\frac{dy_j}{ds_j} \qquad (13.10)$$

with

$$\frac{dy_j}{ds_j}=\frac{d}{ds_j}\left(\frac{1}{1+e^{-s_j}}\right)=y_j\left(1-y_j\right) \qquad (13.11)$$

Therefore,

$$\frac{\partial E(t)}{\partial s_j}=\left(\hat{y}_j-y_j\right)y_j\left(1-y_j\right) \qquad (13.12)$$

and, consequently

$$\frac{\partial E(t)}{\partial w_{ij}}=\frac{\partial E(t)}{\partial s_j}\times\frac{\partial s_j}{\partial w_{ij}}=\frac{\partial E(t)}{\partial s_j}\times y_i^{(1)} \qquad (13.13)$$

Looking at the superscript at the end of eqn.13.13, we say that the error has been propagated back one layer.

Based on our usage of the superscript, we should be able to establish a direct link between parameters of adjacent layers as

$$s_i^{(l)}=\sum_m y_m^{(l+1)}w_{mi}^{(l)}+\theta_i^{(l)} \qquad (13.14)$$

where $w_{mi}^{(l)}$ are the weights and $\theta_i^{(l)}$ is the bias of neuron i in level l ahead of the output layer.

By applying the chain rule of differentiation, to begin with $l=1$,

$$\frac{\partial E(t)}{\partial w_{mi}^{(1)}}=\frac{\partial E(t)}{\partial \hat{y}_i^{(1)}}\cdot\frac{\partial \hat{y}_i^{(1)}}{\partial s_i^{(1)}}\cdot\frac{\partial s_i^{(1)}}{\partial w_{mi}^{(1)}}$$

$$=\frac{\partial E(t)}{\partial \hat{y}_i^{(1)}}\cdot y_i^{(1)}\left(1-y_i^{(1)}\right)y_m^{(2)} \qquad (13.15)$$

Summing over all connections originating from neuron i to the next layer, we have

$$\frac{\partial E(t)}{\partial \hat{y}_i^{(1)}}=\sum_j\frac{\partial E(t)}{\partial s_j}w_{ij}$$

$$=\sum_j\frac{\partial E(t)}{\partial s_j}\cdot\frac{\partial s_j}{\partial \hat{y}_i^{(1)}} \qquad (13.16)$$

Substituting eqn.13.12 in 13.16, and then the result in eqn.13.15 allows us to compute the change in error with respect to the weights $\partial E(t)/\partial w_{mi}^{(1)}$ for the penultimate layer. We then repeat this procedure until $\partial E(t)/\partial w_{mi}^{(l)}$ is computed for all the connections.

Notice that, at each layer, the partial derivatives $\partial E(t)/\partial s_j^{(1)}$ are saved for computations at the next layer $l+1$. However, these partial derivatives are no longer needed once the computations for $l+1$ are done; we simply update the partial derivatives and move ahead to the next pair of adjacent layers.

The process of computing the partial derivatives $\partial E(t)/\partial w_{mi}^{(1)}$ from the output back to the input, linking the inputs to the first hidden layer is called the *backward pass*.

Primarily, there are two approaches to applying the gradient descent method: (i) the periodic updating, and (ii) the continuous updating. In either case, the exemplars are *repeatedly* presented until (hopefully) the algorithm converges. When the exemplars are resent to the network sequentially, an entire pass through all the elements of the training set constitutes an *epoch*. When such an entire pass occurs without error, training is considered complete. As we will discuss some more at a later point, there is no guarantee that this will happen.

In the periodic updating approach, the gradient

$$\frac{\partial E}{\partial w} = \sum_{t=1}^{N} \frac{\partial E(t)}{\partial w} \qquad (13.17)$$

is computed over all the N exemplars, one by one. Here, w is the vector of all the M weights in the network. These weights are updated only once per cycle, after all the training patterns are presented, according to the generalized δ update rule

$$w(t+1) = w(t) - \eta \frac{\partial E}{\partial w} \qquad (13.18)$$

where $\eta > 0$ is a small constant, called the *learning rate*.

The continuous update approach has the weights updated whenever an exemplar is presented. Accordingly, the update rule is

$$w(t+1) = w(t) - \eta \frac{\partial E(t)}{\partial w} \qquad (13.19)$$

The two approaches are essentially the same, and the second one has the advantage that all the partial derivatives $\partial E(t) / \partial w_{mi}^{(l)}$ need not be stored. Looking at these update rules, however intuitive they are, as stochastic partial differential equations, there is no guarantee of convergence to the desired solution. One may think of having η larger, since it is called learning rate, but it has been found to lead to oscillation instead of faster convergence.

An astute reader would have noticed that the bias θ does not appear in the weight updating. A genuine observation indeed. To understand the trick, let us consider two neurons at the input layer—a variable input x and a constant input $x_0 = 1$. If these are multiplied by the weights w_{21} and w_{22} and are fed to a sigmoidal neuron in the next layer, the output would be

$$\frac{1}{1 + e^{-\left(w_{21}x + w_{22} \right)}}$$

We notice an additional weight, w_{22}, that influences the output and, hence, the error. We simply include this new weight in our M-dimensional vector $w \in \mathfrak{R}^M$. At the end of training, the updated w_{22} would be the updated bias.

Example 13.2.2:

Let us develop a neural network to recognize the upper case letters **L** and **T** and their rotated versions. To simplify the presentation, we draw each letter on a grid with 9 squares as shown in Figure 13.10.

Thus, the letters and their rotated versions constitute a set of 8 exemplars. The problem is to

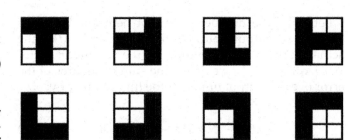

FIGURE 13.10

design a feedforward neural network with one hidden layer so that the network is trained to produce an output value 1 when the input pattern is either **T** or any of its rotated versions, and 0 in the complementary situation.

Since we assume that the letters are fitted to a 3×3 grid, we may have 9 inputs at the input layer—the input is 1 when a square is filled and 0 otherwise. The hidden layer is assumed to have 3 units, and the output layer has only one neuron, as shown in Figure 13.11. We have presented the picture without clutter; we encourage the reader to draw it herself to see the complete set of weights.

All 9 neurons at the input layer are assumed to simply send their 0 or 1 to the three neurons in the hidden layers via appropriately designated weights. In addition, we shall have a 10th neuron with an input 1 to represent the bias in the hidden layer. Likewise, we add a neuron to the hidden layer to take care of the bias in the output layer. Thus, we have $M = 9 \times 3 + 3 + 3 \times 1 + 1 = 34$ weights to be updated, starting from a set of random elements in the interval $(-1, +1)$.

The sigmoid we have been working with so far maps the input to a value in the interval $(0,1)$. To map the input to a value in the interval $(-1,1)$ we define a **tan-sigmoid**, as shown in Figure 13.12:

$$\tan \sigma(z) = 2\sigma(z) - 1 \qquad (13.20)$$

We use the standard sigmoid for the output unit since we detect a **T** or its complement. However, for the hidden layer, we employ the tan-sigmoid.

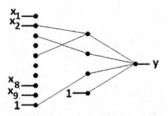

FIGURE 13.11

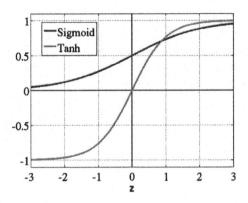

FIGURE 13.12

The eight exemplars are sequentially presented to the network until the error is significantly minimized. The following are the results reported in [14,15].

A set of initial weights and biases have been picked up randomly in the interval $(-1,1)$, and the backpropagation algorithm is applied for 5938 epochs with $\eta = 0.001$, When a satisfactory result is obtained, i.e., when the final sum of squared error is 0.124976, the set of $9 \times 3 = 27$ weights (rounded to 3 decimal places) between the input and the hidden layer are as follows:

−0.943	−0.948	0.538	−0.607	0.254	0.929	−0.596	0.774	0.140
0.742	−0.430	0.716	0.310	−1.288	−0.135	0.431	−0.143	0.535
0.530	−0.437	−0.338	−0.404	1.598	0.134	−0.625	0.399	−0.127

where w_{ij} is the weight from the jth input neuron to the ith hidden neuron. And the 3 biases are as follows:

$$\begin{bmatrix} 1.0021 & -0.3599 & -0.3504 \end{bmatrix}^T$$

Between the hidden layer and the output, the weights are

$$0.2615$$
$$-1.4469$$
$$1.3431$$

and the bias for the output neuron is

$$-0.0564$$

The algorithm was reported to have consumed 8,843,153 floating point operations for the said 5938 epochs. We leave it to the imagination of the reader to think about the computational overhead of the handwritten digit recognition problem and other modern problems, like autonomous driving, which we mentioned in the introduction.

Example 13.2.3:

We continue with the previous example of recognizing handwritten digits from the MNIST data. Figure 13.13 shows a sample of how the digits in the data set look.

The MNIST data comes in two parts. The first part contains 60,000 images to be used as training data. These images are scanned handwriting samples from 250 people, half of whom were US Census Bureau employees and half of whom were high school students. The images are grayscale and 28×28 pixels in size. The second part of the MNIST data set is 10,000 images to be used as test data. Again, these are 28×28 grayscale images. We shall use the test data to evaluate how well our neural network has learned to recognize digits. To make this a good test of performance, the test data was taken from a different set of 250 people than the original training data (albeit still a group split between Census Bureau employees and high school students). This helps give us confidence that our system can recognize digits from people whose writing it didn't see during training.

In the present case, as one would expect, each training input x_i is a $28 \times 28 = 784$-dimensional vector; each entry in the vector represents the gray value for a single pixel in the image. We'll denote the corresponding desired output by y_i, a 10-dimensional vector. For example, if a particular training image, x, depicts a 7, then

$$y = (0,0,0,0,0,0,0,1,0,0)^T$$

is the desired output from the network.

What we are looking at is an algorithm that lets us find weights and biases so that the output from the network approximates y for all training inputs x. The aim of our training algorithm will be to minimize the sum of the squared error in eqn.13.6. From this perspective, a few remarks are in place. Our goal is to find a set of weights and a set of biases, which minimize the quadratic function in eqn.13.6. While this is a well-posed problem, it has got a lot of distracting structure as currently posed—the interpretation of w and b as weights and biases, the σ function lurking in the background, the choice of network architecture, MNIST, and so on. It turns out that we can understand a tremendous amount by ignoring most of that structure and just concentrating on the minimization aspect. In other words, we are going to forget all about the specific form of the error function, the connection to neural networks, and so on. Instead, we're going to imagine that we've

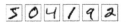

FIGURE 13.13

simply been given a function of many variables and we want to minimize that function.

Furthermore, there are a number of challenges in applying the gradient descent rule. Looking back at the quadratic error function, when the number of training inputs is very large, this can take a long time and, thus, learning occurs slowly. We might also have to live with poor accuracies in the 20–50% range. If you work a bit harder, with a lot more data, more hidden layers, different initial weights, and so on, you can get up over 50%. But to get much higher accuracies, it helps to use established machine learning algorithms, such as the support vector machine (SVM).

Finally, when it comes to implementing our network to classify digits, today Python comes in very handy for those who wish to really do hands-on work. Of course, MATLAB does a more professional job. We will not discuss these issues in this book.

13.2.5.1 Issues in Implementation

For all nice properties, such as continuity and differentiability, we have considered the sigmoid (either log- or tan-). We need to examine the consequences, too, and we should be able weigh the pros and cons of the choice of sigmoid versus some other activation function, as well as the backpropagation algorithm versus another algorithm.

1. Since the sigmoid reaches 0 or 1 only asymptotically, the desired output of the network must be from the set $\{\varepsilon_1, 1 - \varepsilon_2\}$. But this demands a trade-off—if the εs are larger, then the range of weights would be smaller, and perhaps we get the convergence a bit earlier, however, the sensitivity of the network to noise in the input patterns could be higher. It may be shown that $1 - \varepsilon_1 - \varepsilon_2$ indicates the noise tolerance of a given neuron.

2. While the quadratic error function, once again, has nice mathematical properties, there is a possibility of this function getting stuck at a local minimum, fooling us to take it as real minimum; there is also another possibility wherein the weights begin oscillating, i.e. the update partial differential equation (PDE) begins giving oscillatory solutions near the global minimum. Hence, it is good to stop the update at $E < \varepsilon$. Further, the number of iterations should be capped a priori since there is no guarantee that the algorithm would converge and, consequently, the program never terminates.

3. We must be conscious not to choose identical initial weights to begin with—this would keep the change of weights constant over iterations, and hence, the updated weights also remain same. It is, therefore, imperative that we choose to begin with small, random weights; this is usually referred to as *symmetry breaking*.

4. Referring to Example 13.2.2 once again, if the input patterns are binary, then $\{\alpha, 1\}$ is preferred over $\{0,1\}$, where α is a small positive number, say 0.1, or even −1. This choice usually leads to faster convergence in almost all cases.

5. The back propagation algorithm generally slows down by an order of magnitude every time a layer is added to the network (more about this in a later part of the chapter on deep networks). This problem may be solved by selecting either a different algorithm, or a different network architecture that is not necessarily nice layered. Many variants of the back propagation algorithm are possible, say, making the learning rate η either gradually decay.

13.2.6 Unsupervised and Reinforcement Learning

In both the perceptron, and the sigmoidal neuron, we have used the correct output for the current input for learning. This type of learning is called **supervised learning**. A good question could be—what happens if the correct output is either partially or indirectly available or completely unavailable?

When the correct output is not available, and still we are able to learn, we call it **unsupervised learning**. Here, the system is presented with only a set of inputs; the system is not given any external indication as to what the correct responses should be nor whether the generated responses are right or wrong. Basically, unsupervissed learning aims at finding a certain kind of regularity in the data represented by the inputs. Roughly speaking, regularity means that much less data re actually required to approximately describe or classify the exemplars than the amount of data in the exemplars. Examples exploiting data regularity include vector quantization for data compression and Principal Component Analysis, for dimension reduction.

A principal component analysis identifies M orthogonal directions representing, in descending order, the largest variance in the data, where M is less than or equal to the dimension of the data. This amounts, mathematically, to finding the eigenvectors of the covariance matrix

$$C = E\left[(x - \bar{x})(x - \bar{x})^T \right]$$

of the real data vector x. The input and output are assumed to be related by a linar transformation $y = Wx, W \in \Re^{M \times N}$, and objective of principal component analysis is to reconstruct the N-dimensional input x *optimally* from the M-dimensional output y, when $N > M$.

Reinforcement learning is somewhere between supervised learning, where the desired output is available, and unsupervissed learning, when the system gets no feedback at all on how it is performing. In reinforcement learnin the system receives a feedback that tells the system whether its output response is right or wrong, but no information on what the right output should be is provided. Therefore, the system must employsome random search strategy so that the space of plausible and rational choices is searached until a correct answer is found.

Reinforcement learning is usually involved in exploring a new environment when some knowledge (or subjective feeling) about the right response to environmental inuts is available. The system receives an input from the environment and produces an output as rsponse. Subseqnently, it receives a reward or a penalty from the environment. The system, thus, learns from a sequence of such interactions.

The difficulty with reinforcement learning stems not only from the possibility that the precise information about the error is unavailable but also from the likelihood that the reward or penalty may be implemented only after many action steps. An example is a chess game. The only knowledge available is that the game is either won or lost; a loss may be due to errors at several of the earlier moves. In order to learn where and when errors are made, straegies should be identified to reward the right moves and penalize the wrong ones.

A random search component is necessary for reinforcement learning. A well-accepted approach is to constsruct a model of the environment. the model is just an auxiliary learning network that uses the input from the environment, the output from the main learning system, and a reinforcement signal to generate estimates of the desired outputs for the main learning network.

A clever way is to first train the auxiliary network to minimize the square of the difference between the real environmental signal r and the estimated reinforcement signal $\hat{r}$; thus, making it a supervised learning problem. If the auxiliary network is a good model of the environment, the real reward r will be maximum when the estimated reward $\hat{r}$ is maximum. Therefore, at the second stage, the two networks are trained together to maximize the total reward over all the input patterns. This again becomes a supervised learning problem—the weights in hte main network are updated according to gradient descent to maximize the output $\hat{r}$ of the auxilary network.

13.2.7 Radial Basis Networks

The **radial basis function networks** (RBF networks) are a paradigm of neural networks, which was developed considerably later than that of perceptrons. Like perceptrons, the RBF networks are built in layers. But in this case, they have exactly three layers, i.e. only one single layer of hidden neurons. The input layer again does not participate in information processing. The output neurons do little more than adding all input values and returning the sum.

Hidden neurons are called as RBF neurons. As propagation function, each hidden neuron calculates a norm that represents the distance between the input to the network and the so-called position of the neuron (center). This is inserted into a radial activation function that calculates and outputs the activation of the neuron.

Definition 1

The center c_h of an RBF neuron h is the point in the input space where the RBF neuron is located. In general, the closer the input vector is to the center vector of an RBF neuron, the higher is its activation.

Definition 2

The RBF neurons h have a propagation function that determines the distance between the center c_h of a neuron and the input vector x. This distance represents the network input. Then, the network input is sent through a radial basis function that returns the activation or the output of the neuron.

Definition 3

An RBF network has exactly three layers in the following order: The input layer consisting of input neurons, the hidden layer (also called RBF layer) consisting of RBF neurons and the output layer consisting of RBF output neurons. Each layer is completely linked with the following one, in a feedforward topology. The connections between input layer and RBF layer are unweighted, i.e. they only transmit the input. The connections between RBF layer and output layer are weighted. A bias neuron is not used in RBF networks.

The set of input neurons shall be represented by I, the set of hidden neurons by H, and the set of output neurons by O. Therefore, the inner neurons are called radial basis neurons because from their definition follows directly that all input vectors with the same distance from the center of a neuron also produce the same output value.

13.2.7.1 Information Processing of an RBF Network

An RBF network receives the input by means of the unweighted connections. Then the input vector is sent through a norm so that the result is a scalar. This scalar

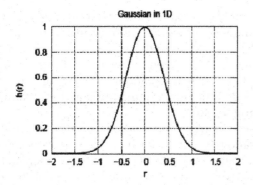

FIGURE 13.14

(which, by the way, can only be positive due to the norm) is processed by a radial basis function, for example by a Gaussian shown in Figure 13.14.

The output values of the different neurons of the RBF layer or of the different Gaussian bells are added within the third layer: basically, in relation to the whole input space, Gaussian bells are added here.

Suppose that we have a second, a third, and a fourth RBF neuron and therefore four differently located centers. Each of these neurons now measures another distance from the input to its own center and de facto provides different values, even if the Gaussian bell is the same. Since these values are finally simply accumulated in the output layer, one can easily see that any surface can be shaped by dragging, compressing, and removing Gaussian bells and, subsequently, accumulating them. Here, the parameters for the superposition of the Gaussian bells are in the weights of the connections between the RBF layer and the output layer.

Since we use a norm to calculate the distance between the input vector and the center of a neuron h, we have different choices. Often, the Euclidian norm is chosen to calculate the distance. From the definition of a norm directly follows that the distance can only be positive. Strictly speaking, we only use the positive part of the activation function. By the way, activation functions other than the Gaussian bell are also possible; normally, functions that are monotonically decreasing over the interval $[0,\infty]$ are chosen.

The output y of an RBF neuron Ω results from combining the functions of an RBF neuron to

$$y_\Omega = \sum_h w_{h,\Omega} \cdot f(\|x - c_h\|) \tag{13.21}$$

And, if set N contains $|N|$ training exemplars, then we have $|N|$ functions of the above form, i.e. one function for each training sample.

Next, let us assume that the widths $\sigma_1,...,\sigma_k$, the centers $c_1,...,c_k$, and the training samples $\hat{y}_i$ including the teaching input x_i are given. We are then looking for the weights w_h with $|H|$ weights for one output neuron. Thus, our problem can be seen as a system of equations since the only thing we want to change at the moment are the weights. Consequently, this demands a distinction of cases concerning the number of training samples $|N|$ and the number of RBF neurons $|H|$:

1. The sizes $|N|$ and $|H|$ are same. In this case, using the previous equation 13.21, the transformation is

$$Y = MW \tag{13.22}$$

where Y is the vector desired output, M is the matrix of the outputs of all $|H|$ neurons to $|N|$ training inputs, and W is the vector of desired weights. Since M is square in this case, we may attempt to invert it and compute the weights as

$$W = M^{-1}Y$$

2. There are more training samples than the neurons, i.e. $|N| > |H|$. In this case, we go with the least squares estimate to obtain

$$W = \left[M^T M\right]^{-1} M^T Y$$

3. The third case that can be enumerated, in principle, does not arise in practice, and hence, we do not discuss about it.

It is easy to see that the multiple output neurons (in the third layer) can be trivially dealt with at no extra computational effort.

But, the good news is only so far so good. The least squares solution does not guarantee that the output vector corresponds to the desired output, even though the calculation is mathematically correct; our computers can only provide us with (nonetheless very good) approximations of the pseudo-inverse matrices. This means that we also get only approximations of the correct weights (maybe with a lot of accumulated numerical errors) and, therefore, only an approximation (maybe very rough or even unrecognizable) of the desired output. Thus, even if we have enough computing power to analytically determine a weight vector, we should use it only as an initial value for our learning process, which leads us to the real training methods, and we resort to the generalized δ rule.

We generally perform the whole training process in three phases. In the first phase, we obtain an approximate set of weights using the analytical computation using the pseudo inverse. In the second phase, we refine the results using the δ rule. The second phase is further split into two phases—online and then offline—for faster and accurate convergence.

As already indicated, the training of an RBF network is not just obtaining an optimal set of weights between the hidden and the output layer. We need to take a look at the possibility of varying the widths σ and the center c, which is not always trivial.

One might suspect if it is any obvious that the approximation accuracy of RBF networks can be increased by *adapting* the widths and positions of the Gaussian bells in the input space to the problem that needs to be approximated. There are several methods to deal with this.

1. Fixed selection: The centers and widths can be selected in a fixed manner and regardless of the training exemplars—this is what we have assumed until now.

2. Conditional, fixed selection: Again centers and widths are selected fixedly, but we have previous knowledge about the functions to be approximated and comply with it.

3. Adaptive to the learning process: This is definitely the most elegant variant, but certainly the most challenging one, too.

We will not discuss these in this book, and the reader is suggested several good references.

We close this part on radial basis function networks comparing them to the multilayer networks.

1. Input dimension: We must be careful with RBF networks in high-dimensional functional spaces, since the network could very quickly require huge memory storage and computational effort. Here, a multilayer network would cause less problems.

2. Center selection: Selecting the centers c for RBF networks is still a major problem. Please use any previous knowledge you have when applying them. Such problems do not occur with the multilayer networks.

3. Output dimension: The advantage of RBF networks is that the training is not much influenced when the output dimension of the network is high. On the other hand, for a multilayered network, a learning procedure such as backpropagation will be very time-consuming.

4. Extrapolation: An advantage as well as a disadvantage of RBF networks is the lack of extrapolation capability: An RBF network returns the result 0 far away from the centers of the RBF

layer. However, unlike the multilayered network, the RBF network is capable of using this to tell us, "I do not know," which could be an advantage.

5. Lesion tolerance: For the output of a multilayered network, it is no so important if a weight or a neuron is missing. It will only worsen a little in total. If a weight or a neuron is missing in an RBF network, then large parts of the output remain practically uninfluenced. But one part of the output is heavily affected because a Gaussian bell is directly missing. Thus, we can choose between a strong local error for lesion and a weak but global error.

6. Spread: Here, the multilayered network has a clear advantage, since RBF networks are used considerably less often, the multilayered networks seem to have a considerably longer tradition and they are working too good.

13.2.8 Recurrent Neural Networks

Up until now, we've been discussing neural networks where the output from one layer is used as input to the next layer, the hidden layer(s), and then to the output layer. Such networks are called **feedforward neural networks**. This means there are no loops in the network—information is always fed forward, never fed back. If we did have loops, we would end up with situations where the input to the σ function depends on the output. This makes life quite complicated and we simply avoid them.

However, there are other models of artificial neural networks in which feedback loops are possible. These models are called **recurrent neural networks**. The idea in these models is to have neurons that fire for some limited duration of time, before becoming quiescent. That firing can stimulate other neurons, which may fire a little while later and for a limited duration. That causes still more neurons to fire, and so over time we get a cascade of neurons firing. Loops don't cause problems in such a model, since a neuron's output only affects its input at some later time, not instantaneously.

Recurrent neural nets have been less influential than feedforward networks, in part because the learning algorithms for recurrent nets are (at least to date) less powerful. But recurrent networks are still extremely interesting. They're much closer in spirit to how our brains work than feedforward networks. Also, it's possible that recurrent networks can solve important problems that can only be solved with great difficulty by feedforward networks. However, to limit our scope, in this book we're going to concentrate on the more widely-used feedforward networks.

13.2.9 Towards Deep Learning

While our basic neural networks do give us impressive performance, that performance is somewhat mysterious. The weights and biases in the network were discovered automatically. And that means we do not immediately have an explanation of how the network does what it does. Can we find some way to understand the principles by which our network is classifying handwritten digits? And, given such principles, can we do better?

To address these questions, let's think back to the example of autonomous driving we briefly discussed at the start of the chapter. Suppose we want to determine what would a good human driver does.

The end result is a network that breaks down such a very complicated question into very simple questions answerable at the level of, say, single pixels. It does this through a series of many layers, with early layers answering very simple and specific questions about the input image, and later layers building up a hierarchy of ever more complex and abstract concepts. Networks with this kind of many-layer structure—two or more hidden layers—are called **deep neural networks**. Researchers in the 1980s and 1990s tried using stochastic gradient descent and backpropagation to train deep networks. Unfortunately, except for a few special architectures, they didn't have much luck. The networks would learn, but very slowly, and in practice often too slowly to be useful.

Since 2006, a set of techniques has been developed that enable learning in deep neural nets. These deep learning techniques are based on stochastic gradient descent and backpropagation, but also introduce new ideas. These techniques have enabled much deeper (and larger) networks to be trained—people now routinely train networks with 5–10 hidden layers. And, it turns out that these perform far better on many problems than shallow neural networks, i.e. networks with just a single hidden layer. The reason, of course, is the ability of deep nets to build up a complex hierarchy of concepts. It is a bit like the way conventional programming languages use modular design and ideas about abstraction to enable the creation of complex computer programs. Comparing a deep network to a shallow network is a bit like comparing a programming language with the ability to make function calls to a stripped down language with no ability to make such calls. Abstraction takes a different form in neural networks than it does in conventional programming, but it's just as important.

Much of the ongoing research aims to better understand the challenges that can occur when training deep networks. In this smaller part of the book we, in no way, attempt to comprehensively summarize that work, we nevertheless thought of citing a couple of good papers, to give the reader the flavor of some of the questions.

As a first example, in 2010 Glorot and Bengio[2] found evidence suggesting that the use of sigmoid activation functions can cause problems training deep networks. In particular, they found evidence that the use of sigmoids will cause the activations in the final hidden layer to saturate near 0, early in training, substantially slowing down learning. They have also suggested some alternative activation functions, which appear not to suffer as much from this saturation problem.

As a second example, in 2013 Sutskever, Martens, Dahl and Hinton[3] studied the impact on deep learning of both the random weight initialization and the momentum schedule in momentum-based stochastic gradient descent. In both cases, making good choices made a substantial difference in the ability to train deep networks.

Thus, there is a significant role played by the choice of activation function, the way weights are initialized, and even details of how learning by gradient descent is implemented. Of course, choice of network architecture and other hyper-parameters is also important. Thus, many factors can play a role in making deep networks hard to train, and understanding all those factors is still a subject of ongoing research.

13.2.10 Summary

In recent years, there has been a great deal of effort to design feedback control systems that mimic the functions of living biological systems. There has been great interest recently in controllers that work independent of a particular mathematical model of the plant, but attempt to understand the dynamics of the plant at higher cognitive levels, so to say; for instance, what could have been a differential equation governing the recognintion of Ts and Ls, in spite of rotations, in Example 13.2.2? There is a sound argument behind this—today, when no system is purely mechanical or electrical or hydraulic, how do we obtain a meaningful mathematical model? On the other hand, as the plant is operated over a long period of time can we extract some meaning out of the data? Techniques include artificial neural networks, fuzzy logic control, which mimics linguistic and reasoning functions (next section), and evolutionary algorithms (the last part of this chapter). By now, the theory and applications of these nonlinear network structures in feedback control have been well documented. We briefly present the working principles of these techniques in this chapter, but Chapters 14

---

[2] "Understanding the difficulty of training deep feedforward neural networks," by Xavier Glorot and Yoshua Bengio (2010)

[3] "On the importance of initialization and momentum in deep learning," by Ilya Sutskever, James Martens, George Dahl and Geoffrey Hinton (2013)

through 16 have been dedicated to develop and examine the case studies in control applications. Many researchers have contributed to the development of a firm foundation for analysis and design of neural networks in control system applications.

This section attempted to show how ANNs fulfill the promise of providing model-free learning controllers for a class of nonlinear systems, in the sense that a structural or parameterized model of the system dynamics is not needed. It will be shown, in the rest of the book, that as uncertainty about the controlled system increases or as one desires to consider human user inputs at higher levels of abstraction, the ANN controllers acquire more and more structure, eventually acquiring a deep hierarchical structure that resembles some of the elegant architectures proposed by computer science engineers using high-level design approaches based on theories, such as cognitive linguistics, reinforcement learning, psychological theories, adaptive critics, or optimal dynamic programming.

13.3 Fuzzy Logic

Multiple-valued logic has a long history. Aristotle raised the question of whether all valid propositions can only be assigned the logical values true or false. The first attempts at formulating a multiple-valued logic were made by logicians, such as Charles Sanders Pierce at the end of the nineteenth and beginning of the twentieth century. The first well-known system of multiple valued logic was introduced by the Polish logician Jan Lukasiewicz in the 1920s. By defining a third truth value, Lukasiewicz created a system of logic that was later axiomatized by other researchers. From 1930 onwards, renowned mathematicians such as Gödel, Brouwer, and von Neumann continued work on developing an alternative system of logic that could be used in mathematics or physics. In their investigations, they considered the possibility of an infinite number of truth values.

Fuzzy logic, as formulated by Zadeh in 1965, is a multiple-valued logic with a continuum of truth values. The term fuzzy logic really refers more to a whole family of possible logic theories that vary in the definition of their logical operators. First attempts to use fuzzy logic for control systems were extensively examined by Mamdanis group in England in the 1970s. Since then, fuzzy controllers have left the research laboratories and are used in industrial and consumer electronics. Over the last few years interest in fuzzy controllers has increased dramatically. Some companies already offer microchips with hardwired fuzzy operators and fuzzy inference rules.

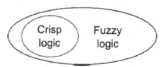

FIGURE 13.15

Fuzzy Logic is based on the theory of fuzzy sets, which is, in some sense, a generalization of the classical set theory. While an exposition to the classical set theory leaves one unambiguous, fuzzy logic leads to quite a good amount of debate. Perhaps, *extension* is a better word than *generalization*. Nevertheless, to make a quick and tangible exposition to fuzzy logic, we take it that the classical set theory is a subset of the theory of fuzzy sets, as Figure 13.15 shows.

The following definition and the example throw more light on the idea of fuzzy sets.

Definition 4:

Let X be a set. A fuzzy subset A of X is characterized by a *membership function*:

$$\mu_A:X \to [0,1] \tag{13.23}$$

where we normalize the peak magnitude of the function to 1, consistent with any other elementary signal we consider, e.g. unit step function.

Example 13.3.1:

When we, for instance, avail an online service, such as booking an air ticket, we are prompted to rate the service. In most cases, we will be suggested to rate the service as either poor, good, or excellent. From our perspective, it is likely that the services are neither poor nor excellent. Accordingly, mapping our perspective, we may come up with the subsets, say, poor, good, excellent, and the membership function as shown in Figure 13.16.

That this function is a Gaussian is by sheer intuition. Other shapes of the membership function, such as sigmoid, hyperbolic, tangent, and exponential, may be chosen arbitrarily by following expert advice or statistical patterns. Moreover, the user has the freedom to either expand or compress the Gaussian, or any other chosen shape.

In sharp contrast, had it been the classical—Yes or No type—set theory, we would rate the services as either "good" or "no good" (one is a complement of the other) and certainly no notion of anything in between or beyond. In other words, $\mu_A:X \to \{0,1\}$, and not an interval as before.

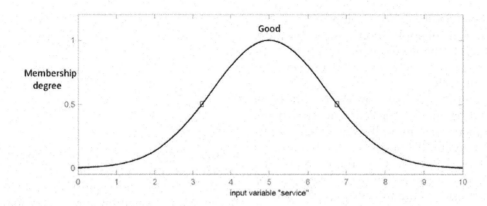

FIGURE 13.16

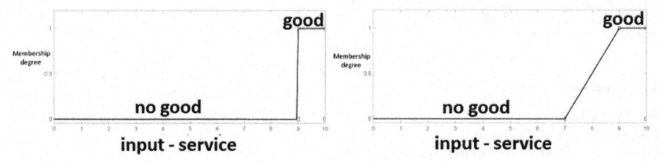

FIGURE 13.17

On a 0–10 scale, we show the contrast between the classical, also known as crisp, set theoretic notion and the fuzzy set theoretic notion in Figure 13.17.

Fuzzy set have a number of properties, some of the important ones enumerated below. Typically, the reader would see these definitions as extended versions of the crisp sets and subsets.

Definition 5:

The height of A, denoted $h(A)$, corresponds to the upper bound of the co-domain of its membership function:

$$h(A) = \sup\{\mu_A(x):x \in X\} \qquad (13.24)$$

Definition 6:

The set A is said to be **normalised** if and only if $h(A) = 1$.

In practice, it is extremely rare to work on non-normalised fuzzy sets.

Definition 7:

The **support** of A is the set of elements of X belonging to at least some A (i.e. the membership degree of x is strictly positive). In other words, the support is the set

$$supp(A) = \{x \in X:\mu_A(x) > 0\} \qquad (13.25)$$

Definition 8:

The **kernel** of A is the set of elements of X belonging entirely to A. In other words, the kernel is

$$ker(A) = \{x \in X:\mu_A(x) = 1\} \qquad (13.26)$$

We may readily observe that, by defintion, $ker(A) \subset supp(A)$.

Definition 9:

An α-**cut** of A is the classical subset of elements with a membership degree greater than or equal to α:

$$\alpha - cut(A) = \{x \in X : \mu_A(x) \geq \alpha\} \qquad (13.27)$$

The above definitions may be illustrated in Figure 13.18.

13.3.1 The Linguistic Variables

The idea behind the definition of a membership function discussed above is to provide us with a definition of fuzzy systems in natural language.

Definition 10:

Let V be a variable, X the range of values of the variable, and T_V a finite or infinite set of fuzzy sets. A linguistic variable corresponds to the triple (V, X, T_V).

Example 13.3.2:

Figures 13.19 through 13.21 show some examples of linguistic variables and the corresponding triples. The following remarks are important.

1. When we define the fuzzy sets of linguistic variables, we do not have to exhaustively define the linguistic variables. Instead, we only define a few fuzzy subsets that will be useful later in definition of the rules that we apply it. For instance, the service is rated among poor, good, and excellent; if it is useful we can also accommodate subsets such as "extremely poor," "average," "wonderful," and the like either at the beginning, in the middle, or at the end.

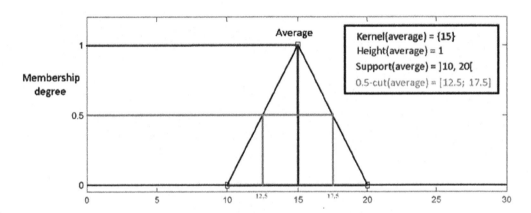

FIGURE 13.18

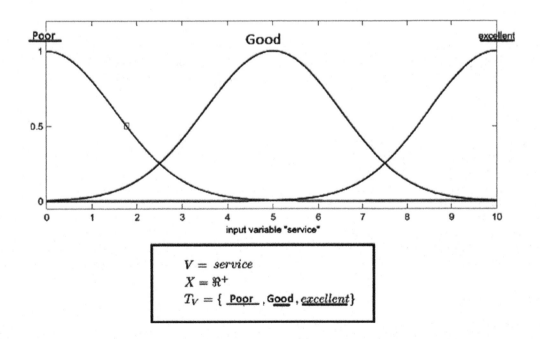

FIGURE 13.19

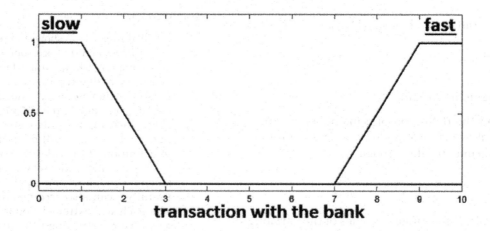

FIGURE 13.20

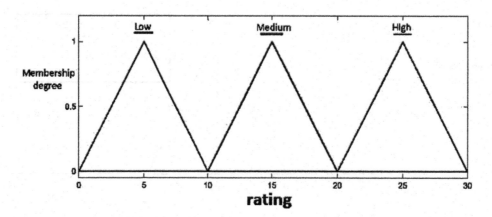

FIGURE 13.21

2. It is rather obvious that the highest rating 10, on the 0–10, scale has a height shorter than, say, 8. For instance, in Figure 13.2, the height of "good," with a numerical rating of 5, is 1.0 while that of "excellent" with a numerical rating 10 is almost 0.0. As shown in Figure 13.7, if we define the other two sets "poor" and "excellent," then the height of "excellent," with the numerical rating of 10, can be shown to be 1.0 since it makes *more* sense to call 10 as *excellent* in the respective set, whereas it is an outlier in the set "good." By the same token, in the interval defining "excellent," the membership of "good" is shown to decrease monotonically.

13.3.2 The Fuzzy Operators

Recall that once the sets are definied, in the classical theory, we move on to operations such as complement, union, and intersection, on the sets. We extend the same

in order to easily manipulate fuzzy sets. Here are the two sets of operators for the complement (NOT), the intersection (AND), and union (OR) most commonly used:

Name	AND $\mu_{A \cap B}$	OR $\mu_{A \cup B}$	NOT μ_{A^c}
Zadeh	$\min\left(\mu_A(x), \mu_B(x)\right)$	$\min\left(\mu_A(x), \mu_B(x)\right)$	$1 - \mu_A(x)$
Probabilistic	$\mu_A(x) \times \mu_B(x)$	$\mu_A(x) + \mu_B(x)$ $- \mu_A(x) \times \mu_B(x)$	$1 - \mu_A(x)$

With these definitions, fuzzy operators satisfy the properties of commutativity, distributivity, and associativity classics. However, there are two notable exceptions:

1. The law of excluded middle is contradicted:

$$A \cup A^c \neq X, \quad \text{i.e.,} \quad \mu_{A \cup A^c} \neq 1 \qquad (13.28)$$

2. An element can belong to A and A^c at the same time:

$$A \cap A^c \neq \phi, \quad \text{i.e.,} \quad \mu_{A \cap A^c} \neq 0 \qquad (13.29)$$

These elements correspond to the set $supp(A) - ker(A)$.

13.3.3 Reasoning with Fuzzy Sets

In classical logic, the arguments are of the form:

$$\textit{If } p \textit{ then } q \qquad (13.30)$$

where p and q are truth statements such as

$$p\text{:it is an aeroplane} \quad q\text{:it flies} \qquad (13.31)$$

At any point of time, each of these statements is either *true* or *false*, and the connecting words—*if* and *then*—suggest an implication. The truth of the implication can be verified using a Boolean enumeration in a truth table. Moreover, we can establish a clear negation—if it does not fly, then it is not an aeroplane. The arguments need not be restricted to just a pair of statements. We observe that the statement is either true or false once and for all universally.

Reasoning in fuzzy logic, better known as approximate reasoning, is based on fuzzy rules that may be expressed in a similar manner; however, the statements are no longer crisp. Hence, one would expect the membership functions to play a significant role. For instance, consider:

$$\begin{aligned} &\text{If the service is } \textit{excellent,} \\ &\text{then rating is } \textit{very high} \end{aligned} \qquad (13.32)$$

and the meaning of "excellent" and "very high" are understood from the respective fuzzy sets, and they might vary from person to person.

To determine the degree of truth of the proposition fuzzy "rating is very high," we need a definition of the fuzzy implication. Interestingly, like other fuzzy operators, there is no single definition of the fuzzy implication, and we have a variety of definitions to choose from. Here are two popular definitions of fuzzy implication:

Name	Truth Value
Mamdani	$\min(\mu_A(x), \mu_B(x))$
Larsen	$\mu_A(x) \times \mu_B(x)$

In all likelihood, these two definitions do not match the readers intuitive extension of the notion of classical implication. And, there are other definitions that do present an extension but they are not as popular. We prefer rigor to intuition.

The antecedent p and the consequent q come from different universes of discourse; therefore, we might expect that the new universe of discourse that describes the relationship between p and q is the Cartesian product of all pairs (p_i, q_i), where $p_i \in p$, and $q_i \in q$. If R is a fuzzy relation, then R is a fuzzy subset of $p \times q$, with every pair (p_i, q_i) having as associated membership $\propto_R(p_i, q_i)$. Just to illustrate the point let us consider

$$\begin{aligned} p = \text{small numbers over} \\ \text{a universe } U_1 = \{1, 2\} \end{aligned} \qquad (13.33)$$

and

$$\begin{aligned} q = \text{big numbers over} \\ \text{a universe } U_2 = \{1, 2, 3\} \end{aligned} \qquad (13.34)$$

Let us further assume that the membership values for the elements of the fuzzy set p are the ordered pairs of the form $(p_i, \mu(p_i))$ and likewise for the elements of the fuzzy set q:

$$\begin{aligned} &\text{for } p : (1,1),(2,0.6) \quad \text{and} \\ &\text{for } q : (1,0.1),(2,0.6),(3,1) \end{aligned} \qquad (13.35)$$

Then, $R = p \times q$ may be expressed as the following matrix:

$$R = \begin{array}{c|ccc} & 1 & 2 & 3 \\ \hline 1 & 0.1 & 0.6 & 1 \\ 2 & 0.1 & 0.6 & 0.6 \end{array} \qquad (13.36)$$

where the membership values of the members of R are taken as $\min\{\mu_p(p_i), \mu_q(q_j)\}$, following Mamdani's formula.

Fuzzy relationships can be, like the classical ones, chained together as well. If R is a fuzzy relation from p to q, and S is from q to r, then the composition is

$$R \circ S \in p \times r \qquad (13.37)$$

and this allows us to treat compound fuzzy conditional statements. For instance, we may arrive at decisions such as "If p is *small* then q is *large* else q is *not very large*." This kind of inferences lead naturally to an extension where fuzzy rules are applied to control systems where several of them are used to match observed conditions and consequently to recommend an alternative action.

The following three examples illustrate the fuzzy—reasoning process.

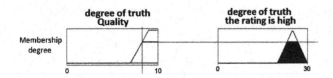

FIGURE 13.22

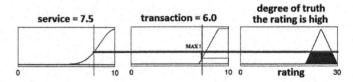

FIGURE 13.23

Example 13.3.3:

Let us consider the fuzzy rule:

$$\text{If the service is } excellent, \\ \text{then rating is } very\ high \qquad (13.38)$$

where, in Figure 13.5, the service is *excellent* with a score of 8.5 on a 0–10 scale. Using Mamdani's impllication, we get the fuzzy rating corresponding to *high* as shown in Figure 13.22.

What do we do with the dark shaded region? For now let us just see it as a fuzzy set, rather a subset, wherein our decision lies. How we extract our real decision from this fuzzy set is the topic of the next part.

In general, the result of applying fuzzy rules depends on the following factors:

1. The definition of the fuzzy implication chosen,
2. The definition of the membership function of the fuzzy set of the proposition located at the conclusion of the fuzzy rule,
3. The degree of validity of propositions located premise.

As we have defined the fuzzy operators AND, OR, and NOT, the premise of a fuzzy rule may well be formed from a combination of fuzzy propositions. All the rules of a fuzzy system is called the **decision matrix**. Here is a typical decision matrix, with reference to make use of a website to book an air ticket:

If	Then
the service is good	the rating is average
the service is bad	
OR the transaction with the bank is slow	the rating is low
the service is excellent	
OR the payment is done fast	the rating is high

Example 13.3.4:

In this example, we brought in another proposition, the speed of bank transaction that would also influence the rating.

Figure 13.23 illustrates what we get for the fuzzy rule 3 in the decision matrix above:

$$\text{If } the\ service\ is\ excellent \text{ OR} \\ the\ transaction\ is\ fast, \qquad (13.39) \\ \text{then } the\ rating\ is\ high$$

where the service is given a score of 7.5 out of 10 and the transaction is 6.5 out of 10, if we choose the Mamdani implication and the translation of OR by MAX.

Example 13.3.5:

We will now apply all the 3 rules from our decision matrix. However, we will get three fuzzy sets for the *rating*, and we will aggregate them by the operator MAX, the method that is quite popular.

Figure 13.24 shows this aggregation. Once again, what do we do with the dark shaded area?

The inference we have, i.e. the dark shaded area, is the fuzzified final decision of the rating we wanted to give to the service and the transaction together. We need to convert this back to the real world, and we discuss this next.

13.3.4 The Defuzzification

The transformation from the fuzzy world to the real world is, once again, done in many ways, each having its own advantages. In this part, we will briefly introduce the idea without getting into too many methods. In particular, we will discuss (i) the method of the mean of maxima (MoM) and (ii) the method of center of gravity (CoG).

The MoM, as one might guess from the name, defuzzifies the decision, such as the overall rating, as the average of the abscissas of the maxima of the fuzzy set resulting from the aggregation of the implication results. The average is computed using the following formula.

$$\text{Decision} = \frac{\int_S y \cdot dy}{\int_S dy} \qquad (13.40)$$

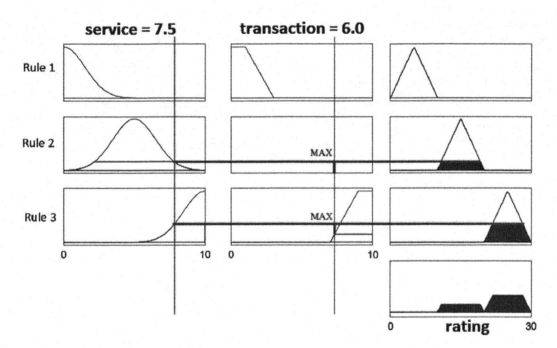

FIGURE 13.24

where

$$S = \left\{ y_m \in R, \mu(y_m) = \sup_{y \in R}(\mu(y)) \right\} \qquad (13.41)$$

and R is the fuzzy set resulting from the aggregation of the implication results.

For the fuzzy decision in blue in Figure 13.10, applying MoM suggests that the overall rating is 25.0, as shown in Figure 13.25.

The second method, the CoG, is more commonly used. It defines the output as corresponding to the abscissa of the center of gravity of the surface of the membership function characterizing the fuzzy set resulting from the aggregation of the implication results. The computation of CoG is given by

$$\text{Decision} = \frac{\int_S y \cdot \mu(y)dy}{\int_S \mu(y)dy} \qquad (13.42)$$

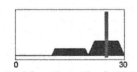

Decision: the rating is 25.0

FIGURE 13.25

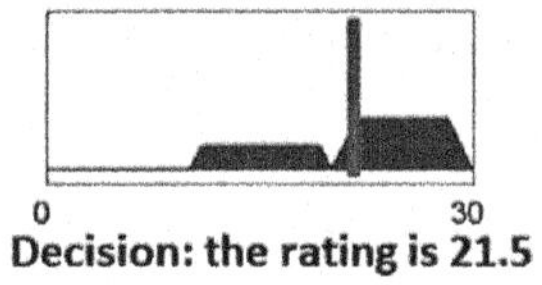

0   30
Decision: the rating is 21.5

FIGURE 13.26

And, this method suggests the overall rating for the airline booking as 21.5, as shown in Figure 13.26.

Technically speaking, this method avoids discontinuities that could appear in the MoM method, but it is more complex and computationally expensive. Certainly, there are a host of other methods that are as effective, yet computationally cheaper. We direct the reader to the references at the end of this chapter. We close this part by merely stating that the choice of the defuzzification method can have a significant effect on the final decision.

Example 13.3.6:

A thermostat in the water heater is an excellent example to introduce, as well as elaborate, the decision process. Likewise, the reader is free to generate several day-to-day examples such as adding the right amount of sweetener to a hot cup of coffee. We shall map all such phenomena into the following. Let us assume that there is a car on a track that is L meters long. And, we are interested in either *pushing* or *pulling*, obviously with a certain force (measured in Newtons) the car to bring it to the center of the track, say, in minimum time. We are interested in this example, rather than brewing coffee, because we can readily identify the mechanical quantitite—position and velocity—and comfortably understand the dynamics of the corresponding second-order system. All we need to do is to compute the right amount of force to place the car at the center of the track.

With reference to the center, let us define the fuzzy set for position x as shown in Figure 13.27.

Assuming that the center is at the origin of our coordinate system, −2 m to −1 m is certainly to the *Left* of the center, and −1 m to the center is "linearly" losing the property of being left, while the *Middle* is linearly gaining the property of being the middle; the opposite is true on the positive x-axis.

Similarly, we may describe the car's velocity in terms of *moving left*, or *standing still*, or *moving right*, as shown in Figure 13.28.

Finally, the force in Newtons may also be described in terms of *pull*, or *no force*, or *push*, assuming *push* pushes away the car from the origin, and *pull* pulls the car towards the origin. Figure 13.29 describes this.

Next, we may prescribe the following rules, by experience, or equivalently, by intuition.

1. If the car is near the *middle* and is *standing still* then apply *no force*
2. If *left* then *push*
3. If *right* then *pull*
4. If *middle* then *no force*
5. If *moving left* then *push*
6. If *standing still* then *no force*
7. If *moving right* then *pull*
8. If *left* AND *moving left* then *push*
9. If *right* AND *moving right* then *pull*

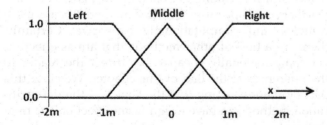

FIGURE 13.27

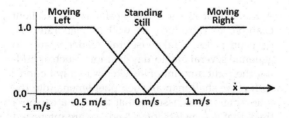

FIGURE 13.28

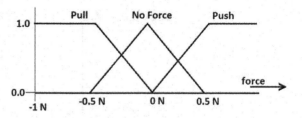

FIGURE 13.29

This set of nine rules is only a suggestion; the imaginative reader may be cleverer than the authors.

For any input (initial) condition defined by the state x_0, $\dot{x}_0$, we can determine, first, the fuzzy outcome for each rule, i.e. the fuzzy sets tht define the action to be taken on each rule. Then we aggregate all the results from each rule, and then, finally, defuzzify the fuzzy set to arrive at a single value.

Let us now work out for the initial condition: $x_0 = -0.5$ m, and $\dot{x}_0 = 0$, i.e. the car is standing still slightly towards the left of the center. Using minimum for AND and Mamdani's rule for implilcation, we proceed as follows.

Given the membership values for $x = -0.5$ in *middle* (with reference to rule 1) and $\dot{x} = 0$ in *standing still*, and applying Mamdani's minimum rule, we get the membership value of the antecedent as $\min\{0.5,1\} = 0.5$. Going rule by rule, we get the values of the antecedents as:

1. Rule 1: 0.5
2. Rule 2: 0.5
3. Rule 3: 0.0
4. Rule 4: 0.5
5. Rule 5: 0.0
6. Rule 6: 1.0
7. Rule 7: 0.0
8. Rule 8: 0.0
9. Rule 9: 0.0

The resulting fuzzy sets using Mamdani's rule are as follows:

Rule	μ_p	Consequent	Result
1	0.5	$\{(-1,0),(0,1),(1,0)\}$	$\{(-1,0),(0,0.5),(1,0)\}$
2	0.5	$\{(-1,0),(0,0),(1,1)\}$	$\{(-1,0),(0,0),(1,0.5)\}$
3	0.0	$\{(-1,1),(0,0),(1,0)\}$	$\{(-1,0),(0,0),(1,0)\}$
4	0.5	$\{(-1,0),(0,1),(1,0)\}$	$\{(-1,0),(0,0.5),(1,0)\}$
5	0.0	$\{(-1,0),(0,0),(1,1)\}$	$\{(-1,0),(0,0),(1,0)\}$
6	1.0	$\{(-1,0),(0,1),(1,0)\}$	$\{(-1,0),(0,1),(1,0)\}$
7	0.0	$\{(-1,1),(0,0),(1,0)\}$	$\{(-1,0),(0,0),(1,0)\}$
8	0.0	$\{(-1,0),(0,0),(1,1)\}$	$\{(-1,0),(0,0),(1,0)\}$
9	0.0	$\{(-1,1),(0,0),(1,0)\}$	$\{(-1,0),(0,0),(1,0)\}$

We now aggregate all of the results from each rule, by simply taking the maximum membership for each state in the consequent, across all of the results of each rule. Thus, we get the *fuzzy decision*:

$$\{(-1,0),(0,1),(1,0.5)\} \tag{13.43}$$

This is the fuzzy set that descries the membership of each possible action of pullling, no force, or pushing the car given the initial condition and the implications of the rules that were framed.

And, finally let us defuzzify using the CoG method to arrive at the real decision:

$$\text{Decision} = \frac{\sum_{i=1}^{3} y_i \times \mu(y_i)}{\sum_{i=1}^{3} \mu(y_i)} = \frac{-1 \times 0 + 0 \times 1 + 1 \times 0.5}{0 + 1 + 0.5} \tag{13.44}$$

$$= \frac{1}{3} \text{Newtons}$$

We understand, and justify intuitively, that a gentle *push* by 0.3333 N to the right, from its initial condition, would bring the car to the center of the track.

The fuzzy inference requires an iterative adjustment of the rules and the membership functions that comprise those rules; arriving at a concrete decision is almost impossible without a few trials and errors. However, we emphasize that we will be able to make a pragmatic, read as computationally simpler, decision using fuzzy rules than we would otherwise using Boolean logic.

13.3.4.1 Some Remarks

In the definitions, we have seen that the designer of a fuzzy system must make a number of important choices. These choices are based mainly on the advice of the expert or statistical analysis of past data, in particular to define the membership functions and the decision matrix. The methodology of fuzzy logic allows to build inference systems in which decisions are without discontinuities, exible and nonlinear, i.e. closer to human behavior than classical logic is. In addition, the rules of the decision matrix are expressed in natural language. This has many advantages, such as include knowledge of a non-expert computer system at the heart of decision making model or finer aspects of natural language.

13.3.5 Type II Fuzzy Systems and Control

Despite having a name that carries the connotation of uncertainty, research has shown that there are limitations in the ability of the fuzzy systems to model and minimize the effect of uncertainties. This is because, at the core, the fuzzy system is certain in the sense that its membership grades are crisp values. To overcome such limitations, the idea that the membership functions themselves can be fuzzy, has been recently attracting interests. For obvious reasons, the nomenclature has moved from type-1 fuzzy systems to **type-2 fuzzy systems**. A further improvement is sought in the **Interval Type-2 fuzzy systems** (IT2FS) and are currently the most widely used for their reduced computational cost.

An example of an IT2FS, $\tilde{X}$, is shown in Figure 13.30.

We observe that, unlike the basic T1FS, whose membership for each x is a number, the membership of an IT2FS is an interval. For example, the membership of number 3 is [0.25,1]. And, an IT2FS is bounded from the above and below by two T1FSs, $\underline{X}$ and $\overline{X}$, which are called **upper MF** (UMF) and **lower MF** (LMF), respectively. The area between $\underline{X}$ and $\overline{X}$ is called as **the footprint of uncertainty** (FoU).

IT2FSs are particularly useful when it is difficult to determine the exact MF, or in modeling the diverse opinions from different individuals. These membership functions can be constructed from surveys, or using optimization algorithms.

In Figure 13.31 we show the schematic diagram of an IT2FLS. It is similar to its T1 counterpart, with the major difference being that at least one of the FSs in the rule base is an IT2FS. Hence, the outputs of the inference engine are IT2FSs, and a type-reducer is needed to convert them into a T1FS before defuzzification can be carried out.

Example 13.3.7:

Consider an IT2FLS that has two inputs, x_1 and x_2, and one output y. Each input domain consists of two IT2FSs, shown as the shaded areas in Figure 13.32.

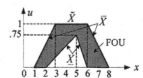

FIGURE 13.30

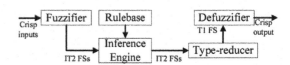

FIGURE 13.31

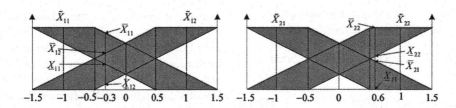

FIGURE 13.32

The rulebase has the following four rules:

R1: If x_1 is $\tilde{X}_{11}$ and x_2 is $\tilde{X}_{21}$, then y is Y_1.
R2: If x_1 is $\tilde{X}_{11}$ and x_2 is $\tilde{X}_{22}$, then y is Y_2.
R3: If x_1 is $\tilde{X}_{12}$ and x_2 is $\tilde{X}_{21}$, then y is Y_3.
R4: If x_1 is $\tilde{X}_{12}$ and x_2 is $\tilde{X}_{22}$, then y is Y_4.

The complete rulebase and the corresponding consequents are given in the following table

	$\tilde{X}_{21}$	$\tilde{X}_{22}$
$\tilde{X}_{11}$	$Y_1 = \left[\underline{y}_1, \overline{y}_1\right]$	$Y_2 = \left[\underline{y}_2, \overline{y}_2\right]$
	$= [-1.0, -0.9]$	$= [-0.6, -0.4]$
$\tilde{X}_{12}$	$Y_3 = \left[\underline{y}_3, \overline{y}_3\right]$	$Y_4 = \left[\underline{y}_4, \overline{y}_4\right]$
	$= [0.4, 0.6]$	$= [0.9, 1.0]$

Typical computations in an IT2FLS involve the following steps, more or less in the same pattern as before:

1. Compute the membership of each of the inputs x_i on the respective sets X_i.
2. Compute the aggregate decision rule.
3. Perform type-reduction; there are many such methods, and the most commonly used one is the center-of-sets, a generalization of CoG:

$$Y_{CoS}(x) = \bigcup_n \frac{\sum_{n=1}^{N} f^n y_n}{\sum_{n=1}^{N} f^n} = \left[y_l, y_r\right] \qquad (13.45)$$

If the reader can imagine a family of polygons for which we get a left CoG and a right CoG, the following formulae, based on the Karnik-Mendel (KM) algorithms make sense:

$$y_l = \min_{k \in [1, N-1]} \frac{\sum_{n=1}^{k} \overline{f}^n \underline{y}_n + \sum_{n=k+1}^{N} \underline{f}^n \underline{y}_n}{\sum_{n=1}^{k} \overline{f}^n + \sum_{n=k+1}^{N} \underline{f}^n} = \frac{\sum_{n=1}^{L} \overline{f}^n \underline{y}_n + \sum_{n=L+1}^{N} \underline{f}^n \underline{y}_n}{\sum_{n=1}^{L} \overline{f}^n + \sum_{n=L+1}^{N} \underline{f}^n}$$

$$y_r = \max_{k \in [1, N-1]} \frac{\sum_{n=1}^{k} \underline{f}^n \overline{y}_n + \sum_{n=k+1}^{N} \overline{f}^n \overline{y}_n}{\sum_{n=1}^{k} \underline{f}^n + \sum_{n=k+1}^{N} \overline{f}^n} = \frac{\sum_{n=1}^{R} \underline{f}^n \overline{y}_n + \sum_{n=R+1}^{N} \overline{f}^n \overline{y}_n}{\sum_{n=1}^{R} \underline{f}^n + \sum_{n=R+1}^{N} \overline{f}^n} \qquad (13.46)$$

Here, L and R are called as the switch points and are determined by the range

$$\underline{y}_L \le y_l \le \underline{y}_{L+1} \qquad (13.47)$$

$$\overline{y}_R \le y_r \le \overline{y}_{R+1} \qquad (13.48)$$

and the sequences $\{\underline{y}_n\}$ and $\{\overline{y}_n\}$ have been sorted in ascending order.
4. Finally, the defuzzified output is

$$y = \frac{y_l + y_r}{2} \qquad (13.49)$$

With the input $\mathbf{x}' = (-0.3, 0.6)$, the decision intervals of the four IT2FSs are

$$\left[\mu_{\underline{X}_{11}}(x_1'), \mu_{\overline{X}_{11}}(x_1')\right] = [0.4, 0.9]$$

$$\left[\mu_{\underline{X}_{12}}(x_1'), \mu_{\overline{X}_{12}}(x_1')\right] = [0.1, 0.6]$$

$$\left[\mu_{\underline{X}_{21}}(x_1'), \mu_{\overline{X}_{21}}(x_1')\right] = [0.0, 0.45]$$

$$\left[\mu_{\underline{X}_{22}}(x_1'), \mu_{\overline{X}_{22}}(x_1')\right] = [0.55, 1.0]$$

The decision intervals of the 4 rules are:

Rule	Decision Interval	Consequent
1	$\left[\underline{f}^1,\overline{f}^1\right]=[0.4\times0,0.9\times0.45]$	$\left[\underline{y}^1,\overline{y}^1\right]$
	$=[0,0.405]$	$=[-1,-0.9]$
2	$\left[\underline{f}^2,\overline{f}^2\right]=[0.4\times0.55,0.9\times1.0]$	$\left[\underline{y}^2,\overline{y}^2\right]$
	$=[0.22,0.9]$	$=[-0.6,-0.4]$
3	$\left[\underline{f}^3,\overline{f}^3\right]=[0.1\times0,0.6\times0.45]$	$\left[\underline{y}^3,\overline{y}^3\right]$
	$=[0,0.27]$	$=[0.4,0.6]$
4	$\left[\underline{f}^4,\overline{f}^4\right]=[0.1\times0.55,0.6\times1.0]$	$\left[\underline{y}^4,\overline{y}^4\right]$
	$=[0.055,0.6]$	$=[0.9,1.0]$

From the KM algorithms, we find that $L=1$ and $R=3$. Hence,

$$y_l = \frac{\overline{f}^1\underline{y}_1+\underline{f}^2\underline{y}_2+\underline{f}^3\underline{y}_3+\underline{f}^4\underline{y}_4}{\overline{f}^1+\underline{f}^2+\underline{f}^3+\underline{f}^4}$$

$$=\frac{0.405\times-1+0.22\times-0.6+0.0\times0.4+0.055\times0.9}{0.405+0.22+0.0+0.055}$$

$$=-0.7169$$

$$y_r = \frac{\underline{f}^1\overline{y}_1+\underline{f}^2\overline{y}_2+\underline{f}^3\overline{y}_3+\overline{f}^4\overline{y}_4}{\underline{f}^1+\underline{f}^2+\underline{f}^3+\overline{f}^4}$$

$$=\frac{0.0\times-0.9+0.22\times-0.4+0\times0.6+0.6\times1.0}{0.0+0.22+0+0.6}$$

$$=0.6244$$

Finally, the crisp output of the IT2FLS is:

$$y=\frac{y_l+y_r}{2}=-0.0463 \tag{13.50}$$

13.3.5.1 MATLAB Implementation

A MATLAB function, *IT2FLS.m*, is available for computing the output of an IT2FLS given its rulebase and inputs. It uses the most efficient algorithm to compute y_l and y_r. Typically, each IT2FS is represented by a 9-point vector $(p_1,\ldots,p_9)$ as shown in Figure 13.33.

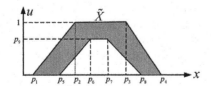

FIGURE 13.33

13.3.6 Summary

A challenging question is: What are the fundamental differences between IT2FLSs and the type 1 systems? Once the fundamental differences are clear, we can better understand the advantages of IT2FLSs and, hence, better make use of them.

In the literature there has been considerable effort on answering this challenging and fundamental question. Some important arguments are as follows:

1. An IT2FS can better model intra-personal (i.e. the uncertainty a person has about the word) and inter-personal (i.e. the uncertainty that a group of people have about the word) uncertainties, which are intrinsic to natural language, because the membership grade of an IT2FS is an interval instead of a crisp number.

2. Using IT2FSs to represent the inputs and outputs will result in the reduction of the rule-base when compared to using the T1FSs, as the ability of the FoU to represent more uncertainties enables one to cover the input/output domains with fewer FSs. This makes it easier to construct the rule-base using expert knowledge and also increases robustness.

3. IT2 reasoning is more adaptive and it can realize more complex input-output relationships, which cannot be achieved by T1FSs. Karnik and Mendel pointed out that an IT2FS can be thought of as a collection of many different embedded T1FSs.

4. IT2FSs have a novelty that does not exist in traditional T1FSs. It has been shown that in an IT2FSs different membership grades from the same IT2FS can be used in different rules, whereas for traditional T1FSs the same membership grade from the same T1FS is always used in different rules. This again implies that an IT2FS is more complex than a T1FS and it cannot be implemented by a T1FS using the same rulebase.

5. There are two more fundamental differences uncovered recently: (i) *Adaptiveness*, meaning that the embedded T1 fuzzy sets used to compute the bounds of the type-reduced interval change as input changes, and (ii) *novelty*, meaning that the upper and lower membership functions of the same IT2 fuzzy set may be used simultaneously in computing each bound of the type-reduced interval. T1FSs do not have these properties.

13.4 Genetic Algorithms and Other Nature Inspired Methods

As the reader has been exposed to the optimal control problems in section II of this book, a piece of good news is that we have a rigorous framework like state space methods, LQR/LQG, and Robust control. Nevertheless, this very framework may be compared to a huge playground wherein a game may be played in virtually an infinite number of ways; a game of chess could be a better illustration—the placement of the chessmen and the rules of the game are deceptively simple, but the possible moves explode combinatorially. Technically speaking, we are bowled by the question: How do we arrive at the best solution. Thus, the optimization problems in real world are often very challenging to solve, and many applications have to deal with what are well-known as **NP-hard** problems, which loosely mean that the solutions are not *always* found in a predictable amount of time, yet there is no guarantee that the optimal solution can be arrived at. In fact, for these class of NP problems, there are no efficient algorithms at all. All we hope is to attempt at a solution by several trials and errors using various techniques. In addition, new algorithms are being continuously developed to see if they can cope with the computational challenges.

In the current literature, there are about 40 different algorithms. It is indeed task to classify these algorithms systematically. However, we give a broad classification, based on the inspiration from Nature, with a motivation to provide the reader a direction to catch up with the literature. One would have already heard of almost all the names in different contexts. Among these new algorithms, many algorithms such as particle swarm optimization, cuckoo search and firefly algorithm, have gained popularity due to the promise they have shown in providing solutions to some of the important problems. In this part, of the chapter we first provide a very brief classification followed by an introduction to the well established genetic algorithms, and particle swarm optimization.

Nature has inspired many researchers in many ways; thus, it is a rich source of inspiration. Apart from the development of an airplane inspired by birds, the nose of the fastest bullet train has been inspired by the beak of Kingfisher bird, which dives into water smoothly minimizing micro-pressure By far, the majority of nature-inspired algorithms are based on some successful characteristics of biological systems. Therefore, the largest fraction of nature-inspired algorithms are biology-inspired, or bio-inspired for short.

Among bio-inspired algorithms, a special class of algorithms have been developed by drawing inspiration from **swarm intelligence** (SI). Therefore, some of the bio-inspired algorithms can be called swarm-intelligence based. SI-based algorithms are among the most popular and widely used. There are many reasons for such popularity, one of the reasons is that SI-based algorithms usually sharing information among multiple agents, so that self-organization, co-evolution and learning during iterations may help to provide the high efficiency of most SI-based algorithms. Another reason is that multiple agent can be parallelized easily so that large-scale optimization becomes more practical from the implementation point of view.

The classical particle swarm optimization (PSO) uses the swarming behavior of fish and birds, while firefly algorithm (FA) uses the flashing behavior of swarming fireflies. Cuckoo search (CS) is based on the brooding parasitism of some cuckoo species, while bat algorithm uses the echolocation of foraging bats. Ant colony optimization uses the interaction of social insects (e.g. ants), while the class of bee algorithms are all based on the foraging behavior of honey bees.

Obviously, SI-based algorithms belong to a wider class of algorithms, called bio-inspired algorithms. In fact, bio-inspired algorithms form a majority of all nature-inspired algorithms. From the set theory point of view,

SI-based algorithms $\subset$ bio-inspired $\subset$ nature inspired

It is necessary to mention here that many bio-inspired algorithms do not use directly the swarming behavior. Therefore, it is better to call them bio-inspired, but not SI-based. For example, genetic algorithms are bio-inspired, but not SI-based.

And, not all nature inspired algorithms are bio-inspired; possibly, one might think of physical sciences as the inspiration for algorithms such as Central Force Optimization, Electro-Magnetism Optimization, Galaxy-based search algorithm, River formation dynamics, and the more popular Simulated Annealing. We do not cover any of these in this book.

13.4.1 Genetic Algorithms

Genetic Algorithms (GAs) represent one branch of the bio-inspired category called **evolutionary computation**, in that they imitate the biological processes of reproduction and natural selection to solve for the *fittest* solutions. Like in evolution, a GA's processes are random, nevertheless we have to set the level of randomization and, hence, the level of control. It is this handle that puts these algorithms higher above than the normal random search strategy, and yet the GAs require no extra information about the given problem. Moreover, it is this feature that allows them to find solutions to problems in spite of a lack of continuity, derivatives, linearity, or other "nice" mathematical features.

191

The GAs are designed to simulate a biological process, therefore, much of the relevant terminology is borrowed from biology. However, the ideas are pretty intuitive, and one need not refer to huge biological texts to understand the terms. The basic components common to almost all genetic algorithms are:

1. A **fitness function** for optimization, such as the sum of the squared errors
2. A **population of chromosomes**
3. Selection of which chromosomes will reproduce
4. Crossover to produce **next generation** of chromosomes
5. **Random mutation** of chromosomes in new generation

The term *chromosome* refers to a numerical value, or values, that represent a candidate solution to the problem that the GA is trying to solve. Each candidate solution is encoded as an array of parameter values. If a problem has N dimensions, then typically each chromosome is encoded as an N-element array:

$$\text{chromosome} = \left[p_1, p_2, \ldots, p_N \right]$$

where each p_i is a particular value of the ith parameter. It is up to the creator of the genetic algorithm to devise how to translate the sample space of candidate solutions into chromosomes. One approach is to convert each parameter value into a bit string (sequence of 1s and 0s), then concatenate the parameters end-to-end like genes in a DNA strand to create the chromosomes.

A GA begins with a randomly chosen assortment of chromosomes, the initial population, which serves as the first generation. Then each chromosome in the population is evaluated by the fitness function to test how well it solves the problem at hand. So far, it is very much like the forward pass in the back propagation algorithm. What makes the difference, is as follows. A selection operator chooses some of the chromosomes for reproduction based on a probability distribution defined by the user. The fitter a chromosome is, the more likely it is to be selected. For example, if f is a non-negative fitness function, then the probability that chromosome C_i is chosen to reproduce might be

$$P(C_i) = \left| \frac{f(C_i)}{\sum_{j \in \text{population}} f(C_j)} \right| \quad (13.51)$$

The selection operator chooses chromosomes with replacement, and hence, the same chromosome can be chosen more than once. The *crossover operator* swaps a subsequence of two of the chosen chromosomes to create two offsprings. For example, if the parent chromosomes

[11010111001000] & [01011101010010]

are crossed-over after the 4th bit, then the offsprings are

[01010111001000] & [11011101010010]

The *mutation operator* randomly flips individual bits in the new chromosomes. Typically mutation happens with a very low probability, say 0.001. Some algorithms implement the mutation operator before the selection and crossover operators; this is a matter of preference. At first glance, the mutation operator may seem unnecessary; in fact, it plays an important role. Selection and crossover maintain the genetic information of fitter chromosomes, but these chromosomes are only fitter relative to the current generation. This can cause the algorithm to converge too quickly and lose "potentially useful genetic material (1s or 0s at particular locations)." This is to, mildly, warn the reader that the algorithm can get stuck at a local optimum before finding the global optimum. The mutation operator helps protect against this problem by maintaining diversity in the population, but it can also make the algorithm converge more slowly.

Typically, the selection, crossover, and mutation process continues until the number of offsprings is the same as the initial population, so that the second generation is composed entirely of new offsprings and the first generation is completely replaced.

Now the second generation is tested by the fitness function, and the cycle repeats. It is a common practice to record the chromosome with the highest fitness (along with its fitness value) from each generation, or the "best-so-far" chromosome. This is, loosely, akin to memoization in dynamic programming. GAs are iterated until the fitness value of the best-so-far chromosome stabilizes and does not change for many generations. This means the algorithm has converged to a solution(s). The whole process of iterations is called a run. At the end of each run there is usually at least one chromosome that is a highly fit solution to the original problem.

Depending on how the algorithm is written, this could be the most fit of all the best-so-far chromosomes, or the most fit of the final generation. The "performance" of a genetic algorithm depends highly on the method used to encode candidate solutions into chromosomes and the *particular criterion for success*, or what the fitness function is actually measuring. Other important details are the probability of crossover, the probability of mutation, the size of the population, and the number of iterations. These values are usually tuned after assessing the algorithm's performance on a few trial runs.

Continuing the analogy of natural selection in biological evolution, the fitness function is like the habitat to which organisms adapt. It is the only step in the algorithm that determines how the chromosomes will change over time, and can mean the difference between finding the optimal solution and finding no solutions at all. Kinnear[4] explains that the fitness function is the only chance that you have to communicate your intentions to the powerful process that genetic programming represents. Moreover, the fitness function must be more sensitive than just detecting what is a "good" chromosome versus a "bad" chromosome: it needs to accurately score the chromosomes based on a range of fitness values, so that a somewhat complete solution can be distinguished from a more complete solution. It is important to consider which partial solutions should be favored over other partial solutions because that will determine the direction in which the whole population moves.

Before we go for an illustrative example, we shall have the algorithm as follows.

1. Generate an initial population of candidate solutions, called individuals.

2. Calculate the fitness of each individual in the current population.

3. Sect some number of the individuals, preferably, with highest fitness to be the parents of the next generation.

4. Pair up the selected parents, with some chance of random mutations, and the offspring enter the new population.

5. Go To Step 2

The above steps are repeated for some number of generations.

Example 13.4.1:

Let us look at this simple example and determine

$$\max_{x} f(x) = -\frac{x^2}{10} + 3x$$

It is elementary that $f(x)$ is maximum when $x = 15$ with $f(15) = 22.5$. We shall illustrate how we genetically evolve to this solution.

We will first assume that x varies between 0 and 31, and we will encode these as binary chromosomes: 00000 to 11111. Then, we will randomly pick up the first generation of population, say 10, among the possible 32; this is easily done by

---

[4] KE Kinnear, "A Perspective on the Work in this Book," In K. E. Kinnear (Ed.), *Advances in Genetic Programming* (pp. 3–17). Cambridge, MA: MIT Press, 1994.

tossing a coin 5 times. With this we will tabulate some preliminary computations as follows:

Chromosome C_i	Initial Population	x	$f(x)$ Fitness	Selection Probability
0	01001	11	20.9	0.1416
1	10001	17	22.1	0.1497
2	11110	30	0	0
3	10110	22	17.6	0.1192
4	01110	14	22.4	0.1518
5	11010	26	10.4	0.0705
6	11000	12	21.6	0.1463
7	00011	3	8.1	0.0549
8	00010	2	5.6	0.0379
9	10001	9	18.9	0.1280

With the initial population,

$$\sum_{j} f(x) = 147.6, \quad \text{and} \quad P(C_i) = \frac{f(x_i)}{\sum_{j} f(x_j)}$$

and these probabilities are put in the last column. One would have noticed that, among the randomly chosen initial population, the function has a maximum of 22.4.

Next, we randomly choose 5 pairs of the initial chromosomes; we do not need to consider all the 10 from the initial population to make 5 pairs. We give the following table, but the reader is encouraged to experiment with any other choice of the 5 pairs to produce the required 10 offsprings.

Mating Pairs	Crossover Point	New Population	x	$f(x)$ Fitness
01100	2nd	01010	10	20
11010	bit	11100	28	5.6
10001	3rd	10001	17	22.1
01001	bit	01001	0	18.9
10110	1st	10110	22	17.6
01110	bit	01110	14	22.4
01110	4th	01111	15	22.5
01001	bit	01000	8	17.6
00011	4th	01010	10	20
11010	bit	11011	27	8.1

And, the reader would have detected that the maximum value, 22.5, has evolved. We once again emphasize that the purpose of this example is only to illustrate the computations; the reader has to work on this simple example and observe the patterns—strings of 0s and 1s—evolving towards the solution.

The GA in the previous example is simple, but versions of the same have been applied in many scientific and engineering areas, as well as fine arts. Just to mention a few of them, the General

Electric (GE) has used the GAs for automating parts of aircraft design, Texas Instruments (TI) for chip design, computer-animated horses for the movie *The Lord of the Rings: The Return of the King*, analysis of credit card data, and financial forecasting.

Example 13.4.2:

We close this part of the chapter with an interesting connection, being pursued for over two decades, between the ANNs and the GAs. In their review, Haruna et al. [16] have reported that the GAs have been extensively used to optimize weights, for topology, to select features, for training and to enhance interpretation. Typically, in a GA, a population of networks is considered. The complete set of weights for each of the networks is encoded as a binary string. Each weight string has a different evolution path. The fitness function is used to evaluated the fitness of each weight string. The weight string giving less classification error is considered as fitter. Successive strings are constructed by performing mutation and crossover on old strings. Since no differentiation is required, neither the fitness function nor the transfer characteristics of the units need be differentiable.

13.4.2 Particle Swarm Optimization

Particle swarm optimization (PSO) is another most influential technique applied to almost every area of optimization. It was developed by Kennedy and Eberhart in 1995 based on swarm behavior in Nature, such as fish and bird schooling. In the past two decades, there are a score of PSO variants investigated extensively.

The PSO uses, instead of the mutation/crossover idea, real-number randomness and global communication among the swarm particles. Therefore, it is also easier to implement because there is no encoding or decoding of the parameters into binary strings as with those in genetic algorithms where real-number strings can also be used. Many new algorithms that are based on swarm intelligence may have drawn inspiration from different sources, but they have some similarity to some of the components that are used in PSO. In this sense, PSO pioneered the basic ideas of swarm-intelligence based computation.

The PSO algorithm searches the space of an objective function by adjusting the trajectories of individual agents, called particles, as the piecewise paths formed by positional vectors in a quasi-stochastic manner. The movement of a swarming particle consists of two major components: a stochastic component and a deterministic component. Each particle is attracted toward the position of the current global best g^* and its own

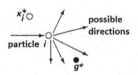

FIGURE 13.34

best location x_i^* in history, while at the same time it has a tendency to move randomly. When a particle finds a location that is better than any previously found locations, updates that location as the new current best for particle i. There is a current best for all n particles at any time t during iterations. The aim is to find the global best among all the current best solutions until the objective no longer improves or after a certain number of iterations.

The movement of particles is schematically represented in Figure 13.34, where $x_i^*(t)$ is the current best for particle i, and

$$g^* \approx \min\{f(x_i)\} \text{ for } i = 1, 2, \ldots, n \qquad (13.52)$$

is the current global best at t.

Let x_i and v_i be the position vector and velocity for particle i, respectively. The new velocity vector is determined by the following formula:

$$
\begin{aligned}
&v_i(t+1) \\
&= v_i(t) + \alpha\theta_1\left(g^* - x_i(t)\right) + \beta\theta_2\left(x_i^*(t) - x_i(t)\right)
\end{aligned}
\qquad (13.53)
$$

where θ_1 and θ_2 are two random vectors, and each entry takes the values between 0 and 1. The parameters α and β are the learning parameters or acceleration constants, which can typically be taken as, say, $\alpha \approx \beta \approx 2$.

The initial locations of all particles should distribute relatively uniformly so that they can sample over most regions, which is especially important for multimodal problems. The initial velocity of a particle can be taken as zero, that is, $v_i(0) = 0$. The new position can then be updated by

$$x_i(t+1) = x_i(t) + v_i(t+1) \qquad (13.54)$$

Although v_i can be any values, it is usually bounded in some range $[0, v_{max}]$.

There are many variants that extend the standard PSO algorithm, and the most noticeable improvement is probably to use an inertia function ϕ so that eqn.13.53 is replaced by

$$
\begin{aligned}
&v_i(t+1) \\
&= \phi v_i(t) + \alpha\theta_1\left(g^* - x_i(t)\right) + \beta\theta_2\left(x_i^*(t) - x_i(t)\right)
\end{aligned}
\qquad (13.55)
$$

where ϕ takes the values between 0 and 1, typically 0.5. This is equivalent to introducing a virtual mass to stabilize the motion of the particles; thus, the algorithm is expected to converge more quickly.

The essential steps of the particle swarm optimization can be summarized as the following pseudo code:

Objective function $f(x)$, $x = [x_1, \ldots, x_d]^T$

Initialize locations x_i and velocity v_i of the n particles

Find g^* from $\min\{f(x_1), \ldots, f(x_n)\}$ at $t = 0$

while [criterion]

    for loop over all n particles and all d dimensions

    Generate new velocity $v_i(t+1)$ using eqn. 13.53

    Calculate new locations $x_i(t+1) = x_i(t) + v_i(t+1)$

    Evaluate objective functions at new locations $x_i(t+1)$

    Find the current best for each particle x_i

    end for

Find the current global best g

Update $t = t + 1$

end while

Output the final results x_i and g^*

13.4.2.1 Accelerated PSO

A simplified version that could accelerate the convergence of the algorithm is to use only the global best. The so-called accelerated particle swarm optimization (APSO) was developed by Xin-She Yang in 2008. In the APSO, the velocity vector is generated by a simpler formula

$$v_i(t+1) = v_i(t) + \alpha\left(\theta - \frac{1}{2}\right) + \beta\left(g^* - x_i(t)\right) \quad (13.56)$$

where θ is a random variable with values from 0 to 1, and the shift $1/2$ is purely out of convenience. We can also use a standard normal distribution $\alpha\theta(t)$, where $\theta(t)$ is drawn from $N(0,1)$ to replace the second term, so that

$$v_i(t+1) = v_i(t) + \beta\left(g^* - x_i(t)\right) + \alpha\theta(t) \quad (13.57)$$

where $\theta(t)$ can be drawn from a Gaussian distribution or any other suitable distributions. The update of the position is simply

$$x_i(t+1) = x_i(t) + v_i(t+1)$$

$$= (1-\beta)x_i(t) + g^* + \alpha\theta(t) \quad (13.58)$$

Typical values for the accelerated PSO are α in the range 0.1 to 0.4 and β in the range 0.1–0.7. It is worth pointing out that the parameters α and β should, in general, be related to the scales of the independent variables x_i and the search domain.

Interestingly, this simplified APSO has been shown to have global convergence. A further improvement to the accelerated PSO is to reduce the randomness as iterations proceed. This means that we can use a monotonically decreasing function such as

$$\alpha = \alpha_0 e^{-\gamma t} \quad (13.59)$$

or

$$\alpha = \alpha_0 \gamma^t, \ 0 < \gamma < 1 \quad (13.60)$$

where α_0 is chosen in the range 0.5–1.0. Here t is the number of iterations or time steps. $0 < \gamma < 1$ is a control parameter. Obviously, other non-increasing function forms $\alpha(t)$ can also be used; in addition, these parameters should be fine-tuned to suit your optimization problems of interest.

13.4.3 Summary

PSO shares many similarities with GAs. Both the techniques begin with a group of a randomly generated population and utilize a fitness value to evaluate the population. They update the population and search for the optimum with random techniques. However, the main difference between the PSO approach compared to GAs is that PSO does not have genetic operators such as crossover and mutation. Particles update themselves with the internal velocity and they have a memory that is important to the algorithm. Compared with GAs, the information sharing mechanism in PSO is significantly different. In PSO, only the best particle gives out the information to others. It is a one-way information sharing mechanism, and the evolution only looks for the best solution. Compared to GAs, all the particles tend to converge to the best solution quickly even in the local version in most cases; moreover, PSO is easy to implement and there are few parameters to adjust.

In terms of computational effort, in many cases such as process modeling, the GA approach is faster, although it should be noted that neither algorithm takes what can be considered an unacceptably long time to determine their results. With respect to that accuracy of model parameters, the GA determines values that are closer to the known values than does the PSO. Finally, the GA seems to arrive at its final parameter values in fewer generations than the PSO.

13.5 Chapter Summary

When we say *traditional* optimization methods, we are referring to three main types: calculus-based, exhaustive search, and random. Calculus-based optimization methods come in two categories: direct and indirect. The direct method *jumps onto* the objective function and follows the direction of the gradient towards a local maximum or minimum value. This is also known as the gradient descent (or, if you prefer, gradient ascent) method. The indirect method takes the gradient of the objective function, sets it equal to zero, then solves the set of equations that results. Although these calculus-based techniques have been studied extensively and improved, they still have insurmountable problems. First, they only search for local optima, which renders them useless if we do not know the neighborhood of the global optimum or if there are other local optima nearby (and we usually do not know). Second, these methods require the existence of derivatives, and this is virtually never the case in practical applications. However pessimistic this paragraph would have read, the ANNs, largely belonging to this class, have a way of their own in solving several practical problems.

Exhaustive search algorithms perform, as the name itself suggests, an exhaustive search. Evidently, these algorithms require a finite search space or a discretized infinite search space of possible values for the objective function. Then, they test *every single value* in the space, one at a time, to find the maximum or minimum. While this method is intuitive, it is the least efficient of all optimization algorithms. In practical problems, the search spaces are too vast to test every possibility one at a time would demand a virtually infinite amount of time. Again, with certain linear-like interpolations (particularly with the case of triangular membership functions),

fuzzy logic attempts to solve problems, virtually searching exhaustively.

Random search algorithms became increasingly popular as people realized the shortcomings of calculus-based and exhaustive search algorithms. This style of algorithm randomly chooses some representative sampling from the search space and finds the optimal value in that sampling. While faster than an exhaustive search, this method can be expected to do no better than an exhaustive search. Using this type of algorithm means that we leave it up to *chance* whether we will be somewhat near the optimal solution or miles away from it.

Genetic algorithms have many advantages over these traditional methods. Unlike calculus-based methods, genetic algorithms progress from a population of candidate solutions instead of a single value. This greatly reduces the likelihood of finding a local optimum instead of the global optimum. Genetic algorithms do not require extra information, say the derivatives, that is unrelated to the values of the possible solutions themselves. The only mechanism that guides their search is the numerical fitness value of the candidate solutions, based on the creator's definition of fitness. This allows them to function when the search space is noisy, nonlinear, and derivatives do not even exist. This also makes them widely applicable.

A lot more can be said about these techniques but they are not a panacea. There are control parameters involved in these meta-heuristics, and appropriately setting these parameters is a key point for success. In general, the possibility of utilizing hybrid approaches should be considered. Additionally, for both approaches, the major issue in implementation lies in the selection of an appropriate objective function.

We present several case studies that make use of ANNs, Fuzzy Logic, GAs, and GAs in combination with ANNs and Fuzzy Logic, and PSO in the next three chapters.

14

ANN-Based Control Systems

The use of ANNs in control systems was first proposed in 1989 by Werbos followed by Narendra in 1990. ANN-based control basically has two quite strong implications: approximate dynamic programming, which uses an ANN to approximately solve a given optimal control problem, and then another ANN for closed loop feedback control. Ever since, many researchers have contributed to the development of these fields. For more on this, see the Historical Development section at the end of this chapter.

Borrowing from the seminal work of Narendra and Parthasarathy, here we present several ANN-based feedback control topologies.

Solid lines denote control signal flow loops while dashed lines denote tuning loops. There are basically two sorts of feedback control topologies—indirect and direct control, as shown in Figures 14.1 and 14.2, respectively.

However, the direct control is more efficient and involves directly tuning the parameters of an adjustable ANN controller. There is a topology that has two neural network controllers, but as one would expect, it turns out to be computationally expensive.

The challenge in using an ANN for feedback control purposes is to select a suitable control system structure and then to demonstrate using mathematically acceptable techniques how the weights of the ANN can be tuned to ensure the desired closed loop behavior. In this chapter, we will briefly present some of the success stories reported in the literature [7–23].

14.1 Applications of Radial Basis Function Neural Networks

In the standard Radial Basis Function (RBF) network, the estimations are expressed as a linear combination of M Gaussian Basis functions:

$$y_s(x) = We^{\frac{(x-\mu)^T (x-\mu)}{2\sigma^2}} \tag{14.1}$$

where $x \in \Re^n$ is the input vector, the parameters μ and σ are the basis center and width, respectively. In the standard implementation, the hidden layer neurons are a priori statically allocated on a uniform grid that covers the whole input space and only the weights w_{ij} are updated. This approach requires an exponentially increasing number of basis functions versus the dimension of the input space.

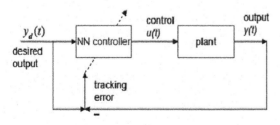

FIGURE 14.1

FIGURE 14.2

14.1.1 Fully Tuned Extended Minimal Resource Allocation Network RBF

In order to avoid the dimensionality problems generated by standard RBF, Platt proposed a sequential learning technique for RBFNs (Platt, 1991). The resulting architecture was called the resource allocating network (RAN) and has proven to be suitable for online modeling of non-stationary processes with only an incremental growth in model complexity. The RAN learning algorithm proceeds as follows: At each sampling instant, some units are added if the following three criteria are met:

1. Current estimation error criteria, where the error must be greater than a pre-defined threshold:

$$e(t) = y(t) - \hat{y}(t) \geq E_1$$

2. Novelty criteria, where the nearest center distance must be greater than a threshold:

$$\inf_j \| x(t) - \mu_j(t) \| \geq E_2$$

3. Windowed mean error criteria, where windowed mean error must be greater than a threshold:

$$\frac{1}{T} \sum_{i=0}^{T} [y(t - T + i) - \hat{y}(t - T + i)] \geq E_3$$

If one or more of the above criteria are not met, the existing network parameters (the centers, the weights, and the variances) are adjusted using a suitable online learning algorithm, such as the extended Kalman filter (EKF). To avoid an excessive increase of the Network size, a pruning strategy can also be applied. When this happens, the network is called Minimal RAN (MRAN) (Lu et al. 1997). The sequential learning algorithm for MRAN can be summarized as follows.

1. Obtain an input and calculate the network output and the corresponding errors, elaborated above.
2. Create a new RBF center if all the three inequalities hold.
3. If the inequality on the nearest center distance is not met, adjust the weights and widths of the existing RBF network using EKF.
4. In addition, a pruning strategy is adopted:
   a. If a center's normalized contribution to the output for a certain number of consecutive inputs is found to be below a threshold value, that center is pruned.
   b. The dimensions of the corresponding matrix are adjusted and the next input is evaluated.

In the MRAN algorithm, the parameters of the network (including all the hidden neurons' centers, widths, and weights) have to be updated in every step. This causes the size of the matrices to become large as the hidden neurons increase and the RBF network structure becomes more complex computationally, which directly results in a large computation load and limits the use of MRAN for real-time implementations.

Based on this analysis, an extension to MRAN called Extended MRAN (EMRAN) has been proposed. The focus is to reduce the computation load of the MRAN and to realize a scheme for fast online identification. For this purpose, a winner neuron strategy is incorporated into the conventional MRAN algorithm. The key idea in the EMRAN algorithm is that in every step, only those parameters that are related to the selected winner neuron are updated by EKF. EMRAN attempts to reduce the online computation time considerably and avoids overflow of the memory, retaining at the same time the positive characteristics of MRAN, namely a smaller number of hidden neurons and lower approximation error. This strategy implies a significant reduction of the number of parameters being updated online with just a small performance degradation compared to MRAN (Lu et al. 1997).

A number of successful applications of MRAN in different areas, such as pattern classification, function approximation, and time series prediction, have been reported. Now we present an application in the aerospace[1] area, where a former colleague of the first author of this book and a former doctoral student of the second author of this book has worked extensively.

Example 14.1: Autolanding of Fixed Wing Aircraft

Fault tolerant control has generated a great deal of interest due to its potential applications for both manned and unmanned aircraft. Landing is the most critical flight phase for any air vehicle. In the case of commercial aircraft, most accidents have occurred during the approach and landing phases, although this accounts for only a small portion of the total flight. Further, almost half of these accidents were due to severe wind turbulences during landing and the other half due to stuck actuator failures. The probability of the actuator fault during the landing phase is higher compared to that during the cruise phase because the control surfaces of the aircraft are commanded more rapidly and frequently during the landing phase. The failure of the control surfaces during the landing phase causes more fatal accidents than during the cruise phase because there is limited time and space for recovery due to the proximity of the aircraft to the ground. Recourse to autolanding ensures cost reduction, repeatability, and safety. The autolanding controller should be robust enough for various disturbances, such as wind turbulence, unpredictable gusts near the ground, and control surface failures.

---

[1] Shaik Ismail, *Fault-Tolerant Autolanding Controllers Using Diagonally Dominant Backstepping and Neural-Sliding Mode Augmentation*, Ph.D. Thesis, National Institute of Technology, Tiruchirappalli, India, 2014.

At present, a majority of the autolanding controllers are classical PID controllers, which perform well under normal landing conditions but fail under external disturbances (like wind shear and turbulence) and aircraft component and actuator failure conditions. Hence, fault-tolerant control systems (FTCS) for landing are being extensively investigated. Neural networks, with their ability to approximate nonlinear functions and capability for online learning, provide a fast mechanism for adapting the aircraft control systems to unknown actuator failures, structural damage, and wind disturbances. Hence, these are increasingly being used for fault tolerant flight control.

Using a six degrees of freedom (6 DOF) simulation model of a typical modern high performance aircraft with independent left and right elevator and aileron control surfaces, several neural-aided and neuro-fuzzy controllers have been reported in open literature for autolanding of the aircraft under unknown actuator failures and external wind disturbances. The conventional PID, LQR, H_∞, and H_2 controllers were used as baseline controllers for training the neural networks online. A glaring deficiency of these controllers is the presence of gaps in the fault tolerance range or envelope of the controllers. In the current application, the baseline control structure is the nonlinear dynamic inversion (NDI)-based control, also designed by the same authors.

14.1.2 Autolanding Problem Formulation

A high performance fighter aircraft model chosen for the present study has conventional control surfaces, but with independent left and right elevator and aileron controls. The CFD method was used to generate additional aerodynamic data for the split elevator and aileron control surfaces.[2] The two elevators can be moved together or in differential mode (−25 to +25 degrees). The deflection range for the independent ailerons is −20 to +20 degrees, and for the rudder it is −30 to +30 degrees.

The hydraulic actuators are modeled as first order lags with a time constant of 50 ms and a rate limit of 60 deg/sec. In the present study six types of actuator failures have been considered:

1. Failure of left elevator alone,
2. Failure of either left or right aileron alone,
3. Combined failure of left elevator and left aileron,

4. Combined failure of left elevator and right aileron,
5. Combined failure of both the ailerons,
6. Failure of rudder alone.

Failure of both the elevators is not considered because this case is, in general, not recoverable.

Failure of actuators can occur at any time during the flight, but the effect is acute at the time of landing. In the present study, failures were injected just before the two critical stages of the landing flight: level turn and descent. Further, the failed control surfaces can be stuck at any value within the permissible range of deflections.

The autolanding trajectory chosen for this study is shown in Figure 14.3.

The trajectory consists of 8 segments: wings-level flight at 600m altitude, two coordinated level turns, glide slope descent, and finally the flare maneuver and touchdown on the runway. Two level turns are included in the trajectory so that the first turn segment can be used for online training of neural networks and actuator failures are injected just before the second turn.

The wind disturbances are assumed to be present along all the axes throughout the landing mission and are modeled on Dryden spectrum, as shown in Figure 14.4.

The desired touchdown point of aircraft under normal operating conditions is:

$$x = 0, \ y = 0 \ \text{and} \ z = 0 \text{ m}$$

Since the ideal touchdown cannot be achieved under unknown actuator failures, some safety and performance criteria need to be satisfied as specified below:

- x-distance and y-distance:

$$-100 \text{ m} \le x \le 400 \text{ m}, \ -5 \text{ m} \le y \le 5 \text{ m}$$

- to restrict the landing area to a rectangle of 500×10 m, also called a "pill-box,"
- Total velocity: $V_T \ge 60$ m/sec to prevent stall,

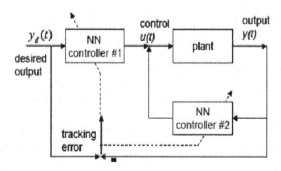

FIGURE 14.3

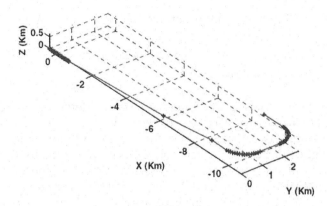

FIGURE 14.4

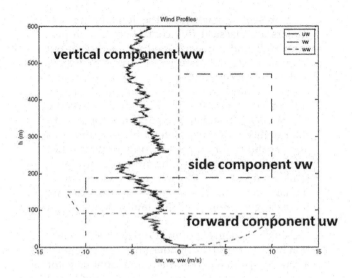

FIGURE 14.5

- Sink rate: $\dot{h} \geq -2$ m/sec to prevent landing gear damage,
- Bank angle: $\phi \leq 10$ degrees to prevent wing tips touching the ground,
- Heading angle error: $\psi \leq 15$ degrees to prevent excessive side loads on landing gear.

It is obvious that all the possible actuator stuck positions can not be accommodated by any controller because in some cases the resulting moments cannot be trimmed out for the landing maneuver. That is, a steady level turn or wings level descent may not be possible. Thus, the full range of hard over positions must be checked for the feasible subset. The feasible range is computed by trimming the aircraft model with the control surfaces in failed positions. If trim is achieved then that particular failed position belongs to the feasible region. In the present work, the feasible region for autolanding is the union of the following trim computations:

- Region of level flight trim: body axis rates and flight path angle $p = q = r = \gamma = 0$, 6 DOF accelerations = 0,
- Region of level descent trim: body axis rates $p = q = r = 0$, flight path angle $\gamma = -6$ degrees, 6 DOF accelerations = 0,
- Region of level turning trim: bank angle $\phi = 40$ degrees, 6 DOF accelerations = 0.

A generic architecture for autolanding controllers is shown in Figure 14.5. The controller comprises two parts—a tracking command generator that generates the commands based on trajectory deviations and a feedback controller (FC) that accepts the command signals as its inputs. The FC is designed to make the closed-loop system stable along the desired trajectory and to meet the performance requirements with or without actuator failures.

The tracking command generator determines the offset of the aircraft from the desired ground track for each segment of flight, which can be approximated by straight lines or arcs of circles. The reference commands consist of Altitude (h_{ref}), Velocity (V_{ref}), Cross Distance from the desired track vector (δ_{ref}), and the angular error of the aircraft velocity vector from the desired track vector (ψ_{ref}).

Nonlinear dynamic inversion with full state feedback is used to develop the baseline architecture called a full state feedback controller (FSFC). The architecture of the baseline FSFC is shown in Figures 14.6 and 14.7. High gains are used in the inner loops: $K_q = -105$ deg/rad/sec in the pitch rate loop, $K_{ps} = -25$ deg/rad/sec in the stability axis roll rate loop, and $K_{rs} = -180$ deg/rad/sec in the stability axis yaw rate loop to increase the robustness of the controller to actuator failures. The outer loop bandwidth is kept low to ensure that high frequency commands are not generated in response to disturbances due to severe winds.

The design begins by considering the three axes for feedback control. The roll rate and yaw rate have been transformed into stability axes for the purpose of developing the control laws based on dynamic inversion. The details of controller design are given in [24]. Successful decoupling between the longitudinal and lateral-directional axes is achieved by the use of stability axis roll, pitch, and yaw rates. The decoupling is further improved if the angle of attack used in the body to stability axis transformation is the 1-g trim value. Multiple surface redundancy is used effectively to enhance the ability of the controller to handle failures like both ailerons failing. The robustness of the baseline FSFC controller to aileron and rudder failures is enhanced by using the independent elevator control

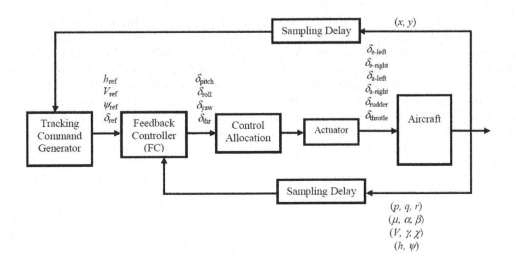

FIGURE 14.6

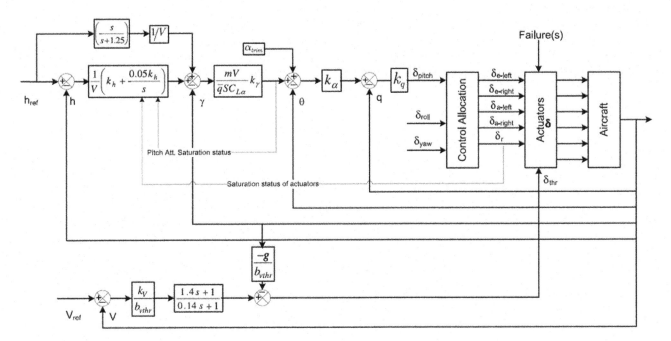

FIGURE 14.7

surfaces in differential mode for the control of roll and yaw responses of the aircraft, as shown below:

$$
\begin{bmatrix} \delta_{e-left} \\ \delta_{e-right} \\ \delta_{a-left} \\ \delta_{a-right} \\ \delta_{rudder} \end{bmatrix} = \begin{bmatrix} 1.0 & -0.75 & -0.27 \\ 1.0 & 0.76 & 0.27 \\ 0.0 & -1.0 & 0.0 \\ 0.0 & 1.0 & 0.0 \\ 0.0 & 1.66 & 1.0 \end{bmatrix} \begin{bmatrix} \delta_{pitch} \\ \delta_{roll} \\ \delta_{yaw} \end{bmatrix}
$$

The gain entry of 1.66 is the aileron to rudder interconnect. The entries ±0.75 represent the use of elevators in differential mode to achieve control in the roll axis. This gain allows us to tolerate the additional failure case where both ailerons have failed. Similarly, the gain entries ±0.27 are intended to create differential elevator control in response to demand for control in the yaw axis and enhance the ability of the FSFC to handle rudder failures.

A few comments are in order here. Firstly, the structure of this controller is based on nonlinear dynamic inversion, where a few key stability and control derivatives are used, and other derivatives do not play a significant role. The cascade structure of the controller is designed with well separated time constants in the inner and outer loops.

The principal aim of the autolanding controller is to reject winds and follow the predefined inertial trajectory for autolanding. Therefore, it seems appropriate to replace whenever possible the air-data feedback of angle of attack and sideslip with physically similar inertial signals. In the case of angle of attack, we have considered the pitch attitude for feedback. However, in case of the sideslip, placement of a low pass filter with a first order time constant of 30 ms in the feedback loop seemed to be the most appropriate solution.

Next, the failure cases for the FSFC were examined and it was found that the performance degrades due to the saturation of the rudder actuator due to the aileron to rudder gain. Therefore, the aileron to rudder interconnect gain was reduced to a value of 1.2 from its desired value of 1.66. In fact, when a detailed study was conducted to identify the reasons for the "gaps" in the fault-tolerance envelope, it was found that these are mainly due to position/rate saturation of the actuators. Position saturation results in open loop response due to loss of regulatory action. On the other hand, rate saturation of the actuators causes a significant reduction in the phase bandwidth of an actuator when the amplitude of oscillation is increased. This is illustrated in Figure 14.8. It can be seen that at about 10% of the actuator amplitude, the frequency response of the actuator is close to the desired transfer function (20 rad/sec bandwidth), whereas at about 50% of the amplitude of oscillation, the effective bandwidth is about half of the desired value due to rate limiting in the actuator.

The inputs to the neural controller consist of desired attitude rates (p_d, q_d, r_d), the stability axis angular rates (p_s, q_d, r_s), the angle of attack and side-slip angle (α, β), and control surface deflections (δ_e, δ_a, δ_r). The desired rates are taken from the FSFC. The output is comprised of neural controller outputs for the five aerodynamic control surfaces that tend to drive the differences between the desired and actual control surface signals to zero.

The first step in the evaluation of fault-tolerance capabilities of any autolanding controller, using 6 DOF simulations, is the examination of its feasibility regions for various types of actuator failures.

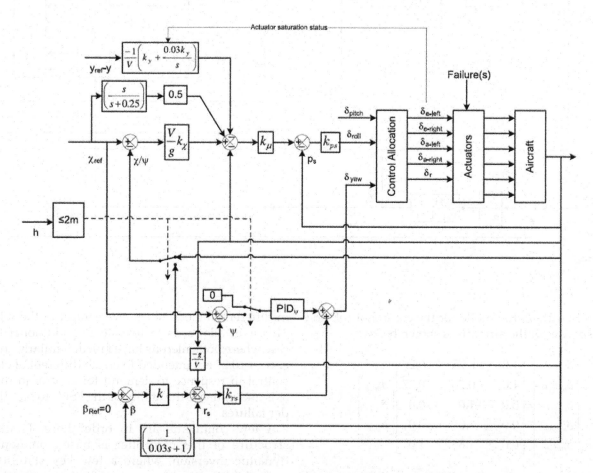

FIGURE 14.8

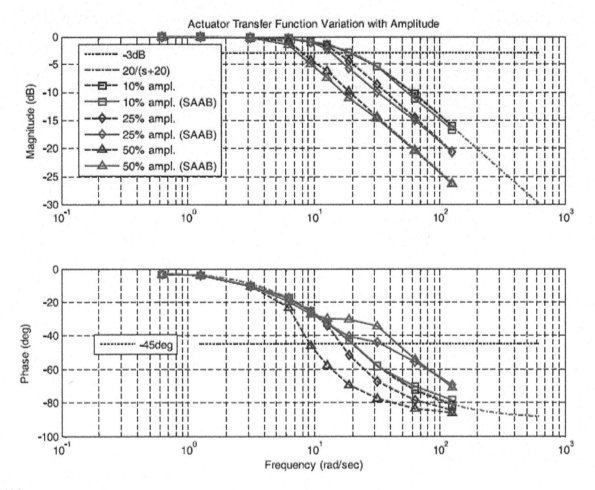

FIGURE 14.9

The case of simultaneous failure of three actuators is also studied. It is assumed that the left elevator, the right aileron, and the rudder are stuck at 2 degrees, −2 degrees, and 4 degrees, respectively, during level turn. The longitudinal and lateral-directional responses are shown in Figure 14.9.

The sideslip is about 3 degrees subsequent to failure and remains at this value during touchdown, as well. There are multiple solutions for landing in cross-winds, particularly for aircraft such as ours, which has a stable dutch-roll mode. At one extreme, we have the zero sideslip solution with non-zero bank angle (i.e. crabbed landing), while at the other end we can also have a non-zero sideslip with zero bank angle. In this particular case, with three simultaneous control surfaces stuck, the controller is able to find a solution that is in between these extremes. Similarly, the velocity shows significant variations due to the severe vertical winds (representing a microburst) particularly just before touchdown. That the controller is able to handle this case attests to its robustness with respect to unknown failures and severe winds.

Some of the major findings of this research work on combining nonlinear dynamic inversion with radial basis neural networks are as follows:

1. The gains in the innermost rotational equation loop should be chosen as high as possible to handle hardover control surface failures while at the same time low enough to prevent nonlinear actuator limits getting excited (position and rate limiting). Outer loop trajectory following gains are a compromise between a low value (to ensure that the flight controller does not create large inner loop demands and saturate the actuators when it encounters severe winds) and high value (to ensure closed loop performance in terms of accurate trajectory following is achieved).

2. The controller is able to handle severe winds by replacing the air data sensors (e.g. angle of attack) by the feedback of an equivalent inertial signal (e.g. pitch attitude). Similarly, sharp changes in sideslip due to severe winds are prevented from

affecting controller performance by introducing a low pass filter in the feedback path.

3. Multiple redundancies in control surfaces are handled by design of suitable interconnect gains. Wherever actuator rate limiting is a concern, exact control decoupling can be sacrificed to achieve better fault tolerance envelope.

4. Actuator failure tolerance range maps, also called feasibility maps, show that the proposed controller is more robust than the other controllers reported in open literature for the same autolanding problem. The proposed controller can handle more types of actuator failures than the previous controllers.

Example 14.2: ANN-based Backstepping Control

In this example, we show that backstepping can be applied in conjunction with the ANNs. Two NN will be required in this controller. This is the work of C. Kwan (adopted from [25]), who has also applied backstepping to NN control of flexible joint robot arms, induction motors, and a large class of nonlinear systems.

The dynamics of an n-link rigid robot arm with no actuator shaft compliance and non-negligible motor electrical dynamics are given by

$$M(q)\ddot{q} + V_m(q,\dot{q})\dot{q} + F(\dot{q}) + G(q) + \tau_d = K_T i \qquad (14.2)$$

$$L\frac{di}{dt} + R(i,\dot{q}) + \tau_e = u_e \qquad (14.3)$$

with $q(t) \in \Re^n$ the joint variable, $i(t) \in \Re^n$ the motor armature currents, K_T a diagonal electromechanical conversion matrix, L a matrix of electrical inductances, $R(i,\dot{q})$ representing both electrical resistance and back emf, $\tau_d(t)$ and $\tau_e(t)$ the mechanical and electrical disturbances, and motor terminal voltage vector $u_e(t) \in \Re^n$ the control input voltage.

We next define the tracking error

$$e = q_d - q \qquad (14.4)$$

with $q_d(t)$ the desired robot arm trajectory, and the filtered tracking error

$$r = \dot{e} + \Lambda e \qquad (14.5)$$

with $\Lambda > 0$ a diagonal design matrix. Differentiating $r(t)$ and using the dynamics we find

$$M(q)\dot{r} = -V_m r + F_1(X_1) + \tau_d - K_T i \qquad (14.6)$$

where the unknown nonlinear robot function is

$$F_1(X_1) = M(q)(\ddot{q}_d + \Lambda\dot{e}) + V_m(q,\dot{q})(\dot{q}_d + \Lambda e) + F(\dot{q}) + G(q) \qquad (14.7)$$

and, for instance, one may select

$$X_1 = \begin{bmatrix} e \\ \dot{e} \\ q_d \\ \dot{q}_d \\ \ddot{q}_d \end{bmatrix}$$

Let us assume the availability of $i_d(t)$, a value of the current $i(t)$ that stabilizes the dynamics, and let the corresponding error term be

$$v = i_d - i$$

so that the dynamics are rewritten as follows:

$$M(q)\dot{r} = -V_m r + F_1 + \tau_d - K_T(i_d - v) \qquad (14.8)$$

The dynamics of the electrical sub-system would be

$$L\frac{dv}{dt} = \underbrace{L\frac{d}{dt}(i_d) + R(i,\dot{q}) + \tau_e}_{\overset{\Delta}{=} F_2(X_2)} - u_e \qquad (14.9)$$

where F_2 is the unknown nonlinear motor function.

We must notice that, since $i_d(t)$ will turn out to be a complex function of F_1 and $r(t)$, the function F_2 is complicated. Further, we need the derivative of $i_d(t)$. For X_2, one may select

$$X_2 = \begin{bmatrix} i \\ e \\ \dot{e} \\ q_d \\ \dot{q}_d \\ \ddot{q}_d \end{bmatrix}$$

We now need to explore how to select $i_d(t)$ and $u_e(t)$ so that the error dynamics are stable, yielding tracking of the desired robot arm trajectory $q_d(t)$. The matrix K_T is unknown, hence, we assume the following scalar bounds

$$K_{B_1} < \| K_T \| < K_{B_2} \qquad (14.10)$$

The interesting part is to assume that there exists *two* networks that can approximate the unknown nonlinear functions F_1 and F_2 on a compact set. Thus,

$$F_1(X_1) = W_1^T \varphi_1(X_1) + \varepsilon_1 \qquad (14.11)$$

$$F_2(X_2) = W_2^T \varphi_2(X_2) + \varepsilon_2 \qquad (14.12)$$

with W_i the tunable ANN weights, $\varphi_i(\cdot)$ the activation functions, and ε_i the ANN functional reconstruction errors that are assumed bounded. The neural networks could be, for simplicity, using radial basis functions.

The estimates for F_1 and F_2 are given by

$$\hat{F}_1(X_1) = \hat{W}_1^T \varphi_1(X_1) \qquad (14.13)$$

$$\hat{F}_2(X_2) = \hat{W}_2^T \varphi_2(X_1) \qquad (14.14)$$

with $\hat{W}_i$ the current values of the ANN weights as provided by the tuning algorithms.

We will next *select* the desirable value of armature current as

$$i_d = \frac{1}{K_{B_1}} \left(\hat{W}_1^T \varphi_1(X_1) + K_r r + v_i \right) \qquad (14.15)$$

with $K_r > 0$ a gain matrix and v_i a robustifing term to be defined.

Likewise, we will select the control input as

$$u_e = \hat{W}_2^T \varphi_2(X_2) + K_v v \qquad (14.16)$$

with $K_v > 0$ a gain matrix.

This is clearly the feedback linearization motivated design. We summarize the ANN-based backstepping control that provides stable tracking performance as follows:

Control Signal	$u_e = \hat{W}_2^T \varphi_2(X_2) + K_v v$
Auxiliary Signal	$i_d = \dfrac{1}{K_{B_1}} \left(\hat{W}_1^T \varphi_1(X_1) + K_r r + v_i \right)$
ANN Weight Updating	$\dot{\hat{W}}_1 = \Gamma_1 \varphi_1 r^T - c_1 \Gamma_1 \lVert \zeta \rVert \hat{W}_1$
	$\dot{\hat{W}}_2 = \Gamma_2 \varphi_2 r^T - c_2 \Gamma_2 \lVert \zeta \rVert \hat{W}_2$
	where $\zeta = [r^T \quad v^T]^T$
Design Parameters	Γ_i positive definite matrices
	$c_i > 0$ small constants
Tracking Error	$e = q_d - q$
	$r = \dot{e} + \Lambda e, \Lambda > 0$
	$v = i_d - i$
Robustifying Signal	$v_i(t) = \dfrac{r\rho^2}{\lVert r \rVert \rho + \varepsilon_v}$
	where $\rho = \lVert \hat{W}_1^T \varphi_1 \rVert, \varepsilon_v > 0$ and small

The ANN backstepping controller is shown in Figure 14.10. The stability proof hinges on using a suitable Lyapunov function that weights both the tracking errors and the ANN weight estimation errors. It can be shown that the tracking error $r(t)$—and $v(t)$—can be made arbitrarily small by increasing the gains K_r and K_v.

The ANN backstepping controller provides advantages, including fast online weight tuning, no persistence

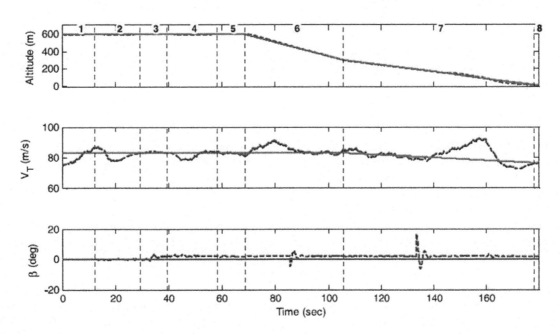

FIGURE 14.10

of excitation, and no need to compute any regression matrices. In contrast, the nonlinear function F_2 depends on the derivative i_d, which is a function of $\hat{F}_1$. In backstepping using standard adaptive control techniques, one must determine a regression matrix for F_1 and then differentiate it to find a regression matrix for F_2. All this is unnecessary using the ANN approach.

Initialization of the ANN weights at zero amounts to setting $\hat{F}_1 = 0 = \hat{F}_2$ and looking back at the corresponding equations for u_e and i_d—one would immediately observe a PD tracking controller plus extra terms in the robust term $v_i(t)$ and the armature current $i(t)$. This holds the system in stable tracking until the ANNs begin to learn, improving the closed loop performance.

Lastly, care must be exercised so that the initial tracking errors are in a certain set of allowable initial conditions This initial condition set depends on the speed of the desired trajectory and the size of the compact sets over which the approximation properties hold. And, the approximation accuracy may be increased with increasing the number of hidden layer neurons in the ANN, essentially moving towards deep neural networks.

14.2 Optimal Control Using Artificial Neural Network

We have, till now, discussed the design of neural network controllers for tracking and stabilization based on control theory techniques including dynamic inversion, feedback linearization, and backstepping. Rigorous neural controller design algorithms may be given in terms of Lyapunov energy-based techniques, passivity, and so on.

Nonlinear optimal control design provides a very powerful theory that is applicable for systems in any form. Solving the so-called Hamilton-Jacobi (HJ) equations will directly yield a controller with guaranteed properties in terms of stability and performance for any sort of nonlinear system. Unfortunately, the HJ equations are difficult to solve and may not even have analytic solutions for general nonlinear systems. In the special case of *linear* optimal control, solution techniques are available based on, for instance, Riccati equation techniques, and that theory provides a cornerstone of control design for aerospace systems, vehicles, and industrial plants. It would be valuable to have tractable controller design techniques for general nonlinear systems. Indeed, it has been shown that neural networks afford computationally effective techniques for solving general HJ equations.

14.2.1 Neural Network LQR Control Using the Hamilton-Jacobi-Bellman Equation

In work by Abu-Khalaf and Lewis (2004), it has been shown how to solve the Hamilton-Jacobi-Bellman (HJB) that appears in optimal control for general nonlinear systems by a successive approximation (SA) technique based on neural networks. Rigorous results have been proven and a computationally effective scheme for nearly optimal controller design was provided based on ANN. This technique allows one to consider general affine nonlinear systems of the form

$$\dot{x} = f(x) + g(x)u \tag{14.17}$$

To give internal stability and good closed loop performance, one may select the L_2 norm performance index:

$$V(x(0)) = \int_0^\infty [Q(x) + u^T R u] dt \tag{14.18}$$

with matrix R positive definite and $Q(x)$ generally selected as a norm. It is desired to select the control input $u(t)$ to minimize the cost $V(x)$. Under suitable assumptions of detectability, this guarantees that the states and controls are bounded and hence that the closed loop system is stable.

An infinitesimal equivalent to the cost is given by

$$0 = \frac{\partial V^T}{\partial x}(f + gu) + Q + u^T R u \equiv H\left(x, \frac{\partial V}{\partial x}, u\right) \tag{14.19}$$

which defines the Hamiltonian function $H(.)$ and the co-state as the cost gradient $\partial V / \partial x$.

This is a nonlinear Lyapunov equation, known as the "generalized HJB equation." Differentiating with respect to the control u to find a minimum yields:

$$u(x) = -\frac{1}{2} R^{-1} g^T(x) \frac{\partial V(x)}{\partial x} \tag{14.20}$$

Substituting this into the previous equation yields the HJB equation of optimal control:

$$0 = \frac{\partial V^T}{\partial x} f + Q - \frac{1}{4} \frac{\partial V^T}{\partial x} g(x) R^{-1} g(x) \frac{\partial V}{\partial x} \tag{14.21}$$

The boundary condition for this equation is $V(0) = 0$. Solving this equation yields the optimal value function $V(x)$ and the optimal control may be computed from the cost gradient.

This procedure will give the optimal control in feedback form for any nonlinear system. Unfortunately, the HJB equation cannot be solved for most nonlinear systems.

Therefore, one may use a successive approximation (SA) approach wherein equations 14.19 and 14.20 are iterated to determine sequences V_i, u_i. The initial stabilizing control u_0 used in equation 14.19 to find V_0 is easily determined using, for example, the LQR for the linearization of equation 14.17. It has been shown by Saridis (1979) that the SA converges to the optimal solution V^*, u^* of the HJB. Let the region of asymptotic stability (RAS) of the optimal solution be Ω^* and the RAS at iteration i be Ω_i. Then, in fact, it has been shown that:

1. u_i is stabilizing for all i,
2. $V_i \to V^*, u_i \to u^*, \Omega_i \to \Omega^*$ uniformly,
3. $V_i(x) \geq V_{i+1}(x)$, that is the value function decreases,
4. $\Omega_i \leq \Omega_{i+1}$, that is the RAS increases.

In fact, Ω^* is the largest RAS of any other admissible control law.

It is difficult to solve equations 14.19 and 14.20 as required for the SA method just given. Beard and Saridis (1997) showed how to implement the SA algorithm using Galerkin Approximation to solve the nonlinear Lyapunov equation. This method is computationally intensive, since it requires the evaluation of numerous integrals. Abu-Khalaf and Lewis (2004) showed how to use ANN to compute the SA solution at each iteration. This yields a computationally effective method for determining nearly optimal controls for a general class of nonlinear constrained input systems. The value function at each iteration is approximated using an ANN by

$$V(x) \approx V(x, w_j) = w_j^T \sigma(x) \qquad (14.22)$$

with w_j the ANN weights and $\sigma(x)$ a basis set of activation functions. To satisfy the initial condition $V_i(0) = 0$ and the symmetry requirements on $V(x)$, the activation functions were selected as a basis of even polynomials in x. Then the parameterized nonlinear Lyapunov equation becomes

$$0 = w_j^T \nabla \sigma(x)\big(f + g u_j\big) + Q + u_j^T R u_j \qquad (14.23)$$

with u_i the current control value. Evaluating this equation with a high enough number of sample values of x, it can easily be solved for the weights using, for instance, least-squares. The sample values of x must satisfy a condition known as persistence of excitation in order to obtain a unique least-squares solution for the weights. The number of samples selected must be greater than the number of ANN weights. Then, the next iteration value of the control is given by:

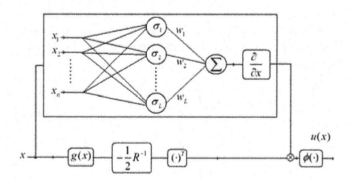

FIGURE 14.11

$$u_{i+1}(x) = -\frac{1}{2} R^{-1} g^T(x) \nabla \sigma^T(x) w_i \qquad (14.24)$$

Using a Sobolev space setting, it was shown that under certain mild assumptions, the ANN solution converges in the mean to a suitably close approximation of the optimal solution. Moreover, if the initial ANN weights are selected to yield an admissible control, then the control is admissible (which implies stable) at each iteration. The control given by this approach is shown in Figure 14.11.

It is a feedback control in terms of a nonlinear neural network. This approach has also been given for constrained input systems, such as industrial and aircraft actuator systems.

14.2.2 Neural Network H_∞ Control Using the Hamilton-Jacobi-Isaacs Equation

Many systems contain unknown disturbances, and the optimal control approach just given may not be effective. In this case, one may use the H_∞ design procedure as follows.

Consider the dynamical system in Figure 14.12, where $u(t)$ is an action or control input, $d(t)$ is a disturbance, $y(t)$ is the measured output, and $z(t)$ is a performance output with

$$\| z \|_2 = h^T h + \| u \|_2$$

$$\dot{x} = f(x) + g(x)u + k(x)d$$
$$y = x$$
$$z = \psi(x, u)$$
$$u = l(y)$$

FIGURE 14.12

Here, we take full state feedback $y = x$ and desire to determine the action or control $u(t) = u(x(t))$ such that, under the worst disturbance, one has the L_2 gain bounded by a prescribed γ so that

$$\frac{\int_0^\infty \|z\|^2\,dt}{\int_0^\infty \|d\|^2\,dt} = \frac{\int_0^\infty \left(h^T h + \|u\|^2\right)dt}{\int_0^\infty \|d\|^2\,dt} \le \gamma^2 \quad (14.25)$$

This is a *differential game* with two players and can be confronted by defining the utility

$$r(x,u,d) = h^T(x)h(x) + \|u(t)\|^2 - \gamma^2 \|d(t)\|^2 \quad (14.26)$$

and the long term value (cost-to-go)

$$V(x(t)) = \int_0^\infty r(x,u,d)dt \quad (14.27)$$

The optimal value is given by

$$V^*(x(t)) = \min_u \max_d \int_0^\infty r(x,u,d)dt \quad (14.28)$$

The optimal control and worst-case disturbance are given by the stationarity conditions as

$$u^*(x(t)) = -\frac{1}{2}g^T(x)\frac{\partial V^*}{\partial x} \quad (14.29)$$

$$d^*(x(t)) = \frac{1}{2\gamma^2}k^T(x)\frac{\partial V^*}{\partial x} \quad (14.30)$$

If the min-max and max-min solutions are the same, then a saddle point exists and the game has a unique solution. Otherwise, we consider the min-max solution, which confers a slight advantage to the action input $u(t)$.

The infinitesimal equivalent to equation 14.27 is found using Leibniz's formula to be

$$0 = \dot V + r(x,u,d) = \left(\frac{\partial V}{\partial x}\right)^T \dot x + r(x,u,d) \equiv H\left(x,\frac{\partial V}{\partial x},u,d\right) \quad (14.31)$$

with $V(0) = 0$, where $H(x,\lambda,u,d)$ is the Hamiltonian with $\lambda(t)$ the co-state, and $\dot x = f(x) + g(x)u + k(x)d$. This is a nonlinear Lyapunov equation.

Substituting u^* and d^* into equation 14.31 yields the nonlinear Hamilton-Jacobi-Isaacs (HJI) equation

$$0 = \left(\frac{\partial V^*}{\partial x}\right)^T f + h^T h - \frac{1}{4}\left(\frac{\partial V^*}{\partial x}\right)^T gg^T\left(\frac{\partial V^*}{\partial x}\right) + \frac{1}{4\gamma^2}\left(\frac{\partial V^*}{\partial x}\right)^T kk^T\left(\frac{\partial V^*}{\partial x}\right) \quad (14.32)$$

whose solution provides the optimal value V^*, and hence the solution to the min-max differential game.

Unfortunately, this equation cannot generally be solved. In Abu-Khalaf et al. (2004), it has been shown that the following two-loop successive approximation policy iteration algorithm has very desirable properties like those delineated above for the LQR case. First, one finds a stabilizing control for zero disturbance. Then one iterates equations 14.30 and 14.31 until convergence with respect to the disturbance. Now one selects an improved control using equation 14.29. The procedure repeats until convergence of both loops. Note that it is easy to select the initial stabilizing control u_0 by setting $d(t) = 0$ and using LQR design on the linearized system dynamics.

To implement this algorithm practically, one may approximate the value at each step using a one-tunable-layer neural network as

$$V(x) \approx V\left(x,w_{ij}\right) = w_{ij}^T(x)\sigma(x) \quad (14.33)$$

with $\sigma(x)$ a basis set of activation functions. The disturbance iteration is in index i and the control iteration is in index j. Then the parameterized nonlinear Lyapunov equation 14.31 becomes

$$0 = w_{ij}^T \nabla \sigma(x)\dot x + r\left(x,u_j,d_i\right)$$
$$= w_{ij}^T \nabla \sigma(x)F\left(x,u_j,d_i\right) + h^T h + \|u_j\|^2 - \gamma^2 \|d_i\|^2 \quad (14.34)$$

which can easily be solved for the weights using, for example least-squares. Then, on disturbance iterations, the next disturbance is given by

$$d_{i+1}(x) = -\frac{1}{2}R^{-1}k^T \nabla \sigma^T(x)w_{ij} \quad (14.35)$$

and on control iterations, the improved control is given by

$$u_{j+1}(x) = -\frac{1}{2}R^{-1}g^T \nabla \sigma^T(x)w_{ij} \quad (14.36)$$

This algorithm is shown to converge to the approximately optimal H_∞ solution. This yields a neural network feedback controller as shown in the figure above for the LQR case.

14.3 Historical Development

We close this chapter with a glimpse of historical development. The use of ANNs in feedback control systems was first proposed by Werbos in 1989.

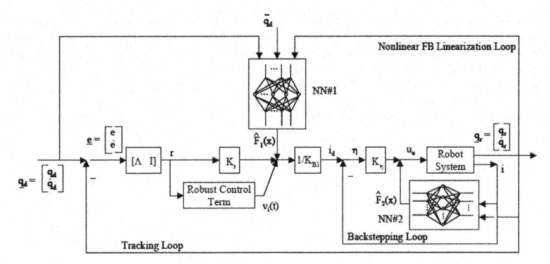

FIGURE 14.13
Feedback linearization and backstepping using ANN.

Since then, ANN-based control has been studied extensively. Eventually, ANNs have become a part of the mainstream of control theory as a natural extension of adaptive control to systems that are *nonlinear* in the tunable parameters. The reader may find several review papers in the references that highlight a host of difficulties to be addressed for closed loop control applications, which are fundamentally different from open loop applications, such as classification and image processing. The basic problems that still qualify as open problems include weight initialization for feedback stability, determining the gradients needed for backpropagation tuning, determining what to backpropagate, obviating the need for preliminary off-line tuning, modifying backprop so that it tunes the weights forward through time, and several other related problems. These issues have since been addressed by many approaches.

Initial work was in NN for system identification and identification-based indirect control. In closed loop control applications, it is necessary to show the stability of the tracking error as well as boundedness of the ANN weight estimation errors.

The approximation properties of ANNs by Cybenko in 1989 and by Park and Sandberg in 1991 are basic to their feedback controls applications. Based on this and analysis of the error dynamics, various modifications to backprop were presented that guaranteed closed loop stability as well as weight error boundedness. These are akin to terms added in adaptive control to make algorithms robust to high-frequency unmodeled dynamics. Back in 1992, Sanner and Slotine used radial basis functions in control and showed how to select the ANN basis functions. Lewis in 1995 used back propagation with an e-mod term.

Multilayer NN were rigorously used for discrete time control by Jagannathan and Lewis in 1996. Most stability results on ANN-based control have been local in nature, and global stability has been treated by Annaswamy (1997). In 2001, Calise used ANNs in conjunction with dynamic inversion to control aircraft and missiles. Feedback linearization and backstepping using ANN has been a favorite with many for the past three decades. Applications of NN control have been extended to partial differential equation systems by Padhi, Balakrishnan, and Randolph (2001). ANNs have been used for control of stochastic systems, receding horizon controllers, and hybrid discrete event NN controllers.

Computational complexity makes ANNs with many hidden layer neurons difficult to implement. However, the emerging deep neural networks have busted this myth to a large extent and people now routinely train networks with 5 to 10 hidden layers. And, it turns out that these perform far better on many problems than shallow neural networks, such as networks with just a single hidden layer (Figure 14.13).

Appendix 14A: Adaptive Neurocontrol

We have seen in Chapter 13 that the single hidden layer feedforward neural network (FFNN) has important characteristics: (i) a single layer of N hidden layer neuronal elements; and (ii) model of each neuron incorporates a nonlinear activation function that are two types: (a) sigmoidal function and (b) radial basis function; and (c) feedforward connectivity. This FFNN maps n-dimensional input into m-dimensional output: $x \rightarrow NN(x)$, assuming that the state variable and the function NN(.) are in the space of real variables. In this appendix, we discuss the design of MRAC (Chapter 11), for which the (system's) uncertainty is modeled using an RBF artificial neural network, then the adaptive control laws are derived using the Lyapunov's normalized energy functional.

We have the following variables defined [A14A.1] as:

$$\text{Matrix of inner layer weights: } V = (v_1, v_2, ..., v_N) \in R^{n \times N} \quad (14A.1)$$

$$\text{Matrix of outer layer weights: } W = (w_1, w_2, ..., w_m) \in R^{N \times m} \quad (14A.2)$$

$$\text{Vector of output biases: } c \in R^m; \text{ thresholds } \theta \in R^N \quad (14A.3)$$

Then, we have the following functional relationships

$$\text{Sigmoidal: } NN(x) = W^T \sigma(V^T x + \theta) + b \quad (14A.4)$$

$$\text{RBF: } NN(x) = W^T \begin{bmatrix} \varphi(\|x - C_1\|) \\ : \\ \varphi(\|x - C_N\|) \end{bmatrix} + b = W^T \Phi(x) + b \quad (14A.5)$$

In the matrix form, we have

$$NN(x) = W^T \sigma \left[V^T \begin{pmatrix} x \\ 1 \end{pmatrix} \right] + c \quad (14A.6)$$

The vector of hidden layer sigmoids is given as

$$\sigma(V^T x + \theta) = (\sigma(v_1^T x + \theta_1)...\sigma(v_N^T x + \theta_N))^T \quad (14A.7)$$

The kth output is then given as

$$NN_k(x) = w_1^T \sigma(v_k^T x + \theta_k) + c_k$$
$$= \sum_{j=1}^{N} w_{jk} \sigma \left(\sum_{i=1}^{n} v_{ik} x_i + \theta_k \right) + c_k \quad (14A.8)$$

The sigmoidal NN has some important properties: (i) universal approximation: large class of functions can be approximated by single layer NNs within any given tolerance on compacted domains, (ii) a large class of functions can be approximated using linear combinations of shifted and scaled sigmoids, (iii) the NN approximation error decreases as the number of hidden layer neurons, N increases, (iv) the inclusion of biases and thresholds in the NN weight matrices simplifies bookkeeping, and (v) function approximation

using the NN sigmoidal elements is to determine connection between weight W and V. In case of the RBF, we have the following definitions and functional relationships:

$$\text{Matrix form: } NN(x) = W^T \Phi(x) + b \quad (14A.9)$$

$$\text{The vector of RBFs: } \Phi(x) = \begin{bmatrix} e^{-\frac{\|x - C_1\|^2}{2\sigma_1^2}} ... e^{-\frac{\|x - C_N\|^2}{2\sigma_N^2}} \end{bmatrix}^T \quad (14A.10)$$

In (14A.10), C is the matrix of RBF centers, σ is the vector of RBF widths, W is the matrix of output weights, and in (14A.9), b is the vector of output biases. Then, the kth output is given as

$$NN_k(x) = w_k^T \Phi(x) + b_k = \left(\sum_{j=1}^{N} w_{jk} + e^{-\frac{\|x - C_j\|^2}{2\sigma_j^2}} \right) + b_k \quad (14A.11)$$

In case of the RBF, the function approximation is to determine output weights W, centers C, and the widths, σ.

14A.1 n-th Order System with Matched Uncertainties

An FFNN can be used in the structure of MRAC (Chapter 11): (i) in off-line/online approximation of uncertain nonlinearities in the system dynamics: (a) modeling errors, (b) battle damage, and (c) control failures; (ii) start with fixed width RBF NNs, linear in unknown parameters; and (iii) generalize to using the sigmoidal NNs. We have the system dynamics as

$$\dot{x} = Ax + B(u - \Lambda f(x)), \Lambda = diag(\lambda_1, ..., \lambda_m),$$
$$\text{constant unknown matrices} \quad (14A.12)$$

In (14A.12), B is known constant matrix. The uncertainty is represented as

$$f(x) = \Theta^T \Phi(x) + \varepsilon_f(x); \Theta \text{ is the matrix of}$$
unknown parameters,
$$\Phi(x) \text{ is vector of } N \text{ fixed RBFs:} (\varphi_1(x)...\varphi_N(x))^T, \quad (14A.13)$$
$$\varepsilon_f(x) \text{ is function approximation tolerance}$$

We assume that, with the unknown output weights and widths, the RBF NN approximates the nonlinearity within a given tolerance:

$$\|\varepsilon_f(x)\| = \|f(x) - \Theta^T\Phi(x)\| \leq \varepsilon \qquad (14A.14)$$

The RBF NN estimator is given as

$$\hat{f}(x) = \hat{\Theta}^T\Phi(x) \qquad (14A.15)$$

The estimator error is given as

$$\begin{aligned} NN(x) - f(x) &= (\hat{\Theta} - \Theta)^T\Phi(x) - \varepsilon_f(x) \\ &= \Delta\Theta^T\Phi(x) - \varepsilon_f(x) \end{aligned} \qquad (14A.16)$$

The stable reference model is given as

$$\dot{x}_m = A_m x_m + B_m r, \; A_m \text{ is Hurwitz} \qquad (14A.17)$$

Our control goal is specified as bounded tracking

$$\lim_{t\to\infty}\|x(t) - x_m(t)\| \leq \varepsilon_x \qquad (14A.18)$$

For the MRAC design using FFNN, we can choose N and the vector of widths σ, and this can be done offline for inclusion of any a priori knowledge about the uncertainty, then the MRAC is designed and the CL performance is evaluated. One can repeat the previous two steps if needed for finer tuning of the modeling and control. The form of the feedback control law is chosen [A14A.1] as:

$$u = \hat{K}_x^T x + \hat{K}_r^T r + \hat{\Theta}^T\Phi(x) \qquad (14A.19)$$

The CL dynamics using (14A.12), (14A.16), and (14A.19) are given as

$$\dot{x} = (A + B\Lambda\hat{K}_x^T)x + B\Lambda(\hat{K}_r^T r + \Lambda\Theta^T\Phi(x) - \varepsilon_f(x)) \qquad (14A.20)$$

The desired dynamics are given as (14A.17), and from (14A.20) and (14A.17) we say that there exist ideal gains (K_x, K_r) such that we have the following conditions

$$A + B\Lambda K_x^T = A_m; \; B\Lambda K_r^T = B_m \qquad (14A.21)$$

As such, the knowledge of the ideal gain is not required. We have for the equality (14A.21) the following expressions, from (14A.17), (14A.20), and (14A.21) after simplification

$$\begin{aligned} A + B\Lambda\hat{K}_x^T - A_m = B\Lambda\Delta K_x^T; \quad \Delta K_x^T = (\hat{K}_x - K_x)^T \\ B\Lambda\hat{K}_r^T - B_m = B\Lambda\Delta K_r^T; \quad \Delta K_x^T = (\hat{K}_r - K_r)^T \end{aligned} \qquad (14A.22)$$

Since the tracking error is defined as $e(t) = x(t) - x_m(t)$, we have the error dynamics (difference between the system states and the desired states) obtained as

$$\begin{aligned} \dot{e}(t) &= \dot{x}(t) - \dot{x}_m(t) = (A + B\Lambda\hat{K}_x^T)x \\ &\quad + B\Lambda(\hat{K}_r^T r + \Delta\Theta^T\Phi(x) - \varepsilon_f(x)) - A_m x_m - B_m r \pm A_m x \\ &= A_m(x - x_m) + (A + B\Lambda\hat{K}_x^T - A_m)x + B\Lambda(\hat{K}_r - K_r)^T r \\ &\quad + B\Lambda(\Delta\Theta^T\Phi(x) - \varepsilon_f(x)) \\ &= A_m e(t) + B\Lambda(\Delta K_x^T x + \Delta K_r^T r + \Delta\Theta^T\Phi(x) - \varepsilon_f(x)) \end{aligned} \qquad (14A.23)$$

Now, we can use Lyapunov function as follows

$$\begin{aligned} V(e, \nabla K_x, \nabla K_r, \Delta\Theta) &= e^T P e + trace\{\nabla K_x^T\Gamma_x^{-1}K_x|\Lambda|\} \\ &\quad + trace\{\nabla K_r^T\Gamma_r^{-1}K_r|\Lambda|\} \\ &\quad + trace\{\nabla\Theta^T\Gamma_\Theta^{-1}\Theta|\Lambda|\} \end{aligned} \qquad (14A.24)$$

In (14A.24), all the $\Gamma(.) > 0$ and are symmetric PD matrices. The matrix P is the symmetric PD solution of the algebraic Lyapunov equation $PA + A^T P = -Q$, and Q is symmetric PD matrix. Now, the time derivative of the Lyapunov function is obtained as [A14A.1]

$$\begin{aligned} \dot{V} &= \dot{e}^T P e + e^T P \dot{e} + 2trace\{\nabla K_x^T\Gamma_x^{-1}\dot{\hat{K}}_x|\Lambda|\} \\ &\quad + 2trace\{\nabla K_r^T\Gamma_r^{-1}\dot{\hat{K}}_r|\Lambda|\} \\ &\quad + 2trace\{\nabla\Theta^T\Gamma_\Theta^{-1}\dot{\hat{\Theta}}|\Lambda|\} \end{aligned} \qquad (14A.25)$$

$$\begin{aligned} &= \{A_m e + B\Lambda(\Delta K_x^T x + \Delta K_r^T r + \Delta\Theta^T\Phi(x) - \varepsilon_f(x))\}^T P e \\ &\quad + e^T P\{A_m e(t) + B\Lambda(\Delta K_x^T x + \Delta K_r^T r + \Delta\Theta^T\Phi(x) - \varepsilon_f(x))\} \\ &\quad + 2trace\{\nabla K_x^T\Gamma_x^{-1}\dot{\hat{K}}_x|\Lambda|\} + 2trace\{\nabla K_r^T\Gamma_r^{-1}\dot{\hat{K}}_r|\Lambda|\} \\ &\quad + 2trace\{\nabla\Theta^T\Gamma_\Theta^{-1}\dot{\hat{\Theta}}|\Lambda|\} \end{aligned} \qquad (14A.26)$$

$$\begin{aligned} &= e^T(A_m P + PA_m)e + 2e^T PB\Lambda(\Delta K_x^T x + \Delta K_r^T r \\ &\quad + \Delta\Theta^T\Phi(x) - \varepsilon_f(x)) \\ &\quad + 2trace\{\nabla K_x^T\Gamma_x^{-1}\dot{\hat{K}}_x|\Lambda|\} + 2trace\{\nabla K_r^T\Gamma_r^{-1}\dot{\hat{K}}_r|\Lambda|\} \\ &\quad + 2trace\{\nabla\Theta^T\Gamma_\Theta^{-1}\dot{\hat{\Theta}}|\Lambda|\} \end{aligned} \qquad (14A.27)$$

$$\begin{aligned} \dot{V} &= -e^T Q e - 2e^T PB\Lambda\varepsilon_f(x) \\ &\quad + 2e^T PB\Lambda\Delta K_x^T x + 2trace\{\nabla K_x^T\Gamma_x^{-1}\dot{\hat{K}}_x|\Lambda|\} \\ &\quad + 2e^T PB\Lambda\Delta K_r^T r + 2trace\{\nabla K_r^T\Gamma_r^{-1}\dot{\hat{K}}_r|\Lambda|\} \\ &\quad + 2e^T PB\Lambda\Delta\Theta^T\Phi(x) + 2trace\{\nabla\Theta^T\Gamma_\Theta^{-1}\dot{\hat{\Theta}}|\Lambda|\} \end{aligned} \qquad (14A.28)$$

Using the trace identity, we obtain the following

$$\dot{V} = -e^T Q e - 2e^T PB\Lambda \varepsilon_f(x)$$

$$+ 2 trace\{\Delta K_x^T(\Gamma_x^{-1}\dot{\hat{K}}_x + xe^T PB\,\mathrm{sgn}(\Lambda))|\Lambda|\}$$

$$+ 2 trace\{\Delta K_r^T(\Gamma_r^{-1}\dot{\hat{K}}_r + re^T PB\,\mathrm{sgn}(\Lambda))|\Lambda|\} \quad (14A.29)$$

$$+ 2 trace\{\Delta \Theta^T(\Gamma_\Theta^{-1}\dot{\hat{\Theta}} + \Phi(x)e^T PB\,\mathrm{sgn}(\Lambda))|\Lambda|\}$$

The adaptive parameters K's(.) and Θ(.) should be chosen such that time derivative in (14A.29) becomes negative definite, and all the parameters remain bounded for all future times. So, we chose the following adaptive laws

$$\dot{\hat{K}}_x = -\Gamma_x xe^T PB\,\mathrm{sgn}(\Lambda)$$

$$\dot{\hat{K}}_r = -\Gamma_r re^T PB\,\mathrm{sgn}(\Lambda) \quad (14A.30)$$

$$\dot{\hat{\Theta}} = -\Gamma_\Theta \Phi(x)e^T PB\,\mathrm{sgn}(\Lambda)$$

Then, the time derivative of the Lyapunov function is

$$\dot{V} = -e^T Q e - 2e^T PB\Lambda \varepsilon_f(x) \le -\lambda_{min}\{Q\}\cdot\|e\|^2$$
$$+ 2\|e\|\|PB\|\lambda_{max}\{\Lambda\}\cdot\|\varepsilon\| \quad (14A.31)$$

Thus, under the condition of (14A.31), the adaptive control laws of (14A.30) are valid. One can keep the adaptive parameters bounded by the following modifications

σ-modification:

$$\dot{\hat{K}}_x = -\Gamma_x(xe^T PB + \sigma_x\hat{K}_x)\mathrm{sgn}(\Lambda)$$

$$\dot{\hat{K}}_r = -\Gamma_r(re^T PB + \sigma_r\hat{K}_r)\mathrm{sgn}(\Lambda) \quad (14A.32)$$

$$\dot{\hat{\Theta}} = -\Gamma_\Theta(\Phi(x)e^T PB + \sigma_\Theta\hat{\Theta})\mathrm{sgn}(\Lambda)$$

e-modification:

$$\dot{\hat{K}}_x = -\Gamma_x(xe^T PB + \sigma_x\|e^T PB\|\hat{K}_x)\mathrm{sgn}(\Lambda)$$

$$\dot{\hat{K}}_r = -\Gamma_r(re^T PB + \sigma_r\|e^T PB\|\hat{K}_r)\mathrm{sgn}(\Lambda) \quad (14A.33)$$

$$\dot{\hat{\Theta}} = -\Gamma_\Theta(\Phi(x)e^T PB + \sigma_\Theta\|e^T PB\|\hat{\Theta})\mathrm{sgn}(\Lambda)$$

So, the added damping to the adaptive laws is controlled by the parameters σ's(.) > 0, and there is a trade-off between adaptation rate and the damping.

Reference

A14A.1 Lavretsky, E. Adaptive control: Introduction, overview, and applications. PPT slides, Robust and Adaptive Control Workshop. http://web.iitd.ac.in/~sbhasin/docs/IEEE_WorkShop_Slides_Lavretsky_Adaptive_Control.pdf, accessed July 2017.

Appendix 14B: Neural Network-Based Fault Tolerant Control: Lyapunov Stability Analysis

When a fault is detected, an FTC (fault tolerant computer) tries to compensate the effect of the fault by adding an auxiliary signal to the normal control (see Appendix 7B, Example 7B.3), and this additional control loop can influence the stability of the entire control system. Here, stability of such an FTC, based on Lyapunov's method (A14B.1) is concisely studied. Such an FTC system uses a state space model of the system and nonlinear state observer, both structured/modeled as ANNs [A14B.1].

A14B.1 Model Formulations

A nonlinear dynamic system is represented as

$$x(k+1) = f(x(k), u(k)) + g(x(k), u(k)) \quad (14B.1)$$

In (14B.1), f(.) is a plant that is working under the normal conditions, and g(.) represents a fault-affecting mechanism. Since the state vector is not always available, one can use the model of the plant in normal condition, and use an observer for estimation of these states using the same/similar model of the plant

$$\bar{x}(k+1) = \bar{f}(\bar{x}(k), u(k)) \quad (14B.2)$$

The state observer is given as

$$\hat{x}(k+1) = \hat{f}(\hat{x}(k), u(k), y(k)) \quad (14B.3)$$

In (14B.3), y(.) is the output of the plant. Using the preceding models, we can approximate the unknown fault as

$$\hat{g} = \hat{f}(\hat{x}(k), u(k), y(k)) - \bar{f}(\bar{x}(k), u(k)) \quad (14B.4)$$

Now, the effect of fault can be compensated or eliminated by application of an adequate auxiliary input based on the estimated fault function g, and the augmented control law is given as

$$u_c(k) = u(k) + u_a(k) \qquad (14B.5)$$

The nonlinear state space model (for the plant [14B.2]) is formulated as the structure of an ANN:

$$\bar{x}(k+1) = \bar{f}(\bar{x}(k), u(k)); \; \bar{y}(k) = H\bar{x}(k) \qquad (14B.6)$$

In (14B.6), the state vector is the output of the hidden layer (of an ANN), the nonlinear function represents hidden layer, and H denotes the synaptic weights between the hidden and output neurons. The FTC scheme requires the estimation of the state vector of the plant, for which the following neural model (for the nonlinear observer) can be used

$$\hat{x}(k+1) = \hat{f}(\hat{x}(k), u(k), e(k)); \; \hat{y}(k) = H\hat{x}(k) \qquad (14B.7)$$

In (14B.7), $e(k)$ is the error between the output of the nonlinear observer, and measured output of the system; here again the nonlinear function represents the hidden layer. Thus, one can determine the unknown fault-function, g by using the estimated state of the system from (14B.7), and (14B.42).

A14B.2 Control Law

Here, idea is to choose the additional control signal to compensate the fault effect. It is easier to attempt to solve an equivalent linear problem. So, we assume that nominal system model is linear and rewrite the system, (14B.1); and using (14B.5) as

$$x(k+1) = Ax(k) + B(u(k) + u_a(k)) + g(x(k), u(k)) \qquad (14B.8)$$

From (14B.8), we see that the fault can be completely compensated. If the fault model of the system is very close to the normal model, then we get

$$Bu_a(k) + g(x(k), u(k)) = 0; \; u_a(k) = -B^{-1}g(x(k), u(k)) \qquad (14B.9)$$

Since the fault function is unknown, one can use the estimate of it. Then using (14B.4), we obtain

$$u_a(k) = -B^{-1}\{\hat{f}(\hat{x}(k), u(k), y(k)) - \bar{f}(\bar{x}(k), u(k))\} \qquad (14B.10)$$

The discrete time state space form of the PI controller is

$$x_r(k+1) = x_r(k) + K_i e_r(k); \; u_r(k) = x_r(k) + K_p e_r(k) \qquad (14B.11)$$

In (14B.11), K's are the integral and proportional gains of the controller, $e_r(.)$ is the regulation error $e_r(k) = y_r(k) - y(k)$, $x_r(.)$ is the controller state, and $y_r(.)$ is the reference signal. The regulation error is equivalently written as: $e_r(k) = y_r(k) - \bar{y}(k) = y_r(k) - H\bar{x}(k)$; and the controller from is written as

$$x_r(k+1) = x_r(k) + K_i(y_r(k) - H\bar{x}(k));$$
$$u_r(k) = x_r(k) + K_p(y_r(k) - H\bar{x}(k)) \qquad (14B.12)$$

The total component of the control law is now given, after substituting for $u(k) (= u_r(k))$, and (14B.10) into (14B.5) as

$$
\begin{aligned}
u_c(k) &= u(k) + u_a(k) \\
&= x_r(k) + K_p(y_r(k) - H\bar{x}(k)) \\
&\quad - B^{-1}\{\hat{f}(\hat{x}(k), u(k), y(k)) - \bar{f}(\bar{x}(k), u(k))\}
\end{aligned}
\qquad (14B.13)
$$

Since the form of (14B.13) is quite complicated, the non-linearities in (14B.13) are linearized. Thus, in order to be able to use the control law, the following *NN* model (for the plant) is used and linearized:

$$\bar{x}(k+1) = h(W^x \bar{x}(k) + W^u u(k)) \qquad (14B.14)$$

$$\bar{x}(k+1) = A\bar{x}(k) + Bu(k); \; \bar{y}(k) = H\bar{x}(k) \qquad (14B.15)$$

In (14B.15), we have $A = h'W^x$, $B = h'W^u$, and h' is the first derivative of the nonlinear activation function of the hidden (layer) neurons. In fact, W^x is the matrix of recurrent links, and W^u is the input weight matrix. In the similar way, the NN nonlinear model of the observer is used, and then linearized as

$$
\begin{aligned}
\hat{x}(k+1) &= h(W^x \hat{x}(k) + W^u u(k) + \hat{y}(k)) \\
&= h(W^x \hat{x}(k) + W^u u(k) + H\hat{x}(k))
\end{aligned}
\qquad (14B.16)
$$

$$\hat{x}(k+1) = A\hat{x}(k) + Bu(k) + D\hat{x}(k) \qquad (14B.17)$$

In (14B.16), the linearized matrices A and B are the same as in (14B.14) and matrix $D = h'W^x H$. Now, we use these linearized versions (14B.15) and (14B.17) in (14B.13) to simplify the part {.} as

$$
\begin{aligned}
&\{\hat{f}(\hat{x}(k), u(k), y(k)) - \bar{f}(\bar{x}(k), u(k))\} \\
&= A\hat{x}(k) + Bu(k) + D\hat{x}(k) - [A\bar{x}(k) + Bu(k)] \qquad (14B.18) \\
&= A(\hat{x}(k) - \bar{x}(k)) + D\hat{x}(k)
\end{aligned}
$$

Substituting this simplification in (14B.13), we obtain the total controller form as

$$u_c(k) = u(k) + u_a(k)$$

$$= x_r(k) + K_p(y_r(k) - H\overline{x}(k)) \tag{14B.19}$$

$$- B^{-1}\{A(\hat{x}(k) - \overline{x}(k)) + D\hat{x}(k)\}$$

To get the final form of the NN-based form of the state model of the plant, we substitute (14B.19) in (14B.14)

$$\overline{x}(k+1) = h(W^x\overline{x}(k) + W^u u(k))$$

$$= h(W^x\overline{x}(k) + W^u\{x_r(k)$$

$$+ K_p(y_r(k) - H\overline{x}(k)) \tag{14B.20}$$

$$- B^{-1}\{A(\hat{x}(k) - \overline{x}(k)) + D\hat{x}(k)\}$$

Simplifying (14B.20), we get

$$\overline{x}(k+1) = h(W^x\overline{x}(k) + W^u x_r(k) + W^u K_p y_r(k)$$

$$- W^u K_p H\overline{x}(k) - W^u B^{-1} A\hat{x}(k) \tag{14B.21}$$

$$+ W^u B^{-1}\overline{x}(k) - W^u B^{-1} D\hat{x}(k))$$

$$\overline{x}(k+1) = h((W^x - W^u K_p H + W^u B^{-1})\overline{x}(k)$$

$$- W^u B^{-1} A\hat{x}(k) - W^u B^{-1} D\hat{x}(k) \tag{14B.22}$$

$$+ W^u x_r(k) + W^u K_p y_r(k))$$

$$\overline{x}(k+1) = h((W^x - W^u K_p H + W^u B^{-1})\overline{x}(k)$$

$$- W^u B^{-1}(A\hat{x}(k) + D\hat{x}(k)) \tag{14B.23}$$

$$+ W^u x_r(k) + W^u K_p y_r(k))$$

$$\overline{x}(k+1) = h((W^x - W^u K_p H + W^u B^{-1})\overline{x}(k)$$

$$- W^u B^{-1}(A+D)\hat{x}(k) + W^u x_r(k) \tag{14B.24}$$

$$+ W^u K_p y_r(k))$$

In a similar way, we should obtain the NN-based form of the state-space representation for the nonlinear observer as follows:

$$\hat{x}(k+1) = h(W^x\hat{x}(k) + W^u u_c(k) + W^\varepsilon\varepsilon(k));$$

$$\varepsilon(k) = \hat{y}(k) - \overline{y}(k) = H(\hat{x}(k) - \overline{x}(k))$$

$$= h(W^x\hat{x}(k) + W^\varepsilon H(\hat{x}(k) - \overline{x}(k)) \tag{14B.25}$$

$$+ W^u[x_r(k) + K_p(y_r(k) - H\overline{x}(k))$$

$$- B^{-1}\{A(\hat{x}(k) - \overline{x}(k)) + D\hat{x}(k)\}])$$

Simplifying (14B.25), we obtain

$$\hat{x}(k+1) = h(W^x\hat{x}(k) + W^u u_c(k) + W^\varepsilon\varepsilon(k))$$

$$= h(W^x\hat{x}(k) + W^\varepsilon H\hat{x}(k) - W^\varepsilon H\overline{x}(k)$$

$$+ W^u x_r(k) + W^u K_p y_r(k) - W^u K_p H\overline{x}(k) - W^u B^{-1} A\hat{x}(k)$$

$$+ W^u B^{-1}\overline{x}(k) - W^u B^{-1} D\hat{x}(k)) \tag{14B.26}$$

$$\hat{x}(k+1) = h((-W^\varepsilon H - W^u K_p H + W^u B^{-1})\overline{x}(k)$$

$$+ (W^x - W^u B^{-1}(A+D) + W^\varepsilon H)\hat{x}(k) \tag{14B.27}$$

$$+ W^u x_r(k) + W^u K_p y_r(k))$$

The state equation for the controller (14B.11) can be rewritten in the form

$$x_r(k+1) = x_r(k) + K_i(y_r(k) - H\overline{x}(k))$$

$$= -K_i H\overline{x}(k) + x_r(k) + K_i y_r(k) \tag{14B.28}$$

A14B.3 Lyapunov Stability Analysis

Now, for the stability analysis, we should consider the three systems, as outlined in the preceding section: (i) the NN model of the plant, (ii) NN model of the observer, and (iii) the state space model of the PI controller; in a composite manner with a combined state vector as $x(k) = [\overline{x}(k) \; \hat{x}(k) \; x_r(k)]^T$, as follows

$$x(k+1) = h(A_c x(k) + B_c) \tag{14B.29}$$

The matrices A_c and B_c in (14B.29) are different form the matrices A and B; the latter are the linearization matrices of the NN models of the plant/observer. In (14B.29), we have the following equivalences:

$$A_c = \begin{bmatrix} W^x - W^u K_p H + W^u B^{-1} & W^u B^{-1}(A+D) & W^u \\ -W^\varepsilon H - W^u K_p H + W^u B^{-1} & W^x - W^u B^{-1}(A+D) + W^\varepsilon H & W^u \\ -K_i H & 0 & 1 \end{bmatrix}$$

$$B_c = \begin{bmatrix} W^u K_p y_r(k) \\ W^u K_p y_r(k) \\ K_i y_r(k) \end{bmatrix}$$

$$\tag{14B.30}$$

Since the composite system (14B.30) is a bit involved, it would be worthwhile to make some transformations without making any approximations. Let us have the following linear transformation:

$$v(k) = A_c x(k) + B_c; \ v(k + 1) = A_c x(k + 1) + B_c \quad \text{(14B.31)}$$

Substituting (14B.29) in (14B.31), we obtain

$$v(k+1) = A_c x(k+1) + B_c = A_c h(A_c x(k) + B_c) + B_c$$

$$v(k+1) = A_c h(v(k)) + B_c \quad \text{(14B.32)}$$

In (14B.32), we have used (14B.31), and we see that the matrices A_c and B_c have come out of the nonlinear $h(.)$, the NN activation function. Now, let us introduce the equivalent coordinate transformation as

$$z(k) = v(k) - v^*(k);$$

$$v^*(k + 1) = A_c h(v^*(k)) + B_c \quad \text{(14B.33)}$$

The composite system can be finally transformed as

$$z(k + 1) = A_c g_c(z(k)) \quad \text{(14B.34)}$$

In (14B.34), g_c is a new function, which is obtained as follows:
From (14B.33), we obtain

$$z(k + 1) = v(k + 1) - v^*(k + 1) \quad \text{(14B.35)}$$

Substituting from (14B.32) and (14B.33) in (14B.35), we obtain

$$z(k + 1) = A_c h(v(k)) + B_c - A_c h(v^*(k)) - B_c \quad \text{(14B.36)}$$

Substituting for $v(k)$ from (14B.33), we obtain

$$z(k + 1) = A_c h(z(k) + v^*(k)) - A_c h(v^*(k)) \quad \text{(14B.37)}$$

$$z(k + 1) = A_c[h(z(k) + v^*(k)) - h(v^*(k))] \quad \text{(14B.38)}$$

Comparing (14B.34) and (14B.38), we see that g_c is given as

$$g_c(z(k)) = h(z(k) + v^*(k)) - h(v^*(k)) \quad \text{(14B.39)}$$

So, in the final analysis, we have the transformations of the state variables as: $x(.) \to v(.) \to z(.)$, and we see that this $v^*(.)$ in (14B.33) is considered as the equilibrium point of the system. Also, in the final transformed composite system, the nonlinear function h has come out of A_c, and this would make the stability analysis a bit easier. Now, we define a Lyapunov normalized energy function (LEF) for the composite system as

$$V(z) = z^T P z \quad \text{(14B.40)}$$

The finite difference of the LEF is obtained as

$$\Delta V(z(k)) = V(z(k + 1)) - V(z(k));$$

$$V(z(k + 1)) = z(k + 1)^T P(z(k + 1)) \quad \text{(14B.41)}$$

Substituting from (14B.34) in (14B.41), we get

$$\Delta V(z(k)) = z(k+1)^T P(z(k + 1)) - V(z(k)) \quad \text{(14B.42)}$$

$$= (A g_c(z(k)))^T P\{A g_c(z(k))\} - z^T(k) P z(k) \quad \text{(14B.43)}$$

$$= g_c(z(k))^T A^T P A(g_c(z(k))) - z^T(k) P z(k) \quad \text{(14B.44)}$$

We see from (14B.39) that, in effect, the nonlinear function $g_c(.)$ is the NN-nonlinear activation function h only, and is operating over the composite state vector, and its equilibrium point. So, it is bounded as

$$|g_c(z(k))| \le |(z(k))| \quad \text{(14B.45)}$$

Taking this fact into account, we can rewrite (14B.44) as

$$\Delta V(z(k)) \le (z(k))^T A^T P A(z(k)) - z^T(k) P z(k) \quad \text{(14B.46)}$$

$$\le (z(k))^T [A^T P A - P] z(k) \quad \text{(14B.47)}$$

Since (14B.47) is quadratic, we see that if $[A^T P A - P] < 0$; then ΔV is negative definite, the LEF would continuously decrease, and the system will be stable. Further study and analysis of $[A^T P A - P] < 0$, by substituting for A_c from (14B.30) would yield specific conditions on the ranges of the NN-weight matrices: W^x, W^u, and $W\varepsilon$ as well as permissible values of the PI controller gains: K_p, and K_i, since A_c is specified in terms of these parameters.

Reference

A14B.1 Czajkowski, A., Patan, K., and Korbicz, J. Stability analysis of the neural network based fault tolerant control for the boiler unit; In *Artificial Intelligence and Soft Computing: 11th International Conference*, ICAISC-2012, Zakopane, Poland, April 29–May 3, 2012, Proceedings, Part II pp. 548–556; https://www.researchgate.net/publication/259633138_Stability_Analysis_of_the_Neural_Network_Based_Fault_Tolerant_Control_for_the_Boiler_Unit, accessed December 2018.

15

Fuzzy Control Systems

Fuzzy logic happened to find a prominent place in control systems for a variety of reasons, both rigorous as well as matter-of-fact. If it is not an exaggeration to say many a consumer product is a control system, chances are that fuzzy logic is the principal ingredient.

Fuzzy control systems use a collection of fuzzy conditional statements—if (·) then (·)—derived from a database that is expected to encompass an almost complete working of the plant. This modern paradigm of control system design is based on interpolative and approximate reasoning [26–29]. Interestingly, the controllers do not depend on any mathematical model, prompting us to call this paradigm as model-free, yet typically any nonlinear function can be handled with the desired precision.

Assumptions are rather a rule than an exception in any design, and fuzzy control systems are designed on the following:

1. The plant is observable and controllable. It may sound dubious to say, on one hand, that it is a model-free paradigm, and on the other, that the plant has certain mathematical structure. But what we emphasize is that the plant is equipped with an adequate number of sensors and actuators that give us the required numbers, if not the equations.

2. The data from the sensors and the actuators, coupled with operator experience and engineering comon sense, can be put in the form of a vast set of *lingustic* "if (·) then (·)" statements, which in turn allow us to make a set of rules for an inference.

3. With whatever is available from the previous two assumptions, a solution exists. We do not have a stringent requirement that it is optimal in a rigorous sense.

4. The decisions we make using lingustic statements can be translated into the right technical data to be presented to the plant.

15.1 Simple Examples

The following example opens up the paradigm.

Example 15.1:

In the classical paradigm, the output of the plant from the sensors is fedback and is compared to the reference input. In the fuzzy paradigm, the analog signals, sampled at appropriate intervals, would be scaled and normalized and then compared. The error is then *fuzzified*—e.g. negative small (NS), positive medium (PM), and so on—using the database from the operator experience. For all good reasons, we consider the rate of change of error $\dot{e}$, too. Based on this linguistic description of the state of the plant, we make a decision, for example

if the error is NS **and** the error-rate is PM

then the control is NS $\qquad$ (15.1)

The reader will have quickly noticed that the decision is also a linguistic statement. These fuzzy decisions are first *defuzzified*—i.e. the right numbers $\hat{u}$ for the control actions are determined. Then they are descaled and denormalized to u before they are fed to the actuators. Figure 15.1 shows this schematic.

Assume that the output of a plant is the pressure of a vessel, say a boiler. The measurements out of the sensors are precise, i.e. crisp, but the linguistic variables defining the error and error rate in pressure, and that of the decided actuator input, are all fuzzy. Since physical systems cannot work on fuzzy linguistic commands, the fuzzy control signal must be converted back to crisp quantities using defuzzification techniques discussed in an earlier chapter.

The above description may be elaborated into the following steps:
1. We assume that normalized error is in the interval $[-e_m, e_m]$ and is defined by seven linguistic variables—negative

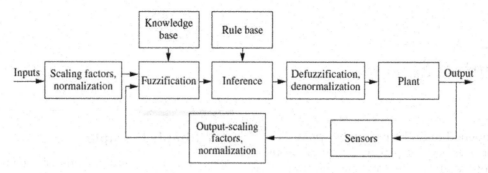

FIGURE 15.1
Schematic of a fuzzy control system.

large (NL), negative medium (NM), negative small (NS), zero (Z), positive small (PS), positive medium (PM), positive large (PL). As discussed in Chapter 13, these variables belong to, not necessarily disjoint, subintervals of the original interval. Likewise, for the purpose of illustration, we assume that the normalized error rate and the normalized output have also been defined by sets of seven and linguistic variables, respectively. Tables 15.1 and 15.2 below show these definitions.

For illustrative purposes, the sampling of normalized error in the interval [−1,1] has been coarse; in practice it can be fine. The reader will have also noticed that an error of 0.0, for instance, could mean something in terms of negative as well as positive error. And there is no need for symmetry in the assignment. Depending upon the context, for instance, an error of +1.0 may have a non-zero weightage, say, 0.02 for the lingustic variable PS. However, symmetry allows us to draw good looking triangles and trapeziums as the membership functions described in Chapter 13.

2. Gathered from the earlier plant operation data logs, we may formulate the contol rules of Table 15.3 as follows:

if the error is NS (row-wise)

and the error-rate is PM (column-wise)

then the control is PS (*i*th row *j*th column)

For a finer control action, Tables 15.3 and 15.4 could be quite large, with more subintervals and fuzzy variables.

3. The fuzzy decisions, such as the ones given in Table 15.3, must map into precise ones, using the defuzzification rules of Chapter 13. This is shown in Table 15.4.

TABLE 15.1

Fuzzy Assignment of Error

Error	NL	NM	NS	Z	PS	PM	PL
−1.0	1	0.75	0	0	0	0	0
−0.5	0	1	0.6	0	0	0	0
.0	0	0	0.2	1	0.1	0	0
+0.5	0	0	0	0	0.5	1	0.2
.0	0	0	0	0	0	0.85	1

TABLE 15.2

Fuzzy Assignment of Error Rate

Error	NL	NM	NS	Z	PS	PM	PL
−1.0	1	0.7	0.4	0	0	0	0
−0.5	0	1	0.55	0.1	0	0	0
.0	0	0	0.1	1	0.3	0	0
+0.5	0	0	0	0	0.7	1	0.2
.0	0	0	0	0	0.5	0.8	1

TABLE 15.3

Decision Rules for Fuzzy Output

$e↓/\dot{e}→$	NL	NM	NS	Z	PS	PM	PL
NL	PL	PL	PL	PS	PS	PS	Z
NM	PL	PL	PS	PS	PS	Z	Z
NS	PL	PS	PS	PS	Z	Z	Z
Z	Z	Z	Z	Z	Z	Z	Z
PS	Z	Z	NS	NS	NS	NL	NL
PM	Z	Z	PS	NS	NS	NL	NL
PL	Z	NS	NS	NS	NL	NL	NL

TABLE 15.4

Defuzzified Assignment of the Output

output	−1	−0.5	0	0.5	1.0
NL	1	0.5	0	0	0
NS	0	0.5	0.75	0	0
Z	0	0	03	1	0.3
PS	0	0	0.4	1	0.5
PL	0	0	0	0.85	1

Example 15.2: A Toy Aircraft Landing Problem

We will present a simulation result of the landing dynamics of an aircraft. The desired profile is shown in Figure 15.2.

The desired downward velocity is thus chosen to be parabolic with the altitude; it could have been linear as well. As the aircraft descends, the expected velocity gets smaller and smaller, approaching zero as the aircraft touches the land gently to avoid damage.

The two state variables for this simulation will be the height above ground, h, and the vertical velocity of the aircraft, v. The control output will be a force that, when applied to the aircraft, will alter its height (h) and velocity (v). Looking at the system as a mass m, moving with a velocity v, having a momentum $p = mv$ would give us the following intuitive update rules:

$$v_{i+1} = v_i + \frac{f_i \cdot \Delta t(=1)}{m(=1)}$$

$$= v_i + f_i$$

$$h_{i+1} = h_i + v_i \cdot \Delta t(=1)$$

$$= h_i + v_i$$

where, for the sake of quick programming, we have assumed that the mass is 1 unit and the update interval Δt is also set to 1 unit. These two control equations define the new value of the state variables v and h in response to the control input and the previous state variable values.

Next, we construct membership functions for h, v, and the control force f as follows:

1. Define simple triangular membership functions for state variables as shown in Figures 15.3 and 15.4.
2. Define a membership function for the control output, as shown in Figure 15.5.
3. Define the rules and summarize them in Table 15.5. The values in the table, of course, are the control outputs.
4. Define the initial conditions, and conduct a simulation for four cycles. Since our objective is to control the aircraft's landing, we will assume the following initial conditions for the simulations:
   - Initial height, $h_0 = 1000$ ft,
   - Initial velocity, $v_0 = -20$ fps,
   - The control force f_0 to be computed to get to the next set of state variables $h(1)$ and $v(1)$.

Using the update difference equations above for each cycle, we defuzzify using the centroid method and obtain $f_0 = 5.8$. This is the output

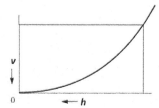

FIGURE 15.2

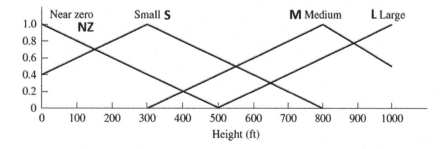

FIGURE 15.3

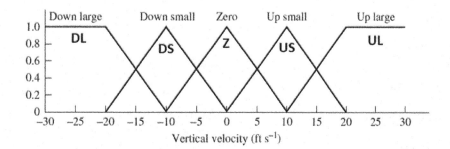

FIGURE 15.4

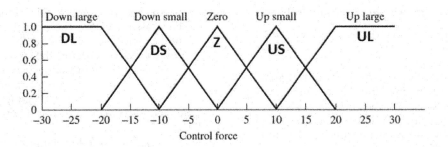

FIGURE 15.5

TABLE 15.5

Fuzzy Rules for Example 15.2

$h{\downarrow}/v{\rightarrow}$	DL	DS	Z	US	UL
L	Z	DS	DL	DL	DL
M	US	Z	DS	DL	DL
S	UL	US	Z	DS	DL
NZ	UL	UL	Z	DS	D

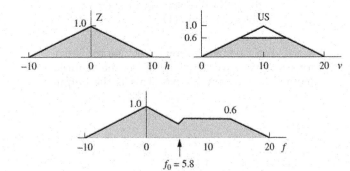

FIGURE 15.6
Truncated consequents and union of fuzzy consequent for cycle 1.

force computed from the initial conditions, and the results for cycle 1 appear in Figure 15.6.

We may now update the state variables and the output for the next cycle:

$$h_1 = 1000 - 20 = 980$$

$$v_1 = -20 + 5.8 = -14.2$$

Using these, we find the centroid to be $f_1 = -0.5$, as shown in Figure 15.7.

At the end of two cycles, the updates are

$$h_2 = 980 - 14.2 = 965.8$$

$$v_2 = -14.20 - 0.5 + 5.8 = -14.7$$

For the third cycle, we get

$$h_3 = 965.8 + (-14.7) = 951.1$$

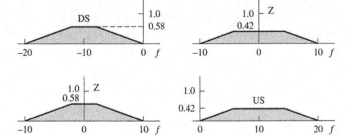

FIGURE 15.7
Truncated consequents for cycle 2.

$$v_3 = -14.7 + (-0.4) = -15.1$$

And for the last cycle of this simulation, we get

$$h_4 = 951.1 + (-15.1) = 936.0$$

$$v_4 = -15.1 + 0.3 = -14.8$$

Looking at the progression of the simulation, the descent velocity versus altitude appears to be reasonably following our desired parabolic curve shown in Figure 15.2.

Clearly, the design, which demands several iterations, is a cakewalk using a computer and all we need is an intuitive, though modest, beginning. We may assume cubic or higher order polynomials for the desired velocity profile, and still we would be able to achieve design with an appropriate scaling of the parameters as well as fine tuning on the time-step Δt.

15.2 Industrial Process Control Case Study

In this part we present a case that demonstrates the flexibility and reasonable accuracy of a typical application in fuzzy control. We have moved out of the classical fuzzy control and incorporated the second generation fuzzy logic, the type II fuzzy logic, capable of also handling uncertainties, thus allowing us to design real robust control systems. We describe some of the experimental equipment and the performance analysis conducted thereupon under the supervision of the second author.

Example 15.3:

In countless industrial process applications, maintaining the liquid level is of significant interest. The three-tank hybrid system, shown in Figure 15.2, is a highly uncertain system that represents various industrial processes quite reasonably.

Therefore, the study for a three-tank level control system has an essential theoretical significance and practical importance. The purpose of the three-tank system is to track the reference point and stabilize the level in the tanks with minimum settling time and fewer oscillations. For properties like varying dynamics, interactions, nonlinearity, and large time delay, the three-tank hybrid system is considered a benchmark problem. Due to the occurrence of both continuous and discrete dynamics of the system, it is also called a *hybrid system*. The presence of the ON-OFF solenoid valves brings discrete dynamics, and continuous dynamics are present in the system inherently. In recent years, because of its combination of continuous and discrete dynamics, it has attracted researchers to develop progressively better control strategies for the three-tank system.

A linear quadratic regulator (LQR) is typically used to determine the state feedback gain for a closed loop system. This is an optimal regulator, by which the open loop poles can be relocated to get a stable system with optimal control and minimum cost. Fuzzy logic controllers have also been tried and much literature is already available.

Nevertheless, the performance of such classical and even the type I fuzzy systems can be inadequate in the presence of noise and uncertainty in the understanding of the dynamics of the plant. Continuing with our arguments from Chapter 13, the fuzzy logic system can be affected by uncertainty in three ways:

1. Different people can interpret different meaning from the words used in antecedents and consequents.
2. Polling done by a group of experts can lead to different consequents for an identical rule.
3. Presence of corrupted training and measurement data owing to faulty instruments.

The type II fuzzy systems, hereafter referred to as T2FLCs for brevity, are capable of outperforming the former type I fuzzy systems since their fuzzy sets are characterized by membership functions, which themselves are fuzzy.

Type II fuzzy sets consist of the well known *footprint of uncertainty*, a bounded region of the uncertainty in the primary membership grades. It is the union of all primary membership grades that makes it generally more robust. The interval type II fuzzy logic system (IT2FLS), a special case of the T2FLCs, is most extensively used for its minimal computational cost. In this case study, we have separately designed an LQR controller and an IT2FLC to regulate the level of a three-tank system by manipulating the inflow rate. In addition, we have imposed a requirement on minimal oscillations. We have simulated and implemented both the controllers on the laboratory scale three-tank and analysed the robustness with and without the presence of disturbance in the system.

Referring to Figure 15.8, the system contains three water tanks that can be filled with two independent pumps acting on the outer tanks 1 and 2. The pumps on the left and the right have liquid flow rates Q_1 and Q_2, respectively, and they are assumed to be continuously manipulated from 0 to a maximum flow Q_{max}. Interaction in the system is caused by upper and lower pipes, and the flow over these pipes can be modulated using valves. The heights h_1, h_2, and h_3 are the liquid levels of each tank and can be measured using differential pressure transmitter (DPT) sensors. The nominal outflow from the system is located at the bottom of tanks 1 and 3.

Here we present a simplified physical description of the system. From the conservation of mass in the tanks we obtain the differential equations:

$$\frac{d}{dt}h_1 = \frac{1}{A}\left(Q_1 - Q_{13v1} - Q_{13v3}Q_{L1}\right)$$

$$\frac{d}{dt}h_2 = \frac{1}{A}\left(Q_2 - Q_{23v2} - Q_{23v23}\right)$$

$$\frac{d}{dt}h_3 = \frac{1}{A}\left(Q_{13v1} + Q_{13v13} + Q_{23v2} + Q_{23v23} - Q_{N3}\right) \quad (15.2)$$

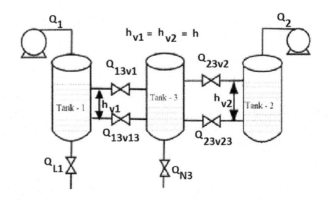

FIGURE 15.8

where the Qs denote flows and A is the is cross-sectional area of each of the tanks (assumed to be the same). All the variables that appear in the mathematical model may be summerized as follows:

1	h_i	Water level in tank i, $=1,2,3$
2	Q_i	Inflow rate through pump i, $=1,2$
3	Q_{ijvk}	Flow between tank i and tank j through valve k
4	Q_{L1} and Q_{N3}	Outflow from tanks 1 and 3

By Torricelli's law, the flow through a lower valve $V_{i3}, i = 1,2$ is

$$Q_{i3vi3} = V_{i3}a_zS_{i3}sgn(h_i - h_3)\sqrt{|2g(h_i - h_3|)} \quad (15.3)$$

$$\approx k_{i3}V_{i3}(h_i - h_3). \quad (15.4)$$

$$\text{where } k_{i3} = a_zS_{i3}\sqrt{\frac{2g}{h_{max}}}$$

We approximate the nonlinearity in equation 15.3 with the right hand side of equation 15.4.

The flow through the valve $V_{L1}(V_{N3})$ is obtained by setting $h_1(h_3)$ in place of $(h_i - h_3)$ in equation 15.3, and through the upper valves V_i by setting $\max\{h_v, h_j\}$ in place of $h_j, j = 1,2,3$. And, by linear approximation we have

$$Q_{L1} \approx k_{L1}V_{L1}h_1 \quad (15.5)$$

$$Q_{N3} \approx k_{N3}V_{N3}h_3 \quad (15.6)$$

$$\text{where } k_* = a_zS_*\sqrt{\frac{2g}{h_{max} - h_v}}$$

In addition, $h_i, i = 1,2,3$ and $Q_j, j = 1,2$ must satisfy the operating constraints

$$0 \le h_i \le h_{max} \quad \text{and} \quad 0 \le Q_j \le Q_{max}$$

For the simulation we have taken the parameters in Table 15.6.

TABLE 15.6

Parameters for the Three-Tank System

Inner diameter of all tanks	15 cm
Height of each tank	100 cm
Overflow height of tanks, h_v	95 cm
Inter-connecting pipes inner diameter	1.25 cm
Orifice diameter of tanks	1.25 cm
k_{13}	0.60
k_{23}	0.72
k_{L1}	0.855
k_{N3}	0.1839

15.2.1 Results

Now we present the results from simulation (using MATLAB) with IT2FLC as well as a comparison between IT2FLC and the LQR-based controller from the experimental study. We considered three conditions to assess the performance of the controllers: (i) When higher interaction is not in effect, (ii) When all the interactions are in effect, and (iii) When the middle tank (tank 3) has been introduced to direct inflow of liquid from a redundant pump, thus acting as a disturbance for tank 1 and tank 2 and resulting in a change in the dynamics of the system. In our simulations, we have considered conditions (ii) and (iii) together.

Figure 15.9 shows the simulation results for set-point tracking of the levels of tank 1 and tank 2 for the

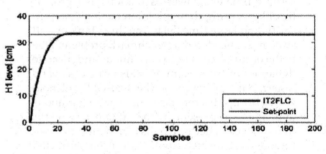

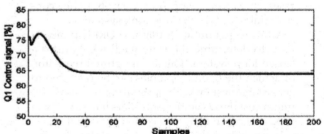

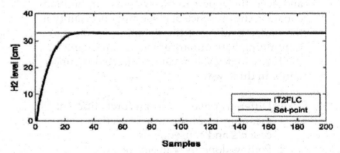

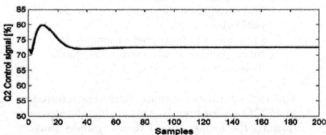

FIGURE 15.9

three-tank system for condition (i). We have arbitrarily taken 33 cm as reference. The control signal generated by IT2FLC for changing and manipulating the inflow rate has also been plotted in Figure 15.9.

Figure 15.10 presents the results of conditions (ii) and (iii). To ensure maximum possible interaction, the reference level is taken as 50 cm. First, we let the controller track the setpoint and bring the system to steady state. Once steady state is reached, at the 200th sample, we introduce a disturbance by switching on the redundant pump, which provides direct inflow to the middle tank (tank 3) at a flow rate of $63 \text{cm}^3/\text{sec}$. This causes the system to deviate from the steady state. Nevertheless, the IT2FLC takes the right action to regulate the level providing disturbance rejection.

While conducting the simulation, characteristics such as uncertainties, noise, and transport delay were not modelled to optimize the fuzzy controller parameters. Instead, the capability of the IT2FLC to handle uncertainties and noises has been determined by investigating the performance on the experimental setup, and comparing the same with an LQR controller. The sampling time taken for entire experiments was 5 sec. The experimental results are shown in Figure 15.11 for the setpoint tracking of the levels of tanks 1 and 2 for condition (i). The LQR controller has been designed based on various models developed for several conditions. However, in the present case,

the response is oscillatory. On the other hand, IT2FLC provides much better response.

The experiment is conducted for conditions (ii) and (iii) together and the results are presented in Figures 15.12 and 15.13. First, we have given the setpoint as 50 cm to the system and let the controller track the level and reach to steady state. At the steady state, at the 200th sample, we have introduced a disturbance by switching on the redundant pump. Despite the change in dynamics, the control action brought back the steady state.

The step responses indicate that both IT2FLC and the LQR controller are capable of providing satisfactory performance in practice, though the presence of IT2FLC is a shade better in the set up we have created. To complete the comparison, we have found that the integral square errors (ISEs) for the IT2FLC were found to be 43.74% (for h_1) and 37.71% (for h_2) less than that of the LQR controller for condition (i), while the respective figures were 24.53% and 23.44% for conditions (ii) and (iii).

Apart from the success story of this particular case study, it is generally felt that the Interval Type II Fuzzy System (IT2FS) appears to be attractive because of its capability to handle uncertainties and noise. Needless to say, there is a necessity to go beyond simulations and test the controller in a real-time setting. One has to weigh between an enormous amount of effort to identify and model such nonlinear systems as a three-tank hybrid system and designing a IT2FLC from a linearized version of the first principles.

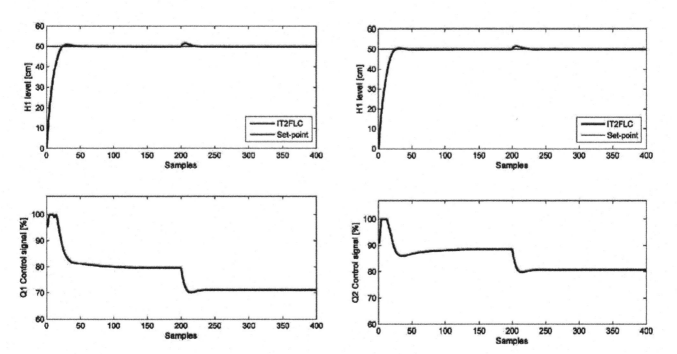

FIGURE 15.10

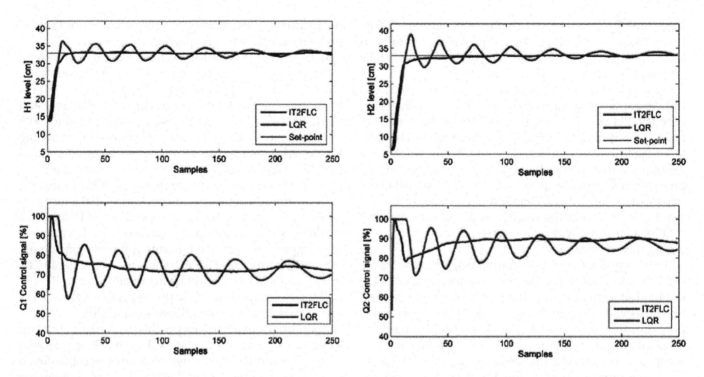

FIGURE 15.11

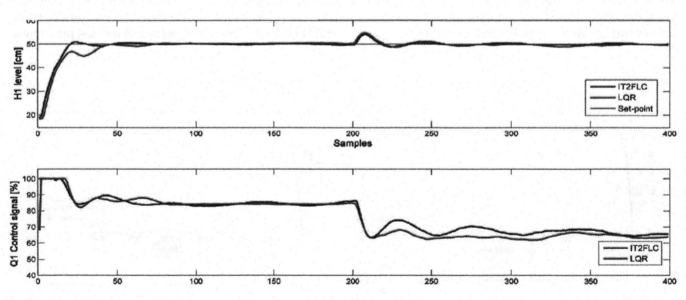

FIGURE 15.12

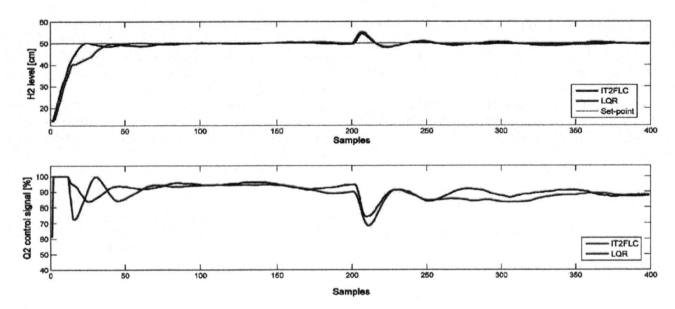

FIGURE 15.13

15.3 Chapter Summary

Consumers prefer gadgets based on their ability to streamline their routine and mundane tasks, giving the consumer time to put toward more creative tasks. There is no doubt that fuzzy logic is being incorporated worldwide in appliances to accomplish these goals, primarily in the control mechanisms designed to make them work. Products with fuzzy logic monitor user-defined settings, for instance autofocus in DSLR cameras, and then automatically set the equipment to function at the user's preferred level for a given task. The literature abounds in papers and books pertaining to fuzzy control systems.

One can readily talk about newer generations of fuzzy logic controllers being based on the integration of conventional and fuzzy controllers. Fuzzy clustering techniques have also been used to extract the linguistic rules from numerical experimental data. In general, the trend is toward the compilation and fusion of different forms of knowledge representation for the best possible identification and control of ill-defined complex systems. Advanced fuzzy controllers use adaptation capabilities to tune the vertices or supports of the membership functions or to add and delete rules to optimize the performance and compensate for the effects of any internal or external perturbations. Learning fuzzy systems try to learn the membership functions or the rules. In addition, principles of genetic algorithms, for example, have been used to fine-tune the rather coarse symmetrical triangular membership functions.

A fuzzy controller is often, but not always, simpler to write, especially if the engineer understands the physics of the system. A fuzzy control is often faster than a PID control, and with little or no overshoot. But, sometimes fuzzy control can get complicated if the membership functions are difficult to determine. This is where we keep our fingers crossed in discussing the stability of fuzzy control systems.

Appendix 15A: Fuzzy Logic in Flight Control-Gain Scheduling

The fuzzy logic (FL) is emerging as a new paradigm for nonlinear modeling, especially to represent a certain kind of uncertainty with more rigor. In fact, FL has made tremendous strides in the last four/five decades. The FL-based systems have the features [A15A.1]: (i) they are based on multi-valued logic as against the bi-valued (crisp) logic, (ii) they do not have any specific architecture like a neural network or even any conventional one, (iii) they are based on certain rules that need to be a priori specified and are mainly obtained from a domain/human expert, (iv) it is a machine intelligent approach in which desired behavior can be specified by the rules in which an expert's (design engineer's) experience can be captured/accommodated, but it is not considered as a machine learning approach, since there is no such learning like in artificial neural networks (ANNs), (v) FL-system deals with approximate reasoning in uncertain situations where truth is a matter of degree, and (vi) fuzzy system is based on the computational mechanism (algorithm) with which decisions can be inferred despite incomplete/imprecise knowledge, it is the process of inference engine, also called FIS (fuzzy inference system). FL-based control is suited to multivariable and non-linear processes/dynamics. The measured plant variables are first fuzzified, then the inference engine/ FIS is invoked. Finally, the results are defuzzified to convert the composite membership function of the output into a single crisp value, since such signals/ outputs provide connectivity to the output world, this then specifies the desired control action. The heuristic fuzzy control does not require deep knowledge of the controlled process. In fact, the FL-based control is a non-model-based approach. The heuristic knowledge of the control policy/law should be known a priori. The FL can be used to augment the flight control system: (i) to approximately replicate some of the ways a pilot might respond to an aircraft that is not behaving as expected due to a damage or failure, say of a control surface and/or wing, (ii) to incorporate the sophisticated nonlinear strategies based on pilot's or system design engineer's experience and intelligence within the control law, (iii) to do adaptive fuzzy gain scheduling (GS) using the fuzzy relationships between the scheduling variables (SVs) and controller parameters, and (iv) FL-based adaptive tuning of Kalman filter for adaptive estimation/control. As such conventional control systems are not regarded as the intelligent system; however, if in the conventional control strategy, some logic is utilized, then it can be a start of the intelligent control system. Fuzzy logic provides this possibility via If...Then... rule. However, the paradigms of ANNs, FL, and GAs are better called the part of the soft computing technology, since for artificial intelligent (AI) some more facets of uses of (symbolic) language, learning, supervision, and automated decision making might be required (see Appendix IVA for a brief exposure on artificial intelligence, intelligent systems, and intelligent control). The FL-controller is suited to keep the output variables between the specified limits/ ranges and also to keep the control actuation, i.e. the control input and the related variables, between limits/ ranges. In fact, the FL deals with vagueness rather than the (classical and usual) uncertainty. The FL in control would be useful: (a) if the system dynamics are slow and/or nonlinear, (b) when models of the system are not available, and (c) competent human operators/ domain experts, to derive the expert rules, are available. Thus, the rule-based FL system can model any continuous function or system and the quality of fuzzy approximation would depend on the quality of the

rules utilized. Primarily, these rules can be formed by the experts and, if required, the ANNs can be used to learn the rules from the empirical data (by using an optimization method, say e.g. GA). The latter systems are called the ANFIS (adaptive neuro-fuzzy inference system). The basic unit of the fuzzy system-cum-approximation is: "If...Then... rule." A fuzzy variable is one whose values can be considered labels of fuzzy sets: temperature→fuzzy variable→linguistic values such as low, medium, normal, high, very high, etc., leading to membership values (on the universe of discourse, e.g. degree Celsius); thus, the temperature (state) variable is fuzzified and provided with a range, rather than having a crisp value. The dependence of a linguistic variable on another variable is described by means of a fuzzy conditional statement:

$$R: \text{If } S_1 \text{ (is true), then } S_2 \text{ (is true) Or } S_1 \rightarrow S_2. \quad (15A.1)$$

For example: (i) if the load is small then the torque is very high, ii) if the error is negative-large, then output is −ve large, etc. The composite conditional statement would be:
R_1: If S_1 then (If S_2 then S_3) is equivalent to:

$$R_1: \text{If } S_1 \text{ then } R_2 \text{ AND } R_2: \text{If } S_2 \text{ then } S_3. \quad (15A.2)$$

The number of rules could be large (30) or, for more complex process control plant, 60 to 80 rules might be required; however, for small tasks, 5 to 10 rules might be sufficient like for an intelligent washing machine [A15A.1]. Then a fuzzy algorithm is formed by combining 2 or 3 fuzzy conditional statements: "If speed error is negative large, then (if change in speed error is NOT (negative large Or negative medium) then change in fuel is positive large),..., Or,..., Or,..." The knowledge necessary to control a plant is specified as a set of linguistic rules of the form: If (cause) then (effect). These are the rules that the new human-operators are trained to control a plant/process and the set of rules constitute the "knowledge base" of the system. However, all the rules necessary to control a plant might not be elicited or even known. It is, therefore, essential to use some technique capable of inferring the control action from available rules:

Forward fuzzy reasoning (Generalized Modus Ponens):

Premise 1: x is A',
Premise 2: If x is A, then y is B,

Then the Consequence: y is B'.

This is the forward data driven inference used in all fuzzy controllers, i.e. given the cause infer the effect. The directly related to backwards goal-driven inference mechanism, i.e. infer the cause that lead to a particular effect:

Generalized Modus Tollens:

Premise 1: y is B',
Premise 2: If x is A then y is B,

Then the consequence: x is A'.

Reasons for use of fuzzy logic in control are: (i) for a complex system, math model of the plant is hard to obtain, (ii) fuzzy control is model free approach, (iii) human experts can provide linguistic descriptions about the system and control instructions, and (iv) fuzzy controllers provide a systematic way to incorporate the knowledge of human experts (and their reasoning). FL-based approach reduces the design iterations and simplifies the design complexity. It can reduce the hardware cost and improve the control performance, especially for the nonlinear control systems. The assumptions invoked in a FL-control system are: (a) the basic plant is observable and controllable, (b) expert linguistic rules are available/or formulated using engineering common sense, intuition or an analytical model, (c) a solution exists, (d) one is seeking a good enough (approximate reasoning) solution and not necessarily the optimal one, (e) goal is to design a controller to the best of our knowledge and within an acceptable precision range, and (f) the problem of stability/optimality could remain as open issue. The heuristic knowledge-based control does not require thorough and deep knowledge of the controlled plant. This approach is popular in industry and manufacturing environment where such knowledge is lacking. This approach is case dependent and does not resolve the issue of overall system's stability and performance automatically. The model-based fuzzy control (MBFC) combines the FL and modern/optimal control methods, especially the known model of the plant can be used. This approach is useful for control of high-speed trains, in robotics, helicopters, and flight control.

In this Appendix, we consider the use of FL for gain scheduling. In the classical GS control, the control gains are computed as a function of some variable, e.g. the dynamic pressure at flight conditions. The fuzzy gain scheduling (FGS) is a special case of the MBFC and uses linguistic rules and fuzzy reasoning to determine the control laws at various flight conditions. The issues of

the stability, pole placement, and closed loop dynamic behavior are resolved by using the conventional and modern control approaches. The Tagaki and Sugeno (TS) controller is signified by a set of fuzzy rules that specify a relation between the present state of the dynamic system to its math model and the corresponding control law with a general rule→ R: If (state) Then (fuzzy plant/process model) AND (fuzzy control law). Let the dynamic system (A15A.1) be given as

$$\dot{x}(t) = f(x,u); \; x_0 \qquad (15A.3)$$

The fuzzy state variables for the discrete case are

$$\phi X_{ij} = \sum \mu_{\varphi X_{ij}}(x)/x \qquad (15A.4)$$

Each element of the crisp state variable is fuzzified and in each fuzzy region of fuzzy state space, the local plant model is defined

$$R_s^i : if \; x = \phi x^i \; then \; \dot{x}^i = f_i(x^i, u^i) \qquad (15A.5)$$

These rules can be given in terms of the elements of the crisp process state. The process is specified by the state space model with degrees of fulfillment of the local models of the process employing Mamdani rule.

$$\dot{x}^i = \mu_s^i(x) f_i(x,u)$$
$$= \min(\mu_{\phi X_1}^i(x_1),...,\mu_{\phi X_n}^i(x_n)) f_i(x,u) \qquad (15A.6)$$

The fuzzy open loop model is given as

$$\dot{x} = \sum w_s^i(x) f_i(x,u) \qquad (15A.7)$$

In (15A.7), $w_s^i(x) = \frac{\mu_s^i(x)}{\sum \mu_s^i}$ is the normalized degrees of fulfillment. The fuzzy control law is defined

$$R_s^i : if \; x = \phi x^i \; then \; u = g_i(x) \qquad (15A.8)$$

As the development of the fuzzy open loop/process model, the control law is defined as

$$u = \sum w_c^j(x) g_j(x) \qquad (15A.9)$$

In (15A.9), $w(j,c)$ are the control weights. In general, an aircraft's dynamic characteristics would change with Mach number and/or altitude (or dynamic pressure). The GS is used for selecting the appropriate filters' (feedback controllers) characteristics (gain and time constants) so that the performance of the aircraft/control system is acceptable despite the change in its dynamics. The concept is illustrated by the following example [A15A.2].

Example 15A.1:

Use the system process rules as (A15A.1):

$$R^o: if\ x_d = 0,\ then\ \dot{x} = f_1(x,u) = -0.4x + 0.4u$$

$$R^1: if\ x_d = 1,\ then\ \dot{x} = f_2(x,u) = -2.5x + 2.5u$$

$$(15A.10)$$

Then,

1. Obtain step responses of these two systems/process rules and compare the results.
2. Use the fuzzy GS controller concept with the following rules

$$R^o: if\ x_d = 0,\ then\ u_1 = g_1(x,x_d) = k_1(x - x_d)$$
$$R^1: if\ x_d = 1,\ then\ u_2 = g_2(x,x_d) = k_2(x - x_d)$$
$$(15A.11)$$

The state feedback gains used are: $k_1 = 0.5$ and $k_2 = 0.92$. Then, the overall fuzzy process model is given by the weighted sum

$$\dot{x} = w_1\{-0.4(x - x_d) + 0.4u\}$$
$$+ w_2\{-2.5(x - x_d) + 2.5u\}$$
$$(15A.12)$$

Design the (15A.11) fuzzy GS and obtain the responses of the final system.

3. Compare the responses of the designed system with the system: $\dot{x} = -0.2x + 0.2u$
4. Plot all the results: step responses, Bode diagrams of the open loop and closed loop systems.

From (15A.10), we see that the process time constants are respectively: 2.5 s. and 0.4 s. The step input responses of these processes are shown in Figure 15A.1. It is obvious that the second process rule signifies the faster process. The fading of the slow plant dynamics to the fast one is represented by the fuzzy membership functions plotted in Figure 15A.2, as altitude increases. The fuzzy control law is given as:

$$u = 0.5w_1(x - x_d) + 0.92w_2(x - x_d) \quad (15A.13)$$

The fuzzy controller is realized, and the unit step responses are shown in Figure 15A.3.

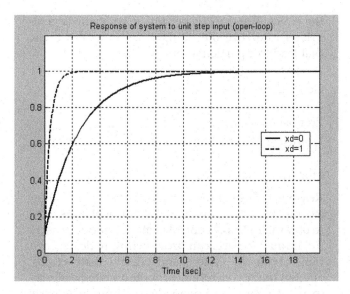

FIGURE 15A.1
Responses for two processes: at low altitude $x_d = 0$ (slow dynamics, -); at high altitude $x_d = 1$ (fast dynamics, --). (From Raol, J.R. and Singh, J., *Flight Mechanics Modelling and Analysis*, Taylor & Francis Informa Group, Boca Raton, FL, 2009.)

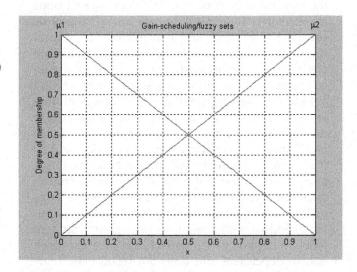

FIGURE 15A.2
GS-fuzzy membership functions: transition from slow dynamics to fast dynamics as the altitude increases. (From Raol, J.R. and Singh, J., *Flight Mechanics Modelling and Analysis*, Taylor & Francis Informa Group, Boca Raton, FL, 2009.)

The comparison of the time responses with the invariant system is shown in Figure 15A.4. The Bode-frequency responses for the open loop and closed loop systems are shown in Figures 15A.5 and 15A.6. The fuzzy controller meets the requirements at both the ends.

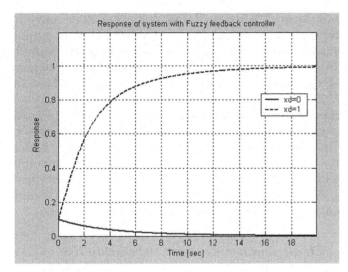

FIGURE 15A.3
Responses of the system for two processes: at low altitude $x_d = 0$ (slow dynamics, -); at high altitude $x_d = 1$ (fast dynamics, --) with fuzzy controllers. (From Raol, J.R. and Singh, J., *Flight Mechanics Modelling and Analysis*, Taylor & Francis Informa Group, Boca Raton, FL, 2009.)

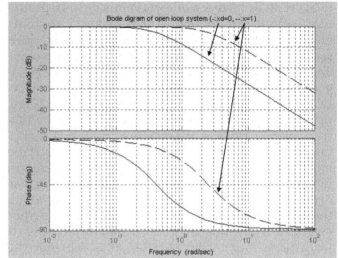

FIGURE 15A.5
Bode-frequency responses of the system for two processes: at low altitude $x_d = 0$ (- slow dynamics); at high altitude $x_d = 1$ (-- fast dynamics). (From Raol, J.R. and Singh, J., *Flight Mechanics Modelling and Analysis*, Taylor & Francis Informa Group, Boca Raton, FL, 2009.)

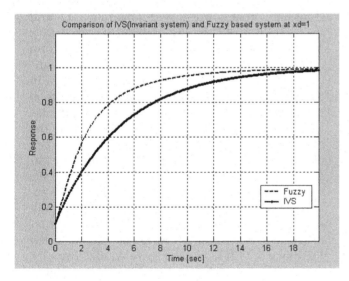

FIGURE 15A.4
The step responses of the fuzzy controlled (--) and the invariant system (-). (From Raol, J.R. and Singh, J., *Flight Mechanics Modelling and Analysis*, Taylor & Francis Informa Group, Boca Raton, FL, 2009.)

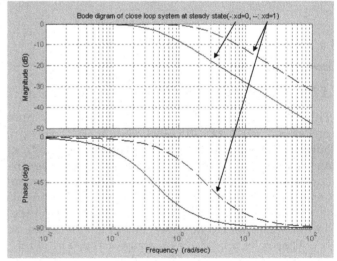

FIGURE 15A.6
CL Bode-frequency responses of the system for two processes: at low altitude $x_d = 0$ (- slow dynamics); at high altitude $x_d = 1$ (-- fast dynamics). (From Raol, J.R. and Singh, J., *Flight Mechanics Modelling and Analysis*, Taylor & Francis Informa Group, Boca Raton, FL, 2009.)

References

A15A.1 King R.E., *Computational Intelligence in Control Engineering*, Marcel Dekker Publications, New York, 1999.

A15A.2 Raol, J. R., and J. Singh. *Flight Mechanics Modelling and Analysis*. Taylor & Francis Informa Group, Boca Raton, FL, 2009.

Appendix 15B: Interval Type II Fuzzy Logic (IT2FL)-Based Pilot's Situation Assessment

Here, we study an application of interval type II fuzzy logic (IT2FL) for handling an aviation pilot's situation assessment in an aviation scenario of an airlane monitoring, which is a critical requirement to avert an occurrence of certain mishaps. In particular, airlane information is provided to auto-pilots and automatic landing systems, and continual decision making is incorporated based on the sensor and airlane data within an automated aircraft navigation system. Incorporation of human intelligence, i.e. at the base level expert's experience/knowledge, in avionics control systems can be now done by FL's If ... Then ... rules. A study of interval type II fuzzy logic-based decision scheme (IT2FLDS) for airlane monitoring is discussed and realized using an interval type II Mamdani model. This IT2FLDS was designed to ascertain if an aircraft is flying along the designated airlane or not, and it was further extended to include flight level data of an aircraft for airlane monitoring (A15B.1). It is felt that the present study of the application of IT2FL system can be easily extended to design and build control algorithms for similar and related applications, because the procedure has been now established, and validated using simulations (A15B.1), which were carried using MATLAB/Simulink, and IT2FL-system toolbox (A15B.2).

A15B.1 The IT2FLDS for Airlane Monitoring

This IT2FLDS primarily consists of blocks for: (i) input processing, (ii) fuzzy inference engine (FIE), and (iii) output processing, Figure A15B.1. The sensor data from an aircraft act as the inputs: aircraft speed, bearing, aircraft class, and altitude. Some IT2FL-based fuzzy memberships functions are used to map these sensor inputs to the FIE block that applies type II fuzzy reasoning, and obtains fuzzy output set, in fact this is based on If ...Then rules provided as is done for Type I FL (T1FL), and the same or similar rules can be/were used. In fact, the T2FSs are the ones whose (secondary) membership (functions) grades are T1FSs. The output block consists of a type II reducer and defuzzifier, the type reducer

transforms T2FSs into T1FSs, and defuzzifier converts type reduced output set to precise or crisp output values using centroid-based defuzzification method. The mathematical and other developmental details of the IT2FL-based inference engine are provided in (A15B.1). Appropriate (fuzzy) rules, (trapezoidal interval type II) membership functions and an FIE are designed to monitor if an aircraft is flying along the airlane or not. The initial experiment was conducted to prove better uncertainty handling capabilities of IT2FLDS compared with T1FLDS, then the model used for the Experiment 1 (E1) was further extended to handle additional sensor data (to represent altitude or flight level) for the E2.

E1: Air-lane monitoring with uncertainties

The experimental Simulink model of IT2FLDS is shown in Figure A15B.2, and the simulation data used are given in Table A15B.1 (A15B.3). Constant velocity of aircraft is considered with the model represented as $x(k+1) = Fx(k) + Gw(k)$, where F is the state transition matrix, G is the process noise gain matrix, and w is the zero mean Gaussian process noise with covariance Q, and $k = (1,2,3,4\cdots,30)$ is the index or scan value. The values of Q, F, and G were obtained from (A15B.3). Absolute difference (between the aircraft under test and airlane along the y-axis) in terms of distance and bearing are the inputs to the IT2FLDS model. Class/type of aircraft, i.e. the civilian, fighter, etc., was considered as the third input. Every aircraft of a particular type/class has a predetermined flight plan, and the airlane is established on basis of the flight plan. We have considered a civilian aircraft. The random-noise generated was controlled using the signal-to-noise ratio (SNR) parameter. The rules similar to (A15B.3_bookmark27) were considered in the FIE block of IT2FLDS, see Figure A15B.2. Initially, a scenario with SNR = 100 dB was considered (~no noise case). The outputs obtained using T1FLDS and IT2FLDS were stored. The positions of airlane (at x-pos = 3000 m) and aircraft along

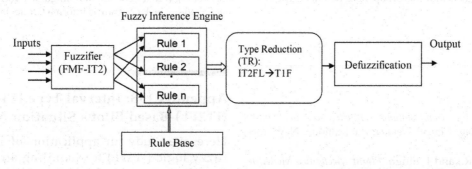

FIGURE A15B.1
An IT2FL-FIE/FIS.

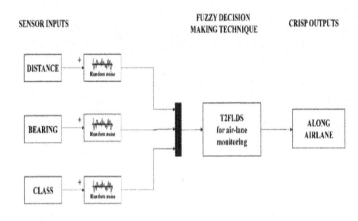

FIGURE A15B.2
IT2FLDS Simulink block-Experiment E1. (Adapted from Lakshmi, S. and Raol, J.R., *IET Intell. Transp. Sy.*, 12, 860–867, 2018.)

the x- and y-axes use in IT2FLDS are shown in Figure A15B.3. The bearing results at various data points, k of aircraft and airlane obtained with respect to the origin (0, 0) are shown in Figure A15B.4. Similar results for this experiment reported in (A15B.3) for T1FLDS, establishing uniformity of simulation environment considered. The crisp output obtained from IT2FLDS is as shown in Figure A15B.5, and shows the decision output of IT2FLDS, whether the aircraft is flying along the airlane or not, this output is true/yes if fuzzy output value is ≥ 0.5 and false/no if value is below 0.5. The decision-making results obtained using IT2FLDS and T1FLDS (based on standard fuzzy operator/implication functions) are shown in Table A15B.2, and these results

TABLE A15B.1

The Data for Simulations E-1 and E-2

Parameter	Simulation Values–Experiment 1 (A15B.3)	Simulation Values–Experiment 2
Initial state $X = \begin{bmatrix} x & \overset{n}{x} & y & \overset{n}{y} \end{bmatrix}$	$X = \begin{bmatrix} 2990\,\text{m} & 0\,\text{m} & 0\,\text{m/s} & 332\,\text{m/s} \end{bmatrix}$	$X = \begin{bmatrix} 361.67\,\text{ft} & 361.67\,\text{ft} & 567\,\text{knots} & 567\,\text{knots} \end{bmatrix}$
No of sensor data	3	4
Type of sensors	*Speed, Bearing and aircraft class*	*Speed, Bearing, aircraft class and altitude*
Sensor update rate	1 Hz	1 Hz
Air-lane location	$x = 3000\,\text{m}$ along y axis	$z = 29,000\,\text{ft}$ along z axis
Simulation time	30 s	30 s
Number of rules	2	3

Source: Lakshmi, S. and Raol, J.R., *IET Intell. Transp. Sy.*, 12, 860–867, 2018.

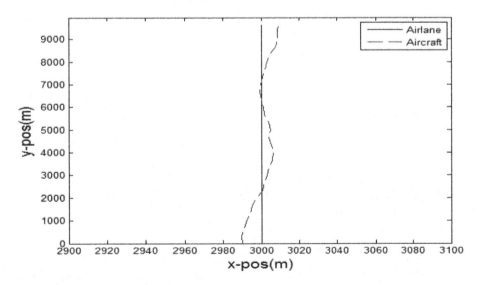

FIGURE A15B.3
x, y positions of airlane and aircraft in yaw plane. (Adapted from Lakshmi, S. and Raol, J.R., *IET Intell. Transp. Sy.*, 12, 860–867, 2018.)

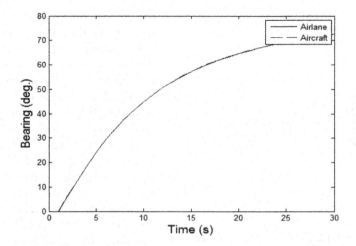

FIGURE A15B.4
Bearing angle of airlane, aircraft with respect to (0, 0). (Adapted from Lakshmi, S. and Raol, J.R., *IET Intell. Transp. Sy.*, 12, 860–867, 2018.)

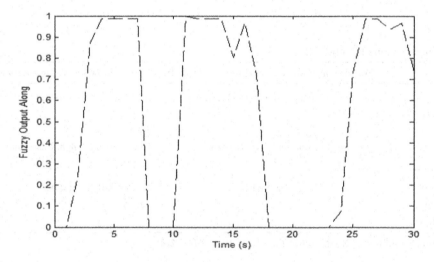

FIGURE A15B.5
IT2FLDS output for airlane decision-E1. (Adapted from Lakshmi, S. and Raol, J.R., *IET Intell. Transp. Sy.*, 12, 860–867, 2018.)

for IT2FLDS are similar to those of T1FLDS for algebraic sum (AS), fuzzy implication function (A15B.3). Now, considering uniform parameters, SNR of random noise generator block was varied to generate various cases (SNR = 90, 80, 70, 50, and 10 dB) to induce uncertainties in sensor input data provided to fuzzy decision systems (for IT2FLDS and T1FLDS). Initial results obtained for airlane decision (SNR = 100 dB) for IT2FLDS and T1FLDS were stored. For each case, output for airlane decision making by IT2FLDS and T1FLDS was noted. The output (decision) errors observed were quantified using: mean square error (MSE), root MSE (RMSE), peak SNR (PSNR), and maximum error (MAX ERROR), which are shown in Table A15B.3. For cases 1, 2, and 3, the IT2FLDS exhibits minor errors in the decision; whereas, for cases 4 and 5, these errors are relatively large, but

in absolute terms these are yet small. In T1FLDS, minor errors are observed in cases 1 and 2 and increase for cases 3, 4, and 5. The performance improvements, due to the use of IT2FLDS, are given in Table A15B.4; thus, T2FL exhibits better uncertainty handling capability.

E2: airlane monitoring considering speed, bearing, class, and flight level

The altitude to be maintained (generally, in feet) by an aircraft is known as flight level. The Aviation Rulemaking Committee (ARC) of the FAA (A15B.4) has defined eight phases of a flight, Figure A15B.6. Airlane discipline is a way to reduce airlane mishaps, and height/altitude of an aircraft is an important factor to be considered for this. An aircraft spends 57% of total flight time in cruising altitude, or phase

TABLE A15B.2

Decision Results-E1

	T1FLDS (AS) (A15B.3)		IT2FLDS	
Time (s)	Defuzzified Output	Decision	Defuzzified Output	Decision
1	0	No	0	No
3	1	Yes	0.874	Yes
5	1	Yes	0.987	Yes
7	1	Yes	0.987	Yes
9	0	No	0	No
11	1	Yes	1.	Yes
13	1	Yes	0.987	Yes
15	1	Yes	0.802	Yes
17	1	Yes	0.725	Yes
19	0	No	0	No
21	0	No	0	No
23	0.	No	0	No
25	1	Yes	0.73	Yes
27	1	Yes	0.987	Yes
29	1	Yes	0.966	Yes

Source: Lakshmi, S. and Raol, J.R., *IET Intell. Transp. Sy.*, 12, 860–867, 2018.

TABLE A15B.3

Airlane Decision Errors Observed for IT2FLDS and T1FLDS (A15B.3)-E1

Case	SNR-dB	MSE		RMSE		PSNR		MAX ERROR	
		IT2FLDS	T1FLDS	IT2FLDS	T1FLDS	IT2FLDS	T1FLDS	IT2FLDS	T1FLDS
1	90	2.16E−09	2.07E−08	6.57E−05	2.03E−04	1.35E+02	1.25E+02	1.99E−04	8.12E−04
2	80	3.75E−08	3.58E−07	2.74E−04	8.46E−04	1.22E+02	1.13E+02	8.31E−04	3.30E−03
3	70	4.34E−07	4.13E−06	9.32E−04	2.80E−03	1.12E+02	1.02E+02	2.80E−03	1.14E−02
4	50	1.36E−05	3.79E−04	5.20E−03	2.75E−02	9.68E+01	8.23E+01	1.98E−02	1.07E−01
5	10	1.70E−02	4.75E−02	1.84E−01	3.09E−01	6.58E+01	6.14E+01	2.70E−01	4.98E−01

Source: Lakshmi, S. and Raol, J.R., *IET Intell. Transp. Sy.*, 12, 860–867, 2018.

TABLE A15B.4

Airlane Decision Improvement; IT2FLDS and T1FLDS (A15B.3)-E1

Case	SNR-dB	MSE (%) Reduction	RMSE (%) Reduction	PSNR (%) Improvement	MAX ERROR Reduction (%)
1	90	89.55	67.68	7.85	75.44
2	80	89.54	67.65	8.70	74.83
3	70	89.49	66.73	9.59	75.44
4	50	96.40	81.09	17.53	81.48
5	10	64.21	40.26	7.29	45.79

Source: Lakshmi, S. and Raol, J.R., *IET Intell. Transp. Sy.*, 12, 860–867, 2018.

five, and we have considered this phase for our simulation. The application of IT2FLDS for airlane monitoring considers aircraft distance, bearing, altitude, and class; the Simulink block for which is shown in Figure A15B.7, with the simulation parameters as in Table A15B.1; E-2.

For distance, bearing, and class sensor inputs, the MFs were adopted from E1. The rules for FIE used are:

Rule 1: IF aircraft has same bearing as airlane AND if it close to airlane, THEN aircraft is flying along airlane.

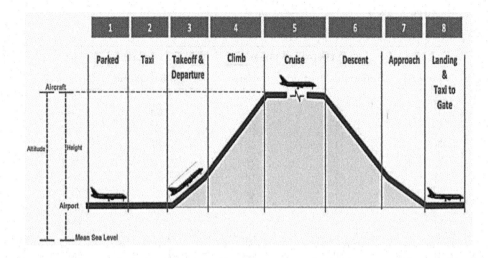

FIGURE A15B.6
Phases of a typical flight. (From FAA; Portable Electronic Devices, Aviation Rulemaking Committee Report, 30 September 2013; https://www.faa.gov/about/ initiatives/ped/ media / ped _ arc_ final_ report.pdf, accessed December 2016).

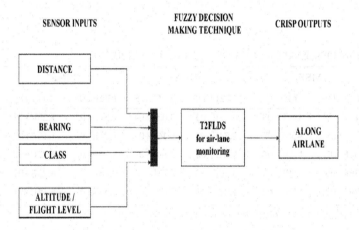

FIGURE A15B.7
IT2FLDS Simulink block: airlane monitoring (data: distance, bearing, aircraft class and altitude)-E2. (Adapted from Lakshmi, S. and Raol, J.R., *IET Intell. Transp. Sy.*, 12, 860–867, 2018.)

Rule 2: IF aircraft class is civilian, THEN possibility of aircraft flying along airlane is high. Rule 3: IF altitude of aircraft is within permissible levels, THEN aircraft is flying along airlane. The T2TMF (type II trapezoidal membership function) and If…THEN…rule for altitude data were designed in accordance with reduced vertical separation minimum specifications (A15B.4). The locations of the aircraft and airlane, in the yaw plane, are shown in Figure A15B.8; in the pitch plane, in Figure A15B.9; and in 3D-space (pitch + yaw), in Figure A15B.10. It can be seen that, in the yaw plane, matching of the aircraft and airlane is good. The difference between aircraft trajectory and airlane is evident in pitch plane and 3D space representation. Bearing of airlane/aircraft, as seen from origin (0,0) and at various

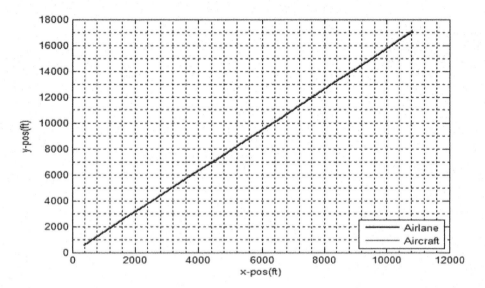

FIGURE A15B.8

x, y positions of airlane, aircraft in yaw plane-E2. (Adapted from Lakshmi, S. and Raol, J.R., *IET Intell. Transp. Sy.*, 12, 860–867, 2018.)

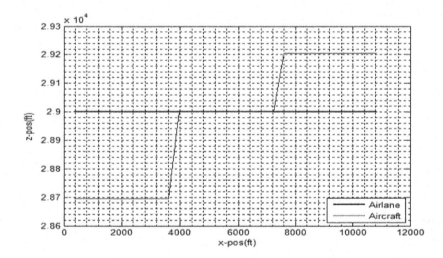

FIGURE A15B.9

x, z positions of airlane and aircraft in pitch plane-E2. (Adapted from Lakshmi, S. and Raol, J.R., *IET Intell. Transp. Sy.*, 12, 860–867, 2018.)

data points, is shown in Figure A15B.11; and the altitude of airlane/aircraft is shown in Figure A15B.12.

It has been established that the inclusion of flight level (altitude) data for airlane decision making in IT2FLDS gives more realistic results compared to similar T1FLDS system. Also, the use of the IT2FL decision (making-cum-fusion) system for the assessment of an aviation pilot's situation assessment is a novel application. The mathematical details of IT2FL are given in Chapter 3 of (A15B.5).

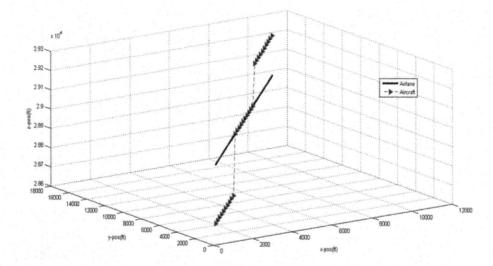

FIGURE A15B.10
x, y, z positions of airlane, aircraft; pitch and yaw plane-E2. (Adapted from Lakshmi, S. and Raol, J.R., *IET Intell. Transp. Sy.*, 12, 860–867, 2018.)

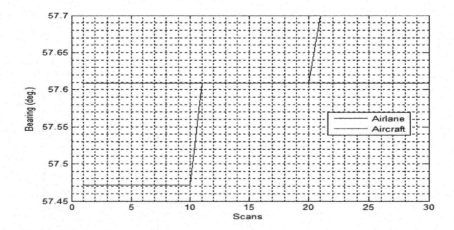

FIGURE A15B.11
Bearing angle of airlane, aircraft with respect to (0, 0)-E2. (Adapted from Lakshmi, S. and Raol, J.R., *IET Intell. Transp. Sy.*, 12, 860–867, 2018.)

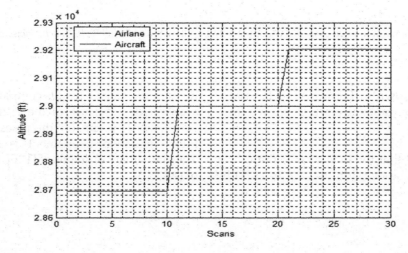

FIGURE A15B.12
Altitude in feet of airlane and aircraft-E2. (Adapted from Lakshmi, S. and Raol, J.R., *IET Intell. Transp. Sy.*, 12, 860–867, 2018.)

References

A15B.1 Lakshmi, S., and Raol, J. R. Interval type-2 fuzzy-logic-based decision fusion system for air-lane monitoring. *IET Intel. Trans. Sys.*, Vol. 12, No. 8, pp. 860–867, 2018.

A15B.2 Castro, J.R., Castillo, O., Melin, P. et al. Building fuzzy inference systems with a new interval type-2 fuzzy logic toolbox, in *Transactions on Computational Science I (LNCS)*, Springer, Berlin,Germany, Vol. 4750, pp. 104–114, 2008.

A15B.3 Raol, J.R., and Kashyap, S.K. Decision fusion using fuzzy logic type 1 in two aviation scenarios, *Jl. Aerosp. Sci. Technol. (Aero, Soc. India)*, Vol. 65, No. 3, pp. 273–286, 2013.

A15B.4 Anon. FAA: Portable Electronic Devices, Aviation Rulemaking Committee Report, https://www.faa.gov/about/initiatives/ped/media/ped_arc_final_report.pdf; and DGCA (Director General of Civil Aviation, India): Requirements for implementation of reduced vertical separation minimum (RVSM), Civil aviation requirements Section 8–aircraft operations series 'S', part II, issue I; 2013; http://dgca.nic.in/cars/D8S-S2.pdf, accessed December 2016.

A15B.5 Raol, J. R. *Data Fusion Mathematics; Theory and Practice*. Taylor & Francis Informa Group, Boca Raton, FL, 2015.

16

Nature Inspired Optimization for Controller Design

Recent applications of intelligent control include without restrictions: transportation systems, medical, biomedical and biological systems, aerospace, automation, biotechnology, mechatronics, manufacturing, process control, power systems, energy and smart grid, agriculture, environmental systems, robotics and autonomous systems, and economics and business systems; the list is endless. This last chapter of the book presents glimpses of recent high quality research results regarding not only the theoretic development in integration of computational intelligence theories and control techniques, but also related effective applications to some new and useful physical systems [30–37].

16.1 Control Application in Light Energy Efficiency

We begin with a sligthly different application, compared to the robotics and aerospace problems in earlier chapters. Due to shortage in energy resources and the greenhouse effect, energy savings have become one of the most interesting and challenging research area. Several control strategies have been proposed to increase energy saving in offices. One such methodology is the intelligent lighting control systems. These are based on intelligent algorithms to control the lamps as a function of various specifications. Most of these strategies consider the lighting control problem as a constrained minimization problem. However, it has been shown that particle swarm optimization (PSO) might be a suitable method for the lighting control problem. Even further, it was shown that this was not a significant improvement. In what follows we describe the most recent work of Copot et al. [38], where they have modeled tuning the PID controller parameters, K_P, K_I, K_D as a multi-objective optimization problem and employed PSO to obtain the optimal parameters of the PID controller for a light control system.

The focus of this research is to regulate the light amount—a combination of natural and artificial lighting—in a room at a constant level, irrespective of the disturbances from outside, such as weather conditions. Thus, a control system for closed loop regulation of the light amount in building rooms is designed.

The main benefits would be a higher level of comfort and a continuous saving of energy. The obtained results verify that the multi-objective particle swarm optimization (MOPSO) is able to perform appropriately in complex systems, such as light control environment.

16.1.1 A Control Systems Perspective

Lighting accounts for a significant amount of the primary energy use worldwide in office buildings. And, perhaps it is equally natural to conserve the energy by reducing the amount of wasted energy in office lighting. Apart from the naive approach to replace low energy incandescent lamps with high energy efficiency fluorescent lamps, with a straight savings up to 40%, one may find in the literature several studies that investigated the energy saving in lighting systems from different perspectives; for instance, energy savings by controlling light in commercial space, either private office or open office, and control decisions, such as multilevel switching, manual switching, daylight harvesting, and occupancy sensors. A comprehensive literature review and analysis for energy saving from several types of lighting control can be found in [39].

Recall that we have made a comment, earlier in Chapter 1, that industrial acceptance for modern controllers would be higher if the innovation is centered around PIDs. Copot et al. [38] have proposed a PSO algorithm for the design of a PID control strategy. This method tunes the PID controller in loop with the given plant using an optimization algorithm to minimize some cost function. In contrast to several previous studies, which considered the problem as a single objective problem, their work employs a PSO algorithm MOPSO with an accelerated update methodology to calculate the optimal parameters of PID controllers. The MOPSO algorithm is deployed to tune PID controllers considering desired criteria, which are overshoot, robustness to variation in the gain of the plant, settling time, and rise time.

The control system for lighting, like any other control system, can be seen to consist of three main elements: (i) a decision making element (controller), (ii) sensors to supply information to the controller, and (iii) switches or variable controls in series with the supply to the luminaries and capable of being remotely controlled. To model and identify the lighting system, the supply

FIGURE 16.1
(a,b) Full and Partial interactions among cubicles.

voltage to the lamp dimmer circuits should vary step-wise and the resulting response, measured by the light sensors, is assumed to have been recorded with a data acquisition board. This model is then used to implement and validate the light system simulator.

The lighting system considered in their simulator has a MIMO (multiple-input multiple-output) configuration that consists of 8 zones, in two circumstances: (i) full interaction and (ii) partial interaction between them as shown in Figure 16.1. The amount of interaction is determined by the presence of full-way or half-way delimiter walls between the work cubicles. In this way, the reflection of light on the working area is altered. It is assumed that each of the 8 zones in the room has its own light sensor and its own separately controlled bank of lamps.

The linear model from the dimmer voltage to the light meter for one room was found to be

$$G(s) = \frac{13.6}{s(s^2 + 11.8s + 32.8)} \qquad (16.1)$$

assuming that the window is closed and that there were no other disturbances present.

An intuitive idea, which was confirmed in an earlier work by one of the authors, that the partial interaction between the cubicles would lead to higher savings, has been considered in the reported work on PID controller tuning.

The closed loop control system, with the optimal set of controller gains, $X = \{K_P, K_I, K_D\}$, is expected to meet the standard specifications on the overshoot M_p, the settling time T_s, the rise time T_r, and disturbance rejection. Thus, there are *three* objective functions:

$$J_1(X) = f_1(M_p), J_2(X) = f_2(t_s - t_r),$$
$$\text{and } J_3(X) = f_3(\text{disturbance rejection}) \qquad (16.2)$$

The block diagram of the PSO-based closed loop system is given in Figure 16.2.

Going by the PSO methodology of Chapter 13, the particle is a three dimensional one, and the PSO algorithm assignsarbitrary values to X, computes the objective functions, and then updates the gains in the direction of minimizing the objective functions. More formally, the following composite objective function is considered:

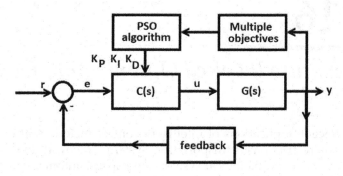

FIGURE 16.2

$$J(X) = \sum_{i=1}^{3} \alpha_i J_i(X) \qquad (16.3)$$

The authors have initially chosen $\alpha_1 = 0.35$, $\alpha_2 = 0.50$, $\alpha_3 = 0.15$ as the nominal values. Subsequently, for the disturbance rejection part, they have been varied by 35%. At the end of MOPS optimization, the following were the gains obtained: $K_P = 0.4283$ $K_I = 1.4068$ $K_D = 0.2209$

The step response has been reported to have an overshoot less than 20%, settling time less than 5 sec when there was no disturbance, and an almost identical performance when the α_i parameters have been randomly changed in the range of 35% of their nominal values.

The inference was that the MOPSO-PID controller reacts very fast and tracks the reference input remarkably well. The controller was demonstrated to maintain a constant level of the office lighting system for both partial- and full-interaction cases.

16.2 PSO Aided Fuzzy Control System

This is a case study reported by Oh et al. [40] which reports the design methodology of an optimized fuzzy controller with the aid of PSO for a ball and beam system, a well-known control engineering experimental setup, which consists of a servo motor, beam, and ball. This is widely used in several universities as it exhibits a number of interesting and challenging properties when being considered from the control perspective. The ball and beam system determines the position of the ball through the control of a servo motor. The displacement change of the position of the ball leads to the change of the angle of the beam, which determines the position angle of a servo motor. The fixed membership function design of a type 1-based fuzzy logic controller (FLC) leads to the difficulty of rule-based control design when representing the linguistic nature of knowledge. On the other hand, a type 2 FLC exhibits some robustness by comparison. The research goes a step forward and fine tunes the type 2 FLC using PSO.

To evaluate performance of each controller, the authors have considered controller characteristic parameters, such as maximum overshoot, rise time, settling time, and a steady state error.

As we have presented in Chapters 13 and 15, Type-2 FLCs have been used successfully in many applications, such as process control, time-series forecasting, communication and networks, decision making, data and survey processing, word modeling, and phoneme recognition. The ongoing challenge for advanced system control has resulted in a diversity of design methodologies and detailed algorithms. One of the difficulties in controlling complex systems is to come up with a suite of optimized control parameters, such as linguistic control rules, scaling factors, and membership functions of the fuzzy controller. With this regard, it is difficult to assume that the given expert's knowledge captured in the form of the fuzzy controller may immediately lead to the optimal control. To improve the performance of the controller through adjusting control rules and membership functions, one has to exploit some effective optimization vehicle, such as particle swarm optimization.

Oh et al. [40] have compared a conventional optimized type 1 fuzzy cascade controller with the optimized interval type 2 fuzzy cascade controller in application to the ball and beam system. Owing to the advantages of the PSO, they have considered it as a viable alternative. The comparative analysis involves two tuning cases, such as scaling factor (SF) tuning and fuzzy rule (SF + R) tuning.

The main advantages of the fuzzy cascade control structure become particularly visible when the first control unit contains significant nonlinearities that limit the overall control performance. Furthermore, the controller of the fuzzy cascade structure attenuates the disturbance as well as improves dynamic properties of the control loop. The structure of the fuzzy cascade controller consists of an outer controller and an inner controller. The setpoint value of the inner controller corresponds to the position angle of a servo motor and is given as a reference value, which enters to the inner controller as the inner controller of the control cascade. The overall control architecture is shown in Figure 16.3. Notice the similarity of this architecture to the lead-lag architecture of Chapter 1.

Let us now proceed to describe the fuzzy cascade controller. The commonly known ball and beam system is also referred to as a structure that realizes a task of "balancing a ball on a beam." The problem is generally linked to real-world control problems, such as horizontally stabilizing an airplane during landing and in

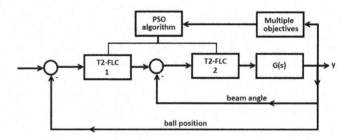

FIGURE 16.3

turbulent airflow and the balance problem dealing with goods to be carried by a moving robot. There are two degrees of freedom in this system. One is the ball rolling up and down the beam and the other is the beam itself rotating through its connected axis.

The control goal is to govern the position of the ball by applying a suitable voltage level to a servo motor. We maintain a steady state of the ball by adjusting an angle of the beam through the movement of the servo motor. The position of the ball is obtained by measuring the voltage at the steel rod while the angle of servo motor is recorded by the position of the encoder. It is difficult to directly control the velocity and acceleration of the ball due to the friction coefficient between the ball and beam, as well as serial linkage (connection) between the individual elements of the ball and beam system. The nonlinear representation of the system can be simplified by deriving a transfer function of the linearized equations of the system. Typically, the result would be an unstable 4th order system of the form

$$G(s) = \frac{K}{s^4 + \alpha s^3} \qquad (16.4)$$

In their paper the authors have reported to have used 150 number of particles, and the number of generations is equal to 300. In the PSO(SF), the objective function in the type 2 fuzzy cascade also has been optimized quite fast by being reduced from 10.14 (initial generation) to 8.17 (reported after 30 generations) and then slowly reduced reaching the value of 8.00 (obtained after 300 generations). In the PSO(SF + Rule), the objective function in the type 2 fuzzy cascade has also been reduced significantly, reaching the values of 9.66 (initial generation), 7.95 (after 30 generations) and 7.76 (after 300 generations). The objective function chosen was the Integral of Absolute Error (IAE) criterion, and the PSO has been reported to have arrived at a value of 8.0044 for PSO using SF, and 7.7639 for PSO using SF + R. The parameters of the type 2 fuzzy cascade controller based on PSO, however, converge quite rapidly and this might also be attributed to the efficient computing realized by the PSO.

After setting the values of the scaling factors, the consequent interval sets of type 2 FLC have been tuned and it was shown that the type 2 fuzzy cascade controller based on PSO(SF) reaches the setpoint quickly. The same pattern of behavior was observed for the type 2 fuzzy cascade controller based on PSO(SF + R). The following were the numerical details presented in their work.

Spec	PSO (SF)	PSO (SF + R)
M_p	3.04	0
t_r	0.45 sec	0.50 sec
t_s	1.36 sec	1.31 sec
e_{ss}	0	0

There was a report of the performance of the optimized type 2 fuzzy cascade controller including three sources of 5% white noise—two immediately after the type 2 FLCs, and one in the feedback loop, with reference to Figure 16.3. In conclusion, type 2 cascaded FLC was found to exhibit robust characteristics when compared with the type 1 fuzzy cascade controller. The type 2 fuzzy controller's characteristic has an excellent consistency because membership function of type 2 fuzzy's characteristic includes the uncertainty. And also, it can include linguistic rules wider than the type 1 fuzzy controller because the inference result values come out the intervals.

16.3 Genetic Algorithms (GAs) Aided Semi-Active Suspension System

Suppression of vibration in passive suspensions depends on the spring stiffness, damping coefficient, and car mass. Due to the fact that they cannot satisfy the comfort requirement in different road conditions, more and more interest is being devoted to active and semi-active suspension in both academia and industry. Over the last few years, various active and semi-active strategies have been proposed to improve performance and remove the restrictions of passive suspensions. The first controlled suspension to appear was the active suspension system, which requires a power source to generate suspension forces according to some prescribed criterion. With such a strategy, ride comfort and track holding ability can be enhanced simultaneously. Another strategy uses basically dissipative elements to modulate the control force actively. This semi-active strategy appears more attractive

since it needs no additional energy. In this section, we describe the semi-active strategy presented by Zhang et al. [41]. In the next section, we will look into the active control of a suspension system.

The objective of semi-active control for a semi-active suspension system is to improve the ride comfort and holding ability under different road conditions. Many kinds of control approaches have been applied to semi-active suspension systems. The fuzzy logical controller has been applied and emerged as another methodology for semi-active suspension systems to deal with the nonlinearities associated with the actuator dynamics, shock absorbers, suspension springs, etc. An important point in utilizing the FLC for a semi-active suspension system is to determine the control rules, such that the satisfactory performance under different road profiles can be achieved. The conventional fuzzy control rules are established by the knowledge and experience of expertise or a skilled operator; in other words, the rules may not be optimal. It also faces a challenge in the development of fuzzy rules. Many studies have the ability to solve this problem for a conventional FLC. However, the GAs are found to be more suitable to deal with the problem of lacking experience or knowledge than other methods—in particular, when the phenomena being analyzed are describable in terms of rules for action and learning processes. Figure 16.4 presents the schematic of the fuzzy control system of a semi-active suspension.

It consists of a suspension system model, a fuzzy logic controller, genetic algorithm program, and an adjustable damper, where u is the control forces of the damper. In this study, GA is used to produce the rules of the fuzzy controller, and the damper modulates the force u, so as to make the vehicle body accelerator approach minimal value.

The multibody system is applied to a broad variety of engineering problems from aerospace to civil engineering, from vehicle design to micromechanical analysis, and from robotics to biomechanics. The requirements for more complex models and the fast development of

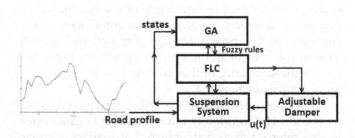

FIGURE 16.4

more and more powerful computers led to the development of multibody system dynamics.

In what follows we discuss the multibody model of a full car semi-active suspension system [41]. A seven degrees of freedom (DOF) full-car suspension multibody model is created using ADAMS. This model is composed of a double-wishbone front suspension (inferior and superior wishbone arms, the steering gear connecting rod, and the wheel hub), a trailing-arm rear suspension, a body, four tires and suspension test rig, six revolute joints, six spherical joints, four translational joints, nine fixed joints, two hooke joints, four inplane primitive joints, and six motions. The typical parameters for the model are as follows:

Sprung mass, in Kg	2010
Moment of Inertia I_{xx}	1.06E+009
Moment of Inertia I_{yy}	2.28E+009
Moment of Inertia I_{zz}	2.18E+009
Tire Mass, in Kg	50
Length of the Superior Wishbone Arms, in m	0.35
Length of the Inferior Wishbone Arms, in m	0.5
Length of the Wheel Hub, in m	0.33
Vehicle Speed in m/sec	25 & 35

In this work, an automated design of the fuzzy rules using GA is presented. The two input variables are defined as

$$E_v = (0 - v)/K_1 = -v / K_1, \quad \text{and} \quad E_a = a / K_2$$

where E_v is the velocity error, E_2 is the accelerator, a is the maximum accelerator of the vehicle, v is the maximum velocity of the vehicle, and $K_i (i = 1, 2)$ are the scaling factors of the inputs. The output of the FLC is presented by

$$u = C/K_3$$

where C is the coefficient of the damper and K_3 is the scaling factors of the output.

The scaling factors $K_i (i = 1, 2)$ are chosen in such a way as to convert the two inputs within the universe of discourse and activate the rule base effectively, whereas K_3 is selected such that it activates the system to generate the desired output. All these scaling factors are chosen based on trial and error.

Triangular membership functions are chosen for both inputs and output. The fuzzy sets of two inputs are defined as

$$\{NB, NM, NS, ZE, PS, PM, PB\}$$

The fuzzy sets of the output are defined as

$$\{C1, C2, C3, C4, C5, C6, C7\}$$

Triangular membership functions of 7 linguistic variables are used for both inputs and output, as shown below:

	NB	NM	NS	Z	PS	PM	PB
NB	7	6	6	5	6	6	7
NM	6	5	4	4	4	5	6
NS	4	3	2	2	2	3	4
Z	1	1	1	1	1	1	1
PS	4	4	3	2	3	4	4
PM	6	5	4	4	4	5	6
PB	7	6	6	5	6	6	7

Sine the binary representation traditionally used in GAs has some drawbacks when applied to multidimensional problems, the authors have used real coded genes or floating point representation to overcome this problem. It uses an integer set $\{1, 2, 3, 4, 5, 6, 7\}$ to represent the fuzzy set $\{C1, C2, C3, C4, C5, C6, C7\}$, respectively. The decision making logic constructed by 49 rules and the fuzzy set of the output variable are selected to form the chromosome for the GA. In this case study, a chromosome is thus obtained from the decision table by the code string of GA illustrated by the 49 integers, such as

$$\{7665667654445643222341111111443234465444567665667\}$$

The performance index of ride quality α can be calculated by the ADAMS software automatically. The fitness function in the GA can be selected as $F = 1 / RMS(\alpha)$.

Recall that the GAs are based on a heuristic ascension method (selection and crossover) and a semi-random exploration method (mutations). Each new generation inherits properties from the best solutions of the precedent GAs. After some generations, the population evolution terminates, and the best individual represents the optimal solution. However, the process of traditional GA cycle, offspring individuals are potentially mated or mutated using the crossover operator or mutation operator depending on the crossover rate P_C and the mutation rate P_M, respectively. Too large a P_C destroys too great a deal of good schemata, and too small a P_C reduces the reproduction of individuals. Since this can restrain the algorithm global search ability and result in unsatisfactory solutions, the parameters of GAs are all set according to the experiments and subjectivity. In order to deal with this kind

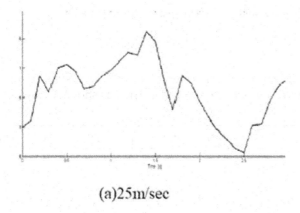

(a)25m/sec

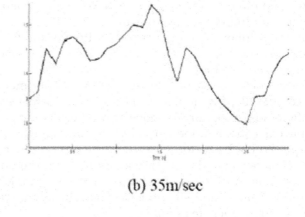

(b) 35m/sec

FIGURE 16.5
(a) 25 m/sec and (b) 35 m/sec.

of problem, the authors in [41] have proposed the following improvement. At first, the program initializes and creates a population of N_P individuals. The initial generation will then be evaluated. Following this evaluation, crossover and mutation operators are applied to the same parent population unconditionally; also, a mutation operator makes random mutations of the N_C. The population of the reproduced individuals then has size $3 \times N_M$. Then, the fitness of the new individual is calculated. Until now, the offspring individuals for the next generation are generated. Next, both the parent individuals N_P and the offspring individuals $3 \times N_M$ are compared via their fitness value to get a new population N_P. The weaker individuals may be eliminated.

The authors considered the population size, crossover, mutation probability, and termination generation as 10, 1, 1, and 50, respectively. Further, the vehicle is assumed to run at the speed of 25 and 35 m/sec and the random road profile is shown in Figure 16.5.

To accurately mimic a driving car, a relevant road profile needs to be formed. According to the ISO 8608 norm, a road profile can be mathematically composed, based on the assumption that a given road has equal statistical properties everywhere along a section to be classified. In other words, the road surface is a combination of a large number of longer and shorter periodic bumps with different amplitudes. Another input parameter for the road profile formulation is the road roughness factor, varying from 1 to 8, with 1 being a high quality (smooth) road surface like an asphalt layer and a road roughness factor of 8 represents very

poor road quality, as in roadway layers consisting of cobblestones.

In their study, ADAMS Version 2005 and MATLAB Version 7.0 were utilized to model this semi-active suspension system with a fuzzy controller-based GA. The ADAMS software performs the integration of dynamic vehicle model and MATLAB executes the GA-based fuzzy control system. During the simulation, both packages take simultaneous time steps. Each package independently computes the solution for its respective models.

With the optimized fuzzy control input using the improved GA, three independent runs were performed for each speed. The simulation values are depicted as follows:

Speed	Simulations	RMS Acc.	Percent of Improvement	Max. ⌊Acc.⌋	Percent of Improvement
25 m/sec	Run 1	45.873	83.9	314.801	56.0
	Run 2	45.873	83.9	314.799	56.0
	Run 3	45.873	83.9	314.799	56.0
35 m/sec	Run 1	47.059	86.4	176.031	79.2
	Run 2	46.999	86.1	180.570	79.3
	Run 3	46.999	86.1	180.570	79.3

The results tabulated above include the percentage of improvement with reference to the passive suspension system. Simulation results suggest that the optimal fuzzy control suspension registered more than 80% improvement in ride performance compared to the passive system.

Applying the GA evolution procedure to the suspension system, after 50 generations, the rule tables for the fuzzy controllers evolve as follows:

	NB	NM	NS	Z	PS	PM	PB
NB	5	6	7	5	5	5	6
NM	2	1	4	7	5	3	5
NS	5	5	4	1	6	6	6
Z	3	3	6	1	6	2	6
PS	5	1	4	6	4	7	3
PM	5	6	2	2	5	2	4
PB	6	6	7	6	7	6	5

For speed 25 m/sec

	NB	NM	NS	Z	PS	PM	PB
NB	7	6	5	5	7	6	5
NM	2	6	5	3	5	4	3
NS	6	4	2	5	4	7	4
Z	1	4	4	1	1	5	6
PS	1	4	6	5	5	5	3
PM	5	3	5	5	6	7	1
PB	7	5	6	6	6	7	5

For speed 35 m/sec

Simulation results suggest that a modest ride comfort improvement is observed when compared to the passive suspension. The optimal fuzzy control can bring about a possible solution to the ride comfort.

16.4 GA Aided Active Suspension System

In this section we will describe another case study where the suspension systems are designed to be active, which requires a power source to generate suspension forces according to some prescribed criterion. With such a strategy, ride comfort and track holding ability can be enhanced simultaneously. For further ride comfort improvements, active suspension systems can adjust the system energy to control the vibration of the vehicle body, leading to increased ride comfort. We do not debate here whether semi-active is better or active suspension is better; we just describe the design methodologies, which were hugely benefited by the intelligent control techniques.

In recent years, active suspension control technologies have become an extensive research topic, hence these systems have significant influence on the vehicle's subjective ride comfort impression. In this section, we describe the recent results presented in [42]. Fully active suspension systems have the ability to actively control the vertical movement of the vehicle body, relative to the wheels. This is achieved by applying an independent force on the suspension, accomplished hydraulically or electrically.

Haemers et al. [42] designed an optimal control for a full-car electromechanical active suspension. For this purpose, an active car suspension lab setup representing a full-car suspension system was built, based on the widely available theoretical full-car active suspension model [43]. This new approach allows one to accurately simulate a driving car without the need of two degrees of freedom for every wheel. Instead, it uses one fixed base and one moveable platform supported by six active rods, leading to a more convenient setup. Kinematic and dynamic analysis have been performed in order to assure system behavior that matches typical full-car dynamics.

We need to bring to the readers attention ISO 2631 norm according to which, driver comfort is quantified by the acceleration levels in the three principal axes of translation (vertical, longitudinal, and lateral). As the active car suspension lab setup kinematically doesn't allow for longitudinal and lateral movement, the control objective is to minimize the vertical accelerations in Z-axis, thus maximizing driver comfort. For simplification, a full-car active suspension model is often reduced to a half-car or even a quartercar model, meaning only one actively controlled spring-damper-wheel system is examined. This results in less degrees of freedom and an overall simpler structure, but subsystem interactions are neglected. Many control optimisation techniques have been carried out in order to obtain an adequate control for both full-car, half-car, and quarter-car active suspension models. For example, there were H_∞ and reduced order H_∞ controller designs available in the literature.

The full-car suspension arrangement is represented as a linearised seven degrees of freedom system. It consists of a single mass m_s representing the car body (or chassis) connected to four wheel masses—front left, front right, rear left, and rear right—at each corner. The vehicle body mass is free to heave (z translation), pitch (angular displacement θ), and roll (angular displacement ψ). The suspensions between vehicle body and wheel masses are modelled as a linear viscous damper in parallel with a spring element, while the tire elasticities are modelled as simple linear springs without damping. An active car suspension is equipped with the ability to impose a force on the wheel masses relative to the vehicle body mass, modelled as the forces f. For simplicity, all pitch and roll angles are assumed to be small. The accompanying motion equations can be found in [43]. From these motion equations, a dynamic state space representation can be extracted.

The state space mode has four states—the heave position z, heave velocity $\dot{z}$, pitch angular position θ, and velocity $\dot{\theta}$. There are five inputs—the actuator forces on each of the four wheel rods, and the gravity g; these are considered as two sets of inputs (force control inputs u_1 to each of the actuators of the suspension, and the collection of disturbances u_2, primarily due to gravity but, predefined by the road profile). To minimize the objective function, the accelerations in the upward z direction. An integral control has been used in addition to the state feedback control law to make the steady state error zero. Together, the task is to design a Proportional Integral (PI) controller. Let us enumerate a little more so that the optimization process becomes clear. Since there are 4 states and 4 inputs, the state feedback control matrix is 4×4. Likewise, the integral control matrix has size 4×2, as there are 2 references z and θ. Thus, there are 24 parameters, quite a good number to be determined, and this is where the genetic algorithms become handy.

As driver comfort is maximized for minimal acceleration in z-axis, the GA's objective funcion is to minimize the Root Mean Square (RMS) value (according to ISO 2631 norm) of the central body's acceleration in z-axis. Also, nonlinear constraints can be taken into account by the GA. The schematic of the algorithm is shown in Figure 16.6.

After passing through 10 generations and calculating for about 60 minutes, the genetic algorithm was reported to have successfully determined the MIMO controller's 24 optimal feedback gains K_0 and K_i. The resulting performance of this MIMO controller is compared to a full-car passive suspension. It was found that the vehicle's body acceleration is dramatically reduced, resulting in a much more comfortable ride. The driver comfort quantified by the RMS value of the acceleration changed from 3.8975 for passive control to 0.0530 in the case of the active control with the MIMO controller, meaning that there was a tremendous increase in driver comfort.

From these results, it can be concluded that a GA aided PI state feedback hybrid controller can be successfully designed without the need for further manual fine-tuning.

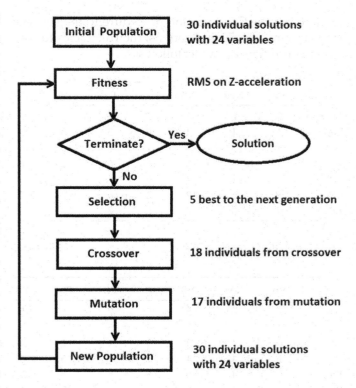

FIGURE 16.6
Adapted from Haemers, M. et al., *IFAC PapersOnLine*, 51, 1–6, 2018.)

16.5 Training ANNs Using GAs

In this part, we present the work of Montana and Davis [44]. The work is a bit dated, but we find it relevant in the current context, wherein the results of this work might be an inspiration to develop algorithms for deep neural networks.

Montana and Davis have reported that while multi-layered feedforward neural networks possess a number of properties that make them particularly suited to complex pattern classification problems, nevertheless, their application to some real-world problems has been hampered by the lack of a training algorithm that reliably finds a nearly globally optimal set of weights in a relatively short time. They proposed that GAs, which are primarily good at exploring a large and complex space, provide an intelligent way to find values close to the global optimum, and hence, they are well suited to the problem of training feedforward networks. In their work, they have described a set of experiments performed on data from a sonar image classification problem.

The performance degradation in multilayer feedforward neural networks appears to stem from the fact that complex spaces have nearly global minima, which are sparse among the local minima. Gradient search techniques tend to get trapped at local minima. With a high enough gain (or momentum), backpropagation can escape these local minima. However, it is not always assured and when the nearly global minima are well hidden among the local minima, backpropagation can end up bouncing between local minima without much overall improvement.

We now briefly discuss the main idea of employing a GA to optimize the weights in a given neural network. The five components of a GA are used in the following manner.

1. Chromosome Encoding: The weights (and biases) in the neural network are encoded as a list of real numbers.

2. Evaluation Function: First, the weights on the chromosome are assigned to the links in a network of a given architecture, and then the network is run over the training set of examples. Finally, the sum of the squares of the errors is returned.

3. Initialization Procedure: The weights of the initial members of the population are chosen at random with an a priori known probability distribution. This is different from the initial probability distribution of the weights usually used in backpropagation, which is a uniform distribution between −1.0 and 1.0. The choice of probability distribution reflects the empirical observation by researchers that optimal solutions tend to contain weights with small absolute values but can have weights with arbitrarily large absolute values. This leads to seeking the initial population with genetic material, which allows the genetic algorithm to explore the range of all possible solutions but tends to favor those solutions which are most likely a priori.

4. Operators: A large number of different types of genetic operators have been created and the goal of the simulation/experiment is to find out how different operators perform in different situations and thus to be able to select a good set of operators for the final algorithm. The operators can be grouped into the three basic categories: mutations, crossovers, and gradients.

We next discuss each of the operators individually, one category at a time.

a. **Unbiased-Mutate-Weights:** For each entry in the chromosome, this operator will find a weight with fixed probability $p = 0.1$ and replace it with a random value chosen from the initialization probability distribution.

b. **Biased-Mutate-Weights:** For each entry in the chromosome, this operator will find a weight with fixed probability $p = 0.1$ and add to it a random value chosen from the initialization probability distribution. We expect biased mutation to be better than unbiased mutation for the following reason. Right from the start of a run, parents are chosen, which tend to be better than average. Therefore, the weight settings in these parents tend to be better than random settings. Hence, biasing the probability distribution by the present value of the weight should give better results than a probability distribution centered on zero.

c. **Mutate-Nodes:** This operator selects n non-input nodes of the network, which the parent chromosome represents. For each of the ingoing links to these n nodes, the operator adds to the links of a weight a random value from the initialization probability distribution. It then encodes this new network on the child chromosome. The intuition here is that the ingoing links to a node form a logical subgroup of all the links in terms of the operation of the network. By confining its changes to a small number of these subgroups, it will make its improvements more likely to result in a good evaluation.

d. **Mutate-Weakest-Nodes:** The concept of node strength is different from the concept of error used in backpropagation. For example, a node can have zero error if all its output links are set to zero, but such a node is not contributing anything positive to the network and is thus not a strong node. We define the strength of a hidden node in a feedforward network as the difference between the evaluation of the network intact and the evaluation of the network with that node with its output links set to zero.

e. **Crossover-Weights:** This operator puts a value into each position of the child's chromosome by randomly selecting one of the two parents and using the value in the same position on that parent's chromosome.

f. **Crossover-Nodes:** For each node in the network encoded by the child chromosome, this operator selects one of the two parents' networks and finds the corresponding node in this network. It then puts the weight of each ingoing link to the parent's node into the corresponding link of the child's network. The intuition here is that networks succeed because of the synergism between their various weights, and this synergism is greatest among weights from ingoing links to the same node. Therefore, as genetic material gets passed around, these logical subgroups should stay together.

g. **Crossover-Features:** Different nodes in a neural network perform different roles. For a fully connected, layered network, the role that a given node can play depends only on which layer it is in and not on its position in that layer. In fact, we can exchange the role of two nodes A and B in the same layer of a network as follows. Loop over all nodes C connected (by either an ingoing or outgoing link) to A (and thus also to B). Exchange the weight on the link between C and A with that on the link between C and B. Ignoring the internal structure—the new network is identical to the old network, i.e. given the same inputs they will produce the same outputs.

The child produced by the previously discussed crossovers is greatly affected by the internal structures of the parents. The crossover features operator reduces this dependence on internal structure by doing the following. For each node in the first parent's network, it tries to find a node in the second parent's network that is playing the same role by showing a number of inputs to both networks and comparing the responses of different nodes. It then rearranges the second parent's network so nodes playing the same role are in the same position. At this point, it forms a child in the same way as crossover nodes. The greatest improvement gained from this operator over the other crossover operators should come at the beginning of a run before all the members of a population start looking alike.

h. **HillClimb:** This operator calculates the gradient for each member of the training set and sums them together to get a total gradient. It then normalizes this gradient by dividing by the magnitude. The child is obtained from the parent by taking a step in the direction determined by the normalized gradient of size step size, where step size is a parameter that adapts throughout the run in the following way. If the evaluation of the child is worse than the parent's, step size is multiplied by the parameter step-size-decay = 0.4; if the child is better than the parent, step size is multiplied by step-size-expand = 1.4.

This operator differs from backpropagation in the following ways: (i) weights are adjusted only after calculating the gradient for all members of the training set, and (ii) the gradient is normalized so that the step size is not proportional to the size of the gradient.

5. Parameter Settings: There are a number of parameters whose values can greatly influence the performance of the algorithm. Except where stated otherwise, we kept these constant across runs. We now discuss some of the important parameters individually.

a. **Parent-Scalar:** This parameter determines with what probability each individual is chosen as a parent. The second best individual is "Parent-Scalar" times as likely as the best to be chosen, the third-best is "Parent-Scalar" times as likely as the second-best, and so on. The value was usually linearly interpolated between 0.92 and 0.89 over the course of a run.

b. **Operator-Probabilities:** This list of parameters determines with what probability each operator in the operator pool is selected. These values were usually initialized so that the operators all had equal probabilities of selection. An adaptation mechanism changes these probabilities over the course of a run to reflect the performance of the operators, increasing the probability of selection for operators that are doing well and decreasing it for operators performing poorly. This saves the user from having to hand-tune these probabilities.

c. **Population-Size:** This self-explanatory parameter was usually set to 50.

In their pioneering work, Montana and Davis claim that through this fusion, they have demonstrated a real-world application of a genetic algorithm to a large and complex problem. Moreover, they claim that they have introduced a new type of training algorithm that is likely to outperform the backpropagation algorithm. The work described here only touches the surface of the potential for using genetic algorithms to train neural networks. In the realm of feedforward networks, there are a host of other operators with which one might experiment. As a general purpose optimization tool, genetic algorithms should be applicable to any type of neural network (and not just feedforward networks whose nodes have smooth transfer functions) for which an evaluation function can be derived. The existence of genetic algorithms for training could aid in the development of other types of neural networks.

16.6 Chapter Summary

Genetic algorithms and their cousins, like PSO algorithms, are capable of searching the entire solution space with more likelihood of finding the global optimum than conventional optimization methods. Indeed, conventional methods usually require the objective function to be well behaved (such as continuity), whereas the generational nature of GAs can tolerate noisy and discontinuous function evaluations. However, there are a number of considerations commonly arising in control engineering problems. A few of them are briefly touched upon below.

Representation

Continuous decision variables may be handled either directly through real-valued representations and the appropriate genetic operators or by using binary representation schemes and standard genetic operators. In the case of binary representations, real values can be approximated to the necessary degree with a fixed point binary scheme. In most control problems however, it is the relative precision of the parameters that is significant rather than absolute precision. In such cases, the logarithm of the parameter may be encoded reducing the number of bits and hence memory usage.

Scale

The concept of fitness is central to all GA approaches. Given that many optimization problems are characterised by a real-valued objective function, these values must be converted into a non-negative fitness value if they are to be handled correctly by the GA. Early work on GAs concentrated on the use of offsetting objective function values so that selection could be based directly on an individuals performance within a population. The use of scaling retains an individual's relative performance and also attempts to bias the selective pressure toward better individuals, yet allowing relatively unfit individuals the potential

to reproduce. Alternatively, by discarding the relative differences between the individual's raw performance and only considering them on their rank in a population, a constant selective pressure may be applied throughout the evolutionary process. Offsetting and scaling can result in more and more individuals receiving fitnesses with relatively small differences as the population converges. The rank-based methods maintain a constant selective pressure towards good individuals throughout the convergence process and is claimed to bring a number of other advantages.

Constraints

Most control engineering problems are subject to constraints, and GAs handle constraints in a number of ways. The most efficient and direct method is to embed these constraints in the coding of the individuals. Where this is not possible, penalty functions may be used to ensure that invalid individuals have a fitness that reflects that they are low performers. However, appropriate penalty functions are not always easy to design for a given problem and may affect the efficiency of the search.

Adaptation

The vast majority of applications of GAs have concentrated on their use as a function optimizer. However, GAs have been shown to be well suited to tracking time-varying systems (Dasgupta and McGregor, 1992), i.e. one in which the optimum fitness or fitness criterion change over time. The GAs have the advantage over many conventional methods of being able to respond to such changes by exploiting the diversity of the individuals in the current population. If there is insufficient diversity in the population, then new material can be readily introduced by replacing some individuals with randomly initialised individuals.

Parallelism

Apart from the obvious benefits in execution time, parallel GAs have a higher degree of robustness, can be made fault-tolerant, and typically require fewer function evaluations to reach optimal solutions. By employing a distributed population structure with local selection and reproduction and some form of genetic mobility, the performance of the GA may be enhanced over one where the population is treated globally. These benefits may be realised even when the GA is implemented on a sequential machine.

Lastly, the application of GAs to control engineering can be broadly classified into two main areas: off-line design and analysis and on-line adaptation and tuning. In off-line applications, a GA can be employed as a search and optimization engine—for example, to select suitable control laws for a known plant to satisfy given performance criteria or to search for optimal parameter settings for a particular controller structure. In online adaptation, GAs may be used as a learning mechanism to identify characteristics of unknown or non-stationary systems or for adaptive controller tuning for known or unknown plants. However, it was agreed upon in the literature that the off-line implementations are a particularly rewarding application area for GAs. In control systems design, and multiobjective optimization in particular, GAs offer the opportunity to address problems that where hitherto not susceptible to efficient solution by conventional methods. Such problems may be addressed directly through GAs and, by including the control engineer in the optimization process as a decision maker, afford the opportunity for the designer to guide the search while learning about the problems' trade-offs. There has been a rapid explosion in research in the past few years where learning systems, including artificial neural networks, classifier systems, and adaptive fuzzy controllers, may be coupled with GAs to expediate their learning.

Appendix 16A: Genetic Algorithm-Based Tuning of PI Controllers

Traditionally, for last several decades, the control-design of industrial processes/plants has been performed using the conventional PID control methods, which is due to its simplicity, low cost of design and robust performance in a wide range of operating situations; nearly 90% controllers are PID based [A16A.1]. However, design/tuning of a PID controller is largely intuitive, and could be difficult, if multiple objectives are to be achieved; of course, there are auto-tuning, adaptive, and compensation schemes-based PID controllers. In simple PID controllers, it is difficult to generate a derivative term, which can be used to reduce oscillations, due to use of high proportional gain; however, it can also enhance any random noise/error in the signal, and this gets reflected in the controller output. In many cases, the integral action (PI controller) makes the steady state (speed) error zero. Genetic Algorithm

(GA) is an efficient search technique that manipulates the coding of a parameter set to search a near optimal, in fact sub-optimal (and yet robust) solution by the process of cooperation and competition among the potential solutions. GA is very useful for industrial applications, because it can handle problems with nonlinear constraints, multiple objectives and dynamic components. GA has two main elements: (i) the encoding scheme represents the possible solutions to the problem, individual parameters can be encoded in some alphabets like binary strings/real numbers/vectors, and a population pool of chromosomes is installed and set to a random value; and (ii) the evaluation function; in each cycle of GA-evaluation, a subsequent generation is created from the chromosomes in the current population, and the cycle of evaluation is repeated until a termination criterion is reached, i.e. a fitness value. In this appendix, GA-based tuning of PI controllers is briefly discussed [A16A.1].

A16A.1 PI Control Design Procedure

In a proportional-integral (PI) controller, P depends on the present error, and I depends on the accumulation of past errors; then, a weighted sum of these two is used to adjust the plant via a control element, by tuning these two parameters in the PI controller algorithm. The TFs of the PI controller for continuous and discrete time systems (CTS, DTS) are given as

$$G_c(s) = K_p + \frac{K_i}{s}; \quad G_c(z) = K_p + K_i T\left(\frac{z}{z-1}\right);$$
(16A.1)

$$T = \text{sampling interval}$$

A simple procedure is: (i) SET K_p, starting with $K_p = 0$ and $K_i = 0$; increase K_p until the output starts overshooting and reducing the settling time significantly; (ii) SET K_i; Increase K_i until the final error is equal to zero.

Algorithm for the Design of PI Controller

First, let us have the closed loop transfer function (CLTFs) of the systems (CTSTF and DTSTF) with controller as

$$TF(s) = G_{cl}(s) = \frac{G(s)G_c(s)}{1 + G(s)G_c(s)};$$

$$TF(z) = G_{cl}(z) = \frac{G(z)G_c(z)}{1 + G(z)G_c(z)}$$
(16A.2)

The usual performance criteria for the controller design are specified as: (i) settling time ≤ 3 sec.; (ii) peak overshoot ≤ 2%, and (iii) steady state error ≤ 1%. Normal steps for the design of PI controller are: (i) read the open loop transfer function (OLTF) of the given higher order system; (ii) form the CLTF; (iii) obtain the step response of closed loop control system (CLCS); (iv) check/tally the response for the required specifications, if not satisfactory, obtain a reduced order model of the system, and design a new controller for the reduced order model; (v) obtain the initial values of the parameter K_p and K_i by pole-zero cancellation; (vi) cascade the controller with reduced order model and get the CLCS's response with these initial values; (vii) find the optimum values for the controller parameters which satisfy the required specifications; (viii) with the optimum values, cascade this controller with the original system; (ix) obtain the CLCS's response of the system with the controller; and (x) if the specifications are met then design is done, otherwise tune the parameters till these meet the required specifications.

A16A.1.1 Existing Tuning Methods

The conventional tuning methods are: (i) Ziegler-Nichols (Z-N) method, (ii) magnitude optimum (MO) method and (iii) symmetric optimum (SO) tuning method.

   i. **Z-N method**: In this very useful method, the tuning formula is obtained when the plant model is given by a first order plus dead time (FOPDT):

$$G(s) = \frac{ke^{-s\tau}}{1+sT_c} ; \ T_{tc} = \text{time constant} \quad (16A.3)$$

To extract the parameters in (16A.3), one can use the step response from which the parameters, k, τ, and T_{tc} (or $a = k\tau/T_{tc}$) can be extracted, see Figure A16A.1a; and with these, the Ziegler–Nichols formula of Table A16A.1 can be used to get the controller parameters.

Also, if the frequency response experiment can be performed, the crossover frequency ω_c and the gain K_c can be determined from the Nyquist plot, see Figure A16A.1b, and one can use Table A16A.1; with $T_c = 2p\omega_c$. The demerits of Z-N method area: (i) the appropriate values of τ, and T can be obtained only if the curve obtained is S-shaped, otherwise deviations in values might occur, (ii) further fine tuning is needed, (iii) controller settings being aggressive, large overshoot and oscillatory responses might occur, (iv) poor performance for plants with a dominant delay, (v) closed loop sensitive to parameter variations, and (vi) parameters of the step response may be hard to determine due to measurement noise.

  ii. **Magnitude optimum (MO) and symmetric optimum (SO) methods:** These are loop shaping tuning methods. The controller is able to shape the available OLTF of the system in a desired manner. The requirements are specified for the CLTF $Gcl(s)$:

$$G_{cl}(s) = 0; \ \lim_{\omega \to} \left(\frac{d^n(G_{cl}(j))}{d\omega^n} \right) = 0; \quad (16A.4)$$

for as many n as possible

Let the desired OLTF be

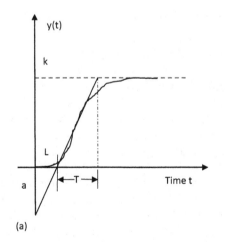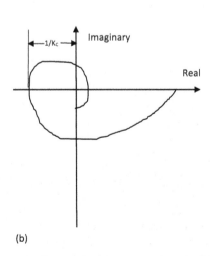

(a)   (b)

FIGURE A16A.1

(a) Step response of a typical first order system, (b) Frequency response. (Adapted from Anon, http://shodhganga.inflibnet.ac.in/bitstream/10603/16159/8/08_chapter%203.pdf.)

TABLE A16A.1

The Z-N Tuning Formulae

Controller	Using Step Response			Using Frequency Response		
	K_p	T_i	T_d	K_p	T_i	T_d
P	$1/a$	–	–	$0.5K_c$		
PI	$0.9/a$	3τ	–	$0.4K_c$	$0.8T_c$	
PID	$1.2/a$	2τ	$\tau/2$	$0.6K_c$	$0.5T_c$	$0.12T_c$

$$G_{ol}(s) = \frac{\omega_n^2}{s(s + 2\xi\omega_n)} \quad (16A.5)$$

The PI controller is used when it is possible to approximate the model of the plant with the TF:

$$G(s) = \frac{K}{(1 + T_1 s)(1 + T_2 s)}; \quad T_2 < T_1 \quad (16A.6)$$

By setting the damping ratio as 0.707, the PI parameters can be calculated as

$$Kp = \frac{T_I}{2KT_2}; \quad T_I = T_1; \quad \omega_n = 0.707/T_2 \quad (16A.7)$$

The dominant pole is canceled by the zero of the PI controller and the closed loop dynamics are determined with the smaller time constant T_2. The MO design method optimizes the CLTF $G_{cl}(s)$ between the reference input and the output signal. It often cancels the plant poles by the controller zeros that can lead to poor performance in response to load disturbance.

The objective of the SO method is to obtain an OLTF

$$G_{ol}(s) = \frac{b\omega_c^2\left(s + \dfrac{\omega_c}{b}\right)}{s^2(s + b\omega_c)} \quad (16A.8)$$

In (16A.8), the parameter a is related to the phase margin (PM) of the control system as

$$PM = \theta = 2b\tan\left(\frac{b-1}{b+1}\right); \quad b = \frac{1 + \sin\theta}{\cos\theta} \quad (16A.9)$$

The method maximizes the *PM* and leads to symmetrical phase and amplitude characteristics. In (16A.8), there is a (phase) lead TF that supplies phase at the cross over frequency. The common choice of the parameters is: $b = 2$ and gives the PM as 37 deg. The SO method can give a good

response to a load disturbance, but would give large overshoot, which can be reduced by using a 2 DOF controller or with a pre-filter. Because, the controller tuning methods are available for second order system, for higher order models, model order reduction is required; the main aim of model order reduction is to design a controller of lower order that should effectively control the original higher order system. There are two approaches: (i) obtain the controller on the basis of reduced order plant model, the plant-reduction approach is computationally simpler, since it deals with lower order model/controller, but errors are introduced in the design process because the reduction is carried out at the early stages; and (ii) the controller is designed for the higher order system, then, in the controller reduction approach, error propagation is minimized as the design process is carried out at the final stages of reduction, but this introduces computational complexity, because higher order models are used. In this approach, the CL response (Bode-frequency response) of higher order controller with original system is reduced, and a lower order TF is fitted, so that in the required range of frequency, both the responses match closely.

A16A.2 GA-Based Approach

A GA can be used to reduce the mismatch between the given higher order and the resulting reduced order models. GA is used for the following: (i) to reduce the error value between given higher order and obtained reduced order model; and (ii) to determine the optimized value of PI controller parameters namely K_p and K_i.

A16A.2.1 Initialization of the Parameters

The GA-based procedure for PID Controller tuning is: (i) initialize the population (K_p, K_i); (ii) give the specifications as: settling time ≤ 3 sec., and peak overshoot ≤ 2%; (iii) if generations > max., then stop, otherwise continue; (iv) reproduction, (v) crossover, (vi) mutation, (vii)—generation = generation + 1, (viii) go to step (ii). For GA one needs to select the parameters: (i) population size, (ii) bit length of the chromosome to represent the information, (iii) number of iterations, (iv) type of cross over and mutation; the typical values of which are given taken as: (i) population type: double vector, (ii) population size: 10, (iii) length of chromosome: 6, (iv) number of generations: 100, (v) crossover: single point, (vi) crossover function: intermediate, (vii) crossover probability:1.0, and (viii) uniform type mutation. In each generation, the genetic operators (reproduction, crossover, with a fraction of 0.8, and mutation) are applied to selected individuals (K_p and K_i) from the current population and a new population is created. The arithmetic crossover generates an offspring as a component-wise linear combination of the parents in latter phases of evolution. It is desirable to keep individuals intact and to use an adaptively changing crossover rate, higher rates in early phases and a lower rate at the end of the GA. In the beginning, mutation operators resulting in bigger jumps in the search space are preferred, but, when the solution is approaching, a mutation operator leading to slighter shifts, in the search space should be preferred. An example is illustrated here, and the detailed results including graphs/plots can be found in [A16A.1].

A16A.3 Numerical Example

Consider a 4th order system TF

$$G(s) = \frac{14s^3 + 248s^2 + 900s + 1200}{s^4 + 18s^3 + 102s^2 + 180s + 120} \quad (16A.10)$$

The reduced 2nd order TF of the system is

$$G_r(s) = \frac{11.98s + 12.53}{s^2 + 2.138s + 1.253}$$

The controller parameters are determined from the Z-N tuning method in Table A16A.2.

The GA tuning of the reduced order transfer function results in: $Kd = 1.00404$, $Kp = 22.84908$ and $Ki = 25.6977$. Various results are summarized in Tables A16A.3 and A16A.4.

It is has been observed that [A16A.1]: (i) the conventional Ziegler-Nichols PI controller tuning method has peak overshoot more than the GA-based method, (ii) The GA-based PI controller reveals shorter settling time and rise time, (iii) produces an output which is quite ahead of that of PI in the rise time analysis, and (iv) set point tracking/disturbance rejection has been obtained in the GA-PI controller.

TABLE A16A.2

Parameters Selected by Z-N Method

Controller	K_p	K_i	K_d
P	7.32	0	0
PI	6.59	158	0
PID	8.78	35.16	0.055

TABLE A16A.3

Comparison of Time Domain Specifications: Higher Order System (HOS)

Specifications	HOS	Z-N with PI Controller	Z-N with PID Controller	GA with PID Controller	GA with PI Controller
Rise time (sec.)	1.69	0.0157	0.0285	0.122	0.0069
Peak-overshoot (%)	0.433	13	1.4	0.0405	0.061
Settling time (sec.)	2.55	0.107	0.12	0.398	0.0124
Peak time (sec.)	3.98	0.041	0.0815	0.97	0.0266
Peak amplitude	10	1.13	1.01	1	1

Source: Anon, http://shodhganga.inflibnet.ac.in/bitstream/10603/16159/8/08_chapter % 203.pdf.

TABLE A16A.4

Comparison of Time Domain Specifications: Lower Order System (LOS)

Specifications	ROS	Z-N with PI Controller	Z-N with PID Controller	GA with PID Controller	GA with PI Controller
Rise time (sec.)	1.69	0.0178	0.0312	0.109	0.00803
Peak-overshoot (%)	0.433	14.4	2.06	0.0116	0.0075
Settling time (sec.)	2.55	0.12	0.243	0.189	0.0143
Peak time (sec.)	3.98	0.0477	0.106	0.41	0.0383
Peak amplitude	10	1.14	1.02	1	1

Source: Anon, http://shodhganga.inflibnet.ac.in/bitstream/10603/16159/8/08_chapter % 203.pdf.

References

A16A.1 Anon. http://shodhganga.inflibnet.ac.in/ bitstream/10603/16159/8/08_chapter%203. pdf, accessed December 2018.

Appendix 16B: Helicopter Control Design Process Using Genetic Algorithm

A military helicopter has to satisfy certain so called handling qualities (HQ) requirements (ADS-33; U.S. Army Aviation and Troop Command, 1994), which specify the allowable cross-coupling and the required frequency responses of the helicopter for the specific mission [A16B.1]. A helicopter is characterized by the instability, the higher degrees of freedom (9 to 15 state variables, or even more), and the high degree of coupling between some of these state variables. Mainly two approaches have been followed for the design of a controller of a helicopter: (i) eigen-structure assignment [A16B.2], and (ii) H$_2$, H$_\infty$ and μ synthesis. The HI (H-infinity/H$_\infty$) method can give robust designs to uncertainties and reliable algorithms; however, it depends on the proper selection of the weighting functions (Chapter 8 for HI design approach). These weighting functions can shape the sensitivity functions, and are obtained by analysis of uncertainties present in the system, and from frequency- and / or time-domain specifications. With the weighing functions, the design can be formulated as a multi-objective optimization problem, in which both time- and frequency-domain specifications are considered simultaneously. The GA's can be used to solve complicated nonlinear optimization problems.

A16B.1 Mathematical Aspects of the Helicopter and HQ Requirements

The LTI mathematical model was obtained from a fully nonlinear flight dynamic model of the Sikorsky UH-60A helicopter using a numerical linearization about a trimmed flight condition at 30-kn forward flight [A16B.1]. The basic model with 25 states consists of: (i) fuselage rigid body modes, (ii) flap degrees of freedom for main rotor and tail rotor, and (iii) first harmonic dynamic inflow for the main rotor and tail rotor dynamic inflow; and the model was further augmented

with an 8-state model of the actuator dynamics (of four identical, uncoupled, second TFs. In fact, the highest order model with 33 states had been used for performance evaluation of the design. A reduced-order model with nearly 17 states was used as the nominal design model. The aim of rotorcraft control system design was to achieve level 1 ADS-33D HQ performance-specifications/requirements [A16B.3], and these specifications are divided into hover/low speed and forward flight regimes, and the helicopter in the present case was a utility aircraft and falls under the "moderate maneuvering" MTE category. Satisfaction of the design requirements is assumed to be for an attitude command attitude hold (ACAH) response-type performing the "target acquisition and tracking mission task element" in a low-speed flight with a useable cue environment (UCE) of 1 and divided attention operations [A16B.1].

A16B.2 H$_\infty$ Design with Genetic Algorithm

A continuous time LTI system with TF $G(s)$, and the forward loop compensator as $K(s)$ is considered with a unity feedback, as a tracking control problem. The variables are: (i) closed loop error e, i.e. tracking error, (ii) reference input r, (iii) u as the control input to the plant, and (iv) y as the (feedback) output of the plant affected by the disturbance d. A generalized system is constructed using the frequency-dependent weighting functions: $W_1(s)$ for tracking error signal $e(.)$, $W_2(s)$ for the control input $u(.)$, and $W_3(s)$ for the output $y(.)$. The CLTF of this system is given as

$$G_{cl} = \begin{bmatrix} W_1 S \\ W_2 KS \\ W_3 GKS \end{bmatrix} = \begin{bmatrix} W_1 S \\ W_2 R \\ W_3 T \end{bmatrix} \quad (16B.1)$$

In (16B.1), $S(s)$ is the sensitivity function, and $T(s)$ the complementary sensitivity function (see Chapter 8). The idea is to find a rational TF controller that stabilizes the CLCS and satisfies the criterion

$$\|G_{cl}\|_\infty \le \gamma; \quad \gamma > \gamma_0, \gamma_0 = \min\|G_{cl}\|_\alpha \quad (16B.2)$$

To guarantee the good ability of rejecting the disturbance and tracking, low-pass filters are used on the diagonal of W_1; whereas the W_2 and W_3 are high-passed. The following inequality needs to be satisfied

$$[\bar{\sigma}(W_1^{-1}(j\omega)) + \bar{\sigma}(W_3^{-1}(j\omega))] \ge 1 \quad (16B.3)$$

In the mixed sensitivity control design problem, the weighting matrices are considered as the design parameters, then the HI design problem is treated as a minimization problem with the inequality constraints:

$$\begin{aligned} &\underset{W_1,W_2,W_3}{\text{minimize}}\,\psi\,(W_1,W_2,W_3); \\ &\text{subject to } \bar{\sigma}(W_1^{-1}(j\omega))+\bar{\sigma}(W_3^{-1}(j\omega))] \geq 1 \end{aligned} \tag{16B.4}$$

We also have

$$\gamma_0(W_1,W_2,W_3) \leq \varepsilon_0;\ \phi_i(W_1,W_2,W_3) \leq \varepsilon_i;\ i=1,2,3,...,n \tag{16B.5}$$

In (16B.4), ψ is the performance index to be minimized, and in (16B.5), ϕ_i's are the indices that denote rise time, overshoot, and BW; which are to be kept below/or within certain bounds/ranges as specified. The proper weighting matrices can be found, and simultaneously the controller can be synthesized by solving (16B.4) and (16B.5). The constrained problem is usually a non-convex, non-smooth, and multi-objective with several conflicting design criteria, it is difficult to solve it by using the conventional approach. This requires an efficient numerical search algorithm like a multi-objective genetic algorithm.

A16B.2.1 Chromosome Coding

The structures of the weighing matrices are selected as

$$W_1 = diag(w_{11},w_{12},...,w_{1m})$$
$$W_2 = diag(w_{21},w_{32},...,w_{2m}) \tag{16B.6}$$
$$W_3 = diag(w_{31},w_{32},...,w_{3m})$$

In (16B.6), the individual weights (their TFs; since these are also considered filters) are selected as

$$w_{1i} = \frac{k_{1i}\left(1+s\dfrac{1}{\omega_{1i}}\right)}{\left(1+s\dfrac{1}{\omega_{2i}}\right)};\ w_{3i} = \frac{k_{3i}\left(1+s\dfrac{1}{\omega_{3i}}\right)}{\left(1+s\dfrac{1}{\omega_{4i}}\right)}; \tag{16B.7}$$

$$w_{2i} = k_{2i};\ i=1,2,...,m$$

In (16B.7), m is the dimension of the output vector. Each parameter in (16B.7) is coded by a binary string, and a chromosome or an individual is generated by joining all the strings in series, which is defined by

$$g_r = \{\omega_{11},\omega_{12},...,\omega_{4m},k_{11},k_{12},...,k_{3m}\} \tag{16B.8}$$

The algorithms for the design process are given [A16B.1] as follows:

Algorithm 1: to compute the objective function

Step1: For each individual, generate the corresponding weights

Step2: If the constraint (16B.3) is satisfied, then

a. Synthesize the controller $K(s)$ based using the HI method, and compute γ_0;

b. If $\gamma_0 < \varepsilon_0$, compute the performance indices of the CLCS; compute the objective function as

$$f = \psi + \sum_{i=1}^{n} r_i;\ r_i\begin{cases} 0, & \phi_i \leq \varepsilon_i \\ (\varepsilon_i - \phi_i)^2, & \phi_i < \varepsilon_i \end{cases} \tag{16B.9}$$

c. If $\gamma_0 > \varepsilon_0$, then the objective function is defined as $f = n\gamma_0$.

Step3: If $[\bar{\sigma}(W_1^{-1}(j\omega))+\bar{\sigma}(W_3^{-1}(j\omega))] < 1$, then take the function value as $f \sim 10^6$.

A linear ranking approach may be used to convert the objective function to the fitness value.

Algorithm 2: to optimize the weighting matrices for HI controller

Step1: For the nominal plant $G(s)$, determine y and ϕ_i.

Step2: Determine ε_0 and ε_i in terms of the requirements on the performance.

Step3: Determine the structures of W_1, W_2 and W_3, and the search domains of the parameters.

Step4: Select the control parameters for GA: population size, crossover rate, mutation rate, etc.

Step5: Randomly generate the first population.

Step6: Compute the objective value and assign the fitness value to each chromosome.

Step7: Start the process of GA search: (a) perform selection operation using elitist and tournament selection strategies, (b) perform crossover and mutation and generate new individuals, and (c) compute the objective values of the new individuals.

Step8: Terminate if the condition (e.g. the number of generation exceeds the given maximum D); otherwise go to Step 7.

A16B.3 Helicopter Flight Control Design-Objective Function

The reference input vector was defined to be the pilot commands in heave, pitch, roll, and yaw:

$$r(t) = [\delta_c \ \delta_e \ \delta_a \ \delta_p]^T = \text{in}[heave, pitch, roll, yaw] \quad (16\text{B}.10)$$

The control input vector to the plant (to helicopter dynamics) consists of the swashplate tilt and main and tail rotor collective:

$$u(t) = [\theta_0 \ \theta_{1s} \ \theta_{1c} \ \theta_{0T}]^T \quad (16\text{B}.11)$$

The pitch rate q and roll rate p are included in the ACAH design to help stabilize the plant, as rate feedback, the feedback (output) is defined by

$$y(t) = [\dot{H}\theta + k_q q\varphi + k_p pr]^T; \ k_p = k_q = 0.1; \ H= \text{altitude/height} \quad (16\text{B}.12)$$

Now, the objective functions can be chosen for level 1 ADS-33D as:

1. Attitude-quickness response is described by error functions produced by on-axis step responses of the helicopter. The flight control system is a 4-input and 4-output system. Let $y_{ji}(t)$ ($i = 1, 2, 3, 4; j = 1, 2, 3, 4$) denote the outputs of the CLCS to unit step inputs $r_i(t) = 1(t)$ ($i = 1, 2, 3, 4$), then error functions are specified:

$$e_i(t) = \int_0^{T_0} [r_i - y_{ii}(t)]^2 dt; \ e_{ji}(t) = \int_0^{T_0} [0 - y_{ji}(t)]^2 dt; \ j \neq i \quad (16\text{B}.13)$$

In (16B.13), $y_{ii}(.)$ and $y_{ji}(.)$ are the same and cross axes responses, and T_0 is the time constant (>transient time), is taken as 5 sec. Thus, the PI to be minimize is specified as

$$\psi = k_e \max_i(e_i); \ k_e = 0.5 \quad (16\text{B}.14)$$

2. The degree of inter-axis coupling is reflected by the off-axis response errors, e_{ji}, then

$$\phi_1 = \max_{j \neq i}(e_{ji}); \ \varepsilon_1 = 0.05 \quad (16\text{B}.15)$$

3. The collective climb rate (to step command input) should be like a delayed first order response

$$\phi_2 = \max_t(y_{11}(t)); \ \varepsilon_2 = 1 \quad (16\text{B}.16)$$

4. Denote the BWs for the pitch, roll, and yaw axes as: $\omega_{b\theta}, \omega_{b\varphi}, \omega_{b\psi}$ and $\tau_\theta, \tau_\varphi, \tau_\psi$ as the time delays these axes. Then, other control system performance indices are specified (16B.5) as [A16B.1]:

$$\phi_3 = -\omega_{b\theta}, \qquad \varepsilon_3 = -2$$

$$\phi_4 = \tau_\theta - 0.1\omega_{b\theta}, \quad \varepsilon_4 = -0.05$$

$$\phi_5 = -\omega_{b\varphi}, \qquad \varepsilon_5 = -2.5$$

$$\phi_6 = \tau_\varphi, \qquad \varepsilon_6 = 0.16 \qquad (16\text{B}.17)$$

$$\phi_7 = -\omega_{b\psi} \qquad \varepsilon_7 = -3.5$$

$$\phi_8 = \tau_\psi - 0.11\omega_{b\theta}, \ \varepsilon_8 = -0.225$$

5. For the pole placement requirement, we specify that there are l number of pairs of complex poles: $(\zeta_i, \omega_{di}); \ i = 1, 2, ..., l; \ \omega_{di} \geq 0.5; \ \zeta_i \geq 0.35$, the damping ratio, then the performance index is specified as

$$\phi_9 = -\min_i(\zeta_i); \ \varepsilon_i = -0.35 \quad (16\text{B}.18)$$

Certain parameters for the GA are chosen as

$$R_1 = [10^2, 10^3], R_2 = [10^{-3}, 10^{-2}], R_3 = [10^{-3}, 10^{-2}]$$

$$R_4 = [10^2, 10^3], R_5 = [50, 10^3], \quad R_6 = [10^{-2}, 10], R_7 = [0.01, 10]$$

$$k_{11} = k_{12} = k_{13} = k_{14}; \quad k_{31} = k_{32} = k_{33} = k_{34} = 1 / k_{11}$$

$$(16\text{B}.19)$$

The other parameters are: population size, N = 31, cross-over rate = 0.85, mutation rate = 0.001, generation termination $D = 40$, and tournament size = 2. With these parameters and using Algorithms 1 and 2, the satisfactory weighing matrices were found, and the controller design based on H-infinity and GA was achieved with $\gamma_0 = 0.5$. The stability robustness test was satisfied in the entire range of frequency, the dynamic responses to the pilots commands were satisfactory (simulated with the original 33rd order model of the helicopter), and the design also met the ADS-33D HQ requirements for level 1 [A16B.1], wherein the detailed results with plots can be found.

References

A16B.1 Dai, Ji-yang, and Mao, Jian-qin-. Robust flight controller design for helicopters based on genetic algorithms, *Aeronauticaet Astronautica Sinica*, Vol. 22, No. 5, pp. 471–473,2001;http://hkxb.buaa.edu.cn/EN/Y2001/V22/I5/471; also, 15th Triennial World Congress of IFAC, Barcelona, Spain, 2002; (Elsevier-IFAC Publications; IFAC Proceedings Volumes, Vol. 35, No. 1, pp 73–78, 2002. https://ac.els-cdn.com/S1474667015387681/1-s2.0-S1474667015387681-main.pdf?_tid=8d5031e4-01dc-45c1-ace5-8d3beb1eb885&acdnat=15-44599470_df9366857e-3a3f91774f0db2ec9c9003, accessed December 2018.

A16B.2 Srinathkumar, S. *Eigenstructure Control Algorithms: Applications to Aircraft/Rotorcraft Handling Qualities Design.* IET/IEE Control Engineering, London, UK, 2011.

A16B.3 Anon. Directorate for Engineering, *Aeronautical Design Standard: Handling QualitiesRequirements for Military Rotorcraft (ADS-33D)*, U. S. Army Aviation and Troop Command, St. Louis, MO, 1994.

Appendix IVA: Artificial Intelligence, Intelligent Systems, and Intelligent Control

We briefly dwell on certain aspects and features of artificial intelligence (AI), some area of AI, intelligent systems (IS), and intelligent control (IC). Since we believe we are an intelligent species, (as an outcome of long drawn evolutionary process, say millions of years), we would like to eventually build intelligent-systems-based technology around us to help us in carrying out ever increasing complex and sophisticated tasks to facilitate our useful and happier living on this planet. Now that we have invented and used computers, mainly digital, we further want them to think, since we are also a thinking (natural-) machine. Hence, we would like to have the automation of activities that are associated with our own thinking, like learning, and decision making. An attempt in the realm of AI is an art of creating (mostly computing smart) machines that perform functions that require intelligence (of humans). In order to build AI-based systems, we need to study our own mental faculties, may be through the use of computational models. Hence, one important facet of the pursuit of AI is the study of computations that make it possible to perceive, reason, and act. Figure AIVA.1 depicts several aspects of AI and areas of AI: search, logic, knowledge representation, machine learning, planning, visualization, building expert systems, and utilization for a desired application, e.g. robotics.

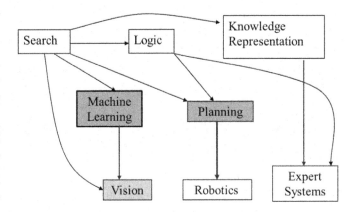

FIGURE AIVA.1
Main ingredients of Artificial intelligence (AI) and areas of AI.

AIVA.1 Concept of AI

AI is a field of study that seeks to explain, emulate (human's) intelligent behaviour in terms of computational processes, it is also considered as a branch of computing (computer) science that is concerned with the automation of intelligent behaviour (AIVA.1). The four quadrants of AI (AIQ) are depicted in Figure AIVA.2.

Let us pick up the element (2,1) of the AIQ which signifies: systems that act like humans. Here, by systems, we mean: (i) an algorithm (like an agent), (ii) a SW program, (iii) HW system (a smart device/sensor/computer), and/or (iv) any switching mechanism that has a logic. The main aspect here being how to make the computers (in fact any system that has some logic and computing mechanism), to do things at which, in the present/current times, the humans are better (in some qualitative or quantitative manner) than other species.

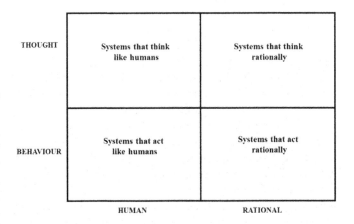

FIGURE AIVA.2
Four quadrant of AI: each element (*i*, *j*) of the matrix signifies an import aspect of AI.

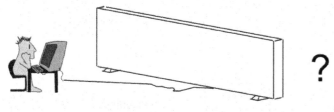

FIGURE AIVA.3
A system, i.e. a computer, that thinks like a human.

When certain tasks are performed by the machines, then we want them to perform the way we perform these tasks using our intelligence. This is highlighted by the so called Turing test, see Figure AIVA.3; assume you have entered a room that has a computer terminal, and within a fixed period of time, you type what you want, and study the replies. At the other end, there is either a human, or a computer. If it is a computer, and at the end of the period of time, if you cannot reliably ascertain whether it is a system/computer, or a human, then the system can be deemed to be an intelligent one. Here, the idea is to achieve human-level performance in all cognitive tasks. These include: (i) natural language processing for communication with human, (ii) knowledge representation to store processed information effectively and efficiently, (iii) automated reasoning to retrieve and answer questions using the stored information-cum-knowledge (ICK), and (iv) machine learning (ML) to adapt to new circumstances. The complete Turing test includes two more aspects: (a) computer vision to perceive objects, e.g. seeing, and (b) robotics-to move objects, i.e. acting.

Let us now pick the element (1,1) of the AIQ which signifies: systems that think like humans. Here, the emphasis for the systems is to think like we think. This requires cognitive modelling, that is the humans as observed from 'inside' (mainly this involves study of our biological neural network). We can do this by introspection vs. psychological experiments. The idea, thus in cognitive science is to make computers think, i.e. the machines with minds, relatively, in nearly complete and literal sense. Here, we call upon the automation of activities that humans associate with their own thinking: decision making, problem solving, and learning.

Let us now pick the element (1,2) of the AIQ which signifies: systems that think rationally. Here, the emphasis for the systems is to think rationally. Ironically, humans are not always 'rational' in their thinking. As we know from the evolutionary nature, that all (or almost all) biological species are neither kind nor cruel, they are just indifferent, i.e. often we observe that these are cruel to the members of the other species, and kind to their own members. In fact, these species do not think rationally or irrationally. But, with the highest level of intelligence, the humans can think irrationally also, since the very fact of ability of rational thinking in humans, can produce an irrational thought as a by-product (under existential pressure, difficult times, under fear, over-sentimentalism, over-reactions, knee-jerk responses). The rationality can be defined in terms of logic, but logic cannot express every aspect of act. The uncertainty is difficult to rationalize. Of course, now fuzzy logic can be used to define a kind of uncertainty that is called vagueness. The logical approach would need more computational time, and programming, since it also needs 'guidance'. The demand here is to study the mental faculties by using computational models, and evolve the computational mechanisms that make machines perceive, reason, and act.

Let us now pick the element (2,2) of the AIQ which signifies: systems that act rationally. Here, the emphasis for the systems is to act rationally, which signifies doing the right thing. This has to be in terms of maximizing the goal function (i.e. achievement), given certain amount of information. Here, it is not very important whether the AI-system replicates a human thought process, makes the same decisions as humans do, or uses purely a logical reasoning. The latter is a part of the rational agent, and does not necessarily cover all of the rationality. This is because, sometimes a logic cannot reason a correct inference. In such cases, a specific-in domain human ICK is used. This element of AIQ suggests that one should study the AI as a rational agent, because it is logic+domain knowledge. The rational agent (RA) is an entity that perceives and acts. In an abstract sense, RA is a function from percept histories (past actions, experiences) to actions: f: P→A. So, here we seek an agent with best performance for a given environment and task, e.g. design a best program for given machine resources.

So, AI can be in plain term stated as: produced by human art or an effort rather than originally done naturally, and is the ability to acquire knowledge and use it. Thus, in AI, we develop the ideas that enable the computers to be intelligent.

AIVA.2 Foundation of AI

The foundation of AI has major constituents: (i) philosophy, (ii) psychology, (iii) computer engineering, (iv) control theory and cybernetics, and (v) linguistics. The origin of AI could be that, the study of human intelligence began with no formal expression, and the ideas were initiated about mind that it can pave the way for the machines to act as humans, and perform their internal operations as well. The mathematics imparts a formalism to the three main aspects of AI: (i) computation, (ii) logic, and (iii) probability (which can be replaced or aided by the fuzzy logic). The computation leads to analysis of the problems that can be made amenable to be computing. The logic provides some reasoning, like If,…, Then,…, etc. Probability assigns a degree of belief to handle uncertainty in AI. Then, the decision theory combines probability and utility theory to arrive at some final actions. The fuzzy logic assigns a degree of 'belongingness' and permits to incorporate human experts' knowledge. In psychology, we ask questions: how humans think and act? It is a study of human reason and acting, and thus provides reasoning models for AI. Thus, in this field of study, the humans and other animals can be thought of as information processing/ gathering/giving machines to support the developmental ideas for AI. In the areas of computer engineering, we work to develop efficient computers, algorithms, time-sharing concepts, object oriented programs, and concurrent processing, and even build the parallel computing machines/algorithms based on the architectures of recurrent neural networks. The control theory and cybernetics try to answer the question of how can artefacts operate under their own control? These artefacts tend/try to adjust their actions to do better for the environment over time, based on an objective function/performance criterion, and feedback from the environment. These aspects are also applicable to language, vision, and planning. As for the linguistics, for understanding natural languages for the purpose of use in AI, different approaches can be adopted from the linguist work: syntactic, semantic, and knowledge representation.

Some merits of AI are: (i) more powerful, more useful computers have/can come out, (ii) the interfaces have improved, (iii) very difficult and new problems are being solved, and (iv) information is handled in better ways, and relieves the information overload, because it is converted into knowledge. The demerits of AI are: (i) costs are increased, (ii) the SW development is slow/ expensive, and (iii) experienced programmers are a few.

AIVA.3 Knowledge Representation and Learning for/in AI

Acquiring information (and data), learning (something required), and knowledge representation and reasoning (KRR) are most important aspects in AI. The point is that if we are going to act in a rational manner, then we should be able to describe that environment and draw some inferences by representing the acquired information in a useable form. Questions in the domain of KRR are: (i) how to describe the existing knowledge, (ii) how

to do it concisely, (iii) how do we do it rightly, (iv) how do we generate new knowledge, and (v) how do we deal with the uncertainty (or uncertain knowledge, i.e. the one with large dispersion). Since learning from the presented data/information is very important in AI, we dwell more on this facet.

AIVA.3.1 Learning

The idea here is to generate new facts/concepts/inferences from the old/previous facts. We must be able to distinguish different situations from in the new environments. So, if a system is going to (rationally) act, then it must be able to change its actions with new experience, i.e. act like a learned person. This requires that to generate an intelligent behaviour, the system has to interact with the environment. The properly developed intelligent systems are expected to: (a) accept inputs from sensors (sensory inputs, sound, vision,…), (b) interact with humans (to understand language, recognize speech, generate text/speech/graphics), and (c) modify the environment. Learning: (i) is essential for unknown environment (since the designer does not have omniscience/all-pervading knowledge), (ii) is useful as a method of constructing/developing a system (an agent is exposed to reality, new environment), and (iii) modifies the agents' decision making approaches to improve the performance of the AI systems. Thus, the learning brings about: (i) a persistent change in performance due to interaction with the environment, (ii) a permanent change in knowledge/behaviour due to experience, and (iii) an enduring change in capacity due to practice.

AIVA.3.2 Theory and Approaches of Learning

Theory of learning encompasses: (a) behaviourism, (b) cognitivism, (c) social learning aspects, (d) social constructivism, (e) multiple intelligences, and (f) brain-based learning. Learning is defined as an outward expression (showing off/announcing, displaying skills), and focusses on observable behaviour. It is context dependent, and is based on classical/operant conditioning: reflexes, and feedback/reinforcement. The cognitivism grew in response to behaviourism, on the concept that the knowledge is stored (cognitively) as symbols. Thus, then the learning becomes a process of connecting symbols in a meaningful and memorable ways. So, the study of learning focused on the mental processes that facilitate symbol-connections. The cognitive learning theory is based on: (i) discovery learning, and (ii) meaningful verbal learning. The features of the discovery learning are: (i) one can learn anything, if it is presented in terms that he/she can understand, and (ii) confront the learner with problems and help find the solutions. The meaningful verbal learning has a feature: any new material is presented in a systematic way, and then, it is connected to existing structures in a meaningful way, the latter aspect also has to be defined. If learner has difficulty with a new presented fact/data, then one needs to go back to the concrete anchors, i.e. advance organisers, and then provide a discovery approach to the learner, and the results of the learning would be enhanced. The cognitive learning has some criticism: (i) I/O models are mechanistic/deterministic, (ii) does not account adequately for individuality, and (iii) a little emphasis on affective characteristics.

The aspects/features in/of/for brain-based learning (BBL) that grew out of neuroscience, and constructivism, are: (i) brain is a parallel processor (biological neural networks are massively parallel and naturally adaptive circuits/systems), (ii) entire body is learning (from five sensors: seeing/vision, hearing/acoustic, taste, smell, and touch), (iii) it is a search for meaning and patterning, (iv) emotions are critical, (v) the processing is in parts as well as in wholes, (vi) focused attention and peripheral perception, (vii) it is conscious as well as unconscious (in the sense of unawareness) process, (viii) each brain is unique, and has several types of memories. The BBL provides opportunities for group learning and for self-expression and making personal connections to the content; it has to take into account the regular effect of (multi-sensory) environmental changes. The main problems of the BBL are: (i) research is conducted by neuroscientists, and lesser by the educational/teaching researchers, and (ii) lack of understating of the brain itself makes BBL questionable.

As far as knowledge in learning is concerned, for logical learning, we do not need a priori knowledge, but then we need to incorporate any available knowledge, otherwise effort of the previous knowledge is a waste. So, we need to have the approach of inductive learning in the logical setting. The approach is to find a hypothesis that explains the classifications of the examples given their descriptions: hypothesis * description → classification. In a learning process one can use feedback: (i) supervised learning, correct answers for each example, (ii) unsupervised learning, correct answers not given, and (iii) reinforcement learning, occasional rewards.

Function learning is a learning of a function from the examples, say 'f' is a target/desired function: for pair $(x, f(x))$, find a hypothesis 'h' such that 'h' is nearly 'f', given a training set of examples, the curve fitting, as seen from Figure AIVA.4, if 'h' agrees with all the examples presented, then 'h' is consistent.

In computational learning theory we are seeking general laws that govern learning, and the features are: (i) sample (examples) complexity, how many sets we need to learn a hypothesis successfully, (ii) computational complexity needed to learn, and (iii) mistakes in learning, errors of fitting the hypothesis to the function. In the case of statistical learning, we need to include the agents to handle uncertainty by using probability/decision theory/fuzzy logic, since these uncertainties abound in nature and in the data/examples presented to the learner. The Bayesian learning calculates probability of each hypothesis given the data, and makes predictions based on this, i.e. computes the a posteriori probability, from which some recursive solutions are required to be found. This learning is based on all the hypotheses that are weighted by their probabilities, and then the predictions are made. Since the problem space is very large, one resorts to approximations. The approaches are: (i) maximum a posteriori (MAP), which needs a priori probability, and (ii) maximum likelihood (ML). Here, a priori probability is uniform (=1) and, hence, not required. This means that it is not relevant if we are using a large amount of data, we get best fit to the data. Often ML is regarded as a non-Bayesian statistical learning method, since a priori probability is not taken into account, yet produces reasonable results even if all the hypotheses are of the same complexity.

The full Bayesian learning gives best prediction, but it is complex process and often intractable. The MAP selects a single best hypothesis and the prior is still used. The ML assumes a uniform prior. The process is: (i) choose parameterized family of models to describe the data, (ii) write down the likelihood of the data as a function of parameters, (iii) take the derivative of the log-likelihood wrt to each parameter, and (iv) determine the parameters values such that the derivatives are zero; leading to ML estimates.

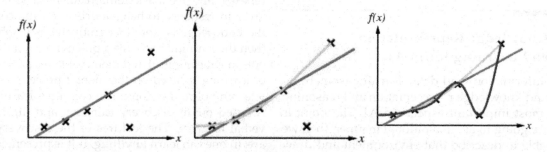

FIGURE AIVA.4
The learning by induction: curve fitting, learning a function from the presented data (for the third case a better hypothesis/model is required for the given data).

Most applications of AI are/could be for: (i) autonomous planning and scheduling (autonomous rovers, telescope scheduling), (ii) analysis of data from space-based missions, (iii) image-guided surgery, (iv) image analysis and enhancement, (v) autonomous vehicle control, (vi) pedestrian detection, (vii) games, (viii) bioinformatics (gene expression data analysis, prediction of protein structures), (ix) text classification/document sorting (web pages, e-mails, news articles), and (x) video image classification (music composition, picture drawing, natural language processing).

AIVA.4 Intelligent Systems

Intelligent systems (ISs) are computationally intensive paradigms that are nature-inspired, yet mathematically sound and are increasingly finding applications in many fields of science and engineering. The so called artificial-intelligent approaches, including soft computing (artificial neural networks (ANNs), fuzzy logic (FL), and genetic algorithms (GAs)) and systems employ computers to emulate various faculties of human intelligence and biological functions (AIVA.2). The intelligent systems are suited for: (i) search and optimization, (ii) pattern recognition and matching, (iii) planning and uncertainty management, and (iv) control, and adaptation. In essence, the intelligence is often identified/specified/signified in terms of (a) competence (ability to do a task in a better way), (b) expertise (real good working knowledge of carrying out a task), (c) talent (much more than average ability in many things), (d) schooling (training), (e) IQ (intelligent quotient-cognitive intelligence), and (f) social interaction (mutual behavioural aspects). In an alternative way, it is signified as: (i) cognitive intelligence, (ii) emotional intelligence, and (iii) spiritual intelligence. However, from the computational point-of-view, the intelligence is characterized by its: (a) flexibility (expanding and accommodating ability), (b) adaptability (meet the changing environment), (c) memory (storage capacity, remembering from the experience), (d) learning (new strategies), (e) temporal dynamics (adopting to time dimension, time varying aspect), (f) reasoning (thinking and decision making ability), and (g) the ability to manage uncertain/imprecise information; it is in the latter aspect that the soft computing, especially fuzzy logic technology plays a very significant role. Many important features of such ISs are: (i) information gathering (collecting data and acquiring the knowledge), (ii) understanding (assimilating), (iii) making inferences (deciphering), and (iv) applying it to understand and solve new problems efficiently. In AI we have: (i) rationally think as humans

do, and (ii) rationally act as humans act. An IS includes intelligent behaviour as seen in nature as a whole: (a) biological genetics (distributed information), (b) evolution (survivability), (c) chaos (structured randomness), and (d) natural adaptation (sufficability). The idea in IS, supported by AI is to find solutions that are better than their predecessors, and not necessarily the optimal ones. In a simpler way the IS should mimic some aspects of intelligence exhibited by nature: (i) learning (from the presented data/information), (ii) adaptability (meet the demands of changes in the system), (iii) robustness across problem domains (assured performance in the presence of the uncertainty), (iv) improving efficiency (more output than the input over time and/or space), (v) information compression (data to knowledge), and (vi) extrapolated reasoning (AIVA.2). Another new or novel perceptive for the AI should be: make an attempt to build algorithms, software, machines, and/or systems that perceive, think, reason out, make decision and act rationally in the sense of achieving 'probably approximately correct algorithm/performance/perfect behaviour' in the areas of problem solving and learning (AIVA.3). Then gradually approach towards the perfection by ever refining these artefacts. This perspective of AI is practically achievable and would give a definite measure of evaluating the performance of the AI systems and one can calibrate the nearness of it to the human's thinking and acting; that it gives a means of measure in the sense of achieving the higher and higher probability in the sense of/for lesser and lesser approximation, higher probability for more and more correct computational algorithm: for learning, fuzzy logic based inferences, and decision making.

AIVA.4.1 ISs in Aerospace Engineering

Some novel applications of IS are: (i) spacecraft autonomy, (ii) aircraft control (fault diagnosis, fuzzy gain scheduling), (iii) modelling, (iv) airfoil design, (v) satellite operations, (vi) missile design, and (vii) vehicle health management. Using ISs, complex/sophisticated ideas can be implemented and tested with rapid development cycles. The role of ISs in aerospace engineering is: (a) function as intelligent assistants to augment human expertise (e.g. pilots' situation assessment), (b) act as a substitute for human expertise in endeavours that save cost, time, and life.

AIVA.4.2 ISs for Modelling

Intelligent systems provide two aspects for modelling: (i) generalization, implies that the model could be used not only to represent just the data gathered but the knowledge that the data represents, and (ii) robustness, is the system's ability to perform within certain bounds of its nominal performance in-spite of uncertainty, that is however bounded. The techniques of ANNs, FL, GA, and expert systems can be used by aerospace engineers for modelling. Knowledge representation, in general contains syntax and semantics, and includes (AIVA.2):

- Mathematical equations (ARMA, state space models)

- Rule-based/expert systems/inferencing system (based on reasoning) consists of: (a) a knowledge base (a set of rules/known facts), acquired data (derived facts/data), and an inference engine.

- Fuzzy modelling consists of: (i) system I/O variables, (ii) fuzzy membership functions, (iii) fuzzy implication functions, (iv) aggregation functions, and (v) defuzzification. Attractive features of this are: (a) reduced design complexity, (b) rapid prototyping, (c) flexibility, (d) simplicity, and (e) cost effectiveness.

- Neural models: ANNs are brain-inspired connectionist models that consist of many similar linear and nonlinear computational elements called neurons connected in parallel and complex patterns. These simple neurons have the ability to perform tasks such as memory recall, pattern recognition, and learning.

- Tree structure: an algorithm for placing and locating objects in a database; it finds data by repeatedly making choices at decision points called nodes.

Intelligent modelling tools are used for operator behaviour modelling, flight test data modelling, and aerodynamic modelling for design.

AIVA.4.3 Improved Robustness Using Fuzzification

The technique of the fuzzy granularization combined with the knowledge available through an existing linear dynamic controller provides an outer fuzzy shell to existing control techniques.

The fuzzification of the inputs and outputs might lead to better robustness to system uncertainties via the manipulation of the distance between the membership functions.

AIVA.4.4 Intelligent Systems for Search

Search techniques are employed in combinatorial optimization, and consist of: (i) an initial state, (ii) a set of operators, (iii) application of the operators produces a sequence of new states called the path, and (iv) a goal test is defined and tested to determine when a new state becomes the desired goal state. A general combinatorial task involves deciding: (a) the specifications of those entities (what type of aircraft, the opponents aircraft descriptions, identification number of aircraft), and (b) the way in which those entities are brought together (various elementary tactics formations with their relative positions). A few search techniques used by the ISs are (AIVA.2):

- Golden section: the interval is reduced by throwing out regions that do not contain the optimum.

- Breadth-first search: search a state space by constructing a tree consisting of a set of leaves and branches, the algorithm defines a way to move through the tree structure.

- Depth-first search: before going deeper, the algorithm follows a single branch of the tree down as many levels as possible until a solution/dead-end is reached.

- Heuristic dynamic programming (HDP): based on an attempt to approximate the Bellman equation (see Chapter 6), variations of HDP can be used in solving combinatorial optimization problems.

- A*Search: combines the cost estimate $J(n)$ of traversing from step n to the goal state and $U(n)$ which is the known path cost from the start node to step n (AIVA.3).

- Genetic algorithm: a type of evolutionary computation that uses a genetic/evolutionary metaphor.

For aerospace applications, intelligent search techniques are used for: (a) searching a design space, (b) searching for an optimal scheduling and planning schemes, and (c) solving combinatorial optimization problems.

AIVA.4.5 Search for Optimal Air Combat Tactics

In case of a four-airplane division of two elements, each element consists of two aircraft, treated as a unit, and this hierarchical concept can be used to develop a GA-based approach to search for optimal air-combat tactics. The tactics implementation proceeds as:

- Specify a set of commonly-used element, the division formations, the underlying tactical maneuvers, and attack tactics.

- Formulate a set of principles for aggregating the small formation tactics for large engagements, and implement a method for doing this in the GA algorithm.

- Use the resultant formation tactics to drive the engagement, and use the performance metric generator, and search for the best engagement tactics so as to optimize these metrics.

AIVA.4.6 IS for Design

In aerospace systems, the design is interdisciplinary: wing (aerodynamic), propulsion (engine control), and automatic control (autopilot, stability augmentation) designs might need extensive redesign (design iterations) for arriving at an integrated design. ISs can help in many ways: (i) neural networks and fuzzy logic for representation of known data/designer expertise (mathematical modelling), and (ii) genetic algorithms as optimization procedures for searching through large spaces with several constraints (of systems' dynamics and stability), and subsystem interactions. For example, to maximize engine performance under degraded conditions, a fault tolerant engine control scheme can be applied, for which an engine performance estimator using a combination of a genetic algorithm (GA) and RBFNN for the implementation can be designed. The usual input selection, executed by simple inspection of data files, can be replaced by automatic inspection using GAs. The performance can be evaluated by an objective function (PI), the error is defined as the difference between desired performance and the current design's performance.

AIVA.5 Intelligent Control

The two important aspects to intelligent control (IC) are: (i) the 'intelligence' to analyze the changing environment, i.e. the analytical ability to comprehend/react to the changing environment; and (ii) the resources to respond to the changing environment, i.e. the physical components of the system that are necessary to react to the environment. One application is: maintain performance under any of the situations: (a) loss of control due to failure, (b) aircraft characteristics change due to damage (C.G., or mass/inertia), and/or c) environmental effects due to wind and turbulence. The IC encompasses: (i) conventional control-optimal, robust, stochastic, linear, and nonlinear, and (ii) fuzzy, genetic, and neuro-control technologies.

AIVA.5.1 Levels of Intelligent Control

An IC design is stated as: given the dynamic system

$$x(t + 1) = f(x(t), u(t), t) + w; \qquad \text{(IVA.1)}$$

In (IVA.1), w is an unknown disturbance. A set of goals is generated:

$$x_g(t + 1) = g(x_g(t), x(t), t) \qquad \text{(IVA.2)}$$

A performance measure is defined as:

$$J(t + 1) = \sum_{t=0}^{T} F(x_g(t), x(t), u(t), t) \qquad \text{(IVA.3)}$$

A planning function is specified as

$$P(t + 1) = p(x(t), P(t), t, \nu) \qquad \text{(IVA.4)}$$

In (IVC.4), ν is system faults and emergencies. The intelligent controller needs to arrive at a control, $u(t)$, such that the system: (i) is locally stable (includes requirements of handling quality of the system), (ii) follows closely the desired path (closeness defined by a PI), (iii) constantly optimizes long-term and short-term goals, and (iv) reacts to changing environments by properly adapting the planning functionality (AIVA.2). A practical way to accommodate the IC is to have various levels of capabilities for self-improvement (using some adaptive mechanism, and switching based on some logic, say fuzzy logic), which is also an important goal of human intelligence. The IC is defined with various levels of intelligence (control) so that it will accommodate easily the contributions from: cognitive science, computer hardware, sensors/actuators, learning theory, and newer control architectures. For various tasks in the levels of self-improvement, the designer of control systems can use the combination of the existing control methods augmented by the soft-computing based paradigms, many of which have been discussed in the presence volume.

AIVA.5.2 Level Self-Improvement of Description

Some main levels for IC and Self-improvement are briefly stated here (AIVA.2), in all or some of which ANNs, FL, and/or GA-based approaches can be used appropriately (see Appendixes of Chapters 14 through 16):

Level 0 IC-a robust controller: Tracking Error (TE) with robust feedback control tends to zero, self-improvement of TE is an important goal of many control techniques. To achieve this, one designs robust feedback controllers with constant gains that improve the error (i.e. reduce the errors) as time goes to infinity.

Level 1 IC-an adaptive controller: TE + Control Parameters (CP), robust feedback control with adaptive control parameters (error tends to zero for non-nominal operations); feedback control is self-improving of control parameters towards the goal of achieving better tracking error or some error oriented goal, is the next level in intelligent control.

Level 2 IC-an optimal controller: TE + CP + Performance Measure (PM), optimal control: robust, adaptive feedback control that minimizes or maximizes a utility function over time. Self-improvement of an estimate of the performance error towards the goal of optimization of an utility function over time (error tends to zero and a PI is optimized) is the next level in IC.

Level 3 IC-a planning controller: TE + CP + PM + Planning function: Level 2 + the ability to plan ahead of time for uncertain situations, simulate, and model uncertainties. This Level 3 IC includes self-improvement of planning functions for contingency, emergencies, and faults. These planning functions could be static for Level 2 but should be self-improving for Level 3.

AIVA.6 Intelligent Flight Control-A Brief Note

In most flight control systems for aerospace vehicles one finds extensive/extended application of the classical control methods (Chapters 1 through 4), and/or the so called modern control approaches (Chapters 5 through 12), in selective cases. The aim of design and development of complex need-based/sophisticated control for flight (-dynamics) is: (a) to improve the otherwise poor static stability or low damping in dynamics for a particular axis, (b) to provide enhanced stability to inherently unstable (or relaxed stability) aircraft/dynamics, (c) to improve the pilot-aircraft interactions/coupling, (d) to improve the safety and reliability of the aircraft's functions (in presence of certain expected/limited failures), and (e) to reduce the workload of the pilot in handling the complex flight missions/operations and tasks, e.g. use of an autopilot. The conventional methods have been very successfully applied to the problems of flight control, guidance and navigation for variety of aircrafts in the last few decades. However, recently there has been an upward trend of applications of artificial neural networks (ANNs), fuzzy logic/system (FL/S), and GA's to the design of flight control systems for aircraft, missiles, UAVs (unmanned aerial vehicles) and MAVs (micro/mini air vehicles), along with use of allied and supportive technologies: (a) parameter/state estimation, (b) H-infinity control/filtering, (c) object oriented programming (OOP), and (d) handling quality evaluation of pilot-aircraft interaction. We feel that some more rigorous blending and cross-fertilization of these approaches with conventional/classical approaches can emerge as do-able, useful and robust and yet intelligent

methods for design, evaluation and synthesis of flight control systems [AIVA.3]. The ANNs, FLSs and GAs that are called the major components of the soft computing (SC) are crucial subsystems/ingredients of a successfully built AI system. An AI system would use all the SC paradigms in an integrated way of providing: (i) the structure (ANNs), (ii) incorporation of knowledge (FLSs), and (iii) optimization (GAs) of the strategies.

AIVA.6.1 ANN Based Systems for Flight Control

ANNs are emerging paradigms for solving complex/involved problems in sciences and engineering and have salient features: (a) they mimic the behaviour of human brain, (b) they have massively parallel architecture/structures, (c) they can be represented by adaptive circuits with input channel/s, weights (parameters/coefficients), one or two hidden layers and output channel/s with some nonlinearities, (d) the weights can be tuned to obtain optimum performance of the ANN – be it modelling of a dynamic system or nonlinear curve fitting, (e) it requires to use training algorithm to determine the weights (where the acquired knowledge is stored), (f) they can have feedback type arrangement within the neuronal structure, (g) the trained network can be used for predicting the behaviour of the dynamic system that is modelled using the ANN, (h) the ANN can be easily coded/validated using standard software procedures, (i) optimally structured ANNs can be hard-wired/firm-wired and embedded into a chip for practical applications – this will be the generalisation of the erstwhile analogue circuits-cum-computers, and (j) then the neural network based system can be truly termed as a new generation powerful and massively parallel computer. An aerodynamic model (used for design of flight control laws) could be highly nonlinear and dependent on many physical variables, which might be difficult to model and to have an explicit math-model; the difference/s between the math-model/s and the real system/s' data would cause performance degradation of the flight control system. To alleviate this problem, an ANN can be used and the weights of the NW are so adjusted as to compensate for the effect of the (initial) modelling errors. The ways in which ANNs can be used for control augmentation are: (i) the normal control can be aided by RBFNN controller for online learning to represent the local inverse dynamics of an aircraft, (ii) to compensate for uncertainty without explicitly identifying changes in the aircraft model, (iii) the ANN nonlinearity, the sigmoidal function can be made adaptive and used in the desired dynamics block of the flight controller (say MRAC), (iv) the learning ability can be incorporated into the gain scheduling of the controller, and (v) sensor and/or actuator failure detection and management. Major merits would be: (i) the controller becomes more robust and more insensitive to the plant parameter variations, and (ii) the online learning ability would be useful in handling certain unexpected behaviour.

AIVA.6.2 FLS for Flight Control

The FLSs have the salient features: (a) they are based on multivalued logic, (b) they do not have any specific architecture like ANN, (c) they are based on certain rules which need to be a priori specified, (d) fuzzy logic is a machine intelligent approach in which desired behaviour can be specified in rules in which an expert's (design engineer's) experience can be captured/incorporated, (e) it deals with approximate reasoning in uncertain situations where truth is a matter of degree, (f) it is based on the computational mechanism (algorithm) with which decisions can be inferred, in-spite of the incomplete knowledge, the process of inference engine can be used for decision making and control. FL based control is suited to multivariable and non-linear processes and can achieve desired control action. The heuristic fuzzy control does not require deep knowledge of the controlled process, but, the control policy should be known a priori; however, an improvement to this type of controller can be made by an application of (math-) model-based fuzzy system, wherein the model of the controlled process is assumed to be known and used. This will require to assure the stability of the FL-based flight control system. FLS can be used to augment the flight control system in certain ways: (a) FL will approximately duplicate some of the ways a pilot might respond to an aircraft (useful in case of a damage or failure, (b) the rules can be used to create a nonlinear response with low damping for large errors and high damping for small error, (c) to incorporate the complex nonlinear strategies based on pilot or system design engineer's experience and intelligence within the control law, (d) adaptive fuzzy gain scheduling (AGS), and (e) FL based adaptive tuning of Kalman filter (KF) for adaptive estimation/control.

A brief application review is presented in Table AIVA.1, and is just an indicative, and not exhaustive of various possibilities [AIVA.5 through AIVA.9].

AIVA.6.3 Associated Technologies for Flight Control

Certain associated technologies (for which certain paradigms of soft computing can be used) for flight control also play a key role in understanding the behaviour of flight control systems and are briefly highlighted next

System modelling, identification and parameter estimation: Math-modelling of aerospace vehicles is very important since many applications require such information in the form of aerodynamic (stability and control) derivatives that appear in the model: (a) to explain aerodynamic, stability and control behaviour of the flight vehicle, (b) for design of flight control systems, and (c) for high fidelity flight simulators. Determination of aerodynamic derivatives of an aircraft with a fly-by wire control system (FBWCS for unstable aircraft) poses new situations for identification and parameter estimation problem: when identification experiments are performed with an aircraft operating in the closed loop, the feedback introduces correlations among the I/O variables. This correlation causes the identifiability problems that would render estimates of some derivatives uncertain and unclear. Also, due to the feedback action, that would be constantly trying to generate controlled responses, the measured responses would not display the modes of the (original) vehicle, causing an identifiability issue. Also, the complexity of the problem increases when the basic aircraft is unstable, due to the fact that the numerical integration of the state model (in the estimation program) leads to numerical divergence. The problem is further compounded if these data are noisy. This is a fertile field for application of ANNs and FLSs based methods for system identification, parameter estimation and state estimation for; (a) modelling pilot's control activity and behaviour in several mission-oriented tasks to quantitatively evaluate his/her workload and supervisory effectiveness in a FBW aircraft, and (b) modelling of the aircraft from control requirement point of view can also be done using the system identification methods.

H-Infinity control and filtering: In the HI (H-infinity) concept, the minimum of the worst case error (i.e. the least of the maximum error) is sought in order to arrive at the robust (suboptimal) solution to control/filtering problem. Specific studies useful to flight control are: (i) concept of gain and phase margin, (ii) H-infinity filtering for generalizations of the

TABLE AIVA.1

A Brief Note on Applications of ANNs and FLS to Flight Control

Approach	Aircraft and/or Problem Addressed	Inferences Drawn	References
FFNN-online learning	F-8; an autopilot mode to track the pilot's throttle/pitch rate command		AIVA.5
RBFNN (radial basis function NN-variable Gaussian functions	F-8; online learning to represent the local inverse dynamics	Compact ANN structure; small tracking error	AIVA.6
ANN	F-16; to compensate for the effect of the aerodynamic modelling error	The errors exponentially converge to a compact set	AIVA.7
FL	Guided missile; adaptive fuzzy gain scheduling	High degree of automation; better terminal guidance performance; less control effort; a smaller miss distance.	AIVA.8
ANN; adaptive nonlinearity	U.S. Navy; U.S. AF & NASA program on X-36/Xv-15	compensates an uncertainty without explicitly identifying changes in the aircraft model	AIVA.9
FL	E-6A aircraft in a limited way; simulation mode for F-18 automatic carrier landing system	allows the incorporation of complex nonlinear strategies based on pilot's and/or design engineer's intelligence in the control law	AIVA.9

conventional/KF algorithms, and (iii) combination of H-2 (Kalman filter) and HI filtering algorithms.

The object oriented programming (OOP) in control system design: The OOP has a good potential application to the design of flight control system, and the merits are: (a) use of data modelling concept, (b) encapsulation of data and functions, and (c) OOP makes software decomposition like the decomposition of a dynamical system. The 'objects' provide a much more natural basis for a good CACSD (computer aided control system design/SW tool) than the conventionally used simple data types and functions. The aspect is that the OO-paradigm can be used in the field of CACSD, i.e. the package would provide a set of data structures adequate to describe the 'objects' that are found in control system design; as for example, the fundamental 'objects' are 'system' and 'results': a model in the form of a set of equations is a 'system' and the time and frequency responses are a 'result'.

A synergy of a typical use of ANN-FL based systems for flight control with some associated technologies is depicted in Figure AIVA.5.

AIVA.6.4 Recent Intelligent Flight Control System

The NASA's intelligent flight control system (IFCS) (AIVA.10) is, perhaps, one of the latest developments in this direction; to effectively optimize aircraft performance in normal as well as failure conditions. It is designed to incorporate self-learning ANN into the flight control software (SW) that enables the pilot to maintain the control and to safely land the aircraft, in case the test aircraft suffers a failure of a control surface/a damage to the airframe. The IFCS is an integration of the ANN technology and the state-of-the-art control strategies/algorithms that identify the changes in the aircraft stability/control characteristics in case of failure; it responds to such changes and immediately adjusts to maintain the best possible performance of the aircraft-flight. The adaptive ANN-SW learns the new changed flight characteristics all in real time, and helps the pilot regain the control of the aircraft and/or prevent an accident; the ANN strategy is directly applied to FCS-feedback errors. The ANNs are also used/validated to organize/map the aerodynamic changes and these are then fed to FCS, which in turn uses this information to stabilize the aircraft. In some cases, the ANN based strategy can be used in the direct adaptive mode to continuously apply the required corrections to FCS and also learn by observing the effects on the performance of the aircraft behaviour.

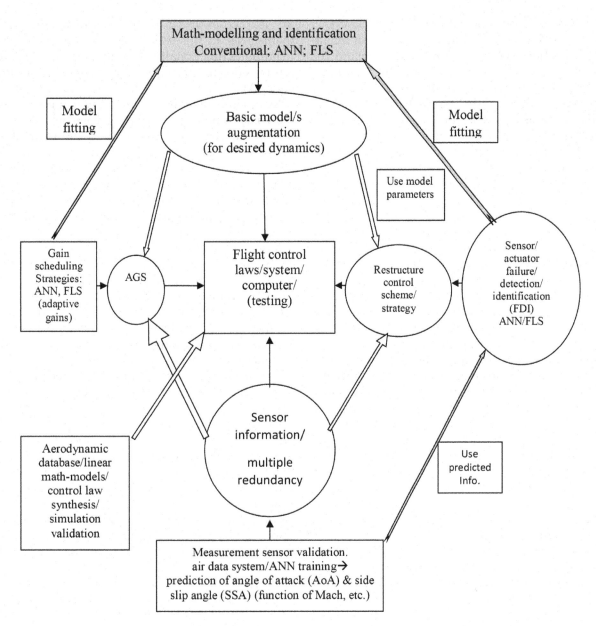

FIGURE AIVA.5
A concept of ANN-FLS augmented, aided intelligent-flight control system: simulation and evaluation. (From Raol, J.R. and Singh, J., *Flight Mechanics Modelling and Analysis*, CRC Press, Boca Raton, FL, 2009).

AIVA.7 Classical vs Intelligent Control

We can now summarize some important differences between the classical and intelligent control. For developing a classical control system (CCS, including modern control) to control a plant (process, dynamic system, or an agent), the designer constructs a *mathematical model* (MM) of the system. This MM contains all (or should contain most) of the dynamics of the plant that affect controlling the very plant [AIVA.11]. This form of control can be called the mathematicians' approach, because the designer should mathematically model the plant that is to be controlled. Whereas, for developing an intelligent control system (ICS) to control an agent (plant), the designer *inputs the system behaviour* and the ICS *abstractly models* the plant. This type of control is called the lazy man's approach, because the designer doesn't need to know the internal dynamics of the plant. It must be emphasized here, that in many cases, the plant may be too complex to be modelled. Of course, for general systems and intelligent systems (based on AI), there are many more differences than highlighted here from the control-point-of-view.

One can use the notion of 'shifting' intelligence: (i) for CCS, the designer must model the system to be controlled and therefore the intelligence/knowledge lies with her/him; the intelligence is shifted towards the designer, and (ii) for ICS, the software abstractly models the system and therefore the intelligence/knowledge lies with it; the intelligence is shifted towards the software. However, it must be noted here that the designer must have some knowledge of the plant, since he is designing the system, ultimately; it is that s/he just doesn't need (more) enough to develop an accurate model of the plant.

Since we have studied classical control (and modern control) in the three sections of the present volume, here we just capture the main aspects of ICS; which is that the system that has to be controlled does not have to be rigidly modelled. The designer inputs the appropriate stimuli to the ICS, and evaluate its performance (on its outputs), the ICS itself develops a model (now it is not a MM in the sense of the classical sense), of the plant to be controlled; of course, for the ICS the modelling is accomplished by ANNs, and FL.

Humans perform many complex tasks without knowing exactly how they do these, often in a very simple way. The types of ICS-approaches popularly studied and applied are: (i) ANNs, (ii) fuzzy logic, (iii) genetic algorithms/programming, (iv) reinforcement learning, and (v) support vector machines. Now, let us dwell on how some of these are called the ingredients of the intelligent systems, and help build AI-based machines/computers and control systems.

First, let us take the ANNs. These are so constructed as to mimic some (functional) behaviour of the so called biological neural networks (BNNs). The functional structure of an ANN is such that these are several simple neuronal units/circuits connected in parallel, and these are in millions; the knowledge is stored in the input weights (synapses) of the neurons, and adjusting these weights gives the neurons the ability to store different information. The ANN is presented with the data/examples and trained to learn, say patterns. Once, it is trained well, it is able to recognize the patterns and works or acts as humans act for the same job. Similarly, ANNs can be trained to do many complex tasks, and the learned ANNs can then perform these tasks by themselves. In this way ANNs are relatively close(-er) to human's functionality, and can perform the tasks that humans do, and thus, the ANNs are called an ingredient of AI.

Now, let us take the case of FL which is modelled on the basis of how the reasoning goes on in the human brain; the human's reasoning is approximate, non-quantitative, and non-binary, i.e. it is multi-valued. The FL's inference process is devised such that it is able to incorporate the heuristic knowledge of the human expert (about the process/plant/environment), and also fuzzifies the input variables and takes into account the ranges of the variables rather than the crisp values, i.e. it assign a degree of 'belongingness' of a variable to certain (fuzzy) sets. Thus the FL works with imprecise information, due to uncertainty, and still makes decisions like humans do in the similar situations.

Genetic programming (GP) is derived from GA, except that the output of GP is another program. The GA is based on the evolutionary-nature's strategy of mating-crossover-reproduction-retaining of the fitter off-springs and propagation, and repeating the process; over a period of time (from a few seconds/days to thousands of years, depending up on the type of species), the process produces evolutionary stable strategies (ESS) for the survival of the species. Now, in this process of EP (evolutionary process/programming), there is no 'explicit' intelligence seen working, but as we can conjecture, there should be some 'implicit' intelligence working or playing role, without which such an awesome task cannot be performed. Thus, GA/GP are safely considered as a part of AI. The GA in fact is very simple, and yet very powerful in solving optimization problems, and can lead robust solutions, if not optimal ones. The idea is to create new (GP) programs to solve a control problem based on the ones that work best now. The steps to implementing GP are [AIVA.11]: (i) to generate a random group of functions and terminals, each random group will constitute a computer program, functions are operators such as +, −, *, /, etc. Terminals are the I/Os to the problem; (ii) to execute each program and assign a number (called a fitness value) to each program to see how well it did in solving the problem; (iii) to create a new population via mutation, crossover and the most fit programs, mutation involves randomly changing functions and terminals of a program.

The reinforcement learning involves presenting an agent (a robot) with a set of action choices, (initially equally rewarding). The agent makes choices, based on a behavioural policy, and receives rewards or punishments via a reward function based on the choices made. In the process, the rewards/punishments affect how the agent will make its choices in future encounters, and thus it learns what choices are rewarding and which are not.

References

AIVA.1   Russell, S. J., and Norvig, P. *Artificial Intelligence-A Modern Approach*. Prentice Hall, Englewood Cliffs, NJ, 1995.

AIVA.2   Krisha Kumar, K. Intelligent Systems for Aerospace Engineering-An Overview. https://ti.arc.nasa.gov/m/pub-archive/364h/0364%20(Krishna).pdf.

AIVA.3   Raol, J. R., and Ajith, K. Gopal (Eds.). *Mobile Intelligent Autonomous Systems*. CRC Press, Boca Raton, FL, 2013.

AIVA.4   Raol, J. R. Intelligent flight control-A brief survey. *CADFEJL*, Vol. 1, No.1, pp. 2–9, 2017.

AIVA.5   Sadhukhan D., and Feteih S. F8 neurocontroller based on dynamic inversion. *J. Guid. Control Dyn.*, Vol. 19, No.1, pp. 150–156, 1996.

AIVA.6   Li, Y., Sundararajan, N., and Saratchandran, P. Stable neuro-flight controller using fully tuned radial basis function neural networks. *J. Guid. Control Dyn.*, Vol. 24, No. 4, pp. 665–674, 2001.

AIVA.7   Lee, T., and Kim, Y. Non-linear adaptive flight control using backstopping and neural networks controller. *J. Guid. Control Dyn.*, Vol. 24, No. 4, pp. 675–682, 2001.

AIVA.8   Lin, C. L., and Su, H. W. Adaptive fuzzy gain scheduling in guidance system design. *J. Guid. Control Dyn.*, Vol. 24, No. 4, pp. 683–692, 2001.

AIVA.9   Steingerg, M. L. Comparison of intelligent, adaptive and non-linear flight control laws. *J. Guid. Control Dyn.*, Vol. 24, No. 4, pp. 693–699, 2001.

AIVA.10   Gibbs, Y. (Eds.), NASA Armstrong fact sheet: Intelligent flight control system, March 2014. https://www.nasa.gov/centers/armstrong/news/FactSheets/ FS-076-DFRC.html. Accessed February 2017.

AIVA.11   Smith, B. Classical vs intelligent control. Special topics in robotics, Memorial University, Newfoundland, Canada, 2002. https://www.engr.mun.ca~baxter/Publications/ClassicalvsIntelligentControl.pdf, accessed December, 2018.

Exercises for Section IV

1. Design a network of perceptrons to classify the following three-dimensional patterns:

    Class A:{x} = {(0,0,0), (1,1,1)}

    and Class B:{x} = {(0,0,1), (0,1,1)}

2. Design a suitable multilayer feedforward network capable of plotting the function:

    $f(x_1,x_2) = \cos(2\pi x_1)\cos(2\pi x_2)$, $0 \le x_i \le 0.5$, $i = 1,2$

    by using as exemplars the values of the function evaluated over, say, 441 sample points in the x_1–x_2 plane. You may select one or two hidden layers, each having between 5 and 10 units. The network should have the potential for interpolating between the sample points so that a continuous map from the space of inputs to the output is well approximated. Comment on the success or lack of it in reaching your goals.

3. A critical goal during training is to make the hidden layer small enough to improve the generalization performance and large enough to permit learning. Justify this statement and analyze how an optimal number of hidden nodes may be arrived at (in the sense that both larger and smaller numbers of hidden nodes produce larger mean-squared errors).

4. In the lines of example 15.2, conduct a simulation for the cruise control system of a car with speed v (between 0 to 100 kmph) and angle of banking θ (between $\pm 10°$) of the road as inputs and the throttle position (0 to 10) as the output. Owing to the mechanical nature of the system, you may assume that the speed and throttle are related in discrete time $k = 0,1,2,\cdots$ as follows:

    $$T_k = m(v_{k+1}-v_k) + \kappa v_k mg \sin(\theta_k)\theta_k$$

    where κ is the friction coefficient. You may rewrite this equation to show the update on the velocity as

    $$v_{k+1} = k_1 v_k + \begin{bmatrix} k_1 & k_2 \end{bmatrix} \begin{bmatrix} T_k \\ \theta_k \end{bmatrix}$$

    where κ_i may be obtained from κ_1, m, and $mg \sin(\theta_k)$.

    To cruise at a nominal speed of 50 kmph, assume suitably scaled values of κ_1 and m, for instance, $\kappa_1/m = 0.1$, and triangular membership functions for v and θ. Construct a suitable set of fuzzy rules and conduct at least 4 cycles of simulation using the initial conditions $v_0 = 55$ kmph, and $\theta_0 = -5°$.

5. Let us consider the problem of level control in a cylindrical tank, of height h and cross-sectional area A. If f_i and f_o are the liquid *in-* and *out-* flow rates, respectively, we wish to maintain the liquid level at h_s. Basic engineering intuition tells us the following:

    • The rate of change in the liquid level, obviously, is proportional to the difference in the in- and out-flows, i.e.

    $$\frac{dh}{dt} = \frac{f_i - f_o}{A}$$

    the larger the area the slower the change.

- Liquid flows into the tank through a valve, at a nominal rate F, and hence, in-flow may be controlled handling the valve:

$$\Delta f_i = \frac{F - f_i}{F}\%$$

- The error between the desired and actual level levels, an index of *performance*, is

$$e = \frac{h_s - h}{h_s}\%$$

The control logic is clear now:

If $e < -10\%$ then $\Delta f_i = 5\%$

If $-1\% < e < 1\%$ then $\Delta f_i = 0\%$

If $e > 10\%$ then $\Delta f_i = -5\%$

Let the initial values be

$$f_i = F = f_o = 0.3, h = 2, A = 3, h_s = 1$$

after appropriate scaling.

At $t = 0$, $e = 1$, let the in-flow be $f_i = 0.4$. At the next time instant, say at $t = 0.5$, $\delta h = (0.4-0.3)/3 = 0.03$, and hence, $h = 0.03$ and $e = (1-0.03)/1 = 7\%$

With the above information, design a fuzzy controller, using a weighted average defuzzification, and conduct a four-cycle simulation of this system.

Use GAs or PSO for the following problems:

6. An apparently simple function to start with:

$$N f(x) = {}^{x}i x^2_i, \ x \in [-10,10]$$

$$i = 1$$

The global minimum is, obviously, zero and is located at the origin of the N-dimensional space.

7. Obtain the global maximum of the following well-known Easom function:

$$f(x) = -\cos(x)e^{-(x-\pi)2} \ x \in [-10,10]$$

8. Let us now consider the two-dimensional Easom function:

$$f(x) = -\cos(x_1)\cos(x_2)e$$

$$-((x_1-\pi)_2+(x_2-\pi)_2) \ xi \in [-100,100]$$

Show that the global minimum of this function is located at $x^* = (-\pi, \pi)$ and $f(x^*) = -1$.

9. Show that the following Hansen's function:

$$f(x) = \sum_{i=0}^{4}(i+1)\cos(ix_1 + i + 1)$$

$$+ \sum_{j=0}^{4}(j+2)\cos((j+2)x_2 + j + 1) \text{ with } x_i \in [-10,10]$$

has multiple global minima located at $x^* = \{(7.589893, 7.708314), (7.589893, 1.425128), (7.589893, 4.858057), (1.306708, 7.708314), (1.306708, 4.858057), (4.976478, 4.858057), (4.976478, 1.425128), (4.976478, 7.708314)\}$

10. Let us now look at the so-called ripple function

$$f(x) = \sum_{i=1}^{2}-e^{-2\ln 2\left(\frac{x_i-0.1}{0.8}\right)^2}\left(\sin^6(5\pi x_i)+0.1\cos^2(500\pi x_i)\right)$$

subject to $0 \leq x_i \leq 1$. It has one global minimum and 252,004 local minima.

The global form of the function consists of 25 holes, which form a 5×5 regular grid. Additionally, the whole function landscape is full of small ripples caused by a high-frequency cosine function that creates a large number of local minima.

References for Section IV

1. Ross, T. *Fuzzy Logic with Engineering Applications*. 3rd ed., John Wiley & Sons, Chichester, UK, 2010.
2. Mendel, J. M. *Uncertain Rule-Based Fuzzy Logic Systems: Introduction and New Directions*. Prentice-Hall, Upper Saddle River, NJ, 2001.
3. Castillo, O., and Melin, P. *Type-2 Fuzzy Logic: Theory and Applications*. Springer-Verlag, Berlin, Germany, 2008.
4. Mendel, J. M. et al. Interval type-2 fuzzy logic systems made simple. *IEEE Trans. Fuzzy Sys.*, Vol. 14, No. 6, pp. 808–821, 2006.
5. Aisbett, J., Rickard, J. T., and Morgenthaler, D. Type-2 fuzzy sets as functions on spaces. *IEEE Trans. Fuzzy Sys.*, Vol. 18, No. 4, pp. 841–844, 2010.
6. Goldberg, D. E. *Genetic Algorithms in Search, Optimization, and Machine Learning*.Addison-Wesley, Reading, MA, 1989.
7. Haupt, R. L., and Haupt, S. E. *Practical Genetic Algorithms* (2nd ed.). John Wiley & Sons, Hoboken, NJ, 2004.

8. Mitchell, M. *An Introduction to Genetic Algorithms.* MIT Press, Cambridge, UK, 1996.

9. Simon, D. *Evolutionary Optimization Algorithms: Biologically-Inspired and Population-Based Approaches to Computer Intelligence.* John Wiley & Sons, Hoboken, NJ, 2013.

10. Dernoncourt, F. Introduction to fuzzy logic, Tutorial, Copyright© 2013 – Franck Dernoncourt, http://francky.me.

11. Leekwijck, W. V., and Kerre, E. E. Defuzzication: Criteria and classication. *Fuzzy Sets Sys.*, Vol. 108, No. 2, pp. 159–178, 1999.

12. Madau, D. et al. Influence value defuzzication method. *Proc. Fifth IEEE Int. Conf. Fuzzy Syst.*, Vol. 3, pp. 1819–1824, 1996.

13. Zadeh, L. **Fuzzy sets**. *Inf. Control*, Vol. 8, No. 3, pp. 338–353, 1965.

14. Bose, N. K., and Liang, P. *Neural Network Fundamentals with Graphs, Algorithms, and Applications,* McGraw-Hill, New York, 1996.

15. Hagan, M. T. et al., An introduction to the use of neural networks in control systems. *Int. J. Robust Nonlinear Control*, Vol. 12, pp. 959–985, 2002.

16. Haruna, C. et al. Neural networks optimization through genetic algorithm searches: A review. *Appl. Math. Inf. Sci.* Vol. 11, No. 6, pp. 1543–1564, 2017.

17. Ismail, S., Pashilkar, A. A., **Ayyagari, R.,** and Sundararajan, N. Diagonally dominant backstepping Autopilot for aircraft with unknown actuator failures and severe winds. *Aeronaut. J.*, Vol. 118, No. 1207, 2014.

18. Ismail, S., Pashilkar, A. A., **Ayyagari, R.,** and Sundararajan, N. Improved neural-aided sliding mode controller for autolanding under actuator failures and severe winds. *Elsevier J. Aerospace Sc. Technology*, Vol. 33, No. 1, pp. 55–64, 2014.

19. Ismail, S., Pashilkar, A. A., and **Ayyagari**, R. Phase Compensation & anti-windup design for neural-aided sliding mode fault-tolerant autoland controller. In *Proceedings of IEEE International Conference on Cognitive Computing and Information Processing* (2015 CCIP), Noida, India, 3–5 March 2015.

20. Pashilkar, A. A., Ismail, S., **Ayyagari**, R., and Sundarrajan, N. Design of a nonlinear dynamic inversion controller for trajectory following and maneuvering for fixed wing aircraft, *Presented at IEEE Symposium on Computational Intelligence for Security and Defense Applications (2013 CISDA)*, Nanyung Technological University, Singapore, 16–19 April 2013.

21. Ismail, S., Pashilkar, A. A., and **Ayyagari, R.** Guaranteed stability and improved performance against actuator failure using neural aided sliding mode controller for autolanding tasks. *Presented at the IFAC Workshop on Embedded Guidance Navigation and Control in Aerospace (EGNCA),* IISc Bangalore, India, 11th to 15th February 2012.

22. Lu, Y. W., Sundararajan, N., and Saratchandran, P. A sequential learning scheme for function approximation and using minimal radial basis neural networks. *Neural Comput.*, Vol. 9, pp. 1–18, 1997.

23. Lu, Y. W., Sundararajan, N., and Saratchandran, P. A sequential minimal radial basis function (RBF) neural network learning algorithm. *IEEE Tr. Neural Networks*, Vol. 9, No. 2, pp. 308–318, 1998.

24. Pashilkar, A. A., Ismail, S., **Ayyagari**, R., and Sundarrajan, N. Improved autolanding controller for aircraft encountering unknown actuator failures. *Presented at IEEE Symposium on Computational Intelligence for Security and Defense Applications (2013 CISDA)*, Nanyung Technological University, 16–19 April 2013.

25. Lewis, F. L. et al., *Neural Network Control of Robot Manipulators and Nonlinear Systems.* Taylor & Francis Group, London, UK, 1999.

26. Liang, Q., and Mendel, J. M. Interval type-2 fuzzy logic systems: Theory and design. *IEEE Tr. Fuzzy Sys.*, Vol. 8, No. 5, pp. 535–550, 2000.

27. Wu, D. On Interval Type-2 Fuzzy Sets and Systems private communication.

28. Hagras, H. Type-2 FLCs: A new generation of fuzzy controllers. *IEEE Comp. Intell. Mag.*, Vol. 2, pp. 30–43, 2007.

29. Sahu, H., and **Ayyagari**, R. Interval fuzzy Type-II controller for the level control of a three tank system. In *Proceedings of Fourth International Conference on Advances in Control & Optimization of Dynamical Systems*, NIT Tiruchirappalli, India, 2016. accessed http://www.ifac-papersonline.net, pp. 561–566.

30. Mitchell, M. *Complexity: A Guided Tour.* Oxford University Press, New York, 2009.

31. Nazaruddin, Y. Y. et al. PSO based PID controller for quadrotor with virtual sensor. *IFAC PapersOnLine*, Vol. 51, No. 4, pp. 358–363. 2018.

32. Erguzel, T. T. Fuzzy controller parameter optimization using genetic algorithm for a real time controlled system, *Proceedings of the World Congress on Engineering 2013*, Vol. II, WCE 2013, July 3–5, 2013, London, U.K.

33. Tahmasebi. P., and Hezarkhani, A. A hybrid neural networks-fuzzy logic-genetic algorithm for grade estimation, *Comput. Geosci.*, Vol. 42, pp. 18–27, 2012, (Elsevier).

34. Corus, D., and Oliveto, P. S. Standard steady state genetic algorithms can hillclimb faster than mutation-only evolutionary algorithms. *IEEE Tr. Evol. Comp.* Vol. 22, No. 5, pp. 720–732.

35. Rizk, Y. et al. Decision making in multiagent systems: A survey, *IEEE Tr. Cogn. Develop. Sys.*, Vol. 10, No. 3, pp. 514–529, 2018.

36. Si, W. et al. An improved control algorithm for lighting systems by using PSO, *IEE J. Trans. Elec., Inform. Sys.*, Vol. 133, pp. 1501–1508, 2013.

37. Wang, H., Mustafa, G. I., and Tian, Y. Model-free-fractional-order sliding mode control for an active vehicle suspension system. *Advances Engg. Software*, Vol. 115, pp. 452–461, 2018.

38. Copot, C. et al. PID based particle swarm optimization in offices light control. *IFAC PapersOnLine*, Vol. 51, No. 4, pp. 382–387, 2018.

39. Alison, W. et al. Lighting controls in commercial buildings, leukos. *J. Illum. Eng. Soc. North America*, Vol. 8, No. 3, pp. 61–180, 2012.

40. Oh, S., Jang, H., and Pedrycz, W. A comparative experimental study of type-1/type-2 fuzzy cascade controller based on genetic algorithms and particle swarm optimization. *Expert Sys. Appl.*, Vol.38, pp. 11217–11229, 2011 (Elsevier).

41. Zhang, J. et al. Optimal design of fuzzy logic controller for multi-body model of semi-active suspension based on genetic algorithm. *Proc. fif*th *IEEE Conf. Indust. Elec. Appl.*, pp. 1478–1483, 2010.

42. Haemers, M. et al. Proportional-integral state-feedback controller optimization for a full-car active suspension setup using a genetic algorithm. *IFAC PapersOnLine*, Vol. 51, No. 4, pp. 1–6, 2018.

43. Ahmad, A. Active suspension control based on a full-vehicle model. *IOSR J. Electr. Electron. Engg.* (IOSR JEEE), Vol. 9, No. 2, pp. 6–18, 2014.

44. Montana, D. J., and Davis, L. Training feedforward neural networks using genetic algorithms. *Proceedings of the 11th International Joint Conference on Artificial Intelligence*, Vol. 1, Morgan Kaufmann Publishers, San Francisco, CA, pp. 762–767, 1989.

Section V

System Theory and Control Related Topics

In this last section of the book, we briefly discuss topics of controllability, observability, identifiability, and estimability, which are fundamental system-theoretic results that are often necessary and essential conditions for control and estimation methods. Also, the topic of stochastic processes and calculus is briefly treated, which helps understand the stochastic control and the estimation theory. The latter is needed for the development of mathematical models of the plants/processes from the measurement-data of these processes. In most cases, the analysis, design, and synthesis of the controllers is based on these mathematical models. The Lyapunov stability results are very useful in analysis and design of certain control approaches, such as MRAC, besides the stability study of nonlinear dynamic systems. A brief note on game theory-based control, control-based game, and decentralized/centralized control is given. Also, some MATLAB-based additional illustrative examples for certain control systems are given.

Appendix A: Controllability, Observability, Identifiability, and Estimability

In classical/conventional control system theory, the I/O and various error signals are mainly used and the analysis/design of these systems is carried out using TFs for linear systems, and for nonlinear systems one uses the describing functions. Here, the unique aspect is that the study is based on the I/O relation of the underlying dynamic system, and internal charateristics/behavior of these systems are not generally taken into account (AA.1) through (AA.3). Of course, the approach has a limitation, when it is extended to the time varying, highly nonlinear, and multiple-input multiple-output (MIMO) systems. Modern control engineering (optimal, model predictive, robust, digital, and adaptive controls) is based on the concept of state space. Also, the modern estimation and filtering methods are mainly based on the mathematical models (MMs) that are based on the state space concept. This approach perhaps stems from the field of classical dynamics: the phase plane method is a two-dimensional state space method, and the former is often used for analysis of the nonlinear control systems. A state of a system is the smallest collection of (state) variables such that the (current) knowledge of the state variables together with the input to the system completely determines the behavior of the system for any future time ($t > t(0)$). Some other questions of: (i) the existence of the optimal control solutions, (ii) whether a dynamic system's parameters are identifiable or not, and (iii) whether system's states are observable/estimable or not, can be answered using the concepts of controllability, observability, identifiability, and estimability. These concepts are as such applicable to the MMs of the linear systems and do not directly apply to the actual systems; however, as the models of the systems are presumed to be accurate representations of the underlying systems, the concepts then can be said to apply to the actual systems (also). For study of these aspects, it is assumed that the basic dynamic system is generally stable and it possesses BIBO stability. In general, a linear system is controllable (observable) if all the components (state variables) of the state vector are controllable (observable). As such, we need control and observe all the state variables; in some cases, it is enough to take care of only the offending states, i.e. only of the unstable components of the state vector, which leads to the concept of stabilizability and detectability. A linear system is stabilizable (detectable), if all the unstable modes are controllable (observable).

A.1 Controllability and Observability

Intuitively, any given system must be controllable/observable if we want to do something with the system under the specified control input to that system. Since basically the concepts of controllability and observability are related to linear and linearized systems (of algebraic equations, and here for the solution of these equations, some rank-conditions are useful), the tests of controllability and observability can be connected to the rank tests of certain matrices from the mathematical models of these system; these matrices are known as controllability and observability matrices.

A.1.1 Controllability of Discrete and Continuous Time Systems

Let a linear discrete time-invariant system be given as

$$x(k+1) = \phi x(k) + Bu(k)$$
$$y(k) = Hx(k) + Du(k) \qquad (A.1)$$

For the controllability, one should be able to transfer the system from its initial state $x(0)$ to any desired final state $x(t_f) = x_f$ in a finite time; i.e. be able to find a control input sequence $u(0)$, $u(1)$,..., $u(n-1)$, such that we obtain $x(t_f) = x_f$. Thus, by definition, the linear discrete time system (1) is controllable iff, the rank (CoM) is n:

$$CoM(\phi, B) = [B \ \phi B ... \phi^{n-1} B] \qquad (A.2)$$

If the transition matrix in (A.1) has distinct eigenvalues and is in the diagonal (or Jordan) canonical form, the system (in fact, its state model) is controllable iff and all the rows of B are nonzero. If the transition matrix has multiple order eigenvalues and is in the Jordan canonical form, the system is controllable iff, each Jordan block corresponds to one distinct eigenvalue, and the elements of B corresponding to the last row of each Jordan block are not all zero. The system of (A.1) is completely output controllable iff, the *OCo* matrix has rank p:

$$OCo = [D\ HB\ H\varphi B\ H\varphi^2 B...H\phi^{n-1}B] \quad (A.3)$$

Let a linear continuous time-invariant system be given as

$$\dot{x}(t) = Ax(t) + Bu(t) \quad (A.4)$$

The system of equation (A.4) is controllable iff, the rank (CoM) is n:

$$\text{CoM}\ (A,B) = [B\ AB\ ...\ A^{n-1}B] \quad (A.5)$$

It must be noted that the B vector/matrix in (A.1) and (A.4) would, in general, have different numerical values as their elements. The controllability is invariant under similarity transformation if this transformation matrix is nonsingular. There are three aspects of state controllability, the application of an input transfers: (a) any state to any state, (b) any state to zero state, this is controllability to the origin, and (c) zero state to any state, this is referred as controllability from the origin, or reachability. For the CTS (continuous time system), (A.4), these three aspects are equivalent; however, for DTS (discrete time system), the aspects (a) and (c) are equivalent but not the aspect (b).

A.1.2 Observability of Discrete and Continuous Time Systems

We want to learn (almost) everything about the dynamic behavior of the state space variables, i.e. the states, $x(t)$, by using only the information from the output measurements: from the initial state of the system, we must know about (all) the states at all the other times in future; hence, the concept of observability. A linear continuous time-invariant system without input is given as

$$\dot{x}(t) = Ax(t);\ y(t) = Cx(t) \quad (A.6)$$

One can see from (A.6) that the knowledge of $x(0)$ is sufficient to determine $x(t)$ at any time instant; hence, the system, (A.6) is observable iff, the observability matrix has full rank, n:

$$\text{ObM}(A,C) = \begin{bmatrix} C \\ CA \\ ... \\ CA^{n-1} \end{bmatrix} \quad (A.7)$$

Let the linear discrete time system be given as

$$x(k+1) = \phi x(k);\ \text{with unknown initial} \quad (A.8)$$
$$\text{state } x(0);\ y(k) = Hx(k)$$

We can easily see from (A.8) that if we know $x(0)$, then the recursion in (A.8) itself gives us the complete knowledge about the state variables at any discrete time instant in the future. We can use the measurements as in (A.8) to determine the unknown $x(0)$; hence, the system, (A.8) is observable iff, the observability matrix has rank equal to n:

$$\text{ObM}(\phi,H) = \begin{bmatrix} H \\ H\phi \\ ... \\ H\phi^{n-1} \end{bmatrix} \quad (A.9)$$

The system's observability is also invariant under similarity transformation if the transformation matrix is nonsingular. Also, if the state space model is in observable canonical form, then the system is observable; and if A (or the transition matrix) has distinct eigenvalues and is in diagonal canonical form, the state model is observable iff, none of the columns of (C), H contain zeros. If the transition matrix has multiple order eigenvalues and is in Jordan canonical form, then the state model is observable, iff: (i) each Jordan block corresponds to one distinct eigenvalue, and ii) the elements of H that correspond to first column of each Jordan block are not all zero. The phase variable canonical form is both controllable and observable: in such case, one state variable is directly measured, and since other state variables are connected in a chain fashion, they are known in turn and the system is observable. Also, since all the state variables are directly affected by the control input, the system is controllable. In the modal (diagonal) canonical form all the eigenvalues appear on the diagonal of the system coefficient matrix (A or ϕ), so we get n completely decoupled first order system. In this case, if all the elements of the control input coefficient vector $u(.)$ and all the elements of the output coefficient vector (C, H) are nonzero, then the system is controllable and observable. Also, the pair (A, B) is controllable iff, the pair (A^T, B^T) is observable. If there is a pole-zero cancellation (in the TF), the state space model will be either uncontrollable or unobservable or both. If a continuous time system is sampled, its controllability or observability would depend on the sampling period T and the location of the eigenvalues of A: (a) loss of controllability and/or observability can occur only in the presence of oscillatory modes of the system, (b) a sufficient condition for the discrete model with sampling period T to be controllable is:

$$\text{Re}[\lambda_i - \lambda_j] = 0; \; |\text{Im}[\lambda_i - \lambda_i]| \neq \frac{2\pi m}{T}, \text{for } m = 1, 2, 3, \ldots \quad \text{(A.10)}$$

(c) the condition in (b) is also a necessary condition for a single input case, and (d) if a continuous time system is not controllable or observable, then its discrete time version, with any sampling period, is not controllable or observable.

A.2 Identifiability and Estimability

The concepts of identifiability and estimability are very important for the system identification and estimation-cum-filtering problems, in general. The determination of the numerical values of the parameters of a system from its I/O data (i.e. the parameter estimation problem) is often called the identification problem. In fact, the identifiability problem is given a MM of the system and the specific I/O data, and we would like the parameters of the model to be uniquely determined. The identifiability problem is concerned with the existence of the unique solutions. Mainly, the identifiability is the appropriate determination of the form of the MM to the empirical data. The identification-cum-parameter estimation problem (often run hand-in-hand) requires systematic setting up of: (i) an experiment and data gathering procedures, (ii) choosing a suitable estimation algorithm, and (iii) analysis of these gathered data. System identification, estimation, and filtering are important steps in the modeling of dynamic systems for control, prediction, and simulation applications, and are data-dependent algorithms based on some error criteria: LS (least squares) or MS (mean squares).

A.2.1 Identifiablility

For the given experimental conditions, the best model is determinable by using some identification method. For the models based on difference/differential equations, TFs, impulse responses, and the state space systems, a system is identifiable if it is completely controllable and the input signal $u(.)$ is persistently exciting; also, we assume that the input signal $u(.)$ and the noise signal $v(.)$ are uncorrelated. Let the system be represented by S and a class of mathematical models denoted by M; this class is parameterized by β, the parameter vector that contains the unknown parameter of the system, a quadratic loss function J, and an identification method denoted as I. The identification

environment E. A linear, sable, and constant coefficient dynamic system S are said to be identifiable in the deterministic sense if the parameter vector β and the initial state $x(0)$ are uniquely determinable from a set of finite number N of the (undisturbed) measurements of control input $u(t)$ and $y(t)$. A linear stable/constant coefficient dynamic system is identifiable if the sequence of estimates $\hat{\beta}(N)$ converges to β in a stochastic case. A linear, stable and constant coefficient dynamic system is structurally identifiable if the loss function J has a minimum. The system is structurally identifiable if the optimization problem has a unique solution. A linear, stable, and constant coefficient dynamic system is system identifiable under given experimental situation: M, I, E, if $\hat{\beta}(N) \rightarrow \beta(S, M)$ with probability 1 as $N \rightarrow \infty$. Here, $\beta(S, M)$ is the set of all the parameters that give models that describe the system without error in the mean square (MS) sense. A linear, stable, and constant coefficient dynamic system is strongly system identifiable under given I and E, if it is system identifiable for all M, such that the parameter set $\beta(S, M)$ is nonempty. A linear, stable, and constant coefficient dynamic system is parameter identifiable under given M, I, and E, if it is system identifiable and the parameter set $\beta(S, M)$ consists of only one element of this set. Interestingly the concepts of system and parameter identifiablility contain the experimental conditions explicitly. A parameter vector β is structurally and locally identifiable if for any solution $\hat{\beta}$, and the solution is unique in some neighborhood of $\hat{\beta}$. Consider the linear model of (6) in a form to make the parameters to be estimated explicit

$$Z = x\beta + v \quad \text{(A.11)}$$

The parameter vector is identifiable if the knowledge of the mean $E\{z\}$ gives us β. The parameterization β is identifiable if for any β_1 and β_2, $f(\beta_1) = f(\beta_2)$ implies $\beta_1 = \beta_2$. In a linear model setting, we have $f(\beta) = x\beta$, and let $x^T x$ be a nonsingular matrix, then $x\beta_1 = x\beta_2$ implies the

$$\beta_1 = (x^T x)^{-1} x^T x \beta_1 = (x^T x)^{-1} x^T x \beta_2 = \beta_2 \quad \text{(A.12)}$$

Hence, the parameter vector β is identifiable.

A.2.2 Estimability

If suppose that a linear combination of the measured data z (A.11) is available and that it has the expected value $L\{\beta\}$, then a linear combination of the parameters $L\{\beta\}$ is estimable. In that sense, any linear combination of the data z, for example Kz, would have expectation

$E\{Kz\} = K.x\beta$, and thus, the expected value of any linear combination of the data z is equal to that same linear combination of the rows of x multiplied by β; hence $L\{\beta\}$ is estimable iff, there is a linear combination of the rows of x that is equal to L, i.e. iff, there is a K such that $L = Kx$. Interestingly, then, the rows of x form a generating set from which any estimable L can be constructed. Now, because the row space of x is the same as the row space of x^Tx, the rows of x^Tx also form a generating set from which all estimable L's can be constructed. This is also true for the rows of $(x^Tx)^{-1}x^Tx$ since these form a generating set for L. Now, once an estimable L can be determined, $L\{\beta\}$ can be estimated by computing Lb, where $b = (x^Tx)^{-1}x^Tz$; and from the general theory of linear models, the unbiased estimator Lb is the best linear unbiased estimator of $L\{\beta\}$ in the sense of having the minimum variance (also in the sense of the LS, least squares) and as ML estimator, if the residuals are Gaussian.

References

AA.1 Cellier, F. E. Lectures notes on digital control, University of Arizona, USA, 2015. https://www.inf.ethz.ch/personal/cellier/Lect/DC/Lect_dc_index.html. (dc_1 to dc_29); accessed December 2017.

AA.2 Kar, I., and Somanath, M. Introduction to Digital Control. Lecture Notes; Modules 1 to 11. Deptartment of Electronics and Electrical Engineering, Indian Institute of Technology (IIT) Guwahati, Guwahati, Assam, India. https://nptel.ac.in/courses/108103008/PDF/module1/m1_lec1.pdf; (https://nptel.ac.in/downloads/108103008/); accessed November 2017.

AA.3 Raol, J. R., Girija Gopalratnam, and B. Twala. *Nonlinear Filtering: Concepts and Engineering Applications.* Taylor & Francis Group, Boca Raton, FL, 2017.

Appendix B: Stochastic Processes and Stochastic Calculus-Brief Treatment

A very brief exposure of the theory of probability and stochastic processes/calculus that support the treatment of equations that involve noise (random/stochastic signals or disturbances/phenomena), which in turn provide mathematical basis and framework to stochastic control and linear/nonlinear filtering, is given here (AB.1) through (AB.3).

B.1 Probability Theory

We often encounter discrete or real valued (continuous) random variables (DRVs/CRVs); for the DRVs, we can simply assign a probability to every possible outcome (a value) of a RV; then, taking expectations is easy. In the case of real valued RVs (CRVs), one would work with probability densities:

$$\Pr(X \in [a,b]) = \int_a^b p_X(x)dx \; ; \quad E\{X\} = \int_{-\infty}^{\infty} xp_X(x)dx \quad \text{(B.1)}$$

In (B.1), Pr is the probability, p is the probability density function (pdf) of the CRV x, and E stands for the mathematical expectation. If we do not know the pdf p_X, then we use the uniform distribution, in that case all the values are equally weighted.

B.1.1 Probability Spaces, Events, and Some Properties

In a set Ω, every element $\omega \in \Omega$ symbolizes one possible occurrence of an event. For this occurrence in the event space, we need to specify what yes-no questions make real sense, then the sensible questions are put in a set F, which is a collection of subsets of Ω (also called a topology space), the latter is a larger sample space; not every such F would qualify, because of the requirement to make some sense. All these sensible questions then lead to a set F that is called σ–algebra (sigma-algebra). An element A of this algebra set F is called an F-measurable set, or an event, and set F is closed under countable interactions. A probability is assigned to every event A in F, by the axioms of the probability: a number $P(A)$, called the probability of the event A (the occurrence of the event A, in the sample space Ω, defined earlier),

and is chosen such that it satisfies the following conditions; the axioms of probability: (a) $P(A) \geq 0$, (b) $P(\Omega) = 1$, and (c) if $AB = \{0\}$, then $P(A + B) = P(A) + P(B)$. Thus, the probability measure, the triple (Ω, F, P), is a map (mapping) $P{:}F \to [0,1]$. If the sample space Ω consists of a non-countable infinity of elements, then its probabilities cannot be determined in terms of the probabilities of the elementary events: for example, if the space Ω is the set of all real numbers (CRVs), its subsets can be considered as the sets of points on the real line, then it is impossible to define probabilities to all subjects of Ω so as to satisfy the axioms, because there are infinite numbers on the real line. In this case, the events, all intervals $x_1 \leq x < x_2$ are considered on the real line and their countable unions and interactions as a probability space. Then, probabilities to the events $x \leq x_i$ is assignable, and all other probabilities can be obtained from the axioms of probability. Suppose that $p(x)$ is a pdf such that

$$\int_{-\infty}^{\infty} p(x)dx = 1, \quad p(x) \geq 0 \quad \text{(B.2)}$$

The probability of the event $\{x \leq x_i\}$ is defined as

$$P(x \leq x_i) = \int_{-\infty}^{x_i} p(x)dx \quad \text{(B.3)}$$

Probability of another event $x_1 < x \leq x_2$ consisting of all points in the interval (x_1, x_2) is given by

$$P(x_1 < x \leq x_2) = \int_{x_1}^{x_2} p(x)dx \quad \text{(B.4)}$$

If pdf $p(x)$ is bounded, then the integral in equation (B.4) tends to zero as $(x_1 \to x_2)$, signifying that the probability of the event $\{x_2\}$ consisting of the single outcome x_2 is zero for every x_2. For CRVs, the probability of a single outcome as a particular event cannot be defined, i.e. we cannot say what is the probability of x_1; in most cases the equality is not valid. For the probability space(Ω, F, P), we have: (a) $A \in F \Rightarrow P(A^c) = 1 - P(A)$, (b) $A, B \in F, A \subset B \Rightarrow P(A) \leq P(B)$, here, A^c is the complement of A, i.e. NOT(A); (c) $\{A_n\} \subset F$ countable $\Rightarrow P(\cup_n A_n) \leq \sum_n P(A_n)$, (d) $A_1 \subset A_2 \subset ... \in F \Rightarrow \lim_{n\to\infty} P(A_n) = P(\cup_n A_n)$, and (e) $A_1 \supset A_2 \supset ... \in F \Rightarrow \lim_{n\to\infty} P(A_n) = P(\cap_n A_n)$.

B.1.2 Random Variables and Their Probability Aspects

If (Ω, F, P) describes all possible happenings of the (random) system and their probabilities, then RVs describe concrete observations that we can make on/from such a system. Assume a measurement sensor returns an element, i.e. an observation, in some set S: then, the outcome of such an observation is described by specifying what value it takes for every possible fate of the system $\omega \in \Omega$; and a function of time $f(t, \omega)$, real or complex is assigned, i.e. we have a family of functions, one for each ω. This family is a stochastic process (or a function), and is a function of two variables ω, a sample in the space, and t, time: (i) fix ω, then $f(t, \omega) = f^{\omega}(t)$ is a real function of time; for each outcome ω, there corresponds a function of time, called a realization, or a sample function of the stochastic process; (ii) fix t, then $f(t, \omega) = f_t(\omega)$ is a family of RVs depending upon the parameter t. A random process can be regarded as a family of realizations of RVs. In order to specify a stochastic process, one has to provide the probability, pdf, of occurrence of the various realizations. Now, if $f(t)$ is a real stochastic process, then its cumulative (probability) distribution function (CDF) is given by

$$F_c(x, t) = P(f(t) \leq x) \qquad (B.5)$$

We also have $F_c(-\infty) = 0; F_c(\infty) = 1$. Then, the pdf corresponding to the CDF $F_c(x,t)$ is given by

$$p(x, t) = \frac{\partial F_c(x, t)}{\partial x} \qquad (B.6)$$

For two instances, t_1 and t_2, we have random variables $f(t_1)$ and $f(t_2)$, and their joint CDF is

$$F_c(x_1, x_2; t_1, t_2) = P(f(t_1) \leq x_1, f(t_2) \leq x_2) \qquad (B.7)$$

The joint pdf is

$$p(x_1, x_2; t_1, t_2) = \frac{\partial^2 F_c(x_1, x_2; t_1, t_2)}{\partial x_1 \partial x_2} \qquad (B.8)$$

It might happen that $F_c(x)$ in (B.5) and (B.6) might not have a derivative for every x. For CRVs, we consider that the number of points where $p(x)$ does not exist is a countable set, and the number of points of discontinuity are relatively few; (B.2) holds true, and if Δx is sufficiently small, then

$$P(x_1 \leq x \leq x_1 + \Delta x) = p(x)\Delta x \qquad (B.9)$$

Then, from (B.9) one gets

$$p(x) = \lim_{\Delta x \to 0} \frac{P(x_1 \leq x \leq x_2 + \Delta x)}{\Delta x} \qquad (B.10)$$

The discrete RVs (DRVs) have a CDF $F_c(x)$ that resembles a staircase, and its pdf (B.4) cannot be obtained. In such cases, we make the use of Dirac delta function $\delta(x)$ and define the pdf as

$$p(x) = \sum_i P_i \delta(x - x_i) \qquad (B.11)$$

$$P_i = P(x = x_i) = F_c(x_i) - F_c(x_i^-) \qquad (B.12)$$

The right-hand side of equation (B.12) is the differential height of CDF (values) at x_i.

B.1.3 Conditional Probability

In linear and nonlinear stochastic estimation/filtering, one is interested in determination of (unknown) states of a dynamic system, given the measurements that are affected by a stochastic process, and hence, the determination of the states is conditioned on knowing the measurements. The conditional probability determines the probability of an event A, given the fact that event B (related to even A) has already occurred. If event A does not depend on event B, then the events are independent. The probability (is affected because the event B has already occurred) of occurrence of event A is determined by the conditional probability

$$P(A/B) = \frac{P(A, B)}{P(B)} \qquad (B.13)$$

In (B.13), $P(A,B)$ is the joint PDF (probability or probability distribution function). The conditional pdf is

$$p(x_2, t_2 / x_1, t_1) = \frac{p(x_1, x_2; t_1, t_2)}{p(x_1, t_1)} \qquad (B.14)$$

Then the relation between the joint and conditional probability densities follows easily

$$p(x_1, x_2; t_1, t_2) = p(x_2, t_2 / x_1, t_1) p(x_1, t_1) \qquad (B.15)$$

B.2 Brownian Motion, Wiener Process and White Noise

The most famous example of observable fluctuations in a physical system is the so-called Brownian motion (BM) that is the random and incessant movement of pollen grain/s suspended in a fluid: (i) the motion of the pollen particles is incessant and highly irregular, and the path appears to have no tangent at any point, on an observable scale, the path appears non-differentiable (although it is more like continuous) since it has notches everywhere, (ii) the particles appear to move independently of one another, even when they approach each other closely, (iii) the motion is not affected by the molecular composition and mass density of the particles, (iv) as viscosity of the solvent is decreased, the motion becomes more active, (v) with the decrease of the particle radius, the motion becomes more active, and (vi) with the increase in the ambient temperature (more kinetic energy), the motion becomes more active. The limiting motion of the pollen particle as N (the number of molecules of the fluid per unit time) tends to infinity is known as BM, and its limiting stochastic process x_t, with $\sigma^2 = 1$ (variance) is known as the Wiener process (WP).

B.2.1 Random Walk and Wiener Process

Suppose in an experiment of the tossing of a fair coin an infinite time, we assume that the tossings occur every T seconds and after each tossing a step of length s is taken to the right if the "heads" show up, and to the left if the "tails" show up. In a 2-DoF graph, the movement can be drawn as up and down, respectively, and would look like discrete steps. This process starts at $t = 0$, and location at time t is a staircase function, with the discontinuities at the points $t = n$. This is a discrete time stochastic process $x(t)$ whose samples, $x(t, \omega)$ depend on the particular sequence of heads and tails (a random phenomenon). Suppose that at the first n tossings we observe k heads and $n - k$ tails. Thus, the random walk consists of k steps to the right and $n - k$ steps to the left. The limit form of the random walk as $n \rightarrow \infty$ or $T \rightarrow 0$ is called the Wierner Process (WP). It is also a Gaussian process, and it is a random process with independent increments.

B.2.2 Brownian Motion and Wiener Process

The position of a particle in BM can be modeled as a random process that satisfies

$$m\frac{d^2x(t)}{dt^2} + \gamma\frac{dx(t)}{dt} + \kappa\,x(t) = F_r(t) \qquad (B.16)$$

The forcing input function is the collision force, and m is the mass of the particle. The force $F_r(t)$ is a fluctuating force; hence, a random function, and is considered as a normal white noise with zero mean and a power spectrum. Now, if $\kappa \neq 0$ (bound motion), then the particle reaches the steady state and $x(t)$ is a stationary process; and if $\kappa = 0$, then $x(t)$ is an unbounded non-stationary process that would approach the WP as $t \rightarrow \infty$, may be the WP is defined as such.

B.2.3 Wiener Process and White Noise

From (B.16) with $\kappa = 0$, the velocity of a Brownian particle is a stationary process:

$$m\frac{dv(t)}{dt} + \gamma\,v(t) = F_r(t) \qquad (B.17)$$

This process $v(t)$ (the velocity of $x(.)$) is a normal process with zero mean. Now, the position $x(t)$ of the particle can be written as the integral of the its velocity:

$$x(t) = \int_0^t v(t)dt \qquad (B.18)$$

Here, $x(0) = 0$. Because of the linear operation, $x(t)$ is also a normal process. We see that from (B.17), by neglecting the acceleration term, we get:

$$\gamma\frac{dx(t)}{dt} \cong F_r(t) \qquad (B.19)$$

$$x(t) \cong \frac{1}{\gamma}\int_0^t F_r(t)dt \qquad (B.20)$$

Here, we know that $F_r(t)$ is a white noise process (also denoted as $v(t)$). So, for the condition $\kappa = 0$, the $x(t)$ is WP process, given by the integral of (B.20). Hence, we can define WP as the integral of normal white noise process with zero mean as

$$W(t) = \int_0^t v(t)dt; \quad R_v(\tau) = \alpha\delta(\tau) \qquad (B.21)$$

It follows that the WP $W(t)$ is a normal process with zero mean with independent increments:

$$R_w(t_1, t_2) = \alpha\min(t_1, t_2) \qquad (B.22)$$

In a nutshell, the WP is a stochastic process with the properties: (i) $W_0 = 0$; (ii) W_t has stationary and independent increments; (iii) it has normal distribution with zero mean and variance t; and (iv) it has continuous sample paths.

B.2.4 White Noise

We can model a signal received by

$$z(k) = y(k) + v(k) \qquad (B.23)$$

Here, $v(k)$ are i.i.d (independent and identically distributed) random variables with zero mean. We can assume that every disturbance is itself generated by several independent small effects, by the central limit theorem (CLT), then $v(k)$ would be Gaussian RVs; we say that $v(k)$ is a discrete time white noise or white Gaussian noise (WGN). Discrete time white noise is easy to conceive and represent. In another way, we can define (continuous time) white noise as a process such that its integral is a WP by (B.20). Also, the white noise is defined as the process that is uncorrelated in time, and the latter can be considered as the intrinsic property of the white noise. In that case, the white noise is a regular stochastic process that is totally unpredictable. The white noise also has the following property

$$E\{v(t)v(\tau)\} = R\delta(t-\tau) \qquad (B.24)$$

From (B.24), the noise $v(t)$ with a Dirac-delta correlation function is called white noise because one can easily show that the fluctuation spectrum of the Dirac-delta is a constant

$$\text{Fourier Transform }\{\delta(t)\} = \int_{-\infty}^{\infty} \delta(t)e^{j\omega t}dt = 1 \qquad (B.25)$$

Thus, in the spectrum, all the frequencies (infinite number of them) are equally represented, in analogy with white light that contains all frequencies of visible light. Because the correlation function for white noise is delta-correlated, it is not integrable in the ordinary sense. Hence, some interpretations are needed beyond the ordinary rule of calculus: (i) Ito and/or (ii) Stratonovich. We now assume that the continuous time signal $y(t)$ is transmitted and is affected by white noise, and that measured process $z(t)$ is obtained

$$z(t) = y(t) + v(t) \qquad (B.26)$$

As such, (B.26) is not very meaningful due to the presence of white noise, $v(t)$, and one works with the integrated version, using (B.20):

$$Z(t) = \int_0^t z(\tau)d\tau = \int_0^t y(\tau)d\tau + \int_0^t v(\tau)d\tau \qquad (B.27)$$

$$= \int_0^t y(\tau)d\tau + W(t) \qquad (B.28)$$

Now, (B.28) uses WP and $W(t)$, which is mathematically meaningful; the process $Z(t)$ would contain the same information as $z(t)$; to estimate signal $y(t)$ (from the observations $z(t)$), we use $Z(t)$ instead; the latter is a mathematically well-posed problem. We can also use the concept of band-limited white noise (BLWN) knowing very well that the dynamic system's bandwidth is much smaller than the bandwidth of the BLWN, thereby the system sees the noise as if its bandwidth is very large/infinite. The fact that white noise is

independent at different times has some good consequence; for example, dynamic systems driven by white noise have the Markov property.

B.3 Moments of a Stochastic Process

The mean (or the mathematical expectation, E) of the stochastic process (we do not distinguish between the random variable $f_t(\omega)$, and its stochastic process realization, x) is defined by

$$\text{mean }\{x(t)\} = E\{x(t)\} = \int_{-\infty}^{\infty} x(t)p(x,t)dx \qquad (B.29)$$

The correlation function is given by

$$\text{correlation funcion }\{x(t_1,t_2)\} = E\{x(t_1)x(t_2)\}$$

$$= \int_{-\infty}^{\infty} \int_{-\infty}^{\infty} x_1 x_2 p(x_1,x_2;t_1,t_2)dx_1 dx_2 \qquad (B.30)$$

In terms of the conditional pdf

$$E\{x(t_1)x(t_2)\} = \int_{-\infty}^{\infty} \int_{-\infty}^{\infty} x_1 x_2 [p(x_2,t_2 / x_1,t_1) \times p(x_1,t_1)]dx_1 dx_2 \qquad (B.31)$$

In term of the conditional average, replacing the joint pdf by the conditional one we have

$$E\{x(t_1)x(t_2)\} = \int_{-\infty}^{\infty} x_1 E\{x_2 / x_1\} \, p(x_1,t_1)dx_1 \qquad (B.32)$$

$$E\{x(t_2 / t_1)\} = E\{x_2 / x_1\} = \int_{-\infty}^{\infty} x_2 p(x_2,t_2 / x_1,t_1)dx_2 \qquad (B.33)$$

For a discrete random process mathematical expectation is defined as

$$E\{X\} = \sum_{k=1}^{N} x(k)P(X = x(k)) \qquad (B.34)$$

Often, it makes sense to think of the conditional mean (defined based on conditional pdf) as a true function of the measurements: $f(z(k)) = E(X/Z = z(k))$; $E(X/Z)$ is called the conditional expectation of X given Z. This is a basic notion of an estimator/estimation: $E(X/Z)$ is a good estimate of X given Z, because it is determined by taking the mean values of X based on the measurements of Z. The same notion is applicable to discrete time random variables also. Then, the covariance is

$$\text{covariance } \{(x(t_1, t_2)\} = E\{(x(t_1) - E\{x(t_1)\})(x(t_2) - E\{x(t_2)\})\} \quad (B.35)$$

$$\text{covariance } \{x(t) - \hat{x}(t)\} = E\{(x(t) - \hat{x}(t))(x(t) - \hat{x}(t))^T\} \quad (B.36)$$

In (B.36), "cap" denotes the estimation of the state variable x, and the state is assumed to be $n \times 1$ vector. It could be replaced by the parameter vector, in case of a parameter estimation problem. The variance of the stochastic process is given as

$$\sigma^2 = E\{(x - E\{x\})^2\}$$

$$= \int_{-\infty}^{\infty} (x - E\{x\})^2 p(x)\,dx; \quad \text{for continuous time} \quad (B.37)$$

$$= \sum_k (x(k) - E\{x(k)\})^2 p(k); \quad \text{for discrete time}$$

The moments, in general, of a stochastic process can be written as

$$m_n = E\{x^n\} = \int_{-\infty}^{\infty} x^n p(x)\,dx; \quad \text{for continuous time}$$

$$= \sum_k (x^n(k)p(k); \quad \text{for discrete time} \quad (B.38)$$

From (B.38), we can see that for n = 0, m_0 = 1, and for n = 1, m_1 = E{x}. The characteristic function, the Fourier transform of the probability density p(x), of a stochastic process is defined as

$$\Phi(\omega) = E\{e^{i\omega x}\} = \int_{-\infty}^{\infty} e^{i\omega x} p(x)\,dx \quad (B.39)$$

B.4 Markov Process

The most commonly used stochastic descriptor has the property that the future state of a system is determined by the present state, and not any states that have occurred in the past. Markov property states that the conditional probability of $x(t)$ given information up until, say time $s < t$, depends only on $x(s)$, and the process is defined by the fact that the conditional pdf has the following property

$$p(x_i, t_i \,/\, x_1, ..., x_{i-1}; t_1, ..., t_{i-1}) = p(x_i, t_i \,/\, x_{i-1}; t_{i-1}) \quad (B.40)$$

This means that the cpdf (conditional probability density function) at t_n given the value at x_{i-1} at t_{i-1} is not affected by the values at the earlier times, and the process is without memory. We observe that the Markov process (MP) is fully determined by $p(x_1, t_1)$ and $p(x_2, t_2 \,/\, x_1, t_1)$; the entire hierarchy can be constructed for these pdfs:

$$p(x_1, x_2, x_3; t_1, t_2, t_3) = p(x_3, t_3 \,/\, x_2, t_2) p(x_2, t_2 \,/\, x_1, t_1) p(x_1, t_1) \quad (B.41)$$

Equation (B.41) looks analogous to differential equations: the output pdf on the left-hand side is generated with a propagator, the conditional pdf, carrying the system forward in time and starting with the initial pdf, the right most term. A large class of Markov processes with continuous sample paths can be obtained as the solution of an appropriate SDE. In a filtering problem, we can approximate our signal process by a discrete time finite-state MP, called Markov chains.

B.5 Stochastic Differential Equations

As such, we would like to think of stochastic differential equations (SDE) as ordinary differential equations (ODE) driven by, say, white noise. Now, as we have seen, the BM is an abstraction of random walk process and has the property that its increments are independent. Also, the direction and the magnitude of each change of the process is completely random and independent of the previous changes. One way to think of BM as the solution of the following SDE

$$\frac{dB(t)}{dt} = v(t) \quad (B.42)$$

Here, $v(t)$ is a white noise process. The SDEs can be used for modeling dynamic phenomena, which are affected by random noise processes:

$$\frac{dx}{dt} = f(x, t) + g(x, t)v(t) \quad (B.43)$$

Because of $v(.)$, the solution of (B.43) would be a stochastic process; hence, we seek the statistics of the solutions over all the realizations. The first term, $f(x,t)$ is the drift term, and the second term $g(x,t)$ (either vector or matrix) is the diffusion or the dispersion term that determines how the noise $v(.)$ enters the dynamic system, and it is modeled as Gaussian and white noise (WGN).

B.5.1 Heuristic Solution of Linear SDEs

We consider the linear time-invariant SDE

$$\frac{dx(t)}{dt} = Ax(t) + Gv(t): \quad x(0) \sim N(x_{m0}, P_0) \quad \text{(B.44)}$$

Here, A and G are constant matrices, and $v(t)$ is white noise process with zero mean and the spectral density as Q, the initial condition for the solution is also specified. If $v(.)$ is deterministic and continuous process, we obtain the solution of equation (B.44) from the linear system theory as

$$x(t) = \exp(At)x(0) + \int_0^t \exp(A(t-\tau))Gv(\tau)d\tau \quad \text{(B.45)}$$

We can now see if the solution is also valid if $v(.)$ is a white noise process, because the SDE is linear, and the solution process $x(t)$ is Gaussian. Since the white noise $v(.)$ has zero mean, taking expectations on both the sides of (B.45) we obtain

$$x_m(t) = E\{x(t)\} = \exp(At)x_{m0} \quad \text{(B.46)}$$

This is the expected value of the SDE solutions over all the realizations of noise. The mean function is denoted as $x_m(t) = E\{x(t)\}$; the covariance of the solution is given by

$$P(t) = \text{cov}(x(t)) = E\left\{(x(t) - x_m(t))(x(t) - x_m(t))^T\right\}$$

$$= \exp(At)P_0\exp(At)^T + \int_0^t \exp(A(t-\tau))GQG^T\exp(A(t-\tau))$$

$$\text{(B.47)}$$

We can differentiate the mean (B.46) and covariance (B.47) solutions and simplify to obtain, because of the exponentials,

$$\frac{dx_m(t)}{dt} = Ax_m(t) \quad \text{(B.48)}$$

$$\frac{dP(t)}{dt} = AP(t) + P(t)A^T + GQG^T \quad \text{(B.49)}$$

Though (B.48) and (B.49) are correct equations for the mean and covariance of the solution process $x(t)$, in general, one must be careful in extrapolation of results to stochastic settings.

B.5.2 Heuristic Solution of Nonlinear SDEs

Consider the nonlinear differential equation

$$\frac{dx}{dt} = f(x,t) + g(x,t)v(t) \quad \text{(B.50)}$$

Here, f and g are nonlinear functions, and $v(.)$ is a white noise with spectral density Q. However, because of the basic requirement of the continuity of the righthand side and differentiability, we have difficulty getting the solution of (B.50), especially since the functions are nonlinear. We can have the following iteration

$$\hat{x}_t(k+1) = \hat{x}_t(k) + f(\hat{x}_t(k),k)\Delta t + g(\hat{x}_t(k),k)\Delta B(k) \quad \text{(B.51)}$$

Here, $\Delta B(k)$ is a Gaussian RV with variance as $Q\Delta t$. We can simulate trajectories from SDEs, and the result converges to the true solution in the limit as $\Delta t \to 0$. In the case of the SDEs, the convergence is a bit more involved because of path-wise approximations, which correspond to approximating the solution with fixed $v(t)$. These are strong solutions and give rise to a strong order of convergence; however, we can approximate the pdf or the moments of the solutions, then these give rise to weak solutions and weak order of convergence.

References

AB.1 Sarkka, S. Applied stochastic differential equations. Written material for the course held in Autumn 2012, Version 1.0, November, 2012. http://users.aalto.fi/~ssarkka/course_s2012/pdf/sde_course_booklet_2012.pdf; accessed May 2015.

AB.2 Handel Ramon van. Stochastic calculus, filtering and stochastic control. Lectures notes, May, 2007. https://www.princeton.edu/~rvan/acm217/ACM217.pdf; accessed April 2015.

AB.3 Raol, J. R., Girija Gopalratnam, and B. Twala. *Nonlinear Filtering: Concepts and Engineering Applications.* CRC Press, Taylor & Francis Group, Boca Raton, FL, 2017.E

Appendix C: Lyapunov Stability Theory Results

The Lyapunov's direct (second) method (LDM) is based on the generalization of the energy concept. The Lyapunov energy function(-al), LEF, $V(x)$, is such that it is positive definite, has continuous partial derivative, and its time derivative should be negative semi-definite/definite, if the dynamic system to which this LEF belongs is to be (locally)stable/asymptotically stable. The LEF is a bowl-shaped quadratic function and $V(x(t))$ always moves down the bowl, and the system's states move across the contour curves of the bowl towards toward the origin, Ref. [5] of Section III. The Lyapunov stability analysis is very useful for establishing the convergence of estimators/observers that are used in the state feedback control systems. Also, these results are useful, in general, for establishing the stability of control systems, especially if one used artificial neural networks (ANNs) or fuzzy logic (FL) for control of dynamic systems.

C.1 Motivational Example

Let a nonlinear dynamic system be described as

$$\dot{x} = f(x,t); \ x \in R^n, \ t \in R; \ R \text{ is real space};$$
$$\dot{x} = f(x) \text{ autonomous system} \qquad (C.1)$$

For system (C.1), if $x(t) = x_e$ and if it remains the same for all future times, then it is an equilibrium state: $0 = f(x)$; e.g. we can get the equilibrium points of a pendulum from its state space model. The pendulum differential equation-based model is:

$$ml^2\ddot{\theta} + b\dot{\theta} + mgl\sin(\theta) = 0; \ m = \text{mass}, \ l = \text{length} \qquad (C.2)$$

The state space model for (C.2) is given as

$$\dot{x}_1 = x_2$$
$$\dot{x}_2 = -\frac{b}{ml^2}x_2 - \frac{g}{l}\sin(x_1) \qquad (C.3)$$

The equilibrium points for the pendulum are

$$0 = x_2$$
$$0 = -\frac{b}{ml^2}x_2 - \frac{g}{l}\sin(x_1) \rightarrow x_2 = 0; \ \sin(x_1) = 0 \qquad (C.4)$$

This leads to

$$x_e = \begin{pmatrix} \pi k \\ 0 \end{pmatrix}, \ k = 0, \pm 1, \pm 2, \ldots \qquad (C.5)$$

The autonomous LTI (linear-time invariant) system, $\dot{x} = Ax$, has a single equilibrium point, i.e. the origin, if the system matrix A is non-singular; otherwise, it has an infinite number of such points in the null space of A: $Ax_e = 0$. The trajectories of this LTI are given as

$$x(t) = e^{A(t-t_0)}x(t_0) \qquad (C.6)$$

If all the eigenvalues of A are in the LHS (left hand s) plane, then the trajectories, (C.6) converge to the origin exponentially. Behavior of the original system can be studied in the neighborhood of the origin, or by perturbing and seeing that the system eventually comes back to its steady state. The latter can be done by studying the dynamics of the perturbed system, meaning study the dynamics of the error: the stability of the motion error $e(t) = x(t) - x^*(t)$; these dynamics being given as

$$\dot{e}(t) = f(x^*(t) + e(t)) - f(x^*(t)) \equiv g(e,t); \ e(0) = \delta x_0 \qquad (C.7)$$

Lyapunov stability means that the dynamic system trajectory can be kept arbitrarily close to the origin by starting sufficiently close to it:

$$\{\|x(0)\| < r; r > 0\} \Rightarrow \{\|x(t)\| < R; \forall R > 0; \forall t \geq 0\} \qquad (C.8)$$

The domain $R >$ domain r and R includes the r. Asymptotic stability means that the equilibrium is stable, and the states started close to 0 actually converge to 0 as time t goes to infinity; equilibrium point that is stable but not asymptotically stable is called marginally stable. It is defined as

$$\{\|x(0)\| < r; r > 0\} \Rightarrow \left\{\lim_{t \to \infty}\|x(t)\| = 0\right\} \qquad (C.9)$$

The state vector of an exponentially stable system converges to the origin faster than an exponential function, and exponential stability implies asymptotic stability:

$$\exists r, \alpha, \gamma > 0 \ \{\|x(0)\| < r; t > 0\} :: \{\|x(t)\| \leq \alpha \|x(0)\| e^{-\gamma t}\} \qquad (C.10)$$

If asymptotic (exponential) stability holds for any initial states, the equilibrium point is called globally asymptotically (exponentially) stable; and the LTI systems are either exponentially stable, marginally stable, or unstable. Stability is always global; local stability notion is needed only for nonlinear systems.

C.1.1 Lyapunov's First Method

For the system (C.1), it is assumed that $f(x)$ is continuously differentiable. We have the linearized dynamics as

$$\dot{x} = \frac{\partial f(x)}{\partial x}|_{x=0} x + h.o.t \cong Ax \quad (C.11)$$

In (C.11), if the matrix A is Hurwitz, then the equilibrium is locally asymptotically stable. If A has at least one eigenvalue in the RHS plane, then the equilibrium is unstable. If A has at least one eigenvalue on the imaginary axis, then inference is unclear from the linear approximation.

C.1.2 Lyapunov's Second (Direct) Method (LDM)

It is based on the (intuitive) fact that the total energy of a dynamic system is continuously dissipated, then the system, whether it is linear or nonlinear, should settle down eventually to an equilibrium. However, if this does not happen, then the system is unstable; meaning some energy is continuously pumped in and the system's states keep growing without any bounds. Thus, the idea is to study a single scalar energy function that is related to the states of the n-dimensional system and see how this functional reduces as the time lapses. Let us illustrate this by the example of the nonlinear spring mass damper (SMD) system, the equation for which is given as

$$m\ddot{x} + b\dot{x}|\dot{x}| + k_0 x + k_1 x^3 = 0 \quad (C.12)$$

For (C.12) to check the stability definitions is not easy, and the linearized system would be marginally stable, if at all. The total mechanical energy of the system is given as

$$V(x) = \frac{1}{2}m\dot{x}^2 + \int_0^x (k_0 x + k_1 x^3)dx = \frac{1}{2}m\dot{x}^2 + \frac{1}{2}k_0 x^2 + \frac{1}{4}k_1 x^4 \quad (C.13)$$

The rate of the change of the energy is given as

$$\dot{V}(x) = m\dot{x}\ddot{x} + (k_0 x + k_1 x^3)\dot{x} = \dot{x}(-b\dot{x}|\dot{x}|) = -b|\dot{x}|^3 \leq 0 \quad (C.14)$$

We see from (C.14) that the energy of the system is dissipated until the mass settles down; $\dot{x} = 0$. Lyapunov stability theorem states that the scalar function $V(x) > 0$ (i.e. it is PD, positive definite) with continuous partial derivatives, is such if its time derivative is negative semi-definite, $\dot{V}(x) \leq 0$, then the equilibrium point "0" is stable; if the time derivative is (locally) negative definite $\dot{V}(x) < 0$, then the stability is asymptotically stable. If LEF is radially unbounded, i.e. $\lim_{\|x\|\to\infty} V(x) = \infty$, the origin is globally asymptotically stable. Local stability of the pendulum is ascertained as follows. The differential equation is given as

$$\ddot{\theta} + \dot{\theta} + \sin(\theta) = 0; \ x = [\theta \ \dot{\theta}]^T \quad (C.15)$$

The LEF is chosen as

$$V(x) = (1 - \cos(\theta)) + \frac{\dot{\theta}^2}{2} \quad (C.16)$$

The LEF represents the total energy of the pendulum and is locally PD. Its time derivative is given as

$$\dot{V}(x) = \frac{\partial V(x)}{\partial \theta}\dot{\theta} + \frac{\partial V(x)}{\partial \dot{\theta}}\ddot{\theta} = \dot{\theta}\sin(\theta) + \dot{\theta}\ddot{\theta} = -\dot{\theta}^2 \leq 0 \quad (C.17)$$

Since the time derivative of the LEF is negative semi-definite, the pendulum system is locally stable. An example of the asymptotic stability is given as follows. Let the system dynamic be given as

$$\dot{x}_1 = x_1(x_1^2 + x_2^2 - 2) - 4x_1 x_2^2$$
$$\dot{x}_2 = x_2(x_1^2 + x_2^2 - 2) - 4x_1^2 x_2 \quad (C.18)$$

LEF is chosen as

$$V(x_1, x_2) = x_1^2 + x_2^2 \quad (C.19)$$

The time derivative of the LEF is obtained as

$$\dot{V}(x_1, x_2) = 2(x_1^2 + x_2^2)(x_1^2 + x_2^2 - 2) < 0 \quad (C.20)$$

The time derivative is negative definite in the two-dimensional ball specified by $x_1^2 + x_2^2 < 2$; hence, the system (C.18) is (locally) asymptotically stable. Let us now take a first order nonlinear system:

$$\dot{x} = -c(x); \ xc(x) > 0 \quad (C.21)$$

The LEF is taken as $V(x) = x^2$, which is globally PD. The time derivative of $V(x)$ is obtained:

$$\dot{V}(x) = 2x\dot{x} = -2xc(x) < 0 \quad (C.22)$$

The system (C.21) is globally asymptotically stable.

C.1.3 Extension of the Concept of Lyapunov Function

If the time derivative of the LEF is only negative semi-definite, then it is still possible to draw inferences on the asymptotic stability. Invariant set theorems (ISTs), due to La Salle, extend the concept of Lyapunov function. The LF for LTI systems $\dot{x} = Ax$ can be chosen as $V(x) = x^T P x$; P is PD, and the quadratic LF P is, hence, PD. Here, $V(.)$ is the normalized LEF. The time derivative of $V(x(t))$ is given as

$$\dot{V}(x) = \dot{x}^T P x + x^T P \dot{x} = x^T (A^T P + PA) x = -x^T Q x < 0 \quad (C.23)$$

In (C.23), the matrix Q is symmetric PD, and we have the Lyapunov equation, LE:

$$A^T P + PA = -Q \quad (C.24)$$

The procedure to check the stability now is simplified as: (i) choose a symmetric PD Q, (ii) solve the Lyapunov equation (C.24) to obtain P, and (iii) check if P is PD or not. Thus, an LTI system is (globally exponentially) stable if, and only if, for any symmetric PD matrix Q, the unique matrix solution P of the LE is symmetric PD. In most cases, Q can be chosen as a diagonal matrix. The ISTs of La Salle provide asymptotic stability analysis tools for autonomous systems with a negative semi-definite time derivative of a LF. Barbalat's Lemma extends Lyapunov stability analysis to nonautonomous systems also, like MRAC. The Lemma is: if a differentiable function $f(t)$ has a finite limit as $t \to \infty$, and if $\dot{f}(t)$ is uniformly continuous, then

$$\lim_{t \to \infty} \dot{f}(t) = 0 \quad (C.25)$$

The condition of uniform continuity of a function is difficult to verify directly. Hence, a simple sufficient condition is: if derivative is bounded, then the function is uniformly continuous. Let the LTI be given as

$$\dot{x} = Ax + Bu; \quad y = Cx, \text{ control } u \text{ is bounded.} \quad (C.26)$$

Since u is bounded and the system is stable, then x is bounded, and the time derivative of output is bounded:

$$\dot{y} = C\dot{x} = C(Ax + Bu) \quad (C.27)$$

Thus, using the Barbalat's Lemma, one can conclude that the output is uniformly continuous in time. We also have Lyapunov-like Lemma: if a scalar function $V(x, t)$ satisfies the conditions: (i) function is lower bounded and (ii) its time derivative along the system's state trajectories is negative semi-definite and uniformly continuous in time, then $\lim_{t \to \infty} \dot{V}(t) = 0$. This Lemma provides theoretical foundations for stable adaptive control design; as an example, we consider the following error dynamics:

$$\dot{e} = -e + \theta w(t); \quad \dot{\theta} = -ew(t) \quad (C.28)$$

In (C.28), e is the tracking error, θ is the parameter error, and $w(.)$ is a bounded continuous function. Now, let the LF be given as

$$V(e, \theta) = e^2 + \theta^2 \quad (C.29)$$

The time derivative of the LF is obtained as

$$\dot{V}(e, \theta) = 2e(-e + \theta w) + 2\theta(-ew) = -2e^2 \le 0 \quad (C.30)$$

As a result, e and θ are both bounded. Also, we have $\ddot{V}(e, \theta) = -4e(-e + \theta w)$ that is bounded, and hence, $\dot{V}(e, \theta)$ is uniformly continuous and

$$\lim_{t \to \infty} \dot{V}(e, \theta) = \lim_{t \to \infty} (-2e^2) = 0 \Rightarrow \lim_{t \to \infty} e(t) \quad (C.31)$$

The example with the result, as in (C.31), indicates a stable adaptation.

C.2 Lyapunov Stability Analysis for a Pilot's Situation Assessment in a Control Framework

In various aviation scenarios, the decision making and decision fusion are very important aspects for a pilot's situation assessment (SA) and activity (control task); and here, the decision making and the system's dynamics can be assumed to be represented as a closed loop control system. In this case for the controller part, the interval type 2 fuzzy logic (IT2FL) is used. Then, derivation of the Lyapunov stability result for IT2FL for the pilot's SA considering aircraft dynamics and decision making as a composite control system and combination of four type 1 FL (T1FL) systems is established, the latter representing equivalently the IT2FL. Also, some interesting and novel inferences on the derived conditions, which are based on non-FL based control system approach, are made, and as a special case, the results for type 1 FL (T1FL) can be readily obtained (AC.1).

C.2.1 A Brief Description of the Problem

One main aim of decision-making-cum-decision fusion (DMDF) would be to take one final action in a surveillance volume that is being covered by the pilot while flying an airplane for a specific mission. The SA aids in decision making to: (i) avoid collision with any closely flying aircraft, (ii) decipher the intentions of the enemy, and (iii) communicate with other friendly aircraft. The application of FL at the higher levels of DMDF is a proper choice for a precise decision making: if one output of SA is "aircraft is non-friendly and targeting tank," then it can be interpreted really as a decision that the aircraft is non-friendly and it is targeting a tank. The inference methods from artificial intelligence (AI)-cum-soft computing can be used (say FL) since the fusion of symbolic information would require reasoning and inference in the presence of modeling and/or data uncertainty; the latter is modeled by FL. In IT2FL, the primary MF itself has the MF that is another T1FL for each value of the input variable; hence, IT2FL provides an additional design degree of freedom (DOF) in fuzzy logic systems. In that case, the resulting IT2FL can potentially provide better parameterization to represent the real problem with uncertainty (thus, a third dimension compared to the T1FL is attained) and enhance the performance compared to T1FL. The IT2FL consists of: (i) fuzzifier (with T2FL MFs), inference mechanism/engine (FIS, If... Then...rules, and FIF for operations on fuzzy sets that are characterized by MFs), (ii) the Type Reducer (TR, not present in T1FL), and (iii) finally the defuzzification operation (DFO); the latter two are the O/P, output processor of a IT2FLS. For our purpose: (a) pilot's SA, (b) aircraft (plant) dynamics, and (c) decision making are considered as a composite control system that is controlled or regulated by using IT2FL. Hence, it is of great importance to ascertain stability of the combined control system, since the augmenting sub-systems are inherently nonlinear. The stability of the combined control system is studied by using LEF, and the IT2FL is considered as a logical combination of four T1FL systems (AC.1). Interestingly, the obtained condition for the assured stability is also heuristically obtained by using the non-fuzzy logic-based control system. It is then conjectured that if certain basic assumptions on control inputs are satisfied, then the LEF-based stability result for the control system is also the result for the non-FL-based control system, including the pilot's SA.

C.2.2 Pilot's Situation Assessment as a Fuzzy Logic-Based Control System

It is desirable to assist the pilot by employing an intelligent decision support system; hence, the use of fuzzy logic is a good choice. The problem is represented as a combined FL-based control and composite dynamic system; the latter consists of the pilot's command input based on the decision making and the aircraft's (plant)

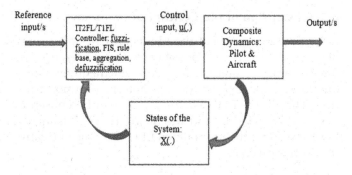

FIGURE AC.1
Combined FL-based pilot's SA (FIS-fuzzy inference system).

dynamics, as shown in Figure AC.1 as a nonlinear state-feedback control system, in which X is the universe of discourse. Consider a single input n-th order composite nonlinear system that represents the state space equations of the controlled process, including the pilot's activity and the aircraft dynamics:

$$\dot{x} = f(x) + g(x)u; \quad x(t(0)) = x_0 \qquad (C.32)$$

When FL-based control is used, then one needs to specify the rule base: the i-th fuzzy control rule in the rule base of the T-S FLC (fuzzy logic control) base is of the form (AC.2):

$$\text{Rule i: If } x_1 \text{ is } X_{i,1} \text{ AND } x_2 \text{ is } X_{i,2} \text{ AND...AND } x_n \text{ is } X_{i,n};$$
$$\text{Then } u = u_i(x), i = 1 \text{ to } r, r \in N \qquad (C.33)$$

Thus, r is the total number of rules; the $X_{i,1}, X_{i,2}, ..., X_{i,n}$ are fuzzy sets that describe the linguistics terms of the input variables $x_k, k = 1$ to n; $u = u_i(x)$ is the control signal of rule i; and the function AND is a t-norm. The u_i can be a single value or a function of the state vector x. Each fuzzy rule generates the firing strength defined as

$$\alpha_i(x) = \text{AND}(\mu_{i,1}(x_1), \mu_{i,2}(x_2), ..., \mu_{i,n}(x_n)) \in [0,1], \forall x \in X,$$
$$i = 1 \text{ to } r \qquad (C.34)$$

C.2.3 IT2FL as a Combination of T1FL

Next, the IT2FL-based controller is considered and viewed as an equivalent combination of four T1FL systems (AC.1), as in Figure AC.2 (to be viewed in conjunction with Figure AC.3) for the sake of the Lyapunov stability analysis: the four blocks of Figure AC.2 are based on the decomposition of T2FMFs into 4 T1FMFs, as shown in Figure AC.3. It is assumed that for any $x \in X$ there exists among all rules, at least one $\alpha_i(x)$

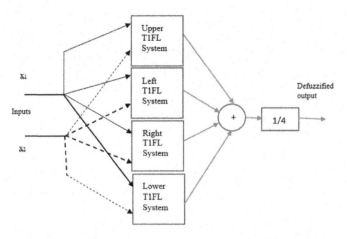

FIGURE AC.2
T2FL system simplified: The controller output is the average of the four outputs of the embedded upper, left, right, and lower T1FL systems. (From Ibrahim Abdel Fattah Abdel Hameed Ibrahim. New applications and developments of fuzzy systems. Thesis, Doctor of Philosophy, Korea University, Seoul, South Korea, February 2010.)

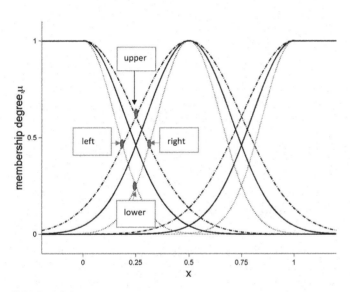

FIGURE AC.3
Decomposition of T2MFs into 4 T1MFs. (From Ibrahim Abdel Fattah Abdel Hameed Ibrahim. New applications and developments of fuzzy systems. Thesis, Doctor of Philosophy, Korea University, Seoul, South Korea, February 2010.)

$\in (0,1]$, $i = 1,...,r$; and the control input u is a function of $\alpha_i(x)$ and u_i. Using the weighted sum defuzzification method, the output of the FLC is obtained (based on Figure AC.2) as

$$u = k \frac{\sum_{i=1}^{r} \alpha_i u_i}{\sum_{i=1}^{r} \alpha_i} \qquad (C.35)$$

In (C.35), k is equal to ¼. Certain assumptions used in the stability analysis are (AC.2):

a. If for any input $x_0 \in X$, the firing strength $\alpha_i(x_0)$ corresponding to the fuzzy rule i is zero, then, that fuzzy rule i, for $i = 1,...,r$, is considered as an inactive rule for that input; otherwise, it is considered as an active fuzzy rule.

b. An active region of the fuzzy rule i is defined as a set $X_i^A = \{x \in X \ / \ \alpha_i(x) \neq 0\}, i = 1,...,r$.

With the assumptions (a) and (b), (C.35) can be written as

$$u(x(0)) = k \frac{\sum_{i=1, \alpha_i \neq 0}^{r} \alpha_i(x(0)) u_i(x(0))}{\sum_{i=1, \alpha_i \neq 0}^{r} \alpha_i(x(0))} \qquad (C.36)$$

C.2.4 Lyapunov Stability Analysis

The LEF considered is $V(x) = x^T P x$; it is PD, since P is taken as a PD; P is also assumed to be independent of time and is a weighting matrix; hence, V is a normalized LEF. The LEF, V is regarded as a bounded function of the state trajectories that are governed by (C.32) in order to obtain the stable and bounded CL system dynamics/ responses. Also, it is assumed that V has continuous partial derivatives, at least of first order. In the Lyapunov stability analysis, one has to obtain the derivative of V with respect to time and establish that this derivative function is negative definite, based on the constraint of the system dynamics (C.32); and in the present case (C.36), since FL-based control strategy is used. Thus, one has the following expression for the time derivative of V (for simplicity dependence on x is often not mentioned) [AC.2]:

$$\dot{V}(t) = \dot{x}^T P x + x^T P \dot{x} \qquad (C.37)$$

Next, substitute (C.32) in (C.37) to obtain

$$\dot{V}(t) = (f(x) + g(x)u)^T P x + x^T P(f(x) + g(x)u) \qquad (C.38)$$

Considering time being the control input as scalar quantity without loss of generality, one obtains

$$\dot{V}(t) = f^T(x)Px + g^T(x)u \cdot Px + x^T Pf(x) + x^T Pg(x)u \qquad (C.39)$$

$$\dot{V}(t) = f^T(x)Px + x^T Pf(x) + g^T(x)Pxu + x^T Pg(x)u \qquad (C.40)$$

$$\dot{V}(t) = [f^T(x)Px + x^T Pf(x)] + [g^T(x)Px + x^T Pg(x)]u$$
$$= F(x) + B(x)u \qquad (C.41)$$

544
Appendix C

From (C.41), it can be seen that one has to establish that the time derivative of V is negative definite. To establish this, the following definitions/assumptions on the sets, functions, and input are required to be specified (AC.2):

1. $B^0 = \{x \in X \mid B(x) = 0\}$; $B^+ = \{x \in X \mid B(x) > 0\}$; and $B^- = \{x \in X \mid B(x) < 0\}$.

2. The dynamic system is described by (C.32) with $x = 0 \in R^n$ as an equilibrium point.

3. The LEF $V{:}R^n{\to}R$; $V(x) = x^T P x$, $P \in R^{n\times n}$ is PD and assumed bounded, as discussed earlier to obtain consistent results with the stability theory of dynamic system (BIBO stability, bounded input, bounded output). In the present case of the nonlinear dynamic system of (C.32), the BIBO stability theory is considered to be applicable to a good and proper approximation of the nonlinear system or to the linearized nonlinear system. Also, $V(0) = 0$, $V(x) > 0$, $\forall x \neq 0$.

4. $F(x) \leq 0$, $\forall x \in B^0$

5. $u_i(x) \leq -kF(x)/B(x)$ for $x \in X_i^A \cap B^+$; and $u_i(x) \geq -kF(x)/B(x)$ for $x \in X_i^A \cap B^-, I = 1,\ldots, r.$

6. The set $\{x \in X \mid \dot{V}(x) = 0\}$ contains no state trajectories except the trivial one, $x(t) = 0$ for $t \geq 0$.

If all the foregoing assumptions are valid, then the combined and composite closed loop system with the T-S FLC will be globally asymptotically stable in the sense of Lyapunov at the origin. In (AC.2), only T1FL is considered; however, here the result is extended to IT2FL, which can be established as follows:

1. For $B(x(0))$ strictly positive,

$$u_i(x(0)) \leq -k\frac{F(x(0))}{B(x(0))}$$

$$\to u(x(0)) = k\frac{\sum_{i=1,\alpha_i\neq 0}^{r}\alpha_i(x(0)u_i(x(0))}{\sum_{i=1,\alpha_i\neq 0}^{r}\alpha_i(x(0))}$$

$$\leq k\frac{-k\frac{F(x(0))}{B(x(0))}\sum_{i=1,\alpha_i\neq 0}^{r}\alpha_i(x(0)}{\sum_{i=1,\alpha_i\neq 0}^{r}\alpha_i(x(0))} = -\frac{F(x(0))}{B(x(0))} \quad (C.42)$$

$$\to \dot{V}(x(0)) = F(x(0)) + B(x(0))u(x(0)) \leq F(x(0))$$

$$+ B(x(0))\left\{-\frac{F(x(0))}{B(x(0))}\right\} = 0.$$

Hence,

$$u_i(x(0)) \leq -k\frac{F(x(0))}{B(x(0))} \to \dot{V}(x(0)) \leq 0. \quad (C.43)$$

2. For $B(x(0))$ strictly negative,

$$u_i(x(0)) \geq -k\frac{F(x(0))}{B(x(0))}$$

$$\to u(x(0)) = k\frac{\sum_{i=1,\alpha_i\neq 0}^{r}\alpha_i(x(0)u_i(x(0))}{\sum_{i=1,\alpha_i\neq 0}^{r}\alpha_i(x(0))} \quad (C.44)$$

Hence,

$$u_i(x(0)) \geq -k\frac{F(x(0))}{B(x(0))} \to \dot{V}(x(0)) \leq 0. \quad (C.45)$$

3. For $x(0) \in B^0$,

$$\dot{V}(x(0)) = F(x(0)) + B(x(0))u(x(0)) = F(x(0)) \leq 0. \quad (C.46)$$

Thus, for all the three cases considered, the time derivative of the LEF is negative definite; hence, the asymptotic stability results for the combined IT2FL-based control and the pilot's situation assessment-cum-aircraft plant dynamics are established.

C.2.5 Further Stability Analysis

The stability result (AC.1), when T1FL is considered, can be obtained when $k = 1$, from the results of Section C.2.4. Hence, the asymptotic stability results for the combined T1FL-based control and the pilot's situation assessment-cum-aircraft plant dynamics are established. It is ascertained here that the pilot's situation assessment was not considered in (AC.1) and (AC.2). Further, one can consider the following time derivative of the LEF to satisfy the stability condition

$$\dot{V}(t) = [f^T(x)Px + x^T Pf(x)] + [g^T(x)Px + x^T Pg(x)]u \leq 0. \quad (C.47)$$

From (C.47), one can obtain for input u,

$$u \leq -\frac{\|f\|\|P\|\|x\| + \|x\|\|P\|\|f\|}{\|g\|\|P\|\|x\| + \|x\|\|P\|\|g\|} = -\frac{2\|f\|\|P\|\|x\|}{2\|g\|\|P\|\|x\|} \quad (C.48)$$

$$\leq -\frac{\|f\|}{\|g\|} \quad (C.49)$$

The condition in (C.49) is equivalent to the assumption 5) of Section C.2.4:

$$u_i(x(0)) \leq -k \frac{F(x(0))}{B(x(0))} \tag{C.50}$$

It is easily ascertained that the condition in (C.49) is obtained based on no assumption that the controller is a fuzzy logic based. However, it is based on the assumptions of boundedness of the nonlinear functions of (C.32) and the boundedness of the weighting matrix P. However, even if P is unbounded, the condition (C.49) would still be valid (AC.1). The main aspect is that the control input should be bounded. For linear dynamic system, the condition (C.49) yields

$$u \leq -\frac{\|Ax\|}{\|Bx\|} = -\frac{\|A\|\|x\|}{\|B\|\|x\|}$$
$$\leq -\frac{\|A\|}{\|B\|} \tag{C.51}$$

Again, we see that the condition in (C.51) is equivalent to the assumption v) of Section 3, and (C.50). These foregoing conditions and observations can be corroborated based on the following fundamental aspect of the dynamic system. Let the linear dynamic system be considered, for the sake of simplicity,

$$\dot{x} = Ax + Bu \tag{C.52}$$

When the system reaches the steady state, the time derivative of the state is zero and (C.52) becomes

$$Ax + Bu = 0 \tag{C.53}$$

From (C.53), one can obtain the following result

$$u \leq -\frac{\|A\|\|x\|}{\|B\|} \tag{C.54}$$

The condition in (C.54) is the same as (C.51), except the norm of x because the norm obviously cancels out in (C.51). Thus, the fundamental condition for the stability, apart from the inherent stability of the dynamic system being controlled, is on the boundedness of the input as specified by all the conditions in (C.33), (C.35), (C.36), (C.39), (C.50), (C.51), and (C.54), which indicate and signify the same result in different ways. Thus, the observations and the inferences made in this paper are novel in terms of the interpretation of the Lyapunov stability result in the context of FL-based control system.

Thus, the Lyapunov method can be used to derive the asymptotic stability condition when the pilot's situation assessment is viewed as composite control system with the aircraft dynamics and the controller is based on interval type 2 FL (IT2FL). The condition for the case when the controller is T1FL is easily obtained as the special case of the T2FL, and the condition is similar to one obtained in the literature. Thus, using the LEF, it has been established that the type 2 (as well as type 1) fuzzy logic-based controllers used in the pilot's SA in conjunction with nonlinear dynamic plant lead to asymptotic stability results that are further supported by some new intuitive inferences deducted from the basic theory of control system. The performance of the IT2FL decision system has been evaluated by numerical simulations carried out in MATLAB/SIMULINK and is presented in Ref. (AC.3).

References

AC1. Ibrahim, A.F.A.H.I. . New applications and developments of fuzzy systems. Thesis, Doctor of Philosophy, Korea University, Seoul, South Korea, February. 2010.

AC2. Precup, R. E., Tomescu, M. L., and St. Preitl. Fuzzy logic control system stability analysis based on Lyapunov's direct method. *Int. J of Computers, Communications, & Control*, Vol. IV, No. 4, pp. 415–426, 2009.

AC3. Lakshmi, S., and Raol, J. R. Lyapunov stability analysis for pilot's situation assessment within control system framework. *Control Data Fusion e-J.*, Vol. 1, No. 4, pp. 02–09, 2017s.

Appendix D: Game Theory and Decentralized/Centralized Control

We briefly discuss game theory in comparison to the (general) control theory; and compare centralized and decentralized control strategies. The control-oriented games have the features: (i) adaptive strategy in games, (ii) state space approach, and (iii) man-machine games (MMG) (AD.1). In the game-based controls we have: (i) consensus of MAS (multi-agent system), (ii) distributed coverage of graphs, and (iii) congestion games. Some forms of the present control and game theories mainly were evolved and/or originated with the work of Norbert Wiener (AD.2), and John von Neumann (AD.3).

When competition in the world increases, or becomes essential, the growing and expanding companies and organizations should become: (i) more flexible to accommodate changes in policies and procedures; (ii) be adaptive to the effects/changes of the environmental/systems' own dynamical aspects; and (iii) more responsive to meet the demands and targets to remain up-to-date; however, this would be very difficult in the presence of complexities and uncertainties like: (a) absence of workers/reduced work force, (b) machine failures/outages, and (c) changes in customer/s' behaviors (an inevitable situation). Interestingly enough, colonies of ants, bees, wasps, and termites (all of Nature's social insects) exhibit flexibility: (a) adaptive behavior that is desired/required by the general industry, (b) achieve it without any central planning/management, and (c) coordinate many insects/workers (AD.4); hence, these societies can provide important models for industrial logistics/work procedures and ideas for learning and controls, which one can utilize for many engineering technology development. We have discussed some aspects of differential game/s in Chapter 6, in the context of optimal control, i.e. a strategy that a player should/would employ.

D.1 Game Theory

A normal non-cooperative game $G = (N,S,J)$ has: (i) player, $N = \{1,2,...,n\}$; (ii) strategy, $S_i = \{1,2,...,k_i\}$, $I = 1,...,n$; a situation (profile) $S = \prod_{i=1}^{n} S_i$; and (iii) payoff/cost functional,

$$c_j(s) = S \to \mathbb{R}, \; j = 1,...,n; \; \text{payoff, } c = \{c_1,...c_n\} \quad \text{(D.1)}$$

A situation $s = (x_1^*,...,x_n^*) \in S$ is called a Nash equilibrium if the inequality holds true

$$c_j(x_1^*,...,x_j^*,...,x_n^*) \geq c_j(x_1^*,...,x_j,...,x_n^*) \quad \text{(D.2)}$$

In a cooperative game, a transferable utility game G consists of three things: (i) n players $N: = \{p_1,...,p_n\} = \{1,...,m\}$; (ii) subsets $\{S \mid S \in 2^N\}$, each is named a coalition, $S = 0$ is an empty coalition, and $S = N$ is a full coalition; and (iii) $v : 2^N \to \mathbb{R}$ is called the characteristic function; $v(S)$ is the worth of S, this means that the profit, i.e. the cost $c : 2^N \to \mathbb{R}$, of coalition S, $v(0) = 0$.

If in a game G with $P = \{p_1, p_2,...,p_n\}$; $R = \{p_i \in P \mid p_i$ has a right hand glove$\}$; $L = \{p_i \in P \mid p_i$ has a left hand glove$\}$.

Let $S \in 2^P$, a single glove (0.01), a pair of gloves (1), then:

$$v(S) = \min\{|S \cap L|, |S \cap R| + 0.01[n - 2\min\{|S \cap L|, |S \cap R|\}]\} \quad \text{(D.3)}$$

For a given cooperative game $G(N,v)$, $x \in \mathbb{R}^n$ is an imputation if

$$x_i \geq v(\{i\}), \; i = 1,...,n; \; \sum_{i=1}^{N} x_i = v(N) \quad \text{(D.4)}$$

For an evolutionary (non-cooperative) game, we have the assumptions: (i) these are finitely or infinitely repeated: $G \to G^N$, or $G \to G^\infty$, and (ii) the dynamics of the strategies are given as (AD.1):

$$x_1(k+1) = f_1(x_1(k),...,x_n(k),...,x_1(1),...,x_n(1))$$

$$x_2(k+1) = f_2(x_1(k),...,x_n(k),...,x_1(1),...,x_n(1))$$

$$\vdots$$

$$x_n(k+1) = f_n(x_1(k),...,x_n(k),...,x_1(1),...,x_n(1)) \quad \text{(D.5)}$$

$$x_i \in D_{t_i}, \text{and} f : \prod_{j=1}^{n} D_{t_j}^k \to D_{t_i}, i = 1,...,n$$

For an NEG (a networked evolutionary game), $((N,E),G,\pi)$ consists of: (i) a network/graph (N,E); (ii) an FNG (fundamental network game), G, such that if $(i,j) \in E$, then i and j play FNG with strategies $x_i(k)$ and $x_j(k)$, respectively; and

(iii) use of a currently available local information-based strategy for updating a gaming rule. In an asocial world, the cooperation can start based on reciprocity, and there are certain rules for the evolution of cooperation.

D.1.1 Comparison of Game with Control

The common point for both game and control is the purpose of actions, i.e. an individual intends to manipulate the object of the game provided to her; e.g. a game-device or a system for controlling. However, there are certain differences: (i) the object for the control is some machine (industrial plant/process, aircraft, ship, robot, etc.), and as such an "intelligent thing" (meaning an "intelligence") is not an object of direct control; for a game, an object is an "intelligent thing," and has an ability in anti-control, i.e. an opponent does not want to be controlled and opposes any control from the opposite player; and (ii) for the control, the goal is optimization (for optimal control) or any other control philosophy, and for a game, the aim is to achieve Nash equilibrium. For control, we have the following basics (Chapter 6) repeated here for clarity. Consider a linear system

$$\dot{x} = Ax + Bu \tag{D.6}$$

In an optimal control problem, we need to determine a control law, $u(t)$, such that the following cost function is minimized

$$\min_u J = \min_u \int_0^\infty [x^T R x + u^T Q u] dt \tag{D.7}$$

Then the solution to the optimal control problem results into the control law

$$u^*(t) = -Q^{-1}B^T P x(t) \tag{D.8}$$

In (D.8), the matrix P satisfies the algebraic matrix Riccati equation

$$PA + A^T P = R - PBQ^{-1}B^T P = 0 \tag{D.9}$$

In game theory, we can consider a linear system as

$$\dot{x} = Ax + B_1 u_1 + B_2 u_2 \tag{D.10}$$

The aim is to minimize the cost function as

$$\min_{u_i} J_i = \min_{u_i} \int_0^\infty [x^T R_i x + u^T Q_i u] dt, \ i = 1, 2 \tag{D.11}$$

Here, we use the Nash equilibrium

$$u_1^* = -Q_1^{-1}B_1^T P_1 x; \ u_2^* = -Q_2^{-1}B_2^T P_2 x; \ P_i > 0, \ i = 1, 2 \tag{D.12}$$

For this case, the following coupled algebraic Riccati equations need to be satisfied

$$\begin{aligned} & P_1(A - B_2 Q_2^{-1} B_2^T P_2) + \\ & (A - B_2 Q_2^{-1} B_2^T P_2)^T P_1 + R_1 - P_1 B_1 Q_1^{-1} B_1^T P_1 = 0 \\ & P_2(A - B_1 Q_1^{-1} B_1^T P_1) + \\ & (A - B_1 Q_1^{-1} B_1^T P_1)^T P_2 + R_2 - P_2 B_2 Q_2^{-1} B_2^T P_2 = 0 \end{aligned} \tag{D.13}$$

In a nonlinear case, it becomes a problem of differential games (DGs), and one has to solve coupled Hamilton-Jacobi-Bellman equation (Chapter 6).

As a part of cross discipline between control and game, we have: (a) control theory $\rightarrow$ game theory leading to control-oriented games, and (b) game theory $\rightarrow$ control theory leading to game-based controls.

D.1.2 Control-Oriented Games

In here, we have three aspects: (i) learning control in games, (ii) state space approach, and (iii) man-machine games. In the first aspect, the learning control in games, we have: (i) strategy in Rock-Paper-Scissors game and (ii) convolutional neural network leading to optimal action. Consider the Rock-Paper-Scissors game: (a) game theory$\rightarrow$ the game properties: zero-sum; pure harmonic game; Nash equilibrium: (1/3,1/3,1/3);..., and b) control theory (player's perspective): How to win?

In the state space approach, we recall here the definition of a networked evolutionary game, section AD.1, and add the definition: (i) (N,E) is called a graph, where N is the set of nodes and E ($N \times N$) the set of edges, (ii) $U_d(i) = \{j | \text{there is a path connecting } i, j \text{ with length} \leq d\}$, and (iii) if $(i,j) \in E$ implies $(j,i) \in E$, the graph is undirected; otherwise, it is directed. A network is homogeneous if each node has same degree, for undirected graph; and in-degree and out-degree for the directed graph. The FNG is defined as: (i) a normal game with two players is called a fundamental network game if $S_1 = S_2 := S_0 = \{1,2,...,t\}$, and (ii) an FNG is symmetric if $c_{1,2}(x,y) = c_{2,1}(y,x), \ \forall x, y \in S_0$. Then, overall payoff (cost function) is given as

$$c_i(k) = \sum_{j \in U(i) \setminus i} c_{ij}(k), \ i \in N \tag{D.14}$$

We have the strategy updating rule (SUR) for a NEG, denoted by π, and it is a set of mappings:

$$x_i(k+1) = f_i(\{x_j(k), c_j(k) | j \in U(i)\}), k \geq 0, i \in N \tag{D.15}$$

In (D.15), f_i could be a probabilistic mapping, and when the network is homogeneous, $f_i, i \in N$ are the same. Also, since $c_j(k)$ depends on $x_t(k)$, $t \in U(j)$, it follows that $x_i(k+1)$ depends on

$$x_j(k), \ j \in U_2(i), \text{ so we have}$$
$$x_i(k+1) = f_i(\{x_j(k) \mid j \in U_2(i)\}), \ i \in N \quad \text{(D.16)}$$

There are certain aspects for further study: (i) convergence of network evolutionary game, (ii) strategical equivalence, and (iii) evolutionary stable strategy (ESS). The latter is in the spirit of Nature's biological evolutionary process (due to which the species stabilize their existence and continue to pervade for much longer times) and we can then presume that Nature's ESS is an unintended/unintentional self-evolving "game," which is not purposefully designed as a game but it appears so, such things in the theory of evolution are called "designoids," as opposed to the man-made design, such as the intentional design of an aircraft.

In man-machine games, we have the following model, for n machines v/s m players (AD.1):

$$m_1(k+1) = f_1(m_1(k), ..., m_n(k), h_1(k), ..., h_m(k))$$
$$m_2(k+1) = f_2(m_1(k), ..., m_n(k), h_1(k), ..., h_m(k))$$
$$\vdots \quad \text{(D.17)}$$
$$m_n(k+1) = f_n(m_1(k), ..., m_n(k), h_1(k), ..., h_m(k))$$

The goal function is given as

$$\max_{h(k) \in D_p} \sum_{k=1}^{N} \lambda^k c_h(k), \ 0 < \lambda < 1 \quad \text{(D.18)}$$

In this case for pure strategy, the optimal solution appears on a cycle, so the idea is to find the best cycle by using the concepts of: (i) optimization and identification in non-equilibrium dynamical games and/or (ii) optimal control of logical control networks. For mixed strategy, we have (i) dynamic programming (DP) for $N < \infty$ (Chapter 6) and (ii) DP + receding horizon control for $N = \infty$, for the latter, one follows the receding horizon-based feedback optimization for mix-valued logical networks.

D.1.3 Potential Games

Consider a finite game $G = (N,S,C)$. G is a positive game if there exists a function $P : S \to \mathbb{R}$, called the potential function, such that for every $i \in N$ and for every $s^{-i} \in S^{-i}$ and $\forall x, y \in S_i$, we have

$$c_i(x, s^{-i}) - c_i(y, s^{-i}) = P(x, s^{-i}) - P(y, s^{-i}), \ i = 1, ..., n \quad \text{(D.19)}$$

There are certain fundamental properties that are now briefly stated: (i) if G is a potential game, then the potential function P is unique up to a constant number; if P_1 and P_2 are two potential functions, then $P_1 - P_2 = c_0 \in \mathbb{R}$; (ii) every finite potential game possesses a pure Nash equilibrium and sequential or cascading leads to a Nash equilibrium; and (iii) G is a potential game if, and only if, there exist $d_i(x_1, ..., \hat{x}_i, ..., x_n)$, which is independent of x_i, such that

$$c_i(x_1, ..., x_n) = P(x_1, ..., x_n) + d_i(x_1, ..., \hat{x}_i, ..., x_n), \ i = 1, ..., n \quad \text{(D.20)}$$

D.1.4 Game-Based Controls

Game-based controls include: (i) control of power systems, (ii) multi-objective optimization, and (iii) robust optimization/control all via game theory. In game-based controls, there are: (i) consensus of MAS, (ii) distributed coverage of graphs, and (iii) congestion games (AD.1).

For the consensus of MAS, let us have the network graph: $(N,E(k)):N = \{1,2,...,n\}$ with varying topology: $E(k)$. The model of MAS is taken as

$$a_i(k+1) = f_i(a_j(k) \mid j \in U(i)), \ i = 1, ..., n \quad \text{(D.21)}$$

We also have the set of strategies $a_i \in \in A_i \subset \mathbb{R}^n$, $i = 1, ..., n$. The potential function is given as

$$P(a) = -\sum_{i \in N} \sum_{j \in U(i)} \frac{\|a_j - a_i\|}{2} \quad \text{(D.22)}$$

The payoff functions are:

$$c_i(a) = -\sum_{j \in U(i)} \|a_j - a_i\|, \ i = 1, ..., n \quad \text{(D.23)}$$

Then, $\max_{a \in A} P(a) \to$ consensus.

In distributed coverage of graphs, in an unknown connected graph, mobile agents are initially arbitrarily deployed, and as usual potential and payoff functions are defined, the number of covered nodes is asymptotically maximized in probability.

In a congestion game, N number of players, M number of facilities, and S number of strategies are defined, and facility cost is also specified. Again, as the usual payoff and potential functions are defined. In effect, every congestion game is a potential game, and every finite potential game is isometric to a congestion game. Other game-based control related aspects are scheduling-allocation in power/energy distributed systems and cooperative game leading to control strategies. The cross-fertilizations of the two fields can usher into a new discipline: (i) in a game theory one can use the state space approach and learning-control; and (ii) in control theory, one can imbibe the features of control of MASs, distributed graph covering (for distributed control), and congestion control.

D.1.5 Differential Game-Based Tidal Power Generation Process Control

A control method based on differential game theory is studied for tidal power process control (AD.5). In the generation of the tidal power, the potential and kinetic energies of sea water are converted into electric energy through the water-turbine generator set. Tidal energy is produced by the periodic movement of sea water under the influence of the tidal force due to the rotation of the earth: (i) in the rising tide, the water level rises rapidly, and the kinetic energy is changing to the potential energy, and (ii) in the falling tide, the water level drops, and the potential energy changes to kinetic energy; the sum of conversion of these energies is the tidal energy. For the generation of power, a water-turbine is driven by the drop of sea water to generate electricity. In the process of utilization of tidal resource, considering the characteristics of intermittency and fluidity of the sea tide, the virtual opponent is introduced; each generator is controlled by the game of the cost function and the other generators to control the direct current (DC) bus voltage, so as to achieve the Nash equilibrium. The generation strategy of each independent tidal power generation unit (TPGU) is optimal and minimizes its cost function; and all the optimal control strategies can keep the DC bus voltage stable. When the Nash equilibrium is reached, no participant will change the strategy actively, because if a participant does change the control strategy, the cost function would increase. Hence, all the participants must maintain their own strategy, and the entire tidal power system would reach a stable state. The cost function is given as

$$J_j(d_j^*, d^{-j}) \le J_j(d_j, d^{-j}); \quad d_j^*(j = 1, 2, ..., N) \quad \text{(D.24)}$$

In (D.24), $d^*(.)$ is the power generation control strategy when the TPGU j reaches Nash equilibrium. The d^{-j} is the power generation control strategy (PGCS) of other power generators other than the TPGUs. The Nash equilibrium (NE) theory is used in the control method of the tidal power generation process (TPGP) based on differential game (DG) theory, by which the cost function of each TPGU minimum assumes there are N power generation devices (PGD) in the tidal power generation system, and for each such device,

there are $N-1$ competitors. However, since there is no communication line, every PGD cannot understand the number of competitors and any other information, and a concept of virtual rivals is introduced, and other PGDs, except themselves, are regarded as an equivalent virtual competitor (AD.5). Any other one or several PGDs modify the control strategy, which can be regarded as the virtual generation device to modify its own control strategy. Thus, in this virtual competitor method, the game problem of N participants is simplified into the game problem of two participants thereby greatly simplifying the problem. Assume N TPGUs supply power to the DC bus via the converter, and super capacitor is used as energy storage device for power generation system. Let i_{jt} represent the output current of the j-th tidal PGD at the time t, then:

$$i_{1t} + i_{2t} + ... + i_N t = i_{Ct} + i_{ot} \quad \text{(D.25)}$$

In (D.25), i_{Ct} is the charging current of the energy storage device, and the i_{ot} is the load current of the TPGU. The total current can be obtained as

$$\sum_{j=1}^{N} i_{1t} = E_a \frac{d_j u_{ot}}{dt} + \frac{u_{ot}}{J_j} \quad \text{(D.26)}$$

The DC bus voltage u_{ot} is obtained as

$$u_{ot} = u_{jt} \left(\sum_{j=1}^{N} i_{jt} \right); \quad u_{jt} = u_{ot} + \varepsilon_{jt} \quad \text{(D.27)}$$

In (D.27), we have u_{jt} as the DC bust voltage at the time t detected by the j-th TPGU and ε_{jt} is the detection error of the TPGU. Also, it must be recognized that the DC bus voltage in the process of TPG (D.27), is determined by the parallel current sources. For a TPG device in the game to reach the Nash equilibrium state, the cost function of the j-th PGD us defined as

$$J_{jt}(u, i) = A u_{jt} \Delta i_{jt} + B \sum_{j=1}^{t} (k_{j1} u_r - k_{j2} u_{jm} - k_{j3} i_{jm})^2 + E(u_r - u_{jt})^2$$

$$\text{(D.28)}$$

In (D.28), A, B, E, and $k(..)$'s are nonnegative constants, $u(.)$ and $i(.)$ are the DC bus voltage and the output current of each PGD, and u_r is the DC bus voltage setting value of the tidal power generation system. The first term (D.28) is the cost of the game of control of the tidal power generation process (when this term is positive, then the output power of the unit is increased, if the term is negative, the power is decreased), and Δi_{jt} is the change in the current. The second term is a penalty function for the deviation between the set value and the output voltage of the historical time and the load compensation. The third term is the control error of the DC bus voltage. By minimizing the cost function $J(..)$ of each TPGD, the control strategy of the tidal power circuit is obtained. By using the optimization process, the reference value of the output current of the TPG process can be obtained, from (D.28) as

$$i_{jr}(t + 1) = i_{jr}(t) + \Delta i_j(t)$$

$$= i_{jr}(t) + \frac{2Bk_{j2}}{A}(k_{j1}u_r - k_{j2}u_j(t) - k_{j3}i_j(t)) \quad \text{(D.29)}$$

$$+ \frac{2E(u_r - u_j(t))}{A}$$

In (D.29), $i_{jr}(t + 1)$ is the objective reference current of the j-th TPGD at the time $t + 1$, and this is equal to the sum of the reference current $i_{jr}(t)$ at the previous time and the change in the current at the present time. After this current value is obtained, the current control loop enables the output current to track the reference current. Each PGD detects local bus voltage and output current data/information. Since the information of other PGD is not known, as these devices with control strategy based on the DGT are rational, each generator can pursue its best interest by minimizing its cost function. Finally, the result of the game is that the whole TPGS reaches the Nash equilibrium state, i.e. the DC bus voltage of the device is stable at the setting point, and the output power of each generating device is evenly divided.

Results: The numerical results of power generation flow characteristics (by two methods) are given in Tables AD.1 and AD.2. It can be seen that the game theory-based control of the tidal power control process makes good use of tidal levels and has higher power generation flow, efficiency, and power, as seen

in Table AD.1. Also, in the case of the small tidal level, the large power generation flow and the power can be obtained (AD.5).

Although, we feel that the solution (D.29) looks quite like the conventional one in optimization theory, we have included this piece of work for the importance it carries in the area of power generation in an alternative way and is a potential application of modern control methods discussed in this volume.

D.2 Decentralized/Centralized Control Design-Optimization Procedure

Fighter aircrafts often perform high angle-of-attack (HAOA) maneuvers and have short take-off and landing (STOL) capability; they have more control surfaces than conventional aerodynamic rudder, e.g. direct force control and thrust vectoring nozzle, and these control surfaces might have been distributed in/on the airframe and propulsion systems (APS). Hence, interestingly, and yet very difficult problem, integrated flight/propulsion control (IFPC) is to effectively coordinate these control surfaces to achieve the commands from the on-board flight computer and, at the same time, reduce the pilot workload. In a centralized approach to IFPC, a global-level controller to coordinate the APSs is designed by considering all the interconnections in the aircraft to yield the optimal design. This invariably would yield a high-order controller, which is difficult to implement and validate. In a decentralized control approach, the airframe and propulsion subsystem controllers are designed in a hierarchical decentralized structure resulting in controllers with lower order and, yet, meeting the implementation requirements. The approaches to decentralized controller design are: (i) to directly impose an upper triangular structure on the centralized

TABLE AD.1

Values of the Power Generation Flow Characteristics: DGT Based Control

TPGU Running time (in Hr.)	1	2	3	4	5	6
Tidal level (m)	5	6	7	8	9	10
Power generation flow (kW/h)	1200	1300	1550	1780	2058	3069
Efficiency (%)	75.8	78.9	80.1	83.7	86.5	92.4
Power (kW)	152.9345	173.6548	192.8816	214.5697	235.6874	293.1953

TABLE AD.2

Values of the Power Generation Flow Characteristics: Conventional Method

TPGU Running time	1	2	3	4	5	6
Tidal level(m)	5	6	7	8	9	10
Power generation flow (kW/h)	700	900	950	1120	1257	1538
Efficiency (%)	76.2	78.3	80.2	83.1	86.4	93.2
Power (kW)	102.3654	113.2478	132.0658	141.1024	152.3970	176.053

controller and using Riccati-like equations, solve for the decentralized controller (for high order systems, this is computationally intensive), (ii) to first design a centralized controller (for the airframe/engine integrated system), then partition the centralized controller into decentralized airframe and propulsion sub-controllers (with a specific interconnection). The latter approach leads to a very highly iterative design, and for the full envelope of an aircraft, the design effort will be very large (AD.6). Here, the design of decentralized controllers is considered in the optimization framework with use of GA (AD.6).

D.2.1 Flight-Propulsion Integrated Model

The model consists of an integrated airframe and propulsion state space representation of a modern fighter aircraft powered by a turbofan engine and equipped with 2D thrust vectoring nozzle. The vehicle dynamics are linearized at a flight condition representative of the STOL approach to landing task (airspeed 61.73 m/sec, flight path angle −3°) (AD.6):

$$\dot{x} = Ax + Bu; \quad y = Cx + Du$$

$$x = [u, w, q, \theta, h, N_2, N_{25}, P_6, T_{41}]^T$$

$$u = [W_f, A_{78}, A_8, \delta_{tv}]^T \qquad \text{(D.30)}$$

$$y = [V, q_v, N_{2P}, R, F_x, F_z, T_M]^T$$

In (D.30), we have:

u- aircraft's body axis forward velocity (m/sec); w- its vertical axis velocity (m/sec); q- pitch rate (°/s); θ- pitch angle (°); h- altitude (m); N_2- engine fan speed (rad/min); N_{25}- core compressor speed (rad/min); P_6- engine mixing plane pressure (kPa); T_{41}- engine high pressure turbine blade temperature (Kelvin); W_f - engine main burner fuel flow rate (kg/h); A_{78}- thrust reverser port area (cm²); A_8- main nozzle throat area (cm²); δ_{tv}- nozzle thrust vectoring angle (°); V- aircraft airspeed (m/sec); $q_v = q + 0.1\theta$ - pitch variable; N_{2P}- engine fan speed (% of maximum allowable rad/min at operating condition); R- engine pressure ratio; F_x, and F_z- total nozzle forces (in Nw) in the x-, and z-axis direction; T_M- total nozzle pitching moment (N.m). q_v is used in rate command attitude hold (RCAH) control to track low-frequency pitch rate command and high-frequency pitch angle command; and N_{2P}, R can be used as the scheduling variable for the baseline engine control.

D.2.2 Centralized Design

A GA-based mixed sensitivity HI controller design method is used for the centralized control system for

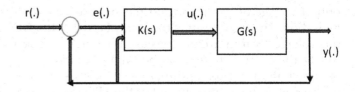

FIGURE AD.1
Centralized control system.

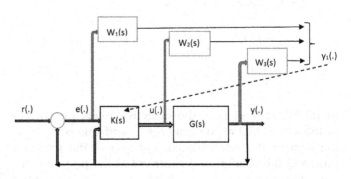

FIGURE AD.2
Mixed sensitivity synthesis.

an IAPM (integrated airframe/propulsion model) $G(s)$, Figures AD.1, and AD.2, and controller $K = [K_e \ K_y]$. The 2-DOF control structure allows simultaneous design of command tracking/loop shaping system and is also based on requirements of common flight control system to meet the desired piloted HQ (AD.7). The CLTF for the centralized control system T_c is given as:

$$T_c = [I + G(K_e - K_y)]^{-1}GK_e \qquad \text{(D.31)}$$

From Figure AD.2, we have the following relationship among the variables:

$$\begin{bmatrix} y_1 \\ y_2 \end{bmatrix} = P(s)\begin{bmatrix} r \\ u \end{bmatrix};$$

$$e = r - y; \ y_1 = [W_1(s)e \ W_2(s)u \ W_e(s)y]^T; \ y_2 = [e, y]^T \qquad \text{(D.32)}$$

$$P(s) = \begin{bmatrix} W_1(s) & -W_1(s)G(s) \\ 0 & W_2(s) \\ 0 & W_3(s)G(s) \\ \hline I & -G(s) \\ 0 & G(s) \end{bmatrix} \qquad \text{(D.33)}$$

The HI cost function, in terms of the CLTF from $r(s)$ to $y_1(s)$, to find the controller $K(s)$ is given as:

$$\min \left\| T_{y_1 r}(s) \right\|_\infty, \text{ OR } \left\| T_{y_1 r}(s) \right\|_\infty < \gamma; \left\| T_{y_1 r}(s) \right\|_\infty, \tag{D.34}$$

$$= \sup_\omega (\bar{\sigma}\{T_{y_1 r}(j\omega)\})$$

The optimization problem is expressed, in terms of fitness function of GA, as:

$$\min_{W_1, W_2, W_3} \psi_1(W_1, W_2, W_3) = \sum_{i=1}^{m} k_{1i} \varphi_{1i} \tag{D.35}$$

In (D.35), we have the multiple performance indices with their corresponding weights, $k(..)$.

The constraints are:

$$\bar{\sigma}(W_1^{-1}(s)) + \bar{\sigma}(W_3^{-1}(s))] \geq 1; \ \gamma_0(W_1, W_2, W_3) \leq \varepsilon_0;$$
$$\varepsilon_{i1} \leq \phi_i(W_1, W_2, W_3) \leq \varepsilon_{i2}; \tag{D.36}$$

In (D.36), γ_0 is the norm of the CLTF $T_{y_1} r(s)$; and ϕ_i are the performance functions of rise time, overshoot, and BW, and so on. The desired velocity response should be well-damped with minimal overshoot and a rapid settling time, and a 90% rise time to a step input command in 5 sec. The desired control BWs are: 1 rad/sec for V loop, 5 rad/sec for q_v loop, 5 rad/sec from N_{2P} loop, and 10 rad/sec for the R loop. The weighting functions W_1 and W_3 were selected as first order TFs (filters) to provide good frequency response shaping without increasing the order of controller, and W_2 was selected as scalar. In the fitness function of GA, some tips can be useful; e.g. when the BW of q_v loop has achieved 4.9 rad/sec, (a gap of 0.1 rad/sec to desired BW), GA may be stopped to optimize this index; and can increase the bound to 5.5 rad/sec for the solution to satisfy desired BW.

The 17th order augmented plant consists of 9th order integrated model, first order sensitivity/complementary sensitivity weighting function for four controlled variables. The optimal weighting functions and the 17th order centralized controller were found by GA. From the Bode-frequency and step input command responses of the closed loop system, it was found that the system could track velocity commands to the desired BW with small response in velocity to other commands; and the command tracking bandwidths of centralized CLCS met the HQ specifications.

D.2.3 Decentralized Design

A very suitable control structure for the IFPC problem is a hierarchical one with the flight controller generating commands for the aircraft control surfaces and the propulsion subsystem, see Figure AD.3; the subscripts/superscripts are: $a \rightarrow$ airframe, $e \rightarrow$ engine (propulsion system); and e are the corresponding errors; and the intermediate variables y_{ea} represent propulsion system quantities (e) that affect the airframe (a), e.g. propulsion

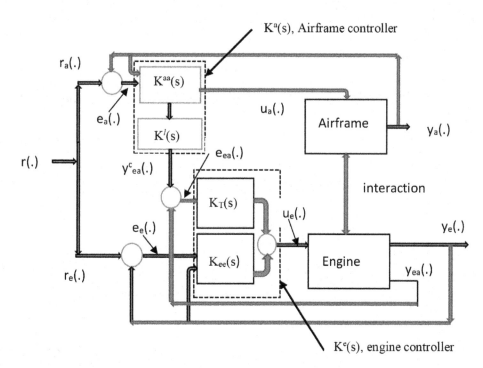

FIGURE AD.3
Decentralized control of two systems.

system generated forces and moments. We assume that the centralized controller $K(s)$ is obtained, with the following variables:

$$u = K(s) \begin{bmatrix} e \\ y \end{bmatrix}; \; u = \begin{bmatrix} u_a \\ u_e \end{bmatrix}; \; e = \begin{bmatrix} e_a \\ e_e \end{bmatrix}; \; y = \begin{bmatrix} y_a \\ y_e \end{bmatrix} \quad (D.37)$$

The aim is to determine the decentralized airframe and engine sub-controllers, $K^a(s)$, and $K^e(s)$ such that the closed loop performance/robustness (with these controllers) match with the centralized controller $K(s)$. Let the controlled plant $G'(s)$ be:

$$\begin{bmatrix} y_a \\ y_e \\ y_{ea} \end{bmatrix} = \begin{bmatrix} G_{aa}(s) & G_{ae}(s) \\ G_{ea}(s) & G_{ee}(s) \\ G^a_{ea}(s) & G^e_{ea}(s) \end{bmatrix} \begin{bmatrix} u_a \\ u_e \end{bmatrix} = G'(s) \begin{bmatrix} u_a \\ u_e \end{bmatrix} \quad (D.38)$$

D.2.3.1 The Assignments of the Plant's I/Os

The thrust vectoring nozzle δ_{tv} is primarily a pitch effector, thus $u_a = [\delta_{tv}]$, $\mathbf{y}_a = [V, q_v]^T$; the main fuel flow rate W_f, thrust reverser port area A_{78}, and main nozzle thrust area A_8 mainly affect engine's operation, thus $\mathbf{u}_e = [W_f, A_{78}, A_8]^T$, $\mathbf{y}_e = [N_{2P}, R]^T$. One can choose the forces and moments as the interface variables to control the coupling of engine to airframe. It was found from the centralized control loop design analysis that the changes of F_z and T_M were mainly affected by thrust nozzle angle δ_{tv}, while F_x was to track the step velocity command. Since F_x is the most critical variable, which needs to control, it was retained as the interface variable.

D.2.3.2 Airframe Sub-Controller Design

Flight sub-controller: the configuration for this design is shown in Figure AD.4. In the figure, $K(a,a)$ is flight controller with the TF from $[e_a, y_a]^T$ to $[u_a, y^c_{ea}]^T$; the order of this flight controller is 21 (centralized controller + 4th order engine

subsystem). For this design, the centralized controller's order was not reduced.

Lead filter: The filter $K^l(s)$ is designed to compensate for the delay of engine subsystem tracking, y^c_{ea} command due to the engine combustion process: high lead could result in saturation of the engine actuators, and the low lead would need large y^c_{ea} tracking BW. Since the controller $K_T(s)$ provides decoupled tracking of y^c_{ea}, the filter was taken as

$$K^l(s) = \text{diag}\left\{ \frac{s+a}{a} \; \frac{b}{s+b} \right\}; \; a < b; \; a = 10, \text{ and } b = 30;$$
$$\text{BW} = 30 \text{ rad/sec.} \quad (D.39)$$

Thus, the airframe sub-controller $K^a(s)$ consists of $K^{aa}(s)$ and the lead filter $K^l(s)$.

D.2.3.3 Engine Sub-Controller Design

1. Interface variable tracking controller, $K^T(s)$ is designed to track y^c_{ea} commands and provides decoupling between $r_e(t)$ and y^c_{ea} responses, the control loop is shown in Figure AD.5.

2. The design of $K_{ee}(s)$: this is regarded as reduced order (10th order) approximation of the $K_{ee}(s)$ block (17th order) of the centralized controller, which is partitioned as

$$K(s) = \begin{bmatrix} K_{aa}(s) & K_{ae}(s) \\ K_{ea}(s) & K_{ee}(s) \end{bmatrix} \quad (D.40)$$

The engine sub-controller $K^e(s)$ consists of $K^T(s)$ and $K^{ee}(s)$. The CLTF of the decentralized design $T_d(s)$ can be obtained as per the I/O connections shown in Figure AD.3.

D.2.3.4 The Optimization of the Decentralized Control

Once the centralized controller is properly designed, both of $K^a(s)$ and K_{ee} are determined; and the controller

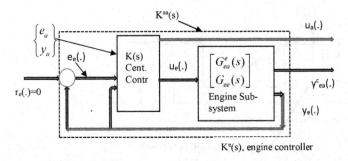

FIGURE AD.4
Airframe sub-controller.

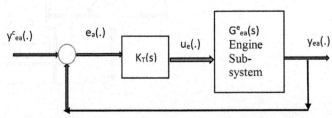

FIGURE AD.5
Control loop to determine the interface variable-tracking controller.

K_T is the part that is tuned in the decentralized control CLTF T_d. The point of the controller reduction (with stability criteria) is that the closed loop stability is assured, and the closed loop performance degradation is limited, if $||T_d - T_c||_\infty$ is sufficiently small. Also, the CLTF, T_d does not really make the error, $\bar{\sigma}(T_d - T_c)(j\omega)$ sufficiently small uniformly over all the frequencies of the interest. The decentralized control CLCS is stable if

$$\left\| T_d - T_c \right\|_\infty < 1 \tag{D.41}$$

In (D.41), $\left\| T_d - T_c \right\|_\infty < 1$ denotes the maximum value of $\bar{\sigma}(T_d - T_c)(j\omega)$ over all the frequencies. The decentralized controller design problem in the optimization framework is stated as:

$$\min_{K_T} \left\| T_d(j\omega) - T_c(j\omega) \right\|_\infty \tag{D.42}$$

The error is expressed as

$$\min_{W_S, W_C, W_T} \psi_2(W_S, W_C, W_T) = \frac{1}{m} \sum_1^m [\bar{\sigma}(T_d(j\omega) - T_c(j\omega))]^2; \tag{D.43}$$

m = no. of frequency points.

The K_T should meet the requirements of tracking interface variable; tracking bandwidth, rise time, and overshoot. Previously, it was found that the higher BW of K_T had resulted into a smaller error for CLCS; the BW was chosen to be as the BW of engine actuator, 30 rad/sec. As in the case of the fitness function of the centralized controller (for 4th order engine subsystem), the sensitivity weight W_S, the complementary weight W_T, and the control weight W_C were the design parameters. Then GA is used to search these optimal weighting functions with the fitness function as:

$$f_2 = k_\psi \psi_2 + \sum_{j=1}^{n_2} k_{2i} \varphi_{2i} \tag{D.44}$$

The useful frequency band of the aircraft is $(\omega_1, \omega_2) = (0.01, 100)(\text{rad/sec})$ with 20 frequency points/decade resulting in 81 points. For the full order decentralized control CLCS, $\left\| T_{fd} - T_c \right\|_\infty = 0.1$, and $\left\| T_{rd} - T_c \right\|_\infty = 0.2$, is for the reduced order decentralized control CLCS.

Remarks: It was found that: (i) the full and reduced order decentralized sub-controllers maintained the performance of tracking velocity command (without overshoot), and the decoupling of pitch variable, engine fan speed, and engine pressure ratio (as by the centralized controller); (ii) the maximum value (over frequency) of the structured singular value was 1.1226 with the centralized controller, 1.1226 with the full order decentralized sub-controllers, and 1.1082 with the reduced order decentralized sub-controllers; and (iii) their stability margins SM = $1/\mu_{max}$ were 0.8908, 0.8907 and 0.9024, respectively; thus, the optimization design approach in the H_∞ control setting uses genetic algorithm for the design of the decentralized control maintained the robustness characteristics of the centralized control [AD.6].

References

AD.1 Cheng, D. Control theory vs game theory. Institute of Systems Science, Academy of Mathematics and Systems Science, Chinese Academy of Sciences, April 2015; http://cnc.sjtu.edu.cn/Shanghai2015-C.pdf; accessed October, 2018.

AD.2 Wiener, N. *Cybernetics, Control and Communication in the Animal and the Machine.* Hermann & Cambridge. Press, Paris, France, 1948.

AD.3 J. von Neumann, and Morgenstern, O. *Theory of Games and Economic Behaviour.* Princeton University Press, Princeton, NJ, 1944.

AD.4 Anderson, C., and John J. Bartholdi, J. J. III. Centralized versus decentralized control in manufacturing: Lessons from social insects, pp. 92–105 in *"Complexity and Complex Systems in Industry,"* Proceedings, University of Warwick, 19th–20th September 2000, (McCarthy, I. P. and Rakotobe-Joel, T., Eds.). The University of Warwick, London, UK 652 p.

AD.5 Xu, H., and Wang, D. Tidal power generation process control based on differential game theory. *J Coast Res.,* No. 83, pp. 959–963, 2018.

AD.6 Huang, Y., Dai, J., and Peng, C. Optimization design of decentralized control for complex decentralized systems. https://export.arxiv.org/ftp/arxiv/papers/1809/1809.00596. pdf; accessed December 2018.

AD.7 Raol, J. R., and Singh, J. *Flight Mechanics Modelling and Analysis.* Taylor & Francis Informa Group, Boca Raton, FL, 2008.

Appendix E: Examples of Control of Dynamic Systems

In this Appendix, we present illustrative examples of PID tuner, temperature control in a heat exchanger, control process with long dead time-Smith predictor, and digital servo control of hard disk drive, all based on MATLAB. Where feasible, the results and the plots for this Appendix are generated by using the MATLAB scripts assembled from (AE.1) through (AE.4), (and rerun by the first author of the current volume).

Example E.1: Here, the aim is to design a proportional-integral (PI) controller with reasonably good disturbance/rejection performance using the PID Tuner tool and an ISA-PID controller for both good reference tracking/disturbance rejection (AE.1). The results for this example can be generated by running the MATLAB script "**piddrtuner.m**." The PID Tuner with initial PID design can be launched. The plant math model used is given as

$$G(s) = \frac{6(s+5)e^{-s}}{(s+1)(s+2)(s+3)(s+4)} \quad \text{(AE.1)}$$

The PID Tuner will design a PI controller in parallel form for the plant $G(s)$. This PID Tuner automatically designs an initial PI controller and clicking on the "Show parameters" button will display the controller gains and performance metrics. One can see the step response plot: the settling time is ~12 sec and the overshoot is ~6.3%, as shone in Figure AE.1. Now, the tuning of PID for disturbance rejection can be done: (i) assume that a step disturbance occurs at the plant input and the aim of the controller is to reject the disturbance quickly, and (ii) it is also expected that the reference tracking would be degraded as disturbance rejection performance improves. Now, click "Add Plot," select "Input disturbance rejection," and click "Add" to plot the input disturbance step response. See Figure AE.2: the peak deviation is about 1 and it settles to less than 0.1 in about 9 sec. You can tile the plots to show both the reference tracking and input disturbance responses, in a side by side manner. You can move the response time slider to the right to increase the response speed (open loop bandwidth). The gain in the "Controller parameters" table first increases and then decreases, with the maximum value occurring at 0.3. When K_i is 0.3, the peak deviation is reduced to 0.9 (about 10% improvement) and it settles to less than 0.1 in about 6.7 sec (about 25%

improvement). Because of increased BW, the step reference tracking response becomes more oscillatory, and the overshoot exceeds 15%, which is normally unacceptable. However, this type of performance trade-off between reference tracking and disturbance rejection would exist, because a single PID controller is not able to satisfy both design goals simultaneously. Now, click "Export" to export the designed PI controller to the MATLAB workspace. The controller is represented by a PID object, and one can create an ISA-PID controller. One can manually create

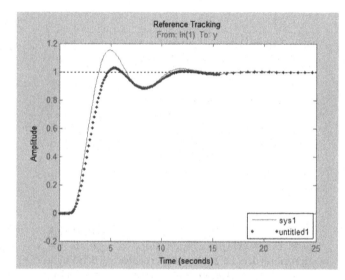

FIGURE AE.1
PID tuner, step response: Reference tracking (Example AE.1).

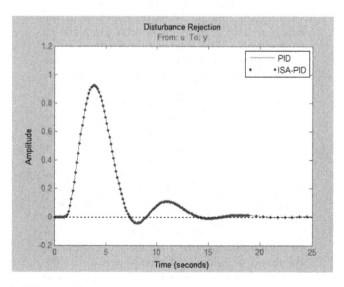

FIGURE AE.2
PID tuner: Disturbance rejection (Example AE.2).

the same PI controller in MATLAB workspace by using the "pid" command and directly specify the K_p and K_i gains obtained from the parameter table of the PID Tuner. An easy proposition to make a PI controller perform well for reference tracking/disturbance rejection is to upgrade it to an ISA-PID controller, which improves reference tracking by additional tuning parameters "b" that allows independent control of the impact of the reference signal on the proportional action. In the ISA-PID structure, there is a feedback controller C and a feedforward (pre-) filter F: C is a regular PI controller in parallel form that can be represented by a PID object:

$$C(s) = pid(K_p, K_i) = K_p + \frac{K_i}{s}; \quad F(s) = \frac{bK_p s + K_i}{K_p s + K_i} \quad \text{(AE.2)}$$

The ISA-PID has two inputs, r and y, and one output u (and y that is fed-back), and set-point weight b is a real number between 0 and 1; when it decreases, the overshoot in the reference tracking response is reduced. The reference tracking response with ISA-PID controller has much less overshoot because set-point weight b reduces overshoot, see Figure AE.1; whereas the disturbance rejection responses are the same because set-point weight b only affects reference tracking, Figure AE.2.

Example E.2: The aim is to design feedback and feedforward compensators to regulate the temperature in a chemical reactor (AE.2). In a chemical reactor/heat exchanger (the stirring tank), the top inlet pipe/duct delivers liquid to be mixed in the tank, and the liquid must be maintained at a constant temperature by varying the amount of steam supplied to the heat exchanger, from its bottom pipe via its control valve. The temperature fluctuations of the inlet flow are considered as the main source of disturbances to this process. One can use the measured data to model the heat exchanger dynamics. To derive a first-order-plus-dead-time model of the heat exchanger characteristics, one can inject a step disturbance in the valve voltage V and record the effect on the tank temperature T over time, see Figure AE.3. The results for this example can be generated by running the script "**tcheatexchanger.**" The values t_1 and t_2, in Figure AE.3 are the times where the response attains 28.3% and 63.2% of its final value, from which the time constant *tau* and dead time *theta* for the heat exchanger are computed. By comparing the first-order-plus-dead-time response with the measured response, as seen in Figure AE.3, one can see that the experimental/measured data and the model/simulated responses are in good

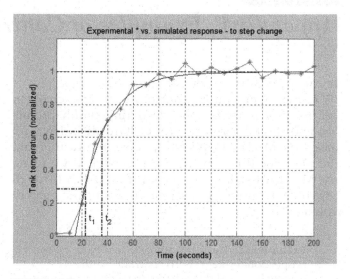

FIGURE AE.3
The tank temperature measured and simulated responses (Example AE.2).

agreement. The model fitted to these measured data ('*') is given as

$$G_p(s) = \frac{e^{(-14.7s)}}{21.3s + 1} \quad \text{(AE.3)}$$

The G_p models the effect on the tank temperature due to the steam valve opening (due in turn a change in the voltage that drive the valve). A similar bump test experiment could be conducted to estimate the first-order response to a step disturbance in inflow temperature, the model for which is given as

$$G_d(s) = \frac{e^{(-35s)}}{25s + 1} \quad \text{(AE.4)}$$

To regulate the tank temperature T around a given set-point T_{sp}, one can use the feedback architecture to control the valve opening (voltage V): (i) use the PI control in the forward loop before the heat exchanger model (this is not the direct feedforward controller), and (ii) simple feedback loop around the heat exchanger. The PI controller is given as

$$C(s) = K_c \left(1 + \frac{1}{\tau_c s} \right) \quad \text{(AE.5)}$$

The ITAE formulae are used to decide the values for the controller parameters. Now, one can close the loop and obtain the response to the step change in the set-point temperature, see Figure AE.4, from which it is seen that the response is fast with some overshoot. Next, we obtain the Bode diagram of the plant and

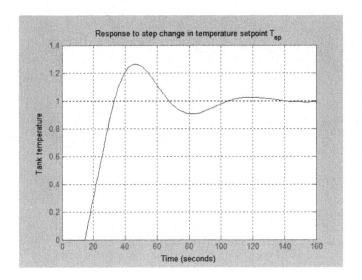

FIGURE AE.4
The tank temperature response to step change in the set-point temperature (Example AE.2).

the controller as the OL (loop TF) as shown in Figure AE.5, and it is seen that the GM seems low. One can reduce the proportional gain to 0.9. We can see from Figure AE.6 that the stability is improved at the expense of the performance, this is seen from Figure AE.7, and the rise time is increased. Now we can use the direct feedforward

controller before the heat exchanger. Since the changes in the inflow temperature are the main problem of temperature variations in the tank, to mitigate such disturbances, one can use the feedforward control instead of the feedback control. In such a case, the conventional feedback loop is not used; here, the feedforward controller F uses measurements of the inflow temperature to adjust the steam valve opening (voltage V); and feedforward control anticipates and pre-empts the effect of inflow temperature changes even as the disturbance is affecting the plant output. The computation shows that the overall transfer from temperature disturbance d to tank temperature T is given as

$$T(s) = [G_p(s)F(s) + G_d(s)]d(s); \; G_p(s)F(s) + G_d(s) = 0 \qquad (AE.6)$$

If we want it so the disturbance does not affect the temperature, then we should meet the condition of (AE.6), for which we should have

$$F(s) = -\frac{G_d(s)}{G_p(s)} = -\frac{(21.3s+1)e^{-20.3s}}{25s+1} \qquad (AE.7)$$

Due to modeling inaccuracies, exact disturbance rejection cannot/would not happen, but feedforward control will help minimize temperature fluctuations due to inflow disturbances. Now,

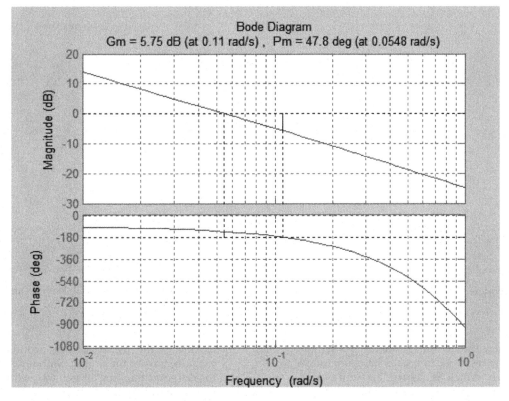

FIGURE AE.5
The Bode plot (of G_p*C) with gain margins with $K_c = 1.2341$ (Example AE.2).

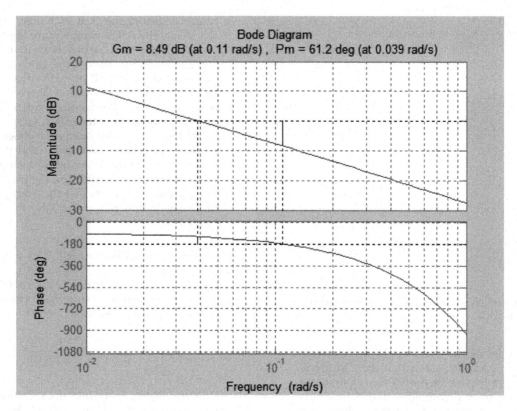

FIGURE AE.6
The Bode plot (of G_p*C) with gain margins with $K_c = 0.9$ (Example AE.2).

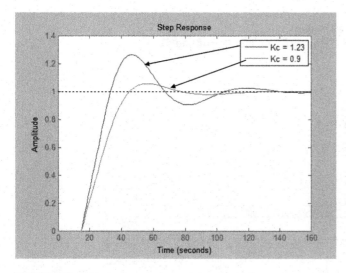

FIGURE AE.7
The tank temperature responses with two type of gains (Example AE.2).

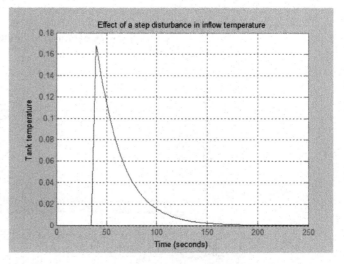

FIGURE AE.8
The tank temperature responses with feedforward (F)*plant (G_p) and G_d (Example AE.2).

increase the ideal feedforward delay by 5 sec and simulate the response to a step change in inflow temperature, which is shown in Figure AE.8. Now, one can combine the feedforward and feedback control for the temperature control: feedback control (with PI controller from error block and the feedback loop around the heat exchanger from the output to the input) is good for set-point tracking in general, while feedforward control (no PI controller and no looping around, just direct feed-controller from the disturbance) can help with rejection of measured

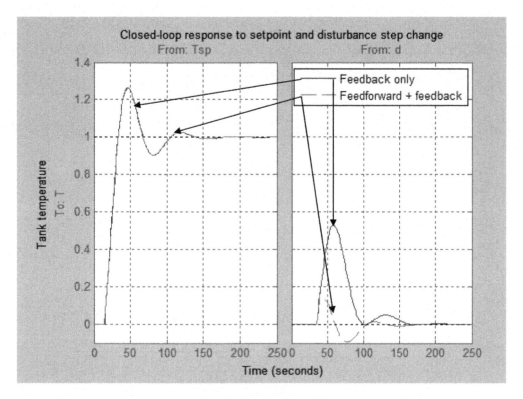

FIGURE AE.9
The CL responses of T for the set-point and disturbance change (Example AE.2).

disturbances. One can use the "connect" to build the corresponding closed-loop model from Tsp, d (inputs: set-point temperature and the disturbance, and to the output temperature) to *T*: (i) first name the input and output channels of each block, (ii) then let "connect" automatically wire the diagram, and (iii) to compare the closed loop responses with and without feedforward control, calculate the corresponding CLTF for the feedback-only configuration. Also, compare the two designs: the two designs have identical performance for set-point tracking, yet, the additional feedforward control is obviously beneficial for disturbance rejection as seen from Figure AE.9. This is also supported by the CL Bode plot, as is seen from Figure AE.10.

Example E.3: The aim is to see the limitations of PI control for processes with long dead time, and the benefits of a control strategy called "Smith Predictor" (SP) [AE.3]. The results and plots for this example can be generated by running the script **cpldeadtsmith.m.** The plant model P (open loop response) is a first order TF with time constant of 40.2 sec, and with the dead time of 93.9 sec:

$$P(s) = \frac{Y(s)}{U(s)} = \frac{5.6e^{-93.9s}}{40.2s+1} = e^{-93.9s}G_p(s) = e^{-93.9s}\frac{5.6}{40.2s+1}$$

$$(AE.8)$$

The delay (93.9 sec) in the system is more than twice the time constant (40.2 sec) of the basic process, and this is typical of many chemical processes. The response of the plant is shown in Figure AE.11. A PI control is a commonly used approach in the process control industry. One can use the "pidtune" to design a PI controller with OL BW at 0.006 rad/sec, with $K_p = 0.0501$ and $T_i = 47.3$. To evaluate the performance, we close the feedback loop and simulate the responses to step changes in the reference signal y_{sp} (basically reference signal r) and output disturbance signal d. Due to the delay in the feedback path, it is necessary to convert P or C_{pi} to the state space representation using the steady-state (SS) command. The C_{pi} (PI controller) is given as

$$C_{pi}(s) = K_p\left(1 + \frac{1}{T_i}\frac{1}{s}\right); K_p = 0.0501, T_i = 47.3 \qquad (AE.9)$$

We see from Figure AE.12 that the CL response (after PI controller has been incorporated) has tolerable and, yet, somewhat sluggish overshoot because it settles in ~600 sec. Increasing the proportional gain K_p speeds up the response; however, it increases overshoot and quickly leads to instability, as is seen in Figure AE.13, indicating that the performance of the PI controller is highly limited due to the presence of the long dead time. The reason being that the PI controller has

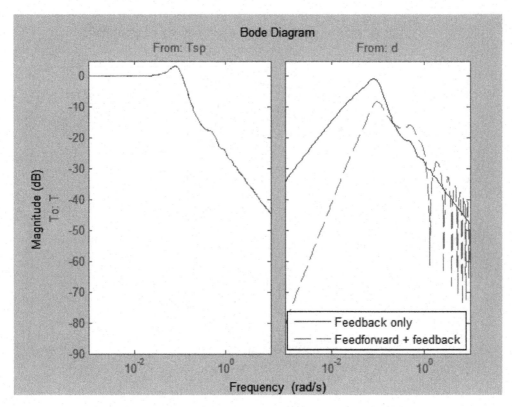

FIGURE AE.10
The CL Bode plot (Example AE.2).

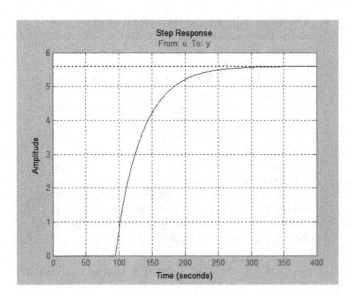

FIGURE AE.11
Step response of the process with the dead time (Example AE.3).

no knowledge of the dead time and reacts in a "haste" when the actual output y does not match the desired set-point y_{sp}. From our experience of using a shower, we know that we get hot and cold water in rapid succession, because we adjust the knobs in a great hurry, so a better way is to wait for a change in a particular temperature setting

to take effect before making more adjustments; and once it is learned what knob setting delivers the comfortable temperature, we can get the correct temperature in just the time it takes the shower to react. This is an intuitive control strategy learned from bad/good experience in the real life. This "optimal" control strategy is the main idea behind the Smith Predictor scheme, which uses an internal model G_p to predict the delay-free response y_p of the process; it then compares this prediction y_p with the desired set-point y_{sp} to decide what adjustments are needed (control u). In order to prevent drifting and reject external disturbances, the Smith Predictor compares the actual process output with a prediction y_1 that takes the dead time into account. The difference $d_y = y - y_1$ is fed back through a filter F and it contributes to the overall error signal e. Utilizing the Smith Predictor scheme requires: (i) a model G_p of the process dynamics (only original plant without the dead time) and an estimate *tau* of the process dead time, and (ii) adequate settings for the compensator and filter dynamics, C and F. The F is a first-order filter with a 20 sec time constant to capture low-frequency disturbances. For the controller C, redesign the PI controller with the overall plant seen by the PI controller, which incorporates the dynamics from P (original plant with dead time included), G_p (the original process model without dead time), F (filter),

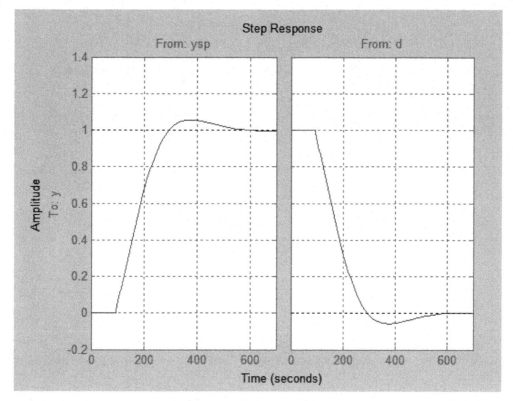

FIGURE AE.12
Step response of the process (with the dead time) and PI controller (Example AE.3).

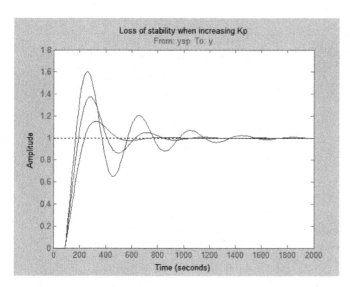

FIGURE AE.13
Step response of the process (& the dead time) with PI controller and increased gain (Example AE.3).

and dead time (only dead time model, D_p). Due to utilization of the Smith Predictor control, one is able to increase the OL BW to achieve faster response and increase the phase margin to reduce the overshoot. This new controller is similar to (AE.9), but with $K_p = 0.574$, $T_i = 40.2$. Now we can compare the performance of the PI controller and

the Smith Predictor. First derive the CLTF from y_{sp}, d to y for the Smith Predictor structure, and to simplify the task of connecting all the blocks involved, name all their I/O channels and use "connect" to do the "wiring." Compare the step responses (Smith Predictor '-', and PI controller '--') as depicted in Figure AE.14, from which it can be seen that the Smith Predictor obtains a much faster response with no overshoot, which is also seen in the closed loop Bode diagram, Figure AE.15, the Smith Predictor obtains the higher BW. In many real-life situations, the internal model is an approximation of the true process dynamics; hence, it is important to see how robust the Smith Predictor is to uncertainty on the process dynamics and the dead time. Let us consider two perturbed plant math models that represent the range of uncertainty on the process parameters. The Bode diagrams of the nominal and the perturbed models are shown in Figure AE.16. Now, to study this aspect, gather the nominal and perturbed models into an array of process models, rebuild the CLTFs for the PI and the Smith predictor, and generate the CL responses, which are shown in Figure AE.17. We see that both the designs are sensitive to model mismatch, as is also supported by the respective CL Bode diagrams, see Figure AE.18. It would be now advisable to improve the robustness. First check the stability margins for the inner and outer loops.

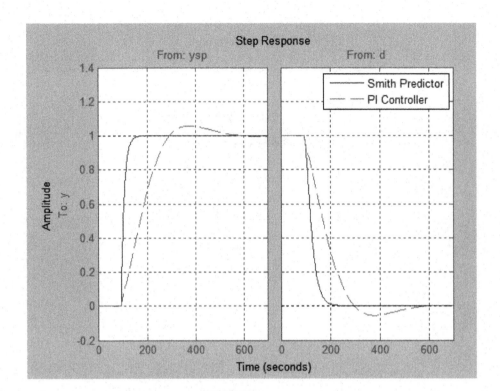

FIGURE AE.14
Step responses of the CL systems with Smith Predictor (-) and PI controller (--); (Example AE.3).

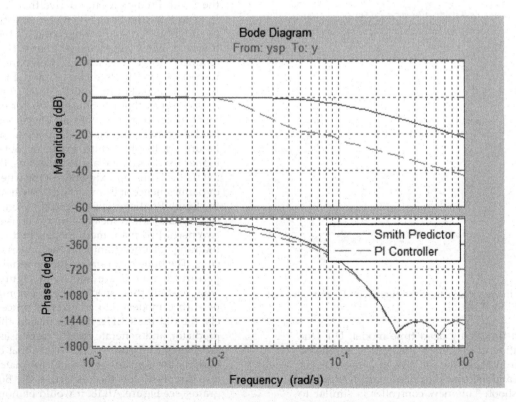

FIGURE AE.15
The CL Bode diagrams for the Smith Predictor, and PI controlled plant (P); (Example AE.3).

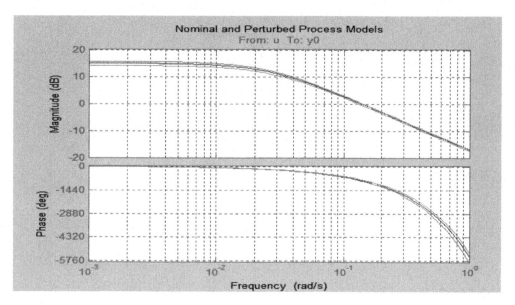

FIGURE AE.16

The Bode diagrams of the nominal and its two perturbed models (Example AE.3).

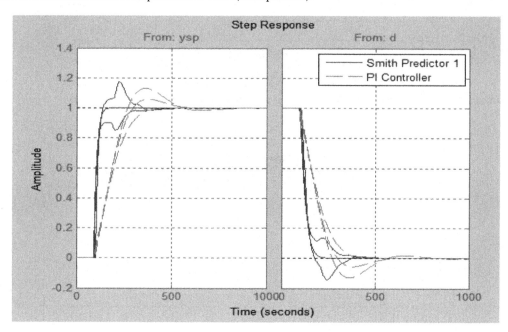

FIGURE AE.17

Step responses of the CL systems with SP and PI controllers: model mismatch (Example AE.3).

The inner loop has the (loop) OL TF as $C.G_p$; and the margins are shown in Figure AE.19 and are satisfactory.

Now, we focus on the outer loop: use "connect" to derive the OLTF L from y_{sp} to d_p with the inner loop closed. The Bode plot of the same is shown in Figure AE.20, which is essentially zero, and this is to be expected when the process and prediction models match nearly perfectly. Next, we obtain the stability margins for the outer loop and use one perturbed model, P_1. The results are shown in Figure AE.21, from which we can see that the gain

curve has a hump near 0.04 rad/sec, which has lowered the GM and would increase the hump in the CL step response. To mitigate this effect, use a filter F that rolls off earlier and more quickly:

$$F(s) = \frac{(1+10s)}{(1+100s)} \qquad \text{(AE.10)}$$

By doing this, we see that the GM has improved near the 0.04 rad/sec phase crossing, see Figure AE.22. Next, generate the CL responses with the modified filter. This design provides

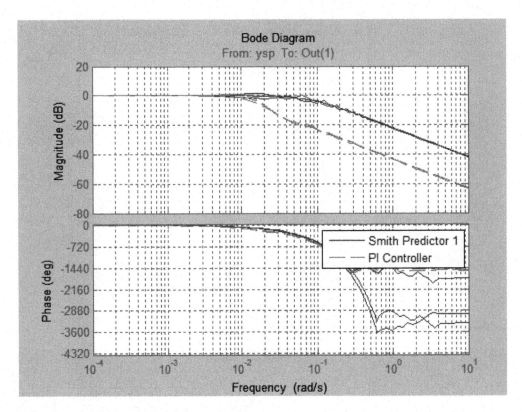

FIGURE AE.18
The CL Bode diagrams for the Smith Predictor, and PI controlled plant with perturbed models (Example AE.3).

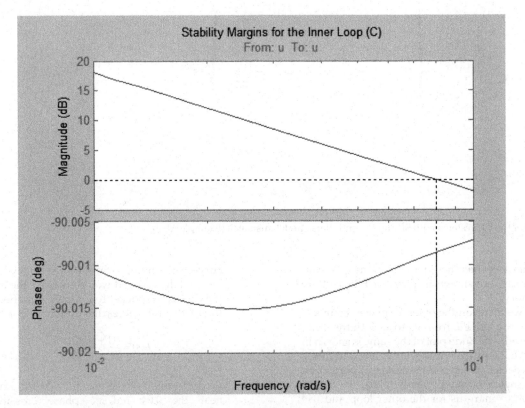

FIGURE AE.19
Inner loop stability margins (Example AE.19).

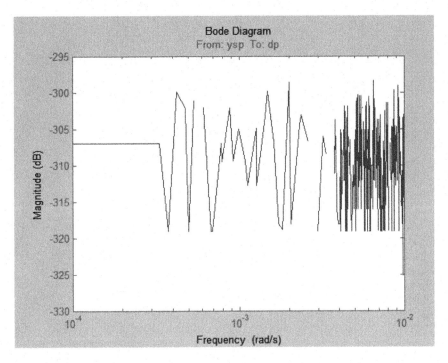

FIGURE AE.20
Open loop TF L, from y_{sp} to d_p (Example AE.3).

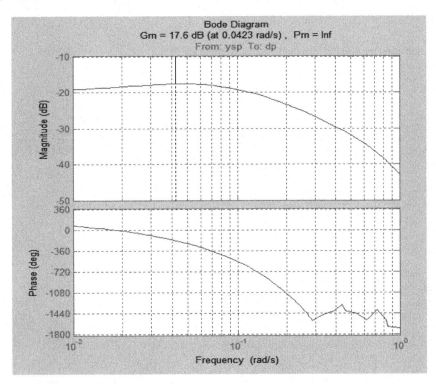

FIGURE AE.21
Open loop TF L, with one perturbed model (P_1) from ysp to d_p (Example AE.3).

more consistent performance, but a slightly slower nominal response as seen in Figure AE.23. We can further improve the design for disturbance rejection. The formulae for the CLTF from d to y show that the optimal choice of the filter F is $F(s) = e^{\tau s}$, the delay being the internal model's dead time. An approximation suggested is $F(s) = e^{\tau s} \approx \frac{(1+B(s))}{(1+B(s)e^{-\tau s})}$; $B(s)$ is a low-pass filter with the same time constant as the internal model $Gp(s)$. The PI controller is redesigned; the

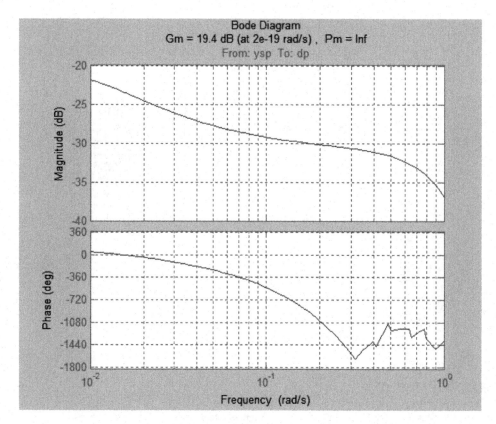

FIGURE AE.22
Open loop TF L, with one perturbed model (P1) with modified F (Example AE.3).

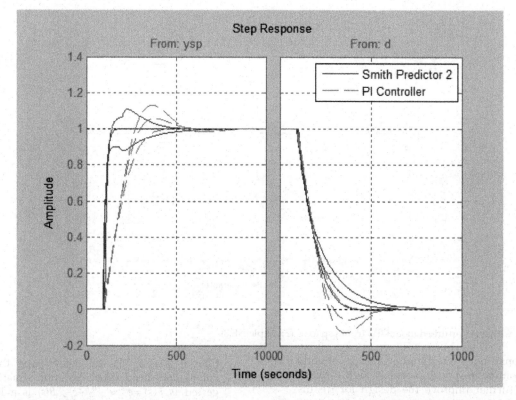

FIGURE AE.23
The CL step responses with the modified design (new F filter) (Example AE.3).

CL step responses are compared in Figure AE.24 from which it is seen that this design speeds up disturbance rejection at the cost of slower set-point tracking.

Example E.4: The aim to use Control System Toolbox™ (of MATLAB) to design a digital servo controller for a disk drive read/write head (AE.4). The head-disk assembly (HDA) and actuators are modeled by a 10th order TFs including two rigid-body modes and the first four resonances. The model input is the current i_c driving the voice coil motor, and the output is the position error signal (PES, in % of track width); the model includes a small delay. The drive model is given as

$$G(s) = G_r(s)G_f(s); \; G_r(s) = \exp\{(-1e-5)s\} \frac{1e6}{s(s+12.5)}$$

$$G_f(s) = \sum_{i=1}^{4} \frac{\omega_i(a_i s + b_i \omega_i)}{s + 2\zeta_i \omega_i s + \omega_i^2} \tag{AE.11}$$

The coupling coefficients and the damping and the natural frequencies (Hz) for the dominant flexible modes are:

$$(a_i, b_i, \zeta_i, \omega_i) \mid_{i=1:4} = (0.0000115, -0.00575, 0.05, 70);$$

$$(0, 0.023, 0.005, 2200);$$

$$= (0, 0.8185, 0.05, 4000);$$

$$(0.0273, 0.1642, 0.005, 9000) \tag{AE.12}$$

From the data in (AE.12), a nominal model of the head assembly is constructed. The results for this example can be generated by running the script **dscharddisk.m.** The Bode diagram of the head assembly is shown in Figure AE.25. Servo control is used to keep the read/write head "on track." It is a digital controller $C(z)$ and designed to maintain the PES (offset from the track center) close to zero. For the present case, the disturbance is a step variation d in the input current i_c. The task is to design this compensator with adequate disturbance rejection performance. The controller is in the feedback path, and there is a ZOH, the signal from which goes back to input and the loop error signal drives the plant, $G(s)$, i.e. the position error signal is fed back after the on/off switch ($Ts = 7e-5$ sec; 14.2 kHz) to the $C(z)$. The design specifications are: (i) OL gain > 20 dB at 100 Hz; (ii) BW > 800 Hz; (iii) GM > 10 dB; (iv) PM > 45°;

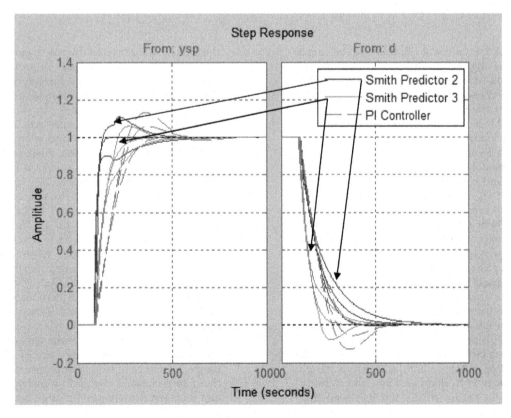

FIGURE AE.24
The CL step responses with the modified PI controller (another new F filter) (Example AE.3).

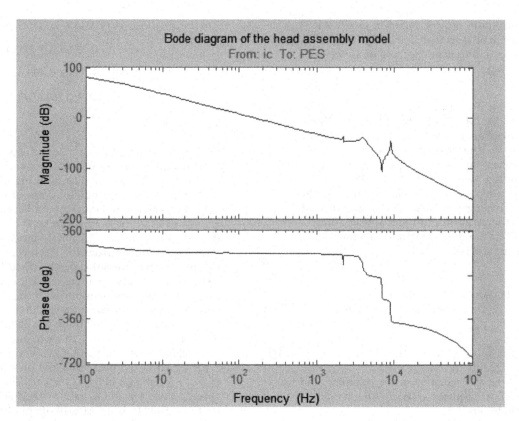

FIGURE AE.25
Magnitude and phase of the head assembly (Example AE.4).

and (v) Peak CL gain < 4 dB. Because the servo controller is digital, the design is done in the discrete time domain. Hence, the HDA model is discretized using c2d and the ZOH. The Bode diagrams of the CT and DT models are shown in Figure AE.26. For the controller design, one starts with a pure integrator; $1/(z–1)$, in order to ensure zero SS error. You can plot the root locus of the OL model G_d. C, i.e. discrete plant and the controller, and zoom around $z = 1$. The root locus is shown in Figure AE.27. Due to the two poles at $z = 1$, the servo loop is unstable for all positive gains; hence, to stabilize the feedback loop, first add a pair of zeros near $z = 1$. The new root locus is shown in Figure AE.28. Then, adjust the loop gain by clicking on the locus and dragging the black square inside the unit circle. This gain is displayed in the data marker, and the gain of ~ 50 stabilizes the loop; hence, set $C_1 = 50*C$. Simulate the CL response to a step disturbance in current i_c. This disturbance is smoothly rejected, but the PES is large. The head deviates from track center by 45% of track width, see Figure AE.29. Next, check the OL Bode plot and the stability margins as in Figure AE.30, and it is seen that

the gain at 100 Hz is 15 dB only (spec. is 20 dB), and the GM is 7 dB. So, an option is to add a notch filter near 4000 Hz resonance to make it possible to have higher low-frequency gain, see Figure AE.31. Now, one can safely double the loop gain, and from Figure AE.32, it is seen that the stability margins and gain at 100 Hz are within the specifications. Also, one can see that the step disturbance rejection has greatly improved. The PES is below 20% of the track width, see Figure AE.33. Also, one can check if the 3dB gain specification on $T = G_dC/(1 + G_dC)$ and the CL sensitivity is met, see Figure AE.34. Next, the robustness analysis is done for variations in the damping and natural frequencies of the second and third flexible modes: $\zeta_2 = 0.005 \pm 50\%; \zeta_3 = 0.05 \pm 50\%; \omega_2 = 2200 + 10\%; \omega_3 = 4000+ 20\%$. An array of 16 models related to all the possible combinations of extremal values is generated, and these models are discretized at once and the effect of the parameter variations on the OL response is seen, see Figure AE.35. Finally, the performance of step disturbance rejection for these 16 models is checked as in Figure AE.36, and it is seen that the servo design is robust.

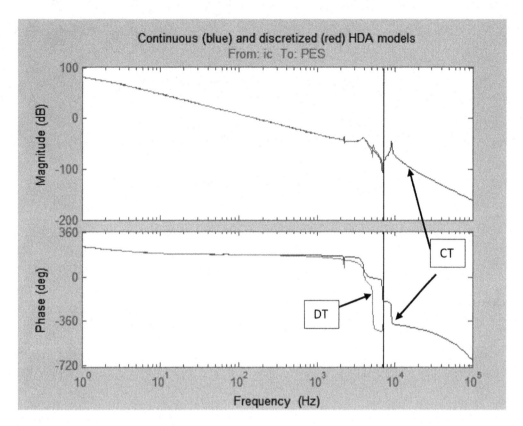

FIGURE AE.26
The Bode diagrams of the CT and DT models (Example AE.4).

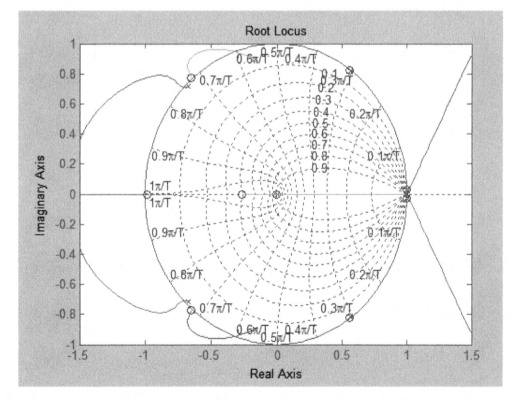

FIGURE AE.27
Root locus of the OL model $G_d C$, i.e. discrete plant and the controller, and zoom around $z = 1$ (Example AE.4).

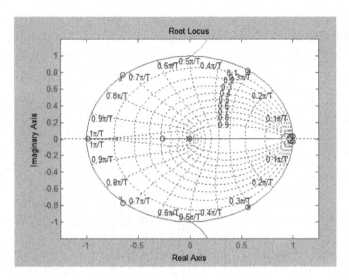

FIGURE AE.28

The root locus: stabilize the feedback loop, added a pair of zeros near $z = 1$; (Example AE.4).

FIGURE AE.29

The rejection of the disturbance (Example AE.4).

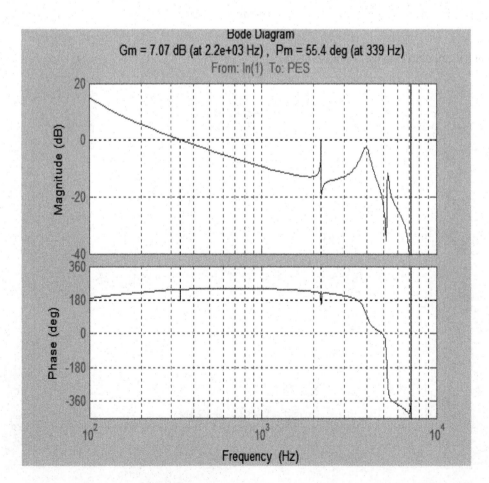

FIGURE AE.30

The OL Bode plot and the stability margins; (Example AE.4).

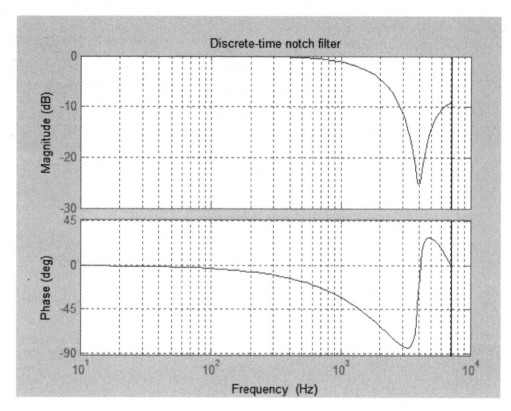

FIGURE AE.31
Bode plot of DT notch filter (Example AE.4).

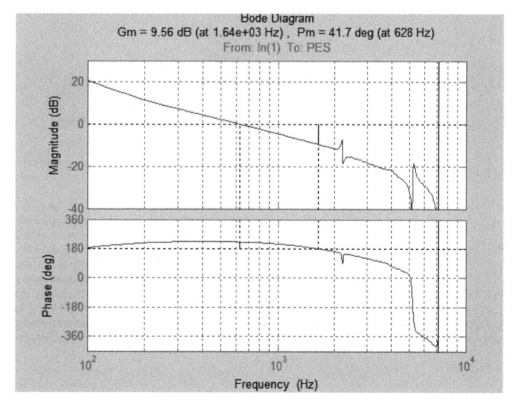

FIGURE AE.32
Frequency response after adding the notch filter (Example AE.4).

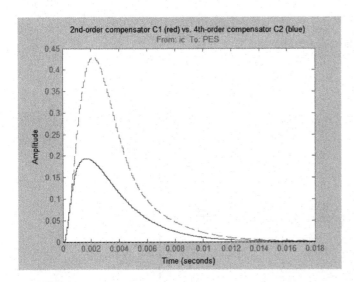

FIGURE AE.33
Improvement in the (step) rejection of the disturbance (Example AE.4).

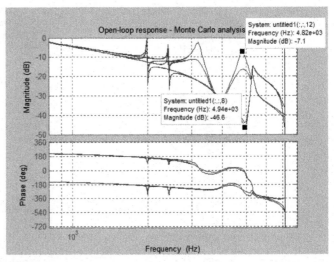

FIGURE AE.35
The effect of the parameter variations on the OL response (Example AE.4).

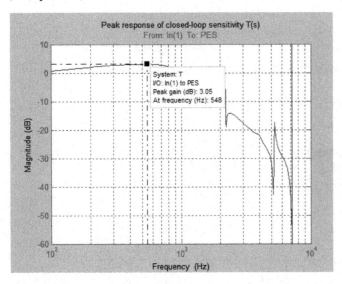

FIGURE AE.34
Improvement in the Bode-frequency (rejection) of the disturbance (Example AE.4).

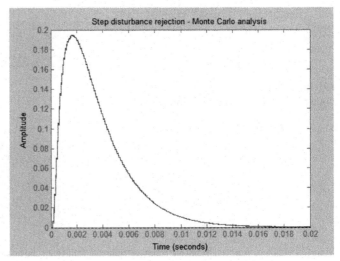

FIGURE AE.36
The performance of step disturbance rejection for the 16 models (Example AE.4).

AE.2 Temperature Control in a Heat Exchanger. https://in.mathworks.com/help/control/examples/temperature-control-in-a-heat-exchanger.html, accessed August 2018.

AE.3 Control of Processes with Long Dead Time: The Smith Predictor. https://in.mathworks.com/help/control/examples/control-of-processes-with-long-dead-time-the-smith-predictor.html, accessed August 2018.

AE.4 Digital Servo Control of a Hard-Disk Drive. https://in.mathworks.com/help/control/examples/digital-servo-control-of-a-hard-disk-drive.html, accessed August 2018.

References

AE.1 Designing PID for Disturbance Rejection with PID Tuner. https://in.mathworks.com/help/control/examples/designing-pid-for-disturbance-rejection-with-pid-tuner.html, accessed August 2018.

Appendix F: Examples of Aircraft Control

Here, we present two examples based on MATLAB for the design of a digital pitch control and yaw damper for an aircraft. The results and the plots for these examples are generated by using the MATLAB scripts assembled from (AF.1), (AF.3), and rerun by the first author of the current volume.

Example F.1: The aim is to design a digital pitch control for the aircraft, which would permit the aircraft to operate at a high angle-of-attack (HAOA) with minimal pilot workload (AF.1), (AF.2). One can generate the results and plots for this example by running the script "**designhaoapmc.m**." In a Simulink model of the aircraft, the control systems in the Controllers block can be switched in the model. This would allow one to see the analog response, then one can switch to a design created using the Control System Toolbox's LTI (linear time-invariant) objects. A controller is incorporated that is a discrete version of the analog design; and this implementation is similar to the algorithm that would go into an on-board flight computer. The math model can be linearized: (i) in the Control and Estimation Tools Manager (CETM) that can be launched from "slexAircraftPitchControlAutop," (ii) with the CETM, you can select an operating point and click the Linearize Model button, (iii) an LTI Viewer can be created showing a step plot of the linearization, (iv) to browse around the LTI Viewer, right click on the graph window to see your options, and (v) for help, type "help slcontrol" or "help ltiview." By uncommenting certain statements in the MATLAB script, the model parameters can be displayed in the command window. There are three types of LTI objects to develop a linear model: (a) state space (SS), (b) TF, and (c) Zero-Pole-Gain (ZPG) objects. The "contap" is a state space object. One can get the other types, one at time, by using other commands. When you create the object in MATLAB, you can manipulate it using *, +, −, etc., which is called "overloading" the MATLAB operators. You can try creating an object of your own choice to see what happens when adding, multiplying, and so forth with the "contap" object. To see what is stored in the LTI object, type "get(contap)" or "contap. InputName." The LTI object is used to design the digital autopilot to replace the analog autopilot; the analog system is coded into the LTI object "contap" (CONtinuous AutoPilot); the first attempt at creating a digital autopilot will use a ZOH with T = 0.1 sec, the discrete object maintains the type (ss, tf, or zpk). One can see

from Bode plot in Figure AF.1, that the systems do not match in phase angle from 3 rad/sec to the half sample frequency (the vertical black line) for the pilot stick input and the AOA (angle of attack) sensor; this design has poor response compared to the analog system. Use the Simulink model and start the simulation making sure you can see the scope windows. When the simulation is on, double-click the manual switch labeled Analog or Digital. One can use the Tustin transformation; it does better, see Figure AF.2. The simulation uses the LTI object as designed. Now, look in the Controllers subsystem by using the browser or by double-clicking the icon; the LTI block picks up an LTI object from the workspace. Use "discap1," the Tustin discretization of the analog design. Though the Tustin transform performs better than the ZOH, use of T = 0.1 sec, appears to be too slow for the discrete system to track the performance of the analog system at half the sample frequency. Hence, use for the transformation T = 0.05 sec, and the result of the improvement can be seen from Figure AF.3.

Example F.2: The aim is to design a yaw damper for a 747® jet aircraft using the classical control design features in Control System Toolbox™ (AF.3). A simplified trim (hence a linear) math model of the aircraft dynamics during its cruise flight can be represented in the state space form

$$\frac{dx}{dt} = Ax + Bu; \ y = Cx + Du \qquad (AF.1)$$

The four states used are: (i) beta (angle of slide slip/AOSS, in rad), (ii) phi (bank/roll angle, in rad), (iii) yaw rate (rad/sec), and (iv) roll rate (rad/sec). The two inputs are: (i) the rudder (pedals, rad) and (ii) aileron deflections (rad). The results and plots of this example can be generated by running the script **yawdamper.m**. The mathematical model (MM) data of the aircraft (at trim) are given in the script. You can uncomment a certain statement to see the parameters of the aircraft model. From these data, a state space model of the aircraft is created. The MM of the aircraft has a pair of lightly damped poles, see Figure AF.4a. Referring to the theory of the flight mechanics (AF.2), we ascertain that these modes correspond to the Dutch roll. One can right-click and select "grid" so that the damping/natural frequency values are plotted. One can see the lightly damped modes in

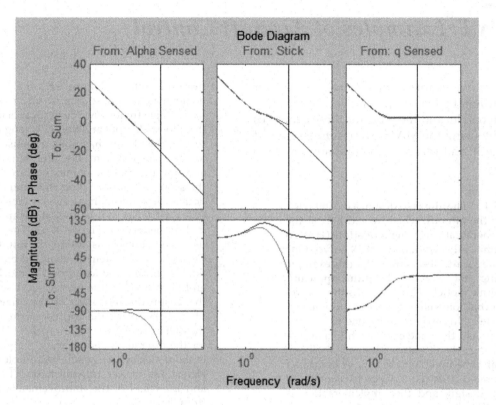

FIGURE AF.1
Frequency responses comparing analog and ZOH controllers.

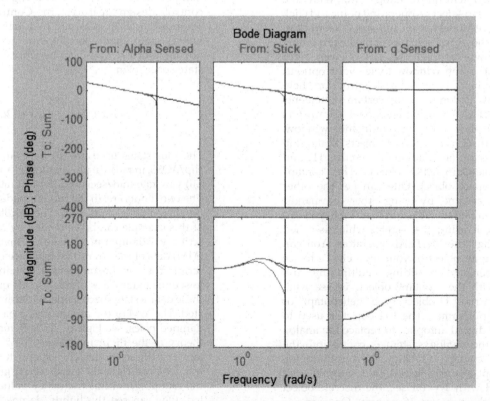

FIGURE AF.2
Frequency responses comparing analog and $T = 0.1$ sec Tustin controllers.

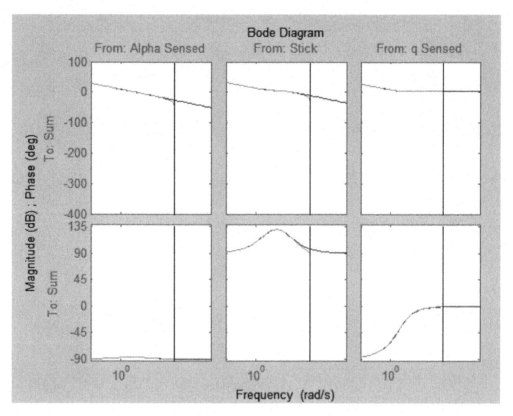

FIGURE AF.3
Frequency responses comparing analog and $T = 0.05$ sec Tustin controllers.

Figure AF.4b as well as from the impulse response in Figure AF.5; also, it can be seen in Figure AF.6a over a smaller time span of 20 sec. One can also expand only the plot from aileron to bank angle (phi) by right-clicking on the chosen plot, then select on the (2,2) entry in "I/O Selector." As can be seen from Figure AF.6b, the aircraft is oscillating around a non-zero bank angle and turns in response to an aileron impulse. One needs to design a compensator that would enhance the damping of these poles.

The yaw damper is designed using yaw rate as the sensed/measured output and rudder angle as the input. The frequency response of this I/O pair is shown in Figure AF.7, from which it can be seen that the rudder has a lot of authority around the lightly damped Dutch roll mode (1 rad/sec). So, the idea is to provide a damping ratio > 0.5 with natural frequency < 1.0 rad/sec. The simplest compensator is a gain. One can use the root locus to select a proper value of the feedback gain, see Figure AF.8 (see the code for this, here we use "rlocusplot(sys11)"), which shows that we might need a positive feedback; hence we get another plot as in Figure AF.9a (we use "rlocusplot(-sys11)"), and this seems better. By using the data curser on the plot (blue curve/semi-circle), we can see that best achievable CL damping is nearly 0.45 for a

gain of $K = 2.85$, see Figure AF.9b. Using this gain, we can close the SISO feedback loop and study the impulse response, see Figure AF.10. It is seen that the CL response is quite good. Next, close the loop around the complete MIMO model and study the response from the aileron. The feedback loop involves input 1 and output 1 of the plant. We can see from Figure AF.11 that the yaw rate is well damped. Yet, moving the aileron, the system does not continue to bank like a normal aircraft, see Figure AF.12, and the spiral mode seems to have been over-stabilized. This is typically a very slow mode that allows the aircraft to bank and turn without constant aileron input; interestingly, many pilots are used to this behavior and will not like a design that does not fly normally. So, we need to make sure that the spiral mode does not shift farther into the LHS plane when we close the loop. One can use a washout filter:

$$H(s) = \frac{ks}{s+a}; \quad a = 0.2, k = 1 \qquad \text{(AF.2)}$$

One can use the SISO tool (sisotool) and graphically tune the parameters k and a to determine a proper and adequate combination. Now, one can connect the washout filter in cascade with the design model and use the root locus to

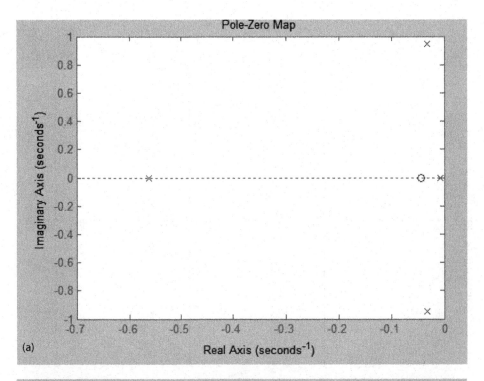

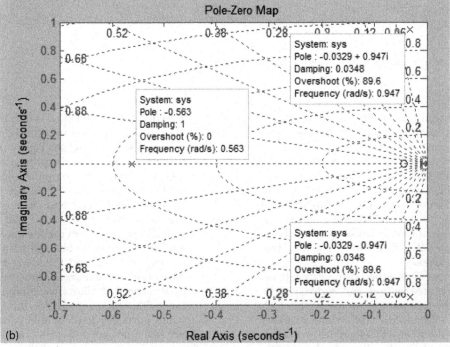

FIGURE AF.4

(a) The aircraft pole-zero disposition (b) The aircraft's lightly damped modes (0.038).

determine the filter gain k, see Figure AF.13. The best damping is 0.305 for $k = 2.34$. Now, close the loop with MIMO model, and study the impulse response, Figure AF.14. We can see that the washout filter has restored the normal bank-and-turn behavior of the aircraft. This can be ascertained from Figure AF.15, the impulse response from aileron to bank angle. Though not perfect, the present design has enhanced the damping (ratio) and, at the same time, would allow the pilot to fly the aircraft normally, as she/he is used to.

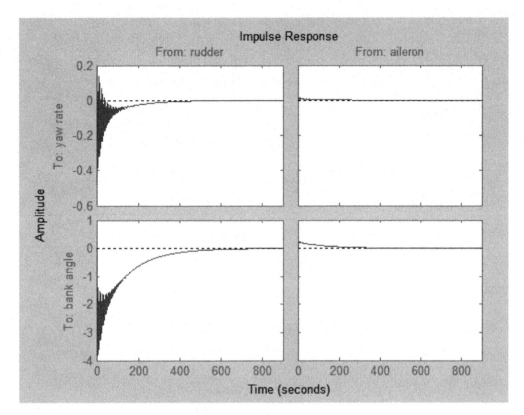

FIGURE AF.5
Impulse response of the aircraft showing the effect of lightly damped modes.

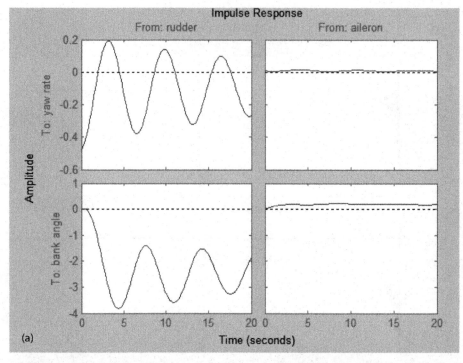

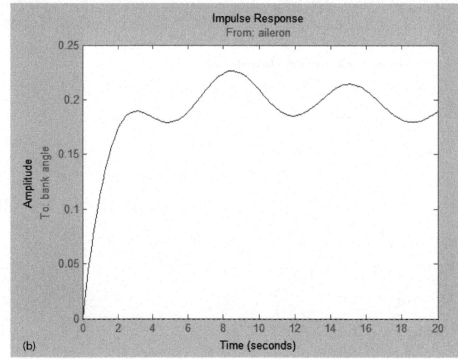

FIGURE AF.6

(a) The impulse responses over a smaller span of 20 sec. (b) The impulse response from ailron to bank angle.

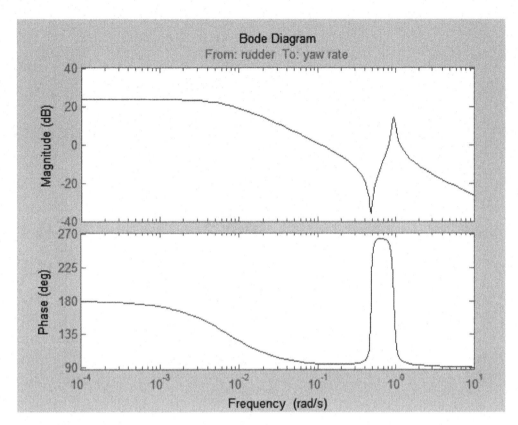

FIGURE AF.7
The Bode plot of rudder to yaw rate of the aircraft.

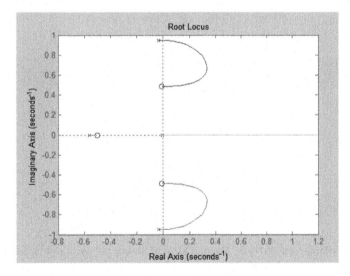

FIGURE AF.8
The root locus of the aircraft dynamics (with rlocusplot(sys11)).

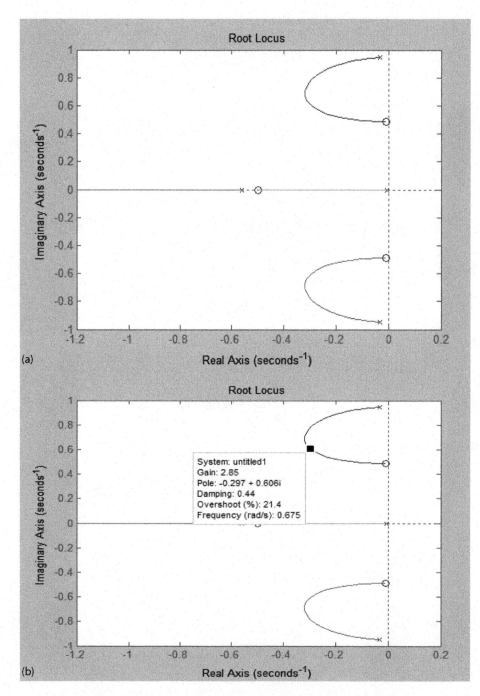

FIGURE AF.9
(a) The root locus of the aircraft (with rlocusplot(-sys11)) (b) The root locus of the aircraft (with rlocusplot(-sys11)) to see the achievable damping (0.45).

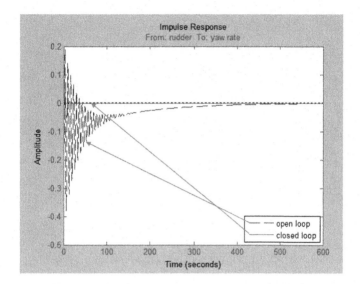

FIGURE AF.10
The impulse responses for open and closed loop.

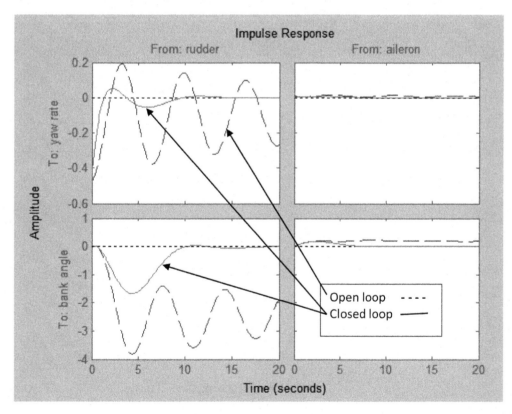

FIGURE AF.11
The impulse responses: The bank-and-turn is not like a normal aircraft behavior.

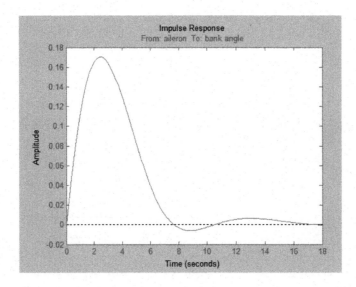

FIGURE AF.12
The impulse response: The spiral mode is over-stabilized.

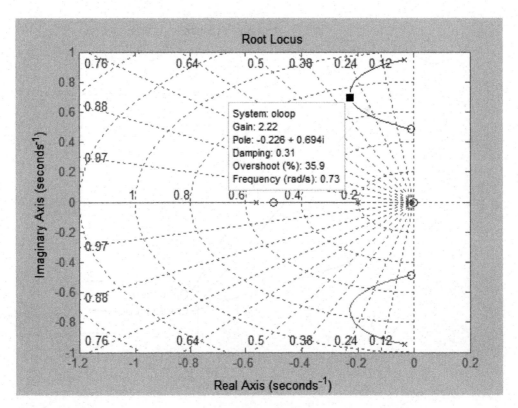

FIGURE AF.13
The root locus to determine the washout filter gain.

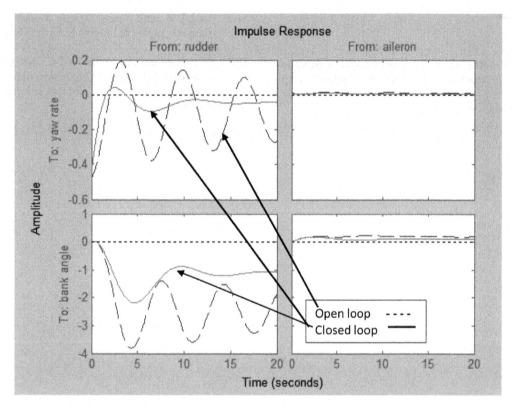

FIGURE AF.14
Impulse responses: Improvement using the washout filter; the normal bank-and-turn behavior of the aircraft (see the arrows).

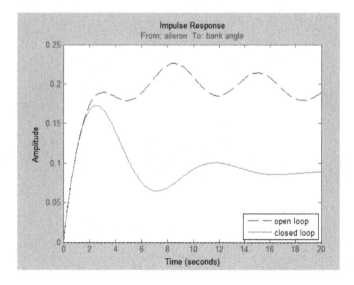

FIGURE AF.15
The impulse response: From aileron to bank angle, improvement using the washout filter; the normal bank-and-turn behavior of the aircraft.

References

AF.1   https://in.mathworks.com/examples/simulink/mw/
       simulink_product-slexAircraftPitch ControlExample-
       designing-a-high-angle-of-attack-pitch-mode-control,
       accessed August 2018.

AF.2   Raol, J. R., and Singh, J. *Flight Mechanics Modelling and
       Analysis.* Taylor & Francis Informa Group, Boca Raton,
       FL, 2008.

AF.3   Yaw Damper Design for a 747® Jet Aircraft. https://
       in.mathworks.com/help/control/examples/yaw-
       damper-design-for-a-747-jet-aircraft.html, accessed
       August 2018.

Appendix G: Additional Topics on Control and Summary

We present here brief descriptions of several other approaches and certain different aspects of the concepts of control theory and design, some of which are the extension/s of the approaches presented in the Chapters/Appendixes of the present volume. We finally present a brief summary/highlights (not necessarily in the same order as the chapters in the book), with some additional insights (on stability and performance) of several control concepts [AG.1–AG.11].

G.1 Control Design Using Polynomial Matrix Descriptions

For the design of a controller of a linear system, polynomial matrix techniques can be used as an alternative to state space methods, for (i) dynamics assignment—a generalization of eigenvalue placement, (ii) deadbeat control—finding a control law such that the state of a discrete time system is driven to the origin as fast as possible (in a minimal number of time-steps), and (iii) H_2 optimal control—a stabilizing control law is sought that minimizes the H_2 norm of some transfer function. Although the control problems are formulated in the state space setting, these can be solved in the polynomial setting. These matrices arise naturally in modeling of the physical systems [AG.1]. The dynamic systems in mechanics, acoustics, and linear stability flows in fluid dynamics can be modeling as second order vector differential equation in the form of matrices:

$$A_2\ddot{x}(t) + A_1\dot{x}(t) + A_0 x(t) = 0 \qquad (AG.1)$$

Based on the application of Laplace Transform to (AG.1), we can get the following characteristic matrix polynomial

$$A(s) = A_2 s^2 + A_1 s + A_0 \qquad (AG.2)$$

In (AG.2), A_0 is the symmetric negative definite stiffness matrix, A_1 is the anti-symmetric damping matrix, and A_2 is the symmetric positive definite (PD) matrix. This is a model of a gyroscopic system. In fact, many control design problems for industrial systems attempt to solve mathematical equations involving polynomial matrices; these were formulated for scalar plants and involved manipulations on scalar polynomials/scalar rational

functions. Presently, most of the results are available both in the state space and polynomial frameworks.

G.1.1 Polynomial Matrix Approach

The problems are formulated in the state space, and then solved by the polynomial methods, so as to better understand the connection between the two approaches.

G.1.1.1 Dynamic Assignment

Consider a discrete time system as

$$x(k+1) = \varphi x(k) + Bu(k); \ \varphi - \text{the transition matrix} \qquad (AG.3)$$

Let the state feedback be given as

$$u(k) = -Kx(k) \qquad (AG.4)$$

The aim in state feedback is to determine K, so that the closed loop system, $(\varphi - BK)$ has the desired eigenvalues. In the polynomial framework it is assigning the characteristic polynomial, $\det(zI - \varphi + BK)$. In this case, the pair (φ, B) must be reachable.

An involved problem of the eigenstructure assignment of the closed loop system matrix, in the polynomial framework is captured by the similarity invariants of matrix $(\varphi - BK)$, i.e. the polynomials that appear in the Smith diagonal form of the polynomial matrix, $zI - \varphi + BK$.

Let (φ, B) be a reachable pair with reachability indices $k_1 \geq ... \geq k_m$; and let $c_1(z),...,c_p(z)$ be monic polynomials such that the second one divides the first one, and so on; and $\sum_{i=1}^{p} \deg\{c_i(z)\} = n$; then the feedback matrix exists such that the closed loop matrix $(\varphi - BK)$ has similarity invariants $c_i(z)$, iff:

$$\sum_{i=1}^{p} \deg\{c_i(z)\} \geq \sum_{i=1}^{k} k_i; \quad k = 1,...,p. \qquad (AG.5)$$

The result of (AG.5) signifies that one can place the eigenvalues at arbitrary desired locations, but the structure of these is limited. One cannot split it into many repeated eigenvalues. Next, a procedure to assign a set of invariant polynomials by static state feedback method is discussed. Let us write the following factorization form

$$(zI - \varphi)^{-1} B = N_R(z) D_R^{-1}(z) \qquad (AG.6)$$

In (AG.6), $N(.,.)$ and $D(.,.)$ are the right coprime polynomial matrices. Now, form a column-reduced polynomial matrix $C(z)$ with invariant polynomials $c_1(z),...,c_p(z)$, which has the same degrees as $D_R(z)$. Then, we to solve the following equation

$$X_L D_R(z) + Y_L N_R(z) = C(z); \quad K = X_L^{-1} Y_L \quad \text{(AG.7)}$$

The equation (AG.7) is a Diophantine polynomial equation; and there is always a constant solution X_L, and Y_L to (AG.7).

Example G.1

Let the matrices (φ, B) be given as

$$\varphi = \begin{bmatrix} 1 & 0 & 0 & 1 \\ 0 & 0 & 1 & 0 \\ 0 & 0 & 0 & 1 \\ 1 & 0 & 0 & 0 \end{bmatrix}, \quad B = \begin{bmatrix} 0 & 1 \\ 0 & 0 \\ 1 & 1 \\ 0 & 0 \end{bmatrix} \quad \text{(AG.8)}$$

Let us chose the polynomials as

$$c_1(z) = z^3 - z^2, \quad c_2(z) = z \quad \text{(AG.9)}$$

We obtain the following polynomial matrices

$$D_R(z) = \begin{bmatrix} z^2 & z \\ 0 & z^2 - z - 1 \end{bmatrix}, \quad N_R(z) = \begin{bmatrix} 0 & z \\ 1 & 1 \\ z & z \\ 0 & 1 \end{bmatrix} \quad \text{(AG.10)}$$

The matrices in (AG.10) satisfy (AG.6). The following inequalities are also satisfied:

$$\deg c_1(z) \geq 2; \quad \deg c_1(z) + \deg c_2(z) \geq 4 \quad \text{(AG.11)}$$

Then, we have the following matrices

$$C(z) = \begin{bmatrix} z^2 & 0 \\ z & z^2 - z \end{bmatrix}, \quad X_L \begin{bmatrix} z^2 & z \\ 0 & z^2 - z - 1 \end{bmatrix}$$

$$+ Y_L \begin{bmatrix} 0 & z \\ 1 & 1 \\ z & z \\ 0 & 1 \end{bmatrix} = \begin{bmatrix} z^2 & 0 \\ z & z^2 - z \end{bmatrix} \quad \text{(AG.12)}$$

The constant solution and the feedback gain are obtained as

$$X_L = \begin{bmatrix} 1 & 0 \\ 0 & 1 \end{bmatrix}; \quad Y_L = \begin{bmatrix} -1 & 0 & 0 & 0 \\ -1 & 0 & 1 & 1 \end{bmatrix}; \quad K = Y_L \quad \text{(AG.13)}$$

G.1.1.2 Deadbeat Regulation

We consider the problem as in (AG.3) and (AG.4). The idea here is that every initial state x(0) is driven to the origin in the shortest possible time, and this is closely related to discrete time controllability, i.e. the system pair (φ, B) must be controllable. The Diophantine equation is given as

$$(I - \varphi z^{-1}) X_R(z^{-1}) + B z^{-1} Y_R(z^{-1}) = I; \ X_R(z^{-1}) \text{ is nonsingular}$$

$$K = Y_R(z^{-1}) X_R^{-1}(z^{-1})$$

$$\text{(AG.14)}$$

In (AG.14), K is the deadbeat feedback gain. From the foregoing development, we see that the deadbeat problem is a special case of the dynamic assignment; and this means that the deadbeat regulator assigns all the closed loop eigenvalues at the origin.

Example G.2

Let the matrices (φ, B) be given as

$$\varphi = \begin{bmatrix} 0 & 1 & 0 & 0 \\ 0 & 0 & 1 & 0 \\ 0 & 0 & 0 & 1 \\ 0 & 0 & 1 & 0 \end{bmatrix}, \quad B = \begin{bmatrix} 1 & 0 \\ 0 & 0 \\ 0 & 0 \\ 0 & 1 \end{bmatrix} \quad \text{(AG.15)}$$

The (φ, B) in (AG.15) is controllable. The Diophantine equation (AG.14) is obtained as

$$\begin{bmatrix} 1 - z^{-1} & 0 & 0 \\ 0 & 1 & -z^{-1} & 0 \\ 0 & 0 & 1 & -z^{-1} \\ 0 & 0 & -z^{-1} & 1 \end{bmatrix} X_R(z^{-1}) + \begin{bmatrix} z^{-1} & 0 \\ 0 & 0 \\ 0 & 0 \\ 0 & z^{-1} \end{bmatrix} Y_R(z^{-1}) = I \quad \text{(AG.16)}$$

This equation has a minimum-column-degree solution pair

$$X_R(z^{-1}) = \begin{bmatrix} 1 & 0 & 0 & 0 \\ 0 & 1 & z^{-1} & z^{-2} \\ 0 & 0 & 1 & z^{-1} \\ 0 & 0 & 0 & 1 \end{bmatrix}; \quad Y_R(z^{-1}) = \begin{bmatrix} 0 & 1 & z^{-1} & z^{-2} \\ 0 & 0 & 1 & z^{-1} \end{bmatrix} \quad \text{(AG.17)}$$

The feedback gain and the closed loop dynamics are given as

$$K = \begin{bmatrix} 0 & 1 & 0 & 0 \\ 0 & 0 & 1 & 0 \end{bmatrix};$$

$$x(z^{-1}) = \sum_k z^{-k}(\varphi - BK)^k x(0) = \begin{bmatrix} 1 & 0 & 0 & 0 \\ 0 & 1 & z^{-1} & z^{-2} \\ 0 & 0 & 1 & z^{-1} \\ 0 & 0 & 0 & 1 \end{bmatrix} x(0) \quad (AG.18)$$

From (AG.18), it is observed that every initial condition $x(0)$ is driven to the origin in three steps, since $(\varphi - BK)^k = 0$, for $k \geq 3$.

G.1.1.3 *H₂ Optimal Control*

In the H_2 optimal control problem, the transfer matrix of a linear system attains a minimum norm in the Hardy space H_2. Let us take a continuous time plant as

$$\dot{x}(t) = Ax(t) + B_1 v(t) + B_2 u(t)$$

$$z(t) = C_1 x(t) + D_{11} v(t) + D_{12} u(t) \quad (AG.19)$$

$$y(t) = C_2 x(t) + D_{21} v(t) + D_{22} u(t)$$

The plant TF matrix is obtained as

$$\begin{bmatrix} P_{11}(s) & P_{12}(s) \\ P_{21}(s) & P_{22}(s) \end{bmatrix} = \begin{bmatrix} C_1 \\ C_2 \end{bmatrix} (sI - A)^{-1} \begin{bmatrix} B_1 & B_2 \end{bmatrix} + \begin{bmatrix} D_{11} & D_{12} \\ D_{21} & D_{22} \end{bmatrix} (AG.20)$$

With $K(s)$ as the controller TF matrix, we have the control system TF matrix from $v(t)$ to $z(t)$ obtained as

$$G(s) = P_{11}(s) + P_{12}(s)[I - K(s)P_{22}(s)]^{-1}K(s)P_{21}(s) \quad (AG.21)$$

In the H_2 problem the following cost function is minimized

$$\int \text{trace}\{G'(-s)G(s)\}ds \quad (AG.22)$$

In (AG.22), the integral is a contour integral, first covering the imaginary axis, and then an infinite semi-circle in the left half s-plane. The following assumptions are made:

(A, B_1) is stabilizabe, (C_1, A) is detectable; $D_{11} = 0$

(A, B_2) is controllabe, (C_2, A) is observable; $D_{22} = 0$

$D'_{12} C_1 = 0$; $D'_{12} D_{12} = I$

$B_1 D'_{21} = 0$; $D_{21} D'_{21} = I$

$\qquad\qquad\qquad\qquad\qquad\qquad (AG.23)$

Now, let us write the following in terms of factors

$$C_2(sI - A)^{-1} = D_L^{-1}(s)N_L(s) \quad (AG.24)$$

In (AG.25), the $N_L(.)$, and $D_L(.)$ are the left coprime polynomial matrices, and $D_L(s)$ is row-reduced; also, we can have

$$(sI - A)^{-1}B_2 = N_R(s)D_R^{-1}(s) \quad (AG.25)$$

In (AG.25), $N_R(.)$ and $D_R(.)$ are the right coprime polynomial matrices such that $D_R(s)$ is column-reduced. Next, we introduce the following matrix

$$C_L(s)C'_L(-s) = [N_L(s)B_1 + D_L(s)D_{21}]$$
$$[N_L(-s)B_1 + D_L(-s)D_{21}]' \quad (AG.26)$$

In (AG.26), $C_L(s)$ is a square polynomial matrix with Hurwitz determinant (i.e. all its roots have strictly negative real parts). Similarly we have $C_R(s)$ matrix with Hurwitz determinant:

$$C'_R(-s)C_R(s) = [C_1 N_R(-s) + D_{12}D_R(-s)]'$$
$$[C_1 N_R(s) + D_{12}D_R(s)] \quad (AG.27)$$

The matrices $C_L(s)$, and $C_R(s)$ in (AG.26) and (AG.27) are the spectral factors, and can be uniquely determined up to right and left orthogonal factors. Then, we have the unique constant solution X_R, and Y_R to the polynomial Diophantine equation

$$D_L(s)X_R + N_L(s)Y_R = C_L(s) \quad (AG.28)$$

Then, the feedback gain matrix is given as

$$K_R = Y_R X_R^{-1} \quad (AG.29)$$

Similarly we have

$$X_L D_R(s) + Y_L N_R(s) = C_R(s); \quad K_L = X_L^{-1} Y_L \quad (AG.30)$$

Finally the unique H_2 controller feedback gain matrix is given as

$$K(s) = -K_L(sI - A + K_R C_2 + B_2 K_L)^{-1} K_R \quad (AG.31)$$

From (AG.28), and (AG.30) we see that the H_2 optimal problem is solved in two stages as dynamic assignment problem, since the assigned optimal invariant polynomials are obtained by the solutions of two polynomial matrix spectral factorization problems.

Example G.3:

The dynamics are given as

$$A = \begin{bmatrix} 1 & 1 \\ 0 & 0 \end{bmatrix}; B_1 = \begin{bmatrix} \sqrt{2} & 0 & 0 \\ 0 & 1 & 0 \end{bmatrix}; B_2 = \begin{bmatrix} 0 \\ 1 \end{bmatrix};$$

$$C_1 = \begin{bmatrix} 2 & 0 \\ 0 & 0 \end{bmatrix}; C_2 = \begin{bmatrix} 1 & 0 \end{bmatrix}; D_{12} = \begin{bmatrix} 0 \\ 1 \end{bmatrix}; D_2 = \begin{bmatrix} 0 & 0 & 1 \end{bmatrix} \tag{AG.32}$$

We have the following MFD (matrix fraction description) by extracting a minimal polynomial of a left null-space

$$[N_L(s)\ D_L(s)] \begin{bmatrix} sI - A \\ -C_2 \end{bmatrix} = 0;\ N_L(s) = [s\ 1],\ D_L(s) = s^2 \tag{AG.33}$$

Similarly for the right null-space we have

$$[sI - A - B_2] \begin{bmatrix} N_R(s) \\ D_R(s) \end{bmatrix} = 0;\ N_R(s) = \begin{bmatrix} 1 \\ s \end{bmatrix},\ D_R(s) = s^2 \tag{AG.34}$$

Performing the spectral factorizations as suggested in (AG.26), and (AG.27) we obtain the following for this example:

$$C_L(s)C_L'(-s) = [\sqrt{2}s\ 1\ s^2] \begin{bmatrix} -\sqrt{2}s \\ 1 \\ s^2 \end{bmatrix} = 1 - 2s^2 + s^4$$

$$= (s-1)^2(s+1)^2 \tag{AG.35}$$

The Hurwitz spectral factor: $C_L(s) = s^2 + 2s + 1$

$$= (s+1)^2$$

And,

$$C_R'(-s)C_R(s) = [2\ s^2] \begin{bmatrix} 2 \\ s^2 \end{bmatrix} = s^4 + 4$$

$$= (s^2 - 2s + 2)(s^2 + 2s + 2) \tag{AG.36}$$

The Hurwitz spectral factor: $C_R(s) = s^2 + 2s + 2$

$$= (s+1+i)(s+1-i)$$

We next, solve the Diophantine equation (AG.28) to obtain

$$[s^2\ s\ 1] \begin{bmatrix} X_R \\ Y_R \end{bmatrix} = s^2 + 2s + 1;\ X_R = 1;\ Y_R = \begin{bmatrix} 2 \\ 1 \end{bmatrix};\ K_R = \begin{bmatrix} 2 \\ 1 \end{bmatrix} \tag{AG.37}$$

Similarly solving the equation (AG.30) we obtain

$$[X_L\ Y_L] \begin{bmatrix} s^2 \\ s \\ 1 \end{bmatrix} = s^2 + 2s + 2;\ X_L = 1;\ Y_L = \begin{bmatrix} 2 & 2 \end{bmatrix};\ K_L = \begin{bmatrix} 2 & 2 \end{bmatrix} \tag{AG.38}$$

Finally, utilizing (AG.31), building its right MFD by extracting a minimal polynomial basis of a right-null space, we obtain the H_2 optimal controller with the TF

$$K(s) = \frac{6s + 2}{s^2 + 4s + 7} \tag{AG.39}$$

G.2 Advanced Topics in Artificial Intelligence (AI)-Based Intelligent Control (IC)

Artificial Intelligence (AI) (coined by John McCarthy in 1956) and machine learning (ML) (coined by Arthur Samuel) are two fields that have made great strides in last few decades: AI covers search, planning, reasoning, learning, natural language processing, perception, vision, and decision making/decision fusion; and ML tries to achieve AI through training and learning with its basis in statistical learning theory [AG.2]. The tools available in AI-ML seem to be very successful in learning complex functions and discovering intricate structures in high-dimensional data (data mining), and complicated network systems; with superior performance in image/speech recognition and are being applied in/for: drug discovery, particle physics, astronomy, and biomedicine [AG.2]. There are several aspects of the advancements in AI-ML for applications to general systems/control and related fields [AG.2]: (i) learning of high dimensional sparse sets (sparse models/sparse graphs) of parameters from measurements/observations (e. g. from planetary systems, multiple and comprehensive sensing systems, environmental systems/events); (ii) online/sequential learning, and decision making, the algorithms can assure asymptotic optimality; and (iii) in the context of modeling: (a) in learning mixtures of Gaussian models, the discovery of structural and latent variables using spectral/tensor approaches is very useful, and (b) application of optimization using ML (maximum likelihood) learning for non-convex network models, and stochastic gradient algorithms for deep learning.

Interestingly, the AI-ML approach, strengthened by the advancement in: (i) neuroscience, (ii) cognitive science, (iii) reinforcement learning, and (iv) deep learning, has made great progress in championship games demonstrating human-level control. However, in the domain of control it is still required to address the problems of control of large, complex, and distributed dynamical systems in presence of high levels of uncertainties and rapid changes in the system control's environments. As for the utilization of AI-ML based approaches for solving complex control problems, and the already well established control methods (robust/adaptive control, stochastic control, system identification, MPC, decentralized/distributed control, and agent based control) to strengthen the AI-ML approach, there are several potential possibilities: (a) reinforcement learning and stochastic learning, (b) sensorimotor neural systems and control, and (c) the free-energy principle/unified brain theory (neuronal perceptions) and control. One interesting confluence may be explored: the control addresses the problem of characterizing the minimal information required about a plant to control it; whereas statistical learning theory delves into obtaining information and theoretic limits of the achievable accuracy in modeling of/from the data.

Several more possibilities are: (i) to utilize the background foundational knowledge (systems-related concepts, available model structures, estimation/filtering methods) from control theory/practice for strengthening the AI-ML methods; (ii) exploration/utilization of the relationship between the spectral methods for ML and the classical model reduction approaches; (iii) utilization of unstructured models (hidden Markov models, deep networks) for designing algorithms for data-driven decision making; (iv) development of newer insights for the combination of neuroscience, cognitive science, reinforcement learning, and in general AI to build new architectures for more versatile, intelligent and adaptive controllers, e.g. neural prosthetics (to compensate for spinal cord injuries), which can embed biologically-sound signal representations that would describe high level objectives and translate these into specific control actions; (v) generate and provide smart services using cognitive science, and game theory; (vi) studying and designing artificial agents from the perspective of control theory; (vii) to find out non-convex optimization solutions to the saddle point problem; this a critical limitation in optimization and training of deep neural networks; and (viii) utilization/application of the concepts of robust and adaptive stochastic controls to make the AI-ML algorithms more robust in classification and recognition [AG.2].

A few simple applications of the hybrid intelligent systems for control, wherein some AI-based ISs' paradigms are used for realizing some control goals, are discussed next.

G.2.1 Hybrid Intelligent Systems for Control

We have studied some intelligent systems (ISs) and their applications in Section IV (and Appendix IVA) of this book. Here, we study hybridization of some basic intelligent systems. The intelligent systems include artificial neural networks (ANNs), fuzzy logic-based systems (FLS) and genetic algorithms (GAs); each of which has certain properties of ability of learning (FL with ANN), modeling, classifying, obtaining empirical rules, solving optimizing tasks; to varying degrees, appropriate to a specific kind of task [AG.3]. The combination/hybridization of these ISs, including neuro-fuzzy system (NFS), fuzzy-GA system (FGS), and neuro-GA (NGS) systems, and other such hybrid systems can be exploited in applications to modeling and control of dynamic systems, thereby deriving the merits of each individual method and eliminating certain demerits. Often, we need to acquire information from various sources, and combine it to generate the required knowledge, for which certain combinations of the sub-systems of ISs would be needed and this can be obtained at various levels and degrees as seen from Table AG.1 [AG.3]:

The hybrid systems can be realized and utilized as: (i) in NFS, the fuzzy system's parameters are encoded in several layers of an ANN, and the ANN is trained with the available system's data and the parameters can be adapted leading to ANFIS (adaptive neuro-fuzzy inference system). This way the FLS is able to represent the given data and has captured the knowledge from the data. Thus, the trained (with the help of ANN) ANFIS can be utilized for prediction. In FGS, the FLS's parameters are optimized using a GA. In NGS, the ANN's parameters are optimized using a GA.

TABLE AG.1

Available Benefits from ISs

	FLS	**ANN**	**GA's**
Knowledge representation	Good	~Bad	~Bad
Uncertainty tolerance	Good	Good	Good
Imprecision tolerance	Good	Good	Good
Adaptability	~Bad	Good	Good
Learning ability	Bad	Good	Good
Explain-ability	Good	~Bad	~Bad
Knowledge discovery/data mining	~Bad	Good	~Good
Maintainability	~Good	Good	~Good

Source: Didekova, Z. and Kajan, S. Applications of intelligent hybrid systems in MATLAB, http://dsp.vscht.cz/konference_matlab/MATLAB09/prispevky/021_didekova.pdf, accessed December 2018.

G.2.2 GA Based Approach

When combined with an ANN, the required parameters in the chromosomes of GA can be the connections in the ANN, the numerical values of weights/coefficients and/or biases in the ANN modeling topology. The required parameters are the ones that define (fuzzy) membership functions, and/or the basis of the (fuzzy) If ... Then rules. The criterion functions used are:

i) for modelling: $J = \sum_{i=1}^{N} e^2(i) = \sum_{i=1}^{N} (y_m(i) - y_p(i))^2$

$$(AG.40)$$

ii) for control: $J = \sum_{i=1}^{N} e^2(i) = \sum_{i=1}^{N} (r(i) - y(i))^2$

In (AG.40), $y_m(.)$ is the measurements/data from the system, and $y_p(.)$ is the predicted output from the model; and $r(i)$ the reference input to the system, and $y(.)$ is the controlled output; $e(.)$ for the modeling is called residual error (or innovations sequence), and for the control it is a control error; and N is the number of data/patterns. Since, in a GA we are searching for best fitness function, the criteria of (AG.40) may be appropriately transformed/translated to the fitness function.

G.2.3 ANFIS

A neuro-fuzzy system represents connection of numerical data and linguistic representation of the provided information, this subsequently gets as knowledge built-up. The structure of an ANFIS is similar to a multilayer ANN. It has I/O layers, and three hidden layers for membership functions, and (fuzzy) rules. The encoding of the FLS can be in terms of Mamdani or Sugeno fuzzy inference system (FIS). The parameters of the FLS are determined by the ANN's training procedure. A typical Sugeno rule is given as:

IF x_1 is A_1 AND x_2 is A_2 .... AND x_m is A_m
THEN $y = f(x_1, x_2, ..., x_m)$ (AG.41)

A typical small portion of the ANFIS code (based on MATLAB) is shown below:

```
data = [x y z];
in_fis = genfis1(data, [7 7],
char('gbellmf','gbellmf'));
epoch_n = 200;
out_fis = anfis(data, in_fis, epoch_n);
outfissim = evalfis([x y], out_fis);
```

The code has the following details: (i) the ANFIS structure is created by using **genfis1.m,** and uses say, 7

Bell-shaped membership functions; (ii) then, the **anfis.m** uses a hybrid learning algorithm to identify the parameters of the membership function, with single output. The trained ANFIS then can be tested on the test-data.

G.2.4 Fuzzy PI Controller with Optimization Using GA

In such a FGS, the inputs to the fuzzy PI controller are control error e (and derivative of control error), and the output from the controller is the control input $u(.)$ (and its derivative). Each input and output can have 5 triangular membership functions, the shapes of which are optimized. For determining these FLS parameters, appropriate tuning parameters of GA should be chosen (say, size of population as 30; and number of generations as 200).

G.2.5 Neuro-Predictive Controller with Optimization Using GA

In this case, the plant's behavior can be captured by an MLP (multilayer perceptron) ANN, with three layers. The design parameters for the optimization are: number of neurons in the input and hidden layers, and their interconnections. A learning rule with back-propagation with Levenberg-Marquardt modification can be used. The model predictive control (MPC) can be based on the receding horizon technique, and the ANN model predicts the plant response over a specified time horizon, Figure AG.1; and these predictions are used by a numerical optimization algorithm to determine the control signal $u(k)$ in each control period k, which minimizes a performance criterion J over the specific horizon:

$$J = \sum_{i=N_1}^{N_2} [r(k+i) - \hat{y}(k+i)]^2 + \gamma \sum_{i=1}^{N_u} [\Delta u(k+i-1)]^2$$

$$u(k) = u(k-1) + \Delta u(k);$$

$$(AG.42)$$

to determine $\Delta u(k)$ to minimize J

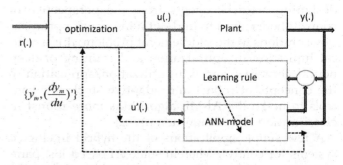

FIGURE AG.1
Neuro-predictive controller.

In (AG.42), N_1, and N_2 are lower/upper output prediction horizons, and N_u is the control horizon. The number of inputs, i.e. the past samples $y(k-1), y(k-2),...,y(k-n)$; and $u(k), u(k-1),..., u(k-m)$; and the number of neurons in the hidden layer are decided by the design with some experience. GA's are used as a search procedure for the determination of the neural model architecture and for the optimization of the predictive controller parameters.

G.2.5.1 Neural Model Structure Optimization

For modeling of the (nonlinear) system, one can chose a neural structure with 6 inputs, and 10 hidden neuronal elements. Then, the interconnection map ("1" for the connection and "0" for no connection) of the neural network is coded into the chromosomes of the GA. One can choose the length of the chromosome as 70 genes. Any unrealizable neural network structure is delated/not used further. The following cost function can be used in GA for searching an optimal neural model structure:

$$J = MSE + (1-\alpha) \cdot \left(\frac{w}{w_n}\right) \cdot 10^{-5} + \alpha \cdot \left(\frac{n}{n_n}\right) \cdot 10^{-5} \quad \text{(AG.43)}$$

In (AG.43), MSE is mean square error of the neural model; α–a constant = 0.7; w-number of weighted interconnections between the input and hidden layers; w_n-maximum number of such connections; n-number of neurons in the network; and n_n-maximum number of neurons in the network.

G.2.5.2 Neural Controller Optimization

The next design step would be to search for the predictive controller cost function coefficients N_1, N_2, N_u, and γ; which can be done by using (another) GA, and the fitness consists of the closed loop (simulation and) performance index for the control error $e(.)$:

$$ISE = \int_0^T (e(t))^2 \, dt \quad \text{(AG.44)}$$

The minimization of (AG.44) would determine the controller performance.

G.3 Receding Horizon Control (MPC)

In general, the idea would be to design the control input sequence $u(t,$ or $k; k_f = 1,2,..., k_{ch})$, using the assumed/available mathematical model of the plant and a set of constraints. If the model errors and/or the disturbances exist, then these inputs will not generate the desired response. The MPC (also called receding horizon control), is a strategy for a closed loop control to compensate these errors, which is to use the knowledge of the system at time k; then use this knowledge to design an input sequence:

$$u(k / k), u(k+1 / k), u(k+2 / k),..., u(k+N / k)$$

This is done over a finite horizon, N. If a fraction of the sequence is used then it is called a receding horizon control, otherwise it is called fixed/finite horizon control. A fixed horizon optimization (MPC) leads to a control signal/sequence of: $u(.) = \{u(k), ..., u(k+N)\}$. This signal begins at the current time instant k, and ends at some future time instant $k + N$. This would suffer from two problems: (i) the fixed choice of the control signal will not be useful, if something unusual/unexpected happens at some future time in the (future) interval, $\{k, k+N1\}$; because that situation was not predicted/included in the model; (ii) as one approaches the final time, $k + N$, the control law is sort of "gives up trying," because there is not enough time to keep trying to achieve something useful to reduce the cost function [AG.4]. The fraction of the input sequence, is usually the first step; then the process is repeated from time $k + 1$ at the state $x(k + 1)$. Thus, it is clear from this that the control algorithm is based on solving the optimization problem numerically at each step. The merits of the MPC are: (i) accounts the system's constraints explicitly, and (ii) can handles nonlinear and time varying plants, the latter is so because the controller is explicitly a function of the model, and this model can be modified in real time. In the receding horizon (control, it is also known as MPC, or vice versa), the procedure followed is:

1. At time k, and for the current state $x(k)$, solve an optimal control problem for a fixed (future) interval, $\{k, k+N\}$ with the current and future constraints;

2. Apply only the first step in the resulting optimal control sequence;

3. Measure the state reached at time $k + 1$; and

4. Repeat the fixed horizon optimization at the time $k + 1$ over the future interval $\{k + 1, k + N\}$, starting from the present state, $x(k + 1)$.

In the absence of any disturbance, the state measured at the step c), will be the same as the one predicted by the model. It is safer to use the measured state rather than the predicted one; however, in a practical case, one uses an observer to estimate the state $x(k + 1)$ at time $k + 1$; in that case we assume that the full state vector is measured or available. For LTI, if the model and the cost function are time-invariant, then the same input $u(k)$ will result when the state takes the same value. Thus, in particular, the receding optimization is a time-invariant state feedback control strategy. We need to solve the following problem at the current time and for the current state:

$$J_{N,opt}(x) \triangleq \min J_N(\{x(k)\}, \{u(k)\}); \ x(k+1) =$$
$$f(x(k), u(k)), \text{ for } k = 0, 1, 2, ..., N-1.$$
(AG.45)

$$J_{N,opt}(x) \triangleq \min J_N(\{x(k)\}, \{u(k)\})$$
$$= \min[W_N(x(N)) + \sum_{k=0}^{N-1} W_{x,u}(x(k), u(k))]$$
(AG.46)

We should reassert here that the control signals and the states are the members of the finite real sets; and the sequences $\{u(k)\} = [u(0),...,u(N-1)]$, and $\{x(k)\} = [x(0), ..., x(N)]$, which satisfy these constraints and the state equation of the dynamic system, (AG.45) are the feasible sequences, and such a pair is called a feasible solution to the receding horizon optimization problem. In (AG.46), the function W_N is the terminal (state) weighting matrix, and $W_{x,u}$ is the per-stage weighting matrix. We assume that f, and $W(.)$'s are the continuous functions of their arguments. The set of all $u(.)$ is compact, and the set of all $x(.)$ is closed set. Since N is finite, the assumptions made here are sufficient to ensure the existence of a minimum of the cost function by the Weierstrass theorem. The typical choices of the weighing functions are:

$$W_N(x) = x^T P x; \ W_{x,u}(x) = x^T R x + u^T Q u \quad \text{(AG.47)}$$

In (AG.47), P, and R, are symmetric and positive semi-definite matrices, and Q is symmetric and PD matrix. One, can use more general functions in the form of p-norm as

$$W_N(x) = \|Px\|_p; \ \text{and} \ W_{x,u}(x) = \|Rx\|_p$$
$$+ \|Qu\|_p; \ p = 1, 2, \infty$$
(AG.48)

Now, denote the minimizing control sequence (that is as such a function of the current state, $x(k)$) as

$$\hat{u}_{x(k)} = \{\hat{u}(0), \hat{u}(1), ..., \hat{u}(N-1)\} \quad \text{(AG.49)}$$

Then, the control signal applied to the plant is $u(k) = \hat{u}(0)$. Then, we move on one instant and repeat the procedure leading to the receding horizon control strategy: we solve the finite/fixed horizon optimization problem, and extract only the first component of the optimal control sequence and use this as a control law for the given system. Thus, we obtain (implicitly) the time-invariant control policy $K_N(x) = \hat{u}(0)$; though every time the first control input used might be different. However, it requires that the minimizer is unique, which is obtainable, if the cost function is convex. If the system model is nonlinear, then the cost function might be non-convex, then we would get only locally optimal solution.

Let the plant dynamics for a LTI system be given as

$$x(k+1) = \varphi x(k) + Bu(k); \ y(k) = Hx(k)$$

The cost is defined as

$$J = \sum_{0}^{N} \{y(k+j/k)^T R_y y(k+j/k)$$
$$+ u(k+j/k)^T Q u(k+j/k)\} + \text{terminal cost}$$

If $N \to \infty$, and there are no additional constraints on the variable y or u, then this is the LQR formulation. The problem, in general, might require solving it for a very long input sequence. This would not be advisable, if the model errors and/or the disturbance are present, which would be the case normally. The longer plans have more degrees of freedom but take more time to compute; hence, design is done for small N so at least some required goal is met. It is assumed that the full state is available, otherwise one has to use an estimator. Also, the corresponding controller should be applied immediately. If the constraints are not active, then it looks like a state feedback: $x(k/k) = -Kx(k)$, then the performance is checked by using the closed loop eigenvalues. If the constraints are active, then the stability depends on the terminal cost and its constraints.

G.4 Hybrid Dynamic Systems and Control Aspects

A hybrid system (HS), in the context of system/control theory and practice, is a combination of a continuous time and a discrete time dynamical system. In general context, such a system is heterogeneous in composition/nature and is a combination of processes with distinct features and characteristics; and generate signals that are both continuous (a set of real numbers) and discrete (the set of symbols: $\{a, b, c\}$) in nature; and their dependencies would be on time, which also would be continuous (t) and discrete, $(k = 1, 2, ..., N)$. Also, the signals could be time driven, and/or event driven. Some examples are: (i) manufacturing systems, (ii) communication networks, (iii) autopilot, (iv) propulsion engine, (v) computer synchronization, (vi) traffic control, (vii) chemical processes, (viii) embedded control systems, and (iv) hierarchically organized complex combination of discrete planning and continuous control algorithms in autonomous-intelligent systems [AG.5].

G.4.1 Background

In order to take into consideration the relations/interaction of the continuous and discrete parts of the hybrid system, it is important to develop models that accurately describe their dynamic behavior. A manufacturing process is composed of the event-driven dynamics of the parts moving among different machines and the time-driven dynamics of the processes within particular machines. An accurate model of a four-stroke gasoline engine has a natural hybrid representation: (i) the power train and air dynamics are continuous time processes, and (ii) the pistons have four modes of operation that correspond to the stroke they are in, their behavior is represented as a discrete event process. A study of hybrid systems is essential in designing sequential supervisory controllers for continuous systems, and it is central in designing intelligent control systems. An example of a hybrid control system is a switching system where the behavior can be described by a finite number of dynamical models that are typically sets of differential or difference equations, together with a set of rules for switching among these models. The furnace and air conditioner, along with the heat flow characteristics form a continuous time system (to be controlled), and the thermostat is a simple asynchronous discrete-event system. In order to avoid dealing directly with a set of nonlinear equations one can work with sets of simpler equations (that are linear), and switch among these simpler models. Whenever the behavior of a computer program depends on values of continuous variables within that program (continuous time clocks), one needs hybrid system approach to guarantee correctness of the program. Certain classes of hybrid systems have been studied as variable structure control (VSC), sliding mode control (SMC), and bang-bang control (BBC).

G.4.2 Modeling of Systems

The continuous and discrete time signals are studied in the time domain or in transform domain using Fourier, Laplace, and z transforms. The real numbers are such that between any two there is always a third real number, and our measurements provide us only with a finite number of decimal digits. In fact, the real numbers (do not really exist in nature, but they) represent an idealization that help us understand phenomena ranging from

the motion of planets to the behavior of atoms; thus, the real numbers have retained their usefulness on scales smaller than one hundredth of the classical diameter of subatomic particles (electron, proton) and are possibly valid down to the quantum gravity scale (twenty orders of magnitude smaller than such a particle). The modeling of hybrid systems can be done using: (i) differential/difference equations, (ii) finite automata, and (iii) Petri nets. In the description of an RLC circuit, one can use nonlinear differential equations to take into account the parasitic effects of the capacitors. For certain systems: production lines in manufacturing, computer networks, and traffic systems, their evolution in time depends on complex interactions of the timing of various discrete events; and such discrete event dynamical systems are modeled by discrete models, such as finite automata.

G.4.3 Supervisory Control of Hybrid System

The plant is approximated by a discrete event system (DES) and the design is carried out in the discrete domain. The hybrid control system (HCS) consists of a continuous (state, variable) system to be controlled, and a discrete event controller connected to the plant via an interface in a feedback configuration, Figure AG.2. It is assumed that the dynamic behavior of the plant is representation by a set of known nonlinear ordinary differential equations

$$\dot{x}(t) = f(x(t), u(t)) \qquad \text{(AG.50)}$$

The plant contains all continuous components of the hybrid control system (any conventional continuous controllers), a clock if time and synchronous operations are to be modeled; the controller is an event driven, asynchronous DES that is described by a finite state automaton. The generator is a partitioning of the state space. The actuator's signal is a piecewise continuous

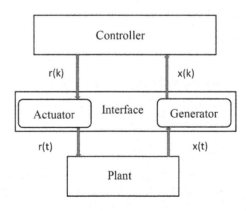

FIGURE AG.2
Hybrid system in supervisory control.

command signal—a staircase signal. The partition of the state space is determined by physical constraints and it is fixed and given. In such a HCS, the plant taken together with the actuator and generator, behaves like a DES; it accepts symbolic inputs via the actuator and produces symbolic outputs via the generator. This is somewhat analogous to a continuous time plant, with a ZOH and a sampler, and "look" like a discrete time system. From the DES controller's point of view, it is the DES plant model that is controlled, and this plant model is an approximation of the actual system, and its behavior is an abstraction of the system's behavior. Hence, the future behavior of the actual continuous system cannot be determined uniquely from the DES plant states/input. Thus, in the supervisory control, all the possible future behaviors of the continuous plant are incorporated into the DES plant model.

A dynamical system can be described as a triple $(T, X, f/F)$ with T, the time axis, S the signal space, and f as the set of all functions $f{:}T \to X$, the behavior; the latter consists of all the pairs of plant/control symbols that it can generate. A necessary condition for the DES plant model, as a valid approximation of the continuous plant, is that the behavior of the continuous plant model is contained in the behavior of the DES plant model; then the aim of the controller is to restrict the behavior of the DES plant model in order to satisfy the control specifications. So, the requirement is to find a discrete abstraction, F_d that is an approximation *pf* the behavior F_c of the continuous plant; and design a supervisor so that the behavior of the CLCS meets the specifications. Interestingly, the DES's behavior is nondeterministic.

Example G.4: Thermostat/Furnace System

Assume that the thermostat is set at 70° Fahrenheit (°F). If the room temperature falls below 70°, the furnace starts and remains on until the room temperature reaches ~72°. At 72°, the furnace shuts off. When the furnace is on, it produces a constant amount of heat per time. The plant in the thermostat/furnace hybrid control system is the furnace and room and is modeled as [AG.5]:

$$\dot{x}(t) = 0.0042(T_0 - x(t)) + 0.1u(t) \quad \text{(AG.51)}$$

In (AG.51), x is the plant state, the temperature of the room in °F; $u(t)$ is the voltage of the furnace control circuit, and T_0 is the outside temperature. The thermostat partitions the state space of the plant as

$$h_1(x) = x - 75; \quad h_2(x) = 70 - x \quad \text{(AG.52)}$$

The 1st hypersurface (h_1) detects when the temperature state exceeds 72°F; and the second one (h_2), when the temperature falls below 70°F. The associated functions are

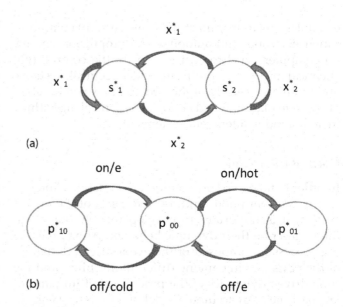

FIGURE AG.3
(a) The DES controller (b) The DES plant.

$\alpha_i(x) = x_i^*$; $i = 1,2$; and two plant symbols are:
x_1^*, x_2^*. $\qquad$ (AG.53)

The DES plant and the controller are shown in Figure AG.3. The output function of the controller is defined as:

$$\phi(s_1^*) = \hat{u}_1 \Leftrightarrow off; \quad \phi(s_2^*) = \hat{u}_2 \Leftrightarrow on \quad \text{(AG.54)}$$

The actuator operates as

$$\varphi(\hat{u}_1) = 0; \quad \varphi(\hat{u}_2) = 12 \quad \text{(AG.55)}$$

In (AG.55), the constants for the controller inputs depend on the given data. For the labeling of the arcs, the convention is: the controller symbols, which enable the transition followed by a "/," and then the plant symbols, which can be generated by the transition; and the two of the transitions are labeled with null symbols, e; this reflects the fact that nothing might actually happen at these transitions, i.e. when the controller receives a null symbol it remains in the same state and reissues the current controller symbol. This is equivalent to the controller doing nothing, but it serves to keep all the symbolic sequences, $s^*, p^*, \dots$, in phase with each other.

G.4.4 Stability and Design of Hybrid Systems

The hybrid dynamical systems are continuous and/or discrete time subsystems and a rule that determines the switching between them; and also arise when switching controllers (any suitable of many) are used to achieve stability and improve performance: computer disk drives, constrained mechanical systems, switching power converters, and automotive powertrain applications. Such a hybrid system is modeled as

$$\dot{x}(t) = f(x(t), y(t), u(t)); \quad y(t^+) = \delta(x(t), y(t)) \qquad \text{(AG.56)}$$

In (AG.56), $x(t)$ is the continuous state, $y(t) \rightarrow \{/k:1, 2, ..., N\}$ is the discrete state that indexes the sub-systems $f_{y(t)}$, $u(t)$ is a continuous control signal, or an external reference/disturbance signal to the continuous sub-system, and δ is the switching law that specifies the logical and/or DES dynamics.

Example G.5: Car with an Automatic Transmission

The model of the car dynamics is given as [AG.5]:

$$\dot{v}(t) = -\frac{k}{m}v^2 sign(v) - g\sin(\theta) + \frac{G_{y(t)}}{m}T;$$

$$\omega = G_{y(t)}v \qquad \text{(AG.57)}$$

In (AG.57), m is the mass of the car, v its velocity, θ the road angle, G_i, $i = 1,2,3,4$, are the transmission gear ratios (normalized by the radius of the wheels, r), ω is the angular velocity of the motor/engine, and T is the torque generated by the car engine. The discrete state transition function that determines the switching between the gears is given as [AG.5]:

$$y(t^+) = \begin{cases} i+1, & \text{if } y(t) = i \neq 4 \text{ and } v = \frac{1}{G_i}\omega_{high} \\ i, & \text{if } y(t) = i \geq 2 \text{ and } v = \frac{1}{G_{i+1}}\omega_{low} \end{cases} \qquad \text{(AG.58)}$$

The discrete state transition finite automaton is depicted in Figure AG.4.

For hybrid systems, the stability analysis is carried out by employing multiple Lyapunov functions (MLFs), so that a single piecewise continuous

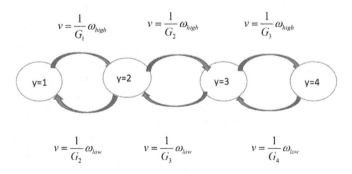

FIGURE AG.4
Finite automaton—the switching of the gears.

and piecewise differential Lyapunov function can be composed. Let us have the autonomous form of the hybrid system model as

$$\dot{x}(t) = f(x(t), y(t)) = f_{y(t)}(x(t)) \qquad \text{(AG.59)}$$

It is assumed that there are only finite number of switches in a bounded time interval. Now, consider the family of Lyapunov-like functions: $\{V_i, i = 1,2, ..., N\}$, with each $V(.)$ associated with the subsystems $f_i(x)$, and these functions are positive definite. Now, if the following is true, then the system is stable in the sense of Lyapunov:

$$V_i(x(t_{i,k})) \leq V_i(x(t_{i,k-1})) \qquad \text{(AG.60)}$$

In (AG.60), $t(i, k)$ is the k-th time instant at which the subsystem f_i is switched in. The condition in (AG.60) is a sufficient condition for the stability, and interestingly it provides a procedure for switching between subsystems to achieve a stable trajectory: a strategy it to choose that subsystem that causes maximal descent of a particular energy function.

Example G.6: Linear Switched System

A linear system is described as

$$\dot{x}(t) = A_{y(t)}x(t); \quad y(t^+) = \delta(x(t), y(t)) \qquad \text{(AG.61)}$$

Even for this restricted class of hybrid systems, there are two possibilities: (i) switched system can be unstable even if all the basic subsystems are stable, and (ii) it is possible to stabilize linear switched system even if all the subsystems are unstable. Let the hybrid system be given as in (AG.60), with $y(i)$, $i = 1, 2$, and the following matrices

$$A_1 = \begin{bmatrix} -1 & -100 \\ 10 & -1 \end{bmatrix}; A_2 = \begin{bmatrix} -1 & 10 \\ -100 & -1 \end{bmatrix} \qquad \text{(AG.62)}$$

The systems of (AG.62) have the eigenvalues as $\lambda_{1,2} = -1 \pm \sqrt{1000}$ and are stable. However, the switched systems using A_1 in the second and 4th quadrants and A_2 in the 1st and 3rd quadrants are unstable, have limit cycles.

It is important to have the conditions that assure the stability of a switched system for any switching signal, for which two approaches are feasible [AG.5]: (i) a supervisor determines which controller is to be connected in closed loop with the plant at each instant of time; and the stability of the switched system is ensured by keeping each controller in the loop long enough for the transient effects to dissipate; (ii) to demonstrate the stability for any switching signals, by ensuring that the matrices A_i share a common quadratic Lyapunov function, and the stability is ensured in the sense of Lyapunov, as is done usually in case of other systems (see Appendix C). The application of certain theoretical results to hybrid systems is accomplished usually using a linear matrix inequality (LMI) formulation for constructing a set of Lyapunov-like functions; and the existence of a solution to the LMI problem is a sufficient condition and guarantees that the hybrid system is stable. The procedure is to partition the state space into some R-regions that are defined by quadratic forms, and for this some physical insight, and a good understanding of the LMI problem are required. Let R_i denote a region where one searches for a Lyapunov function $V_i = x^T P_i x$ that satisfies:

$$\dot{V}_i(x) = \left[\frac{\partial V_i(x)}{\partial x} \right] A_i x = x^T (A_i^T P_i + P_i A_i) x \leq 0 \quad \text{(AG.63)}$$

The idea in (AG.63) is to find matrices $P_i > 0$ (i.e. PD), so that the condition in (AG.63) is satisfied. There are two steps: (i) the region R_i must be expressed by the quadrant $x^T Q_i x \geq 0$; and (ii) a technique called S-procedure is applied to replace a constrained stability condition to a condition without a constraint. Introduce a new unknown variable $\lambda \geq 0$ to obtain the unconstrained problem as

$$A_i^T P_i + P_i A_i + \lambda Q_i \leq 0 \quad \text{(AG.64)}$$

This unconstraint problem can be solved by LMI tools, and the solution will be also a solution to the constraint problem.

G.4.5 Hybrid Automata

Hybrid automata provide a general modeling representation for the analytical specification and algorithmic analysis of hybrid systems. A hybrid automaton is a finite state machine/mechanism with a set of real-valued variables. The state of the automaton changes either instantaneously via a discrete transition or, as time elapses, via a continuous activity.

Example G.7: Hybrid Automaton of a Thermostat

The topology is shown in Figure AG.5. When the heater is off, the room temperature falls according to $\dot{x}(t) = -Kx$, and when it is on (i.e. the control is on), the temperature rises as per $\dot{x}(t) = K(h - x)$. Initially, the temperature is $x = 72$ and the heater is off, the heater will go on as soon as the temperature falls at 70° (as per the condition $x = 70$); and when the heater is on the temperature is rising and when it reaches 75° (as per the condition $x = 75$), the heater will go off and the temperature will start falling again; this control strategy

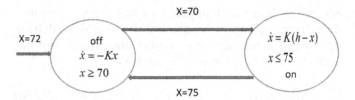

FIGURE AG.5
The hybrid automaton of a thermostat.

guarantees that the room temperature will remain between 70° and 75° [AG.5].

A hybrid automaton consists of a finite set of real-valued variables: $X = \{x_1, \ldots, x_n\}$ and a labeled/directed graph (V, E), V is a finite set of vertices and E is a set of directed arcs/edges between vertices; and the graph models the discrete (event) portion of the hybrid system, see Figure AG.5 (with two vertices and two edges; and the arc labeled as $x = 72$ is for the initialization of the system). The vertices V represent continuous activities labeled with constraints on the derivatives of the variables in X. Specifically, a vertex v (of V) called a control mode or location has the features; (i) a flow condition/activity described by a differential equation, and when the hybrid automaton is in control mode v, the variables x_i change according the flow condition, e.g. in the thermostat automaton, the flow condition $\dot{x}(t) = K(h - x)$ of the control mode "on" ensures that the temperature is rising while the heater is on; and (ii) an invariant condition, inv(v) assigns to each control mode a region of real domain R, the hybrid automaton may reside in control mode v only while the condition inv(v) is true, e.g. the invariant condition $x \leq 75$ of the control mode "on" ensures that the heater must go off when the temperature rises at 75°.

An edge e (of E) is control switch or transition labeled with guarded assignment in the variables in X; it is enabled, if the associated guard is true and its execution modifies the values of the variables according to the assignment; e.g. for the thermostat automaton, the control switch from control mode "on" to "off" is described by the condition $x = m$. A state $\sigma = (v, x)$ consists of a control location v, and a valuation x of the variables in X; and it can change either by a discrete and instantaneous transition or by a time delay. A discrete transition changes both the control location/the real valued variables, whereas a time delay changes only the values of the variables in X according to the flow condition. A run of a hybrid automaton H is a finite of infinite sequence of states, $\sigma_i = (v_i, x_i)$ with flow conditions f_i:

$$\rho : \sigma_0 \xrightarrow[f_0]{t_0} \sigma_1 \xrightarrow[f_1]{t_1} \sigma_2 \xrightarrow[f_2]{t_2} \cdots \quad \text{(AG.65)}$$

In (AG.65), we have: (i) $f_i(0) = x_i$, (ii) $f_i(t)$ (of inv(v_i)) for all t: $0 \leq t \leq t_i$, and (iii) σ_{i+1} is a transition successor of $\sigma_i' = (v_i, f_i(t_i))$, and σ_i' is a time successor of σ_i. A hybrid automaton is said to be nonzeno if it cannot prevent time for diverging, then only finitely many transitions can be executed in every bounded time interval. The hybrid automaton is linear if its flow conditions, invariants and transition relations can be defined by linear expressions. An interesting special case of a linear hybrid automaton is a timed automaton; wherein each continuous variable increases uniformly with time (with slope 1) and can be considered as a clock; a discrete transition either resets the clock or leaves it unchanged. Another case is of a rectangular automaton: if the flow conditions are independent of the control modes, and the variables are pairwise independent; the invariant condition and the transition relation are described by linear predicates that also correspond to n-dimensional rectangles. These characterize an exact boundary between the decidability and undecidability of verification problems of hybrid automata. The main decision problem concerning the analysis/verification of hybrid systems is the reachability, and the concept is almost similar or the same as is for the linear systems in terms of the reachability of a state. In general, the reachability problem is undecidable even for very restricted classes of hybrid automata. The undecidability of the reachability problem is an obstacle in the analysis/controller synthesis for linear hybrid automata. Controllability of hybrid integrator systems is defined with respect to a pair of regions of the hybrid state space: it is controllable with respect to (R_1, R_2) if there exists an acceptable trajectory that drives the state (v, x) from R_1 to R_2. The methodologies for synthesizing controllers for nonlinear hybrid automata based on a game theoretical framework have been also developed; the computation of the (predecessor) operators is carried out using an appropriate Hamilton-Jacobi-Bellman equation (Chapter 6). Although, many important problems related to hybrid automata are very difficult, there are efficient algorithms for large classes of systems; and many practical applications can be modeled accurately enough by simple hybrid models [AG.5].

G.5 Some Different Ideas in Control

Often for solving any (newer) control problem, we are guided by the well-established or familiar methods. We also, generally do not look deeper into the dynamics of the system that we are trying to control beyond the modeling of the plant, and sometimes its environment.

As we are aware there are certain interesting aspects to look into during/in this process of modeling and applications of the control methods to some engineering problems [AG.6]: (i) strong coupling between subsystems, (ii) mutually interacting relationships (beyond the cause-effect), and (iii) need of accurate and non-isolated models. Thus, an idea is to base the control (philosophy) on: (a) energy, (b) dissipation, and (c) interaction. In the classical setting, usually the overall system is considered as "closed and isolated;" and here coupling with other systems is difficult; the interactions, in general is given less priority. The latter is justified on the basis of time-scale separation between naturally interacting systems, or two/three interacting processes, like in an adaptive control, for the system's dynamics, the control (horizon) action, and the system identification. In the classical/modern control, the system model and the controller are considered as the signal processors, and the design specifications are given in terms of signals: tracking, disturbance, and robustness to model uncertainty.

An alternative (and perhaps supplementary) concept it to view the system/control as an open and energy processing devices/systems/multi-ports (multiple ports). The main aspect is that of "passivity" and it is: (i) a restatement of energy conservation, and (ii) a natural generalization to nonlinear systems, of positivity of matrices and phase-shift of LTI systems, i.e. it is sign preserving property; the property is natural since the mechanical systems and parameter estimators define passive maps. So, the passivity-based control (PBC) is looked-upon from energy processing point-of-view: (a) the plant is viewed as energy-transportation multi-port system/devices; (b) the physical systems satisfy the generalized energy-conservation, i.e. stored energy = supplied energy + dissipation; (c) the control strategy is to preserve the energy-conservation property but with the desired energy and dissipation functions, i.e. desired stored energy = new supplied energy + desired dissipation; or PBC = energy shaping + damping assignment; and (d) for general systems the goal is to achieve passivation.

The possible formulations are: (i) state feedback either to passivize or to change the energy function, and the dissipation; (ii) control by interaction (CBI); the plant and controller are energy-transformation devices, whose energy is added up; and (iii) decompose the system into passive, or passifiable, sub-systems and design PBC's for each of them.

The merits of PBC are [AG.6]:

1. Energy and dissipation are additive: (i) applicable to nonlinear systems, (ii) suitable to handle interconnections of open systems, and (iii) model of uncertainty, say friction is naturally captured;

2. The performance can be handled because of shaping and dissipation;

3. The structure of the system can be effectively utilized to incorporate physical knowledge and to provide physical interpretations to the control action;

4. CBI is applicable to multi-domain systems;

5. There is a geometrical characterization of:
   (i) power-conserving interconnections, and
   (ii) passifiable nonlinear systems.

A multi-port passive system is shown in Figure AG.6. The plant Σ, and the controller Σ_c with the states x, and x_c are cyclo-passive:

$$\dot{H} = u^T y - d; \quad \dot{H}_c = u_c^T y_c - d_c;$$

$$d's \geq 0, \text{ are dissipations} \tag{AG.66}$$

The interconnecting sub-system Σ_I is power-preserving, Figure AG.6;

$$y^T u + y_c^T u_c = y^T v; \quad \Leftarrow u = -y_c + v, \, u_c = y \tag{AG.67}$$

The interconnected system satisfies the following equation

$$\dot{H} + \dot{H}_c = v^T y - d - d_c;$$

$$H(x) + H_c(x_c) \Rightarrow \text{new energy}, \tag{AG.68}$$

$$d + d_c \Rightarrow \text{new dissipation}$$

In (AG.68), one needs to know how to affect x, and the energy functions are coupled via the generation of invariant spaces. An alternative is to make $H_c(x, x_c)$. Then, the idea is to restrict the motion to a subspace of (x, x_c). Define the set

$$S_k \triangleq \{(x, x_c) \mid x_c = F(x) + k\} \tag{AG.69}$$

In (AG.68), k is determined by the initial conditions of the controllers. In S_0 we have

$$H_d(x) \triangleq H(x) + H_c(F(x)) \tag{AG.70}$$

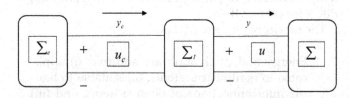

FIGURE AG.6
Multi-port passive system.

The problem is represented as

$$J(x, x_c) \triangleq F(x) - x_c \tag{AG.71}$$

Determine $F(.)$ so that the set S is invariant

$$\frac{d\{J(x, x_c)\}}{dt}\Big|_{J=0} \equiv 0 \tag{AG.72}$$

The satisfaction of (AG.72) would require solving a resultant partial differential equation.

A suitable port-Hamiltonian (PH) model of a physical system can be described as:

$$\sum_{(u,y)} : \begin{cases} x = [I_m(x) - R(x)]\nabla H + g(x)u \\ y = g^T(x)\nabla H \end{cases} \tag{AG.73}$$

In (AG.73) we have: (i) $u^T y$ has units of power, (ii) I_m is negatively symmetric interconnection matrix, it specifies the internal power-conserving structure (exchanges/oscillations between potential and kinetic energies, Kirchhoff's laws, transformers); (iii) R is a symmetric (≥ 0, i.e. PD) matrix and signifies friction, resistors, etc.); and (iv) g is an input matrix. The PH systems are cyclo-passive:

$$\dot{H} = -\nabla H^T R \nabla H + u^T y \tag{AG.74}$$

The power preserving interconnection of a PH system is also a PH system.

Several interesting and further results on: (i) mechanical systems (under-actuated mechanical system, pendulum systems, VTOL aircraft, bilateral tele-operators); (ii) electrical systems (PI control of power converters, transient stability of power systems); (iii) electro-mechanical systems (energy shaping of a MEMS actuator, induction motor); (iv) energy management (dynamic energy router); and (v) wind speed estimation in windmill systems are given in [AG.6]:

G.6 Summary of Important Concepts of Control Theory

Here, we briefly summarize (a short revisit highlighting) various basic concepts of control theory and design, both classical and optimal/modern.

In control theory our primary aim is to analyze a given (open loop and/or closed loop) system, and then see if its stability and/or performance need to be improved. Our secondary aim is to proceed with designing a suitable controller for the plant to meet the required specifications. This controller will be a feed-forward and/or feedback, both, or any other combination. The specifications are given by the user/engineer of the control system, or are decided by the designer based on the task that the closed loop system is to perform. These specifications could be in time domain and/or frequency domain. This process is called the control design and configuration process, the latter can be called the control synthesis process. However, the control design and synthesis terms are used often interchangeably.

G.6.1 Modeling and Analysis

For the analysis we use either time domain or frequency domain methods. We mainly need a mathematical model of the system, and/or the time and/or frequency domain data (from any experiments performed on the original system). The system for which the analysis and/or the controller is to be designed, is often called a process, or a plant. There are various types of mathematical forms/structures available for modeling of the plant; often this form is well known for certain systems, like aerospace vehicles (e.g. flight mechanics model based on the equations of motion of an aircraft, which in turn are based on Newtonian mechanics). Then one can proceed with the determination of the unknown parameters of the model structure, which is done using the experimental data and a parameter estimation method. In case the model structure is not known, one can use a system identification (-cum-parameter estimation) procedure and determine the structure as well as the coefficients of the model (especially the time series models). So, then at this stage we have a reasonably good model of the plant available and this could be in the form of: (i) TFs (or describing functions for nonlinear systems), (ii) linear or nonlinear differential/difference equations, (iii) state space-based, and/or (iv) time series models. For linear systems, one can obtain the unique I/O TF from the states space model; however, the state space model obtained from a TF would not be unique, since the state space describes also the internal structure of the plant model, whereas the TF is the I/O model, which is naturally unique. Then, based on the available mathematical model, a suitable analysis method can be used for characterizing the behavior of the dynamic system.

There are several aspects to be studied in connection with the analysis of the (feedback) control system: (a) sensitivity of the plant TF with respect to a change in its parameter/s, (ii) disturbance rejection, (iii) stability, and (iv) I/O behavior of the system.

G.6.1.1 Sensitivity

The TF characteristics might change when one or more parameters of the TF change due to wear and tear of the components, or the aging of the same. This can be evaluated by sensitivity of the closed loop TF, which is defined as

$$S = \frac{\Delta T(s)}{T(s)} \frac{b}{\Delta b} \tag{AG.75}$$

The sensitivity function is defined as

$$S_b^T = \lim_{\Delta b \to 0} \frac{\Delta T(s)}{\Delta b} \frac{b}{T(s)} = \frac{\partial T(s)}{\partial b} \frac{b}{T(s)}; \tag{AG.76}$$
$$OR \quad S_b^T = \frac{\partial W}{\partial b} \frac{b}{W}$$

The idea in the disturbance rejection is to minimize its effect on the system's performance. One can see that the overall CLCS TF with the disturbance as one of the inputs is given as [AG.7]:

$$G_{cl}(s) = \frac{G_c(s)G_p(s)}{1 + G_c(s)G_p(s)H(s)} R(s)$$
$$+ \frac{G_d(s)}{1 + G_c(s)G_p(s)H(s)} D(s) \tag{AG.77}$$

In (AG.77), $G_c(s)$ is the controller TF, $G_d(s)$ the disturbance TF, and $H(s)$ is the feedback TF. So, for the disturbance rejection, one needs to minimize the complete second term in (AG.79). This should be done at least in the significant range of the required BW of the system. One can increase the loop gain $G_c(s)G_p(s)$, without of course increasing the $G_d(s)$, and choosing the compensator properly. The disturbance magnitude itself can be reduced if feasible, or if it can be measured, then one can use feed forward compensator. The denominator of (AG.79) equated to zero is called the characteristic equation of the system.

G.6.1.2 Stability

Next, important aspect is to evaluate the stability of the system for which one can study its impulse (or step) response with the input signal $u(t)$. We have studied various aspects and definitions of the stability in Sections I and III, for continuous and discrete time systems. The BIBO stability should be valid for all the bounded input signals; in terms of an impulse response, the (convolution) integral should be finite. The stability can also be evaluated by using TFs, in terms of its poles and zeros. Also, the eigenvalues of the system, i.e. of the coefficient matrix A, can be evaluated and the stability determined (assuming there is no pole-zero cancellation, i.e. the system is a minimal realization). The stability can be assessed by using several stability criteria/techniques: (i) Routh-Hurwitz, (ii) root locus, (iii) Nyquist, and (iv) Bode plots. For a LTI system, all the poles of the CLCS should lie in the LHS plane of the complex plane (S-plane). This also is determined from the characteristic equation that its roots should lie in the LHS plane; also meaning that the real roots and the real parts of the complex roots should lie in the LHS plane. In case of a general nonlinear system, one can use computer simulation for studying the time responses and assess the stability characteristics. From the set of criteria based on root locus, Nyquist plot, and Bode plot; one can also obtain the information on the settling time, and the relative stability, i.e. how much away is the system from the stability (instability) boundary. The criterion at (i) only tells the absolute stability of the system (it is stable or unstable), except when the criterion is modified. In terms of poles and zeros we have: (i) the complex pole pair would give growing or decaying oscillations, (ii) real pole will give exponentially growing or decaying responses, and (iii) the position of zeros determine the amplitude and the phase of the response.

There are certain performance specifications defined in terms of gain margin (GM), and phase margin (PM) and these are made clear from a Bode diagram (Bode-frequency response) or plot of the (open) loop TF of the plant. Often, 20 dB GM, and 45° of PM are required, and this provides sufficient distance of the system from the verge of instability; this means that if the gain of the system is increased by 20 dB, and/or the phase lag is increased up to 45°, then the system has just reached the boundary and the system is ready to slip into the instability, and in the case of the Nyquist criteria, this critical point is taken as $-1\pm j0$ on the Nyquist (polar) plot, and signifies as the unity gain and the phase angle of $-180°$. A root locus is a plot of the system characteristic (in fact again the GH(s) plot), and usually the feedback gain is varied. By plotting the root locus (in MATLAB) for a given system, a designer can figure out how much GM and PM are available for a particular parametric curve, since one can see the movement of the roots when the gain is varied.

The discrete time systems are described by z-transforms (analogous to Laplace transform), and difference equations, which can be equivalently put in the form of state space forms. In fact, the z-transform can be said as the generalized discrete time Fourier transform. Interestingly, the following equivalence is easily established:

$$F(z) = Z\{f(k)\} = \sum_{k=0}^{\infty} f(k)e^{-i\omega k}\big|_{e^{i\omega}=z} = \sum_{k=0}^{\infty} f(k)z^{-k};$$

(AG.78)

z is a complex variable

For discrete time systems, the stability is considered with reference to the unit circle, since the continuous time s-plane is equivalently translated to z-plane, wherein the LHS plane is compacted inside the unit circle, and the RHS plane is outside of this circle. If the poles are farther inside the unit circle, then the stability is higher. Also, as in the case of a continuous time state space model, the stability can be judged from the eigenvalues of the system, if the magnitude of all the eigenvalues is less than 1, then the system is stable. Interestingly the I/O description of a discrete time system from its state space formulation is given as

$$G(z) = H(zI - \varphi)^{-1}B + D$$

(AG.79)

This is exactly similar in form and detail to the CTTF, with the appropriate system's coefficient matrices.

G.6.2 Feedback Control

For a control design broadly at the top level one of the two requirements are posed: (i) a regulator (in which case a physical variable is kept at a constant value, even in the presence of perturbation), and (ii) in a tracking or servomechanism a physical variable is required to be followed or track a desired time function. In this case a compensator, also called a controller or even a filter is designed to provide satisfactory behavior of the closed loop control system (CLCS).

G.6.2.1 Classical

In a classical idea, a portion of the plant output is fed back and (negatively) added to the original control input; thus, the error signal drives the original plant, which enhances the stability and the performance of the CLCS. The open loop gain is the magnitude of the plant TF, $|G(s)|$; and the closed loop gain is the magnitude of the CLCS TF, $|G(s)/(1 + G(s)H(s))|$.

In a proportional feedback, the system's differential equation remains the same, but, the term related to the natural frequency is altered as is seen from the following equations:

$$\ddot{x} + \zeta\dot{x} + \omega_n^2 x = u; \ u_{fb} = -K_P x(t); \ u = u + u_{fb}$$
$$\ddot{x} + \zeta\dot{x} + (\omega_n^2 + K_P)x = u$$

(AG.80)

One would get the CLTF as

$$G_{cl}(s) = \frac{1}{s^2 + \zeta s + (\omega_n^2 + K_P)}$$

(AG.81)

One can easily see the effect of the proportional feedback, which increases the BW of the system, but the overall gain will reduce slightly, this can be ascertained by plotting the Bode frequency response.

We can now see the effect of the integral feedback, which adds an extra state:

$$u_{fb}(t) = -K_I \int_0^t x(\tau)d\tau; \quad \ddot{x} + \zeta\dot{x} + \omega_n^2 x = u - K_I \int_0^t x(\tau)d\tau$$

$$\dddot{x} + \zeta\ddot{x} + \omega_n^2\dot{x} + K_I x = \dot{u} \qquad \text{(AG.82)}$$

$$G_{cl}(s) = \frac{s}{s(s^2 + \zeta s + \omega_n^2) + K_I}$$

One can see from (AG.82) that the closed loop gain at $s = 0$ (steady state gain or DC gain) is zero, it means that the integral feedback would reject the constant disturbance. Also, we can see that for large s (i.e. higher frequencies), the closed loop TF approaches the open loop TF, and the Bode diagram would be almost the same as the open loop Bode for this range of frequencies.

Now, we can see the effect of the derivative feedback as follows:

$$u_{fb}(t) = -K_D\dot{x}(t); \quad \ddot{x} + (\zeta + K_D)\dot{x} + \omega_n^2 x = u$$

$$G_{cl}(s) = \frac{1}{s^2 + (\zeta + K_D)s + \omega_n^2} \qquad \text{(AG.83)}$$

It is seen from (AG.83) that the damping of the system is enhanced/improved, and the BW remains the same, the response is flattened. But, use of differentiation could enhance the effect of noise.

Now, one can combine the effects of the three types of the feedbacks, the control signal would be $u_{fb} = u_P + u_I + u_D$, such a controller is called PID controller, and one can see that one needs to tune the PID controller gains for achieving the desired stability and (tracking) error performance of the CLCS, then at this stage the design process depends on the intuition and the designer's engineering judgement, keeping in mind the design goal, so it becomes an "art." For simple and lower order systems, this approach is easy, and the design can be done with intuition.

Also, there are lead, lag, and lead-lag compensators that are explicitly put in the forward and/or feedback paths/loops that can be utilized to improve phase lead, time response, and BW of the system: if the system is "slow," one can use the lead compensator, and if the system is "fast," one can use lag compensator. In most cases, these are first order TFs, and are also called filters.

For large and complex system this design procedure becomes a difficult process: higher order systems, more numbers of complex real poles, and so on; however,

such a problem can be solved in an easier way by the full state feedback control, which is a part of the modern control approach.

G.6.2.2 State Feedback Control

Here, we consider the state feedback control, before we consider the optimal control, because this directly provides connection with PID control philosophy. In a state feedback we have the following formulation:

$$\dot{x}(t) = Ax(t) + Bu(t); \quad y(t) = Cx(t);$$
$$\text{and } u(t) = -Kx(t) \qquad \text{(AG.84)}$$

Thus, for the proportional control we have $u_P = -K_P x$; and for the derivative feedback we have

$$u_D = -K_D\dot{x}(t) \qquad \text{(AG.85)}$$

With both the proportional and derivative feedback we will have

$$u_{fb} = -[K_P \ K_D]\begin{bmatrix} x_1 \\ x_2 \end{bmatrix} = -K\bar{x}; \ x_1 = x, \text{ and } x_2 = \dot{x}_1 = \dot{x} \text{ (AG.86)}$$

From (AG.84), we obtain the closed loop system as

$$\dot{x}(t) = (A - BK)x(t) \qquad \text{(AG.87)}$$

From (AG.87), we can see that by choosing appropriate state feedback gain matrix, we can alter the eigensystem of the closed loop system and place these as desired. Since, for the minimal system, the eigenvalues are the poles of the closed loop system, this is also a pole placement/assignment problem. So, the PID control, the state feedback and the pole placement are very closely connected philosophies.

G.6.3 Nonlinear Control

The approaches of analysis and design so far studied can all be applied to nonlinear systems, but in a limited sense, i.e. these would be applicable to only the local linearizations of the nonlinear systems. This also means that one might need to design controllers at all these locally linearized systems, and then use some switching mechanism to bring in a particular controller gains'/filters' when another linearized system is to be controlled. This is also a part of gain scheduling design. Often such approaches are used for design of flight controllers of an aircraft, and performance validation is carried out using extensive nonlinear simulations (in piloted ground-based/air-based flight [called in-flight] simulators). For gain scheduling either table look up or curve fitting methods are used. As of now this approach is found to work well for an inherently unstable (relaxed static stability) fighter aircraft that is being developed in the country. The analysis and the designs would work reasonably well, if the nonlinear systems work within the domain of their approximating linearization; however, when there are exceedances, the behavior of the overall CLCS might not be acceptable. In such cases adaptive and/or nonlinear control methods with some formal approaches might be needed for the design: (i) input-state feedback linearization, (ii) input-output feedback linearization, (iii) basic concepts of zero dynamics, (iv) partial feedback linearization, and (v) integrator back-stepping [AG.8].

G.6.3.1 *Input-State Feedback Linearization*

The conventional linearization is valid only for small operating domain near equilibriums, and it is useful for piecewise linear systems. This linearization is not suited to dynamic environments. Let us take a second order nonlinear system [AG.8]:

$$\dot{x}_1 = -x_1 - x_1^3 + x_2; \quad \dot{x}_2 = x_1 - x_2 + bu;$$
$$\text{equilibrium point} = (0,0) \tag{AG.88}$$

The linearized model of (AG.88) is valid only for $||x_1|| < 1$. We observe that the model of the nonlinear system contains an unmatched nonlinearity, i.e. the cubic term present in the first equation is not spanned by the control input u. The system of (AG.88) is then transformed as follows, with new state defined and using (AG.88):

$$z_1 = x_1, z_2 = -x_1^3 + x_2;$$
$$\dot{z}_1 = -z_1 + z_2; \dot{z}_2 = 2z_1^3 - 3z_1^2 z_2 + z_1 - z_2 + bu \tag{AG.89}$$

One can see that all the nonlinear terms (in the second equation) are matched by the control input. Then, the idea is to design a control law u, so that all the nonlinear entries of the second equation are canceled out. This results in

$$u = \frac{1}{b}[-2z_1^3 + 3z_1^2 z_2 + v]$$

$$\dot{z}_1 = -z_1 + z_2; \quad \dot{z}_2 = z_1 - z_2 + v$$

$$\begin{bmatrix} \dot{z}_1 \\ \dot{z}_2 \end{bmatrix} = \begin{bmatrix} -1 & 1 \\ 1 & -1 \end{bmatrix} \begin{bmatrix} z_1 \\ z_2 \end{bmatrix} + \begin{bmatrix} 0 \\ 1 \end{bmatrix} v; \tag{AG.90}$$

$$|(sI - A)| = 0 \Rightarrow poles(-2,0)$$

The poles of the new system are located at $(-2,0)$. If we want the closed loop system to have damping ratio as 0.7, and the natural frequency as 5 rad/s, then the poles should be placed at $(-3.5 \pm 3.75j)$, and the control input can be selected as

$$v = -[19.999 \quad 5] \begin{bmatrix} z_1 \\ z_2 \end{bmatrix} \tag{AG.91}$$

Finally, in the terms of the original states as in (AG.88), and (AG.89), we get the control law as

$$u = \frac{1}{b}[-19.999x_1 - 5x_2 + 3x_1^3 + 3x_1^2 x_2 - 3x_1^5] \tag{AG.92}$$

The demerits of the input-state feedback linearization are: (a) all the states would not be available for the measurements and, hence, the feedback, (b) more error will be induced if more measurements are used, and (c) if only output form the system is available, then input-state feedback linearization cannot be implemented.

G.6.3.2 *Input-Output Feedback Linearization*

Let us now take another system as:

$$\dot{x}_1 = \cos(x_2) + (x_2 + 1)x_3$$
$$\dot{x}_2 = x_1^3 + x_3 \tag{AG.93}$$
$$\dot{x}_3 = x_1^2 + u; \quad y = x_1$$

Since we have only output measurable, we need to obtain an explicit relation between input (u) and output signals. Hence, differentiating equation for y twice we obtain

$$\ddot{y} = (x_2 + 1)u + f_1(\overline{x}); \quad f_1(\overline{x}) = (x_1^3 + x_3)(x_3 - \sin(x_2))$$
$$+ (x_2 + 1)x_1^2 \tag{AG.94}$$

We can use the following control input signal to cancel the nonlinear terms in (AG.94):

$$u = \frac{1}{x_2 + 1}(v - f_1(\overline{x})); \quad \ddot{y} = v \tag{AG.95}$$

We see from (AG.95) that a double integrator relation is obtained between the input v and the output y. For tracking control law, we have the following:

$$v = y_d - k_1 e - k_2 \dot{e}; \quad \ddot{e} + k_2 \dot{e} + k_1 e = 0; \quad e = y - y_d \tag{AG.96}$$

Here, for positive values of the coefficients, one should/ can assure stable error dynamics so that a perfect tracking is obtained. For an n-th order system, if r number of differentiations are needed in order to obtain an explicit relation between input u and the output y, then, the system is said to have a relative degree of r. Often, a part of the system dynamics is rendered unobservable at the time of I/O linearization, and this part is called the internal dynamics of the original system. Then, the stability of such dynamics can be evaluated by using the concept of zero dynamics.

G.6.3.3 Concept of Zero Dynamics

Let us study the 3rd order TF [AG.8]:

$$\frac{y(s)}{u(s)} = \frac{1}{s^3 + 6s^2 + 11s + 6};$$

$$\begin{bmatrix} \dot{x}_1 \\ \dot{x}_2 \\ \dot{x}_3 \end{bmatrix} = \begin{bmatrix} 0 & 1 & 0 \\ 0 & 0 & 1 \\ -6 & -11 & -6 \end{bmatrix} \begin{bmatrix} x_1 \\ x_2 \\ x_3 \end{bmatrix} + \begin{bmatrix} 0 \\ 0 \\ 1 \end{bmatrix} u; \quad y = x_1 \quad \text{(AG.97)}$$

By differentiating the output y, three times we obtain

$$\dddot{y} = -6x_1 - 11x_2 - 6x_3 + u;$$
$$\text{the relative degree } r = 3 \quad \text{(AG.98)}$$

The relative degree indicates the excess number of poles over the number of zeros; hence, the order of the internal dynamics in this case is 3–3 = 0. Let us append one zero to (AG.97):

$$\frac{y(s)}{u(s)} = \frac{s+4}{s^3 + 6s^2 + 11s + 6};$$

$$\begin{bmatrix} \dot{x}_1 \\ \dot{x}_2 \\ \dot{x}_3 \end{bmatrix} = \begin{bmatrix} 0 & 1 & 0 \\ 0 & 0 & 1 \\ -6 & -11 & -6 \end{bmatrix} \begin{bmatrix} x_1 \\ x_2 \\ x_3 \end{bmatrix} + \begin{bmatrix} 0 \\ 0 \\ 1 \end{bmatrix} u; \quad y = 4x_1 + x_2 \quad \text{(AG.99)}$$

Let us differentiate y of (AG.99) two times to obtain

$$\ddot{y} = -6x_1 - 11x_2 - 2x_3 + u;$$
$$\text{the relative degree } r = 2. \quad \text{(AG.100)}$$

In this case the order of the internal dynamics is $3 - 2 = 1$. Let us use the control law as:

$$u = 6x_1 + 11x_2 + 2x_3 + v; \quad \ddot{y} = v \quad \text{(AG.101)}$$

Now, we consider the output y as the state of the system, and new control signal v, as the input we obtain

$$z_1 = y; z_2 = \dot{y}$$
$$\dot{z}_1 = z_2 \quad \text{(AG.102)}$$
$$\dot{z}_2 = v$$

From (AG.102), we see that this is a second order model (two first order), and the input u considered in (AG.101) converts the original system into second order system; equivalently, it is now a cascade combination of reduced order system and internal dynamics. From (AG.99), we can observe that the internal dynamics can be represented as

$$\dot{x}_1 = x_2 = -4x_1 + y \quad \text{(AG.103)}$$

The stability of the internal dynamics can be studied by the location of the zero in the s-plane.

Now, consider yet another system as;

$$\dot{x}_1 = x_1 x_2 - x_1$$
$$\dot{x}_2 = 2x_1 x_2 - 3x_2 + u; \quad y = x_2 \quad \text{(AG.104)}$$

By first order differentiation of output y, we get

$$\dot{y} = 2x_1 x_2 - 3x_2 + u \quad \text{(AG.105)}$$

We consider the input u as:

$$u = -2x_1 x_2 + 3x_2 + v; \quad \dot{y} = v \quad \text{(AG.106)}$$

The internal dynamics from (AG.104) is obtained as: $\dot{x}_1 = x_1 y - x_1$. The zero dynamics can be determined by using the concept of the input that makes the output zero. In the present case, the zero output implies $y = 0$, and $\dot{y} = 0$; this also implies that $x_2 = 0$; hence

$$\dot{x}_1 = -x_1$$
$$\dot{x}_2 = u; \dot{y} = 0 \Rightarrow \dot{x}_2 = 0 \Rightarrow u = 0 \quad \text{(AG.107)}$$

Thus, the zero dynamics equation is obtained as $\dot{x}_1 = -x_1$

G.6.3.4 Partial Feedback Linearization

In case of the under-actuated mechanical systems (UMS), the conventional linearization does not work. A concept of partial feedback linearization to at least linearize the model "somewhat" is useful. Let us study the Lagrangian model for a generic UMS case [AG.8]:

$$m_{11}(x)\ddot{x}_1 + m_{12}(x)\ddot{x}_2 + h_1(x, y) = 0$$
$$m_{21}(x)\ddot{x}_1 + m_{22}(x)\ddot{x}_2 + h_2(x, y) = B(x)w \quad \text{(AG.108)}$$

Now, we substitute for the first state, $x(..,1)$ from the 1st equation of (AG.108)) into the second equation of (AG.108), and then choose the following control law w:

$$w = B^{-1}(x)[\{h_2(x, y) - m_{21}(x)m_{11}^{-1}(x)h_1(x, y)\}$$
$$+ \{m_{22}(x) - m_{21}(x)m_{11}^{-1}(x)m_{12}(x)\}u] \quad \text{(AG.109)}$$

When we substitute this in the active joint equation, we obtain

$$\ddot{x}_2 = u \tag{AG.110}$$

Next substituting (AG.110) in the 1st equation of (AG.108), we obtain the following equations for the passive joints:

$$\ddot{x}_1 = -\frac{m_{12}(x)u}{m_{11}(x)} - \frac{h_1(x,y)}{m_{11}(x)}$$

$$\dot{x}_1 = y_1$$

$$\dot{y}_1 = g(x)u + f(x,y); \quad f(x,y) = -m_{11}^{-1}(x)h_1(x,y); \tag{AG.111}$$

$$g(x) = -m_{11}^{-1}(x)m_{12}(x)$$

$$\dot{x}_2 = y_2$$

$$\dot{y}_2 = u$$

The partial feedback linearization as discussed in the foregoing is known as co-located feedback linearization, and results in a linear model for the active joints. Next, we define a new control law as

$$v = g(x)u + f(x,y)$$ and obtain the following equations

$$\dot{x}_1 = y_1$$
$$\dot{y}_1 = v$$
$$\dot{x}_2 = y_2 \tag{AG.112}$$
$$\dot{y}_2 = \frac{v-f}{g}$$

When the passive joints, as in (AG.112) are linearized, it is called non-co-located linearization.

G.6.3.5 Integrator Back-Stepping

Consider the system as

$$\dot{x} = -\sin x - x^3 + z; \quad \dot{z} = u \tag{AG.113}$$

Design aim is to find a control law u that ensures the regulation of the state variables $x(t)$, and $z(t)$ for all the initial conditions; $x = 0$, $z = 0$. We can have the control law ($z = -c_1x + \sin(x)$), but z is a state variable and not the control input. We can then specify the desired value of z as

$$z_{des} = -c_1 + \sin x \triangleq \alpha_s(x) \tag{AG.114}$$

Then, the error is defined as:

$$e = z - z_{des} = z - \alpha_s(x) = z + c_1x - \sin x \tag{AG.115}$$

In such a case the z is called a virtual control; and the desired value of z, i.e. $\alpha_s(x)$ are the stabilizing functions. Using the state equation (AG.113), and the error definition, (AG.115), we can have the following dynamics [AG.8]:

$$\dot{x} = -\sin x - x^3 + (z + c_1x - \sin x)$$
$$-c_1x + \sin x = -c_1x - x^3 + e \tag{AG.116}$$

$$\dot{e} = \dot{z} - \dot{\alpha}_s = \dot{z} + (c_1x - \cos x)\dot{x}$$
$$= u + (c_1x - \cos x)(-c_1x - x^3 + e) \tag{AG.117}$$

In (AG.117) the signal, $-d(\alpha_s(x))/dt$, is called the backstepping signal, and it serves the purpose of the feedback control law inside the block and it can be given through an integrator, as $-\alpha_s(x)$.

Now, we use Lyapunov's method of stability analysis, and determine a suitable control law. The Lypunov energy functional is specified as

$$V_a(x,e) = V(x) + \frac{1}{2}e^2 = \frac{1}{2}x^2 + \frac{1}{2}e^2 \tag{AG.118}$$

As usual this function is a PD and in order that the designed control system is stable, the time derivative of this function should be negative (semi) definite. We obtain the expressions for the time derivative as follows, by substituting appropriate expressions from (AG.116), and (AG.117) in the time derivative expression of (AG.118):

$$\dot{V}_a(x,e) = x(-c_1x - x^3 + e)$$
$$+ e(u + (c_1x - \cos x)(-c_1x - x^3 + e)) \tag{AG.119}$$

$$= -c_1x^2 - x^4 + e[x + u + (c_1x - \cos x)(-c_1x - x^3 + e)] \tag{AG.120}$$

A choice of the control can be made as

$$u = -c_2e - x - (c_1x - \cos x)(-c_1x - x^3 + e) \tag{AG.121}$$

Then, substitute (AG.121) in (AG.120) to obtain the time derivative as

$$\dot{V}_a = -c_1x^2 - x^4 - c_2e^2 \tag{AG.122}$$

We can easily see that if the constants c's are positive, then since, all other variables appear as squared quantities, the time derivative of the Lypunov function is −ve; thus, the system designed is (asymptotically) stable.

G.6.4 Optimization, Dynamic Programming and Calculus of Variation

We capture and revisit several aspects of techniques leading to the theory of optimal control, then revisit briefly the concepts of optimal control [AG.9].

G.6.4.1 Nonlinear Optimization (Unconstraint)

Here, the aim is to minimize a nonlinear function F(x), then $x^* = \arg\{\min_x F(x)\}$. A strong minimum is when the objective function increases locally in all the directions. In a weak minimum, the objective function remains the same in some directions, and increases locally in other directions. If the first order derivative $g = \left(\frac{\partial F}{\partial x}\right)^T | x = x^* = 0$, then this is necessary and sufficient condition for a point to be a stationary point. Also, it is a necessary but not the sufficient condition for the function F(x) to be a minimum. The stationary point could be either maximum or a saddle point. The sufficient condition for the minimum is that the Hessian matrix (G_k; with its elements as the second order partial differentiations of the function F(x) with respect to all the components of the vector x) is PD; if it is positive semidefinite (PSD), then it gives the necessary condition. Typically the minimization problem is solved by using an iterative algorithm: $x^*(k + 1) = x^*(k) + \alpha_k p_k$ = previous estimate + scalar (tuning parameter/step size) (into) search direction. One search direction is that of the well know steepest descent, a first order gradient method: $x^*(k + 1) = x * (k) - \alpha_k g_k$. This ensures that the cost function gradually decreases:

$$g_k^T p_k < 0; \quad p_k = -g_k \qquad (AG.123)$$

With the given search direction, one needs to tune the step size so that the process is fast, accurate and easy. One can do the line search using (AG.123), then if the local minimum is bracketed/near; then one can use the methods: (i) Golden section search, (ii) Bisection, and (iii) polynomial approximations. The first too methods have linear convergence. For polynomial search the function is approximated as a quadratic/cubic in the interval, and the minimum of this polynomial is used as the estimate of the local minimum.

The second order methods for the search of the optimum are faster than the line search methods:

$$p_k = -G_k^{-1} g_k \qquad (AG.124)$$

Since, the function F(x) might not be quadratic always, to get the minimum solution we need iteration of the search method. If the gradients (formulas) are not explicitly available, then one can use the finite difference methods to obtain these. One can use quasi-Newton approximation also.

G.6.4.2 Nonlinear Optimization (with Constraints)

This problem is posed as optimization of some function such that some other function of the same independent variables is to be satisfied either with equality (or inequality) constraints:

$$\min_y F(y), \ f(y) = 0; \quad \min_u F(x,u), \ f(x,u) = 0 \qquad (AG.125)$$

A direct procedure is: (i) solve for x in terms of u from the constraint, and (ii) substitute in F(x, u) and solve for u as for the unconstraint optimization. This will work well if the f is linear. A general approach for nonlinear functions is to use the method of Lagrange multiplier (LM) λ [AG.9]:

$$L(x,u,\lambda) = F(x,u) + \lambda^T f(x,u) \qquad (AG.126)$$

$$dL = \frac{\partial L}{\partial x} dx + \frac{\partial L}{\partial u} du \qquad (AG.127)$$

The first term in (AG.126) is obtained as:

$$\frac{\partial L}{\partial x} \triangleq \frac{\partial F}{\partial x} + \lambda^T \frac{\partial f}{\partial x} \equiv 0 \Rightarrow \lambda^T = -\frac{\partial F}{\partial x}\left(\frac{\partial f}{\partial x}\right)^{-1};$$

$$\frac{\partial F}{\partial y} + \lambda^T \frac{\partial f}{\partial y} = 0 \qquad (AG.128)$$

Also, we have from the constraint equation

$$df = \frac{\partial f}{\partial x} dx + \frac{\partial f}{\partial u} du \equiv 0 \Rightarrow dx = -\left(\frac{\partial f}{\partial x}\right)^{-1} \frac{\partial f}{\partial u} u \qquad (AG.129)$$

The permissible cost variations are

$$dF = \frac{\partial F}{\partial x} dx + \frac{\partial F}{\partial u} du$$

$$= \left(-\frac{\partial F}{\partial x}\left(\frac{\partial f}{\partial x}\right)^{-1} \frac{\partial f}{\partial u} + \frac{\partial F}{\partial u}\right) du \qquad (AG.130)$$

$$= \left(\frac{\partial F}{\partial u} + \lambda^T \frac{\partial f}{\partial u} +\right) du = \frac{\partial L}{\partial u} du$$

Thus, we obtain the expression for the gradient of the cost function with respect to u, while satisfying the constraint, and this should be zero, and we get the following (necessary) conditions to solve the constraint optimization problem:

$$\frac{\partial L}{\partial x} = 0; \quad \frac{\partial L}{\partial u} = 0; \quad \frac{\partial L}{\partial \lambda} = f(x,u) = 0 \qquad (AG.131)$$

The constraint solution will be a point of tangency of the constant cost curves and the constraint function. This also means that the gradient of the cost function (normal to the constant cost curve) $\frac{\partial F}{\partial y}$ must lie in the space spanned by the constraint gradients $\frac{\partial f}{\partial y}$. It is possible to express the cost gradient as a linear combination of the constraint gradient. Often, the improvements in the cost cannot be done without violating the constraints.

We can have the optimization problem with inequality constraint

$$\min_y F(y), \ f(y) \le 0 \qquad \text{(AG.132)}$$

For this problem all the constraints are not active. The problem can be partitioned as

$$\frac{\partial F}{\partial y} = -\underbrace{\sum_i \lambda_i \frac{\partial f_i}{\partial y}}_{active} - \underbrace{\sum_j \lambda_j \frac{\partial f_j}{\partial y}}_{inactive}; \qquad \text{(AG.133)}$$

$$\lambda_j = 0, \text{ for inactive constraints}$$

For the inequality constraint problem we have

$$\frac{\partial L}{\partial y} = 0; \quad \lambda_i \frac{\partial L}{\partial \lambda_i} = 0, \text{ for } \forall i \qquad \text{(AG.134)}$$

For active constraints we have:

$$\frac{\partial L}{\partial \lambda_i} = f_i = 0, \ \lambda_i \ge 0 \qquad \text{(AG.135)}$$

For inactive constraints we have:

$$\frac{\partial L}{\partial \lambda_i} = f_i < 0, \ \lambda_i = 0 \qquad \text{(AG.136)}$$

The essence of the Kuhn-Tucker theorem in nonlinear programming is captured in (AG.135), and (AG.136). The relationship between the changes in the cost function and the constraints is captured as follows, at the solution point:

$$\frac{\partial L}{\partial y} = 0 \Rightarrow \frac{\partial F}{\partial y} = -\lambda^T \frac{\partial f}{\partial y}$$

$$In \cos t \qquad \Delta F = \frac{\partial F}{\partial y} \Delta y$$

$$\text{Constraint} \quad \Delta f = \frac{\partial f}{\partial y} \Delta y \qquad \text{(AG.137)}$$

$$\Delta F = -\lambda^T \frac{\partial f}{\partial y} \Delta y = -\lambda^T \Delta f \Rightarrow \frac{dF}{df} = -\lambda^T$$

From (AG.137), it is clear the constraint-sensitivity of the cost function is directly related to the Lagrange multiplier, LM. If the constraint is specified as $f \le c$; then the condition is the same as in (AG.138): $\frac{\partial F}{\partial c} = -\lambda^T$.

G.6.4.3 Dynamic Programming

The idea in the DP is based on the principle of optimality (PO): assume that the optimal solution (between A-B-C) for a problem passes through an intermediate point (x_1, t_1), B, then the optimal solution to the same problem starting at (x_1, t_1), B must be the continuous of the same path (B-C), it means this new solution is also optimal. The PO leads to the DP, the numerical solution procedure for solving multi-stage decision making problems, and can be used to design a control law. It has been found in the case of the classical shortest path problems: (i) robot navigation, and (ii) aircraft path problems, which going forward from point A to B, the destination, the cost, and computation time required are higher, compared to going backwards from point B to A, in the latter the principle of optimality is invoked.

For more general control problems we have the assumptions/features: (i) some structure of the systems states and control inputs, like bounded but also discretized; (ii) grid the time/state and quantized control inputs; (iii) the discrete time problem leads to discrete LQR; and (iv) the continuous time problem leads to calculus of variation, which in turn leads to continuous time LQR.

For a classical control problem we have to minimize

$$J = \phi(x(t_f)) + \int_{t_0}^{t_f} F(x(t), u(t), t)dt; \ \dot{x} = f(x, u, t); \qquad \text{(AG.138)}$$

$$x(t_0) = fixed, t_f = fixed$$

Solution approach: (i) develop a grid over (i.e. discretize) space/time and look at possible final states, and evaluate final costs; and (ii) go backward one step in time and consider all the probable ways of completing the solution. To evaluate the cost of a control action, the integral in the cost has to be approximated. The procedure is illustrated as follows [AG.9]:

1. Apply the control $u^{ij}(k)$ to move to state x^j at time $t(k+1) = t(k) + \Delta t$; from the state x^i at time $t(k)$.

2. The cost is approximated:

$$\int_{t_k}^{t_{k+1}} F(x(t), u(t), t)dt \approx F(x^i(k), u^{ij}(k), t(k))\Delta t \qquad \text{(AG.139)}$$

3. Solve the system equations approximately for getting the control inputs directly:

$$x^j(k+1) \approx x^i(k) + f(x^i(k), u^{ij}(k), t(k))\Delta t$$

$$f(x^i(k), u^{ij}(k), t(k)) = \frac{x^j(k+1) - x^i(k)}{\Delta t} \qquad \text{(AG.140)}$$

4. The control input can be easily obtained if the system is affine:

$$\dot{x} = f(x,t) + g(x,t)u \qquad (AG.141)$$

From (AG.142) we obtain using (AG.140):

$$u^{ij}(k) = g^{-1}(x^i(k),$$

$$t(k)) \begin{bmatrix} \dfrac{x^j(k+1) - x^i(k)}{\Delta t} \\ \\ -f(x^i(k), t(k)) \end{bmatrix} \qquad (AG.142)$$

5. Since we know the optimal path from each new terminal point $x^j(k + 1)$, one can establish the optimal path to take from $x^i(k)$:

$$J^*(x^i(k), t(k)) = \min_{x^j(k+1)} [\Delta J(x^i(k),$$

$$x^j(k+1)) + J^*(x^j(k+1)) \qquad (AG.143)$$

Thus, for each $x^i(k)$, the result is: (i) best $x^i(k + 1)$ to pick, since it gives the lowest cost, and (ii) the control input required to obtain this best cost.

6. Work backward in time till we reach the $x(t(0))$, and only one value of x is permitted specified by the given initial condition.

There could be bounds on the useable/applicable control input (on its magnitude), due to certain state transitions that would not be permitted, as is the case with certain constraints on the state.

The problem of DP also can be extended

$$J = \phi(x(t_f), t_f) + \int_{t_0}^{t_f} F(x(t), u(t), t)dt; \\ \psi(x(t_f), t_f) = 0 \qquad (AG.144)$$

This would give a group of points that satisfy the terminal constraints, evaluate the cost, and then proceed backwards form there. For higher dimensional problem, it might also be necessary to quantize the control inputs as well, like grid of, now a state vector (that contains more components). If there are N_t points in time and N_x quantized states of dimension n, then the points to consider are $N = N_t (N_x)^n$, and this is known as the curse of dimensionality.

In most cases, the DP must be solved numerically, yet, a few cases can be solved analytically, like the discrete LQR. Here, the aim is to minimize the cost

$$J = \frac{1}{2} x^T(N)Wx(N)$$

$$+ \frac{1}{2} \sum_{k=0}^{N-1} \{x^T(k)Rx(k) + u^T(k)Qu(k)\} \qquad (AG.145)$$

$$x(k+1) = \varphi(k)x(k) + B(k)u(k);$$

$$W = I \text{ for the present} \qquad (AG.146)$$

In (AG.145), R, Q, and W are the respective Gramian-weight matrices of the states, inputs, and the final state. The application of the DP procedures results in the following solution/algorithm to the LQR problem (for implementation on a digital computer):

i. $$P(N) = I; \text{ or } P(N) = W$$

ii. $$F(k) = [Q + B^T(k)P(k+1)B(k)]^{-1} \qquad (AG.147)$$

$$B^T(k)P(k+1)\varphi(k)$$

iii. $$P(k) = R + F^T(k)QF(k) + [\varphi(k) - B(k)F(k)]^T \qquad (AG.148)$$

$$P(k+1)[\varphi(k) - B(k)F(k)]$$

iv. $$u^*(k) = -F(k)x^*(k);$$

$$x^*(k+1) = \varphi(k)x^*(k) + B(k)u^*(k) \qquad (AG.149)$$

The steps (ii–iv), i.e. the recursions should be evaluated from N–1 to 0. One can see that the matrices $P(.)$, and $F(k)$ are independent of the state and can be pre-computed and stored. The iterations (ii) and (iii) can be combined as:

$$P(k) = R + \varphi^T(k)\{P(k+1) - P(k+1)B(k)$$

$$[Q + B^T(k)P(k+1)B(k)]^{-1} \cdot \qquad (AG.150)$$

$$B^T(k)P(k+1)\}\varphi(k)$$

For the LTI problem (the system is stabilizable, and the uncontrollable modes are stable) we obtain the following steady equation that is discrete form of the well-known algebraic Riccati equation (DARE):

$$P_s = R + \varphi^T\{P_s - P_sB[Q + B^TP_sB]^{-1}B^TP_s\}\varphi \qquad (AG.151)$$

The DARE can be solved numerically. From (AG.147), we can obtain following insights into the functioning of the algorithm/controller:

$$F(N-1) = [Q + B^T(N-1)P(N)B(N-1)]^{-1} \qquad (AG.152)$$

$$B^T(N-1)P(N)\varphi(N-1)$$

For the scalar case (AG.153) reduces to

$$F(N-1) = \frac{B(N-1)P(N)\varphi(N-1)}{Q+B^2(N-1)P(N)} \qquad \text{(AG.153)}$$

If there is very high weighting, W on the terminal state, and $P(N)$ is very large, then we have

$$F(N-1) \Rightarrow \frac{B(N-1)P(N)\varphi(N-1)}{B^2(N-1)P(N)} = \frac{\varphi(N-1)}{B(N-1)} = \frac{\varphi}{B} \quad \text{(AG.154)}$$

Then the terminal state is

$$x(N) = (\varphi - BF)x(N-1) = (\varphi - B\frac{\varphi}{B})x(N-1) = 0. \quad \text{(AG.155)}$$

This is a nilpotent controller. If the control penalty is set as very small, then $Q \to 0$ (R/Q is large), then also we obtain the same condition as in (AG.155), i.e. if the state is penalized, but not the controller (the weight on Q is less), then the controller will put large efforts to make the state x small; and if there are no limits on the controller, then it will drive $x(1) = 0$, regardless of what is $x(0)$.

G.6.4.4 Dynamic Programming in Continuous Time

The exact solution for DP in continuous time will result in a nonlinear partial differential equation (PDE) and is called the Hamilton-Jacobi-Bellman (HJB) equation. The procedure is to split the time interval $[t, t_f]$ into $[t, t + \Delta t]$ and $[t + \Delta t, t_f]$; and the interest is in the case when $\Delta t \to 0$. Then, identify the optimal cost-to-go $J^*(x(t + \Delta t), t + \Delta t)$; determine the "stage cost" in time $[t, t + \Delta t]$. Combine these to find the best strategy from time t. Then the result is manipulated into HJB equation. Interestingly, the HJB is a necessary and sufficient condition for optimality, but in general, it is difficult to solve this equation and, hence, numerical methods are used. If we specialize the problem to a linear system with a quadratic cost function, then it will result into continuous time LQR:

$$\dot{x}(t) = A(t)x(t) + B(t)u(t)$$

$$J = \frac{1}{2}x^T(t_f)Sx(t_f) + \frac{1}{2}\int_{t_0}^{t_f}\{x^T(t)Rx(t) + u^T(t)Qu(t)\}dt \quad \text{(AG.156)}$$

In (AG.156), S is the Gramian-weight matrix of the final state $x(t_f)$. We assume that the terminal time is fixed, and there are no bounds on u. The Hamiltonian is defined as [AG.9]:

$$H(x, u, J_x^*, t) = \frac{1}{2}[x^T(t)Rx(t) + u^T(t)Qu(t)]$$
$$+ J_x^*(x(t), t)[A(t)x(t) + B(t)u(t)] \quad \text{(AG.157)}$$

Then, we have

$$\frac{\partial H}{\partial u} = u^T(t)Q + J_x^*(x(t), t)B(t) = 0;$$

$$u^*(t) = -Q^{-1} + B^T J_x^{*T}(x(t), t) \quad \text{(AG.158)}$$

$$\frac{\partial^2 H}{\partial u^2} = Q > 0$$

With the expression of the control law introduced into (AG.157), and suspending the argument t from some variables for the sake of simplicity:

$$H = \frac{1}{2}x^T(t)Rx(t) + J_x^*(x(t), t)A(t)x(t)$$
$$\qquad\qquad\qquad\qquad\qquad\qquad \text{(AG.159)}$$
$$- \frac{1}{2}J_x^*(x(t), t)B(t)Q^{-1}B^T(t)J_x^{*T}(x(t), t)$$

We see that the boundary condition for this PDE is $J^*(x(t_f), t_f) = \frac{1}{2}x^T(t_f)Wx(t_f)$, so we can assume that $J^*(x(t), t) = \frac{1}{2}x^T(t)P(t)x(t)$; $P(.)$ is symmetric. So, we have the following conditions

$$\frac{\partial J^*}{\partial x} = x^T(t_f)P(t); \quad \frac{\partial J^*}{\partial t} = \frac{1}{2}x^T(t)\dot{P}(t)x(t) \quad \text{(AG.160)}$$

Now, we can use the HJB equation:

$$-J_t^*(x(t), t) = \min_{u(t)\in U} H(x(t), u(t), J_x^*, t) \quad \text{(AG.161)}$$

Now using the expressions from (AG.160), and (AG.159) in (AG.161), we obtain after the simplification:

$$-\dot{P}(t) = P(t)A + A^TP(t) + R - P(t)BQ^{-1}B^TP(t); \qquad \text{(AG.162)}$$
$$P(t_f) = S$$

If the solution of the differential Riccati equation is obtained as $P(t)$, then the HJB equation is satisfied, and the resulting control is optimal, then we have

$$u^*(t) = -Q^{-1}B^TP(t)x(t) = -K_{sf}x(t) \quad \text{(AG.163)}$$

This solution is a state feedback control. We can also, get the steady state differential Riccati equation as

$$PA + A^TP + R - PBQ^{-1}B^TP = 0 \quad \text{(AG.164)}$$

This equation is known as (control) ARE (CARE); and if a constant steady state solution to the differential Riccati equation exists, then it is a PSD (positive semi-definite), and symmetric solution of the CARE. If a steady state solution, P_s exists to the differential Riccati equation, then the CLCS using the feedback control $u(t) = -Q^{-1}B^TP_sx(t) = -K_{sf}x(t)$ is asymptotically stable (if the original dynamic system is stabilizable, and detectable). Also, the steady state control minimizes the infinite horizon cost function $\lim\limits_{t_f \to \infty} J, \forall H \geq 0$. Also, the P_s is PD, if and only if the system is stabilizable, and completely observable. Some insights can be gained by using the scalar LQR problem:

$$\dot{x} = ax + bu; \quad J = \int_0^\infty (Rx^2(t) + Qu^2(t))dt \quad \text{(AG.165)}$$

Then, the steady state gain and the CLCS dynamics that are stable:

$$K_s = \frac{a + \sqrt{a^2 + b^2 R/Q}}{b}; \quad \text{(AG.166)}$$

$$\dot{x} = (a - bK_s)x = -\sqrt{a^2 + b^2 R/Q}\,x$$

The observations are:

1. If $R/Q \to \infty$, $A_{cl} = -|b|\sqrt{R/Q}$, smaller Q yields faster response (good damping is important, controls the state excursion and, hence, damping);

2. If $R/Q \to 0$ (controls the actuator/control effort, and it is expensive), $K \sim (a+|a|)/b$;

   a. If $a < 0 (OL\ stable)$, $K \approx 0$, and $A_{cl} = a - bK \approx a$; and

   b. If $a > 0 (OL\ unstable)$,

   $$K \approx 2a/b, \text{ and } A_{cl} = a - bK \approx -a.$$

A good thumb rule for selection of the weighing matrices is to normalize the appropriate and related signals.

G.6.4.5 Calculus of Variation

It is an alternative approach to solve general optimization problems for continuous time systems, i.e. variational calculus. The cost function used is

$$J = \phi(x(t_f)) + \int_{t_0}^{t_f} F(x(t), u(t), t)dt \quad \text{(AG.167)}$$

$$\dot{x} = f(x, u, t); \quad x(t_0), t_0 \text{ are given}; \quad \psi(x(t_f), t_f) = 0$$

For the case of functions (studied in DP, LQR), we evaluated the gradient and equated to zero to find the stationary points; and higher order derivatives were studied to determine if it were a maximum or a minimum. We define something equivalent to the differential of a function called a variation of a functional, first the increment of a functional:

$$\Delta J(x(t), \delta x(t)) = J(x(t) + \delta x(t)) - J(x(t)) \quad \text{(AG.168)}$$

A variation of the functional is a linear approximation of this increment:

$$\Delta J(x(t), \delta x(t)) = \delta J(x(t) + \delta x(t)) + HOT.;$$

$$\text{higher order terms} \quad \text{(AG.169)}$$

The first term in (AG.169) is linear in $\delta x(t)$. The fundamental theorem of the calculus of variations: for a differentiable functional $J(x)$, if x^* is an extremal value, then the variation of J must vanish on x^*:

$$\delta J(x(t) + \delta x(t)) = 0, \ \forall \delta x \quad \text{(AG.170)}$$

The variation is computed as follows. Let the cost be given as

$$J(x(t)) = \int_{t_0}^{t_f} F(x(t))dt \quad \text{(AG.171)}$$

Then we have

$$\delta J(x(t) + \delta x) = \int_{t_0}^{t_f} \left\{ \frac{\partial F(x(t))}{\partial x(t)} \right\} \delta x\, dt + F(x(t_f))\delta t_f$$
$$- F(x(t_0))\delta t_0$$
$$\text{(AG.172)}$$

$$= \int_{t_0}^{t_f} F_x \delta x\, dt + F(x(t_f))\delta t_f - F(x(t_0))\delta t_0$$

For a cost functional,

$$J(x(t)) = \int_{t_0}^{t_f} F(x(t), \dot{x}(t), t)dt \quad \text{(AG.173)}$$

the process yields the following Euler equation:

$$\frac{\partial F(x(t), \dot{x}(t), t)}{\partial x} - \frac{d}{dt}\left(\frac{\partial F(x(t), \dot{x}(t), t)}{\partial \dot{x}} \right) = 0 \quad \text{(AG.174)}$$

We now consider some cases of terminal conditions.

1. *Free terminal time*:

   The cost function is the same as (AG.173), and $t(0)$, and $x(t(0))$ are fixed, and the terminal time t_f is free. Interestingly, independent of the terminal constraint, the extremal solution has to satisfy the same Euler equation:

   $$F_x(x^*(t), \dot{x}^*(t), t) - \frac{d}{dt} F_{\dot{x}}(x^*(t), \dot{x}^*(t), t) = 0 \quad \text{(AG.175)}$$

2. *The t_f and $x(t_f)$ are free but unrelated:*

The condition at (AG.175), and the following to be satisfied:

$$F(x^*(t_f), \dot{x}^*(t_f), t_f) - F_{\dot{x}}(x^*(t_f), \dot{x}^*(t_f), t_f)\dot{x}^*(t_f) = 0 \quad \text{(AG.176)}$$

$$F_{\dot{x}}(x^*(t_0), \dot{x}^*(t_0), t_0) = 0$$

This is a TPBV problem, since the conditions are at t_0 and t_f.

3. *The t_f and $x(t_f)$ are free but related, $x(t_f) = \Theta(t_f)$.*

The condition at (AG.175), and the following to be satisfied:

$$F_{\dot{x}}(x^*(t_f), \dot{x}^*(t_f), t_f)\left[\frac{d\Theta}{dt}(t_f) - \dot{x}^*(t_f) \right]$$
$$+ F(x^*(t_f), \dot{x}^*(t_f), t_f) = 0 \quad \text{(AG.177)}$$

This is the Transversality condition.

4. *The corner conditions, when the solutions are not smooth:*

Here, $x(t)$ is continuous, but there are discontinuities in $\dot{x}(t)$. This will happen, if there are state constraints, or there are jumps in the control signal. Here, t_0, t_f, $x(t_0)$, and $x(t_f)$ are fixed. The necessary conditions are:

$$F_x - \frac{d}{dt}(F_{\dot{x}}) = 0, \ t \in (t_0, t_f)$$

$$F_{\dot{x}}(t_1^-) = F_{\dot{x}}(t_1^+) \quad \text{(AG.178)}$$

$$F(t_1^-) - F_{\dot{x}}(t_1^-)\dot{x}(t_1^-) = F(t_1^+) - F_{\dot{x}}(t_1^+)\dot{x}(t_1^+)$$

The last two of (AG.178) are the Weierstrass-Erdmann conditions.

G.6.4.6 *Calculus of Variation with Constraints*

The cost function is the same as (AG.173) and the constraint is given as $f(x(t), \dot{x}(t), t) = 0$.

Let us have, which seems similar to Hamiltonian in DP problem:

$$g(x(t), \dot{x}(t), t) \equiv F(x(t), \dot{x}(t), t) + \lambda^T(t)f(x(t), \dot{x}(t), t) \quad \text{(AG.179)}$$

The constraint problem can be solved if the following generalized Equations are satisfied:

$$\frac{\partial g(x(t), \dot{x}(t), t)}{\partial x} - \frac{d}{dt}\left(\frac{\partial g(x(t), \dot{x}(t), t)}{\partial \dot{x}} \right) = 0 \quad \text{(AG.180)}$$

$$f(x(t), \dot{x}(t), t) = 0$$

The t_f free and $x(t_f)$ given as: $\psi(x(t_f), t_f) = 0$

$$J(x(t), t) = \phi(x(t_f), t_f)$$
$$+ \int_{t_0}^{t_f} F(x(t), \dot{x}(t), t)dt; \ \psi(x(t_f), t_f) = 0 \quad \text{(AG.181)}$$

Using the Lagrange multiplier we obtain

$$J_a(x(t), \lambda, t) = \phi(x(t_f), t_f) + \lambda^T \psi(x(t_f), t_f)$$
$$+ \int_{t_0}^{t_f} F(x(t), \dot{x}(t), t)dt \quad \text{(AG.182)}$$

The necessary conditions are:

$$F_x - \frac{d}{dt}(F_{\dot{x}}) = 0$$

$$\phi_x(t_f) + F_{\dot{x}}(t_f) = 0 \quad \text{(AG.183)}$$

$$\phi_{t_f} + F(t_f) - F_{\dot{x}}(t_f)\dot{x}(t_f) = 0; \ \phi \equiv \phi(x(t_f), \lambda, t_f)$$

In (AG.183), we have

$$\phi(x(t_f), \lambda, t_f) = \phi(x(t_f), t_f) + \lambda^T \psi(x(t_f), t_f) \quad \text{(AG.184)}$$

If t_f is fixed, and $x(t_f)$ is free, then there is no ψ, and λ, then ϕ is simplified.

G.6.5 **Optimal Control**

Let the problem be specified as

$$J = \phi(x(t_f), t_f)$$
$$+ \int_{t_0}^{t_f} F(x(t), u(t), t)dt; \ \dot{x}(t) = f(x(t), u(t), t) \quad \text{(AG.185)}$$

1. $x(t_0)$, and t_0 are fixed; t_f is free, $x(t_f)$ are fixed or free.

So, the augmented cost function is

$$J_a = \phi(x(t_f), t_f)$$
$$+ \int_{t_0}^{t_f} [F(x(t), u(t), t) + \lambda^T \{f(x(t), u(t), t) - \dot{x}\}]dt \quad \text{(AG.186)}$$

The Hamiltonian is given as

$$H(x, u, \lambda, t) = F(x(t), u(t), t) + \lambda^T f(x(t), u(t), t) \quad \text{(AG.187)}$$

The necessary conditions for $\delta J_a = 0$ are as follows

$$\dot{x} = f(x, u, t)$$

$$\dot{\lambda} = -H_x^T \qquad \text{(AG.188)}$$

$$H_u = 0$$

With the boundary condition:

$$\phi_{t_f} + F + \lambda^T f = \phi_{t_f} + H(t_f) = 0 \qquad \text{(AG.189)}$$

If $x_i(t_f)$ is fixed, then $x_i(t_f) = x_{i_f}$

If $x_i(t_f)$ is free, then $\lambda_i(t_f) = \dfrac{\partial \phi}{\partial x_i}(t_f)$

We can see the symmetry in the differential equations of the necessary conditions, (AG.188):

$$\dot{x} = f(x, u, t) = \left(\frac{\partial H}{\partial \lambda} \right)^T$$

$$\dot{\lambda} = -H_x^T = -\left(\frac{\partial H}{\partial x} \right)^T = -\left(\frac{\partial (F + \lambda^T f)}{\partial x} \right)^T \quad \text{(AG.190)}$$

$$= -\left(\frac{\partial f}{\partial x} \right)^T \lambda - \left(\frac{\partial F}{\partial x} \right)^T$$

From (AG.190), it is observed that the dynamics of the co-state are the linearized system dynamics.

2. t_f free and $\psi(x(t_f), t_f) = 0$, the more general terminal conditions

   For the cost function as in (AG.185), define the following augmented functions:

   Hamiltonian: $H(x, u, \lambda, t) = F(x(t), u(t), t) + \lambda^T f(x(t), u(t), t)$

   Augmented general terminal condition: $\phi(x(t_f), \mu, t_f) = \phi(x(t_f), t_f) + \mu^T \psi(x(t_f), t_f)$

   Then, we obtain the following necessary conditions for the optimal control

$$\dot{x} = f(x, u, t)$$

$$\dot{\lambda} = -H_x^T$$

$$H_u = 0$$

$$H(t_f) + \phi_{t_f}(t_f) = 0; \ \lambda(t_f) = \phi_x^T(t_f) = \left[\frac{\partial \phi}{\partial x}(t_f) \right]^T \quad \text{(AG.191)}$$

G.6.5.1 The LQR Variational Solution

The plant is given as

$$\dot{x}(t) = Ax(t) + Bu(t), \ x(t_0) = x_0$$

$$y(t) = Cx(t) \qquad \text{(AG.192)}$$

The cost function is given as

$$J = \int_{t_0}^{t_f} [y^T(t)\Re y(t) + u^T(t)Qu(t)]dt$$

$$+ x^T(t_f)Sx(t_f); \ R = C^T\Re C \qquad \text{(AG.193)}$$

In (AG.193), The matrices, R, $\Re$, Q and S are the respective Gramians (or the weight matrices) for the variables x, y, u, and $x(t_f)$, and presently they do not have any stochastic definitions and interpretations. The Hamiltonian is defined as

$$H = \frac{1}{2}[x^T(t)Rx(t) + u^T(t)Qu(t)]dt + \lambda^T(Ax(t) + B(t)) \quad \text{(AG.194)}$$

Applying the necessary conditions we obtain

$$\dot{x}(t) == Ax(t) + Bu(t)$$

$$\dot{\lambda}(t) = -\frac{\partial H^T}{\partial x} = -Rx(t) - A^T\lambda(t); \ \lambda(t_f) = Sx(t_f) \quad \text{(AG.195)}$$

$$\frac{\partial H}{\partial u} = 0 = Qu(t) + B^T\lambda(t) \Rightarrow u^* = -Q^{-1}B^T\lambda(t)$$

Now, we can substitute for u in (AG.192) to obtain

$$\dot{x}(t) = Ax(t) - BQ^{-1}B^T\lambda(t) \qquad \text{(AG.196)}$$

The adjoint/co-state equation is given as

$$\dot{\lambda}(t) = -Rx(t) - A^T\lambda(t) = -C^T\Re C - A^T\lambda(t) \quad \text{(AG.197)}$$

We can now combine (AG.196), and (AG.197) to obtain

$$\begin{bmatrix} \dot{x}(t) \\ \dot{\lambda}(t) \end{bmatrix} = \begin{bmatrix} A & -BQ^{-1}B^T \\ -C^T\Re C & -A^T \end{bmatrix} \begin{bmatrix} x(t) \\ \lambda(t) \end{bmatrix} \quad \text{(AG.198)}$$

The block matrix in (AG.199) is called the Hamiltonian matrix and defines the coupled closed loop dynamics for the state and co-state variables. Interestingly, though the dynamics are coupled, the initial condition of $x(t)$ is known as $x(t_0)$; and for the co-state it is known at the terminal time as is seen from (AG.195), and this leads to a TPBVP. This problem is solved as follows:

Define $\lambda(t) = P(t)x(t)$, then solve (AG.198) by obtaining the transition matrix, and use this to relate $x(t)$ to $x(t_f)$, and $\lambda(t_f)$; find $\lambda(t)$ in terms of $x(t_f)$, and then eliminate $x(t_f)$ to obtain the co-state in terms of $x(t)$. From this now we can find the equation for $P(t)$ by using

$$\dot{\lambda}(t) = \dot{P}(t)x(t) + P(t)\dot{x}(t) \qquad (AG.199)$$

Now substituting in (AG.199), the required equations from (AG.198), and then simplifying we obtain

$$\dot{P}(t) = -A^T P(t) - P(t)A - C^T \Re C + P(t)BQ^{-1}B^T P(t) \quad (AG.200)$$

In (AG.200), we have the matrix differential Riccati equation. The optimal value of $P(t)$ can be found by solving this equation from t_f with $P(t_f) = S$, in backwards direction. The control signal then is obtained as

$$\hat{u}(t) = -Q^{-1}B^T P(t)x(t) = -K(t)x(t) \qquad (AG.201)$$

So, the control law is again the state feedback control. We see that the coupled closed loop dynamics are coupled as in (AG.198), and the Hamiltonian matrix defines these; hence, the closed loop poles can be found by determining the eigenvalues of this Hamiltonian matrix, i.e. solve the characteristic equation: $\det(sI - H) = 0$.

For a SISO system, one can relate the closed loop poles to a symmetric root locus (SRL) for the TF

$$G(s) = C(sI - A)^{-1}B = \frac{N(s)}{D(s)} \qquad (AG.202)$$

The closed loop poles for the design of (AG.198) are given by the LHS plane roots of

$$D(s)D(-s) + \Re Q^{-1} N(s)N(-s) = 0 \qquad (AG.203)$$

This can be plotted using the standard root locus rules, and the locus drawn is symmetric wrt to both the real and imaginary axes.

Interpretation of the design of (AG.198) and the SRL are briefly given here:

1. For a SISO, define $\frac{\Re}{Q} = 1/r$ and consider r as a tuning parameter. Then, when $r \to \infty$, the control cost is high, and we have from (AG.203), $D(s)D(-s) = 0$; thus, the closed loop poles are the stable roots of the open loop system (which are already in the LHS plane), and this is the reflection about the $j\omega$ axis of the unstable open loop poles.

2. When $r \to 0$, then it is a low cost control situation, and from (AG.2020, we have $N(s)N(-s) = 0$, here, the closed loop go to m finite system that are in the LHS plane (or the reflections of the system zeros in the LHS plane), and the system zeros at infinity (there are $n-m$). The poles that tend to infinity do via very specific paths that form a Butterworth pattern.

3. The LQR/SRL selects the closed poles that try a balance between the system errors (lower Q) and the control effort (higher Q); one can see that in the design iteration with tuning of Q (r), the poles move along the SRL, and the design can focus on the system's performance aspects.

4. The LQR also exhibits very good stability margins. Let us represent the problem in a simplified form as;

$$J = \int_0^\infty (y^T y + \rho u^T u)dt; \ \dot{x} = Ax + Bu; y = Cx; R = C^T C;$$
$$(AG.204)$$
$$\text{since } \Re = I.$$

The TFs are given as:

$$\text{Loop TF: } L(s) = K(sI - A)^{-1}B \qquad (AG.205)$$

$$\text{Cost TF: } C_j(s) = C(sI - A)^{-1}B \qquad (AG.206)$$

The relationship between the open loop cost $C_j(s)$ and the closed loop return difference $I + L(s)$ is given as

$$[I + L(-s)]^T [I + L(s)] = 1 + \frac{1}{\rho} C_j^T(-s)C_j(s) \quad (AG.207)$$

This is known as the Kalman frequency domain equality. In terms of the magnitude (AG.207) becomes

$$|1 + L(j\omega)|^2 = 1 + \frac{1}{\rho}|C_j(j\omega)|^2 \geq 1 \qquad (AG.208)$$

The Nyquist plot of $L(j\omega)$ will always be outside the unit circle that is centred at $(-1,0)$. We can recall the Nyquist stability theorem here: if the loop TF $L(s)$ has P number of poles in the RHS plane, then for closed loop stability, the locus of $L(j\omega)$ ($\omega{:}-\infty$ to ∞) must encircle the critical point $(-1,0)$, P times in the counter clockwise direction. If the open loop is stable, then, any change in the gain, K yields a stable closed loop system. For, and unstable system, the change in the gain, of $(K/2$ to $\infty)$ results in a stable closed loop system and gain margin would be $(1/2, \infty)$, and the phase margin of at least $\pm 60°$.

G.6.5.2 Properties of Optimal Control

If $F = F(x, u)$, and $f = f(x, u)$ do not explicitly depend of time t, then the Hamiltonian H is at least piecewise constant:

$$H = F(x,u) + \lambda^T f(x,u) \qquad \text{(AG.209)}$$

From (AG.209), by total differentiation we obtain

$$\frac{dH}{dt} = \frac{\partial H}{\partial t} + \left(\frac{\partial H}{\partial x}\right)\frac{dx}{dt} + \left(\frac{\partial H}{\partial u}\right)\frac{du}{dt} + \left(\frac{\partial H}{\partial \lambda}\right)\frac{d\lambda}{dt} \qquad \text{(AG.210)}$$

$$= H_x f + H_u \dot{u} + H_\lambda \dot{\lambda}$$

Using the standard Euler equation (in terms of Hamiltonian) we get

$$\frac{dH}{dt} = H_x f + H_u \dot{u} + H_\lambda \dot{\lambda}$$

$$= -\dot{\lambda}^T f + f^T \dot{\lambda} + H_u \dot{u} \qquad \text{(AG.211)}$$

$$= H_u \dot{u} = (0)\dot{u} = 0$$

From (AG.211), we see that the Hamiltonian is at least piecewise constant.

For free final time problem, we have from the transversality condition $\phi_t + H(t_f) = 0$, if ϕ is not a function of time t, then again H is constant. In case the solution has a corner (not induced by an intermediate state variable constraint), then H, λ, and H_u all are continuous across the corner.

If there is an interior point state constraint $N(x(t_1), t_1) = 0$, then by using the LM method and calculus of variation we obtain the following conditions, showing the discontinuity of the co-state and the Hamiltonian:

$$\lambda^T(t_1^-) = \lambda^T(t_1^+) + \upsilon^T N_x(t_1)$$

$$H(t_1^-) = H(t_1^+) - \upsilon^T N_x(t_1) \qquad \text{(AG.212)}$$

G.6.5.3 Constrained Optimal Control (on the Input u(t))

Let the modified cost function, with the LM that allows to incorporate the constraint V, of the control input $u(t)$, be given as

$$J_a = \phi(x(t_f), t_f) + \int_{t_0}^{t_f} [H - \lambda^T \dot{x} + \mu^T V] dt; \qquad \text{(AG.213)}$$

$$V(x, u, t) \le 0; \; u(t) \in U$$

We obtain the following necessary conditions

$$\dot{x} = f(x, u, t)$$

$$\dot{\lambda} = -(H_a)_x^T \qquad \text{(AG.214)}$$

$$(H_a)_u = 0; \quad \mu_i V_i(x, u, t) = 0$$

In (AG.214), the augmented Hamiltonian is defined as

$$H_a = F + \lambda^T(t) f + \mu^T(t) V \qquad \text{(AG.215)}$$

G.6.5.4 Pontryagin's Minimum Principle

We have the normal situation of $H_u = 0$ (from the calculus of variation), if the control constraints are not active, but with the constraint on control, at the constraint boundary we could have H_u different from zero, and it could be that $H_u > 0$ or $H_u < 0$, and this would depend on the direction (i.e. sing) of the permissible δ_u. The stronger condition would be that H must be minimized over the set of all possible u. So, the necessary condition of $H_u = 0$ should be replaced by

$$u^*(t) = \arg\left\{\min_{u(t) \in U} H(x, u, \lambda, t)\right\} \qquad \text{(AG.216)}$$

Thus, the problem is to find minimizing control inputs $u(t)$ given the constraints. This is the problem of Pontryagin's minimum principle (often, this is known as the Pontryagin's maximum principle, depending upon the context of the cost to be maximized), and it would handle the "edges" as well and apply to all the constrained control problems.

Consider the minimum time-fuel problem:

$$\min J = \int_0^{t_f} (1 + b|u(t)|) dt \qquad \text{(AG.217)}$$

The aim is to propel the state to the origin with minimum cost. In the cost function (AG.217), the use of the integration of $|u(t)|$ sums up the fuel used (e.g. useful in spacecraft problems); whereas in the usual optimal and variational control problem, we use integral of $u^2(t)$, which sums up the power used. The dynamics are expressed as:

$$\text{Define, } x_1 = y, \; x_2 = \dot{y} \Rightarrow \dot{x}_1 = x_2, \dot{x}_2 = u \quad \text{(AG.218)}$$

This results in

$$H(u) = b|u| + \lambda_2 u \qquad \text{(AG.219)}$$

The application of the PMP with the constraints suggest that we find u that minimizes H(u), and the resulting control law is given as

$$u(t) = \begin{cases} -u_m, & b < \lambda_2(t) \\ 0, & -b < \lambda_2(t) < b \\ u_m, & \lambda_2(t) < -b \end{cases} \quad \text{(AG.220)}$$

The solution has a Bang-off-Bang nature. As a special case when $b \rightarrow 0$ (fuel problem is deleted), the solution of the minimum time problem is that the control signal is just Bang-Bang. The coasting is fuel efficient, but it takes a long time.

Now, consider the minimum time problem:

The aim is to determine the control input sequence

$$M_i^- \leq u_i(t) \leq M_i^+ \quad \text{(AG.221)}$$

So that the following system is driven from an arbitrary state x_0 to the origin to minimize the maneuver time

$$\dot{x} = A(x,t) + B(x,t)u; \min J = \int_0^{t_f} dt \quad \text{(AG.222)}$$

The Hamiltonian is formed as

$$H = 1 + \lambda^T(t)[A(x,t) + B(x,t)u]$$
$$= 1 + \lambda^T(t)A(x,t) + \sum_{i=1}^m \lambda^T(t)b_i(x,t)u_i(t) \quad \text{(AG.223)}$$

Using PMP, we get the Bang-Bang control law:

$$u_i(t) = \begin{cases} M_i^+, & \text{if } \lambda^T(t)b_i(x,t) < 0 \\ M_i^-, & \text{if } \lambda^T(t)b_i(x,t) > 0 \end{cases} \quad \text{(AG.224)}$$

Next, one has to solve for the co-state:

$$\dot{\lambda} = -H_x^T = -\left(\frac{\partial A}{\partial x} + \frac{\partial B}{\partial x}u\right)^T \lambda \quad \text{(AG.225)}$$

Some properties are: (i) if the real parts of the eigenvalues of $A \leq 0$, then an optimal control exists, (ii) muse eliminate unstable plants (from this statement) since the control is unbounded, (iii) if an external control exists, then it is unique,

the satisfaction of the PMP is both necessary and sufficient for time-optimal control of a LTI system.

Now, consider the minimum fuel problem:

Let the system be given as in (AG.222), and the control constraint as in (AG.221). Define the cost function as

$$\min J = \int_0^{t_f} \left(\sum_{i=1}^m c_i |u_i(t)| \right) dt \quad \text{(AG.226)}$$

The use of PMP results in the following control law:

$$u_i^*(t) = \begin{cases} M_i^-, & \text{if } c_i < \lambda^T(t)b_i \\ 0, & \text{if } -c_i < \lambda^T(t)b_i < c_i \\ M_i^+, & \text{if } \lambda^T(t)b_i < -c_i \end{cases} \quad \text{(AG.227)}$$

Consider the minimum energy problem:

With the fixed final time and terminal constraints

$$\min J = \frac{1}{2} \int_0^{t_f} (u^T Q u) dt; \dot{x} = A(x,t) + B(x,t)u \quad \text{(AG.228)}$$

$$H = \frac{1}{2} u^T Q u + \lambda^T[A(x,t) + B(x,t)u] \quad \text{(AG.229)}$$

Without constraints on the control we have: $u = -Q^{-1}B^T\lambda(t)$. With bounded controls we have to solve

$$u^*(t) = \arg\left\{ \min_{u(t) \in U} \left[\frac{1}{2} u^T Q u + \lambda^T B(x,t)u \right] \right\} \quad \text{(AG.230)}$$

The control inputs are

$$u_i(t) = \begin{cases} M_i^-, & \text{if } \tilde{u}_i < M_i^- \\ \tilde{u}_i, & \text{if } M_i^- < \tilde{u}_i < M_i^+ \\ M_i^+, & \text{if } M_i^+ < \tilde{u}_i \end{cases}; \tilde{u}_i = -Q_{ii}^{-1}\lambda^T b_i \quad \text{(AG.231)}$$

G.6.6 Stochastic Optimal Control

The aim is to design optimal compensators for systems with noisy observations. We first assume that the states are perfectly measurable. The system dynamics are given as

$$\dot{x}(t) = Ax(t) + Bu(t) + Gw(t) \quad \text{(AG.232)}$$

In (AG.232), $w(t)$ is Gaussian white noise with zero mean and the covariance $R_w = E\{w(t)w^T(t)\}$. The initial conditions are also random variables; $E\{x(t_0)\} = 0$, and $X_0 = E\{x(t_0)x^T(t_0)\}$. The modified cost function is given as

$$J_s = E\left\{\begin{array}{l} \dfrac{1}{2}x^T(t_f)Sx(t_f) \\[2mm] +\dfrac{1}{2}\displaystyle\int_{t_0}^{t_f}(x^T(t)Rx(t)+u^T(t)Qu(t))dt \end{array}\right\}; \quad \text{(AG.233)}$$

In (AG.233), the matrices S, R, and Q are still the Gramian-weight matrices. Because the noise process is white, the correlation times-scales are very short as compared to the system dynamics. Also, as we know, the system state $x(t)$ encompasses all the past information about the system. Here, the case of the optimal controller is identical to the deterministic case. If the noise process is not white, then we need to incorporate a model of this process, also called a shaping filter, which will be modeling/capturing the spectral content (temporal correlation) of the noise. The shaping filter will be another dynamical system with its states as x_s.

G.6.6.1 Disturbance Feedforward

Consider stochastic LQR, We can modify the state weighting matrix as

$$R_a = \begin{bmatrix} R & 0 \\ 0 & 0 \end{bmatrix}; u = -[K \ K_s]\begin{bmatrix} x \\ x_s \end{bmatrix} \quad \text{(AG.234)}$$

In (AG.234), the solution is same as in the case of the deterministic case, but requires that the shaping filter states be available, which are anyway fictitious and need to be estimated. The gain K is independent of the properties of the disturbance. However, the control inputs based on the filter's states will tend to improve the performance of the system, i.e. providing disturbance feedforward (control). The specific initial conditions do not affect the LQR controller; however, they do affect the cost-to-go from the initial time, and this cost:

$$J_s = \frac{1}{2}trace\left\{P(t_0)X(t_0)+\int_{t_0}^{t_f}(P(t)GR_wG^T)dt\right\};$$

$$X = E\{x(t)x^T(t)\}; X_0 \quad\quad\quad\quad \text{(AG.235)}$$

$$= X(t_0) = X(0) = E\{x(t_0)x^T(t_0)\}$$

We can see from (AG.236) that the first term is the same as the cost-to-go from the uncertain initial condition.

The second is the increase in the cost due to the presence of the noise in the system. When the t_0 is 0, and the $t_f \to \infty$, then the performance would be infinite. The cost can be modified to consider the time-average:

$$J_s = \lim_{t_f \to \infty}\frac{1}{t_f - t_0}J_s \quad \text{(AG.236)}$$

There is no effect on the necessary conditions, since it is still a fixed end time problem; the initial conditions are irrelevant. For LTI system with stationary process noise and a well-posed time-invariant control problem, i.e. $u(t) = -K_s x(t)$, we obtain the following (in steady state, s

$$J_s = \frac{1}{2}trace\{[R + K_s^T QK_s]X_s\}; X_s = \lim_{t \to \infty}X(t)$$

$$\equiv \frac{1}{2}trace\{P_s GR_w G^T\} \quad \text{(AG.237)}$$

G.6.6.2 Full Control Problem

The aim is to design an optimal controller with incomplete and noisy measurements. Let us have the system as

$$\dot{x} = Ax + Bu + Gw; z = C_z x; y = C_y x + v; \quad \text{(AG.238)}$$
$$w \sim N(0, R_w), v \sim N(0, R_n)$$

$$X_{e0} = E\{(x_0(t)-\bar{x}_0(t))(x_0(t)-\bar{x}_0(t))^T\};$$
$$x_0(t) \sim N(\bar{x}_0, X_{e0}); R_v = E\{v(t)v^T(t)\}. \quad \text{(AG.239)}$$

The cost function is defined as

$$J = E\left\{\begin{array}{l} \dfrac{1}{2}x^T(t_f)Sx(t_f) \\[2mm] +\dfrac{1}{2}\displaystyle\int_{t_0}^{t_f}[z^T(t)R_z z(t)+u^T(t)Qu(t)]dt \end{array}\right\} \quad \text{(AG.240)}$$

In (AG. 240), R_z is the Gramian-weight matrix for the variable z. The idea is to determine the control input as a function of the output that minimizes the cost function J; this is a stochastic optimal output feedback problem:

$$u(t) = f[y(\tau), t_0 \le \tau \le t]; \quad t_0 \le t \le t_f \quad \text{(AG.241)}$$

The result is the LQG controller that would use the appropriate LQE (linear quadratic estimator) to obtain the optimal state estimates of $x(t)$ from the noisy measurements $y(t)$, using the observer gain $L(t)$, and use LQR to obtain the optimal feedback control $u(t) = -Kx(t)$, and this is also the use of the separation principle.

The regulator is given as

$$u(t) = -K(t)\hat{x}(t)$$
$$\dot{P}(t) = -A^T P(t) - P(t)A - C_z^T R_z C \qquad \text{(AG.242)}$$
$$+ P(t)BQ^{-1}B^T P(t); \ P(t_f) = S$$

The estimator is given as

$$\dot{\hat{x}} = A\hat{x} + Bu + L(t)(y(t) - C_y\hat{x}(t)),$$
$$\hat{x}(t_0) = \bar{x}_0 \text{ and } X_e(t_0) = X_{e0} \qquad \text{(AG.243)}$$

$$\dot{X}_e(t) = AX_e(t) + X_e(t)A^T + GR_wG^T$$
$$- X_e(t)C_y^T R_v^{-1} C_y X_e(t); \qquad \text{(AG.244)}$$
$$L(t) = X_e(t)C_y^T R_v^{-1}$$

The compact form of the compensator is then obtained as

$$\dot{x}_c = A_c x_c + B_c y; \ u = -C_c x_c; \ x_c = \hat{x}$$
$$A_c = A - BK(t) - L(t)C_y \qquad \text{(AG.245)}$$
$$B_c = L(t); \ C_c = K(t)$$

This controller design is valid for SISO as well as MIMO systems, and for time-varying plants also; and if we use the steady state regulator and estimator gains for an LTI system, the compensator will be constant.

G.6.6.2.1 *The Infinite Horizon LQG*

We use the LTI plant, and use the time-averaged cost as

$$\bar{J} = \lim_{\substack{t_0 \to -\infty \\ t_f \to \infty}} \frac{1}{t_f - t_0} J \qquad \text{(AG.246)}$$

In (AG.246), the $-\infty$ is for the initial time (t_0) in order to settle down the estimator. Here, the optimality conditions are still the same. Then, in the final the steady state solutions are given as

$$K_s = Q^{-1}B^T P_s \text{ and } L_s = X_{es}C_y^T R_v^{-1} \qquad \text{(AG.247)}$$

The performance can be evaluated by using the steady state solutions of (AG.242) and (AG.244); however, the cost (AG.246) has to account for the estimation error, since the latter would degrade the performance of the overall system (CLCS). There are two types of errors: (i) regulation error and (ii) estimation error.

G.6.6.2.2 *Some Properties and Interpretations*

1. $\bar{J}_s = Trace\{P_s GR_w G^T + X_{es}K_s^T Q K_s\}$ (AG.248)

2. Even if there is no controller activity, the performance is lower bounded by the estimation error, X_{es}, similarly in case of the regulator error, the performance is lower bounded by it.

3. Thus, there is no point in making one faster than the other, any further increase in the authority of one over the other, brings diminishing returns. So, it is prudent to make the performance shares from the both comparable.

4. Due to the application and the validation of the separation theorem, the closed loop poles of the system consists of the union of the regulator and the estimator (observer) poles. Thus, these designs can be done independently and the combination will obtain the CLCS stability.

5. In a servo (tacking) approach, one can feedback $e(t) = r(t) - y(t)$, instead of only the output $y(t)$.

6. The performance optimality is more than just the application of the separation principle. It can be established/evaluated as follows: the cost is rewritten in terms of the estimator states and the dynamics, the designing a stochastic LQR with full state feedback. The final cost will be

$$J = E\left\{ \frac{1}{2} \int_{t_0}^{t_f} [\hat{x}^T(t)R_x\hat{x}(t) + u^T(t)Qu(t)]dt \right\}$$
$$+ \frac{1}{2}\int_{t_0}^{t_f} [trace\{R_x X_e\}]dt \qquad \text{(AG.249)}$$

In (AG.249), since the last term is independent of $u(t)$, we need to determine control $u(t)$ to minimize only the first term. We have some more considerations as equation (AG.243) with the innovations process (or in fact the observer/estimator residuals) as

$$r(t) = y(t) - C_y x(t); \ r(t) \sim N(0, R_v + C_y X_e C_y^T) \qquad \text{(AG.250)}$$

Then (AG.243) is rewritten as

$$\dot{\hat{x}} = A\hat{x} + Bu + L(t)r(t) \qquad \text{(AG.251)}$$

Thus, (AG.251) is (like) an LTI with $r(t)$ as a process noise. Thus, the problem is to minimize the cost in (AG.249), subject to the dynamics in (AG.251). Interestingly, the solution is independent of the driving process noise: $u(t) = -K(t)\hat{x}(t)$; and the gain is obtained from LQR with A, B, R, and Q, as in the original problem. This combination of LQE/LQR will give performance optimal result.

G.6.6.3 LQG Robustness

When the combination of optimal estimator and the optimal regulator are used to design a controller the strategy is the LQG design: (i) can be designed using the principle of separation, and (ii) the stability of the CLCS is guaranteed. Then the analyst has to focus on the performance only. The designed controller must be able to tolerate some modeling errors: (i) because of the use of linearized models, (ii) some parameters are not known accurately, and (iii) some higher frequency dynamics are ignored. The eigenvalues can be evaluated for the designed CLCS, but it is very difficult to predict how the closed loop poles will shift as a function of errors in the plant model, since the perturbed closed loop eigenvalues are not the union of the regulator and the estimator poles. One can do the parametric analysis by perturbing the dynamics and then evaluating the closed loop eigenvalues. In frequency domain, one can use the "closeness" of the loop TF L(s) to the critical point as a measure of "closeness" to changing the number of encirclements of the root locus of $L(s)$. Form this we can see that the perturbations can destabilize the system. Let us have the $G_n(s)$ as the nominal model, $L_n(s) = G_n(s)G_c(s)$; then, the large nominal errors when $L_n \sim -1$, should not be allowed to occur.

Hence, the LQG is a good way to design a controller for a nominal system. However, if there are perturbations in the plant, then for the actual (disturbed) plant the stability/performance cannot be guaranteed. One should employ the sensitivity plots (sensitivity TF) for analysis to study this aspect. One can use the lower controller BW, but then BW is sometimes required to stabilize the system. Such limitations prompt the control design engineers to go for alternative design methods like H_∞, and μ-synthesis.

G.6.7 Multivariable Frequency Response

In the MIMO control problems the system $G(s)$ is described by a $p \times m$ TF matrix. The TFM is still given as $G(s) = C(sI–A)^{-1}B + D$. If we plot the individual TFs, then the coupling between various interconnected sub-systems is ignored, and it does not shade any light on the complete behavior of the system. Then, basic idea would be to restrict all the inputs to the same frequency, and then see how the system responds to this frequency, then study the changes in the responses with the frequency. For the MIMO systems, the SVD (singular value decomposition) analysis is found to be effective, i.e. obtain the SVD of the matrix $G(s)$ at each frequency $s = j\omega$. The singular values give a detailed description of how a matrix (i.e. the system G) acts on

an input vector at a particular frequency. We consider the inputs/outputs as follows:
Inputs:

$$u = u_c e^{j\omega t}; \Rightarrow y = G(s)|_{s=j\omega} u; \Rightarrow y = y_c e^{j\omega t} \quad \text{(AG.252)}$$

Thus, we need to analyse only the $y_c = G(j\omega)u_c$. One can use the following steps:

1. Fix the size of $||u_c||_2 = 1$ of the input, and study effect on the largeness of the output.
2. Analyze the relation $y_c = G_\omega u_c$; $G_\omega = G(s = j\omega)$.
3. Define the maximum and minimum amplifications as

$$\bar{\sigma} = \max_{\|u_c\|_2=1} \|y_c\|_2; \quad \underline{\sigma} = \min_{\|u_c\|_2=1} \|y_c\|_2 \quad \text{(AG.253)}$$

Since the $G(s)$ changes with the frequency, the singular values (norms) will also change with frequency. The SVD gives the singular values and the associated eigenvectors, the latter ones determine the phases of the each of the components of the input vector u_c. Thus, the MIMO frequency responses are the plots of the maximum and minimum singular values.

G.6.8 H_∞ Control

The H-infinity (HI) control philosophy is based on the HI norm. The H_2 norm is concerned with the overall response, whereas the H_∞ (HI) is concerned with the peaks in the frequency response. Interestingly, the property $\|GH\|_\infty \le \|G\|_\infty \cdot \|H\|_\infty$ is valid for HI norm, but it is not so in general for the $||GH||_2$. The main idea in the HI controller design is to determine a stabilizing controller such that the peaks in the TFM (response) of interests are kept under certain number:

$$\left|\max_\omega \bar{\sigma}(T(j\omega))\right| \equiv \|T(s)\|_\infty < \lambda; \lambda \text{ is a small constant} \quad \text{(AG.254)}$$

For H_2 norm:

In contrast, for the H_2 control we are looking for a design of a stabilizing controller such that $||T(s)||_2$ is reduced as much as possible. In fact, the H_2 control and LQG are the same designs.

Let us assume that we have $G(s) = C(sI − A)^{-1}B + D$, and the real parts of the eigenvalues of A are all negative; hence, the system $G(s)$ is stable. Then, the H_2 norm requires a strictly proper system, i.e. $D = 0$:

$$\dot{x} = Ax + Bu; \; y = Cx \quad \text{(AG.255)}$$

Then, with the observability and the controllability Gramians defined as

$$A^T P_0 + P_0 A + C^T C = 0 \Leftrightarrow P_0 = \int_0^\infty (e^{A^T t} C^T C e^{At}) dt \tag{AG.256}$$

$$A P_c + P_c A^T + BB^T = 0 \Leftrightarrow P_c = \int_0^\infty (e^{At} BB^T e^{A^T t}) dt$$

Then we have for the norm of $G(s)$

$$\|G(s)\|_2^2 = trace\{B^T P_0 B\} = trace\{C P_c C^T\} \tag{AG.257}$$

For H_∞ norm:
Define the Hamiltonian matrix as

$$H = \begin{bmatrix} A + B(\gamma^2 I - D^T D)^{-2} D^T C & B(\gamma^2 I - D^T D)^{-1} B^T \\ -C^T (I + D(\gamma^2 I - D^T D)^{-2} D^T) C & -(A + B(\gamma^2 I - D^T D)^{-2} D^T C)^T \end{bmatrix} \tag{AG.258}$$

Interestingly enough, the following is true:

$\|G(s)\|_\infty < \gamma$, iff $\bar\sigma(D) < \gamma$ and H has no eigenvalues on the $j\omega$-axis. One keeps tuning the γ and find its smallest value for the maximum value of the $G(s)$ to be less than this constant. A bisection search algorithm may be used for searching the value of γ.

G.6.8.1 Riccati Equation Test

If $D = 0$, then the Hamiltonian matrix simplifies to

$$H = \begin{bmatrix} A & \dfrac{1}{\gamma^2} BB^T \\ -C^T C & -A^T \end{bmatrix} \tag{AG.259}$$

Another test is that there exists a X (some symmetric matrix), so that the following are true:

$$A^T X + XA + C^T C + \frac{1}{\gamma^2} XBB^T X = 0;$$

$$A + \frac{1}{\gamma^2} BB^T X \text{ is stable} \tag{AG.260}$$

One can compare the ARE (AG.260), with the one that we got in the case of an LQR control design

$$A^T P + PA + C^T C - \frac{1}{\rho} PBB^T P = 0;$$

(A, B, C) is stabilizable/detectable $\tag{AG.261}$

There is a sign difference, in the quadratic term in the case of ARE of (AG.260), and the solutions of the both could behave quite differently.

G.6.8.2 H_∞ Synthesis

For this design, the systems dynamics are represented in the generalized form with two inputs (w and u), and the two outputs (z and y); here the output y is fed back with the designed controller: $u = G_c(s)y$. These variables are: z-performance output, w-disturbance or the reference input, y-sensor outputs (measurements/observations), and u-actuator input (i.e. control law). The generalises plant is defined as

$$P(s) = \begin{bmatrix} P_{zw}(s) & P_{zu}(s) \\ P_{yw}(s) & P_{yu}(s) \end{bmatrix} \tag{AG.262}$$

We can then get the following CLTF between z and w as:

$$P_{zwcl}(s) = P_{zw}(s) + P_{zu}(s) G_c(s)(I - P_{yu}(s) G_c)^{-1} P_{yw}(s) \tag{AG.263}$$
$$= F_l(P, G_c)$$

This is called a (lower) LFT (linear fractional transformation), and the aim of the design is to determine $G_c(s)$ to stabilize the CLCS, such that HI norm of the LFT: $\|F_l(P, G_c)\|_\infty < \gamma$. One can use the bisection method that is also called γ-iteration to find the smallest value of γ to satisfy the LFT inequality. For our system (AG.238), we have the following generalized plant description:

$$P(s) = \begin{bmatrix} P_{zw}(s) & P_{zu}(s) \\ P_{yw}(s) & P_{yu}(s) \end{bmatrix} = \begin{bmatrix} A & |B_w & B \\ \hline C_z & |0 & D_{zu} \\ C_y & |D_{yw} & 0 \end{bmatrix} \tag{AG.264}$$

In (AG.264), we have the following assumptions: (i) (A, B, C_y), and (A, B_w, C_z) are stabilizable/detectable, (ii) $D_{zu}^T [C_z \ D_{zu}] = [0 \ I]$, and (iii) $\begin{bmatrix} B_w \\ D_{yw} \end{bmatrix} D_{yw}^T = \begin{bmatrix} 0 \\ I \end{bmatrix}$; these are essential and would simplify the design. Then, there would exist a stabilizing controller $G_c(s)$ that would satisfy the condition on the LFT norm: $\|F_l(P(s), G_c(s))\|_\infty < \gamma$ iff the following are true:

1. There exists $X \geq 0$ that solves the following ARE

$$A^T X + XA + C_z^T C_z + X(\gamma^{-2} B_w B_w^T - BB^T) X = 0; \tag{AG.265}$$

$$Real\{\lambda_i[A + (\gamma^{-2} B_w B_w^T - BB^T) X]\} < 0, \ \forall i$$

2. There exists $Y \geq 0$ that solves the following ARE

$$AY + YA^T + B_w^T B_w + Y(\gamma^{-2} C_z^T C_z - C_y^T C_y)Y = 0;$$

$$\text{Real}\{\lambda_i[A + Y(\gamma^{-2} C_z^T C_z - C_y^T C_y)]\} < 0, \ \forall i$$

(AG.266)

3. $\rho(XY) < \gamma^2$; the spectral radius is defined as

$$\rho(A) = \max_i |\lambda_i(A)| \qquad \text{(AG.267)}$$

From the foregoing solution for X, and Y we obtain the following compensator:

$$G_c(s) = \begin{bmatrix} A + (\gamma^{-2} B_w B_w^T - BB^T)X - ZYC_y^T C_y & ZYC_y^T \\ -B^T X & 0 \end{bmatrix};$$

$$Z = (I - \gamma^{-2} YX)^{-1}$$

(AG.268)

One can put the LQG (H_2) design also into the generalized framework as is done for the HI synthesis.

From the central controller we obtain the following dynamics:

$$\dot{\hat{x}} = A\hat{x} + B_w(\gamma^{-2} B_w^T X)\hat{x} + Bu + L(y - C_y \hat{x}); \ L = ZYC_y^T \quad \text{(AG.269)}$$

The form in (AG.269) is similar to the one that would have been obtained from a Kalman filter or observe (for state estimation) for the LQG design. However, there is an additional (second) term in (AG.269), which is the estimate of the worst-case disturbance to the system. Also, the separation rule as such does not exist for the H_∞ controller.

G.6.9 Adaptive Control/Learning Systems

The adaptive control theory/systems can be put in two classes: (i) model-based, and (ii) data-driven adaptive control. In (i) the controllers are fully or partially based on a given model of the system; i.e. all or part of the controller is based on a model of the system. In the fully model-based adaptive control, both the controller and the adaptation filters are based on the model [AG.10]: MRAC, MM adaptive switching control, concurrent adaptive control, adaptive regulation or disturbance rejection, l_1 adaptive control, speed-gradient-based control, passivity-based adaptive control, composite (or combined) adaptive control, retrospective cost adaptive control, set-theoretic-based adaptive controller, and robust adaptive control. In case of (ii) the controllers are fully based on learning from direct interaction with the environment: data-driven extremum seekers (ES), model-free reinforcement learning (RL), genetic algorithms, neural network and deep learning algorithms, kernel function-based parameterization, particle filters, and model-free iterative learning control (ILC) [AG.10]. In the

partially model-based/learning-based sub-class, the controller is based on a model of the system, but the adaptation filters are data-driven, say by learning based on machine learning algorithms; the adaptation layer is based on direct interaction with the environment/data: ES-based modular adaptive control, Gaussian-process modular adaptive control, NN-based modular adaptive control, model-based RL control, model-based approximate/adaptive dynamic programming (ADP) control, and model-based neurodynamic programming (NDP) control [AG.10]. In fact, the fully data-driven methods can learn (or expected to learn) the best control policies by direct interaction with the plant without any prior knowledge about the model of the system. This flexibility requires extensive measurements, probing of the system/data collection, and high computational power.

An important point is that of the stability, and performance assurances that are easy to achieve and analyze in the fully model-based (and the learning-based) methods; but not in the fully data-driven approaches. The issues and challenges for the data-driven approaches, in study and application are: (i) hybrid dynamical systems, (ii) the learning algorithms often rely on a proper choice of weight functions, (iii) choice of the bases functions in ANN algorithms. These fully data-driven approaches would benefit from the theoretical tools in dynamical systems theory, as well as, nonlinear and robust control theory. A feature of adaptive control is asymptotic convergence to zero for tracking errors but not for parameters, but without persistent excitation the equilibrium point is not unique; of course depending on the design goals, parameter convergence to true values may not be required. Thus, the aspects of stability, performance, and convergence are interesting and challenging problems to be studied in the context of fully data-driven adaptive control systems. One approach of data-driven algorithm (though, based on the conventional and not on the soft-computing method) is discussed next.

G.6.9.1 Online Learning for Inverse Dynamics

Most control designs need mathematical models of the dynamics of the plants that are often difficult to obtain for the complex systems. An alternative to modeling is the approach of machine learning algorithm for supervised learning of this complex behavior. An online incremental learning algorithm based on Gaussian process (GP) learning is briefly discussed here. The approach does not need any knowledge of the structure of the dynamics of the plant.

The dynamics of say, a robotic manipulator characterizes the relationship between its motion (position, velocity and acceleration), and the joint torques that cause these motions [AG.11]:

$$M(q)\ddot{q} + C(q,\dot{q}) + G(q) = T_o + T_a \qquad \text{(AG.270)}$$

In (AG.270), q is the joint position vector, $M(q)$ is the inertia matrix, $C(.,.)$ is the centripetal, and Coriolis torque vector, $G(.)$ is the gravity loading vector, T_o is the torque, and T_a is the additional torque due to forces that are not included in the rigid body, i.e. friction and contact forces. In the inverse dynamics approach, the model-based controllers apply the joint space dynamics to cancel the nonlinear and coupling effects that are due to the manipulator. The control signal is computed as

$$u = \hat{M}(q)\ddot{q} + \hat{C}(q,\dot{q}) + \hat{G}(q) \qquad (\text{AG.271})$$

Normal adaptive control schemes need an adequate knowledge of the structure of the dynamic model and are prone to effects of the unmodeled dynamics. The data-driven approaches seem to be very suitable for such applications, in which the inverse dynamics relationship of a manipulator is learnt. In such an approach, a "model" is inferred that describes the observed data as closely as possible; some methods are: (a) global methods: (i) Gaussian process regression (GPR), and (ii) support vector regression (SVR); and (b) locally weighted projection regression (LWPR). As we know, the GP is characterized by its mean and covariance, which provide the priors for the Bayesian inference. This GP prior is updated with the training data to yield the posterior GP, i.e. the parameters of the mean and covariance, known as the hyper-parameters; and is done by selecting these parameters that maximize the probability of observing the training data. The SPGP (sparse pseudo-input GP) is an approximation of the full procedure of GP. The SPGP is the process of simultaneously finding an active set of point locations for "pseudo-inputs," denoted by X while learning the hyper-parameters of the GP in a joint optimization scheme. The sparse online Gaussian process (SOGP) method reduces the computational load of GPR by keeping track of a sparse set of inputs named Basis Vectors (BV); and incorporates a method of incrementally processing data.

G.6.9.2 The Sparse Online GP Algorithm

The aim is to find the function f that maps the inputs X to their target values y:

$$y \sim N(0, K(X,X) + \sigma_n^2 I) \qquad (\text{AG.272})$$

In (AG.272), $K(.,.)$ is the covariance matrix that is composed of the covariances, $k(x, x')$ that are evaluated at all the pairs of the training points. For learning the inverse dynamics of a robot manipulator, the X are set to the joint angles, velocities and accelerations $(q, \dot{q}, \ddot{q})$ and y is the joint torque T_o as in (AG.270). The covariance function is given as

$$k(x, x') = \theta_1 e^{(-\theta_2 \|x - x'\|^2)} \qquad (\text{AG.273})$$

The hyper-parameters, in (AG.273), can be learned for a particular data set by maximizing the log likelihood using an optimization method. The joint distribution of the output and the predicted function is given by

$$\begin{bmatrix} y \\ f^*(x^*) \end{bmatrix} \sim N\left(0, \begin{bmatrix} K(X,X) + \sigma_n^2 I & k(X, x^*) \\ k(x^*, X) & k(x^*, x^*) \end{bmatrix} \right) \qquad (\text{AG.274})$$

In (AG.274), $f^*(.)$ is a prediction for a new input vector x^*. In SOGP, the posterior mean and covariance are treated as linear combination of the prior covariance functions [AG.11]:

$$\bar{f}_N = \bar{f}_0 + \sum_{i=1}^{N} K_0(x, x_i)\alpha_N(i) = \bar{f}_0 + \alpha_N^T k_x \qquad (\text{AG.275})$$

$$K(X,X) = K_0(x, x') + k_x^T C_N^{GP} k_{x'}$$

In (AG.275), the mean, the covariance parameters, and the kernel function are denoted, respectively as:

$$\alpha_N = [\alpha_N(1), ..., \alpha_N(t)]^T ; \quad C_N^{GP} = \{C_N(ij)\}_{i,j=1} ;$$
$$k_x = [K_0(x_1, x), ..., K_0(x_N, x)]^T \qquad (\text{AG.276})$$

The kernel functions are centred on each combination of the training points. The parameters in (AG.275) can be updated iteratively as [AG.11]:

$$\alpha_{N+1} = T_{N+1}(\alpha_N) + p_{N+1}(s_{N+1})$$
$$C_{N+1}^{GP} = U_{N+1}(C_N^{GP}) + r_{N+1}(s_{N+1} s_{N+1}^T) \qquad (\text{AG.277})$$
$$s_{N+1} = T_{N+1}(C_N^{GP} k_{N+1}) + e_{N+1}$$

In (AG.278), $e_{N+1} = [0, ..., 1]^T$, the operators $T(.)$, and $U(.)$ extend an N-dim vector and $N \times N$-dim matrix to an $N + 1$ dimensions, by appending a zero to the end of the vector, and a zero row and column to the matrix. The scalars $p(.)$ and $r(.)$ are computed as

$$p_{N+1} = \frac{\partial}{\partial f_N(x_N + 1)} \ln \langle P(y_{N+1} \mid f_N(x_N + 1)) \rangle_N$$

$$\qquad (\text{AG.278})$$

$$r_{N+1} = \frac{\partial^2}{\partial f_N^2(x_N + 1)} \ln \langle P(y_{N+1} \mid f_N(x_N + 1)) \rangle_N$$

In (AG.278), the angular parenthesis denotes the averages with respect to the GP at the N-th iteration. In (AG.276), the dimension increases with each data point added. However, the following is used to avoid this increase in the dimension (as a linear combination of the covariance functions):

$$K(x, x_{N+1}) = \sum_{i=1}^{N} e_{N+1} K_0(x, x_i) \qquad (AG.279)$$

An approximation of the solution for (AG.279) can be obtained by minimizing the following error measure:

$$\left\| K_0(x, x_{N+1}) - \sum_{i=1}^{N} e_{N+1}(i) K_0(x, x_i) \right\|^2 \qquad (AG.280)$$

The solution is given as

$$\hat{e}_{N+1} = K_N^{-1} k_{N+1}; \quad K_N = \{K_0(x_i, x_j)\}_{i,j=1} \qquad (AG.281)$$

Using this solution we get the following equations:

$$\hat{K}_0(x, x_{N+1}) = \sum_{i=1}^{N} \hat{e}_{N+1}(i) K_0(x, x_i)$$
$$\varepsilon_{N+1} = K_0(x_{N+1}, x_{N+1}) - k_{N+1}^T K_N^{-1} k_{N+1} \qquad (AG.282)$$

In the 1st equation of (AG.282), we have the orthogonal projection $\hat{K}_0(x, x_{N+1})$ of the $K_0(x, x_{N+1})$ on the linear span of the functions $K_0(x, x_i)$. The second equation is the residual error from the projection (the 1st equation), and it determines whether or not the current input x_N will be included in the BV set. If the input yields a high value of the error, the second equation, then it means that the new data cannot be described adequately using the linear combinations of the existing data points and it should be added to the BV set. The SOGP regression algorithm is given in Table AG.2.

G.6.9.3 Adaptation

Model (AG.270) of the robot manipulator provides a prior knowledge to the learning controller, and this would improve the generalization performance. For the SOGP regression, a zero mean function for the GP is assumed. To incorporate a prior knowledge, one can set the mean function in (AG.272) equal to the available model of the system, this will bias the system to the a priori knowledge. Yet, the covariance function, $k(.,.)$ requires the adaptation to the data that are observed; hence, the hyper-parameters of the covariance function should also be optimized to the incoming training set/data. For this a nonlinear conjugate gradient descent (NCGD) method can be used. This technique minimizes the negative log marginal likelihood with respect to the hyper-parameters. The initial guess values can be obtained by performing the optimization on a batch of the training data that might have been collected from the (experimental or real) runs of the (robotic) system. An NCGD algorithm is describe in [AG.11].

The main points of the application of the SOGP-regression and the NCGD algorithms to a 6-DOF Puma 560 robot study using the Robotics Toolbox (RTB) and the open-source SPGP and SOGP (incorporating the a-priori knowledge and incremental updating) [AG.11] are:

TABLE AG.2

The Sparse Online Gaussian Process Regression Algorithm

Steps	Computations
1. For a new data pair, (y_{N+1}, x_{N+1}), do the following	
2. Compute p_{N+1}, and r_{N+1} as per (AG.278), $\hat{e}_{N+1}$, and ε_{N+1} as per (AG.281), and (AG.282)	
3. If $\varepsilon_{N+1} <$ tolerance, then	
4. Compute α and C^{GP} (without extending their size)	
$\alpha_{N+1} = \alpha_N + p_{N+1}(s_{N+1})$	
5. $C_{N+1}^{GP} = C_N^{GP} + r_{N+1}(s_{N+1} s_{N+1}^T)$	
$s_{N+1} = C_N^{GP} k_{N+1} + \hat{e}_{N+1}$	
6. else	
7. Compute α and C^{GP} as follows	
8. $\alpha_{N+1} = T_{N+1}(\alpha_N) + p_{N+1}(s_{N+1})$	
$C_{N+1}^{GP} = U_{N+1}(C_N^{GP}) + r_{N+1}(s_{N+1} s_{N+1}^T)$	
$s_{N+1} = T_{N+1}(C_N^{GP} k_{N+1}) + e_{N+1}$	
9. Add the current input to x_{N+1} to BV set	
10. end if	
11. If size of BV set $> M_{BV}$ then	
12. Compute $\varepsilon_{N+1} = K_0(x_{N+1}, x_{N+1}) - k_{N+1}^T K_N^{-1} k_{N+1}$ as per (AG.282)	
13. Remove the lowest scoring one	
14. end if	

1. The effects of Coloumb and viscous friction were simulated with the friction constants obtained from the defaults for the Puma 560, and to simulate the effects of uncertainty in the knowledge of the inertial parameters, a 10% percent error was introduced

2. Tracking performance of the controller was evaluated on a "star-like" asterisk pattern (X), i.e. the asterisk trajectory, which is a difficult trajectory due to the high components of velocity and acceleration, and it normally needs a model-based control for good tracking accuracy. For the present case, inverse kinematics was used to convert this trajectory from task to joint space.

3. In the case with the full knowledge of the model, given sufficient data, the learning controllers were able to outperform the CT (computed torque) method by learning the nonlinear behavior of Coloumb and viscous friction.

4. With a partial knowledge of the model, the algorithms were initialized with only the gravity loading vector $G(q)$ of the model. The learning controllers compensated for friction and outperformed the CT controller; however, it took longer to achieve the same tracking performance as in the case of full knowledge. By incorporating a priori knowledge, the algorithms performed reasonably well from the start.

G.7 Concluding Remarks

In this appendix, we have briefly summarized several control philosophies and concepts as a small projection of the main topics discussed in five sections of the book. Several other supporting aspects of the control, modeling and system identification-cum-parameter estimation, and state estimation, filtering, and observers can be found in the (companion) volumes listed as Ref. [AG.12–AG.14].

References

AG.1 Henrion, D., and Sebek, M. Controller design using polynomial matrix description. https://homepages.laas.fr/henrion/Papers/6-43-13-16.pdf, accessed December 2018.

AG.2 Khargonekar, P. Advancing systems and control research in the era of ML and AI. Annual review in control, Elsevier JL, Vol. 45, pp. 1–4, 2018. https://www.Sciencedirect.com/science/article/pii/S1367578818300269; https://faculty.sites.uci.edu/khargonekar/files/2018/04/Control_ML_AI_Final.pdf, accessed December 2018.

AG.3 Didekova, Z., and Kajan, S. Applications of intelligent hybrid systems in MATLAB. http://dsp.vscht.cz/konference_matlab/MATLAB09/prispevky/021_didekova.pdf, accessed December 2018.

AG.4 Seron, M., M. Receding horizon control. Centre of Complex Dynamic Systems and Control. The University of Newcastle, Australia. http://www.eng.newcastle.edu.au/eecs/cdsc/books/cce/Slides/RecedingHorizonControl.pdf, accessed December 2018.

AG.5 Antsaklis, P. J., and Koutsoukos, X. D. Hybrid systems control. https://www3.nd.edu/~isis/techreports/isis-2001-003.pdf, accessed December 2018.

AG.6 Ortega, R. A "new" control theory to face the challenges of modern technology. PPTs., Supelec, pp. 1–91, January 2012; http://webpages.lss.supelec.fr/ortega/sli_eeci12.pdf, accessed December 2018.

AG.7 Simrock, S. Tutorial on control theory. ICALEPCS, WTC Grenoble, France, 2011. https://accelconf.web.cern.ch/accelconf/icalepcs2011/talks/tutmukp01_talk.pdf, accessed December 2018.

AG.8 Anon. A few illustrations of the basic concepts of nonlinear control. https://slideplayer.com/slide/12336729/, accessed December 2018.

AG.9 Anon. Principles of optimal control. MIT OpenCourseWare, Spring, USA, 2008. http://ocw.mit.edu, accessed December 2018.

AG.10 Ariyur, K., Benosman, M., and Black, W. Adaptive and learning Systems: Historic ambitions, current reality, and future potential, August, 2018. http://blog.ifac-control.org/adaptive-control/adaptive-and-learning-systems-historic-ambitions-current-reality-and-future-potential/, accessed December 2018.

AG.11 Sun de la Cruz, J., Owen, W., and Kulic, D. Online learning of inverse dynamics via Gaussian process regression, 2012 IEEE/RSJ International Conference on Intelligent Robots and Systems, October 7–12, 2012. Vilamoura, Algarve, Portugal; https://www.academia.edu/2818566/Online_learning_of_inverse_dynamics_via_Gaussian_Process_Regression, accessed December 2018.

AG.12 Raol, J. R., Girija, G., and J. Singh. *Modelling and Parameter Estimation of Dynamic Systems*. The IET/IEE Control Series Book, Vol. 65, London, UK 2004.

AG.13 Raol, J. R., and J. Singh. *Flight Mechanics Modelling and Analysis.* Taylor & Francis Informa Group, Boca Raton, FL, 2009.

AG.14 Raol, J. R., Girija, G., and B. Twala. *Nonlinear Filtering: Concepts and Engineering Applications*. Taylor & Francis Informa Group, Boca Raton, FL, 2017.

Index

Note: Page numbers in italic and bold refer to figures and tables, respectively.